Algebra and Trigonometry

EIGHTH EDITION

Ron Larson
The Pennsylvania State University, The Behrend College

Prepared by

Ron Larson
The Pennsylvania State University, The Behrend College

BROOKS/COLE
CENGAGE Learning

Australia • Brazil • Japan • Korea • Mexico • Singapore • Spain • United Kingdom • United States

BROOKS/COLE
CENGAGE Learning

For product information and technology assistance, contact us at **Cengage Learning Customer & Sales Support, 1-800-354-9706**

For permission to use material from this text or product, submit all requests online at **www.cengage.com/permissions** Further permissions questions can be emailed to **permissionrequest@cengage.com**

ISBN-13: 978-1-4390-4851-1
ISBN-10: 1-4390-4851-7

Brooks/Cole
20 Channel Center Street
Boston, MA 02210
USA

Cengage Learning is a leading provider of customized learning solutions with office locations around the globe, including Singapore, the United Kingdom, Australia, Mexico, Brazil, and Japan. Locate your local office at: **www.cengage.com/global**

Cengage Learning products are represented in Canada by Nelson Education, Ltd.

To learn more about Brooks/Cole, visit **www.cengage.com/brookscole**

Purchase any of our products at your local college store or at our preferred online store **www.CengageBrain.com**

Printed in the United States of America
2 3 4 5 6 7 14 13 12 11

CONTENTS

Preface

This *Student Solutions Manual* for *Algebra and Trigonometry*, Eighth Edition, is a supplement to the textbook by Ron Larson. Like the previous edition's *Student Solutions Guide*, this manual includes solutions to every exercise in the text with all essential algebraic steps included.

We would like to thank Dianna L. Zook for her contributions to the prior editions of this supplement.

PART I

CHAPTER P
Prerequisites

C H A P T E R   P
Prerequisites

Section P.1   Review of Real Numbers and Their Properties

- You should know the following sets.
  - (a) The set of real numbers includes the rational numbers and the irrational numbers.
  - (b) The set of rational numbers includes all real numbers that can be written as the ratio p/q of two integers, where $q \neq 0$.
  - (c) The set of irrational numbers includes all real numbers which are not rational.
  - (d) The set of integers: $\{\ldots, -3, -2, -1, 0, 1, 2, 3, \ldots\}$
  - (e) The set of whole numbers: $\{0, 1, 2, 3, 4, \ldots\}$
  - (f) The set of natural numbers: $\{1, 2, 3, 4, \ldots\}$

- The real number line is used to represent the real numbers.

- Know the inequality symbols.
  - (a) $a < b$ means a is less than b.
  - (b) $a \leq b$ means a is less than or equal to b.
  - (c) $a > b$ means a is greater than b.
  - (d) $a \geq b$ means a is greater than or equal to b.

Interval Notation	*Inequality Notation*	*Graph*	*Type*
$[a, b]$	$a \leq x \leq b$		Bounded and Closed
(a, b)	$a < x < b$		Bounded and Open
$[a, b)$	$a \leq x < b$		Bounded
$(a, b]$	$a < x \leq b$		Bounded
$[a, \infty)$	$x \geq a$		Unbounded
(a, ∞)	$x > a$		Unbounded
$(-\infty, b]$	$x \leq b$		Unbounded
$(-\infty, b)$	$x < b$		Unbounded
$(-\infty, \infty)$	$-\infty < x < \infty$		Unbounded

- You should know that $|a| = \begin{cases} a, & \text{if } a \geq 0 \\ -a, & \text{if } a < 0 \end{cases}$.

- Know the properties of absolute value.
  - (a) $|a| \geq 0$
  - (b) $|-a| = |a|$
  - (c) $|ab| = |a||b|$
  - (d) $\left|\dfrac{a}{b}\right| = \dfrac{|a|}{|b|}, b \neq 0$

—continued—

- The distance between a and b on the real line is $d(a, b) = |b - a| = |a - b|$.

- You should be able to identify the terms in an algebraic expression.

- You should know and be able to use the basic rules of algebra.

- **Commutative Property**

  (a) Addition: $a + b = b + a$ (b) Multiplication: $a \cdot b = b \cdot a$

- **Associative Property**

  (a) Addition: $(a + b) + c = a + (b + c)$ (b) Multiplication: $(ab)c = a(bc)$

- **Identity Property**

  (a) Addition: 0 is the identity; $a + 0 = 0 + a = a$. (b) Multiplication: 1 is the identity; $a \cdot 1 = 1 \cdot a = a$.

- **Inverse Property**

  (a) Addition: $-a$ is the additive inverse of a; $a + (-a) = -a + a = 0$.

  (b) Multiplication: $1/a$ is the multiplicative inverse of a, $a \neq 0$; $a(1/a) = (1/a)a = 1$.

- **Distributive Property**

  (a) $a(b + c) = ab + ac$ (b) $(a + b)c = ac + bc$

- **Properties of Negation**

  (a) $(-1)a = -a$ (b) $-(-a) = a$ (c) $(-a)b = a(-b) = -ab$

  (d) $(-a)(-b) = ab$ (e) $-(a + b) = (-a) + (-b) = -a - b$

- **Properties of Equality**

  (a) If $a = b$, then $a \pm c = b \pm c$. (b) If $a = b$, then $ac = bc$.

  (c) If $a \pm c = b \pm c$, then $a = b$. (d) If $ac = bc$ and $c \neq 0$, then $a = b$.

- **Properties of Zero**

  (a) $a \pm 0 = a$ (b) $a \cdot 0 = 0$ (c) $0 \div a = 0/a = 0, a \neq 0$

  (d) $a/0$ is undefined. (e) If $ab = 0$, then $a = 0$ or $b = 0$.

- **Properties of Fractions** $(b \neq 0, d \neq 0)$

  (a) Equivalent Fractions: $a/b = c/d$ if and only if $ad = bc$.

  (b) Rule of Signs: $-a/b = a/-b = -(a/b)$ and $-a/-b = a/b$

  (c) Equivalent Fractions: $a/b = ac/bc, c \neq 0$

  (d) Addition and Subtraction

   1. Like Denominators: $(a/b) \pm (c/b) = (a \pm c)/b$ 2. Unlike Denominators: $(a/b) \pm (c/d) = (ad \pm bc)/bd$

  (e) Multiplication: $(a/b) \cdot (c/d) = (ac)/bd$

  (f) Division: $(a/b) \div (c/d) = (a/b) \cdot (d/c) = (ad)/(bc)$ if $c \neq 0$.

1. rational

3. origin

5. composite

7. variables; constants

9. coefficient

11. $-9, -\frac{7}{2}, 5, \frac{2}{3}, \sqrt{2}, 0, 1, -4, 2, -11$

    (a) Natural numbers: $5, 1, 2$

    (b) Whole numbers: $0, 5, 1, 2$

    (c) Integers: $-9, 5, 0, 1, -4, 2, -11$

    (d) Rational numbers: $-9, -\frac{7}{2}, 5, \frac{2}{3}, 0, 1, -4, 2, -11$

    (e) Irrational numbers: $\sqrt{2}$

13. $2.01, 0.666\ldots, -13, 0.010110111\ldots, 1, -6$

    (a) Natural numbers: 1

    (b) Whole numbers: 1

    (c) Integers: $-13, 1, -6$

    (d) Rational numbers: $2.01, 0.666\ldots, -13, 1, -6$

    (e) Irrational numbers: $0.010110111\ldots$

15. $-\pi, -\frac{1}{3}, \frac{6}{3}, \frac{1}{2}\sqrt{2}, -7.5, -1, 8, -22$

    (a) Natural numbers: $\frac{6}{3}, 8$

    (b) Whole numbers: $\frac{6}{3}, 8$

    (c) Integers: $\frac{6}{3}, -1, 8, -22$

    (d) Rational numbers: $-\frac{1}{3}, \frac{6}{3}, -7.5, -1, 8, -22$

    (e) Irrational numbers: $-\pi, \frac{1}{2}\sqrt{2}$

17. (a)

    (b)

    (c)

    (d)

19. $\frac{5}{8} = 0.625$

21. $\frac{41}{333} = 0.\overline{123}$

23. $-2.5 < 2$

25. $-4 > -8$

27. $\frac{3}{2} < 7$

29. $\frac{5}{6} > \frac{2}{3}$

31. (a) The inequality $x \le 5$ denotes the set of all real numbers less than or equal to 5.

    (b)

    (c) The interval is unbounded.

33. (a) The inequality $x < 0$ denotes the set of all real numbers less than zero.

    (b)

    (c) The interval is unbounded.

35. (a) The interval $[4, \infty)$ denotes the set of all real numbers greater than or equal to 4.

    (b)

    (c) The interval is unbounded.

37. (a) The inequality $-2 < x < 2$ denotes the set of all real numbers greater than -2 and less than 2.

    (b)

    (c) The interval is bounded.

39. (a) The inequality $-1 \le x < 0$ denotes the set of all real numbers greater than or equal to -1 and less than zero.

    (b)

    (c) The interval is bounded.

41. (a) The interval $[-2, 5)$ denotes the set of all real numbers greater than or equal to -2 and less than 5.

    (b)

    (c) The interval is bounded.

43. $y \ge 0; [0, \infty)$

45. $-2 < x \le 4; (-2, 4]$

47. $10 \le t \le 22; [10, 22]$

49. $W > 65; (65, \infty)$

51. $|-10| = -(-10) = 10$

53. $|3 - 8| = |-5| = -(-5) = 5$

55. $|-1| - |-2| = 1 - 2 = -1$

57. $\dfrac{-5}{|-5|} = \dfrac{-5}{-(-5)} = \dfrac{-5}{5} = -1$

59. If $x < -2$, then $x + 2$ is negative.

So, $\dfrac{|x + 2|}{x + 2} = \dfrac{-(x + 2)}{x + 2} = -1$.

61. $|-3| > -|-3|$ because $3 > -3$.

63. $-5 = -|5|$ since $-5 = -5$.

65. $-|-2| = -|2|$ because $-2 = -2$.

67. $d(126, 75) = |75 - 126| = 51$

69. $d\left(-\tfrac{5}{2}, 0\right) = \left|0 - \left(-\tfrac{5}{2}\right)\right| = \tfrac{5}{2}$

71. $d\left(\tfrac{16}{5}, \tfrac{112}{75}\right) = \left|\tfrac{112}{75} - \tfrac{16}{5}\right| = \tfrac{128}{75}$

73. $d(x, 5) = |x - 5|$ and $d(x, 5) \le 3$, so $|x - 5| \le 3$.

75. $d(y, 0) = |y - 0| = |y|$ and $d(y, 0) \ge 6$, so $|y| \ge 6$.

77. $d(57, 236) = |57 - 236| = 179$ mi

79.

Budgeted Expense, b	Actual Expense, a	$|a - b|$	0.05b
$112,700	$113,356	$656	$0.05(112,700) = \$5635$

Because $656 < $5635 but $656 > $500, the actual expense does not pass the "budget variance test."

81.

Budgeted Expense, b	Actual Expense, a	$|a - b|$	0.05b
$37,640	$37,335	$305	$0.05(37,640) = \$1882$

Because $305 < $500 and $305 < $1882, the actual expense passes the "budget variance test."

83.

Receipts	Expenditures	$|Receipts - Expenditures|$
$1453.2	$1560.6	$|1453.2 - 1560.6| = \$107.4$ billion

85.

Receipts	Expenditures	$|Receipts - Expenditures|$
$2025.5	$1789.2	$|2025.5 - 1789.2| = \$236.3$ billion

87.

Receipts	Expenditures	$|Receipts - Expenditures|$
$1880.3	$2293.0	$|1880.3 - 2293.0| = \$412.7$ billion

89. $7x + 4$

Terms: $7x, 4$

Coefficient: 7

91. $\sqrt{3}x^2 - 8x - 11$

Terms: $\sqrt{3}x^2, -8x, -11$

Coefficients: $\sqrt{3}, -8$

93. $4x^3 + \dfrac{x}{2} - 5$

Terms: $4x^3, \dfrac{x}{2}, -5$

Coefficients: $4, \dfrac{1}{2}$

95. $4x - 6$

(a) $4(-1) - 6 = -4 - 6 = -10$

(b) $4(0) - 6 = 0 - 6 = -6$

97. $x^2 - 3x + 4$

(a) $(-2)^2 - 3(-2) + 4 = 4 + 6 + 4 = 14$

(b) $(2)^2 - 3(2) + 4 = 4 - 6 + 4 = 2$

99. $\dfrac{x+1}{x-1}$

(a) $\dfrac{1+1}{1-1} = \dfrac{2}{0}$

Division by zero is undefined.

(b) $\dfrac{-1+1}{-1-1} = \dfrac{0}{-2} = 0$

101. $x + 9 = 9 + x$

Commutative Property of Addition

103. $\dfrac{1}{(h+6)}(h+6) = 1, h \neq -6$

Multiplicative Inverse Property

105. $2(x+3) = 2x + 6$

Distributive Property

107. $1 \cdot (1+x) = 1 + x$

Multiplicative Identity Property

109. $x + (y + 10) = (x + y) + 10$

Associative Property of Addition

111. $3(t-4) = 3 \cdot t - 3 \cdot 4$

Distributive Property

113. $\dfrac{3}{16} + \dfrac{5}{16} = \dfrac{8}{16} = \dfrac{1}{2}$

115. $\dfrac{5}{8} - \dfrac{5}{12} + \dfrac{1}{6} = \dfrac{15}{24} - \dfrac{10}{24} + \dfrac{4}{24} = \dfrac{9}{24} = \dfrac{3}{8}$

117. $12 \div \dfrac{1}{4} = 12 \cdot \dfrac{4}{1} = 12 \cdot 4 = 48$

119. $\dfrac{2x}{3} - \dfrac{x}{4} = \dfrac{8x}{12} - \dfrac{3x}{12} = \dfrac{5x}{12}$

121. (a) Because $A > 0, -A < 0$.

The expression is negative.

(b) Because $B < A, B - A < 0$.

The expression is negative.

123. (a)

n	1	0.5	0.01	0.0001	0.000001
$5/n$	5	10	500	50,000	5,000,000

(b) The value of $5/n$ approaches infinity as n approaches 0.

125. True. Because $b < 0$, $a - b$ subtracts a negative number from (or adds a positive number to) a positive number. The sum of two positive numbers is positive.

127. False. If $a < b$, then $\dfrac{1}{a} > \dfrac{1}{b}$, where $a \neq b \neq 0$.

129. (a) No. $|u + v| \neq |u| + |v|$ if u is positive and v is negative or vice versa.

(b) No. $|u + v| \leq |u| + |v|$

They are equal when u and v have the same sign. If they differ in sign, $|u + v|$ is less than $|u| + |v|$.

131. The only even prime number is 2, because its factors are itself and 1.

133. Yes, if a is a negative number, then $-a$ is positive. So, $|a| = -a$ if a is negative.

Section P.2   Exponents and Radicals

- You should know the properties of exponents.

(a) $a^1 = a$

(b) $a^0 = 1, a \neq 0$

(c) $a^m a^n = a^{m+n}$

(d) $a^m/a^n = a^{m-n}, a \neq 0$

(e) $a^{-n} = 1/a^n = (1/a)^n, a \neq 0$

(f) $(a^m)^n = a^{mn}$

(g) $(ab)^n = a^n b^n$

(h) $(a/b)^n = a^n/b^n, b \neq 0$

(i) $(a/b)^{-n} = (b/a)^n, a \neq 0, b \neq 0$

(j) $|a^2| = |a|^2 = a^2$

- You should be able to write numbers in scientific notation, $c \times 10^n$, where $1 \leq c < 10$ and n is an integer.

- You should be able to use your calculator to evaluate expressions involving exponents.

—*continued*—

■ You should know the properties of radicals.

(a) $\sqrt[n]{a^m} = \left(\sqrt[n]{a}\right)^m$, $a > 0$

(b) $\sqrt[n]{a} \cdot \sqrt[n]{b} = \sqrt[n]{ab}$

(c) $\dfrac{\sqrt[n]{a}}{\sqrt[n]{b}} = \sqrt[n]{\dfrac{a}{b}}$, $b \neq 0$

(d) $\sqrt[m]{\sqrt[n]{a}} = \sqrt[mn]{a}$

(e) $\left(\sqrt[n]{a}\right)^n = a$

(f) For n even, $\sqrt[n]{a^n} = |a|$.

For n odd, $\sqrt[n]{a^n} = a$.

(g) $a^{1/n} = \sqrt[n]{a}$

(h) $a^{m/n} = \left(\sqrt[n]{a}\right)^m = \sqrt[n]{a^m}$, $a \geq 0$

■ You should be able to simplify radicals.

(a) All possible factors have been removed from the radical sign.

(b) All fractions have radical-free denominators.

(c) The index for the radical has been reduced as far as possible.

■ You should be able to use your calculator to evaluate radicals.

---

1. exponent; base

3. square root

5. index; radicand

7. like radicals

9. rationalizing

11. (a) $3^2 \cdot 3 = 3^3 = 27$

(b) $3 \cdot 3^3 = 3^4 = 81$

13. (a) $\left(3^3\right)^0 = 1$

(b) $-3^2 = -9$

15. (a) $\dfrac{3}{3^{-4}} = 3 \cdot 3^4 = 3^5 = 243$

(b) $48(-4)^{-3} = \dfrac{48}{(-4)^3} = -\dfrac{48}{64} = -\dfrac{3}{4}$

17. (a) $2^{-1} + 3^{-1} = \dfrac{1}{2} + \dfrac{1}{3} = \dfrac{3}{6} + \dfrac{2}{6} = \dfrac{5}{6}$

(b) $\left(2^{-1}\right)^{-2} = 2^{(-1)(-2)} = 2^2 = 4$

19. $(-4)^3\left(5^2\right) = (-64)(25) = -1600$

21. $\dfrac{3^6}{7^3} = \dfrac{729}{343} \approx 2.125$

23. When $x = 2$,

$-3x^3 = -3(2)^3 = -24$.

25. When $x = 10$,

$6x^0 = 6(10)^0 = 6(1) = 6$.

27. When $x = -3$,

$2x^3 = 2(-3)^3 = 2(-27) = -54$.

29. When $x = -\dfrac{1}{2}$,

$-20x^2 = -20\left(-\dfrac{1}{2}\right)^2 = -20\left(\dfrac{1}{4}\right) = -5$.

31. (a) $(-5z)^3 = (-5)^3 z^3 = -125z^3$

(b) $5x^4\left(x^2\right) = 5x^{4+2} = 5x^6$

33. (a) $6y^2\left(2y^0\right)^2 = 6y^2(2 \cdot 1)^2 = 6y^2(4) = 24y^2$

(b) $\dfrac{3x^5}{x^3} = 3x^{5-3} = 3x^2$

35. (a) $\dfrac{7x^2}{x^3} = 7x^{2-3} = 7x^{-1} = \dfrac{7}{x}$

(b) $\dfrac{12(x+y)^3}{9(x+y)} = \dfrac{4}{3}(x+y)^{3-1} = \dfrac{4}{3}(x+y)^2$

---

8 Chapter P Prerequisites

37. (a) $\left[\left(x^2y^{-2}\right)^{-1}\right]^{-1} = \left(x^{-2}y^2\right)^{-1}$

$= x^2y^{-2}$

$= \dfrac{x^2}{y^2}$

(b) $\left(\dfrac{a^{-2}}{b^{-2}}\right)\left(\dfrac{b}{a}\right)^3 = \left(\dfrac{a^{-2}}{b^{-2}}\right)\left(\dfrac{b^3}{a^3}\right)$

$= \left(a^{-2}\right)\left(b^2\right)\left(b^3\right)\left(a^{-3}\right)$

$= a^{-5}b^5$

$= \dfrac{b^5}{a^5}$

39. (a) $(x+5)^0 = 1,\ x \neq -5$

(b) $\left(2x^2\right)^{-2} = \dfrac{1}{\left(2x^2\right)^2} = \dfrac{1}{4x^4}$

41. (a) $\left(-2x^2\right)^3\left(4x^3\right)^{-1} = \dfrac{-8x^6}{4x^3} = -2x^3$

(b) $\left(\dfrac{x}{10}\right)^{-1} = \dfrac{10}{x}$

43. (a) $3^n \cdot 3^{2n} = 3^{n+2n} = 3^{3n}$

(b) $\left(\dfrac{a^{-2}}{b^{-2}}\right)\left(\dfrac{b}{a}\right)^3 = \left(\dfrac{b^2}{a^2}\right)\left(\dfrac{b^3}{a^3}\right) = \dfrac{b^5}{a^5}$

45. $10,250.4 = 1.02504 \times 10^4$

47. $-0.000125 = -1.25 \times 10^{-4}$

49. $57,300,000 = 5.73 \times 10^7$ square miles

51. $0.0000899 = 8.99 \times 10^{-5}$ gram per cubic centimeter

53. $1.25 \times 10^5 = 125,000$

55. $-2.718 \times 10^{-3} = -0.002718$

57. $1.5 \times 10^7 = 15,000,000$ degrees Celsius

59. $9.0 \times 10^{-5} = 0.00009$ meter

61. (a) $\left(2.0 \times 10^9\right)\left(3.4 \times 10^{-4}\right) = 6.8 \times 10^5$

(b) $\left(1.2 \times 10^7\right)\left(5.0 \times 10^{-3}\right) = 6.0 \times 10^4$

63. (a) $750\left(1 + \dfrac{0.11}{365}\right)^{800} \approx 954.448$

(b) $\dfrac{67,000,000 + 93,000,000}{0.0052} = 30,769,230,769.2$

$\approx 3.077 \times 10^{10}$

65. (a) $\sqrt{9} = 3$

(b) $\sqrt[3]{\dfrac{27}{8}} = \dfrac{\sqrt[3]{27}}{\sqrt[3]{8}} = \dfrac{3}{2}$

67. (a) $32^{-3/5} = \dfrac{1}{32^{3/5}} = \dfrac{1}{\left(\sqrt[5]{32}\right)^3} = \dfrac{1}{(2)^3} = \dfrac{1}{8}$

(b) $\left(\dfrac{16}{81}\right)^{-3/4} = \left(\dfrac{81}{16}\right)^{3/4} = \left(\sqrt[4]{\dfrac{81}{16}}\right)^3 = \left(\dfrac{3}{2}\right)^3 = \dfrac{27}{8}$

69. (a) $\left(-\dfrac{1}{64}\right)^{-1/3} = (-64)^{1/3} = \sqrt[3]{-64} = -4$

(b) $\left(\dfrac{1}{\sqrt{32}}\right)^{-2/5} = \left(\sqrt{32}\right)^{2/5} = \sqrt[5]{\left(\sqrt{32}\right)^2} = \sqrt[5]{32} = 2$

71. (a) $\sqrt{57} \approx 7.550$

(b) $\sqrt[5]{-27^3} = (-27)^{3/5} \approx -7.225$

73. (a) $(-12.4)^{-1.8} \approx -0.011$

(b) $\left(5\sqrt{3}\right)^{-2.5} \approx 0.005$

75. (a) $\sqrt{4.5 \times 10^9} \approx 67,082.039$

(b) $\sqrt[3]{6.3 \times 10^4} \approx 39.791$

77. (a) $\left(\sqrt[5]{2}\right)^5 = 2^{5/5} = 2^1 = 2$

(b) $\sqrt[5]{96x^5} = \sqrt[5]{32x^5 \cdot 3}$

$= 2\sqrt[5]{3}x$

79. (a) $\sqrt{20} = \sqrt{4 \cdot 5}$

$= \sqrt{4}\sqrt{5} = 2\sqrt{5}$

(b) $\sqrt[3]{128} = \sqrt[3]{64 \cdot 2}$

$= \sqrt[3]{64}\sqrt[3]{2} = 4\sqrt[3]{2}$

81. (a) $\sqrt{72x^3} = \sqrt{36x^2 \cdot 2x}$

$= 6x\sqrt{2x}$

(b) $\sqrt{\dfrac{18^2}{z^3}} = \dfrac{\sqrt{18^2}}{\sqrt{z^2 \cdot z}} = \dfrac{18}{z\sqrt{z}} = \dfrac{18\sqrt{z}}{z^2}$

83. (a) $\sqrt[3]{16x^5} = \sqrt[3]{8x^3 \cdot 2x^2} = 2x\sqrt[3]{2x^2}$

(b) $\sqrt{75x^2y^{-4}} = \sqrt{\dfrac{75x^2}{y^4}} = \dfrac{\sqrt{25x^2 \cdot 3}}{\sqrt{y^4}} = \dfrac{5|x|\sqrt{3}}{y^2}$

85. (a) $2\sqrt{50} + 12\sqrt{8} = 2\sqrt{25 \cdot 2} + 12\sqrt{4 \cdot 2} = 2\left(5\sqrt{2}\right) + 12\left(2\sqrt{2}\right) = 10\sqrt{2} + 24\sqrt{2} = 34\sqrt{2}$

(b) $10\sqrt{32} - 6\sqrt{18} = 10\sqrt{16 \cdot 2} - 6\sqrt{9 \cdot 2} = 10\left(4\sqrt{2}\right) - 6\left(3\sqrt{2}\right) = 40\sqrt{2} - 18\sqrt{2} = 22\sqrt{2}$

87. (a) $5\sqrt{x} - 3\sqrt{x} = 2\sqrt{x}$

(b) $-2\sqrt{9y} + 10\sqrt{y} = -2\left(3\sqrt{y}\right) + 10\sqrt{y}$

$\quad = -6\sqrt{y} + 10\sqrt{y} = 4\sqrt{y}$

89. (a) $3\sqrt{x+1} + 10\sqrt{x+1} = 13\sqrt{x+1}$

(b) $7\sqrt{80x} - 2\sqrt{125x} = 7\sqrt{16 \cdot 5x} - 2\sqrt{25 \cdot 5x}$

$\quad = 7\left(4\sqrt{5x}\right) - 2\left(5\sqrt{5x}\right)$

$\quad = 28\sqrt{5x} - 10\sqrt{5x}$

$\quad = 18\sqrt{5x}$

91. $\sqrt{5} + \sqrt{3} \approx 3.968$ and $\sqrt{5+3} = \sqrt{8} \approx 2.828$

So, $\sqrt{5} + \sqrt{3} > \sqrt{5+3}$.

93. $\sqrt{3^2 + 2^2} = \sqrt{9+4} = \sqrt{13} \approx 3.606$

So, $5 > \sqrt{3^2 + 2^2}$.

95. $\dfrac{1}{\sqrt{3}} = \dfrac{1}{\sqrt{3}} \cdot \dfrac{\sqrt{3}}{\sqrt{3}} = \dfrac{\sqrt{3}}{3}$

97. $\dfrac{5}{\sqrt{14} - 2} = \dfrac{5}{\sqrt{14} - 2} \cdot \dfrac{\sqrt{14} + 2}{\sqrt{14} + 2} = \dfrac{5\left(\sqrt{14} + 2\right)}{\left(\sqrt{14}\right)^2 - (2)^2} = \dfrac{5\left(\sqrt{14} + 2\right)}{14 - 4} = \dfrac{5\left(\sqrt{14} + 2\right)}{10} = \dfrac{\sqrt{14} + 2}{2}$

99. $\dfrac{\sqrt{8}}{2} = \dfrac{\sqrt{4 \cdot 2}}{2} = \dfrac{2\sqrt{2}}{2} = \dfrac{\sqrt{2}}{1} \cdot \dfrac{\sqrt{2}}{\sqrt{2}} = \dfrac{2}{\sqrt{2}}$

101. $\dfrac{\sqrt{5} + \sqrt{3}}{3} = \dfrac{\sqrt{5} + \sqrt{3}}{3} \cdot \dfrac{\sqrt{5} - \sqrt{3}}{\sqrt{5} - \sqrt{3}} = \dfrac{5 - 3}{3\left(\sqrt{5} - \sqrt{3}\right)} = \dfrac{2}{3\left(\sqrt{5} - \sqrt{3}\right)}$

Radical Form	*Rational Exponent Form*

103. $\sqrt{2.5}$, Given $\qquad\qquad (2.5)^{1/2}$, Answer

105. $\sqrt[4]{81} = 3$, Answer $\qquad\qquad 81^{1/4}$, Given

107. $\sqrt[3]{-216} = -6$, Given $\qquad\qquad (-216)^{1/3} = -6$, Answer

109. $\sqrt[4]{81^3} = 27$, Given $\qquad\qquad 81^{3/4} = 27$, Answer

111. $\dfrac{\left(2x^2\right)^{3/2}}{2^{1/2}x^4} = \dfrac{2^{3/2}\left(x^2\right)^{3/2}}{2^{1/2}x^4} = \dfrac{2^{3/2}|x|^3}{2^{1/2}x^4} = 2^{3/2 - 1/2}|x|^{3-4} = 2^1|x|^{-1} = \dfrac{2}{|x|}$

113. $\dfrac{x^{-3} \cdot x^{1/2}}{x^{3/2} \cdot x^{-1}} = \dfrac{x^{1/2} \cdot x^1}{x^{3/2} \cdot x^3} = x^{1/2 + 1 - 3/2 - 3} = x^{-3} = \dfrac{1}{x^3}, x > 0$

115. (a) $\sqrt[4]{3^2} = 3^{2/4} = 3^{1/2} = \sqrt{3}$

(b) $\sqrt[6]{(x+1)^4} = (x+1)^{4/6} = (x+1)^{2/3} = \sqrt[3]{(x+1)^2}$

117. (a) $\sqrt{\sqrt{32}} = \left(32^{1/2}\right)^{1/2} = 32^{1/4} = \sqrt[4]{32} = \sqrt[4]{16 \cdot 2} = 2\sqrt[4]{2}$

(b) $\sqrt{\sqrt[4]{2x}} = \left(\left(2x\right)^{1/4}\right)^{1/2} = \left(2x\right)^{1/8} = \sqrt[8]{2x}$

119. $T = 2\pi\sqrt{\dfrac{2}{32}} = 2\pi\sqrt{\dfrac{1}{16}} = 2\pi\left(\dfrac{1}{4}\right) = \dfrac{\pi}{2} \approx 1.57$ seconds

121. $t = 0.03\left[12^{5/2} - \left(12 - h\right)^{5/2}\right], 0 \le h \le 12$

(a)

h (in centimeters)	t (in seconds)
0	0
1	2.93
2	5.48
3	7.67
4	9.53
5	11.08
6	12.32
7	13.29
8	14.00
9	14.50
10	14.80
11	14.93
12	14.96

(b) As h approaches 12, t approaches
$0.03\left(12^{5/2}\right) = 8.64\sqrt{3} \approx 14.96$ seconds.

123. True. When dividing variables, you subtract exponents.

125. $1 = \dfrac{a^m}{a^m} = a^{m-m} = a^0, a \neq 0$

127. No. A number is in scientific notation only when there is one nonzero digit to the left of the decimal point.

129. No. Rationalizing the denominator produces a number equivalent to the original fraction; squaring does not.

Section P.3   Polynomials and Special Products

- Given a polynomial in x, $a_nx^n + a_{n-1}x^{n-1} + \ldots + a_1x + a_0$, where $a_n \neq 0$, and n is a nonnegative integer, you should be able to identify the following.

  (a) Degree: n

  (b) Terms: $a_nx^n, a_{n-1}x^{n-1}, \ldots, a_1x, a_0$

  (c) Coefficients: $a_n, a_{n-1}, \ldots, a_1, a_0$

  (d) Leading coefficient: a_n

  (e) Constant term: a_0

- You should be able to add and subtract polynomials.

- You should be able to multiply polynomials by the Distributive Properties.

- You should be able to multiply two binomials by the FOIL Method.

—continued—

- Given a polynomial in x, $a_nx^n + a_{n-1}x^{n-1} + \ldots + a_1x + a_0$, where $a_n \neq 0$, and n is a nonnegative integer, you should be able to identify the following.

    (a) Degree: n (b) Terms: a_nx^n, $a_{n-1}x^{n-1}$, ..., a_1x, a_0

    (c) Coefficients: a_n, a_{n-1}, ..., a_1, a_0 (d) Leading coefficient: a_n

    (e) Constant term: a_0

- You should be able to add and subtract polynomials.

- You should be able to multiply polynomials by the Distributive Properties.

- You should be able to multiply two binomials by the FOIL Method.

- You should know the special binomial products.

    (a) $(u + v)(u - v) = u^2 - v^2$ (b) $(u \pm v)^2 = u^2 \pm 2uv + v^2$

    (c) $(u \pm v)^3 = u^3 \pm 3u^2v + 3uv^2 \pm v^3$

- You should be able to factor out all common factors, the first step in factoring.

- You should be able to factor the following special polynomial forms.

    (a) $u^2 - v^2 = (u + v)(u - v)$ (b) $u^2 \pm 2uv + v^2 = (u \pm v)^2$

    (c) $u^3 \pm v^3 = (u \pm v)(u^2 \mp uv + v^2)$

- You should be able to factor by grouping.

- You should be able to factor some trinomials by grouping.

---

1. n; a_n; a_0

3. monomial; binomial; trinomial

5. First terms; Outer terms; Inner terms; Last terms

6. (c) The sum and difference of same terms

7. (a) A binomial sum squared

8. (b) A binomial difference squared

9. (d) 12 is a polynomial of degree zero.

10. (e) $-3x^5 + 2x^3 + x$ is a polynomial of degree five.

11. (b) $1 - 2x^3 = -2x^3 + 1$ is a binomial with leading coefficient -2.

12. (a) $3x^2$ is a monomial of positive degree.

13. (f) $\frac{2}{3}x^4 + x^2 + 10$ is a trinomial with leading coefficient $\frac{2}{3}$.

14. (c) $x^3 + 3x^2 + 3x + 1$ is a third-degree polynomial with leading coefficient 1.

15. Sample answer: $-2x^3 + 4x^2 - 3x + 20$

17. Sample answer: $-15x^4 + 1$

19. (a) Standard form: $-\frac{1}{2}x^5 + 14x$

     (b) Degree: 5

         Leading coefficient: $-\frac{1}{2}$

     (c) Binomial

21. (a) Standard form: $-3x^4 + x^2 - 4$

     (b) Degree: 4

         Leading coefficient: -3

     (c) Trinomial

23. (a) Standard form: $-x^6 + 3$

     (b) Degree: 6

         Leading coefficient: -1

     (c) Binomial

25. (a) Standard form: 3

(b) Degree: 0

Leading coefficient: 3

(c) Monomial

27. (a) Standard form: $-4x^5 + 6x^4 + 1$

(b) Degree: 5

Leading coefficient: -4

(c) Trinomial

29. (a) Standard form: $4x^3 y$

(b) Degree: 4 (add the exponents on x and y)

Leading coefficient: 4

(c) Monomial

31. $2x - 3x^3 + 8$ *is* a polynomial.

Standard form: $-3x^3 + 2x + 8$

41. $\left(15x^2 - 6\right) - \left(-8.3x^3 - 14.7x^2 - 17\right) = 15x^2 - 6 + 8.3x^3 + 14.7x^2 + 17$

$$= 8.3x^3 + \left(15x^2 + 14.7x^2\right) + \left(-6 + 17\right)$$

$$= 8.3x^3 + 29.7x^2 + 11$$

43. $5z - \left[3z - \left(10z + 8\right)\right] = 5z - \left(3z - 10z - 8\right)$

$$= 5z - 3z + 10z + 8$$

$$= \left(5z - 3z + 10z\right) + 8$$

$$= 12z + 8$$

45. $3x\left(x^2 - 2x + 1\right) = 3x\left(x^2\right) + 3x\left(-2x\right) + 3x\left(1\right)$

$$= 3x^3 - 6x^2 + 3x$$

47. $-5z\left(3z - 1\right) = -5z\left(3z\right) + \left(-5z\right)\left(-1\right)$

$$= -15z^2 + 5z$$

55. $\left(7x^3 - 2x^2 + 8\right) + \left(-3x^2 - 4\right) = \left(7x^3 - 3x^3\right) + \left(-2x^2\right) + \left(8 - 4\right)$

$$= 4x^3 - 2x^2 + 4$$

57. $\left(5x^2 - 3x + 8\right) - \left(x - 3\right) = 5x^2 - 3x + 8 - x + 3$

$$= 5x^2 + \left(-3x - x\right) + \left(8 + 3\right)$$

$$= 5x^2 - 4x + 11$$

59. $\left(x + 7\right)\left(2x + 3\right) = 2x^2 + 3x + 14x + 21$ FOIL

$$= 2x^2 + 17x + 21$$

33. $\dfrac{3x + 4}{x} = 3 + \dfrac{4}{x} = 3 + 4x^{-1}$ is *not* a polynomial

because it includes a term with a negative exponent.

35. $y^2 - y^4 + y^3$ *is* a polynomial.

Standard form: $-y^4 + y^3 + y^2$

37. $\left(6x + 5\right) - \left(8x + 15\right) = 6x + 5 - 8x - 15$

$$= \left(6x - 8x\right) + \left(5 - 15\right)$$

$$= -2x - 10$$

39. $-\left(t^3 - 1\right) + \left(6t^3 - 5t\right) = -t^3 + 1 + 6t^3 - 5t$

$$= \left(-t^3 + 6t^3\right) - 5t + 1$$

$$= 5t^3 - 5t + 1$$

49. $\left(1 - x^3\right)\left(4x\right) = 1\left(4x\right) - x^3\left(4x\right)$

$$= 4x - 4x^4$$

$$= -4x^4 + 4x$$

51. $\left(1.5t^2 + 5\right)\left(-3t\right) = \left(1.5t^2\right)\left(-3t\right) + \left(5\right)\left(-3t\right)$

$$= -4.5t^3 - 15t$$

53. $-2x\left(0.1x + 17\right) = \left(-2x\right)\left(0.1x\right) + \left(-2x\right)\left(17\right)$

$$= -0.2x^2 - 34x$$

61. Multiply: $x^2 + 2x + 3$

$\quad\quad\quad\quad\underline{x^2 - 2x + 3}$

$\quad\quad\quad\quad x^4 + 2x^3 + 3x^2$

$\quad\quad\quad\quad\quad\quad -2x^3 - 4x^2 - 6x$

$\quad\quad\quad\quad\quad\quad\quad\quad\quad 3x^2 + 6x + 9$

$\quad\quad\quad\quad\overline{x^4 + 0x^3 + 2x^2 + 0x + 9} = x^4 + 2x^2 + 9$

63. $(x + 3)(x + 4) = x^2 + 4x + 3x + 12$ \quad FOIL

$\quad\quad\quad\quad\quad = x^2 + 7x + 12$

65. $(3x - 5)(2x + 1) = 6x^2 + 3x - 10x - 5$ \quad FOIL

$\quad\quad\quad\quad\quad = 6x^2 - 7x - 5$

67. $(x + 10)(x - 10) = x^2 - 10^2 = x^2 - 100$

69. $(x + 2y)(x - 2y) = x^2 - (2y)^2$

$\quad\quad\quad\quad\quad = x^2 - 4y^2$

71. $(2x + 3)^2 = (2x)^2 + 2(2x)(3) + 3^2$

$\quad\quad\quad\quad = 4x^2 + 12x + 9$

73. $(x + 1)^3 = x^3 + 3x^2(1) + 3x(1^2) + 1^3$

$\quad\quad\quad\quad = x^3 + 3x^2 + 3x + 1$

75. $(2x - y)^3 = (2x)^3 - 3(2x)^2 y + 3(2x)y^2 - y^3$

$\quad\quad\quad\quad = 8x^3 - 12x^2 y + 6xy^2 - y^3$

77. $(4x^3 - 3)^2 = (4x^3)^2 - 2(4x^3)(3) + (3)^2$

$\quad\quad\quad\quad = 16x^6 - 24x^3 + 9$

79. Multiply: $x^2 - x + 1$

$\quad\quad\quad\quad\underline{x^2 + x + 1}$

$\quad\quad\quad\quad x^4 - x^3 + x^2$

$\quad\quad\quad\quad\quad\quad x^3 - x^2 + x$

$\quad\quad\quad\quad\quad\quad\quad\quad x^2 - x + 1$

$\quad\quad\quad\quad\overline{x^4 - 0x^3 + x^2 + 0x + 1} = x^4 + x^2 + 1$

81. $(-x^2 + x - 5)(3x^2 + 4x + 1)$

$\quad$ Multiply: $-x^2 + x - 5$

$\quad\quad\quad\quad\quad\underline{3x^2 + 4x + 1}$

$\quad\quad\quad\quad -3x^4 + 3x^3 - 15x^2$

$\quad\quad\quad\quad\quad\quad -4x^3 + 4x^2 - 20x$

$\quad\quad\quad\quad\quad\quad\quad\quad -x^2 + x - 5$

$\quad\quad\quad\quad\overline{-3x^4 - x^3 - 12x^2 - 19x - 5}$

83. $[(m - 3) + n][(m - 3) - n] = (m - 3)^2 - n^2$

$\quad\quad\quad\quad\quad = m^2 - 6m + 9 - n^2$

$\quad\quad\quad\quad\quad = m^2 - n^2 - 6m + 9$

85. $[(x - 3) + y]^2 = (x - 3)^2 + 2y(x - 3) + y^2$

$\quad\quad\quad\quad = x^2 - 6x + 9 + 2xy - 6y + y^2$

$\quad\quad\quad\quad = x^2 + 2xy + y^2 - 6x - 6y + 9$

87. $(2r^2 - 5)(2r^2 + 5) = (2r^2)^2 - 5^2 = 4r^4 - 25$

89. $\left(\frac{1}{4}x - 5\right)^2 = \left(\frac{1}{4}x\right)^2 - 2\left(\frac{1}{4}x\right)(5) + (-5)^2$

$\quad\quad\quad\quad = \frac{1}{16}x^2 - \frac{5}{2}x + 25$

91. $\left(\frac{1}{5}x - 3\right)\left(\frac{1}{5}x + 3\right) = \left(\frac{1}{5}x\right)^2 - (3)^2 = \frac{1}{25}x^2 - 9$

93. $(2.4x + 3)^2 = (2.4x)^2 + 2(2.4x)(3) + (3)^2$

$\quad\quad\quad\quad = 5.76x^2 + 14.4x + 9$

95. $(1.5x - 4)(1.5x + 4) = (1.5x)^2 - 4^2$

$\quad\quad\quad\quad = 2.25x^2 - 16$

97. $5x(x + 1) - 3x(x + 1) = 2x(x + 1)$

$\quad\quad\quad\quad = 2x^2 + 2x$

99. $(u + 2)(u - 2)(u^2 + 4) = (u^2 - 4)(u^2 + 4)$

$\quad\quad\quad\quad = u^4 - 16$

101. $(\sqrt{x} + \sqrt{y})(\sqrt{x} - \sqrt{y}) = (\sqrt{x})^2 - (\sqrt{y})^2$

$\quad\quad\quad\quad = x - y$

103. $(x - \sqrt{5})^2 = x^2 - 2(x)(\sqrt{5}) + (\sqrt{5})^2$

$\quad\quad\quad\quad = x^2 - 2\sqrt{5}x + 5$

105. (a) Profit = Revenue − Cost

$\quad\quad$ Profit $= 95x - (73x + 25,000)$

$\quad\quad\quad\quad = 95x - 73x - 25,000$

$\quad\quad\quad\quad = 22x - 25,000$

$\quad$ (b) For $x = 5000$:

$\quad\quad$ Profit $= 22(5000) - 25,000$

$\quad\quad\quad\quad = 110,000 - 25,000$

$\quad\quad\quad\quad = \$85,000$

107. (a)
$$500(1 + r)^2 = 500(r + 1)^2$$
$$= 500(r^2 + 2r + 1)$$
$$= 500r^2 + 1000r + 500$$

(b)

r	$2\frac{1}{2}\%$	3%	4%	$4\frac{1}{2}\%$	5%
$500(1 + r)^2$	\$525.31	\$530.45	\$540.80	\$546.01	\$551.25

(c) As r increases, the amount increases.

109. (a)
$$V = l \cdot w \cdot h = (26 - 2x)(18 - 2x)(x)$$
$$= 2(13 - x)(2)(9 - x)(x)$$
$$= 4x(-1)(x - 13)(-1)(x - 9)$$
$$= 4x(x - 13)(x - 9)$$
$$= 4x^3 - 88x^2 + 468x$$

(b)

x (cm)	1	2	3
V (cm^3)	384	616	720

111. (a) Area of shaded region = Area of outer rectangle − Area of inner rectangle
$$A = 2x(2x + 6) - x(x + 4)$$
$$= 4x^2 + 12x - x^2 - 4x$$
$$= 3x^2 + 8x$$

(b) Area of shaded region = Area of outer triangle − Area of inner triangle
$$A = \frac{1}{2}(9x)(12x) - \frac{1}{2}(6x)(8x)$$
$$= 54x^2 - 24x^2$$
$$= 30x^2$$

(c) Area of shaded region = Area of outer triangle − Area of inner triangle
$$A = \frac{1}{2}(x + 2)(5x) - \frac{1}{2}(x + 1)(3x)$$
$$= \frac{5}{2}x^2 + 5x - \frac{3}{2}x^2 - \frac{3}{2}x$$
$$= x^2 + \frac{7}{2}x$$

OR

Area of shaded region = Area of trapezoid
$$A = \frac{1}{2}(2x)\left[(x + 2) + (x + 1)\right]$$
$$= x(2x + 3)$$
$$= 2x^2 + 3x$$

Note: $x = \frac{1}{2}$ is the only x-value yielding a triangle with the given dimensions.

(d) Area of shaded region = Area of rectangle − Area of triangle
$$A = (3x + 10)(x + 6) - \frac{1}{2}(3x + 10)(x + 6)$$
$$= \frac{1}{2}(3x + 10)(x + 6)$$
$$= \frac{1}{2}(3x^2 + 28x + 60)$$
$$= \frac{3}{2}x^2 + 14x + 30$$

113. Area $=$ length $\times$ width $= (2x + 14)(22) = (2x)(22) + (14)(22) = 44x + 308$

115. (a) Estimates will vary. Actual safe loads for $x = 12$:

$$S_6 = \left(0.06(12)^2 - 2.42(12) + 38.71\right)^2 = 335.2561 \quad \text{(using a calculator)}$$

$$S_8 = \left(0.08(12)^2 - 3.30(12) + 51.93\right)^2 = 568.8225 \quad \text{(using a calculator)}$$

Difference in safe loads $= 568.8225 - 335.2561 = 233.5664$ pounds

(b) The difference in safe loads decreases in magnitude as the span increases.

117. $(x + 1)(x + 4) = x^2 + 4x + x + 4$

This illustrates the Distributive Property.

119. False. $\left(4x^2 + 1\right)(3x + 1) = 12x^3 + 4x^2 + 3x + 1$

121. Because $x^m x^n = x^{m+n}$, the degree of the product is $m + n$.

123. $(x - 3)^2 \neq x^2 + 9$

The student did not remember the middle term when squaring the binomial. The correct method for squaring the binomial is:

$$(x - 3)^2 = (x)^2 - 2(x)(3) + (3)^2 = x^2 - 6x + 9$$

125. No; $\left(x^2 + 1\right) + \left(-x^2 + 3\right) = 4$, which is not a second-degree polynomial.

127. $(x + y)^2 \neq x^2 + y^2$

Let $x = 3$ and $y = 4$.

$$(3 + 4)^2 = (7)^2 = 49$$
$$3^2 + 4^2 = 9 + 16 = 25$$

$>$ Not Equal

If either x or y is zero, then $(x + y)^2$ would equal $x^2 + y^2$.

Section P.4   Factoring Polynomials

1. factoring

3. factoring by grouping

5. $80 = 2 \cdot 2 \cdot 2 \cdot 2 \cdot 5$

$280 = 2 \cdot 2 \cdot 2 \cdot 5 \cdot 7$

Greatest common factor: $2 \cdot 2 \cdot 2 \cdot 5 = 40$

7. $12x^2y^3 = 2 \cdot 2 \cdot 3 \cdot x \cdot x \cdot y \cdot y \cdot y$

$18x^2y = 2 \cdot 3 \cdot 3 \cdot x \cdot x \cdot y$

$24x^3y^2 = 2 \cdot 2 \cdot 2 \cdot 3 \cdot x \cdot x \cdot x \cdot y \cdot y$

Greatest common factor: $2 \cdot 3 \cdot x \cdot x \cdot y = 6x^2y$

9. $4x + 16 = 4(x + 4)$

11. $2x^3 - 6x = 2x\left(x^2 - 3\right)$

13. $3x(x - 5) + 8(x - 5) = (x - 5)(3x + 8)$

15. $(x + 3)^2 - 4(x + 3) = (x + 3)\left[(x + 3) - 4\right]$

$\qquad\qquad\qquad\qquad\quad = (x + 3)(x - 1)$

17. $\frac{1}{2}x + 4 = \frac{1}{2}x + \frac{8}{2} = \frac{1}{2}(x + 8)$

19. $\frac{1}{2}x^3 + 2x^2 - 5x = \frac{1}{2}x^3 + \frac{4}{2}x^2 - \frac{10}{2}x$

$\qquad\qquad\qquad\quad = \frac{1}{2}x\left(x^2 + 4x - 10\right)$

21. $\frac{2}{3}x(x - 3) - 4(x - 3) = \frac{2}{3}x(x - 3) - \frac{12}{3}(x - 3)$

$\qquad\qquad\qquad\qquad\qquad = \frac{2}{3}(x - 3)(x - 6)$

23. $x^2 - 81 = x^2 - 9^2 = (x + 9)(x - 9)$

25. $48y^2 - 27 = 3\left(16y^2 - 9\right)$

$\qquad\qquad\quad = 3\left((4y)^2 - 3^2\right)$

$\qquad\qquad\quad = 3(4y + 3)(4y - 3)$

27. $16x^2 - \frac{1}{9} = (4x)^2 - \left(\frac{1}{3}\right)^2 = \left(4x + \frac{1}{3}\right)\left(4x - \frac{1}{3}\right)$

29. $(x-1)^2 - 4 = (x-1)^2 - (2)^2$
$\qquad = [(x-1) + 2][(x-1) - 2]$
$\qquad = (x+1)(x-3)$

31. $9u^2 - 4v^2 = (3u)^2 - (2v)^2 = (3u + 2v)(3u - 2v)$

33. $x^2 - 4x + 4 = x^2 - 2(2)x + 2^2 = (x-2)^2$

35. $4t^2 + 4t + 1 = (2t)^2 + 2(2t)(1) + 1^2 = (2t+1)^2$

37. $25y^2 - 10y + 1 = (5y)^2 - 2(5y)(1) + 1^2 = (5y-1)^2$

39. $9u^2 + 24uv + 16v^2 = (3u)^2 + 2(3u)(4v) + (4v)^2$
$\qquad = (3u + 4v)^2$

41. $x^2 - \frac{4}{3}x + \frac{4}{9} = x^2 - 2(x)\left(\frac{2}{3}\right) + \left(\frac{2}{3}\right)^2 = \left(x - \frac{2}{3}\right)^2$

43. $4x^2 - \frac{4}{3}x + \frac{1}{9} = (2x)^2 - 2(2x)\left(\frac{1}{3}\right) + \left(\frac{1}{3}\right)^2 = \left(2x - \frac{1}{3}\right)^2$

45. $x^3 - 8 = x^3 - 2^3 = (x-2)(x^2 + 2x + 4)$

47. $y^3 + 64 = y^3 + 4^3 = (y+4)(y^2 - 4y + 16)$

49. $x^3 - \frac{8}{27} = (x)^3 - \left(\frac{2}{3}\right)^3 = \left(x - \frac{2}{3}\right)\left(x^2 + \frac{2}{3}x + \frac{4}{9}\right)$

51. $8t^3 - 1 = (2t)^3 - 1^3 = (2t - 1)(4t^2 + 2t + 1)$

53. $u^3 + 27v^3 = u^3 + (3v)^3 = (u + 3v)(u^2 - 3uv + 9v^2)$

55. $(x+2)^3 - y^3 = (x+2)^3 - (y)^3$
$\qquad = (x + 2 - y)\left[(x+2)^2 + y(x+2) + y^2\right]$
$\qquad = (x + 2 - y)(x^2 + 4x + 4 + xy + 2y + y^2)$

57. $x^2 + x - 2 = (x+2)(x-1)$

59. $s^2 - 5s + 6 = (s-3)(s-2)$

61. $20 - y - y^2 = -(y^2 + y - 20) = -(y+5)(y-4)$

63. $x^2 - 30x + 200 = (x-20)(x-10)$

65. $3x^2 - 5x + 2 = (3x-2)(x-1)$

67. $5x^2 + 26x + 5 = (5x+1)(x+5)$

69. $-9z^2 + 3z + 2 = -(9z^2 - 3z - 2)$
$\qquad = -(3z-2)(3z+1)$

71. $x^3 - x^2 + 2x - 2 = x^2(x-1) + 2(x-1)$
$\qquad = (x-1)(x^2 + 2)$

73. $2x^3 - x^2 - 6x + 3 = x^2(2x-1) - 3(2x-1)$
$\qquad = (2x-1)(x^2 - 3)$

75. $6 + 2x - 3x^3 - x^4 = 2(3+x) - x^3(3+x)$
$\qquad = (3+x)(2 - x^3)$

77. $6x^3 - 2x + 3x^2 - 1 = 2x(3x^2 - 1) + 1(3x^2 - 1)$
$\qquad = (3x^2 - 1)(2x + 1)$

79. $a \cdot c = (3)(8) = 24$. Rewrite the middle term, $10x = 6x + 4x$, because $(6)(4) = 24$ and $6 + 4 = 10$.
$3x^2 + 10x + 8 = 3x^2 + 6x + 4x + 8$
$\qquad = 3x(x+2) + 4(x+2)$
$\qquad = (x+2)(3x+4)$

81. $a \cdot c = (6)(-2) = -12$. Rewrite the middle term, $x = 4x - 3x$, because $4(-3) = -12$ and $4 + (-3) = 1$.
$6x^2 + x - 2 = 6x^2 + 4x - 3x - 2$
$\qquad = 2x(3x+2) - 1(3x+2)$
$\qquad = (2x-1)(3x+2)$

83. $a \cdot c = (15)(2) = 30$. Rewrite the middle term, $-11x = -6x - 5x$, because $(-6)(-5) = 30$ and $(-6) + (-5) = -11$.
$15x^2 - 11x + 2 = 15x^2 - 6x - 5x + 2$
$\qquad = 3x(5x-2) - 1(5x-2)$
$\qquad = (3x-1)(5x-2)$

85. $6x^2 - 54 = 6(x^2 - 9) = 6(x+3)(x-3)$

87. $x^3 - x^2 = x^2(x-1)$

89. $x^3 - 16x = x(x^2 - 16) = x(x + 4)(x - 4)$

91. $x^2 - 2x + 1 = (x - 1)^2$

93. $1 - 4x + 4x^2 = (1 - 2x)^2$

95. $2x^2 + 4x - 2x^3 = -2x(-x - 2 + x^2)$
$$= -2x(x^2 - x - 2)$$
$$= -2x(x + 1)(x - 2)$$

97. $\frac{1}{81}x^2 + \frac{2}{9}x - 8 = \frac{1}{81}x^2 + \frac{18}{81}x - \frac{648}{81}$
$$= \frac{1}{81}(x^2 + 18x - 648)$$
$$= \frac{1}{81}(x + 36)(x - 18)$$

99. $3x^3 + x^2 + 15x + 5 = x^2(3x + 1) + 5(3x + 1)$
$$= (3x + 1)(x^2 + 5)$$

101. $x^4 - 4x^3 + x^2 - 4x = x(x^3 - 4x^2 + x - 4)$
$$= x\left[x^2(x - 4) + (x - 4)\right]$$
$$= x(x - 4)(x^2 + 1)$$

103. $2x^3 + x^2 - 8x - 4 = x^2(2x + 1) - 4(2x + 1)$
$$= (2x + 1)(x^2 - 4)$$
$$= (2x + 1)(x + 2)(x - 2)$$

105. $\frac{1}{4}x^3 + 3x^2 + \frac{3}{4}x + 9 = \frac{1}{4}x^3 + \frac{12}{4}x^2 + \frac{3}{4}x + \frac{36}{4}$
$$= \frac{1}{4}(x^3 + 12x^2 + 3x + 36)$$
$$= \frac{1}{4}\left[x^2(x + 12) + 3(x + 12)\right]$$
$$= \frac{1}{4}(x + 12)(x^2 + 3)$$

107. $(t - 1)^2 - 49 = (t - 1)^2 - (7)^2$
$$= \left[(t - 1) + 7\right]\left[(t - 1) - 7\right]$$
$$= (t + 6)(t - 8)$$

109. $(x^2 + 8)^2 - 36x^2 = (x^2 + 8)^2 - (6x)^2$
$$= \left[(x^2 + 8) - 6x\right]\left[(x^2 + 8) + 6x\right]$$
$$= (x^2 - 6x + 8)(x^2 + 6x + 8)$$
$$= (x - 4)(x - 2)(x + 4)(x + 2)$$

111. $5x^3 + 40 = 5(x^3 + 8)$
$$= 5(x^3 + 2^3)$$
$$= 5(x + 2)(x^2 - 2x + 4)$$

113. $5(3 - 4x)^2 - 8(3 - 4x)(5x - 1) = (3 - 4x)\left[5(3 - 4x) - 8(5x - 1)\right]$
$$= (3 - 4x)[15 - 20x - 40x + 8]$$
$$= (3 - 4x)(23 - 60x)$$

115. $7(3x + 2)^2(1 - x)^2 + (3x + 2)(1 - x)^3 = (3x + 2)(1 - x)^2\left[7(3x + 2) + (1 - x)\right]$
$$= (3x + 2)(1 - x)^2(21x + 14 + 1 - x)$$
$$= (3x + 2)(1 - x)^2(20x + 15)$$
$$= 5(3x + 2)(1 - x)^2(4x + 3)$$

117. $3(x - 2)^2(x + 1)^4 + (x - 2)^3(4)(x + 1)^3 = (x - 2)^2(x + 1)^3\left[3(x + 1) + 4(x - 2)\right]$
$$= (x - 2)^2(x + 1)^3(3x + 3 + 4x - 8)$$
$$= (x - 2)^2(x + 1)^3(7x - 5)$$

119. $5(x^6 + 1)^4(6x^5)(3x + 2)^3 + 3(3x + 2)^2(3)(x^6 + 1)^5 = 3(x^6 + 1)^4(3x + 2)^2\left[10x^5(3x + 2) + 3(x^6 + 1)\right]$

$$= 3(x^6 + 1)^4(3x + 2)^2(30x^6 + 20x^5 + 3x^6 + 3)$$

$$= 3(x^6 + 1)^4(3x + 2)^2(33x^6 + 20x^5 + 3)$$

$$= 3\left[(x^2)^3 + 1\right]^4(3x + 2)^2(33x^6 + 20x^5 + 3)$$

$$= 3\left[(x^2 + 1)(x^4 - x^2 + 1)\right]^4(3x + 2)^2(33x^6 + 20x^5 + 3)$$

$$= 3(x^2 + 1)^4(x^4 - x^2 + 1)^4(3x + 2)^2(33x^6 + 20x^5 + 3)$$

121. $a^2 - b^2 = (a + b)(a - b)$

Matches model (b).

123. $a^2 + 2a + 1 = (a + 1)^2$

Matches model (a).

125. $3x^2 + 7x + 2 = (3x + 1)(x + 2)$

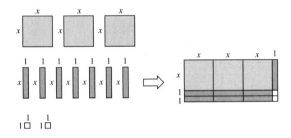

127. $2x^2 + 7x + 3 = (2x + 1)(x + 3)$

129. $A = \pi(r + 2)^2 - \pi r^2$

$$= \pi\left[(r + 2)^2 - r^2\right]$$

$$= \pi\left[r^2 + 4r + 4 - r^2\right]$$

$$= \pi(4r + 4)$$

$$= 4\pi(r + 1)$$

131. $A = 8(18) = 4x^2 = 4(36 - x^2) = 4(6 - x)(6 + x)$

133. $x^4(4)(2x + 1)^3(2x) + (2x + 1)^4(4x^3) = 2x^3(2x + 1)^3\left[4x^2 + 2(2x + 1)\right]$

$$= 2x^3(2x + 1)^3(4x^2 + 4x + 2)$$

$$= 4x^3(2x + 1)^3(2x^2 + 2x + 1)$$

135. $(2x - 5)^4(3)(5x - 4)^2(5) + (5x - 4)^3(4)(2x - 5)^3(2)$

$$= (2x - 5)^3(5x - 4)^2\left[15(2x - 5) + 8(5x - 4)\right]$$

$$= (2x - 5)^3(5x - 4)^2(30x - 75 + 40x - 32)$$

$$= (2x - 5)^3(5x - 4)^2(70x - 107)$$

137. $\dfrac{(5x - 1)3 - (3x + 1)(5)}{(5x - 1)^2} = \dfrac{-8}{(5x - 1)^2}$

139. For $x^2 + bx - 15$ to be factorable, b must equal $m + n$ where $mn = -15$.

Factors of −15	Sum of factors
$(15)(-1)$	$15 + (-1) = 14$
$(-15)(1)$	$-15 + 1 = -14$
$(3)(-5)$	$3 + (-5) = -2$
$(-3)(5)$	$-3 + 5 = 2$

The possible b-values are $14, -14, -2,$ or 2.

141. For $x^2 + bx + 50$ to be factorable, b must equal $m + n$ where $mn = 50$.

Factors of 50	Sum of factors
$(1)(50)$	$1 + 50 = 51$
$(-1)(-50)$	$-1 + (-50) = -51$
$(5)(10)$	$5 + 10 = 15$
$(-5)(-10)$	$-5 + (-10) = -15$
$(2)(25)$	$2 + 25 = 27$
$(-2)(-25)$	$-2 + (-25) = -27$

The possible b-values are $-51, 51, -15, 15, -27, 27$.

143. For $2x^2 + 5x + c$ to be factorable, the factors of $2c$ must add up to 5.

Possible c-values	$2c$	Factors of $2c$ that add up to 5
2	4	$(1)(4) = 4$ and $1 + 4 = 5$
3	6	$(2)(3) = 6$ and $2 + 3 = 5$
-3	-6	$(6)(-1) = -6$ and $6 + (-1) = 5$
-7	-14	$(7)(-2) = -14$ and $7 + (-2) = 5$
-12	-24	$(8)(-3) = -24$ and $8 + (-3) = 5$

These are a few possible c-values. There are *many* correct answers.

If $c = 2$: $2x^2 + 5x + 2 = (2x + 1)(x + 2)$

If $c = 3$: $2x^2 + 5x + 3 = (2x + 3)(x + 1)$

If $c = -3$: $2x^2 + 5x - 3 = (2x - 1)(x + 3)$

If $c = -7$: $2x^2 + 5x - 7 = (2x + 7)(x - 1)$

If $c = -12$: $2x^2 + 5x - 12 = (2x - 3)(x + 4)$

145. For $3x^2 - x + c$ to be factorable, the factors of $3c$ must add up to -1.

Possible c-values	$3c$	Factors of $3c$ must add up to -1
-2	-6	$(2)(-3) = -6$ and $2 + (-3) = -1$
-4	-12	$(3)(-4) = -12$ and $3 + (-4) = -1$
-10	-30	$(5)(-6) = -30$ and $5 + (-6) = -1$

These are a few possible c-values. There are *many* correct answers.

If $c = -2$: $3x^2 - x - 2 = (3x + 2)(x - 1)$

If $c = -4$: $3x^2 - x - 4 = (3x - 4)(x + 1)$

If $c = -10$: $3x^2 - x - 10 = (3x + 5)(x - 2)$

147. (a) $V = \pi R^2 h - \pi r^2 h$

$= \pi h\left(R^2 - r^2\right)$

$= \pi h(R - r)(R + r)$

(b) The average radius is $\dfrac{R + r}{2}$.

The thickness of the tank is $R - r$.

$V = \pi h(R - r)(R + r) = 2\pi\left(\dfrac{R + r}{2}\right)(R - r)h$

$= 2\pi(\text{average radius})(\text{thickness})h$

(c) $V = \pi h\left(4 - 3\dfrac{2}{3}\right)\left(4 + 3\dfrac{2}{3}\right) = \pi h\left(\dfrac{1}{3}\right)\left(7\dfrac{2}{3}\right) = \dfrac{25\pi h}{9}$ ft^3

Number of bags: $\dfrac{V}{3/5} = \dfrac{5}{3}V = \dfrac{5}{3}\left(\dfrac{23\pi h}{9}\right) = \dfrac{115\pi}{27}$

$\approx 13.38h$ bags

(d)

h (feet)	Number of bags
$\frac{1}{2}$	6.69
1	13.38
$\frac{3}{2}$	20.07
2	26.76
$\frac{5}{2}$	33.45
3	40.14
$\frac{7}{2}$	46.83
4	53.52
$\frac{9}{2}$	60.21
5	66.90
$\frac{11}{2}$	73.59
6	80.29

149. True. $a^2 - b^2 = (a + b)(a - b)$

151. $9x^2 - 9x - 54 = 9\left(x^2 - x - 6\right) = 9(x + 2)(x - 3)$

The error on the problem in the book was that 3 factored out of the first binomial but not out of the second binomial.

$(3x + 6)(3x - 9) = 3(x + 2)(3)(x - 3) = 9(x + 2)(x - 2)$

153. $x^{2n} - y^{2n} = \left(x^n\right)^2 - \left(y^n\right)^2 = \left(x^n + y^n\right)\left(x^n - y^n\right)$

This is not completely factored unless $n = 1$.

For $n = 2$: $\left(x^2 + y^2\right)\left(x^2 - y^2\right) = \left(x^2 + y^2\right)(x + y)(x - y)$

For $n = 3$: $\left(x^3 + y^3\right)\left(x^3 - y^3\right) = (x + y)\left(x^2 - xy + y^2\right)(x - y)\left(x^2 + xy + y^2\right)$

For $n = 4$: $\left(x^4 + y^4\right)\left(x^4 - y^4\right) = \left(x^4 + y^4\right)\left(x^2 + y^2\right)(x + y)(x - y)$

155. Answers will vary. Sample answer: $x^2 - 3$

157. $u^6 - v^6 = \left(u^3\right)^2 - \left(v^3\right)^2$

$= \left(u^3 + v^3\right)\left(u^3 - v^3\right)$

$= \left[(u + v)\left(u^2 - uv + v^2\right)\right]\left[(u - v)\left(u^2 + uv + v^2\right)\right]$

$= (u + v)(u - v)\left(u^2 + uv + v^2\right)\left(u^2 - uv + v^2\right)$

$x^6 - 1 = (x + 1)(x - 1)\left(x^2 + x + 1\right)\left(x^2 - x + 1\right)$

$x^6 - 64 = x^6 - 2^6 = (x + 2)(x - 2)\left(x^2 + 2x + 4\right)\left(x^2 - 2x + 4\right)$

Section P.5   Rational Expressions

- You should be able to find the domain of a rational expression.

- You should know that a rational expression is the quotient of two polynomials.

- You should be able to simplify rational expressions by reducing them to lowest terms. This may involve factoring both the numerator and the denominator.

- You should be able to add, subtract, multiply, and divide rational expressions.

- You should be able to simplify complex fractions.

- You should be able to simplify expressions with negative or fraction exponents.

1. domain

3. complex

5. equivalent

7. The domain of the polynomial $3x^2 - 4x + 7$ is the set of all real numbers.

9. The domain of the polynomial $4x^3 + 3$, $x \geq 0$ is the set of nonnegative real numbers because the polynomial is restricted to that set.

11. The domain of $\dfrac{1}{3 - x}$ is the set of all real numbers x such that $x \neq 3$.

13. The domain of $\dfrac{x^2 - 1}{x^2 - 2x + 1} = \dfrac{(x + 1)(x - 1)}{(x - 1)(x - 1)}$ is the set of all real numbers x such that $x \neq 1$.

15. The domain of $\dfrac{x^2 - 2x - 3}{x^2 - 6x + 9} = \dfrac{(x - 3)(x + 1)}{(x - 3)(x - 3)}$ is the set of all real numbers x such that $x \neq 3$.

17. The domain of $\sqrt{x + 7}$ is the set of all real numbers x such that $x \geq -7$.

19. The domain of $\sqrt{2x - 5}$ is the set of all real numbers x such that $x \geq \frac{5}{2}$.

21. The domain of $\dfrac{1}{\sqrt{x - 3}}$ is the set of all real numbers x such that $x > 3$.

23. $\dfrac{5}{2x} = \dfrac{5(3x)}{(2x)(3x)} = \dfrac{5(3x)}{6x^2}$, $x \neq 0$

The missing factor is $3x$, $x \neq 0$.

25. $\dfrac{15x^2}{10x} = \dfrac{5x(3x)}{5x(2)} = \dfrac{3x}{2}$, $x \neq 0$

27. $\dfrac{3xy}{xy + x} = \dfrac{x(3y)}{x(y + 1)} = \dfrac{3y}{y + 1}$, $x \neq 0$

29. $\dfrac{4y - 8y^2}{10y - 5} = \dfrac{-4y(2y - 1)}{5(2y - 1)}$
$= -\dfrac{4y}{5}$, $y \neq \dfrac{1}{2}$

31. $\dfrac{x - 5}{10 - 2x} = \dfrac{x - 5}{-2(x - 5)}$
$= -\dfrac{1}{2}$, $x \neq 5$

33. $\dfrac{y^2 - 16}{y + 4} = \dfrac{(y + 4)(y - 4)}{y + 4}$
$= y - 4$, $y \neq -4$

35. $\dfrac{x^3 + 5x^2 + 6x}{x^2 - 4} = \dfrac{x(x + 2)(x + 3)}{(x + 2)(x - 2)} = \dfrac{x(x + 3)}{x - 2}$, $x \neq -2$

37. $\dfrac{y^2 - 7y + 12}{y^2 + 3y - 18} = \dfrac{(y - 3)(y - 4)}{(y + 6)(y - 3)} = \dfrac{y - 4}{y + 6}$, $y \neq 3$

39. $\dfrac{2 - x + 2x^2 - x^3}{x^2 - 4} = \dfrac{(2 - x) + x^2(2 - x)}{(x + 2)(x - 2)}$
$= \dfrac{(2 - x)(1 + x^2)}{(x + 2)(x - 2)}$
$= \dfrac{-(x - 2)(x^2 + 1)}{(x + 2)(x - 2)}$
$= -\dfrac{x^2 + 1}{x + 2}$, $x \neq 2$

41. $\dfrac{z^3 - 8}{z^2 + 2z + 4} = \dfrac{(z - 2)(z^2 + 2z + 4)}{z^2 + 2z + 4} = z - 2$

43. $\dfrac{5x^3}{2x^3 + 4} = \dfrac{5x^3}{2(x^3 + 2)}$

There are no common factors so this expression cannot be simplified. In this case, factors of terms were incorrectly cancelled.

45.

x	0	1	2	3	4	5	6
$\dfrac{x^2 - 2x - 3}{x - 3}$	1	2	3	Undef.	5	6	7
$x + 1$	1	2	3	4	5	6	7

The expressions are equivalent except at $x = 3$.

47. $\dfrac{\pi r^2}{(2r)^2} = \dfrac{\pi r^2}{4r^2} = \dfrac{\pi}{4}, r \neq 0$

49. $\dfrac{5}{x - 1} \cdot \dfrac{x - 1}{25(x - 2)} = \dfrac{1}{5(x - 2)}, x \neq 1$

51. $\dfrac{r}{r - 1} \div \dfrac{r^2}{r^2 - 1} = \dfrac{r}{r - 1} \cdot \dfrac{r^2 - 1}{r^2}$

$= \dfrac{r(r + 1)(r - 1)}{r^2(r - 1)}$

$= \dfrac{r + 1}{r}, r \neq 1, r \neq 0$

53. $\dfrac{t^2 - t - 6}{t^2 + 6t + 9} \cdot \dfrac{t + 3}{t^2 - 4} = \dfrac{(t - 3)(t + 2)(t + 3)}{(t + 3)^2(t + 2)(t - 2)}$

$= \dfrac{t - 3}{(t + 3)(t - 2)}, t \neq -2$

55. $\dfrac{x^2 - 36}{x} \div \dfrac{x^3 - 6x^2}{x^2 + x} = \dfrac{x^2 - 36}{x} \cdot \dfrac{x^2 + x}{x^3 - 6x^2}$

$= \dfrac{(x + 6)(x - 6)}{x} \cdot \dfrac{x(x + 1)}{x^2(x - 6)}$

$= \dfrac{(x + 6)(x + 1)}{x^2}, x \neq 6, -1$

57. $6 - \dfrac{5}{x + 3} = \dfrac{6(x + 3)}{(x + 3)} - \dfrac{5}{x + 3}$

$= \dfrac{6(x + 3) - 5}{x + 3}$

$= \dfrac{6x + 18 - 5}{x + 3}$

$= \dfrac{6x + 13}{x + 3}$

59. $\dfrac{5}{x - 1} + \dfrac{x}{x - 1} = \dfrac{5 + x}{x - 1} = \dfrac{x + 5}{x - 1}$

61. $\dfrac{3}{x - 2} + \dfrac{5}{2 - x} = \dfrac{3}{x - 2} - \dfrac{5}{x - 2} = -\dfrac{2}{x - 2}$

63. $\dfrac{4}{2x + 1} - \dfrac{x}{x + 2} = \dfrac{4(x + 2)}{(2x + 1)(x + 2)} - \dfrac{x(2x + 1)}{(x + 2)(2x + 1)}$

$= \dfrac{4x + 8 - 2x^2 - x}{(x + 2)(2x + 1)}$

$= \dfrac{-2x^2 + 3x + 8}{(x + 2)(2x + 1)}$

65. $\dfrac{1}{x^2 - x - 2} - \dfrac{x}{x^2 - 5x + 6} = \dfrac{1}{(x - 2)(x + 1)} - \dfrac{x}{(x - 2)(x - 3)}$

$= \dfrac{(x - 3) - x(x + 1)}{(x + 1)(x - 2)(x - 3)} = \dfrac{x - 3 - x^2 - x}{(x + 1)(x - 2)(x - 3)}$

$= \dfrac{-x^2 - 3}{(x + 1)(x - 2)(x - 3)} = -\dfrac{x^2 + 3}{(x + 1)(x - 2)(x - 3)}$

67. $-\dfrac{1}{x} + \dfrac{2}{x^2 + 1} + \dfrac{1}{x^3 + x} = \dfrac{-(x^2 + 1)}{x(x^2 + 1)} + \dfrac{2x}{x(x^2 + 1)} + \dfrac{1}{x(x^2 + 1)}$

$= \dfrac{-x^2 - 1 + 2x + 1}{x(x^2 + 1)} = \dfrac{-x^2 + 2x}{x(x^2 + 1)} = \dfrac{-x(x - 2)}{x(x^2 + 1)}$

$= -\dfrac{x - 2}{x^2 + 1} = \dfrac{2 - x}{x^2 + 1}, x \neq 0$

69. $\dfrac{x + 4}{x + 2} - \dfrac{3x - 8}{x + 2} = \dfrac{(x + 4) - (3x - 8)}{x + 2}$

$= \dfrac{x + 4 - 3x + 8}{x + 2} = \dfrac{-2x + 12}{x + 2} = \dfrac{-2(x - 6)}{x + 2}$

The error was incorrect subtraction in the numerator.

71. $\dfrac{\left(\dfrac{x}{2}-1\right)}{(x-2)}=\dfrac{\left(\dfrac{x}{2}-\dfrac{2}{2}\right)}{\left(\dfrac{x-2}{1}\right)}=\dfrac{x-2}{2}\cdot\dfrac{1}{x-2}=\dfrac{1}{2},\,x\neq 2$

73. $\dfrac{\left[\dfrac{x^2}{(x+1)^2}\right]}{\left[\dfrac{x}{(x+1)^3}\right]}=\dfrac{x^2}{(x+1)^2}\cdot\dfrac{(x+1)^3}{x}$

$=x(x+1),\,x\neq -1,0$

75. $\dfrac{\left(\sqrt{x}-\dfrac{1}{2\sqrt{x}}\right)}{\sqrt{x}}=\dfrac{\left(\sqrt{x}-\dfrac{1}{2\sqrt{x}}\right)}{\sqrt{x}}\cdot\dfrac{2\sqrt{x}}{2\sqrt{x}}$

$=\dfrac{2x-1}{2x},\,x>0$

77. $x^5-2x^{-2}=x^{-2}\left(x^7-2\right)=\dfrac{x^7-2}{x^2}$

79. $x^2\left(x^2+1\right)^{-5}-\left(x^2+1\right)^{-4}=\left(x^2+1\right)^{-5}\left[x^2-\left(x^2+1\right)\right]$

$=-\dfrac{1}{\left(x^2+1\right)^5}$

81. $2x^2(x-1)^{1/2}-5(x-1)^{-1/2}=(x-1)^{-1/2}\left[2x^2(x-1)^1-5\right]=\dfrac{2x^3-2x^2-5}{(x-1)^{1/2}}$

83. $\dfrac{3x^{1/3}-x^{-2/3}}{3x^{-2/3}}=\dfrac{3x^{1/3}-x^{-2/3}}{3x^{-2/3}}\cdot\dfrac{x^{2/3}}{x^{2/3}}=\dfrac{3x^1-x^0}{3x^0}=\dfrac{3x-1}{3},\,x\neq 0$

85. $\dfrac{\left(\dfrac{1}{x+h}-\dfrac{1}{x}\right)}{h}=\dfrac{\left(\dfrac{1}{x+h}-\dfrac{1}{x}\right)}{h}\cdot\dfrac{x(x+h)}{x(x+h)}$

$=\dfrac{x-(x+h)}{hx(x+h)}$

$=\dfrac{-h}{hx(x+h)}$

$=-\dfrac{1}{x(x+h)},\,h\neq 0$

87. $\dfrac{\left(\dfrac{1}{x+h-4}-\dfrac{1}{x-4}\right)}{h}=\dfrac{\left(\dfrac{1}{x+h-4}-\dfrac{1}{x-4}\right)}{h}\cdot\dfrac{(x-4)(x+h-4)}{(x-4)(x+h-4)}$

$=\dfrac{(x-4)-(x+h-4)}{h(x-4)(x+h-4)}$

$=\dfrac{-h}{h(x-4)(x+h-4)}$

$=-\dfrac{1}{(x-4)(x+h-4)},\,h\neq 0$

89. $\dfrac{\sqrt{x+2}-\sqrt{x}}{2}=\dfrac{\sqrt{x+2}-\sqrt{x}}{2}\cdot\dfrac{\sqrt{x+2}+\sqrt{x}}{\sqrt{x+2}+\sqrt{x}}$

$=\dfrac{(x+2)-x}{2\left(\sqrt{x+2}+\sqrt{x}\right)}$

$=\dfrac{2}{2\left(\sqrt{x+2}+\sqrt{x}\right)}$

$=\dfrac{1}{\sqrt{x+2}+\sqrt{x}}$

91. $\dfrac{\sqrt{t+3}-\sqrt{3}}{t}=\dfrac{\sqrt{t+3}-\sqrt{3}}{t}\cdot\dfrac{\sqrt{t+3}+\sqrt{3}}{\sqrt{t+3}+\sqrt{3}}$

$=\dfrac{(t+3)-3}{t\left(\sqrt{t+3}+\sqrt{3}\right)}$

$=\dfrac{t}{t\left(\sqrt{t+3}+\sqrt{3}\right)}$

$=\dfrac{1}{\sqrt{t+3}+\sqrt{3}}$

93. $\dfrac{\sqrt{x+h+1}-\sqrt{x+1}}{h} = \dfrac{\sqrt{x+h+1}-\sqrt{x+1}}{h} \cdot \dfrac{\sqrt{x+h+1}+\sqrt{x+1}}{\sqrt{x+h+1}+\sqrt{x+1}}$

$$= \dfrac{(x+h+1)-(x+1)}{h\left(\sqrt{x+h+1}+\sqrt{x+1}\right)}$$

$$= \dfrac{h}{h\left(\sqrt{x+h+1}+\sqrt{x+1}\right)}$$

$$= \dfrac{1}{\sqrt{x+h+1}+\sqrt{x+1}}, \; h \neq 0$$

95. Probability $= \dfrac{\text{Shaded area}}{\text{Total area}}$

$$= \dfrac{x(x/2)}{x(2x+1)}$$

$$= \dfrac{x/2}{2x+1} \cdot \dfrac{2}{2}$$

$$= \dfrac{x}{2(2x+1)}$$

97. (a) $\dfrac{1}{50}$ minute

(b) $x\left(\dfrac{1}{50}\right) = \dfrac{x}{50}$ minutes

(c) $120\left(\dfrac{1}{50}\right) = \dfrac{12}{5} = 2.4$ minutes

99. (a) $r = \dfrac{\left[\dfrac{24(48 \cdot 475 - 20{,}000)}{48}\right]}{\left(20{,}000 + \dfrac{48 \cdot 475}{12}\right)} \approx 0.0639$ or 6.39%

(b) $r = \dfrac{\left[\dfrac{24(NM-P)}{N}\right]}{\left(P + \dfrac{NM}{12}\right)}$

$$= \dfrac{\left[\dfrac{24(NM-P)}{N}\right]}{\left(\dfrac{12P+NM}{12}\right)}$$

$$= \dfrac{288(NM-P)}{N(12P+NM)}$$

Using the simplified formula the annual interest rate is still 6.39%.

101. $T = 10\left(\dfrac{4t^2 + 16t + 75}{t^2 + 4t + 10}\right)$

(a)

t	0	2	4	6	8	10	12	14	16	18	20	22
T	75°	55.9°	48.3°	45°	43.3°	42.3°	41.7°	41.3°	41.1°	40.9°	40.7°	40.6°

(b) T is approaching 40°.

103. False. In order for the simplified expression to be equivalent to the original expression, the domain of the simplified expression needs to be restricted. If n is even, $x \neq \pm 1$. If n is odd, $x \neq 1$.

105. Completely factor the numerator and the denominator. A rational expression is in simplest form if there are no common factors in the numerator and the denominator other than ± 1.

Section P.6 The Rectangular Coordinate System and Graphs

- You should be able to use the point-plotting method of graphing.

- You should be able to find x- and y-intercepts.
  (a) To find the x-intercepts, let $y = 0$ and solve for x.
  (b) To find the y-intercepts, let $x = 0$ and solve for y.

—*continued*—

■ You should be able to test for symmetry.

  (a) To test for x-axis symmetry, replace y with $-y$.

  (b) To test for y-axis symmetry, replace x with $-x$.

  (c) To test for origin symmetry, replace x with $-x$ and y with $-y$.

■ You should know the standard equation of a circle with center (h, k) and radius r:

$$(x - h)^2 + (y - k)^2 = r^2$$

---

1. (a) v   horizontal real number line

   (b) vi vertical real number line

   (c) i   point of intersection of vertical axis and horizontal axis

   (d) iv four regions of the coordinate plane

   (e) iii directed distance from the y-axis

   (f) ii directed distance from the x-axis

3. Distance Formula

5. $A: (2, 6)$, $B: (-6, -2)$, $C: (4, -4)$, $D: (-3, 2)$

7.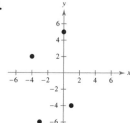

9.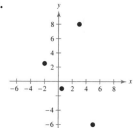

11. $(-3, 4)$

13. $(-5, -5)$

15. $x > 0$ and $y < 0$ in Quadrant IV.

17. $x = -4$ and $y > 0$ in Quadrant II.

19. $y < -5$ in Quadrant III or IV.

21. $(x, -y)$ is in the second Quadrant means that (x, y) is in Quadrant III.

23. (x, y), $xy > 0$ means x and y have the same signs. This occurs in Quadrant I or III.

25.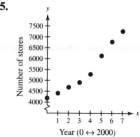

27. $d = \left|5 - (-3)\right| = 8$

29. $d = \left|2 - (-3)\right| = 5$

31. $d = \sqrt{(x_2 - x_1)^2 + (y_2 - y_1)^2}$

    $= \sqrt{(3 - (-2))^2 + (-6 - 6)^2}$

    $= \sqrt{(5)^2 + (-12)^2}$

    $= \sqrt{25 + 144}$

    $= 13$ units

33. $d = \sqrt{(x_2 - x_1)^2 + (y_2 - y_1)^2}$

    $= \sqrt{(-5 - 1)^2 + (-1 - 4)^2}$

    $= \sqrt{(-6)^2 + (-5)^2}$

    $= \sqrt{36 + 25}$

    $= \sqrt{61}$ units

35. $d = \sqrt{(x_2 - x_1)^2 + (y_2 - y_1)^2}$

$= \sqrt{\left(2 - \dfrac{1}{2}\right)^2 + \left(-1 - \dfrac{4}{3}\right)^2}$

$= \sqrt{\left(\dfrac{3}{2}\right)^2 + \left(-\dfrac{7}{3}\right)^2}$

$= \sqrt{\dfrac{9}{4} + \dfrac{49}{9}}$

$= \sqrt{\dfrac{277}{36}}$

$= \dfrac{\sqrt{277}}{6}$ units

37. $d = \sqrt{(x_2 - x_1)^2 + (y_2 - y_1)^2}$

$= \sqrt{(-12.5 - (-4.2))^2 + (4.8 - 3.1)^2}$

$= \sqrt{(-8.3)^2 + (1.7)^2}$

$= \sqrt{68.89 + 2.89}$

$= \sqrt{71.78}$

≈ 8.47 units

39. (a) The distance between $(0, 2)$ and $(4, 2)$ is 4.

The distance between $(4, 2)$ and $(4, 5)$ is 3.

The distance between $(0, 2)$ and $(4, 5)$ is

$\sqrt{(4 - 0)^2 + (5 - 2)^2} = \sqrt{16 + 9} = \sqrt{25} = 5.$

(b) $4^2 + 3^2 = 16 + 9 = 25 = 5^2$

41. (a) The distance between $(-1, 1)$ and $(9, 1)$ is 10.

The distance between $(9, 1)$ and $(9, 4)$ is 3.

The distance between $(-1, 1)$ and $(9, 4)$ is

$\sqrt{(9 - (-1))^2 + (4 - 1)^2} = \sqrt{100 + 9} = \sqrt{109}.$

(b) $10^2 + 3^2 = 109 = \left(\sqrt{109}\right)^2$

43. $d_1 = \sqrt{(4 - 2)^2 + (0 - 1)^2} = \sqrt{4 + 1} = \sqrt{5}$

$d_2 = \sqrt{(4 + 1)^2 + (0 + 5)^2} = \sqrt{25 + 25} = \sqrt{50}$

$d_3 = \sqrt{(2 + 1)^2 + (1 + 5)^2} = \sqrt{9 + 36} = \sqrt{45}$

$\left(\sqrt{5}\right)^2 + \left(\sqrt{45}\right)^2 = \left(\sqrt{50}\right)^2$

45. $d_1 = \sqrt{(1 - 3)^2 + (-3 - 2)^2} = \sqrt{4 + 25} = \sqrt{29}$

$d_2 = \sqrt{(3 + 2)^2 + (2 - 4)^2} = \sqrt{25 + 4} = \sqrt{29}$

$d_3 = \sqrt{(1 + 2)^2 + (-3 - 4)^2} = \sqrt{9 + 49} = \sqrt{58}$

$d_1 = d_2$

47. (a)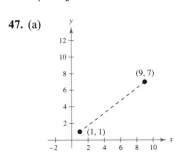

(b) $d = \sqrt{(9 - 1)^2 + (7 - 1)^2}$

$= \sqrt{64 + 36} = 10$

(c) $\left(\dfrac{9 + 1}{2}, \dfrac{7 + 1}{2}\right) = (5, 4)$

49. (a)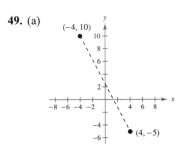

(b) $d = \sqrt{(4 + 4)^2 + (-5 - 10)^2}$

$= \sqrt{64 + 225} = 17$

(c) $\left(\dfrac{4 - 4}{2}, \dfrac{-5 + 10}{2}\right) = \left(0, \dfrac{5}{2}\right)$

51. (a)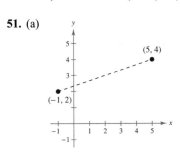

(b) $d = \sqrt{(5 + 1)^2 + (4 - 2)^2}$

$= \sqrt{36 + 4} = 2\sqrt{10}$

(c) $\left(\dfrac{-1 + 5}{2}, \dfrac{2 + 4}{2}\right) = (2, 3)$

53. (a)

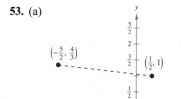

(b) $d = \sqrt{\left(\dfrac{1}{2} + \dfrac{5}{2}\right)^2 + \left(1 - \dfrac{4}{3}\right)^2}$

$= \sqrt{9 + \dfrac{1}{9}} = \dfrac{\sqrt{82}}{3}$

(c) $\left(\dfrac{-(5/2) + (1/2)}{2}, \dfrac{(4/3) + 1}{2}\right) = \left(-1, \dfrac{7}{6}\right)$

55. (a)

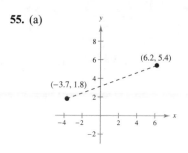

(b) $d = \sqrt{\left(6.2 + 3.7\right)^2 + \left(5.4 - 1.8\right)^2}$

$= \sqrt{98.01 + 12.96} = \sqrt{110.97}$

(c) $\left(\dfrac{6.2 - 3.7}{2}, \dfrac{5.4 + 1.8}{2}\right) = (1.25, 3.6)$

67. (a) Cost during Super Bowl XXXVIII $(2004) \approx \$2,302,000$

Cost during Super Bowl XXXIV $(2000) \approx \$2,100,000$

Increase $= \$2,302,000 - \$2,100,000 = \$202,000$

Percent increase $= \dfrac{\$202,000}{\$2,100,000} \approx 0.096$ or 9.6%

(b) Cost during Super Bowl XLII $(2008) = \$2,700,000$

Cost during Super Bowl XXXIV $(2000) = \$2,100,000$

Increase $= \$2,700,000 - \$2,100,000 = \$600,000$

Percent increase $= \dfrac{\$600,000}{\$2,100,000} \approx 0.286$ or 28.6%

69. The number of performers elected each year seems to be nearly steady except for the middle years. Five performers will be elected in 2010.

57. $d = \sqrt{120^2 + 150^2}$

$= \sqrt{36,900}$

$= 30\sqrt{41}$

≈ 192.09

The plane flies about 192 kilometers.

59. midpoint $= \left(\dfrac{x_1 + x_2}{2}, \dfrac{y_1 + y_2}{2}\right)$

$= \left(\dfrac{2003 + 2007}{2}, \dfrac{4174 + 4656}{2}\right)$

$= (2005, 4415)$

In 2005, the sales for Big Lots were about $4415 million.

61. $(-2 + 2, -4 + 5) = (0, 1)$

$(2 + 2, -3 + 5) = (4, 2)$

$(-1 + 2, -1 + 5) = (1, 4)$

63. $(-7 + 4, -2 + 8) = (-3, 6)$

$(-2 + 4, 2 + 8) = (2, 10)$

$(-2 + 4, -4 + 8) = (2, 4)$

$(-7 + 4, -4 + 8) = (-3, 4)$

65. The highest price of milk is approximately $3.87. This occurred in 2007.

71. midpoint $= \left(\dfrac{x_1 + x_2}{2}, \dfrac{y_1 + y_2}{2}\right)$

$= \left(\dfrac{1999 + 2007}{2}, \dfrac{19,805 + 28,857}{2}\right)$

$= (2003, 24,331)$

In 2003, the sales for the Coca-Cola Company were about $24,331 million.

73. (a)

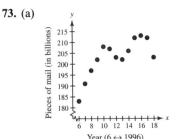

(b) The greatest decrease occurred in 2008.

(c) Answers will vary. Sample answer: Technology now enables us to transport information in ways other than by mail. The Internet is one example.

77. The midpoint of the given line segment is $\left(\dfrac{x_1 + x_2}{2}, \dfrac{y_1 + y_2}{2}\right)$.

The midpoint between (x_1, y_1) and $\left(\dfrac{x_1 + x_2}{2}, \dfrac{y_1 + y_2}{2}\right)$ is $\left(\dfrac{x_1 + \frac{x_1 + x_2}{2}}{2}, \dfrac{y_1 + \frac{y_1 + y_2}{2}}{2}\right) = \left(\dfrac{3x_1 + x_2}{4}, \dfrac{3y_1 + y_2}{4}\right)$.

The midpoint between $\left(\dfrac{x_1 + x_2}{2}, \dfrac{y_1 + y_2}{2}\right)$ and (x_2, y_2) is $\left(\dfrac{\frac{x_1 + x_2}{2} + x_2}{2}, \dfrac{\frac{y_1 + y_2}{2} + y_2}{2}\right) = \left(\dfrac{x_1 + 3x_2}{4}, \dfrac{y_1 + 3y_2}{4}\right)$.

So, the three points are

$\left(\dfrac{3x_1 + x_2}{4}, \dfrac{3y_1 + y_2}{4}\right), \left(\dfrac{x_1 + x_2}{2}, \dfrac{y_1 + y_2}{2}\right)$, and $\left(\dfrac{x_1 + 3x_2}{4}, \dfrac{y_1 + 3y_2}{4}\right)$.

75. Because $x_m = \dfrac{x_1 + x_2}{2}$ and $y_m = \dfrac{y_1 + y_2}{2}$ we have:

$$2x_m = x_1 + x_2 \qquad 2y_m = y_1 + y_2$$
$$2x_m - x_1 = x_2 \qquad 2y_m - y_1 = y_2$$

So, $(x_2, y_2) = (2x_m - x_1, 2y_m - y_1)$.

79.

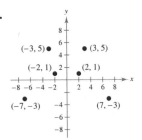

(a) The point is reflected through the y-axis.

(b) The point is reflected through the x-axis.

(c) The point is reflected through the origin.

81. False, you would have to use the Midpoint Formula 15 times.

83. No. It depends on the magnitude of the quantities measured.

85. Use the Midpoint Formula to prove the diagonals of the parallelogram bisect each other.

$$\left(\dfrac{b + a}{2}, \dfrac{c + 0}{2}\right) = \left(\dfrac{a + b}{2}, \dfrac{c}{2}\right)$$
$$\left(\dfrac{a + b + 0}{2}, \dfrac{c + 0}{2}\right) = \left(\dfrac{a + b}{2}, \dfrac{c}{2}\right)$$

Review Exercises for Chapter P

1. $\left\{11, -14, -\frac{8}{9}, \frac{5}{2}, \sqrt{6}, 0.4\right\}$

(a) Natural numbers: 11

(b) Whole numbers: 0, 11

(c) Integers: 11, −14

(d) Rational numbers: $11, -14, -\frac{8}{9}, \frac{5}{2}, 0.4$

(e) Irrational numbers: $\sqrt{6}$

3. (a) $\frac{5}{6} = 0.8\overline{3}$

(b) $\frac{7}{8} = 0.875$

$\frac{5}{6} < \frac{7}{8}$

5. $x \leq 7$

The set consists of all real numbers less than or equal to 7.

7. $d(-74, 48) = |48 - (-74)| = 122$

9. $d(x, 7) = |x - 7|$ and $d(x, 7) \geq 4$, thus $|x - 7| \geq 4$.

11. $d(y, -30) = |y - (-30)| = |y + 30|$ and
$d(y, -30) < 5$, thus $|y + 30| < 5$.

13. $12x - 7$

(a) $12(0) - 7 = -7$

(b) $12(-1) - 7 = -19$

15. $-x^2 + x - 1$

(a) $-(1)^2 + 1 - 1 = -1$

(b) $-(-1)^2 + (-1) - 1 = -3$

17. $2x + (3x - 10) = (2x + 3x) - 10$

Illustrates the Associative Property of Addition

19. $0 + (a - 5) = a - 5$

Illustrates the Additive Identity Property

21. $(t^2 + 1) + 3 = 3 + (t^2 + 1)$

Illustrates the Commutative Property of Addition

23. $|-3| + 4(-2) - 6 = 3 - 8 - 6 = -11$

25. $\dfrac{5}{18} \div \dfrac{10}{3} = \dfrac{\cancel{5}}{\cancel{18}} \cdot \dfrac{\cancel{3}}{\cancel{10}} = \dfrac{1}{12}$

27. $6[4 - 2(6 + 8)] = 6[4 - 2(14)]$
$= 6[4 - 28]$
$= 6(-24)$
$= -144$

29. $\dfrac{x}{5} + \dfrac{7x}{12} = \dfrac{12x}{60} + \dfrac{35x}{60} = \dfrac{47x}{60}$

31. (a) $3x^2(4x^3)^3 = 3x^2(64x^9) = 192x^{11}$

(b) $\dfrac{5y^6}{10y} = \dfrac{y^{6-1}}{2} = \dfrac{y^5}{2}, \quad y \neq 0$

33. (a) $(-2z)^3 = -8z^3$

(b) $\dfrac{(8y)^0}{y^2} = \dfrac{1}{y^2}$

35. (a) $\dfrac{a^2}{b^{-2}} = a^2 b^2$

(b) $(a^2 b^2)(3ab^{-2}) = 3a^{2+1}b^{4-2} = 3a^3 b^2$

37. (a) $\dfrac{(5a)^{-2}}{(5a)^2} = (5a)^{-2-2} = (5a)^{-4} = \dfrac{1}{(5a)^4} = \dfrac{1}{625a^4}$

(b) $\dfrac{4(x^{-1})^{-3}}{4^{-2}(x^{-1})^{-1}} = \dfrac{4^{1-(-2)}x^3}{x} = 4^3 x^{3-1} = 64x^2$

39. $501{,}500{,}000 = 5.015 \cdot 10^8$

41. $4.84 \times 10^8 = 484{,}000{,}000$

43. (a) $\sqrt[3]{27^2} = \left(\sqrt[3]{27}\right)^2 = (3)^2 = 9$

(b) $\sqrt{49^3} = \left(\sqrt{49}\right)^3 = (7)^3 = 343$

45. (a) $\left(\sqrt[3]{216}\right)^3 = \left(\sqrt[3]{6}\right)^3 = (6)^3 = 216$

(b) $\sqrt[4]{32^4} = \left(\sqrt[4]{32}\right)^4 = 32$

47. (a) $\sqrt{50} - \sqrt{18} = \sqrt{25 \cdot 2} - \sqrt{9 \cdot 2}$
$= 5\sqrt{2} - 3\sqrt{2}$
$= 2\sqrt{2}$

(b) $2\sqrt{32} + 3\sqrt{72} = 2\sqrt{16 \cdot 2} + 3\sqrt{36 \cdot 2}$
$= 2(4\sqrt{2}) + 3(6\sqrt{2})$
$= 8\sqrt{2} + 18\sqrt{2}$
$= 26\sqrt{2}$

49. These are not like terms. Radicals cannot be combined by addition or subtraction unless the index and the radicand are the same.

51. $\dfrac{3}{4\sqrt{3}} = \dfrac{3}{4\sqrt{3}} \cdot \dfrac{\sqrt{3}}{\sqrt{3}} = \dfrac{3\sqrt{3}}{4(3)} = \dfrac{\sqrt{3}}{4}$

53. $\dfrac{1}{2-\sqrt{3}} = \dfrac{1}{2-\sqrt{3}} \cdot \dfrac{2+\sqrt{3}}{2+\sqrt{3}} = \dfrac{2+\sqrt{3}}{2^2 - \left(\sqrt{3}\right)^2} = \dfrac{2+\sqrt{3}}{4-3} = \dfrac{2+\sqrt{3}}{1} = 2+\sqrt{3}$

55. $\dfrac{\sqrt{7}+1}{2} = \dfrac{\sqrt{7}+1}{2} \cdot \dfrac{\sqrt{7}-1}{\sqrt{7}-1} = \dfrac{\left(\sqrt{7}\right)^2 - 1^2}{2\left(\sqrt{7}-1\right)} = \dfrac{7-1}{2\left(\sqrt{7}-1\right)} = \dfrac{6}{2\left(\sqrt{7}-1\right)} = \dfrac{3}{\sqrt{7}-1}$

57. $16^{3/2} = \sqrt{16^3} = \left(\sqrt{16}\right)^3 = (4)^3 = 64$

59. $\left(3x^{2/5}\right)\left(2x^{1/2}\right) = 6x^{2/5+1/2} = 6x^{9/10}$

61. Standard form: $-11x^2 + 3$
Degree: 2
Leading coefficient: -11

63. Standard form: $-12x^2 - 4$
Degree: 2
Leading coefficient: -12

65. $-\left(3x^2 + 2x\right) + \left(1 - 5x\right) = -3x^2 - 2x + 1 - 5x$
$= -3x^2 - 7x + 1$

67. $2x\left(x^2 - 5x + 6\right) = (2x)\left(x^2\right) + (2x)(-5x) + (2x)(6)$
$= 2x^3 - 10x^2 + 12x$

69. $\left(2x^3 - 5x^2 + 10x - 7\right) + \left(4x^2 - 7x - 2\right) = 2x^3 + \left(-5x^2 + 4x^2\right) + (10x - 7x) + (-7 - 2) = 2x^3 - x^2 + 3x - 9$

71. $(3x - 6)(5x + 1) = 15x^2 + 3x - 30x - 6$
$= 15x^2 - 27x - 6$

73. $(2x - 3)^2 = (2x)^2 - 2(2x)(3) + 3^3$
$= 4x^2 - 12x + 9$

75. $\left(3\sqrt{5} + 2\right)\left(3\sqrt{5} - 2\right) = \left(3\sqrt{5}\right)^2 - 2^2$
$= 9(5) - 4$
$= 41$

77. $2500(1 + r)^2 = 2500(r + 1)^2$
$= 2500\left(r^2 + 2r + 1\right)$
$= 2500r^2 + 5000r + 2500$

79. Area $= (x + 12)(x + 16)$
$= x^2 + 16x + 12x + 192$
$= x^2 + 28x + 192$ square feet

81. $x^3 - x = x\left(x^2 - 1\right) = x(x + 1)(x - 1)$

83. $25x^2 - 49 = (5x)^2 - 7^2 = (5x + 7)(5x - 7)$

85. $x^3 - 64 = x^3 - 4^3 = (x - 4)\left(x^2 + 4x + 16\right)$

87. $2x^2 + 21x + 10 = (2x + 1)(x + 10)$

89. $x^3 - x^2 + 2x - 2 = x^2(x - 1) + 2(x - 1)$
$= (x - 1)\left(x^2 + 2\right)$

91. The domain of $\dfrac{1}{x + 6}$ is the set of all real numbers except $x = -6$.

93. $\dfrac{x^2 - 64}{5(3x + 24)} = \dfrac{(x + 8)(x - 8)}{5 \cdot 3(x + 8)} = \dfrac{x - 8}{15}, \quad x \neq -8$

95. $\dfrac{x^2 - 4}{x^4 - 2x^2 - 8} \cdot \dfrac{x^2 + 2}{x^2} = \dfrac{\left(x^2 - 4\right)\left(x^2 + 2\right)}{\left(x^2 - 4\right)\left(x^2 + 2\right)x^2}$
$= \dfrac{1}{x^2}, \quad x \neq \pm 2$

97. $\dfrac{1}{x - 1} + \dfrac{1 - x}{x^2 + x + 1} = \dfrac{x^2 + x + 1 + (1 - x)(x - 1)}{(x - 1)\left(x^2 + x + 1\right)}$
$= \dfrac{x^2 + x + 1 + x - 1 - x^2 + x}{(x - 1)\left(x^2 + x + 1\right)}$
$= \dfrac{3x}{(x - 1)\left(x^2 + x + 1\right)}$

99. $\dfrac{\left(\dfrac{3a}{a^2} - 1\right)}{\left(\dfrac{a}{x} - 1\right)} = \dfrac{\left(\dfrac{3a}{a^2 - x}\right)}{\left(\dfrac{a - x}{x}\right)} = \dfrac{3a}{1} \cdot \dfrac{x}{a^2 - x} \cdot \dfrac{x}{a - x}$
$= \dfrac{3ax^2}{\left(a^2 - x\right)(a - x)}$

101. $\dfrac{\dfrac{1}{2(x+h)} - \dfrac{1}{2x}}{h} = \dfrac{\dfrac{x-(x+6)}{2x(x+h)}}{h}$

$\qquad = \dfrac{-h}{2x(x+h)} \cdot \dfrac{1}{h}$

$\qquad = \dfrac{-1}{2x(x+h)}, \quad h \neq 0$

103.

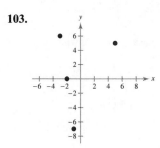

105. $x > 0$ and $y = -2$ in Quadrant IV.

107. (a)

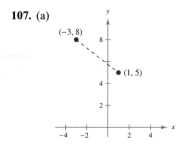

(b) $d = \sqrt{(-3-1)^2 + (8-5)^2} = \sqrt{16+9} = 5$

(c) Midpoint: $\left(\dfrac{-3+1}{2}, \dfrac{8+5}{2}\right) = \left(-1, \dfrac{13}{2}\right)$

109. (a)

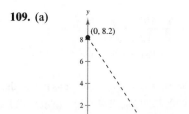

(b) $d = \sqrt{(5.6-0)^2 + (0-8.2)^2}$

$\qquad = \sqrt{31.36 + 67.24} = \sqrt{98.6}$

(c) Midpoint: $\left(\dfrac{0+5.6}{2}, \dfrac{8.2+0}{2}\right) = (2.8, 4.1)$

111. $(4-4, 8-8) = (0, 0)$

$\quad (6-4, 8-8) = (2, 0)$

$\quad (4-4, 3-8) = (0, -5)$

$\quad (6-4, 3-8) = (2, -5)$

113. $(2000, 2.17), (2008, 10.38)$

$\left(\dfrac{2000+2008}{2}, \dfrac{2.17+10.38}{2}\right) = (2004, 6.275)$

In 2004, sales were approximately $6.275 billion.

115. False, $(a+b)^2 = a^2 + 2ab + b^2 \neq a^2 + b^2$

There is also a cross-product term when a binomial sum is squared.

117. $\dfrac{ax-b}{b-ax} = -1$ for all real nonzero numbers a and b

except for $x = \dfrac{b}{a}$. This makes the expression undefined.

Problem Solving for Chapter P

1. (a) Men's

Minimum Volume: $V = \dfrac{4}{3}\pi(55)^3 \approx 696{,}910 \text{ mm}^3$

Maximum Volume: $V = \dfrac{4}{3}\pi(65)^3 \approx 1{,}150{,}347 \text{ mm}^3$

Women's

Minimum Volume: $V = \dfrac{4}{3}\pi\left(\dfrac{95}{2}\right)^3 \approx 448{,}921 \text{ mm}^3$

Maximum Volume: $V = \dfrac{4}{3}\pi(55)^3 \approx 696{,}910 \text{ mm}^3$

(b) Men's

Minimum density: $\dfrac{7.26}{1{,}150{,}347} \approx 6.31 \times 10^{-6} \text{ kg/mm}^3$

Maximum density: $\dfrac{7.26}{696{,}910} \approx 1.04 \times 10^{-5} \text{ kg/mm}^3$

Women's

Minimum density: $\dfrac{4.00}{696{,}910} \approx 5.74 \times 10^{-6} \text{ kg/mm}^3$

Maximum density: $\dfrac{4.00}{448{,}921} \approx 8.91 \times 10^{-6} \text{ kg/mm}^3$

(c) No. The weight would be different. Cork is much lighter than iron so it would have a much smaller density.

3. One gold ball: $\dfrac{1.6 \times 10^7}{1.58 \times 10^8} \approx 0.101$ pound

$0.101(16) = 1.62$ ounces

5. To say that a number has *n* significant digits means that the number has *n* digits with the leftmost non-zero digit and ending with the rightmost non-zero digit. For example; 28,000, 1.400, 0.00079 each have two significant digits.

7. $r = 1 - \left(\dfrac{3225}{12,000}\right)^{1/4} \approx 0.280$ or 28%

9.

Perimeter: $P = 2l + w + \pi r$

$13.14 = 2l + 2 + \pi(1)$

$l = \dfrac{13.14 - 2 - \pi}{2} \approx 4$ feet

Amount of glass = Area of window

$A = lw + \dfrac{1}{2}\pi r^2$

$= (4)(2) + \dfrac{1}{2}\pi(1)^2$

≈ 9.57 square feet

11. $y_1 = 2x\sqrt{1 - x^2} - \dfrac{x^2}{\sqrt{1 - x^2}} \quad y_2 = \dfrac{2 - 3x^2}{\sqrt{1 - x^2}}$

When $x = 0, y_1 = 0$. When $x = 0, y_2 = 2$.

Thus, $y_1 \neq y_2$.

$y_1 = \dfrac{2x\sqrt{1 - x^2}}{1} - \dfrac{x^2}{\sqrt{x - x^2}}$

$= \dfrac{2x\sqrt{1 - x^2}}{1} \cdot \dfrac{\sqrt{1 - x^2}}{\sqrt{1 - x^2}} - \dfrac{x^3}{\sqrt{1 - x^2}}$

$= \dfrac{2x(1 - x^2) - x^3}{\sqrt{1 - x^2}}$

$= \dfrac{2x - 2x^3 - x^3}{\sqrt{1 - x^2}}$

$= \dfrac{2x - 3x^3}{\sqrt{1 - x^2}}$

$= \dfrac{x(2 - 3x^2)}{\sqrt{1 - x^2}}$

Let $y_2 = \dfrac{x(2 - 3x^2)}{\sqrt{1 - x^2}}$. Then $y_1 = y_2$.

13. (a) $(1, -2)$ and $(4, 1)$

The points of trisection are:

$\left(\dfrac{2(1) + 4}{3}, \dfrac{2(-2) + 1}{3}\right) = (2, -1)$

$\left(\dfrac{1 + 2(4)}{3}, \dfrac{-2 + 2(1)}{3}\right) = (3, 0)$

(b) $(-2, -3)$ and $(0, 0)$

The points of trisection are:

$\left(\dfrac{2(-2) + 0}{3}, \dfrac{2(-3) + 0}{3}\right) = \left(-\dfrac{4}{3}, -2\right)$

$\left(\dfrac{-2 + 2(0)}{3}, \dfrac{-3 + 2(0)}{3}\right) = \left(-\dfrac{2}{3}, -1\right)$

Practice Test for Chapter P

1. Evaluate $\dfrac{|-42| - 20}{15 - |-4|}$.

2. Simplify $\dfrac{x}{z} - \dfrac{z}{y}$.

3. The distance between x and 7 is no more than 4. Use absolute value notation to describe this expression.

4. Evaluate $10(-x)^3$ for $x = 5$.

5. Simplify $\left(-4x^3\right)\left(-2x^{-5}\right)\left(\frac{1}{16}x\right)$.

6. Change 0.0000412 to scientific notation.

7. Evaluate $125^{2/3}$.

8. Simplify $\sqrt[4]{64x^7 y^9}$.

9. Rationalize the denominator and simplify $\dfrac{6}{\sqrt{12}}$.

10. Simplify $3\sqrt{80} - 7\sqrt{500}$.

11. Simplify $\left(8x^4 - 9x^2 + 2x - 1\right) - \left(3x^3 + 5x + 4\right)$.

12. Multiply $(x - 3)\left(x^2 + x - 7\right)$.

13. Multiply $\left[(x - 2) - y\right]^2$.

14. Factor $16x^4 - 1$.

15. Factor $6x^2 + 5x - 4$.

16. Factor $x^3 - 64$.

17. Combine and simplify $-\dfrac{3}{x} + \dfrac{x}{x^2 + 2}$.

18. Combine and simplify $\dfrac{x - 3}{4x} \div \dfrac{x^2 - 9}{x^2}$.

19. Simplify $\dfrac{1 - \left(\dfrac{1}{x}\right)}{1 - \dfrac{1}{1 - \left(\dfrac{1}{x}\right)}}$.

20. (a) Plot the points $(-3, 6)$ and $(5, -1)$,

    (b) find the distance between the points, and

    (c) find the midpoint of the line segment joining the points.

CHAPTER 1
Equations, Inequalities, and Mathematical Modeling

C H A P T E R 1
Equations, Inequalities, and Mathematical Modeling

Section 1.1   Graphs of Equations

You should know the following important facts about lines.

- The graph of $y = mx + b$ is a straight line. It is called a linear equation in two variables.
  - (a) The slope (steepness) is m.
  - (b) The y-intercept is $(0, b)$.

- The slope of the line through (x_1, y_1) and (x_2, y_2) is
  $$m = \frac{y_2 - y_1}{x_2 - x_1} = \frac{\text{change in } y}{\text{change in } x} = \frac{\text{rise}}{\text{run}}.$$

- (a) If $m > 0$, the line rises from left to right.
  - (b) If $m = 0$, the line is horizontal.
  - (c) If $m < 0$, the line falls from left to right.
  - (d) If m is undefined, the line is vertical.

- Equations of Lines
  - (a) Slope-Intercept Form: $y = mx + b$
  - (b) Point-Slope Form: $y - y_1 = m(x - x_1)$
  - (c) Two-Point Form: $y - y_1 = \dfrac{y_2 - y_1}{x_2 - x_1}(x - x_1)$
  - (d) General Form: $Ax + By + C = 0$
  - (e) Vertical Line: $x = a$
  - (f) Horizontal Line: $y = b$

- Given two distinct nonvertical lines
  $L_1: y = m_1 x + b_1$ and $L_2: y = m_2 x + b_2$
  - (a) L_1 is parallel to L_2 if and only if $m_1 = m_2$ and $b_1 \neq b_2$.
  - (b) L_1 is perpendicular to L_2 if and only if $m_1 = -1/m_2$.

1. solution or solution point

3. intercepts

5. circle; (h, k); r

7. $y = \sqrt{x + 4}$

(a) $(0, 2)$: $2 \overset{?}{=} \sqrt{0 + 4}$

$\qquad 2 = 2$

Yes, the point *is* on the graph.

(b) $(5, 3)$: $3 \overset{?}{=} \sqrt{5 + 4}$

$\qquad 3 \overset{?}{=} \sqrt{9}$

$\qquad 3 = 3$

Yes, the point *is* on the graph.

9. $y = x^2 - 3x + 2$

(a) $(2, 0)$: $(2)^2 - 3(2) + 2 \overset{?}{=} 0$

$$4 - 6 + 2 \overset{?}{=} 0$$

$$0 = 0$$

Yes, the point *is* on the graph.

(b) $(-2, 8)$: $(-2)^2 - 3(-2) + 2 \overset{?}{=} 8$

$$4 + 6 + 2 \overset{?}{=} 8$$

$$12 \neq 8$$

No, the point *is not* on the graph.

11. (a) $(2, 3)$: $3 \overset{?}{=} |2 - 1| + 2$

$$3 \overset{?}{=} 1 + 2$$

$$3 = 3$$

Yes, the point *is* on the graph.

(b) $(-1, 0)$: $0 \overset{?}{=} |-1 - 1| + 2$

$$0 \overset{?}{=} 2 + 2$$

$$0 \neq 4$$

No, the point *is not* on the graph.

13. (a) $(3, -2)$: $(3)^2 + (-2)^2 \overset{?}{=} 20$

$$9 + 4 \overset{?}{=} 20$$

$$13 \neq 20$$

No, the point *is not* on the graph.

(b) $(-4, 2)$: $(-4)^2 + (2)^2 \overset{?}{=} 20$

$$16 + 4 \overset{?}{=} 20$$

$$20 = 20$$

Yes, the point *is* on the graph.

15. $y = -2x + 5$

x	-1	0	1	2	$\frac{5}{2}$
y	7	5	3	1	0
(x, y)	$(-1, 7)$	$(0, 5)$	$(1, 3)$	$(2, 1)$	$\left(\frac{5}{2}, 0\right)$

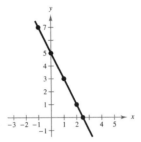

17. $y = x^2 - 3x$

x	-1	0	1	2	3
y	4	0	-2	-2	0
(x, y)	$(-1, 4)$	$(0, 0)$	$(1, -2)$	$(2, -2)$	$(3, 0)$

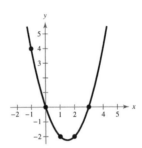

19. x-intercept: $(3, 0)$

y-intercept: $(0, 9)$

21. x-intercept: $(-2, 0)$

y-intercept: $(0, 2)$

23. x-intercept: $(1, 0)$

y-intercept: $(0, 2)$

25. $x^2 - y = 0$

$(-x)^2 - y = 0 \Rightarrow x^2 - y = 0 \Rightarrow y$-axis symmetry

$x^2 - (-y) = 0 \Rightarrow x^2 + y = 0 \Rightarrow$ No x-axis symmetry

$(-x)^2 - (-y) = 0 \Rightarrow x^2 + y = 0 \Rightarrow$ No origin symmetry

27. $y = x^3$

$\quad y = (-x)^3 \Rightarrow y = -x^3 \Rightarrow$ No y-axis symmetry

$\quad -y = x^3 \Rightarrow y = -x^3 \Rightarrow$ No x-axis symmetry

$\quad -y = (-x)^3 \Rightarrow -y = -x^3 \Rightarrow y = x^3 \Rightarrow$ Origin symmetry

29. $y = \dfrac{x}{x^2 + 1}$

$\quad y = \dfrac{-x}{(-x)^2 + 1} \Rightarrow y = \dfrac{-x}{x^2 + 1} \Rightarrow$ No y-axis symmetry

$\quad -y = \dfrac{x}{x^2 + 1} \Rightarrow y = \dfrac{-x}{x^2 + 1} \Rightarrow$ No x-axis symmetry

$\quad -y = \dfrac{-x}{(-x)^2 + 1} \Rightarrow -y = \dfrac{-x}{x^2 + 1} \Rightarrow y = \dfrac{x}{x^2 + 1} \Rightarrow$ Origin symmetry

31. $xy^2 + 10 = 0$

$\quad (-x)y^2 + 10 = 0 \Rightarrow -xy^2 + 10 = 0 \Rightarrow$ No y-axis symmetry

$\quad x(-y)^2 + 10 = 0 \Rightarrow xy^2 + 10 = 0 \Rightarrow x$-axis symmetry

$\quad (-x)(-y)^2 + 10 = 0 \Rightarrow -xy^2 + 10 = 0 \Rightarrow$ No origin symmetry

33.

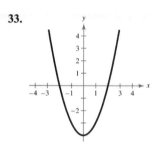

35.

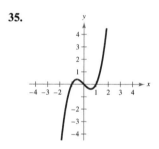

37. $y = -3x + 1$

$\quad x$-intercept: $\left(\tfrac{1}{3}, 0\right)$

$\quad y$-intercept: $(0, 1)$

No symmetry

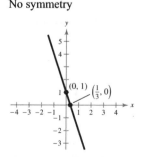

39. $y = x^2 - 2x$

$\quad x$-intercepts: $(0, 0), (2, 0)$

$\quad y$-intercept: $(0, 0)$

No symmetry

x	-1	0	1	2	3
y	3	0	-1	0	3

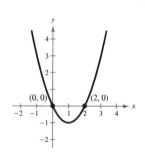

41. $y = x^3 + 3$

x-intercept: $\left(\sqrt[3]{-3}, 0\right)$

y-intercept: $(0, 3)$

No symmetry

x	-2	-1	0	1	2
y	-5	2	3	4	11

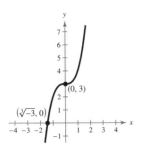

43. $y = \sqrt{x - 3}$

x-intercept: $(3, 0)$

y-intercept: none

No symmetry

x	3	4	7	12
y	0	1	2	3

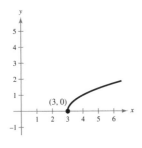

45. $y = |x - 6|$

x-intercept: $(6, 0)$

y-intercept: $(0, 6)$

No symmetry

x	-2	0	2	4	6	8	10
y	8	6	4	2	0	2	4

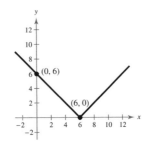

47. $x = y^2 - 1$

x-intercept: $(-1, 0)$

y-intercepts: $(0, -1), (0, 1)$

x-axis symmetry

x	-1	0	3
y	0	± 1	± 2

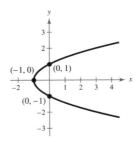

49. $y = 5 - \frac{1}{2}x$

Intercepts: $(10, 0), (0, 5)$

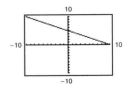

51. $y = x^2 - 4x + 3$

Intercepts: $(3, 0), (1, 0), (0, 3)$

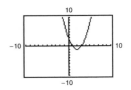

53. $y = \dfrac{2x}{x-1}$

Intercept: $(0, 0)$

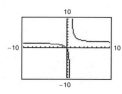

55. $y = \sqrt[3]{x} + 2$

Intercepts: $(-8, 0), (0, 2)$

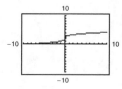

57. $y = x\sqrt{x+6}$

Intercepts: $(0, 0), (-6, 0)$

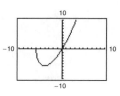

59. $y = |x + 3|$

Intercepts: $(-3, 0), (0, 3)$

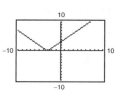

61. Center: $(0, 0)$; Radius: 4

$(x - 0)^2 + (y - 0)^2 = 4^2$

$x^2 + y^2 = 16$

63. Center: $(2, -1)$; Radius: 4

$(x - 2)^2 + (y - (-1))^2 = 4^2$

$(x - 2)^2 + (y + 1)^2 = 16$

65. Center: $(-1, 2)$; Solution point: $(0, 0)$

$(x - (-1))^2 + (y - 2)^2 = r^2$

$(0 + 1)^2 + (0 - 2)^2 = r^2 \Rightarrow 5 = r^2$

$(x + 1)^2 + (y - 2)^2 = 5$

67. Endpoints of a diameter: $(0, 0), (6, 8)$

Center: $\left(\dfrac{0+6}{2}, \dfrac{0+8}{2}\right) = (3, 4)$

$(x - 3)^2 + (y - 4)^2 = r^2$

$(0 - 3)^2 + (0 - 4)^2 = r^2 \Rightarrow 25 = r^2$

$(x - 3)^2 + (y - 4)^2 = 25$

69. $x^2 + y^2 = 25$

Center: $(0, 0)$, Radius: 5

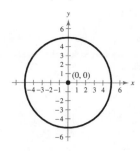

71. $(x - 1)^2 + (y + 3)^2 = 9$

Center: $(1, -3)$, Radius: 3

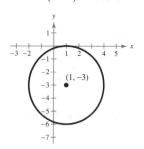

73. $\left(x - \frac{1}{2}\right)^2 + \left(y - \frac{1}{2}\right)^2 = \frac{9}{4}$

Center: $\left(\frac{1}{2}, \frac{1}{2}\right)$, Radius: $\frac{3}{2}$

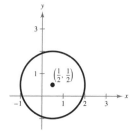

75. $y = 500{,}000 - 40{,}000t, \ 0 \le t \le 8$

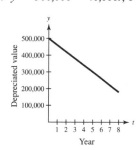

77. (a)

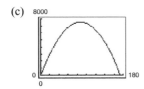

(b) $2x + 2y = \frac{1040}{3}$

$2y = \frac{1040}{3} - 2x$

$y = \frac{520}{3} - x$

$A = xy = x\left(\frac{520}{3} - x\right)$

(c)

8000

180

0

0

(d) When $x = y = 86\frac{2}{3}$ yards, the area is a maximum of $7511\frac{1}{9}$ square yards.

(e) A regulation NFL playing field is 120 yards long and $53\frac{1}{3}$ yards wide. The actual area is 6400 square yards.

79. (a)

100

100

0

0

The model fits the data very well.

Answers will vary.

(b) graphically: 75.4 years

algebraically: $y = -0.0025(90)^2 + 0.574(90) + 44.25$

$y = 75.66$ years

(c) $76 = -0.0025t^2 + 0.574t + 44.25$

$0 = -0.0025t^2 + 0.574t - 31.75$

$t = \dfrac{-0.574 \pm \sqrt{(0.574)^2 - 4(-0.0025)(-31.75)}}{2(-0.0025)}$

$t = \dfrac{-0.574 \pm \sqrt{0.011976}}{-0.005}$

$t \approx 92.9$, or about 93 years

In the year $1900 + 93 = 1993$ the life expectancy was approximately 76 years.

(d) Year $= 2015 \rightarrow t = 115$

$y = -0.0025(115)^2 + 0.574(115) + 44.25$

$y \approx 77.2$

The model predicts the life expectancy to be 77.2 years in 2015.

(e) Answers will vary. Sample answer: No; the model predicts that life expectancies after 2035 will decrease but this is not likely.

81. $y = ax^2 + bx^3$

(a) $y = a(-x)^2 + b(-x)^3$

$\quad = ax^2 - bx^3$

To be symmetric with respect to the *y*-axis; *a* can be any non-zero real number, *b* must be zero.

(b) $-y = a(-x)^2 + b(-x)^3$

$\quad -y = ax^2 - bx^3$

$\quad y = -ax^2 + bx^3$

To be symmetric with respect to the origin; *a* must be zero, *b* can be any non-zero real number.

Section 1.2   Linear Equations in One Variable

- You should know how to solve linear equations.

  $ax + b = 0, a \neq 0$

- An identity is an equation whose solution consists of every real number in its domain.

- To solve an equation you can:

  (a) Add or subtract the same quantity from both sides.

  (b) Multiply or divide both sides by the same nonzero quantity.

  (c) Remove all symbols of grouping and all fractions.

  (d) Combine like terms.

  (e) Interchange the two sides.

- Check the answer!

- A "solution" that does not satisfy the original equation is called an extraneous solution.

- Be able to find intercepts algebraically.

1. equation

3. identities; conditional

5. extraneous

7. $5x - 3 = 3x + 5$

(a) $5(0) - 3 \overset{?}{=} 3(0) + 5$

$\quad -3 \neq 5$

$\quad x = 0$ *is not* a solution.

(b) $5(-5) - 3 \overset{?}{=} 3(-5) + 5$

$\quad -28 \neq -10$

$\quad x = -5$ *is not* a solution.

(c) $5(4) - 3 \overset{?}{=} 3(4) + 5$

$\quad 17 = 17$

$\quad x = 4$ *is* a solution.

(d) $5(10) - 3 \overset{?}{=} 3(10) + 5$

$\quad 47 \neq 35$

$\quad x = 10$ *is not* a solution.

9. $3x^2 + 2x - 5 = 2x^2 - 2$

(a) $3(-3)^2 + 2(-3) - 5 \overset{?}{=} 2(-3)^2 - 2$

$\quad 27 - 6 - 5 \overset{?}{=} 18 - 2$

$\quad 16 = 16$

$\quad x = -3$ *is* a solution.

(b) $3(1)^2 + 2(1) - 5 \overset{?}{=} 2(1)^2 - 2$

$\quad 3 + 2 - 5 \overset{?}{=} 2 - 2$

$\quad 0 = 0$

$\quad x = 1$ *is* a solution.

(c) $3(4)^2 + 2(4) - 5 \overset{?}{=} 2(4)^2 - 2$

$\quad 48 + 8 - 5 \overset{?}{=} 32 - 2$

$\quad 51 \neq 30$

$\quad x = 4$ *is not* a solution.

(d) $3(-5)^2 + 2(-5) - 5 \overset{?}{=} 2(-15)^2 - 2$

$\quad 75 - 10 - 5 \overset{?}{=} 50 - 2$

$\quad 60 \neq 16$

$\quad x = -5$ *is not* a solution.

11. $\dfrac{5}{2x} - \dfrac{4}{x} = 3$

(a) $\dfrac{5}{2(-1/2)} - \dfrac{4}{(-1/2)} \overset{?}{=} 3$

$\quad\quad\quad -5 + 8 \overset{?}{=} 3$

$\quad\quad\quad\quad\quad 3 = 3$

$x = -\dfrac{1}{2}$ *is* a solution.

(b) $\dfrac{5}{2(4)} - \dfrac{4}{4} \overset{?}{=} 3$

$\quad\quad \dfrac{5}{8} - 1 \overset{?}{=} 3$

$\quad\quad\quad -\dfrac{3}{8} = 3$

$x = 4$ *is not* a solution.

(c) $\dfrac{5}{2(0)} = \dfrac{4}{0}$ is undefined.

$x = 0$ *is not* a solution.

(d) $\dfrac{5}{2(1/4)} - \dfrac{4}{1/4} \overset{?}{=} 3$

$\quad\quad 10 - 16 \overset{?}{=} 3$

$\quad\quad\quad -6 = 3$

$x = \dfrac{1}{4}$ *is not* a solution.

13. $3 + \dfrac{1}{x+2} = 4$

(a) $x = -1$

$3 + \dfrac{1}{-1+2} \overset{?}{=} 4$

$3 + \dfrac{1}{1} \overset{?}{=} 4$

$3 + 1 \overset{?}{=} 4$

$4 = 4$

$x = -1$ *is* a solution.

(b) $x = -2$

$3 + \dfrac{1}{-2+2} \overset{?}{=} 4$

$3 + \dfrac{1}{0} \overset{?}{=} 4$

Division by zero is undefined.

$x = -2$ *is not* a solution.

(c) $x = 0$

$3 + \dfrac{1}{0+2} \overset{?}{=} 4$

$3\dfrac{1}{2} \neq 4$

$x = 0$ *is not* a solution.

(d) $x = 5$

$3 + \dfrac{1}{5+2} \overset{?}{=} 4$

$3\dfrac{1}{7} \neq 4$

$x = 5$ *is not* a solution.

15. $\sqrt{3x-2} = 4$

(a) $\sqrt{3(2)-2} \overset{?}{=} 4$

$\sqrt{7} \neq 4$

$x = 3$ *is not* a solution.

(b) $\sqrt{3(2)-2} \overset{?}{=} 4$

$\sqrt{4} \neq 4$

$x = 2$ *is not* a solution.

(c) $\sqrt{3(9)-2} \overset{?}{=} 4$

$\sqrt{25} \neq 4$

$x = 9$ *is not* a solution.

(d) $\sqrt{3(-6)-2} \overset{?}{=} 4$

$\sqrt{-20} \neq 4$

$x = -6$ *is not* a solution.

17. $6x^2 - 11x - 35 = 0$

   (a) $6\left(-\frac{5}{3}\right)^2 - 11\left(-\frac{5}{3}\right) - 35 \overset{?}{=} 0$

        $\frac{50}{3} + \frac{55}{3} - \frac{105}{3} \overset{?}{=} 0$

               $0 = 0$

     $x = -\frac{5}{3}$ *is* a solution.

   (b) $6\left(-\frac{2}{7}\right)^2 - 11\left(\frac{7}{2}\right) - 35 \overset{?}{=} 0$

        $\frac{24}{49} + \frac{154}{49} - \frac{1715}{49} \overset{?}{=} 0$

            $-\frac{1537}{49} \neq 0$

     $x = -\frac{2}{7}$ is *not* a solution.

   (c) $6\left(\frac{7}{2}\right)^2 - 11\left(\frac{7}{2}\right) - 35 \overset{?}{=} 0$

        $\frac{147}{2} - \frac{77}{2} - \frac{70}{2} \overset{?}{=} 0$

               $0 = 0$

     $x = \frac{7}{2}$ is *not* a solution.

   (d) $6\left(\frac{5}{3}\right)^2 - 11\left(\frac{5}{3}\right) - 35 \overset{?}{=} 0$

        $\frac{50}{3} - \frac{55}{3} - \frac{105}{3} \overset{?}{=} 0$

            $-\frac{110}{3} \neq 0$

     $x = \frac{5}{3}$ is *not* a solution.

19. $2(x - 1) = 2x - 2$ is an *identity* by the Distributive Property. It is true for all real values of x.

21. $-6(x - 3) + 5 = -2x + 10$ is *conditional*. There are real values of x for which the equation is not true.

23. $4(x + 1) - 2x = 4x + 4 - 2x = 2x + 4 = 2(x + 2)$

This is an *identity* by simplification. It is true for all real values of x.

25. $(x - 4)^2 - 11 = x^2 - 8x + 16 - 11 = x^2 - 8x + 5$

Thus, $x^2 - 8x + 5 = (x - 4)^2 = 11$ is an *identity* by simplification. It is true for all real values of x.

27. $3 + \dfrac{1}{x + 1} = \dfrac{4x}{x + 1}$ is *conditional*. There are real values of x for which the equation is not true.

29. $2(x - 1) = 2x - 1$ is *conditional*. There are real values of x for which the equation is not true.

31.

$4x + 32 = 83$	Original equation
$4x + 32 - 32 = 83 - 32$	Subtract 32 from each side.
$4x = 51$	Simplify.
$\dfrac{4x}{4} = \dfrac{51}{4}$	Divide each side by 4.
$x = \dfrac{51}{4}$	Simplify.

33.

$$x + 11 = 15$$
$$x + 11 - 11 = 15 - 11$$
$$x = 4$$

35.

$$7 - 2x = 25$$
$$7 - 7 - 2x = 25 - 7$$
$$-2x = 18$$
$$\frac{-2x}{-2} = \frac{18}{-2}$$
$$x = -9$$

37.

$$3x - 5 = 2x + 7$$
$$3x - 2x - 5 = 2x - 2x + 7$$
$$x - 5 = 7$$
$$x - 5 + 5 = 7 + 5$$
$$x = 12$$

39. $4y + 2 - 5y = 7 - 6y$

$$4y - 5y + 2 = 7 - 6y$$
$$-y + 2 = 7 - 6y$$
$$-y + 6y + 2 = 7 - 6y + 6y$$
$$5y + 2 = 7$$
$$5y + 2 - 2 = 7 - 2$$
$$5y = 5$$
$$\frac{5y}{5} = \frac{5}{5}$$
$$y = 1$$

41. $x - 3(2x + 3) = 8 - 5x$

$$x - 6x - 9 = 8 - 5x$$
$$-5x - 9 = 8 - 5x$$
$$-5x + 5x - 9 = 8 - 5x + 5x$$
$$-9 \neq 8$$

No solution

43.
$$\frac{5x}{4} + \frac{1}{2} = x - \frac{1}{2}$$
$$4\left(\frac{5x}{4} + \frac{1}{2}\right) = 4\left(x - \frac{1}{2}\right)$$
$$4\left(\frac{5x}{4}\right) + 4\left(\frac{1}{2}\right) = 4(x) - 4\left(\frac{1}{2}\right)$$
$$5x + 2 = 4x - 2$$
$$x = -4$$

45.
$$\frac{3}{2}(z + 5) - \frac{1}{4}(z + 24) = 0$$
$$4\left[\frac{3}{2}(z + 5) - \frac{1}{4}(z + 24)\right] = 4(0)$$
$$4\left(\frac{3}{2}\right)(z + 5) - 4\left(\frac{1}{4}\right)(z + 24) = 4(0)$$
$$6(z + 5) - (z + 24) = 0$$
$$5z = -6$$
$$z = -\frac{6}{5}$$

47. $0.25x + 0.75(10 - x) = 3$
$$0.25x + 7.5 - 0.75x = 3$$
$$-0.50x + 7.5 = 3$$
$$-0.50x = -4.5$$
$$x = 9$$

49.
$$\frac{3x}{8} - \frac{4x}{3} = 4 \qquad \text{or} \qquad \frac{3x}{8} - \frac{4x}{3} = 4$$
$$\frac{9x}{24} - \frac{32x}{24} = 4 \qquad\qquad 24\left(\frac{3x}{8} - \frac{4x}{3}\right) = 24(4)$$
$$-\frac{23x}{24} = 4 \qquad\qquad 9x - 32x = 96$$
$$\qquad\qquad -23x = 96$$
$$-\frac{23x}{24}\left(-\frac{24}{23}\right) = 4\left(-\frac{24}{23}\right) \qquad x = -\frac{96}{23}$$
$$x = -\frac{96}{23}$$

The second method is easier. The fractions are eliminated in the first step.

51.
$$\frac{2x}{5} + 5x = \frac{4}{3} \qquad \text{or} \qquad \frac{2x}{5} + 5x = \frac{4}{3}$$
$$\frac{2x}{5} + \frac{2x}{5} = \frac{4}{3} \qquad\qquad 15\left(\frac{2x}{5} + 5x\right) = 15\left(\frac{4}{3}\right)$$
$$\frac{27x}{5} = \frac{4}{3} \qquad\qquad 6x + 75x = 20$$
$$\qquad\qquad 81x = 20$$
$$\frac{27x}{5}\left(\frac{5}{27}\right) = \frac{4}{3}\left(\frac{5}{27}\right) \qquad x = \frac{20}{81}$$
$$x = \frac{20}{81}$$

The second method is easier. The fractions are eliminated in the first step.

53. $x + 8 = 2(x - 2) - x$
$$x + 8 = 2x - 4 - x$$
$$x + 8 = x - 4$$
$$8 \neq -4$$

Contradiction; no solution

55.
$$\frac{100 - 4x}{3} = \frac{5x + 6}{4} + 6$$
$$12\left(\frac{100 - 4x}{3}\right) = 12\left(\frac{5x + 6}{4}\right) + 12(6)$$
$$4(100 - 4x) = 3(5x + 6 + 72)$$
$$400 - 16x = 15x + 18 + 72$$
$$-31x = -310$$
$$x = 10$$

57.
$$\frac{5x - 4}{5x + 4} = \frac{2}{3}$$
$$3(5x - 4) = 2(5x + 4)$$
$$15x - 12 = 10x + 8$$
$$5x = 20$$
$$x = 4$$

59.
$$10 - \frac{13}{x} = 4 + \frac{5}{x}$$
$$\frac{10 - 13}{x} = \frac{4x + 5}{x}$$
$$10x - 13 = 4x + 5$$
$$6x = 18$$
$$x = 3$$

61.
$$3 = 2 + \frac{2}{z + 2}$$
$$3(z + 2) = \left(2 + \frac{2}{z + 2}\right)(z + 2)$$
$$3z + 6 = 2z + 4 + 2$$
$$z = 0$$

63.
$$\frac{x}{x + 4} + \frac{4}{x + 4} + 2 = 0$$
$$\frac{x + 4}{x + 4} + 2 = 0$$
$$1 + 2 = 0$$
$$3 \neq 0$$

Contradiction; no solution

65. $\dfrac{2}{(x-4)(x-2)} = \dfrac{1}{x-4} + \dfrac{2}{x-2}$ Multiply both sides by $(x-4)(x-2)$.

$$2 = 1(x-2) + 2(x-4)$$
$$2 = x - 2 + 2x - 8$$
$$2 = 3x - 10$$
$$12 = 3x$$
$$4 = x$$

A check reveals that $x = 4$ is an extraneous solution—it makes the denominator zero. There is no real solution.

67. $\dfrac{1}{x-3} + \dfrac{1}{x+3} = \dfrac{10}{x^2-9}$

$\dfrac{1}{x-3} + \dfrac{1}{x+3} = \dfrac{10}{(x+3)(x-3)}$ Multiply both sides by $(x+3)(x-3)$.

$$1(x+3) + 1(x-3) = 10$$
$$2x = 10$$
$$x = 5$$

69. $\dfrac{3}{x^2-3x} + \dfrac{4}{x} = \dfrac{1}{x-3}$

$\dfrac{3}{x(x-3)} + \dfrac{4}{x} = \dfrac{1}{x-3}$ Multiply both sides by $(x-1)(3x+1)$.

$$3 + 4(x-3) = x$$
$$3 + 4x - 12 = x$$
$$3x = 9$$
$$x = 3$$

A check reveals that $x = 3$ is an extraneous solution since it makes the denominator zero, so there is no solution.

71. $(x+2)^2 + 5 = (x+3)^2$

$$x^2 + 4x + 4 + 4 = x^2 + 6x + 9$$
$$4x + 9 = 6x + 9$$
$$-2x = 0$$
$$x = 0$$

73. $(x+2)^2 - x^2 = 4(x+1)$

$$x^2 + 4x + 4 - x^2 = 4x + 4$$
$$4 = 4$$

The equation is an identity; every real number is a solution.

75. $y = 2(x-1) - 4$

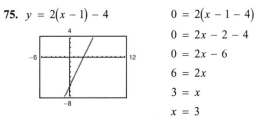

$$0 = 2(x - 1 - 4)$$
$$0 = 2x - 2 - 4$$
$$0 = 2x - 6$$
$$6 = 2x$$
$$3 = x$$
$$x = 3$$

The x-intercept is at 3. The solution of $0 = 2(x-1) - 4$ and the x-intercept of $y = 2(x-1) - 4$ are the same. They are both $x = 3$.

The x-intercept is $(3, 0)$.

77. $y = 20 - (3x - 10)$

The x-intercept is at 10. The solution of $0 = 2 - (3x - 10)$ and the x-intercept of $y = 20 - (3x - 10)$ are the same. They are both $x = 10$. The x-intercept is $(10, 0)$.

$0 = 20 - (3x - 10)$
$0 = 20 - 3x + 10$
$0 = 30 - 3x$
$3x = 30$
$x = 10$

79. $y = -38 + 5(9 - x)$

The x-intercept is at $\frac{7}{5}$. The solution of $0 = -38 + 5(9 - x)$ and the x-intercept of $y = -38 + 5(9 - x)$ are the same. They are both $x = \frac{7}{5}$. The x-intercept is $\left(\frac{7}{5}, 0\right)$.

$0 = -38 + 5(9 - x)$
$0 = -38 + 45 - 5x$
$0 = 7 - 5x$
$5x = 7$
$x = \frac{7}{5}$

81. $y = 12 - 5x$

x-intercept: $0 = 12 - 5x \Rightarrow 5x = 12 \Rightarrow x = \frac{12}{5}$

y-intercept: $y = 12 - 5(0) \Rightarrow y = 12$

The x-intercept is $\left(\frac{12}{5}, 0\right)$ and the y-intercept is $(0, 12)$.

83. $y = -3(2x + 1)$

x-intercept: $0 = -3(2x + 1) \Rightarrow 0 = 2x + 1 \Rightarrow x = -\frac{1}{2}$

y-intercept: $y = -3(2(0) + 1) \Rightarrow y = -3$

The x-intercept is $\left(-\frac{1}{2}, 0\right)$ and the y-intercept is $(0, -3)$.

85. $2x + 3y = 10$

x-intercept: $2x + 3(0) = 10 \Rightarrow 2x = 10 \Rightarrow x = 5$

y-intercept: $2(0) + 3y = 10 \Rightarrow 3y = 10 \Rightarrow y = \frac{10}{3}$

The x-intercept is $(5, 0)$ and the y-intercept is $\left(0, \frac{10}{3}\right)$.

87. $\dfrac{2x}{5} + 8 - 3y = 0 \Rightarrow 2x + 40 - 15y = 0$ Multiply both sides by 5.

x-intercept: $2x + 40 - 15(0) = 0 \Rightarrow 2x + 40 = 0 \Rightarrow x = -20$

y-intercept: $2(0) + 40 - 15y = 0 \Rightarrow 40 - 15y = 0 \Rightarrow y = \dfrac{40}{15} = \dfrac{8}{3}$

The x-intercept is $(-20, 0)$ and the y-intercept is $\left(0, \frac{8}{3}\right)$.

89. $4y - 0.75x + 1.2 = 0$

x-intercept: $4(0) - 0.75x + 1.2 = 0 \Rightarrow -0.75 + 1.2 = 0 \Rightarrow x = \dfrac{1.2}{0.75} = 1.6$

y-intercept: $4y - 0.75(0) + 1.2 = 0 \Rightarrow 4y + 1.2 = 0 \Rightarrow y = \dfrac{-1.2}{4} = -0.3$

The x-intercept is $(1.6, 0)$ and the y-intercept is $(0, -0.3)$.

91. The student solved correctly to find $x = 2$, but forgot to check the solution. In the original equation, $x = 2$ yields a denominator of zero, which is undefined.
So, $x = 2$ is an extraneous solution, and the equation has no solution.

93. $0.275x + 0.725(500 - x) = 300$
$0.275x + 362.5 - 0.725x = 300$
$-0.45x = -62.5$
$x = \dfrac{62.5}{0.45} \approx 138.889$

95. $\dfrac{2}{7.398} - \dfrac{4.405}{x} = \dfrac{1}{x}$ Multiply both sides by $7.398x$.

$$2x - (4.405)(7.398) = 7.398$$
$$2x = (4.405)(7.398) + 7.398$$
$$2x = (5.405)(7.398)$$
$$x = \dfrac{(5.405)(7.398)}{2} \approx 19.993$$

97. $4(x+1) - ax = x + 5$

$$4x + 4 - ax = x + 5$$
$$3x - ax = 1$$
$$x(3-a) = 1$$
$$x = \dfrac{1}{3-a},\ a \neq 3$$

99. $6x + ax = 2x + 5$

$$4x + ax = 5$$
$$x(4+a) = 5$$
$$x = \dfrac{5}{4+a},\ a \neq -4$$

101. $19x + \dfrac{1}{2}ax = x + 9$

$$18x + \dfrac{1}{2}ax = 9 \quad \text{Multiply both sides by 2.}$$
$$36x + ax = 18$$
$$x(36+a) = 18$$
$$x = \dfrac{18}{36+a},\ a \neq -36$$

103. $-2ax + 6(x+3) = -4x + 1$

$$-2ax + 6x + 18 = -4x + 1$$
$$-2ax + 10x + 18 = 1$$
$$-2ax + 10x = -17$$
$$x(-2a+10) = -17$$
$$x = \dfrac{-17}{-2a+10} = \dfrac{-17}{10-2a},\ a \neq 5$$

105. $471 = 2\pi(25) + 2\pi(5h)$

$$471 = 50\pi + 10\pi h$$
$$471 - 50\pi = 10\pi h$$
$$h = \dfrac{471 - 50\pi}{10\pi} = \dfrac{471 - 50(3.14)}{10(3.14)} = 10$$
$$h = 10 \text{ feet}$$

107. (a) Female: $y = 0.432x - 10.44$

For $y = 16$: $\quad 16 = 0.432x - 10.44$
$$26.44 = 0.432x$$
$$\dfrac{26.44}{0.432} = x$$
$$x \approx 61.2 \text{ inches}$$

(b) Male: $y = 0.449x - 12.15$

For $y = 19$: $\quad 19 = 0.449x - 12.15$
$$31.15 = 0.449x$$
$$69.4 \approx x$$

Yes, it is likely that both bones came from the same person because the estimated height of a male with a 19-inch thigh bone is 69.4 inches.

(c)

Height x	Female Femur Length	Male Femur Length
60	15.48	14.79
70	19.80	19.28
80	24.12	23.77
90	28.44	28.26
100	32.76	32.75
110	37.08	37.24

The lengths of the male and female femurs are approximately equal when the lengths are 100 inches.

(d) $0.432x - 10.44 = 0.449x - 12.15$
$$1.71 = 0.017x$$
$$x \approx 100.59 \text{ inches}$$

It is unlikely that a female would be over 8 feet tall, so if a femur of this length was found, it most likely belonged to a very tall male.

109. $y = -7.69t + 1480.7, -4 \le t \le 7$

(a)

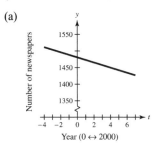

y-intercept: $(0, 1480.7)$

(b) Let $t = 0$: $y = -7.69(0) + 1480.7 = 1480.7$

x-intercept: $(0, 1480.7)$

(c) $1373 = -7.69t + 1480.7$

$-107.7 = -7.69t$

$t = \dfrac{-107.9}{-7.69} \approx 14$

This corresponds to 2014. This answer seems reasonable. Explanations will vary.

111. $10,000 = 0.32m + 2500$

$7500 = 0.32m$

$\dfrac{7500}{0.32} = m$

$m = 23,437.5$ miles

113. False. $x(3 - x) = 10 \Rightarrow 3x - x^2 = 10$

This is a quadratic equation. The equation cannot be written in the form $ax + b = 0$.

115. $2(x - 3) + 1 = 2x - 5$

$2x - 6 + 1 = 2x - 5$

$2x - 5 = 2x - 5$

False. The equation is an identity, so every real number is a solution.

117.
$$2 - \frac{1}{x - 2} = \frac{3}{x - 2}$$
$$(x - 2)\left(2 - \frac{1}{x-2}\right) = (x-2)\left(\frac{3}{x-2}\right)$$
$$2(x - 2) - 1 = 3$$
$$2x - 4 - 1 = 3$$
$$2x - 5 = 3$$
$$2x = 8$$
$$x = 4$$

False. $x = 4$ is a solution.

119. (a)

x	-1	0	1	2	3	4
$3.2x - 5.8$	-9	-5.8	-2.6	0.6	3.8	7

(b) Since the sign changes from negative at 1 to positive at 2, the root is somewhere between 1 and 2.

$1 < x < 2$

(c)

x	1.5	1.6	1.7	1.8	1.9	2
$3.2x - 5.8$	-1	-0.68	-0.36	-0.04	0.28	0.6

(d) Since the sign changes from negative at 1.8 to positive at 1.9, the root is somewhere between 1.8 and 1.9.

$1.8 < x < 1.9$

To improve accuracy, evaluate the expression at subintervals within this interval and determine where the sign changes.

121. (a)

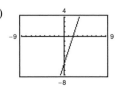

(b) *x*-intercept: $(2, 0)$

(c) The *x*-intercept is the solution of the equation $3x - 6 = 0$.

Section 1.3   Modeling with Linear Equations

- You should be able to set up mathematical models to solve problems.

- You should be able to translate key words and phrases.
  - (a) Equality: Equals, equal to, is, are, was, will be, represents
  - (b) Addition: Sum, plus, greater, increased by, more than, exceeds, total of
  - (c) Subtraction: Difference, minus, less than, decreased by, subtracted from, reduced by, the remainder
  - (d) Multiplication: Product, multiplied by, twice, percent of
  - (e) Division: Quotient, divided by, ratio, per
  - (f) Consecutive: Next, subsequent

- You should know the following formulas:
  - (a) Perimeter:
    1. Square: $P = 4s$
    2. Rectangle: $P = 2l + 2w$
    3. Circle: $C + 2\pi r$
    4. Triangle: $P = a + b + c$
  - (b) Area:
    1. Square: $A = s^2$
    2. Rectangle: $A = lw$
    3. Circle: $A = \pi r^2$
    4. Triangle: $A = \left(\dfrac{1}{2}\right)bh$
  - (c) Volume:
    1. Cube: $V = s^3$
    2. Rectangular solid: $V = lwh$
    3. Cylinder: $V = \pi r^2 h$
    4. Sphere: $V = \left(\dfrac{4}{3}\right)\pi r^3$
  - (d) Simple Interest: $I = Prt$
  - (e) Compound Interest: $A = P\left(1 + \dfrac{r}{n}\right)^{nt}$
  - (f) Distance: $d = rt$
  - (g) Temperature: $F = \dfrac{9}{5}C + 32$

- You should be able to solve word problems. Study the examples in the text carefully.

1. mathematical modeling

3. $A = \pi r^2$

5. $V = s^3$

7. $A = P\left(1 + \dfrac{r}{12}\right)^{12t}$

9. $x + 4$

The sum of a number and 4
A number increased by 4

11. $\dfrac{u}{5}$

The ratio of a number and 5
The quotient of a number and 5
A number divided by 5

13. $\dfrac{y - 4}{5}$

The difference of a number and 4 is divided by 5.
A number decreased by 4 is divided by 5.

15. $-3(b + 2)$

The product of -3 and the sum of a number and 2.
Negative 3 is multiplied by a number increased by 2.

17. $\dfrac{4(p-1)}{p}$

The product of 4 and the difference of a number and 1 is divided by the number.
A number decreased by 1 is multiplied by 4 and divided by the number.

19. *Verbal Model:* $(\text{Sum}) = (\text{first number}) + (\text{second number})$

   Labels: $\text{Sum} = S$, first number $= n$, second number $= n + 1$

   Expression: $S = n + (n + 1) = 2n + 1$

21. *Verbal Model:* $\text{Product} = (\text{first odd integer})(\text{second odd integer})$

   Labels: $\text{Product} = P$, first odd integer $= 2n - 1$, second odd integer $= 2n - 1 + 2 = 2n + 1$

   Expression: $P = (2n - 1)(2n + 1) = 4n^2 - 1$

23. *Verbal Model:* $(\text{distance}) = (\text{rate}) \cdot (\text{time})$

   Labels: Distance $= d$, rate $= 55$ mph, time $= t$

   Expression: $d = 55t$

25. *Verbal Model:* $(\text{Amount of acid}) = 20\% \cdot (\text{amount of solution})$

   Labels: Amount of acid (in gallons) $= A$, amount of solution (in gallons) $= x$

   Expression: $A = 0.20x$

27. *Verbal Model:*   Perimeter $= 2(\text{width}) + 2(\text{length})$

   Labels: Perimeter $= P$, width $= x$, length $= 2(\text{width}) = 2x$

   Expression: $P = 2x + 2(2x) = 6x$

29. *Verbal Model:* $(\text{Total cost}) = (\text{unit cost})(\text{number of units}) + (\text{fixed cost})$

   Labels: Total cost $= C$, unit cost $= \$40$, number of units $= x$, fixed cost $= \$2500$

   Expression: $C = 40x + 2500$

31. *Verbal Model:*   Thirty percent of the list price L

   Expression: $0.30L$

33. *Verbal Model:*   Percent of 672 that is represented by the number N

   Expression: $N = p(672)$, p is in decimal form.

35.

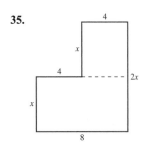

Area $=$ Area of top rectangle $+$ Area of bottom rectangle

   $A = 4x + 8x = 12x$

37. *Verbal Model:*   Sum = (first number) + (second number)

   Labels: Sum = 525, first number = n, second number = $n + 1$

   Equation: $525 = n + (n + 1)$

   $525 = 2n + 1$

   $524 = 2n$

   $n = 262$

   Answer: First number = $n = 262$, second number = $n + 1 = 263$

39. *Verbal Model:*   Difference = (one number) − (another number)

   Labels: Difference = 148, one number = $5x$, another number = x

   Equation: $148 = 5x - x$

   $148 = 4x$

   $x = 37$

   $5x = 185$

   Answer: The two numbers are 37 and 185.

41. *Verbal Model:*   Product = (smaller number) · (larger number) = (smaller number)^2 − 5

   Labels: Smaller number = n, larger number = $n + 1$

   Equation: $n(n + 1) = n^2 - 5$

   $n^2 + n = n^2 - 5$

   $n = -5$

   Answer: Smaller number = $n = -5$, larger number = $n + 1 = -4$

43. $x = $ percent · number

   $= (30\%)(45)$

   $= 0.30(45)$

   $= 13.5$

45. $432 = $ percent · 1600

   $432 = p(1600)$

   $\frac{432}{1600} = p$

   $p = 0.27 = 27\%$

47. $12 = \frac{1}{2}\% \cdot$ number

   $12 = 0.005x$

   $\frac{12}{0.005} = x$

   $x = 2400$

49. *Verbal Model:*   (first paycheck) + (second paycheck) = total

   Labels: second paycheck = x, first paycheck = $0.85x$, total = \$1125

   Equation: $0.85x + x = 1125$

   $1.85x = 1125$

   $x \approx 608.11$

   $0.85x \approx 516.89$

The first salesperson's weekly paycheck is \$516.89 and the second salesperson's weekly paycheck is \$608.11.

51. *Verbal Model:* Loan payments $= 32\%$ (Annual income)

    Labels: Loan payments $= \$15{,}125.50$, Annual income $= I$

    Equation:
$$15{,}125.50 = 0.32I$$
$$I \approx 47{,}267.19$$

The family's annual income is $\$47{,}267.19$.

53. *Verbal Model:* (2007 price) $=$ (percent increase)(2000 price) $+$ (2000 price)

    Labels: 2007 price $= \$2.80$, percent increase $= p$, 2000 price $= \$1.51$

    Equation:
$$2.8 = 1.51p + 1.51$$
$$1.29 = 1.51p$$
$$0.854 \approx p$$

The percent increase in the price of a gallon of regular unleaded gasoline was about 85.4%.

55. *Verbal Model:* (2007 price) $=$ (percent increase)(2000 price) $+$ (2000 price)

    Labels: 2007 price $= \$2.23$, percent increase $= p$, 2000 price $= \$1.63$

    Equation:
$$2.23 = 1.63p + 1.63$$
$$0.6 = 1.63p$$
$$0.368 \approx p$$

The percent increase in the price of a pound of 100% ground beef was about 36.8%.

57. (a)

  (b) $l = 1.5w$
$$P = 2l + 2w = 2(1.5w) + 2w = 5w$$

  (c) $25 = 5w$
$$5 = w$$
Width: $w = 5$ meters
Length: $l = 1.5w = 7.5$ meters
Dimensions: 7.5 meters $\times$ 5 meters

59. *Verbal Model:* $\text{Average} = \dfrac{(\text{test \#1}) + (\text{test \#2}) + (\text{test \#3}) + (\text{test \#4})}{4}$

    Labels: Average $= 90$, test #1 $= 87$, test #2 $= 92$, test #3 $= 84$, test #4 $= x$

    Equation: $90 = \dfrac{87 + 92 + 84 + x}{4}$

    Answer: You must score 97 or better on test #4 to earn an A for the course.

61. $\text{Rate} = \dfrac{\text{distance}}{\text{time}} = \dfrac{50 \text{ kilometers}}{\dfrac{1}{2} \text{ hour}} = 100 \text{ kilometers/hour}$

     $\text{Total time} = \dfrac{\text{total distance}}{\text{rate}} = \dfrac{500 \text{ kilometers}}{100 \text{ kilometers/hour}} = 5 \text{ hours}$

63. *Verbal Model:* $(\text{Distance}) = (\text{rate})(\text{time}_1 + \text{time}_2)$

   Labels: $\text{Distance} = 2 \cdot 200 = 400 \text{ miles, rate} = 2,$

   $$\text{time}_1 = \frac{\text{distance}}{\text{rate}_1} = \frac{200}{55} \text{ hours}$$

   $$\text{time}_2 = \frac{\text{distance}}{\text{rate}_2} = \frac{200}{40} \text{ hours}$$

   Equation: $400 = r\left(\dfrac{200}{55} + \dfrac{200}{40}\right)$

   $$400 = r\left(\frac{1600}{440} + \frac{2200}{440}\right) = \frac{3800}{440}r$$

   $$46.3 \approx r$$

The average speed for the round trip was approximately 46.3 miles per hour.

65. *Verbal Model:* $(\text{Distance}) = (\text{rate})(\text{time})$

   Labels: $\text{Distance} = 1.5 \times 10^{11} \text{ meters}$

   $\text{Rate} = 3.0 \times 10^{8} \text{ meters per second}$

   $\text{Time} = t$

   Equation: $1.5 \times 10^{11} = \left(3.0 \times 10^{8}\right)t$

   $$500 = t$$

Light from the sun travels to the Earth in 500 seconds or approximately 8.33 minutes.

67. *Verbal Model:* $\dfrac{(\text{height of building})}{(\text{length of building's shadow})} = \dfrac{(\text{height of stake})}{(\text{length of stake's shadow})}$

   Labels: Let h = height of the building in feet.

   Equation: $\dfrac{h \text{ feet}}{87 \text{ feet}} = \dfrac{4 \text{ feet}}{1/3 \text{ foot}}$

   $$\frac{h}{87} = \frac{4}{1/3}$$

   $$\frac{1}{3}h = 348$$

   $$h = 1044 \text{ feet}$$

The Chrysler Building is 1044 feet tall.

69. (a)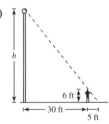

(b) *Verbal Model:* $\dfrac{(\text{height of pole})}{(\text{height of pole's shadow})} = \dfrac{(\text{height of person})}{(\text{height of person's shadow})}$

   Labels: Height of pole = h, height of pole's shadow = $30 + 5 = 35$ feet, height of person = 6 feet, height of person's shadow = 5 feet

   Equation: $\dfrac{h}{35} = \dfrac{6}{5}$

   $$h = \frac{6}{5} \cdot 35 = 42$$

The pole is 42 feet tall.

71. *Verbal Model:* $\left(\text{Interest from } 4\frac{1}{2}\%\right) + \left(\text{interest from } 5\%\right) = \left(\text{total interest}\right)$

    Labels: Amount invested at $4\frac{1}{2}\% = x$, amount invested at $5\% = 12{,}000 - x$, interest from $4\frac{1}{2}\% = x(0.045)$,

    interest from $5\% = (12{,}000 - x)(0.05)$, total annual interest $= \$580$

    Equation:
$$0.045x + 0.05(12{,}000 - x) = 580$$
$$0.045x + 600 - 0.05x = 580$$
$$-0.005x = -20$$
$$x = 4000$$

The smallest amount the can be invested at 5% is $12{,}000 - 4000 = \$8000$.

73. *Verbal Model:* $\left(\text{Profit from dogwood trees}\right) + \left(\text{profit from red maple trees}\right) = \left(\text{total profit}\right)$

    Labels: Inventory of dogwood trees $= x$, inventory of red maple trees $= 40{,}000 - x$,

    profit from dogwood trees $= 0.25x$, profit from red maple trees $= 0.17(40{,}000 - x)$,

    total profit $= 0.20(40{,}000) = 8000$

    Equation:
$$0.25x + 0.17(40{,}000) = 8000$$
$$0.25x + 6800 - 0.17x = 8000$$
$$0.08x = 1200$$
$$x = 15{,}000$$

The amount invested in dogwood trees was $\$15{,}000$ and the amount invested in red maple trees was $40{,}000 - 15{,}000 = \$25{,}000$.

75. (a) *Verbal Model:* $\left(\begin{array}{c}\text{Amount in}\\\text{Solution 1}\end{array}\right) + \left(\begin{array}{c}\text{amount in}\\\text{Solution 2}\end{array}\right) = \left(\begin{array}{c}\text{amount in}\\\text{final solution}\end{array}\right)$

    Labels: Amount in Solution 1 $= 0.10x$, amount in Solution 2 $= 0.30(100 - x)$,

    amount in final solution $= 0.25(100)$

    Equation:
$$0.10x + 0.30(100 - x) = 0.25(100)$$
$$0.10x + 30 - 0.30x = 25$$
$$-0.20x = -5$$
$$x = 25$$

Use 25 gallons of Solution 1 and $100 - 25 = 75$ gallons of Solution 2.

  (b) *Verbal Model:* $\left(\begin{array}{c}\text{Amount in}\\\text{Solution 1}\end{array}\right) + \left(\begin{array}{c}\text{amount in}\\\text{Solution 2}\end{array}\right) = \left(\begin{array}{c}\text{amount in}\\\text{final solution}\end{array}\right)$

    Labels: Amount in Solution 1 $= 0.25x$, amount in Solution 2 $= 0.50(5 - x)$,

    amount in final solution $= 0.30(5)$

    Equation:
$$0.25x + 0.50(5 - x) = 1.5$$
$$0.25x + 2.5 - 0.50x = 1.5$$
$$-0.25x = -1$$
$$x = 4$$

Use 4 liters of Solution 1 and $5 - 4 = 1$ gallons of Solution 2.

©2011 Cengage Learning. All Rights Reserved. May not be scanned, copied or duplicated, or posted to a publicly accessible website, in whole or in part.

(c) *Verbal Model:* $\left(\begin{array}{c}\text{Amount in}\\\text{Solution 1}\end{array}\right) + \left(\begin{array}{c}\text{amount in}\\\text{Solution 2}\end{array}\right) = \left(\begin{array}{c}\text{amount in}\\\text{final solution}\end{array}\right)$

   Labels: Amount in Solution 1 = $0.15x$, amount in Solution 2 = $0.45(10 - x)$,

   amount in final solution = $0.30(10)$

   Equation: $0.15x + 0.45(10 - x) = 0.30(10)$

   $0.15x + 4.5 - 0.45x = 3$

   $-0.30x = -1.5$

   $x = 5$

   Use 5 quarts of Solution 1 and $10 - 5 = 5$ quarts of Solution 2.

(d) *Verbal Model:* $\left(\begin{array}{c}\text{Amount in}\\\text{Solution 1}\end{array}\right) + \left(\begin{array}{c}\text{amount in}\\\text{Solution 2}\end{array}\right) = \left(\begin{array}{c}\text{amount in}\\\text{final solution}\end{array}\right)$

   Labels: Amount in Solution 1 = $0.70x$, amount in Solution 2 = $0.90(25 - x)$,

   amount in final solution = $0.75(25)$

   Equation: $0.70x + 0.90(25 - x) = 0.75(25)$

   $0.70x + 22.5 - 0.90x = 18.75$

   $-0.20x = -3.75$

   $x = 18.75$

   Use 18.75 gallons of Solution 1 and $25 - 18.75 = 6.25$ quarts of Solution 2.

77. *Verbal Model:* $\left(\begin{array}{c}\text{Amount of gasoline}\\\text{in Solution 1}\end{array}\right) + \left(\begin{array}{c}\text{Amount of gasoline}\\\text{in Solution 2}\end{array}\right) = \left(\begin{array}{c}\text{Amount of gasoline}\\\text{in final solution}\end{array}\right)$

   Labels: Amount of gasoline in Solution 1 = $\frac{22}{33}(2)$, amount of gasoline in Solution 2 = x,

   amount of gasoline in final solution = $\frac{40}{41}(2 + x)$

   Equation: $\frac{64}{33} + x = \frac{40}{41}(2 + x)$

   $\frac{64}{33} + x = \frac{80}{41} + \frac{40}{41}x$ Multiply both sides by $33(41)$.

   $2624 + 1353x = 2640 + 1320x$

   $33x = 16$

   $x \approx 0.48$

   Approximately 0.48 gallon of gasoline must be added.

79. *Verbal Model:*   (Fixed costs) + (variable cost per unit)(number of units) = (total cost)

   Labels: Fixed costs = \$14,000, variable cost per unit = \$12.75, number of units = x, total cost = \$110,000

   Equation: $14,000 + 12.75x = 110,000$

   $12.75x = 96,000$

   $x \approx 7529$

   The company can produce no more than 7529 units with \$110,000 budgeted for monthly costs.

81. $A = \frac{1}{2}bh$

   $2A = bh$

   $\frac{2A}{b} = h$

83. $S = C + RC$

   $S = C(1 + R)$

   $\frac{S}{1 + R} = C$

85. $V = \frac{4}{3}\pi a^2 b$

$\frac{3}{4}V = \pi a^2 b$

$\frac{(3/4)V}{\pi a^2} = b$

$\frac{3V}{4\pi a^2} = b$

87. $h = v_0 t + \frac{1}{2}at^2$

$h - v_0 t = \frac{1}{2}at^2$

$2(h - v_0 t) = at^2$

$\frac{2(h - v_0 t)}{t^2} = a$

89. $C = \dfrac{1}{\dfrac{1}{C_1} + \dfrac{1}{C_2}}$

$\dfrac{1}{C} - \dfrac{1}{C_2} = \dfrac{1}{C_1}$

$\dfrac{C_2 - C}{CC_2} = \dfrac{1}{C_1}$

$\dfrac{CC_2}{C_2 - C} = C_1$

91. $L = a + (n - 1)d$

$L - a = nd - d$

$L - a + d = nd$

$\dfrac{L - a + d}{d} = n$

93. $W_1 x = W_2(L - x)$

$50x = 75(10 - x)$

$50x = 750 - 75x$

$125x = 750$

$x = 6$ feet from 50-pound child

95. $V = \frac{4}{3}\pi r^3$

$5.96 = \frac{4}{3}\pi r^3$

$17.88 = 4\pi r^3$

$\dfrac{17.88}{4\pi} = r^3$

$r = \sqrt[3]{\dfrac{4.47}{\pi}} \approx 1.12$ inches

97. $C = \frac{5}{9}(F - 32)$

When $F = 64.4°$, $\frac{5}{9}(64.4 - 32) = 18°$C.

99. $C = \frac{9}{5}C + 32$

When $C = 50°$, $\frac{9}{5}(50) + 32 = 122°$F.

101. False, it should be written as $\dfrac{z^3 - 8}{z^2 - 9}$.

103. $ax + b = 0 \Rightarrow x = -\dfrac{b}{a}$

(a) If $ab > 0$, then a and b have the same sign and $x = -b/a$ is negative.

(b) If $ab < 0$, then a and b have opposite signs and $x = -b/a$ is positive.

105. Answers will vary. Sample answer: $x + 7 = 4$

Section 1.4 Quadratic Equations and Applications

- You should be able to solve a quadratic equation by factoring, if possible.
- You should be able to solve a quadratic equation of the form $u^2 = d$ by extracting square roots.
- You should be able to solve a quadratic equation by completing the square.
- You should know and be able to use the Quadratic Formula: For $ax^2 + bx + c = 0$, $a \neq 0$,

$$x = \frac{-b \pm \sqrt{b^2 - 4ac}}{2a}.$$

—continued—

- You should be able to determine the types of solutions of a quadratic equation by checking the discriminant $b^2 - 4ac$.

  (a) If $b^2 - 4ac > 0$, there are two distinct real solutions. The graph has two x-intercepts.

  (b) If $b^2 - 4ac = 0$, there is one repeated real solution. The graph has one x-intercept.

  (c) If $b^2 - 4ac < 0$, there is no real solution. The graph has no x-intercepts.

- You should be able to use your calculator to solve quadratic equations.
- You should be able to solve applications involving quadratic equation. Study the examples in the text carefully.

---

1. quadratic equation

3. factoring; square roots; completing; square; Quadratic Formula

5. position equation

7. $2x^2 = 3 - 5x$

General form: $2x^2 + 5x - 3 = 0$

9. $(x - 3)^2 = 3$

$x^2 - 6x + 9 = 3$

General form: $x^2 - 6x + 6 = 0$

11. $\frac{1}{5}(3x^2 - 10) = 12x$

$3x^2 - 10 = 60x$

General form: $3x^2 - 60x - 10 = 0$

13. $6x^2 + 3x = 0$

$3x(2x + 1) = 0$

$3x = 0$ or $2x + 1 = 0$

$x = 0$ or $x = -\frac{1}{2}$

15. $x^2 - 2x - 8 = 0$

$(x - 4)(x + 2) = 0$

$x - 4 = 0$ or $x + 2 = 0$

$x = 4$ or $x = -2$

17. $x^2 + 10x + 25 = 0$

$(x + 5)^2 = 0$

$x + 5 = 0$

$x = -5$

19. $3 + 5x - 2x^2 = 0$

$(3 - x)(1 + 2x) = 0$

$3 - x = 0$ or $1 + 2x = 0$

$x = 3$ or $x = -\frac{1}{2}$

21. $x^2 + 4x = 12$

$x^2 + 4x - 12 = 0$

$(x + 6)(x - 2) = 0$

$x + 6 = 0$ or $x - 2 = 0$

$x = -6$ or $x = 2$

23. $\frac{3}{4}x^2 + 8x + 20 = 0$

$4\left(\frac{3}{4}x^2 + 8x + 20\right) = 4(0)$

$3x^2 + 32x + 80 = 0$

$(3x + 20)(x + 4) = 0$

$3x + 20 = 0$ or $x + 4 = 0$

$x = -\frac{20}{3}$ or $x = -4$

25. $x^2 = 49$

$x = \pm 7$

27. $x^2 = 11$

$x = \pm\sqrt{11}$

29. $3x^2 = 81$

$x^2 = 27$

$x = \pm 3\sqrt{3}$

31. $(x - 12)^2 = 16$

$x - 12 = \pm 4$

$x = 12 \pm 4$

$x = 16$ or $x = 8$

33. $(x + 2)^2 = 14$

$x + 2 = \pm\sqrt{14}$

$x = -2 \pm \sqrt{14}$

---

35. $(2x - 1)^2 = 18$

$$2x - 1 = \pm\sqrt{18}$$

$$2x = 1 \pm 3\sqrt{2}$$

$$x = \frac{1 \pm 3\sqrt{2}}{2}$$

37. $(x - 7)^2 = (x + 3)^2$

$$x - 7 = \pm(x + 3)$$

$$x - 7 = x + 3 \quad \text{or} \quad x - 7 = -x - 3$$

$$-7 \neq 3 \qquad \text{or} \qquad 2x = 4$$

$$x = 2$$

The only solution of the equation is $x = 2$.

39. $x^2 + 4x - 32 = 0$

$$x^2 + 4x = 32$$

$$x^2 + 4x + 2^2 = 32 + 2^2$$

$$(x + 2)^2 = 36$$

$$x + 2 = \pm 6$$

$$x = -2 \pm 6$$

$$x = 4 \quad \text{or} \quad x = -8$$

41. $x^2 + 6x + 2 = 0$

$$x^2 + 6x = -2$$

$$x^2 + 6x + 3^2 = -2 + 3^2$$

$$(x + 3)^2 = 7$$

$$x + 3 = \pm\sqrt{7}$$

$$x = -3 \pm \sqrt{7}$$

43. $9x^2 - 18x = -3$

$$x^2 - 2x = -\frac{1}{3}$$

$$x^2 - 2x + 1^2 = -\frac{1}{3} + 1^2$$

$$(x - 1)^2 = \frac{2}{3}$$

$$x - 1 = \pm\sqrt{\frac{2}{3}}$$

$$x = 1 \pm \sqrt{\frac{2}{3}}$$

$$x = 1 \pm \frac{\sqrt{6}}{3}$$

45. $7 + 2x - x^2 = 0$

$$-x^2 + 2x + 7 = 0$$

$$x^2 - 2x - 7 = 0$$

$$x^2 - 2x = 7$$

$$x^2 - 2x + (-1)^2 = 7 + (-1)^2$$

$$(x - 1)^2 = 8$$

$$x - 1 = \pm 2\sqrt{2}$$

$$x = 1 \pm 2\sqrt{2}$$

47. $2x^2 + 5x - 8 = 0$

$$2x^2 + 5x = 8$$

$$x^2 + \frac{5}{2}x = 4$$

$$x^2 + \frac{5}{2}x + \left(\frac{5}{4}\right)^2 = 4 + \left(\frac{5}{4}\right)^2$$

$$\left(x + \frac{5}{4}\right)^2 = \frac{89}{16}$$

$$x + \frac{5}{4} = \pm\frac{\sqrt{89}}{4}$$

$$x = -\frac{5}{4} \pm \frac{\sqrt{89}}{4}$$

$$x = \frac{-5 \pm \sqrt{89}}{4}$$

49. $\dfrac{1}{x^2 + 2x + 5} = \dfrac{1}{x^2 + 2x + 1^2 - 1^2 + 5}$

$$= \dfrac{1}{(x + 1)^2 + 4}$$

51. $\dfrac{4}{x^2 + 4x - 3} = \dfrac{4}{x^2 + 4x + 4 - 4 - 3}$

$$= \dfrac{4}{(x + 2)^2 - 7}$$

53. $\dfrac{1}{4x^2 + 4x + 9} = \dfrac{\frac{1}{4}}{x^2 + x + \frac{9}{4}}$

$$= \dfrac{\frac{1}{4}}{x^2 + x + \left(\frac{1}{2}\right)^2 + \frac{9}{4} - \left(\frac{1}{2}\right)^2}$$

$$= \dfrac{\frac{1}{4}}{\left(x + \frac{1}{2}\right)^2 + 2}$$

55. $\dfrac{1}{\sqrt{6x - x^2}} = \dfrac{1}{\sqrt{-1\left(x^2 - 6x + 3^2 - 3^2\right)}}$

$$= \dfrac{1}{\sqrt{-1\left[(x - 3)^2 - 9\right]}}$$

$$= \dfrac{1}{\sqrt{-(x - 3)^2 + 9}}$$

$$= \dfrac{1}{\sqrt{9 - (x - 3)^2}}$$

57. (a) $y = (x + 3)^2 - 4$

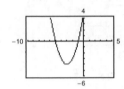

(b) The x-intercepts are $(-1, 0)$ and $(-5, 0)$.

(c) $0 = (x + 3)^2 - 4$

$$4 = (x + 3)^2$$

$$\pm\sqrt{4} = x + 3$$

$$-3 \pm 2 = x$$

$$x = -1 \ \text{ or } \ x = -5$$

(d) The x-intercepts of the graphs are solutions of the equation $0 = (x + 3)^2 - 4$.

59. (a) $y = 1 - (x - 2)^2$

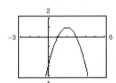

(b) The x-intercepts are $(1, 0)$ and $(3, 0)$.

(c) $0 = 1 - (x - 2)^2$

$$(x - 2)^2 = 1$$

$$x - 2 = \pm 1$$

$$x = 2 \pm 1$$

$$x = 3 \ \text{ or } \ x = 1$$

(d) The x-intercepts of the graphs are solutions of the equation $0 = 1 - (x - 2)^2$.

61. (a) $y = -4x^2 + 4x + 3$

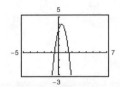

(b) The x-intercepts are $\left(-\frac{1}{2}, 0\right)$ and $\left(\frac{3}{2}, 0\right)$.

(c) $0 = -4x^2 + 4x + 3$

$$4x^2 - 4x = 3$$

$$4\left(x^2 - x\right) = 3$$

$$x^2 - x = \tfrac{3}{4}$$

$$x^2 - x + \left(\tfrac{1}{2}\right)^2 = \tfrac{3}{4} + \left(\tfrac{1}{2}\right)^2$$

$$\left(x - \tfrac{1}{2}\right)^2 = 1$$

$$x - \tfrac{1}{2} = \pm\sqrt{1}$$

$$x = \tfrac{1}{2} \pm 1$$

$$x = \tfrac{3}{2} \ \text{ or } \ x = -\tfrac{1}{2}$$

(d) The x-intercepts of the graphs are solutions of the equation $0 = -4x^2 + 4x + 3$.

63. (a) $y = x^2 + 3x - 4$

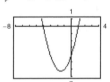

(b) The x-intercepts are $(-4, 0)$ and $(1, 0)$.

(c) $0 = x^2 + 3x - 4$

$$0 = (x + 4)(x - 1)$$

$$x + 4 = 0 \quad \text{or} \quad x - 1 = 0$$

$$x = -4 \quad \text{or} \qquad x = 1$$

(d) The x-intercepts of the graphs are solutions of the equation $0 = x^2 + 3x - 4$.

65. $2x^2 - 5x + 5 = 0$

$$b^2 - 4ac = (-5)^2 - 4(2)(5) = -15 < 0$$

No real solution

67. $2x^2 - z - 1 = 0$

$$b^2 - 4ac = (-1)^2 - 4(2)(-1) = 9 > 0$$

Two real solutions

69. $\frac{1}{3}x^2 - 5x + 25 = 0$

$b^2 - 4ac = (-5)^2 - 4\left(\frac{1}{3}\right)(25) = -\frac{25}{3} < 0$

No real solution

71. $0.2x^2 + 1.2x - 8 = 0$

$b^2 - 4ac = (1.2)^2 - 4(0.2)(-8) = 7.84 > 0$

Two real solutions

73. $2x^2 + x - 1 = 0$

$x = \dfrac{-b \pm \sqrt{b^2 - 4ac}}{2a}$

$= \dfrac{-1 \pm \sqrt{1^2 - 4(2)(-1)}}{2(2)}$

$= \dfrac{-1 \pm 3}{4} = \dfrac{1}{2}, -1$

75. $16x^2 + 8x - 3 = 0$

$x = \dfrac{-b \pm \sqrt{b^2 - 4ac}}{2a}$

$= \dfrac{-8 \pm \sqrt{8^2 - 4(16)(-3)}}{2(16)}$

$= \dfrac{-8 \pm 16}{32} = \dfrac{1}{4}, -\dfrac{3}{4}$

77. $2 + 2x - x^2 = 0$

$-x^2 + 2x + 2 = 0$

$x = \dfrac{-b \pm \sqrt{b^2 - 4ac}}{2a}$

$= \dfrac{-2 \pm \sqrt{2^2 - 4(-1)(2)}}{2(-1)}$

$= \dfrac{-2 \pm 2\sqrt{3}}{-2} = 1 \pm \sqrt{3}$

79. $x^2 + 12x + 16 = 0$

$x = \dfrac{-b \pm \sqrt{b^2 - 4ac}}{2a}$

$= \dfrac{-12 \pm \sqrt{12^2 - 4(1)(16)}}{2(1)}$

$= \dfrac{-12 \pm 4\sqrt{5}}{2}$

$= -6 \pm 2\sqrt{5}$

81. $x^2 + 8x - 4 = 0$

$x = \dfrac{-b \pm \sqrt{b^2 - 4ac}}{2a}$

$= \dfrac{-8 \pm \sqrt{8^2 - 4(1)(-4)}}{2(1)}$

$= \dfrac{-8 \pm 4\sqrt{5}}{2} = -4 \pm 2\sqrt{5}$

83. $12x - 9x^2 = -3$

$-9x^2 + 12x + 3 = 0$

$x = \dfrac{-b \pm \sqrt{b^2 - 4ac}}{2a}$

$= \dfrac{-12 \pm \sqrt{12^2 - 4(-9)(3)}}{2(-9)}$

$= \dfrac{-12 \pm 6\sqrt{7}}{-18} = \dfrac{2}{3} \pm \dfrac{\sqrt{7}}{3}$

85. $9x^2 + 30x + 25 = 0$

$x = \dfrac{-b \pm \sqrt{b^2 - 4ac}}{2a}$

$= \dfrac{-30 \pm \sqrt{30^2 - 4(9)(25)}}{2(9)}$

$= \dfrac{-30 \pm 0}{18} = -\dfrac{5}{3}$

87. $4x^2 + 4x = 7$

$4x^2 + 4x - 7 = 0$

$x = \dfrac{-b \pm \sqrt{b^2 - 4ac}}{2a}$

$= \dfrac{-4 \pm \sqrt{4^2 - 4(4)(-7)}}{2(4)}$

$= \dfrac{-4 \pm 8\sqrt{2}}{8} = -\dfrac{1}{2} \pm \sqrt{2}$

89. $28x - 49x^2 = 4$

$-49x^2 + 28x - 4 = 0$

$x = \dfrac{-b \pm \sqrt{b^2 - 4ac}}{2a}$

$= \dfrac{-28 \pm \sqrt{28^2 - 4(-49)(-4)}}{2(-49)}$

$= \dfrac{-28 \pm 0}{-98} = \dfrac{2}{7}$

91.
$$8t = 5 + 2t^2$$
$$-2t^2 + 8t - 5 = 0$$
$$t = \frac{-b \pm \sqrt{b^2 - 4ac}}{2a}$$
$$= \frac{-8 \pm \sqrt{8^2 - 4(-2)(-5)}}{2(-2)}$$
$$= \frac{-8 \pm 2\sqrt{6}}{-4} = 2 \pm \frac{\sqrt{6}}{2}$$

93.
$$(y - 5)^2 = 2y$$
$$y^2 - 12y + 25 = 0$$
$$y = \frac{-b \pm \sqrt{b^2 - 4ac}}{2a}$$
$$= \frac{-(-12) \pm \sqrt{(-12)^2 - 4(1)(25)}}{2(1)}$$
$$= \frac{12 \pm 2\sqrt{11}}{2} = 6 \pm \sqrt{11}$$

95.
$$\frac{1}{2}x^2 + \frac{3}{8}x = 2$$
$$4x^2 + 3x = 16$$
$$4x^2 + 3x - 16 = 0$$
$$x = \frac{-b \pm \sqrt{b^2 - 4ac}}{2a}$$
$$= \frac{-3 \pm \sqrt{3^2 - 4(4)(-16)}}{2(4)}$$
$$= \frac{-3 \pm \sqrt{265}}{8} = -\frac{3}{8} \pm \frac{\sqrt{265}}{8}$$

97. $5.1x^2 - 1.7x - 3.2 = 0$
$$x = \frac{1.7 \pm \sqrt{(-1.7)^2 - 4(5.1)(-3.2)}}{2(5.1)}$$
$$x \approx 0.976, -0.643$$

99. $-0.067x^2 - 0.852x + 1.277 = 0$
$$x = \frac{-(-0.852) \pm \sqrt{(-0.852)^2 - 4(-0.067)(1.277)}}{2(-0.067)}$$
$$x \approx -14.071, 1.355$$

101. $422x^2 - 506x - 347 = 0$
$$x = \frac{506 \pm \sqrt{(-506)^2 - 4(422)(-347)}}{2(422)}$$
$$x \approx 1.687, -0.488$$

103. $12.67x^2 + 31.55x + 8.09 = 0$
$$x = \frac{-31.55 \pm \sqrt{(31.55)^2 - 4(12.67)(8.09)}}{2(12.67)}$$
$$x \approx 2.200, -0.290$$

105. $x^2 - 2x - 1 = 0$ Complete the square.
$$x^2 - 2x = 1$$
$$x^2 - 2x + 1^2 = 1 + 1^2$$
$$(x - 1)^2 = 2$$
$$x - 1 = \pm\sqrt{2}$$
$$x = 1 \pm \sqrt{2}$$

107. $(x + 3)^2 = 81$ Extract square roots.
$$x + 3 = \pm 9$$
$$x + 3 = 9 \quad \text{or} \quad x + 3 = -9$$
$$x = 6 \quad \text{or} \qquad x = -12$$

109. $x^2 - x - \frac{11}{4} = 0$ Complete the square.
$$x^2 - x = \frac{11}{4}$$
$$x^2 - x + \left(\frac{1}{2}\right)^2 = \frac{11}{4} + \left(\frac{1}{2}\right)^2$$
$$\left(x - \frac{1}{2}\right)^2 = \frac{12}{4}$$
$$x - \frac{1}{2} = \pm\sqrt{\frac{12}{4}}$$
$$x = \frac{1}{2} \pm \sqrt{3}$$

111. $(x + 1)^2 = x^2$ Extract square roots.
$$x^2 = (x + 1)^2$$
$$x = \pm(x + 1)$$
For $x = +(x + 1)$:
$$0 \neq 1 \quad \text{No solution}$$
For $x = -(x + 1)$:
$$2x = -1$$
$$x = -\frac{1}{2}$$

113. (a) $w(w + 14) = 1632$

(b) $w^2 + 14w - 1632 = 0$
$$(w + 48)(w - 34) = 0$$
$$w = -48 \quad \text{or} \quad w = 34$$

Because w must be greater than zero, $w = 34$ feet and the length is $w + 14 = 48$ feet.

115. $S = x^2 + 4xh$

$108 = x^2 + 4x(3)$

$0 = x^2 + 12x - 108$

$0 = (x + 18)(x - 6)$

$x = -18$ or $x = 6$

Because x must be positive, $x = 6$ inches.

The dimensions of the box are 6 inches $\times$ 6 inches $\times$ 3 inches.

117. $(200 - 2x)(100 - 2x) = \dfrac{1}{2}(100)(200)$

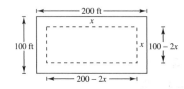

$20,000 - 600x + 4x^2 = 10,000$

$4x^2 - 600x + 10,000 = 0$

$4(x^2 - 150x + 2500) = 0$

Thus, $a = 1$, $b = -150$, and $c = 2500$.

$x = \dfrac{150 \pm \sqrt{(-150)^2 - 4(1)(2500)}}{2(1)} \approx \dfrac{150 \pm 111.8034}{2}$

$x \approx \dfrac{150 + 111.8034}{2} \approx 130.902$ feet, not possible since the lot is only 100 feet wide

$x \approx \dfrac{150 - 111.8034}{2} \approx 19.098$ feet

The first landscaper must go around the lot $\dfrac{19.098 \text{ feet}}{24 \text{ inches}} = \dfrac{19.098 \text{ feet}}{2 \text{ feet}} \approx 9.5$ times.

119. (a) $s = -16t^2 + v_0 t + s_0$

$s = -16t^2 + 0t + 25,000 = 0$

$-16t^2 + 25,000 = 0$

$t^2 = 1562.5$

$t = \sqrt{1562.5}$

≈ 39.5 seconds

(b) *Verbal model:* $(\text{distance}) = (\text{rate}) \cdot (\text{time})$

Labels: Distance $= d$, rate $= 500$ miles per hour, time $= \dfrac{39.5}{3600} \approx 0.011$ hour

Equation: $d \approx 500(0.011) = 5.5$ miles

The supply package will travel about 5.5 miles horizontally.

121. (a) $s = -16t^2 + v_0t + s_0$

$$v_0 = 100 \text{ mph} = \frac{(100)(5280)}{3600} = 146\frac{2}{3} \text{ ft/sec}$$

$$s_0 = 6\frac{1}{4} \text{ feet}$$

$$s = -16t^2 + 146\frac{2}{3}t + 6\frac{1}{4}$$

(b) When $t = 3$: $s(3) = 302.25$ feet

When $t = 4$: $s(4) \approx 336.92$ feet

When $t = 5$: $s(5) \approx 339.58$ feet

During the interval $3 \le t \le 5$, the baseball's speed decreased due to gravity.

(c) The ball hits the ground when $s = 0$.

$$-16t^2 + 146\frac{2}{3}t + 6\frac{1}{4} = 0$$

By the Quadratic Formula, $t \approx -0.042$ or $t \approx 9.209$. Assuming that the ball is not caught and drops to the ground, it will be in the air for approximately 9.209 seconds.

123. $P = 0.0103t^2 + 0.119t + 5.55, 1 \le t \le 8$

(a)

t	1	2	3	4	5	6	7	8
P	5.68	5.83	6.00	6.19	6.40	6.63	6.89	7.16

The average admission price reached or surpassed $6.50 sometime in 2005.

(b) $0.0103t^2 + 0.119t + 5.55 = 6.50$

$0.0103t^2 + 0.119t - 0.95 = 0$

$$t = \frac{-b \pm \sqrt{b^2 - 4ac}}{2a}$$

$$= \frac{-0.119 \pm \sqrt{(0.119)^2 - 4(0.0103)(-0.95)}}{2(0.0103)}$$

$$= \frac{-0.119 \pm \sqrt{0.053301}}{0.0206}$$

$$\approx -16.98, 5.43$$

Because the domain is $1 \le t \le 8$, $t \approx 5.43$ is the only solution. The average admission price reached or surpassed $6.50 sometime in 2005.

(c) For 2014, let $t = 14$.

$P(14) \approx \$9.23$

Answers will vary.

125. (a) *Model:* $(\text{winch})^2 + (\text{distance to dock})^2 = (\text{length of rope})^2$

Labels: winch $= 15$, distance to dock $= x$, length of rope $= l$

Equation: $15^2 + x^2 = l^2$

(b) When $l = 75$: $15^2 + x^2 = 75^2$

$$x^2 = 5625 - 225 = 5400$$

$$x = \sqrt{5400} = 30\sqrt{5} \approx 73.5$$

The boat is approximately 73.5 feet from the dock when there is 75 feet of rope out.

127. $x^2 + x^2 = 9^2$ (Pythagorean Theorem)

$$2x^2 = 81$$

$$x^2 = \frac{81}{2}$$

$$x = \sqrt{\frac{81}{2}} = \frac{9}{\sqrt{2}} = \frac{9\sqrt{2}}{2} \approx 6.36$$

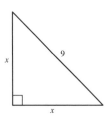

Each leg of the right triangle is about 6.36 centimeters.

129. $x(20 - 0.0002x) = 500,000$

$$0 = 0.0002x^2 - 20x + 500,000$$

$$0 = x^2 - 100,000 + 2,500,000,000$$

$$0 = (x - 50,000)^2$$

$$x = 50,000 \text{ units}$$

131. $0.125x^2 + 20x + 500 = 14,000$

$0.125x^2 + 20x - 13,500 = 0$

$$x = \frac{-20 + \sqrt{20^2 - 4(0.125)(-13,500)}}{2(0.125)}$$

Using the positive value for x, we have

$$x = \frac{-20 + \sqrt{7150}}{0.25} \approx 258 \text{ units.}$$

133. $800 + 0.04x + 0.002x^2 = 1680$

$0.002x^2 + 0.04x - 880 = 0$

$$x = \frac{-0.04 \pm \sqrt{(0.04)^2 - 4(0.002)(-880)}}{2(0.002)}$$

$$= \frac{-0.04 \pm \sqrt{7.0416}}{0.004}$$

Choosing the positive value for x, we have

$$x = \frac{-0.04 + \sqrt{7.0416}}{0.004} \approx 653 \text{ units.}$$

135. $D = 0.032t^2 + 0.21t + 5.6, 0 \le t \le 8$

(a)

t	0	1	2	3	4	5	6	7	8
D	5.600	5.842	6.148	6.518	6.952	7.450	8.012	8.638	9.328

The public debt reached or surpassed $7 trillion sometime in 2004.

(b) $0.032t^2 + 0.21t + 5.6 = 7$

$0.032t^2 + 0.21t - 1.4 = 0$

$$t = \frac{-b \pm \sqrt{b^2 - 4ac}}{2a} = \frac{-0.21 \pm \sqrt{(0.21)^2 - 4(0.032)(-1.4)}}{2(0.032)} = \frac{-0.21 \pm \sqrt{0.2233}}{0.064} \approx -10.66, 4.10$$

Because the domain is $1 \le t \le 8$, $t \approx 4.10$ is the only solution. The public debt reached or surpassed $7 trillion sometime in 2004.

Using the *intersect* feature of a graphing utility, you obtain $t \approx 4.10$ when $y = 7$, which verifies the result from part (a).

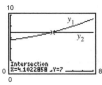

(c) For 2014, let $t = 14$.

$D(14) \approx \$14.812$ billion

Answers will vary.

137. (a)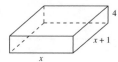

$V = 1024 \text{ ft}^3$, so: $\qquad 4x(x + 1) = 1024$

$$4x^2 + 4x - 1024 = 0$$
$$x = 15.5078 \text{ ft}$$
$$x = 16.5078 \text{ ft}$$

(b) 1 ft³ weighs approximately 63.5 lbs, so
1024 ft³ × 62.4 lbs/ft³ = 63,897.6 lbs.

(c) $\dfrac{1024 \text{ ft}^3}{0.13368 \text{ ft}^3/\text{gal}} = 7660.08 \text{ gal}$

At 5 gallons per minute:

$\dfrac{7660.08 \text{ gal}}{5 \text{ gals/min}} = 1532.02 \text{ min} = 25.5 \text{ hours}$

139. False

$$b^2 - 4ac = (-1)^2 - 4(-3)(-10) < 0,$$

so, the quadratic equation has no real solutions.

141. The student should have subtracted $15x$ from both sides so that the equation is equal to zero. By factoring out an x, there are two solutions, $x = 0$ and $x = 6$.

143. $3(x + 4)^2 + (x + 4) - 2 = 0$

(a) Let $u = x + 4$

$$3u^2 + u - 2 = 0$$
$$(3u - 2)(u + 1) = 0$$

$3u - 2 = 0 \qquad$ or $\qquad u + 1 = 0$
$\quad u = \frac{2}{3} \qquad\qquad\qquad u = -1$
$x + 4 = \frac{2}{3} \qquad\qquad x + 4 = -1$
$\quad x = -\frac{10}{3} \quad$ or $\qquad x = -5$

(b) $3(x^2 + 8x + 16) + (x + 4) - 2 = 0$

$$3x^2 + 24x + 48 + x + 4 - 2 = 0$$
$$3x^2 + 25x + 50 = 0$$
$$(3x + 10)(x + 5) = 0$$

$3x + 10 = 0 \qquad$ or $\qquad x + 5 = 0$
$\quad x = -\frac{10}{3} \qquad\qquad\qquad x = -5$

(c) The method of part (a) reduces the number of algebraic steps.

145. Sample answer: $(x - 3)(x - (-5)) = 0$

$$(x - 3)(x + 5) = 0$$
$$x^2 + 2x - 15 = 0$$

147. 8 and 14

One possible equation is:

$$(x - 8)(x - 14) = 0$$
$$x^2 - 22x + 112 = 0$$

Any non-zero multiple of this equation would also have these solutions.

149. $1 + \sqrt{2}$ and $1 - \sqrt{2}$

One possible equation is:

$$\left[x - \left(1 + \sqrt{2}\right)\right]\left[x - \left(1 - \sqrt{2}\right)\right] = 0$$
$$\left[(x - 1) - \sqrt{2}\right]\left[(x - 1) + \sqrt{2}\right] = 0$$
$$(x - 1)^2 - \left(\sqrt{2}\right)^2 = 0$$
$$x^2 - 2x + 1 - 2 = 0$$
$$x^2 - 2x - 1 = 0$$

Any non-zero multiple of this equation would also have these solutions.

151. (a) The discriminant is positive because the graph has two x-intercepts.

$$b^2 = 4ac = (-2)^2 - 4(1)(0) = 4$$

(b) The discriminant is zero because the graph has one x-intercept.

$$b^2 = 4ac = (-2)^2 - 4(1)(1) = 0$$

(c) The discriminant is negative because the graph has no x-intercepts.

$$b^2 = 4ac = (-2)^2 - 4(1)(2) = -4$$

In part (c), if the linear term was $2x$, the discriminant would still be zero and the equation would have no solutions.

In part (c), if the constant term was -2, the discriminant would be positive and the equation would have two solutions.

153. (a) $S = \dfrac{-b + \sqrt{b^2 - 4ac}}{2a} + \dfrac{-b - \sqrt{b^2 - 4ac}}{2a}$

$\qquad = \dfrac{-2a}{2a}$

$\qquad = -\dfrac{b}{a}$

(b) $P = \left(\dfrac{-b + \sqrt{b^2 - 4ac}}{2a}\right)\left(\dfrac{-b - \sqrt{b^2 - 4ac}}{2a}\right)$

$\qquad = \dfrac{b^2 - \left(b^2 - 4ac\right)}{4a^2}$

$\qquad = \dfrac{4ac}{4a^2}$

$\qquad = \dfrac{c}{a}$

Section 1.5   Complex Numbers

- Standard form: $a + bi$

  If $b = 0$, then $a + bi$ is a real number.

  If $a = 0$ and $b \neq 0$, then $a + bi$ is a pure imaginary number.

- Equality of Complex Numbers: $a + bi = c + di$ if and only if $a = c$ and $b = d$

- Operations on complex numbers

  (a) Addition: $(a + bi) + (c + di) = (a + c) + (b + d)i$

  (b) Subtraction: $(a + bi) - (c + di) = (a - c) + (b - d)i$

  (c) Multiplication: $(a + bi)(c + di) = (ac - bd) + (ad + bc)i$

  (d) Division: $\dfrac{a + bi}{c + di} = \dfrac{a + bi}{c + di} \cdot \dfrac{c - di}{c - di} = \dfrac{ac + bd}{c^2 + d^2} + \dfrac{bc - ad}{c^2 + d^2}i$

- The complex conjugate of $a + bi$ is $a - bi$:

  $(a + bi)(a - bi) = a^2 + b^2$

- The additive inverse of $a + bi$ is $-a - bi$.

- $\sqrt{-a} = \sqrt{a}\,i$ for $a > 0$.

---

1. (a) iii

   (b) i

   (c) ii

3. principal square

5. $a + bi = -12 + 7i$

   $a = -12$

   $b = 7$

7. $(a - 1) + (b + 3)i = 5 + 8i$

   $a - 1 = 5 \Rightarrow a = 6$

   $b + 3 = 8 \Rightarrow b = 5$

9. $8 + \sqrt{-25} = 8 + 5i$

11. $2 - \sqrt{-27} = 2 - \sqrt{27}\,i$

              $= 2 - 3\sqrt{3}\,i$

13. $\sqrt{-80} = 4\sqrt{5}\,i$

15. $14 = 14 + 0i = 14$

17. $-10i + i^2 = -10i - 1 = -1 - 10i$

19. $\sqrt{-0.09} = \sqrt{0.09}\,i$

           $= 0.3i$

21. $(7 + i) + (3 - 4i) = 10 - 3i$

23. $(9 - i) - (8 - i) = 1$

25. $\left(-2 + \sqrt{-8}\right) + \left(5 - \sqrt{-50}\right) = -2 + 2\sqrt{2}\,i + 5 - 5\sqrt{2}\,i$

                               $= 3 - 3\sqrt{2}\,i$

27. $13i - (14 - 7i) = 13i - 14 + 7i$

                 $= -14 + 20i$

29. $-\left(\frac{3}{2} + \frac{5}{2}i\right) + \left(\frac{5}{3} + \frac{11}{3}i\right) = -\frac{3}{2} - \frac{5}{2}i + \frac{5}{3} + \frac{11}{3}i$

                             $= -\frac{9}{6} - \frac{15}{6}i + \frac{10}{6} + \frac{22}{6}i$

                             $= \frac{1}{6} + \frac{7}{6}i$

31. $(1 + i)(3 - 2i) = 3 - 2i + 3i - 2i^2$

                  $= 3 + i + 2 = 5 + i$

33. $12i(1 - 9i) = 12i - 108i^2$

               $= 12i + 108$

               $= 108 + 12i$

35. $\left(\sqrt{14} + \sqrt{10}\,i\right)\left(\sqrt{14} - \sqrt{10}\,i\right) = 14 - 10i^2$

                                 $= 14 + 10 = 24$

37. $(6 + 7i)^2 = 36 + 84i + 49i^2$

              $= 36 + 84i - 49$

              $= -13 + 84i$

39. $(2 + 3i)^2 + (2 - 3i)^2 = 4 + 12i + 9i^2 + 4 - 12i + 9i^2$

$$= 4 + 12i - 9 + 4 - 12i - 9$$

$$= -10$$

41. The complex conjugate of $9 + 2i$ is $9 - 2i$.

$$(9 + 2i)(9 - 2i) = 81 - 4i^2$$

$$= 81 + 4$$

$$= 85$$

43. The complex conjugate of $-1 - \sqrt{5}i$ is $-1 + \sqrt{5}i$.

$$(-1 - \sqrt{5}i)(-1 + \sqrt{5}i) = 1 - 5i^2$$

$$= 1 + 5 = 6$$

45. The complex conjugate of $\sqrt{-20} = 2\sqrt{5}i$ is $-2\sqrt{5}i$.

$$(2\sqrt{5}i)(-2\sqrt{5}i) = -20i^2 = 20$$

47. The complex conjugate of $\sqrt{6}$ is $\sqrt{6}$.

$$(\sqrt{6})(\sqrt{6}) = 6$$

49. $\dfrac{3}{i} \cdot \dfrac{-i}{-i} = \dfrac{-3i}{-i^2} = -3i$

51. $\dfrac{2}{4 - 5i} = \dfrac{2}{4 - 5i} \cdot \dfrac{4 + 5i}{4 + 5i}$

$$= \dfrac{2(4 + 5i)}{16 + 25}$$

$$= \dfrac{8 + 10i}{41}$$

$$= \dfrac{8}{41} + \dfrac{10}{41}i$$

53. $\dfrac{5 + i}{5 - i} \cdot \dfrac{(5 + i)}{(5 + i)} = \dfrac{25 + 10i + i^2}{25 - i^2}$

$$= \dfrac{24 + 10i}{26}$$

$$= \dfrac{12}{13} + \dfrac{5}{13}i$$

55. $\dfrac{9 - 4i}{i} \cdot \dfrac{-i}{-i} = \dfrac{-9i + 4i^2}{-i^2} = -4 - 9i$

57. $\dfrac{3i}{(4 - 5i)^2} = \dfrac{3i}{16 - 40i + 25i^2}$

$$= \dfrac{3i}{-9 - 40i} \cdot \dfrac{-9 + 40i}{-9 + 40i}$$

$$= \dfrac{-27i + 120i^2}{81 + 1600}$$

$$= \dfrac{-120 - 27i}{1681}$$

$$= -\dfrac{120}{1681} - \dfrac{27}{1681}i$$

59. $\dfrac{2}{1 + i} - \dfrac{3}{1 - i} = \dfrac{2(1 - i) - 3(1 + i)}{(1 + i)(1 - i)}$

$$= \dfrac{2 - 2i - 3 - 3i}{1 + 1}$$

$$= \dfrac{-1 - 5i}{2}$$

$$= -\dfrac{1}{2} - \dfrac{5}{2}i$$

61. $\dfrac{i}{3 - 2i} + \dfrac{2i}{3 + 8i} = \dfrac{i(3 + 8i) + 2i(3 - 2i)}{(3 - 2i)(3 + 8i)}$

$$= \dfrac{3i + 8i^2 + 6i - 4i^2}{9 + 24i - 6i - 16i^2}$$

$$= \dfrac{4i^2 + 9i}{9 + 18i + 16}$$

$$= \dfrac{-4 + 9i}{25 + 18i} \cdot \dfrac{25 - 18i}{25 - 18i}$$

$$= \dfrac{-100 + 72i + 225i - 162i^2}{625 + 324}$$

$$= \dfrac{62 + 297i}{949}$$

$$= \dfrac{62}{949} + \dfrac{297}{949}i$$

63. $\sqrt{-6} \cdot \sqrt{-2} = (\sqrt{6}i)(\sqrt{2}i)$

$$= \sqrt{12}i^2$$

$$= (2\sqrt{3})(-1)$$

$$= -2\sqrt{3}$$

65. $(\sqrt{-15})^2 = (\sqrt{15}i)^2 = 15i^2 = -15$

67. $(3 + \sqrt{-5})(7 - \sqrt{-10}) = (3 + \sqrt{5}i)(7 - \sqrt{10}i)$

$$= 21 - 3\sqrt{10}i + 7\sqrt{5}i - \sqrt{50}i^2$$

$$= (21 + \sqrt{50}) + (7\sqrt{5} - 3\sqrt{10})i$$

$$= (21 + 5\sqrt{2}) + (7\sqrt{5} - 3\sqrt{10})i$$

69. $x^2 - 2x + 2 = 0; a = 1, b = -2, c = 2$

$$x = \frac{-(-2) \pm \sqrt{(-2)^2 - 4(1)(2)}}{2(1)}$$

$$= \frac{2 \pm \sqrt{-4}}{2}$$

$$= \frac{2 \pm 2i}{2}$$

$$= 1 \pm i$$

71. $4x^2 + 16x + 17 = 0; a = 4, b = 16, c = 17$

$$x = \frac{-16 \pm \sqrt{(16)^2 - 4(4)(17)}}{2(4)}$$

$$= \frac{-16 \pm \sqrt{-16}}{8}$$

$$= \frac{-16 \pm 4i}{8}$$

$$= -2 \pm \frac{1}{2}i$$

73. $4x^2 + 16x + 15 = 0; a = 4, b = 16, c = 15$

$$x = \frac{-16 \pm \sqrt{(16)^2 - 4(4)(15)}}{2(4)}$$

$$= \frac{-16 \pm \sqrt{16}}{8} = \frac{-16 \pm 4}{8}$$

$$x = -\frac{12}{8} = -\frac{3}{2} \text{ or } x = -\frac{20}{8} = -\frac{5}{2}$$

75. $\frac{3}{2}x^2 - 6x + 9 = 0$ Multiply both sides by 2.

$3x^2 - 12x + 18 = 0$

$$x = \frac{-(-12) \pm \sqrt{(-12)^2 - 4(3)(18)}}{2(3)}$$

$$= \frac{12 \pm \sqrt{-72}}{6}$$

$$= \frac{12 \pm 6\sqrt{2}i}{6} = 2 \pm \sqrt{2}i$$

77. $1.4x^2 - 2x - 10 = 0$ Multiply both sides by 5.

$7x^2 - 10x - 50 = 0$

$$x = \frac{-(-10) \pm \sqrt{(-10)^2 - 4(7)(-50)}}{2(7)}$$

$$= \frac{10 \pm \sqrt{1500}}{14}$$

$$= \frac{10 \pm 10\sqrt{15}}{14}$$

$$= \frac{5}{7} \pm \frac{5\sqrt{15}}{7}$$

79. $-6i^3 + i^2 = -6i^2i + i^2$

$$= -6(-1)i + (-1)$$

$$= 6i - 1$$

$$= -1 + 6i$$

81. $-14i^5 = -14i^2i^2i = -14(-1)(-1)(i) = -14i$

83. $\left(\sqrt{-72}\right)^3 = \left(6\sqrt{2}i\right)^3$

$$= 6^3\left(\sqrt{2}\right)^3 i^3$$

$$= 216\left(2\sqrt{2}\right)i^2i$$

$$= 432\sqrt{2}(-1)i$$

$$= -432\sqrt{2}i$$

85. $\frac{1}{i^3} = \frac{1}{i^2i} = \frac{1}{-i} = \frac{1}{-i} \cdot \frac{i}{i} = \frac{i}{-i^2} = i$

87. $(3i)^4 = 81i^4 = 81i^2i^2 = 81(-1)(-1) = 81$

89. (a) $z_1 = 9 + 16i, z_2 = 20 - 10i$

(b) $\frac{1}{z} = \frac{1}{z_1} + \frac{1}{z_2} = \frac{1}{9 + 16i} + \frac{1}{20 - 10i} = \frac{20 - 10i + 9 + 16i}{(9 + 16i)(20 - 10i)} = \frac{29 + 6i}{340 + 230i}$

$z = \left(\frac{340 + 230i}{29 + 6i}\right)\left(\frac{29 - 6i}{29 - 6i}\right) = \frac{11,240 + 4630i}{877} = \frac{11,240}{877} + \frac{4630}{877}i$

91. (a) $2^4 = 16$

(b) $(-2)^4 = 16$

(c) $(2i)^4 = 2^4 i^4 = 16 i^2 i^2 = 16(-1)(-1) = 16$

(d) $(-2i)^4 = (-2)^4 i^4 = 16 i^2 i^2 = 16(-1)(-1) = 16$

93. False, if $b = 0$ then $a + bi = a - bi = a$.

That is, if the complex number is real, the number equals its conjugate.

95. False.

$$i^{44} + i^{150} - i^{74} - i^{109} + i^{61} = \left(i^2\right)^{22} + \left(i^2\right)^{75} - \left(i^2\right)^{37} - \left(i^2\right)^{54} i + \left(i^2\right)^{30} i$$
$$= (-1)^{22} + (-1)^{75} - (-1)^{37} - (-1)^{54} i + (-1)^{30} i$$
$$= 1 - 1 + 1 - i + i = 1$$

97. $i = i$

$i^2 = -1$

$i^3 = -i$

$i^4 = 1$

$i^5 = i^4 i = i$

$i^6 = i^4 i^2 = -1$

$i^7 = i^4 i^3 = -i$

$i^8 = i^4 i^4 = 1$

$i^9 = i^4 i^4 i = i$

$i^{10} = i^4 i^4 i^2 = -1$

$i^{11} = i^4 i^4 i^3 = -i$

$i^{12} = i^4 i^4 i^4 = 1$

The pattern $i, -1, -i, 1$ repeats. Divide the exponent by 4.

If the remainder is 1, the result is i.

If the remainder is 2, the result is -1.

If the remainder is 3, the result is $-i$.

If the remainder is 0, the result is 1.

99. $\sqrt{-6}\sqrt{-6} = \sqrt{6}i\sqrt{6}i = 6i^2 = -6$

101. $\left(a_1 + b_1 i\right) + \left(a_2 + b_2 i\right) = \left(a_1 + a_2\right) + \left(b_1 + b_2\right)i$

The complex conjugate of this sum is
$\left(a_1 + a_2\right) - \left(b_1 + b_2\right)i.$

The sum of the complex conjugates is
$\left(a_1 - b_1 i\right) + \left(a_2 - b_2 i\right) = \left(a_1 + a_2\right) - \left(b_1 + b_2\right)i.$

So, the complex conjugate of the sum of two complex numbers is the sum of their complex conjugates.

Section 1.6   Other Types of Equations

- You should be able to solve certain types of nonlinear or nonquadratic equations by rewriting them in a form in which you can factor, extract square roots, complete the square, or use the Quadratic Formula.
- For equations involving radicals or rational exponents, raise both sides to the same power.
- For equations that are of the quadratic type, $au^2 + bu + c = 0$, $a \neq 0$, use either factoring, the Quadratic Formula, or completing the square.
- For equations with fractions, multiply both sides by the least common denominator to clear the fractions.
- For equations involving absolute value, remember that the expression inside the absolute value can be positive or negative.
- Always check for extraneous solutions.

1. polynomial

3. quadratic type

5. $6x^4 - 14x^2 = 0$

$$2x^2(3x^2 - 7) = 0$$

$$2x^2 = 0 \Rightarrow x = 0$$

$$3x^2 - 7 = 0 \Rightarrow x = \pm\frac{\sqrt{21}}{3}$$

7. $x^4 - 81 = 0$

$$(x^2 + 9)(x + 3)(x - 3) = 0$$

$$x^2 + 9 = 0 \Rightarrow x = \pm 3i$$

$$x + 3 = 0 \Rightarrow x = -3$$

$$x - 3 = 0 \Rightarrow x = 3$$

9. $x^3 + 512 = 0$

$$x^3 + 8^3 = 0$$

$$(x + 8)(x^2 - 8x + 64) = 0$$

$$x + 8 = 0 \Rightarrow x = -8$$

$$x^2 - 8x + 64 = 0 \Rightarrow x = 4 \pm 4\sqrt{3}\,i$$

11. $5x^3 + 3 - x^2 + 45x = 0$

$$5x(x^2 + 6x + 9) = 0$$

$$5x(x + 3)^2 = 0$$

$$5x = 0 \Rightarrow x = 0$$

$$x + 3 = 0 \Rightarrow x = -3$$

13. $x^3 - 3x^2 - x + 3 = 0$

$$x^2(x - 3) - (x - 3) = 0$$

$$(x - 3)(x^2 - 1) = 0$$

$$(x - 3)(x + 3)(x - 1) = 0$$

$$x - 3 = 0 \Rightarrow x = 3$$

$$x + 1 = 0 \Rightarrow x = -1$$

$$x - 1 = 0 \Rightarrow x = 1$$

15. $x^4 - x^3 + x - 1 = 0$

$$x^3(x - 1) + (x - 1) = 0$$

$$(x - 1)(x^3 + 1) = 0$$

$$(x - 1)(x + 1)(x^2 - x + 1) = 0$$

$$x - 1 = 0 \Rightarrow x = 1$$

$$x + 1 = 0 \Rightarrow x = -1$$

$$x^2 - x + 1 = 0 \Rightarrow x = \frac{1}{2} \pm \frac{\sqrt{3}}{2}i$$

17. $x^4 - 4x^2 + 3 = 0$

$$(x^2 - 3)(x^2 - 1) = 0$$

$$(x + \sqrt{3})(x - \sqrt{3})(x + 1)(x - 1) = 0$$

$$x + \sqrt{3} = 0 \Rightarrow x = -\sqrt{3}$$

$$x - \sqrt{3} = 0 \Rightarrow x = \sqrt{3}$$

$$x + 1 = 0 \Rightarrow x = -1$$

$$x - 1 = 0 \Rightarrow x = 1$$

19. $4x^4 - 65x^2 + 16 = 0$

$$(4x^2 - 1)(x^2 - 16) = 0$$

$$(2x + 1)(2x - 1)(x + 4)(x - 4) = 0$$

$$2x + 1 = 0 \Rightarrow x = -\tfrac{1}{2}$$

$$2x - 1 = 0 \Rightarrow x = \tfrac{1}{2}$$

$$x + 4x = 0 \Rightarrow x = -4$$

$$x - 4 = 0 \Rightarrow x = 4$$

21. $x^6 + 7x^3 - 8 = 0$

$$(x^3 + 8)(x^3 - 1) = 0$$

$$(x + 2)(x^2 - 2x + 4)(x - 1)(x^2 + x + 1) = 0$$

$$x + 2 = 0 \Rightarrow x = -2$$

$$x^2 - 2x + 4 = 0 \Rightarrow x = 1 \pm \sqrt{3}\,i$$

$$x - 1 = 0 \Rightarrow x = 1$$

$$x^2 + x + 1 = 0 \Rightarrow x = -\frac{1}{2} \pm \frac{\sqrt{3}}{2}i$$

23. $\dfrac{1}{x^2} + \dfrac{8}{x} + 15 = 0$

$1 + 8x + 15x^2 = 0$

$(1 + 3x)(1 + 5x) = 0$

$1 + 3x = 0 \Rightarrow x = -\dfrac{1}{3}$

$1 + 5x = 0 \Rightarrow x = -\dfrac{1}{5}$

25. $2\left(\dfrac{x}{x+2}\right)^2 - 3\left(\dfrac{x}{x+2}\right) - 2 = 0$

$2x^2 - 3x(x+2) - 2(x+2)^2 = 0$

$2x^2 - 3x^2 - 6x - 2x^2 - 8x - 8 = 0$

$-3x^2 - 14x - 8 = 0$

$3x^2 + 14x + 8 = 0$

$(3x + 2)(x + 4) = 0$

$3x + 2 = 0 \Rightarrow x -\dfrac{2}{3}$

$x + 4 = 0 \Rightarrow x = -4$

27. $2x + 9\sqrt{x} = 5$

$2x + 9\sqrt{x} - 5 = 0$

$2\left(\sqrt{x}\right)^2 + 9\sqrt{x} - 5 = 0$

$\left(2\sqrt{x} - 1\right)\left(\sqrt{x} + 5\right) = 0$

$\sqrt{x} = \tfrac{1}{2} \Rightarrow x = \tfrac{1}{4}$

$\left(\sqrt{x} = -5 \text{ is not a solution.}\right)$

29. $3x^{1/3} + 2x^{2/3} = 5$

$2x^{2/3} + 3x^{1/3} - 5 = 0$

$2\left(x^{1/3}\right)^2 + 3x^{1/3} - 5 = 0$

$\left(2x^{1/3} + 5\right)\left(x^{1/3} - 1\right) = 0$

$2x^{1/3} + 5 = 0 \Rightarrow x^{1/3} = -\tfrac{5}{2} \Rightarrow x = \left(-\tfrac{5}{2}\right)^3 = -\tfrac{125}{8}$

$x^{1/3} - 1 = 0 \Rightarrow x^{1/3} = 1 \Rightarrow x = (1)^3 = 1$

31. $y = x^3 - 2x^2 - 3x$

(a)

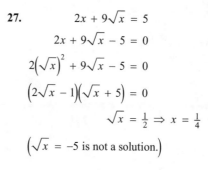

(b) x-intercepts: $(-1, 0), (0, 0), (3, 0)$

(c) $0 = x^3 - 2x^2 - 3x$

$0 = x(x + 1)(x - 3)$

$x = 0$

$x + 1 = 0 \Rightarrow x = -1$

$x - 3 = 0 \Rightarrow x = 3$

(d) The x-intercepts of the graph are the same as the solutions of the equation.

33. $y = x^4 - 10x^2 + 9$

(a)

(b) x-intercepts: $(\pm 1, 0), (\pm 3, 0)$

(c) $0 = x^4 - 10x^2 + 9$

$0 = \left(x^2 - 1\right)\left(x^2 - 9\right)$

$0 = (x + 1)(x - 1)(x + 3)(x - 3)$

$x + 1 = 0 \Rightarrow x = -1$

$x - 1 = 0 \Rightarrow x = 1$

$x + 3 = 0 \Rightarrow x = -3$

$x - 3 = 0 \Rightarrow x = 3$

(d) The x-intercepts of the graph are the same as the solutions of the equation.

35. $\sqrt{3x} - 12 = 0$

$\qquad \sqrt{3x} = 12$

$\qquad 3x = 144$

$\qquad x = 48$

37. $\sqrt{x - 10} - 4 = 0$

$\qquad \sqrt{x - 10} = 0$

$\qquad x - 10 = 16$

$\qquad x = 26$

39. $\sqrt[3]{2x + 5} + 3 = 0$

$\qquad \sqrt[3]{2x + 5} = -3$

$\qquad 2x + 5 = -27$

$\qquad 2x = -32$

$\qquad x = -16$

41. $-\sqrt{26 - 11x} + 4 = x$

$\qquad 4 - x = \sqrt{26 - 11x}$

$\qquad 16 - 8x + x^2 = 26 - 11x$

$\qquad x^2 + 3x - 10 = 0$

$\qquad (x + 5)(x - 2) = 0$

$\qquad x + 5 = 0 \Rightarrow x = -5$

$\qquad x - 2 = 0 \Rightarrow x = 2$

43. $\sqrt{x + 1} = \sqrt{3x + 1}$

$\qquad x + 1 = 3x + 1$

$\qquad -2x = 0$

$\qquad x = 0$

45. $\sqrt{x} - \sqrt{x - 5} = 1$

$\qquad \sqrt{x} = 1 + \sqrt{x - 5}$

$\qquad \left(\sqrt{x}\right)^2 = \left(1 + 2\sqrt{x - 5}\right)^2$

$\qquad x = 1 + 2\sqrt{x - 5} + x - 5$

$\qquad 4 = 2\sqrt{x - 5}$

$\qquad 2 = \sqrt{x - 5}$

$\qquad 4 = x - 5$

$\qquad 9 = x$

47. $\sqrt{x + 5} + \sqrt{x - 5} = 10$

$\qquad \sqrt{x + 5} = 10 - \sqrt{x - 5}$

$\qquad \left(\sqrt{x + 5}\right)^2 = \left(10 - \sqrt{x - 5}\right)^2$

$\qquad x + 5 = 100 - 20\sqrt{x - 5} + x - 5$

$\qquad -90 = -20\sqrt{x - 5}$

$\qquad 9 = 2\sqrt{x - 5}$

$\qquad 81 = 4(x - 5)$

$\qquad 81 = 4x - 20$

$\qquad 101 = 4x$

$\qquad \frac{101}{4} = x$

49. $\sqrt{x + 2} - \sqrt{2x - 3} = -1$

$\qquad \sqrt{x + 2} = \sqrt{2x - 3} - 1$

$\qquad \left(\sqrt{x + 2}\right)^2 = \left(\sqrt{2x - 3} - 1\right)^2$

$\qquad x + 2 = 2x - 3 - 2\sqrt{2x - 3} + 1$

$\qquad x + 2 = 2x - 2 - 2\sqrt{2x - 3}$

$\qquad -x + 4 = -2\sqrt{2x - 3}$

$\qquad x - 4 = 2\sqrt{2x - 3}$

$\qquad (x - 4)^2 = \left(2\sqrt{2x - 3}\right)^2$

$\qquad x^2 - 8x + 16 = 4(2x - 3)$

$\qquad x^2 - 8x + 16 = 8x - 12$

$\qquad x^2 - 16x + 28 = 0$

$\qquad (x - 2)(x - 14) = 0$

$\qquad x - 2 = 0 \Rightarrow x = 2$, extraneous

51. $(x - 5)^{3/2} = 8$

$\qquad (x - 5)^3 = 8^2$

$\qquad x - 5 = \sqrt[3]{64}$

$\qquad x = 5 + 4 = 9$

53. $(x + 3)^{2/3} = 8$

$\qquad (x + 3)^2 = 8^3$

$\qquad x + 3 = \pm\sqrt{8^3}$

$\qquad x + 3 = \pm\sqrt{512}$

$\qquad x = -3 \pm 16\sqrt{2}$

55. $\left(x^2 - 5\right)^{3/2} = 27$

$\left(x^2 - 5\right)^3 = 27^2$

$x^2 - 5 = \sqrt[3]{27^2}$

$x^2 = 5 + 9$

$x^2 = 14$

$x = \pm\sqrt{14}$

57. $3x(x - 1)^{1/2} + 2(x - 1)^{3/2} = 0$

$(x - 1)^{1/2}\left[3x + 2(x - 1)\right] = 0$

$(x - 1)^{1/2}(5x - 2) = 0$

$(x - 1)^{1/2} = 0 \Rightarrow x - 1 = 0 \Rightarrow x = 1$

$5x - 2 = 0 \Rightarrow x = \frac{2}{5}$, extraneous

59. $y = \sqrt{11x - 30} - x$

(a)

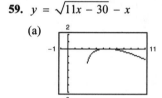

(b) x-intercepts: $(5, 0), (6, 0)$

(c) $0 = \sqrt{11x - 30} - x$

$x = \sqrt{11x - 30}$

$x^2 = 11x - 30$

$x^2 - 11x + 30 = 0$

$(x - 5)(x - 6) = 0$

$x - 5 = 0 \Rightarrow x = 5$

$x - 6 = 0 \Rightarrow x = 6$

(d) The x-intercepts of the graph are the same as the solutions of the equation.

61. $y = \sqrt{7x + 36} - \sqrt{5x + 16} - 2$

(a)

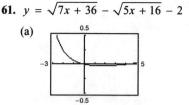

(b) x-intercepts: $(0, 0), (4, 0)$

(c) $0 = \sqrt{7x + 26} - \sqrt{5x + 16} - 2$

$-\sqrt{7x + 36} = -\sqrt{5x + 16} - 2$

$\sqrt{7x + 36} = 2 + \sqrt{5x + 16}$

$\left(\sqrt{7x + 36}\right)^2 = \left(2 + \sqrt{5x + 16}\right)^2$

$7x + 36 = 4 + 4\sqrt{5x + 16} + 5x + 16$

$7x + 36 = 5x + 20 + 4\sqrt{5x + 16}$

$2x + 16 = 4\sqrt{5x + 16}$

$x + 8 = 2\sqrt{5x + 16}$

$(x + 8)^2 = \left(2\sqrt{5x + 16}\right)^2$

$x^2 + 16x + 64 = 4(5x + 16)$

$x^2 + 16x + 64 = 20x + 64$

$x^2 - 4x = 0$

$x(x - 4) = 0$

$x = 0$

$x - 4 = 0 \Rightarrow x = 4$

(d) The x-intercepts of the graph are the same as the solutions of the equation.

63.

$x = \frac{3}{x} + \frac{1}{2}$

$(2x)(x) = (2x)\left(\frac{3}{x}\right) + (2x)\left(\frac{1}{2}\right)$

$2x^2 = 6 + x$

$2x^2 - x - 6 = 0$

$(2x + 3)(x - 2) = 0$

$2x + 3 = 0 \Rightarrow x = -\frac{3}{2}$

$x - 2 = 0 \Rightarrow x = 2$

65.
$$\frac{1}{x} - \frac{1}{x+1} = 3$$
$$x(x+1)\frac{1}{x} - x(x+1)\frac{1}{x+1} = x(x+1)(3)$$
$$x + 1 - x = 3x(x+1)$$
$$1 = 3x^2 + 3x$$
$$0 = 3x^2 + 3x - 1$$
$$a = 3, b = 3, c = -1$$
$$x = \frac{-3 \pm \sqrt{(3)^2 - 4(3)(-1)}}{2(3)} = \frac{-3 \pm \sqrt{21}}{6}$$

67. $\dfrac{30 - x}{x} = x$
$$30 - x = x^2$$
$$0 = x^2 + x - 30$$
$$0 = (x + 6)(x - 5)$$
$$x + 6 = 0 \Rightarrow x = -6$$
$$x - 5 = 0 \Rightarrow x = 5$$

69.
$$\frac{x}{x^2 - 4} + \frac{1}{x + 2} = 3$$
$$(x + 2)(x - 2)\frac{x}{x^2 - 4} + (x + 2)(x - 2)\frac{1}{x + 2} = 3(x + 2)(x - 2)$$
$$x + x - 2 = 3x^2 - 12$$
$$3x^2 - 2x - 10 = 0$$
$$a = 3, b = -2, c = -10$$
$$x = \frac{-(-2) \pm \sqrt{(-2)^2 - 4(3)(-10)}}{2(3)}$$
$$= \frac{2 \pm \sqrt{124}}{6} = \frac{2 \pm 2\sqrt{31}}{6} = \frac{1 \pm \sqrt{31}}{3}$$

71. $|2x - 5| = 11$
$$2x - 5 = 11 \Rightarrow x = 8$$
$$-(2x - 5) = 11 \Rightarrow x = -3$$

73. $|x| = x^2 + x - 24$

First equation:
$$x = x^2 + x - 24$$
$$x^2 - 24 = 0$$
$$x^2 = 24$$
$$x = \pm 2\sqrt{6}$$

Second equation:
$$-x = x^2 + x - 24$$
$$x^2 + 2x - 24 = 0$$
$$(x + 6)(x - 4) = 0$$
$$x + 6 = 0 \Rightarrow x = -6$$
$$x - 4 = 0 \Rightarrow x = 4$$

Only $x = 2\sqrt{6}$ and $x = -6$ are solutions of the original equation. $x = -2\sqrt{6}$ and $x = 4$ are extraneous.

75. $|x + 1| = x^2 - 5$

First equation:
$$x + 1 = x^2 - 5$$
$$x^2 - x - 6 = 0$$
$$(x - 3)(x + 2) = 0$$
$$x - 3 = 0 \Rightarrow x = 3$$
$$x + 2 = 0 \Rightarrow x = -2$$

Second equation:
$$-(x + 1) = x^2 - 5$$
$$-x - 1 = x^2 - 5$$
$$x^2 + x - 4 = 0$$
$$x = \frac{-1 + \sqrt{17}}{2}$$

Only $x = 3$ and $x = \dfrac{-1 - \sqrt{17}}{2}$ are solutions of the original equation. $x = -2$ and $x = \dfrac{-1 + \sqrt{17}}{2}$ are extraneous.

77. $y = \dfrac{1}{x} - \dfrac{4}{x-1} - 1$

(a)

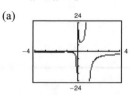

(b) x-intercept: $(-1, 0)$

(c) $0 = \dfrac{1}{x} - \dfrac{4}{x-1} - 1$

$0 = (x-1) - 4x - x(x-1)$

$0 = x - 1 - 4x - x^2 + x$

$0 = -x^2 - 2x - 1$

$0 = x^2 + 2x + 1$

$0 = (x+1)^2$

$x + 1 = 0 \Rightarrow x = -1$

(d) The x-intercept of the graph is the same as the solution of the equation.

79. $y = |x+1| - 2$

(a)

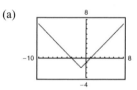

(b) x-intercepts: $(1, 0), (-3, 0)$

(c) $0 = |x+1| - 2$

$2 = |x+1|$

$x + 1 = 2 \qquad$ or $\quad -(x+1) = 2$

$x = 1 \qquad\quad$ or $\quad\; -x - 1 = 2$

$-x = 3$

$x = -3$

(d) The x-intercepts of the graph are the same as the solutions of the equation.

81. $3.2x^4 - 1.5x^2 - 2.1 = 0$

$x^2 = \dfrac{1.5 \pm \sqrt{1.5^2 - 4(3.2)(-2.1)}}{2(3.2)}$

Using the positive value for x^2, we have

$x = \pm\sqrt{\dfrac{1.3 + \sqrt{29.13}}{6.4}} \approx \pm 1.038.$

83. $7.08x^6 + 4.15x^3 - 9.6 = 0$

$a = 7.8, b = 4.15, c = -9.6$

$x^3 = \dfrac{-4.15 \pm \sqrt{(4.15)^2 - 4(7.08)(-9.6)}}{2(7.08)}$

$= \dfrac{-4.15 \pm \sqrt{2.89.0945}}{14.16}$

$x = \sqrt[3]{\dfrac{-4.15 + \sqrt{289.0945}}{14.16}} \approx 0.968$

$x = \sqrt[3]{\dfrac{-4.15 - \sqrt{289.0945}}{14.16}} \approx -1.143$

85. $1.8x - 6\sqrt{x} - 5.6 = 0$ Given equation

$1.8\left(\sqrt{x}\right)^2 - 6\sqrt{x} - 5.6 = 0$

Use the Quadratic Formula with $a = 1.8, b = -6$, and $c = -5.6$.

$\sqrt{x} = \dfrac{6 \pm \sqrt{36 - 4(1.8)(-5.6)}}{2(1.8)} \approx \dfrac{6 \pm 8.7361}{3.6}$

Considering only the positive value for $\sqrt{x}$, we have:

$\sqrt{x} \approx 4.0934$

$x \approx 16.756.$

87. $4x^{2/3} + 8x^{1/3} + 3.6 = 0$

$a = 4, b = 8, c = 3.6$

$x^{1/3} = \dfrac{-8 \pm \sqrt{8^2 - 4(4)(3.6)}}{2(4)}$

$x = \left[\dfrac{-8 + \sqrt{6.4}}{8}\right]^3 \approx -0.320$

$x = \left[\dfrac{-8 - \sqrt{6.4}}{8}\right]^3 \approx -2.280$

89. $-4, 7$

Sample answer: $(x - (-4))(x - 7) = 0$

$(x + 4)(x - 7) = 0$

$x^2 - 3x - 28 = 0$

91. $-\frac{7}{3}$ and $\frac{6}{7}$

One possible equation is:

$x = -\frac{7}{3} \Rightarrow 3x = -7 \Rightarrow 3x + 7$ is a factor.

$x = \frac{6}{7} \Rightarrow 7x = 6 \Rightarrow 7x - 6$ is a factor.

$(3x + 7)(7x - 6) = 0$

$21x^2 + 31x - 42 = 0$

Any non-zero multiple of this equation would also have these solutions.

93. $\sqrt{3}, -\sqrt{3}$, and 4

One possible equation is:

$\left(x - \sqrt{3}\right)\left(x - \left(-\sqrt{3}\right)\right)(x - 4) = 0$

$\left(x - \sqrt{3}\right)\left(x + \sqrt{3}\right)(x - 4) = 0$

$\left(x^2 - 3\right)(x - 4) = 0$

$x^3 - 4x^2 - 3x + 12 = 0$

Any non-zero multiple of this equation would also have these solutions.

95. $i, -i$

Sample answer: $(x - i)(x - (-i)) = 0$

$(x - i)(x + i) = 0$

$x^2 - i^2 = 0$

$x^2 + 1 = 0$

97. $-1, 1, i$, and $-i$

One possible equation is:

$(x - (-1))(x - 1)(x - i)(x - (-i)) = 0$

$(x + 1)(x - 1)(x - i)(x + i) = 0$

$\left(x^2 - 1\right)\left(x^2 + 1\right) = 0$

$x^4 - 1 = 0$

Any non-zero multiple of this equation would also have these solutions.

99. Let $x =$ the number of students in the original group. Then, $\dfrac{1700}{x} =$ the original cost per student.

When six more students join the group, the cost per student becomes $\dfrac{1700}{x} - 7.50$.

Model: (Cost per student) $\cdot$ (Number of students) = Total cost

$$\left(\frac{1700}{x} - 7.5\right)(x + 6) = 1700$$

$$(3400 - 15x)(x + 6) = 3400x \quad \text{Multiply both sides by } 2x \text{ to clear fraction.}$$

$$-15x^2 - 90x + 20,400 = 0$$

$$x = \frac{90 \pm \sqrt{(-90)^2 - 4(-15)(20,400)}}{2(-15)} = \frac{90 \pm 1110}{-30}$$

Using the positive value for x we conclude that the original number was $x = 34$ students.

101. *Model:* $\text{Time} = \dfrac{\text{Distance}}{\text{Rate}}$

Labels: Let $x =$ average speed of the plane. Then we have a travel time of $t = 145/x$. If the average speed is increased by 40 mph, then

$$t - \dfrac{12}{60} = \dfrac{145}{x + 40}$$

$$t = \dfrac{145}{x + 40} + \dfrac{1}{5}.$$

Now, we equate these two equations and solve for x.

Equation: $\dfrac{145}{x} = \dfrac{145}{x + 40} + \dfrac{1}{5}$

$$145(5)(x + 40) = 145(5)x + x(x + 40)$$

$$725x + 29{,}000 = 725x + x^2 + 40x$$

$$0 = x^2 + 40x - 29{,}000$$

Using the positive value for x found by the Quadratic Formula, we have $x \approx 151.5\,\text{mph}$ and $x + 40 = 191.5$ mph. The airspeed required to obtain the decrease in travel time is 191.5 miles per hour.

103.
$$A = P\left(1 + \dfrac{r}{n}\right)^{nt}$$

$$3052.49 = 2500\left(1 + \dfrac{r}{12}\right)^{(12)(5)}$$

$$1.220996 = \left(1 + \dfrac{r}{12}\right)^{60}$$

$$(1.220996)^{1/60} = 1 + \dfrac{r}{12}$$

$$\left[(1.220996)^{1/60} - 1\right](12) = r$$

$$r \approx 0.04 = 4\%$$

105. $D = 431{,}61 + 121.8\sqrt{t},\ 8 \le t \le 16$

(a) $431.61 + 121.8\sqrt{t} = 875$

$$121.8\sqrt{t} = 443.39$$

$$t = \left(\dfrac{443.39}{121.8}\right)^2$$

$$\approx 13.25$$

The number of medical doctors reached 875,000 during 2003.

(b) $431.61 + 121.8\sqrt{t} = 1000$

$$121.8\sqrt{t} = 568.39$$

$$t = \left(\dfrac{568.39}{121.8}\right)^2$$

$$\approx 21.78$$

The model predicts the number of medical doctors will reach 1,000,000 during 2011. Answers will vary.

107. $T = 75.82 - 2.11x + 43.51\sqrt{x},\ 5 \le x \le 40$

(a)

Absolute Pressure, x	Temperature, T
5	162.56
10	192.31
15	212.68
20	228.20
25	240.62
30	250.83
35	259.38
40	266.60

(b) $T = 212°$ when $x \approx 15$ pounds per square inch.

(c) $212 = 75.82 - 2.11x + 43.51\sqrt{x}$

$$0 = -2.11x + 43.51\sqrt{x} - 136.18$$

By the Quadratic Formula, we have

$$\sqrt{x} \approx 16.77928 \Rightarrow x \approx 281.333$$

$$\sqrt{x} \approx 3.84787 \Rightarrow x \approx 14.806.$$

Since x is restricted to $5 \le x \le 40$, let $x = 14.806$ pounds per square inch.

(d)

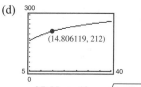

(14.806119, 212)

109. $37.55 = 40 - \sqrt{0.01x + 1}$

$$\sqrt{0.01x + 1} = 2.45$$

$$0.01x + 1 = 6.0025$$

$$0.01x = 5.0025$$

$$x = 500.25$$

Rounding x to the nearest whole unit yields $x \approx 500$ units.

111. *Model:* $\left(\begin{array}{c}\text{Distance from}\\\text{home to 1st}\end{array}\right)^2 + \left(\begin{array}{c}\text{distance from}\\\text{1st to 2nd}\end{array}\right)^2 = \left(\begin{array}{c}\text{distance from}\\\text{home to 2nd}\end{array}\right)^2$

Labels: Distance from home to 1st $= x$, distance from 1st to 2nd $= x$, distance from home to 2nd $= 127.5$

Equation: $x^2 + x^2 = (127.5)^2$

$$2x^2 = 16{,}256.25$$

$$x^2 = \frac{16{,}256.25}{2}$$

$$x = \pm\sqrt{8128.125} \approx \pm 90$$

The distance between bases is approximately 90 feet.

113. $S = 8\pi\sqrt{64 + h^2}$

(a)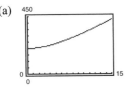

When $S = 350$, $h \approx 11.4$.

(b)

h	8	9	10	11	12	13
S	284.3	302.6	321.9	341.8	362.5	383.6

$S = 350$ when h is between 11 and 12 inches.

(c)

$$350 = 8\pi\sqrt{64 + h^2}$$

$$(350)^2 = \left(8\pi\sqrt{64 + h^2}\right)^2$$

$$122{,}500 = 64\pi^2\left(64 + h^2\right)$$

$$64 + h^2 \approx 193.935$$

$$h^2 \approx 129.935$$

$$h \approx 11.4$$

(d) Solving graphically or numerically yields an approximate solution. An exact solution is obtained algebraically.

115. *Model:* $\left(\begin{array}{c}\text{Portion done by}\\\text{first person}\end{array}\right) + \left(\begin{array}{c}\text{portion done by}\\\text{second person}\end{array}\right) = \left(\begin{array}{c}\text{work}\\\text{done}\end{array}\right)$

Labels: Work done $= 1$, rate of first person $= \dfrac{1}{r}$, time worked by first person $= 12$, rate of second person $= \dfrac{1}{r+3}$, time worked by second person $= 12$

Equation:

$$\frac{12}{r} + \frac{12}{r+3} = 1$$

$$r(r+3)\frac{12}{r} + r(r+3)\frac{12}{r+3} = r(r+3)$$

$$12r + 36 + 12r = r^2 + 3r$$

$$0 = r^2 - 21r - 36$$

$$r = \frac{-(-21) \pm \sqrt{(-21)^2 - 4(1)(-36)}}{2(1)} = \frac{21 \pm \sqrt{585}}{2}$$

$$r \approx 23 \quad \text{(Choose the positive value for } r\text{.)}$$

It would take approximately 23 hours and 26 hours individually.

117.
$$v = \sqrt{\frac{gR}{\mu s}}$$
$$v^2 = \frac{gR}{\mu s}$$
$$v^2 \mu s = gR$$
$$\frac{v^2 \mu s}{R} = g$$

119. False—See Example 7 on page 133.

121. True. There is no value to satisfy this equation.
$$\sqrt{x + 10} - \sqrt{x - 10} = 0$$
$$\sqrt{x + 10} = \sqrt{x - 10}$$
$$x + 10 = x - 10$$
$$10 \neq -10$$

123. The distance between $(1, 2)$ and $(x, -10)$ is 13.
$$\sqrt{(x - 1)^2 + (-10 - 2)^2} = 13$$
$$(x - 1)^2 + (-12)^2 = 13^2$$
$$x^2 - 2x + 1 + 144 = 169$$
$$x^2 - 2x - 24 = 0$$
$$(x + 4)(x - 6) = 0$$
$$x + 4 = 0 \Rightarrow x = -4$$
$$x - 6 = 0 \Rightarrow x = 6$$

Both $(-4, -10)$ and $(6, -10)$ are a distance of 13 from $(1, 2)$.

129. $20 + \sqrt{20 - 1} = b$
$$\sqrt{20 - a} = b - 20$$
$$20 - a = b^2 - 40b + 400$$
$$-a = b^2 - 40b + 380$$
$$a = -b^2 + 40b - 380$$

This formula gives the relationship between a and b. From the original equation we know that $a \leq 20$ and $b \geq 20$. Choose a b value, where $b \geq 20$ and then solve for a, keeping in mind that $a \leq 20$.

Some possibilities are: $b = 20, a = 20$
$$b = 21, a = 19$$
$$b = 22, a = 16$$
$$b = 23, a = 11$$
$$b = 24, a = 4$$
$$b = 25, a = -5$$

125. The distance between $(0, 0)$ and $(8, y)$ is 17.
$$(8 - 0)^2 + (y - 0)^2 = 17$$
$$(8)^2 + (y)^2 = 17^2$$
$$64 + y^2 = 289$$
$$y^2 = 225$$
$$y = \pm\sqrt{225}$$
$$= \pm15$$

Both $(8, 15)$ and $(8, -15)$ are a distance of 17 from $(0, 0)$.

127. $9 + |9 - a| = b$
$$|9 - a| = b - 9$$
$$9 - a = b - 9 \quad \text{or} \quad 9 - a = -(b - 9)$$
$$-a = b - 18 \qquad 9 - a = -b + 9$$
$$a = 18 - b \qquad -a = -b$$
$$a = b$$

Thus, $a = 18 - b$ *or* $a = b$. From the original equation we know that $b \geq 9$.

Some possibilities are: $b = 9, a = 9$
$$b = 10, a = 8 \ or \ a = 10$$
$$b = 11, a = 7 \ or \ a = 11$$
$$b = 12, a = 6 \ or \ a = 12$$
$$b = 13, a = 5 \ or \ a = 13$$
$$b = 14, a = 4 \ or \ a = 14$$

Section 1.7 Linear Inequalities in One Variable

- You should know the properties of inequalities.

  (a) Transitive: $a < b$ and $b < c$ implies $a < c$.

  (b) Addition: $a < b$ and $c < d$ implies $a + c < b + d$.

  (c) Adding or Subtracting a Constant: $a \pm c < b \pm c$ if $a < b$.

  (d) Multiplying or Dividing a Constant: For $a < b$,

      1. If $c > 0$, then $ac < bc$ and $\dfrac{a}{c} < \dfrac{b}{c}$.

      2. If $c < 0$, then $ac > bc$ and $\dfrac{a}{c} > \dfrac{b}{c}$.

- You should be able to solve absolute value inequalities.

  (a) $|x| < a$ if and only if $-a < x < a$.

  (b) $|x| > a$ if and only if $x < -a$ or $x > a$.

1. solution set

3. negative

5. double

7. Interval: $[0, 9)$

   (a) Inequality: $0 \le x \le 9$

   (b) The interval is bounded.

9. Interval: $[-1, 5]$

   (a) Inequality: $-1 \le x \le 5$

   (b) The interval is bounded.

11. Interval: $(11, \infty)$

   (a) Inequality: $x > 11$

   (b) The interval is unbounded.

23. $5x - 12 > 0$

13. Interval: $(-\infty, -2)$

   (a) Inequality: $x < -2$

   (b) The interval is unbounded.

15. $x < 3$

   Matches (b).

17. $-3 < x \le 4$

   Matches (e).

19. $|x| < 3 \Rightarrow -3 < x < 3$

   Matches (f).

21. $-1 \le x \le \frac{5}{2}$

   Matches (g).

(a) $x = 3$

$5(3) - 12 \overset{?}{>} 0$

$3 > 0$

Yes, $x = 3$ *is*
a solution.

(b) $x = -3$

$5(-3) - 12 \overset{?}{>} 0$

$-27 \not> 0$

No, $x = -3$ *is not*
a solution.

(c) $x = \frac{5}{2}$

$5\left(\frac{5}{2}\right) - 12 \overset{?}{>} 0$

$\frac{1}{2} > 0$

Yes, $x = \frac{5}{2}$ *is*
a solution.

(d) $x = \frac{3}{2}$

$5\left(\frac{3}{2}\right) - 12 \overset{?}{>} 0$

$-\frac{9}{2} \not> 0$

No, $x = \frac{3}{2}$ *is not*
a solution.

25. $0 < \dfrac{x-2}{4} < 2$

(a) $x = 4$

$0 \overset{?}{<} \dfrac{4-2}{4} \overset{?}{<} 2$

$0 < \dfrac{1}{2} < 2$

Yes, $x = 4$ *is*

a solution.

(b) $x = 10$

$0 \overset{?}{<} \dfrac{10-2}{4} \overset{?}{<} 2$

$0 < 2 \not< 2$

No, $x = 10$ *is not*

a solution.

(c) $x = 0$

$0 \overset{?}{<} \dfrac{0-2}{4} \overset{?}{<} 2$

$0 \not< -\dfrac{1}{2} < 2$

No, $x = 0$ *is not*

a solution.

(d) $x = \dfrac{7}{2}$

$0 \overset{?}{<} \dfrac{(7/2)-2}{4} \overset{?}{<} 2$

$0 < \dfrac{3}{8} < 2$

Yes, $x = \dfrac{7}{2}$ *is*

a solution.

27. $|x - 10| \geq 3$

(a) $x = 13$

$|13 - 10| \overset{?}{\geq} 3$

$3 \geq 3$

Yes, $x = 13$ *is*

a solution.

(b) $x = -1$

$|-1 - 10| \overset{?}{\geq} 3$

$11 \geq 3$

Yes, $x = -1$ *is*

a solution.

(c) $x = 14$

$|14 - 10| \overset{?}{\geq} 3$

$4 \geq 3$

Yes, $x = 14$ *is*

a solution.

(d) $x = 9$

$|9 - 10| \overset{?}{\geq} 3$

$1 \not\geq 3$

No, $x = 9$ *is not*

a solution.

29. $4x < 12$

$\dfrac{1}{4}(4x) < \dfrac{1}{4}(12)$

$x < 3$

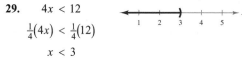

31. $-2x > -3$

$-\dfrac{1}{2}(-2x) < \left(-\dfrac{1}{2}\right)(-3)$

$x < \dfrac{3}{2}$

33. $x - 5 \geq 7$

$x \geq 12$

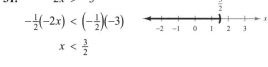

35. $2x + 7 < 3 + 4x$

$-2x < -4$

$x > 2$

37. $2x - 1 \geq 1 - 5x$

$7x \geq 2$

$x \geq \dfrac{2}{7}$

39. $4 - 2x < 3(3 - x)$

$4 - 2x < 9 - 3x$

$x < 5$

41. $\dfrac{3}{4}x - 6 \leq x - 7$

$-\dfrac{1}{4}x \leq -1$

$x \geq 4$

43. $\dfrac{1}{2}(8x + 1) \geq 3x + \dfrac{5}{2}$

$4x + \dfrac{1}{2} \geq 3x + \dfrac{5}{2}$

$x \geq 2$

45. $3.6x + 11 \geq -3.4$

$3.6x \geq 14.4$

$x \geq -4$

47. $1 < 2x + 3 < 9$

$-2 < 2x < 6$

$-1 < x < 3$

49. $-8 \leq 1 - 3(x - 2) < 13$

$-8 \leq 1 - 3x + 6 < 13$

$-8 \leq 7 - 3x < 13$

$-15 \leq -3x < 6$

$5 \geq x > -2$

51. $-4 < \dfrac{2x - 3}{3} < 4$

$-12 < 2x - 3 < 12$

$-9 < 2x < 15$

$-\dfrac{9}{2} < x < \dfrac{15}{2}$

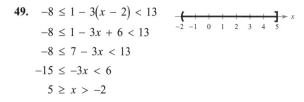

53. $\dfrac{3}{4} > x + 1 > \dfrac{1}{4}$

$-\dfrac{1}{4} > x > -\dfrac{3}{4}$

$-\dfrac{3}{4} < x < -\dfrac{1}{4}$

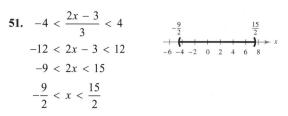

55. $3.2 \leq 0.4x - 1 \leq 4.4$

$4.2 \leq 0.4x \leq 5.4$

$10.5 \leq x \leq 13.5$

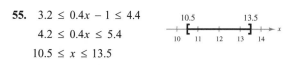

57. $|x| < 5$

$-5 < x < 5$

59. $\left|\dfrac{x}{2}\right| > 1$

$\dfrac{x}{2} < -1$ or $\dfrac{x}{2} > 1$

$x < -2$ $\qquad x > 2$

61. $|x - 5| < -1$

No solution. The absolute value of a number cannot be less than a negative number.

63. $|x - 20| \le 6$

$-6 \le x - 20 \le 6$

$14 \le x \le 26$

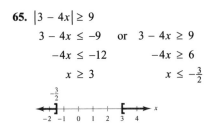

65. $|3 - 4x| \ge 9$

$3 - 4x \le -9$ or $3 - 4x \ge 9$

$-4x \le -12 \qquad -4x \ge 6$

$x \ge 3 \qquad\qquad x \le -\dfrac{3}{2}$

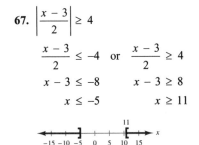

67. $\left|\dfrac{x - 3}{2}\right| \ge 4$

$\dfrac{x - 3}{2} \le -4$ or $\dfrac{x - 3}{2} \ge 4$

$x - 3 \le -8 \qquad x - 3 \ge 8$

$x \le -5 \qquad\quad x \ge 11$

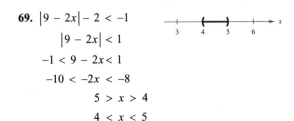

69. $|9 - 2x| - 2 < -1$

$|9 - 2x| < 1$

$-1 < 9 - 2x < 1$

$-10 < -2x < -8$

$5 > x > 4$

$4 < x < 5$

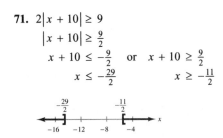

71. $2|x + 10| \ge 9$

$|x + 10| \ge \dfrac{9}{2}$

$x + 10 \le -\dfrac{9}{2}$ or $x + 10 \ge \dfrac{9}{2}$

$x \le -\dfrac{29}{2} \qquad\qquad x \ge -\dfrac{11}{2}$

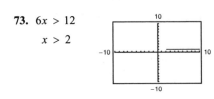

73. $6x > 12$

$x > 2$

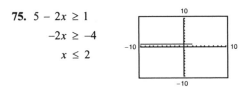

75. $5 - 2x \ge 1$

$-2x \ge -4$

$x \le 2$

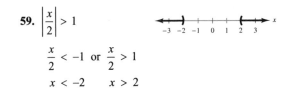

77. $4(x - 3) \le 8 - x$

$4x - 12 \le 8 - x$

$5x \le 20$

$x \le 4$

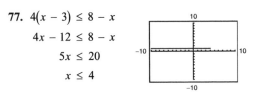

79. $|x - 8| \le 14$

$-14 \le x - 8 \le 14$

$-6 \le x \le 22$

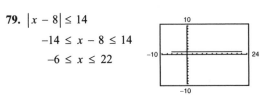

81. $2|x + 7| \ge 13$

$|x + 7| \ge \dfrac{13}{2}$

$x + 7 \le -\dfrac{13}{2}$ or $x + 7 \ge \dfrac{13}{2}$

$x \le -\dfrac{27}{2} \qquad\qquad x \ge -\dfrac{1}{2}$

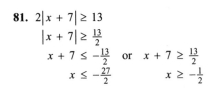

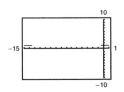

83. $y = 2x - 3$

(a) $\qquad y \ge 1$

$2x - 3 \ge 1$

$2x \ge 4$

$x \ge 2$

(b) $\qquad y \le 0$

$2x - 3 \le 0$

$2x \le 3$

$x \le \dfrac{3}{2}$

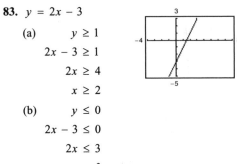

85. $y = -\dfrac{1}{2}x + 2$

(a) $\quad 0 \le y \le 3$

$0 \le -\dfrac{1}{2}x + 2 \le 3$

$-2 \le -\dfrac{1}{2}x \le 1$

$4 \ge x \ge -2$

(b) $\qquad y \ge 0$

$-\dfrac{1}{2}x + 2 \ge 0$

$-\dfrac{1}{2}x \ge -2$

$x \le 4$

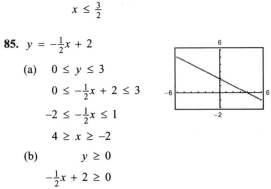

87. $y = |x - 3|$

(a) $y \le 2$

$|x - 3| \le 2$

$-2 \le x - 3 \le 2$

$1 \le x \le 5$

(b) $y \ge 4$

$|x - 3| \ge 4$

$x - 3 \le -4$ or $x - 3 \ge 4$

$x \le -1$ or $x \ge 7$

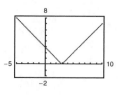

89. $x - 5 \ge 0$

$x \ge 5$

$[5, \infty)$

91. $x + 3 \ge 0$

$x \ge -3$

$[-3, \infty)$

93. $7 - 2x \ge 0$

$-2x \ge -7$

$x \le \frac{7}{2}$

$\left(-\infty, \frac{7}{2}\right]$

95. $|x - 10| < 8$

All real numbers within 8 units of 10

97. The midpoint of the interval $[-3, 3]$ is 0. The interval represents all real numbers x no more than 3 units from 0.

$|x - 0| \le 3$

$|x| \le 3$

99. The graph shows all real numbers at least 3 units from 7.

$|x - 7| \ge 3$

101. All real numbers within 10 units of 12

$|x - 12| < 10$

103. All real numbers more than 4 units from -3

$|x - (-3)| > 4$

$|x + 3| > 4$

105. $\$4.10 \le E \le \$4.25$

107. $p \le 45\%$

109. $r = 220 - A = 220 - 20 = 200$ beats per minute

$0.50(200) \le r \le 0.85(200)$

$100 \le r \le 170$

The target heartrate is at least 100 beats per minute and at most 170 beats per minute.

111. $9.00 + 0.75x > 13.50$

$0.75x > 4.50$

$x > 6$

You must produce at least 6 units each hour in order to yield a greater hourly wage at the second job.

113. $1000\big(1 + r(2)\big) > 1062.50$

$1 + 2r > 1.0625$

$2r > 0.0625$

$r > 0.03125$

$r > 3.125\%$

115. $R > C$

$115.95x > 95x + 750$

$20.95x > 750$

$x \ge 35.7995$

$x \ge 36$ units

117. Let x = number of dozens of doughnuts sold per day.

Revenue: $R = 4.50x$

Cost: $C = 2.75x + 220$

Profit: $P = R - C$

$= 4.50x - (2.75x + 220)$

$= 1.75x - 220$

$60 \le 1.75x - 220 \le 270$

$280 \le 1.75x \le 490$

$160 \le x \le 280$

The daily sales vary between 160 and 280 dozen doughnuts per day.

119. (a) $y = 0.067x - 5.638$

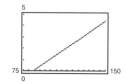

(b) From the graph you see that $y \ge 3$ when $x \ge 129$.

Algebraically:

$3 \le 0.067x - 5.638$

$8.638 \le 0.067x$

$x \ge 129$

IQ scores are not a good predictor of GPAs. Other factors include study habits, class attendance, and attitude.

121. $S = 1.09t + 30.9$

(a) $32.5 \leq 1.09t + 30.9 \leq 42$

$1.6 \leq 1.09t \leq 11.1$

$1.47 \leq t \leq 10.18$

The average salary was between \$32,500 and \$42,000 between the years 1991 and 2000.

(b) $1.09t + 30.9 > 54$

$1.09t > 23.1$

$t > 21.19$

The average salary will exceed \$54,500 sometime during 2011.

123. $\left| s - 10.4 \right| \leq \frac{1}{16}$

$-\frac{1}{16} \leq s - 10.4 \leq \frac{1}{16}$

$-0.0625 \leq s - 10.4 \leq 0.0625$

$10.3375 \leq s \leq 10.4625$

Because $A = s^2$,

$(10.3375)^2 \leq \text{area} \leq (10.4625)^2$

$106.864 \text{ in.}^2 \leq \text{area} \leq 109.464 \text{ in.}^2.$

125. $\frac{1}{10}(\$2.09) \approx \$0.21$

You might have been undercharged or overcharged by \$0.21.

127. $\left| \dfrac{t - 15.6}{1.9} \right| < 1$

$-1 < \dfrac{t - 15.6}{1.9} < 1$

$-1.9 < t - 15.6 < 1.9$

$13.7 < t < 17.5$

Two-thirds of the workers could perform the task in the time interval between 13.7 minutes and 17.5 minutes.

129. $\left| h - 50 \right| \leq 30$

$-30 \leq h - 50 \leq 30$

$20 \leq h \leq 80$

The minimum relative humidity is 20 and the maximum is 80.

131. False. If c is negative, then $ac \geq bc$.

133. $\left| x - a \right| \geq 2$ Matches (b).

$x - a \leq -2$

$x \leq a - 2$ or

$x - a \geq 2$

$x \geq a + 2$

135. Sample answer: $x > 5$

Section 1.8   Other Types of Inequalities

- You should be able to solve inequalities.

  (a) Find the critical number.

    1. Values that make the expression zero

    2. Values that make the expression undefined

  (b) Test one value in each test interval on the real number line resulting from the critical numbers.

  (c) Determine the solution intervals.

1. positive; negative

3. zeros; undefined values

5. $x^2 - 3 < 0$

(a) $x = 3$

$(3)^2 - 3 \overset{?}{<} 0$

$6 \not< 0$

No, $x = 3$ *is not* a solution.

(b) $x = 0$

$(0)^2 - 3 \overset{?}{<} 0$

$-3 < 0$

Yes, $x = 0$ *is* a solution.

(c) $x = \frac{3}{2}$

$\left(\frac{3}{2}\right)^2 - 3 \overset{?}{<} 0$

$-\frac{3}{4} < 0$

Yes, $x = \frac{3}{2}$ *is* a solution.

(d) $x = -5$

$(-5)^2 - 3 \overset{?}{<} 0$

$22 \not< 0$

No, $x = -5$ *is not* a solution.

7. $\dfrac{x+2}{x-4} \ge 3$

(a) $x = 5$

$\dfrac{5+2}{5-4} \overset{?}{\ge} 3$

$7 \ge 3$

Yes, $x = 5$ *is*

a solution.

(b) $x = 4$

$\dfrac{4+2}{4-4} \overset{?}{\ge} 3$

$\dfrac{6}{0}$ is undefined.

No, $x = 4$ *is not*

a solution.

(c) $x = -\dfrac{9}{2}$

$\dfrac{-\dfrac{9}{2}+2}{-\dfrac{9}{2}-4} \overset{?}{\ge} 3$

$\dfrac{5}{17} \overset{?}{\ge} 3$

No, $x = -\dfrac{9}{2}$ *is not*

a solution.

(d) $x = \dfrac{9}{2}$

$\dfrac{\dfrac{9}{2}+2}{\dfrac{9}{2}-4} \overset{?}{\ge} 3$

$13 \ge 3$

Yes, $x = \dfrac{9}{2}$ *is*

a solution.

9. $3x^2 - x - 2 = (3x+2)(x-1)$

$3x + 2 = 0 \Rightarrow x = -\dfrac{2}{3}$

$x - 1 = 0 \Rightarrow x = 1$

The key numbers are $-\dfrac{2}{3}$ and 1.

11. $\dfrac{1}{x-5} + 1 = \dfrac{1 + 1(x-5)}{x-5}$

$= \dfrac{x-4}{x-5}$

$x - 4 = 0 \Rightarrow x = 4$

$x - 5 = 0 \Rightarrow x = 5$

The key numbers are 4 and 5.

13. $x^2 < 9$

$x^2 - 9 < 0$

$(x+3)(x-3) < 0$

Key numbers: $x = \pm 3$

Test intervals: $(-\infty, -3), (-3, 3), (3, \infty)$

Test: Is $(x+3)(x-3) < 0$?

Interval	x-Value	Value of $x^2 - 9$	Conclusion
$(-\infty, -3)$	-4	7	Positive
$(-3, 3)$	0	-9	Negative
$(3, \infty)$	4	7	Positive

Solution set: $(-3, 3)$

15. $(x+2)^2 \le 25$

$x^2 + 4x + 4 \le 25$

$x^2 + 4x - 21 \le 0$

$(x+7)(x-3) \le 0$

Key numbers: $x = -7, x = 3$

Test intervals: $(-\infty, -7), (-7, 3), (3, \infty)$

Test: Is $(x+7)(x-3) \le 0$?

Interval	x-Value	Value of $(x+7)(x-3)$	Conclusion
$(-\infty, -7)$	-8	$(-1)(-11) = 11$	Positive
$(-7, 3)$	0	$(7)(-3) = -21$	Negative
$(3, \infty)$	4	$(11)(1) = 11$	Positive

Solution set: $[-7, 3]$

17. $x^2 + 4x + 4 \ge 9$

$x^2 + 4x - 5 \ge 0$

$(x+5)(x-1) \ge 0$

Key numbers: $x = -5, x = 1$

Test intervals: $(-\infty, -5), (-5, 1), (1, \infty)$

Test: Is $(x+5)(x-1) \ge 0$?

Interval	x-Value	Value of $(x+5)(x-1)$	Conclusion
$(-\infty, -5)$	-6	$(-1)(-7) = 7$	Positive
$(-5, 1)$	0	$(5)(-1) = -5$	Negative
$(1, \infty)$	2	$(7)(1) = 7$	Positive

Solution set: $(-\infty, -5] \cup [1, \infty)$

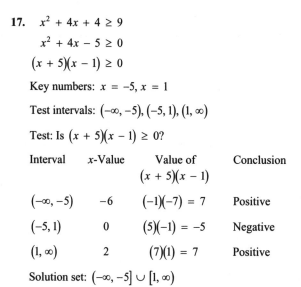

19.
$$x^2 + x < 6$$
$$x^2 + x - 6 < 0$$
$$(x + 3)(x - 2) < 0$$

Key numbers: $x = -3, x = 2$

Test intervals: $(-\infty, -3), (-3, 2), (2, \infty)$

Test: Is $(x + 3)(x - 2) < 0$?

Interval	x-Value	Value of $(x + 3)(x - 2)$	Conclusion
$(-\infty, -3)$	-4	$(-1)(-6) = 6$	Positive
$(-3, 2)$	0	$(3)(-2) = -6$	Negative
$(2, \infty)$	3	$(6)(1) = 6$	Positive

Solution set: $(-3, 2)$

21.
$$x^2 + 2x - 3 < 0$$
$$(x + 3)(x - 1) < 0$$

Key numbers: $x = -3, x = 1$

Test intervals: $(-\infty, -3), (-3, 1), (1, \infty)$

Test: Is $(x + 3)(x - 1) < 0$?

Interval	x-Value	Value of $(x + 3)(x - 1)$	Conclusion
$(-\infty, -3)$	-4	$(-1)(-5) = 5$	Positive
$(-3, 1)$	0	$(3)(-1) = -3$	Negative
$(1, \infty)$	2	$(5)(1) = 5$	Positive

Solution set: $(-3, 1)$

23.
$$3x^2 - 11x > 20$$
$$3x^2 - 11x - 20 > 0$$
$$(3x + 4)(x - 5) > 0$$

Key numbers: $x = 5, x = -\frac{4}{3}$

Test intervals: $\left(-\infty, -\frac{4}{5}\right), \left(-\frac{4}{3}, 5\right), (5, \infty)$

Test: Is $(3x + 4)(x - 5) > 0$?

Interval	x-Value	Value of $(3x + 4)(x - 5)$	Conclusion
$\left(-\infty, -\frac{4}{3}\right)$	-3	$(-5)(-8) = 40$	Positive
$\left(-\frac{4}{3}, 5\right)$	0	$(4)(-5) = -20$	Negative
$(5, \infty)$	6	$(22)(1) = 22$	Positive

Solution set: $\left(-\infty, -\frac{4}{3}\right) \cup (5, \infty)$

25.
$$x^2 - 3x - 18 > 0$$
$$(x + 3)(x - 6) > 0$$

Key numbers: $x = -3, x = 6$

Test intervals: $(-\infty, -3), (-3, 6), (6, \infty)$

Test: Is $(x + 3)(x - 6) > 0$?

Interval	x-Value	Value of $(x + 3)(x - 6)$	Conclusion
$(-\infty, -3)$	-4	$(-1)(-10) = 10$	Positive
$(-3, 6)$	0	$(3)(-6) = -18$	Negative
$(6, \infty)$	7	$(10)(1) = 10$	Positive

Solution set: $(-\infty, -3) \cup (6, \infty)$

27. $x^3 - 3x^2 - x > -3$

$x^3 - 3x^2 - x + 3 > 0$

$x^2(x - 3) - (x - 3) > 0$

$(x - 3)(x^2 - 1) > 0$

$(x - 3)(x + 1)(x - 1) > 0$

Key numbers: $x = -1, x = 1, x = 3$

Test intervals: $(-\infty, -1), (-1, 1), (1, 3), (3, \infty)$

Test: Is $(x - 3)(x + 1)(x - 1) > 0$?

Interval	x-Value	Value of $(x - 3)(x + 1)(x - 1)$	Conclusion
$(-\infty, -1)$	-2	$(-5)(-1)(-3) = -15$	Negative
$(-1, 1)$	0	$(-3)(1)(-1) = 3$	Positive
$(1, 3)$	2	$(-1)(3)(1) = -3$	Negative
$(3, \infty)$	4	$(1)(5)(3) = 15$	Positive

Solution set: $(-1, 1) \cup (3, \infty)$

29. $4x^2 - 4x + 1 \le 0$

$(2x - 1)^2 \le 0$

Key number: $x = \frac{1}{2}$

Test intervals: $\left(-\infty, \frac{1}{2}\right), \left(\frac{1}{2}, \infty\right)$

Test: Is $(2x - 1)^2 \le 0$?

Interval	x-Value	Value of $(2x - 1)^2$	Conclusion
$\left(-\infty, \frac{1}{2}\right)$	0	$(-1)^2 = 1$	Positive
$\left(\frac{1}{2}, \infty\right)$	1	$(1)^2 = 1$	Positive

Solution set: $x = \frac{1}{2}$

31. $4x^3 - 6x^2 < 0$

$2x^2(2x - 3) < 0$

Key numbers: $x = 0, x = \frac{3}{2}$

Test intervals: $(-\infty, 0) \Rightarrow 2x^2(2x - 3) < 0$

$\left(0, \frac{3}{2}\right) \Rightarrow 2 \Rightarrow 2x^2(2x - 3) < 0$

$\left(\frac{3}{2}, \infty\right) \Rightarrow 2x^2(2x - 3) > 0$

Solution set: $(-\infty, 0) \cup \left(0, \frac{3}{2}\right)$

33. $x^3 - 4x \ge 0$

$x(x + 2)(x - 2) \ge 0$

Key numbers: $x = 0, x = \pm 2$

Test intervals: $(-\infty, -2) \Rightarrow x(x + 2)(x - 2) < 0$

$(-2, 0) \Rightarrow x(x + 2)(x - 2) > 0$

$(0, 2) \Rightarrow x(x + 2)(x - 2) < 0$

$(2, \infty) \Rightarrow x(x + 2)(x - 2) > 0$

Solution set: $[-2, 0] \cup [2, \infty)$

35. $(x - 1)^2(x + 2)^3 \ge 0$

Key numbers: $x = 1, x = -2$

Test intervals: $(-\infty, -2) \Rightarrow (x - 1)^2(x + 2)^3 < 0$

$(-2, 1) \Rightarrow (x - 1)^2(x + 2)^3 > 0$

$(1, \infty) \Rightarrow (x - 1)^2(x + 2)^3 > 0$

Solution set: $[-2, \infty)$

37. $y = -x^2 + 2x + 3$

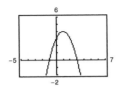

(a) $y \le 0$ when $x \le -1$ or $x \ge 3$.

(b) $y \ge 3$ when $0 \le x \le 2$.

39. $y = \frac{1}{8}x^3 - \frac{1}{2}x$

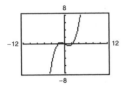

(a) $y \ge 0$ when $-2 \le x \le 0$ or $2 \le x < \infty$.

(b) $y \le 6$ when $x \le 4$.

41. $\dfrac{4x - 1}{x} > 0$

Key numbers: $x = 0$, $x = \dfrac{1}{4}$

Test intervals: $\left(-\infty, 0\right), \left(0, \frac{1}{4}\right), \left(\frac{1}{4}, \infty\right)$

Test: Is $\dfrac{4x - 1}{x} > 0$?

Interval	x-Value	Value of $\dfrac{4x-1}{x}$	Conclusion
$\left(-\infty, 0\right)$	-1	$\dfrac{-5}{-1} = 5$	Positive
$\left(0, \frac{1}{4}\right)$	$\dfrac{1}{8}$	$\dfrac{-\frac{1}{2}}{\frac{1}{8}} = -4$	Negative
$\left(\frac{1}{4}, \infty\right)$	1	$\dfrac{3}{1} = 3$	Positive

Solution set: $\left(-\infty, 0\right) \cup \left(\frac{1}{4}, \infty\right)$

43. $\dfrac{3x - 5}{x - 5} \ge 0$

Key numbers: $x = \dfrac{5}{3}$, $x = 5$

Test intervals: $\left(-\infty, \frac{5}{3}\right), \left(\frac{5}{3}, 5\right), \left(5, \infty\right)$

Test: Is $\dfrac{3x - 5}{x - 5} \ge 0$?

Interval	x-Value	Value of $\dfrac{3x-5}{x-5}$	Conclusion
$\left(-\infty, \frac{5}{3}\right)$	0	$\dfrac{-5}{-5} = 1$	Positive
$\left(\frac{5}{3}, 5\right)$	2	$\dfrac{6-5}{2-5} = -\dfrac{1}{3}$	Negative
$\left(5, \infty\right)$	6	$\dfrac{18-5}{6-5} = 13$	Positive

Solution set: $\left(-\infty, \dfrac{5}{3}\right] \cup \left(5, \infty\right)$

45.

$$\dfrac{x + 6}{x + 1} - 2 < 0$$

$$\dfrac{x + 6 - 2(x + 1)}{x + 1} < 0$$

$$\dfrac{4 - x}{x + 1} < 0$$

Key numbers: $x = -1$, $x = 4$

Test intervals: $\left(-\infty, -1\right) \Rightarrow \dfrac{4 - x}{x + 1} < 0$

$\left(-1, 4\right) \Rightarrow \dfrac{4 - x}{x + 1} > 0$

$\left(4, \infty\right) \Rightarrow \dfrac{4 - x}{x + 1} < 0$

Solution set: $\left(-\infty, -1\right) \cup \left(4, \infty\right)$

47.
$$\frac{2}{x + 5} > \frac{1}{x - 3}$$

$$\frac{2}{x + 5} - \frac{1}{x - 3} > 0$$

$$\frac{2(x - 3) - 1(x + 5)}{(x + 5)(x - 3)} > 0$$

$$\frac{x - 11}{(x + 5)(x - 3)} > 0$$

Key numbers: $x = -5, x = 3, x = 11$

Test intervals: $(-\infty, -5) \Rightarrow \dfrac{x - 11}{(x + 5)(x - 3)} < 0$

$(-5, 3) \Rightarrow \dfrac{x - 11}{(x + 5)(x - 3)} > 0$

$(3, 11) \Rightarrow \dfrac{x - 11}{(x + 5)(x - 3)} < 0$

$(11, \infty) \Rightarrow \dfrac{x - 11}{(x + 5)(x - 3)} > 0$

Solution set: $(-5, 3) \cup (11, \infty)$

49.
$$\frac{1}{x - 3} \le \frac{9}{4x + 3}$$

$$\frac{1}{x - 3} - \frac{9}{4x + 3} \le 0$$

$$\frac{4x + 3 - 9(x - 3)}{(x - 3)(4x + 3)} \le 0$$

$$\frac{30 - 5x}{(x - 3)(4x + 3)} \le 0$$

Key numbers: $x = 3, x = -\dfrac{3}{4}, x = 6$

Test intervals: $\left(-\infty, -\dfrac{3}{4}\right) \Rightarrow \dfrac{30 - 5x}{(x - 3)(4x + 3)} > 0$

$\left(-\dfrac{3}{4}, 3\right) \Rightarrow \dfrac{30 - 5x}{(x - 3)(4x + 3)} < 0$

$(3, 6) \Rightarrow \dfrac{30 - 5x}{(x - 3)(4x + 3)} > 0$

$(6, \infty) \Rightarrow \dfrac{30 - 5x}{(x - 3)(4x + 3)} < 0$

Solution set: $\left(-\dfrac{3}{4}, 3\right) \cup [6, \infty)$

51.
$$\frac{x^2 + 2x}{x^2 - 9} \le 0$$

$$\frac{x(x + 2)}{(x + 3)(x - 3)} \le 0$$

Key numbers: $x = 0, x = -2, x = \pm 3$

Test intervals: $(-\infty, -3) \Rightarrow \dfrac{x(x + 2)}{(x + 3)(x - 3)} > 0$

$(-3, -2) \Rightarrow \dfrac{x(x + 2)}{(x + 3)(x - 3)} < 0$

$(-2, 0) \Rightarrow \dfrac{x(x + 2)}{(x + 3)(x - 3)} > 0$

$(0, 3) \Rightarrow \dfrac{x(x + 2)}{(x + 3)(x - 3)} < 0$

$(3, \infty) \Rightarrow \dfrac{x(x + 2)}{(x + 3)(x - 3)} > 0$

Solution set: $(-3, -2] \cup [0, 3)$

53.
$$\frac{3}{x - 1} + \frac{2x}{x + 1} > -1$$

$$\frac{3(x + 1) + 2x(x - 1) + 1(x + 1)(x - 1)}{(x - 1)(x + 1)} > 0$$

$$\frac{3x^2 + x + 2}{(x - 1)(x + 1)} > 0$$

Key numbers: $x = -1, x = 1$

Test intervals: $(-\infty, -1) \Rightarrow \dfrac{3x^2 + x + 2}{(x - 1)(x + 1)} > 0$

$(-1, 1) \Rightarrow \dfrac{3x^2 + x + 2}{(x - 1)(x + 1)} < 0$

$(1, \infty) \Rightarrow \dfrac{3x^2 + x + 2}{(x - 1)(x + 1)} > 0$

Solution set: $(-\infty, -1) \cup (1, \infty)$

55. $y = \dfrac{3x}{x - 2}$

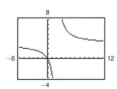

(a) $y \le 0$ when $0 \le x < 2$.

(b) $y \ge 6$ when $2 < x \le 4$.

57. $y = \dfrac{2x^2}{x^2 + 4}$

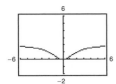

(a) $y \ge 1$ when $x \le -2$ or $x \ge 2$.

This can also be expressed as $|x| \ge 2$.

(b) $y \le 2$ for all real numbers x.

This can also be expressed as $-\infty < x < \infty$.

59. $4 - x^2 \ge 0$

$(2 + x)(2 - x) \ge 0$

Key numbers: $x = \pm 2$

Test intervals: $(-\infty, -2) \Rightarrow 4 - x^2 < 0$

$\qquad\qquad\qquad (-2, 2) \Rightarrow 4 - x^2 > 0$

$\qquad\qquad\qquad (2, \infty) \Rightarrow 4 - x^2 < 0$

Domain: $[-2, 2]$

61. $x^2 - 9x + 20 \ge 0$

$(x - 4)(x - 5) \ge 0$

Key numbers: $x = 4, x = 5$

Test intervals: $(-\infty, 4), (4, 5), (5, \infty)$

Interval	x-Value	Value of $(x - 4)(x - 5)$	Conclusion
$(-\infty, 4)$	0	$(-4)(-5) = 20$	Positive
$(4, 5)$	$\frac{9}{2}$	$\left(\frac{1}{2}\right)\left(-\frac{1}{2}\right) = -\frac{1}{4}$	Negative
$(5, \infty)$	6	$(2)(1) = 2$	Positive

Domain: $(-\infty, 4] \cup [5, \infty)$

63. $\dfrac{x}{x^2 - 2x - 35} \ge 0$

$\dfrac{x}{(x + 5)(x - 7)} \ge 0$

Key numbers: $x = 0, x = -5, x = 7$

Test intervals: $(-\infty, -5) \Rightarrow \dfrac{x}{(x + 5)(x - 7)} < 0$

$\qquad\qquad (-5, 0) \Rightarrow \dfrac{x}{(x + 5)(x - 7)} > 0$

$\qquad\qquad (0, 7) \Rightarrow \dfrac{x}{(x + 5)(x - 7)} < 0$

$\qquad\qquad (7, \infty) \Rightarrow \dfrac{x}{(x + 5)(x - 7)} > 0$

Domain: $(-5, 0] \cup (7, \infty)$

65. $0.4x^2 + 5.26 < 10.2$

$0.4x^2 - 4.94 < 0$

$0.4(x^2 - 12.35) < 0$

Key numbers: $x \approx \pm 3.51$

Test intervals: $(-\infty, -3.51), (-3.51, 3.51), (3.51, \infty)$

Solution set: $(-3.51, 3.51)$

67. $-0.5x^2 + 12.5x + 1.6 > 0$

Key numbers: $x \approx -0.13, x \approx 25.13$

Test intervals: $(-\infty, -0.13), (-0.13, 25.13), (25.13, \infty)$

Solution set: $(-0.13, 25.13)$

69. $\dfrac{1}{2.3x - 5.2} > 3.4$

$\dfrac{1}{2.3x - 5.2} - 3.4 > 0$

$\dfrac{1 - 3.4(2.3x - 5.2)}{2.3x - 5.2} > 0$

$\dfrac{-7.82x + 18.68}{2.3x - 5.2} > 0$

Key numbers: $x \approx 2.39, x \approx 2.26$

Test intervals: $(-\infty, 2.26), (2.26, 2.39), (2.39, \infty)$

Solution set: $(2.26, 2.39)$

71. $s = -16t^2 + v_0t + s_0 = -16t^2 + 160t$

(a) $-16t^2 + 160t = 0$

$-16t(t - 10) = 0$

$t = 0, t = 10$

It will be back on the ground in 10 seconds.

(b) $\qquad -16t^2 + 160t > 384$

$-16t^2 + 160t - 384 > 0$

$-16(t^2 - 10t + 24) > 0$

$t^2 - 10t + 24 < 0$

$(t - 4)(t - 6) < 0$

Key numbers: $t = 4, t = 6$

Test intervals: $(-\infty, 4), (4, 6), (6, \infty)$

Solution set: 4 seconds $< t <$ 6 seconds

73. $2L + 2W = 100 \Rightarrow W = 50 - L$

$LW \geq 500$

$L(50 - L) \geq 500$

$-L^2 + 50L - 500 \geq 0$

By the Quadratic Formula you have:

Key numbers: $L = 25 \pm 5\sqrt{5}$

Test: Is $-L^2 + 50L - 500 \geq 0$?

Solution set: $25 - 5\sqrt{5} \leq L \leq 25 + 5\sqrt{5}$

$\qquad\qquad$ 13.8 meters $\leq L \leq$ 36.2 meters

75. $R = x(75 - 0.0005x)$ and $C = 30x + 250,000$

$P = R - C$

$= (75x - 0.0005x^2) - (30x + 250,000)$

$= -0.0005x^2 + 45x - 250,000$

$\qquad\qquad\qquad\qquad P \geq 750,000$

$-0.0005x^2 + 45x - 250,000 \geq 750,000$

$-0.0005x^2 + 45x - 1,000,000 \geq 0$

Key numbers: $x = 40,000, x = 50,000$ (These were obtained by using the Quadratic Formula.)

Test intervals:
$(0, 40,000), (40,000, 50,000), (50,000, \infty)$

The solution set is $[40,000, 50,000]$ or

$40,000 \leq x \leq 50,000.$ The price per unit is

$p = \dfrac{R}{x} = 75 - 0.0005x.$

For $x = 40,000, p = \$55.$ For $x = 50,000,$

$p = \$50.$ So, for $40,000 \leq x \leq 50,000,$

$\$50.00 \leq p \leq \$55.00.$

77. (a) and (c)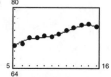

The model fits the data well.

(b) $N \approx -0.00412t^4 + 0.1705t^3 - 2.538t^2$

$\qquad\qquad + 16.55t + 31.5$

(d) 2003 to 2006

(e) No. The degree of the model is even and the leading coefficient is negative, so as t increases, N will continue to decrease toward $-\infty$.

The model decreases sharply after 2006.

79. $\qquad \dfrac{1}{R} = \dfrac{1}{R_1} + \dfrac{1}{2}$

$2R_1 = 2R + RR_1$

$2R_1 = R(2 + R_1)$

$\dfrac{2R_1}{2 + R_1} = R$

Because $R \geq 1,$

$\dfrac{2R_1}{2 + R_1} \geq 1$

$\dfrac{2R_1}{2 + R_1} - 1 \geq 0$

$\dfrac{R_1 - 2}{2 + R_1} \geq 0.$

Because $R_1 > 0,$ the only key number is $R_1 = 2.$ The inequality is satisfied when $R_1 \geq 2$ ohms.

81. True

$x^3 - 2x^2 - 11x + 12 = (x + 3)(x - 1)(x - 4)$

The test intervals are $(-\infty, -3), (-3, 1), (1, 4),$ and $(4, \infty).$

83. $x^2 + bx + 4 = 0$

    (a) To have at least one real solution, $b^2 - 4ac \geq 0$.

$$b^2 - 4(1)(4) \geq 0$$

$$b^2 - 16 \geq 0$$

Key numbers: $b = -4, b = 4$

Test intervals: $(-\infty, -4) \Rightarrow b^2 - 16 > 0$

$$(-4, 4) \Rightarrow b^2 - 16 < 0$$

$$(4, \infty) \Rightarrow b^2 - 16 > 0$$

Solution set: $(-\infty, -4] \cup [4, \infty]$

    (b) $b^2 - 4ac \geq 0$

Key numbers: $b = -2\sqrt{ac}, b = 2\sqrt{ac}$

Similar to part (a), if $a > 0$ and $c > 0$,

$b \leq -2\sqrt{ac}$ or $b \geq 2ac$.

85. $3x^2 + bx + 10 = 0$

    (a) To have at least one real solution, $b^2 - 4ac \geq 0$.

$$b^2 - 4(3)(10) \geq 0$$

$$b^2 - 120 \geq 0$$

Key numbers: $b = -2\sqrt{30}, b = 2\sqrt{30}$

Test intervals: $\left(-\infty, -2\sqrt{30}\right) \Rightarrow b^2 - 120 > 0$

$$\left(-2\sqrt{30}, 2\sqrt{30}\right) \Rightarrow b^2 - 120 < 0$$

$$\left(2\sqrt{30}, \infty\right) \Rightarrow b^2 - 120 > 0$$

Solution set: $\left(-\infty, -2\sqrt{30}\right] \cup \left[2\sqrt{30}, \infty\right]$

    (b) $b^2 - 4ac \geq 0$

Similar to part (a), if $a > 0$ and $c > 0$,

$b \leq -2\sqrt{ac}$ or $b \geq 2ac$.

87. 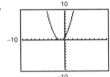 For part (b), the y-values that are less than or equal to 0 occur only at $x = -1$.

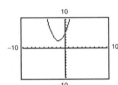 For part (c), there are no y-values that are less than 0.

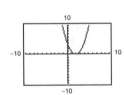 For part (d), the y-values that are greater than 0 occur for all values of x except 2.

Review Exercises for Chapter 1

1. $y = -4x + 1$

x	-2	-1	0	1	2
y	9	5	1	-3	-7

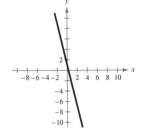

3. x-intercepts: $(1, 0), (5, 0)$

y-intercept: $(0, 5)$

5. $y = -4x + 1$

Intercepts: $\left(\frac{1}{4}, 0\right), (0, 1)$

$y = -4(-x) + 1 \Rightarrow y = 4x + 1 \Rightarrow$ No y-axis symmetry

$-y = -4x + 1 \Rightarrow y = 4x - 1 \Rightarrow$ No x-axis symmetry

$-y = -4(-x) + 1 \Rightarrow y = -4x - 1 \Rightarrow$ No origin symmetry

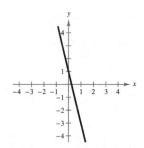

7. $y = 7 - x^2$

Intercepts: $\left(\pm\sqrt{7}, 0\right), (0, 7)$

$y = 7 - (-x)^2 \Rightarrow y = 7 - x^2 \Rightarrow y$-axis symmetry

$y = 7 - x^2 \Rightarrow y = -7 + x^2 \Rightarrow$ No x-axis symmetry

$-y = 7 - (-x)^2 \Rightarrow y = -7 + x^2 \Rightarrow$ No origin symmetry

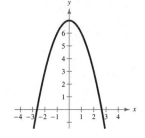

9. $y = x^3 + 3$

Intercepts: $\left(-\sqrt[3]{3}, 0\right), (0, 3)$

$y = (-x)^3 + 3 \Rightarrow y = -x^3 + 3 \Rightarrow$ No y-axis symmetry

$-y = x^3 + 3 \Rightarrow y = -x^3 - 3 \Rightarrow$ No x-axis symmetry

$-y = (-x)^3 + 3 \Rightarrow y = x^3 - 3 \Rightarrow$ No origin symmetry

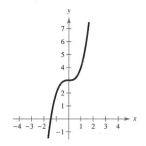

11. $y = -|x| - 2$

Intercept: $(0, -2)$

$y = -|-x| - 2 \Rightarrow y = -|x| - 2 \Rightarrow y$-axis symmetry

$-y = -|x| - 2 \Rightarrow y = |x| + 2 \Rightarrow$ No x-axis symmetry

$-y = -|-x| - 2 \Rightarrow y = |x| + 2 \Rightarrow$ No origin symmetry

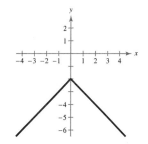

13. $x^2 + y^2 = 9$

Center: $(0, 0)$

Radius: 3

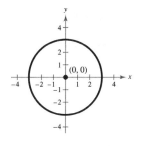

15. $(x + 2)^2 + y^2 = 16$

$(x - (-2))^2 + (y - 0)^2 = 4^2$

Center: $(-2, 0)$

Radius: 4

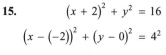
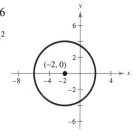

17. Endpoints of a diameter: $(0, 0)$ and $(4, -6)$

Center: $\left(\dfrac{0 + 4}{2}, \dfrac{0 + (-6)}{2}\right) = (2, -3)$

Radius: $r = \sqrt{(2 - 0)^2 + (-3 - 0)^2} = \sqrt{4 + 9} = \sqrt{13}$

Standard form: $(x - 2)^2 + \left(y - (-3)\right)^2 = \left(\sqrt{13}\right)^2$

$$(x - 2)^2 + (y + 3)^2 = 13$$

19. (a) $R = 0.123t^2 + 0.43t + 20.0,\ 8 \le t \le 17$

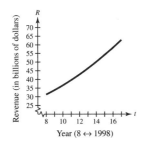

Year (8 ↔ 1998)

(b) The revenue was 50 billion dollars sometime in 2003.

21.
$$6 - (x - 2)^2 = 2 + 4x - x^2$$
$$6 - (x^2 - 4x + 4) = 2 + 4x - x^2$$
$$2 + 4x - x^2 = 2 + 4x - x^2$$
$$0 = 0 \quad \text{Identity}$$

All real numbers are solutions.

23. $-x^3 + x(7 - x) + 3 = x(-x^2 - x) + 7(x + 1) - 4$

$$-x^3 + 7x - x^2 + 3 = -x^3 - x^2 + 7x + 7 - 4$$
$$-x^3 - x^2 + 7x + 3 = -x^3 - x^2 + 7x + 3$$
$$0 = 0 \quad \text{Identity}$$

All real numbers are solutions.

25.
$$8x - 5 = 3x + 20$$
$$5x = 25$$
$$x = 5$$

27. $2(x + 5) - 7 = 3(x - 2)$
$$2x + 10 - 7 = 3x - 6$$
$$2x + 3 = 3x - 6$$
$$-x = -9$$
$$x = 9$$

29.
$$\dfrac{x}{5} - 3 = \dfrac{2x}{2} + 1$$
$$5\left(\dfrac{x}{5} - 3\right) = (x + 1)5$$
$$x - 15 = 5x + 5$$
$$-4x = 20$$
$$x = -5$$

31. $y = 3x - 1$

x-intercept: $0 = 3x - 1 \Rightarrow x = \frac{1}{3}$

y-intercept: $y = 3(0) - 1 \Rightarrow y = -1$

The x-intercept is $\left(\frac{1}{3}, 0\right)$ and the y-intercept is $(0, -1)$.

33. $y = 2(x - 4)$

x-intercept: $0 = 2(x - 4) \Rightarrow x = 4$

y-intercept: $y = 2(0 - 4) \Rightarrow y = -8$

The x-intercept is $(4, 0)$ and the y-intercept is $(0, -8)$.

35. $y = -\dfrac{1}{2}x + \dfrac{2}{3}$

x-intercept: $0 = -\dfrac{1}{2}x + \dfrac{2}{3} \Rightarrow x = \dfrac{2/3}{1/2} = \dfrac{4}{3}$

y-intercept: $y = -\dfrac{1}{2}(0) + \dfrac{2}{3} \Rightarrow y = \dfrac{2}{3}$

The x-intercept is $\left(\dfrac{4}{3}, 0\right)$ and the y-intercept is $\left(0, \dfrac{2}{3}\right)$.

37. $244.92 = 2(3.14)(3)^2 + 2(3.14)(3)h$
$$244.92 = 56.52 + 18.84h$$
$$188.40 = 18.84h$$
$$10 = h$$

The height is 10 inches.

39. *Verbal Model:*   September's profit + October's profit = 689,000

     Labels: Let x = September's profit. Then $x + 0.12x$ = October's profit.

     Equation: $x + (x + 0.12x) = 689{,}000$

$$2.12x = 689{,}000$$

$$x = 325{,}000$$

$$x + 0.12x = 364{,}000$$

Answer: September profit: 325,000, October profit: $364,000

41. Let x = the number of original investors.

Each person's share is $\dfrac{90{,}000}{x}$. If three more people invest, each person's share is $\dfrac{90{,}000}{x + 3}$.

Since this is $2500 less than the original cost, we have:

$$\frac{90{,}000}{x} - 2500 = \frac{90{,}000}{x + 3}$$

$$90{,}000(x + 3) - 2500x(x + 3) = 90{,}000x$$

$$90{,}000x + 270{,}000 - 2500x^2 - 7500x = 90{,}000x$$

$$-2500x^2 - 7500x + 270{,}000 = 0$$

$$-2500(x^2 + 3x - 108) = 0$$

$$-2500(x + 12)(x - 9) = 0$$

$x = -12$, extraneous or $x = 9$

There are currently nine investors.

43. Let x = the number of liters of pure antifreeze.

$$30\% \text{ of } (10 - x) + 100\% \text{ of } x = 50\% \text{ of } 10$$

$$0.30(10 - x) + 1.00x = 0.50(10)$$

$$3 - 0.30x + 1.00x = 5$$

$$0.70x = 2$$

$$x = \frac{2}{0.70} = \frac{20}{7} = 2\frac{6}{7} \text{ liters}$$

45. $V = \dfrac{1}{3}\pi r^2 h$

$$3V = \pi r^2 h$$

$$\frac{3V}{\pi r^2} = h$$

47. $15 + x - 2x^2 = 0$

$$0 = 2x^2 - x - 15$$

$$0 = (2x + 5)(x - 3)$$

$$2x + 5 = 0 \Rightarrow x = -\tfrac{5}{2}$$

$$x - 3 = 0 \Rightarrow x = 3$$

49. $6 = 3x^2$

$$2 = x^2$$

$$\pm\sqrt{2} = x$$

51. $(x + 13)^2 = 25$

$$x + 13 = \pm 5$$

$$x = -13 \pm 5$$

$$x = -18 \text{ or } x = -8$$

53. $x^2 + 12x + 25 = 0$

$$x = \frac{-12 \pm \sqrt{12^2 - 4(1)(25)}}{2(1)}$$

$$= \frac{-12 \pm 2\sqrt{11}}{2}$$

$$= -6 \pm \sqrt{11}$$

55. $-2x^2 - 5x + 27 = 0$

$2x^2 + 5x - 27 = 0$

$x = \dfrac{-5 \pm \sqrt{5^2 - 4(2)(-27)}}{2(2)}$

$= \dfrac{-5 \pm \sqrt{241}}{4}$

57. $M = 500x(20 - x)$

(a) $500x(20 - x) = 0$ when $x = 0$ feet and $x = 20$ feet.

(b)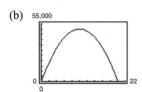

(c) The bending moment is greatest when $x = 10$ feet.

59. $4 + \sqrt{-9} = 4 + 3i$

61. $i^2 + 3i = -1 + 3i$

63. $(7 + 5i) + (-4 + 2i) = (7 - 4) + (5i + 2i) = 3 + 7i$

65. $6i(5 - 2i) = 30i - 12i^2 = 12 + 30i$

67. $\dfrac{6 - 5i}{i} = \dfrac{6 - 5i}{i} \cdot \dfrac{-i}{-i}$

$= \dfrac{-6i + 5i^2}{-i^2}$

$= -5 - 6i$

69. $\dfrac{4}{2 - 3i} + \dfrac{2}{1 + i} = \dfrac{4}{2 - 3i} \cdot \dfrac{2 + 3i}{2 + 3i} + \dfrac{2}{1 + i} \cdot \dfrac{1 - i}{1 - i}$

$= \dfrac{8 + 12i}{4 + 9} + \dfrac{2 - 2i}{1 + 1}$

$= \dfrac{8}{13} + \dfrac{12}{13}i + 1 - i$

$= \left(\dfrac{8}{13} + 1\right) + \left(\dfrac{12}{13}i - i\right)$

$= \dfrac{21}{13} - \dfrac{1}{13}i$

71. $3x^2 + 1 = 0$

$3x^2 = -1$

$x^2 = -\dfrac{1}{3}$

$x = \pm\sqrt{-\dfrac{1}{3}}$

$= \pm\sqrt{\dfrac{1}{3}}i = \pm\dfrac{\sqrt{3}}{3}i$

73. $x^2 - 2x + 10 = 0$

$x^2 - 2x + 1 = -10 + 1$

$(x - 1)^2 = -9$

$x - 1 = \pm\sqrt{-9}$

$x = 1 \pm 3i$

75. $5x^4 - 12x^3 = 0$

$x^3(5x - 12) = 0$

$x^3 = 0$ or $5x - 12 = 0$

$x = 0$ or $\quad x = \dfrac{12}{5}$

77. $x^4 - 5x^2 + 6 = 0$

$(x^2 - 2)(x^2 - 3) = 0$

$x^2 - 2 = 0 \quad$ or $\quad x^2 - 3 = 0$

$x^2 = 2 \qquad\qquad x^2 = 3$

$x = \pm\sqrt{2} \qquad\quad x = \pm\sqrt{3}$

79. $\sqrt{2x + 3} + \sqrt{x - 2} = 2$

$\left(\sqrt{2x + 3}\right)^2 = \left(2 - \sqrt{x - 2}\right)^2$

$2x + 3 = 4 - 4\sqrt{x - 2} + x - 2$

$x + 1 = -4\sqrt{x - 2}$

$(x + 1)^2 = \left(-4\sqrt{x - 2}\right)^2$

$x^2 + 2x + 1 = 16(x - 2)$

$x^2 - 14x + 33 = 0$

$(x - 3)(x - 11) = 0$

$x = 3$, extraneous or $x = 11$, extraneous

No solution

81. $(x - 1)^{2/3} - 25 = 0$

$(x - 1)^{2/3} = 25$

$(x - 1)^2 = 25^3$

$x - 1 = \pm\sqrt{25^3}$

$x = 1 \pm 125$

$x = 126$ or $x = -124$

83. $\dfrac{5}{x} = 1 + \dfrac{3}{x + 2}$

$5(x + 2) = 1(x)(x + 2) + 3x$

$5x + 10 = x^2 + 2x + 3x$

$10 = x^2$

$\pm\sqrt{10} = x$

85. $|x - 5| = 10$

$x - 5 = -10$ or $x - 5 = 10$

$x = -5 \qquad x = 15$

87. $\qquad |x^2 - 3| = 2x$ or $\qquad x^2 - 3 = 2x$

$x^2 - 2x - 3 = 0 \qquad x2 + 2x - 3 = 0$

$(x - 3)(x + 1) = 0 \qquad (x + 3)(x - 1) = 0$

$x = 3$ or $x = -1 \qquad x = -3$ or $x = 1$

The only solutions of the original equation are $x = 3$ or $x = 1$. $(x = 3$ and $x = -1$ are extraneous.)

89. $\qquad 29.95 = 42 - \sqrt{0.001x + 2}$

$-12.05 = -\sqrt{0.001x + 2}$

$\sqrt{0.001x + 2} = 12.05$

$0.001x + 2 = 145.2025$

$0.001x = 143.2025$

$x = 143{,}202.5$

$\approx 143{,}203$ units

91. Interval: $(-7, 2]$

Inequality: $-7 < x \leq 2$

The interval is bounded.

93. Interval: $(-\infty, -10]$

Inequality: $x \leq -10$

The interval is unbounded.

95. $3(x + 2) + 7 < 2x - 5$

$3x + 13 < 2x - 5$

$x < -18$

97. $4(5 - 2x) \leq \frac{1}{2}(8 - x)$

$20 - 8x \leq 4 - \frac{1}{2}x$

$-\frac{15}{2}x \leq -16$

$x \geq \frac{32}{15}$

$\left[\frac{32}{15}, \infty\right)$

99. $|x - 3| > 4$

$x - 3 < -4$ or $x - 3 > 4$

$x < -1$ or $\qquad x > 7$

$(-\infty, -1) \cup (7, \infty)$

101. If the side is 19.3 cm, then with the possible error of 0.5 cm we have:

$18.8 \leq$ side ≤ 19.8

353.44 cm$^2 \leq$ area ≤ 392.04 cm^2

103. $x^2 - 6x - 27 < 0$

$(x + 3)(x - 9) < 0$

Critical numbers: $x = -3, x = 9$

Test intervals: $(-\infty, -3), (-3, 9), (9, \infty)$

Test: Is $(x + 3)(x - 9) < 0$?

By testing an x-value in each test interval in the inequality, we see that the solution set is $(-3, 9)$.

105. $\qquad 6x^2 + 5x < 4$

$6x^2 + 5x - 4 < 0$

$(3x + 4)(2x - 1) < 0$

Critical numbers: $x = -\frac{4}{3}, x = \frac{1}{2}$

Test intervals: $\left(-\infty, -\frac{4}{3}\right), \left(-\frac{4}{3}, \frac{1}{2}\right), \left(\frac{1}{2}, \infty\right)$

Test: Is $(3x + 4)(2x - 1) < 0$?

By testing an x-value in each test interval in the inequality, we see that the solution set is $\left(-\frac{4}{3}, \frac{1}{2}\right)$.

107. $\qquad \frac{2}{x + 1} \leq \frac{3}{x - 1}$

$\frac{2(x - 1) - 3(x + 1)}{(x + 1)(x - 1)} \leq 0$

$\frac{2x - 2 - 3x - 3}{(x + 1)(x - 1)} \leq 0$

$\frac{-(x + 5)}{(x + 1)(x - 1)} \leq 0$

Critical numbers: $x = -5$

Test intervals: $(-5, -1), (-1, 1), (1, \infty)$

Test: Is $\dfrac{-(x + 5)}{(x + 1)(x - 1)} \leq 0$?

By testing an x-value in each test interval in the inequality, we see that the solution set is $[-5, -1) \cup (1, \infty)$.

109. $5000(1 + r)^2 > 5500$

$$(1 + r)^2 > 1.1$$

$$1 + r > 1.0488$$

$$r > 0.0488$$

$$r > 4.9\%$$

111. False

$$\sqrt{-18}\sqrt{-2} = \left(\sqrt{18}i\right)\left(\sqrt{2}i\right) = \sqrt{36}i^2 = -6$$

$$\sqrt{(-8)(-2)} = \sqrt{36} = 6$$

113. Rational equations, equations involving radicals, and absolute value equations, may have "solutions" that are extraneous. So checking solutions, in the original equations, is crucial to eliminate these extraneous values.

Problem Solving for Chapter 1

1.

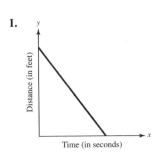

3. (a) $A = \pi ab$

$a + b = 20 \Rightarrow b = 20 - a$, thus:

$$A = \pi a(20 - a)$$

(b)

a	4	7	10	13	16
A	64π 0	91π	100π	91π	64π

(c)

$$300 = \pi a(20 - a)$$

$$300 = 20\pi a - \pi a^2$$

$$\pi a^2 - 20\pi a + 300 = 0$$

$$a = \frac{20\pi \pm \sqrt{(-20\pi)^2 - 4\pi(300)}}{2\pi}$$

$$= \frac{20\pi \pm \sqrt{400\pi^2 - 1200\pi}}{2\pi}$$

$$= \frac{20\pi \pm 20\sqrt{\pi(\pi - 3)}}{2\pi}$$

$$= 10 \pm \frac{10}{\pi}\sqrt{\pi(\pi - 3)}$$

$$a \approx 12.123 \text{ or } a \approx 7.877$$

(d)

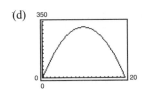

(e)   The a-intercepts occur at $a = 0$ and $a = 20$. Both yield an area of 0. When $a = 0$, $b = 20$ and you have a vertical line of length 40. Likewise when $a = 20$, $b = 0$ and you have a horizontal line of length 40. They represent the minimum and maximum values of a.

(f)   The maximum value of A is $100\pi \approx 314.159$. This occurs when $a = b = 10$ and the ellipse is actually a circle.

5. $h = \left(\sqrt{h_0} - \dfrac{2\pi d^2 \sqrt{3}}{lw} t \right)^2$

$l = 60'', \; w = 30'', \; h_0 = 25'', \; d = 2''$

$h = \left(5 - \dfrac{8\pi \sqrt{3}}{1800} t \right)^2 = \left(5 - \dfrac{\pi \sqrt{3}}{225} t \right)^2$

(a) $\quad 12.5 = \left(5 - \dfrac{\pi \sqrt{3}}{225} t \right)^2$

$\sqrt{12.5} = 5 - \dfrac{\pi \sqrt{3}}{225} t$

$t = \dfrac{225}{\pi \sqrt{3}} \left(5 - \sqrt{12.5} \right) \approx 60.6 \text{ seconds}$

(b) $\quad 0 = \left(\sqrt{12.5} - \dfrac{\pi \sqrt{3}}{225} t \right)^2$

$t = \dfrac{225 \sqrt{12.5}}{\pi \sqrt{3}} \approx 146.2 \text{ seconds}$

(c) The speed at which the water drains decreases as the amount of the water in the bathtub decreases.

7. (a) $5, 12,$ and $13; 8, 15,$ and 17

$\qquad 7, 24,$ and 25

(b) $5 \cdot 12 \cdot 13 = 780$ which is divisible by 3, 4, and 5.

$8 \cdot 15 \cdot 17 = 2040$ which is divisible by 3, 4, and 5.

$7 \cdot 24 \cdot 25 = 4200$ which is also divisible by 3, 4, and 5.

(c) Conjecture: If $a^2 + b^2 = c^2$ where $a, b,$ and c are positive integers, then abc is divisible by 60.

13. (a) $c = 1$

The terms are: $i, -1 + i, -i, -1 + i, -i, -1 + i, -i, -1 + i, -i, \ldots$

The sequence is bounded so $c = i$ *is* in the Mandelbrot Set.

(b) $c = -2$

The terms are: $1 + i, 1 + 3i, -7 + 7i, 1 - 97i, -9407 - 1931, \ldots$

The sequence is unbounded so $c = 1 + i$ *is not* in the Mandelbrot Set.

(c) $c = -2$

The terms are: $-2, 2, 2, 2, 2, \ldots$

The sequence is bounded so $c = -2$ *is* in the Mandelbrot Set.

15. $y = x^4 - x^3 - 6x^2 + 4x + 8 = (x - 2)^2 (x + 1)(x + 2)$

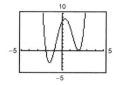

From the graph you see that $x^4 - x^3 - 6x^2 + 4x + 8 > 0$ on the intervals $(-\infty, -2) \cup (-1, 2) \cup (2, \infty)$.

9. $ax^2 + bx + c = 0$

$x^2 + \dfrac{b}{a} x + \dfrac{c}{A} = 0$

$x_1 + x_2 = -\dfrac{b}{a}$

$x_1 \cdot x_2 = \dfrac{c}{a}$

11. (a) $z_m = \dfrac{1}{z}$

$= \dfrac{1}{1 + i} = \dfrac{1}{1 + i} \cdot \dfrac{1 - i}{1 - i}$

$= \dfrac{1 - i}{2} = \dfrac{1}{2} - \dfrac{1}{2} i$

(b) $z_m = \dfrac{1}{z}$

$= \dfrac{1}{3 - i} = \dfrac{1}{3 - i} \cdot \dfrac{3 + i}{3 + i}$

$= \dfrac{3 + i}{10} = \dfrac{3}{10} + \dfrac{1}{10} i$

(c) $z_m = \dfrac{1}{z}$

$= \dfrac{1}{-2 + 8i}$

$= \dfrac{1}{-2 + 8i} \cdot \dfrac{-2 - 8i}{-2 - 8i}$

$= \dfrac{-2 - 8i}{68} = -\dfrac{1}{34} - \dfrac{2}{17} i$

Practice Test for Chapter 1

1. Graph $3x - 5y = 15$.

2. Graph $y = \sqrt{9 - x}$.

3. Solve $5x + 4 = 7x - 8$.

4. Solve $\dfrac{x}{3} - 5 = \dfrac{x}{5} + 1$.

5. Solve $\dfrac{3x + 1}{6x - 7} = \dfrac{2}{5}$.

6. Solve $(x - 3)^2 + 4 = (x + 1)^2$.

7. Solve $A = \frac{1}{2}(a + b)h$ for a.

8. 301 is what percent of 4300?

9. Cindy has $6.05 in quarter and nickels. How many of each coin does she have if there are 53 coins in all?

10. Ed has $15,000 invested in two fund paying $9\frac{1}{2}\%$ and 11% simple interest, respectively. How much is invested in each if the yearly interest is $1582.50?

11. Solve $28 + 5x - 3x^2 = 0$ by factoring.

12. Solve $(x - 2)^2 = 24$ by taking the square root of both sides.

13. Solve $x^2 - 4x - 9 = 0$ by completing the square.

14. Solve $x^2 + 5x - 1 = 0$ by the Quadratic Formula.

15. Solve $3x^2 - 2x + 4 = 0$ by the Quadratic Formula.

16. The perimeter of a rectangle is 1100 feet. Find the dimensions so that the enclosed area will be 60,000 square feet.

17. Find two consecutive even positive integers whose product is 624.

18. Solve $x^3 - 10x^2 + 24x = 0$ by factoring.

19. Solve $\sqrt[3]{6 - x} = 4$.

20. Solve $(x^2 - 8)^{2/5} = 4$.

21. Solve $x^4 - x^2 - 12 = 0$.

22. Solve $4 - 3x > 16$.

23. Solve $\left| \dfrac{x - 3}{2} \right| < 5$.

24. Solve $\dfrac{x + 1}{x - 3} < 2$.

25. Solve $|3x - 4| \geq 9$.

CHAPTER 2
Functions and Their Graphs

CHAPTER 2
Functions and Their Graphs

Section 2.1   Linear Equations in Two Variables

You should know the following important facts about lines.

- The graph of $y = mx + b$ is a straight line. It is called a linear equation in two variables.

  (a) The slope (steepness) is m.

  (b) The y-intercept is $(0, b)$.

- The slope of the line through (x_1, y_1) and (x_2, y_2) is

  $$m = \frac{y_2 - y_1}{x_2 - x_1} = \frac{\text{change in } y}{\text{change in } x} = \frac{\text{rise}}{\text{run}}.$$

- (a) If $m > 0$, the line rises from left to right.

  (b) If $m = 0$, the line is horizontal.

  (c) If $m < 0$, the line falls from left to right.

  (d) If m is undefined, the line is vertical.

- Equations of Lines

  (a) Slope-Intercept Form: $y = mx + b$

  (b) Point-Slope Form: $y - y_1 = m(x - x_1)$

  (c) Two-Point Form: $y - y_1 = \dfrac{y_2 - y_1}{x_2 - x_1}(x - x_1)$

  (d) General Form: $Ax + By + C = 0$

  (e) Vertical Line: $x = a$

  (f) Horizontal Line: $y = b$

- Given two distinct nonvertical lines

  $L_1: y = m_1 x + b_1$ and $L_2: y = m_2 x + b_2$

  (a) L_1 is parallel to L_2 if and only if $m_1 = m_2$ and $b_1 \neq b_2$.

  (b) L_1 is perpendicular to L_2 if and only if $m_1 = -1/m_2$.

---

1. linear

3. parallel

5. rate or rate of change

7. general

9. (a) $m = \frac{2}{3}$. Because the slope is positive, the line rises.
   Matches L_2.

   (b) m is undefined. The line is vertical. Matches L_3.

   (c) $m = -2$. The line falls. Matches L_1.

11.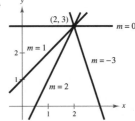

13. Two points on the line: $(0, 0)$ and $(4, 6)$

   $\text{Slope} = \dfrac{y_2 - y_1}{x_2 - x_1} = \dfrac{6}{4} = \dfrac{3}{2}$

15. Two points on the line: $(0, 8)$ and $(2, 0)$

Slope $= \dfrac{y_2 - y_1}{x_2 - x_1} = \dfrac{-8}{2} = -4$

17. $y = 5x + 3$

Slope: $m = 5$

y-intercept: $(0, 3)$

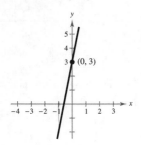

19. $y = -\frac{1}{2}x + 4$

Slope: $m = -\frac{1}{2}$

y-intercept: $(0, 4)$

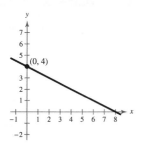

21. $5x - 2 = 0$

$x = \frac{2}{5}$, vertical line

Slope: undefined

No y-intercept

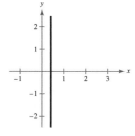

23. $7x + 6y = 30$

$y = -\frac{7}{6}x + 5$

Slope: $m = -\frac{7}{6}$

y-intercept: $(0, 5)$

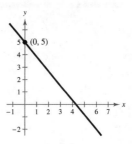

25. $y - 3 = 0$

$y = 3$, horizontal line

Slope: $m = 0$

y-intercept: $(0, 3)$

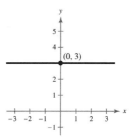

27. $x + 5 = 0$

$x = -5$

Slope: undefined (vertical line)

No y-intercept

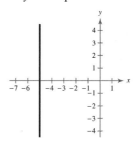

29. $m = \dfrac{0 - 9}{6 - 0} = \dfrac{-9}{6} = -\dfrac{3}{2}$

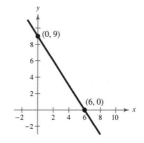

31. $m = \dfrac{6-(-2)}{1-(-3)} = \dfrac{8}{4} = 2$

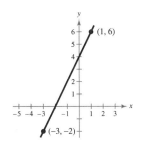

33. $m = \dfrac{-7-(-7)}{8-5} = \dfrac{0}{3} = 0$

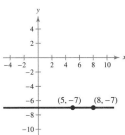

35. $m = \dfrac{4-(-1)}{-6-(-6)} = \dfrac{5}{0}$

m is undefined.

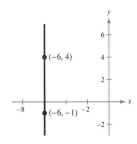

37. $m = \dfrac{-\dfrac{1}{3}-\left(-\dfrac{4}{3}\right)}{-\dfrac{3}{2}-\dfrac{11}{2}} = -\dfrac{1}{7}$

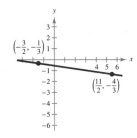

39. $m = \dfrac{1.6-3.1}{-5.2-4.8} = \dfrac{-1.5}{-10} = 0.15$

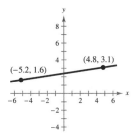

41. Point: $(2,1)$, Slope: $m = 0$

Because $m = 0$, y does not change. Three points are $(0,1), (3,1),$ and $(-1,1)$.

43. Point: $(5,-6)$, Slope: $m = 1$

Because $m = 1$, y increases by 1 for every one unit increase in x. Three points are $(6,-5), (7,-4),$ and $(8,-3)$.

45. Point: $(-8,1)$, Slope is undefined.

Because m is undefined, x does not change. Three points are $(-8,0), (-8,2),$ and $(-8,3)$.

47. Point: $(-5,4)$, Slope: $m = 2$

Because $m = 2 = \dfrac{2}{1}$, y increases by 2 for every one unit increase in x. Three additional points are $(-4,6), (-3,8),$ and $(-2,10)$.

49. Point: $(7,-2)$, Slope: $m = \dfrac{1}{2}$

Because $m = \dfrac{1}{2}$, y increases by 1 unit for every two unit increase in x. Three additional points are $(9,-1), (11,0),$ and $(13,1)$.

51. Point: $(0,-2)$; $m = 3$

$$y + 2 = 3(x-0)$$
$$y = 3x - 2$$

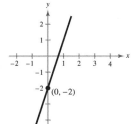

53. Point: $(-3, 6)$; $m = -2$

$$y - 6 = -2(x + 3)$$
$$y = -2x$$

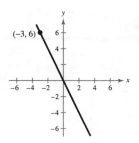

55. Point: $(4, 0)$; $m = -\frac{1}{3}$

$$y - 0 = -\frac{1}{3}(x - 4)$$
$$y = -\frac{1}{3}x + \frac{4}{3}$$

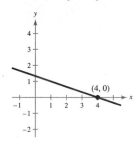

57. Point: $(2, -3)$; $m = -\frac{1}{2}$

$$y - (-3) = -\frac{1}{2}(x - 2)$$
$$y + 3 = -\frac{1}{2}x + 1$$
$$y = -\frac{1}{2}x - 2$$

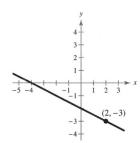

59. Point: $(6, -1)$; m is undefined.

Because the slope is undefined, the line is a vertical line.

$$x = 6$$

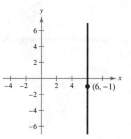

61. Point: $\left(4, \frac{5}{2}\right)$; $m = 0$

$$y - \frac{5}{2} = 0(x - 4)$$
$$y - \frac{5}{2} = 0$$
$$y = \frac{5}{2}$$

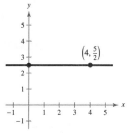

63. Point: $(-5.1, 1.8)$; $m = 5$

$$y - 1.8 = 5(x - (-5.1))$$
$$y = 5x + 27.3$$

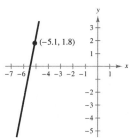

65. $(5, -1), (-5, 5)$

$$y + 1 = \frac{5 + 1}{-5 - 5}(x - 5)$$

$$y = -\frac{3}{5}(x - 5) - 1$$

$$y = -\frac{3}{5}x + 2$$

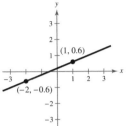

67. $(-8, 1), (-8, 7)$

Because both points have $x = -8$, the slope is undefined, and the line is vertical.

$$x = -8$$

69. $\left(2, \frac{1}{2}\right), \left(\frac{1}{2}, \frac{5}{4}\right)$

$$y - \frac{1}{2} = \frac{\frac{5}{4} - \frac{1}{2}}{\frac{1}{2} - 2}(x - 2)$$

$$y = -\frac{1}{2}(x - 2) + \frac{1}{2}$$

$$y = -\frac{1}{2}x + \frac{3}{2}$$

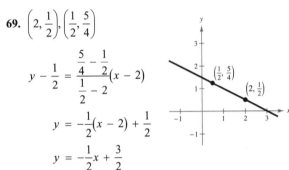

71. $\left(-\frac{1}{10}, -\frac{3}{5}\right), \left(\frac{9}{10}, -\frac{9}{5}\right)$

$$y - \left(-\frac{3}{5}\right) = \frac{-\frac{9}{5} - \left(-\frac{3}{5}\right)}{\frac{9}{10} - \left(-\frac{1}{10}\right)}\left(x - \left(-\frac{1}{10}\right)\right)$$

$$y = -\frac{6}{5}\left(x + \frac{1}{10}\right) - \frac{3}{5}$$

$$y = -\frac{6}{5}x - \frac{18}{28}$$

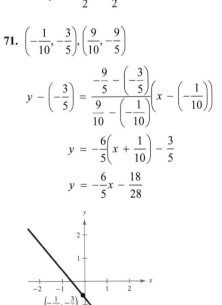

73. $(1, 0.6), (-2, -0.6)$

$$y - 0.6 = \frac{-0.6 - 0.6}{-2 - 1}(x - 1)$$

$$y = 0.4(x - 1) + 0.6$$

$$y = 0.4x + 0.2$$

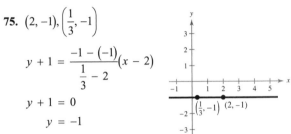

75. $(2, -1), \left(\frac{1}{3}, -1\right)$

$$y + 1 = \frac{-1 - (-1)}{\frac{1}{3} - 2}(x - 2)$$

$$y + 1 = 0$$

$$y = -1$$

The line is horizontal.

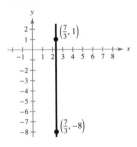

77. $\left(\frac{7}{3}, -8\right), \left(\frac{7}{3}, 1\right)$

$$m = \frac{1 - (-8)}{\frac{7}{3} - \frac{7}{3}} = \frac{9}{0} \text{ and is undefined.}$$

$$x = \frac{7}{3}$$

The line is vertical.

79. $L_1: y = \frac{1}{3}x - 2$

$$m_1 = \frac{1}{3}$$

$$L_2: y = \frac{1}{3}x + 3$$

$$m_2 = \frac{1}{3}$$

The lines are parallel.

81. $L_1: y = \frac{1}{2}x - 3$

$\quad m_1 = \frac{1}{2}$

$\quad L_2: y = -\frac{1}{2}x + 1$

$\quad m_2 = -\frac{1}{2}$

The lines are neither parallel nor perpendicular.

83. $L_1: (0, -1), (5, 9)$

$\quad m_1 = \dfrac{9+1}{5-0} = 2$

$\quad L_2: (0, 3), (4, 1)$

$\quad m_2 = \dfrac{1-3}{4-0} = -\dfrac{1}{2}$

The lines are perpendicular.

85. $L_1: (3, 6), (-6, 0)$

$\quad m_1 = \dfrac{0-6}{-6-3} = \dfrac{2}{3}$

$\quad L_2: (0, -1), \left(5, \dfrac{7}{3}\right)$

$\quad m_2 = \dfrac{\frac{7}{3}+1}{5-0} = \dfrac{2}{3}$

The lines are parallel.

87. $4x - 2y = 3$

$\quad y = 2x - \frac{3}{2}$

Slope: $m = 2$

(a) $(2, 1),\ m = 2$

$\quad y - 1 = 2(x - 2)$

$\quad\quad y = 2x - 3$

(b) $(2, 1),\ m = -\frac{1}{2}$

$\quad y - 1 = -\frac{1}{2}(x - 2)$

$\quad\quad y = -\frac{1}{2}x + 2$

89. $3x + 4y = 7$

$\quad y = -\frac{3}{4}x + \frac{7}{4}$

Slope: $m = -\frac{3}{4}$

(a) $\left(-\frac{2}{3}, \frac{7}{8}\right),\ m = -\frac{3}{4}$

$\quad y - \frac{7}{8} = -\frac{3}{4}\left(x - \left(-\frac{2}{3}\right)\right)$

$\quad\quad y = -\frac{3}{4}x + \frac{3}{8}$

(b) $\left(-\frac{2}{3}, \frac{7}{8}\right),\ m = \frac{4}{3}$

$\quad y - \frac{7}{8} = \frac{4}{3}\left(x - \left(-\frac{2}{3}\right)\right)$

$\quad\quad y = \frac{4}{3}x + \frac{127}{72}$

91. $y + 3 = 0$

$\quad y = -3$

Slope: $m = 0$

(a) $(-1, 0),\ m = 0$

$\quad y = 0$

(b) $(-1, 0),\ m$ is undefined.

$\quad x = -1$

93. $x - 4 = 0$

$\quad x = 4$

Slope: m is undefined.

(a) $(3, -2),\ m$ is undefined.

$\quad x = 3$

(b) $(3, -2),\ m = 0$

$\quad y = -2$

95. $x - y = 4$

$\quad y = x - 4$

Slope: $m = 1$

(a) $(2.5, 6.8),\ m = 1$

$\quad y - 6.8 = 1(x - 2.5)$

$\quad\quad y = x + 4.3$

(b) $(2.5, 6.8),\ m = -1$

$\quad y - 6.8 = (-1)(x - 2.5)$

$\quad\quad y = -x + 9.3$

97. $\dfrac{x}{2} + \dfrac{y}{3} = 1$

$\quad 3x + 2y - 6 = 0$

99. $\dfrac{x}{-1/6} + \dfrac{y}{-2/3} = 1$

$6x + \dfrac{3}{2}y = -1$

$12x + 3y + 2 = 0$

101. $\dfrac{x}{c} + \dfrac{y}{c} = 1,\, c \neq 0$

$x + y = c$

$1 + 2 = c$

$3 = c$

$x + y = 3$

$x + y - 3 = 0$

103. (a) $y = 2x$

(b) $y = -2x$

(c) $y = \dfrac{1}{2}x$

(b) and (c) are perpendicular.

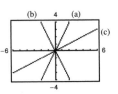

105. (a) $y = -\dfrac{1}{2}x$

(b) $y = -\dfrac{1}{2}x + 3$

(c) $y = 2x - 4$

(a) and (b) are parallel. (c) is perpendicular to (a) and (b).

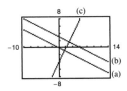

107. Set the distance between $(4, -1)$ and (x, y) equal to the distance between $(-2, 3)$ and (x, y).

$$\sqrt{(x - 4)^2 + \left[y - (-1)\right]^2} = \sqrt{\left[x - (-2)\right]^2 + (y - 3)^2}$$

$$(x - 4)^2 + (y + 1)^2 = (x + 2)^2 + (y - 3)^2$$

$$x^2 - 8x + 16 + y^2 + 2y + 1 = x^2 + 4x + 4 + y^2 - 6y + 9$$

$$-8x + 2y + 17 = 4x - 6y + 13$$

$$0 = 12x - 8y - 4$$

$$0 = 4(3x - 2y - 1)$$

$$0 = 3x - 2y - 1$$

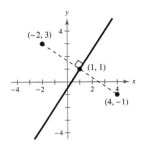

This line is the perpendicular bisector of the line segment connecting $(4, -1)$ and $(-2, 3)$.

109. Set the distance between $\left(3, \dfrac{5}{2}\right)$ and (x, y) equal to the distance between $(-7, 1)$ and (x, y).

$$\sqrt{(x - 3)^2 + \left(y - \dfrac{5}{2}\right)^2} = \sqrt{\left[x - (-7)\right]^2 + (y - 1)^2}$$

$$(x - 3)^2 + \left(y - \dfrac{5}{2}\right)^2 = (x + 7)^2 + (y - 1)^2$$

$$x^2 - 6x + 9 + y^2 - 5y + \dfrac{25}{4} = x^2 + 14x + 49 + y^2 - 2y + 1$$

$$-6x - 5y + \dfrac{61}{4} = 14x - 2y + 50$$

$$-24x - 20y + 61 = 56x - 8y + 200$$

$$80x + 12y + 139 = 0$$

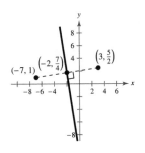

This line is the perpendicular bisector of the line segment connecting $\left(3, \dfrac{5}{2}\right)$ and $(-7, 1)$.

111. (a) $m = 135$. The sales are increasing 135 units per year.

(b) $m = 0$. There is no change in sales during the year.

(c) $m = -40$. The sales are decreasing 40 units per year.

113. (a)   greatest increase = largest slope

$$\left(18, 97{,}486\right), \left(16, 90{,}260\right)$$

$$m_1 = \frac{97{,}486 - 90{,}260}{18 - 16} = 3{,}613$$

The salary increased the greatest amount between 2006 and 2008.

least increase = smallest slope

$$\left(14, 86{,}160\right), \left(12, 83{,}944\right)$$

$$m_2 = \frac{86{,}160 - 83{,}944}{14 - 12} = 1108$$

The salary increased the least amount between 2002 and 2004.

(b) $m = \dfrac{97{,}486 - 69{,}277}{18 - 6} = \dfrac{9403}{12}$ or 2350.75

(c)   The average salary of a senior high school principal increased $2350.75 per year between 1996 and 2008.

115. $y = \frac{6}{100}x$

$y = \frac{6}{100}(200) = 12$ feet

117. $\left(10, 2540\right), m = -125$

$V - 2540 = -125\left(t - 10\right)$

$V - 2540 = -125t + 1250$

$V = -125t + 3790,\ 5 \le t \le 10$

119.   The V-intercept measures the value of the molding machine at the time of purchase $\left(\text{when } t = 0\right)$.

The slope measures the amount the value of the machine depreciates per year.

121.   Using the points $\left(0, 875\right)$ and $\left(5, 0\right)$, where the first coordinate represents the year t and the second coordinate represents the value V, you have

$$m = \frac{0 - 875}{5 - 0} = -175$$

$V = -175t + 875,\ 0 \le t \le 5.$

123.   Sales price = List price − 20% discount

$S = L - 0.20L$

$S = 0.80L$

125. $W = 0.07S + 2500$

127. $\left(17, 1.46\right), \left(9, 1.21\right)$

$$m = \frac{1.46 - 1.21}{17 - 9} = 0.03125$$

$y - 1.46 = 0.03125\left(t - 17\right)$

$y = 0.03125t + 0.92875$

For 2012, use $t = 22$.

$y = 0.03125\left(22\right) + 0.92875$

$y \approx \$1.62$

For 2014, use $t = 24$.

$y = 0.03125\left(24\right) + 0.92875$

$y \approx \$1.68$

129. (a) $\left(0, 40{,}571\right), \left(8, 44{,}112\right)$

$$m = \frac{44{,}112 - 40{,}571}{8 - 0} = 442.625$$

$y - 40{,}571 = 442.625\left(t - 0\right)$

$y = 442.625t + 40{,}571$

(b)   For 2010, use $t = 10$.

$y = 442.625\left(10\right) + 40{,}571$

$y = 44997.25$

$y \approx 44{,}997$ students

For 2015, use $t = 15$.

$y = 442.625\left(15\right) + 40{,}571$

$y = 47{,}210.375$

$y \approx 47{,}210$ students

(c) $m = 442.625$; Each year, enrollment increases by about 443 students.

131. (a) Cost = cost of operation (per hr) + cost of operator (per hour) + cost of machine

$$C = 6.50t + 11.50t + 42,000$$

$$C = 18t + 42,000$$

(b) Revenue = charge per hour

$$R = 30t$$

(c) $P = R - C$

$$P = 30t - (18t + 42,000)$$

$$P = 12t - 42,000$$

(d) $\quad 0 = 12t - 42,000$

$$\frac{42,000}{12} = \frac{12t}{12}$$

$$t = 3500.$$

To break even the company must use the equipment for 3500 hours.

133. (a)

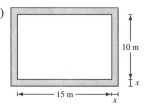

(b) $y = 2(15 + 2x) + 2(10 + 2x) = 8x + 50$

(c)

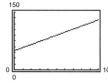

(d) Because $m = 8,$ each 1-meter increase in x will increase y by 8 meters.

135. (a) and (b)

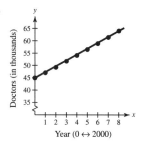

(c) Answers will vary.

Sample answer: Using technology:
$y = 2.3726x + 44.658$

Using 2 points to estimate:

$(3, 51.7), (4, 54.1)$

$$m = \frac{54.1 - 51.7}{4 - 3} = 2.4$$

$$y - y_1 = m(x - x_1)$$

$$y - 51.7 = 2.4(x - 3)$$

$$y = 2.4x + 44.5$$

(d) Answers will vary.

Sample answer: The slope describes the change (increase or decrease) in the number of osteopathic doctors per year.
The y-intercept describes the number of doctors in the year 2000 (when $x = 0$).

(e) The model is a good fit to the data.

(f) Sample answer: For 2012, use $x = 12.$

$$y = 2.3726(12) + 44.658$$

$$y \approx 73.1$$

In the year 2012, there will be approximately 73.1 thousand osteopathic doctors.

137. False. The slope with the greatest magnitude corresponds to the steepest line.

139. Using the Distance Formula, we have

$AB = 6, BC = \sqrt{40}$, and $AC = 2$. Since

$6^2 + 2^2 = \left(\sqrt{40}\right)^2$, the triangle is a right triangle.

141. No. The slope cannot be determined without knowing the scale on the *y*-axis. The slopes will be the same if the scale on the *y*-axis of (a) is $2\frac{1}{2}$ and the scale on the *y*-axis of (b) is 1. Then the slope of both is $\frac{5}{4}$.

143. The line $y = 4x$ rises most quickly.

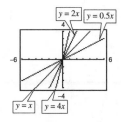

The line $y = -4x$ falls most quickly.

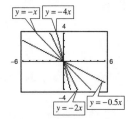

The greater the magnitude of the slope (the absolute value of the slope), the faster the line rises or falls.

145. No, the slopes of two perpendicular lines have opposite signs. (Assume that neither line is vertical or horizontal.)

Section 2.2   Functions

- Given a set or an equation, you should be able to determine if it represents a function.

- Know that functions can be represented in four ways: verbally, numerically, graphically, and algebraically.

- Given a function, you should be able to do the following.
  - (a) Find the domain and range.
  - (b) Evaluate it at specific values.

- You should be able to use function notation.

1. domain; range; function

3. independent; dependent

5. implied domain

7. Yes, the relationship is a function. Each domain value is matched with exactly one range value.

9. No, the relationship is not a function. The domain values are each matched with three range values.

11. Yes, it does represent a function. Each input value is matched with exactly one output value.

13. No, it does not represent a function. The input values of 10 and 7 are each matched with two output values.

15. (a) Each element of *A* is matched with exactly one element of *B*, so it does represent a function.

(b) The element 1 in *A* is matched with two elements, −2 and 1 of *B*, so it does not represent a function.

(c) Each element of *A* is matched with exactly one element of *B*, so it does represent a function.

(d) The element 2 in *A* is not matched with an element of *B*, so the relation does not represent a function.

17. Each is a function. For each year there corresponds one and only one circulation.

19. $x^2 + y^2 = 4 \Rightarrow y = \pm\sqrt{4 - x^2}$

No, *y is not* a function of *x*.

21. $x^2 + y = 4 \Rightarrow y = 4 - x^2$

Yes, *y is* a function of *x*.

23. $2x + 3y = 4 \Rightarrow y = \frac{1}{3}(4 - 2x)$

Yes, y *is* a function of x.

25. $(x + 2)^2 + (y - 1)^2 = 25$

$$y = \pm\sqrt{25 - (x + 2)^2} + 1$$

No, y *is not* a function of x.

27. $y^2 = x^2 - 1 \Rightarrow y = \pm\sqrt{x^2 - 1}$

No, y *is not* a function of x.

29. $y = \sqrt{16 - x^2}$

Yes, y *is* a function of x.

31. $y = |4 - x|$

Yes, y *is* a function of x.

33. $x = 14$

No, this *is not* a function of x.

35. $y + 5 = 0$

$$y = -5 \quad \text{or} \quad y = 0x - 5$$

Yes, y *is* a function of x.

37. $f(x) = 2x - 3$

(a) $f(1) = 2(1) - 3 = -1$

(b) $f(-3) = 2(-3) - 3 = -9$

(c) $f(x - 1) = 2(x - 1) - 3 = 2x - 5$

39. $V(r) = \frac{4}{3}\pi r^3$

(a) $V(3) = \frac{4}{3}\pi(3)^3 = \frac{4}{3}\pi(27) = 36\pi$

(b) $V\left(\frac{3}{2}\right) = \frac{4}{3}\pi\left(\frac{3}{2}\right)^3 = \frac{4}{3}\pi\left(\frac{27}{8}\right) = \frac{9}{2}\pi$

(c) $V(2r) = \frac{4}{3}\pi(2r)^3 = \frac{4}{3}\pi(8r^3) = \frac{32}{3}\pi r^3$

41. $g(t) = 4t^2 - 3t + 5$

(a) $g(2) = 4(2)^2 - 3(2) + 5$

$\qquad = 15$

(b) $g(t - 2) = 4(t - 2)^2 - 3(t - 2) + 5$

$\qquad\qquad = 4t^2 - 19t + 27$

(c) $g(t) - g(2) = 4t^2 - 3t + 5 - 15$

$\qquad\qquad\qquad = 4t^2 - 3t - 10$

43. $f(y) = 3 - \sqrt{y}$

(a) $f(4) = 3 - \sqrt{4} = 1$

(b) $f(0.25) = 3 - \sqrt{0.25} = 2.5$

(c) $f(4x^2) = 3 - \sqrt{4x^2} = 3 - 2|x|$

45. $q(x) = \dfrac{1}{x^2 - 9}$

(a) $q(0) = \dfrac{1}{0^2 - 9} = -\dfrac{1}{9}$

(b) $q(3) = \dfrac{1}{3^2 - 9}$ is undefined.

(c) $q(y + 3) = \dfrac{1}{(y + 3)^2 - 9} = \dfrac{1}{y^2 + 6y}$

47. $f(x) = \dfrac{|x|}{x}$

(a) $f(2) = \dfrac{|2|}{2} = 1$

(b) $f(-2) = \dfrac{|-2|}{-2} = -1$

(c) $f(x - 1) = \dfrac{|x - 1|}{x - 1} = \begin{cases} -1, & \text{if } x < 1 \\ 1, & \text{if } x > 1 \end{cases}$

49. $f(x) = \begin{cases} 2x + 1, & x < 0 \\ 2x + 2, & x \geq 0 \end{cases}$

(a) $f(-1) = 2(-1) + 1 = -1$

(b) $f(0) = 2(0) + 2 = 2$

(c) $f(2) = 2(2) + 2 = 6$

51. $f(x) = \begin{cases} 3x - 1, & x < -1 \\ 4, & -1 \leq x \leq 1 \\ x^2, & x > 1 \end{cases}$

(a) $f(-2) = 3(-2) - 1 = -7$

(b) $f\left(-\frac{1}{2}\right) = 4$

(c) $f(3) = 3^2 = 9$

53. $f(x) = x^2 - 3$

$f(-2) = (-2)^2 - 3 = 1$

$f(-1) = (-1)^2 - 3 = -2$

$f(0) = (0)^2 - 3 = -3$

$f(1) = (1)^2 - 3 = -2$

$f(2) = (2)^2 - 3 = 1$

x	-2	-1	0	1	2
$f(x)$	1	-2	-3	-2	1

55. $h(t) = \frac{1}{2}|t + 3|$

$h(-5) = \frac{1}{2}|-5 + 3| = 1$

$h(-4) = \frac{1}{2}|-4 + 3| = \frac{1}{2}$

$h(-3) = \frac{1}{2}|-3 + 3| = 0$

$h(-2) = \frac{1}{2}|-2 + 3| = \frac{1}{2}$

$h(-1) = \frac{1}{2}|-1 + 3| = 1$

t	-5	-4	-3	-2	-1
$h(t)$	1	$\frac{1}{2}$	0	$\frac{1}{2}$	1

57. $f(x) = \begin{cases} -\frac{1}{2}x + 4, & x \le 0 \\ (x - 2)^2, & x > 0 \end{cases}$

$f(-2) = -\frac{1}{2}(-2) + 4 = 5$

$f(-1) = -\frac{1}{2}(-1) + 4 = 4\frac{1}{2} = \frac{9}{2}$

$f(0) = -\frac{1}{2}(0) + 4 = 4$

$f(1) = (1 - 2)^2 = 1$

$f(2) = (2 - 2)^2 = 0$

x	-2	-1	0	1	2
$f(x)$	5	$\frac{9}{2}$	4	1	0

59. $15 - 3x = 0$

$3x = 15$

$x = 5$

61. $\dfrac{3x - 4}{5} = 0$

$3x - 4 = 0$

$x = \dfrac{4}{3}$

63. $x^2 - 9 = 0$

$x^2 = 9$

$x = \pm 3$

65. $x^3 - x = 0$

$x(x^2 - 1) = 0$

$x(x + 1)(x - 1) = 0$

$x = 0, x = -1, \ \text{or} \ \ x = 1$

67. $f(x) = g(x)$

$x^2 = x + 2$

$x^2 - x - 2 = 0$

$(x - 2)(x + 1) = 0$

$x - 2 = 0 \quad x + 1 = 0$

$x = 2 \qquad x = -1$

69. $f(x) = g(x)$

$x^4 - 2x^2 = 2x^2$

$x^4 - 4x^2 = 0$

$x^2(x^2 - 4) = 0$

$x^2(x + 2)(x - 2) = 0$

$x^2 = 0 \Rightarrow x = 0$

$x + 2 = 0 \Rightarrow x = -2$

$x - 2 = 0 \Rightarrow x = 2$

71. $f(x) = 5x^2 + 2x - 1$

Because $f(x)$ is a polynomial, the domain is all real numbers x.

73. $h(t) = \dfrac{4}{t}$

The domain is all real numbers t except $t = 0$.

75. $g(y) = \sqrt{y - 10}$

Domain: $y - 10 \ge 0$

$y \ge 10$

The domain is all real numbers y such that $y \ge 10$.

77. $g(x) = \dfrac{1}{x} - \dfrac{3}{x + 2}$

The domain is all real numbers x except $x = 0, x = -2$.

79. $f(s) = \dfrac{\sqrt{s-1}}{s-4}$

Domain: $s - 1 \geq 0 \Rightarrow s \geq 1$ and $s \neq 4$

The domain consists of all real numbers s, such that $s \geq 1$ and $s \neq 4$.

81. $f(x) = \dfrac{x-4}{\sqrt{x}}$

The domain is all real numbers x such that $x > 0$ or $(0, \infty)$.

83. $f(x) = x^2$

$f(-2) = (-2)^2 = 4$

$f(-1) = (-1)^2 = 1$

$f(0) = 0^2 = 0$

$f(1) = 1^2 = 1$

$f(2) = 2^2 = 4$

$\{(-2, 4), (-1, 1), (0, 0), (1, 1), (2, 4)\}$

85. $f(x) = |x| + 2$

$f(-2) = |-2| + 2 = 4$

$f(-1) = |-1| + 2 = 3$

$f(0) = |0| + 2 = 2$

$f(1) = |1| + 2 = 3$

$f(2) = |2| + 2 = 4$

$\{(-2, 4), (-1, 3), (0, 2), (1, 3), (2, 4)\}$

87. $A = s^2$ and $P = 4s \Rightarrow \dfrac{P}{4} = s$

$A = \left(\dfrac{P}{4}\right)^2 = \dfrac{P^2}{16}$

89. (a)

Height, x	Volume, V
1	484
2	800
3	972
4	1024
5	980
6	864

The volume is maximum when $x = 4$ and $V = 1024$ cubic centimeters.

(b)

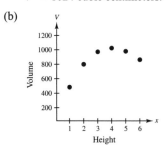

V is a function of x.

(c) $V = x(24 - 2x)^2$

Domain: $0 < x < 12$

91. $A = \dfrac{1}{2}bh = \dfrac{1}{2}xy$

Because $(0, y), (2, 1),$ and $(x, 0)$ all lie on the same line, the slopes between any pair are equal.

$\dfrac{1-y}{2-0} = \dfrac{0-1}{x-2}$

$\dfrac{1-y}{2} = \dfrac{-1}{x-2}$

$y = \dfrac{2}{x-2} + 1$

$y = \dfrac{x}{x-2}$

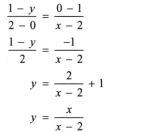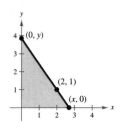

So, $A = \dfrac{1}{2}x\left(\dfrac{x}{x-2}\right) = \dfrac{x^2}{2(x-2)}$.

The domain of A includes x-values such that $x^2/[2(x-2)] > 0$. By solving this inequality, the domain is $x > 2$.

93. $y = -\frac{1}{10}x^2 + 3x + 6$

$y(30) = -\frac{1}{10}(30)^2 + 3(30) + 6 = 6$ feet

If the child holds a glove at a height of 5 feet, then the ball *will* be over the child's head because it will be at a height of 6 feet.

95. $p(t) = \begin{cases} 1.011t^2 - 12.38t + 170.5, & 8 \le t \le 13 \\ -6.950t^2 + 222.55t - 1557.6, & 14 \le t \le 17 \end{cases}$

1998: Use $t = 8$ and find $p(8)$. $p(8) = 1.011(8)^2 - 12.38(t) + 170.5 = 136.164$ thousand $= \$136{,}164$

1999: Use $t = 9$ and find $p(9)$. $p(9) = 1.011(9)^2 - 12.38(9) + 170.5 = 140.971$ thousand $= \$140{,}971$

2000: Use $t = 10$ and find $p(10)$. $p(10) = 1.011(10)^2 - 12.38(10) + 170.5 = 147.800$ thousand $= \$147{,}800$

2001: Use $t = 11$ and find $p(11)$. $p(11) = 1.011(11)^2 - 12.38(11) + 170.5 = 156.651$ thousand $= \$156{,}651$

2002: Use $t = 12$ and find $p(12)$. $p(12) = 1.011(12)^2 - 12.38(12) + 170.5 = 167.524$ thousand $= \$167{,}524$

2003: Use $t = 13$ and find $p(13)$. $p(13) = 1.011(13)^2 - 12.38(13) + 170.5 = 180.419$ thousand $= \$180{,}419$

2004: Use $t = 14$ and find $p(14)$. $p(14) = -6.950(14)^2 + 222.55(14) - 1557.6 = 195.900$ thousand $= \$195{,}900$

2005: Use $t = 15$ and find $p(15)$. $p(15) = -6.950(15)^2 + 222.55(15) - 1557.6 = 216.900$ thousand $= \$216{,}900$

2006: Use $t = 16$ and find $p(16)$. $p(16) = -6.950(16)^2 + 222.55(16) - 1557.6 = 224.000$ thousand $= \$224{,}000$

2007: Use $t = 17$ and find $p(17)$. $p(17) = -6.950(17)^2 + 222.55(17) - 1557.6 = 217.200$ thousand $= \$217{,}200$

97. (a) Cost = variable costs + fixed costs

$C = 12.30x + 98{,}000$

(b) Revenue = price per unit × number of units

$R = 17.98x$

(c) Profit = Revenue − Cost

$P = 17.98x - (12.30x + 98{,}000)$

$P = 5.68x - 98{,}000$

99. (a) $R = n(\text{rate}) = n\big[8.00 - 0.05(n - 80)\big], n \ge 80$

$R = 12.00n - 0.05n^2 = 12n - \dfrac{n^2}{20} = \dfrac{240n - n^2}{20}, n \ge 80$

(b)

n	90	100	110	120	130	140	150
$R(n)$	\$675	\$700	\$715	\$720	\$715	\$700	\$675

The revenue is maximum when 120 people take the trip.

101. (a)

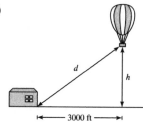

(b) $(3000)^2 + h^2 = d^2$

$h = \sqrt{d^2 - (3000)^2}$

Domain: $d \ge 3000$ (because both $d \ge 0$ and $d^2 - (3000)^2 \ge 0$)

103.
$$f(x) = x^2 - x + 1$$
$$f(2 + h) = (2 + h)^2 - (2 + h) + 1$$
$$= 4 + 4h + h^2 - 2 - h + 1$$
$$= h^2 + 3h + 3$$
$$f(2) = (2)^2 - 2 + 1 = 3$$
$$f(2 + h) - f(2) = h^2 + 3h$$
$$\frac{f(2 + h) - f(2)}{h} = \frac{h^2 + 3h}{h} = h + 3, h \neq 0$$

105.
$$f(x) = x^3 + 3x$$
$$f(x + h) = (x + h)^3 + 3(x + h)$$
$$= x^3 + 3x^2h + 3xh^2 + h^3 + 3x + 3h$$
$$\frac{f(x + h) - f(x)}{h} = \frac{(x^3 + 3x^2h + 3xh^2 + h^3 + 3x + 3h) - (x^3 + 3x)}{h}$$
$$= \frac{h(3x^2 + 3xh + h^2 + 3)}{h}$$
$$= 3x^2 + 3xh + h^2 + 3, h \neq 0$$

107.
$$g(x) = \frac{1}{x^2}$$
$$\frac{g(x) - g(3)}{x - 3} = \frac{\frac{1}{x^2} - \frac{1}{9}}{x - 3}$$
$$= \frac{9 - x^2}{9x^2(x - 3)}$$
$$= \frac{-(x + 3)(x - 3)}{9x^2(x - 3)}$$
$$= -\frac{x + 3}{9x^2}, x \neq 3$$

109.
$$f(x) = \sqrt{5x}$$
$$\frac{f(x) - f(5)}{x - 5} = \frac{\sqrt{5x} - 5}{x - 5}$$

111. By plotting the points, we have a parabola, so $g(x) = cx^2$. Because $(-4, -32)$ is on the graph, you have $-32 = c(-4)^2 \Rightarrow c = -2$. So, $g(x) = -2x^2$.

113. Because the function is undefined at 0, we have $r(x) = c/x$. Because $(-4, -8)$ is on the graph, you have $-8 = c/-4 \Rightarrow c = 32$. So, $r(x) = 32/x$.

115. False. The equation $y^2 = x^2 + 4$ is a relation between x and y. However, $y = \pm\sqrt{x^2 + 4}$ does not represent a function.

117. False. The range is $[-1, \infty)$.

119. $f(x) = \sqrt{x - 1}$ Domain: $x \geq 1$

$g(x) = \dfrac{1}{\sqrt{x - 1}}$ Domain: $x > 1$

The value 1 may be included in the domain of $f(x)$ as it is possible to find the square root of 0. However, 1 cannot be included in the domain of $g(x)$ as it causes a zero to occur in the denominator which results in the function being undefined.

121. No; x is the independent variable, f is the name of the function.

123. (a) Yes. The amount that you pay in sales tax will increase as the price of the item purchased increases.

(b) No. The length of time that you study the night before an exam does not necessarily determine your score on the exam.

Section 2.3   Analyzing Graphs of Functions

- You should be able to determine the domain and range of a function from its graph.
- You should be able to use the vertical line test for functions.
- You should be able to find the zeros of a function.
- You should be able to determine when a function is constant, increasing, or decreasing.
- You should be able to approximate relative minimums and relative maximums from the graph of a function.
- You should know that f is
    - (a) odd if $f(-x) = -f(x)$.
    - (b) even if $f(-x) = f(x)$.

1. ordered pairs

3. zeros

5. maximum

7. odd

9. Domain: $(-\infty, -1] \cup [1, \infty)$

Range: $[0, \infty)$

11. Domain: $[-4, 4]$

Range: $[0, 4]$

13. Domain: $(-\infty, \infty)$; Range: $[-4, \infty)$

   (a) $f(-2) = 0$

   (b) $f(-1) = -1$

   (c) $f\left(\frac{1}{2}\right) = 0$

   (d) $f(1) = -2$

15. Domain: $(-\infty, \infty)$; Range: $(-2, \infty)$

   (a) $f(2) = 0$

   (b) $f(1) = 1$

   (c) $f(3) = 2$

   (d) $f(-1) = 3$

17. $y = \frac{1}{2}x^2$

A vertical line intersects the graph at most once, so y *is* a function of x.

19. $x - y^2 = 1 \Rightarrow y = \pm\sqrt{x - 1}$

y is *not* a function of x. Some vertical lines intersect the graph twice.

21. $x^2 = 2xy - 1$

A vertical line intersects the graph at most once, so y *is* a function of x.

23.
$$f(x) = 2x^2 - 7x - 30$$
$$2x^2 - 7x - 30 = 0$$
$$(2x + 5)(x - 6) = 0$$
$$2x + 5 = 0 \quad \text{or} \quad x - 6 = 0$$
$$x = -\tfrac{5}{2} \quad \text{or} \quad x = 6$$

25.
$$f(x) = \frac{x}{9x^2 - 4}$$
$$\frac{x}{9x^2 - 4} = 0$$
$$x = 0$$

27.
$$f(x) = \tfrac{1}{2}x^3 - x$$
$$\tfrac{1}{2}x^3 - x = 0$$
$$x^3 - 2x = 2(0)$$
$$x(x^2 - 2) = 0$$
$$x = 0 \quad \text{or} \quad x^2 - 2 = 0$$
$$x^2 = 2$$
$$x = \pm\sqrt{2}$$

29. $f(x) = 4x^3 - 24x^2 - x + 6$
$$4x^3 - 24x^2 - x + 6 = 0$$
$$4x^2(x - 6) - 1(x - 6) = 0$$
$$(x - 6)(4x^2 - 1) = 0$$
$$(x - 6)(2x + 1)(2x - 1) = 0$$
$$x - 6 = 0, \quad 2x + 1 = 0, \quad 2x - 1 = 0$$
$$x = 6, \quad x = -\tfrac{1}{2}, \quad x = \tfrac{1}{2}$$

31. $f(x) = \sqrt{2x} - 1$

$$\sqrt{2x} - 1 = 0$$
$$\sqrt{2x} = 1$$
$$2x = 1$$
$$x = \frac{1}{2}$$

33. (a)

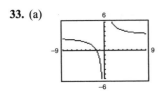

Zero: $x = -\dfrac{5}{3}$

   (b) $f(x) = 3 + \dfrac{5}{x}$

$$3 + \frac{5}{x} = 0$$
$$3x + 5 = 0$$
$$x = -\frac{5}{3}$$

35. (a)

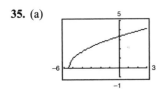

Zero: $x = -\dfrac{11}{2}$

   (b) $f(x) = \sqrt{2x + 11}$

$$\sqrt{2x + 11} = 0$$
$$2x + 11 = 0$$
$$x = -\frac{11}{2}$$

37. (a)

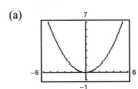

Zero: $x = \dfrac{1}{3}$

   (b) $f(x) = \dfrac{3x - 1}{x - 6}$

$$\frac{3x - 1}{x - 6} = 0$$
$$3x - 1 = 0$$
$$x = \frac{1}{3}$$

39. $f(x) = \frac{3}{2}x$

The function is increasing on $(-\infty, \infty)$.

41. $f(x) = x^3 - 3x^2 + 2$

The function is increasing on $(-\infty, 0)$ and $(2, \infty)$ and decreasing on $(0, 2)$.

43. $f(x) = |x + 1| + |x - 1|$

The function is increasing on $(1, \infty)$.

The function is constant on $(-1, 1)$.

The function is decreasing on $(-\infty, -1)$.

45. $f(x) = \begin{cases} x + 3, & x \le 0 \\ 3, & 0 < x \le 2 \\ 2x + 1, & x > 2 \end{cases}$

The function is increasing on $(-\infty, 0)$ and $(2, \infty)$.

The function is constant on $(0, 2)$.

47. $f(x) = 3$

   (a)

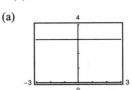

Constant on $(-\infty, \infty)$

   (b)

x	-2	-1	0	1	2
$f(x)$	3	3	3	3	3

49. $g(s) = \dfrac{s^2}{4}$

   (a)

Decreasing on $(-\infty, 0)$; Increasing on $(0, \infty)$

   (b)

s	-4	-2	0	2	4
$g(s)$	4	1	0	1	4

51. $f(t) = -t^4$

(a)

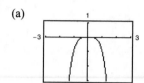

Increasing on $(-\infty, 0)$; Decreasing on $(0, \infty)$

(b)

t	-2	-1	0	1	2
$f(t)$	-16	-1	0	-1	-16

53. $f(x) = \sqrt{1 - x}$

(a)

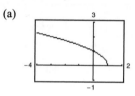

Decreasing on $(-\infty, 1)$

(b)

x	-3	-2	-1	0	1
$f(x)$	2	$\sqrt{3}$	$\sqrt{2}$	1	0

55. $f(x) = x^{3/2}$

(a)

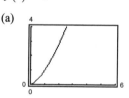

Increasing on $(0, \infty)$

(b)

x	0	1	2	3	4
$f(x)$	0	1	2.8	5.2	8

57. $f(x) = (x - 4)(x + 2)$

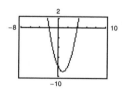

Relative minimum: $(1, -9)$

59. $f(x) = -x^2 + 3x - 2$

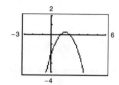

Relative maximum: $(1.5, 0.25)$

61. $f(x) = x(x - 2)(x + 3)$

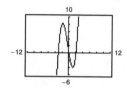

Relative minimum: $(1.12, -4.06)$

Relative maximum: $(-1.79, 8.21)$

63. $g(x) = 2x^3 + 3x^2 - 12x$

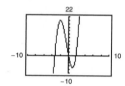

Relative minimum: $(1, -7)$

Relative maximum: $(-2, 20)$

65. $h(x) = (x - 1)\sqrt{x}$

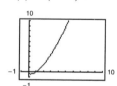

Relative minimum: $(0.33, -0.38)$

67. $f(x) = 4 - x$

$f(x) \geq 0$ on $(-\infty, 4]$

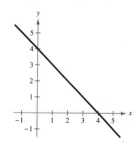

69. $f(x) = 9 - x^2$

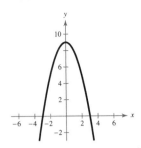

$f(x) \geq 0$ on $[-3, 3]$

71. $f(x) = \sqrt{x - 1}$

$f(x) \geq 0$ on $[1, \infty)$

$\sqrt{x - 1} \geq 0$

$x - 1 \geq 0$

$x \geq 1$

$[1, \infty)$

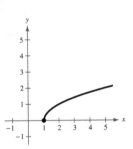

73. $f(x) = -(1 + |x|)$

$f(x)$ is never greater than 0. $(f(x) < 0$ for all $x.)$

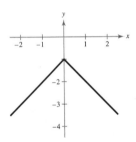

75. $f(x) = -2x + 15$

$$\frac{f(3) - f(0)}{3 - 0} = \frac{9 - 15}{3} = -2$$

The average rate of change from $x_1 = 0$ to $x_2 = 3$ is -2.

77. $f(x) = x^2 + 12x - 4$

$$\frac{f(5) - f(1)}{5 - 1} = \frac{81 - 9}{4} = 18$$

The average rate of change from $x_1 = 1$ to $x_2 = 5$ is 18.

79. $f(x) = x^3 - 3x^2 - x$

$$\frac{f(3) - f(1)}{3 - 1} = \frac{-3 - (-3)}{2} = 0$$

The average rate of change from $x_1 = 1$ to $x_2 = 3$ is 0.

81. $f(x) = -\sqrt{x - 2} + 5$

$$\frac{f(11) - f(3)}{11 - 3} = \frac{2 - 4}{8} = -\frac{1}{4}$$

The average rate of change from $x_1 = 3$ to $x_2 = 11$ is $-\frac{1}{4}$.

83. $f(x) = x^6 - 2x^2 + 3$

$f(-x) = (-x)^6 - 2(-x)^2 + 3$

$\qquad = x^6 - 2x^2 + 3$

$\qquad = f(x)$

The function is even. y-axis symmetry.

85. $g(x) = x^3 - 5x$

$g(-x) = (-x)^3 - 5(-x)$

$\qquad = -x^3 + 5x$

$\qquad = -g(x)$

The function is odd. Origin symmetry.

87. $h(x) = x\sqrt{x + 5}$

$h(-x) = (-x)\sqrt{-x + 5}$

$\qquad = -x\sqrt{5 - x}$

$\qquad \neq h(x)$

$\qquad \neq -h(x)$

The function is neither odd nor even. No symmetry.

89. $f(s) = 4s^{3/2}$

$\qquad = 4(-s)^{3/2}$

$\qquad \neq f(s)$

$\qquad \neq -f(s)$

The function is neither odd nor even. No symmetry.

91.

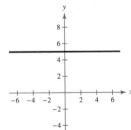

The graph of $f(x) = 5$ is symmetric to the y-axis, which implies $f(x)$ is even.

$f(-x) = 5$

$\qquad = f(x)$

The function is even.

93. $f(x) = 3x - 2$

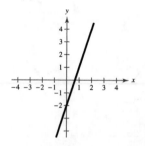

The graph displays no symmetry, which implies $f(x)$ is neither odd nor even.

$$f(-x) = 3(-x) - 2$$
$$= -3x - 2$$
$$\neq f(x)$$
$$\neq -f(x)$$

The function is neither even nor odd.

95. $h(x) = x^2 - 4$

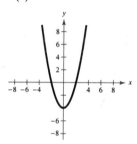

The graph displays y-axis symmetry, which implies $h(x)$ is even.

$$h(-x) = (-x)^2 - 4 = x^2 - 4 = h(x)$$

The function is even.

97. $f(x) = \sqrt{1 - x}$

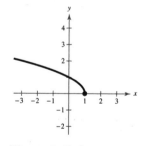

The graph displays no symmetry, which implies $f(x)$ is neither odd nor even.

$$f(-x) = \sqrt{1 - (-x)}$$
$$= \sqrt{1 + x}$$
$$\neq f(x)$$
$$\neq -f(x)$$

The function is neither even nor odd.

99. $f(x) = |x + 2|$

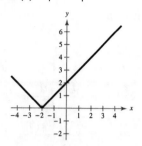

The graph displays no symmetry, which implies $f(x)$ is neither odd nor even.

$$f(-x) = |-x + 2| \neq f(x) \neq -f(x)$$

The function is neither even nor odd.

101. $h = \text{top} - \text{bottom}$
$$= \left(-x^2 + 4x - 1\right) - 2$$
$$= -x^2 + 4x - 3$$

103. $h = \text{top} - \text{bottom}$
$$= \left(4x - x^2\right) - 2x$$
$$= 2x - x^2$$

105. $L = \text{right} - \text{left}$
$$= \tfrac{1}{2}y^2 - 0 = \tfrac{1}{2}y^2$$

107. $L = \text{right} - \text{left}$
$$= 4 - y^2$$

109. $L = -0.294x^2 + 97.744x - 664.875,\ 20 \leq x \leq 90$

(a)
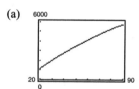

(b) $L = 2000$ when $x \approx 29.9645 \approx 30$ watts.

111. (a) For the average salaries of college professors, a scale of $10,000 would be appropriate.

(b) For the population of the United States, use a scale of 10,000,000.

(c) For the percent of the civilian workforce that is unemployed, use a scale of 1%.

113. (a)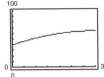

(b) Average rate of change $= \dfrac{r(t_2) - r(t_1)}{t_2 - t_1}$

$= \dfrac{r(35) - r(0)}{t_2 - t_1}$

$= \dfrac{63.975 - 39.3}{35 - 0}$

$= 0.705$

The average rate of change from 1970 to 2005 is 0.705 per year. The enrollment rate of children in preschool has slowly been increasing each year.

115. $s_0 = 6, v_0 = 64$

(a) $s = -16t^2 + 64t + 6$

(b)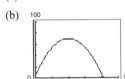

(c) $\dfrac{s(3) - s(0)}{3 - 0} = \dfrac{54 - 6}{3} = 16$

(d) The slope of the secant line is positive.

(e) $s(0) = 6, m = 16$

Secant line: $y - 6 = 16(t - 0)$

$y = 16t + 6$

(f)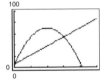

117. $v_0 = 120, s_0 = 0$

(a) $s = -16t^2 + 120t$

(b)

(c) $\dfrac{s(5) - s(3)}{5 - 3} = \dfrac{200 - 216}{2} = -8$

(d) The slope of the secant line is negative.

(e) $s(5) = 200, m = -8$

Secant line: $y - 200 = -8(t - 5)$

$y = -8t + 240$

(f)

119. $v_0 = 0, s_0 = 120$

(a) $s = -16t^2 + 120$

(b)

(c) $\dfrac{s(2) - s(0)}{2 - 0} = \dfrac{56 - 120}{2} = -32$

(d) The slope of the secant line is negative.

(e) $s(0) = 120, m = -32$

Secant line: $y - 120 = -32(t - 0)$

$y = -32t + 120$

(f)

121. False. The function $f(x) = \sqrt{x^2 + 1}$ has a domain of all real numbers.

123. (a) Even. The graph is a reflection in the *x*-axis.

(b) Even. The graph is a reflection in the *y*-axis.

(c) Even. The graph is a vertical translation of *f*.

(d) Neither. The graph is a horizontal translation of *f*.

125. $\left(-\frac{3}{2}, 4\right)$

    (a) If f is even, another point is $\left(\frac{3}{2}, 4\right)$.

    (b) If f is odd, another point is $\left(\frac{3}{2}, -4\right)$.

127. $(4, 9)$

    (a) If f is even, another point is $(-4, 9)$.

    (b) If f is odd, another point is $(-4, -9)$.

129. $(x, -y)$

    (a) $(-x, -y)$

    (b) $(-x, y)$

131. (a) $y = x$

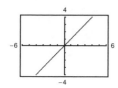

    (b) $y = x^2$

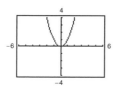

    (c) $y = x^3$

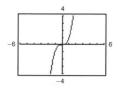

    (d) $y = x^4$

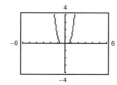

    (e) $y = x^5$

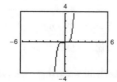

    (f) $y = x^6$

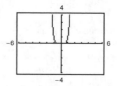

All the graphs pass through the origin. The graphs of the odd powers of x are symmetric with respect to the origin and the graphs of the even powers are symmetric with respect to the y-axis. As the powers increase, the graphs become flatter in the interval $-1 < x < 1$.

133. Average rate of change $= \dfrac{s(t_2) - s(t_1)}{t_2 - t_1}$

$$= \dfrac{s(9) - s(0)}{9 - 0}$$

$$= \dfrac{540 - 0}{9 - 0}$$

$$= 60 \text{ feet per second.}$$

As the time traveled increases, the distance increases rapidly, causing the average speed to increase with each time increment. From $t = 0$ to $t = 4$, the average speed is less than from $t = 4$ to $t = 9$. Therefore, the overall average from $t = 0$ to $t = 9$ falls below the average found in part (b).

135. Answers will vary.

Section 2.4 A Library of Parent Functions

- You should be able to identify and graph the following types of functions:

  (a) Linear functions like $f(x) = ax + b$

  (b) Squaring functions like $f(x) = x^2$

  (c) Cubic functions like $f(x) = x^3$

  (d) Square root functions like $f(x) = \sqrt{x}$

  (e) Reciprocal functions like $f(x) = \dfrac{1}{x}$

  (f) Constant functions like $f(x) = c$

  (g) Absolute value functions like $f(x) = |x|$

  (h) Step and piecewise-defined functions like $f(x) = [\![x]\!]$

- You should be able to determine the following about these parent functions:

  (a) Domain and range

  (b) x-intercept(s) and y-intercept

  (c) Symmetries

  (d) Where it is increasing, decreasing, or constant

  (e) If it is odd, even or neither

  (f) Relative maximums and relative minimums

---

1. $f(x) = [\![x]\!]$

   (g) greatest integer function

3. $f(x) = \dfrac{1}{x}$

   (h) reciprocal function

5. $f(x) = \sqrt{x}$

   (b) square root function

7. $f(x) = |x|$

   (f) absolute value function

9. $f(x) = ax + b$

   (d) linear function

11. (a) $f(1) = 4, f(0) = 6$

   $(1, 4), (0, 6)$

   $m = \dfrac{6 - 4}{0 - 1} = -2$

   $y - 6 = -2(x - 0)$

   $\qquad y = -2x + 6$

   $f(x) = -2x + 6$

   (b)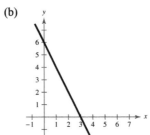

13. (a) $f(5) = -4, f(-2) = 17$

$(5, -4), (-2, 17)$

$m = \dfrac{17 - (-4)}{-2 - 5} = \dfrac{21}{-7} = -3$

$y - (-4) = -3(x - 5)$

$y + 4 = -3x + 15$

$y = -3x + 11$

$f(x) = -3x + 11$

(b)

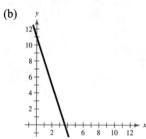

15. (a) $f(-5) = -1, f(5) = -1$

$(-5, -1), (5, -1)$

$m = \dfrac{-1 - (-1)}{5 - (-5)} = \dfrac{0}{10} = 0$

$y - (-1) = 0(x - (-5))$

$y + 1 = 0$

$y = -1$

$f(x) = -1$

(b)

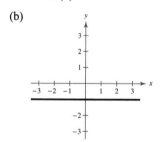

17. (a) $f\left(\dfrac{1}{2}\right) = -6, f(4) = -3$

$\left(\dfrac{1}{2}, -6\right), (4, -3)$

$m = \dfrac{-3 - (-6)}{4 - (1/2)} = \dfrac{3}{7/2} = \dfrac{6}{7}$

$y - (-3) = \dfrac{6}{7}(x - 4)$

$y + 3 = \dfrac{6}{7}x - \dfrac{24}{7}$

$y = \dfrac{6}{7}x - \dfrac{45}{7}$

$f(x) = \dfrac{6}{7}x - \dfrac{45}{7}$

(b)

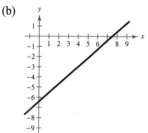

19. $f(x) = 0.8 - x$

21. $f(x) = -\dfrac{1}{6}x - \dfrac{5}{2}$

23. $g(x) = -2x^2$

25. $f(x) = 3x^2 - 1.75$

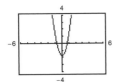

27. $f(x) = x^3 - 1$

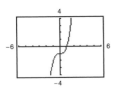

29. $f(x) = (x - 1)^3 + 2$

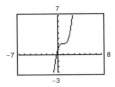

31. $f(x) = 4\sqrt{x}$

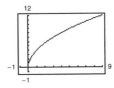

33. $g(x) = 2 - \sqrt{x + 4}$

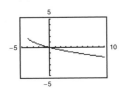

35. $f(x) = -\dfrac{1}{x}$

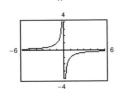

37. $h(x) = \dfrac{1}{x + 2}$

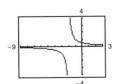

39. $g(x) = |x| - 5$

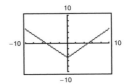

41. $f(x) = |x + 4|$

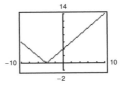

43. $f(x) = [\![x]\!]$
   (a) $f(2.1) = 2$
   (b) $f(2.9) = 2$
   (c) $f(-3.1) = -4$
   (d) $f\left(\dfrac{7}{2}\right) = 3$

45. $h(x) = [\![x + 3]\!]$
   (a) $h(-2) = [\![1]\!] = 1$
   (b) $h\left(\dfrac{1}{2}\right) = [\![3.5]\!] = 3$
   (c) $h(4.2) = [\![7.2]\!] = 7$
   (d) $h(-21.6) = [\![-18.6]\!] = -19$

47. $h(x) = [\![3x - 1]\!]$
   (a) $h(2.5) = [\![6.5]\!] = 6$
   (b) $h(-3.2) = [\![-10.6]\!] = -11$
   (c) $h\left(\dfrac{7}{3}\right) = [\![6]\!] = 6$
   (d) $h\left(-\dfrac{21}{3}\right) = [\![-22]\!] = -22$

49. $g(x) = 3[\![x - 2]\!] + 5$
   (a) $g(-2.7) = 3[\![-4.7]\!] + 5 = 3(-5) + 5 = -10$
   (b) $g(-1) = 3[\![-3]\!] + 5 = 3(-3) + 5 = -4$
   (c) $g(0.8) = 3[\![-1.2]\!] + 5 = 3(-2) + 5 = -1$
   (d) $g(14.5) = 3[\![12.5]\!] + 5 = 3(12) + 5 = 41$

51. $g(x) = -[\![x]\!]$

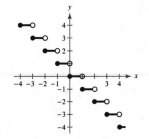

53. $g(x) = [\![x]\!] - 2$

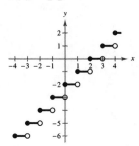

55. $g(x) = [\![x + 1]\!]$

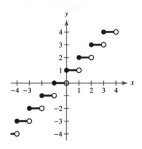

57. $f(x) = \begin{cases} 2x + 3, & x < 0 \\ 3 - x, & x \geq 0 \end{cases}$

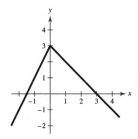

59. $f(x) = \begin{cases} \sqrt{4 + x}, & x < 0 \\ \sqrt{4 - x}, & x \geq 0 \end{cases}$

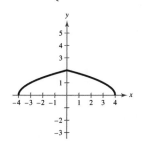

61. $f(x) = \begin{cases} x^2 + 5, & x \leq 1 \\ -x^2 + 4x + 3, & x > 1 \end{cases}$

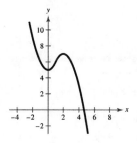

63. $h(x) = \begin{cases} 4 - x^2, & x < -2 \\ 3 + x, & -2 \leq x < 0 \\ x^2 + 1, & x \geq 0 \end{cases}$

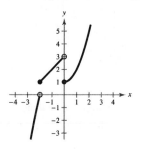

65. $s(x) = 2\left(\frac{1}{4}x - \left[\!\left[\frac{1}{4}x\right]\!\right]\right)$

(a)

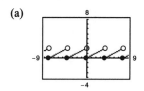

(b) Domain: $(-\infty, \infty)$

   Range: $[0, 2)$

(c) Sawtooth pattern

67. $h(x) = 4\left(\frac{1}{2}x - \left[\!\left[\frac{1}{2}x\right]\!\right]\right)$

(a)

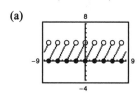

(b) Domain: $(-\infty, \infty)$

   Range: $[0, 4)$

(c) Sawtooth pattern

69. (a)

(b) $C(9.25) = 23.40 + 3.75[\![9.25]\!]$

$\qquad = 23.40 + 3.75(9)$

$\qquad = 57.15$

It costs $57.15 to mail a 9.25 pound package.

71. (a) $W(30) = 14(30) = 420$

$\qquad W(40) = 14(40) = 560$

$\qquad W(45) = 21(45 - 40) + 560 = 665$

$\qquad W(50) = 21(50 - 40) + 560 = 770$

(b) $W(h) = \begin{cases} 14h, & 0 < h \le 45 \\ 21(h - 45) + 630, & h > 45 \end{cases}$

73. (a)

The domain of $f(x) = -1.97x + 26.3$ is $6 < x \le 12$. One way to see this is to notice that this is the equation of a line with negative slope, so the function values are decreasing as x increases, which matches the data for the corresponding part of the table. The domain of $f(x) = 0.505x^2 - 1.47x + 6.3$ is then $1 \le x \le 6$.

(b) $f(5) = 0.505(5)^2 - 1.47(5) + 6.3$

$\qquad = 0.505(25) - 7.35 + 6.3 = 11.575$

$f(11) = -1.97(11) + 26.3 = 4.63$

These values represent the revenue in thousands of dollars for the months of May and November, respectively.

(c) The model values are very close to the actual values.

Month, x	1	2	3	4	5	6	7	8	9	10	11	12
Revenue, y	5.2	5.6	6.6	8.3	11.5	15.8	12.8	10.1	8.6	6.9	4.5	2.7
Model, $f(x)$	5.3	5.4	6.4	8.5	11.6	15.7	12.5	10.5	8.6	6.6	4.6	2.7

75. False. The vertical line $x = 2$ has an x-intercept at the point $(2, 0)$ but does not have a y-intercept. The horizontal line $y = 3$ has a y-intercept at the point $(0, 3)$ but does not have an x-intercept.

Section 2.5   Transformations of Functions

■ You should know the basic types of transformations.

Let $y = f(x)$ and let c be a positive real number.

1.	$h(x) = f(x) + c$	Vertical shift c units upward
2.	$h(x) = f(x) - c$	Vertical shift c units downward
3.	$h(x) = f(x - c)$	Horizontal shift c units to the right
4.	$h(x) = f(x + c)$	Horizontal shift c units to the left
5.	$h(x) = -f(x)$	Reflection in the x-axis
6.	$h(x) = f(-x)$	Reflection in the y-axis
7.	$h(x) = cf(x), c > 1$	Vertical stretch
8.	$h(x) = cf(x), 0 < c < 1$	Vertical shrink
9.	$h(x) = f(cx), c > 1$	Horizontal shrink
10.	$h(x) = f(cx), 0 < c < 1$	Horizontal stretch

---

1. rigid

3. nonrigid

5. vertical stretch; vertical shrink

7. (a) $f(x) = |x| + c$ Vertical shifts

   $c = -1$: $f(x) = |x| - 1$ 1 unit down

   $c = 1$: $f(x) = |x| + 1$ 1 unit up

   $c = 3$: $f(x) = |x| + 3$ 3 units up

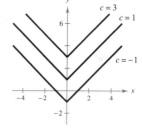

   (b) $f(x) = |x - c|$ Horizontal shifts

   $c = -1$: $f(x) = |x + 1|$ 1 unit left

   $c = 1$: $f(x) = |x - 1|$ 1 unit right

   $c = 3$: $f(x) = |x - 3|$ 3 units right

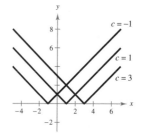

   (c) $f(x) = |x + 4| + c$ Horizontal shift four units left and a vertical shift

   $c = -1$: $f(x) = |x + 4| - 1$ 1 unit down

   $c = 1$: $f(x) = |x + 4| + 1$ 1 unit up

   $c = 3$: $f(x) = |x + 4| + 3$ 3 units up

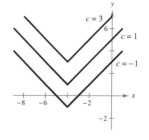

9. (a) $f(x) = [\![x]\!] + c$ Vertical shifts

     $c = -2$: $f(x) = [\![x]\!] - 2$ 2 units down

     $c = 0$: $f(x) = [\![x]\!]$ Parent function

     $c = 2$: $f(x) = [\![x]\!] + 2$ 2 units up

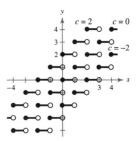

(b) $f(x) = [\![x + c]\!]$ Horizontal shifts

     $c = -2$: $f(x) = [\![x - 2]\!]$ 2 units right

     $c = 0$: $f(x) = [\![x]\!]$ Parent function

     $c = 2$: $f(x) = [\![x + 2]\!]$ 2 units left

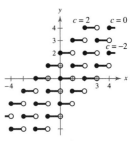

(c) $f(x) = [\![x - 1]\!] + c$ Horizontal shift 1 unit right and a vertical shift

     $c = -2$: $f(x) = [\![x - 1]\!] - 2$ 2 units down

     $c = 0$: $f(x) = [\![x - 1]\!]$

     $c = 2$: $f(x) = [\![x - 1]\!] + 2$ 2 units up

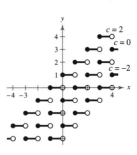

11. (a) $y = f(x) + 2$

Vertical shift 2 units upward

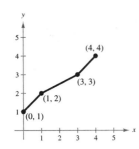

(b) $y = f(x - 2)$

Horizontal shift 2 units to the right

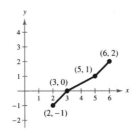

(c) $y = 2f(x)$

Vertical stretch
(each y-value is multiplied by 2)

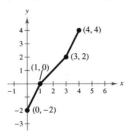

(d) $y = -f(x)$

Reflection in the x-axis

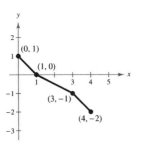

(e) $y = f(x + 3)$

Horizontal shift 3 units to the left

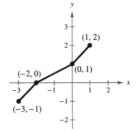

(f) $y = f(-x)$

Reflection in the y-axis

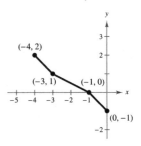

(g) $y = f\left(\frac{1}{2}x\right)$

Horizontal stretch
(each *x*-value is multiplied by 2)

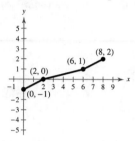

13. (a) $y = f(x) - 1$

Vertical shift 1 unit downward

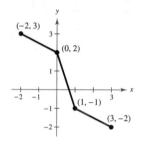

(b) $y = f(x - 1)$

Horizontal shift 1 unit to the right

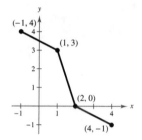

(c) $y = f(-x)$

Reflection in the *y*-axis

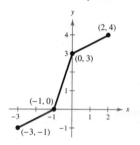

(d) $y = f(x + 1)$

Horizontal shift 1 unit to the left

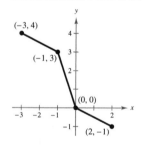

(e) $y = -f(x - 2)$

Reflection in the *x*-axis and a
horizontal shift 2 units to the right

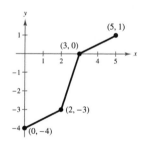

(f) $y = \frac{1}{2}f(x)$

Vertical shrink
$\left(\text{each } y\text{-value is multiplied by } \frac{1}{2}\right)$

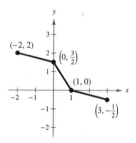

(g) $y = f(2x)$

Horizontal shrink
$\left(\text{each } x\text{-value is multiplied by } \frac{1}{2}\right)$

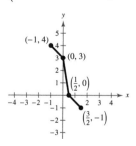

15. Parent function: $f(x) = x^2$

    (a) Vertical shift 1 unit downward

$$g(x) = x^2 - 1$$

    (b) Reflection in the *x*-axis, horizontal shift 1 unit to the left, and a vertical shift 1 unit upward

$$g(x) = -(x + 1)^2 + 1$$

    (c) Reflection in the *x*-axis, horizontal shift 2 units to the right, and a vertical shift 6 units upward

$$g(x) = -(x - 2)^2 + 6$$

    (d) Horizontal shift 5 units to the right and a vertical shift 3 units downward

$$g(x) = (x - 5)^2 - 3$$

17. Parent function: $f(x) = |x|$

    (a) Vertical shift 5 units upward

$$g(x) = |x| + 5$$

    (b) Reflection in the *x*-axis and a horizontal shift 3 units to the left

$$g(x) = -|x + 3|$$

    (c) Horizontal shift 2 units to the right and a vertical shift 4 units downward

$$g(x) = |x - 2| - 4$$

    (d) Reflection in the *x*-axis, horizontal shift 6 units to the right, and a vertical shift 1 unit downward

$$g(x) = -|x - 6| - 1$$

19. Parent function: $f(x) = x^3$

Horizontal shift 2 units to the right

$$y = (x - 2)^3$$

21. Parent function: $f(x) = x^2$

Reflection in the *x*-axis

$$y = -x^2$$

23. Parent function: $f(x) = \sqrt{x}$

Reflection in the *x*-axis and a vertical shift 1 unit upward

$$y = -\sqrt{x} + 1$$

25. $g(x) = 12 - x^2$

    (a) Parent function: $f(x) = x^2$

    (b) Reflection in the *x*-axis and a vertical shift 12 units upward

    (c)

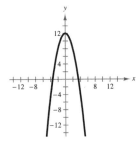

    (d) $g(x) = 12 - f(x)$

27. $g(x) = x^3 + 7$

    (a) Parent function: $f(x) = x^3$

    (b) Vertical shift 7 units upward

    (c)

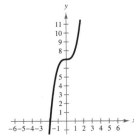

    (d) $g(x) = f(x) + 7$

29. $g(x) = \frac{2}{3}x^2 + 4$

    (a) Parent function: $f(x) = x^2$

    (b) Vertical shrink of two-thirds, and a vertical shift 4 units upward

    (c)

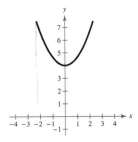

    (d) $g(x) = \frac{2}{3}f(x) + 4$

31. $g(x) = 2 - (x + 5)^2$

(a) Parent function: $f(x) = x^2$

(b) Reflection in the x-axis, horizontal shift 5 units to the left, and a vertical shift 2 units upward

(c)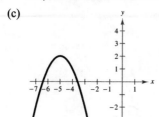

(d) $g(x) = 2 - f(x + 5)$

33. $g(x) = 3 + 2(x - 4)^2$

(a) Parent function: $f(x) = x^2$

(b) Horizontal shift 4 units to the right, vertical shift 3 units upward, vertical stretch

(c)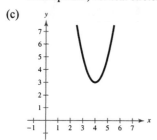

(d) $g(x) = 3 + 2f(x - 4)$

35. $g(x) = \sqrt{3x}$

(a) Parent function: $f(x) = \sqrt{x}$

(b) Horizontal shrink by $\frac{1}{3}$

(c)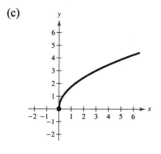

(d) $g(x) = f(3x)$

37. $g(x) = (x - 1)^3 + 2$

(a) Parent function: $f(x) = x^3$

(b) Horizontal shift 1 unit to the right and a vertical shift 2 units upward

(c)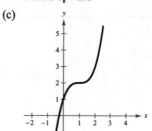

(d) $g(x) = f(x - 1) + 2$

39. $g(x) = 3(x - 2)^3$

(a) Parent function: $f(x) = x^3$

(b) Horizontal shift 2 units to the right, vertical stretch (each y-value is multiplied by 3).

(c)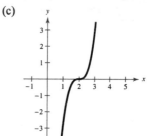

(d) $g(x) = 3f(x - 2)$

41. $g(x) = -|x| - 2$

(a) Parent function: $f(x) = |x|$

(b) Reflection in the x-axis; vertical shift 2 units downward

(c)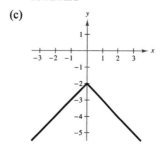

(d) $g(x) = -f(x) - 2$

43. $g(x) = -|x + 4| + 8$

   (a) Parent function: $f(x) = |x|$

   (b) Reflection in the *x*-axis; horizontal shift 4 units to the left; and a vertical shift 8 units upward

   (c)

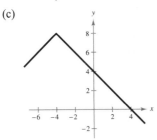

   (d) $g(x) = -f(x + 4) + 8$

45. $g(x) = -2|x - 1| - 4$

   (a) Parent function: $f(x) = |x|$

   (b) Horizontal shift one unit to the right, vertical stretch, reflection in the *x*-axis, vertical shift four units downward

   (c)

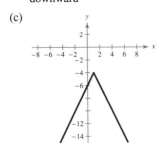

   (d) $g(x) = -2f(x - 1) - 4$

47. $g(x) = 3 - [\![x]\!]$

   (a) Parent function: $f(x) = [\![x]\!]$

   (b) Reflection in the *x*-axis and a vertical shift 3 units upward

   (c)

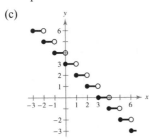

   (d) $g(x) = 3 - f(x)$

49. $g(x) = \sqrt{x - 9}$

   (a) Parent function: $f(x) = \sqrt{x}$

   (b) Horizontal shift 9 units to the right

   (c)

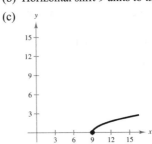

   (d) $g(x) = f(x - 9)$

51. $g(x) = \sqrt{7 - x} - 2$ or $g(x) = \sqrt{-(x - 7)} - 2$

   (a) Parent function: $f(x) = \sqrt{x}$

   (b) Reflection in the *y*-axis, horizontal shift 7 units to the right, and a vertical shift 2 units downward

   (c)

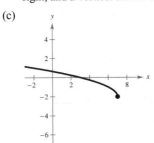

   (d) $g(x) = f(7 - x) - 2$

53. $g(x) = \sqrt{\frac{1}{2}x} - 4$

   (a) Parent function: $f(x) = \sqrt{x}$

   (b) Horizontal stretch (each *x*-value is multiplied by 2) and a vertical shift 4 units down

   (c)

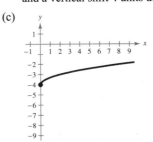

   (d) $g(x) = f\left(\frac{1}{2}x\right) - 4$

55. $f(x) = (x - 3)^2 - 7$

57. $f(x) = x^3$ moved 13 units to the right

$$g(x) = (x - 13)^3$$

59. $f(x) = -|x| + 12$

61. $f(x) = \sqrt{x}$ moved 6 units to the left and reflected in both the *x*- and *y*-axes

$$g(x) = -\sqrt{-x + 6}$$

63. $f(x) = x^2$

   (a) Reflection in the *x*-axis and a vertical stretch (each *y*-value is multiplied by 3)

$$g(x) = -3x^2$$

   (b) Vertical shift 3 units upward and a vertical stretch (each *y*-value is multiplied by 4)

$$g(x) = 4x^2 + 3$$

65. $f(x) = |x|$

   (a) Reflection in the *x*-axis and a vertical shrink (each *y*-value is multiplied by $\frac{1}{2}$)

$$g(x) = -\frac{1}{2}|x|$$

   (b) Vertical stretch (each *y*-value is multiplied by 3) and a vertical shift 3 units downward

$$g(x) = 3|x| - 3$$

67. Parent function: $f(x) = x^3$

Vertical stretch (each *y*-value is multiplied by 2)

$$g(x) = 2x^3$$

69. Parent function: $f(x) = x^2$

Reflection in the *x*-axis, vertical shrink (each *y*-value is multiplied by $\frac{1}{2}$)

$$g(x) = -\frac{1}{2}x^2$$

71. Parent function: $f(x) = \sqrt{x}$

Reflection in the *y*-axis, vertical shrink (each *y*-value is multiplied by $\frac{1}{2}$)

$$g(x) = \frac{1}{2}\sqrt{-x}$$

73. Parent function: $f(x) = x^3$

Reflection in the *x*-axis, horizontal shift 2 units to the right and a vertical shift 2 units upward

$$g(x) = -(x - 2)^3 + 2$$

75. Parent function: $f(x) = \sqrt{x}$

Reflection in the *x*-axis and a vertical shift 3 units downward

$$g(x) = -\sqrt{x} - 3$$

77. (a) $g(x) = f(x) + 2$

Vertical shift 2 units upward

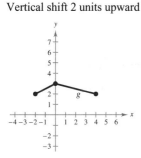

    (b) $g(x) = f(x) - 1$

Vertical shift 1 unit downward

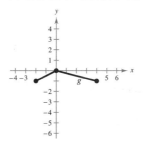

    (c) $g(x) = f(-x)$

Reflection in the *y*-axis

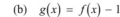

    (d) $g(x) = -2f(x)$

Reflection in the *x*-axis and a vertical stretch (each *y*-value is multiplied by 2)

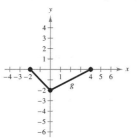

    (e) $g(x) = f(4x)$

Horizontal shrink (each *x*-value is multiplied by $\frac{1}{4}$)

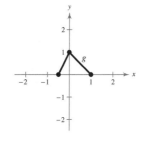

    (f) $g(x) = f(\frac{1}{2}x)$

Horizontal stretch (each *x*-value is multiplied by 2)

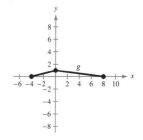

79. (a) Vertical stretch of 128.0 and vertical shift 527 units up

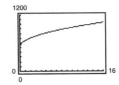

(b) Average rate of change $= \dfrac{M(t_2) - M(t_1)}{t_2 - t_1}$

$= \dfrac{M(16) - M(0)}{16 - 0}$

$= \dfrac{1039 - 527}{16}$

$= 32$

Each year, the total number of miles driven by vans, pick-ups, and SUVs increases by an average of 32 billion miles.

(c) $M = 527 + 128.0\sqrt{t + 10}$; The graph is shifted 10 units to the left.

(d) 2012: Use $t = 12 \rightarrow M(12) = 527 + 128.0\sqrt{12 + 10}$

≈ 1127.37

In the year 2012, vans, pickups, and SUVs will drive about 1127 billion miles.

Answers will vary.

81. False. $y = f(-x)$ is a reflection in the *y*-axis.

83. True; because $|x| = |-x|$, the graphs of $f(x) = |x| + 6$ and $f(x) = |-x| + 6$ are identical.

85. (a) The profits were only $\frac{3}{4}$ as large as expected:

$g(t) = \frac{3}{4}f(t)$

(b) The profits were $10,000 greater than predicted:

$g(t) = f(t) + 10,000$

(c) There was a two-year delay: $g(t) = f(t - 2)$

87. $y = f(x + 2) - 1$

Horizontal shift 2 units to the left and a vertical shift 1 unit downward

$(0, 1) \rightarrow (0 - 2, 1 - 1) = (-2, 0)$

$(1, 2) \rightarrow (1 - 2, 2 - 1) = (-1, 1)$

$(2, 3) \rightarrow (2 - 2, 3 - 1) = (0, 2)$

89. No. $g(x) = -x^4 - 2$. Yes. $h(x) = -(x - 3)^4$

Section 2.6   Combinations of Functions: Composite Functions

■ Given two functions, *f* and *g*, you should be able to form the following functions (if defined):

1.   Sum: $(f + g)(x) = f(x) + g(x)$

2.   Difference: $(f - g)(x) = f(x) - g(x)$

3.   Product: $(fg)(x) = f(x)g(x)$

4.   Quotient: $(f/g)(x) = f(x)/g(x), g(x) \neq 0$

5.   Composition of *f* with *g*: $(f \circ g)(x) = f(g(x))$

6.   Composition of *g* with *f*: $(g \circ f)(x) = g(f(x))$

1. addition; subtraction; multiplication; division

3. $g(x)$

5.

x	0	1	2	3
f	2	3	1	2
g	-1	0	$\frac{1}{2}$	0
$f + g$	1	3	$\frac{3}{2}$	2

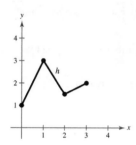

7.

x	-2	0	1	2	4
f	2	0	1	2	4
g	4	2	1	0	2
$f + g$	6	2	2	2	6

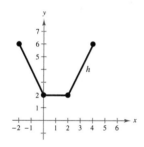

9. $f(x) = x + 2, g(x) = x - 2$

(a) $(f + g)(x) = f(x) + g(x)$
$= (x + 2) + (x - 2)$
$= 2x$

(b) $(f - g)(x) = f(x) - g(x)$
$= (x + 2) - (x - 2)$
$= 4$

(c) $(fg)(x) = f(x) \cdot g(x)$
$= (x + 2)(x - 2)$
$= x^2 - 4$

(d) $\left(\dfrac{f}{x}\right)(x) = \dfrac{f(x)}{g(x)} = \dfrac{x + 2}{x - 2}$

Domain: all real numbers x except $x = 2$

11. $f(x) = x^2, g(x) = 4x - 5$

(a) $(f + g)(x) = f(x) + g(x)$
$= x^2 + (4x - 5)$
$= x^2 + 4x - 5$

(b) $(f - g)(x) = f(x) - g(x)$
$= x^2 - (4x - 5)$
$= x^2 - 4x + 5$

(c) $(fg)(x) = f(x) \cdot g(x)$
$= x^2(4x - 5)$
$= 4x^3 - 5x^2$

(d) $\left(\dfrac{f}{g}\right)(x) = \dfrac{f(x)}{g(x)}$
$= \dfrac{x^2}{4x - 5}$

Domain: all real numbers x except $x = \dfrac{5}{4}$

13. $f(x) = x^2 + 6, g(x) = \sqrt{1 - x}$

(a) $(f + g)(x) = f(x) + g(x) = x^2 + 6 + \sqrt{1 - x}$

(b) $(f - g)(x) = f(x) - g(x) = x^2 + 6 - \sqrt{1 - x}$

(c) $(fg)(x) = f(x) \cdot g(x) = (x^2 + 6)\sqrt{1 - x}$

(d) $\left(\dfrac{f}{g}\right)(x) = \dfrac{f(x)}{g(x)} = \dfrac{x^2 + 6}{\sqrt{1 - x}} = \dfrac{(x^2 + 6)\sqrt{1 - x}}{1 - x}$

Domain: $x < 1$

15. $f(x) = \dfrac{1}{x}, g(x) = \dfrac{1}{x^2}$

(a) $(f + g)(x) = f(x) + g(x) = \dfrac{1}{x} + \dfrac{1}{x^2} = \dfrac{x + 1}{x^2}$

(b) $(f - g)(x) = f(x) - g(x) = \dfrac{1}{x} - \dfrac{1}{x^2} = \dfrac{x - 1}{x^2}$

(c) $(fg)(x) = f(x) \cdot g(x) = \dfrac{1}{x}\left(\dfrac{1}{x^2}\right) = \dfrac{1}{x^3}$

(d) $\left(\dfrac{f}{g}\right)(x) = \dfrac{f(x)}{g(x)} = \dfrac{1/x}{1/x^2} = \dfrac{x^2}{x} = x$

Domain: all real numbers x except $x = 0$

For Exercises 17–27, $f(x) = x^2 + 1$ and $g(x) = x - 4$.

17. $(f + g)(2) = f(2) + g(2) = (2^2 + 1) + (2 - 4) = 3$

19. $(f - g)(0) = f(0) - g(0)$
$= (0^2 + 1) - (0 - 4)$
$= 5$

21. $(f - g)(3t) = f(3t) - g(3t)$

$$= \left[(3t)^2 + 1\right] - (3t - 4)$$

$$= 9t^2 - 3t + 5$$

23. $(fg)(6) = f(6)g(6)$

$$= (6^2 + 1)(6 - 4)$$

$$= 74$$

25. $\left(\dfrac{f}{g}\right)(5) = \dfrac{f(5)}{g(5)} = \dfrac{5^2 + 1}{5 - 4} = 26$

27. $\left(\dfrac{f}{g}\right)(-1) - g(3) = \dfrac{f(-1)}{g(-1)} - g(3)$

$$= \dfrac{(-1)^2 + 1}{-1 - 4} - (3 - 4)$$

$$= -\dfrac{2}{5} + 1 = \dfrac{3}{5}$$

29. $f(x) = \frac{1}{2}x, \ g(x) = x - 1$

$(f + g)(x) = \frac{3}{2}x - 1$

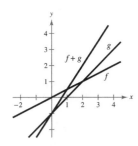

31. $f(x) = x^2, \ g(x) = -2x$

$(f + g)(x) = x^2 - 2x$

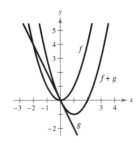

33. $f(x) = 3x, \ g(x) = -\dfrac{x^3}{10}$

$(f + g)(x) = 3x - \dfrac{x^3}{10}$

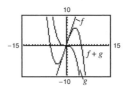

For $0 \le x \le 2$, $f(x)$ contributes most to the magnitude.

For $x > 6$, $g(x)$ contributes most to the magnitude.

35. $f(x) = 3x + 2, \ g(x) = -\sqrt{x + 5}$

$(f + g)x = 3x - \sqrt{x + 5} + 2$

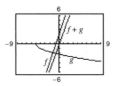

For $0 \le x \le 2$, $f(x)$ contributes most to the magnitude.

For $x > 6$, $f(x)$ contributes most to the magnitude.

37. $f(x) = x^2, \ g(x) = x - 1$

(a) $(f \circ g)(x) = f(g(x)) = f(x - 1) = (x - 1)^2$

(b) $(g \circ f)(x) = g(f(x)) = g(x^2) = x^2 - 1$

(c) $(g \circ g)(x) = g(g(x)) = g(x - 1) = x - 2$

39. $f(x) = \sqrt[3]{x - 1}, \ g(x) = x^3 + 1$

(a) $(f \circ g)(x) = f(g(x))$

$$= f(x^3 + 1)$$

$$= \sqrt[3]{(x^3 + 1) - 1}$$

$$= \sqrt[3]{x^3} = x$$

(b) $(g \circ f)(x) = g(f(x))$

$$= g(\sqrt[3]{x - 1})$$

$$= (\sqrt[3]{x - 1})^3 + 1$$

$$= (x - 1) + 1 = x$$

(c) $(g \circ g)(x) = g(g(x))$

$$= g(x^3 + 1)$$

$$= (x^3 + 1)^3 + 1$$

$$= x^9 + 3x^6 + 3x^3 + 2$$

41. $f(x) = \sqrt{x + 4}$ Domain: $x \geq -4$

$g(x) = x^2$ Domain: all real numbers x

(a) $(f \circ g)(x) = f(g(x)) = f(x^2) = \sqrt{x^2 + 4}$

Domain: all real numbers x

(b) $(g \circ f)(x) = g(f(x))$

$= g(\sqrt{x + 4}) = (\sqrt{x + 4})^2 = x + 4$

Domain: $x \geq -4$

43. $f(x) = x^2 + 1$ Domain: all real numbers x

$g(x) = \sqrt{x}$ Domain: $x \geq 0$

(a) $(f \circ g)(x) = f(g(x))$

$= f(\sqrt{x})$

$= (\sqrt{x})^2 + 1$

$= x + 1$

Domain: $x \geq 0$

(b) $(g \circ f)(x) = g(f(x)) = g(x^2 + 1) = \sqrt{x^2 + 1}$

Domain: all real numbers x

45. $f(x) = |x|$ Domain: all real numbers x

$g(x) = x + 6$ Domain: all real numbers x

(a) $(f \circ g)(x) = f(g(x)) = f(x + 6) = |x + 6|$

Domain: all real numbers x

(b) $(g \circ f)(x) = g(f(x)) = g(|x|) = |x| + 6$

Domain: all real numbers x

47. $f(x) = \dfrac{1}{x}$ Domain: all real numbers x except $x = 0$

$g(x) = x + 3$ Domain: all real numbers x

(a) $(f \circ g)(x) = f(g(x)) = f(x + 3) = \dfrac{1}{x + 3}$

Domain: all real numbers x except $x = -3$

(b) $(g \circ f)(x) = g(f(x)) = g\left(\dfrac{1}{x}\right) = \dfrac{1}{x} + 3$

Domain: all real numbers x except $x = 0$

49. (a) $(f + g)(3) = f(3) + g(3) = 2 + 1 = 3$

(b) $\left(\dfrac{f}{g}\right)(2) = \dfrac{f(2)}{g(2)} = \dfrac{0}{2} = 0$

51. (a) $(f \circ g)(2) = f(g(2)) = f(2) = 0$

(b) $(g \circ f)(2) = g(f(2)) = g(0) = 4$

53. $h(x) = (2x^2 + 1)^2$

One possibility: Let $f(x) = x^2$ and $g(x) = 2x + 1$,

then $(f \circ g)(x) = h(x)$.

55. $h(x) = \sqrt[3]{x^2 - 4}$

One possibility: Let $f(x) = \sqrt[3]{x}$ and $g(x) = x^2 - 4$,

then $(f \circ g)(x) = h(x)$.

57. $h(x) = \dfrac{1}{x + 2}$

One possibility: Let $f(x) = 1/x$ and $g(x) = x + 2$,

then $(f \circ g)(x) = h(x)$.

59. $h(x) = \dfrac{-x^2 + 3}{4 - x^2}$

One possibility: Let $f(x) = \dfrac{x + 3}{4 + x}$ and $g(x) = -x^2$,

then $(f \circ g)(x) = h(x)$.

61. (a) $T(x) = R(x) + B(x) = \frac{3}{4}x + \frac{1}{15}x^2$

(b)

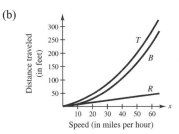

(c) $B(x)$; As x increases, $B(x)$ increases at a faster rate.

63. (a) $c(t) = \dfrac{b(t) - d(t)}{p(t)} \times 100$

(b) $c(5)$ represents the percent change in the population due to births and deaths in the year 2005.

65. (a) $(N + M)(t) = N(t) + M(t)$
$$= 0.227t^3 - 4.11t^2 + 14.6t + 544$$

This represents the combined Navy and Marine personnel (in thousands) from 2000 to 2007, where $t = 0$ corresponds to 2000.

$(N + M)(0) = 544$ thousand

$(N + M)(6) = 532.67$ thousand

$(N + M)(12) = 519.62$ thousand

(b) $(N - M)(t) = N(t) - M(t)$
$$= 0.157t^3 - 3.65t^2 + 11.2t + 200$$

This represents the difference between the number of Navy personnel (in thousands) and the number of Marine personnel from 2000 to 2007, where $t = 0$ corresponds to 2000.

$(N - M)(0) = 200$ thousand

$(N - M)(6) = 169.71$ thousand

$(N - M)(12) = 80.10$ thousand

67. $(B - D)(t) = B(t) - D(t) = -0.197t^3 + 10.17t^2 - 128t + 2043$

This represents the number of births (in millions) more than the number of deaths in the United States from 1990 to 2006, where $t = 0$ corresponds to 1990.

69. (a) T is a function of t since for each time t there corresponds one and only one temperature T.

(b) $T(4) \approx 60°; T(15) \approx 72°$

(c) $H(t) = T(t - 1)$; All the temperature changes would be one hour later.

(d) $H(t) = T(t) - 1$; The temperature would be decreased by one degree.

(e) The points at the endpoints of the individual functions that form each "piece" appear to be $(0, 60), (6, 60), (7, 72), (20, 72), (21, 60),$ and $(24, 60)$. Note that the value $t = 24$ is chosen for the last ordered pair because that is when the day ends and the cycle starts over.

From $t = 0$ to $t = 6$: This is the constant function $T(t) = 60$.

From $t = 6$ to $t = 7$: Use the points $(6, 60)$ and $(7, 72)$.

$$m = \frac{72 - 60}{7 - 6} = 12$$

$y - 60 = 12(x - 6) \Rightarrow y = 12x - 12,$ or $T(t) = 12t - 12$

From $t = 7$ to $t = 20$: This is the constant function $T(t) = 72$.

From $t = 20$ to $t = 21$: Use the points $(20, 72)$ and $(21, 60)$.

$$m = \frac{72 - 60}{20 - 21} = -12$$

$y - 60 = -12(x - 21) \Rightarrow y = -12x + 312,$ or $T(t) = -12t + 312$

From $t = 21$ to $t = 24$: This is the constant function $T(t) = 60$.

A piecewise-defined function is $T(t) = \begin{cases} 60, & 0 \le t \le 6 \\ 12t - 12, & 6 < t < 7 \\ 72, & 7 \le t \le 20 \\ -12t + 312, & 20 < t < 21 \\ 60, & 21 \le t \le 24 \end{cases}$.

71. $(A \circ r)(t) = A(r(t)) = A(0.6t) = \pi(0.6t)^2 = 0.36\pi t^2$

$A \circ r$ represents the area of the circle at the time t.

73. (a) $N(T(t)) = N(3t + 2)$

$$= 10(3t + 2)^2 - 20(3t + 2) + 600$$

$$= 10(9t^2 + 12t + 4) - 60t - 40 + 600$$

$$= 90t^2 + 60t + 600$$

$$= 30(3t^2 + 2t + 20), \quad 0 \le t \le 6$$

This represents the number of bacteria in the food as a function of time.

(b) Use $t = 0.5$.

$$N(T(0.5)) = 30\left(3(0.5)^2 + 2(0.5) + 20\right)$$

$$= 652.5$$

After half an hour, there will be about 653 bacteria.

(c) $30(3t^2 + 2t + 20) = 1500$

$$3t^2 + 2t + 20 = 50$$

$$3t^2 + 2t - 30 = 0$$

By the Quadratic Formula, $t \approx -3.513$ or 2.846.

Choosing the positive value for t, you have $t \approx 2.846$ hours.

75. (a) $f(g(x)) = f(0.03x) = 0.03x - 500,000$

(b) $g(f(x)) = g(x - 500,000) = 0.03(x - 500,000)$

$g(f(x))$ represents your bonus of 3% of an amount over $500,000.

77. False. $(f \circ g)(x) = 6x + 1$ and $(g \circ f)(x) = 6x + 6$

79. Let O = oldest sibling, M = middle sibling, Y = youngest sibling.

Then the ages of each sibling can be found using the equations:

$O = 2M$

$M = \frac{1}{2}Y + 6$

(a) $O(M(Y)) = 2\left(\frac{1}{2}(Y) + 6\right) = 12 + Y$; Answers will vary.

(b) Oldest sibling is 16: $O = 16$

Middle sibling: $O = 2M$

$$16 = 2M$$

$$M = 8 \text{ years old}$$

Youngest sibling: $M = \frac{1}{2}Y + 6$

$$8 = \frac{1}{2}Y + 6$$

$$2 = \frac{1}{2}Y$$

$$Y = 4 \text{ years old}$$

81. Let $f(x)$ and $g(x)$ be two odd functions and define $h(x) = f(x)g(x)$. Then

$h(-x) = f(-x)g(-x)$

$$= \left[-f(x)\right]\left[-g(x)\right] \quad \text{because } f \text{ and } g \text{ are odd}$$

$$= f(x)g(x)$$

$$= h(x).$$

So, $h(x)$ is even.

Let $f(x)$ and $g(x)$ be two even functions and define $h(x) = f(x)g(x)$. Then

$h(-x) = f(-x)g(-x)$

$$= f(x)g(x) \quad \text{because } f \text{ and } g \text{ are even}$$

$$= h(x).$$

So, $h(x)$ is even.

83. (a) $g(x) = \frac{1}{2}\big[f(x) + f(-x)\big]$

To determine if $g(x)$ is even, show

$g(-x) = g(x).$

$g(-x) = \frac{1}{2}\big[f(-x) + f(-(-x))\big]$

$= \frac{1}{2}\big[f(-x) + f(x)\big]$

$= \frac{1}{2}\big[f(x) + f(-x)\big]$

$= g(x) \checkmark$

$h(x) = \frac{1}{2}\big[f(x) - f(-x)\big]$

To determine if $h(x)$ is odd show $h(-x) = -h(x).$

$h(-x) = \frac{1}{2}\big[f(-x) - f(-(-x))\big]$

$= \frac{1}{2}\big[f(-x) - f(x)\big]$

$= -\frac{1}{2}\big[f(x) - f(-x)\big]$

$= -h(x) \checkmark$

(b) Let $f(x) = a$ function

$f(x) = $ even function + odd function.

Using the result from part (a) $g(x)$ is an even function and $h(x)$ is an odd function.

$f(x) = g(x) + h(x)$

$= \frac{1}{2}\big[f(x) + f(-x)\big] + \frac{1}{2}\big[f(x) - f(-x)\big]$

$= \frac{1}{2}f(x) + \frac{1}{2}f(-x) + \frac{1}{2}f(x) - \frac{1}{2}f(-x)$

$= f(x) \checkmark$

(c) $f(x) = x^2 - 2x + 1$

$f(x) = g(x) + h(x)$

$g(x) = \frac{1}{2}\big[f(x) + f(-x)\big]$

$= \frac{1}{2}\big[x^2 - 2x + 1 + (-x)^2 - 2(-x) + 1\big]$

$= \frac{1}{2}\big[x^2 - 2x + 1 + x^2 + 2x + 1\big]$

$= \frac{1}{2}\big[2x^2 + 2\big] = x^2 + 1$

$h(x) = \frac{1}{2}\big[f(x) - f(-x)\big]$

$= \frac{1}{2}\big[x^2 - 2x + 1 - \big((-x)^2 - 2(-x) + 1\big)\big]$

$= \frac{1}{2}\big[x^2 - 2x + 1 - x^2 - 2x - 1\big]$

$= \frac{1}{2}\big[-4x\big] = -2x$

$f(x) = (x^2 + 1) + (-2x)$

$k(x) = \frac{1}{x + 1}$

$k(x) = g(x) + h(x)$

$g(x) = \frac{1}{2}\big[k(x) + k(-x)\big]$

$= \frac{1}{2}\bigg[\frac{1}{x + 1} + \frac{1}{-x + 1}\bigg]$

$= \frac{1}{2}\bigg[\frac{1 - x + x + 1}{(x + 1)(1 - x)}\bigg]$

$= \frac{1}{2}\bigg[\frac{2}{(x + 1)(1 - x)}\bigg]$

$= \frac{1}{(x + 1)(1 - x)}$

$= \frac{-1}{(x + 1)(x - 1)}$

$h(x) = \frac{1}{2}\big[k(x) - k(-x)\big]$

$= \frac{1}{2}\bigg[\frac{1}{x + 1} - \frac{1}{1 - x}\bigg]$

$= \frac{1}{2}\bigg[\frac{1 - x - (x + 1)}{(x + 1)(1 - x)}\bigg]$

$= \frac{1}{2}\bigg[\frac{-2x}{(x + 1)(1 - x)}\bigg]$

$= \frac{-x}{(x + 1)(1 - x)}$

$= \frac{x}{(x + 1)(x - 1)}$

$k(x) = \bigg(\frac{-1}{(x + 1)(x - 1)}\bigg) + \bigg(\frac{x}{(x + 1)(x - 1)}\bigg)$

Section 2.7   Inverse Functions

- Two functions f and g are inverses of each other if $f(g(x)) = x$ for every x in the domain of g and $g(f(x)) = x$ for every x in the domain of f.

- A function f has an inverse function if and only if no **horizontal** line crosses the graph of f at more than one point.

- The graph of f^{-1} is a reflection of the graph of f about the line $y = x$.

- Be able to find the inverse of a function, if it exists.
    1. Use the Horizontal Line Test to see if f^{-1} exists.
    2. Replace $f(x)$ with y.
    3. Interchange x and y and solve for y.
    4. Replace y with $f^{-1}(x)$.

---

1. inverse

3. range; domain

5. one-to-one

7. $f(x) = 6x$

$$f^{-1}(x) = \frac{x}{6} = \frac{1}{6}x$$

$$f(f^{-1}(x)) = f\left(\frac{x}{6}\right) = 6\left(\frac{x}{6}\right) = x$$

$$f^{-1}(f(x)) = f^{-1}(6x) = \frac{6x}{6} = x$$

9. $f(x) = x + 9$

$$f^{-1}(x) = x - 9$$

$$f(f^{-1}(x)) = f(x - 9) = (x - 9) + 9 = x$$

$$f^{-1}(f(x)) = f^{-1}(x + 9) = (x + 9) - 9 = x$$

11. $f(x) = 3x + 1$

$$f^{-1}(x) = \frac{x - 1}{3}$$

$$f(f^{-1}(x)) = f\left(\frac{x - 1}{3}\right) = 3\left(\frac{x - 1}{3}\right) + 1 = x$$

$$f^{-1}(f(x)) = f^{-1}(3x + 1) = \frac{(3x + 1) - 1}{3} = x$$

13. $f(x) = \sqrt[3]{x}$

$$f^{-1}(x) = x^3$$

$$f(f^{-1}(x)) = f(x^3) = \sqrt[3]{x^3} = x$$

$$f^{-1}(f(x)) = f^{-1}(\sqrt[3]{x}) = \left(\sqrt[3]{x}\right)^3 = x$$

15. The inverse is a line through $(-1, 0)$. Matches graph (c).

17. The inverse is half a parabola starting at $(1, 0)$. Matches graph (a).

19. $(f \circ g)(x) = f(g(x)) = f\left(-\dfrac{2x + 6}{7}\right) = -\dfrac{7}{2}\left(-\dfrac{2x + 6}{7}\right) - 3 = x + 3 - 3 = x$

$(g \circ f)(x) = g(f(x)) = g\left(-\dfrac{7}{2}x - 3\right) = -\dfrac{2\left(-\dfrac{7}{2}x - 3\right) + 6}{7} = \dfrac{-(-7x)}{7} = x$

21. $(f \circ g)(x) = f(g(x)) = f(\sqrt[3]{x - 5}) = \left(\sqrt[3]{x - 5}\right)^3 + 5 = x - 5 + 5 = x$

$(g \circ f)(x) = g(f(x)) = g(x^3 + 5) = \sqrt[3]{x^3 + 5 - 5} = \sqrt[3]{x^3} = x$

23. $f(x) = 2x, g(x) = \dfrac{x}{2}$

(a) $f(g(x)) = f\left(\dfrac{x}{2}\right) = 2\left(\dfrac{x}{2}\right) = x$

$g(f(x)) = g(2x) = \dfrac{2x}{2} = x$

(b)

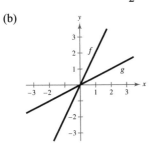

25. $f(x) = 7x + 1, g(x) = \dfrac{x-1}{7}$

(a) $f(g(x)) = f\left(\dfrac{x-1}{7}\right) = 7\left(\dfrac{x-1}{7}\right) + 1 = x$

$g(f(x)) = g(7x+1) = \dfrac{(7x+1)-1}{7} = x$

(b)

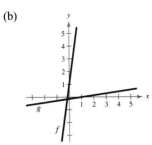

31. $f(x) = 9 - x^2, x \ge 0; g(x) = \sqrt{9-x}, x \le 9$

(a) $f(g(x)) = f\left(\sqrt{9-x}\right), x \le 9 = 9 - \left(\sqrt{9-x}\right)^2 = x$

$g(f(x)) = g(9-x^2), x \ge 0 = \sqrt{9-(9-x^2)} = x$

(b)

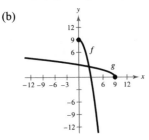

27. $f(x) = \dfrac{x^3}{8}, g(x) = \sqrt[3]{8x}$

(a) $f(g(x)) = f\left(\sqrt[3]{8x}\right) = \dfrac{\left(\sqrt[3]{8x}\right)^3}{8} = \dfrac{8x}{8} = x$

$g(f(x)) = g\left(\dfrac{x^3}{8}\right) = \sqrt[3]{8\left(\dfrac{x^3}{8}\right)} = \sqrt[3]{x^3} = x$

(b)

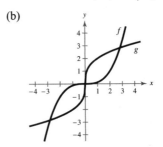

29. $f(x) = \sqrt{x-4}, g(x) = x^2 + 4, x \ge 0$

(a) $f(g(x)) = f(x^2+4), x \ge 0$

$= \sqrt{(x^2+4)-4} = x$

$g(f(x)) = g\left(\sqrt{x-4}\right)$

$= \left(\sqrt{x-4}\right)^2 + 4 = x$

(b)

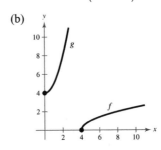

33. $f(x) = \dfrac{x-1}{x+5}$, $g(x) = -\dfrac{5x+1}{x-1}$

(a) $f(g(x)) = f\left(-\dfrac{5x+1}{x-1}\right) = \dfrac{\left(-\dfrac{5x+1}{x-1} - 1\right)}{\left(-\dfrac{5x+1}{x-1} + 5\right)} \cdot \dfrac{x-1}{x-1} = \dfrac{-(5x+1) - (x-1)}{-(5x+1) + 5(x-1)} = \dfrac{-6x}{-6} = x$

$g(f(x)) = g\left(\dfrac{x-1}{x+5}\right) = -\dfrac{\left[5\left(\dfrac{x-1}{x+5}\right) + 1\right]}{\left[\dfrac{x-1}{x+5} - 1\right]} \cdot \dfrac{x+5}{x+5} = -\dfrac{5(x-1) + (x+5)}{(x-1) - (x+5)} = -\dfrac{6x}{-6} = x$

(b)

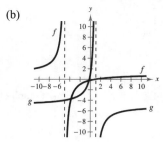

35. No, $\{(-2, -1), (1, 0), (2, 1), (1, 2), (-2, 3), (-6, 4)\}$ does not represent a function. -2 and 1 are paired with two different values.

37.

x	-2	0	2	4	6	8
$f^{-1}(x)$	-2	-1	0	1	2	3

39. Yes, because no horizontal line crosses the graph of f at more than one point, f *has* an inverse.

41. No, because some horizontal lines cross the graph of f twice, f *does not* have an inverse.

43. $g(x) = \dfrac{4 - x}{6}$

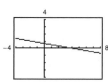

g passes the Horizontal Line Test, so g *has* an inverse.

45. $h(x) = |x + 4| - |x - 4|$

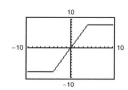

h does not pass the Horizontal Line Test, so h *does not* have an inverse.

47. $f(x) = -2x\sqrt{16 - x^2}$

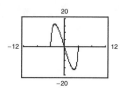

f does not pass the Horizontal Line Test, so f *does not* have an inverse.

49. (a) $f(x) = 2x - 3$ (b)

$y = 2x - 3$

$x = 2y - 3$

$y = \dfrac{x + 3}{2}$

$f^{-1}(x) = \dfrac{x + 3}{2}$

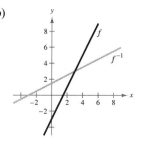

(c) The graph of f^{-1} is the reflection of the graph of f in the line $y = x$.

(d) The domains and ranges of f and f^{-1} are all real numbers.

51. (a) $f(x) = x^5 - 2$ (b)

$y = x^5 - 2$

$x = y^5 - 2$

$y = \sqrt[5]{x + 2}$

$f^{-1}(x) = \sqrt[5]{x + 2}$

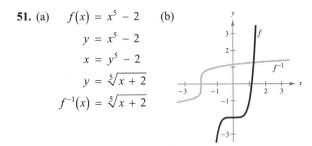

(c) The graph of f^{-1} is the reflection of the graph of f in the line $y = x$.

(d) The domains and ranges of f and f^{-1} are all real numbers.

53. (a) $f(x) = \sqrt{4 - x^2}, 0 \le x \le 2$

$y = \sqrt{4 - x^2}$

$x = \sqrt{4 - y^2}$

$x^2 = 4 - y^2$

$y^2 = 4 - x^2$

$y = \sqrt{4 - x^2}$

$f^{-1}(x) = \sqrt{4 - x^2}, 0 \le x \le 2$

(b)

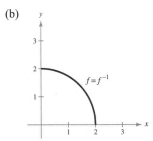

(c) The graph of f^{-1} is the same as the graph of f.

(d) The domains and ranges of f and f^{-1} are all real numbers x such that $0 \le x \le 2$.

55. (a) $f(x) = \dfrac{4}{x}$ (b)

$y = \dfrac{4}{x}$

$x = \dfrac{4}{y}$

$xy = 4$

$y = \dfrac{4}{x}$

$f^{-1}(x) = \dfrac{4}{x}$

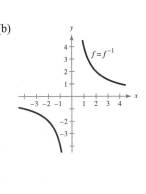

(c) The graph of f^{-1} is the same as the graph of f.

(d) The domains and ranges of f and f^{-1} are all real numbers except for 0.

57. (a) $f(x) = \dfrac{x + 1}{x - 2}$ (b)

$y = \dfrac{x + 1}{x - 2}$

$x = \dfrac{y + 1}{y - 2}$

$x(y - 2) = y + 1$

$xy - 2x = y + 1$

$xy - y = 2x + 1$

$y(x - 1) = 2x + 1$

$y = \dfrac{2x + 1}{x - 1}$

$f^{-1}(x) = \dfrac{2x + 1}{x - 1}$

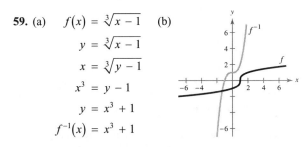

(c) The graph of f^{-1} is the reflection of graph of f in the line $y = x$.

(d) The domain of f and the range of f^{-1} is all real numbers except 2.

The range of f and the domain of f^{-1} is all real numbers except 1.

59. (a) $f(x) = \sqrt[3]{x - 1}$ (b)

$y = \sqrt[3]{x - 1}$

$x = \sqrt[3]{y - 1}$

$x^3 = y - 1$

$y = x^3 + 1$

$f^{-1}(x) = x^3 + 1$

(c) The graph of f^{-1} is the reflection of the graph of f in the line $y = x$.

(d) The domains and ranges of f and f^{-1} are all real numbers.

61. (a)
$$f(x) = \frac{6x + 4}{4x + 5}$$
$$y = \frac{6x + 4}{4x + 5}$$
$$x = \frac{6y + 4}{4y + 5}$$
$$x(4y + 5) = 6y + 4$$
$$4xy + 5x = 6y + 4$$
$$4xy - 6y = -5x + 4$$
$$y(4x - 6) = -5x + 4$$
$$y = \frac{-5x + 4}{4x - 6}$$
$$f^{-1}(x) = \frac{-5x + 4}{4x - 6} = \frac{5x - 4}{6 - 4x}$$

(b)

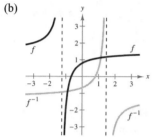

(c) The graph of f^{-1} is the graph of f reflected in the line $y = x$.

(d) The domain of f and the range of f^{-1} is all real numbers except $-\frac{5}{4}$.

The range of f and the domain of f^{-1} is all real numbers except $\frac{3}{2}$.

63. $f(x) = x^4$
$$y = x^4$$
$$x = y^4$$
$$y = \pm\sqrt[4]{x}$$

This does not represent y as a function of x. f does not have an inverse.

65. $g(x) = \frac{x}{8}$
$$y = \frac{x}{8}$$
$$x = \frac{y}{8}$$
$$y = 8x$$

This is a function of x, so g has an inverse.
$$g^{-1}(x) = 8x$$

67. $p(x) = -4$
$$y = -4$$

Because $y = -4$ for all x, the graph is a horizontal line and fails the Horizontal Line Test. p does not have an inverse.

69. $f(x) = (x + 3)^2, x \geq -3 \Rightarrow y \geq 0$
$$y = (x + 3)^2, x \geq -3, y \geq 0$$
$$x = (y + 3)^2, y \geq -3, x \geq 0$$
$$\sqrt{x} = y + 3, y \geq -3, x \geq 0$$
$$y = \sqrt{x} - 3, x \geq 0, y \geq -3$$

This is a function of x, so f has an inverse.
$$f^{-1}(x) = \sqrt{x} - 3, x \geq 0$$

71. $f(x) = \begin{cases} x + 3, & x < 0 \\ 6 - x, & x \geq 0 \end{cases}$

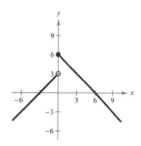

This graph fails the Horizontal Line Test, so f does not have an inverse.

73. $h(x) = -\frac{4}{x^2}$

The graph fails the Horizontal Line Test so h does not have an inverse.

75. $f(x) = \sqrt{2x + 3} \Rightarrow x \geq -\frac{3}{2}, y \geq 0$
$$y = \sqrt{2x + 3}, x \geq -\frac{3}{2}, y \geq 0$$
$$x = \sqrt{2y + 3}, y \geq -\frac{3}{2}, x \geq 0$$
$$x^2 = 2y + 3, x \geq 0, y \geq -\frac{3}{2}$$
$$y = \frac{x^2 - 3}{2}, x \geq 0, y \geq -\frac{3}{2}$$

This is a function of x, so f has an inverse.
$$f^{-1}(x) = \frac{x^2 - 3}{2}, x \geq 0$$

77. $f(x) = (x - 2)^2$

domain of $f: x \geq 2$, range of $f: y \geq 0$

$$f(x) = (x - 2)^2$$
$$y = (x - 2)^2$$
$$x = (y - 2)^2$$
$$\sqrt{x} = y - 2$$
$$\sqrt{x} + 2 = y$$

So, $f^{-1}(x) = \sqrt{x} + 2$.

domain of $f^{-1}: x \geq 0$, range of $f^{-1}: x \geq 2$

79. $f(x) = |x + 2|$

domain of $f: x \geq -2$, range of $f: y \geq 0$

$$f(x) = |x + 2|$$
$$y = |x + 2|$$
$$x = y + 2$$
$$x - 2 = y$$

So, $f^{-1}(x) = x - 2$.

domain of $f^{-1}: x \geq 0$, range of $f^{-1}: y \geq -2$

81. $f(x) = (x + 6)^2$

domain of $f: x \geq -6$, range of $f: y \geq 0$

$$f(x) = (x + 6)^2$$
$$y = (x + 6)^2$$
$$x = (y + 6)^2$$
$$\sqrt{x} = y + 6$$
$$\sqrt{x} - 6 = y$$

So, $f^{-1}(x) = \sqrt{x} - 6$.

domain of $f^{-1}: x \geq 0$, range of $f^{-1}: y \geq -6$

83. $f(x) = -2x^2 + 5$

domain of $f: x \geq 0$, range of $f: y \leq 5$

$$f(x) = -2x^2 + 5$$
$$y = -2x^2 + 5$$
$$x = -2y^2 + 5$$
$$x - 5 = -2y^2$$
$$5 - x = 2y^2$$
$$\sqrt{\frac{5 - x}{2}} = y$$
$$\frac{\sqrt{5 - x}}{\sqrt{2}} \cdot \frac{\sqrt{2}}{\sqrt{2}} = y$$
$$\frac{\sqrt{2(5 - x)}}{2} = y$$

So, $f^{-1}(x) = \frac{\sqrt{-2(x - 5)}}{2}$.

domain of $f^{-1}(x): x \leq 5$, range of $f^{-1}(x): y \geq 0$

85. $f(x) = |x - 4| + 1$

domain of $f: x \geq 4$, range of $f: y \geq 1$

$$f(x) = |x - 4| + 1$$
$$y = x - 3$$
$$x = y - 3$$
$$x + 3 = y$$

So, $f^{-1}(x) = x + 3$.

domain of $f^{-1}: x \geq 1$, range of $f^{-1}: y \geq 4$

In Exercises 87–91, $f(x) = \frac{1}{8}x - 3$, $f^{-1}(x) = 8(x + 3)$, $g(x) = x^3$, $g^{-1}(x) = \sqrt[3]{x}$.

87. $(f^{-1} \circ g^{-1})(1) = f^{-1}(g^{-1}(1))$
$$= f^{-1}(\sqrt[3]{1})$$
$$= 8(\sqrt[3]{1} + 3) = 32$$

89. $(f^{-1} \circ f^{-1})(6) = f^{-1}(f^{-1}(6))$
$$= f^{-1}(8[6 + 3])$$
$$= 8[8(6 + 3) + 3] = 600$$

91. $(f \circ g)(x) = f(g(x)) = f(x^3) = \frac{1}{8}x^3 - 3$

$$y = \frac{1}{8}x^3 - 3$$

$$x = \frac{1}{8}y^3 - 3$$

$$x + 3 = \frac{1}{8}y^3$$

$$8(x + 3) = y^3$$

$$\sqrt[3]{8(x + 3)} = y$$

$$(f \circ g)^{-1}(x) = 2\sqrt[3]{x + 3}$$

In Exercises 93 and 95, $f(x) = x + 4$, $f^{-1}(x) = x - 4$,

$$g(x) = 2x - 5, g^{-1}(x) = \frac{x + 5}{2}.$$

93. $(g^{-1} \circ f^{-1})(x) = g^{-1}(f^{-1}(x))$

$$= g^{-1}(x - 4)$$

$$= \frac{(x - 4) + 5}{2}$$

$$= \frac{x + 1}{2}$$

95. $(f \circ g)(x) = f(g(x))$

$$= f(2x - 5)$$

$$= (2x - 5) + 4$$

$$= 2x - 1$$

$$(f \circ g)^{-1}(x) = \frac{x + 1}{2}$$

Note: Comparing Exercises 93 and 95,

$$(f \circ g)^{-1}(x) = (g^{-1} \circ f^{-1})(x).$$

97. (a) Yes. For each men's European shoe size, there is exactly one men's U.S. shoe size.

(b) $f(11) = 45$

(c) $f^{-1}(43) = 10$ because $f(10) = 43$.

(d) $f(f^{-1}(41)) = f(8) = 41$

(e) $f^{-1}(f(13)) = f^{-1}(47) = 13$

99. (a) Yes.

(b) Given the sales (in millions of dollars) you can determine the year.

(c) $S^{-1}(8430) = 6$ because $S(6) = 8430$.

So, $8430 million in sales occurred in 2006.

(d) No. The function would no longer be one-to-one because the total sales would be $14,532 million in 2007 and 2009.

101. (a) $y = 10 + 0.75x$

$$x = 10 + 0.75y$$

$$x - 10 = 0.75y$$

$$\frac{x - 10}{0.75} = y$$

So, $f^{-1}(x) = \dfrac{x - 10}{0.75}$.

x = hourly wage, y = number of units produced

(b) $y = \dfrac{24.25 - 10}{0.75} = 19$

So, 19 units are produced.

103. False. $f(x) = x^2$ is even and does not have an inverse.

105. Let $(f \circ g)(x) = y$. Then $x = (f \circ g)^{-1}(y)$. Also,

$$(f \circ g)(x) = y \Rightarrow f(g(x)) = y$$

$$g(x) = f^{-1}(y)$$

$$x = g^{-1}(f^{-1}(y))$$

$$x = (g^{-1} \circ f^{-1})(y).$$

Because f and g are both one-to-one

functions, $(f \circ g)^{-1} = g^{-1} \circ f^{-1}$.

107.

x	1	3	4	6
f	1	2	6	7

x	1	2	6	7
$f^{-1}(x)$	1	3	4	6

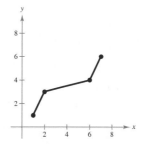

109. This situation could be represented by a one-to-one function if the runner does not stop to rest. The inverse function would represent the time in hours for a given number of miles completed.

111. No. The function oscillates.

113. If $f(x) = k\big(2 - x - x^3\big)$ has an inverse and

$f^{-1}(3) = -2$, then $f(-2) = 3$. So,

$$f(-2) = k\big(2 - (-2) - (-2)^3\big) = 3$$

$$k(2 + 2 + 8) = 3$$

$$12k = 3$$

$$k = \tfrac{3}{12} = \tfrac{1}{4}.$$

So, $k = \tfrac{1}{4}$.

115.

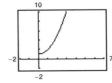

There is an inverse function $f^{-1}(x) = \sqrt{x - 1}$ because the domain of f is equal to the range of f^{-1} and the range of f is equal to the domain of f^{-1}.

Review Exercises for Chapter 2

1. $y = -2x - 7$

Slope: $m = -2 = -\tfrac{2}{1}$

y-intercept: $(0, -7)$

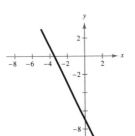

3. $y = 6$

Slope: $m = 0$

y-intercept: $(0, 6)$

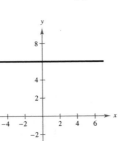

5. $y = -\tfrac{5}{2}x - 1$

Slope: $m = -\tfrac{5}{2}$

y-intercept: $(0, -1)$

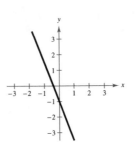

7. $-3x + y = 13$

$y = 3x + 13$

Slope: $m = 3$

y-intercept: $(0, 13)$

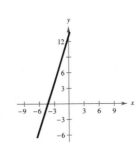

9. $(6, 4), (-3, -4)$

$$m = \frac{4 - (-4)}{6 - (-3)} = \frac{4 + 4}{6 + 3} = \frac{8}{9}$$

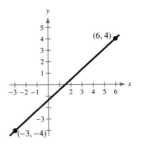

11. $(-4.5, 6), (2.1, 3)$

$$m = \frac{3 - 6}{2.1 - (-4.5)}$$

$$= \frac{-3}{6.6} = -\frac{30}{66} = -\frac{5}{11}$$

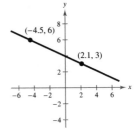

13. $(3, 0), \quad m = \tfrac{2}{3}$

$$y - 0 = \tfrac{2}{3}(x - 3)$$

$$y = \tfrac{2}{3}x - 2$$

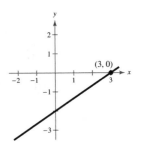

15. $(10, -3)$, $m = -\frac{1}{2}$

$$y - (-3) = -\frac{1}{2}(x - 10)$$

$$y + 3 = -\frac{1}{2}x + 5$$

$$y = -\frac{1}{2}x + 2$$

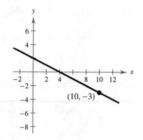

17. $(0, 0)$, $(0, 10)$

$$m = \frac{10 - 0}{0 - 0} = \frac{10}{0}, \text{ undefined}$$

The line is vertical.

$$x = 0$$

19. $(-1, 0)$, $(6, 2)$

$$m = \frac{2 - (0)}{6 - (-1)} = \frac{2}{7}$$

$$y - 0 = \frac{2}{7}(x - (-1))$$

$$y = \frac{2}{7}(x + 1)$$

$$y = \frac{2}{7}x + \frac{2}{7}$$

21. Point: $(3, -2)$

$$5x - 4y = 8$$

$$y = \frac{5}{4}x - 2$$

(a) Parallel slope: $m = \frac{5}{4}$

$$y - (-2) = \frac{5}{4}(x - 3)$$

$$y + 2 = \frac{5}{4}x - \frac{15}{4}$$

$$y = \frac{5}{4}x - \frac{23}{4}$$

(b) Perpendicular slope: $m = -\frac{4}{5}$

$$y - (-2) = -\frac{4}{5}(x - 3)$$

$$y + 2 = -\frac{4}{5}x + \frac{12}{5}$$

$$y = -\frac{4}{5}x + \frac{2}{5}$$

23. $(10, 12{,}500)$, $m = -850$

$$V - 12{,}500 = -850(t - 10)$$

$$V - 12{,}500 = -850t + 8500$$

$$V = -850t + 21{,}000$$

25. $16x - y^4 = 0$

$$y^4 = 16x$$

$$y = \pm 2\sqrt[4]{x}$$

No, y is not a function of x. Some x-values correspond to two y-values.

27. $y = \sqrt{1 - x}$

Yes, the equation represents y as a function of x. Each x-value, $x \leq 1$, corresponds to only one y-value.

29. $f(x) = x^2 + 1$

(a) $f(2) = (2)^2 + 1 = 5$

(b) $f(-4) = (-4)^2 + 1 = 17$

(c) $f(t^2) = (t^2)^2 + 1 = t^4 + 1$

(d) $f(t + 1) = (t + 1)^2 + 1$

$$= t^2 + 2t + 2$$

31. $h(x) = \begin{cases} 2x + 1, & x \leq -1 \\ x^2 + 2, & x > -1 \end{cases}$

(a) $h(-2) = 2(-2) + 1 = -3$

(b) $h(-1) = 2(-1) + 1 = -1$

(c) $h(0) = 0^2 + 2 = 2$

(d) $h(2) = 2^2 + 2 = 6$

33. $f(x) = \sqrt{25 - x^2}$

Domain: $25 - x^2 \geq 0$

$$(5 + x)(5 - x) \geq 0$$

Critical numbers: $x = \pm 5$

Test intervals: $(-\infty, -5), (-5, 5), (5, \infty)$

Test: Is $25 - x^2 \geq 0$?

Solution set: $-5 \leq x \leq 5$

Domain: all real numbers x such that $-5 \leq x \leq 5$, or $[-5, 5]$

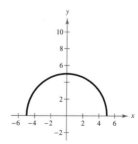

35. $h(x) = \dfrac{x}{x^2 - x - 6}$

$= \dfrac{x}{(x + 2)(x - 3)}$

Domain: All real numbers x except $x = -2, 3$

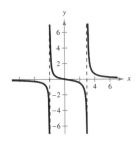

37. $v(t) = -32t + 48$

(a) $v(1) = 16$ feet per second

(b) $0 = -32t + 48$

$t = \dfrac{48}{32} = 1.5$ seconds

(c) $v(2) = -16$ feet per second

39. $f(x) = 2x^2 + 3x - 1$

$\dfrac{f(x + h) - f(x)}{h} = \dfrac{\left[2(x + h)^2 + 3(x + h) - 1\right] - \left(2x^2 + 3x - 1\right)}{h}$

$= \dfrac{2x^2 + 4xh + 2h^2 + 3x + 3h - 1 - 2x^2 - 3x + 1}{h}$

$= \dfrac{h(4x + 2h + 3)}{h}$

$= 4x + 2h + 3, \quad h \neq 0$

41. $y = (x - 3)^2$

A vertical line intersects the graph no more than once, so y *is* a function of x.

43. $x - 4 = y^2$

A vertical line intersects the graph more than once, so y *is not* a function of x.

45. $f(x) = x^2 - 4x - 21$

$x^2 - 4x - 21 = 0$

$(x + 3)(x - 7) = 0$

$x + 3 = 0 \quad$ or $\quad x - 7 = 0$

$x = -3 \qquad\qquad x = 7$

47. $f(x) = \dfrac{8x + 3}{11 - x}$

$\dfrac{8x + 3}{11 - x} = 0$

$8x + 3 = 0$

$x = -\dfrac{3}{8}$

49. $f(x) = x^3 - x^2$

$x^3 - x^2 = 0$

$x^2(x - 1) = 0$

$x^2 = 0 \quad$ or $\quad x - 1 = 0$

$x = 0 \qquad\qquad x = 1$

51. $f(x) = |x| + |x + 1|$

f is increasing on $(0, \infty)$.

f is decreasing on $(-\infty, -1)$.

f is constant on $(-1, 0)$.

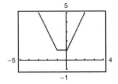

53. $f(x) = -x^2 + 2x + 1$

Relative maximum: $(1, 2)$

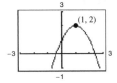

55. $f(x) = x^3 - 6x^4$

Relative maximum: $(0.125, 0.000488) \approx (0.13, 0.00)$

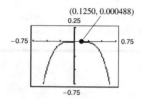

57. $f(x) = -x^2 + 8x - 4$

$$\frac{f(4) - f(0)}{4 - 0} = \frac{12 - (-4)}{4} = 4$$

The average rate of change of f from $x_1 = 0$ to $x_2 = 4$ is 4.

59. $f(x) = 2 - \sqrt{x + 1}$

$$\frac{f(7) - f(3)}{7 - 3} = \frac{(2 - \sqrt{8}) - (2 - 2)}{4}$$

$$= \frac{2 - 2\sqrt{2}}{4} = \frac{1 - \sqrt{2}}{2}$$

The average rate of change of f from $x_1 = 3$ to $x_2 = 7$ is $(1 - \sqrt{2})/2$.

61. $f(x) = x^5 + 4x - 7$

$$f(-x) = (-x)^5 + 4(-x) - 7$$

$$= -x^5 - 4x - 7$$

$$\neq f(x)$$

$$\neq -f(x)$$

Neither even nor odd

63. $f(x) = 2x\sqrt{x^2 + 3}$

$$f(-x) = 2(-x)\sqrt{(-x)^2 + 3}$$

$$= -2x\sqrt{x^2 + 3}$$

$$= -f(x)$$

The function is odd.

65. $f(2) = -6, f(-1) = 3$

Points: $(2, -6), (-1, 3)$

$$m = \frac{3 - (-6)}{-1 - 2} = \frac{9}{-3} = -3$$

$$y - (-6) = -3(x - 2)$$

$$y + 6 = -3x + 6$$

$$y = -3x$$

$$f(x) = -3x$$

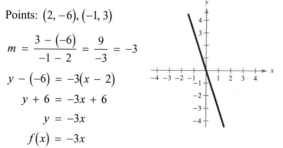

67. $f(x) = x^2 + 5$

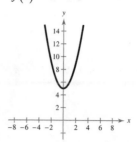

69. $f(x) = -3x^3$

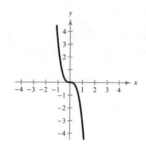

71. $f(x) = -\sqrt{x}$

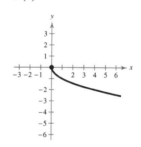

73. $g(x) = \dfrac{3}{x}$

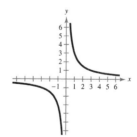

75. $f(x) = [\![x]\!] + 2$

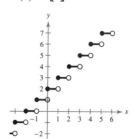

77. $f(x) = \begin{cases} 5x - 3, & x \ge -1 \\ -4x + 5, & x < -1 \end{cases}$

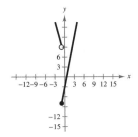

79. Parent function: $f(x) = x^3$

Horizontal shift 4 units to the left and a vertical shift 4 units upward

81. (a) $f(x) = x^2$

(b) $h(x) = x^2 - 9$

Vertical shift 9 units downward

(c)

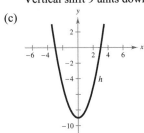

(d) $h(x) = f(x) - 9$

83. $h(x) = -\sqrt{x} + 4$

(a) $f(x) = \sqrt{x}$

(b) Vertical shift 4 units upward, reflection in the *x*-axis

(c)

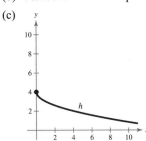

(d) $h(x) = -f(x) + 4$

85. $h(x) = -(x + 2)^2 + 3$

(a) $f(x) = x^2$

(b) Horizontal shift two units to the left, vertical shift 3 units upward, reflection in the *x*-axis.

(c)

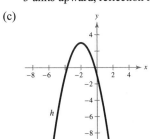

(d) $h(x) = -f(x + 2) + 3$

87. (a) $f(x) = [\![x]\!]$

(b) $h(x) = -[\![x]\!] + 6$

Reflection in the *x*-axis and a vertical shift 6 units upward

(c)

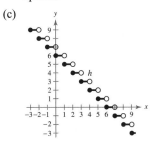

(d) $h(x) = -f(x) + 6$

89. (a) $f(x) = |x|$

(b) $h(x) = -|-x + 4| + 6$

Reflection in both the *x*- and *y*-axes; horizontal shift of 4 unit to the right; vertical shift of 6 units upward

(c)

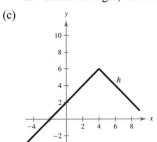

(d) $h(x) = -f(-(x - 4)) + 6 = -f(-x + 4) + 6$

91. (a) $f(x) = [\![x]\!]$

(b) $h(x) = 5[\![x - 9]\!]$

Horizontal shift 9 units to the right and a vertical stretch (each y-value is multiplied by 5)

(c)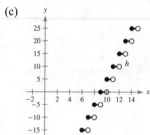

(d) $h(x) = 5f(x - 9)$

93. (a) $f(x) = \sqrt{x}$

(b) $h(x) = -2\sqrt{x - 4}$

Reflection in the x-axis, a vertical stretch (each y-value is multiplied by 2), and a horizontal shift 4 units to the right

(c)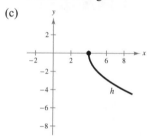

(d) $h(x) = -2f(x - 4)$

95. $f(x) = x^2 + 3, \; g(x) = 2x - 1$

(a) $(f + g)(x) = (x^2 + 3) + (2x - 1) = x^2 + 2x + 2$

(b) $(f - g)(x) = (x^2 + 3) - (2x - 1) = x^2 - 2x + 4$

(c) $(fg)(x) = (x^2 + 3)(2x - 1) = 2x^3 - x^2 + 6x - 3$

(d) $\left(\dfrac{f}{g}\right)(x) = \dfrac{x^2 + 3}{2x - 1}, \;$ Domain: $x \ne \dfrac{1}{2}$

97. $f(x) = \frac{1}{3}x - 3, \; g(x) = 3x + 1$

The domains of $f(x)$ and $g(x)$ are all real numbers.

(a) $(f \circ g)(x) = f(g(x))$

$= f(3x + 1)$

$= \frac{1}{3}(3x + 1) - 3$

$= x + \frac{1}{3} - 3$

$= x - \frac{8}{3}$

Domain: all real numbers

(b) $(g \circ f)(x) = g(f(x))$

$= g\left(\frac{1}{3}x - 3\right)$

$= 3\left(\frac{1}{3}x - 3\right) + 1$

$= x - 9 + 1$

$= x - 8$

Domain: all real numbers

99. $f(x) = x^3 - 4, \; g(x) = \sqrt[3]{x + 7}$

The domains of $f(x)$ and $g(x)$ are all real numbers.

(a) $(f \circ g)(x) = f(g(x))$

$= \left(\sqrt[3]{x + 7}\right)^3 - 4$

$= x + 7 - 4$

$= x + 3$

Domain: all real numbers

(b) $(g \circ f)(x) = g(f(x))$

$= \sqrt[3]{(x^3 - 4) + 7}$

$= \sqrt[3]{x^3 + 3}$

Domain: all real numbers

101. $h(x) = (1 - 2x)^3$

Answer is not unique.

One possibility: Let $f(x) = x^3$ and $g(x) = 1 - 2x$.

$f(g(x)) = f(1 - 2x) = (1 - 2x)^3 = h(x)$.

103. (a) $(r + c)(t) = r(t) + c(t) = 178.8t + 856$

This represents the average annual expenditures for both residential and cellular phone services from 2001 to 2006.

(b)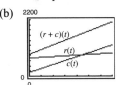

(c) $(r + c)(13) = 178.8(13) + 856 = \$3180.40$

105. $f(x) = 3x + 8$

$y = 3x + 8$

$x = 3y + 8$

$x - 8 = 3y$

$y = \dfrac{x - 8}{3}$

$y = \dfrac{1}{3}(x - 8)$

So $f^{-1}(x) = \dfrac{1}{3}(x - 8)$.

$f\left(f^{-1}(x)\right) = f\left(\dfrac{1}{3}(x - 8)\right) = 3\left(\dfrac{1}{3}(x - 8)\right) + 8 = x - 8 + 8 = x$

$f^{-1}\left(f(x)\right) = f^{-1}(3x + 8) = \dfrac{1}{3}(3x + 8 - 8) = \dfrac{1}{3}(3x) = x$

107. $f(x) = x^3 - 1$

$y = x^3 - 1$

$x = y^3 - 1$

$x + 1 = y^3$

$\sqrt[3]{x + 1} = y$

$f^{-1}(x) = \sqrt[3]{x + 1}$

$f\left(f^{-1}(x)\right) = \left(\sqrt[3]{x + 1}\right)^3 - 1 = x$

$f^{-1}\left(f(x)\right) = \sqrt[3]{(x^3 - 1) + 1} = x$

109. Yes, the function has an inverse because no horizontal lines intersect the graph at more than one point. The function has an inverse.

111. $f(x) = 4 - \frac{1}{3}x$

Yes, the function has an inverse because no horizontal lines intersect the graph at more than one point. The function has an inverse.

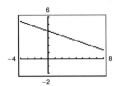

113. $h(t) = \dfrac{2}{t - 3}$

Yes, the function has an inverse because no horizontal lines intersect the graph at more than one point. The function has an inverse.

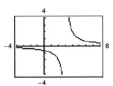

115. (a) $f(x) = \frac{1}{2}x - 3$ (b)

$y = \frac{1}{2}x - 3$

$x = \frac{1}{2}y - 3$

$x + 3 = \frac{1}{2}y$

$2(x + 3) = y$

$f^{-1}(x) = 2x + 6$

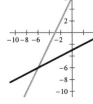

(c) The graph of f^{-1} is the reflection of the graph of f in the line $y = x$.

(d) The domains and ranges of f and f^{-1} are the set of all real numbers.

117. (a) $f(x) = \sqrt{x+1}$

$$y = \sqrt{x+1}$$

$$x = \sqrt{y+1}$$

$$x^2 = y+1$$

$$x^2 - 1 = y$$

$$f^{-1}(x) = x^2 - 1, \ x \geq 0$$

Note: The inverse must have a restricted domain.

(b)

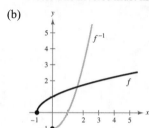

(c) The graph of f^{-1} is the reflection of the graph of f in the line $y = x$.

(d) The domain of f and the range of f^{-1} is $[-1, \infty)$.

The range of f and the domain of f^{-1} is $[0, \infty)$.

119. $f(x) = 2(x-4)^2$ is increasing on $(4, \infty)$.

Let $f(x) = 2(x-4)^2$, $x > 4$ and $y > 0$.

$$y = 2(x-4)^2$$

$$x = 2(y-4)^2, \ x > 0, \ y > 4$$

$$\frac{x}{2} = (y-4)^2$$

$$\sqrt{\frac{x}{2}} = y - 4$$

$$\sqrt{\frac{x}{2}} + 4 = y$$

$$f^{-1}(x) = \sqrt{\frac{x}{2}} + 4, \ x > 0$$

121. False. The graph is reflected in the x-axis, shifted 9 units to the left, then shifted 13 units down ward.

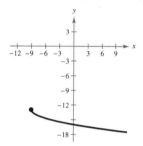

123. A function from a set A to a set B is a relation that assigns to each element x in the set A exactly one element y in the set B.

125. The basic cubic function, $y = x^3$, has the following characteristics. Its domain and range is the set of all real numbers. It is an odd function, thus its graph is symmetric with respect to the origin. It is an increasing function and has an intercept at $(0, 0)$. The cubic function, $f(x) = x^3 - 1$, has the following characteristics. Its domain and range is the set of all real numbers. It is an increasing function with x-intercept at $(1, 0)$ and y-intercept at $(0, -1)$. The graph of $f(x)$ can be obtained by shifting the basic graph of $y = x^3$ downward one unit.

Problem Solving for Chapter 2

1. (a) $W_1 = 0.07S + 2000$

(b) $W_2 = 0.05S + 2300$

(c)

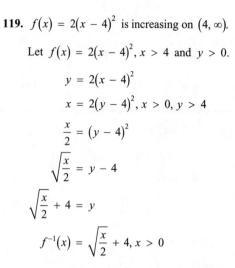

Point of intersection: $(15,000, 3050)$

Both jobs pay the same, $3050, if you sell $15,000 per month.

(d) No. If you think you can sell $20,000 per month, keep your current job with the higher commission rate. For sales over $15,000 it pays more than the other job.

3. (a) Let $f(x)$ and $g(x)$ be two even functions.

Then define $h(x) = f(x) \pm g(x)$.

$h(-x) = f(-x) \pm g(-x)$

$\qquad = f(x) \pm g(x)$ because f and g are even

$\qquad = h(x)$

So, $h(x)$ is also even.

(b) Let $f(x)$ and $g(x)$ be two odd functions.

Then define $h(x) = f(x) \pm g(x)$.

$h(-x) = f(-x) \pm g(-x)$

$\qquad = -f(x) \pm g(x)$ because f and g are odd

$\qquad = -h(x)$

So, $h(x)$ is also odd. $\left(\text{If } f(x) \neq g(x)\right)$

(c) Let $f(x)$ be odd and $g(x)$ be even. Then define $h(x) = f(x) \pm g(x)$.

$h(-x) = f(-x) \pm g(-x)$

$\qquad = -f(x) \pm g(x)$ because f is odd and g is even

$\qquad \neq h(x)$

$\qquad \neq -h(x)$

So, $h(x)$ is neither odd nor even.

5. $f(x) = a_{2n}x^{2n} + a_{2n-2}x^{2n-2} + \cdots + a_2 x^2 + a_0$

$f(-x) = a_{2n}(-x)^{2n} + a_{2n-2}(-x)^{2n-2} + \cdots + a_2(-x)^2 + a_0 = a_{2n}x^{2n} + a_{2n-2}x^{2n-2} + \cdots + a_2 x^2 + a_0 = f(x)$

So, $f(x)$ is even.

7. (a) April 11: 10 hours

April 12: 24 hours

April 13: 24 hours

April 14: $23\dfrac{2}{3}$ hours

Total: $81\dfrac{2}{3}$ hours

(b) Speed $= \dfrac{\text{distance}}{\text{time}} = \dfrac{2100}{81\frac{2}{3}} = \dfrac{180}{7} = 25\dfrac{5}{7}$ mph

(c) $D = -\dfrac{180}{7}t + 3400$

Domain: $0 \le t \le \dfrac{1190}{9}$

Range: $0 \le D \le 3400$

(d)

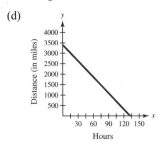

9. (a)–(d) Use $f(x) = 4x$ and $g(x) = x + 6$.

(a) $(f \circ g)(x) = f(x + 6) = 4(x + 6) = 4x + 24$

(b) $(f \circ g)^{-1}(x) = \dfrac{x - 24}{4} = \dfrac{1}{4}x - 6$

(c) $f^{-1}(x) = \dfrac{1}{4}x$

$g^{-1}(x) = x - 6$

(d) $\left(g^{-1} \circ f^{-1}\right)(x) = g^{-1}\!\left(\dfrac{1}{4}x\right) = \dfrac{1}{4}x - 6$

(e) $f(x) = x^3 + 1$ and $g(x) = 2x$

$(f \circ g)(x) = f(2x) = (2x)^3 + 1 = 8x^3 + 1$

$(f \circ g)^{-1}(x) = \sqrt[3]{\dfrac{x - 1}{8}} = \dfrac{1}{2}\sqrt[3]{x - 1}$

$f^{-1}(x) = \sqrt[3]{x - 1}$

$g^{-1}(x) = \dfrac{1}{2}x$

$\left(g^{-1} \circ f^{-1}\right)(x) = g^{-1}\!\left(\sqrt[3]{x - 1}\right) = \dfrac{1}{2}\sqrt[3]{x - 1}$

(f) Answers will vary.

(g) Conjecture: $(f \circ g)^{-1}(x) = \left(g^{-1} \circ f^{-1}\right)(x)$

11. $H(x) = \begin{cases} 1, & x \geq 0 \\ 0, & x < 0 \end{cases}$

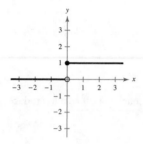

(a) $H(x) - 2$

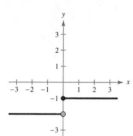

(b) $H(x - 2)$

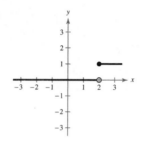

(c) $-H(x)$

(d) $H(-x)$

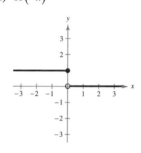

(e) $\frac{1}{2}H(x)$

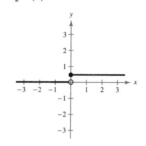

(f) $-H(x - 2) + 2$

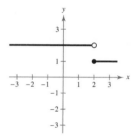

13. $\big(f \circ (g \circ h)\big)(x) = f\big((g \circ h)(x)\big) = f\big(g(h(x))\big) = (f \circ g \circ h)(x)$

$\big((f \circ g) \circ h\big)(x) = (f \circ g)\big(h(x)\big) = f\big(g(h(x))\big) = (f \circ g \circ h)(x)$

15.

x	$f(x)$	$f^{-1}(x)$
-4	—	2
-3	4	1
-2	1	0
-1	0	—
0	-2	-1
1	-3	-2
2	-4	—
3	—	—
4	—	-3

(a)

x	$f\big(f^{-1}(x)\big)$
-4	$f\big(f^{-1}(-4)\big) = f(2) = -4$
-2	$f\big(f^{-1}(-2)\big) = f(0) = -2$
0	$f\big(f^{-1}(0)\big) = f(-1) = 0$
4	$f\big(f^{-1}(4)\big) = f(-3) = 4$

(b)

x	$(f + f^{-1})(x)$
-3	$f(-3) + f^{-1}(-3) = 4 + 1 = 5$
-2	$f(-2) + f^{-1}(-2) = 1 + 0 = 1$
0	$f(0) + f^{-1}(0) = -2 + (-1) = -3$
1	$f(1) + f^{-1}(1) = -3 + (-2) = -5$

(c)

x	$(f \cdot f^{-1})(x)$
-3	$f(-3)f^{-1}(-3) = (4)(1) = 4$
-2	$f(-2)f^{-1}(-2) = (1)(0) = 0$
0	$f(0)f^{-1}(0) = (-2)(-1) = 2$
1	$f(1)f^{-1}(1) = (-3)(-2) = 6$

(d)

x	$\big|f^{-1}(x)\big|$
-4	$\big|f^{-1}(-4)\big| = \big|2\big| = 2$
-3	$\big|f^{-1}(-3)\big| = \big|1\big| = 1$
0	$\big|f^{-1}(0)\big| = \big|-1\big| = 1$
4	$\big|f^{-1}(4)\big| = \big|-3\big| = 3$

Chapter 2 Practice Test

1. Find the equation of the line through $(2, 4)$ and $(3, -1)$.

2. Find the equation of the line with slope $m = 4/3$ and y-intercept $b = -3$.

3. Find the equation of the line through $(4, 1)$ perpendicular to the line $2x + 3y = 0$.

4. If it costs a company \$32 to produce 5 units of a product and \$44 to produce 9 units, how much does it cost to produce 20 units? (Assume that the cost function is linear.)

5. Given $f(x) = x^2 - 2x + 1$, find $f(x - 3)$.

6. Given $f(x) = 4x - 11$, find $\dfrac{f(x) - f(3)}{x - 3}$

7. Find the domain and range of $f(x) = \sqrt{36 - x^2}$.

8. Which equations determine y as a function of x?
   (a) $6x - 5y + 4 = 0$
   (b) $x^2 + y^2 = 9$
   (c) $y^3 = x^2 + 6$

9. Sketch the graph of $f(x) = x^2 - 5$.

10. Sketch the graph of $f(x) = |x + 3|$.

11. Sketch the graph of $f(x) = \begin{cases} 2x + 1, & \text{if } x \geq 0, \\ x^2 - x, & \text{if } x < 0. \end{cases}$

12. Use the graph of $f(x) = |x|$ to graph the following:
    (a) $f(x + 2)$
    (b) $-f(x) + 2$

13. Given $f(x) = 3x + 7$ and $g(x) = 2x^2 - 5$, find the following:
    (a) $(g - f)(x)$
    (b) $(fg)(x)$

14. Given $f(x) = x^2 - 2x + 16$ and $g(x) = 2x + 3$, find $f(g(x))$.

15. Given $f(x) = x^3 + 7$, find $f^{-1}(x)$.

16. Which of the following functions have inverses?

    (a) $f(x) = |x - 6|$

    (b) $f(x) = ax + b, a \neq 0$

    (c) $f(x) = x^3 - 19$

17. Given $f(x) = \sqrt{\dfrac{3 - x}{x}}, 0 < x \leq 3$, find $f^{-1}(x)$.

Exercises 18–20, true or false?

18. $y = 3x + 7$ and $y = \frac{1}{3}x - 4$ are perpendicular.

19. $(f \circ g)^{-1} = g^{-1} \circ f^{-1}$

20. If a function has an inverse, then it must pass both the Vertical Line Test and the Horizontal Line Test.

CHAPTER 3
Polynomial Functions

CHAPTER 3
Polynomial Functions

Section 3.1   Quadratic Functions and Models

You should know the following facts about parabolas.

- $f(x) = ax^2 + bx + c$, $a \neq 0$, is a quadratic function, and its graph is a parabola.

- If $a > 0$, the parabola opens upward and the vertex is the point with the minimum y-value.
  If $a < 0$, the parabola opens downward and the vertex is the point with the maximum y-value.

- The vertex is $\left(-b/2a,\ f(-b/2a) \right)$.

- To find the x-intercepts (if any), solve
  $$ax^2 + bx + c = 0.$$

- The standard form of the equation of a parabola is
  $$f(x) = a(x - h)^2 + k$$
  where $a \neq 0$.
  (a) The vertex is (h, k).
  (b) The axis is the vertical line $x = h$.

---

1. polynomial

3. quadratic; parabola

5. positive; minimum

7. $f(x) = (x - 2)^2$ opens upward and has vertex $(2, 0)$.
Matches graph (e).

8. $f(x) = (x + 4)^2$ opens upward and has vertex $(-4, 0)$.
Matches graph (c).

9. $f(x) = x^2 - 2$ opens upward and has vertex $(0, -2)$.
Matches graph (b).

10. $f(x) = (x + 1)^2 - 2$ opens upward and has vertex $(-1, -2)$. Matches graph (a).

11. $f(x) = 4 - (x - 2)^2 = -(x - 2)^2 + 4$ opens downward and has vertex $(2, 4)$. Matches graph (f).

12. $f(x) = -(x - 4)^2$ opens downward and has vertex $(4, 0)$. Matches graph (d).

13. (a) $y = \frac{1}{2}x^2$ (b) $y = -\frac{1}{8}x^2$ (c) $y = \frac{3}{2}x^2$ (d) $y = -3x^2$

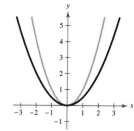

Vertical shrink

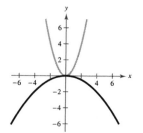

Vertical shrink and reflection in the x-axis

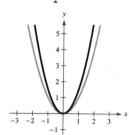

Vertical stretch

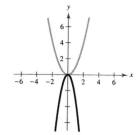

Vertical stretch and reflection in the x-axis

15. (a) $y = (x - 1)^2$

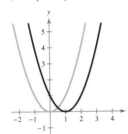

Horizontal shift one unit
to the right

(b) $y = (3x)^2 + 1$

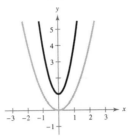

Horizontal shrink
and a vertical shift
one unit upward

(c) $y = \left(\frac{1}{3}x\right)^2 - 3$

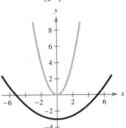

Horizontal stretch and
a vertical shift three
units downward

(d) $y = (x + 3)^2$

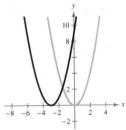

Horizontal shift three
units to the left

17. $f(x) = 1 - x^2$

Vertex: $(0, 1)$

Axis of symmetry: $x = 0$
or the y-axis

Find x-intercepts:

$$1 - x^2 = 0$$
$$1 = x^2$$
$$\pm 1 = x$$

x-intercepts: $(1, 0), (-1, 0)$

19. $f(x) = x^2 + 7$

Vertex: $(0, 7)$

Axis of symmetry: $x = 0$
or the y-axis

Find x-intercepts:

$$x^2 + 7 = 0$$
$$x^2 = -7$$

x-intercepts: none

21. $f(x) = \frac{1}{2}x^2 - 4 = \frac{1}{2}(x - 0)^2 - 4$

Vertex: $(0, -4)$

Axis of symmetry: $x = 0$
or the y-axis

Find x-intercepts:

$$\frac{1}{2}x^2 - 4 = 0$$
$$x^2 = 8$$
$$x = \pm\sqrt{8} = \pm 2\sqrt{2}$$

x-intercepts:
$\left(-2\sqrt{2}, 0\right), \left(2\sqrt{2}, 0\right)$

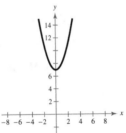

23. $f(x) = (x + 4)^2 - 3$

Vertex: $(-4, -3)$

Axis of symmetry: $x = -4$

Find x-intercepts:

$$0 = (x + 4)^2 - 3$$
$$3 = (x + 4)^2$$
$$\pm\sqrt{3} = x + 4$$
$$-4 \pm \sqrt{3} = x$$

x-intercepts: $\left(-4 + \sqrt{3}, 0\right), \left(-4 - \sqrt{3}, 0\right)$

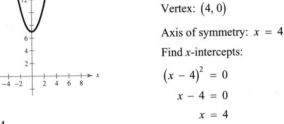

25. $h(x) = x^2 - 8x + 16 = (x - 4)^2$

Vertex: $(4, 0)$

Axis of symmetry: $x = 4$

Find x-intercepts:

$$(x - 4)^2 = 0$$
$$x - 4 = 0$$
$$x = 4$$

x-intercept: $(4, 0)$

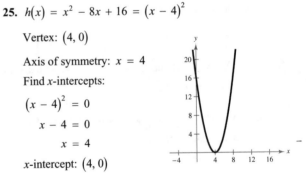

27. $f(x) = x^2 - x + \dfrac{5}{4}$

$$= \left(x^2 - x + \dfrac{1}{4}\right) - \dfrac{1}{4} + \dfrac{5}{4}$$

$$= \left(x - \dfrac{1}{2}\right)^2 + 1$$

Vertex: $\left(\dfrac{1}{2}, 1\right)$

Axis of symmetry: $x = \dfrac{1}{2}$

Find x-intercepts:

$$x^2 - x + \dfrac{5}{4} = 0$$

$$x = \dfrac{1 \pm \sqrt{1 - 5}}{2}$$

Not a real number

No x-intercepts

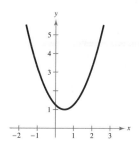

29. $f(x) = -x^2 + 2x + 5$

$$= -\left(x^2 - 2x + 1\right) - (-1) + 5$$

$$= -(x - 1)^2 + 6$$

Vertex: $(1, 6)$

Axis of symmetry: $x = 1$

Find x-intercepts:

$$-x^2 + 2x + 5 = 0$$

$$x^2 - 2x - 5 = 0$$

$$x = \dfrac{2 \pm \sqrt{4 + 20}}{2}$$

$$= 1 \pm \sqrt{6}$$

x-intercepts: $\left(1 - \sqrt{6}, 0\right), \left(1 + \sqrt{6}, 0\right)$

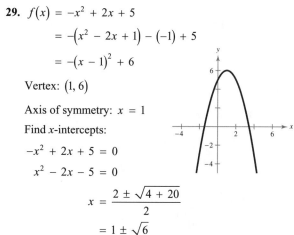

31. $h(x) = 4x^2 - 4x + 21$

$$= 4\left(x^2 - x + \dfrac{1}{4}\right) - 4\left(\dfrac{1}{4}\right) + 21$$

$$= 4\left(x - \dfrac{1}{2}\right)^2 + 20$$

Vertex: $\left(\dfrac{1}{2}, 20\right)$

Axis of symmetry: $x = \dfrac{1}{2}$

Find x-intercepts:

$$4x^2 - 4x + 21 = 0$$

$$x = \dfrac{4 \pm \sqrt{16 - 336}}{2(4)}$$

Not a real number

No x-intercepts

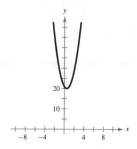

33. $f(x) = \frac{1}{4}x^2 - 2x - 12$

$$= \frac{1}{4}\left(x^2 - 8x + 16\right) - \frac{1}{4}(16) - 12$$

$$= \frac{1}{4}(x - 4)^2 - 16$$

Vertex: $(4, -16)$

Axis of symmetry: $x = 4$

Find x-intercepts:

$$\frac{1}{4}x^2 - 2x - 12 = 0$$

$$x^2 - 8x - 48 = 0$$

$$(x + 4)(x - 12) = 0$$

$$x = -4 \text{ or } x = 12$$

x-intercepts: $(-4, 0), (12, 0)$

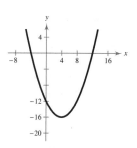

35. $f(x) = -\left(x^2 + 2x - 3\right) = -(x + 1)^2 + 4$

Vertex: $(-1, 4)$

Axis of symmetry: $x = -1$

x-intercepts: $(-3, 0), (1, 0)$

37. $g(x) = x^2 + 8x + 11 = (x + 4)^2 - 5$

Vertex: $(-4, -5)$

Axis of symmetry: $x = -4$

x-intercepts: $(-4 \pm \sqrt{5}, 0)$

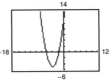

39. $f(x) = 2x^2 - 16x + 31$

$= 2(x - 4)^2 - 1$

Vertex: $(4, -1)$

Axis of symmetry: $x = 4$

x-intercepts: $\left(4 \pm \frac{1}{2}\sqrt{2}, 0\right)$

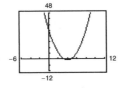

41. $g(x) = \frac{1}{2}(x^2 + 4x - 2) = \frac{1}{2}(x + 2)^2 - 3$

Vertex: $(-2, -3)$

Axis of symmetry: $x = -2$

x-intercepts: $(-2 \pm \sqrt{6}, 0)$

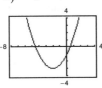

43. $(-1, 4)$ is the vertex.

$y = a(x + 1)^2 + 4$

Because the graph passes through $(1, 0)$,

$0 = a(1 + 1)^2 + 4$

$-4 = 4a$

$-1 = a.$

So, $y = -1(x + 1)^2 + 4 = -(x + 1)^2 + 4.$

45. $(-2, 2)$ is the vertex.

$y = a(x + 2)^2 + 2$

Because the graph passes through $(-1, 0)$,

$0 = a(-1 + 2)^2 + 2$

$-2 = a.$

So, $y = -2(x + 2)^2 + 2.$

47. $(-2, 5)$ is the vertex.

$f(x) = a(x + 2)^2 + 5$

Because the graph passes through $(0, 9)$,

$9 = a(0 + 2)^2 + 5$

$4 = 4a$

$1 = a.$

So, $f(x) = 1(x + 2)^2 + 5 = (x + 2)^2 + 5.$

49. $(1, -2)$ is the vertex.

$f(x) = a(x - 1)^2 - 2$

Because the graph passes through $(-1, 14)$,

$14 = a(-1 - 1)^2 - 2$

$14 = 4a - 2$

$16 = 4a$

$4 = a.$

So, $f(x) = 4(x - 1)^2 - 2.$

51. $(5, 12)$ is the vertex.

$f(x) = a(x - 5)^2 + 12$

Because the graph passes through $(7, 15)$,

$15 = a(7 - 5)^2 + 12$

$3 = 4a \Rightarrow a = \frac{3}{4}.$

So, $f(x) = \frac{3}{4}(x - 5)^2 + 12.$

53. $\left(-\frac{1}{4}, \frac{3}{2}\right)$ is the vertex.

$f(x) = a\left(x + \frac{1}{4}\right)^2 + \frac{3}{2}$

Because the graph passes through $(-2, 0)$,

$0 = a\left(-2 + \frac{1}{4}\right)^2 + \frac{3}{2}$

$-\frac{3}{2} = \frac{49}{16}a \Rightarrow a = -\frac{24}{49}.$

So, $f(x) = -\frac{24}{49}\left(x + \frac{1}{4}\right)^2 + \frac{3}{2}.$

55. $\left(-\frac{5}{2}, 0\right)$ is the vertex.

$f(x) = a\left(x + \frac{5}{2}\right)^2$

Because the graph passes through $\left(-\frac{7}{2}, -\frac{16}{3}\right)$,

$-\frac{16}{3} = a\left(-\frac{7}{2} + \frac{5}{2}\right)^2$

$-\frac{16}{3} = a.$

So, $f(x) = -\frac{16}{3}\left(x + \frac{5}{2}\right)^2.$

57. $y = x^2 - 4x - 5$

x-intercepts: $(5, 0), (-1, 0)$

$0 = x^2 - 4x - 5$

$0 = (x - 5)(x + 1)$

$x = 5 \quad \text{or} \quad x = -1$

59. $f(x) = x^2 - 4x$

x-intercepts: $(0, 0), (4, 0)$

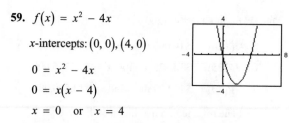

$0 = x^2 - 4x$

$0 = x(x - 4)$

$x = 0$ or $x = 4$

The x-intercepts and the solutions of $f(x) = 0$ are the same.

61. $f(x) = x^2 - 9x + 18$

x-intercepts: $(3, 0), (6, 0)$

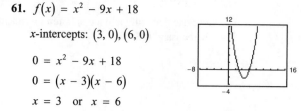

$0 = x^2 - 9x + 18$

$0 = (x - 3)(x - 6)$

$x = 3$ or $x = 6$

The x-intercepts and the solutions of $f(x) = 0$ are the same.

63. $f(x) = 2x^2 - 7x - 30$

x-intercepts: $\left(-\frac{5}{2}, 0\right), (6, 0)$

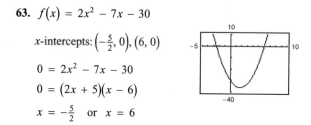

$0 = 2x^2 - 7x - 30$

$0 = (2x + 5)(x - 6)$

$x = -\frac{5}{2}$ or $x = 6$

The x-intercepts and the solutions of $f(x) = 0$ are the same.

65. $f(x) = [x - (-1)](x - 3)$ opens upward

$\quad = (x + 1)(x - 3)$

$\quad = x^2 - 2x - 3$

$g(x) = -[x - (-1)](x - 3)$ opens downward

$\quad = -(x + 1)(x - 3)$

$\quad = -(x^2 - 2x - 3)$

$\quad = -x^2 + 2x + 3$

Note: $f(x) = a(x + 1)(x - 3)$ has x-intercepts $(-1, 0)$ and $(3, 0)$ for all real numbers $a \neq 0$.

67. $f(x) = (x - 0)(x - 10)$ opens upward

$\quad = x^2 - 10x$

$g(x) = -(x - 0)(x - 10)$ opens downward

$\quad = -x^2 + 10x$

Note: $f(x) = a(x - 0)(x - 10) = ax(x - 10)$ has x-intercepts $(0, 0)$ and $(10, 0)$ for all real numbers $a \neq 0$.

69. $f(x) = \left[x - (-3)\right]\left[x - \left(-\frac{1}{2}\right)\right](2)$ opens upward

$\quad = (x + 3)\left(x + \frac{1}{2}\right)(2)$

$\quad = (x + 3)(2x + 1)$

$\quad = 2x^2 + 7x + 3$

$g(x) = -(2x^2 + 7x + 3)$ opens downward

$\quad = -2x^2 - 7x - 3$

Note: $f(x) = a(x + 3)(2x + 1)$ has x-intercepts $(-3, 0)$ and $\left(-\frac{1}{2}, 0\right)$ for all real numbers $a \neq 0$.

71. Let x = the first number and y = the second number. Then the sum is

$x + y = 110 \Rightarrow y = 110 - x$.

The product is $P(x) = xy = x(110 - x) = 110x - x^2$.

$P(x) = -x^2 + 110x$

$\quad = -(x^2 - 110x + 3025 - 3025)$

$\quad = -\left[(x - 55)^2 - 3025\right]$

$\quad = -(x - 55)^2 + 3025$

The maximum value of the product occurs at the vertex of $P(x)$ and is 3025. This happens when $x = y = 55$.

73. Let x = the first number and y = the second number. Then the sum is

$x + 2y = 24 \Rightarrow y = \dfrac{24 - x}{2}$.

The product is $P(x) = xy = x\left(\dfrac{24 - x}{2}\right)$.

$P(x) = \dfrac{1}{2}(-x^2 + 24x)$

$\quad = -\dfrac{1}{2}(x^2 - 24x + 144 - 144)$

$\quad = -\dfrac{1}{2}\left[(x - 12)^2 - 144\right] = -\dfrac{1}{2}(x - 12)^2 + 72$

The maximum value of the product occurs at the vertex of $P(x)$ and is 72. This happens when $x = 12$ and $y = (24 - 12)/2 = 6$. So, the numbers are 12 and 6.

75. $y = -\dfrac{4}{9}x^2 + \dfrac{24}{9}x + 12$

The vertex occurs at $-\dfrac{b}{2a} = \dfrac{-24/9}{2(-4/9)} = 3.$

The maximum height is

$y(3) = -\dfrac{4}{9}(3)^2 + \dfrac{24}{9}(3) + 12 = 16$ feet.

77. $C = 800 - 10x + 0.25x^2 = 0.25x^2 - 10x + 800$

The vertex occurs at $x = -\dfrac{b}{2a} = -\dfrac{-10}{2(0.25)} = 20.$

The cost is minimum when $x = 20$ fixtures.

79. $R(p) = -25p^2 + 1200p$

(a) $R(20) = \$14{,}000$ thousand $= \$14{,}000{,}000$

$R(25) = \$14{,}375$ thousand $= \$14{,}375{,}000$

$R(30) = \$13{,}500$ thousand $= \$13{,}500{,}000$

(b) The revenue is a maximum at the vertex.

$-\dfrac{b}{2a} = \dfrac{-1200}{2(-25)} = 24$

$R(24) = 14{,}400$

The unit price that will yield a maximum revenue of $\$14{,}400$ thousand is $\$24.$

81. (a)

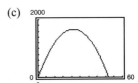

$4x + 3y = 200 \Rightarrow y = \dfrac{1}{3}(200 - 4x) = \dfrac{4}{3}(50 - x)$

$A = 2xy = 2x\left[\dfrac{4}{3}(50 - x)\right] = \dfrac{8}{3}x(50 - x) = \dfrac{8x(50 - x)}{3}$

(b)

x	A
5	600
10	$1066\dfrac{2}{3}$
15	1400
20	1600
25	$1666\dfrac{2}{3}$
30	1600

This area is maximum when $x = 25$ feet and $y = \dfrac{100}{3} = 33\dfrac{1}{3}$ feet.

(c)

This area is maximum when $x = 25$ feet and $y = \dfrac{100}{3} = 33\dfrac{1}{3}$ feet.

(d) $A = \dfrac{8}{3}x(50 - x)$

$= -\dfrac{8}{3}(x^2 - 50x)$

$= -\dfrac{8}{3}(x^2 - 50x + 625 - 625)$

$= -\dfrac{8}{3}\left[(x - 25)^2 - 625\right]$

$= -\dfrac{8}{3}(x - 25)^2 + \dfrac{5000}{3}$

The maximum area occurs at the vertex and is $5000/3$ square feet. This happens when $x = 25$ feet and

$y = \left(200 - 4(25)\right)/3 = 100/3$ feet. The dimensions are $2x = 50$ feet by $33\dfrac{1}{3}$ feet.

(e) They are all identical.

$x = 25$ feet and $y = 33\dfrac{1}{3}$ feet

83. (a)   Revenue $=$ (number of tickets sold)(price per ticket)

Let $y=$ attendance, or the number of tickets sold.

$$m = -100, (20, 1500)$$
$$y - 1500 = -100(x - 20)$$
$$y - 1500 = -100x + 2000$$
$$y = -100x + 3500$$
$$R(x) = (y)(x)$$
$$R(x) = (-100x + 3500)(x)$$
$$R(x) = -100x^2 + 3500x$$

(b)   The revenue is at a maximum at the vertex.

$$-\frac{b}{2a} = \frac{-3500}{2(-100)} = 17.5$$

$$R(17.5) = -100(17.5)^2 + 3500(17.5) = \$30{,}625$$

A ticket price of $17.50 will yield a maximum revenue of $30,625.

85. (a)

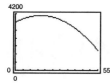

(b)   The maximum annual consumption occurs at the point $(16.9, 4074.813)$.

4075 cigarettes

$1966 \rightarrow t = 16$

The maximum consumption occurred in 1966. After that year, the consumption decreases. It is likely that the warning was responsible for the decrease in consumption.

(c)   Annual consumption per smoker $= \dfrac{\text{Annual consumption in 2005} \; \cdot \; \text{total population}}{\text{total number of smokers in 2005}}$

$$= \frac{1487.9(296{,}329{,}000)}{59{,}858{,}458}$$

$$= 7365.8$$

About 7366 cigarettes per smoker annually

Daily consumption per smoker $= \dfrac{\text{Number of cigarettes per year}}{\text{Number of days per year}}$

$$= \frac{7366}{365}$$

$$\approx 20.2$$

About 20 cigarettes per day

87. True. The equation $-12x^2 - 1 = 0$ has no real solution, so the graph has no x-intercepts.

89. True. The negative leading coefficient causes the parabola to open downward, making the vertex the maximum point on the graph.

91. $f(x) = -x^2 + bx - 75$, maximum value: 25

The maximum value, 25, is the y-coordinate of the vertex.

Find the x-coordinate of the vertex:

$$x = -\frac{b}{2a} = -\frac{b}{2(-1)} = \frac{b}{2}$$

$$f(x) = -x^2 + bx - 75$$

$$f\left(\frac{b}{2}\right) = -\left(\frac{b}{2}\right)^2 + b\left(\frac{b}{2}\right) - 75$$

$$25 = -\frac{b^2}{4} + \frac{b^2}{2} - 75$$

$$100 = \frac{b^2}{4}$$

$$400 = b^2$$

$$\pm 20 = b$$

93. $f(x) = x^2 + bx + 26$, minimum value: 10

The minimum value, 10, is the y-coordinate of the vertex.

Find the x-coordinate of the vertex:

$$x = -\frac{b}{2a} = -\frac{b}{2(1)} = -\frac{b}{2}$$

$$f(x) = x^2 + bx + 26$$

$$f\left(-\frac{b}{2}\right) = \left(-\frac{b}{2}\right)^2 + b\left(-\frac{b}{2}\right) + 26$$

$$10 = \frac{b^2}{4} - \frac{b^2}{2} + 26$$

$$-16 = -\frac{b^2}{4}$$

$$64 = b^2$$

$$\pm 8 = b$$

95. $f(x) = ax^2 + bx + c$

$$= a\left(x^2 + \frac{b}{a}x\right) + c$$

$$= a\left(x^2 + \frac{b}{a}x + \frac{b^2}{4a^2} - \frac{b^2}{4a^2}\right) + c$$

$$= a\left(x + \frac{b}{2a}\right)^2 - \frac{b^2}{4a} + c$$

$$= a\left(x + \frac{b}{2a}\right)^2 + \frac{4ac - b^2}{4a}$$

$$f\left(-\frac{b}{2a}\right) = a\left(\frac{b^2}{4a^2}\right) + b\left(-\frac{b}{2a}\right) + c$$

$$= \frac{b^2}{4a} - \frac{b^2}{2a} + c$$

$$= \frac{b^2 - 2b^2 + 4ac}{4a} = \frac{4ac - b^2}{4a}$$

So, the vertex occurs at $\left(-\dfrac{b}{2a}, \dfrac{4ac - b^2}{4a}\right) = \left(-\dfrac{b}{2a}, f\left(-\dfrac{b}{2a}\right)\right)$.

97. (a)

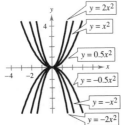

If the a-value is positive, the graph opens upward. If the a-value is negative, the graph opens downward. When the a-value is an integer, the graph is skinnier than $y = \pm x^2$. When the a-value is a fraction, the graph is wider than $y = \pm x^2$.

(b)

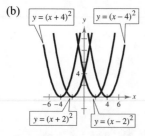

Changing the value of h moves the parabola left or right on the x-axis. When h is positive, the graph moves h units to the left of the origin. When h is negative, the graph moves h units to the right of the origin.

(c)

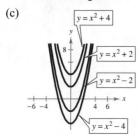

Changing the k-value moves the parabola up or down on the y-axis. When k is positive, the graph moves up k units from the origin. When k is negative, the graph moves down k units from the origin.

99. Yes. A graph of a quadratic equation whose vertex is on the x-axis has only one x-intercept.

Section 3.2   Polynomial Functions of Higher Degree

You should know the following basic principles about polynomials

- $f(x) = a_n x^n + a_{n-1} x^{n-1} + \cdots + a_2 x^2 + a_1 x + a_0, a_n \neq 0,$ is a polynomial function of degree n.

- If f is of odd degree and
    (a) $a_n > 0$, then
        1. $f(x) \to \infty$ as $x \to \infty$.
        2. $f(x) \to -\infty$ as $x \to -\infty$.
    (b) $a_n < 0$, then
        1. $f(x) \to -\infty$ as $x \to \infty$.
        2. $f(x) \to \infty$ as $x \to -\infty$.

- If f is of even degree and
    (a) $a_n > 0$, then
        1. $f(x) \to \infty$ as $x \to \infty$.
        2. $f(x) \to \infty$ as $x \to -\infty$.
    (b) $a_n < 0$, then
        1. $f(x) \to -\infty$ as $x \to \infty$.
        2. $f(x) \to -\infty$ as $x \to -\infty$.

—*continued*—

■ The following are equivalent for a polynomial function.

(a) $x = a$ is a zero of a function.

(b) $x = a$ is a solution of the polynomial equation $f(x) = 0$.

(c) $(x - a)$ is a factor of the polynomial.

(d) $(a, 0)$ is an x-intercept of the graph of f.

■ A polynomial of degree n has at most n distinct zeros and at most $n - 1$ turning points.

■ A factor $(x - a)^k$, $k > 1$, yields a repeated zero of $x = a$ of multiplicity k.

(a) If k is odd, the graph crosses the x-axis at $x = a$.

(b) If k is even, the graph just touches the x-axis at $x = a$.

■ If f is a polynomial function such that $a < b$ and $f(a) \neq f(b)$, then f takes on every value between

$f(a)$ and $f(b)$ in the interval $[a, b]$.

■ If you can find a value where a polynomial is positive and another value where it is negative, then there is at least one real zero between the values.

---

1. continuous

3. x^n

5. (a) solution; (b) $(x - a)$; (c) x-intercept

7. standard

9. $f(x) = -2x + 3$ is a line with y-intercept $(0, 3)$.
Matches graph (c).

10. $f(x) = x^2 - 4x$ is a parabola with intercepts $(0, 0)$ and $(4, 0)$ and opens upward. Matches graph (g).

11. $f(x) = -2x^2 - 5x$ is a parabola with x-intercepts $(0, 0)$ and $\left(-\frac{5}{2}, 0\right)$ and opens downward. Matches graph (h).

12. $f(x) = 2x^3 - 3x + 1$ has intercepts $(0, 1), (1, 0), \left(-\frac{1}{2} - \frac{1}{2}\sqrt{3}, 0\right)$ and $\left(-\frac{1}{2} + \frac{1}{2}\sqrt{3}, 0\right)$.
Matches graph (f).

13. $f(x) = -\frac{1}{4}x^4 + 3x^2$ has intercepts $(0, 0)$ and $\left(\pm 2\sqrt{3}, 0\right)$. Matches graph (a).

14. $f(x) = -\frac{1}{3}x^3 + x^2 - \frac{4}{3}$ has y-intercept $\left(0, -\frac{4}{3}\right)$.
Matches graph (e).

15. $f(x) = x^4 + 2x^3$ has intercepts $(0, 0)$ and $(-2, 0)$.
Matches graph (d).

16. $f(x) = \frac{1}{5}x^5 - 2x^3 + \frac{9}{5}x$ has intercepts $(0, 0), (1, 0), (-1, 0), (3, 0), (-3, 0)$. Matches graph (b).

17. $y = x^3$

(a) $f(x) = (x - 4)^3$

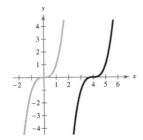

Horizontal shift four units to the right

(b) $f(x) = x^3 - 4$

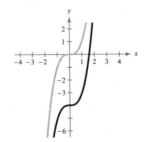

Vertical shift four units downward

(c) $f(x) = -\frac{1}{4}x^3$

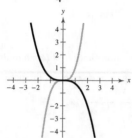

Reflection in the *x*-axis and a vertical shrink
(each *y*-value is multiplied by $\frac{1}{4}$)

(d) $f(x) = (x - 4)^3 - 4$

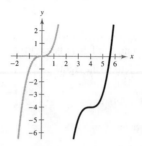

Horizontal shift four units to the right and
vertical shift four units downward

19. $y = x^4$

(a) $f(x) = (x + 3)^4$

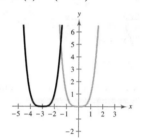

Horizontal shift three units
to the left

(b) $f(x) = x^4 - 3$

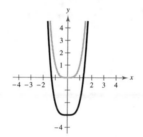

Vertical shift three units
downward

(c) $f(x) = 4 - x^4$

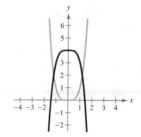

Reflection in the *x*-axis and then
a vertical shift four units upward

(d) $f(x) = \frac{1}{2}(x - 1)^4$

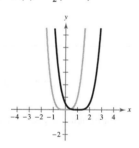
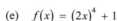

Horizontal shift one unit to
the right and a vertical shrink
(each *y*-value is multiplied by $\frac{1}{2}$)

(e) $f(x) = (2x)^4 + 1$

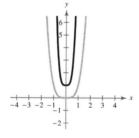

Vertical shift one unit upward
and a horizontal shrink (each
y-value is multiplied by 16)

(f) $f(x) = \left(\frac{1}{2}x\right)^4 - 2$

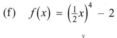
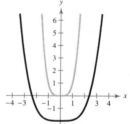

Vertical shift two units downward
and a horizontal stretch (each *y*-value
is multipied by $\frac{1}{16}$)

21. $f(x) = \frac{1}{5}x^3 + 4x$

Degree: 3

Leading coefficient: $\frac{1}{5}$

The degree is odd and the leading coefficient is positive.
The graph falls to the left and rises to the right.

23. $g(x) = 5 - \frac{7}{2}x - 3x^2$

Degree: 2

Leading coefficient: -3

The degree is even and the leading coefficient is
negative. The graph falls to the left and falls to the right.

25. $f(x) = -2.1x^5 + 4x^3 - 2$

Degree: 5

Leading coefficient: -2.1

The degree is odd and the leading coefficient is negative.
The graph rises to the left and falls to the right.

27. $f(x) = 6 - 2x + 4x^2 - 5x^3$

Degree: 3

Leading coefficient: -5

The degree is odd and the leading coefficient is negative.
The graph rises to the left and falls to the right.

29. $f(x) = -\dfrac{3}{4}(t^2 - 3t + 6)$

Degree: 2

Leading coefficient: $-\dfrac{3}{4}$

The degree is even and the leading coefficient is
negative. The graph falls to the left and falls to the right.

31. $f(x) = 3x^3 - 9x + 1;\ g(x) = 3x^3$

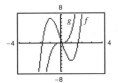

33. $f(x) = -(x^4 - 4x^3 + 16x);\ g(x) = -x^4$

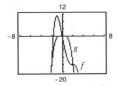

35. $f(x) = x^2 - 36$

(a) $0 = x^2 - 36$

$0 = (x + 6)(x - 6)$

$x + 6 = 0 \qquad x - 6 = 0$

$x = -6 \qquad\ \ x = 6$

Zeros: ± 6

(b) Each zero has a multiplicity of one (odd
multiplicity).

Turning points: 1 (the vertex of the parabola)

(c)

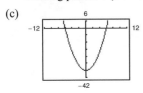

37. $h(t) = t^2 - 6t + 9$

(a) $0 = t^2 - 6t + 9 = (t - 3)^2$

Zero: $t = 3$

(b) $t = 3$ has a multiplicity of 2 (even multiplicity).

Turning points: 1 (the vertex of the parabola)

(c)

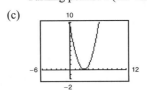

39. $f(x) = \frac{1}{3}x^2 + \frac{1}{3}x - \frac{2}{3}$

(a) $0 = \frac{1}{3}x^2 + \frac{1}{3}x - \frac{2}{3}$

$= \frac{1}{3}(x^2 + x - 2)$

$= \frac{1}{3}(x + 2)(x - 1)$

Zeros: $x = -2,\ x = 1$

(b) Each zero has a multiplicity of 1 (odd multiplicity).

Turning points: 1 (the vertex of the parabola)

(c)

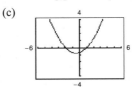

41. $f(x) = 3x^3 - 12x^2 + 3x$

(a) $0 = 3x^3 - 12x^2 + 3x = 3x(x^2 - 4x + 1)$

Zeros: $x = 0,\ x = 2 \pm \sqrt{3}$ (by the Quadratic
Formula)

(b) Each zero has a multiplicity of 1 (odd multiplicity).

Turning points: 2

(c)

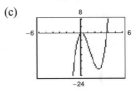

43. $f(t) = t^3 - 8t^2 + 16t$

 (a) $0 = t^3 - 8t^2 + 16t$

   $0 = t(t^2 - 8t + 16)$

   $0 = t(t - 4)(t - 4)$

   $t = 0 \quad t - 4 = 0 \quad t - 4 = 0$

   $t = 0 \qquad t = 4 \qquad t = 4$

   Zeros: $t = 0, t = 4$

 (b) The multiplicity of $t = 0$ is 1 (odd multiplicity).

   The multiplicity of $t = 4$ is 2 (even multiplicity).

   Turning points: 2

 (c)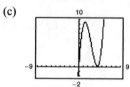

45. $g(t) = t^5 - 6t^3 + 9t$

 (a) $0 = t^5 - 6t^3 + 9t = t(t^4 - 6t^2 + 9) = t(t^2 - 3)^2$

   $= t(t + \sqrt{3})^2(t - \sqrt{3})^2$

   Zeros: $t = 0, t = \pm\sqrt{3}$

 (b) $t = 0$ has a multiplicity of 1 (odd multiplicity).

   $t = \pm\sqrt{3}$ each have a multiplicity of 2 (even multiplicity).

   Turning points: 4

 (c)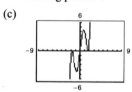

47. $f(x) = 3x^4 + 9x^2 + 6$

 (a) $0 = 3x^4 + 9x^2 + 6$

   $0 = 3(x^4 + 3x^2 + 2)$

   $0 = 3(x^2 + 1)(x^2 + 2)$

   No real zeros

 (b) Turning points: 1

 (c)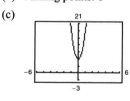

49. $g(x) = x^3 + 3x^2 - 4x - 12$

 (a) $0 = x^3 + 3x^2 - 4x - 12 = x^2(x + 3) - 4(x + 3)$

   $= (x^2 - 4)(x + 3) = (x - 2)(x + 2)(x + 3)$

   Zeros: $x = \pm2, x = -3$

 (b) Each zero has a multiplicity of 1 (odd multiplicity).

   Turning points: 2

 (c)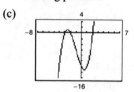

51. $y = 4x^3 - 20x^2 + 25x$

 (a)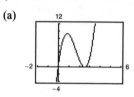

 (b) x-intercepts: $(0, 0), \left(\frac{5}{2}, 0\right)$

 (c) $0 = 4x^3 - 20x^2 + 25x$

   $0 = x(4x^2 - 20x + 25)$

   $0 = x(2x - 5)^2$

   $x = 0, \frac{5}{2}$

 (d) The solutions are the same as the x-coordinates of the x-intercepts.

53. $y = x^5 - 5x^3 + 4x$

 (a)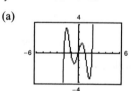

 (b) x-intercepts: $(0, 0), (\pm1, 0), (\pm2, 0)$

 (c) $0 = x^5 - 5x^3 + 4x$

   $0 = x(x^2 - 1)(x^2 - 4)$

   $0 = x(x + 1)(x - 1)(x + 2)(x - 2)$

   $x = 0, \pm1, \pm2$

 (d) The solutions are the same as the x-coordinates of the x-intercepts.

55. $f(x) = (x - 0)(x - 8)$

$\qquad = x^2 - 8x$

Note: $f(x) = ax(x - 8)$ has zeros 0 and 8 for all real numbers $a \neq 0$.

57. $f(x) = (x - 2)(x + 6)$

$\qquad = x^2 + 4x - 12$

Note: $f(x) = a(x - 2)(x + 6)$ has zeros 2 and -6 for all real numbers $a \neq 0$.

59. $f(x) = (x - 0)(x + 4)(x + 5)$

$\qquad = x(x^2 + 9x + 20)$

$\qquad = x^3 + 9x^2 + 20x$

Note: $f(x) = ax(x + 4)(x + 5)$ has zeros $0, -4,$ and -5 for all real numbers $a \neq 0$.

61. $f(x) = (x - 4)(x + 3)(x - 3)(x - 0)$

$\qquad = (x - 4)(x^2 - 9)x$

$\qquad = x^4 - 4x^3 - 9x^2 + 36x$

Note: $f(x) = a(x^4 - 4x^3 - 9x^2 + 36x)$ has zeros $4, -3, 3,$ and 0 for all real numbers $a \neq 0$.

63. $f(x) = \left[x - \left(1 + \sqrt{3}\right)\right]\left[x - \left(1 - \sqrt{3}\right)\right]$

$\qquad = \left[(x - 1) - \sqrt{3}\right]\left[(x - 1) + \sqrt{3}\right]$

$\qquad = (x - 1)^2 - \left(\sqrt{3}\right)^2$

$\qquad = x^2 - 2x + 1 - 3$

$\qquad = x^2 - 2x - 2$

Note: $f(x) = a(x^2 - 2x - 2)$ has zeros $1 + \sqrt{3}$ and $1 - \sqrt{3}$ for all real numbers $a \neq 0$.

65. $f(x) = (x + 3)(x + 3)$

$\qquad = x^2 + 6x + 9$

Note: $f(x) = a(x^2 + 6x + 9), a \neq 0,$ has degree 2 and zero $x = -3$.

67. $f(x) = (x - 0)(x + 5)(x - 1)$

$\qquad = x(x^2 + 4x - 5)$

$\qquad = x^3 + 4x^2 - 5x$

Note: $f(x) = ax(x^2 + 4x - 5), a \neq 0,$ has degree 3 and zeros $x = 0, -5,$ and 1.

69. $f(x) = (x - 0)\left(x - \sqrt{3}\right)\left(x - \left(-\sqrt{3}\right)\right)$

$\qquad = x\left(x - \sqrt{3}\right)\left(x + \sqrt{3}\right) = x^3 - 3x$

Note: $f(x) = a(x^3 - 3x), a \neq 0,$ has degree 3 and zeros $x = 0, \sqrt{3},$ and $-\sqrt{3}$.

71. $f(x) = \left(x - (-5)\right)^2(x - 1)(x - 2) = x^4 + 7x^3 - 3x^2 - 55x + 50$

or $f(x) = \left(x - (-5)\right)(x - 1)^2(x - 2) = x^4 + x^3 - 15x^2 + 23x - 10$

or $f(x) = \left(x - (-5)\right)(x - 1)(x - 2)^2 = x^4 - 17x^2 + 36x - 20$

Note: Any nonzero scalar multiple of these functions would also have degree 4 and zeros $x = -5, 1,$ and 2.

73. $f(x) = x^4(x + 4) = x^5 + 4x^4$

or $f(x) = x^3(x + 4)^2 = x^5 + 8x^4 + 16x^3$

or $f(x) = x^2(x + 4)^3 = x^5 + 12x^4 + 48x^3 + 64x^2$

or $f(x) = x(x + 4)^4 = x^5 + 16x^4 + 96x^3 + 256x^2 + 256x$

Note: Any nonzero scalar multiple of these functions would also have degree 5 and zeros $x = 0$ and -4.

75. $f(x) = x^3 - 25x = x(x + 5)(x - 5)$

(a) Falls to the left; rises to the right

(b) Zeros: $0, -5, 5$

(c)

x	-2	-1	0	1	2
$f(x)$	42	24	0	-24	-42

(d)

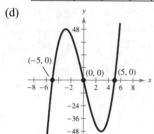

77. $f(t) = \frac{1}{4}(t^2 - 2t + 15) = \frac{1}{4}(t - 1)^2 + \frac{7}{2}$

(a) Rises to the left; rises to the right

(b) No real zeros (no x-intercepts)

(c)

t	-1	0	1	2	3
$f(t)$	4.5	3.75	3.5	3.75	4.5

(d) The graph is a parabola with vertex $\left(1, \frac{7}{2}\right)$.

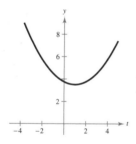

79. $f(x) = x^3 - 2x^2 = x^2(x - 2)$

(a) Falls to the left; rises to the right

(b) Zeros: $0, 2$

(c)

x	-1	0	$\frac{1}{2}$	1	2	3
$f(x)$	-3	0	$-\frac{3}{8}$	-1	0	9

(d)

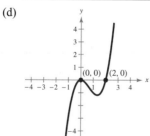

81. $f(x) = 3x^3 - 15x^2 + 18x = 3x(x - 2)(x - 3)$

(a) Falls to the left; rises to the right

(b) Zeros: $0, 2, 3$

(c)

x	0	1	2	2.5	3	3.5
$f(x)$	0	6	0	-1.875	0	7.875

(d)

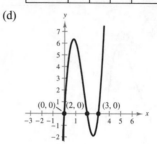

83. $f(x) = -5x^2 - x^3 = -x^2(5 + x)$

(a) Rises to the left; falls to the right

(b) Zeros: $0, -5$

(c)

x	-5	-4	-3	-2	-1	0	1
$f(x)$	0	-16	-18	-12	-4	0	-6

(d)

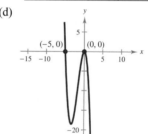

85. $f(x) = x^2(x - 4)$

(a) Falls to the left; rises to the right

(b) Zeros: $0, 4$

(c)

x	-1	0	1	2	3	4	5
$f(x)$	-5	0	-3	-8	-9	0	25

(d)

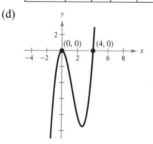

87. $g(t) = -\frac{1}{4}(t-2)^2(t+2)^2$

(a) Falls to the left; falls to the right

(b) Zeros: 2, − 2

(c)

t	−3	−2	−1	0	1	2	3
$g(t)$	$-\frac{25}{4}$	0	$-\frac{9}{4}$	−4	$-\frac{9}{4}$	0	$-\frac{25}{4}$

(d)

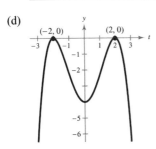

89. $f(x) = x^3 - 16x = x(x-4)(x+4)$

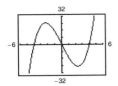

Zeros: 0 of multiplicity 1; 4 of multiplicity 1; and −4 of multiplicity 1

91. $g(x) = \frac{1}{5}(x+1)^2(x-3)(2x-9)$

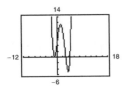

Zeros: −1 of multiplicity 2; 3 of multiplicity 1; $\frac{9}{2}$ of multiplicity 1

93. $f(x) = x^3 - 3x^2 + 3$

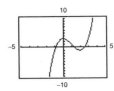

The function has three zeros.
They are in the intervals
$[-1, 0], [1, 2],$ and $[2, 3]$. They
are $x \approx -0.879, 1.347, 2.532.$

x	y
−3	−51
−2	−17
−1	−1
0	3
1	1
2	−1
3	3
4	19

95. $g(x) = 3x^4 + 4x^3 - 3$

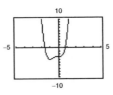

The function has two zeros.
They are in the intervals
$[-2, -1]$ and $[0, 1]$. They
are $x \approx -1.585, 0.779.$

x	y
−4	509
−3	132
−2	13
−1	−4
0	−3
1	4
2	77
3	348

97. (a) Volume $= l \cdot w \cdot h$

height $= x$

length $=$ width $= 36 - 2x$

So, $V(x) = (36 - 2x)(36 - 2x)(x) = x(36 - 2x)^2.$

(b) Domain: $0 < x < 18$

The length and width must be positive.

(c)

Box Height	Box Width	Box Volume, V
1	$36 - 2(1)$	$1[36 - 2(1)]^2 = 1156$
2	$36 - 2(2)$	$2[36 - 2(2)]^2 = 2048$
3	$36 - 2(3)$	$3[36 - 2(3)]^2 = 2700$
4	$36 - 2(4)$	$4[36 - 2(4)]^2 = 3136$
5	$36 - 2(5)$	$5[36 - 2(5)]^2 = 3380$
6	$36 - 2(6)$	$6[36 - 2(6)]^2 = 3456$
7	$36 - 2(7)$	$7[36 - 2(7)]^2 = 3388$

The volume is a maximum of 3456 cubic inches
when the height is 6 inches and the length and width
are each 24 inches. So the dimensions are
$6 \times 24 \times 24$ inches.

(d)

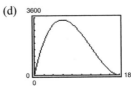

The maximum point on the graph occurs at $x = 6$.
This agrees with the maximum found in part (c).

99. (a) $A = l \cdot w = (12 - 2x)(x) = -2x^2 + 12x$

(b) 16 feet = 192 inches

$V = l \cdot w \cdot h$

$= (12 - 2x)(x)(192)$

$= -384x^2 + 2304x$

(c) Because x and $12 - 2x$ cannot be negative, we have $0 < x < 6$ inches for the domain.

(d)

x	V
0	0
1	1920
2	3072
3	3456
4	3072
5	1920
6	0

When $x = 3$, the volume is a maximum with $V = 3456$ in.^{3}. The dimensions of the gutter cross-section are 3 inches × 6 inches × 3 inches.

(e)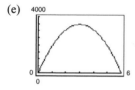

Maximum: (3, 3456)

The maximum value is the same.

(f) No. The volume is a product of the constant length and the cross-sectional area. The value of x would remain the same; only the value of V would change if the length was changed.

101. (a)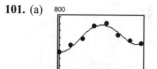

(b) The model fits the data well.

(c) Relative minima: $(0.21, 300.54)$, $(6.62, 410.74)$

Relative maximum: $(3.62, 681.72)$

(d) Increasing: $(0.21, 3.62)$, $(6.62, 7)$

Decreasing; $(0, 0.21)$, $(3.62, 6.62)$

(e) Answers will vary.

103. $G = -0.003t^3 + 0.137t^2 + 0.458t - 0.839, 2 \le t \le 34$

(a)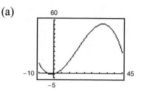

(b) The tree is growing most rapidly at $t \approx 15$.

(c) $y = -0.009t^2 + 0.274t + 0.458$

$-\dfrac{b}{2a} = \dfrac{-0.274}{2(-0.009)} \approx 15.222$

$y(15.222) \approx 2.543$

Vertex $\approx (15.22, 2.54)$

(d) The x-value of the vertex in part (c) is approximately equal to the value found in part (b).

105. False. A fifth-degree polynomial can have at most four turning points.

107. True. A polynomial of degree 7 with a negative leading coefficient rises to the left and falls to the right.

109. $f(x) = x^4$; $f(x)$ is even.

(a) $g(x) = f(x) + 2$

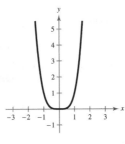

Vertical shift two units upward

$g(-x) = f(-x) + 2$
$\quad\quad\;\; = f(x) + 2$
$\quad\quad\;\; = g(x)$

Even

(b) $g(x) = f(x + 2)$

Horizontal shift two units to the left

Neither odd nor even

(c) $g(x) = f(-x) = (-x)^4 = x^4$

Reflection in the y-axis. The graph looks the same.

Even

(d) $g(x) = -f(x) = -x^4$

Reflection in the x-axis

Even

(e) $g(x) = f\left(\frac{1}{2}x\right) = \frac{1}{16}x^4$

Horizontal stretch

Even

(f) $g(x) = \frac{1}{2}f(x) = \frac{1}{2}x^4$

Vertical shrink

Even

(g) $g(x) = f\left(x^{3/4}\right) = \left(x^{3/4}\right)^4 = x^3, x \geq 0$

Neither odd nor even

(h) $g(x) = (f \circ f)(x) = f(f(x)) = f(x^4) = \left(x^4\right)^4 = x^{16}$

Even

111. (a)

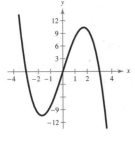

Zeros: 3

Relative minimum: 1

Relative maximum: 1

The number of zeros is the same as the degree and the number of extrema is one less than the degree.

(b)

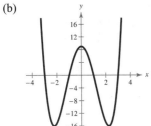

Zeros: 4

Relative minima: 2

Relative maximum: 1

The number of zeros is the same as the degree and the number of extrema is one less than the degree.

(c)

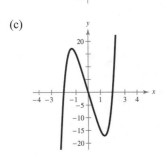

Zeros: 3

Relative minimum: 1

Relative maximum: 1

The number of zeros and the number of extrema are both less than the degree.

Section 3.3   Polynomial and Synthetic Division

You should know the following basic techniques and principles of polynomial division.

- The Division Algorithm (Long Division of Polynomials)

- Synthetic Division

- $f(k)$ is equal to the remainder of $f(x)$ divided by $(x - k)$ (the Remainder Theorem).

- $f(k) = 0$ if and only if $(x - k)$ is a factor of $f(x)$.

---

1. $f(x)$ is the dividend; $d(x)$ is the divisor: $q(x)$ is the quotient: $r(x)$ is the remainder

3. improper

5. Factor

7. $y_1 = \dfrac{x^2}{x + 2}$ and $y_2 = x - 2 + \dfrac{4}{x + 2}$

$$
\begin{array}{r}
x - 2 \\
x + 2 \overline{)x^2 + 0x + 0} \\
\underline{x^2 + 2x} \\
-2x + 0 \\
\underline{-2x - 4} \\
4
\end{array}
$$

So, $\dfrac{x^2}{x + 2} = x - 2 + \dfrac{4}{x + 2}$ and $y_1 = y_2$.

9. $y_1 = \dfrac{x^2 + 2x - 1}{x + 3}$, $y_2 = x - 1 + \dfrac{2}{x + 3}$

(a) and (b)

(c)
$$
\begin{array}{r}
x - 1 \\
x + 3 \overline{)x^2 + 2x - 1} \\
\underline{x^2 + 3x} \\
-x - 1 \\
\underline{-x - 3} \\
2
\end{array}
$$

So, $\dfrac{x^2 + 2x - 1}{x + 3} = x - 1 + \dfrac{2}{x + 3}$ and $y_1 = y_2$.

11.
$$
\begin{array}{r}
2x + 4 \\
x + 3 \overline{)2x^2 + 10x + 12} \\
\underline{2x^2 + 6x} \\
4x + 12 \\
\underline{4x + 12} \\
0
\end{array}
$$

$\dfrac{2x^2 + 10x + 12}{x + 3} = 2x + 4, x \neq 3$

13.
$$
\begin{array}{r}
x^2 - 3x + 1 \\
4x + 5 \overline{)4x^3 - 7x^2 - 11x + 5} \\
\underline{4x^3 + 5x^2} \\
-12x^2 - 11x \\
\underline{-12x^2 - 15x} \\
4x + 5 \\
\underline{4x + 5} \\
0
\end{array}
$$

$\dfrac{4x^3 - 7x^2 - 11x + 5}{4x + 5} = x^2 - 3x + 1, x \neq -\dfrac{5}{4}$

15.
$$
\begin{array}{r}
x^3 + 3x^2 \qquad -1 \\
x + 2 \overline{)x^4 + 5x^3 + 6x^2 - x - 2} \\
\underline{x^4 + 2x^3} \\
3x^3 + 6x^2 \\
\underline{3x^3 + 6x^2} \\
-x - 2 \\
\underline{-x - 2} \\
0
\end{array}
$$

$\dfrac{x^4 + 5x^3 + 6x^2 - x - 2}{x + 2} = x^3 + 3x^2 - 1, x \neq -2$

17.
$$
\begin{array}{r}
x^2 + 3x + 9 \\
x - 3 \overline{)x^3 + 0x^2 + 0x - 27} \\
\underline{x^3 - 3x^2} \\
3x^2 + 0x \\
\underline{3x^2 - 9x} \\
9x - 27 \\
\underline{9x - 27} \\
0
\end{array}
$$

$\dfrac{x^3 - 27}{x - 3} = x^2 + 3x + 9, x \neq 3$

19.

$$
\begin{array}{r}
7 \\
x+2\overline{\smash{\big)}\,7x+\ 3} \\
\underline{7x+14} \\
-11
\end{array}
$$

$$\frac{7x+3}{x+2} = 7 - \frac{11}{x+2}$$

21.

$$
\begin{array}{r}
x \\
x^2+0x+1\overline{\smash{\big)}\,x^3+0x^2+0x-9} \\
\underline{x^3+0x^2+\ x} \\
-x-9
\end{array}
$$

$$\frac{x^3-9}{x^2+1} = x - \frac{x+9}{x^2+1}$$

23.

$$
\begin{array}{r}
2x-8 \\
x^2+0x+1\overline{\smash{\big)}\,2x^3-8x^2+3x-9} \\
\underline{2x^3+0x^2+2x} \\
-8x^2+\ x-9 \\
\underline{-8x^2-0x-8} \\
x-1
\end{array}
$$

$$\frac{2x^3-8x^2+3x-9}{x^2+1} = 2x - 8 + \frac{x-1}{x^2+1}$$

25.

$$
\begin{array}{r}
x+3 \\
x^3-3x^2+3x-1\overline{\smash{\big)}\,x^4+0x^3+0x^2+0x+0} \\
\underline{x^4-3x^3+3x^2-\ x} \\
3x^3-3x^2+\ x+0 \\
\underline{3x^3-9x^2+9x-3} \\
6x^2-8x+3
\end{array}
$$

$$\frac{x^4}{(x-1)^3} = x + 3 + \frac{6x^2-8x+3}{(x-1)^3}$$

27.

$$
\begin{array}{r|rrrr}
5 & 3 & -17 & 15 & -25 \\
  &   & 15 & -10 & 25 \\
\hline
  & 3 & -2 & 5 & 0
\end{array}
$$

$$\frac{3x^3-17x^2+15x-25}{x-5} = 3x^2-2x+5,\ x \neq 5$$

29.

$$
\begin{array}{r|rrrr}
3 & 6 & 7 & -1 & 26 \\
  &   & 18 & 75 & 222 \\
\hline
  & 6 & 25 & 74 & 248
\end{array}
$$

$$\frac{6x^3+7x^2-x+26}{x-3} = 6x^2+25x+74+\frac{248}{x-3}$$

31.

$$
\begin{array}{r|rrrr}
-2 & 4 & 8 & -9 & -18 \\
   &   & -8 & 0 & 18 \\
\hline
   & 4 & 0 & -9 & 0
\end{array}
$$

$$\frac{4x^3+8x^2-9x-18}{x+2} = 4x^2-9,\ x \neq -2$$

33.

$$
\begin{array}{r|rrrr}
-10 & -1 & 0 & 75 & -250 \\
    &   & 10 & -100 & 250 \\
\hline
    & -1 & 10 & -25 & 0
\end{array}
$$

$$\frac{-x^3+75x-250}{x+10} = -x^2+10x-25,\ x \neq -10$$

35.

$$
\begin{array}{r|rrrr}
4 & 5 & -6 & 0 & 8 \\
  &   & 20 & 56 & 224 \\
\hline
  & 5 & 14 & 56 & 232
\end{array}
$$

$$\frac{5x^3-6x^2+8}{x-4} = 5x^2+14x+56+\frac{232}{x-4}$$

37.

$$
\begin{array}{r|rrrrr}
6 & 10 & -50 & 0 & 0 & -800 \\
  &   & 60 & 60 & 360 & 2160 \\
\hline
  & 10 & 10 & 60 & 360 & 1360
\end{array}
$$

$$\frac{10x^4-50x^3-800}{x-6} = 10x^3+10x^2+60x+360+\frac{1360}{x-6}$$

39.

$$
\begin{array}{r|rrrr}
-8 & 1 & 0 & 0 & 512 \\
   &   & -8 & 64 & -512 \\
\hline
   & 1 & -8 & 64 & 0
\end{array}
$$

$$\frac{x^3+512}{x+8} = x^2-8x+64,\ x \neq -8$$

41.

$$
\begin{array}{r|rrrrr}
2 & -3 & 0 & 0 & 0 & 0 \\
  &   & -6 & -12 & -24 & -48 \\
\hline
  & -3 & -6 & -12 & -24 & -48
\end{array}
$$

$$\frac{-3x^4}{x-2} = -3x^3-6x^2-12x-24-\frac{48}{x-2}$$

43.

$$
\begin{array}{r|rrrrr}
6 & -1 & 0 & 0 & 180 & 0 \\
  &   & -6 & -36 & -216 & -216 \\
\hline
  & -1 & -6 & -36 & -36 & -216
\end{array}
$$

$$\frac{180x-x^4}{x-6} = -x^3-6x^2-36x-36-\frac{216}{x-6}$$

45.

$$-\frac{1}{2} \;\Big|\; \begin{array}{cccc} 4 & 16 & -23 & -15 \\ & -2 & -7 & 15 \\ \hline 4 & 14 & -30 & 0 \end{array}$$

$$\frac{4x^3 + 16x^2 - 23x - 15}{x + \dfrac{1}{2}} = 4x^2 + 14x - 30, \; x \neq -\frac{1}{2}$$

47. $f(x) = x^3 - x^2 - 14x + 11, \; k = 4$

$$4 \;\Big|\; \begin{array}{cccc} 1 & -1 & -14 & 11 \\ & 4 & 12 & -8 \\ \hline 1 & 3 & -2 & 3 \end{array}$$

$$f(x) = (x - 4)(x^2 + 3x - 2) + 3$$
$$f(4) = 4^3 - 4^2 - 14(4) + 11 = 3$$

49. $f(x) = 15x^4 + 10x^3 - 6x^2 + 14, \; k = -\frac{2}{3}$

$$-\frac{2}{3} \;\Big|\; \begin{array}{ccccc} 15 & 10 & -6 & 0 & 14 \\ & -10 & 0 & 4 & -\dfrac{8}{3} \\ \hline 15 & 0 & -6 & 4 & \dfrac{34}{3} \end{array}$$

$$f(x) = \left(x + \tfrac{2}{3}\right)\left(15x^3 - 6x + 4\right) + \tfrac{34}{3}$$
$$f\left(-\tfrac{2}{3}\right) = 15\left(-\tfrac{2}{3}\right)^4 + 10\left(-\tfrac{2}{3}\right)^3 - 6\left(-\tfrac{2}{3}\right)^2 + 14 = \tfrac{34}{3}$$

51. $f(x) = x^3 + 3x^2 - 2x - 14, \; k = \sqrt{2}$

$$\sqrt{2} \;\Big|\; \begin{array}{cccc} 1 & 3 & -2 & -14 \\ & \sqrt{2} & 2 + 3\sqrt{2} & 6 \\ \hline 1 & 3 + \sqrt{2} & 3\sqrt{2} & -8 \end{array}$$

$$f(x) = \left(x - \sqrt{2}\right)\left[x^2 + \left(3 + \sqrt{2}\right)x + 3\sqrt{2}\right] - 8$$
$$f\left(\sqrt{2}\right) = \left(\sqrt{2}\right)^3 + 3\left(\sqrt{2}\right)^2 - 2\sqrt{2} - 14 = -8$$

53. $f(x) = -4x^3 + 6x^2 + 12x + 4, \; k = 1 - \sqrt{3}$

$$1 - \sqrt{3} \;\Big|\; \begin{array}{cccc} -4 & 6 & 12 & 4 \\ & -4 + 4\sqrt{3} & -10 + 2\sqrt{3} & -4 \\ \hline -4 & 2 + 4\sqrt{3} & 2 + 2\sqrt{3} & 0 \end{array}$$

$$f(x) = \left(x - 1 + \sqrt{3}\right)\left[-4x^2 + \left(2 + 4\sqrt{3}\right)x + \left(2 + 2\sqrt{3}\right)\right]$$
$$f\left(1 - \sqrt{3}\right) = -4\left(1 - \sqrt{3}\right)^3 + 6\left(1 - \sqrt{3}\right)^2 + 12\left(1 - \sqrt{3}\right) + 4 = 0$$

55. $f(x) = 2x^3 - 7x + 3$

(a) Using the Remainder Theorem:

$$f(1) = 2(1)^3 - 7(1) + 3 = -2$$

Using synthetic division:

$$1 \;\Big|\; \begin{array}{cccc} 2 & 0 & -7 & 3 \\ & 2 & 2 & -5 \\ \hline 2 & 2 & -5 & -2 \end{array}$$

Verify using long division:

$$\begin{array}{r} 2x^2 + 2x - 5 \\ x - 1 \overline{)\; 2x^3 + 0x^2 - 7x + 3} \\ \underline{2x^3 - 2x^2} \\ 2x^2 - 7x \\ \underline{2x^2 - 2x} \\ -5x + 3 \\ \underline{-5x + 5} \\ -2 \end{array}$$

(b) Using the Remainder Theorem:

$$f(-2) = 2(-2)^3 - 7(-2) + 3 = 1$$

Using synthetic division:

$$-2 \;\Big|\; \begin{array}{cccc} 2 & 0 & -7 & 3 \\ & -4 & 8 & -2 \\ \hline 2 & -4 & 1 & 1 \end{array}$$

Verify using long division:

$$\begin{array}{r} 2x^2 - 4x + 1 \\ x + 2 \overline{)\; 2x^3 + 0x^2 - 7x + 3} \\ \underline{2x^3 + 4x^2} \\ -4x^2 - 7x \\ \underline{-4x^2 - 8x} \\ x + 3 \\ \underline{x + 2} \\ 1 \end{array}$$

(c) Using the Remainder Theorem:

$$f\left(\frac{1}{2}\right) = 2\left(\frac{1}{2}\right)^3 - 7\left(\frac{1}{2}\right) + 3 = -\frac{1}{4}$$

Using synthetic division:

$$\frac{1}{2} \begin{array}{|rrrr} 2 & 0 & -7 & 3 \\ & 1 & \frac{1}{2} & -\frac{13}{4} \\ \hline 2 & 1 & -\frac{13}{2} & -\frac{1}{4} \end{array}$$

Verify using long division:

$$x - \frac{1}{2} \overline{\smash{\big)}\,2x^3 + 0x^2 - 7x + 3}$$
quotient $2x^2 + x - \frac{13}{2}$

$$\underline{2x^3 - x^2}$$
$$x^2 - 7x$$
$$\underline{x^2 - \frac{1}{2}x}$$
$$-\frac{13}{2}x + 3$$
$$\underline{-\frac{13}{2}x + \frac{13}{4}}$$
$$-\frac{1}{4}$$

(d) Using the Remainder Theorem:

$$f(2) = 2(2)^3 - 7(2) + 3 = 5$$

Using synthetic division:

$$2 \begin{array}{|rrrr} 2 & 0 & -7 & 3 \\ & 4 & 8 & 2 \\ \hline 2 & 4 & 1 & 5 \end{array}$$

Verify using long division:

$$x - 2 \overline{\smash{\big)}\,2x^3 + 0x^2 - 7x + 3}$$
quotient $2x^2 + 4x + 1$

$$\underline{2x^3 - 4x^2}$$
$$4x^2 - 7x$$
$$\underline{4x^2 - 8x}$$
$$x + 3$$
$$\underline{x - 2}$$
$$5$$

57. $h(x) = x^3 - 5x^2 - 7x + 4$

(a) Using the Remainder Theorem:

$$h(3) = (3)^3 - 5(3)^2 - 7(3) + 4 = -35$$

Using synthetic division:

$$3 \begin{array}{|rrrr} 1 & -5 & -7 & 4 \\ & 3 & -6 & -39 \\ \hline 1 & -2 & -13 & -35 \end{array}$$

Verify using long division:

$$x - 3 \overline{\smash{\big)}\,x^3 - 5x^2 - 7x + 4}$$
quotient $x^2 - 2x - 13$

$$\underline{x^3 - 3x^2}$$
$$-2x^2 - 7x$$
$$\underline{-2x^2 + 6x}$$
$$-13x + 4$$
$$\underline{-13x + 39}$$
$$-35$$

(b) Using the Remainder Theorem:

$$h(2) = (2)^3 - 5(2)^2 - 7(2) + 4 = -22$$

Using synthetic division:

$$2 \begin{array}{|rrrr} 1 & -5 & -7 & 4 \\ & 2 & -6 & -26 \\ \hline 1 & -3 & -13 & -22 \end{array}$$

Verify using long division:

$$x - 2 \overline{\smash{\big)}\,x^3 - 5x^2 - 7x + 4}$$
quotient $x^2 - 3x - 13$

$$\underline{x^3 - 2x^2}$$
$$-3x^2 - 7x$$
$$\underline{-3x^2 + 6x}$$
$$-13x + 4$$
$$\underline{-13x + 26}$$
$$-22$$

(c) Using the Remainder Theorem:

$h(-2) = (-2)^3 - 5(-2)^2 - 7(-2) + 4 = -10$

Using synthetic division:

$$
\begin{array}{r|rrrr}
-2 & 1 & -5 & -7 & 4 \\
   &   & -2 & 14 & -14 \\
\hline
   & 1 & -7 & 7 & -10
\end{array}
$$

Verify using long division:

$$
\begin{array}{r}
x^2 - 7x + 7 \\
x + 2 \overline{)\, x^3 - 5x^2 - 7x + 4} \\
\underline{x^3 + 2x^2} \\
-7x^2 - 7x \\
\underline{-7x^2 - 14x} \\
7x + 4 \\
\underline{7x + 14} \\
-10
\end{array}
$$

(d) Using the Remainder Theorem:

$h(-5) = (-5)^3 - 5(-5)^2 - 7(-5) + 4 = -211$

Using synthetic division:

$$
\begin{array}{r|rrrr}
-5 & 1 & -5 & -7 & 4 \\
   &   & -5 & 50 & -215 \\
\hline
   & 1 & -10 & 43 & -211
\end{array}
$$

Verify using long division:

$$
\begin{array}{r}
x^2 - 10x + 43 \\
x + 5 \overline{)\, x^3 - 5x^2 - 7x + 4} \\
\underline{x^3 + 5x^2} \\
-10x^2 - 7x \\
\underline{-10x^2 - 50x} \\
43x + 4 \\
\underline{43x + 215} \\
-211
\end{array}
$$

59.

$$
\begin{array}{r|rrrr}
2 & 1 & 0 & -7 & 6 \\
  &   & 2 & 4 & -6 \\
\hline
  & 1 & 2 & -3 & 0
\end{array}
$$

$x^3 - 7x + 6 = (x - 2)(x^2 + 2x - 3)$

$\qquad\qquad\quad = (x - 2)(x + 3)(x - 1)$

Zeros: $2, -3, 1$

61.

$$
\begin{array}{r|rrrr}
\frac{1}{2} & 2 & -15 & 27 & -10 \\
            &   & 1 & -7 & 10 \\
\hline
            & 2 & -14 & 20 & 0
\end{array}
$$

$2x^3 - 15x^2 + 27x - 10 = \left(x - \frac{1}{2}\right)(2x^2 - 14x + 20)$

$\qquad\qquad\qquad\qquad\quad = (2x - 1)(x - 2)(x - 5)$

Zeros: $\frac{1}{2}, 2, 5$

63.

$$
\begin{array}{r|rrrr}
\sqrt{3} & 1 & 2 & -3 & -6 \\
         &   & \sqrt{3} & 3 + 2\sqrt{3} & 6 \\
\hline
         & 1 & 2 + \sqrt{3} & 2\sqrt{3} & 0
\end{array}
$$

$$
\begin{array}{r|rrr}
-\sqrt{3} & 1 & 2 + \sqrt{3} & 2\sqrt{3} \\
          &   & -\sqrt{3} & -2\sqrt{3} \\
\hline
          & 1 & 2 & 0
\end{array}
$$

$x^3 + 2x^2 - 3x - 6 = \left(x - \sqrt{3}\right)\left(x + \sqrt{3}\right)(x + 2)$

Zeros: $-\sqrt{3}, \sqrt{3}, -2$

65.

$$
\begin{array}{r|rrrr}
1 + \sqrt{3} & 1 & -3 & 0 & 2 \\
             &   & 1 + \sqrt{3} & 1 - \sqrt{3} & -2 \\
\hline
             & 1 & -2 + \sqrt{3} & 1 - \sqrt{3} & 0
\end{array}
$$

$$
\begin{array}{r|rrr}
1 - \sqrt{3} & 1 & -2 + \sqrt{3} & 1 - \sqrt{3} \\
             &   & 1 - \sqrt{3} & -1 + \sqrt{3} \\
\hline
             & 1 & -1 & 0
\end{array}
$$

$x^3 - 3x^2 + 2 = \left[x - \left(1 + \sqrt{3}\right)\right]\left[x - \left(1 - \sqrt{3}\right)\right](x - 1)$

$\qquad\qquad\quad = (x - 1)\left(x - 1 - \sqrt{3}\right)\left(x - 1 + \sqrt{3}\right)$

Zeros: $1, 1 - \sqrt{3}, 1 + \sqrt{3}$

67. $f(x) = 2x^3 + x^2 - 5x + 2$; Factors: $(x + 2), (x - 1)$

(a)

$$
\begin{array}{r|rrrr}
-2 & 2 & 1 & -5 & 2 \\
   &   & -4 & 6 & -2 \\
\hline
   & 2 & -3 & 1 & 0
\end{array}
$$

$$
\begin{array}{r|rrr}
1 & 2 & -3 & 1 \\
  &   & 2 & -1 \\
\hline
  & 2 & -1 & 0
\end{array}
$$

Both are factors of $f(x)$ because the remainders are zero.

(b) The remaining factor of $f(x)$ is $(2x - 1)$.

(c) $f(x) = (2x - 1)(x + 2)(x - 1)$

(d) Zeros: $\frac{1}{2}, -2, 1$

(e)

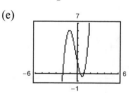

69. $f(x) = x^4 - 4x^3 - 15x^2 + 58x - 40;$

Factors: $(x - 5), (x + 4)$

(a)

```
5 | 1   -4   -15    58   -40
  |       5     5   -50    40
    1    1   -10     8     0
```

```
-4 | 1    1   -10    8
   |     -4    12   -8
     1   -3     2    0
```

Both are factors of $f(x)$ because the remainders are zero.

(b) $x^2 - 3x + 2 = (x - 1)(x - 2)$

The remaining factors are $(x - 1)$ and $(x - 2)$.

(c) $f(x) = (x - 1)(x - 2)(x - 5)(x + 4)$

(d) Zeros: $1, 2, 5, -4$

(e)

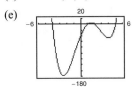

71. $f(x) = 6x^3 + 41x^2 - 9x - 14;$

Factors: $(2x + 1), (3x - 2)$

(a)

```
-1/2 | 6    41    -9   -14
     |      -3   -19    14
       6    38   -28     0
```

```
2/3 | 6    38   -28
    |        4    28
      6    42     0
```

Both are factors of $f(x)$ because the remainders are zero.

(b) $6x + 42 = 6(x + 7)$

This shows that $\dfrac{f(x)}{\left(x + \dfrac{1}{2}\right)\left(x - \dfrac{2}{3}\right)} = 6(x + 7),$

so $\dfrac{f(x)}{(2x + 1)(3x - 2)} = x + 7.$

The remaining factor is $(x + 7)$.

(c) $f(x) = (x + 7)(2x + 1)(3x - 2)$

(d) Zeros: $-7, -\dfrac{1}{2}, \dfrac{2}{3}$

(e)

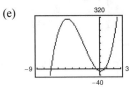

73. $f(x) = 2x^3 - x^2 - 10x + 5;$

Factors: $(2x - 1), \left(x + \sqrt{5}\right)$

(a)

```
1/2 | 2   -1   -10    5
    |      1     0   -5
      2    0   -10    0
```

```
-√5 | 2     0   -10
    |     -2√5    10
      2   -2√5    0
```

Both are factors of $f(x)$ because the remainders are zero.

(b) $2x - 2\sqrt{5} = 2\left(x - \sqrt{5}\right)$

This shows that $\dfrac{f(x)}{\left(x - \dfrac{1}{2}\right)\left(x + \sqrt{5}\right)} = 2\left(x - \sqrt{5}\right),$

so $\dfrac{f(x)}{(2x - 1)\left(x + \sqrt{5}\right)} = x - \sqrt{5}.$

The remaining factor is $\left(x - \sqrt{5}\right)$.

(c) $f(x) = \left(x + \sqrt{5}\right)\left(x - \sqrt{5}\right)(2x - 1)$

(d) Zeros: $-\sqrt{5}, \sqrt{5}, \dfrac{1}{2}$

(e)

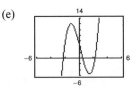

75. $f(x) = x^3 - 2x^2 - 5x + 10$

(a) The zeros of f are $x = 2$ and $x \approx \pm 2.236.$

(b) An exact zero is $x = 2.$

(c)

```
2 | 1   -2   -5    10
  |       2    0   -10
    1    0   -5     0
```

$f(x) = (x - 2)(x^2 - 5)$

$\qquad = (x - 2)\left(x - \sqrt{5}\right)\left(x + \sqrt{5}\right)$

77. $h(t) = t^3 - 2t^2 - 7t + 2$

    (a) The zeros of h are $t = -2, t \approx 3.732, t \approx 0.268$.

    (b) An exact zero is $t = -2$.

    (c)

$$-2 \,\Big|\begin{array}{rrrr} 1 & -2 & -7 & 2 \\ & -2 & 8 & -2 \\ \hline 1 & -4 & 1 & 0 \end{array}$$

        $h(t) = (t + 2)(t^2 - 4t + 1)$

        By the Quadratic Formula, the zeros of
$t^2 - 4t + 1$ are $2 \pm \sqrt{3}$. Thus,

$$h(t) = (t + 2)\Big[t - \big(2 + \sqrt{3}\big)\Big]\Big[t - \big(2 - \sqrt{3}\big)\Big].$$

79. $h(x) = x^5 - 7x^4 + 10x^3 + 14x^2 - 24x$

    (a) The zeros of h are $x = 0, x = 3, x = 4,$
$x \approx 1.414, x \approx -1.414$.

    (b) An exact zero is $x = 4$.

    (c)

$$4 \,\Big|\begin{array}{rrrrr} 1 & -7 & 10 & 14 & -24 \\ & 4 & -12 & -8 & 24 \\ \hline 1 & -3 & -2 & 6 & 0 \end{array}$$

        $h(x) = (x - 4)(x^4 - 3x^3 - 2x^2 + 6x)$

        $= x(x - 4)(x - 3)(x + \sqrt{2})(x - \sqrt{2})$

81. $\dfrac{4x^3 - 8x^2 + x + 3}{2x - 3}$

$$\dfrac{3}{2} \,\Big|\begin{array}{rrrr} 4 & -8 & 1 & 3 \\ & 6 & -3 & -3 \\ \hline 4 & -2 & -2 & 0 \end{array}$$

$$\dfrac{4x^3 - 8x^2 + x + 3}{x - \dfrac{3}{2}} = 4x^2 - 2x - 2 = 2(2x^2 - x - 1)$$

So, $\dfrac{4x^3 - 8x^2 + x + 3}{2x - 3} = 2x^2 - x - 1, x \neq \dfrac{3}{2}$.

83. $\dfrac{x^4 + 6x^3 + 11x^2 + 6x}{x^2 + 3x + 2} = \dfrac{x^4 + 6x^3 + 11x^2 + 6x}{(x + 1)(x + 2)}$

$$-1 \,\Big|\begin{array}{rrrrr} 1 & 6 & 11 & 6 & 0 \\ & -1 & -5 & -6 & 0 \\ \hline 1 & 5 & 6 & 0 & 0 \end{array}$$

$$-2 \,\Big|\begin{array}{rrrr} 1 & 5 & 6 & 0 \\ & -2 & -6 & 0 \\ \hline 1 & 3 & 0 & 0 \end{array}$$

$$\dfrac{x^4 + 6x^3 + 11x^2 + 6x}{(x + 1)(x + 2)} = x^2 + 3x, x \neq -2, -1$$

85. (a) and (b)

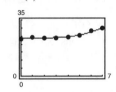

$$A \approx 0.0349t^3 - 0.168t^2 + 0.42t + 23.4$$

    (c)

Year	Actual Value	Estimated Value
0	23.2	23.4
1	24.2	23.7
2	23.9	23.8
3	23.9	24.1
4	24.4	24.6
5	25.6	25.7
6	28.0	27.4
7	29.8	30.1

    (d) $2010 \rightarrow t = 10$

$$10 \,\Big|\begin{array}{rrrr} 0.0349 & -0.168 & 0.42 & 23.4 \\ & 0.349 & 1.81 & 22.3 \\ \hline 0.0349 & 0.181 & 2.23 & 45.7 \end{array}$$

        $A(10) \approx \$45.7$

        In 2010, the amount of money supporting higher education is about \$45.7 billion.

        No, because the model will approach infinity quickly.

87. False. If $(7x + 4)$ is a factor of f, then $-\dfrac{4}{7}$ is a zero of f.

89. True. The degree of the numerator is greater than the degree of the denominator.

91.

$$
\begin{array}{r}
x^{2n} + 6x^n + 9 \\
x^n + 3{\overline{\smash{\big)}\,x^{3n} + 9x^{2n} + 27x^n + 27}} \\
\underline{x^{3n} + 3x^{2n}} \\
6x^{2n} + 27x^n \\
\underline{6x^{2n} + 18x^n} \\
9x^n + 27 \\
\underline{9x^n + 27} \\
0
\end{array}
$$

$$\frac{x^{3n} + 9x^{2n} + 27x^n + 27}{x^n + 3} = x^{2n} + 6x^n + 9, \ x^n \neq -3$$

93. A divisor divides evenly into a dividend if the remainder is zero.

95.

$$
\begin{array}{r|rrrr}
5 & 1 & 4 & -3 & c \\
  &   & 5 & 45 & 210 \\
\hline
  & 1 & 9 & 42 & c + 210
\end{array}
$$

To divide evenly, $c + 210$ must equal zero. So, c must equal -210.

97. If $x - 4$ is a factor of $f(x) = x^3 - kx^2 + 2kx - 8$,

then $f(4) = 0$.

$$f(4) = (4)^3 - k(4)^2 + 2k(4) - 8$$
$$0 = 64 - 16k + 8k - 8$$
$$-56 = -8k$$
$$7 = k$$

99. (a)

$$
\begin{array}{r}
x + 1 \\
x - 1{\overline{\smash{\big)}\,x^2 + 0x - 1}} \\
\underline{x^2 - x} \\
x - 1 \\
\underline{x - 1} \\
0
\end{array}
$$

$$\frac{x^2 - 1}{x - 1} = x + 1, \ x \neq 1$$

(b)

$$
\begin{array}{r}
x^2 + x + 1 \\
x - 1{\overline{\smash{\big)}\,x^3 + 0x^2 + 0x - 1}} \\
\underline{x^3 - x^2} \\
x^2 + 0x \\
\underline{x^2 - x} \\
x - 1 \\
\underline{x - 1} \\
0
\end{array}
$$

$$\frac{x^3 - 1}{x - 1} = x^2 + x + 1, \ x \neq 1$$

(c)

$$
\begin{array}{r}
x^3 + x^2 + x + 1 \\
x - 1{\overline{\smash{\big)}\,x^4 + 0x^3 + 0x^2 + 0x - 1}} \\
\underline{x^4 - x^3} \\
x^3 + 0x^2 \\
\underline{x^3 - x^2} \\
x^2 + 0x \\
\underline{x^2 - x} \\
x - 1 \\
\underline{x - 1} \\
0
\end{array}
$$

$$\frac{x^4 - 1}{x - 1} = x^3 + x^2 + x + 1, \ x \neq 1$$

$$\frac{x^n - 1}{x - 1} = x^{n-1} + x^{n-2} + \cdots + x + 1, \ x \neq 1$$

Section 3.4   Zeros of Polynomial Functions

- You should know that if f is a polynomial of degree $n > 0$, then f has at least one zero in the complex number system.

- You should know the Linear Factorization Theorem.

- You should know the Rational Zero Test.

- You should know shortcuts for the Rational Zero Test. Possible rational zeros $= \dfrac{\text{factors of constant term}}{\text{factors of leading coefficient}}$

  (a) Use a graphing or programmable calculator.
  (b) Sketch a graph.
  (c) After finding a root, use synthetic division to reduce the degree of the polynomial.

- You should know that if $a + bi$ is a complex zero of a polynomial f, with real coefficients, then $a - bi$ is also a complex zero of f.

—continued—

- You should know the difference between a factor that is irreducible over the rationals $\left(\text{such as } x^2 - 7\right)$ and a factor that is irreducible over the reals $\left(\text{such as } x^2 + 9\right)$.

- You should know Descartes's Rule of Signs. (For a polynomial with real coefficients and a non-zero constant term.)
  - (a) The number of positive real zeros of f is either equal to the number of variations in sign of f or is less than that number by an even integer.
  - (b) The number of negative real zeros of f is either equal to the number of variations in sign of $f(-x)$ or is less than that number by an even integer.
  - (c) When there is only one variation in sign, there is exactly one positive (or negative) real zero.

- You should be able to observe the last row obtained from synthetic division in order to determine upper or lower bounds.
  - (a) If the test value is positive and all of the entries in the last row are positive or zero, then the test value is an upper bound.
  - (b) If the test value is negative and the entries in the last row alternate from positive to negative, then the test value is a lower bound. (Zero entries count as positive or negative.)

---

1. Fundamental Theorem of Algebra

3. Rational Zero

5. linear; quadratic; quadratic

7. Descartes's Rule of Signs

9. $f(x) = x(x - 6)^2$

The zeros are: $x = 0, \ x = 6$

11. $g(x) = (x - 2)(x + 4)^3$

The zeros are: $x = 2, \ x = -4$

13. $f(x) = (x + 6)(x + i)(x - i)$

The zeros are: $x = -6, \ x = -i, \ x = i$

15. $f(x) = x^3 + 2x^2 - x - 2$

Possible rational zeros: $\pm 1, \pm 2$

Zeros shown on graph: $-2, -1, 1, 2$

17. $f(x) = 2x^4 - 17x^3 + 35x^2 + 9x - 45$

Possible rational zeros: $\pm 1, \pm 3, \pm 5, \pm 9, \pm 15, \pm 45,$
$\pm \frac{1}{2}, \pm \frac{3}{2}, \pm \frac{5}{2}, \pm \frac{9}{2}, \pm \frac{15}{2}, \pm \frac{45}{2}$

Zeros shown on graph: $-1, \frac{3}{2}, 3, 5$

19. $f(x) = x^3 - 6x^2 + 11x - 6$

Possible rational zeros: $\pm 1, \pm 2, \pm 3, \pm 6$

$$
\begin{array}{r|rrrr}
1 & 1 & -6 & 11 & -6 \\
  &   & 1 & -5 & 6 \\
\hline
  & 1 & -5 & 6 & 0
\end{array}
$$

$x^3 - 6x^2 + 11x - 6 = (x - 1)(x^2 - 5x + 6)$
$= (x - 1)(x - 2)(x - 3)$

So, the rational zeros are 1, 2, and 3.

21. $g(x) = x^3 - 4x^2 - x + 4$
$= x^2(x - 4) - 1(x - 4)$
$= (x - 4)(x^2 - 1)$
$= (x - 4)(x - 1)(x + 1)$

So, the rational zeros are 4, 1, and -1.

23. $h(t) = t^3 + 8t^2 + 13t + 6$

Possible rational zeros: $\pm 1, \pm 2, \pm 3, \pm 6$

$$
\begin{array}{r|rrrr}
-6 & 1 & 8 & 13 & 6 \\
   &   & -6 & -12 & -6 \\
\hline
   & 1 & 2 & 1 & 0
\end{array}
$$

$t^3 + 8t^2 + 13t + 6 = (t + 6)(t^2 + 2t + 1)$
$= (t + 6)(t + 1)(t + 1)$

So, the rational zeros are -1 and -6.

25. $C(x) = 2x^3 + 3x^2 - 1$

Possible rational zeros: $\pm 1, \pm\frac{1}{2}$

$$
\begin{array}{r|rrrr}
-1 & 2 & 3 & 0 & -1 \\
   &   & -2 & -1 & 1 \\
\hline
   & 2 & 1 & -1 & 0
\end{array}
$$

$2x^3 + 3x^2 - 1 = (x + 1)(2x^2 + x - 1)$
$\qquad\qquad\qquad\; = (x + 1)(x + 1)(2x - 1)$
$\qquad\qquad\qquad\; = (x + 1)^2(2x - 1)$

So, the rational zeros are -1 and $\frac{1}{2}$.

27. $f(x) = 9x^4 - 9x^3 - 58x^2 + 4x + 24$

Possible rational zeros:
$\pm 1, \pm 2, \pm 3, \pm 4, \pm 6, \pm 8, \pm 12, \pm 24,$
$\pm\frac{1}{3}, \pm\frac{2}{3}, \pm\frac{4}{3}, \pm\frac{8}{3}, \pm\frac{1}{9}, \pm\frac{2}{9}, \pm\frac{4}{9}, \pm\frac{8}{9}$

$$
\begin{array}{r|rrrrr}
-2 & 9 & -9 & -58 & 4 & 24 \\
   &   & -18 & 54 & 8 & -24 \\
\hline
   & 9 & -27 & -4 & 12 & 0
\end{array}
$$

$$
\begin{array}{r|rrrr}
3 & 9 & -27 & -4 & 12 \\
  &   & 27 & 0 & -12 \\
\hline
  & 9 & 0 & -4 & 0
\end{array}
$$

$f(x) = (x + 2)(x - 3)(9x^2 - 4)$
$\qquad = (x + 2)(x - 3)(3x - 2)(3x + 2)$

So, the rational zeros are $-2, 3, \frac{2}{3}$, and $-\frac{2}{3}$.

29. $z^4 + z^3 + z^2 + 3z - 6 = 0$

Possible rational zeros: $\pm 1, \pm 2, \pm 3, \pm 6$

$$
\begin{array}{r|rrrrr}
1 & 1 & 1 & 1 & 3 & -6 \\
  &   & 1 & 2 & 3 & 6 \\
\hline
  & 1 & 2 & 3 & 6 & 0
\end{array}
$$

$(z - 1)(z^3 + 2z^2 + 3z + 6) = 0$
$(z - 1)(z^2 + 3)(z + 2) = 0$

So, the real zeros are -2 and 1.

31. $2y^4 + 3y^3 - 16y^2 + 15y - 4 = 0$

Possible rational zeros: $\pm\frac{1}{2}, \pm 1, \pm 2, \pm 4$

$$
\begin{array}{r|rrrrr}
1 & 2 & 3 & -16 & 15 & -4 \\
  &   & 2 & 5 & -11 & 4 \\
\hline
  & 2 & 5 & -11 & 4 & 0
\end{array}
$$

$$
\begin{array}{r|rrrr}
1 & 2 & 5 & -11 & 4 \\
  &   & 2 & 7 & -4 \\
\hline
  & 2 & 7 & -4 & 0
\end{array}
$$

$(y - 1)(y - 1)(2y^2 + 7y - 4) = 0$
$(y - 1)(y - 1)(2y - 1)(y + 4) = 0$

So, the real zeros are $-4, \frac{1}{2}$ and 1.

33. $f(x) = x^3 + x^2 - 4x - 4$

(a) Possible rational zeros: $\pm 1, \pm 2, \pm 4$

(b)

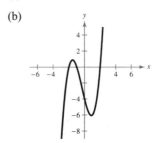

(c) Real zeros: $-2, -1, 2$

35. $f(x) = -4x^3 + 15x^2 - 8x - 3$

(a) Possible rational zeros: $\pm 1, \pm 3, \pm\frac{1}{2}, \pm\frac{3}{2}, \pm\frac{1}{4}, \pm\frac{3}{4}$

(b)

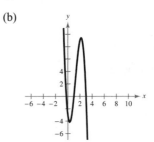

(c) Real zeros: $-\frac{1}{4}, 1, 3$

37. $f(x) = -2x^4 + 13x^3 - 21x^2 + 2x + 8$

(a) Possible rational zeros: $\pm 1, \pm 2, \pm 4, \pm 8, \pm\frac{1}{2}$

(b)
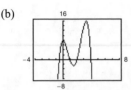

(c) Real zeros: $-\frac{1}{2}, 1, 2, 4$

39. $f(x) = 32x^3 - 52x^2 + 17x + 3$

(a) Possible rational zeros: $\pm 1, \pm 3, \pm\frac{1}{2}, \pm\frac{3}{2}, \pm\frac{1}{4}, \pm\frac{3}{4},$

$$\pm\tfrac{1}{8}, \pm\tfrac{3}{8}, \pm\tfrac{1}{16}, \pm\tfrac{3}{16}, \pm\tfrac{1}{32}, \pm\tfrac{3}{32}$$

(b)
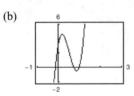

(c) Real zeros: $-\frac{1}{8}, \frac{3}{4}, 1$

41. $f(x) = x^4 - 3x^2 + 2$

(a) $x = \pm 1$, about ± 1.414

(b) An exact zero is $x = 1$.

$$
\begin{array}{r|rrrrr}
1 & 1 & 0 & -3 & 0 & 2 \\
  &   & 1 & 1 & -2 & -2 \\
\hline
  & 1 & 1 & -2 & -2 & 0
\end{array}
$$

(c)
$$
\begin{array}{r|rrrr}
-1 & 1 & 1 & -2 & -2 \\
   &   & -1 & 0 & 2 \\
\hline
   & 1 & 0 & -2 & 0
\end{array}
$$

$$f(x) = (x - 1)(x + 1)(x^2 - 2)$$
$$= (x - 1)(x + 1)(x - \sqrt{2})(x + \sqrt{2})$$

43. $h(x) = x^5 - 7x^4 + 10x^3 + 14x^2 - 24x$

(a) $h(x) = x(x^4 - 7x^3 + 10x^2 + 14x - 24)$

$x = 0, 3, 4$, about ± 1.414

(b) An exact zero is $x = 3$.

$$
\begin{array}{r|rrrrr}
3 & 1 & -7 & 10 & 14 & -24 \\
  &   & 3 & -12 & -6 & 24 \\
\hline
  & 1 & -4 & -2 & 8 & 0
\end{array}
$$

(c)
$$
\begin{array}{r|rrrr}
4 & 1 & -4 & -2 & 8 \\
  &   & 4 & 0 & -8 \\
\hline
  & 1 & 0 & -2 & 0
\end{array}
$$

$$h(x) = x(x - 3)(x - 4)(x^2 - 2)$$
$$= x(x - 3)(x - 4)(x - \sqrt{2})(x + \sqrt{2})$$

45. $f(x) = (x - 1)(x - 5i)(x + 5i)$
$$= (x - 1)(x^2 + 25)$$
$$= x^3 - x^2 + 25x - 25$$

Note: $f(x) = a(x^3 - x^2 + 25x - 25)$, where a is any nonzero real number, has the zeros 1 and $\pm 5i$.

47. If $5 + i$ is a zero, so is its conjugate, $5 - i$.

$$f(x) = (x - 2)(x - (5 + i))(x - (5 - i))$$
$$= (x - 2)(x^2 - 10x + 26)$$
$$= x^3 - 12x^2 + 46x - 52$$

Note: $f(x) = a(x^3 - 12x^2 + 46x - 52)$, where a is any nonzero real number, has the zeros 2 and $5 \pm i$.

49. If $3 + \sqrt{2}i$ is a zero, so is its conjugate, $3 - \sqrt{2}i$.

$$f(x) = (3x - 2)(x + 1)\left[x - \left(3 + \sqrt{2}i\right)\right]\left[x - \left(3 - \sqrt{2}i\right)\right]$$
$$= (3x - 2)(x + 1)\left[(x - 3) - \sqrt{2}i\right]\left[(x - 3) + \sqrt{2}i\right]$$
$$= \left(3x^2 + x - 2\right)\left[(x - 3)^2 - \left(\sqrt{2}i\right)^2\right]$$
$$= \left(3x^2 + x - 2\right)\left(x^2 - 6x + 9 + 2\right)$$
$$= \left(3x^2 + x - 2\right)\left(x^2 - 6x + 11\right)$$
$$= 3x^4 - 17x^3 + 25x^2 + 23x - 22$$

Note: $f(x) = a\left(3x^4 - 17x^3 + 25x^2 + 23x - 22\right)$, where a is any nonzero real number, has the zeros $\frac{2}{3}, -1$, and $3 \pm \sqrt{2}i$.

51. $f(x) = x^4 + 6x^2 - 27$

  (a) $f(x) = (x^2 + 9)(x^2 - 3)$

  (b) $f(x) = (x^2 + 9)(x + \sqrt{3})(x - \sqrt{3})$

  (c) $f(x) = (x + 3i)(x - 3i)(x + \sqrt{3})(x - \sqrt{3})$

53. $f(x) = x^4 - 4x^3 + 5x^2 - 2x - 6$

$$
\begin{array}{r}
x^2 - 2x + 3 \\
x^2 - 2x - 2 \overline{)\, x^4 - 4x^3 + 5x^2 - 2x - 6} \\
\underline{x^4 - 2x^3 - 2x^2} \\
-2x^3 + 7x^2 - 2x \\
\underline{-2x^3 + 4x^2 + 4x} \\
3x^2 - 6x - 6 \\
\underline{3x^2 - 6x - 6} \\
0
\end{array}
$$

  (a) $f(x) = (x^2 - 2x - 2)(x^2 - 2x + 3)$

  (b) $f(x) = (x - 1 + \sqrt{3})(x - 1 - \sqrt{3})(x^2 - 2x + 3)$

  (c) $f(x) = (x - 1 + \sqrt{3})(x - 1 - \sqrt{3})(x - 1 + \sqrt{2}i)(x - 1 - \sqrt{2}i)$

Note: Use the Quadratic Formula for (b) and (c).

55. $f(x) = x^3 - x^2 + 4x - 4$

Because $2i$ is a zero, so is $-2i$.

$$
\begin{array}{r|rrrr}
2i & 1 & -1 & 4 & -4 \\
  &   & 2i & -4-2i & 4 \\
\hline
  & 1 & 2i-1 & -2i & 0
\end{array}
$$

$$
\begin{array}{r|rrr}
-2i & 1 & 2i-1 & -2i \\
  &   & -2i & 2i \\
\hline
  & 1 & -1 & 0
\end{array}
$$

$f(x) = (x - 2i)(x + 2i)(x - 1)$

The zeros of $f(x)$ are $x = 1, \pm 2i$.

Alternate Solution:

Because $x = \pm 2i$ are zeros of $f(x)$,

$(x + 2i)(x - 2i) = x^2 + 4$ is a factor of $f(x)$.

By long division, you have:

$$
\begin{array}{r}
x - 1 \\
x^2 + 0x + 4 \overline{)\, x^3 - x^2 + 4x - 4} \\
\underline{x^3 + 0x^2 + 4x} \\
-x^2 + 0x - 4 \\
\underline{-x^2 + 0x - 4} \\
0
\end{array}
$$

$f(x) = (x^2 + 4)(x - 1)$

The zeros of $f(x)$ are $x = 1, \pm 2i$.

57. $f(x) = 2x^4 - x^3 + 49x^2 - 25x - 25$

Because $5i$ is a zero, so is $-5i$.

$$
\begin{array}{r|rrrrr}
5i & 2 & -1 & 49 & -25 & -25 \\
   &   & 10i & -5i - 50 & -5i + 25 & 25 \\
\hline
   & 2 & -1 + 10i & -1 - 5i & -5i & 0
\end{array}
$$

$$
\begin{array}{r|rrrr}
-5i & 2 & -1 + 10i & -1 - 5i & -5i \\
    &   & -10i & 5i & 5i \\
\hline
    & 2 & -1 & -1 & 0
\end{array}
$$

$$f(x) = (x - 5i)(x + 5i)(2x^2 - x - 1)$$
$$= (x - 5i)(x + 5i)(2x + 1)(x - 1)$$

The zeros of $f(x)$ are $x = \pm 5i, -\frac{1}{2}, 1$.

Alternate Solution:

Because $x = \pm 5i$ are zeros of $f(x)$, $(x - 5i)(x + 5i) = x^2 + 25$ is a factor of $f(x)$.

By long division, you have:

$$
\begin{array}{r}
2x^2 - x - 1 \\
x^2 + 0x + 25 \overline{\smash{\big)}\, 2x^4 - x^3 + 49x^2 - 25x - 25} \\
\underline{2x^4 + 0x^3 + 50x^2} \\
-x^3 - x^2 - 25x \\
\underline{-x^3 + 0x^2 - 25x} \\
-x^2 + 0x - 25 \\
\underline{-x^2 + 0x - 25} \\
0
\end{array}
$$

$$f(x) = (x^2 + 25)(2x^2 - x - 1)$$

The zeros of $f(x)$ are $x = \pm 5i, -\frac{1}{2}, 1$.

59. $g(x) = 4x^3 + 23x^2 + 34x - 10$

Because $-3 + i$ is a zero, so is $-3 - i$.

$$
\begin{array}{r|rrrr}
-3 + i & 4 & 23 & 34 & -10 \\
       &   & -12 + 4i & -37 - i & 10 \\
\hline
       & 4 & 11 + 4i & -3 - i & 0
\end{array}
$$

$$
\begin{array}{r|rrr}
-3 - i & 4 & 11 + 4i & -3 - i \\
       &   & -12 - 4i & 3 + i \\
\hline
       & 4 & -1 & 0
\end{array}
$$

The zero of $4x - 1$ is $x = \frac{1}{4}$. The zeros of $g(x)$ are $x = -3 \pm i, \frac{1}{4}$.

Alternate Solution

Because $-3 \pm i$ are zeros of $g(x)$,

$$[x - (-3 + i)][x - (-3 - i)] = [(x + 3) - i][(x + 3) + i]$$
$$= (x + 3)^2 - i^2$$
$$= x^2 + 6x + 10$$

is a factor of $g(x)$. By long division, you have:

$$
\begin{array}{r}
4x - 1 \\
x^2 + 6x + 10 \overline{\smash{\big)}\, 4x^3 + 23x^2 + 34x - 10} \\
\underline{4x^3 + 24x^2 + 40x} \\
-x^2 - 6x - 10 \\
\underline{-x^2 - 6x - 10} \\
0
\end{array}
$$

$$g(x) = (x^2 + 6x + 10)(4x - 1)$$

The zeros of $g(x)$ are $x = -3 \pm i, \frac{1}{4}$.

61. $f(x) = x^4 + 3x^3 - 5x^2 - 21x + 22$

Because $-3 + \sqrt{2}i$ is a zero, so is $-3 - \sqrt{2}i$, and

$$\left[x - \left(-3 + \sqrt{2}i\right)\right]\left[x - \left(-3 - \sqrt{2}i\right)\right] = \left[(x + 3) - \sqrt{2}i\right]\left[(x + 3) + \sqrt{2}i\right]$$

$$= (x + 3)^2 - \left(\sqrt{2}i\right)^2$$

$$= x^2 + 6x + 11$$

is a factor of $f(x)$. By long division, you have:

$$
\begin{array}{r}
x^2 - 3x + 2 \\
x^2 + 6x + 11 \overline{\smash{\big)}\, x^4 + 3x^3 - 5x^2 - 21x + 22} \\
\underline{x^4 + 6x^3 + 11x^2} \\
-3x^3 - 16x^2 - 21x \\
\underline{-3x^3 - 18x^2 - 33x} \\
2x^2 + 12x + 22 \\
\underline{2x^2 + 12x + 22} \\
0
\end{array}
$$

$$f(x) = \left(x^2 + 6x + 11\right)\left(x^2 - 3x + 2\right)$$

$$= \left(x^2 + 6x + 11\right)(x - 1)(x - 2)$$

The zeros of $f(x)$ are $x = -3 \pm \sqrt{2}i, 1, 2$.

63. $f(x) = x^2 + 36$

$$= (x + 6i)(x - 6i)$$

The zeros of $f(x)$ are $x = \pm 6i$.

65. $h(x) = x^2 - 2x + 17$

By the Quadratic Formula, the zeros of $f(x)$ are

$$x = \frac{2 \pm \sqrt{4 - 68}}{2} = \frac{2 \pm \sqrt{-64}}{2} = 1 \pm 4i.$$

$$f(x) = \left(x - (1 + 4i)\right)\left(x - (1 - 4i)\right)$$

$$= (x - 1 - 4i)(x - 1 + 4i)$$

67. $f(x) = x^4 - 16$

$$= \left(x^2 - 4\right)\left(x^2 + 4\right)$$

$$= (x - 2)(x + 2)(x - 2i)(x + 2i)$$

Zeros: $\pm 2, \pm 2i$

69. $f(z) = z^2 - 2z + 2$

By the Quadratic Formula, the zeros of $f(z)$ are

$$z = \frac{2 \pm \sqrt{4 - 8}}{2} = 1 \pm i.$$

$$f(z) = \left[z - (1 + i)\right]\left[z - (1 - i)\right]$$

$$= (z - 1 - i)(z - 1 + i)$$

71. $g(x) = x^3 - 3x^2 + x + 5$

Possible rational zeros: $\pm 1, \pm 5$

$$
\begin{array}{r|rrrr}
-1 & 1 & -3 & 1 & 5 \\
 & & -1 & 4 & -5 \\
\hline
 & 1 & -4 & 5 & 0
\end{array}
$$

By the Quadratic Formula, the zeros of $x^2 - 4x + 5$

are: $x = \dfrac{4 \pm \sqrt{16 - 20}}{2} = 2 \pm i$

Zeros: $-1, 2 \pm i$

$$g(x) = (x + 1)(x - 2 - i)(x - 2 + i)$$

73. $h(x) = x^3 - x + 6$

Possible rational zeros: $\pm 1, \pm 2, \pm 3, \pm 6$

$$
\begin{array}{r|rrrr}
-2 & 1 & 0 & -1 & 6 \\
   &   & -2 & 4 & -6 \\
\hline
   & 1 & -2 & 3 & 0
\end{array}
$$

By the Quadratic Formula, the zeros of $x^2 - 2x + 3$ are

$$x = \frac{2 \pm \sqrt{4 - 12}}{2} = 1 \pm \sqrt{2}i.$$

Zeros: $-2, 1 \pm \sqrt{2}i$

$$h(x) = (x + 2)\left[x - \left(1 + \sqrt{2}i\right)\right]\left[x - \left(1 - \sqrt{2}i\right)\right]$$
$$= (x + 2)\left(x - 1 - \sqrt{2}i\right)\left(x - 1 + \sqrt{2}i\right)$$

75. $f(x) = 5x^3 - 9x^2 + 28x + 6$

Possible rational zeros:

$$\pm 1, \pm 2, \pm 3, \pm 6, \pm\frac{1}{5}, \pm\frac{2}{5}, \pm\frac{3}{5}, \pm\frac{6}{5}$$

$$
\begin{array}{r|rrrr}
-\frac{1}{5} & 5 & -9 & 28 & 6 \\
   &   & -1 & 2 & -6 \\
\hline
   & 5 & -10 & 30 & 0
\end{array}
$$

By the Quadratic Formula, the zeros of
$5x^2 - 10x + 30 = 5\left(x^2 - 2x + 6\right)$ are

$$x = \frac{2 \pm \sqrt{4 - 24}}{2} = 1 \pm \sqrt{5}i.$$

Zeros: $-\frac{1}{5}, 1 \pm \sqrt{5}i$

$$f(x) = \left[x - \left(-\frac{1}{5}\right)\right](5)\left[x - \left(1 + \sqrt{5}i\right)\right]\left[x - \left(1 - \sqrt{5}i\right)\right]$$
$$= (5x + 1)\left(x - 1 - \sqrt{5}i\right)\left(x - 1 + \sqrt{5}i\right)$$

77. $g(x) = x^4 - 4x^3 + 8x^2 - 16x + 16$

Possible rational zeros: $\pm 1, \pm 2, \pm 4, \pm 8, \pm 16$

$$
\begin{array}{r|rrrrr}
2 & 1 & -4 & 8 & -16 & 16 \\
  &   & 2 & -4 & 8 & -16 \\
\hline
  & 1 & -2 & 4 & -8 & 0
\end{array}
$$

$$
\begin{array}{r|rrrr}
2 & 1 & -2 & 4 & -8 \\
  &   & 2 & 0 & 8 \\
\hline
  & 1 & 0 & 4 & 0
\end{array}
$$

$$g(x) = (x - 2)(x - 2)\left(x^2 + 4\right)$$
$$= (x - 2)^2(x + 2i)(x - 2i)$$

Zeros: $2, \pm 2i$

79. $f(x) = x^4 + 10x^2 + 9$

$$= \left(x^2 + 1\right)\left(x^2 + 9\right)$$
$$= (x + i)(x - i)(x + 3i)(x - 3i)$$

Zeros: $\pm i, \pm 3i$

81. $f(x) = x^3 + 24x^2 + 214x + 740$

Possible rational zeros: $\pm 1, \pm 2, \pm 4, \pm 5, \pm 10, \pm 20, \pm 37,$
$\quad\quad\quad\quad \pm 74, \pm 148, \pm 185, \pm 370, \pm 740$

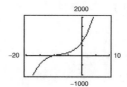

Based on the graph, try $x = -10$.

$$
\begin{array}{r|rrrr}
-10 & 1 & 24 & 214 & 740 \\
    &   & -10 & -140 & -740 \\
\hline
    & 1 & 14 & 74 & 0
\end{array}
$$

By the Quadratic Formula, the zeros of $x^2 + 14x + 74$

are $x = \dfrac{-14 \pm \sqrt{196 - 296}}{2} = -7 \pm 5i.$

The zeros of $f(x)$ are $x = -10$ and $x = -7 \pm 5i$.

83. $f(x) = 16x^3 - 20x^2 - 4x + 15$

Possible rational zeros:

$$\pm 1, \pm 3, \pm 5, \pm 15, \pm\frac{1}{2}, \pm\frac{3}{2}, \pm\frac{5}{2}, \pm\frac{15}{2}, \pm\frac{1}{4}, \pm\frac{3}{4},$$
$$\pm\frac{5}{4}, \pm\frac{15}{4}, \pm\frac{1}{8}, \pm\frac{3}{8}, \pm\frac{5}{8}, \pm\frac{15}{8}, \pm\frac{1}{16}, \pm\frac{3}{16}, \pm\frac{5}{16}, \pm\frac{15}{16}$$

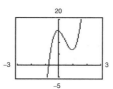

Based on the graph, try $x = -\dfrac{3}{4}$.

$$
\begin{array}{r|rrrr}
-\frac{3}{4} & 16 & -20 & -4 & 15 \\
    &   & -12 & 24 & -15 \\
\hline
    & 16 & -32 & 20 & 0
\end{array}
$$

By the Quadratic Formula, the zeros of
$16x^2 - 32x + 20 = 4\left(4x^2 - 8x + 5\right)$ are

$$x = \frac{8 \pm \sqrt{64 - 80}}{8} = 1 \pm \frac{1}{2}i.$$

The zeros of $f(x)$ are $x = -\dfrac{3}{4}$ and $x = 1 \pm \dfrac{1}{2}i$.

85. $f(x) = 2x^4 + 5x^3 + 4x^2 + 5x + 2$

Possible rational zeros: $\pm 1, \pm 2, \pm\frac{1}{2}$

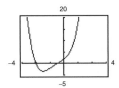

Based on the graph, try $x = -2$ and $x = -\frac{1}{2}$.

$$\begin{array}{r|rrrrr} -2 & 2 & 5 & 4 & 5 & 2 \\ & & -4 & -2 & -4 & -2 \\ \hline & 2 & 1 & 2 & 1 & 0 \end{array}$$

$$\begin{array}{r|rrrr} -\frac{1}{2} & 2 & 1 & 2 & 1 \\ & & -1 & 0 & -1 \\ \hline & 2 & 0 & 2 & 0 \end{array}$$

The zeros of $2x^2 + 2 = 2(x^2 + 1)$ are $x = \pm i$.

The zeros of $f(x)$ are $x = -2$, $x = -\frac{1}{2}$, and $x = \pm i$.

87. $g(x) = 2x^3 - 3x^2 - 3$

Sign variations: 1, positive zeros: 1

$g(-x) = -2x^3 - 3x^2 - 3$

Sign variations: 0, negative zeros: 0

89. $h(x) = 2x^3 + 3x^2 + 1$

Sign variations: 0, positive zeros: 0

$h(-x) = -2x^3 + 3x^2 + 1$

Sign variations: 1, negative zeros: 1

91. $g(x) = 5x^5 - 10x = 5x(x^4 - 2)$

Let $g(x) = x^4 - 2$.

Sign variations: 1, positive zeros: 1

$g(-x) = x^4 - 2$

Sign variations: 1, negative zeros: 1

93. $f(x) = -5x^3 + x^2 - x + 5$

Sign variations: 3, positive zeros: 3 or 1

$f(-x) = 5x^3 + x^2 + x + 5$

Sign variations: 0, negative zeros: 0

95. $f(x) = x^3 + 3x^2 - 2x + 1$

(a) $$\begin{array}{r|rrrr} 1 & 1 & 3 & -2 & 1 \\ & & 1 & 4 & 2 \\ \hline & 1 & 4 & 2 & 3 \end{array}$$

1 is an upper bound.

(b) $$\begin{array}{r|rrrr} -4 & 1 & 3 & -2 & 1 \\ & & -4 & 4 & -8 \\ \hline & 1 & -1 & 2 & -7 \end{array}$$

−4 is a lower bound.

97. $f(x) = x^4 - 4x^3 + 16x - 16$

(a) $$\begin{array}{r|rrrrr} 5 & 1 & -4 & 0 & 16 & -16 \\ & & 5 & 5 & 25 & 205 \\ \hline & 1 & 1 & 5 & 41 & 189 \end{array}$$

5 is an upper bound.

(b) $$\begin{array}{r|rrrrr} -3 & 1 & -4 & 0 & 16 & -16 \\ & & -3 & 21 & -63 & 141 \\ \hline & 1 & -7 & 21 & -47 & 125 \end{array}$$

−3 is a lower bound.

99. $f(x) = 4x^3 - 3x - 1$

Possible rational zeros: $\pm 1, \pm\frac{1}{2}, \pm\frac{1}{4}$

$$\begin{array}{r|rrrr} 1 & 4 & 0 & -3 & -1 \\ & & 4 & 4 & 1 \\ \hline & 4 & 4 & 1 & 0 \end{array}$$

$4x^3 - 3x - 1 = (x-1)(4x^2 + 4x + 1)$
$\qquad = (x-1)(2x+1)^2$

So, the zeros are 1 and $-\frac{1}{2}$.

101. $f(y) = 4y^3 + 3y^2 + 8y + 6$

Possible rational zeros: $\pm 1, \pm 2, \pm 3, \pm 6, \pm\frac{1}{2}, \pm\frac{3}{2}, \pm\frac{1}{4}, \pm\frac{3}{4}$

$$\begin{array}{r|rrrr} -\frac{3}{4} & 4 & 3 & 8 & 6 \\ & & -3 & 0 & -6 \\ \hline & 4 & 0 & 8 & 0 \end{array}$$

$4y^3 + 3y^2 + 8y + 6 = (y + \frac{3}{4})(4y^2 + 8)$
$\qquad = (y + \frac{3}{4})4(y^2 + 2)$
$\qquad = (4y + 3)(y^2 + 2)$

So, the only real zero is $-\frac{3}{4}$.

103. $P(x) = x^4 - \frac{25}{4}x^2 + 9$

$\qquad = \frac{1}{4}\left(4x^4 - 25x^2 + 36\right)$

$\qquad = \frac{1}{4}\left(4x^2 - 9\right)\left(x^2 - 4\right)$

$\qquad = \frac{1}{4}(2x + 3)(2x - 3)(x + 2)(x - 2)$

The rational zeros are $\pm\frac{3}{2}$ and ± 2.

105. $f(x) = x^3 - \frac{1}{4}x^2 - x + \frac{1}{4}$

$\qquad = \frac{1}{4}\left(4x^3 - x^2 - 4x + 1\right)$

$\qquad = \frac{1}{4}\left[x^2(4x - 1) - 1(4x - 1)\right]$

$\qquad = \frac{1}{4}(4x - 1)\left(x^2 - 1\right)$

$\qquad = \frac{1}{4}(4x - 1)(x + 1)(x - 1)$

The rational zeros are $\frac{1}{4}$ and ± 1.

107. $f(x) = x^3 - 1 = (x - 1)\left(x^2 + x + 1\right)$

Rational zeros: $1\,(x = 1)$

Irrational zeros: 0

Matches (d).

108. $f(x) = x^3 - 2$

$\qquad = \left(x - \sqrt[3]{2}\right)\left(x^2 + \sqrt[3]{2}x + \sqrt[3]{4}\right)$

Rational zeros: 0

Irrational zeros: $1\left(x = \sqrt[3]{2}\right)$

Matches (a).

109. $f(x) = x^3 - x = x(x + 1)(x - 1)$

Rational zeros: $3\,(x = 0, \pm 1)$

Irrational zeros: 0

Matches (b).

110. $f(x) = x^3 - 2x$

$\qquad = x\left(x^2 - 2\right)$

$\qquad = x\left(x + \sqrt{2}\right)\left(x - \sqrt{2}\right)$

Rational zeros: $1\,(x = 0)$

Irrational zeros: $2\left(x = \pm\sqrt{2}\right)$

Matches (c).

111. (a)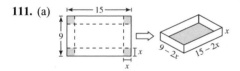

(b) $V = l \cdot w \cdot h = (15 - 2x)(9 - 2x)x$

$\qquad = x(9 - 2x)(15 - 2x)$

Because length, width, and height must be positive, you have $0 < x < \frac{9}{2}$ for the domain.

(c)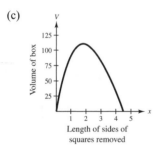

The volume is maximum when $x \approx 1.82$.

The dimensions are: length $\approx 15 - 2(1.82) = 11.36$

$\qquad\qquad\qquad$ width $\approx 9 - 2(1.82) = 5.36$

$\qquad\qquad\qquad$ height $= x \approx 1.82$

$1.82 \text{ cm} \times 5.36 \text{ cm} \times 11.36 \text{ cm}$

(d) $56 = x(9 - 2x)(15 - 2x)$

$\qquad 56 = 135x - 48x^2 + 4x^3$

$\qquad 0 = 4x^3 - 48x^2 + 135x - 56$

The zeros of this polynomial are $\frac{1}{2}, \frac{7}{2},$ and 8.

x cannot equal 8 because it is not in the domain of V. [The length cannot equal -1 and the width cannot equal -7. The product of $(8)(-1)(-7) = 56$ so it showed up as an extraneous solution.]

So, the volume is 56 cubic centimeters when $x = \frac{1}{2}$ centimeter or $x = \frac{7}{2}$ centimeters.

113. $\qquad\qquad\qquad P = -76x^3 + 4830x^2 - 320{,}000,\ 0 \le x \le 60$

$\qquad\qquad 2{,}500{,}000 = -76x^3 + 4830x^2 - 320{,}000$

$76x^3 - 4830x^2 + 2{,}820{,}000 = 0$

The zeros of this equation are $x \approx 46.1,\ x \approx 38.4,$ and $x \approx -21.0$. Because $0 \le x \le 60$, we disregard $x \approx -21.0$. The smaller remaining solution is $x \approx 38.4$. The advertising expense is \$384,000.

115. (a) Current bin: $V = 2 \times 3 \times 4 = 24$ cubic feet

New bin: $V = 5(24) = 120$ cubic feet

$$V(x) = (2 + x)(3 + x)(4 + x) = 120$$

(b) $x^3 + 9x^2 + 26x + 24 = 120$

$x^3 + 9x^2 + 26x - 96 = 0$

The only real zero of this polynomial is $x = 2$. All the dimensions should be increased by 2 feet, so the new bin will have dimensions of 4 feet by 5 feet by 6 feet.

117. $C = 100\left(\dfrac{200}{x^2} + \dfrac{x}{x + 30}\right), x \geq 1$

C is minimum when

$3x^3 - 40x^2 - 2400x - 36000 = 0$.

The only real zero is $x \approx 40$ or 4000 units.

119. $P = R - C = xp - C$

$= x(140 - 0.0001x) - (80x + 150,000)$

$= -0.0001x^2 + 60x - 150,000$

$9,000,000 = -0.0001x^2 + 60x - 150,000$

Thus, $0 = 0.0001x^2 - 60x + 9,150,000$.

$x = \dfrac{60 \pm \sqrt{-60}}{0.0002} = 300,000 \pm 10,000\sqrt{15}i$

Because the solutions are both complex, it is not possible to determine a price p that would yield a profit of 9 million dollars.

121. False. The most complex zeros it can have is two, and the Linear Factorization Theorem guarantees that there are three linear factors, so one zero must be real.

123. $g(x) = -f(x)$. This function would have the same zeros as $f(x)$, so $r_1, r_2,$ and r_3 are also zeros of $g(x)$.

125. $g(x) = f(x - 5)$. The graph of $g(x)$ is a horizontal shift of the graph of $f(x)$ five units of the right, so the zeros of $g(x)$ are $5 + r_1, 5 + r_2,$ and $5 + r_3$.

127. $g(x) = 3 + f(x)$. Because $g(x)$ is a vertical shift of the graph of $f(x)$, the zeros of $g(x)$ cannot be determined.

129. Zeros: $-2, \frac{1}{2}, 3$

$$f(x) = -(x + 2)(2x - 1)(x - 3)$$

$$= -2x^3 + 3x^2 + 11x - 6$$

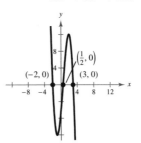

Any nonzero scalar multiple of f would have the same three zeros. Let $g(x) = af(x), a > 0$. There are infinitely many possible functions for f.

131.

133. Because $f(i) = f(2i) = 0$, then i and $2i$ are zeros of f. Because i and $2i$ are zeros of f, so are $-i$ and $-2i$.

$$f(x) = (x - i)(x + i)(x - 2i)(x + 2i)$$

$$= (x^2 + 1)(x^2 + 4)$$

$$= x^4 + 5x^2 + 4$$

135. Because $1 + i$ is a zero of f, so is $1 - i$. From the graph, 1 is also a zero.

$$f(x) = (x - (1 + i))(x - (1 - i))(x - 1)$$

$$= (x^2 - 2x + 2)(x - 1)$$

$$= x^3 - 3x^2 + 4x - 2$$

137. (a) Zeros of $f(x)$: $-2, 1, 4$

(b) The graph touches the x-axis at $x = 1$.

(c) The least possible degree of the function is 4 because there are at least four real zeros (1 is repeated) and a function can have at most the number of real zeros equal to the degree of the function. The degree cannot be odd by the definition of multiplicity.

(d) The leading coefficient of f is positive. From the information in the table, you can conclude that the graph will eventually rise to the left and to the right.

(e) Answers may vary. One possibility is:

$$f(x) = (x - 1)^2 (x - (-2))(x - 4)$$

$$= (x - 1)^2 (x + 2)(x - 4)$$

$$= (x^2 - 2x + 1)(x^2 - 2x - 8)$$

$$= x^4 - 4x^3 - 3x^2 + 14x - 8$$

(f)
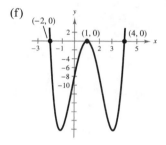

139. (a) $f(x)$ cannot have this graph because it also has a zero at $x = 0$.

(b) $g(x)$ cannot have this graph because it is a quadratic function. Its graph is a parabola.

(c) $h(x)$ is the correct function. It has two real zeros, $x = 2$ and $x = 3.5$, and it has a degree of four, needed to yield three turning points.

(d) $k(x)$ cannot have this graph because it also has a zero at $x = -1$. In addition, because it is only of degree three, it would have at most two turning points.

Section 3.5   Mathematical Modeling and Variation

You should know the following the following terms and formulas.

- Direct variation (varies directly, directly proportional)

  (a) $y = kx$

  (b) $y = kx^n$ (as nth power)

- Inverse variation (varies inversely, inversely proportional)

  (a) $y = k/x$

  (b) $y = k/(x^n)$ (as nth power)

- Joint variation (varies jointly, jointly proportional)

  (a) $z = kxy$

  (b) $z = kx^n y^m$ (as nth power of x and mth power of y)

- k is called the constant of proportionality.

- Least Squares Regression Line $y = ax + b$. Use your calculator or computer to enter the data points and to find the "best-fitting" linear model.

1. variation; regression

3. least squares regression

5. directly proportional

7. directly proportional

9. combined

11.

Year	Actual Number (in thousands)	Model (in thousands)
1992	128,105	127,712
1993	129,200	129,408
1994	131,056	131,104
1995	132,304	132,800
1996	133,943	134,495
1997	136,297	136,191
1998	137,673	137,887
1999	139,368	139,583
2000	142,583	141,279
2001	143,734	142,975
2002	144,863	144,671
2003	146,510	146,367
2004	147,401	148,063
2005	149,320	149,759
2006	151,428	151,454
2007	153,124	153,150

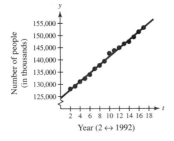

The model is a good fit for the actual data.

13.

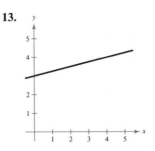

Using the point $(0, 3)$ and $(4, 4)$, $y = \frac{1}{4}x + 3$.

15.

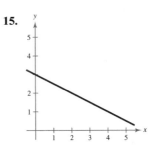

Using the points $(2, 2)$ and $(4, 1)$, $y = -\frac{1}{2}x + 3$.

17. (a)

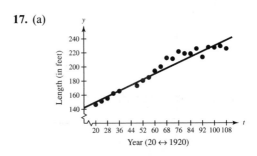

(b) Using the points $(32, 162.3)$ and $(96, 227.7)$:

$$m = \frac{227.7 - 162.3}{96 - 32}$$

$$\approx 1.02$$

$$y - 162.3 = 1.02(t - 32)$$

$$y = 1.02t + 129.66$$

(c) $y \approx 1.01t + 130.82$

(d) The models are similar.

(e) $2012 \rightarrow$ use $t = 112$

Model from part (b):

$$y = 1.02(112) + 129.66 = 243.9 \text{ feet}$$

Model from part (c):

$$y = 1.01(112) + 130.82 = 243.94 \text{ feet}$$

19. (a)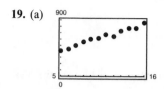

(b) $S \approx 38.3t + 224$

(c)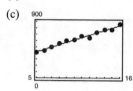

The model is a good fit to the actual data.

(d) $2007 \rightarrow$ use $t = 12$

$$S(12) = 38.3(17) + 224 \approx \$875.1 \text{ million}$$

$2009 \rightarrow$ use $t = 14$

$$S(14) = 38.3(19) + 224 \approx \$951.7 \text{ million}$$

(e) Each year the annual gross ticket sales for Broadway shows in New York City increase by \$38.3 million.

21. The graph appears to represent $y = 4/x$, so y varies inversely as x.

23. $k = 1$

x	2	4	6	8	10
$y = kx^2$	4	16	36	64	100

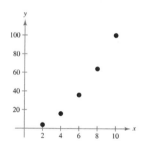

25. $k = \frac{1}{2}$

x	2	4	6	8	10
$y = kx^2$	2	8	18	32	50

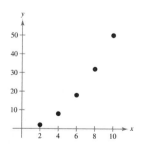

27. $k = 2$

x	2	4	6	8	10
$y = \dfrac{k}{x^2}$	$\dfrac{1}{2}$	$\dfrac{1}{8}$	$\dfrac{1}{18}$	$\dfrac{1}{32}$	$\dfrac{1}{50}$

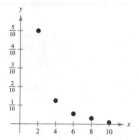

29. $k = 10$

x	2	4	6	8	10
$y = \dfrac{k}{x^2}$	$\dfrac{5}{2}$	$\dfrac{5}{8}$	$\dfrac{5}{18}$	$\dfrac{5}{32}$	$\dfrac{1}{10}$

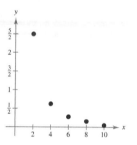

31. $y = \dfrac{k}{x}$

$$1 = \dfrac{k}{5}$$

$$5 = k$$

$$y = \dfrac{5}{x}$$

This equation checks with the other points given in the table.

33. $y = kx$

$$-7 = k(10)$$

$$-\tfrac{7}{10} = k$$

$$y = -\tfrac{7}{10}x$$

This equation checks with the other points given in the table.

35. $y = kx$

$12 = k(5)$

$\frac{12}{5} = k$

$y = \frac{12}{5}x$

37. $y = kx$

$2050 = k(10)$

$205 = k$

$y = 205x$

39. $I = kP$

$113.75 = k(3250)$

$0.035 = k$

$I = 0.035P$

41. $y = kx$

$33 = k(13)$

$\frac{33}{13} = k$

$y = \frac{33}{13}x$

When $x = 10$ inches, $y \approx 25.4$ centimeters.

When $x = 20$ inches, $y \approx 50.8$ centimeters.

43. $y = kx$

$5520 = k(150,000)$

$0.0368 = k$

$y = 0.0368x$

$y = 0.0368(225,000)$

$= \$8280$

The property tax is $8280.

45. $d = kF$

$0.15 = k(265)$

$\frac{3}{5300} = k$

$d = \frac{3}{5300}F$

(a) $d = \frac{3}{5300}(90) \approx 0.05$ meter

(b) $0.1 = \frac{3}{5300}F$

$\frac{530}{3} = F$

$F = 176\frac{2}{3}$ newtons

47. $d = kF$

$1.9 = k(25) \Rightarrow k = 0.076$

$d = 0.076F$

When the distance compressed is 3 inches, we have

$3 = 0.076F$

$F \approx 39.47.$

No child over 39.47 pounds should use the toy.

49. $A = kr^2$

51. $y = \dfrac{k}{x^2}$

53. $F = \dfrac{kg}{r^2}$

55. $P = \dfrac{k}{V}$

57. $F = \dfrac{km_1m_2}{r^2}$

59. $A = \frac{1}{2}bh$

The area of a triangle is jointly proportional to its base and height.

61. $A = \dfrac{\sqrt{3}s^2}{4}$

The area of an equilateral triangle varies directly as the square of one of its sides.

63. $V = \frac{4}{3}\pi r^3$

The volume of a sphere varies directly as the cube of its radius.

65. $r = \dfrac{d}{t}$

Average speed is directly proportional to the distance and inversely proportional to the time.

67. $A = kr^2$

$9\pi = k(3)^2$

$\pi = k$

$A = \pi r^2$

69. $y = \dfrac{k}{x}$

$7 = \dfrac{k}{4}$

$28 = k$

$y = \dfrac{28}{x}$

71. $F = krs^3$

$4158 = k(11)(3)^3$

$k = 14$

$F = 14rs^3$

73. $z = \dfrac{kx^2}{y}$

$6 = \dfrac{k(6)^2}{4}$

$\dfrac{24}{36} = k$

$\dfrac{2}{3} = k$

$z = \dfrac{2/3x^2}{y} = \dfrac{2x^2}{3y}$

75. $d = kv^2$

$0.02 = k\left(\dfrac{1}{4}\right)^2$

$k = 0.32$

$d = 0.32v^2$

$0.12 = 0.32v^2$

$v^2 = \dfrac{0.12}{0.32} = \dfrac{3}{8}$

$v = \dfrac{\sqrt{3}}{2\sqrt{2}} = \dfrac{\sqrt{6}}{4} \approx 0.61 \text{ mi/hr}$

77. $r = \dfrac{kl}{A},\ A = \pi r^2 = \dfrac{\pi d^2}{4}$

$r = \dfrac{4kl}{\pi d^2}$

$66.17 = \dfrac{4(1000)k}{\pi\left(\dfrac{0.0126}{12}\right)^2}$

$k \approx 5.73 \times 10^{-8}$

$r = \dfrac{4\left(5.73 \times 10^{-8}\right)l}{\pi\left(\dfrac{0.0126}{12}\right)^2}$

$33.5 = \dfrac{4\left(5.73 \times 10^{-8}\right)l}{\pi\left(\dfrac{0.0126}{12}\right)^2}$

$\dfrac{33.5\pi\left(\dfrac{0.0126}{12}\right)^2}{4\left(5.73 \times 10^{-8}\right)} = l$

$l \approx 506 \text{ feet}$

79. $W = kmh$

$2116.8 = k(120)(1.8)$

$k = \dfrac{2116.8}{(120)(1.8)} = 9.8$

$W = 9.8mh$

When $m = 100$ kilograms and $h = 1.5$ meters, we have $W = 9.8(100)(1.5) = 1470$ joules.

81. $v = \dfrac{k}{A}$

$v = \dfrac{k}{0.75A} = \dfrac{4}{3}\left(\dfrac{k}{A}\right)$

The velocity is increased by one-third.

83. (a)

Depth (in meters)

(b) Yes, the data appears to be modeled (approximately) by the inverse proportion model.

$4.2 = \dfrac{k_1}{1000}$ $\qquad 1.9 = \dfrac{k_2}{2000}$ $\qquad 1.4 = \dfrac{k_3}{3000}$ $\qquad 1.2 = \dfrac{k_4}{4000}$ $\qquad 0.9 = \dfrac{k_5}{5000}$

$4200 = k_1$ $\qquad 3800 = k_2$ $\qquad 4200 = k_3$ $\qquad 4800 = k_4$ $\qquad 4500 = k_5$

(c) Mean: $k = \dfrac{4200 + 3800 + 4200 + 4800 + 4500}{5} = 4300$, Model: $C = \dfrac{4300}{d}$

(d)

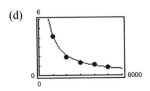

(e) $3 = \dfrac{4300}{d}$

$d = \dfrac{4300}{3} = 1433\dfrac{1}{3}$ meters

85. $y = \dfrac{262.76}{x^{2.12}}$

(a)

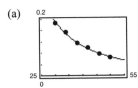

(b) $y = \dfrac{262.76}{(25)^{2.12}}$

≈ 0.2857 microwatts per sq. cm.

87. False. E is jointly proportional (not "directly proportional") to the mass of an object and the square of its velocity.

89. (a) The data shown could be represented by a linear model which would be a good approximation.

(b) The points do not follow a linear pattern. A linear model would be a poor approximation. A quadratic model would be better.

(c) The points do not follow a linear pattern. A linear model would be a poor approximation.

(d) The data shown could be represented by a linear model which would be a good approximation.

91. As one variable increases, the other variable will also increase. Answers will vary.

93. (a) y will change by a factor of one-fourth.

(b) y will change by a factor of four.

Review Exercises for Chapter 3

1. (a) $y = 2x^2$

Vertical stretch

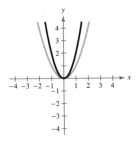

(b) $y = -2x^2$

Vertical stretch and a reflection in the x-axis

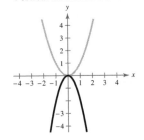

(c) $y = x^2 + 2$

Vertical shift two units upward

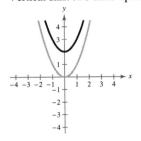

(d) $y = (x + 2)^2$

Horizontal shift two units to the left

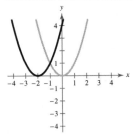

3. $g(x) = x^2 - 2x$

$\quad = x^2 - 2x + 1 - 1$

$\quad = (x - 1)^2 - 1$

Vertex: $(1, -1)$

Axis of symmetry: $x = 1$

$0 = x^2 - 2x = x(x - 2)$

x-intercepts: $(0, 0), (2, 0)$

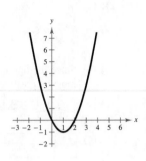

5. $f(x) = x^2 + 8x + 10$

$\quad = x^2 + 8x + 16 - 16 + 10$

$\quad = (x + 4)^2 - 6$

Vertex: $(-4, -6)$

Axis of symmetry: $x = -4$

$0 = (x + 4)^2 - 6$

$(x + 4)^2 = 6$

$x + 4 = \pm\sqrt{6}$

$x = -4 \pm \sqrt{6}$

x-intercepts: $\left(-4 \pm \sqrt{6}, 0\right)$

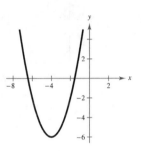

7. $f(t) = -2t^2 + 4t + 1$

$\quad = -2(t^2 - 2t + 1 - 1) + 1$

$\quad = -2\left[(t - 1)^2 - 1\right] + 1$

$\quad = -2(t - 1)^2 + 3$

Vertex: $(1, 3)$

Axis of symmetry: $t = 1$

$0 = -2(t - 1)^2 + 3$

$2(t - 1)^2 = 3$

$t - 1 = \pm\sqrt{\dfrac{3}{2}}$

$t = 1 \pm \dfrac{\sqrt{6}}{2}$

t-intercepts: $\left(1 \pm \dfrac{\sqrt{6}}{2}, 0\right)$

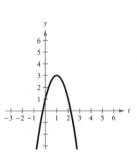

9. $h(x) = 4x^2 + 4x + 13$

$\quad = 4(x^2 + x) + 13$

$\quad = 4\left(x^2 + x + \frac{1}{4} - \frac{1}{4}\right) + 13$

$\quad = 4\left(x^2 + x + \frac{1}{4}\right) - 1 + 13$

$\quad = 4\left(x + \frac{1}{2}\right)^2 + 12$

Vertex: $\left(-\frac{1}{2}, 12\right)$

Axis of symmetry: $x = -\frac{1}{2}$

$0 = 4\left(x + \frac{1}{2}\right)^2 + 12$

$\left(x + \frac{1}{2}\right)^2 = -3$

No real zeros

x-intercepts: none

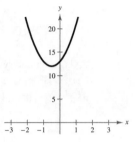

11. $h(x) = x^2 + 5x - 4$

$\quad = x^2 + 5x + \dfrac{25}{4} - \dfrac{25}{4} - 4$

$\quad = \left(x + \dfrac{5}{2}\right)^2 - \dfrac{25}{4} - \dfrac{16}{4}$

$\quad = \left(x + \dfrac{5}{2}\right)^2 - \dfrac{41}{4}$

Vertex: $\left(-\dfrac{5}{2}, -\dfrac{41}{4}\right)$

Axis of symmetry:

$x = -\dfrac{5}{2}$

$0 = x^2 + 5x - 4$

$x = \dfrac{-5 \pm \sqrt{5^2 - 4(1)(-4)}}{2(1)}$

$\quad = \dfrac{-5 \pm \sqrt{41}}{2}$

x-intercepts: $\left(\dfrac{-5 \pm \sqrt{41}}{2}, 0\right)$

206 *Chapter 3 Polynomial Functions*

13. $f(x) = \frac{1}{3}(x^2 + 5x - 4)$

$= \frac{1}{3}\left(x^2 + 5x + \frac{25}{4} - \frac{25}{4} - 4\right)$

$= \frac{1}{3}\left[\left(x + \frac{5}{2}\right)^2 - \frac{41}{4}\right]$

$= \frac{1}{3}\left(x + \frac{5}{2}\right)^2 - \frac{41}{12}$

Vertex: $\left(-\frac{5}{2}, -\frac{41}{12}\right)$

Axis of symmetry: $x = -\frac{5}{2}$

$0 = x^2 + 5x - 4$

$x = \frac{-5 \pm \sqrt{5^2 - 4(1)(-4)}}{2(1)} = \frac{-5 \pm \sqrt{41}}{2}$

x-intercepts: $\left(\frac{-5 \pm \sqrt{41}}{2}, 0\right)$

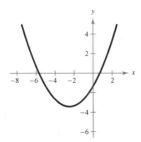

15. Vertex: $(4, 1) \Rightarrow f(x) = a(x - 4)^2 + 1$

Point: $(2, -1) \Rightarrow -1 = a(2 - 4)^2 + 1$

$-2 = 4a$

$-\frac{1}{2} = a$

$f(x) = -\frac{1}{2}(x - 4)^2 + 1$

17. Vertex: $(1, -4) \Rightarrow f(x) = a(x - 1)^2 - 4$

Point: $(2, -3) \Rightarrow -3 = a(2 - 1)^2 - 4$

$1 = a$

$f(x) = (x - 1)^2 - 4$

19. Vertex: $\left(-\frac{3}{2}, 0\right) \Rightarrow f(x) = a\left(x + \frac{3}{2}\right)^2$

Point: $\left(-\frac{9}{2}, -\frac{11}{4}\right) \Rightarrow -\frac{11}{4} = a\left(-\frac{9}{2} + \frac{3}{2}\right)^2$

$-\frac{11}{4} = 9a$

$-\frac{11}{36} = a$

$f(x) = -\frac{11}{36}\left(x + \frac{3}{2}\right)^2$

21. (a) $A = xy = x\left(\frac{8 - x}{2}\right)$ since

$x + 2y - 8 = 0 \Rightarrow y = \frac{8 - x}{2}$.

(b) Since the figure is in the first quadrant and x and y must be positive, the domain of

$A = x\left(\frac{8 - x}{2}\right)$ is $0 < x < 8$.

(c)

x	y	Area
1	$4 - \frac{1}{2}(1)$	$(1)\left[4 - \frac{1}{2}(1)\right] = \frac{7}{2}$
2	$4 - \frac{1}{2}(2)$	$(2)\left[4 - \frac{1}{2}(2)\right] = 6$
3	$4 - \frac{1}{2}(3)$	$(3)\left[4 - \frac{1}{2}(3)\right] = \frac{15}{2}$
4	$4 - \frac{1}{2}(4)$	$(4)\left[4 - \frac{1}{2}(4)\right] = 8$
5	$4 - \frac{1}{2}(5)$	$(5)\left[4 - \frac{1}{2}(5)\right] = \frac{15}{2}$
6	$4 - \frac{1}{2}(6)$	$(6)\left[4 - \frac{1}{2}(6)\right] = 6$

(d)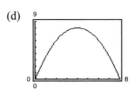

The maximum area of 8 occurs at the vertex when

$x = 4$ and $y = \frac{8 - 4}{2} = 2$.

(e) $A = x\left(\frac{8 - x}{2}\right)$

$= \frac{1}{2}(8x - x^2)$

$= -\frac{1}{2}(x^2 - 8x)$

$= -\frac{1}{2}(x^2 - 8x + 16 - 16)$

$= -\frac{1}{2}\left[(x - 4)^2 - 16\right]$

$= -\frac{1}{2}(x - 4)^2 + 8$

The maximum area of 8 occurs when $x = 4$ and

$y = \frac{8 - 4}{2} = 2$.

23. $R = -10p^2 + 800p$

(a) $R(20) = \$12{,}000$

$R(25) = \$13{,}750$

$R(30) = \$15{,}000$

(b) The maximum revenue occurs at the vertex of the parabola.

$$-\frac{b}{2a} = \frac{-800}{2(-10)} = \$40$$

$R(40) = \$16{,}000$

The revenue is maximum when the price is $40 per unit.

The maximum revenue is $16,000.

Any price greater or less than $40 per unit will not yield as much revenue.

25. $C = 70{,}000 - 120x + 0.055x^2$

The minimum cost occurs at the vertex of the parabola.

Vertex: $-\dfrac{b}{2a} = -\dfrac{-120}{2(0.055)} \approx 1091$ units

About 1091 units should be produced each day to yield a minimum cost.

27. $y = x^3,\ f(x) = -(x-2)^3$

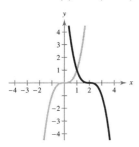

Transformation: Reflection in the *x*-axis and a horizontal shift two units to the right

29. $y = x^4,\ f(x) = 6 - x^4$

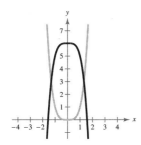

Transformation: Reflection in the *x*-axis and a vertical shift six units upward

31. $y = x^5,\ f(x) = (x-5)^5$

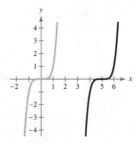

Transformation: Horizontal shift five units to the right

33. $f(x) = -2x^2 - 5x + 12$

The degree is even and the leading coefficient is negative. The graph falls to the left and falls to the right.

35. $g(x) = \frac{3}{4}\left(x^4 + 3x^2 + 2\right)$

The degree is even and the leading coefficient is positive. The graph rises to the left and rises to the right.

37. $f(x) = 3x^2 + 20x - 32$

$0 = 3x^2 + 20x - 32$

$0 = (3x - 4)(x + 8)$

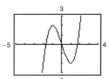

Zeros: $x = \frac{4}{3}$ and $x = -8$,

both of multiplicity 1 (odd multiplicity)

Turning points: 1

39. $f(t) = t^3 - 3t$

$0 = t^3 - 3t$

$0 = t(t^2 - 3)$

Zeros: $t = 0, \pm\sqrt{3}$, all of multiplicity 1 (odd multiplicity)

Turning points: 2

41. $f(x) = -18x^3 + 12x^2$

$0 = -18x^3 + 12x^2$

$0 = -6x^2(3x - 2)$

Zeros: $x = \frac{2}{3}$ of multiplicity 1 (odd multiplicity)

$x = 0$ of multiplicity 2 (even multiplicity)

Turning points: 2

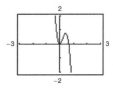

43. $f(x) = -x^3 + x^2 - 2$

(a) The degree is odd and the leading coefficient is negative. The graph rises to the left and falls to the right.

(b) Zero: $x = -1$

(c)

x	-3	-2	-1	0	1	2
$f(x)$	34	10	0	-2	-2	-6

(d)

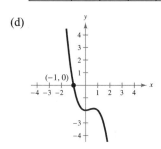

45. $f(x) = x(x^3 + x^2 - 5x + 3)$

(a) The degree is even and the leading coefficient is positive. The graph rises to the left and rises to the right.

(b) Zeros: $x = 0, 1, -3$

(c)

x	-4	-3	-2	-1	0	1	2	3
$f(x)$	100	0	-18	-8	0	0	10	72

(d)

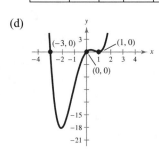

55.

$$
\begin{array}{r}
x^2 - 3x + 2 \\
x^2 + 0x + 2{\overline{\smash{\big)}\,x^4 - 3x^3 + 4x^2 - 6x + 3}} \\
\underline{x^4 + 0x^3 + 2x^2} \\
-3x^3 + 2x^2 - 6x \\
\underline{-3x^3 + 0x^2 - 6x} \\
2x^2 + 0x + 3 \\
\underline{2x^2 + 0x + 4} \\
-1
\end{array}
$$

$$\frac{x^4 - 3x^3 + 4x^2 - 6x + 3}{x^2 + 2} = x^2 - 3x + 2 - \frac{1}{x^2 + 2}$$

47. (a) $f(x) = 3x^3 - x^2 + 3$

x	-3	-2	-1	0	1	2	3
$f(x)$	-87	-25	-1	3	5	23	75

The zero is in the interval $[-1, 0]$.

(b) Zero: $x \approx -0.900$

49. (a) $f(x) = x^4 - 5x - 1$

x	-3	-2	-1	0	1	2	3
$f(x)$	95	25	5	-1	-5	5	65

There are zeros in the intervals $[-1, 0]$ and $[1, 2]$.

(b) Zeros: $x \approx -0.200$, $x \approx 1.772$

51.

$$
\begin{array}{r}
6x + 3 \\
5x - 3{\overline{\smash{\big)}\,30x^2 - 3x + 8}} \\
\underline{30x^2 - 18x} \\
15x + 8 \\
\underline{15x - 9} \\
17
\end{array}
$$

$$\frac{30x^2 - 3x + 8}{5x - 3} = 6x + 3 + \frac{17}{5x - 3}$$

53.

$$
\begin{array}{r}
5x + 4 \\
x^2 - 5x - 1{\overline{\smash{\big)}\,5x^3 - 21x^2 - 25x - 4}} \\
\underline{5x^3 - 25x^2 - 5x} \\
4x^2 - 20x - 4 \\
\underline{4x^2 - 20x - 4} \\
0
\end{array}
$$

$$\frac{5x^3 - 21x^2 - 25x - 4}{x^2 - 5x - 1} = 5x + 4, \; x \neq \frac{5}{2} \pm \frac{\sqrt{29}}{2}$$

57.

$$2 \begin{array}{|rrrrr} 6 & -4 & -27 & 18 & 0 \\ & 12 & 16 & -22 & -8 \\ \hline 6 & 8 & -11 & -4 & -8 \end{array}$$

$$\frac{6x^4 - 4x^3 - 27x^2 + 18x}{x - 2} = 6x^3 + 8x^2 - 11x - 4 - \frac{8}{x - 2}$$

59.

$$8 \begin{array}{|rrrr} 2 & -25 & 66 & 48 \\ & 16 & -72 & -48 \\ \hline 2 & -9 & -6 & 0 \end{array}$$

$$\frac{2x^3 - 25x^2 + 66x + 48}{x - 8} = 2x^2 - 9x - 6, \quad x \neq 8$$

61. $f(x) = 20x^4 + 9x^3 - 14x^2 - 3x$

(a)
$$-1 \begin{array}{|rrrrr} 20 & 9 & -14 & -3 & 0 \\ & -20 & 11 & 3 & 0 \\ \hline 20 & -11 & -3 & 0 & 0 \end{array}$$

Yes, $x = -1$ is a zero of f.

(b)
$$\tfrac{3}{4} \begin{array}{|rrrrr} 20 & 9 & -14 & -3 & 0 \\ & 15 & 18 & 3 & 0 \\ \hline 20 & 24 & 4 & 0 & 0 \end{array}$$

Yes, $x = \tfrac{3}{4}$ is a zero of f.

(c)
$$0 \begin{array}{|rrrrr} 20 & 9 & -14 & -3 & 0 \\ & 0 & 0 & 0 & 0 \\ \hline 20 & 9 & -14 & -3 & 0 \end{array}$$

Yes, $x = 0$ is a zero of f.

(d)
$$1 \begin{array}{|rrrrr} 20 & 9 & -14 & -3 & 0 \\ & 20 & 29 & 15 & 12 \\ \hline 20 & 29 & 15 & 12 & 12 \end{array}$$

No, $x = 1$ is not a zero of f.

63. $f(x) = x^4 + 10x^3 - 24x^2 + 20x + 44$

(a) Remainder Theorem:

$$f(-3) = (-3)^4 + 10(-3)^3 - 24(-3)^2 + 20(-3) + 44$$
$$= -421$$

Synthetic Division:

$$-3 \begin{array}{|rrrrr} 1 & 10 & -24 & 20 & 44 \\ & -3 & -21 & 135 & -465 \\ \hline 1 & 7 & -45 & 155 & -421 \end{array}$$

So, $f(-3) = -421$.

(b) Remainder Theorem:

$$f(-1) = (-1)^4 + 10(-1)^3 - 24(-1)^2 + 20(-1) + 44$$
$$= -9$$

Synthetic Division:

$$-1 \begin{array}{|rrrrr} 1 & 10 & -24 & 20 & 44 \\ & -1 & -9 & 33 & -53 \\ \hline 1 & 9 & -33 & 53 & -9 \end{array}$$

So, $f(-1) = -9$.

65. $f(x) = x^3 + 4x^2 - 25x - 28$; Factor: $(x - 4)$

(a)
$$4 \begin{array}{|rrrr} 1 & 4 & -25 & -28 \\ & 4 & 32 & 28 \\ \hline 1 & 8 & 7 & 0 \end{array}$$

Yes, $(x - 4)$ is a factor of $f(x)$.

(b) $x^2 + 8x + 7 = (x + 7)(x + 1)$

The remaining factors of f are $(x + 7)$ and $(x + 1)$.

(c) $f(x) = x^3 + 4x^2 - 25x - 28$
$$= (x + 7)(x + 1)(x - 4)$$

(d) Zeros: $-7, -1, 4$

(e)
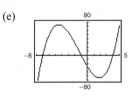

67. $f(x) = x^4 - 4x^3 - 7x^2 + 22x + 24$

Factors: $(x + 2), (x - 3)$

(a)

$$
\begin{array}{r|rrrrr}
-2 & 1 & -4 & -7 & 22 & 24 \\
   &   & -2 & 12 & -10 & -24 \\
\hline
   & 1 & -6 & 5 & 12 & 0
\end{array}
$$

$$
\begin{array}{r|rrrr}
3 & 1 & -6 & 5 & 12 \\
  &   & 3 & -9 & -12 \\
\hline
  & 1 & -3 & -4 & 0
\end{array}
$$

Yes, $(x + 2)$ and $(x - 3)$ are both factors of $f(x)$.

(b) $x^2 - 3x - 4 = (x + 1)(x - 4)$

The remaining factors are $(x + 1)$ and $(x - 4)$.

(c) $f(x) = (x + 1)(x - 4)(x + 2)(x - 3)$

(d) Zeros: $-2, -1, 3, 4$

(e)

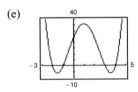

69. $f(x) = 4x(x - 3)^2$

Zeros: $x = 0, 3$

71. $f(x) = x^2 - 11x + 18$

$\quad = (x - 2)(x - 9)$

Zeros: $x = 2, 9$

73. $f(x) = (x + 4)(x - 6)(x - 2i)(x + 2i)$

Zeros: $x = -4, 6, 2i, -2i$

75. $f(x) = -4x^3 + 8x^2 - 3x + 15$

Possible rational zeros:

$\pm 1, \pm 3, \pm 5, \pm 15, \pm\frac{1}{2}, \pm\frac{3}{2}, \pm\frac{5}{2}, \pm\frac{15}{2}, \pm\frac{1}{4}, \pm\frac{3}{4}, \pm\frac{5}{4}, \pm\frac{15}{4}$

77. $f(x) = x^3 + 3x^2 - 28x - 60$

Possible rational zeros:

$\pm 1, \pm 2, \pm 3, \pm 4, \pm 5, \pm 6, \pm 10, \pm 12, \pm 15, \pm 20, \pm 30, \pm 60$

$$
\begin{array}{r|rrrr}
-2 & 1 & 3 & -28 & -60 \\
   &   & -2 & -2 & 60 \\
\hline
   & 1 & 1 & -30 & 0
\end{array}
$$

$x^3 + 3x^2 - 28x - 60 = (x + 2)(x^2 + x - 30)$

$\qquad\qquad\qquad\qquad\quad = (x + 2)(x + 6)(x - 5)$

The zeros of $f(x)$ are $x = -2, x = -6,$ and $x = 5$.

79. $f(x) = x^3 - 10x^2 + 17x - 8$

Possible rational zeros: $\pm 1, \pm 2, \pm 4, \pm 8$

$$
\begin{array}{r|rrrr}
1 & 1 & -10 & 17 & -8 \\
  &   & 1 & -9 & 8 \\
\hline
  & 1 & -9 & 8 & 0
\end{array}
$$

$x^3 - 10x^2 + 17x - 8 = (x - 1)(x^2 - 9x + 8)$

$\qquad\qquad\qquad\qquad\quad = (x - 1)(x - 1)(x - 8)$

$\qquad\qquad\qquad\qquad\quad = (x - 1)^2(x - 8)$

The zeros of $f(x)$ are $x = 1$ and $x = 8$.

81. $f(x) = x^4 + x^3 - 11x^2 + x - 12$

Possible rational zeros: $\pm 1, \pm 2, \pm 3, \pm 4, \pm 6, \pm 12$

$$
\begin{array}{r|rrrrr}
3 & 1 & 1 & -11 & 1 & -12 \\
  &   & 3 & 12 & 3 & 12 \\
\hline
  & 1 & 4 & 1 & 4 & 0
\end{array}
$$

$$
\begin{array}{r|rrrr}
-4 & 1 & 4 & 1 & 4 \\
   &   & -4 & 0 & -4 \\
\hline
   & 1 & 0 & 1 & 0
\end{array}
$$

$x^4 + x^3 - 11x^2 + x - 12 = (x - 3)(x + 4)(x^2 + 1)$

The zeros of $f(x)$ are $x = 3$ and $x = -4$.

83. Because $\sqrt{3}i$ is a zero, so is $-\sqrt{3}i$.

Multiply by 3 to clear the fraction.

$f(x) = 3\left(x - \frac{2}{3}\right)(x - 4)\left(x - \sqrt{3}i\right)\left(x + \sqrt{3}i\right)$

$\quad = (3x - 2)(x - 4)(x^2 + 3)$

$\quad = (3x^2 - 14x + 8)(x^2 + 3)$

$\quad = 3x^4 - 14x^3 + 17x^2 - 42x + 24$

Note: $f(x) = a(3x^4 - 14x^3 + 17x^2 - 42x + 24)$, where a is any real nonzero number, has zeros $\frac{2}{3}, 4,$

and $\pm\sqrt{3}i$.

85. $f(x) = x^3 - 4x^2 + x - 4$, Zero: i

Because i is a zero, so is $-i$.

$$
\begin{array}{r|rrrr}
i & 1 & -4 & 1 & -4 \\
  &   & i & -1-4i & 4 \\
\hline
  & 1 & -4+i & -4i & 0
\end{array}
$$

$$
\begin{array}{r|rrr}
-i & 1 & -4+i & -4i \\
   &   & -i & 4i \\
\hline
   & 1 & -4 & 0
\end{array}
$$

$f(x) = (x - i)(x + i)(x - 4)$

Zeros: $x = \pm i, 4$

87. $g(x) = 2x^4 - 3x^3 - 13x^2 + 37x - 15$, Zero: $2 + i$

Because $2 + i$ is a zero, so is $2 - i$.

$$
\begin{array}{r|rrrrr}
2+i & 2 & -3 & -13 & 37 & -15 \\
    &   & 4+2i & 5i & -31-3i & 15 \\
\hline
    & 2 & 1+2i & -13+5i & 6-3i & 0
\end{array}
$$

$$
\begin{array}{r|rrrr}
2-i & 2 & 1+2i & -13+5i & 6-3i \\
    &   & 4-2i & 10-5i & -6+3i \\
\hline
    & 2 & 5 & -3 & 0
\end{array}
$$

$g(x) = \left[x - (2 + i)\right]\left[x - (2 - i)\right](2x^2 + 5x - 3)$

$\quad = (x - 2 - i)(x - 2 + i)(2x - 1)(x + 3)$

Zeros: $x = 2 \pm i, \frac{1}{2}, -3$

89. $f(x) = x^3 + 4x^2 - 5x$

$\quad = x(x^2 + 4x - 5)$

$\quad = x(x + 5)(x - 1)$

Zeros: $x = 0, -5, 1$

91. $g(x) = x^4 + 4x^3 - 3x^2 + 40x + 208$, Zero: $x = -4$

$$
\begin{array}{r|rrrrr}
-4 & 1 & 4 & -3 & 40 & 208 \\
   &   & -4 & 0 & 12 & -208 \\
\hline
   & 1 & 0 & -3 & 52 & 0
\end{array}
$$

$$
\begin{array}{r|rrrr}
-4 & 1 & 0 & -3 & 52 \\
   &   & -4 & 16 & -52 \\
\hline
   & 1 & -4 & 13 & 0
\end{array}
$$

$g(x) = (x + 4)^2(x^2 - 4x + 13)$

By the Quadratic Formula the zeros of $x^2 - 4x + 13$ are $x = 2 \pm 3i$. The zeros of $g(x)$ are $x = -4$ and $x = 2 \pm 3i$.

$g(x) = (x + 4)^2\left[x - (2 + 3i)\right]\left[x - (2 - 3i)\right]$

$\quad = (x + 4)^2(x - 2 - 3i)(x - 2 + 3i)$

93. $f(x) = x^4 + 2x + 1$

(a)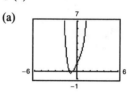

(b) The graph has two x-intercepts, so there are two real zeros.

(c) The zeros are $x = -1$ and $x \approx -0.54$.

95. $h(x) = x^3 - 6x^2 + 12x - 10$

(a)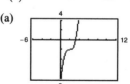

(b) The graph has one x-intercept, so there is one real zero.

(c) $x \approx 3.26$

97. $g(x) = 5x^3 + 3x^2 - 6x + 9$

$g(x)$ has two variations in sign, so g has either two or no positive real zeros.

$g(-x) = -5x^3 + 3x^2 + 6x + 9$

$g(-x)$ has one variation in sign, so g has one negative real zero.

99. $f(x) = 4x^3 - 3x^2 + 4x - 3$

(a)
$$
\begin{array}{r|rrrr}
1 & 4 & -3 & 4 & -3 \\
  &   & 4 & 1 & 5 \\
\hline
  & 4 & 1 & 5 & 2
\end{array}
$$

Because the last row has all positive entries, $x = 1$ is an upper bound.

(b)
$$
\begin{array}{r|rrrr}
-\frac{1}{4} & 4 & -3 & 4 & -3 \\
  &   & -1 & 1 & -\frac{5}{4} \\
\hline
  & 4 & -4 & 5 & -\frac{17}{4}
\end{array}
$$

Because the last row entries alternate in sign, $x = -\frac{1}{4}$ is a lower bound.

101. $V = -0.742t + 13.62$

(a)

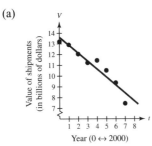

(b) The model is a good fit to the actual data.

103. $D = km$

$4 = 2.5k$

$1.6 = k$

$D = \frac{8}{5}m$ or $D = 1.6m$

In 2 miles:

$D = 1.6(2) = 3.2$ kilometers

In 10 miles:

$D = 1.6(10) = 16$ kilometers

105. $F = ks^2$

If speed is doubled,

$F = k(2s)^2$

$F = 4ks^2$.

So, the force will be changed by a factor of 4.

107. $T = \dfrac{k}{r}$

$3 = \dfrac{k}{65}$

$k = 3(65) = 195$

$T = \dfrac{195}{r}$

When $r = 80$ mph,

$T = \dfrac{195}{80} = 2.4375$ hours

$\approx$ 2 hours, 26 minutes.

109. False. A fourth-degree polynomial can have at most four zeros, and complex zeros occur in conjugate pairs.

111. The maximum (or minimum) value of a quadratic function is located at its graph's vertex. To find the vertex, either write the equation in standard form or use the formula

$$\left(-\frac{b}{2a}, f\left(-\frac{b}{2a}\right) \right).$$

If the leading coefficient is positive, the vertex is a minimum. If the leading coefficient is negative, the vertex is a maximum.

Problem Solving for Chapter 3

1. (a) (i) $g(x) = x^2 - 4x - 12$

$0 = (x - 6)(x + 2)$

Zeros: 6, −2

(ii) $g(x) = x^2 + 5x$

$0 = x(x + 5)$

Zeros: 0, −5

(iii) $g(x) = x^2 + 3x - 10$

$0 = (x + 5)(x - 2)$

Zeros: −5, 2

(iv) $g(x) = x^2 - 4x + 4$

$0 = (x - 2)(x - 2)$

Zeros: 2, 2

(v) $g(x) = x^2 - 2x - 6$

$0 = x^2 - 2x - 6$

By the Quadratic Formula, $x = 1 \pm \sqrt{7}$.

Zeros: $1 \pm \sqrt{7}$

(vi) $g(x) = x^2 + 3x + 4$

$0 = x^2 + 3x + 4$

By the Quadratic Formula, $x = \dfrac{-3 \pm \sqrt{7}i}{2}$.

Zeros: $\dfrac{-3 \pm \sqrt{7}i}{2}$

(b) (i) $f(x) = (x - 2)(x^2 - 4x - 12)$

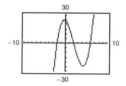

(ii) $f(x) = (x - 2)(x^2 + 5x)$

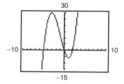

(iii) $f(x) = (x - 2)(x^2 + 3x - 10)$

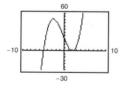

(iv) $f(x) = (x - 2)(x^2 - 4x + 4)$

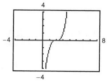

(v) $f(x) = (x - 2)(x^2 - 2x - 6)$

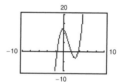

(vi) $f(x) = (x - 2)(x^2 + 3x + 4)$

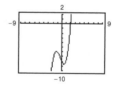

$x = 2$ is an x-intercept of $f(x)$ in all six graphs. All the graphs, except (iii), cross the x-axis at $x = 2$.

(c) (i)   other x-intercepts: $(-2, 0), (6, 0)$

(ii)   other x-intercepts: $(-5, 0), (0, 0)$

(iii)   other x-intercepts: $(-5, 0)$

(iv)   other x-intercepts: No additional x-intercepts

(v)   other x-intercepts: $(-1.6, 0), (3.6, 0)$

(vi)   other x-intercepts: No additional x-intercepts

(d) The x-intercepts of $f(x) = (x - 2)g(x)$ are the same as the zeros of $g(x)$ plus $x = 2$.

3. $f(x) = ax^3 + bx^2 + cx + d$

$$
\begin{array}{r}
ax^2 + (ak + b)x + (ak^2 + bk + c) \\
x - k \overline{)\; ax^3 + bx^2 \quad\;\; + cx \quad\quad\;\; + d} \\
\underline{ax^3 - akx^2} \\
(ak + b)x^2 + cx \\
\underline{(ak + b)x^2 - (ak^2 + bk)x} \\
(ak^2 + bk + c)x + d \\
\underline{(ak^2 + bk + c)x - (ak^3 + bk^2 + ck)} \\
(ak^3 + bk^2 + ck + d)
\end{array}
$$

So, $f(x) = ax^3 + bx^2 + cx + d = (x - k)\left[ax^2 + (ak + b)x + (ak^2 + bx + c)\right] + ak^3 + bk^2 + ck + d$ and

$f(x) = ak^3 + bk^2 + ck + d$. Because the remainder is $r = ak^3 + bk^2 + ck + d$, $f(k) = r$.

5. $V = l \cdot w \cdot h = x^2(x + 3)$

$x^2(x + 3) = 20$

$x^3 + 3x^2 - 20 = 0$

Possible rational zeros: $\pm 1, \pm 2, \pm 4, \pm 5, \pm 10, \pm 20$

$$
\begin{array}{r|rrrr}
2 & 1 & 3 & 0 & -20 \\
  &   & 2 & 10 & 20 \\
\hline
  & 1 & 5 & 10 & 0
\end{array}
$$

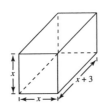

$(x - 2)(x^2 + 5x + 10) = 0$

$x = 2$ or $x = \dfrac{-5 \pm \sqrt{15}i}{2}$

Choosing the real positive value for x we have:

$x = 2$ and $x + 3 = 5$.

The dimensions of the mold are 2 inches $\times$ 2 inches $\times$ 5 inches.

7. False. Since $f(x) = d(x)q(x) + r(x)$, we have $\dfrac{f(x)}{d(x)} = q(x) + \dfrac{r(x)}{d(x)}$.

The statement should be corrected to read $f(-1) = 2$ since $\dfrac{f(x)}{x + 1} = q(x) + \dfrac{f(-1)}{x + 1}$.

9. (a) Slope $= \dfrac{9 - 4}{3 - 2} = 5$

Slope of tangent line is less than 5.

(b) Slope $= \dfrac{4 - 1}{2 - 1} = 3$

Slope of tangent line is greater than 3.

(c) Slope $= \dfrac{4.41 - 4}{2.1 - 2} = 4.1$

Slope of tangent line is less than 4.1.

(d) Slope $= \dfrac{f(2 + h) - f(2)}{(2 + h) - 2} = \dfrac{(2 + h)^2 - 4}{h} = \dfrac{4h + h^2}{h} = 4 + h, h \neq 0$

(e)   Slope $= 4 + h, \quad h \neq 0$

   $4 + (-1) = 3$

   $4 + 1 = 5$

   $4 + 0.1 = 4.1$

The results are the same as in (a)–(c).

(f) Letting h get closer and closer to 0, the slope approaches 4. So, the slope at $(2, 4)$ is 4.

11. Let $x =$ length of the wire used to form the square. Then $100 - x =$ length of wire used to form the circle.

(a) Let $s =$ the side of the square. Then $4s = x \Rightarrow s = x/4$ and the area of the square is $s^2 = (x/4)^2$.

Let $r =$ the radius of the circle. Then

$$2\pi r = 100 - x \Rightarrow r = \frac{100 - x}{2\pi} \text{ and the area of the circle is } \pi r^2 = \pi\left(\frac{100 - x}{2\pi}\right)^2.$$

The combined area is:

$$A(x) = \left(\frac{x}{4}\right)^2 + \pi\left(\frac{100 - x}{2\pi}\right)^2$$

$$= \frac{x^2}{16} + \pi\left(\frac{10{,}000 - 200x + x^2}{4\pi^2}\right)$$

$$= \frac{x^2}{16} + \frac{2500}{\pi} = \frac{50x}{\pi} + \frac{x^2}{4\pi}$$

$$= \left(\frac{1}{16} + \frac{1}{4\pi}\right)x^2 - \frac{50x}{\pi} + \frac{2500}{\pi}$$

$$= \left(\frac{x + 4}{16\pi}\right)x^2 - \frac{50}{\pi}x + \frac{2500}{\pi}$$

(b) Domain: Since the wire is 100 cm, $0 \leq x \leq 100$.

(c) $A(x) = \left(\dfrac{x + 4}{16\pi}\right)x^2 - \dfrac{50}{\pi}x + \dfrac{2500}{\pi}$

$$= \left(\frac{x + 4}{16\pi}\right)\left(x^2 - \frac{800}{\pi + 4}x\right) + \frac{2500}{\pi}$$

$$= \left(\frac{x + 4}{16\pi}\right)\left[x^2 - \frac{800}{\pi + 4}x + \left(\frac{400}{\pi + 4}\right)^2 - \left(\frac{400}{\pi + 4}\right)^2\right] + \frac{2500}{\pi}$$

$$= \left(\frac{x + 4}{16\pi}\right)\left[x - \left(\frac{400}{\pi + 4}\right)\right]^2 - \left(\frac{\pi + 4}{16\pi}\right)\left(\frac{400}{\pi + 4}\right)^2 + \frac{2500}{\pi}$$

$$= \left(\frac{x + 4}{16\pi}\right)\left[x - \left(\frac{400}{\pi + 4}\right)\right]^2 - \frac{10{,}000}{\pi(\pi + 4)} + \frac{2500}{\pi}$$

$$= \left(\frac{x + 4}{16\pi}\right)\left[x - \left(\frac{400}{\pi + 4}\right)\right]^2 + \frac{2500}{\pi + 4}$$

The minimum occurs at the vertex when $x = 400/(\pi + 4) \approx 56$ cm and $A(x) \approx 350$ cm^2. The maximum occurs at one of the endpoints of the domain. When $x = 0$, $A(x) \approx 796$ cm^2. When $x = 100$, $A(x) \approx 625$ cm^2. Thus, the area is maximum when $x = 0$ cm.

(d) Answers will vary. Graph $A(x)$ to see where the minimum and maximum values occur.

Chapter 3 Practice Test

1. Sketch the graph of $f(x) = x^2 - 6x + 5$ and identify the vertex and the intercepts.

2. Find the number of units x that produce a minimum cost C if
$$C = 0.01x^2 - 90x + 15,000.$$

3. Find the quadratic function that has a maximum at $(1, 7)$ and passes through the point $(2, 5)$.

4. Find two quadratic functions that have x-intercepts $(2, 0)$ and $\left(\frac{4}{3}, 0\right)$.

5. Use the leading coefficient test to determine the right and left end behavior of the graph of the polynomial function $f(x) = -3x^5 + 2x^3 - 17$.

6. Find all the real zeros of $f(x) = x^5 - 5x^3 + 4x$.

7. Find a polynomial function with 0, 3, and −2 as zeros.

8. Sketch $f(x) = x^3 - 12x$.

9. Divide $3x^4 - 7x^2 + 2x - 10$ by $x - 3$ using long division.

10. Divide $x^3 - 11$ by $x^2 + 2x - 1$.

11. Use synthetic division to divide $3x^5 + 13x^4 + 12x - 1$ by $x + 5$.

12. Use synthetic division to find $f(-6)$ given $f(x) = 7x^3 + 40x^2 - 12x + 15$.

13. Find the real zeros of $f(x) = x^3 - 19x - 30$.

14. Find the real zeros of $f(x) = x^4 + x^3 - 8x^2 - 9x - 9$.

15. List all possible rational zeros of the function $f(x) = 6x^3 - 5x^2 + 4x - 15$.

16. Find the rational zeros of the polynomial $f(x) = x^3 - \frac{20}{3}x^2 + 9x - \frac{10}{3}$.

17. Write $f(x) = x^4 + x^3 + 5x - 10$ as a product of linear factors.

18. Find a polynomial with real coefficients that has $2, 3 + i$, and $3 - 2i$ as zeros.

19. Use synthetic division to show that $3i$ is a zero of $f(x) = x^3 + 4x^2 + 9x + 36$.

20. Find a mathematical model for the statement, "z varies directly as the square of x and inversely as the square of x and inversely as the square root of y."

C H A P T E R 4
Rational Functions and Conics

C H A P T E R   4
Rational Functions and Conics

Section 4.1   Rational Functions and Asymptotes

- You should know the following basic facts about rational functions.

  (a) A function of the form $f(x) = N(x)/D(x),\ D(x) \neq 0$, where $N(x)$ and $D(x)$ are polynomials, is called a **rational function.**

  (b) The domain of a rational function is the set of all real numbers except those which make the denominator zero.

  (c) If $f(x) = N(x)/D(x)$ is in reduced form, and a is a value such that $D(a) = 0$, then the line $x = a$ is a **vertical asymptote** of the graph of f. $f(x) \to \infty$ or $f(x) \to -\infty$ as $x \to a$.

  (d) The line $y = b$ is a **horizontal asymptote** of the graph of f if $f(x) \to b$ as $x \to \infty$ or $x \to -\infty$.

  (e) Let $f(x) = \dfrac{N(x)}{D(x)} = \dfrac{a_n x^n + a_{n-1}x^{n-1} + \cdots + a_1 x + a_0}{b_m x^m + b_{m-1}x^{m-1} + \cdots + b_1 x + b_0}$, where $N(x)$ and $D(x)$ have no common factors.

    1. If $n < m$, then the x-axis $(y = 0)$ is a horizontal asymptote.

    2. If $n = m$, then $y = \dfrac{a_n}{b_m}$ is a horizontal asymptote.

    3. If $n > m$, then there is no horizontal asymptote.

1. rational functions

3. horizontal asymptote

5. $f(x) = \dfrac{1}{x-1}$

(a) Because the denominator is zero when $x - 1 = 0$, the domain of f is all real numbers except $x = 1$.

(b)

x	0.5	0.9	0.99	0.999
$f(x)$	-2	-10	-100	-1000

x	1.5	1.1	1.01	1.001
$f(x)$	2	10	100	1000

x	-1.5	-1.1	-1.01	-1.001
$f(x)$	-0.4	-0.48	-0.498	-0.4998

x	-0.5	-0.9	-0.99	-0.999
$f(x)$	$-0.6\overline{6}$	-0.53	-0.503	-0.5003

(c) As x approaches 1 from the left, $f(x)$ decreases without bound. As x approaches 1 from the right, $f(x)$ increases without bound.

7. $f(x) = \dfrac{3x^2}{x^2 - 1}$

(a) Because the denominator is zero when $x^2 - 1 = 0$, the domain of f is all real numbers except $x = \pm 1$.

(b)

x	0.5	0.9	0.99	0.999
$f(x)$	-1	-12.79	-147.8	-1498

x	1.5	1.1	1.01	1.001
$f(x)$	5.4	17.29	152.3	1502

x	-1.5	-1.1	-1.01	-1.001
$f(x)$	5.4	17.29	152.3	1502

x	-0.5	-0.9	-0.99	-0.999
$f(x)$	-1	-12.79	-147.8	-1498

(c) As x approaches 1 from the left, $f(x)$ decreases without bound. As x approaches 1 from the right, $f(x)$ increases without bound. As x approaches -1 from the left, $f(x)$ increases without bound. As x approaches -1 from the right, $f(x)$ decreases without bound.

9. $f(x) = \dfrac{4}{x^2}$

Domain: all real numbers except $x = 0$

Vertical asymptote: $x = 0$

Horizontal asymptote: $y = 0$

$\left[\text{Degree of } N(x) < \text{degree of } D(x) \right]$

11. $f(x) = \dfrac{5 + x}{5 - x} = \dfrac{x + 5}{-x + 5}$

Domain: all real numbers except $x = 5$

Vertical asymptote: $x = 5$

Horizontal asymptote: $y = -1$

$\left[\text{Degree of } N(x) = \text{degree of } D(x) \right]$

13. $f(x) = \dfrac{x^3}{x^2 - 1}$

Domain: all real numbers except $x = \pm 1$

Vertical asymptotes: $x = \pm 1$

Horizontal asymptote: None

$\left[\text{Degree of } N(x) > \text{degree of } D(x) \right]$

15. $f(x) = \dfrac{3x^2 + 1}{x^2 + x + 9}$

Domain: All real numbers. The denominator has no real zeros. [Try the Quadratic Formula on the denominator.]

Vertical asymptote: None

Horizontal asymptote: $y = 3$

$\left[\text{Degree of } N(x) = \text{degree of } D(x) \right]$

17. $f(x) = \dfrac{4}{x + 5}$

Vertical asymptote: $y = -5$

Horizontal asymptote: $y = 0$

Matches graph (d).

18. $f(x) = \dfrac{5}{x - 2}$

Vertical asymptote: $x = 2$

Horizontal asymptote: $y = 0$

Matches graph (a).

19. $f(x) = \dfrac{x - 1}{x - 4}$

Vertical asymptote: $x = 4$

Horizontal asymptote: $y = 1$

Matches graph (c).

20. $f(x) = -\dfrac{x + 2}{x + 4}$

Vertical asymptote: $x = -4$

Horizontal asymptote: $y = -1$

Matches graph (b).

21. $g(x) = \dfrac{x^2 - 1}{x + 1} = \dfrac{(x - 1)(x + 1)}{x + 1}$

·The only zero of $g(x)$ is $x = 1$.

$x = -1$ makes $g(x)$ undefined.

23.

$$h(x) = 2 + \dfrac{5}{x^2 + 2}$$

$$0 = 2 + \dfrac{5}{x^2 + 2}$$

$$-2 = \dfrac{5}{x^2 + 2}$$

$$-2x^2 - 4 = 5$$

$$x^2 = -\dfrac{9}{2}$$

No real solution, $h(x)$ has no real zeros.

25. $f(x) = 1 - \dfrac{3}{x - 3}$

$$0 = 1 - \dfrac{3}{x - 3}$$

$$-1 = -\dfrac{3}{x - 3}$$

$$x - 3 = 3$$

$$x = 6$$

The zero is $x = 6$.

27. $g(x) = \dfrac{x^3 - 8}{x^2 + 1}$

$$0 = \dfrac{x^3 - 8}{x^2 + 1}$$

$$0 = x^3 - 8$$

$$8 = x^3$$

$$2 = x$$

The zero is $x = 2$.

29. $f(x) = \dfrac{x - 4}{x^2 - 16} = \dfrac{1}{x + 4}, \; x \neq 4$

Domain: all real numbers x except $x = \pm 4$

Horizontal asymptote: $y = 0$

$\left(\text{Degree of } N(x) < \text{degree of } D(x) \right)$

Vertical asymptote: $x = -4$

(Because $x - 4$ is a common factor of $N(x)$ and $D(x)$, $x = 4$ is not a vertical asymptote of $f(x)$.)

31. $f(x) = \dfrac{x^2 - 1}{x^2 - 2x - 3} = \dfrac{(x+1)(x-1)}{(x+1)(x-3)} = \dfrac{x-1}{x-3}, \; x \neq -1$

Domain: all real numbers x except $x = -1$ and $x = 3$

Horizontal asymptote: $y = 1 \left(\text{Degree of } N(x) = \text{degree of } D(x) \right)$

Vertical asymptote: $x = 3$

(Because $x + 1$ is a common factor of $N(x)$ and $D(x)$, $x = -1$ is not a vertical asymptote of $f(x)$.)

33. $f(x) = \dfrac{x^2 - 3x - 4}{2x^2 + x - 1} = \dfrac{(x+1)(x-4)}{(2x-1)(x+1)} = \dfrac{x-4}{2x-1}, \; x \neq -1$

Domain: all real numbers x except $x = \dfrac{1}{2}$ and $x = -1$

Horizontal asymptote: $y = \dfrac{1}{2}$

$\left(\text{Degree of } N(x) = \text{degree of } D(x) \right)$

Vertical asymptote: $x = \dfrac{1}{2}$

(Because $x + 1$ is a common factor of $N(x)$ and $D(x)$, $x = -1$ is not a vertical asymptote of $f(x)$.)

35. $f(x) = \dfrac{6x^2 + 5x - 6}{3x^2 - 8x + 4} = \dfrac{(3x-2)(2x+3)}{(3x-2)(x-2)} = \dfrac{2x+3}{x-2}, \; x \neq \dfrac{2}{3}$

Domain: all real numbers x except $x = \dfrac{2}{3}$ and $x = 2$

Horizontal asymptote: $y = 2 \left(\text{Degree of } N(x) = \text{degree of } D(x) \right)$

Vertical asymptote: $x = 2$

(Because $3x - 2$ is a common factor of $N(x)$ and $D(x)$, $x = \dfrac{2}{3}$ is not a vertical asymptote of $f(x)$.)

37. $f(x) = \dfrac{x^2 - 4}{x + 2}, \; g(x) = x - 2$

$f(x) = \dfrac{(x+2)(x-2)}{x+2}$

(a) Domain of f: all real numbers except $x = -2$

Domain of g: all real numbers

(b) $f(x) = x - 2$

Because $x + 2$ is a common factor of both the numerator and the denominator of $f(x)$, $x = -2$ is not a vertical asymptote of f. f has no vertical asymptotes.

(c)

x	−4	−3	−2.5	−2	−1.5	−1	0
$f(x)$	−6	−5	−4.5	Undef.	−3.5	−3	−2
$g(x)$	−6	−5	−4.5	−4	−3.5	−3	−2

(d) f and g differ only at $x = -2$, where f is undefined and g is defined.

39. $f(x) = \dfrac{2x-1}{2x^2-x}$, $g(x) = \dfrac{1}{x}$

$$f(x) = \dfrac{2x-1}{x(2x-1)}$$

(a) Domain of f: all real numbers except $x = 0$ and

$$x = \dfrac{1}{2}$$

Domain of g: all real numbers except $x = 0$

(b) $f(x) = \dfrac{1}{x}$

Because $2x - 1$ is a common factor of both the numerator and the denominator of f, $x = 0.5$ is not a vertical asymptote of f. The only vertical asymptote is $x = 0$.

(c)

x	-1	-0.5	0	0.5	2	3	4
$f(x)$	-1	-2	Undef.	Undef.	$\frac{1}{2}$	$\frac{1}{3}$	$\frac{1}{4}$
$g(x)$	-1	-2	Undef.	2	$\frac{1}{2}$	$\frac{1}{3}$	$\frac{1}{4}$

(d) f and g differ only at $x = 0.5$, where f is undefined and g is defined.

41. $C = \dfrac{255p}{100-p}$, $0 \le p < 100$

(a)
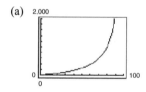

(b) $C(10) = \dfrac{255(10)}{100-10} \approx \$28.33$ million

$C(40) = \dfrac{255(40)}{100-40} = \$170$ million

$C(75) = \dfrac{255(75)}{100-75} = \$765$ million

(c) $C \to \infty$ as $x \to 100$. No, it would not be possible to remove 100% of the pollutants because the model is undefined for $p = 100$.

43. $t = \dfrac{38M + 16{,}965}{10(M + 5000)}$

(a)

M	200	400	600	800	1000
t	0.472	0.596	0.710	0.817	0.916

M	1200	1400	1600	1800	2000
t	1.009	1.096	1.178	1.255	1.328

(b) The model is a good fit to the actual data. The greater the mass, the more time required per oscillation.

(c) For $t = 1.056$ seconds:

$$1.056 = \dfrac{38M + 16{,}965}{10(M + 5000)}$$

$$10.56M + 52{,}800 = 38M + 16{,}965$$

$$35{,}835 = 27.44M$$

$$1306 \approx M$$

The mass is approximately 1306 grams.

45. (a)
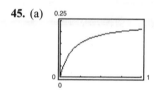

(b) The moth will become satiated at the horizontal asymptote:

$$y = \dfrac{1.568}{6.360} \approx 0.247 \text{ mg}$$

47. False. Polynomial functions do not have vertical asymptotes.

49. $f(x) = 4 - \dfrac{1}{x}$

(a) As $x \to \pm\infty$, $f(x) \to 4$.

(b) As $x \to \infty$, $f(x) \to 4$ but is less than 4.

(c) As $x \to -\infty$, $f(x) \to 4$ but is greater than 4.

51. $f(x) = \dfrac{2x-1}{x-3}$

(a) As $x \to \pm\infty$, $f(x) \to 2$.

(b) As $x \to \infty$, $f(x) \to 2$ but is greater than 2.

(c) As $x \to -\infty$, $f(x) \to 2$ but is less than 2.

53. Vertical asymptote: None $\Rightarrow$ The denominator is not zero for any value of x (unless the numerator is also zero there).

Horizontal asymptote: $y = 2 \Rightarrow$ The degree of the numerator equals the degree of the denominator.

$f(x) = \dfrac{2x^2}{x^2 + 1}$ is one possible function. There are many correct answers.

55. Domain: All real numbers

Example: $f(x) = \dfrac{1}{x^2 + 2}$

Domain: All real numbers except $x = 15$

Example: $f(x) = \dfrac{1}{x - 15}$

There are many correct answers.

Section 4.2   Graphs of Rational Functions

> ■ You should be able to graph $f(x) = \dfrac{N(x)}{D(x)}$, where $N(x)$ and $D(x)$ are polynomials with no common factors.
>
>   (a) Find the x- and y-intercepts.
>
>   (b) Test for symmetry.
>
>   (c) Find any vertical or horizontal asymptotes.
>
>   (d) Plot additional points.
>
>   (e) If the degree of the numerator is one more than the degree of the denominator, use long division to find the slant asymptote.

1. slant asymptote

3. $g(x) = \dfrac{2}{x} + 4$

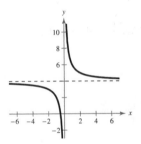

Vertical shift four units upward

5. $g(x) = -\dfrac{2}{x}$

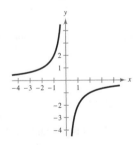

Reflection in the x-axis

7. $g(x) = \dfrac{3}{x^2} - 1$

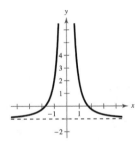

Vertical shift one unit downward

9. $g(x) = \dfrac{3}{(x - 1)^2}$

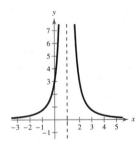

Horizontal shift one unit to the right

11. $g(x) = \dfrac{4}{(x+2)^3}$

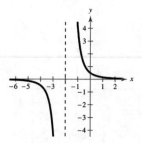

Horizontal shift two units to the left

13. $g(x) = -\dfrac{4}{x^3}$

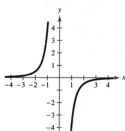

Reflection in the *x*-axis

15. $f(x) = \dfrac{1}{x+2}$

(a) Domain: all real numbers *x* except $x = -2$

(b) *y*-intercept: $\left(0, \dfrac{1}{2}\right)$

(c) Vertical asymptote: $x = -2$

Horizontal asymptote: $y = 0$

(d)

x	-4	-3	-1	0	1
$f(x)$	$-\dfrac{1}{2}$	-1	1	$\dfrac{1}{2}$	$\dfrac{1}{3}$

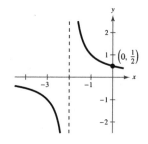

17. $h(x) = \dfrac{-1}{x+4}$

(a) Domain: all real numbers *x* except $x = -4$

(b) *y*-intercept: $\left(0, -\dfrac{1}{4}\right)$

(c) Vertical asymptote: $x = -4$

Horizontal asymptote: $y = 0$

(d)

x	-6	-5	-3	-2	-1	0
$h(x)$	$\dfrac{1}{2}$	1	-1	$-\dfrac{1}{2}$	$-\dfrac{1}{3}$	$-\dfrac{1}{4}$

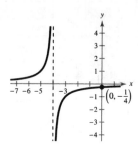

19. $C(x) = \dfrac{7+2x}{2+x}$

(a) Domain: all real numbers *x* except $x = -2$

(b) *x*-intercept: $\left(-\dfrac{7}{2}, 0\right)$

y-intercept: $\left(0, \dfrac{7}{2}\right)$

(c) Vertical asymptote: $x = -2$

Horizontal asymptote: $y = 2$

(d)

x	-4	-3	-1	0	1
$C(x)$	$\dfrac{1}{2}$	-1	5	$\dfrac{7}{2}$	3

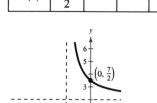

21. $g(x) = \dfrac{1}{x+2} + 2 = \dfrac{2x+5}{x+2}$

    (a) Domain: all real numbers x except $x = -2$

    (b) x-intercept: $\left(-\dfrac{5}{2}, 0\right)$

        y-intercept: $\left(0, \dfrac{5}{2}\right)$

    (c) Vertical asymptote: $x = -2$

        Horizontal asymptote: $y = 2$

    (d)

x	-4	-3	-1	0	1
$g(x)$	$\dfrac{3}{2}$	1	3	$\dfrac{5}{2}$	$\dfrac{7}{3}$

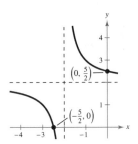

23. $f(x) = \dfrac{x^2}{x^2 + 9}$

    (a) Domain: all real numbers x

    (b) Intercept: $(0, 0)$

    (c) Horizontal asymptote: $y = 1$

    (d)

x	± 1	± 2	± 3
$f(x)$	$\dfrac{1}{10}$	$\dfrac{4}{13}$	$\dfrac{1}{2}$

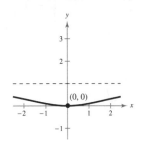

25. $h(x) = \dfrac{x^2}{x^2 - 9}$

    (a) Domain: all real numbers x except $x = \pm 3$

    (b) Intercept: $(0, 0)$

    (c) Vertical asymptotes: $x = \pm 3$

        Horizontal asymptote: $y = 1$

    (d)

x	± 5	± 4	± 2	± 1	0
$h(x)$	$\dfrac{25}{16}$	$\dfrac{16}{7}$	$-\dfrac{4}{5}$	$-\dfrac{1}{8}$	0

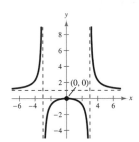

27. $g(s) = \dfrac{4s}{s^2 + 4}$

    (a) Domain: all real numbers s

    (b) Intercept: $(0, 0)$

    (c) Horizontal asymptote: $y = 0$

    (d)

s	-2	-1	0	1	2
$g(s)$	-1	$-\dfrac{4}{5}$	0	$\dfrac{4}{5}$	1

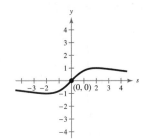

29. $g(x) = \dfrac{4(x+1)}{x(x-4)}$

    (a) Domain: all real numbers x except $x = 0$ and
        $x = 4$

    (b) x-intercept: $(-1, 0)$

    (c) Vertical asymptotes: $x = 0$, $x = 4$

        Horizontal asymptote: $y = 0$

    (d)

x	-2	-1	1	2	3	5	6
$g(x)$	$-\dfrac{1}{3}$	0	$-\dfrac{8}{3}$	-3	$-\dfrac{16}{3}$	$\dfrac{24}{5}$	$\dfrac{7}{3}$

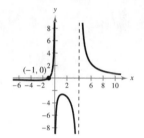

31. $f(x) = \dfrac{2x}{x^2 - 3x - 4} = \dfrac{2x}{(x-4)(x+1)}$

    (a) Domain: all real numbers x except $x = 4$ and
        $x = -1$

    (b) Intercept: $(0, 0)$

    (c) Vertical asymptotes: $x = 4$, $x = -1$

        Horizontal asymptote: $y = 0$

    (d)

x	-3	-2	0	1	2	3	5
$f(x)$	$-\dfrac{3}{7}$	$-\dfrac{2}{3}$	0	$-\dfrac{1}{3}$	$-\dfrac{2}{3}$	$-\dfrac{3}{2}$	$\dfrac{5}{3}$

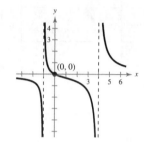

33. $h(x) = \dfrac{x^2 - 5x + 4}{x^2 - 4} = \dfrac{(x-1)(x-4)}{(x+2)(x-2)}$

    (a) Domain: all real numbers x except $x = \pm 2$

    (b) x-intercepts: $(1, 0)$, $(4, 0)$

        y-intercept: $(0, -1)$

    (c) Vertical asymptotes: $x = -2$, $x = 2$

        Horizontal asymptote: $y = 1$

    (d)

x	-4	-3	-1	0	1	3	4
$h(x)$	$\dfrac{10}{3}$	$\dfrac{28}{5}$	$-\dfrac{10}{3}$	-1	0	$-\dfrac{2}{5}$	0

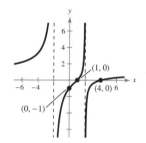

35. $f(x) = \dfrac{6x}{x^2 - 5x - 14} = \dfrac{6x}{(x+2)(x-7)}$

    (a) Domain: all real numbers x except $x = -2$ and $x = 7$

    (b) Intercept: $(0, 0)$

    (c) Vertical asymptotes: $x = -2$, $x = 7$

        Horizontal asymptote: $y = 0$

    (d)

x	-6	-4	0	2	4	6	8	10
$f(x)$	$-\dfrac{9}{13}$	$-\dfrac{12}{11}$	0	$-\dfrac{3}{5}$	$-\dfrac{4}{3}$	$-\dfrac{9}{2}$	$\dfrac{24}{5}$	$\dfrac{5}{3}$

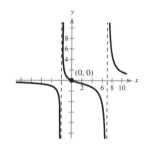

37. $f(x) = \dfrac{2x^2 - 5x - 3}{x^3 - 2x^2 - x + 2} = \dfrac{(2x + 1)(x - 3)}{(x - 2)(x + 1)(x - 1)}$

(a) Domain: all real numbers x except $x = 2$, $x = \pm 1$

(b) x-intercepts: $\left(-\dfrac{1}{2}, 0\right)$, $(3, 0)$

y-intercept: $\left(0, -\dfrac{3}{2}\right)$

(c) Vertical asymptotes: $x = 2$, $x = -1$, and $x = 1$

Horizontal asymptote: $y = 0$

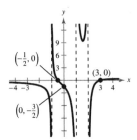

(d)

x	-3	-2	0	$\dfrac{3}{2}$	3	4
$f(x)$	$-\dfrac{3}{4}$	$-\dfrac{5}{4}$	$-\dfrac{3}{2}$	$\dfrac{48}{5}$	0	$\dfrac{3}{10}$

39. $f(x) = \dfrac{x^2 + 3x}{x^2 + x - 6} = \dfrac{x(x + 3)}{(x + 3)(x - 2)} = \dfrac{x}{x - 2}, \quad x \neq -3$

(a) Domain: all real numbers x except $x = -3$ and $x = 2$

(b) Intercept: $(0, 0)$

(c) Vertical asymptote: $x = 2$

Horizontal asymptote: $y = 1$

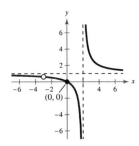

(d)

x	-1	0	1	3	4
$f(x)$	$\dfrac{1}{3}$	0	-1	3	2

41. $f(x) = \dfrac{2x^2 - 5x + 2}{2x^2 - x - 6} = \dfrac{(2x - 1)(x - 2)}{(2x + 3)(x - 2)} = \dfrac{2x - 1}{2x + 3}, \quad x \neq 2$

(a) Domain: all real numbers x except $x = 2$ and $x = -\dfrac{3}{2}$

(b) x-intercept: $\left(\dfrac{1}{2}, 0\right)$

y-intercept: $\left(0, -\dfrac{1}{3}\right)$

(c) Vertical asymptote: $x = -\dfrac{3}{2}$

Horizontal asymptote: $y = 1$

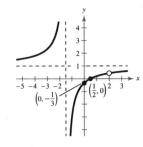

(d)

x	-3	-2	-1	0	1
$f(x)$	$\dfrac{7}{3}$	5	-3	$-\dfrac{1}{3}$	$\dfrac{1}{5}$

43. $f(t) = \dfrac{t^2 - 1}{t - 1} = \dfrac{(t + 1)(t - 1)}{t - 1} = t + 1, \quad t \ne 1$

(a) Domain: all real numbers t except $t = 1$

(b) t-intercept: $(-1, 0)$

  y-intercept: $(0, 1)$

(c) No asymptotes

(d)

t	-3	-2	-1	0	1	2
$f(t)$	-2	-1	0	1	Undef.	3

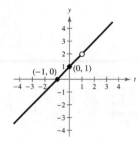

45. $f(x) = \dfrac{x^2 - 1}{x + 1}, \quad g(x) = x - 1$

(a) Domain of f: all real numbers x except $x = -1$

  Domain of g: all real numbers x

(b) $f(x) = \dfrac{(x + 1)(x - 1)}{x + 1} = x - 1$

  Because $x + 1$ is a factor of both the numerator and the denominator of f, $x = -1$ is not a vertical asymptote. f has no vertical asymptotes.

(c)

x	-3	-2	-1.5	-1	-0.5	0	1
$f(x)$	-4	-3	-2.5	Undef.	-1.5	-1	0
$g(x)$	-4	-3	-2.5	-2	-1.5	-1	0

(d)

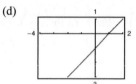

(e) Because there are only a finite number of pixels, the utility may not attempt to evaluate the function where it does not exist.

47. $f(x) = \dfrac{x - 2}{x^2 - 2x}, \quad g(x) = \dfrac{1}{x}$

(a) Domain of f: all real numbers x except $x = 0$ and $x = 2$

  Domain of g: all real numbers x except $x = 0$

(b) $f(x) = \dfrac{x - 2}{x(x - 2)} = \dfrac{1}{x}$

  Because $x - 2$ is a factor of both the numerator and the denominator of f, $x = 2$ is not a vertical asymptote. The only vertical asymptote of f is $x = 0$.

(d)

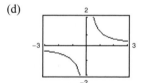

(e) Because there are only a finite number of pixels, the utility may not attempt to evaluate the function where it does not exist.

(c)

x	-0.5	0	0.5	1	1.5	2	3
$f(x)$	-2	Undef.	2	1	$\dfrac{2}{3}$	Undef.	$\dfrac{1}{3}$
$g(x)$	-2	Undef.	2	1	$\dfrac{2}{3}$	$\dfrac{1}{2}$	$\dfrac{1}{3}$

49. $h(x) = \dfrac{x^2 - 9}{x} = x - \dfrac{9}{x}$

    (a) Domain: all real numbers x except $x = 0$

    (b) x-intercepts: $(-3, 0), (3, 0)$

    (c) Vertical asymptote: $x = 0$

       Slant asymptote: $y = x$

    (d)

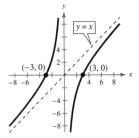

x	-6	-4	-3	-2	2	3	4	6
$h(x)$	$-\dfrac{9}{2}$	$-\dfrac{7}{4}$	0	$\dfrac{5}{2}$	$-\dfrac{5}{2}$	0	$\dfrac{7}{4}$	$\dfrac{9}{2}$

51. $f(x) = \dfrac{2x^2 + 1}{x} = 2x + \dfrac{1}{x}$

    (a) Domain: all real numbers x except $x = 0$

    (b) No intercepts

    (c) Vertical asymptote: $x = 0$

       Slant asymptote: $y = 2x$

    (d)

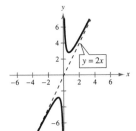

x	-4	-2	2	4	6
$f(x)$	$-\dfrac{33}{4}$	$-\dfrac{9}{2}$	$\dfrac{9}{2}$	$\dfrac{33}{4}$	$\dfrac{73}{6}$

53. $g(x) = \dfrac{x^2 + 1}{x} = x + \dfrac{1}{x}$

    (a) Domain: all real numbers x except $x = 0$

    (b) No intercepts

    (c) Vertical asymptote: $x = 0$

       Slant asymptote: $y = x$

    (d)

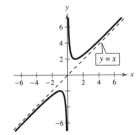

x	-4	-2	2	4	6
$g(x)$	$-\dfrac{17}{4}$	$-\dfrac{5}{2}$	$\dfrac{5}{2}$	$\dfrac{17}{4}$	$\dfrac{37}{6}$

55. $f(t) = \dfrac{t^2 + 1}{t + 5} = -t + 5 - \dfrac{26}{t + 5}$

    (a) Domain: all real numbers t except $t = -5$

    (b) Intercept: $\left(0, -\dfrac{1}{5}\right)$

    (c) Vertical asymptote: $t = -5$

       Slant asymptote: $y = -t + 5$

    (d)

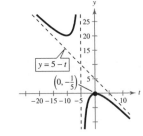

t	-7	-6	-4	-3	0
$f(t)$	25	37	-17	-5	$-\dfrac{1}{5}$

57. $f(x) = \dfrac{x^3}{x^2 - 4} = x + \dfrac{4x}{x^2 - 4}$

    (a) Domain: all real numbers x except $x = \pm 2$

    (b) Intercept: $(0, 0)$

    (c) Vertical asymptotes: $x = \pm 2$

        Slant asymptote: $y = x$

    (d)

x	-6	-4	-1	0	1	4	6
$f(x)$	$-\dfrac{27}{4}$	$-\dfrac{16}{3}$	$\dfrac{1}{3}$	0	$-\dfrac{1}{3}$	$\dfrac{16}{3}$	$\dfrac{27}{4}$

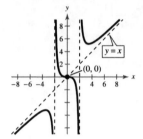

59. $f(x) = \dfrac{x^3 - 1}{x^2 - x} = \dfrac{(x-1)(x^2 + x + 1)}{x(x-1)} = \dfrac{x^2 + x + 1}{x}$

$\qquad = x + 1 + \dfrac{1}{x}, \quad x \neq 1$

    (a) Domain: all real numbers x except $x = 0$ and $x = 1$

    (b) Intercepts: none

    (c) Vertical asymptote: $x = 0$

        Note: $x = 1$ is not a vertical asymptote since it also makes the numerator zero.

        Slant asymptote: $y = x + 1$

    (d)

x	-2	-1	$\dfrac{1}{2}$	2
$f(x)$	$-\dfrac{3}{2}$	-1	$\dfrac{7}{2}$	$\dfrac{7}{2}$

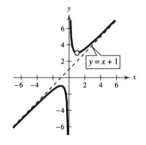

61. $f(x) = \dfrac{x^2 - x + 1}{x - 1} = x + \dfrac{1}{x - 1}$

    (a) Domain: all real numbers x except $x = 1$

    (b) y-intercept: $(0, -1)$

    (c) Vertical asymptote: $x = 1$

        Slant asymptote: $y = x$

    (d)

x	-4	-2	0	2	4
$f(x)$	$-\dfrac{21}{5}$	$-\dfrac{7}{3}$	-1	3	$\dfrac{13}{3}$

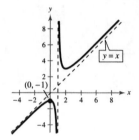

63. $f(x) = \dfrac{2x^3 - x^2 - 2x + 1}{x^2 + 3x + 2}$

$\qquad = \dfrac{(2x - 1)(x + 1)(x - 1)}{(x + 1)(x + 2)}$

$\qquad = \dfrac{(2x - 1)(x - 1)}{x + 2}, \quad x \neq -1$

$\qquad = \dfrac{2x^2 - 3x + 1}{x + 2}$

$\qquad = 2x - 7 + \dfrac{15}{x + 2}, \quad x \neq -1$

    (a) Domain: all real numbers x except $x = -1$ and $x = -2$

    (b) y-intercept: $\left(0, \dfrac{1}{2}\right)$

        x-intercepts: $\left(\dfrac{1}{2}, 0\right), (1, 0)$

    (c) Vertical asymptote: $x = -2$

        Slant asymptote: $y = 2x - 7$

    (d)

x	-4	-3	$-\dfrac{3}{2}$	0	1
$f(x)$	$-\dfrac{45}{2}$	-28	20	$\dfrac{1}{2}$	0

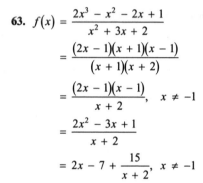

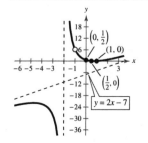

65. $f(x) = \dfrac{x^2 + 5x + 8}{x + 3} = x + 2 + \dfrac{2}{x + 3}$

Domain: all real numbers x except $x = -3$

y-intercept: $\left(0, \dfrac{8}{3}\right)$

Vertical asymptote: $x = -3$

Slant asymptote: $y = x + 2$

Line: $y = x + 2$

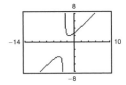

67. $g(x) = \dfrac{1 + 3x^2 - x^3}{x^2} = \dfrac{1}{x^2} + 3 - x = -x + 3 + \dfrac{1}{x^2}$

Domain: all real numbers x except $x = 0$

Vertical asymptote: $x = 0$

Slant asymptote: $y = -x + 3$

Line: $y = -x + 3$

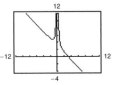

69. $y = \dfrac{x + 1}{x - 3}$

(a) x-intercept: $(-1, 0)$

(b) $\quad 0 = \dfrac{x + 1}{x - 3}$

$\quad 0 = x + 1$

$\quad -1 = x$

71. $y = \dfrac{1}{x} - x$

(a) x-intercepts: $(-1, 0), (1, 0)$

(b) $\quad 0 = \dfrac{1}{x} - x$

$\quad x = \dfrac{1}{x}$

$\quad x^2 = 1$

$\quad x = \pm 1$

73. $y = \dfrac{1}{x + 5} + \dfrac{4}{x}$

(a)

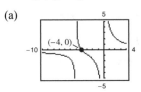

x-intercept: $(-4, 0)$

(b) $\quad 0 = \dfrac{1}{x + 5} + \dfrac{4}{x}$

$\quad -\dfrac{4}{x} = \dfrac{1}{x + 5}$

$\quad -4(x + 5) = x$

$\quad -4x - 20 = x$

$\quad -5x = 20$

$\quad x = -4$

75. $y = x - \dfrac{6}{x - 1}$

(a)

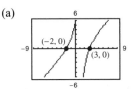

x-intercepts: $(-2, 0), (3, 0)$

(b) $\quad 0 = x - \dfrac{6}{x - 1}$

$\quad \dfrac{6}{x - 1} = x$

$\quad 6 = x(x - 1)$

$\quad 0 = x^2 - x - 6$

$\quad 0 = (x + 2)(x - 3)$

$\quad x = -2, \ x = 3$

77. (a) $0.25(50) + 0.75(x) = C(50 + x)$

$$C = \frac{12.50 + 0.75x}{50 + x} \cdot \frac{4}{4} = \frac{50 + 3x}{4(50 + x)} = \frac{3x + 50}{4(x + 50)}$$

(b) Domain: $x \geq 0$ and $x \leq 1000 - 50$

Thus, $0 \leq x \leq 950$. Using interval notation, the domain is $[0, 950]$.

(c)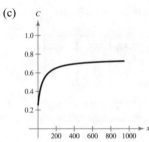

(d) As the tank is filled, the concentration increases more slowly. It approaches the horizontal asymptote of $C = \dfrac{3}{4} = 0.75$.

79. (a) $A = xy$ and

$$(x - 4)(y - 2) = 30$$

$$y - 2 = \frac{30}{x - 4}$$

$$y = 2 + \frac{30}{x - 4} = \frac{2x + 22}{x - 4}$$

Thus, $A = xy = x\left(\dfrac{2x + 22}{x - 4}\right) = \dfrac{2x(x + 11)}{x - 4}$.

(b) Domain: Since the margins on the left and right are each 2 inches, $x > 4$. In interval notation, the domain is $(4, \infty)$.

(c)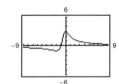

The area is minimum when $x \approx 11.75$ inches and $y \approx 5.87$ inches.

x	5	6	7	8	9	10	11	12	13	14	15
y_1 (Area)	160	102	84	76	72	70	69.143	69	69.333	70	70.909

The area is minimum when x is approximately 12.

81. $f(x) = \dfrac{3(x + 1)}{x^2 + x + 1}$

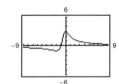

Relative minimum: $(-2, -1)$

Relative maximum: $(0, 3)$

83. $C = 100\left(\dfrac{200}{x^2} + \dfrac{x}{x + 30}\right), x \geq 1$

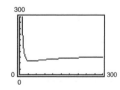

The minimum occurs when $x \approx 40.45$, or 4045 components.

85. (a) Let t_1 = time from Akron to Columbus
and t_2 = time from Columbus back
to Akron.

$$xt_1 = 100 \Rightarrow t_1 = \frac{100}{x}$$

$$yt_2 = 100 \Rightarrow t_2 = \frac{100}{y}$$

$$50(t_1 + t_2) = 200$$

$$t_1 + t_2 = 4$$

$$\frac{100}{x} + \frac{100}{y} = 4$$

$$100y + 100x = 4xy$$

$$25y + 25x = xy$$

$$25x = xy - 25y$$

$$25x = y(x - 25)$$

Thus, $y = \dfrac{25x}{x - 25}$.

(b) Vertical asymptote: $x = 25$

Horizontal asymptote: $y = 25$

(c)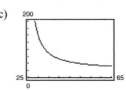

(d)

x	30	35	40	45	50	55	60
y	150	87.5	66.7	56.3	50	45.8	42.9

(e) Sample answer: No. You might expect the average
speed for the round trip to be the average of the
average speeds for the two parts of the trip.

(f) No. At 20 miles per hour you would use more time in
one direction than is required for the round trip at an
average speed of 50 miles per hour.

87. False. There are two distinct branches of the graph.

89. False. The degree of the numerator is two more than the degree of the denominator. To have a slant asymptote,
it has to be exactly one more.

91. $h(x) = \dfrac{6 - 2x}{3 - x} = \dfrac{2(3 - x)}{3 - x}$

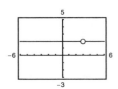

Since $h(x)$ is not reduced and $3 - x$ is a factor of both the numerator and the denominator, $x = 3$ is not a vertical asymptote.

93. Sample answer: If the degree of the numerator of f is exactly one more than the degree of the denominator of f, then the graph
of f has a slant asymptote. To find the equation of a slant asymptote, use long division to divide the numerator by the
denominator. The quotient represents the equation of the slant asymptote.

Section 4.3 Conics

You should know the following basic definitions of conic sections.

■ A parabola is the set of all points (x, y) in a plane that are equidistant from a fixed line (directrix) and a fixed point (focus)
not on the line.

(a) Standard equation with vertex $(0, 0)$ and directrix $y = -p$ (vertical axis): $x^2 = 4py$

(b) Standard equation with vertex $(0, 0)$ and directrix $x = -p$ (horizontal axis): $y^2 = 4px$

(c) The focus lies on the axis p units (directed distance) from the vertex.

— *continued* —

■ An ellipse is the set of all points (x, y) in a plane the sum of whose distances from two distinct fixed points (foci) is constant.

(a) Standard equation of an ellipse with center $(0, 0)$, major axis length $2a$, and minor axis length $2b$:

1. Horizontal major axis: $\dfrac{x^2}{a^2} + \dfrac{y^2}{b^2} = 1$

2. Vertical major axis: $\dfrac{x^2}{b^2} + \dfrac{y^2}{a^2} = 1$

(b) The foci lie on the major axis, c units from the center, where a, b, and c are related by the equation $c^2 = a^2 - b^2$.

(c) The vertices and endpoints of the minor axis are:

1. Horizontal axis: $(\pm a, 0)$ and $(0, \pm b)$

2. Vertical axis: $(0, \pm a)$ and $(\pm b, 0)$

■ A hyperbola is the set of all points (x, y) in a plane the difference of whose distances from two distinct fixed points (foci) is constant.

(a) Standard equation of hyperbola with center $(0, 0)$:

1. Horizontal transverse axis: $\dfrac{x^2}{a^2} - \dfrac{y^2}{b^2} = 1$

2. Vertical transverse axis: $\dfrac{y^2}{a^2} - \dfrac{x^2}{b^2} = 1$

(b) The vertices and foci lie on the transverse axis and are, respectively, a and c units from the center and $b^2 = c^2 - a^2$.

(c) The asymptotes of the hyperbola are:

1. Horizontal transverse axis: $y = \pm \dfrac{b}{a}x$

2. Vertical transverse axis: $y = \pm \dfrac{a}{b}x$

---

1. conic or conic section

3. parabola; directrix; focus

5. axis

7. major axis; center

9. hyperbola; foci

11. $x^2 = 2y$

Parabola opening upward

Not shown

12. $x^2 = -2y$

Parabola opening downward

Matches (c).

13. $y^2 = 2x$

Parabola opening to the right

Matches (e).

14. $y^2 = -2x$

Parabola opening to the left

Matches (a).

15. $9x^2 + y^2 = 9$

$\dfrac{x^2}{1} + \dfrac{y^2}{9} = 1$

Ellipse with vertical major axis

Not shown

16. $x^2 + 9y^2 = 9$

$\dfrac{x^2}{9} + \dfrac{y^2}{1} = 1$

Ellipse with horizontal major axis

Matches (h).

17. $9x^2 - y^2 = 9$

$\dfrac{x^2}{1} - \dfrac{y^2}{9} = 1$

Hyperbola with horizontal transverse axis

Matches (f).

18. $y^2 - 9x^2 = 9$

$\dfrac{y^2}{9} - \dfrac{x^2}{1} = 1$

Hyperbola with vertical transverse axis

Matches (b).

19. $x^2 + y^2 = 25$

Circle with radius 5

Matches (d).

20. $x^2 + y^2 = 16$

Circle with radius 4

Matches (g).

21. $y = \frac{1}{2}x^2$

$x^2 = 2y = 4\left(\frac{1}{2}\right)y;\ p = \frac{1}{2}$

Vertex: $(0, 0)$

Focus: $\left(0, \frac{1}{2}\right)$

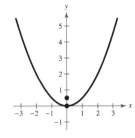

23. $y^2 = -6x$

$y^2 = 4\left(-\frac{3}{2}\right)x;\ p = -\frac{3}{2}$

Vertex: $(0, 0)$

Focus: $\left(-\frac{3}{2}, 0\right)$

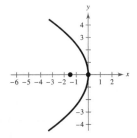

25. $x^2 + 12y = 0$

$x^2 = 4(-3)y;\ p = -3$

Vertex: $(0, 0)$

Focus: $(0, -3)$

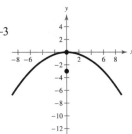

27. Focus: $(-2, 0)$

$y^2 = 4(-2)x$

$y^2 = -8x$

29. Focus: $\left(0, \frac{1}{2}\right)$

$x^2 = 4\left(\frac{1}{2}\right)y$

$x^2 = 2y$

31. Directrix: $y = 1$

$x^2 = 4(-1)y$

$x^2 = -4y$

33. Directrix: $x = -1$

$y^2 = 4(1)x$

$y^2 = 4x$

35. $y^2 = 4px$

$6^2 = 4p(4)$

$36 = 16p$

$p = \frac{9}{4}$

$y^2 = 4\left(\frac{9}{4}\right)x$

$y^2 = 9x$

37. $x^2 = 4py$

$(-2)^2 = 4p\left(\frac{1}{4}\right)$

$4 = p$

$x^2 = 4(4)y$

$x^2 = 16y$

39. $x^2 = 4py$

$3^2 = 4p(6)$

$9 = 24p$

$\frac{3}{8} = p$

$x^2 = 4\left(\frac{3}{8}\right)y$

$x^2 = \frac{3}{2}y$

Focus: $\left(0, \frac{3}{8}\right)$

41. $y^2 = 4px$

$(-3)^2 = 4p(5)$

$9 = 20p$

$\frac{9}{20} = p$

$y^2 = 4\left(\frac{9}{20}\right)x$

$y^2 = \frac{9}{5}x$

Focus: $\left(\frac{9}{20}, 0\right)$

43. $y^2 = 4px,\ p = 1.5$

$y^2 = 4(1.5)x$

$y^2 = 6x$

45. (a)

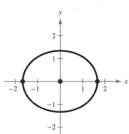

(b) $x^2 = 4py$

$(640)^2 = 4p(152) \Rightarrow p = \dfrac{409,600}{608} = \dfrac{12,800}{19}$

$x^2 = 4\left(\dfrac{12,800}{19}\right)y$

$y = \dfrac{19x^2}{51,200}$

(c)

Distance, x	Height, y
0	0
200	$14\frac{27}{32}$
400	$59\frac{3}{8}$
500	$92\frac{99}{128}$
600	$133\frac{19}{32}$

47. $\dfrac{x^2}{25} + \dfrac{y^2}{16} = 1$

Horizontal major axis

$a = 5, b = 4$

Center: $(0, 0)$

Vertices: $(\pm 5, 0)$

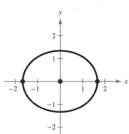

49. $\dfrac{x^2}{25/9} + \dfrac{y^2}{16/9} = 1$

Horizontal major axis

$a = \dfrac{5}{3}, b = \dfrac{4}{3}$

Center: $(0, 0)$

Vertices: $\left(\pm\dfrac{5}{3}, 0\right)$

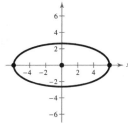

51. $\dfrac{x^2}{36} + \dfrac{y^2}{7} = 1$

Horizontal major axis

$a = 6, b = \sqrt{7}$

Center: $(0, 0)$

Vertices: $(\pm 6, 0)$

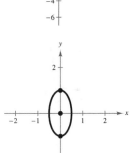

53. $4x^2 + y^2 = 1$

$\dfrac{x^2}{1/4} + y^2 = 1$

Vertical major axis

$a = 1, b = \dfrac{1}{2}$

Center: $(0, 0)$

Vertices: $(0, \pm 1)$

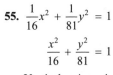

55. $\dfrac{1}{16}x^2 + \dfrac{1}{81}y^2 = 1$

$\dfrac{x^2}{16} + \dfrac{y^2}{81} = 1$

Vertical major axis

$a = 9, b = 4$

Center: $(0, 0)$

Vertices: $(0, \pm 9)$

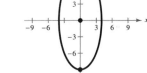

57. Vertices: $(0, \pm 2) \Rightarrow a = 2$

Minor axis of length $2 \Rightarrow b = 1$

Vertical major axis

$\dfrac{x^2}{b^2} + \dfrac{y^2}{a^2} = 1$

$\dfrac{x^2}{1} + \dfrac{y^2}{4} = 1$

59. Vertices: $(\pm 2, 0) \Rightarrow a = 2$

Minor axis of length $3 \Rightarrow b = \dfrac{3}{2}$

Horizontal major axis

$\dfrac{x^2}{a^2} + \dfrac{y^2}{b^2} = 1$

$\dfrac{x^2}{4} + \dfrac{y^2}{9/4} = 1$

61. Vertices: $(\pm 5, 0) \Rightarrow a = 5$

Foci: $(\pm 2, 0) \Rightarrow c = 2$

$b = \sqrt{5^2 - 2^2} = \sqrt{21}$

Horizontal major axis

$\dfrac{x^2}{a^2} + \dfrac{y^2}{b^2} = 1$

$\dfrac{x^2}{25} + \dfrac{y^2}{21} = 1$

63. Foci: $(\pm 5, 0) \Rightarrow c = 5$

Major axis of length $14 \Rightarrow a = 7$

$b = \sqrt{7^2 - 5^2} = \sqrt{24}$

Horizontal major axis

$\dfrac{x^2}{a^2} + \dfrac{y^2}{b^2} = 1$

$\dfrac{x^2}{49} + \dfrac{y^2}{24} = 1$

65. Vertices: $(0, \pm5) \Rightarrow a = 5$

Vertical major axis

$\dfrac{x^2}{b^2} + \dfrac{y^2}{25} = 1$

Passes through $(4, 2)$

$\dfrac{(4)^2}{b^2} + \dfrac{(2)^2}{25} = 1$

$\dfrac{16}{b^2} = \dfrac{21}{25}$

$\dfrac{400}{21} = b^2$

$\dfrac{21x^2}{400} + \dfrac{y^2}{25} = 1$

67. The length of string needed is $2a$ or $2(3) = 6$ feet. The positions of the tacks are $(\pm c, 0)$ where c is given by

$c^2 = a^2 - b^2$

$= 3^2 - 2^2 = 9 - 4 = 5 \Rightarrow c = \pm\sqrt{5}.$

Positions: $(\pm\sqrt{5}, 0)$

69. (a)

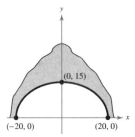

(b) Horizontal major axis

$a = 20, b = 15$

$\dfrac{x^2}{400} + \dfrac{y^2}{225} = 1 \Rightarrow y^2 = 225\left(1 - \dfrac{x^2}{400}\right)$

$y = \pm\sqrt{\dfrac{225}{400}(400 - x^2)}$

$y = \dfrac{3}{4}\sqrt{400 - x^2}$ Top half of ellipse

(c) When $x = 5$:

$y = \dfrac{3}{4}\sqrt{400 - 5^2}$

$y \approx 14.52$ feet

Yes, the truck can clear the tunnel with 0.52 foot clearance.

71. $\dfrac{x^2}{4} + \dfrac{y^2}{1} = 1$

$a = 2, b = 1, c = \sqrt{3}$

Points on the ellipse: $(\pm2, 0), (0, \pm1)$

Length of each latus rectum: $\dfrac{2b^2}{a} = \dfrac{2(1)}{2} = 1$

Additional points: $\left(\sqrt{3}, \pm\dfrac{1}{2}\right), \left(-\sqrt{3}, \pm\dfrac{1}{2}\right)$

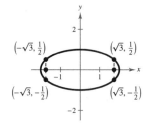

73. $9x^2 + 4y^2 = 36$

$\dfrac{x^2}{4} + \dfrac{y^2}{9} = 1$

$a = 3, b = 2, c = \sqrt{5}$

Points on the ellipse: $(\pm2, 0), (0, \pm3)$

Length of each latus rectum: $\dfrac{2b^2}{a} = \dfrac{2(2)^2}{3} = \dfrac{8}{3}$

Additional points: $\left(\pm\dfrac{4}{3}, -\sqrt{5}\right), \left(\pm\dfrac{4}{3}, \sqrt{5}\right)$

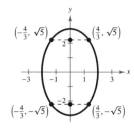

75. $x^2 - y^2 = 1$

$a = 1, b = 1$

Center: $(0, 0)$

Vertices: $(\pm1, 0)$

Asymptotes: $y = \pm x$

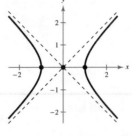

77. $\dfrac{y^2}{1} - \dfrac{x^2}{4} = 1$

$a = 1, b = 2$

Center: $(0, 0)$

Vertices: $(0, \pm1)$

Asymptotes: $y = \pm\dfrac{1}{2}x$

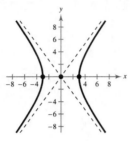

79. $\dfrac{y^2}{49} - \dfrac{x^2}{196} = 1$

$a = 7, b = 14$

Center: $(0, 0)$

Vertices: $(0, \pm7)$

Asymptotes: $y = \pm\dfrac{1}{2}x$

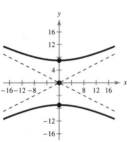

81. $4y^2 - x^2 = 1$

$\dfrac{y^2}{1/4} - \dfrac{x^2}{1} = 0$

$a = \dfrac{1}{2}, b = 1$

Center: $(0, 0)$

Vertices: $\left(0, \pm\dfrac{1}{2}\right)$

Asymptotes: $y = \pm\dfrac{1}{2}x$

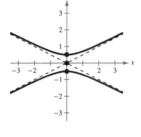

83. $\dfrac{1}{36}y^2 - \dfrac{1}{100}x^2 = 1$

$\dfrac{y^2}{36} - \dfrac{x^2}{100} = 1$

$a = 6, b = 10$

Center: $(0, 0)$

Vertices: $(0, \pm6)$

Asymptotes: $y = \pm\dfrac{3}{5}x$

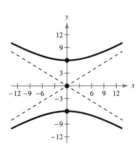

85. Vertices: $(0, \pm2) \Rightarrow a = 2$

Foci: $(0, \pm6) \Rightarrow c = 6$

Vertical transverse axis

$b^2 = c^2 - a^2 = 32$

$\dfrac{y^2}{a^2} - \dfrac{x^2}{b^2} = 1$

$\dfrac{y^2}{4} - \dfrac{x^2}{32} = 1$

87. Vertices: $(\pm1, 0) \Rightarrow a = 1$

Asymptotes: $y = \pm3x$

Horizontal transverse axis

$3 = \dfrac{b}{a} = \dfrac{b}{1} \Rightarrow b = 3$

$\dfrac{x^2}{a^2} - \dfrac{y^2}{b^2} = 1$

$\dfrac{x^2}{1} - \dfrac{y^2}{9} = 1$

89. Foci: $(0, \pm8) \Rightarrow c = 8$

Asymptotes: $y = \pm4x$

Vertical transverse axis

$4 = \dfrac{a}{b} \Rightarrow a = 4b$

$a^2 + b^2 = c^2 \Rightarrow 16b^2 + b^2 = (8)^2$

$b^2 = \dfrac{64}{17} \Rightarrow a^2 = \dfrac{1024}{17}$

$\dfrac{y^2}{a^2} - \dfrac{x^2}{b^2} = 1$

$\dfrac{y^2}{1024/17} - \dfrac{x^2}{64/17} = 1$

$\dfrac{17y^2}{1024} - \dfrac{17x^2}{64} = 1$

91. Vertices: $(0, \pm3) \Rightarrow a = 3$

Vertical transverse axis

$\dfrac{y^2}{9} - \dfrac{x^2}{b^2} = 1$

Point on the graph: $(-2, 5)$

$\dfrac{5^2}{9} - \dfrac{(-2)^2}{b^2} = 1$

$b^2 = \dfrac{9}{4}$

$\dfrac{y^2}{9} - \dfrac{x^2}{9/4} = 1$

93. (a) Vertices: $(\pm 1, 0) \Rightarrow a = 1$

Horizontal transverse axis

$$\frac{x^2}{a^2} - \frac{y^2}{b^2} = 1$$

Point on the graph: $(2, 13)$

$$\frac{2^2}{1^2} - \frac{13^2}{b^2} = 1$$

$$4 - \frac{169}{b^2} = 1$$

$$3b^2 = 169$$

$$b^2 = \frac{169}{3}$$

Thus, we have $\dfrac{x^2}{1} - \dfrac{3y^2}{169} = 1.$

(b) When $y = 5$:

$$x^2 = 1 + \frac{3(5^2)}{169}$$

$$x \approx \sqrt{1 + \frac{75}{169}} \approx 1.2016$$

Width: $2x \approx 2.403$ feet

95. $\dfrac{x^2}{100} - \dfrac{y^2}{4} = 1$

The shortest horizontal distance would be the distance between the center and a vertex of the hyperbola, that is, 10 miles.

97. False. The equation represents a hyperbola.

$$\frac{x^2}{144} - \frac{y^2}{144} = 1$$

99. False. If the graph crossed the directrix, there would exist points nearer the directrix than the focus.

101. (a) $a + b = 20 \Rightarrow b = 20 - a$

$A = \pi ab = \pi a(20 - a)$

(b) $264 = \pi a(20 - a)$

$\pi a^2 - 20\pi a + 264 = 0$

$a \approx 14$ or $a \approx 6$ by the Quadratic Formula

$b = 6$ $b = 14$

Since $a > b$ we choose $a = 14$ and $b = 6$.

$$\frac{x^2}{14^2} + \frac{y^2}{6^2} = 1, \frac{x^2}{196} + \frac{y^2}{36} = 1$$

(c)

a	8	9	10	11	12	13
A	301.6	311.0	314.2	311.0	301.6	285.9

Conjecture: Area is maximum when $a = b = 10$ and the shape is a circle.

(d)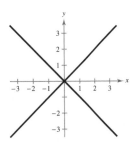

The area is maximum when $a = b = 10$ and the shape is a circle.

103. An ellipse is a circle if $a = b$.

105. $x^2 - y^2 = 0$

$y^2 = x^2$

$y = \pm x$

Two intersecting lines

107. Answers will vary. From the exercise set, we have examples of hyperbolas in optics, art, navigation, and aeronautics.

109. Let (x, y) be such that the sum of the distances from $(-c, 0)$ and $(c, 0)$ is $2a$.

(We are only deriving the form where the major axis is horizontal.)

$$\sqrt{(x + c)^2 + y^2} + \sqrt{(x - c)^2 + y^2} = 2a$$

$$\sqrt{(x + c)^2 + y^2} = 2a - \sqrt{(x - c)^2 + y^2}$$

$$(x + c)^2 + y^2 = 4a^2 - 4a\sqrt{(x - c)^2 + y^2} + (x - c)^2 + y^2$$

$$4a\sqrt{(x - c)^2 + y^2} = 4a^2 + (x - c)^2 - (x + c)^2$$

$$4a\sqrt{(x - c)^2 + y^2} = 4a^2 - 4cx$$

$$a\sqrt{(x - c)^2 + y^2} = a^2 - cx$$

$$a^2\left[(x - c)^2 + y^2\right] = a^4 - 2a^2cx + c^2x^2$$

$$a^2x^2 - 2a^2cx + a^2c^2 + a^2y^2 = a^4 - 2a^2cx + c^2x^2$$

$$a^2x^2 + a^2c^2 + a^2y^2 = a^4 + c^2x^2$$

$$a^2x^2 - c^2x^2 + a^2y^2 = a^4 - a^2c^2$$

$$x^2(a^2 - c^2) + a^2y^2 = a^2(a^2 - c^2)$$

Let $b^2 = a^2 - c^2$. Then $x^2b^2 + a^2y^2 = a^2b^2 \Rightarrow \dfrac{x^2}{a^2} + \dfrac{y^2}{b^2} = 1$.

111. Sample answer: The smaller the distance between the thumbtacks, the more circular the ellipse becomes. The larger the distance between the thumbtacks, the more long and narrow the ellipse becomes.

Section 4.4   Translations of Conics

You should know the following basic facts about conic sections.

■ Parabola with vertex (h, k)

(a) Vertical axis

  1. Standard equation: $(x - h)^2 = 4p(y - k)$

  2. Focus: $(h, k + p)$

  3. Directrix: $y = k - p$

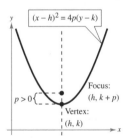

(b) Horizontal axis

  1. Standard equation: $(y - k)^2 = 4p(x - h)$

  2. Focus: $(h + p, k)$

  3. Directrix: $x = h - p$

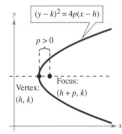

■ Circle with center (h, k) and radius r

  Standard equation: $(x - h)^2 + (y - k)^2 = r^2$

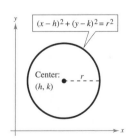

—continued—

- Ellipse with center (h, k)

  (a) Horizontal major axis:

     1. Standard equation: $\dfrac{(x - h)^2}{a^2} + \dfrac{(y - k)^2}{b^2} = 1$

     2. Vertices: $(h \pm a, k)$

     3. Foci: $(h \pm c, k)$

     4. $c^2 = a^2 - b^2$

     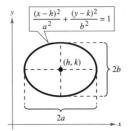

  (b) Vertical major axis:

     1. Standard equation: $\dfrac{(x - h)^2}{b^2} + \dfrac{(y - k)^2}{a^2} = 1$

     2. Vertices: $(h, k \pm a)$

     3. Foci: $(h, k \pm c)$

     4. $c^2 = a^2 - b^2$

     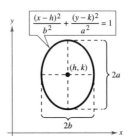

- Hyperbola with center (h, k)

  (a) Horizontal transverse axis:

     1. Standard equation: $\dfrac{(x - h)^2}{a^2} - \dfrac{(y - k)^2}{b^2} = 1$

     2. Vertices: $(h \pm a, k)$

     3. Foci: $(h \pm c, k)$

     4. Asymptotes: $y - k = \pm \dfrac{b}{a}(x - h)$

     5. $c^2 = a^2 + b^2$

     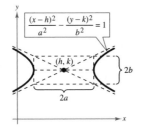

  (b) Vertical transverse axis:

     1. Standard equation: $\dfrac{(y - k)^2}{a^2} - \dfrac{(x - h)^2}{b^2} = 1$

     2. Vertices: $(h, k \pm a)$

     3. Foci: $(h, k \pm c)$

     4. Asymptotes: $y - k = \pm \dfrac{a}{b}(x - h)$

     5. $c^2 = a^2 + b^2$

     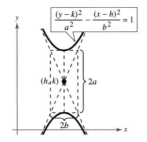

1. Hyperbola with horizontal transverse axis $\Rightarrow \dfrac{(x - h)^2}{a^2} - \dfrac{(y - k)^2}{b^2} = 1$

Matches (b).

2. Ellipse with vertical major axis $\Rightarrow \dfrac{(x - h)^2}{b^2} + \dfrac{(y - k)^2}{a^2} = 1$

Matches (d).

3. Parabola with vertical axis $\Rightarrow (x - h)^2 = 4p(y - k)$

Matches (e).

4. Hyperbola with vertical transverse axis $\Rightarrow \dfrac{(y-k)^2}{a^2} - \dfrac{(x-h)^2}{b^2} = 1$

Matches (c).

5. Ellipse with horizontal major axis $\Rightarrow \dfrac{(x-h)^2}{a^2} + \dfrac{(y-k)^2}{b^2} = 1$

Matches (a).

6. Parabola with horizontal axis $\Rightarrow (y-k)^2 = 4p(x-h)$

Matches (f).

7. $(x+2)^2 + (y-1)^2 = 4$

Center: $(-2, 1)$

The graph has been shifted two units to the left and one unit upward from standard position.

9. $\dfrac{(y+3)^2}{4} - (x-1)^2 = 1$

Center: $(1, -3)$

The graph has been shifted one unit to the right and three units downward from standard position.

11. $\dfrac{(x+4)^2}{9} + \dfrac{(y+2)^2}{16} = 1$

Center: $(-4, -2)$

The graph has been shifted four units to the left and two units downward from standard position.

13. $x^2 + y^2 = 49$

Center: $(0, 0)$

Radius: 7

15. $(x-4)^2 + (y-5)^2 = 36$

Center: $(4, 5)$

Radius: 6

17. $(x-1)^2 + y^2 = 10$

Center: $(1, 0)$

Radius: $\sqrt{10}$

19.
$$x^2 + y^2 - 2x + 6y + 9 = 0$$
$$\left(x^2 - 2x\right) + \left(y^2 + 6y\right) = -9$$
$$\left(x^2 - 2x + 1\right) + \left(y^2 + 6y + 9\right) = -9 + 1 + 9$$
$$(x-1)^2 + (y+3)^2 = 1$$

Center: $(1, -3)$

Radius: 1

21.
$$x^2 + y^2 - 8x = 0$$
$$\left(x^2 - 8x\right) + y^2 = 0$$
$$\left(x^2 - 8x + 16\right) + y^2 = 16$$
$$(x-4)^2 + y^2 = 16$$

Center: $(4, 0)$

Radius: 4

23.
$$4x^2 + 4y^2 + 12x - 24y + 41 = 0$$
$$x^2 + y^2 + 3x - 6y + \tfrac{41}{4} = 0$$
$$\left(x^2 + 3x\right) + \left(y^2 - 6y\right) = -\tfrac{41}{4}$$
$$\left(x^2 + 3x + \tfrac{9}{4}\right) + \left(y^2 - 6y + 9\right) = -\tfrac{41}{4} + \tfrac{9}{4} + 9$$
$$\left(x + \tfrac{3}{2}\right)^2 + (y-3)^2 = 1$$

Center: $\left(-\tfrac{3}{2}, 3\right)$

Radius: 1

25. $(x-1)^2 + 8(y+2) = 0$
$$(x-1)^2 = 4(-2)(y+2); \; p = -2$$

Vertex: $(1, -2)$

Focus: $(1, -4)$

Directrix: $y = 0$

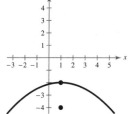

27. $\left(y + \tfrac{1}{2}\right)^2 = 2(x-5)$
$$\left(y + \tfrac{1}{2}\right)^2 = 4\left(\tfrac{1}{2}\right)(x-5); \; p = \tfrac{1}{2}$$

Vertex: $\left(5, -\tfrac{1}{2}\right)$

Focus: $\left(\tfrac{11}{2}, -\tfrac{1}{2}\right)$

Directrix: $x = \tfrac{9}{2}$

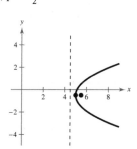

29.
$$y = \tfrac{1}{4}\left(x^2 - 2x + 5\right)$$
$$x^2 - 2x = 4y - 5$$
$$x^2 - 2x + 1 = 4y - 5 + 1$$
$$(x - 1)^2 = 4y - 4$$
$$(x - 1)^2 = 4(1)(y - 1); \ p = 1$$

Vertex: $(1, 1)$

Focus: $(1, 2)$

Directrix: $y = 0$

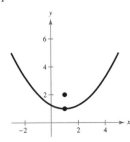

31. $y^2 + 6y + 8x + 25 = 0$
$$y^2 + 6y = -8x - 25$$
$$y^2 + 6y + 9 = -8x - 25 + 9$$
$$(y + 3)^2 = -8x - 16$$
$$(y + 3)^2 = 4(-2)(x + 2); \ p = -2$$

Vertex: $(-2, -3)$

Focus: $(-4, -3)$

Directrix: $x = 0$

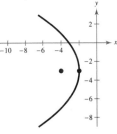

33. Vertex: $(3, 2)$

Focus: $(1, 2)$

Horizontal axis

$$p = 1 - 3 = -2$$
$$(y - 2)^2 = 4(-2)(x - 3)$$
$$(y - 2)^2 = -8(x - 3)$$

35. Vertex: $(0, 4)$

Directrix: $y = 2$

Vertical axis

$$p = 4 - 2 = 2$$
$$(x - 0)^2 = 4(2)(y - 4)$$
$$x^2 = 8(y - 4)$$

37. Focus: $(4, 4)$

Directrix: $x = -4$

Horizontal axis

Vertex: $(0, 4)$

$$p = 4 - 0 = 4$$
$$(y - 4)^2 = 4(4)(x - 0)$$
$$(y - 4)^2 = 16x$$

39. $x^2 = -39{,}204(y - 30{,}000)$

When $y = 0$, $x^2 = 1{,}176{,}120{,}000$,

$x \approx 34{,}295$ feet.

41. (a) $S = -0.355t^2 + 4.33t + 0.7$

(b) t-coordinate of vertex:

$$-\frac{b}{2a} = -\frac{4.33}{2(-0.355)} \approx 6.099$$

y-coordinate of vertex:

$$-0.355(6.099)^2 + 4.33(6.099) + 0.7 \approx 13.903$$

The vertex is about $(6.099, 13.903)$. This means that the model predicts the maximum sales are about $13.903 billion during the year when $t \approx 6$, or 2006.

(c)
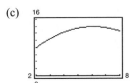

(d)
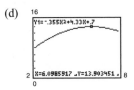

2006

(e)
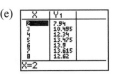

2006

(f) The results are the same. Sample answer: It is helpful to solve problems in multiple ways to verify your results.

43. $\dfrac{(x-1)^2}{9} + \dfrac{(y-5)^2}{25} = 1$

$a = 5, b = 3, c = \sqrt{a^2 - b^2} = 4$

Center: $(1, 5)$

Foci: $(1, 1), (1, 9)$

Vertices: $(1, 0), (1, 10)$

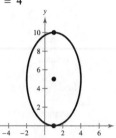

45. $\dfrac{(x+2)^2}{1} + \dfrac{(y+4)^2}{1/4} = 1$

$a = 1, b = \dfrac{1}{2}, c = \sqrt{a^2 - b^2} = \sqrt{\dfrac{3}{4}} = \dfrac{\sqrt{3}}{2}$

Center: $(-2, -4)$

Foci: $\left(-2 \pm \dfrac{\sqrt{3}}{2}, -4\right)$

Vertices: $(-3, -4), (-1, -4)$

47. $9x^2 + 25y^2 - 36x - 50y + 52 = 0$

$9(x^2 - 4x) + 25(y^2 - 2y) = -52$

$9(x^2 - 4x + 4) + 25(y^2 - 2y + 1) = -52 + 9(4) + 25(1)$

$9(x - 2)^2 + 25(y - 1)^2 = 9$

$(x - 2)^2 + \dfrac{(y - 1)^2}{9/25} = 1$

$a = 1, b = \dfrac{3}{5}, c = \sqrt{1 - \dfrac{9}{25}} = \sqrt{\dfrac{16}{25}} = \dfrac{4}{5}$

Horizontal major axis

Center: $(2, 1)$

Foci: $\left(\dfrac{14}{5}, 1\right), \left(\dfrac{6}{5}, 1\right)$

Vertices: $(1, 1), (3, 1)$

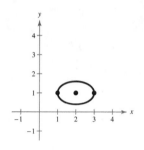

49. $9x^2 + 4y^2 + 36x - 24y + 36 = 0$

$9(x^2 + 4x) + 4(y^2 - 6y) = -36$

$9(x^2 + 4x + 4) + 4(y^2 - 6y + 9) = -36 + 36 + 36$

$9(x + 2)^2 + 4(y - 3)^2 = 36$

$\dfrac{(x + 2)^2}{4} + \dfrac{(y - 3)^2}{9} = 1$

$a = 3, b = 2, c = \sqrt{a^2 - b^2} = \sqrt{5}$

Vertical major axis

Center: $(-2, 3)$

Foci: $\left(-2, 3 \pm \sqrt{5}\right)$

Vertices: $(-2, 0), (-2, 6)$

51. Vertices: $(3, 3), (3, -3)$

Minor axis of length $2 \Rightarrow b = 1$

Center: $(3, 0) \Rightarrow a = 3$

Vertical major axis

$\dfrac{(x - 3)^2}{1} + \dfrac{y^2}{9} = 1$

53. Foci: $(0, 0), (4, 0) \Rightarrow c = 2$

Major axis of length $8 \Rightarrow a = 4$

$b^2 = a^2 - c^2 = 12$

Center: $(2, 0)$

Horizontal major axis

$\dfrac{(x - 2)^2}{16} + \dfrac{y^2}{12} = 1$

55. Center: $(0, 4)$

$a = 2c$

Vertices: $(-4, 4), (4, 4) \Rightarrow a = 4$

$c = 2, b^2 = a^2 - c^2 = 12$

Horizontal major axis

$$\frac{x^2}{16} + \frac{(y - 4)^2}{12} = 1$$

57. Vertices: $(0, 2), (4, 2) \Rightarrow a = 2$

Endpoints of the minor axis:

$(2, 3), (2, 1) \Rightarrow b = 1$

Center: $(2, 2)$

Horizontal major axis

$$\frac{(x - 2)^2}{4} + \frac{(y - 2)^2}{1} = 1$$

59. Vertices: $(\pm 5, 0)$

$$e = \frac{c}{a} = \frac{3}{5}$$

$a = 5, c = 3, b = 4$

$$\frac{x^2}{25} + \frac{y^2}{16} = 1$$

61. $a = 3.67 \times 10^9$

$$e = \frac{c}{a} = 0.249$$

$c = 913,830,000$

Smallest distance: $a - c = 2,756,170,000$ miles

Greatest distance: $a + c = 4,583,830,000$ miles

63. $$\frac{(x - 2)^2}{16} - \frac{(y + 1)^2}{9} = 1$$

$a = 4, b = 3, c = \sqrt{a^2 + b^2} = 5$

Center: $(2, -1)$

Horizontal transverse axis

Vertices: $(6, -1), (-2, -1)$

Foci: $(7, -1), (-3, -1)$

Asymptotes:

$$y = \pm\frac{3}{4}(x - 2) - 1$$

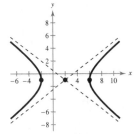

65. $(y + 6)^2 - (x - 2)^2 = 1$

$a = 1, b = 1, c = \sqrt{a^2 + b^2} = \sqrt{2}$

Center: $(2, -6)$

Vertical transverse axis

Vertices: $(2, -7), (2, -5)$

Foci: $\left(2, -6 \pm \sqrt{2}\right)$

Asymptotes:

$y = \pm(x - 2) - 6$

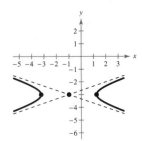

67. $x^2 - 9y^2 + 2x - 54y - 85 = 0$

$(x^2 + 2x) - 9(y^2 + 6y) = 85$

$(x^2 + 2x + 1) - 9(y^2 + 6y + 9) = 85 + 1 - 9(9)$

$(x + 1)^2 - 9(y + 3)^2 = 5$

$$\frac{(x + 1)^2}{5} - \frac{(y + 3)^2}{5/9} = 1$$

$a = \sqrt{5}, b = \frac{\sqrt{5}}{3}, c = \sqrt{a^2 + b^2} = \frac{5\sqrt{2}}{3}$

Center: $(-1, -3)$

Horizontal transverse axis

Vertices: $\left(-1 \pm \sqrt{5}, -3\right)$

Foci: $\left(-1 \pm \frac{5\sqrt{2}}{3}, -3\right)$

Asymptotes: $y = \pm\frac{1}{3}(x + 1) - 3$

69.
$$9x^2 - y^2 - 36x - 6y + 18 = 0$$
$$9(x^2 - 4x) - (y^2 + 6y) = -18$$
$$9(x^2 - 4x + 4) - (y^2 + 6y + 9) = -18 + 9(4) - 9$$
$$9(x - 2)^2 - (y + 3)^2 = 9$$
$$(x - 2)^2 - \frac{(y + 3)^2}{9} = 1$$
$$a = 1, b = 3, c = \sqrt{a^2 + b^2} = \sqrt{10}$$

Center: $(2, -3)$

Horizontal transverse axis

Vertices: $(1, -3), (3, -3)$

Foci: $\left(2 \pm \sqrt{10}, -3\right)$

Asymptotes:
$$y = \pm 3(x - 2) - 3$$

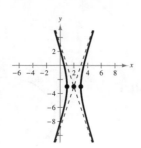

71. Vertices: $(0, 0), (0, 2)$

Foci: $(0, -1), (0, 3)$

Center: $(0, 1)$

Vertical transverse axis
$$a = 1, c = 2, b^2 = c^2 - a^2 = 3$$
$$\frac{(y - 1)^2}{1} - \frac{x^2}{3} = 1$$

73. Vertices: $(2, 0), (6, 0)$

Foci: $(0, 0), (8, 0)$

Center: $(4, 0)$

Horizontal transverse axis
$$a = 2, c = 4, b^2 = c^2 - a^2 = 12$$
$$\frac{(x - 4)^2}{4} - \frac{y^2}{12} = 1$$

75. Vertices: $(2, 3), (2, -3)$

Passes through the point $(0, 5)$

Center: $(2, 0)$

Vertical transverse axis
$$a = 3$$
$$\frac{y^2}{9} - \frac{(x - 2)^2}{b^2} = 1$$
$$\frac{5^2}{9} - \frac{(0 - 2)^2}{b^2} = 1$$
$$b^2 = \frac{9}{4}$$
$$\frac{y^2}{9} - \frac{4(x - 2)^2}{9} = 1$$

77. Vertices: $(0, 2), (6, 2)$

Asymptotes: $y = \frac{2}{3}x, \ y = 4 - \frac{2}{3}x$

Center: $(3, 2)$

Horizontal transverse axis
$$a = 3, b = 2$$
$$\frac{(x - 3)^2}{9} - \frac{(y - 2)^2}{4} = 1$$

79.
$$x^2 + y^2 - 6x + 4y + 9 = 0$$
$$(x^2 - 6x) + (y^2 + 4y) = -9$$
$$(x^2 - 6x + 9) + (y^2 + 4y + 4) = -9 + 9 + 4$$
$$(x - 3)^2 + (y + 2)^2 = 4$$

Circle

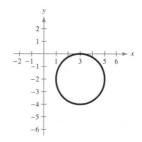

81.
$$y^2 - x^2 + 4y = 0$$
$$(y^2 + 4y + 4) - x^2 = 4$$
$$(y + 2)^2 - x^2 = 4$$
$$\frac{(y + 2)^2}{4} - \frac{x^2}{4} = 1$$

Hyperbola

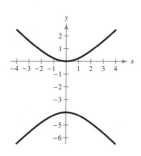

83. $16y^2 + 128x + 8y - 7 = 0$

$$16\left(y^2 + \tfrac{1}{2}y\right) = -128x + 7$$

$$16\left(y^2 + \tfrac{1}{2}y + \tfrac{1}{16}\right) = -128x + 7 + 16\left(\tfrac{1}{16}\right)$$

$$16\left(y + \tfrac{1}{4}\right)^2 = -128x + 8$$

$$\left(y + \tfrac{1}{4}\right)^2 = -8x + \tfrac{1}{2}$$

$$\left(y + \tfrac{1}{4}\right)^2 = -8\left(x - \tfrac{1}{16}\right)$$

Parabola

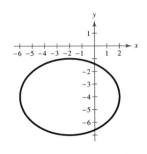

85. $9x^2 + 16y^2 + 36x + 128y + 148 = 0$

$$9\left(x^2 + 4x\right) + 16\left(y^2 + 8y\right) = -148$$

$$9\left(x^2 + 4x + 4\right) + 16\left(y^2 + 8y + 16\right) = -148 + 9(4) + 16(16)$$

$$9(x + 2)^2 + 16(y + 4)^2 = 144$$

$$\frac{(x + 2)^2}{16} + \frac{(y + 4)^2}{9} = 1$$

Ellipse

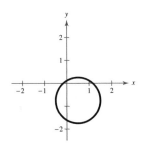

87. $16x^2 + 16y^2 - 16x + 24y - 3 = 0$

$$16\left(x^2 - x\right) + 16\left(y^2 + \tfrac{3}{2}y\right) = 3$$

$$16\left(x^2 - x + \tfrac{1}{4}\right) + 16\left(y^2 + \tfrac{3}{2}y + \tfrac{9}{16}\right) = 3 + 16\left(\tfrac{1}{4}\right) + 16\left(\tfrac{9}{16}\right)$$

$$16\left(x - \tfrac{1}{2}\right)^2 + 16\left(y + \tfrac{3}{4}\right)^2 = 16$$

$$\left(x - \tfrac{1}{2}\right)^2 + \left(y + \tfrac{3}{4}\right)^2 = 1$$

Circle

89. $3x^2 + 2y^2 - 18x - 16y + 58 = 0$

$$3\left(x^2 - 6x\right) + 2\left(y^2 - 8y\right) = -58$$

$$3\left(x^2 - 6x + 9\right) + 2\left(y^2 - 8y + 16\right) = -58 + 27 + 32$$

$$3(x - 3)^2 + 2(y - 4)^2 = 1$$

$$\frac{(x - 3)^2}{1/3} + \frac{(y - 4)^2}{1/2} = 1$$

True. The equation in standard form is

$$\frac{(x - 3)^2}{1/3} + \frac{(y - 4)^2}{1/2} = 1.$$

91. (a) For an ellipse $e = c/a$ and $c^2 = a^2 - b^2$.

$$e = \frac{c}{a} \Rightarrow ea = c$$

$$e^2a^2 = c^2$$

$$e^2a^2 = a^2 - b^2$$

$$b^2 = a^2 - e^2a^2$$

$$b^2 = a^2\left(1 - e^2\right)$$

Thus, by substitution, $\dfrac{(x - h)^2}{a^2} + \dfrac{(y - k)^2}{b^2} = 1$ is

equivalent to $\dfrac{(x - h)^2}{a^2} + \dfrac{(y - k)^2}{a^2\left(1 - e^2\right)} = 1$.

(b)

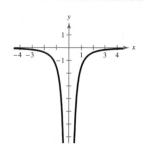

As e approaches 0, the ellipse becomes a circle.

Review Exercises for Chapter 4

1. $f(x) = \dfrac{3x}{x + 10}$

Domain: all real numbers except $x = -10$

3. $f(x) = \dfrac{8}{x^2 - 10x + 24} = \dfrac{8}{(x - 4)(x - 6)}$

Domain: all real numbers except $x = 4$ and $x = 6$

5. $f(x) = \dfrac{4}{x + 3}$

Vertical asymptote: $x = -3$

Horizontal asymptote: $y = 0$

7. $g(x) = \dfrac{x^2}{x^2 - 4}$

Vertical asymptotes: $x = -2, x = 2$

Horizontal asymptote: $y = 1$

9. $h(x) = \dfrac{5x + 20}{x^2 - 2x - 24}$

$$= \dfrac{5(x + 4)}{(x - 6)(x + 4)}$$

$$= \dfrac{5}{x - 6}, \quad x \neq -4$$

Vertical asymptote: $x = 6$

Horizontal asymptote: $y = 0$

11. $\overline{C} = \dfrac{C}{x} = \dfrac{0.5x + 500}{x}, \quad 0 < x$

Horizontal asymptote: $\overline{C} = \dfrac{0.5}{1} = 0.5$

As x increases, the average cost per unit approaches the horizontal asymptote, $\overline{C} = 0.5 = \$0.50$.

13. $f(x) = \dfrac{-3}{2x^2}$

(a) Domain: all real numbers x except $x = 0$

(b) No intercepts

(c) Vertical asymptote: $x = 0$

Horizontal asymptote: $y = 0$

(d)

x	-3	-2	-1	1	2	3
$f(x)$	$-\dfrac{1}{6}$	$-\dfrac{3}{8}$	$-\dfrac{3}{2}$	$-\dfrac{3}{2}$	$-\dfrac{3}{8}$	$-\dfrac{1}{6}$

15. $g(x) = \dfrac{2+x}{1-x} = -\dfrac{x+2}{x-1}$

    (a) Domain: all real numbers x except $x = 1$

    (b) x-intercept: $(-2, 0)$

        y-intercept: $(0, 2)$

    (c) Vertical asymptote: $x = 1$

        Horizontal asymptote: $y = -1$

    (d)

x	-1	0	2	3
$g(x)$	$\dfrac{1}{2}$	2	-4	$-\dfrac{5}{2}$

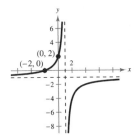

17. $p(x) = \dfrac{5x^2}{4x^2 + 1}$

    (a) Domain: all real numbers x

    (b) Intercept: $(0, 0)$

    (c) Horizontal asymptote: $y = \dfrac{5}{4}$

    (d)

x	-3	-2	-1	0	1	2	3
$p(x)$	$\dfrac{45}{37}$	$\dfrac{20}{17}$	1	0	1	$\dfrac{20}{17}$	$\dfrac{45}{37}$

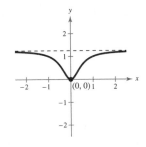

19. $f(x) = \dfrac{x}{x^2 + 1}$

    (a) Domain: all real numbers x

    (b) Intercept: $(0, 0)$

    (c) Horizontal asymptote: $y = 0$

    (d)

x	-2	-1	0	1	2
$f(x)$	$-\dfrac{2}{5}$	$-\dfrac{1}{2}$	0	$\dfrac{1}{2}$	$\dfrac{2}{5}$

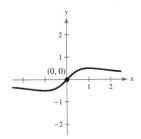

21. $f(x) = \dfrac{-6x^2}{x^2 + 1}$

    (a) Domain: all real numbers x

    (b) Intercept: $(0, 0)$

    (c) Horizontal asymptote: $y = -6$

    (d)

x	± 3	± 2	± 1	0
$f(x)$	$-\dfrac{27}{5}$	$-\dfrac{24}{5}$	-3	0

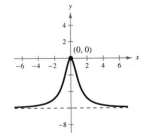

23. $f(x) = \dfrac{6x^2 - 11x + 3}{3x^2 - x}$

$= \dfrac{(3x - 1)(2x - 3)}{x(3x - 1)} = \dfrac{2x - 3}{x}, \; x \ne \dfrac{1}{3}$

(a) Domain: all real numbers x except $x = 0$ and

$x = \dfrac{1}{3}$

(b) x-intercept: $\left(\dfrac{3}{2}, 0\right)$

(c) Vertical asymptote: $x = 0$

Horizontal asymptote: $y = 2$

(d)

x	-2	-1	1	2	3	4
$f(x)$	$\dfrac{7}{2}$	5	-1	$\dfrac{1}{2}$	1	$\dfrac{5}{4}$

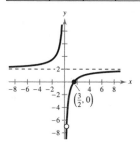

25. $f(x) = \dfrac{2x^3}{x^2 + 1} = 2x - \dfrac{2x}{x^2 + 1}$

(a) Domain: all real numbers x

(b) Intercept: $(0, 0)$

(c) Slant asymptote: $y = 2x$

(d)

x	-2	-1	0	1	2
$f(x)$	$-\dfrac{16}{5}$	-1	0	1	$\dfrac{16}{5}$

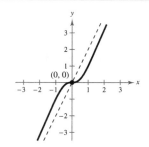

27. $f(x) = \dfrac{x^2 + 3x - 10}{x + 2} = x + 1 - \dfrac{12}{x + 2}$

(a) Domain: all real numbers x except $x = -2$

(b) y-intercept: $(0, -5)$

x-intercepts: $(2, 0), (-5, 0)$

(c) Vertical asymptote: $x = -2$

Slant asymptote: $y = x + 1$

(d)

x	-6	-4	-3	-1	0	1	2	4
$f(x)$	-2	3	10	-12	-5	-2	0	3

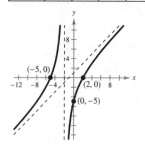

29. $f(x) = \dfrac{3x^3 - 2x^2 - 3x + 2}{3x^2 - x - 4}$

$= \dfrac{(3x - 2)(x + 1)(x - 1)}{(3x - 4)(x + 1)}$

$= \dfrac{(3x - 2)(x - 1)}{3x - 4}$

$= x - \dfrac{1}{3} + \dfrac{2/3}{3x - 4}, \; x \ne -1$

(a) Domain: all real numbers x except

$x = -1$ and $x = \dfrac{4}{3}$

(b) x-intercepts: $(1, 0), \left(\dfrac{2}{3}, 0\right)$

y-intercept: $\left(0, -\dfrac{1}{2}\right)$

(c) Vertical asymptote: $x = \dfrac{4}{3}$

Slant asymptote: $y = x - \dfrac{1}{3}$

(d)

x	-3	-2	0	1	2	3
$f(x)$	$-\dfrac{44}{13}$	$-\dfrac{12}{5}$	$-\dfrac{1}{2}$	0	2	$\dfrac{14}{5}$

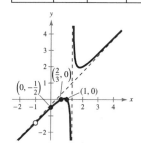

31. $\overline{C} = \dfrac{100{,}000 + 0.9x}{x}, \quad x > 0$

(a)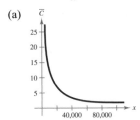

(b) When $x = 1000$,

$$C = \dfrac{100{,}000 + 900}{1000} = \$100.90.$$

When $x = 10{,}000$,

$$C = \dfrac{100{,}000 + 9000}{10{,}000} = \$10.90.$$

When $x = 100{,}000$,

$$C = \dfrac{100{,}000 + 90{,}000}{100{,}000} = \$1.90.$$

(c) The average cost per unit is always greater than $\$0.90$ because $\$0.90$ is the horizontal asymptote of the function.

33. $y = \dfrac{18.47x - 2.96}{0.23x + 1}, \quad 0 < x$

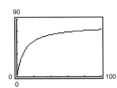

The limiting amount of CO_2 uptake is determined by the

horizontal asymptote, $y = \dfrac{18.47}{0.23} \approx 80.3 \text{ mg/dm}^2/\text{hr.}$

35. $y^2 = -16x$

$y^2 = 4(-4)x$

Parabola

37. $\dfrac{x^2}{64} - \dfrac{y^2}{4} = 1$

Hyperbola

39. $x^2 + 20y = 0$

$\quad x^2 = -20y$

$\quad x^2 = 4(-5)y$

Parabola

41. $\dfrac{y^2}{49} - \dfrac{x^2}{144} = 1$

Hyperbola

43. Vertex: $(0, 0)$

Point: $(3, 6)$

Horizontal axis

$y^2 = 4px$

$6^2 = 4p(3) \Rightarrow p = 3$

$y^2 = 4(3)x$

$y^2 = 12x$

45. Vertex: $(0, 0)$

Focus: $(-6, 0) \Rightarrow p = -6$

Horizontal axis

$y^2 = 4px$

$y^2 = 4(-6)x$

$y^2 = -24x$

47. Vertex: $(0, 0)$

Directrix: $y = -3 \Rightarrow p = 3$

Vertical axis

$x^2 = 4py$

$x^2 = 4(3)y$

$x^2 = 12y$

49. $y = \dfrac{x^2}{200}, \quad -100 \le x \le 100$

Vertex: $(0, 0)$

$x^2 = 200y$

$x^2 = 4(50)y$

$p = 50$

Focus: $(0, 50)$

51. Vertices: $(\pm 9, 0) \Rightarrow a = 9$

Minor axis of length $6 \Rightarrow b = 3$

Center: $(0, 0)$

Horizontal major axis

$\dfrac{x^2}{a^2} + \dfrac{y^2}{b^2} = 1$

$\dfrac{x^2}{81} + \dfrac{y^2}{9} = 1$

53. Center: $(0, 0)$

Vertices: $(0, \pm 6) \Rightarrow a = 6$

Vertical major axis

$$\frac{x^2}{b^2} + \frac{y^2}{36} = 1$$

Using $(2, 2)$:

$$\frac{(2)^2}{b^2} + \frac{(2)^2}{36} = 1$$

$$b^2 = \frac{9}{2}$$

$$\frac{2x^2}{9} + \frac{y^2}{36} = 1$$

55. Foci: $(\pm 14, 0) \Rightarrow c = 14$

Center: $(0, 0)$

Minor axis of length $10 \Rightarrow b = 5$

$$c^2 = a^2 - b^2$$

$$196 = a^2 - 25$$

$$a^2 = 221$$

$$\frac{x^2}{221} + \frac{y^2}{25} = 1$$

57. $a = 5, b = 4, c = \sqrt{a^2 - b^2} = 3$

The foci should be placed 3 feet on either side of the center and have the same height as the pillars.

59. Vertices: $(0, \pm 1)$

Foci: $(0, \pm 5)$

Vertical transverse axis

Center: $(0, 0)$

$a = 1, c = 5$

$b = \sqrt{25 - 1} = \sqrt{24}$

$$\frac{y^2}{a^2} - \frac{x^2}{b^2} = 1$$

$$\frac{y^2}{1} - \frac{x^2}{24} = 1$$

61. Center: $(0, 0)$

Horizontal transverse axis

Vertices: $(\pm 1, 0) \Rightarrow a = 1$

Asymptotes: $y = \pm 2x \Rightarrow \dfrac{b}{a} = 2 \Rightarrow b = 2$

$$\frac{x^2}{1} - \frac{y^2}{4} = 1$$

63. Vertex: $(-8, 8)$

Directrix: $y = 1$

Vertical axis

$p = 8 - 1 = 7$

$(x + 8)^2 = 4(7)(y - 8)$

$(x + 8)^2 = 28(y - 8)$

65. Vertex: $(4, 2)$

Focus: $(4, 0)$

Vertical axis, $p = -2$

$(x - 4)^2 = 4(-2)(y - 2)$

$(x - 4)^2 = -8(y - 2)$

67. Vertices: $(0, 3), (12, 3) \Rightarrow a = 6$

Passes through: $(6, 0)$

Center: $(6, 3)$

Horizontal major axis

$$\frac{(x - 6)^2}{36} + \frac{(y - 3)^2}{b^2} = 1$$

$$\frac{(6 - 6)^2}{36} + \frac{(0 - 3)^2}{b^2} = 1$$

$$\frac{9}{b^2} = 1$$

$$b^2 = 9$$

$$\frac{(x - 6)^2}{36} + \frac{(y - 3)^2}{9} = 1$$

69. Vertices: $(-3, 0), (7, 0) \Rightarrow a = 5$

Foci: $(0, 0), (4, 0) \Rightarrow c = 2$

Horizontal major axis

Center: $(2, 0)$

$b = \sqrt{25 - 4} = \sqrt{21}$

$$\frac{(x - h)^2}{a^2} + \frac{(y - k)^2}{b^2} = 1$$

$$\frac{(x - 2)^2}{25} + \frac{y^2}{21} = 1$$

71. Vertices: $(\pm 6, 7) \Rightarrow$ Center: $(0, 7)$ and $a = 6$

Asymptotes: $y = \pm\dfrac{1}{2}x + 7 \Rightarrow \pm\dfrac{b}{a} = \pm\dfrac{1}{2} = \pm\dfrac{b}{6}$

$\Rightarrow b = 3$

Horizontal transverse axis

$$\frac{x^2}{36} - \frac{(y - 7)^2}{9} = 1$$

73. Vertices: $(-10, 3), (6, 3) \Rightarrow a = 8$

Foci: $(-12, 3), (8, 3) \Rightarrow c = 10 \Rightarrow b^2 = \sqrt{c^2 - a^2} = \sqrt{100 - 64} = 6$

Horizontal transverse axis

Center: $(-2, 3)$

$$\frac{(x + 2)^2}{64} - \frac{(y - 3)^2}{36} = 1$$

75. Foci: $(0, 0), (8, 0) \Rightarrow c = 4$

Asymptotes: $y = \pm 2(x - 4)$

Horizontal transverse axis

Center: $(4, 0)$

$$\frac{b}{a} = 2 \Rightarrow b = 2a$$

$$a^2 + b^2 = c^2$$

$$a^2 + (2a)^2 = 4^2$$

$$a^2 = \frac{16}{5}$$

$$b^2 = \frac{64}{5}$$

$$\frac{(x - h)^2}{a^2} - \frac{(y - k)^2}{b^2} = 1$$

$$\frac{5(x - 4)^2}{16} - \frac{5y^2}{64} = 1$$

77. $x^2 - 6x + 2y + 9 = 0$

$$(x - 3)^2 = -2y$$

$$(x - 3)^2 = 4\left(-\frac{1}{2}\right)y \Rightarrow p = -\frac{1}{2}$$

Parabola

Vertex: $(3, 0)$

Focus: $\left(3, -\frac{1}{2}\right)$

The graph of $x^2 = -2y$ has been shifted to the right three units.

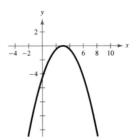

79. $x^2 + y^2 - 2x - 4y + 5 = 0$

$$(x^2 - 2x) + (y^2 - 4y) = -5$$

$$(x^2 - 2x + 1) + (y^2 - 4y + 4) = -5 + 1 + 4$$

$$(x - 1)^2 + (y - 2)^2 = 0$$

Point: $(1, 2)$

Note: This is a degenerate conic, a circle of radius zero.

The graph has been shifted to the right one unit and upward two units from the origin.

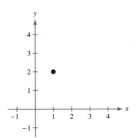

81. $x^2 + 9y^2 + 10x - 18y + 25 = 0$

$$(x^2 + 10x) + 9(y^2 - 2y) = -25$$

$$(x^2 + 10x + 25) + 9(y^2 - 2y + 1) = -25 + 25 + 9$$

$$(x + 5)^2 + 9(y - 1)^2 = 9$$

$$\frac{(x + 5)^2}{9} + (y - 1)^2 = 1$$

Ellipse

Center: $(-5, 1)$

Vertices: $(-8, 1), (-2, 1)$

The graph of $\dfrac{x^2}{9} + y^2 = 1$ has been shifted left five units and upward one unit.

83.
$$9x^2 - y^2 - 72x + 8y + 119 = 0$$
$$9(x^2 - 8x) - (y^2 - 8y) = -119$$
$$9(x^2 - 8x + 16) - (y^2 - 8y + 16) = -119 + 144 - 16$$
$$9(x - 4)^2 - (y - 4)^2 = 9$$
$$\frac{(x - 4)^2}{1} - \frac{(y - 4)^2}{9} = 1$$

Hyperbola

Center: $(4, 4)$

Vertices: $(3, 4), (5, 4)$

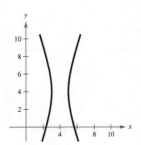

The graph of $x^2 - \dfrac{y^2}{9} = 1$ has been shifted right four units and upward four units from standard position.

85.
$$x^2 = 4p(y - 12)$$
$$(\pm 4)^2 = 4p(10 - 12)$$
$$16 = -8p$$
$$-2 = p$$
$$x^2 = 4(-2)(y - 12)$$
$$x^2 = -8(y - 12)$$

When $y = 0$, we have:

$$x^2 = 96$$
$$x = \pm\sqrt{96} = \pm 4\sqrt{6}$$

At ground level, the width is $2x = 8\sqrt{6} \approx 19.6$ meters.

87. $x^2 + y^2 - 200x - 52{,}500 = 0$

(a) The conic is a circle because x^2 and y^2 have identical coefficients.

(b) $x^2 - 200x + 10{,}000 + y^2 = 52{,}500 + 10{,}000$
$$(x - 100)^2 + y^2 = 62{,}500$$

Center: $(100, 0)$

Radius: 250

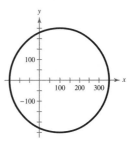

89. True. (See Exercise 79.)

Problem Solving for Chapter 4

1. $f(x) = \dfrac{ax + b}{cx + d}$

Vertical asymptote: $x = -\dfrac{d}{c}$

Horizontal asymptote: $y = \dfrac{a}{c}$

(i) $a > 0, b < 0, c > 0, d < 0$

   Both the vertical asymptote and the horizontal asymptote are positive. Matches graph (d).

(ii) $a > 0, b > 0, c < 0, d < 0$

   Both the vertical asymptote and the horizontal asymptote are negative. Matches graph (b).

(iii) $a < 0, b > 0, c > 0, d < 0$

   The vertical asymptote is positive and the horizontal asymptote is negative. Matches graph (a).

(iv) $a > 0, b < 0, c > 0, d > 0$

   The vertical asymptote is negative and the horizontal asymptote is positive. Matches graph (c).

3. (a)

Age, x	Near point, y
16	3.0
32	4.7
44	9.8
50	19.7
60	39.4

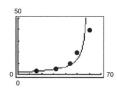

$$y \approx 0.031x^2 - 1.59x + 21.0$$

(b) $\dfrac{1}{y} \approx -0.007x + 0.44$

$$y \approx \frac{1}{-0.007x + 0.44}$$

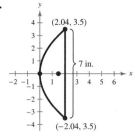

(c)

Age, x	Near point, y	Quadratic Model	Rational Model
16	3.0	3.66	3.05
32	4.7	2.32	4.63
44	9.8	11.83	7.58
50	19.7	19.97	11.11
60	39.4	38.54	50.00

The models are fairly good fits to the data. The quadratic model seems to be a better fit for older ages and the rational model a better fit for younger ages.

(d) For $x = 25$, the quadratic model yields $y \approx 0.625$ inch and the rational model yields $y \approx 3.774$ inches.

(e) The reciprocal model cannot be used to predict the near point for a person who is 70 years old because it results in a negative value $(y \approx -20)$. The quadratic model yields

$$y \approx 63.37 \text{ inches.}$$

5. A hyperbola is the set of all points (x, y) in a plane, the difference of whose distances from two distinct fixed points (foci) is a positive constant.

Using the vertex $(a, 0)$, the distance to $(-c, 0)$ is $a + c$, and the distance to $(c, 0)$ is $(c - a)$.

$$d_2 - d_1 = (a + c) - (c - a) = 2a$$

By definition, $|d_2 - d_1|$ is constant for any point (x, y) on the hyperbola. Thus, $|d_2 - d_1| = 2a$.

7.

Place the vertex at $(0, 0)$. Then, $y^2 = 4px$ and $p = 1.5 \Rightarrow y^2 = 6x$.

When $y = 3.5$, we have $x = \dfrac{(3.5)^2}{6} \approx 2.04$.

The reflector is approximately 2.04 inches deep.

9. Tangent line: $y - y_1 = \dfrac{x_1}{2p}(x - x_1)$

    (a) The slope is $m = \dfrac{x_1}{2p}$.

    (b) For $x^2 = 4y$ or $y = x^2/4$, the endpoints of the chord are $(\pm 2, 1)$.

        The tangent lines are:

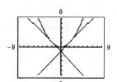

$$y - 1 = \frac{2}{2(1)}(x - 2) \Rightarrow y = x - 1$$

$$y - 1 = \frac{-2}{2(1)}(x + 2) \Rightarrow y = -x - 1$$

        For $x^2 = 8y$ or $y = x^2/8$, the endpoints of the chord are $(\pm 4, 2)$.

        The tangent lines are:

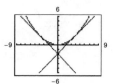

$$y - 2 = \frac{4}{2(2)}(x - 4) \Rightarrow y = x - 2$$

$$y - 2 = \frac{-4}{2(2)}(x + 4) \Rightarrow y = -x - 2$$

        For $x^2 = 12y$ or $y = x^2/12$, the endpoints of the chord are $(\pm 6, 3)$.

        The tangent lines are:

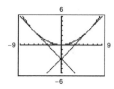

$$y - 3 = \frac{6}{2(3)}(x - 6) \Rightarrow y = x - 3$$

$$y - 3 = \frac{-6}{2(3)}(x + 6) \Rightarrow y = -x - 3$$

        For $x^2 = 16y$ or $y = x^2/16$, the endpoints of the chord are $(\pm 8, 4)$.

        The tangent lines are:

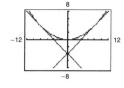

$$y - 4 = \frac{8}{2(4)}(x - 8) \Rightarrow y = x - 4$$

$$y - 4 = \frac{-8}{2(4)}(x + 8) \Rightarrow y = -x - 4$$

11. $Ax^2 + Cy^2 + Dx + Ey + F = 0$

Assume that the conic is *not* degenerate.

(a) $A = C, A \neq 0$

$$Ax^2 + Ay^2 + Dx + Ey + F = 0$$

$$x^2 + y^2 + \frac{D}{A}x + \frac{E}{A}y + \frac{F}{A} = 0$$

$$\left(x^2 + \frac{D}{A}x + \frac{D^2}{4A^2}\right) + \left(y^2 + \frac{E}{A}y + \frac{E^2}{4A^2}\right) = -\frac{F}{A} + \frac{D^2}{4A^2} + \frac{E^2}{4A^2}$$

$$\left(x + \frac{D}{2A}\right)^2 + \left(y + \frac{E}{2A}\right)^2 = \frac{D^2 + E^2 - 4AF}{4A^2}$$

This is a circle with center $\left(-\frac{D}{2A}, -\frac{E}{2A}\right)$ and radius $\frac{\sqrt{D^2 + E^2 - 4AF}}{2|A|}$.

(b) $A = 0$ or $C = 0$ (but not both). Let $C = 0$.

$Ax^2 + Dx + Ey + F = 0$

$$x^2 + \frac{D}{A}x = -\frac{E}{A}y - \frac{F}{A}$$

$$x^2 + \frac{D}{A}x + \frac{D^2}{4A^2} = -\frac{E}{A}y - \frac{F}{A} + \frac{D^2}{4A^2}$$

$$\left(x + \frac{D}{2A}\right)^2 = -\frac{E}{A}\left(y + \frac{F}{E} - \frac{D^2}{4AE}\right)$$

This is a parabola with vertex $\left(-\frac{D}{2A}, \frac{D^2 - 4AF}{4AE}\right)$. $A = 0$ yields a similar result.

(c) $AC > 0 \Rightarrow A$ and C are either both positive or are both negative (if that is the case, move the terms to the other side of the equation so that they are both positive).

$$Ax^2 + Cy^2 + Dx + Ey + F = 0$$

$$A\left(x^2 + \frac{D}{A}x + \frac{D^2}{4A^2}\right) + C\left(y^2 + \frac{E}{C}y + \frac{E^2}{4C^2}\right) = -F + \frac{D^2}{4A} + \frac{E^2}{4C}$$

$$A\left(x + \frac{D}{2A}\right)^2 + C\left(y + \frac{E}{2C}\right)^2 = \frac{CD^2 + AE^2 - 4ACF}{4AC}$$

$$\frac{\left(x + \frac{D}{2A}\right)^2}{\frac{CD^2 + AE^2 - 4ACF}{4A^2C}} + \frac{\left(y + \frac{E}{2C}\right)^2}{\frac{CD^2 + AE^2 - 4ACF}{4AC^2}} = 1$$

Since A and C are both positive, $4A^2C$ and $4AC^2$ are both positive. $CD^2 + AE^2 - 4ACF$ must be positive or the conic is degenerate. Thus, we have an ellipse with center $\left(-D/2A, -E/2C\right)$.

(d) $AC < 0 \Rightarrow A$ and C have opposite signs. Let's assume that A is positive and C is negative. (If A is negative and C is positive, move the terms to the other side of the equation.) From part (c) we have

$$\frac{\left(x + \frac{D}{2A}\right)^2}{\frac{CD^2 + AE^2 - 4ACF}{4A^2C}} + \frac{\left(y + \frac{E}{2C}\right)^2}{\frac{CD^2 + AE^2 - 4ACF}{4AC^2}} = 1.$$

Since $A > 0$ and $C < 0$, the first denominator is positive if $CD^2 + AE^2 - 4ACF < 0$ and is negative if $CD^2 + AE^2 - 4ACF > 0$, since $4A^2C$ is negative. The second denominator would have the *opposite* sign since $4AC^2 > 0$. Thus, we have a hyperbola with center $\left(-D/2A, -E/2C\right)$.

Practice Test for Chapter 4

1. Sketch the graph of $f(x) = \dfrac{x-1}{2x}$ and label all intercepts and asymptotes.

2. Sketch the graph of $f(x) = \dfrac{3x^2 - 4}{x}$ and label all intercepts and asymptotes.

3. Find all the asymptotes of $f(x) = \dfrac{8x^2 - 9}{x^2 + 1}$.

4. Find all the asymptotes of $f(x) = \dfrac{4x^2 - 2x + 7}{x - 1}$.

5. Sketch the graph of $f(x) = \dfrac{x - 5}{(x - 5)^2}$.

6. Find the vertex, focus, and directrix of the parabola $x^2 = 20y$.

7. Find the equation of the parabola with vertex $(0, 0)$ and focus $(7, 0)$.

8. Find the center, foci, and vertices of the ellipse $\dfrac{x^2}{144} + \dfrac{y^2}{25} = 1$.

9. Find the equation of the ellipse with foci $(\pm 4, 0)$ and minor axis of length 6.

10. Find the center, vertices, foci, and asymptotes of the hyperbola $\dfrac{y^2}{144} - \dfrac{x^2}{169} = 1$.

11. Find the equation of the hyperbola with vertices $(\pm 4, 0)$ and asymptotes of $y = \pm\frac{1}{2}x$.

12. Find the equation of the parabola with vertex $(6, -1)$ and focus $(6, 3)$.

13. Find the center, foci, and vertices of the ellipse $16x^2 + 9y^2 - 96x + 36y + 36 = 0$.

14. Find the equation of the ellipse with vertices $(-1, 1)$ and $(7, 1)$ and minor axis of length 2.

15. Find the center, vertices, foci, and asymptotes of the hyperbola $4(x + 3)^2 - 9(y - 1)^2 = 1$.

16. Find the equation of the hyperbola with vertices $(3, 4)$ and $(3, -4)$ and foci $(3, 7)$ and $(3, -7)$.

CHAPTER 5
Exponential and Logarithmic Functions

C H A P T E R   5
Exponential and Logarithmic Functions

Section 5.1   Exponential Functions and Their Graphs

- You should know that a function of the form $f(x) = a^x$, where $a > 0, a \neq 1$, is called an exponential function with base a.

- You should be able to graph exponential functions.

- You should know formulas for compound interest.

  (a) For n compoundings per year: $A = P\left(1 + \dfrac{r}{n}\right)^{nt}$.

  (b) For continuous compoundings: $A = Pe^{rt}$.

---

1. algebraic

3. One-to-One

5. $A = P\left(1 + \dfrac{r}{n}\right)^{nt}$

7. $f(1.4) = (0.9)^{1.4} \approx 0.863$

9. $f(-\pi) = 5^{-\pi} \approx 0.006$

11. $g(x) = 5000(2^x) = 5000(2^{-1.5})$
    ≈ 1767.767

13. $f(x) = 2^x$

  Increasing

  Asymptote: $y = 0$

  Intercept: $(0, 1)$

  Matches graph (d).

15. $f(x) = 2^{-x}$

  Decreasing

  Asymptote: $y = 0$

  Intercept: $(0, 1)$

  Matches graph (a).

17. $f(x) = \left(\frac{1}{2}\right)^x$

x	-2	-1	0	1	2
$f(x)$	4	2	1	0.5	0.25

Asymptote: $y = 0$

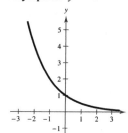

19. $f(x) = 6^{-x}$

x	-2	-1	0	1	2
$f(x)$	36	6	1	0.167	0.028

Asymptote: $y = 0$

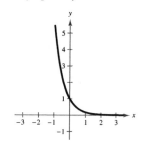

21. $f(x) = 2^{x-1}$

x	-2	-1	0	1	2
$f(x)$	0.125	0.25	0.5	1	2

Asymptote: $y = 0$

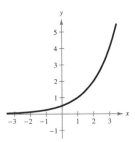

23. $f(x) = 3^x$, $g(x) = 3^x + 1$

Because $g(x) = f(x) + 1$, the graph of g can be obtained by shifting the graph of f one unit upward.

25. $f(x) = 2^x$, $g(x) = 3 - 2^x$

Because $g(x) = 3 - f(x)$, the graph of g can be obtained by reflecting the graph of f in the x-axis and shifting the graph 3 units upward. (**Note:** This is equivalent to shifting the graph of f 3 units upward and then reflecting the graph in the x-axis.)

27. $f(x) = \left(\frac{7}{2}\right)^x$, $g(x) = -\left(\frac{7}{2}\right)^{-x}$

Because $g(x) = -f(-x)$, the graph of g can be obtained by reflecting the graph of f in the x-axis and y-axis.

29. $y = 2^{-x^2}$

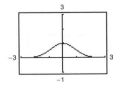

31. $f(x) = 3^{x-2} + 1$

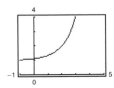

33. $h\left(\frac{3}{4}\right) = e^{-3/4} \approx 0.472$

35. $f(10) = 2e^{-5(10)} \approx 3.857 \times 10^{-22}$

37. $f(6) = 5000e^{0.06(6)} \approx 7166.647$

39. $f(x) = e^x$

x	-2	-1	0	1	2
$f(x)$	0.135	0.368	1	2.718	7.389

Asymptote: $y = 0$

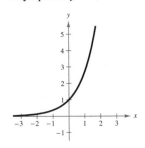

41. $f(x) = 3e^{x+4}$

x	-8	-7	-6	-5	-4
$f(x)$	0.055	0.149	0.406	1.104	3

Asymptote: $y = 0$

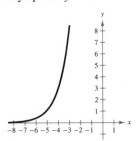

43. $f(x) = 2e^{x-2} + 4$

x	-2	-1	0	1	2
$f(x)$	4.037	4.100	4.271	4.736	6

Asymptote: $y = 4$

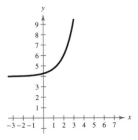

45. $y = 1.08^{-5x}$

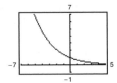

47. $s(t) = 2e^{0.12t}$

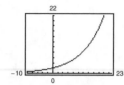

49. $g(x) = 1 + e^{-x}$

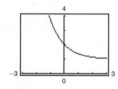

51. $3^{x+1} = 27$

$3^{x+1} = 3^3$

$x + 1 = 3$

$x = 2$

53. $\left(\frac{1}{2}\right)^x = 32$

$\left(\frac{1}{2}\right)^x = \left(\frac{1}{2}\right)^{-5}$

$x = -5$

55. $e^{3x+2} = e^3$

$3x + 2 = 3$

$3x = 1$

$x = \frac{1}{3}$

57. $e^{x^2-3} = e^{2x}$

$x^2 - 3 = 2x$

$x^2 - 2x - 3 = 0$

$(x - 3)(x + 1) = 0$

$x = 3 \quad \text{or} \quad x = -1$

59. $P = \$1500, r = 2\%, t = 10$ years

Compounded n times per year: $A = P\left(1 + \dfrac{r}{n}\right)^{nt} = 1500\left(1 + \dfrac{0.02}{n}\right)^{10n}$

Compounded continuously: $A = Pe^{rt} = 1500e^{0.02(10)}$

n	1	2	4	12	365	Continuous
A	\$1828.49	\$1830.29	\$1831.19	\$1831.80	\$1832.09	\$1832.10

61. $P = \$2500, r = 4\%, t = 20$ years

Compounded n times per year: $A = P\left(1 + \dfrac{r}{n}\right)^{nt} = 2500\left(1 + \dfrac{0.04}{n}\right)^{20n}$

Compounded continuously: $A = Pe^{rt} = 2500e^{0.04(20)}$

n	1	2	4	12	365	Continuous
A	\$5477.81	\$5520.10	\$5541.79	\$5556.46	\$5563.61	\$5563.85

63. $A = Pe^{rt} = 12{,}000e^{0.04t}$

t	10	20	30	40	50
A	\$17,901.90	\$26,706.49	\$39,841.40	\$59,436.39	\$88,668.67

65. $A = Pe^{rt} = 12{,}000e^{0.065t}$

t	10	20	30	40	50
A	\$22,986.49	\$44,031.56	\$84,344.25	\$161,564.86	\$309,484.08

67. $A = 30{,}000e^{(0.05)(25)} \approx \$104{,}710.29$

69. $C(10) = 23.95(1.04)^{10} \approx \$35.45$

71. (a)

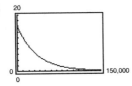

(b)

t	15	16	17	18	19	20	21	22
P	40.19	40.59	40.99	41.39	41.80	42.21	42.62	43.04

t	23	24	25	26	27	28	29	30
P	43.47	43.90	44.33	44.77	45.21	45.65	46.10	46.55

(c) When $t = 38$, $P \approx 50.35$. The population of California will exceed 50 million in 2038.

73. $Q = 16\left(\frac{1}{2}\right)^{t/24,100}$

(a) $Q(0) = 16$ grams

(b) $Q(75,000) \approx 1.85$ grams

(c)

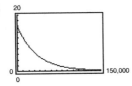

75. (a) $V(t) = 30,500\left(\frac{7}{8}\right)^t$

(b) $V(4) \approx \$17,878.54$

77. True. The line $y = -2$ is a horizontal asymptote for the graph of $f(x) = 10^x - 2$.

79. $f(x) = 3^{x-2}$

$= 3^x 3^{-2}$

$= 3^x\left(\frac{1}{3^2}\right)$

$= \frac{1}{9}(3^x)$

$= h(x)$

So, $f(x) \neq g(x)$, but $f(x) = h(x)$.

81. $f(x) = 16(4^{-x})$ and $f(x) = 16(4^{-x})$

$= 4^2(4^{-x})$ $\qquad = 16(2^2)^{-x}$

$= 4^{2-x}$ $\qquad\qquad = 16(2^{-2x})$

$= \left(\frac{1}{4}\right)^{-(2-x)}$ $\qquad = h(x)$

$= \left(\frac{1}{4}\right)^{x-2}$

$= g(x)$

So, $f(x) = g(x) = h(x)$.

83. $y = 3^x$ and $y = 4^x$

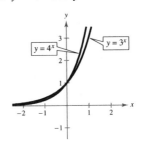

x	-2	-1	0	1	2
3^x	$\frac{1}{9}$	$\frac{1}{3}$	1	3	9
4^x	$\frac{1}{16}$	$\frac{1}{4}$	1	4	16

(a) $4^x < 3^x$ when $x < 0$.

(b) $4^x > 3^x$ when $x > 0$.

85.

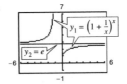

As x increases, the graph of y_1 approaches e, which is y_2.

87. (a)

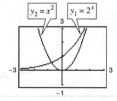

At $x = 2$, both functions have a value of 4. The function y_1 increases for all values of x. The function y_2 is symmetric with respect to the y-axis.

(b)

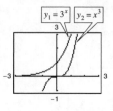

Both functions are increasing for all values of x. For $x > 0$, both functions have a similar shape. The function y_2 is symmetric with respect to the origin.

89. $P = \$3000, r = 6\%, t = 10$ years

$$A = P\left(1 + \frac{r}{n}\right)^{nt} = 3000\left(1 + \frac{0.06}{n}\right)^{10n}$$

(a) $n = 365$

  $A \approx \$5466.09$

(b) $n = 365(24) = 8760$

  $A \approx \$5466.35$

(c) $n = 365(24)(60) = 525{,}600$

  $A \approx \$5466.36$

(d) $n = 365(24)(60)(60) = 31{,}536{,}000$

  $A \approx \$5466.38$

Increasing the number of compoundings does not result in unlimited growth because the balance increases slowly from compounding by the day to compounding by the second.

Section 5.2   Logarithmic Functions and Their Graphs

- You should know that a function of the form $y = \log_a x$, where $a > 0, a \neq 1$, and $x > 0$, is called a logarithm of x to base a.

- You should be able to convert from logarithmic form to exponential form and vice versa.

  $y = \log_a x \Leftrightarrow a^y = x$

- You should know the following properties of logarithms.

  (a) $\log_a 1 = 0$ since $a^0 = 1$.

  (b) $\log_a a = 1$ since $a^1 = a$.

  (c) $\log_a a^x = x$ since $a^x = a^x$.

  (d) $a^{\log_a x} = x$ Inverse Property

  (e) If $\log_a x = \log_a y$, then $x = y$.

- You should know the definition of the natural logarithmic function.

  $\log_e x = \ln x, x > 0$

- You should know the properties of the natural logarithmic function.

  (a) $\ln 1 = 0$ since $e^0 = 1$.

  (b) $\ln e = 1$ since $e^1 = e$.

  (c) $\ln e^x = x$ since $e^x = e^x$.

  (d) $e^{\ln x} = x$ Inverse Property

  (e) If $\ln x = \ln y$, then $x = y$.

- You should be able to graph logarithmic functions.

1. logarithmic

3. natural; e

5. $x = y$

7. $\log_4 16 = 2 \Rightarrow 4^2 = 16$

9. $\log_9 \frac{1}{81} = -2 \Rightarrow 9^{-2} = \frac{1}{81}$

11. $\log_{32} 4 = \frac{2}{5} \Rightarrow 32^{2/5} = 4$

13. $\log_{64} 8 = \frac{1}{2} \Rightarrow 64^{1/2} = 8$

15. $5^3 = 125 \Rightarrow \log_5 125 = 3$

17. $81^{1/4} = 3 \Rightarrow \log_{81} 3 = \frac{1}{4}$

19. $6^{-2} = \frac{1}{36} \Rightarrow \log_6 \frac{1}{36} = -2$

21. $24^0 = 1 \Rightarrow \log_{24} 1 = 0$

23. $f(x) = \log_2 x$

$f(64) = \log_2 64 = 6$ because $2^6 = 64$

25. $f(x) = \log_8 x$

$f(1) = \log_8 1 = 0$ because $8^0 = 1$

27. $g(x) = \log_a x$

$g(a^2) = \log_a a^2$

$\qquad = 2$ by the Inverse Property

29. $f(x) = \log x$

$f\left(\frac{7}{8}\right) = \log\left(\frac{7}{8}\right) \approx -0.058$

31. $f(x) = \log x$

$f(12.5) \approx 1.097$

33. $\log_{11} 11^7 = 7$ because $11^7 = 11^7$

35. $\log_\pi \pi = 1$ because $\pi^1 = \pi$.

37. $f(x) = \log_4 x$

Domain: $(0, \infty)$

x-intercept: $(1, 0)$

Vertical asymptote: $x = 0$

$y = \log_4 x \Rightarrow 4^y = x$

x	$\frac{1}{4}$	1	4	2
$f(x)$	-1	0	1	$\frac{1}{2}$

39. $y = -\log_3 x + 2$

Domain: $(0, \infty)$

x-intercept:

$\quad -\log_3 x + 2 = 0$

$\qquad\qquad 2 = \log_3 x$

$\qquad\qquad 3^2 = x$

$\qquad\qquad 9 = x$

The x-intercept is $(9, 0)$.

Vertical asymptote: $x = 0$

$y = -\log_3 x + 2$

$\log_3 x = 2 - y \Rightarrow 3^{2-y} = x$

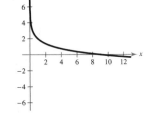

x	27	9	3	1	$\frac{1}{3}$
y	-1	0	1	2	3

41. $f(x) = -\log_6(x + 2)$

Domain: $x + 2 > 0 \Rightarrow x > -2$

The domain is $(-2, \infty)$.

x-intercept:

$\quad 0 = -\log_6(x + 2)$

$\quad 0 = \log_6(x + 2)$

$\quad 6^0 = x + 2$

$\quad 1 = x + 2$

$\quad -1 = x$

The x-intercept is $(-1, 0)$.

Vertical asymptote: $x + 2 = 0 \Rightarrow x = -2$

$\qquad y = -\log_6(x + 2)$

$\qquad -y = \log_6(x + 2)$

$6^{-y} - 2 = x$

x	4	-1	$-1\frac{5}{6}$	$-1\frac{35}{36}$
$f(x)$	-1	0	1	2

43. $y = \log\left(\dfrac{x}{7}\right)$

Domain: $\dfrac{x}{7} > 0 \Rightarrow x > 0$

The domain is $(0, \infty)$.

x-intercept: $\log\left(\dfrac{x}{7}\right) = 0$

$$\dfrac{x}{7} = 10^0$$

$$\dfrac{x}{7} = 1$$

$$x = 7$$

The x-intercept is $(7, 0)$.

Vertical asymptote: $\dfrac{x}{7} = 0 \Rightarrow x = 0$

The vertical asymptote is the y-axis.

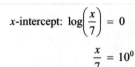

x	1	2	3	4	5
y	−0.85	−0.54	−0.37	−0.24	−0.15

x	6	7	8
y	−0.069	0	0.06

45. $f(x) = \log_3 x + 2$

Asymptote: $x = 0$

Point on graph: $(1, 2)$

Matches graph (c).

The graph of $f(x)$ is obtained by shifting the graph of $g(x)$ upward two units.

47. $f(x) = -\log_3(x + 2)$

Asymptote: $x = -2$

Point on graph: $(-1, 0)$

Matches graph (d).

The graph of $f(x)$ is obtained by reflecting the graph of $g(x)$ in the x-axis and shifting the graph two units to the left.

49. $f(x) = \log_3(1 - x) = \log_3\left[-(x - 1)\right]$

Asymptote: $x = 1$

Point on graph: $(0, 0)$

Matches graph (b).

The graph of $f(x)$ is obtained by reflecting the graph of $g(x)$ in the y-axis and shifting the graph one unit to the right.

51. $\ln \frac{1}{2} = -0.693\ldots \Rightarrow e^{-0.693\ldots} = \frac{1}{2}$

53. $\ln 7 = 1.945\ldots \Rightarrow e^{1.945\ldots} = 7$

55. $\ln 250 = 5.521\ldots \Rightarrow e^{5.521\ldots} = 250$

57. $\ln 1 = 0 \Rightarrow e^0 = 1$

59. $e^4 = 54.598\ldots \Rightarrow \ln 54.598\ldots = 4$

61. $e^{1/2} = 1.6487\ldots \Rightarrow \ln 1.6487\ldots = \frac{1}{2}$

63. $e^{-0.9} = 0.406\ldots \Rightarrow \ln 0.406\ldots = -0.9$

65. $e^x = 4 \Rightarrow \ln 4 = x$

67. $f(x) = \ln x$

$f(18.42) = \ln 18.42 \approx 2.913$

69. $g(x) = 8 \ln x$

$g(0.05) = 8 \ln 0.05 \approx -23.966$

71. $g(x) = \ln x$

$g(e^5) = \ln e^5 = 5$ by the Inverse Property

73. $g(x) = \ln x$

$g(e^{-5/6}) = \ln e^{-5/6} = -\frac{5}{6}$ by the Inverse Property

75. $f(x) = \ln(x - 4)$

Domain: $x - 4 > 0 \Rightarrow x > 4$

The domain is $(4, \infty)$.

x-intercept: $0 = \ln(x - 4)$

$$e^0 = x - 4$$

$$5 = x$$

The x-intercept is $(5, 0)$.

Vertical asymptote: $x - 4 = 0 \Rightarrow x = 4$

x	4.5	5	6	7
$f(x)$	−0.69	0	0.69	1.10

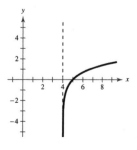

77. $g(x) = \ln(-x)$

Domain: $-x > 0 \Rightarrow x < 0$

The domain is $(-\infty, 0)$.

x-intercept:

$$0 = \ln(-x)$$
$$e^0 = -x$$
$$-1 = x$$

The x-intercept is $(-1, 0)$.

Vertical asymptote: $-x = 0 \Rightarrow x = 0$

x	-0.5	-1	-2	-3
$g(x)$	-0.69	0	0.69	1.10

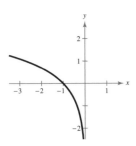

79. $y = \log(x + 9)$

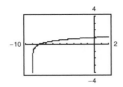

81. $y_1 = \ln(x - 1)$

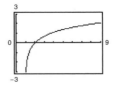

83. $y = \ln x + 8$

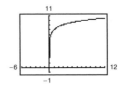

85. $\log_5(x + 1) = \log_5 6$

$$x + 1 = 6$$
$$x = 5$$

87. $\log(2x + 1) = \log 15$

$$2x + 1 = 15$$
$$x = 7$$

89. $\ln(x + 4) = \ln 12$

$$x + 4 = 12$$
$$x = 8$$

91. $\ln(x^2 - 2) = \ln 23$

$$x^2 - 2 = 23$$
$$x^2 = 25$$
$$x = \pm 5$$

93. $t = 16.625 \ln\left(\dfrac{x}{x - 750}\right), \ x > 750$

    (a) When $x = \$897.72$: $t = 16.625 \ln\left(\dfrac{897.72}{897.72 - 750}\right) \approx 30$ years

          When $x = \$1659.24$: $t = 16.625 \ln\left(\dfrac{1659.24}{1659.24 - 750}\right) \approx 10$ years

    (b) Total amounts: $(897.72)(12)(30) = \$323{,}179.20 \approx \$323{,}179$

                      $(1659.24)(12)(10) = \$199{,}108.80 \approx \$199{,}109$

    (c) Interest charges: $323{,}179.20 - 150{,}000 = \$173{,}179.20 \approx \$173{,}179$

                    $199{,}108.80 - 150{,}000 = \$49{,}108.80 \approx \$49{,}109$

    (d) The vertical asymptote is $x = 750$. The closer the payment is to \$750 per month, the longer the length of the mortgage will be. Also, the monthly payment must be greater than \$750.

95. $C = 10.355 - 0.298 + \ln t, 1 \le t \le 6$

(a)

t	1	2	3	4	5	6
C	10.36	9.94	9.37	8.70	7.96	7.15

(b)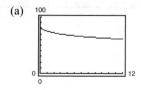

(c) No, the model begins to decrease rapidly, eventually producing negative values.

97. $f(t) = 80 - 17 \log(t + 1), 0 \le t \le 12$

(a)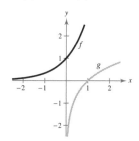

(b) $f(0) = 80 - 17 \log 1 = 80.0$

(c) $f(4) = 80 - 17 \log 5 \approx 68.1$

(d) $f(10) = 80 - 17 \log 11 \approx 62.3$

99. False. Reflecting $g(x)$ about the line $y = x$ will determine the graph of $f(x)$.

101. $f(x) = 3^x, g(x) = \log_3 x$

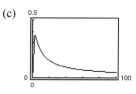

f and g are inverses. Their graphs are reflected about the line $y = x$.

107. $f(x) = \dfrac{\ln x}{x}$

(a)

x	1	5	10	10^2	10^4	10^6
$f(x)$	0	0.322	0.230	0.046	0.00092	0.0000138

(b) As $x \to \infty, f(x) \to 0$.

(c)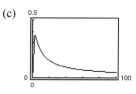

103. $f(x) = e^x, g(x) = \ln x$

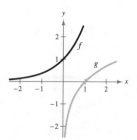

f and g are inverses. Their graphs are reflected about the line $y = x$.

105. $f(x) = 10^x$

x	-2	-1	0	1	2
$f(x)$	$\frac{1}{100}$	$\frac{1}{10}$	1	10	100

$f(x) = \log x$

x	$\frac{1}{100}$	$\frac{1}{10}$	1	10	100
$f(x)$	-2	-1	0	1	2

The domain of $f(x) = 10^x$ is equal to the range of $f(x) = \log x$.

The range of $f(x) = 10^x$ is equal to the domain of $f(x) = \log x$.

$f(x) = 10^x$ and $f(x) = \log x$ are inverses of each other.

109. $y = \log_a x \Rightarrow a^y = x$, so, for example, if $a = -2$, there is no value of y for which $(-2)^y = -4$. If $a = 1$, then every power of a is equal to 1, so x could only be 1. So, $\log_a x$ is defined only for $0 < a < 1$ and $a > 1$.

111. (a) $h(x) = \ln(x^2 + 1)$

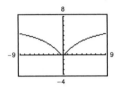

(b) Increasing on $(0, \infty)$

Decreasing on $(-\infty, 0)$

(c) Relative minimum: $(0, 0)$

Section 5.3   Properties of Logarithms

- You should know the following properties of logarithms.

(a) $\log_a x = \dfrac{\log_b x}{\log_b a} \quad \log_a x = \dfrac{\log_{10} x}{\log_{10} a} \quad \log_a x = \dfrac{\ln x}{\ln a}$

(b) $\log_a(uv) = \log_a u + \log_a v \qquad \ln(uv) = \ln u + \ln v$

(c) $\log_a\left(\dfrac{u}{v}\right) = \log_a u - \log_a v \qquad \ln\left(\dfrac{u}{v}\right) = \ln u - \ln v$

(d) $\log_a u^n = n \log_a u \qquad \ln u^n = n \ln u$

- You should be able to rewrite logarithmic expressions using these properties.

1. change-of-base

3. $\dfrac{1}{\log_b a}$

5. $\ln u^n = n \ln u$

This is the Power Property. Matches (a).

7. (a) $\log_5 16 = \dfrac{\log 16}{\log 5}$

(b) $\log_5 16 = \dfrac{\ln 16}{\ln 5}$

9. (a) $\log_{1/5} x = \dfrac{\log x}{\log(1/5)}$

(b) $\log_{1/5} x = \dfrac{\ln x}{\ln(1/5)}$

11. (a) $\log_x \dfrac{3}{10} = \dfrac{\log(3/10)}{\log x}$

(b) $\log_x \dfrac{3}{10} = \dfrac{\ln(3/10)}{\ln x}$

13. (a) $\log_{2.6} x = \dfrac{\log x}{\log 2.6}$

(b) $\log_{2.6} x = \dfrac{\ln x}{\ln 2.6}$

15. $\log_3 7 = \dfrac{\log 7}{\log 3} = \dfrac{\ln 7}{\ln 3} \approx 1.771$

17. $\log_{1/2} 4 = \dfrac{\log 4}{\log(1/2)} = \dfrac{\ln 4}{\ln(1/2)} = -2.000$

19. $\log_9 0.1 = \dfrac{\log 0.1}{\log 9} = \dfrac{\ln 0.1}{\ln 9} \approx -1.048$

21. $\log_{15} 1250 = \dfrac{\log 1250}{\log 15} = \dfrac{\ln 1250}{\ln 15} \approx 2.633$

23. $\log_4 8 = \dfrac{\log_2 8}{\log_2 4} = \dfrac{\log_2 2^3}{\log_2 2^2} = \dfrac{3}{2}$

25. $\log_5 \dfrac{1}{250} = \log_5\left(\dfrac{1}{125} \cdot \dfrac{1}{2}\right)$

$= \log_5 \dfrac{1}{125} + \log_5 \dfrac{1}{2}$

$= \log_5 5^{-3} + \log_5 2^{-1}$

$= -3 - \log_5 2$

27. $\ln\left(5e^6\right) = \ln 5 + \ln e^6$

$\qquad = \ln 5 + 6$

$\qquad = 6 + \ln 5$

29. $\log_3 9 = 2 \log_3 3 = 2$

31. $\log_2 \sqrt[4]{8} = \frac{1}{4} \log_2 2^3 = \frac{3}{4} \log_2 2 = \frac{3}{4}(1) = \frac{3}{4}$

33. $\log_4 16^2 = 2 \log_4 16 = 2 \log_4 4^2 = 2(2) = 4$

35. $\log_2(-2)$ is undefined. -2 is not in the domain of $\log_2 x$.

37. $\ln e^{4.5} = 4.5$

39. $\ln \dfrac{1}{\sqrt{e}} = \ln 1 - \ln \sqrt{e} = 0 - \dfrac{1}{2} \ln e = 0 - \dfrac{1}{2}(1) = -\dfrac{1}{2}$

41. $\ln e^2 + \ln e^5 = 2 + 5 = 7$

43. $\log_5 75 - \log_5 3 = \log_5 \dfrac{75}{3}$

$\qquad = \log_5 25$

$\qquad = \log_5 5^2$

$\qquad = 2 \log_5 5$

$\qquad = 2$

45. $\ln 4x = \ln 4 + \ln x$

47. $\log_8 x^4 = 4 \log_8 x$

49. $\log_5 \dfrac{5}{x} = \log_5 5 - \log_5 x$

$\qquad = 1 - \log_5 x$

51. $\ln \sqrt{z} = \ln z^{1/2} = \dfrac{1}{2} \ln z$

53. $\ln xyz^2 = \ln x + \ln y + \ln z^2$

$\qquad = \ln x + \ln y + 2 \ln z$

55. $\ln z(z-1)^2 = \ln z + \ln(z-1)^2$

$\qquad = \ln z + 2 \ln(z-1), \ z > 1$

57. $\log_2 \dfrac{\sqrt{a-1}}{9} = \log_2 \sqrt{a-1} - \log_2 9$

$\qquad = \dfrac{1}{2} \log_2(a-1) - \log_2 3^2$

$\qquad = \dfrac{1}{2} \log_2(a-1) - 2 \log_2 3, \ a > 1$

59. $\ln \sqrt[3]{\dfrac{x}{y}} = \dfrac{1}{3} \ln \dfrac{x}{y}$

$\qquad = \dfrac{1}{3}[\ln x - \ln y]$

$\qquad = \dfrac{1}{3} \ln x - \dfrac{1}{3} \ln y$

61. $\ln x^2 \sqrt{\dfrac{y}{z}} = \ln x^2 + \ln \sqrt{\dfrac{y}{z}}$

$\qquad = \ln x^2 + \dfrac{1}{2} \ln \dfrac{y}{z}$

$\qquad = \ln x^2 + \dfrac{1}{2}[\ln y - \ln z]$

$\qquad = 2 \ln x + \dfrac{1}{2} \ln y - \dfrac{1}{2} \ln z$

63. $\log_5 \left(\dfrac{x^2}{y^2 z^3}\right) = \log_5 x^2 - \log_5 y^2 z^3$

$\qquad = \log_5 x^2 - \left(\log_5 y^2 + \log_5 z^3\right)$

$\qquad = 2 \log_5 x - 2 \log_5 y - 3 \log_5 z$

65. $\ln \sqrt[4]{x^3(x^2+3)} = \dfrac{1}{4} \ln x^3(x^2+3)$

$\qquad = \dfrac{1}{4}\left[\ln x^3 + \ln(x^2+3)\right]$

$\qquad = \dfrac{1}{4}\left[3 \ln x + \ln(x^2+3)\right]$

$\qquad = \dfrac{3}{4} \ln x + \dfrac{1}{4} \ln(x^2+3)$

67. $\ln 2 + \ln x = \ln 2x$

69. $\log_4 z - \log_4 y = \log_4 \dfrac{z}{y}$

71. $2 \log_2 x + 4 \log_2 y = \log_2 x^2 + \log_2 y^4 = \log_2 x^2 y^4$

73. $\dfrac{1}{4} \log_3 5x = \log_3 (5x)^{1/4} = \log_3 \sqrt[4]{5x}$

75. $\log x - 2 \log(x+1) = \log x - \log(x+1)^2$

$\qquad = \log \dfrac{x}{(x+1)^2}$

77. $\log x - 2 \log y + 3 \log z = \log x - \log y^2 + \log z^3$

$\qquad = \log \dfrac{x}{y^2} + \log z^3$

$\qquad = \log \dfrac{xz^3}{y^2}$

79. $\ln x - \left[\ln(x+1) + \ln(x-1)\right] = \ln x - \ln(x+1)(x-1) = \ln \dfrac{x}{(x+1)(x-1)}$

81. $\frac{1}{3}\left[2\ln(x+3) + \ln x - \ln(x^2 - 1)\right] = \frac{1}{3}\left[\ln(x+3)^2 + \ln x - \ln(x^2 - 1)\right]$

$$= \frac{1}{3}\left[\ln x(x+3)^2 - \ln(x^2 - 1)\right]$$

$$= \frac{1}{3}\ln\frac{x(x+3)^2}{x^2 - 1}$$

$$= \ln\sqrt[3]{\frac{x(x+3)^2}{x^2 - 1}}$$

83. $\frac{1}{3}\left[\log_8 y + 2\log_8(y+4)\right] - \log_8(y-1) = \frac{1}{3}\left[\log_8 y + \log_8(y+4)^2\right] - \log_8(y-1)$

$$= \frac{1}{3}\log_8 y(y+4)^2 - \log_8(y-1)$$

$$= \log_8\sqrt[3]{y(y+4)^2} - \log_8(y-1)$$

$$= \log_8\left(\frac{\sqrt[3]{y(y+4)^2}}{y-1}\right)$$

85. $\log_2\frac{32}{4} = \log_2 32 - \log_2 4 \neq \frac{\log_2 32}{\log_2 4}$

The second and third expressions are equal by Property 2.

87. $\beta = 10\log\left(\frac{I}{10^{-12}}\right) = 10\left[\log I - \log 10^{-12}\right] = 10\left[\log I + 12\right] = 120 + 10\log I$

When $I = 10^{-6}$:

$\beta = 120 + 10\log 10^{-6} = 120 + 10(-6) = 60$ decibels

89. $\beta = 10\log\left(\frac{I}{10^{-12}}\right)$

Difference $= 10\log\left(\frac{10^{-4}}{10^{-12}}\right) - 10\log\left(\frac{10^{-11}}{10^{-12}}\right) = 10\left[\log 10^8 - \log 10\right] = 10(8 - 1) = 10(7) = 70$ dB

91.

x	1	2	3	4	5	6
y	1.000	1.189	1.316	1.414	1.495	1.565
$\ln x$	0	0.693	1.099	1.386	1.609	1.792
$\ln y$	0	0.173	0.275	0.346	0.402	0.448

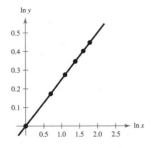

The slope of the line is $\frac{1}{4}$. So, $\ln y = \frac{1}{4}\ln x$.

93.

x	1	2	3	4	5	6
y	2.500	2.102	1.900	1.768	1.672	1.597
$\ln x$	0	0.693	1.099	1.386	1.609	1.792
$\ln y$	0.916	0.743	0.642	0.570	0.514	0.468

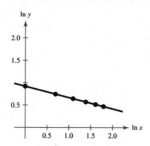

The slope of the line is $-\frac{1}{4}$. So, $\ln y = -\frac{1}{4} \ln x + \ln \frac{5}{2}$.

95.

Weight, x	25	35	50	75	500	1000
Galloping Speed, y	191.5	182.7	173.8	164.2	125.9	114.2
$\ln x$	3.219	3.555	3.912	4.317	6.215	6.908
$\ln y$	5.255	5.208	5.158	5.101	4.835	4.738

Using a graphing utility, $\ln y = -0.14 \ln x + 5.7$.

97. (a)

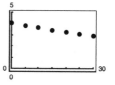

(b) $T - 21 = 54.4(0.964)^t$

$T = 54.4(0.964)^t + 21$

See graph in (a).

(c)

t (in minutes)	T (°C)	$T - 21$ (°C)	$\ln(T - 21)$	$1/(T - 21)$
0	78	57	4.043	0.0175
5	66	45	3.807	0.0222
10	57.5	36.5	3.597	0.0274
15	51.2	30.2	3.408	0.0331
20	46.3	25.3	3.231	0.0395
25	42.5	21.5	3.068	0.0465
30	39.6	18.6	2.923	0.0538

$\ln(T - 21) = -0.037t + 4$

$T = e^{-0.037t + 3.997} + 21$

This graph is identical to T in (b).

(d) $\dfrac{1}{T - 21} = 0.0012t + 0.016$

$T = \dfrac{1}{0.001t + 0.016} + 21$

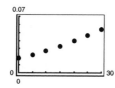
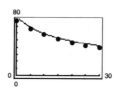

(e) Taking logs of temperatures led to a linear scatter plot because the log function increases very slowly as the x-values increase. Taking the reciprocals of the temperatures led to a linear scatter plot because of the asymptotic nature of the reciprocal function.

99. Let $x = \log_b u$, then $u = b^x$ and $u^n = b^{nx}$.

$\log_b u^n = \log_b b^{nx} = nx = n \log_b u$

101. $f(x) = \ln x$

False, $f(0) \neq 0$ because 0 is not in the domain of $f(x)$.

$f(1) = \ln 1 = 0$

103. False.

$$f(x) - f(2) = \ln x - \ln 2 = \ln \frac{x}{2} \neq \ln(x - 2)$$

105. False.

$$f(u) = 2f(v) \Rightarrow \ln u = 2 \ln v \Rightarrow \ln u = \ln v^2 \Rightarrow u = v^2$$

107. $f(x) = \log_2 x = \dfrac{\log x}{\log 2} = \dfrac{\ln x}{\ln 2}$

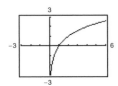

111. $f(x) = \log_{11.8} x$

$= \dfrac{\log x}{\log 11.8} = \dfrac{\ln x}{\ln 11.8}$

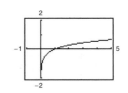

113. $f(x) = \ln \dfrac{x}{2}, \ g(x) = \dfrac{\ln x}{\ln 2}, \ h(x) = \ln x - \ln 2$

$f(x) = h(x)$ by Property 2

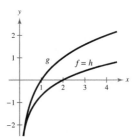

109. $f(x) = \log_{1/2} x$

$= \dfrac{\log x}{\log(1/2)} = \dfrac{\ln x}{\ln(1/2)}$

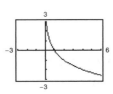

115. $\ln 2 \approx 0.6931, \ \ln 3 \approx 1.0986, \ \ln 5 \approx 1.6094$

$\ln 1 = 0$

$\ln 2 \approx 0.6931$

$\ln 3 \approx 1.0986$

$\ln 4 = \ln(2 \cdot 2) = \ln 2 + \ln 2 \approx 0.6931 + 0.6931 = 1.3862$

$\ln 5 \approx 1.6094$

$\ln 6 = \ln(2 \cdot 3) = \ln 2 + \ln 3 \approx 0.6931 + 1.0986 = 1.7917$

$\ln 8 = \ln 2^3 = 3 \ln 2 \approx 3(0.6931) = 2.0793$

$\ln 9 = \ln 3^2 = 2 \ln 3 \approx 2(1.0986) = 2.1972$

$\ln 10 = \ln(5 \cdot 2) = \ln 5 + \ln 2 \approx 1.6094 + 0.6931 = 2.3025$

$\ln 12 = \ln(2^2 \cdot 3) = \ln 2^2 + \ln 3 = 2 \ln 2 + \ln 3 \approx 2(0.6931) + 1.0986 = 2.4848$

$\ln 15 = \ln(5 \cdot 3) = \ln 5 + \ln 3 \approx 1.6094 + 1.0986 = 2.7080$

$\ln 16 = \ln 2^4 = 4 \ln 2 \approx 4(0.6931) = 2.7724$

$\ln 18 = \ln(3^2 \cdot 2) = \ln 3^2 + \ln 2 = 2 \ln 3 + \ln 2 \approx 2(1.0986) + 0.6931 = 2.8903$

$\ln 20 = \ln(5 \cdot 2^2) = \ln 5 + \ln 2^2 = \ln 5 + 2 \ln 2 \approx 1.6094 + 2(0.6931) = 2.9956$

Section 5.4 Exponential and Logarithmic Equations

- To solve an exponential equation, isolate the exponential expression, then take the logarithm of both sides. Then solve for the variable.

  1. $\log_a a^x = x$

  2. $\ln e^x = x$

- To solve a logarithmic equation, rewrite it in exponential form. Then solve for the variable.

  1. $a^{\log_a x} = x$

  2. $e^{\ln x} = x$

- If $a > 0$ and $a \neq 1$ we have the following:

  1. $\log_a x = \log_a y \Leftrightarrow x = y$

  2. $a^x = a^y \Leftrightarrow x = y$

- Check for extraneous solutions.

1. solve

3. (a) One-to-One

   (b) logarithmic; logarithmic

   (c) exponential; exponential

5. $4^{2x-7} = 64$

   (a) $\quad x = 5$

   $\quad 4^{2(5)-7} = 4^3 = 64$

   Yes, $x = 5$ *is* a solution.

   (b) $\quad x = 2$

   $\quad 4^{2(2)-7} = 4^{-3} = \frac{1}{64} \neq 64$

   No, $x = 2$ *is not* a solution.

7. $3e^{x+2} = 75$

   (a) $\qquad x = -2 + e^{25}$

   $\quad 3e^{\left(-2+e^{25}\right)+2} = 3e^{e^{25}} \neq 75$

   No, $x = -2 + e^{25}$ *is not* a solution.

   (b) $\qquad x = -2 + \ln 25$

   $\quad 3e^{(-2+\ln 25)+2} = 3e^{\ln 25} = 3(25) = 75$

   Yes, $x = -2 + \ln 25$ *is* a solution.

   (c) $\qquad x \approx 1.219$

   $\quad 3e^{1.219+2} = 3e^{3.219} \approx 75$

   Yes, $x \approx 1.219$ *is* a solution.

9. $\log_4(3x) = 3 \Rightarrow 3x = 4^3 \Rightarrow 3x = 64$

   (a) $\qquad x \approx 21.333$

   $\quad 3(21.333) \approx 64$

   Yes, $x \approx 21.333$ *is* an approximate solution.

   (b) $\qquad x = -4$

   $\quad 3(-4) = -12 \neq 64$

   No, $x = -4$ *is not* a solution.

   (c) $\qquad x = \frac{64}{3}$

   $\quad 3\left(\frac{64}{3}\right) = 64$

   Yes, $x = \frac{64}{3}$ *is* a solution.

11. $\ln(2x + 3) = 5.8$

   (a) $\qquad\qquad x = \frac{1}{2}(-3 + \ln 5.8)$

   $\quad \ln\left[2\left(\frac{1}{2}\right)(-3 + \ln 5.8) + 3\right] = \ln(\ln 5.8) \neq 5.8$

   No, $x = \frac{1}{2}(-3 + \ln 5.8)$ *is not* a solution.

   (b) $\qquad\qquad x = \frac{1}{2}\left(-3 + e^{5.8}\right)$

   $\quad \ln\left[2\left(\frac{1}{2}\right)\left(-3 + e^{5.8}\right) + 3\right] = \ln\left(e^{5.8}\right) = 5.8$

   Yes, $x = \frac{1}{2}\left(-3 + e^{5.8}\right)$ *is* a solution.

   (c) $\qquad\qquad x \approx 163.650$

   $\quad \ln\left[2(163.650) + 3\right] = \ln 330.3 \approx 5.8$

   Yes, $x \approx 163.650$ *is* an approximate solution.

13. $4^x = 16$

$4^x = 4^2$

$x = 2$

15. $\left(\frac{1}{2}\right)^x = 32$

$2^{-x} = 2^5$

$-x = 5$

$x = -5$

17. $\ln x - \ln 2 = 0$

$\ln x = \ln 2$

$x = 2$

19. $e^x = 2$

$\ln e^x = \ln 2$

$x = \ln 2$

$x \approx 0.693$

21. $\ln x = -1$

$e^{\ln x} = e^{-1}$

$x = e^{-1}$

$x \approx 0.368$

23. $\log_4 x = 3$

$4^{\log_4 x} = 4^3$

$x = 4^3$

$x = 64$

25. $f(x) = g(x)$

$2^x = 8$

$2^x = 2^3$

$x = 3$

Point of intersection:

$(3, 8)$

27. $f(x) = g(x)$

$\log_3 x = 2$

$x = 3^2$

$x = 9$

Point of intersection:

$(9, 2)$

29. $e^x = e^{x^2-2}$

$x = x^2 - 2$

$0 = x^2 - x - 2$

$0 = (x + 1)(x - 2)$

$x = -1, x = 2$

31. $e^{x^2-3} = e^{x-2}$

$x^2 - 3 = x - 2$

$x^2 - x - 1 = 0$

By the Quadratic Formula

$x \approx 1.618, x \approx -0.618.$

33. $4(3^x) = 20$

$3^x = 5$

$\log_3 3^x = \log_3 5$

$x = \log_3 5 = \dfrac{\log 5}{\log 3}$ or $\dfrac{\ln 5}{\ln 3}$

$x \approx 1.465$

35. $2e^x = 10$

$e^x = 5$

$\ln e^x = \ln 5$

$x = \ln 5 \approx 1.609$

37. $e^x - 9 = 19$

$e^x = 28$

$\ln e^x = \ln 28$

$x = \ln 28 \approx 3.332$

39. $3^{2x} = 80$

$\ln 3^{2x} = \ln 80$

$2x \ln 3 = \ln 80$

$x = \dfrac{\ln 80}{2 \ln 3} \approx 1.994$

41. $5^{-t/2} = 0.20$

$5^{-t/2} = \dfrac{1}{5}$

$5^{-t/2} = 5^{-1}$

$-\dfrac{t}{2} = -1$

$t = 2$

43. $3^{x-1} = 27$

$3^{x-1} = 3^3$

$x - 1 = 3$

$x = 4$

45.
$$2^{3-x} = 565$$
$$\ln 2^{3-x} = \ln 565$$
$$(3-x)\ln 2 = \ln 565$$
$$3\ln 2 - x\ln 2 = \ln 565$$
$$-x\ln 2 = \ln 565 - 3\ln 2$$
$$x\ln 2 = 3\ln 2 - \ln 565$$
$$x = \frac{3\ln 2 - \ln 565}{\ln 2}$$
$$= 3 - \frac{\ln 565}{\ln 2} \approx -6.142$$

47. $8(10^{3x}) = 12$
$$10^{3x} = \frac{12}{8}$$
$$\log 10^{3x} = \log\left(\frac{3}{2}\right)$$
$$3x = \log\left(\frac{3}{2}\right)$$
$$x = \frac{1}{3}\log\left(\frac{3}{2}\right)$$
$$\approx 0.059$$

49. $3(5^{x-1}) = 21$
$$5^{x-1} = 7$$
$$\ln 5^{x-1} = \ln 7$$
$$(x-1)\ln 5 = \ln 7$$
$$x - 1 = \frac{\ln 7}{\ln 5}$$
$$x = 1 + \frac{\ln 7}{\ln 5} \approx 2.209$$

51. $e^{3x} = 12$
$$3x = \ln 12$$
$$x = \frac{\ln 12}{3} \approx 0.828$$

53. $500e^{-x} = 300$
$$e^{-x} = \frac{3}{5}$$
$$-x = \ln\frac{3}{5}$$
$$x = -\ln\frac{3}{5}$$
$$= \ln\frac{5}{3} \approx 0.511$$

55. $7 - 2e^x = 5$
$$-2e^x = -2$$
$$e^x = 1$$
$$x = \ln 1 = 0$$

57. $6(2^{3x-1}) - 7 = 9$
$$6(2^{3x-1}) = 16$$
$$2^{3x-1} = \frac{8}{3}$$
$$\log_2 2^{3x-1} = \log_2\left(\frac{8}{3}\right)$$
$$3x - 1 = \log_2\left(\frac{8}{3}\right) = \frac{\log(8/3)}{\log 2} \text{ or } \frac{\ln(8/3)}{\ln 2}$$
$$x = \frac{1}{3}\left[\frac{\log(8/3)}{\log 2} + 1\right] \approx 0.805$$

59. $e^{2x} - 4e^x - 5 = 0$
$$(e^x + 1)(e^x - 5) = 0$$
$$e^x = -1 \text{ or } e^x = 5$$
(No solution) $x = \ln 5 \approx 1.609$

61. $e^{2x} - 3e^x - 4 = 0$
$$(e^x + 1)(e^x - 4) = 0$$
$$e^x + 1 = 0 \Rightarrow e^x = -1$$
Not possible since $e^x > 0$ for all x.
$$e^x - 4 = 0 \Rightarrow e^x = 4 \Rightarrow x = \ln 4 \approx 1.386$$

63. $\dfrac{500}{100 - e^{x/2}} = 20$
$$500 = 20(100 - e^{x/2})$$
$$25 = 100 - e^{x/2}$$
$$e^{x/2} = 75$$
$$\frac{x}{2} = \ln 75$$
$$x = 2\ln 75 \approx 8.635$$

65. $\dfrac{3000}{2 + e^{2x}} = 2$
$$3000 = 2(2 + e^{2x})$$
$$1500 = 2 + e^{2x}$$
$$1498 = e^{2x}$$
$$\ln 1498 = 2x$$
$$x = \frac{\ln 1498}{2} \approx 3.656$$

67. $\left(1 + \dfrac{0.065}{365}\right)^{365t} = 4$

$\ln\left(1 + \dfrac{0.065}{365}\right)^{365t} = \ln 4$

$365t \ln\left(1 + \dfrac{0.065}{365}\right) = \ln 4$

$t = \dfrac{\ln 4}{365 \ln\left(1 + \dfrac{0.065}{365}\right)} \approx 21.330$

69. $\left(1 + \dfrac{0.10}{12}\right)^{12t} = 2$

$\ln\left(1 + \dfrac{0.10}{12}\right)^{12t} = \ln 2$

$12t \ln\left(1 + \dfrac{0.10}{12}\right) = \ln 2$

$t = \dfrac{\ln 2}{12 \ln\left(1 + \dfrac{0.10}{12}\right)} \approx 6.960$

71. $y_1 = 7$

$y_2 = 2^x$

From the graph $x \approx 2.807$ when $y = 7$.

Algebraically:

$2^x = 7$

$\ln 2^x = \ln 7$

$x \ln 2 = \ln 7$

$x = \dfrac{\ln 7}{\ln 2} \approx 2.807$

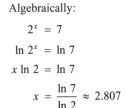

73. $g(x) = 6e^{1-x} - 25$

Algebraically:

$6e^{1-x} = 25$

$e^{1-x} = \dfrac{25}{6}$

$1 - x = \ln\left(\dfrac{25}{6}\right)$

$x = 1 - \ln\left(\dfrac{25}{6}\right)$

$x \approx -0.427$

The zero is $x \approx -0.427$.

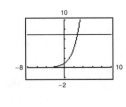

75. $f(x) = 3e^{3x/2} - 962$

Algebraically:

$3e^{3x/2} = 962$

$e^{3x/2} = \dfrac{962}{3}$

$\dfrac{3x}{2} = \ln\left(\dfrac{962}{3}\right)$

$x = \dfrac{2}{3} \ln\left(\dfrac{962}{3}\right)$

$x \approx 3.847$

The zero is $x \approx 3.847$

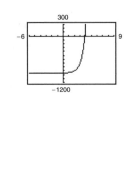

77. $g(t) = e^{0.09t} - 3$

Algebraically:

$e^{0.09t} = 3$

$0.09t = \ln 3$

$t = \dfrac{\ln 3}{0.09}$

$t \approx 12.207$

The zero is $t \approx 12.207$.

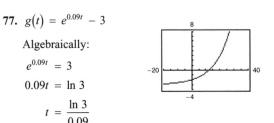

79. $h(t) = e^{0.125t} - 8$

Algebraically:

$e^{0.125t} - 8 = 0$

$e^{0.125t} = 8$

$0.125t = \ln 8$

$t = \dfrac{\ln 8}{0.125}$

$t \approx 16.636$

This zero is $t \approx 16.636$.

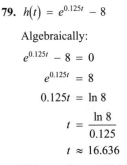

81. $\ln x = -3$

$x = e^{-3} \approx 0.050$

83. $\ln x - 7 = 0$

$\ln x = 7$

$x = e^7 \approx 1096.633$

85. $\ln 2x = 2.4$

$2x = e^{2.4}$

$x = \dfrac{e^{2.4}}{2} \approx 5.512$

87. $\log x = 6$

$x = 10^6$

$= 1,000,000$

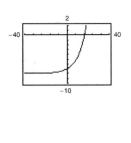

89. $3 \ln 5x = 10$

$\ln 5x = \dfrac{10}{3}$

$5x = e^{10/3}$

$x = \dfrac{e^{10/3}}{5} \approx 5.606$

91. $\ln \sqrt{x + 2} = 1$

$\sqrt{x + 2} = e^1$

$x + 2 = e^2$

$x = e^2 - 2$

≈ 5.389

93. $7 + 3 \ln x = 5$

$3 \ln x = -2$

$\ln x = -\dfrac{2}{3}$

$x = e^{-2/3}$

≈ 0.513

95. $-2 + 2 \ln 3x = 17$

$2 \ln 3x = 19$

$\ln 3x = \dfrac{19}{2}$

$3x = e^{19/2}$

$x = \dfrac{e^{19/2}}{3} \approx 4453.242$

97. $6 \log_3(0.5x) = 11$

$\log_3(0.5x) = \dfrac{11}{6}$

$3^{\log_3(0.5x)} = 3^{11/6}$

$0.5x = 3^{11/6}$

$x = 2(3^{11/6}) \approx 14.988$

99. $\ln x - \ln(x + 1) = 2$

$\ln\left(\dfrac{x}{x + 1}\right) = 2$

$\dfrac{x}{x + 1} = e^2$

$x = e^2(x + 1)$

$x = e^2 x + e^2$

$x - e^2 x = e^2$

$x(1 - e^2) = e^2$

$x = \dfrac{e^2}{1 - e^2} \approx -1.157$

This negative value is extraneous. The equation has no solution.

101. $\ln x + \ln(x - 2) = 1$

$\ln[x(x - 2)] = 1$

$x(x - 2) = e^1$

$x^2 - 2x - e = 0$

$x = \dfrac{2 \pm \sqrt{4 + 4e}}{2}$

$= \dfrac{2 \pm 2\sqrt{1 + e}}{2} = 1 \pm \sqrt{1 + e}$

The negative value is extraneous. The only solution is $x = 1 + \sqrt{1 + e} \approx 2.928$.

103. $\ln(x + 5) = \ln(x - 1) - \ln(x + 1)$

$\ln(x + 5) = \ln\left(\dfrac{x - 1}{x + 1}\right)$

$x + 5 = \dfrac{x - 1}{x + 1}$

$(x + 5)(x + 1) = x - 1$

$x^2 + 6x + 5 = x - 1$

$x^2 + 5x + 6 = 0$

$(x + 2)(x + 3) = 0$

$x = -2 \text{ or } x = -3$

Both of these solutions are extraneous, so the equation has no solution.

105. $\log_2(2x - 3) = \log_2(x + 4)$

$2x - 3 = x + 4$

$x = 7$

107. $\log(x + 4) - \log x = \log(x + 2)$

$\log\left(\dfrac{x + 4}{x}\right) = \log(x + 2)$

$\dfrac{x + 4}{x} = x + 2$

$x + 4 = x^2 + 2x$

$0 = x^2 + x - 4$

$x = \dfrac{-1 \pm \sqrt{17}}{2}$ Quadratic Formula

The negative value is extraneous.

The only solution is $x = \dfrac{-1 + \sqrt{17}}{2} \approx 1.562$.

109. $\log_4 x - \log_4(x - 1) = \dfrac{1}{2}$

$$\log_4\left(\dfrac{x}{x-1}\right) = \dfrac{1}{2}$$

$$4^{\log_4[x/(x-1)]} = 4^{1/2}$$

$$\dfrac{x}{x-1} = 4^{1/2}$$

$$x = 2(x - 1)$$

$$x = 2x - 2$$

$$-x = -2$$

$$x = 2$$

111. $\log 8x - \log\left(1 + \sqrt{x}\right) = 2$

$$\log\dfrac{8x}{1 + \sqrt{x}} = 2$$

$$\dfrac{8x}{1 + \sqrt{x}} = 10^2$$

$$8x = 100\left(1 + \sqrt{x}\right)$$

$$2x = 25\left(1 + \sqrt{x}\right) = 25 + 25\sqrt{x}$$

$$2x - 25 = 25\sqrt{x}$$

$$\left(2x - 25\right)^2 = \left(25\sqrt{x}\right)^2$$

$$4x^2 - 100x + 625 = 625x$$

$$4x^2 - 725x + 625 = 0$$

$$x = \dfrac{725 \pm \sqrt{725^2 - 4(4)(625)}}{2(4)} = \dfrac{725 \pm \sqrt{515{,}625}}{8} = \dfrac{25\left(29 \pm 5\sqrt{33}\right)}{8}$$

$$x \approx 0.866 \text{ (extraneous) or } x \approx 180.384$$

The only solution is $x = \dfrac{25\left(29 + 5\sqrt{33}\right)}{8} \approx 180.384$.

113. $y_1 = 3$

$y_2 = \ln x$

From the graph

$x \approx 20.086$ when $y = 3$.

Algebraically:

$3 - \ln x = 0$

$\ln x = 3$

$x = e^3 \approx 20.086$

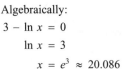

115. $y_1 = 2\ln(x + 3)$

$y_2 = 3$

From the graph, $x \approx 1.482$ when $y = 3$.

Algebraically:

$2\ln(x + 3) = 3$

$\ln(x + 3) = \dfrac{3}{2}$

$x + 3 = e^{3/2}$

$x = e^{3/2} - 3 \approx 1.482$

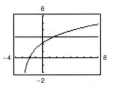

117. (a) $r = 0.05$

$$A = Pe^{rt}$$

$$5000 = 2500e^{0.05t}$$

$$2 = e^{0.05t}$$

$$\ln 2 = 0.05t$$

$$\frac{\ln 2}{0.05} = t$$

$$t \approx 13.86 \text{ years}$$

(b) $r = 0.05$

$$A = Pe^{rt}$$

$$7500 = 2500e^{0.05t}$$

$$3 = e^{0.05t}$$

$$\ln 3 = 0.05t$$

$$\frac{\ln 3}{0.05} = t$$

$$t \approx 21.97 \text{ years}$$

119. (a) $r = 0.025$

$$A = Pe^{rt}$$

$$5000 = 2500e^{0.025t}$$

$$2 = e^{0.025t}$$

$$\ln 2 = 0.025t$$

$$\frac{\ln 2}{0.025} = t$$

$$t \approx 27.73 \text{ years}$$

(b) $r = 0.025$

$$A = Pe^{rt}$$

$$7500 = 2500e^{0.025t}$$

$$3 = e^{0.025t}$$

$$\ln 3 = 0.025t$$

$$\frac{\ln 3}{0.025} = t$$

$$t \approx 43.94 \text{ years}$$

121. $2x^2e^{2x} + 2xe^{2x} = 0$

$$\left(2x^2 + 2x\right)e^{2x} = 0$$

$$2x^2 + 2x = 0 \quad \left(\text{because } e^{2x} \neq 0\right)$$

$$2x(x + 1) = 0$$

$$x = 0, -1$$

123. $-xe^{-x} + e^{-x} = 0$

$$\left(-x + 1\right)e^{-x} = 0$$

$$-x + 1 = 0 \quad \left(\text{because } e^{-x} \neq 0\right)$$

$$x = 1$$

125. $2x \ln x + x = 0$

$$x(2 \ln x + 1) = 0$$

$$2 \ln x + 1 = 0 \quad (\text{because } x > 0)$$

$$\ln x = -\tfrac{1}{2}$$

$$x = e^{-1/2} \approx 0.607$$

127. $\dfrac{1 + \ln x}{2} = 0$

$$1 + \ln x = 0$$

$$\ln x = -1$$

$$x = e^{-1} = \frac{1}{e} \approx 0.368$$

129. $p = 1000\left(1 - \dfrac{5}{5 + e^{-0.001x}}\right)$

(a) When $p = 139.50$:

$$139.50 = 1000\left(1 - \frac{5}{5 + e^{-0.001x}}\right)$$

$$0.1395 = 1 - \frac{5}{5 + e^{-0.001x}}$$

$$\frac{5}{5 + e^{-0.001x}} = 0.8605$$

$$5 = 4.3025 + 0.8605e^{-0.001x}$$

$$0.6975 = 0.8605e^{-0.001x}$$

$$\frac{6975}{8605} = e^{-0.001x}$$

$$\ln \frac{6975}{8605} = \ln e^{-0.001x}$$

$$\ln \frac{6975}{8605} = -0.001x$$

$$x = -\frac{\ln(6975/8605)}{0.001} \approx 210 \text{ coins}$$

(b) When $p = 99.99$:

$$99.99 = 1000\left(1 - \frac{5}{5 + e^{-0.001x}}\right)$$

$$0.09999 = 1 - \frac{5}{5 + e^{-0.001x}}$$

$$\frac{5}{5 + e^{-0.001x}} = 0.90001$$

$$5 = 4.50005 + 0.90001e^{-0.001x}$$

$$0.49995 = 0.90001e^{-0.001x}$$

$$\frac{49{,}995}{90{,}001} = e^{-0.001x}$$

$$\ln \frac{49{,}995}{90{,}001} = \ln e^{-0.001x}$$

$$\ln \frac{49{,}995}{90{,}001} = -0.001x$$

$$x = -\frac{\ln(49{,}995/90{,}001)}{0.001} \approx 588 \text{ coins}$$

131. $V = 6.7e^{-48.1/t}, t \geq 0$

(a)

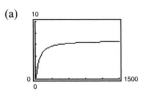

(b) As $t \to \infty, V \to 6.7.$

Horizontal asymptote: $V = 6.7$

The yield will approach 6.7 million cubic feet per acre.

(c) $1.3 = 6.7e^{-48.1/t}$

$\dfrac{1.3}{6.7} = e^{-48.1/t}$

$\ln\left(\dfrac{13}{67}\right) = \dfrac{-48.1}{t}$

$t = \dfrac{-48.1}{\ln(13/67)} \approx 29.3$ years

133. $y = -451 + 444 \ln t, 10 \leq t \leq 17$

$-451 + 444 \ln t = 690$

$444 \ln t = 1141$

$\ln t = \dfrac{1141}{444}$

$t = e^{1141/444} \approx 13$

$t \approx 13$ corresponds to the year 2003.

135. (a) From the graph shown in the textbook you see horizontal asymptotes at $y = 0$ and $y = 100$.

These represent the lower and upper percent bounds; the range falls between 0% and 100%.

(b) Males

$50 = \dfrac{100}{1 + e^{-0.6114(x-69.71)}}$

$1 + e^{-0.6114(x-69.71)} = 2$

$e^{-0.6114(x-69.71)} = 1$

$-0.6114(x - 69.71) = \ln 1$

$-0.6114(x - 69.71) = 0$

$x = 69.71$ inches

Females

$50 = \dfrac{100}{1 + e^{-0.66607(x-64.51)}}$

$1 + e^{-0.66607(x-64.51)} = 2$

$e^{-0.6667(x-64.51)} = 1$

$-0.66607(x - 64.51) = \ln 1$

$-0.66607(x - 64.51) = 0$

$x = 64.51$ inches

137. $y = -3.00 + 11.88 \ln x + \dfrac{36.94}{x}$

(a)

x	0.2	0.4	0.6	0.8	1.0
y	162.6	78.5	52.5	40.5	33.9

(b)

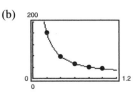

The model seems to fit the data well.

(c) When $y = 30$:

$30 = -3.00 + 11.88 \ln x + \dfrac{36.94}{x}$

Add the graph of $y = 30$ to the graph in part (a) and estimate the point of intersection of the two graphs.

You find that $x \approx 1.20$ meters.

(d) No, it is probably not practical to lower the number of *g*s experienced during impact to less than 23 because the required distance traveled at $y = 23$ is $x \approx 2.27$ meters. It is probably not practical to design a car allowing a passenger to move forward 2.27 meters (or 7.45 feet) during an impact.

139. $\log_a(uv) = \log_a u + \log_a v$

True by Property 1 in Section 3.3.

141. $\log_a(u - v) = \log_a u - \log_a v$

False.

$1.95 \approx \log(100 - 10)$

$\neq \log 100 - \log 10 = 1$

143. Yes, a logarithmic equation can have more than one extraneous solution. See Exercise 103.

145. Yes.

Time to Double	Time to Quadruple
$2P = Pe^{rt}$	$4P = Pe^{rt}$
$2 = e^{rt}$	$4 = e^{rt}$
$\ln 2 = rt$	$\ln 4 = rt$
$\dfrac{\ln 2}{r} = t$	$\dfrac{2\ln 2}{r} = t$

So, the time to quadruple is twice as long as the time to double.

147. $f(x) = \log_a x, g(x) = a^x, a > 1.$

(a) $a = 1.2$ 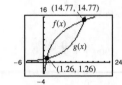   The curves intersect twice: $(1.258, 1.258)$ and $(14.767, 14.767)$

(b) If $f(x) = \log_a x = a^x = g(x)$ intersect exactly once, then

$x = \log_a x = a^x \Rightarrow a = x^{1/x}.$

The graphs of $y = x^{1/x}$ and $y = a$ intersect once for $a = e^{1/e} \approx 1.445.$ Then

$\log_a x = x \Rightarrow \left(e^{1/e}\right)^x = x \Rightarrow e^{x/e} = x \Rightarrow x = e.$

For $a = e^{1/e},$ then curves intersect once at $(e, e).$

(c) For $1 < a < e^{1/e}$ the curves intersect twice. For $a > e^{1/e},$ the curves do not intersect.

Section 5.5   Exponential and Logarithmic Models

- You should be able to solve growth and decay problems.

  (a) Exponential growth if $b > 0$ and $y = ae^{bx}.$

  (b) Exponential decay if $b > 0$ and $y = ae^{-bx}.$

- You should be able to use the Gaussian model

  $y = ae^{-(x-b)^2/c}.$

- You should be able to use the logistic growth model

  $y = \dfrac{a}{1 + be^{-rx}}.$

- You should be able to use the logarithmic models

  $y = a + b \ln x, \ y = a + b \log x.$

1. $y = ae^{bx}; y = ae^{-bx}$

3. normally distributed

5. $y = \dfrac{a}{1 + be^{-rx}}$

7. $y = 2e^{x/4}$

This is an exponential growth model. Matches graph (c).

9. $y = 6 + \log(x + 2)$

This is a logarithmic function shifted up six units and left two units. Matches graph (b).

11. $y = \ln(x + 1)$

This is a logarithmic model shifted left one unit. Matches graph (d).

13. (a) $A = Pe^{rt}$

$\dfrac{A}{e^{rt}} = P$

(b) $A = Pe^{rt}$

$\dfrac{A}{P} = e^{rt}$

$\ln \dfrac{A}{P} = \ln e^{rt}$

$\ln \dfrac{A}{P} = rt$

$\dfrac{\ln(A/P)}{r} = t$

15. Because $A = 1000e^{0.035t}$, the time to double is given by

$2000 = 1000e^{0.035t}$ and you have

$2 = e^{0.035t}$

$\ln 2 = \ln e^{0.035t}$

$\ln 2 = 0.035t$

$t = \dfrac{\ln 2}{0.035} \approx 19.8$ years.

Amount after 10 years: $A = 1000e^{0.35} \approx \$1419.07$

17. Because $A = 750e^{rt}$ and $A = 1500$ when $t = 7.75$, you have

$1500 = 750e^{7.75r}$

$2 = e^{7.75r}$

$\ln 2 = \ln e^{7.75r}$

$\ln 2 = 7.75r$

$r = \dfrac{\ln 2}{7.75} \approx 0.089438 = 8.9438\%.$

Amount after 10 years: $A = 750e^{0.089438(10)} \approx \$1834.37$

19. Because $A = 500e^{rt}$ and $A = \$1505.00$ when $t = 10$, you have

$1505.00 = 500e^{10r}$

$r = \dfrac{\ln(1505.00/500)}{10} \approx 0.110 = 11.0\%.$

The time to double is given by

$1000 = 500e^{0.110t}$

$t = \dfrac{\ln 2}{0.110} \approx 6.3$ years.

21. Because $A = Pe^{0.045t}$ and $A = 10,000.00$ when $t = 10$, you have

$10,000.00 = Pe^{0.045(10)}$

$\dfrac{10,000.00}{e^{0.045(10)}} = P \approx \$6376.28.$

The time to double is given by

$t = \dfrac{\ln 2}{0.045} \approx 15.40$ years.

23. $A = 500,000, r = 0.05, n = 12, t = 10$

$A = P\left(1 + \dfrac{r}{n}\right)^{nt}$

$500,000 = P\left(1 + \dfrac{0.05}{12}\right)^{12(10)}$

$P = \dfrac{500,000}{\left(1 + \dfrac{0.05}{12}\right)^{12(10)}}$

$\approx \$303,580.52$

25. $P = 1000, r = 0.1, A = 2000$

$A = P\left(1 + \dfrac{r}{n}\right)^{nt}$

$2000 = 1000\left(1 + \dfrac{0.1}{n}\right)^{nt}$

$2 = \left(1 + \dfrac{0.1}{n}\right)^{nt}$

(a) $n = 1$

$(1 + 0.1)^t = 2$

$(1.1)^t = 2$

$\ln(1.1)^t = \ln 2$

$t \ln 1.1 = \ln 2$

$t = \dfrac{\ln 2}{\ln 1.1} \approx 7.27$ years

(b) $n = 12$

$\left(1 + \dfrac{0.1}{12}\right)^{12t} = 2$

$\ln\left(\dfrac{12.1}{12}\right)^{12t} = \ln 2$

$12t \ln\left(\dfrac{12.1}{12}\right) = \ln 2$

$12t = \dfrac{\ln 2}{\ln(12.1/12)}$

$t = \dfrac{\ln 2}{12 \ln(12.1/12)} \approx 6.96$ years

(c) $n = 365$

$\left(1 + \dfrac{0.1}{365}\right)^{365t} = 2$

$\ln\left(\dfrac{365.1}{365}\right)^{365t} = \ln 2$

$365t \ln\left(\dfrac{365.1}{365}\right) = \ln 2$

$365t = \dfrac{\ln 2}{\ln(365.1/365)}$

$t = \dfrac{\ln 2}{365 \ln(365.1/365)} \approx 6.93$ years

(d) Compounded continuously

$A = Pe^{rt}$

$2000 = 1000e^{0.1t}$

$2 = e^{0.1t}$

$\ln 2 = \ln e^{0.1t}$

$0.1t = \ln 2$

$t = \dfrac{\ln 2}{0.1} \approx 6.93$ years

27. $3P = Pe^{rt}$

$3 = e^{rt}$

$\ln 3 = rt$

$\dfrac{\ln 3}{r} = t$

r	2%	4%	6%	8%	10%	12%
$t = \dfrac{\ln 3}{r}$ (years)	54.93	27.47	18.31	13.73	10.99	9.16

29. $3P = P(1 + r)^t$

$3 = (1 + r)^t$

$\ln 3 = \ln(1 + r)^t$

$\ln 3 = t \ln(1 + r)$

$\dfrac{\ln 3}{\ln(1 + r)} = t$

r	2%	4%	6%	8%	10%	12%
$t = \dfrac{\ln 3}{\ln(1 + r)}$ (years)	55.48	28.01	18.85	14.27	11.53	9.69

31. Continuous compounding results in faster growth.

$A = 1 + 0.075[\![t]\!]$ and $A = e^{0.07t}$

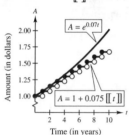

33. $a = 10,\ y = \dfrac{1}{2}(10) = 5,\ t = 1599$

$y = ae^{-bt}$

$5 = 10e^{-b(1599)}$

$0.5 = e^{-1599b}$

$\ln 0.5 = \ln e^{-1599b}$

$\ln 0.5 = -1599b$

$b = -\dfrac{\ln 0.5}{1599}$

Given an initial quantity of 10 grams, after 1000 years, you have

$y = 10e^{-[-(\ln 0.5)/1599](1000)} \approx 6.48$ grams.

35. $a = 2.1,\ y = \dfrac{1}{2}(2.1) = 1.05,\ t = 24{,}100$

$y = ae^{-bt}$

$1.05 = 2.1e^{-b(24{,}100)}$

$0.5 = e^{-24{,}100b}$

$\ln 0.5 = \ln e^{-24{,}100b}$

$\ln 0.5 = -24{,}100b$

$b = -\dfrac{\ln 0.5}{24{,}100}$

Given an initial quantity of 2.1 grams, after 1000 years, you have

$y = 2.1e^{-[-(\ln 0.5)/24{,}100](1000)} \approx 2.04$ grams.

37. $y = 2,\ a = 2(2) = 4,\ t = 5715$

$y = ae^{-bt}$

$2 = 4e^{-b(5715)}$

$0.5 = e^{-5715b}$

$\ln 0.5 = \ln e^{-5715b}$

$\ln 0.5 = -5715b$

$b = -\dfrac{\ln 0.5}{5715}$

Given 2 grams after 1000 years, the initial amount is

$2 = ae^{-[-(\ln 0.5)/5715](1000)}$

$a \approx 2.26$ grams.

39. $y = ae^{bx}$

$1 = ae^{b(0)} \Rightarrow 1 = a$

$10 = e^{b(3)}$

$\ln 10 = 3b$

$\dfrac{\ln 10}{3} = b \Rightarrow b \approx 0.7675$

So, $y = e^{0.7675x}$.

41. $y = ae^{bx}$

$5 = ae^{b(0)} \Rightarrow 5 = a$

$1 = 5e^{b(4)}$

$\dfrac{1}{5} = e^{4b}$

$\ln\left(\dfrac{1}{5}\right) = 4b$

$\dfrac{\ln(1/5)}{4} = b \Rightarrow b \approx -0.4024$

So, $y = 5e^{-0.4024x}$.

43. $P = -18.5 + 92.2e^{0.0282t}$

(a)

Year	1970	1980	1990
Population	73.7	103.74	143.56

Year	2000	2007
Population	196.35	243.24

(b)
$$P = 300$$
$$300 = -18.5 + 92.2e^{0.0282t}$$
$$318.5 = 92.2e^{0.0282t}$$
$$\frac{3185}{922} = e^{0.0282t}$$
$$\ln \frac{3185}{922} = \ln e^{0.0282t}$$
$$\ln \frac{3185}{922} = 0.0282t$$
$$t = \frac{\ln(3185/922)}{0.0282} \approx 44$$

$t \approx 44$ corresponds to the year 2014.

The population will reach 300,000 in 2014.

(c) The model is not valid for long-term predictions of the population because the population will not continue to grow at such a quick rate.

45. $y = 4080e^{kt}$

When $t = 3$, $y = 10,000$:
$$10,000 = 4080e^{k(3)}$$
$$\frac{10,000}{4080} = e^{3k}$$
$$\ln\left(\frac{10,000}{4080}\right) = 3k$$
$$k = \frac{\ln(10,000/4080)}{3} \approx 0.2988$$

When $t = 24$: $y = 4080e^{0.2988(24)} \approx 5,309,734$ hits

47. $P = 346.8e^{kt}$

(a)
$$395 = 346.8e^{k(5)}$$
$$\ln \frac{395}{346.8} = 5k \Rightarrow k \approx 0.02603$$

Because k is positive, the population is increasing.

(b) For 2010, use $t = 10$:

$P \approx 449.910$ thousand $= 449,910$

For 2015, use $t = 15$:

$P \approx 512.447$ thousand $= 512,447$

The results are reasonable because the population is growing at a slow rate.

(c)
$$500 = 346.8e^{0.02603t}$$
$$\ln \frac{500}{346.8} = 0.02603t$$
$$t = \frac{\ln(500/346.8)}{0.02603} \approx 14$$

$t = 14$ corresponds to the year 2014.

The population will reach 500,000 in 2014.

49. $y = ae^{bt}$

When $t = 3$, $y = 100$:   When $t = 5$, $y = 400$:
$$100 = ae^{3b} \qquad\qquad 400 = ae^{5b}$$
$$\frac{100}{e^{3b}} = a$$

Substitute $\frac{100}{e^{3b}}$ for a in the equation on the right.

$$400 = \frac{100}{e^{3b}}e^{5b}$$
$$400 = 100e^{2b}$$
$$4 = e^{2b}$$
$$\ln 4 = 2b$$
$$\ln 2^2 = 2b$$
$$2 \ln 2 = 2b$$
$$\ln 2 = b$$

$$a = \frac{100}{e^{3b}} = \frac{100}{e^{3\ln 2}} = \frac{100}{e^{\ln 2^3}} = \frac{100}{2^3} = \frac{100}{8} = 12.5$$

$$y = 12.5e^{(\ln 2)t}$$

After 6 hours, there are

$$y = 12.5e^{(\ln 2)(6)} = 800 \text{ bacteria.}$$

51. $R = \dfrac{1}{10^{12}}e^{-t/8223}$

(a) $\quad R = \dfrac{1}{8^{14}}$

$\dfrac{1}{10^{12}}e^{-t/8223} = \dfrac{1}{8^{14}}$

$e^{-t/8223} = \dfrac{10^{12}}{8^{14}}$

$-\dfrac{t}{8223} = \ln\left(\dfrac{10^{12}}{8^{14}}\right)$

$t = -8223\ln\left(\dfrac{10^{12}}{8^{14}}\right) \approx 12{,}180$ years old

(b) $\dfrac{1}{10^{12}}e^{-t/8223} = \dfrac{1}{13^{11}}$

$e^{-t/8223} = \dfrac{10^{12}}{13^{11}}$

$-\dfrac{t}{8223} = \ln\left(\dfrac{10^{12}}{13^{11}}\right)$

$t = -8223\ln\left(\dfrac{10^{12}}{13^{11}}\right) \approx 4797$ years old

53. $(0, 23{,}300), (2, 12{,}500)$

(a) $m = \dfrac{12{,}500 - 23{,}300}{2 - 0} = -5400$

$V = -5400t + 23{,}300$

(b) $V = ae^{kt}$

$12{,}500 = 23{,}300e^{k(2)}$

$\ln\dfrac{12{,}500}{23{,}300} = 2k \Rightarrow k \approx -0.311$

$V = 23{,}300e^{-0.311t}$

(c)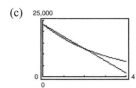

The exponential model depreciates faster in the first 2 years.

(d)

t	1	3
$V = -5400t + 23{,}300$	\$17,900	\$7100
$V = 23{,}300e^{-0.311t}$	\$17,072	\$9166

(e) *Sample answer:* The linear model gives a higher value for the car for the first two years, then the exponential model yields a higher value. If the car is less than two years old, the seller would most likely want to use the linear model and the buyer the exponential model. If it is more than two years old, the opposite is true.

55. $S(t) = 100\left(1 - e^{kt}\right)$

(a) $\quad 15 = 100\left(1 - e^{k(1)}\right)$

$-85 = -100e^{k}$

$\dfrac{85}{100} = e^{k}$

$0.85 = e^{k}$

$\ln 0.85 = \ln e^{k}$

$k = \ln 0.85$

$k \approx -0.1625$

$S(t) = 100\left(1 - e^{-0.1625t}\right)$

(b)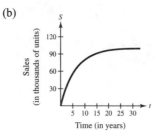

(c) $S(5) = 100\left(1 - e^{-0.1625(5)}\right) \approx 55.625 = 55{,}625$ units

57. $y = 0.0266e^{-(x-100)^2/450}, \ 70 \le x \le 116$

(a)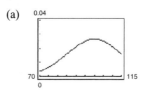

(b) The average IQ score of an adult student is 100.

59. (a) For 1985, use $t = 5$: $y \approx 715$ cell sites

For 2000, use $t = 20$: $y \approx 90{,}880$ cell sites

For 2006, use $t = 26$: $y \approx 199{,}043$ cell sites

(b)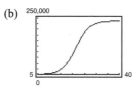

(c) From the graph, the number of cell sites will reach 235,000 in 2014.

(d) $\quad 235{,}000 = \dfrac{237{,}101}{1 + 1950e^{-0.355t}}$

$1 + 1950e^{-0.355t} = \dfrac{237{,}101}{235{,}000}$

$e^{-0.355t} = \dfrac{2101}{458{,}250{,}000}$

$-0.355t = \ln\dfrac{2101}{458{,}250{,}000}$

$t \approx 34.63$

$t \approx 34.63$ corresponds to 2014.

61. $p(t) = \dfrac{1000}{1 + 9e^{-0.1656t}}$

(a) $p(5) = \dfrac{1000}{1 + 9e^{-0.1656(5)}} \approx 203$ animals

(b) $500 = \dfrac{1000}{1 + 9e^{-0.1656t}}$

$1 + 9e^{-0.1656t} = 2$

$9e^{-0.1656t} = 1$

$e^{-0.1656t} = \dfrac{1}{9}$

$t = -\dfrac{\ln(1/9)}{0.1656} \approx 13$ months

(c)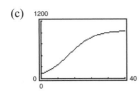

The horizontal asymptotes are $p = 0$ and $p = 1000$.

The asymptote with the larger p-value, $p = 1000$, indicates that the population size will approach 1000 as time increases.

63. $R = \log \dfrac{I}{I_0} = \log I$ because $I_0 = 1$.

(a) $8.5 = \log I \Rightarrow I = 10^{8.5} \approx 316{,}227{,}766$

(b) $5.4 = \log I \Rightarrow I = 10^{5.4} \approx 251{,}189$

(c) $6.1 = \log I \Rightarrow I = 10^{6.1} \approx 1{,}258{,}925$

65. $\beta = 10 \log \dfrac{I}{I_0}$ where $I_0 = 10^{-12}$ watt/m^2.

(a) $\beta = 10 \log \dfrac{10^{-10}}{10^{-12}} = 10 \log 10^2 = 20$ decibels

(b) $\beta = 10 \log \dfrac{10^{-5}}{10^{-12}} = 10 \log 10^7 = 70$ decibels

(c) $\beta = 10 \log \dfrac{10^{-8}}{10^{-12}} = 10 \log 10^4 = 40$ decibels

(d) $\beta = 10 \log \dfrac{1}{10^{-12}} = 10 \log 10^{12} = 120$ decibels

67. $\beta = 10 \log \dfrac{I}{I_0}$

$\dfrac{\beta}{10} = \log \dfrac{I}{I_0}$

$10^{\beta/10} = 10^{\log I/I_0}$

$10^{\beta/10} = \dfrac{I}{I_0}$

$I = I_0 10^{\beta/10}$

% decrease $= \dfrac{I_0 10^{9.3} - I_0 10^{8.0}}{I_0 10^{9.3}} \times 100 \approx 95\%$

69. pH $= -\log\left[H^+\right]$

$-\log\left(2.3 \times 10^{-5}\right) \approx 4.64$

71. $5.8 = -\log\left[H^+\right]$

$-5.8 = \log\left[H^+\right]$

$10^{-5.8} = 10^{\log\left[H^+\right]}$

$10^{-5.8} = \left[H^+\right]$

$\left[H^+\right] \approx 1.58 \times 10^{-6}$ moles per liter

73. $2.9 = -\log\left[H^+\right]$

$-2.9 = \log\left[H^+\right]$

$\left[H^+\right] = 10^{-2.9}$ for the apple juice

$8.0 = -\log\left[H^+\right]$

$-8.0 = \log\left[H^+\right]$

$\left[H^+\right] = 10^{-8}$ for the drinking water

$\dfrac{10^{-2.9}}{10^{-8}} = 10^{5.1}$ times the hydrogen ion concentration of drinking water

75. $t = -10 \ln \dfrac{T - 70}{98.6 - 70}$

At 9:00 A.M. you have:

$t = -10 \ln \dfrac{85.7 - 70}{98.6 - 70} \approx 6$ hours

From this you can conclude that the person died at 3:00 A.M.

77. $u = 120{,}000 \left[\dfrac{0.075t}{1 - \left(\dfrac{1}{1 + 0.075/12} \right)^{12t}} - 1 \right]$

(a)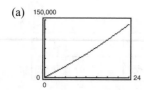

(b) From the graph, $u = \$120{,}000$ when $t \approx 21$ years. It would take approximately 37.6 years to pay $\$240{,}000$ in interest. Yes, it is possible to pay twice as much in interest charges as the size of the mortgage. It is especially likely when the interest rates are higher.

79. False. The domain can be the set of real numbers for a logistic growth function.

81. False. The graph of $f(x)$ is the graph of $g(x)$ shifted upward five units.

83. Answers will vary

Review Exercises for Chapter 5

1. $f(x) = 0.3^x$

$f(1.5) = 0.3^{1.5} \approx 0.164$

3. $f(x) = 2^{-0.5x}$

$f(\pi) = 2^{-0.5(\pi)} \approx 0.337$

5. $f(x) = 7(0.2^x)$

$f(-\sqrt{11}) = 7(0.2^{-\sqrt{11}}) \approx 1456.529$

7. $f(x) = 2^x, g(x) = 2^x - 2$

Because $g(x) = f(x) - 2$, the graph of g can be obtained by shifting the graph of f two units downward.

9. $f(x) = 4^x, g(x) = 4^{-x+2}$

Because $g(x) = f(-x + 2)$, the graph of g can be obtained by reflecting the graph of f in the y-axis and shifting the graph two units to the right. (**Note:** This is equivalent to shifting f two units to the right and then reflecting the graph in the y-axis.)

11. $f(x) = 3^x, g(x) = 1 - 3^x$

Because $g(x) = 1 - f(x)$, the graph of g can be obtained by reflecting the graph of f in the x-axis and shifting the graph one unit upward. (**Note:** This is equivalent to shifting the graph of f one unit upward and then reflecting the graph in the x-axis.)

13. $f(x) = \left(\frac{1}{2}\right)^x$

$g(x) = -\left(\frac{1}{2}\right)^{x+2}$

Because $g(x) = -f(x + 2)$, the graph of g can be obtained by reflecting the graph of f in the x-axis and shifting the graph two units to the left.

15. $f(x) = 4^{-x} + 4$

Horizontal asymptote: $y = 4$

x	−1	0	1	2	3
$f(x)$	8	5	4.25	4.063	4.016

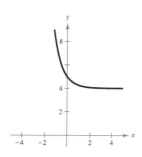

17. $f(x) = 5^{x-2} + 4$

Horizontal asymptote: $y = 4$

x	−1	0	1	2	3
$f(x)$	4.008	4.04	4.2	5	9

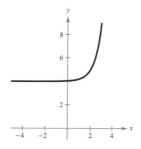

19. $f(x) = \left(\frac{1}{2}\right)^{-x} + 3 = 2^x + 3$

Horizontal asymptote: $y = 3$

x	-2	-1	0	1	2
$f(x)$	3.25	3.5	4	5	7

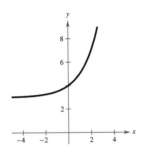

21. $\left(\frac{1}{3}\right)^{x-3} = 9$

$\left(\frac{1}{3}\right)^{x-3} = 3^2$

$\left(\frac{1}{3}\right)^{x-3} = \left(\frac{1}{3}\right)^{-2}$

$x - 3 = -2$

$x = 1$

23. $e^{3x-5} = e^7$

$3x - 5 = 7$

$3x = 12$

$x = 4$

25. $e^8 \approx 2980.958$

27. $e^{-1.7} \approx 0.183$

29. $h(x) = e^{-x/2}$

x	-2	-1	0	1	2
$h(x)$	2.72	1.65	1	0.61	0.37

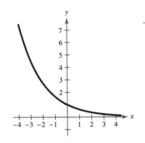

31. $f(x) = e^{x+2}$

x	-3	-2	-1	0	1
$f(x)$	0.37	1	2.72	7.39	20.09

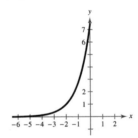

33. $P = \$5000, r = 3\%, t = 10$ years

Compounded n times per year: $A = P\left(1 + \frac{r}{n}\right)^{nt} = 5000\left(1 + \frac{0.03}{n}\right)^{10n}$

Compounded continuously: $A = Pe^{rt} = 5000e^{0.03(10)}$

n	1	2	4	12	365	Continuous
A	\$6719.58	\$6734.28	\$6741.74	\$6746.77	\$6749.21	\$6749.29

35. $F(t) = 1 - e^{-t/3}$

  (a) $F\left(\frac{1}{2}\right) \approx 0.154$

  (b) $F(2) \approx 0.487$

  (c) $F(5) \approx 0.811$

37. $3^3 = 27$

$\log_3 27 = 3$

39. $e^{0.8} = 2.2255\ldots$

$\ln 2.2255\ldots = 0.8$

41. $f(x) = \log x$

$f(1000) = \log 1000 = \log 10^3 = 3$

43. $g(x) = \log_2 x$

$g\left(\frac{1}{4}\right) = \log_2 \frac{1}{4} = \log_2 2^{-2} = -2$

45. $\log_4(x + 7) = \log_4 14$

$\qquad x + 7 = 14$

$\qquad\qquad x = 7$

47. $\ln(x + 9) = \ln 4$

$\qquad x + 9 = 4$

$\qquad\qquad x = -5$

49. $g(x) = \log_7 x \Rightarrow x = 7^y$

Domain: $(0, \infty)$

x-intercept: $(1, 0)$

Vertical asymptote: $x = 0$

x	$\frac{1}{7}$	1	7	49
$g(x)$	-1	0	1	2

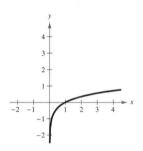

51. $f(x) = 4 - \log(x + 5)$

Domain: $(-5, \infty)$

Because

$4 - \log(x + 5) = 0 \Rightarrow \log(x + 5) = 4$

$\qquad\qquad\qquad\qquad x + 5 = 10^4$

$\qquad\qquad\qquad\qquad\quad x = 10^4 - 5$

$\qquad\qquad\qquad\qquad\qquad = 9995.$

x-intercept: $(9995, 0)$

Vertical asymptote: $x = -5$

x	-4	-3	-2	-1	0	1
$f(x)$	4	3.70	3.52	3.40	3.30	3.22

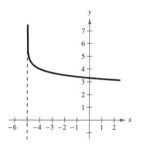

53. $f(x) = \ln x$

(a) $f(22.6) \approx 3.118$

(b) $f(0.98) \approx -0.020$

55. $f(x) = \ln x + 3$

Domain: $(0, \infty)$

$\ln x + 3 = 0$

$\qquad \ln x = -3$

$\qquad\quad x = e^{-3}$

x-intercept: $(e^{-3}, 0)$

Vertical asymptote: $x = 0$

x	1	2	3	$\frac{1}{2}$	$\frac{1}{4}$
$f(x)$	3	3.69	4.10	2.31	1.61

57. $h(x) = \ln(x^2) = 2 \ln|x|$

Domain: $(-\infty, 0) \cup (0, \infty)$

x-intercepts: $(\pm 1, 0)$

Vertical asymptote: $x = 0$

x	± 0.5	± 1	± 2	± 3	± 4
y	-1.39	0	1.39	2.20	2.77

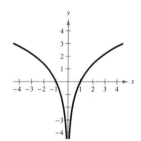

59. $h = 116 \log(a + 40) - 176$

$h(55) = 116 \log(55 + 40) - 176$

$\qquad\quad \approx 53.4$ inches

61. $\log_2 6 = \dfrac{\log 6}{\log 2} \approx 2.585$

$\log_2 6 = \dfrac{\ln 6}{\ln 2} \approx 2.585$

63. $\log_{1/2} 5 = \dfrac{\log 5}{\log(1/2)} \approx -2.322$

$\log_{1/2} 5 = \dfrac{\ln 5}{\ln(1/2)} \approx -2.322$

65. $\log 18 = \log(2 \cdot 3^2)$

$\qquad = \log 2 + 2 \log 3$

$\qquad \approx 1.255$

67. $\ln 20 = \ln(2^2 \cdot 5)$

$\qquad = 2 \ln 2 + \ln 5 \approx 2.996$

69. $\log_5 5x^2 = \log_5 5 + \log_5 x^2$

$\qquad = 1 + 2 \log_5 x$

71. $\log_3 \dfrac{9}{\sqrt{x}} = \log_3 9 - \log_3 \sqrt{x}$

$\qquad = \log_3 3^2 - \log_3 x^{1/2}$

$\qquad = 2 - \dfrac{1}{2} \log_3 x$

73. $\ln x^2 y^2 z = \ln x^2 + \ln y^2 + \ln z$

$\qquad = 2 \ln x + 2 \ln y + \ln z$

75. $\log_2 5 + \log_2 x = \log_2 5x$

77. $\ln x - \dfrac{1}{4} \ln y = \ln x - \ln \sqrt[4]{y} = \ln \dfrac{x}{\sqrt[4]{y}}$

79. $\dfrac{1}{2} \log_3 x - 2 \log_3(y + 8) = \log_3 x^{1/2} - \log_3(y + 8)^2$

$\qquad = \log_3 \sqrt{x} - \log_3(y + 8)^2$

$\qquad = \log_3 \dfrac{\sqrt{x}}{(y + 8)^2}$

81. $t = 50 \log \dfrac{18{,}000}{18{,}000 - h}$

(a) Domain: $0 \le h < 18{,}000$

(b)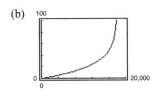

Vertical asymptote: $h = 18{,}000$

(c) As the plane approaches its absolute ceiling, it climbs at a slower rate, so the time required increases.

(d) $50 \log \dfrac{18{,}000}{18{,}000 - 4000} \approx 5.46$ minutes

83. $5^x = 125$

$5^x = 5^3$

$x = 3$

85. $e^x = 3$

$x = \ln 3 \approx 1.099$

87. $\ln x = 4$

$x = e^4 \approx 54.598$

89. $e^{4x} = e^{x^2 + 3}$

$4x = x^2 + 3$

$0 = x^2 - 4x + 3$

$0 = (x - 1)(x - 3)$

$x = 1, x = 3$

91. $2^x - 3 = 29$

$2^x = 32$

$2^x = 2^5$

$x = 5$

93. $25e^{-0.3x} = 12$

Graph $y_1 = 25e^{-0.3x}$ and $y_2 = 12$.

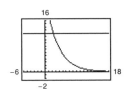

The graphs intersect at $x \approx 2.447$.

95. $\ln 3x = 8.2$

$e^{\ln 3x} = e^{8.2}$

$3x = e^{8.2}$

$x = \dfrac{e^{8.2}}{3} \approx 1213.650$

97. $\ln x - \ln 3 = 2$

$\ln \dfrac{x}{3} = 2$

$e^{\ln(x/3)} = e^2$

$\dfrac{x}{3} = e^2$

$x = 3e^2 \approx 22.167$

99. $\ln \sqrt{x} = 4$

$\dfrac{1}{2} \ln x = 4$

$\ln x = 8$

$x = e^8 \approx 2980.958$

101. $\log_8(x - 1) = \log_8(x - 2) - \log_8(x + 2)$

$$\log_8(x - 1) = \log_8\left(\frac{x - 2}{x + 2}\right)$$

$$x - 1 = \frac{x - 2}{x + 2}$$

$$(x - 1)(x + 2) = x - 2$$

$$x^2 + x - 2 = x - 2$$

$$x^2 = 0$$

$$x = 0$$

Because $x = 0$ is not in the domain of $\log_8(x - 1)$ or of $\log_8(x - 2)$, it is an extraneous solution. The equation has no solution.

103. $\log(1 - x) = -1$

$$1 - x = 10^{-1}$$

$$1 - \frac{1}{10} = x$$

$$x = 0.900$$

105. $2\ln(x + 3) - 3 = 0$

Graph $y_1 = 2\ln(x + 3) - 3$.

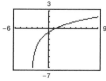

The x-intercept is at $x \approx 1.482$.

107. $6\log(x^2 + 1) - x = 0$

Graph $y_1 = 6\log(x^2 + 1) - x$.

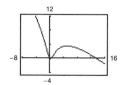

The x-intercepts are at $x = 0$, $x \approx 0.416$, and $x \approx 13.627$.

109. $P = 8500$, $A = 3(8500) = 25,500$, $r = 3.5\%$

$$A = Pe^{rt}$$

$$25,500 = 8500e^{0.035t}$$

$$3 = e^{0.035t}$$

$$\ln 3 = 0.035t$$

$$t = \frac{\ln 3}{0.035} \approx 31.4 \text{ years}$$

111. $y = 3e^{-2x/3}$

Exponential decay model

Matches graph (e).

113. $y = \ln(x + 3)$

Logarithmic model

Vertical asymptote: $x = -3$

Graph includes $(-2, 0)$

Matches graph (f).

115. $y = 2e^{-(x+4)^2/3}$

Gaussian model

Matches graph (a).

117. $y = ae^{bx}$

$$2 = ae^{b(0)} \Rightarrow a = 2$$

$$3 = 2e^{b(4)}$$

$$1.5 = e^{4b}$$

$$\ln 1.5 = 4b$$

$$\frac{\ln 1.5}{4} = b$$

$$b \approx 0.1014$$

So, $y = 2e^{0.1014x}$.

119. (a)

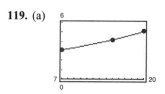

The model fits the data well

(b) $P = 5.5$

$$P = 2.36e^{0.0382t}$$

$$5.5 = 2.36e^{0.0382t}$$

$$\frac{275}{118} = e^{0.0382t}$$

$$\ln\frac{275}{118} = 0.0382t$$

$$t = \frac{\ln(275/118)}{0.0382} \approx 22$$

$t \approx 22$ corresponds to 2022.

Answers will vary.

121. $y = 0.0499e^{-(x-71)^2/128}$, $40 \le x \le 100$

(a) Graph $y_1 = 0.0499e^{-(x-71)^2/128}$.

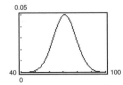

(b) The average test score is 71.

123.
$$\beta = 10 \log\left(\frac{I}{10^{-12}}\right)$$

$$\frac{\beta}{10} = \log\left(\frac{I}{10^{-12}}\right)$$

$$10^{\beta/10} = \frac{I}{10^{-12}}$$

$$I = 10^{\beta/10-12}$$

(a) $\beta = 60$

$I = 10^{60/10-12}$

$= 10^{-6}$ watt/m^2

(b) $\beta = 135$

$I = 10^{135/10-12}$

$= 10^{1.5}$

$= 10\sqrt{10}$ watts/m^2

(c) $\beta = 1$

$I = 10^{1/10-12}$

$= 10^{\frac{1}{10}} \times 10^{-12}$

$\approx 1.259 \times 10^{-2}$ watt/m^2

125. True. By the inverse properties, $\log_b b^{2x} = 2x$.

Problem Solving for Chapter 5

1. $y = a^x$

$y_1 = 0.5^x$

$y_2 = 1.2^x$

$y_3 = 2.0^x$

$y_4 = x$

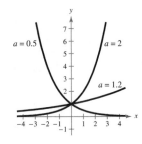

The curves $y = 0.5^x$ and $y = 1.2^x$ cross the line $y = x$. From checking the graphs it appears that $y = x$ will cross $y = a^x$ for $0 \le a \le 1.44$.

3. The exponential function, $y = e^x$, increases at a faster rate than the polynomial $y = x^n$.

5. (a) $f(u + v) = a^{u+v} = a^u \cdot a^v = f(u) \cdot f(v)$

(b) $f(2x) = a^{2x} = \left(a^x\right)^2 = \left[f(x)\right]^2$

7. (a)

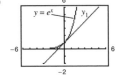

(b)

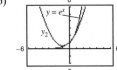

(c)

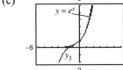

9.
$$f(x) = e^x - e^{-x}$$
$$y = e^x - e^{-x}$$
$$x = e^y - e^{-y}$$
$$x = \frac{e^{2y} - 1}{e^y}$$
$$xe^y = e^{2y} - 1$$
$$e^{2y} - xe^y - 1 = 0$$
$$e^y = \frac{x \pm \sqrt{x^2 + 4}}{2} \quad \text{Quadratic Formula}$$

Choosing the positive quantity for e^y you have

$$y = \ln\left(\frac{x + \sqrt{x^2 + 4}}{2}\right). \text{ So,}$$

$$f^{-1}(x) = \ln\left(\frac{x + \sqrt{x^2 + 4}}{2}\right).$$

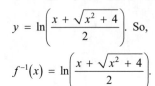

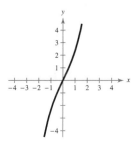

11. Answer (c). $y = 6\left(1 - e^{-x^2/2}\right)$

The graph passes through $(0, 0)$ and neither (a) nor (b) pass through the origin. Also, the graph has y-axis symmetry and a horizontal asymptote at $y = 6$.

13. $y_1 = c_1\left(\frac{1}{2}\right)^{t/k_1}$ and $y_2 = c_2\left(\frac{1}{2}\right)^{t/k_2}$

$$c_1\left(\frac{1}{2}\right)^{t/k_1} = c_2\left(\frac{1}{2}\right)^{t/k_2}$$

$$\frac{c_1}{c_2} = \left(\frac{1}{2}\right)^{(t/k_2 - t/k_1)}$$

$$\ln\left(\frac{c_1}{c_2}\right) = \left(\frac{t}{k_2} - \frac{t}{k_1}\right)\ln\left(\frac{1}{2}\right)$$

$$\ln c_1 - \ln c_2 = t\left(\frac{1}{k_2} - \frac{1}{k_1}\right)\ln\left(\frac{1}{2}\right)$$

$$t = \frac{\ln c_1 - \ln c_2}{\left[(1/k_2) - (1/k_1)\right]\ln(1/2)}$$

15. (a) $y_1 \approx 252.606(1.0310)^t$

(b) $y_2 \approx 400.88t^2 - 1464.6t + 291,782$

(c)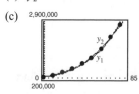

(d) The exponential model is a better fit for the data, but neither would be reliable to predict the population of the United States in 2015. The exponential model approaches infinity rapidly.

17.
$$\left(\ln x\right)^2 = \ln x^2$$
$$\left(\ln x\right)^2 - 2\ln x = 0$$
$$\ln x(\ln x - 2) = 0$$
$$\ln x = 0 \quad \text{or} \quad \ln x = 2$$
$$x = 1 \quad \text{or} \quad x = e^2$$

19. $y_4 = (x - 1) - \frac{1}{2}(x - 1)^2 + \frac{1}{3}(x - 1)^3 - \frac{1}{4}(x - 1)^4$

The pattern implies that

$$\ln x = (x - 1) - \frac{1}{2}(x - 1)^2 + \frac{1}{3}(x - 1)^3 - \frac{1}{4}(x - 1)^4 + \dots.$$

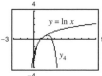

21. $y = 80.4 - 11\ln x$

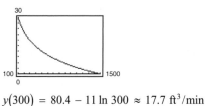

$y(300) = 80.4 - 11\ln 300 \approx 17.7 \text{ ft}^3/\text{min}$

23. (a)

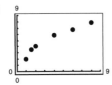

(b) The data could best be modeled by a logarithmic model.

(c) The shape of the curve looks much more logarithmic than linear or exponential.

(d) $y \approx 2.1518 + 2.7044 \ln x$

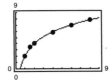

(e) The model is a good fit to the actual data.

25. (a)

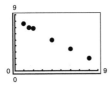

(b) The data could best be modeled by a linear model.

(c) The shape of the curve looks much more linear than exponential or logarithmic.

(d) $y \approx -0.7884x + 8.2566$

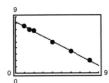

(e) The model is a good fit to the actual data

Chapter 5   Practice Test

1. Solve for x: $x^{3/5} = 8$.

2. Solve for x: $3^{x-1} = \frac{1}{81}$.

3. Graph $f(x) = 2^{-x}$.

4. Graph $g(x) = e^x + 1$.

5. If $5000 is invested at 9% interest, find the amount after three years if the interest is compounded
   (a) monthly.
   (b) quarterly.
   (c) continuously.

6. Write the equation in logarithmic form: $7^{-2} = \frac{1}{49}$.

7. Solve for x: $x - 4 = \log_2 \frac{1}{64}$.

8. Given $\log_b 2 = 0.3562$ and $\log_b 5 = 0.8271$, evaluate $\log_b \sqrt[4]{8/25}$.

9. Write $5 \ln x - \frac{1}{2} \ln y + 6 \ln z$ as a single logarithm.

10. Using your calculator and the change of base formula, evaluate $\log_9 28$.

11. Use your calculator to solve for N: $\log_{10} N = 0.6646$

12. Graph $y = \log_4 x$.

13. Determine the domain of $f(x) = \log_3(x^2 - 9)$.

14. Graph $y = \ln(x - 2)$.

15. True or false: $\dfrac{\ln x}{\ln y} = \ln(x - y)$

16. Solve for x: $5^x = 41$

17. Solve for x: $x - x^2 = \log_5 \frac{1}{25}$

18. Solve for x: $\log_2 x + \log_2(x - 3) = 2$

19. Solve for x: $\dfrac{e^x + e^{-x}}{3} = 4$

20. Six thousand dollars is deposited into a fund at an annual interest rate of 13%. Find the time required for the investment to double if the interest is compounded continuously.

C H A P T E R   6
Trigonometry

C H A P T E R 6
Trigonometry

Section 6.1   Angles and Their Measure

You should know the following basic facts about angles, their measurement, and their applications.

- Types of Angles:
    - (a) Acute: Measure between $0°$ and $90°$.
    - (b) Right: Measure $90°$.
    - (c) Obtuse: Measure between $90°$ and $180°$.
    - (d) Straight: Measure $180°$.

- Two positive angles, α and β are complementary if $\alpha + \beta = 90°$. They are supplementary if $\alpha + \beta = 180°$.

- Two angles in standard position that have the same terminal side are called coterminal angles.

- To convert degrees to radians, use $1° = \pi/180$ radians.

- To convert radians to degrees, use 1 radian $= (180/\pi)°$.

- $1' =$ one minute $= 1/60$ of $1°$.

- $1'' =$ one second $= 1/60$ of $1' = 1/3600$ of $1°$.

- The length of a circular arc is $s = r\theta$ where θ is measured in radians.

- Speed $=$ distance/time

- Angular speed $= \theta/t = s/rt$

---

1. Trigonometry

3. coterminal

5. acute; obtuse

7. radian

9. angular

11.

The angle shown is approximately $210°$.

13.

The angle shown is approximately $-60°$.

15. (a) Since $90° < 130° < 180°$; $130°$ lies in Quadrant II.

    (b) Since $270° < 285° < 360°$; $285°$ lies in Quadrant IV.

17. (a) Since $-180° < -132°50' < -90°$; $-132°\ 50'$ lies in Quadrant III.

    (b) Since $-360° < -336° < -270°$; $-336°$ lies in Quadrant I.

19. (a) $30°$

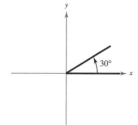

    (b) $150°$

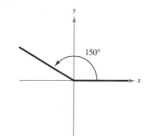

21. (a) 405°

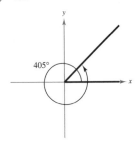

(b) 480°

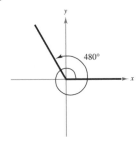

23. (a) Coterminal angles for 45°

$45° + 360° = 405°$

$45° - 360° = -315°$

(b) Coterminal angles for $-36°$

$-36° + 360° = 324°$

$-36° - 360° = -396°$

25. (a) Coterminal angles for 300°

$300° + 360° = 660°$

$300° - 360° = -60°$

(b) Coterminal angles for 740°

$740° - 2(360°) = 20°$

$20° - 360° = -340°$

27. (a) $54°45' = 54° + \left(\frac{45}{60}\right)° = 54.75°$

(b) $-128°30' = -128° - \left(\frac{30}{60}\right)° = -128.5°$

29. (a) $85°18'30'' = \left(85 + \frac{18}{60} + \frac{30}{3600}\right)° \approx 85.308°$

(b) $330°25'' = \left(330 + \frac{25}{3600}\right)° \approx 330.007°$

31. (a) $240.6° = 240° + 0.6(60)' = 240°36'$

(b) $-145.8° = -\left[145° + 0.8(60')\right] = -145°48'$

33. (a) $2.5° = 2° + 0.5(60)' = 2°30'$

(b) $-3.58° = -\left[3° + 0.58(60)'\right]$

$= -[3°34.8']$

$= -\left[3°34' + 0.8(60)''\right]$

$= -3°34'48''$

35. (a) Complement: $90° - 18° = 72°$

Supplement: $180° - 18° = 162°$

(b) Complement: $90° - 85° = 5°$

Supplement: $180° - 85° = 95°$

37. (a) Complement: $90° - 24° = 66°$

Supplement: $180° - 24° = 156°$

(b) Complement: Not possible. 126° is greater than 90°.

Supplement: $180° - 126° = 54°$

39.

The angle shown is approximately 2 radians.

41.

The angle shown is approximately -3 radians.

43.

The angle shown is approximately 1 radian.

45. (a) Because $0 < \frac{\pi}{4} < \frac{\pi}{2}; \frac{\pi}{4}$ lies in Quadrant I.

(b) Because $\pi < \frac{5\pi}{4} < \frac{3\pi}{2}; \frac{5\pi}{4}$ lies in Quadrant III.

47. (a) Because $0 < \frac{\pi}{5} < \frac{\pi}{2}; \frac{\pi}{5}$ lies in Quadrant I.

(b) Because $\pi < \frac{7\pi}{5} < \frac{3\pi}{2}; \frac{7\pi}{5}$ lies in Quadrant III.

49. (a) Because $-\frac{\pi}{2} < -1 < 0; -1$ lies in Quadrant IV.

(b) Because $-\pi < -2 < -\frac{\pi}{2}; -2$ lies in Quadrant III.

51. (a) $\dfrac{\pi}{3}$

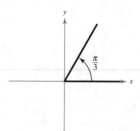

(b) $-\dfrac{2\pi}{3}$

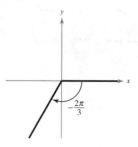

53. (a) $\dfrac{11\pi}{6}$

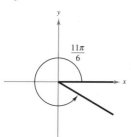

(b) -3

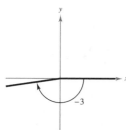

55. (a) Coterminal angles for $\dfrac{\pi}{6}$

$$\dfrac{\pi}{6} + 2\pi = \dfrac{13\pi}{6}$$

$$\dfrac{\pi}{6} - 2\pi = -\dfrac{11\pi}{6}$$

(b) Coterminal angles for $\dfrac{5\pi}{6}$

$$\dfrac{5\pi}{6} + 2\pi = \dfrac{17\pi}{6}$$

$$\dfrac{5\pi}{6} - 2\pi = -\dfrac{7\pi}{6}$$

57. (a) Coterminal angles for $\dfrac{2\pi}{3}$

$$\dfrac{2\pi}{3} + 2\pi = \dfrac{8\pi}{3}$$

$$\dfrac{2\pi}{3} - 2\pi = -\dfrac{4\pi}{3}$$

(b) Coterminal angles for $-\dfrac{\pi}{12}$

$$-\dfrac{\pi}{12} + 2\pi = \dfrac{23\pi}{12}$$

$$-\dfrac{\pi}{12} - 2\pi = -\dfrac{25\pi}{12}$$

59. (a) Coterminal angles for $-\dfrac{9\pi}{4}$

$$-\dfrac{9\pi}{4} + 4\pi = \dfrac{7\pi}{4}$$

$$-\dfrac{9\pi}{4} + 2\pi = -\dfrac{\pi}{4}$$

(b) Coterminal angles for $-\dfrac{2\pi}{15}$

$$-\dfrac{2\pi}{15} + 2\pi = \dfrac{28\pi}{15}$$

$$-\dfrac{2\pi}{15} - 2\pi = -\dfrac{32\pi}{15}$$

61. (a) Complement: $\dfrac{\pi}{2} - \dfrac{\pi}{12} = \dfrac{5\pi}{12}$

Supplement: $\pi - \dfrac{\pi}{12} = \dfrac{11\pi}{12}$

(b) Complement: Not possible; $\dfrac{11\pi}{12}$ is greater than $\dfrac{\pi}{2}$

Supplement: $\pi - \dfrac{11\pi}{12} = \dfrac{\pi}{12}$

63. (a) $30° = 30\left(\dfrac{\pi}{180°}\right) = \dfrac{\pi}{6}$

  (b) $45° = 45\left(\dfrac{\pi}{180°}\right) = \dfrac{\pi}{4}$

65. (a) $-20° = -20\left(\dfrac{\pi}{180°}\right) = -\dfrac{\pi}{9}$

  (b) $-60° = -60\left(\dfrac{\pi}{180°}\right) = -\dfrac{\pi}{3}$

67. (a) $\dfrac{3\pi}{2} = \dfrac{3\pi}{2}\left(\dfrac{180°}{\pi}\right) = 270°$

  (b) $\dfrac{7\pi}{6} = \dfrac{7\pi}{6}\left(\dfrac{180°}{\pi}\right) = 210°$

69. (a) $\dfrac{5\pi}{4} = \dfrac{5\pi}{4}\left(\dfrac{180°}{\pi}\right) = 225°$

  (b) $-\dfrac{7\pi}{3} = -\dfrac{7\pi}{3}\left(\dfrac{180°}{\pi}\right) = -420°$

71. $45° = 45\left(\dfrac{\pi}{180°}\right) \approx 0.785$ radian

73. $-216.35° = -216.35\left(\dfrac{\pi}{180°}\right) \approx -3.776$ radians

75. $532° = 532\left(\dfrac{\pi}{180°}\right) \approx 9.285$ radians

77. $-0.83° = -0.83\left(\dfrac{\pi}{180°}\right) \approx -0.014$ radian

79. $\dfrac{\pi}{7} = \dfrac{\pi}{7}\left(\dfrac{180°}{\pi}\right) \approx 25.714°$

81. $\dfrac{15\pi}{8} = \dfrac{15\pi}{8}\left(\dfrac{180°}{\pi}\right) = 337.500°$

83. $-2 = -2\left(\dfrac{180°}{\pi}\right) \approx -114.592°$

85. $s = r\theta$

  $6 = 5\theta$

  $\theta = \dfrac{6}{5}$ radians

87. $s = r\theta$

  $32 = 7\theta$

  $\theta = \dfrac{32}{7}$ radians

89. $r = 4$ inches, $s = 18$ inches

  $18 = 4\theta$

  $\theta = \dfrac{9}{2}$ radians

91. $s = r\theta$

  $25 = 14.5\theta$

  $\theta = \dfrac{25}{14.5} = \dfrac{50}{29}$ radians

93. $r = 15$ inches, $\theta = 120° = \dfrac{2\pi}{3}$

  $s = r\theta = 15\left(\dfrac{2\pi}{3}\right) = 10\pi$ inches ≈ 31.42 inches

95. $s = r\theta$, θ in radians

  $s = 3(1) = 3$ meters

97. $r = 6$ inches, $\theta = \dfrac{\pi}{3}$

  $A = \dfrac{1}{2}r^2\theta = \dfrac{1}{2}(6)^2\left(\dfrac{\pi}{3}\right) = 6\pi$ in.^{2} ≈ 18.85 in.^{2}

99. $A = \dfrac{1}{2}r^2\theta$

  $A = \dfrac{1}{2}(2.5)^2(225)\left(\dfrac{\pi}{180}\right) \approx 12.27$ square feet

101. $\theta = 41°15'50'' - 32°47'9'' \approx 8.47806° \approx 0.14797$ radian

  $s = r\theta \approx 4000(0.14782) \approx 592$ miles

103. $\theta = \dfrac{s}{r} = \dfrac{450}{6378} \approx 0.071$ radian $\approx 4.04°$

105. $\theta = \dfrac{s}{r} = \dfrac{2.5}{6} = \dfrac{25}{60} = \dfrac{5}{12}$ radian

107. (a)   65 miles per hour $= 65(5280)/60$

$\qquad\qquad\qquad\qquad = 5720$ feet per minute

The circumference of the tire is $C = 2.5\pi$ feet.

The number of revolutions per minute is $r = 5720/2.5\pi \approx 728.3$ rev/minute.

(b) The angular speed is θ/t.

$$\theta = \frac{5720}{2.5\pi}(2\pi) = 4576 \text{ radians}$$

$$\text{Angular speed} = \frac{4576 \text{ radians}}{1 \text{ minute}}$$

$$= 4576 \text{ radians/minute}$$

109. (a)   Angular speed $= \dfrac{(5200)(2\pi) \text{ radians}}{1 \text{ minute}}$

$\qquad\qquad\qquad\;\; = 10,400\pi$ radians per minute

(b)   Linear speed $= \dfrac{\left(\dfrac{7.25}{2}\text{ in.}\right)\left(\dfrac{1 \text{ ft}}{12 \text{ in.}}\right)(5200)(2\pi)\text{ feet}}{1 \text{ minute}}$

$\qquad\qquad\qquad\; = 3141\dfrac{2}{3}\pi$ feet per minute

$\qquad\qquad\qquad\; \approx 164.5$ feet per second

111. (a) $(200)(2\pi) \le$ Angular speed $\le (500)(2\pi)$ radians per minute

Interval: $[400\pi, 1000\pi]$ radians per minute

(b) $(6)(200)(2\pi) \le$ Linear speed $\le (6)(500)(2\pi)$ centimeters per minute

Interval: $[2400\pi, 6000\pi]$ centimeters per minute

113. $A = \dfrac{1}{2}r^2\theta$

$\qquad = \dfrac{1}{2}(15)^2(140°)\left(\dfrac{\pi}{180°}\right)$

$\qquad = 87.5\pi \text{ m}^2$

$\qquad \approx 274.89 \text{ m}^2$

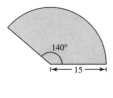

115. $A = \dfrac{1}{2}r^2\theta$

$\qquad = \dfrac{1}{2}(35)^2(140°)\left(\dfrac{\pi}{180°}\right)$

$\qquad \approx 476.39\pi \text{ square meters}$

$\qquad \approx 1496.62 \text{ square meters}$

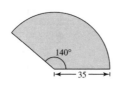

117. False. A measurement of 4π radians corresponds to two complete revolutions from the initial to the terminal side of an angle.

119. False. The terminal side of $-1260°$ lies on the negative x-axis.

121. The speed increases, since the linear speed is proportional to the radius.

123. Since the arc length s is given by $s = r\theta$, if the central angle θ is fixed while the radius r increases, then s increases in proportion to r.

Section 6.2   Right Triangle Trigonometry

■ You should know the right triangle definition of trigonometric functions.

(a) $\sin \theta = \dfrac{\text{opp}}{\text{hyp}}$ (b) $\cos \theta = \dfrac{\text{adj}}{\text{hyp}}$ (c) $\tan \theta = \dfrac{\text{opp}}{\text{adj}}$

(d) $\csc \theta = \dfrac{\text{hyp}}{\text{opp}}$ (e) $\sec \theta = \dfrac{\text{hyp}}{\text{adj}}$ (f) $\cot \theta = \dfrac{\text{adj}}{\text{opp}}$

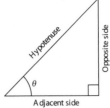

■ You should know the following identities.

(a) $\sin \theta = \dfrac{1}{\csc \theta}$ (b) $\csc \theta = \dfrac{1}{\sin \theta}$ (c) $\cos \theta = \dfrac{1}{\sec \theta}$

(d) $\sec \theta = \dfrac{1}{\cos \theta}$ (e) $\tan \theta = \dfrac{1}{\cot \theta}$ (f) $\cot \theta = \dfrac{1}{\tan \theta}$

(g) $\tan \theta = \dfrac{\sin \theta}{\cos \theta}$ (h) $\cot \theta = \dfrac{\cos \theta}{\sin \theta}$ (i) $\sin^2 \theta + \cos^2 \theta = 1$

(j) $1 + \tan^2 \theta = \sec^2 \theta$ (k) $1 + \cot^2 \theta = \csc^2 \theta$

■ You should know that two acute angles α and β are complementary if $\alpha + \beta = 90°$, and that the cofunctions of complementary angles are equal.

■ You should know the trigonometric function values of 30°, 45°, and 60°, or be able to construct triangles from which you can determine them.

---

1. (i) $\dfrac{\text{hypotenuse}}{\text{adjacent}} = \sec \theta$ (e) $\qquad$ (iv) $\dfrac{\text{adjacent}}{\text{hypotenuse}} = \cos \theta$ (b)

$\quad$ (ii) $\dfrac{\text{adjacent}}{\text{opposite}} = \cot \theta$ (f) $\qquad$ (v) $\dfrac{\text{opposite}}{\text{hypotenuse}} = \sin \theta$ (a)

$\quad$ (iii) $\dfrac{\text{hypotenuse}}{\text{opposite}} = \csc \theta$ (d) $\qquad$ (vi) $\dfrac{\text{opposite}}{\text{adjacent}} = \tan \theta$ (c)

3. Complementary

5. $\text{hyp} = \sqrt{6^2 + 8^2} = \sqrt{36 + 64} = \sqrt{100} = 10$

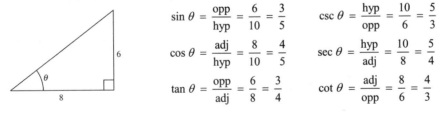

$\sin \theta = \dfrac{\text{opp}}{\text{hyp}} = \dfrac{6}{10} = \dfrac{3}{5}$ $\csc \theta = \dfrac{\text{hyp}}{\text{opp}} = \dfrac{10}{6} = \dfrac{5}{3}$

$\cos \theta = \dfrac{\text{adj}}{\text{hyp}} = \dfrac{8}{10} = \dfrac{4}{5}$ $\sec \theta = \dfrac{\text{hyp}}{\text{adj}} = \dfrac{10}{8} = \dfrac{5}{4}$

$\tan \theta = \dfrac{\text{opp}}{\text{adj}} = \dfrac{6}{8} = \dfrac{3}{4}$ $\cot \theta = \dfrac{\text{adj}}{\text{opp}} = \dfrac{8}{6} = \dfrac{4}{3}$

7. $\text{adj} = \sqrt{41^2 - 9^2} = \sqrt{1681 - 81} = \sqrt{1600} = 40$

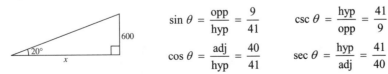

$\sin \theta = \dfrac{\text{opp}}{\text{hyp}} = \dfrac{9}{41}$ $\csc \theta = \dfrac{\text{hyp}}{\text{opp}} = \dfrac{41}{9}$

$\cos \theta = \dfrac{\text{adj}}{\text{hyp}} = \dfrac{40}{41}$ $\sec \theta = \dfrac{\text{hyp}}{\text{adj}} = \dfrac{41}{40}$

$\tan \theta = \dfrac{\text{opp}}{\text{adj}} = \dfrac{9}{40}$ $\cot \theta = \dfrac{\text{adj}}{\text{opp}} = \dfrac{40}{9}$

9.

$$\text{hyp} = \sqrt{15^2 + 8^2} = \sqrt{289} = 17$$

$$\sin \theta = \frac{\text{opp}}{\text{hyp}} = \frac{8}{17} \qquad \csc \theta = \frac{\text{hyp}}{\text{opp}} = \frac{17}{8}$$

$$\cos \theta = \frac{\text{adj}}{\text{hyp}} = \frac{15}{17} \qquad \sec \theta = \frac{\text{hyp}}{\text{adj}} = \frac{17}{15}$$

$$\tan \theta = \frac{\text{opp}}{\text{adj}} = \frac{8}{15} \qquad \cot \theta = \frac{\text{adj}}{\text{opp}} = \frac{15}{8}$$

$$\text{hyp} = \sqrt{7.5^2 + 4^2} = \frac{17}{2}$$

$$\sin \theta = \frac{\text{opp}}{\text{hyp}} = \frac{4}{(17/2)} = \frac{8}{17} \qquad \csc \theta = \frac{\text{hyp}}{\text{opp}} = \frac{(17/2)}{4} = \frac{17}{8}$$

$$\cos \theta = \frac{\text{adj}}{\text{hyp}} = \frac{7.5}{(17/2)} = \frac{15}{17} \qquad \sec \theta = \frac{\text{hyp}}{\text{adj}} = \frac{(17/2)}{7.5} = \frac{17}{15}$$

$$\tan \theta = \frac{\text{opp}}{\text{adj}} = \frac{4}{7.5} = \frac{8}{15} \qquad \cot \theta = \frac{\text{adj}}{\text{opp}} = \frac{7.5}{4} = \frac{15}{8}$$

The function values are the same because the triangles are similar, and corresponding sides are proportional.

11. $\text{adj} = \sqrt{3^2 - 1^2} = \sqrt{8} = 2\sqrt{2}$

$$\sin \theta = \frac{\text{opp}}{\text{hyp}} = \frac{1}{3} \qquad \csc \theta = \frac{\text{hyp}}{\text{opp}} = 3$$

$$\cos \theta = \frac{\text{adj}}{\text{hyp}} = \frac{2\sqrt{2}}{3} \qquad \sec \theta = \frac{\text{hyp}}{\text{adj}} = \frac{3}{2\sqrt{2}} = \frac{3\sqrt{2}}{4}$$

$$\tan \theta = \frac{\text{opp}}{\text{adj}} = \frac{1}{2\sqrt{2}} = \frac{\sqrt{2}}{4} \qquad \cot \theta = \frac{\text{adj}}{\text{opp}} = 2\sqrt{2}$$

$\text{adj} = \sqrt{6^2 - 2^2} = \sqrt{32} = 4\sqrt{2}$

$$\sin \theta = \frac{\text{opp}}{\text{hyp}} = \frac{2}{6} = \frac{1}{3} \qquad \csc \theta = \frac{\text{hyp}}{\text{opp}} = \frac{6}{2} = 3$$

$$\cos \theta = \frac{\text{adj}}{\text{hyp}} = \frac{4\sqrt{2}}{6} = \frac{2\sqrt{2}}{3} \qquad \sec \theta = \frac{\text{hyp}}{\text{adj}} = \frac{6}{4\sqrt{2}} = \frac{3}{2\sqrt{2}} = \frac{3\sqrt{2}}{4}$$

$$\tan \theta = \frac{\text{opp}}{\text{adj}} = \frac{5\sqrt{2}}{4} = \frac{1}{2\sqrt{2}} = \frac{\sqrt{2}}{4} \qquad \cot \theta = \frac{\text{adj}}{\text{opp}} = \frac{4\sqrt{2}}{2} = 2\sqrt{2}$$

The function values are the same since the triangles are similar and the corresponding sides are proportional.

13. Given: $\tan \theta = \dfrac{3}{4} = \dfrac{\text{opp}}{\text{adj}}$

$3^2 + 4^2 = (\text{hyp})^2$

$\text{hyp} = 5$

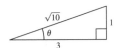

$\sin \theta = \dfrac{\text{opp}}{\text{hyp}} = \dfrac{3}{5}$

$\cos \theta = \dfrac{\text{adj}}{\text{hyp}} = \dfrac{4}{5}$

$\csc \theta = \dfrac{\text{hyp}}{\text{opp}} = \dfrac{5}{3}$

$\sec \theta = \dfrac{\text{hyp}}{\text{adj}} = \dfrac{5}{4}$

$\cot \theta = \dfrac{\text{adj}}{\text{opp}} = \dfrac{4}{3}$

15. Given: $\sec \theta = \dfrac{3}{2} = \dfrac{\text{hyp}}{\text{adj}}$

$(\text{opp})^2 + 2^2 = 3^2$

$\text{opp} = \sqrt{5}$

$\sin \theta = \dfrac{\text{opp}}{\text{hyp}} = \dfrac{\sqrt{5}}{3}$

$\cos \theta = \dfrac{\text{adj}}{\text{hyp}} = \dfrac{2}{3}$

$\tan \theta = \dfrac{\text{opp}}{\text{adj}} = \dfrac{\sqrt{5}}{2}$

$\csc \theta = \dfrac{\text{hyp}}{\text{opp}} = \dfrac{3\sqrt{5}}{5}$

$\cot \theta = \dfrac{\text{adj}}{\text{opp}} = \dfrac{2\sqrt{5}}{5}$

17. Given: $\sin \theta = \dfrac{1}{5} = \dfrac{\text{opp}}{\text{hyp}}$

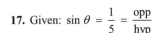

$1^2 + (\text{adj})^2 = 5^2$

$\text{adj} = \sqrt{24} = 2\sqrt{6}$

$\cos \theta = \dfrac{\text{adj}}{\text{hyp}} = \dfrac{2\sqrt{6}}{5}$

$\tan \theta = \dfrac{\text{opp}}{\text{adj}} = \dfrac{\sqrt{6}}{12}$

$\csc \theta = \dfrac{\text{hyp}}{\text{opp}} = 5$

$\sec \theta = \dfrac{\text{hyp}}{\text{adj}} = \dfrac{5\sqrt{6}}{12}$

$\cot \theta = \dfrac{\text{adj}}{\text{opp}} = 2\sqrt{6}$

19. Given: $\cot \theta = 3 = \dfrac{3}{1} = \dfrac{\text{adj}}{\text{opp}}$

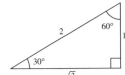

$1^2 + 3^2 = (\text{hyp})^2$

$\text{hyp} = \sqrt{10}$

$\sin \theta = \dfrac{\text{opp}}{\text{hyp}} = \dfrac{\sqrt{10}}{10}$

$\cos \theta = \dfrac{\text{adj}}{\text{hyp}} = \dfrac{3\sqrt{10}}{10}$

$\tan \theta = \dfrac{\text{opp}}{\text{adj}} = \dfrac{1}{3}$

$\csc \theta = \dfrac{\text{hyp}}{\text{opp}} = \sqrt{10}$

$\sec \theta = \dfrac{\text{hyp}}{\text{adj}} = \dfrac{\sqrt{10}}{3}$

21.

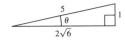

$30° = 30°\left(\dfrac{\pi}{180°}\right) = \dfrac{\pi}{6} \text{ radian}$

$\sin 30° = \dfrac{\text{opp}}{\text{hyp}} = \dfrac{1}{2}$

23.

degree	radian	value
$\sec 45°$	$\dfrac{\pi}{4}$	$\sqrt{2}$

$\sec \dfrac{\pi}{4} = \dfrac{\sqrt{2}}{1} = \sqrt{2}$

25.

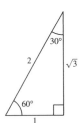

$\cot \theta = \dfrac{\sqrt{3}}{3} = \dfrac{1}{\sqrt{3}} = \dfrac{\text{adj}}{\text{opp}}$

$\theta = 60° = \dfrac{\pi}{3} \text{ radian}$

27.

$\dfrac{\pi}{6} = \dfrac{\pi}{6}\left(\dfrac{180°}{\pi}\right) = 30°$

$\csc \dfrac{\pi}{6} = \dfrac{\text{hyp}}{\text{opp}} = 2$

29.

$$\cot\theta = 1 = \frac{1}{1} = \frac{\text{adj}}{\text{opp}}$$

$$\theta = 45° = 45°\left(\frac{\pi}{180°}\right) = \frac{\pi}{4}$$

31. $\sin 60° = \dfrac{\sqrt{3}}{2}$, $\cos 60° = \dfrac{1}{2}$

(a) $\sin 30° = \cos 60° = \dfrac{1}{2}$

(b) $\cos 30° = \sin 60° = \dfrac{\sqrt{3}}{2}$

(c) $\tan 60° = \dfrac{\sin 60°}{\cos 60°} = \sqrt{3}$

(d) $\cot 60° = \dfrac{\cos 60°}{\sin 60°} = \dfrac{1}{\sqrt{3}} = \dfrac{\sqrt{3}}{3}$

33. $\cos\theta = \dfrac{1}{3}$

(a) $\sin^2\theta + \cos^2\theta = 1$

$$\sin^2\theta + \left(\frac{1}{3}\right)^2 = 1$$

$$\sin^2\theta = \frac{8}{9}$$

$$\sin\theta = \frac{2\sqrt{2}}{3}$$

(b) $\tan\theta = \dfrac{\sin\theta}{\cos\theta} = \dfrac{\dfrac{2\sqrt{2}}{3}}{\dfrac{1}{3}} = 2\sqrt{2}$

(c) $\sec\theta = \dfrac{1}{\cos\theta} = 3$

(d) $\csc\left(90° - \theta\right) = \sec\theta = 3$

35. $\cot\alpha = 5$

(a) $\tan\alpha = \dfrac{1}{\cot\alpha} = \dfrac{1}{5}$

(b) $\csc^2\alpha = 1 + \cot^2\alpha$

$\csc^2\alpha = 1 + 5^2$

$\csc^2\alpha = 26$

$\csc\alpha = \sqrt{26}$

(c) $\cot\left(90° - \alpha\right) = \tan\alpha = \dfrac{1}{5}$

(d) $\sec^2\alpha = 1 + \tan^2\alpha$

$$\sec^2\alpha = 1 + \left(\frac{1}{5}\right)^2$$

$$\sec^2\alpha = \frac{26}{25}$$

$$\sec\alpha = \frac{\sqrt{26}}{5}$$

$$\cos\alpha = \frac{1}{\sec\alpha} = \frac{5\sqrt{26}}{26}$$

37. $\tan\theta\cot\theta = \tan\theta\left(\dfrac{1}{\tan\theta}\right) = 1$

39. $\tan\alpha\cos\alpha = \left(\dfrac{\sin\alpha}{\cos\alpha}\right)\cos\alpha = \sin\alpha$

41. $\left(1 + \sin\theta\right)\left(1 - \sin\theta\right) = 1 - \sin^2\theta = \cos^2\theta$

43. $\left(\sec\theta + \tan\theta\right)\left(\sec\theta - \tan\theta\right) = \sec^2\theta - \tan^2\theta$

$$= \left(1 + \tan^2\theta\right) - \tan^2\theta$$

$$= 1$$

45. $\dfrac{\sin\theta}{\cos\theta} + \dfrac{\cos\theta}{\sin\theta} = \dfrac{\sin^2\theta + \cos^2\theta}{\sin\theta\cos\theta}$

$$= \frac{1}{\sin\theta\cos\theta}$$

$$= \frac{1}{\sin\theta} \cdot \frac{1}{\cos\theta}$$

$$= \csc\theta\sec\theta$$

47. (a) $\tan 23.5° \approx 0.4348$

(b) $\cot 66.5° = \dfrac{1}{\tan 66.5°} \approx 0.4348$

49. (a) $\cos 16°18' = \cos\left(16 + \dfrac{18}{60}\right)° \approx 0.9598$

(b) $\sin 73°56' = \sin\left(73 + \dfrac{56}{60}\right)° \approx 0.9609$

51. Make sure that your calculator is in radian mode.

(a) $\cot \dfrac{\pi}{16} = \dfrac{1}{\tan \dfrac{\pi}{16}} \approx 5.0273$

(b) $\tan \dfrac{\pi}{16} \approx 0.1989$

53. Make sure that your calculator is in radian mode.

(a) $\csc 1 = \dfrac{1}{\sin 1} \approx 1.1884$

(b) $\tan \dfrac{1}{2} \approx 0.5463$

55. (a) $\sin \theta = \dfrac{1}{2} \Rightarrow \theta = 30° = \dfrac{\pi}{6}$

(b) $\csc \theta = 2 \Rightarrow \theta = 30° = \dfrac{\pi}{6}$

57. (a) $\sec \theta = 2 \Rightarrow \theta = 60° = \dfrac{\pi}{3}$

(b) $\cot \theta = 1 \Rightarrow \theta = 45° = \dfrac{\pi}{4}$

59. (a) $\csc \theta = \dfrac{2\sqrt{3}}{3} \Rightarrow \theta = 60° = \dfrac{\pi}{3}$

(b) $\sin \theta = \dfrac{\sqrt{2}}{2} \Rightarrow \theta = 45° = \dfrac{\pi}{4}$

61. $\sin 60° = \dfrac{y}{18}$

$y = 18 \sin 60° = 18 \dfrac{\sqrt{3}}{2} = 9\sqrt{3}$

63.

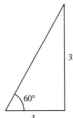

$\tan 60° = \dfrac{32}{x}$

$\sqrt{3} = \dfrac{32}{x}$

$\sqrt{3}x = 32$

$x = \dfrac{32}{\sqrt{3}} = \dfrac{32\sqrt{3}}{3}$

65.

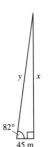

$\tan 82° = \dfrac{x}{45}$

$x = 45 \tan 82°$

Height of the building:

$123 + 45 \tan 82° \approx 443.2$ meters

Distance between friends:

$\cos 82° = \dfrac{45}{y} \Rightarrow y = \dfrac{45}{\cos 82°}$

≈ 323.34 meters

67.

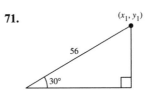

$\sin \theta = \dfrac{1250}{2500} = \dfrac{1}{2}$

$\theta = 30° = \dfrac{\pi}{6}$

69. (a) $\sin 43° = \dfrac{150}{x}$

$x = \dfrac{150}{\sin 43°} \approx 219.9$ ft

(b) $\tan 43° = \dfrac{150}{y}$

$y = \dfrac{150}{\tan 43°} \approx 160.9$ ft

71.

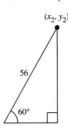

$\sin 30° = \dfrac{y_1}{56}$

$y_1 = (\sin 30°)(56) = \left(\dfrac{1}{2}\right)(56) = 28$

$\cos 30° = \dfrac{x_1}{56}$

$x_1 = \cos 30°(56) = \dfrac{\sqrt{3}}{2}(56) = 28\sqrt{3}$

$(x_1, y_1) = \left(28\sqrt{3}, 28\right)$

$\sin 60° = \dfrac{y_2}{56}$

$y_2 = \sin 60°(56) = \left(\dfrac{\sqrt{3}}{2}\right)(56) = 28\sqrt{3}$

$\cos 60° = \dfrac{x_2}{56}$

$x_2 = (\cos 60°)(56) = \left(\dfrac{1}{2}\right)(56) = 28$

$(x_2, y_2) = \left(28, 28\sqrt{3}\right)$

73. $x \approx 9.397$, $y \approx 3.420$

$$\sin 20° = \frac{y}{10} \approx 0.34$$

$$\cos 20° = \frac{x}{10} \approx 0.94$$

$$\tan 20° = \frac{y}{x} \approx 0.36$$

$$\cot 20° = \frac{x}{y} \approx 2.75$$

$$\sec 20° = \frac{10}{x} \approx 1.06$$

$$\csc 20° = \frac{10}{y} \approx 2.92$$

75. $\sin 60° \csc 60° = 1$

True, $\csc x = \dfrac{1}{\sin x} \Rightarrow \sin 60° \csc 60° = \sin 60°\left(\dfrac{1}{\sin 60°}\right) = 1$

77. $\sin 45° + \sin 45° = 1$

False, $\dfrac{\sqrt{2}}{2} + \dfrac{\sqrt{2}}{2} = \sqrt{2} \neq 1$

79. $\dfrac{\sin 60°}{\sin 30°} = \sin 2°$

False, $\dfrac{\sin 60°}{\sin 30°} = \dfrac{\cos 30°}{\sin 30°} = \cot 30° \approx 1.7321$

$\sin 2° \approx 0.0349$

81. This is true because the corresponding sides of similar triangles are proportional.

83. (a)

θ	0.1	0.2	0.3	0.4	0.5
$\sin \theta$	0.0998	0.1987	0.2955	0.3894	0.4794

(b) In the interval $(0, 0.5]$, $\theta > \sin \theta$.

(c) As $\theta \to 0$, $\sin \theta \to 0$, and $\dfrac{\theta}{\sin \theta} \to 1$.

85.

θ	0°	20°	40°	60°	80°
$\cos \theta$	1	0.94	0.77	0.50	0.17
$\sin(90° - \theta)$	1	0.94	0.77	0.50	0.17

$\cos \theta$ and $\sin(90° - \theta)$ are equal. θ and $90° - \theta$ are complementary angles because $\theta + (90° - \theta) = 90°$.

Section 6.3 Trigonometric Functions of Any Angle

- Know the Definitions of Trigonometric Functions of Any Angle.

  If θ is in standard position, (x, y) a point on the terminal side and $r = \sqrt{x^2 + y^2} \neq 0$, then

$$\sin \theta = \frac{y}{r} \qquad\qquad \csc \theta = \frac{r}{y}, y \neq 0$$

$$\cos \theta = \frac{x}{r} \qquad\qquad \sec \theta = \frac{r}{x}, x \neq 0$$

$$\tan \theta = \frac{y}{x}, x \neq 0 \qquad \cot \theta = \frac{x}{y}, y \neq 0$$

- You should know the signs of the trigonometric functions in each quadrant.

- You should know the trigonometric function values of the quadrant angles $0, \dfrac{\pi}{2}, \pi,$ and $\dfrac{3\pi}{2}$.

- You should be able to find reference angles.

- You should be able to evaluate trigonometric functions of any angle. (Use reference angles.)

- You should know that the period of sine and cosine is 2π.

- You should know which trigonometric functions are odd and even.

  Even: $\cos x$ and $\sec x$

  Odd: $\sin x, \tan x, \cot x, \csc x$

1. reference

3. period

5. (a) $(x, y) = (4, 3)$

$r = \sqrt{16 + 9} = 5$

$\sin \theta = \dfrac{y}{r} = \dfrac{3}{5} \qquad\qquad \csc \theta = \dfrac{r}{y} = \dfrac{5}{3}$

$\cos \theta = \dfrac{x}{r} = \dfrac{4}{5} \qquad\qquad \sec \theta = \dfrac{r}{x} = \dfrac{5}{4}$

$\tan \theta = \dfrac{y}{x} = \dfrac{3}{4} \qquad\qquad \cot \theta = \dfrac{x}{y} = \dfrac{4}{3}$

(b) $(x, y) = (-8, 15)$

$r = \sqrt{64 + 225} = 17$

$\sin \theta = \dfrac{y}{r} = \dfrac{15}{17} \qquad\qquad \csc \theta = \dfrac{r}{y} = \dfrac{17}{15}$

$\cos \theta = \dfrac{x}{r} = -\dfrac{8}{17} \qquad\qquad \sec \theta = \dfrac{r}{x} = -\dfrac{17}{8}$

$\tan \theta = \dfrac{y}{x} = -\dfrac{15}{8} \qquad\qquad \cot \theta = \dfrac{x}{y} = -\dfrac{8}{15}$

7. (a) $(x, y) = \left(-\sqrt{3}, -1\right)$

$r = \sqrt{3 + 1} = 2$

$\sin \theta = \dfrac{y}{r} = -\dfrac{1}{2} \qquad\qquad \csc \theta = \dfrac{r}{y} = -2$

$\cos \theta = \dfrac{x}{r} = -\dfrac{\sqrt{3}}{2} \qquad\qquad \sec \theta = \dfrac{r}{x} = -\dfrac{2\sqrt{3}}{3}$

$\tan \theta = \dfrac{y}{x} = \dfrac{\sqrt{3}}{3} \qquad\qquad \cot \theta = \dfrac{x}{y} = \sqrt{3}$

(b) $(x, y) = (4, -1)$

$r = \sqrt{16 + 1} = \sqrt{17}$

$\sin \theta = \dfrac{y}{r} = -\dfrac{1}{\sqrt{17}} = -\dfrac{\sqrt{17}}{17} \qquad \csc \theta = \dfrac{r}{y} = -\sqrt{17}$

$\cos \theta = \dfrac{x}{r} = \dfrac{4}{\sqrt{17}} = \dfrac{4\sqrt{17}}{17} \qquad \sec \theta = \dfrac{r}{x} = \dfrac{\sqrt{17}}{4}$

$\tan \theta = \dfrac{y}{x} = -\dfrac{1}{4} \qquad\qquad \cot \theta = \dfrac{x}{y} = -4$

9. $(x, y) = (5, 12)$

$r = \sqrt{25 + 144} = 13$

$\sin \theta = \dfrac{y}{r} = \dfrac{12}{13}$ $\csc \theta = \dfrac{r}{y} = \dfrac{13}{12}$

$\cos \theta = \dfrac{x}{r} = \dfrac{5}{13}$ $\sec \theta = \dfrac{r}{x} = \dfrac{13}{5}$

$\tan \theta = \dfrac{y}{x} = \dfrac{12}{5}$ $\cot \theta = \dfrac{x}{y} = \dfrac{5}{12}$

11. $x = -5,\ y = -2$

$r = \sqrt{(-5)^2 + (-2)^2} = \sqrt{29}$

$\sin \theta = \dfrac{y}{r} = \dfrac{-2}{\sqrt{29}} = -\dfrac{2\sqrt{29}}{29}$

$\cos \theta = \dfrac{x}{r} = \dfrac{-5}{\sqrt{29}} = -\dfrac{5\sqrt{29}}{29}$

$\tan \theta = \dfrac{y}{x} = \dfrac{-2}{-5} = \dfrac{2}{5}$

$\csc \theta = \dfrac{r}{y} = \dfrac{\sqrt{29}}{-2} = -\dfrac{\sqrt{29}}{2}$

$\sec \theta = \dfrac{r}{x} = \dfrac{\sqrt{29}}{-5} = -\dfrac{\sqrt{29}}{5}$

$\cot \theta = \dfrac{x}{y} = \dfrac{-5}{-2} = \dfrac{5}{2}$

13. $(x, y) = (-5.4, 7.2)$

$r = \sqrt{29.16 + 51.84} = 9$

$\sin \theta = \dfrac{y}{r} = \dfrac{7.2}{9} = \dfrac{4}{5}$ $\csc \theta = \dfrac{r}{y} = \dfrac{9}{7.2} = \dfrac{5}{4}$

$\cos \theta = \dfrac{x}{r} = -\dfrac{5.4}{9} = -\dfrac{3}{5}$ $\sec \theta = \dfrac{r}{x} = -\dfrac{9}{5.4} = -\dfrac{5}{3}$

$\tan \theta = \dfrac{y}{x} = -\dfrac{7.2}{5.4} = -\dfrac{4}{3}$ $\tan \theta = \dfrac{x}{y} = -\dfrac{5.4}{7.2} = -\dfrac{3}{4}$

15. $\sin \theta > 0 \Rightarrow \theta$ lies in Quadrant I or in Quadrant II.

$\cos \theta > 0 \Rightarrow \theta$ lies in Quadrant I or in Quadrant IV.

$\sin \theta > 0$ and $\cos \theta > 0 \Rightarrow \theta$ lies in Quadrant I.

17. $\sin \theta > 0 \Rightarrow \theta$ lies in Quadrant I or in Quadrant II.

$\cos \theta < 0 \Rightarrow \theta$ lies in Quadrant II or in Quadrant III.

$\sin \theta > 0$ and $\cos \theta < 0 \Rightarrow \theta$ lies in Quadrant II.

19. $\tan \theta < 0$ and $\sin \theta > 0 \Rightarrow \theta$ is in Quadrant II
$\Rightarrow x < 0$ and $y > 0$.

$\tan \theta = \dfrac{y}{x} = \dfrac{15}{-8} \Rightarrow r = 17$

$\sin \theta = \dfrac{y}{r} = \dfrac{15}{17}$ $\csc \theta = \dfrac{r}{y} = \dfrac{17}{15}$

$\cos \theta = \dfrac{x}{r} = -\dfrac{8}{17}$ $\sec \theta = \dfrac{r}{x} = -\dfrac{17}{8}$

$\tan \theta = \dfrac{y}{x} = -\dfrac{15}{8}$ $\cot \theta = \dfrac{x}{y} = -\dfrac{8}{15}$

21. $\sin \theta = \dfrac{y}{r} = \dfrac{3}{5} \Rightarrow x^2 = 25 - 9 = 16$

θ in Quadrant II $\Rightarrow x = -4$

$\sin \theta = \dfrac{y}{r} = \dfrac{3}{5}$ $\csc \theta = \dfrac{r}{y} = \dfrac{5}{3}$

$\cos \theta = \dfrac{x}{r} = -\dfrac{4}{5}$ $\sec \theta = \dfrac{r}{x} = -\dfrac{5}{4}$

$\tan \theta = \dfrac{y}{x} = -\dfrac{3}{4}$ $\cot \theta = \dfrac{x}{y} = -\dfrac{4}{3}$

23. $\cot \theta = \dfrac{x}{y} = -\dfrac{3}{1} = \dfrac{3}{-1}$

$\cos \theta > 0 \Rightarrow \theta$ is in Quadrant IV $\Rightarrow x$ is positive;
$x = 3,\ y = -1,\ r = \sqrt{10}$

$\sin \theta = \dfrac{y}{r} = -\dfrac{\sqrt{10}}{10}$ $\csc \theta = \dfrac{r}{y} = -\sqrt{10}$

$\cos \theta = \dfrac{x}{r} = \dfrac{3\sqrt{10}}{10}$ $\sec \theta = \dfrac{r}{x} = \dfrac{\sqrt{10}}{3}$

$\tan \theta = \dfrac{y}{x} = -\dfrac{1}{3}$ $\cot \theta = \dfrac{x}{y} = -3$

25. $\sec\theta = \dfrac{r}{x} = \dfrac{2}{-1} \Rightarrow y^2 = 4 - 1 = 3$

$\sin\theta < 0 \Rightarrow \theta$ is in Quadrant III $\Rightarrow y = -\sqrt{3}$

$\sin\theta = \dfrac{y}{r} = -\dfrac{\sqrt{3}}{2}$ $\qquad$ $\csc\theta = \dfrac{r}{y} = -\dfrac{2}{\sqrt{3}} = -\dfrac{2\sqrt{3}}{3}$

$\cos\theta = \dfrac{x}{r} = -\dfrac{1}{2}$ $\qquad$ $\sec\theta = \dfrac{r}{x} = -2$

$\tan\theta = \dfrac{y}{x} = \sqrt{3}$ $\qquad$ $\cot\theta = \dfrac{x}{y} = \dfrac{1}{\sqrt{3}} = \dfrac{\sqrt{3}}{3}$

27. $\cot\theta$ is undefined,

$\dfrac{\pi}{2} \le \theta \le \dfrac{3\pi}{2} \Rightarrow y = 0 \Rightarrow \theta = \pi$

$\sin\theta = 0$ $\qquad$ $\csc\theta$ is undefined.

$\cos\theta = -1$ $\qquad$ $\sec\theta = -1$

$\tan\theta = 0$ $\qquad$ $\cot\theta$ is undefined.

29. To find a point on the terminal side of θ, use any point on the line $y = -x$ that lies in Quadrant II. $(-1, 1)$ is one such point.

$x = -1,\ y = 1,\ r = \sqrt{2}$

$\sin\theta = \dfrac{1}{\sqrt{2}} = \dfrac{\sqrt{2}}{2}$ $\qquad$ $\csc\theta = \sqrt{2}$

$\cos\theta = -\dfrac{1}{\sqrt{2}} = -\dfrac{\sqrt{2}}{2}$ $\qquad$ $\sec\theta = -\sqrt{2}$

$\tan\theta = -1$ $\qquad$ $\cot\theta = -1$

31. To find a point on the terminal side of θ, use any point on the line $y = 2x$ that lies in Quadrant III. $(-1, -2)$ is one such point.

$x = -1,\ y = -2,\ y = \sqrt{5}$

$\sin\theta = -\dfrac{2}{\sqrt{5}} = -\dfrac{2\sqrt{5}}{5}$ $\qquad$ $\csc\theta = \dfrac{\sqrt{5}}{-2} = -\dfrac{\sqrt{5}}{2}$

$\cos\theta = -\dfrac{1}{\sqrt{5}} = -\dfrac{\sqrt{5}}{5}$ $\qquad$ $\sec\theta = \dfrac{\sqrt{5}}{-1} = -\sqrt{5}$

$\tan\theta = \dfrac{-2}{-1} = 2$ $\qquad$ $\cot\theta = \dfrac{-1}{-2} = \dfrac{1}{2}$

33. $(x, y) = (-1, 0),\ r = 1$

$\sin\pi = \dfrac{y}{r} = \dfrac{0}{1} = 0$

35. $(x, y) = (0, -1),\ r = 1$

$\sec\dfrac{3\pi}{2} = \dfrac{r}{x} = \dfrac{1}{0} \Rightarrow$ undefined

37. $(x, y) = (0, 1),\ r = 1$

$\sin\dfrac{\pi}{2} = \dfrac{y}{r} = \dfrac{1}{1} = 1$

39. $(x, y) = (-1, 0),\ r = 1$

$\csc\pi = \dfrac{r}{y} = \dfrac{1}{0} \Rightarrow$ undefined

41. $\theta = 160°$

$\theta' = 180° - 160° = 20°$

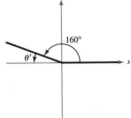

43. $\theta = -125°$

$360° - 125° = 235°$ (coterminal angle)

$\theta' = 235° - 180° = 55°$

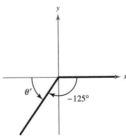

45. $\theta = \dfrac{2\pi}{3}$

$\theta' = \pi - \dfrac{2\pi}{3} = \dfrac{\pi}{3}$

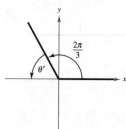

47. $\theta = 4.8$

$\theta' = 2\pi - 4.8\pi \approx 1.4832$

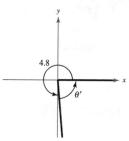

49. $\theta = 225°, \theta' = 45°,$ Quadrant III

$\sin 225° = -\sin 45° = -\dfrac{\sqrt{2}}{2}$

$\cos 225° = -\cos 45° = -\dfrac{\sqrt{2}}{2}$

$\tan 225° = \tan 45° = 1$

51. $\theta = 750°, \theta' = 30°,$ Quadrant I

$\sin 750° = \sin 30° = \dfrac{1}{2}$

$\cos 750° = \cos 30° = \dfrac{\sqrt{3}}{2}$

$\tan 750° = \tan 30° = \dfrac{\sqrt{3}}{3}$

53. $\theta = -150°, \theta' = 30°,$ Quadrant III

$\sin(-150°) = -\sin 30° = -\dfrac{1}{2}$

$\cos(-150°) = -\cos 30° = -\dfrac{\sqrt{3}}{2}$

$\tan(-150°) = \tan 30° = \dfrac{\sqrt{3}}{3}$

55. $\theta = \dfrac{2\pi}{3}, \theta' = \dfrac{\pi}{3}$ in Quadrant II

$\sin \dfrac{2\pi}{3} = \sin \dfrac{\pi}{3} = \dfrac{\sqrt{3}}{2}$

$\cos \dfrac{2\pi}{3} = -\cos \dfrac{\pi}{3} = -\dfrac{1}{2}$

$\tan \dfrac{2\pi}{3} = -\tan \dfrac{\pi}{3} = -\sqrt{3}$

57. $\theta = \dfrac{5\pi}{4}, \theta' = \dfrac{\pi}{4}$ in Quadrant III

$\sin \dfrac{5\pi}{4} = -\sin \dfrac{\pi}{4} = -\dfrac{\sqrt{2}}{2}$

$\cos \dfrac{5\pi}{4} = -\cos \dfrac{\pi}{4} = -\dfrac{\sqrt{2}}{2}$

$\tan \dfrac{5\pi}{4} = \tan \dfrac{\pi}{4} = 1$

59. $\theta = -\dfrac{\pi}{6}, \theta' = \dfrac{\pi}{6},$ Quadrant IV

$\sin\left(-\dfrac{\pi}{6}\right) = -\sin \dfrac{\pi}{6} = -\dfrac{1}{2}$

$\cos\left(-\dfrac{\pi}{6}\right) = \cos \dfrac{\pi}{6} = \dfrac{\sqrt{3}}{2}$

$\tan\left(-\dfrac{\pi}{6}\right) = -\tan \dfrac{\pi}{6} = -\dfrac{\sqrt{3}}{3}$

61. $\theta = \dfrac{11\pi}{4}, \theta' = \dfrac{\pi}{4},$ Quadrant II

$\sin \dfrac{11\pi}{4} = \sin \dfrac{\pi}{4} = \dfrac{\sqrt{2}}{2}$

$\cos \dfrac{11\pi}{4} = -\cos \dfrac{\pi}{4} = -\dfrac{\sqrt{2}}{2}$

$\tan \dfrac{11\pi}{4} = -\tan \dfrac{\pi}{4} = -1$

63. $\theta = \dfrac{9\pi}{4}, \theta' = \dfrac{\pi}{4}$ in Quadrant I

$\sin \dfrac{9\pi}{4} = \sin \dfrac{\pi}{4} = \dfrac{\sqrt{2}}{2}$

$\cos \dfrac{9\pi}{4} = \cos \dfrac{\pi}{4} = \dfrac{\sqrt{2}}{2}$

$\tan \dfrac{9\pi}{4} = \tan \dfrac{\pi}{4} = 1$

65. $\sin \theta = -\dfrac{3}{5}$

$$\sin^2 \theta + \cos^2 \theta = 1$$

$$\cos^2 \theta = 1 - \sin^2 \theta$$

$$\cos^2 \theta = 1 - \left(-\dfrac{3}{5}\right)^2$$

$$\cos^2 \theta = 1 - \dfrac{9}{25}$$

$$\cos^2 \theta = \dfrac{16}{25}$$

$\cos \theta > 0$ in Quadrant IV.

$$\cos \theta = \dfrac{4}{5}$$

67. $\tan \theta = \dfrac{3}{2}$

$$\sec^2 \theta = 1 + \tan^2 \theta$$

$$\sec^2 \theta = 1 + \left(\dfrac{3}{2}\right)^2$$

$$\sec^2 \theta = 1 + \dfrac{9}{4}$$

$$\sec^2 \theta = \dfrac{13}{4}$$

$\sec \theta < 0$ in Quadrant III.

$$\sec \theta = -\dfrac{\sqrt{13}}{2}$$

69. $\cos \theta = \dfrac{5}{8}$

$$\cos \theta = \dfrac{1}{\sec \theta} \Rightarrow \sec \theta = \dfrac{1}{\cos \theta}$$

$$\sec \theta = \dfrac{1}{\dfrac{5}{8}} = \dfrac{8}{5}$$

71. $\sin 10° \approx 0.1736$

73. $\cos(-110°) \approx -0.3420$

75. $\tan 304° \approx -1.4826$

77. $\sec 72° = \dfrac{1}{\cos 72°} \approx 3.2361$

79. $\tan 4.5 \approx 4.6373$

81. $\tan\left(\dfrac{\pi}{9}\right) \approx 0.3640$

83. $\sin(-0.65) \approx -0.6052$

85. $\cot\left(-\dfrac{11\pi}{8}\right) = \dfrac{1}{\tan(-11\pi/8)} \approx -0.4142$

87. (a) $\sin \theta = \dfrac{1}{2} \Rightarrow$ reference angle is $30°$ or $\dfrac{\pi}{6}$ and θ

   is in Quadrant I or Quadrant II.

   Values in degrees: $30°, 150°$

   Values in radian: $\dfrac{\pi}{6}, \dfrac{5\pi}{6}$

(b) $\sin \theta = \dfrac{1}{2} \Rightarrow$ reference angle is $30°$ or $\dfrac{\pi}{6}$ and θ

   is in Quadrant III or Quadrant IV.

   Values in degrees: $210°, 330°$

   Values in radians: $\dfrac{7\pi}{6}, \dfrac{11\pi}{6}$

89. (a) $\csc \theta = \dfrac{2\sqrt{3}}{3} \Rightarrow$ reference angle is $60°$ or $\dfrac{\pi}{3}$ and

   θ is in Quadrant I or Quadrant II.

   Values in degrees: $60°, 120°$

   Values in radians: $\dfrac{\pi}{3}, \dfrac{2\pi}{3}$

(b) $\cot \theta = -1 \Rightarrow$ reference angle is $45°$ or $\dfrac{\pi}{4}$ and θ

   is in Quadrant II or Quadrant IV.

   Values in degrees: $135°, 315°$

   Values in radians: $\dfrac{3\pi}{4}, \dfrac{7\pi}{4}$

91. (a) $\tan \theta = 1 \Rightarrow$ reference angle is $45°$ or $\dfrac{\pi}{4}$ and θ

   is in Quadrant I or Quadrant III.

   Values in degrees: $45°, 225°$

   Values in radians: $\dfrac{\pi}{4}, \dfrac{5\pi}{4}$

(b) $\cot \theta = -\sqrt{3} \Rightarrow$ reference angle is $30°$ or $\dfrac{\pi}{6}$ and

   θ is in Quadrant II or Quadrant IV.

   Values in degrees: $150°, 330°$

   Values in radians: $\dfrac{5\pi}{6}, \dfrac{11\pi}{6}$

93. $\left(\dfrac{\sqrt{2}}{2}, \dfrac{\sqrt{2}}{2}\right)$ corresponds to $t = \dfrac{\pi}{4}$ on the unit circle.

$\sin\dfrac{\pi}{4} = \dfrac{\sqrt{2}}{2}$ since $\sin t = y$.

$\cos\dfrac{\pi}{4} = \dfrac{\sqrt{2}}{2}$ since $\cos t = x$.

$\tan\dfrac{\pi}{4} = 1$ since $\tan t = \dfrac{y}{x}$.

95. $\left(-\dfrac{\sqrt{3}}{2}, \dfrac{1}{2}\right)$ corresponds to $t = \dfrac{5\pi}{6}$ on the unit circle.

$\sin\dfrac{5\pi}{6} = \dfrac{1}{2}$ since $\sin t = y$.

$\cos\dfrac{5\pi}{6} = -\dfrac{\sqrt{3}}{2}$ since $\cos t = x$.

$\tan\dfrac{5\pi}{6} = -\dfrac{\sqrt{3}}{3}$ since $\tan t = \dfrac{y}{x}$.

97. $\left(-\dfrac{1}{2}, -\dfrac{\sqrt{3}}{2}\right)$ corresponds to $t = \dfrac{4\pi}{3}$ on the unit circle.

$\sin\dfrac{4\pi}{3} = -\dfrac{\sqrt{3}}{2}$ since $\sin t = y$.

$\cos\dfrac{4\pi}{3} = -\dfrac{1}{2}$ since $\cos t = x$.

$\tan\dfrac{4\pi}{3} = \sqrt{3}$ since $\tan t = \dfrac{y}{x}$.

99. $t = \dfrac{\pi}{2}, (x, y) = (0, 1)$

$\sin\dfrac{\pi}{2} = 1$ because $\sin t = y$.

$\cos\dfrac{\pi}{2} = 0$ because $\cos t = x$.

$\tan\dfrac{\pi}{2}$ is undefined because $\tan t = \dfrac{y}{x} = \dfrac{1}{0}$.

101. (a) $\sin 5 \approx -1$

(b) $\cos 2 \approx -0.4$

103. (a) $\sin t = 0.25$

$t \approx 0.25$ or 2.89

(b) $\cos t = -0.25$

$t \approx 1.82$ or 4.46

105. (a) New York City:
$N \approx 22.1 \sin(0.52t - 2.22) + 55.01$

Fairbanks: $F \approx 36.6 \sin(0.50t - 1.83) + 25.61$

(b)

Month	New York City	Fairbanks
February	35°	–1°
March	41°	14°
May	63°	48°
June	72°	59°
August	76°	56°
September	69°	42°
November	47°	7°

(c) The periods are about the same for both models, approximately 12 months.

107. $y(t) = 2\cos 6t$

(a) $y(0) = 2\cos 0 = 2$ centimeters

(b) $y\left(\dfrac{1}{4}\right) = 2\cos\left(\dfrac{3}{2}\right) \approx 0.14$ centimeter

(c) $y\left(\dfrac{1}{2}\right) = 2\cos 3 \approx -1.98$ centimeters

109. $I = 5e^{-2t}\sin t$

$I(0.7) = 5e^{-1.4}\sin 0.7 \approx 0.79$

111. False. In each of the four quadrants, the sign of the secant function and the cosine function will be the same since they are reciprocals of each other.

113. $h(t) = f(t)g(t)$

$h(-t) = f(-t)g(-t)$

$= -f(t)g(t)$

$= -h(t)$

Therefore, $h(t)$ is odd.

115. (a)

θ	0°	20°	40°	60°	80°
$\sin\theta$	0	0.342	0.643	0.866	0.985
$\sin(180° - \theta)$	0	0.342	0.643	0.866	0.985

(b) Conjecture: $\sin\theta = \sin(180° - \theta)$

117. $y = \sin x$

Domain: All real numbers x

Range: $[-1, 1]$

Period: 2π

Zeros: $n\pi$

The function is odd.

$y = \cos x$

Domain: All real numbers x

Range: $[-1, 1]$

Period: 2π

Zeros: $n\pi + \dfrac{\pi}{2}$

The function is even.

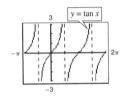

$y = \tan x$

Domain: All real numbers x except $x = n\pi + \dfrac{\pi}{2}$

Range: $(-\infty, \infty)$

Period: π

Zeros: $n\pi$

The function is odd.

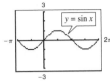

$y = \csc x$

Domain: All real numbers x except $x = n\pi$

Range: $(-\infty, -1] \cup [1, \infty)$

Period: 2π

Zeros: none

The function is odd.

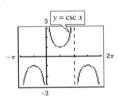

$y = \sec x$

Domain: All real numbers x except $x = n\pi + \dfrac{\pi}{2}$

Range: $(-\infty, -1] \cup [1, \infty)$

Period: 2π

Zeros: none

The function is even.

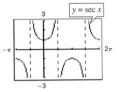

$y = \cot x$

Domain: All real numbers x except $x = n\pi$

Range: $(-\infty, \infty)$

Period: π

Zeros: $n\pi + \dfrac{\pi}{2}$

The function is odd.

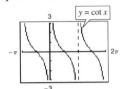

The secant function is similar to the tangent function because they both have vertical asymptotes at

$$x = n\pi + \frac{\pi}{2}.$$

The cotangent function and the cosecant function both have vertical asymptotes at $x = n\pi$. A maximum point on the sine curve corresponds to a relative minimum on the cosecant curve. The maximum points of sine and cosine are interchanged with the minimum points of cosecant and secant. The x-intercepts of the sine and cosine functions become vertical asymptotes of the cosecant and secant functions.

The graphs of sine and cosine may be translated left or right (respectively) to $\pi/2$ to coincide with each other.

119. (a) The points have y-axis symmetry.

(b) $\sin t_1 = \sin(\pi - t_1)$ because they have the same y-value.

(c) $\cos(\pi - t_1) = -\cos t_1$ because the x-values have opposite signs.

Section 6.4   Graphs of Sine and Cosine Functions

- You should be able to graph $y = a \sin(bx - c)$ and $y = a \cos(bx - c)$. (Assume $b > 0$.)

- Amplitude: $|a|$

- Period: $\dfrac{2\pi}{b}$

- Shift: Solve $bx - c = 0$ and $bx - c = 2\pi$.

- Key increments: $\dfrac{1}{4}$ (period)

1. cycle

3. phase shift

5. $y = 2 \sin 5x$

Period: $\dfrac{2\pi}{5}$

Amplitude: $|2| = 2$

7. $y = \dfrac{3}{4} \cos \dfrac{x}{2}$

Period: $\dfrac{2\pi}{1/2} = 4\pi$

Amplitude: $\left|\dfrac{3}{4}\right| = \dfrac{3}{4}$

9. $y = \dfrac{1}{2} \sin \dfrac{\pi x}{3}$

Period: $\dfrac{2\pi}{\pi/3} = 6$

Amplitude: $\left|\dfrac{1}{2}\right| = \dfrac{1}{2}$

11. $y = -4 \sin x$

Period: $\dfrac{2\pi}{1} = 2\pi$

Amplitude: $|-4| = 4$

13. $y = 3 \sin 10x$

Period: $\dfrac{2\pi}{10} = \dfrac{\pi}{5}$

Amplitude: $|3| = 3$

15. $y = \dfrac{5}{3} \cos \dfrac{4x}{5}$

Period: $\dfrac{2\pi}{4/5} = \dfrac{10\pi}{4} = \dfrac{5\pi}{2}$

Amplitude: $\left|\dfrac{5}{3}\right| = \dfrac{5}{3}$

17. $y = \dfrac{1}{4} \sin 2\pi x$

Period: $\dfrac{2\pi}{2\pi} = 1$

Amplitude: $\left|\dfrac{1}{4}\right| = \dfrac{1}{4}$

19. $f(x) = \sin x$

$g(x) = \sin(x - \pi)$

g is a horizontal shift to the right π units of the graph of f (a phase shift).

21. $f(x) = \cos 2x$

$g(x) = -\cos 2x$

g is a reflection in the x-axis of the graph of f.

23. $f(x) = \cos x$

$g(x) = \cos 2x$

The period of f is twice that of g.

25. $f(x) = \sin 2x$

$f(x) = 3 + \sin 2x$

g is a vertical shift three units upward of the graph of f.

27. The graph of g has twice the amplitude as the graph of f. The period is the same.

29. The graph of g is a horizontal shift π units to the right of the graph of f.

31. $f(x) = -2 \sin x$

Period: $\dfrac{2\pi}{b} = \dfrac{2\pi}{1} = 2\pi$

Amplitude: 2

Symmetry: origin

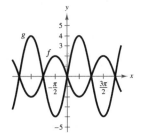

Key points: Intercept   Minimum   Intercept   Maximum   Intercept

$(0, 0)$ $\left(\dfrac{\pi}{2}, -2\right)$ $(\pi, 0)$ $\left(\dfrac{3\pi}{2}, 0\right)$ $(2\pi, 0)$

Because $g(x) = 4 \sin x = (-2)f(x)$, generate key points for the graph of $g(x)$ by multiplying the y-coordinate of each key point of $f(x)$ by -2.

33. $f(x) = \cos x$

Period: $\dfrac{2\pi}{b} = \dfrac{2\pi}{1} = 2\pi$

Amplitude: $|1| = 1$

Symmetry: y-axis

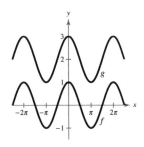

Key points: Maximum   Intercept   Minimum   Intercept   Maximum

$(0, 1)$ $\left(\dfrac{\pi}{2}, 0\right)$ $(\pi, -1)$ $\left(\dfrac{3\pi}{2}, 0\right)$ $(2\pi, 1)$

Because $g(x) = 2 + \cos x = f(x) + 2$, the graph of $g(x)$ is the graph of $f(x)$, but translated upward by two units.
Generate key points of $g(x)$ by adding 2 to the y-coordinate of each key point of $f(x)$.

35. $f(x) = -\dfrac{1}{2} \sin \dfrac{x}{2}$

Period: $\dfrac{2\pi}{b} = \dfrac{2\pi}{1/2} = 4\pi$

Amplitude: $\dfrac{1}{2}$

Symmetry: origin

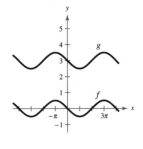

Key points: Intercept   Minimum   Intercept   Maximum   Intercept

$(0, 0)$ $\left(\pi, -\dfrac{1}{2}\right)$ $(2\pi, 0)$ $\left(3\pi, \dfrac{1}{2}\right)$ $(4\pi, 0)$

Because $g(x) = 3 - \dfrac{1}{2} \sin \dfrac{x}{2} = 3 - f(x)$, the graph of $g(x)$ is the graph of $f(x)$, but translated upward by three units.

Generate key points for the graph of $g(x)$ by adding 3 to the y-coordinate of each key point of $f(x)$.

37. $f(x) = 2 \cos x$

Period: $\dfrac{2\pi}{b} = \dfrac{2\pi}{1} = 2\pi$

Amplitude: 2

Symmetry: y-axis

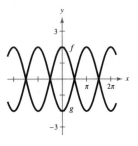

Key points: Maximum   Intercept   Minimum   Intercept   Maximum

$(0, 2)$ $\left(\dfrac{\pi}{2}, 0\right)$ $(\pi, -2)$ $\left(\dfrac{3\pi}{2}, 0\right)$ $(2\pi, 2)$

Because $g(x) = 2 \cos(x + \pi) = f(x + \pi)$, the graph of $g(x)$ is the graph of $f(x)$, but with a phase shift (horizontal translation) of $-\pi$. Generate key points for the graph of $g(x)$ by shifting each key point of $f(x)$ π units to the left.

39. $y = 5 \sin x$

Period: 2π

Amplitude: 5

Key points:

$(0, 0), \left(\dfrac{\pi}{2}, 5\right), (\pi, 0),$

$\left(\dfrac{3\pi}{2}, -5\right), (2\pi, 0)$

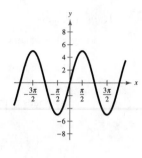

41. $y = \dfrac{1}{3} \cos x$

Period: 2π

Amplitude: $\dfrac{1}{3}$

Key points:

$\left(0, \dfrac{1}{3}\right), \left(\dfrac{\pi}{2}, 0\right), \left(\pi, -\dfrac{1}{3}\right),$

$\left(\dfrac{3\pi}{2}, 0\right), \left(2\pi, \dfrac{1}{3}\right)$

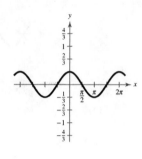

43. $y = \cos \dfrac{y}{2}$

Period $\dfrac{2\pi}{1/2} = 4\pi$

Amplitude: 1

Key points:

$(0, 1), (\pi, 0), (2\pi, -1),$

$(3\pi, 0), (4\pi, 1)$

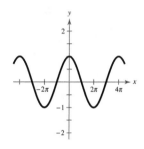

45. $y = \cos 2\pi x$

Period: $\dfrac{2\pi}{2\pi} = 1$

Amplitude: 1

Key points:

$(0, 1), \left(\dfrac{1}{4}, 0\right), \left(\dfrac{1}{2}, -1\right), \left(\dfrac{3}{4}, 0\right)$

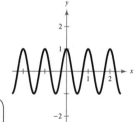

47. $y = -\sin \dfrac{2\pi x}{3}$

Period: $\dfrac{2\pi}{2\pi/3} = 3$

Amplitude: 1

Key points:

$(0, 0), \left(\dfrac{3}{4}, -1\right), \left(\dfrac{3}{2}, 0\right),$

$\left(\dfrac{9}{4}, 1\right), (3, 0)$

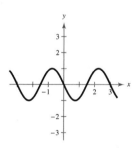

49. $y = \sin\left(x - \dfrac{\pi}{2}\right)$

Period: 2π

Amplitude: 1

Shift: Set $x - \dfrac{\pi}{2} = 0$ and $x - \dfrac{\pi}{2} = 2\pi$

$\qquad x = \dfrac{\pi}{2} \qquad\qquad x = \dfrac{5\pi}{2}$

Key points: $\left(\dfrac{\pi}{2}, 0\right), (\pi, 1), \left(\dfrac{3\pi}{2}, 0\right), (2\pi, -1), \left(\dfrac{5\pi}{2}, 0\right)$

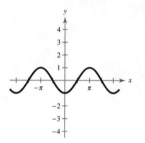

51. $y = 3 \cos(x + \pi)$

Period: 2π

Amplitude: 3

Shift: Set $x + \pi = 0$ and $x + \pi = 2\pi$

$\qquad x = -\pi \qquad\qquad x = \pi$

Key points: $(-\pi, 3), \left(-\dfrac{\pi}{2}, 0\right), (0, -3), \left(\dfrac{\pi}{2}, 0\right), (\pi, 3)$

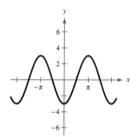

53. $y = 2 - \sin \dfrac{2\pi x}{3}$

Period: $\dfrac{2\pi}{2\pi/3} = 3$

Amplitude: 1

Key points:

$(0, 2), \left(\dfrac{3}{4}, 1\right), \left(\dfrac{3}{2}, 2\right),$

$\left(\dfrac{9}{4}, 3\right), (3, 2)$

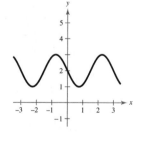

55. $y = 2 + \dfrac{1}{10}\cos 60\pi x$

Period: $\dfrac{2\pi}{60\pi} = \dfrac{1}{30}$

Amplitude: $\dfrac{1}{10}$

Vertical shift two units upward

Key points:

$(0, 2.1), \left(\dfrac{1}{120}, 2\right)\left(\dfrac{1}{60}, 1.9\right),$

$\left(\dfrac{1}{40}, 2\right), \left(\dfrac{1}{30}, 2.1\right)$

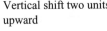

57. $y = 3\cos(x + \pi) - 3$

Period: 2π

Amplitude: 3

Shift: Set $x + \pi = 0$ and $x + \pi = 2\pi$

$x = -\pi$ $\qquad x = \pi$

Key points: $(-\pi, 0), \left(-\dfrac{\pi}{2}, -3\right), (0, -6), \left(\dfrac{\pi}{2}, -3\right), (\pi, 0)$

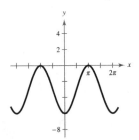

59. $y = \dfrac{2}{3}\cos\left(\dfrac{x}{2} - \dfrac{\pi}{4}\right)$

Period: $\dfrac{2\pi}{1/2} = 4\pi$

Amplitude: $\dfrac{2}{3}$

Shift: $\dfrac{x}{2} - \dfrac{\pi}{4} = 0$ and $\dfrac{x}{2} - \dfrac{\pi}{4} = 2\pi$

$x = \dfrac{\pi}{2}$ $\qquad x = \dfrac{9\pi}{2}$

Key points:

$\left(\dfrac{\pi}{2}, \dfrac{2}{3}\right), \left(\dfrac{3\pi}{2}, 0\right), \left(\dfrac{5\pi}{2}, \dfrac{-2}{3}\right),$

$\left(\dfrac{7\pi}{2}, 0\right), \left(\dfrac{9\pi}{2}, \dfrac{2}{3}\right)$

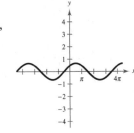

61. $g(x) = \sin(4x - \pi)$

(a) $g(x)$ is obtained by a horizontal shrink of four and a phase shift of $\dfrac{\pi}{4}$; and one cycle of $g(x)$ corresponds to the interval $\left[\dfrac{\pi}{4}, \dfrac{3\pi}{4}\right]$.

(b)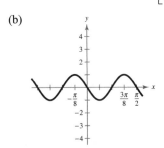

(c) $g(x) = f(4x - \pi)$ where $f(x) = \sin x$.

63. $g(x) = \cos(x - \pi) + 2$

(a) $g(x)$ is obtained by shifting $f(x)$ two units upward and a phase shift of π; and one cycle of $g(x)$ corresponds to the interval $[\pi, 3\pi]$.

(b)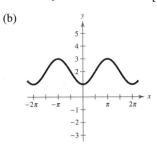

(c) $g(x) = f(x - \pi) + 2$ where $f(x) = \cos x$

65. $g(x) = 2\sin(4x - \pi) - 3$

(a) $g(x)$ is obtained by a horizontal shrink of four, a phase shift of $\dfrac{\pi}{4}$, shifting $f(x)$ three units downward, and has an amplitude of two. One cycle of $g(x)$ corresponds to the interval $\left[\dfrac{\pi}{4}, \dfrac{3\pi}{4}\right]$.

(b)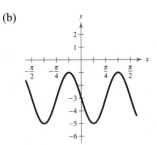

(c) $g(x) = 2f(4x - \pi) - 3$ where $f(x) = \sin x$

67. $y = -2\sin(4x + \pi)$

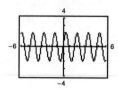

69. $y = \cos\left(2\pi x - \dfrac{\pi}{2}\right) + 1$

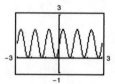

71. $y = -0.1\sin\left(\dfrac{\pi x}{10} + \pi\right)$

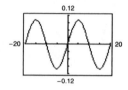

73. $f(x) = a\cos x + d$

Amplitude: $\dfrac{1}{2}[3 - (-1)] = 2 \Rightarrow a = 2$

$3 = 2\cos 0 + d$

$d = 3 - 2 = 1$

$a = 2, d = 1$

75. $f(x) = a\cos x + d$

Amplitude: $\dfrac{1}{2}[8 - 0] = 4$

Reflected in the x-axis: $a = -4$

$0 = -4\cos 0 + d$

$d = 4$

$a = -4, d = 4$

87. $y = 0.85\sin\dfrac{\pi t}{3}$

(a) Time for one cycle $= \dfrac{2\pi}{\pi/3} = 6$ seconds

(b) Cycles per min $= \dfrac{60}{6} = 10$ cycles per minute

(c) Amplitude: 0.85; Period: 6

Key points: $(0, 0), \left(\dfrac{3}{2}, 0.85\right), (3, 0), \left(\dfrac{9}{2}, -0.85\right), (6, 0)$

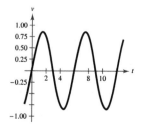

77. $y = a\sin(bx - c)$

Amplitude: $|a| = |3|$

Since the graph is reflected in the x-axis, we have $a = -3$.

Period: $\dfrac{2\pi}{b} = \pi \Rightarrow b = 2$

Phase shift: $c = 0$

$a = -3, b = 2, c = 0$

79. $y = a\sin(bx - c)$

Amplitude: $a = 2$

Period: $2\pi \Rightarrow b = 1$

Phase shift: $bx - c = 0$ when $x = -\dfrac{\pi}{4}$

$(1)\left(-\dfrac{\pi}{4}\right) - c = 0 \Rightarrow c = -\dfrac{\pi}{4}$

$a = 2, b = 1, c = -\dfrac{\pi}{4}$

81. $y_1 = \sin x$

$y_2 = -\dfrac{1}{2}$

In the interval $[-2\pi, 2\pi]$,

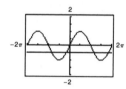

$y_1 = y_2$ when $x = -\dfrac{5\pi}{6}, -\dfrac{\pi}{6}, \dfrac{7\pi}{6}, \dfrac{11\pi}{6}$.

Answers for 83 and 85 are sample answers.

83. $f(x) = 2\sin(2x - \pi) + 1$

85. $f(x) = \cos(2x + 2\pi) - \dfrac{3}{2}$

89. (a) $y = a \cos(bt - c) + d$

Amplitude: $a = \dfrac{1}{2}[\text{max temp} - \text{min temp}] = \dfrac{1}{2}[78.6 - 13.8] = 32.4$

Period: $p = 2[\text{month of max temp} - \text{month of min temp}] = 2[7 - 1] = 12$

$$b = \dfrac{2\pi}{p} = \dfrac{2\pi}{12} = \dfrac{\pi}{6}$$

Because the maximum temperature occurs in the seventh month,

$\dfrac{c}{b} = 7$ so $c \approx 3.67.$

The average temperature is $\dfrac{1}{2}(78.6 + 13.8) = 46.2°$, so $d = 46.2.$

So, $I(t) = 32.4 \cos\left(\dfrac{\pi}{6}t - 3.67\right) + 46.2.$

(b)

The model fits the data well.

(c)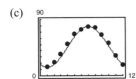

The model fits the data well.

(d) The d value in each model represents the average temperature.

Las Vegas: 80.6°; International Falls: 46.2°

(e) The period of each model is 12. This is what you would expect because the time period is one year (twelve months).

(f) International Falls has the greater temperature variability. The amplitude determines the variability. The greater the amplitude, the greater the temperature varies.

91. $y = 0.001 \sin 880\pi t$

(a) Period: $\dfrac{2\pi}{880\pi} = \dfrac{1}{440}$ seconds

(b) $f = \dfrac{1}{p} = 440$ cycles per second

93. $C = 30.3 + 21.6 \sin\left(\dfrac{2\pi t}{365} + 10.9\right)$

(a) Period $= \dfrac{2\pi}{\dfrac{2\pi}{365}} = 365$

Yes, this is what is expected because there are 365 days in a year.

(b) 30.3 gallons; the average daily fuel consumption is given by the amount of the vertical shift (from 0) which is given by the constant 30.3.

(c)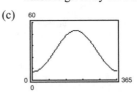

The consumption exceeds 40 gallons per day when $124 < t < 252.$

95. False. The graph of $\sin(x + 2\pi)$ is the graph of $\sin(x)$ translated to the *left* by one period, and the graphs are indeed identical.

97. True.

Because $\cos x = \sin\left(x + \dfrac{\pi}{2}\right)$, $y = -\cos x = -\sin\left(x + \dfrac{\pi}{2}\right)$, and so is a reflection in the x-axis of $y = \sin\left(x + \dfrac{\pi}{2}\right)$.

99.

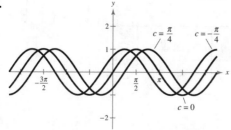

The value of c is the horizontal translation of the graph.

101.

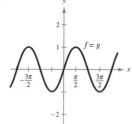

Because the graphs are the same, the conjecture is that

$$\sin(x) = \cos\left(x - \dfrac{\pi}{2}\right).$$

103. (a)

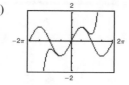

The graphs are nearly the same for $-\dfrac{\pi}{2} < x < \dfrac{\pi}{2}$.

(b)

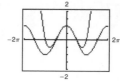

The graphs are nearly the same for $-\dfrac{\pi}{2} < x < \dfrac{\pi}{2}$.

(c) $\sin x \approx x - \dfrac{x^3}{3!} + \dfrac{x^5}{5!} - \dfrac{x^7}{7!}$

$\cos x \approx 1 - \dfrac{x^2}{2!} + \dfrac{x^4}{4!} - \dfrac{x^6}{6!}$

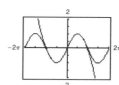 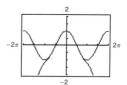

The graphs now agree over a wider range,

$$-\dfrac{3\pi}{4} < x < \dfrac{3\pi}{4}.$$

Section 6.5   Graphs of Other Trigonometric Functions

- You should be able to graph

  $y = a \tan(bx - c)$ $y = a \cot(bx - c)$

  $y = a \sec(bx - c)$ $y = a \csc(bx - c)$

- When graphing $y = a \sec(bx - c)$ or $y = a \csc(bx - c)$ you should first graph $y = a \cos(bx - c)$ or
  $y = a \sin(bx - c)$ because

  (a) The x-intercepts of sine and cosine are the vertical asymptotes of cosecant and secant.

  (b) The maximums of sine and cosine are the local minimums of cosecant and secant.

  (c) The minimums of sine and cosine are the local maximums of cosecant and secant.

- You should be able to graph using a damping factor.

1. odd; origin

3. reciprocal

5. π

7. $(-\infty, -1] \cup [1, \infty)$

9. $y = \sec 2x$

Period: $\dfrac{2\pi}{2} = \pi$

Matches graph (e).

11. $y = \dfrac{1}{2} \cot \pi x$

Period: $\dfrac{\pi}{\pi} = 1$

Matches graph (a).

13. $y = \dfrac{1}{2} \sec \dfrac{\pi x}{2}$

Period: $\dfrac{2\pi}{b} = \dfrac{2\pi}{\pi/2} = 4$

Asymptotes: $x = -1, x = 1$

Matches graph (f).

15. $y = \dfrac{1}{3} \tan x$

Period: π

Two consecutive asymptotes:

$x = -\dfrac{\pi}{2}$ and $x = \dfrac{\pi}{2}$

x	$-\dfrac{\pi}{4}$	0	$\dfrac{\pi}{4}$
y	$-\dfrac{1}{3}$	0	$\dfrac{1}{3}$

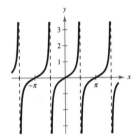

17. $y = -2 \tan 3x$

Period: $\dfrac{\pi}{3}$

Two consecutive asymptotes:

$x = -\dfrac{\pi}{6}, x = \dfrac{\pi}{6}$

x	$-\dfrac{\pi}{3}$	0	$\dfrac{\pi}{3}$
y	0	0	0

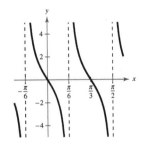

19. $y = -\dfrac{1}{2} \sec x$

Period: 2π

Two consecutive asymptotes:

$x = -\dfrac{\pi}{2}, x = \dfrac{\pi}{2}$

x	$-\dfrac{\pi}{3}$	0	$\dfrac{\pi}{3}$
y	-1	$-\dfrac{1}{2}$	-1

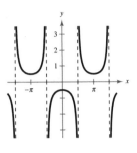

21. $y = \csc \pi x$

Period: $\dfrac{2\pi}{\pi} = 2$

Two consecutive asymptotes:

$x = 0, x = 1$

x	$\dfrac{1}{6}$	$\dfrac{1}{2}$	$\dfrac{5}{6}$
y	2	1	2

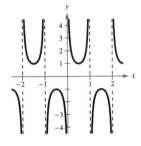

23. $y = \dfrac{1}{2} \sec \pi x$

Period: 2

Two consecutive asymptotes:

$x = -\dfrac{1}{2}, x = \dfrac{1}{2}$

x	-1	0	1
y	$-\dfrac{1}{2}$	$\dfrac{1}{2}$	$-\dfrac{1}{2}$

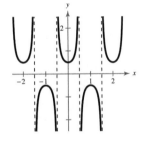

25. $y = \csc \dfrac{x}{2}$

Period: $\dfrac{2\pi}{1/2} = 4\pi$

Two consecutive asymptotes:

$x = 0, x = 2\pi$

x	$\dfrac{\pi}{3}$	π	$\dfrac{5\pi}{3}$
y	2	1	2

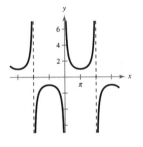

27. $y = 3 \cot 2x$

Period: $\dfrac{\pi}{2}$

Two consecutive asymptotes:

$x = -\dfrac{\pi}{2}, x = \dfrac{\pi}{2}$

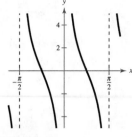

x	$-\dfrac{\pi}{6}$	$-\dfrac{\pi}{8}$	$\dfrac{\pi}{8}$	$\dfrac{\pi}{6}$
y	$-3\sqrt{3}$	-3	3	$3\sqrt{3}$

29. $y = 2 \sec 3x$

Period: $\dfrac{2\pi}{3}$

Two consecutive asymptotes:

$x = -\dfrac{\pi}{6}, x = \dfrac{\pi}{6}$

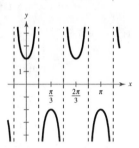

x	$-\dfrac{\pi}{3}$	0	$\dfrac{\pi}{3}$
y	-2	2	-2

31. $y = \tan \dfrac{\pi x}{4}$

Period: $\dfrac{\pi}{\pi/4} = 4$

Two consecutive asymptotes:

$\dfrac{\pi x}{4} = -\dfrac{\pi}{2} \Rightarrow x = -2$

$\dfrac{\pi x}{4} = \dfrac{\pi}{2} \Rightarrow x = 2$

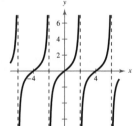

x	-1	0	1
y	-1	0	1

33. $y = 2 \csc(x - \pi)$

Period: 2π

Two consecutive asymptotes:

$x = -\pi, x = \pi$

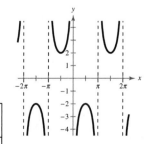

x	$-\dfrac{\pi}{2}$	$\dfrac{\pi}{2}$	$\dfrac{3\pi}{2}$
y	2	-2	-2

35. $y = 2 \sec(x + \pi)$

Period: 2π

Two consecutive asymptotes:

$x = -\dfrac{\pi}{2}, x = \dfrac{\pi}{2}$

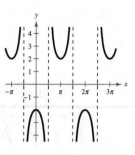

x	$-\dfrac{\pi}{3}$	0	$\dfrac{\pi}{3}$
y	-4	-2	-4

37. $y = \dfrac{1}{4} \csc\left(x + \dfrac{\pi}{4}\right)$

Period: 2π

Two consecutive asymptotes:

$x = -\dfrac{\pi}{4}, x = \dfrac{3\pi}{4}$

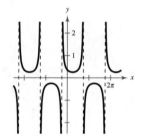

x	$-\dfrac{\pi}{12}$	$\dfrac{\pi}{4}$	$\dfrac{7\pi}{12}$
y	$\dfrac{1}{2}$	$\dfrac{1}{4}$	$\dfrac{1}{2}$

39. $y = \tan \dfrac{x}{3}$

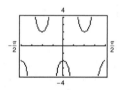

41. $y = -2 \sec 4x = \dfrac{-2}{\cos 4x}$

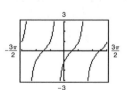

43. $y = \tan\left(x - \dfrac{\pi}{4}\right)$

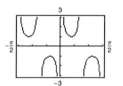

45. $y = -\csc(4x - \pi)$

$y = \dfrac{-1}{\sin(4x - \pi)}$

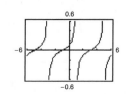

47. $y = 0.1 \tan\left(\dfrac{\pi x}{4} + \dfrac{\pi}{4}\right)$

49. $\tan x = 1$

$$x = -\frac{7\pi}{4}, -\frac{3\pi}{4}, \frac{\pi}{4}, \frac{5\pi}{4}$$

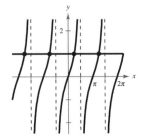

51. $\cot x = -\frac{\sqrt{3}}{3}$

$$x = -\frac{4\pi}{3}, -\frac{\pi}{3}, \frac{2\pi}{3}, \frac{5\pi}{3}$$

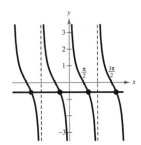

53. $\sec x = -2$

$$x = \frac{2\pi}{3}, \frac{4\pi}{3}, -\frac{2\pi}{3}, -\frac{4\pi}{3}$$

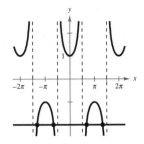

55. $\csc x = \sqrt{2}$

$$x = -\frac{7\pi}{4}, -\frac{5\pi}{4}, \frac{\pi}{4}, \frac{3\pi}{4}$$

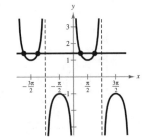

57. $f(x) = \sec x = \dfrac{1}{\cos x}$

$$f(-x) = \sec(-x)$$
$$= \frac{1}{\cos(-x)}$$
$$= \frac{1}{\cos x}$$
$$= f(x)$$

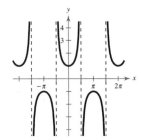

So, $f(x) = \sec x$ is an even function and the graph has y-axis symmetry.

59. $g(x) = \cot x = \dfrac{1}{\tan x}$

$$g(-x) = \cot(-x)$$
$$= \frac{1}{\tan(-x)}$$
$$= -\frac{1}{\tan x}$$
$$= -g(x)$$

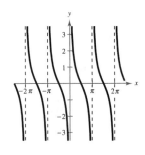

So, $g(x) = \cot x$ is an odd function and the graph has origin symmetry.

61. $f(x) = x + \tan x$

$$f(-x) = (-x) + \tan(-x)$$
$$= -x - \tan x$$
$$= -(x + \tan x)$$
$$= -f(x)$$

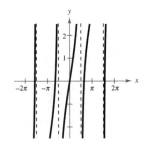

So, $f(x) = x + \tan x$ is an odd function and the graph has origin symmetry.

63. $g(x) = x \csc x = \dfrac{x}{\sin x}$

$$g(-x) = (-x) \csc(-x)$$
$$= \frac{-x}{\sin(-x)}$$
$$= \frac{-x}{-\sin x}$$
$$= \frac{x}{\sin x}$$
$$= x \csc x$$
$$= g(x)$$

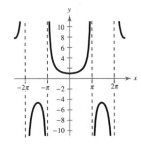

So, $g(x) = x \csc x$ is an even function and the graph has y-axis symmetry.

65. $f(x) = 2 \sin x$

$g(x) = \dfrac{1}{2} \csc x$

(a)

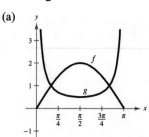

(b) $f > g$ on the interval, $\dfrac{\pi}{6} < x < \dfrac{5\pi}{6}$

(c) As $x \to \pi$, $f(x) = 2 \sin x \to 0$ and

$g(x) = \dfrac{1}{2} \csc x \to +\infty$ because $g(x)$ is the

reciprocal of $f(x)$.

67. $y_1 = \sin x \csc x$ and $y_2 = 1$

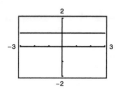

$\sin x \csc x = \sin x \left(\dfrac{1}{\sin x} \right) = 1$, $\sin x \neq 0$

The expressions are equivalent except when $\sin x = 0$ and y_1 is undefined.

69. $y_1 = \dfrac{\cos x}{\sin x}$ and $y_2 = \cot x = \dfrac{1}{\tan x}$

$\cot x = \dfrac{\cos x}{\sin x}$

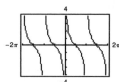

The expressions are equivalent.

71. $y_1 = 1 + \cot^2 x$ and $y_2 = \csc^2 x$

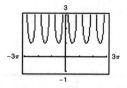

Sample answer:

$1 + \cot^2 x = 1 + \dfrac{\cos^2 x}{\sin^2 x}$

$= \dfrac{\sin^2 x + \cos^2 x}{\sin^2 x}$

$= \dfrac{1}{\sin^2 x}$

$= \csc^2 x$

The expressions are equivalent.

73. $f(x) = |x \cos x|$

Matches graph (d).

As $x \to 0$, $f(x) \to 0$.

75. $g(x) = |x| \sin x$

Matches graph (b).

As $x \to 0$, $g(x) \to 0$.

77. $f(x) = \sin x + \cos\left(x + \dfrac{\pi}{2}\right)$

$g(x) = 0$

$f(x) = g(x)$

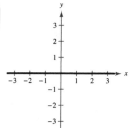

79. $f(x) = \sin^2 x$

$g(x) = \dfrac{1}{2}(1 - \cos 2x)$

$f(x) = g(x)$

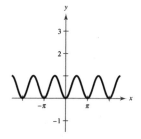

81. $g(x) = e^{-x^2/2} \sin x$

Damping factor: $e^{-x^2/2}$

As $x \to \infty$, $g(x) \to 0$.

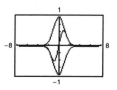

83. $f(x) = 2^{-x/4} \cos \pi x$

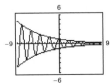

Damping factor: $y = 2^{-x/4}$

As $x \to \infty$, $f(x) \to 0$.

85. $y = \dfrac{6}{x} + \cos x$, $x > 0$

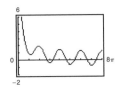

As $x \to 0$, $y \to \infty$.

87. $g(x) = \dfrac{\sin x}{x}$

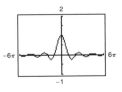

As $x \to 0$, $g(x) \to 1$.

89. $f(x) = \sin \dfrac{1}{x}$

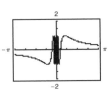

As $x \to 0$, $f(x)$ oscillates between -1 and 1.

91. $\tan x = \dfrac{7}{d}$

$d = \dfrac{7}{\tan x} = 7 \cot x$

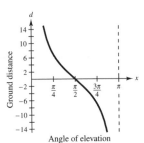

93. (a) Period of $\cos \dfrac{\pi t}{6} = \dfrac{2\pi}{\pi/6} = 12$

Period of $\sin \dfrac{\pi t}{6} = \dfrac{2\pi}{\pi/6} = 12$

The period of $H(t)$ is 12 months.

The period of $L(t)$ is 12 months.

(b) From the graph, it appears that the greatest difference between high and low temperatures occurs in the summer. The smallest difference occurs in the winter.

(c) The highest high and low temperatures appear to occur about half of a month after the time when the sun is northernmost in the sky.

95. (a) $y = \frac{1}{2} e^{-t/4} \cos 4t$

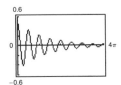

(b) The displacement is a damped sine wave.

$y \to 0$ as t increases.

97. True.

$y = \sec x = \dfrac{1}{\cos x}$

If the reciprocal of $y = \sin x$ is translated $\pi/2$ units to the left, then

$$y = \dfrac{1}{\sin\left(x + \dfrac{\pi}{2}\right)} = \dfrac{1}{\cos x} = \sec x.$$

99. $f(x) = \tan x$

(a) As $x \to \dfrac{\pi}{2}^{+}$, $f(x) \to -\infty$

(b) As $x \to \dfrac{\pi}{2}^{-}$, $f(x) \to \infty$

(c) As $x \to -\dfrac{\pi}{2}^{+}$, $f(x) \to -\infty$

(d) As $x \to -\dfrac{\pi}{2}^{-}$, $f(x) \to \infty$

101. $f(x) = \cot x$

(a) $x \to 0^{+}$, $f(x) \to \infty$

(b) $x \to 0^{-}$, $f(x) \to -\infty$

(c) $x \to \pi^{+}$, $f(x) \to \infty$

(d) $x \to \pi^{-}$, $f(x) \to -\infty$

103. $f(x) = x - \cos x$

(a)

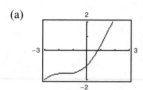

The zero between 0 and 1 occurs at $x \approx 0.7391$.

(b) $x_n = \cos(x_{n-1})$

$x_0 = 1$

$x_1 = \cos 1 \approx 0.5403$

$x_2 = \cos 0.5403 \approx 0.8576$

$x_3 = \cos 0.8576 \approx 0.6543$

$x_4 = \cos 0.6543 \approx 0.7935$

$x_5 = \cos 0.7935 \approx 0.7014$

$x_6 = \cos 0.7014 \approx 0.7640$

$x_7 = \cos 0.7640 \approx 0.7221$

$x_8 = \cos 0.7221 \approx 0.7504$

$x_9 = \cos 0.7504 \approx 0.7314$

$\vdots$

This sequence appears to be approaching the zero of
f: $x \approx 0.7391$.

105. $y_1 = \sec x$

$$y_2 = 1 + \frac{x^2}{2!} + \frac{5x^4}{4!}$$

The graphs appear to coincide on the
interval $-1.1 \le x \le 1.1$.

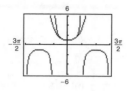

Section 6.6   Inverse Trigonometric Functions

- You should know the definitions, domains, and ranges of $y = \arcsin x$, $y = \arccos x$, and $y = \arctan x$.

Function	Domain	Range
$y = \arcsin x \Rightarrow x = \sin y$	$-1 \le x \le 1$	$-\dfrac{\pi}{2} \le y \le \dfrac{\pi}{2}$
$y = \arccos x \Rightarrow x = \cos y$	$-1 \le x \le 1$	$0 \le y \le \pi$
$y = \arctan x \Rightarrow x = \tan y$	$-\infty < x < \infty$	$-\dfrac{\pi}{2} < x < \dfrac{\pi}{2}$

- You should know the inverse properties of the inverse trigonometric functions.

$\sin(\arcsin x) = x$ and $\arcsin(\sin y) = y, -\dfrac{\pi}{2} \le y \le \dfrac{\pi}{2}$

$\cos(\arccos x) = x$ and $\arccos(\cos y) = y, 0 \le y \le \pi$

$\tan(\arctan x) = x$ and $\arctan(\tan y) = y, -\dfrac{\pi}{2} < y < \dfrac{\pi}{2}$

- You should be able to use the triangle technique to convert trigonometric functions of inverse trigonometric functions into algebraic expressions.

	Function	*Alternative Notation*	*Domain*	*Range*
1.	$y = \arcsin x$	$y = \sin^{-1} x$	$-1 \le x \le 1$	$-\dfrac{\pi}{2} \le y \le \dfrac{\pi}{2}$
3.	$y = \arctan x$	$y = \tan^{-1} x$	$-\infty < x < \infty$	$-\dfrac{\pi}{2} < y < \dfrac{\pi}{2}$

5. $y = \arcsin \dfrac{1}{2} \Rightarrow \sin y = \dfrac{1}{2}$ for $-\dfrac{\pi}{2} \le y \le \dfrac{\pi}{2} \Rightarrow y = \dfrac{\pi}{6}$

7. $y = \arccos \dfrac{1}{2} \Rightarrow \cos y = \dfrac{1}{2}$ for $0 \le y \le \pi \Rightarrow y = \dfrac{\pi}{3}$

9. $y = \arctan \dfrac{\sqrt{3}}{3} \Rightarrow \tan y = \dfrac{\sqrt{3}}{3}$ for $-\dfrac{\pi}{2} < y < \dfrac{\pi}{2} \Rightarrow y = \dfrac{\pi}{6}$

11. $y = \cos^{-1}\left(-\dfrac{\sqrt{3}}{2}\right) \Rightarrow \cos y = -\dfrac{\sqrt{3}}{2}$ for $0 \le y \le \pi \Rightarrow y = \dfrac{5\pi}{6}$

13. $y = \arctan\left(-\sqrt{3}\right) \Rightarrow \tan y = -\sqrt{3}$ for $-\dfrac{\pi}{2} < y < \dfrac{\pi}{2} \Rightarrow y = -\dfrac{\pi}{3}$

15. $y = \arccos\left(-\dfrac{1}{2}\right) \Rightarrow \cos y = -\dfrac{1}{2}$ for $0 \le y \le \pi \Rightarrow y = \dfrac{2\pi}{3}$

17. $y = \sin^{-1} -\dfrac{\sqrt{3}}{2} \Rightarrow \sin y = -\dfrac{\sqrt{3}}{2}$ for $-\dfrac{\pi}{2} \le y \le \dfrac{\pi}{2} \Rightarrow y = -\dfrac{\pi}{3}$

19. $y = \tan^{-1} 0 \Rightarrow \tan y = 0$ for $-\dfrac{\pi}{2} < y < \dfrac{\pi}{2} \Rightarrow y = 0$

21. $f(x) = \sin x$
$g(x) = \arcsin x$
$y = x$

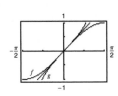

23. $\arccos 0.37 = \cos^{-1}(0.37) \approx 1.19$

25. $\arcsin(-0.75) = \sin^{-1}(-0.75) \approx -0.85$

27. $\arctan(-3) = \tan^{-1}(-3) \approx -1.25$

29. $\sin^{-1} 0.31 = \sin^{-1} 0.31 \approx 0.32$

31. $\arccos(-0.41) = \cos^{-1}(-0.41) \approx 1.99$

33. $\arctan 0.92 = \tan^{-1} 0.92 \approx 0.74$

35. $\arcsin \frac{7}{8} = \sin^{-1}\left(\frac{7}{8}\right) \approx 1.07$

37. $\tan^{-1}\left(\frac{19}{4}\right) \approx 1.36$

39. $\tan^{-1}\left(-\sqrt{372}\right) \approx -1.52$

41. $\arctan\left(-\sqrt{3}\right) = -\dfrac{\pi}{3}$
$\tan\left(-\dfrac{\pi}{6}\right) = -\dfrac{\sqrt{3}}{3}$
$\tan\left(\dfrac{\pi}{4}\right) = 1$

43. $\tan \theta = \dfrac{x}{4}$
$\theta = \arctan \dfrac{x}{4}$

45. $\sin \theta = \dfrac{x+2}{5}$
$\theta = \arcsin\left(\dfrac{x+2}{5}\right)$

47. $\cos \theta = \dfrac{x+3}{2x}$
$\theta = \arccos \dfrac{x+3}{2x}$

49. $\sin(\arcsin 0.3) = 0.3$

51. $\cos\left[\arccos(-0.1)\right] = -0.1$

53. $\arcsin(\sin 3\pi) = \arcsin(0) = 0$

   Note: 3π is not in the range of the arcsine function.

55. Let $y = \arctan \dfrac{3}{4}$.

$\tan y = \dfrac{3}{4}, 0 < y < \dfrac{\pi}{2}$,

$\sin\left(\arctan \dfrac{3}{4}\right) = \sin y = \dfrac{3}{5}$

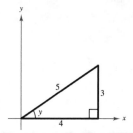

57. Let $y = \tan^{-1} 2$,

$\tan y = 2 = \dfrac{2}{1}, 0 < y < \dfrac{\pi}{2}$,

$\cos\left(\tan^{-1} 2\right) = \cos y = \dfrac{1}{\sqrt{5}} = \dfrac{\sqrt{5}}{5}$.

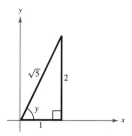

59. Let $y = \arcsin \dfrac{5}{13}$,

$\sin y = \dfrac{5}{13}, 0 < y < \dfrac{\pi}{2}$,

$\cos\left(\arcsin \dfrac{5}{13}\right) = \cos y = \dfrac{12}{13}$.

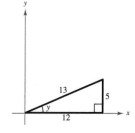

61. Let $y = \arctan\left(-\dfrac{3}{5}\right)$,

$\tan y = -\dfrac{3}{5}, -\dfrac{\pi}{2} < y < 0$,

$\sec\left[\arctan\left(-\dfrac{3}{5}\right)\right] = \sec y = \dfrac{\sqrt{34}}{5}$.

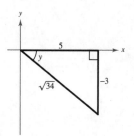

63. Let $y = \arccos\left(-\dfrac{2}{3}\right)$.

$\cos y = -\dfrac{2}{3}, \dfrac{\pi}{2} < y < \pi$,

$\sin\left[\arccos\left(-\dfrac{2}{3}\right)\right] = \sin y = \dfrac{\sqrt{5}}{3}$.

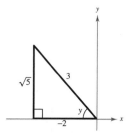

65. Let $u = \cos^{-1} \dfrac{\sqrt{3}}{2}$.

$\cos u = \dfrac{\sqrt{3}}{2}, 0 < u < \dfrac{\pi}{2}$,

$\csc\left[\cos^{-1} \dfrac{\sqrt{3}}{2}\right] = \csc u = 2$.

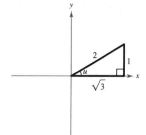

67. Let $y = \arctan x$.

$$\tan y = x = \frac{x}{1},$$

$$\cot(\arctan x) = \cot y = \frac{1}{x}$$

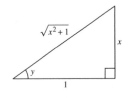

69. Let $y = \arcsin(2x)$.

$$\sin y = 2x = \frac{2x}{1},$$

$$\cos(\arcsin 2x) = \cos y = \sqrt{1 - 4x^2}$$

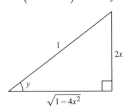

71. Let $y = \arccos x$.

$$\cos y = x = \frac{x}{1},$$

$$\sin(\arccos x) = \sin y = \sqrt{1 - x^2}$$

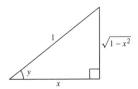

73. Let $y = \arccos\left(\dfrac{x}{3}\right)$.

$$\cos y = \frac{x}{3},$$

$$\tan\left(\arccos\frac{x}{3}\right) = \tan y = \frac{\sqrt{9 - x^2}}{x}$$

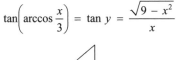
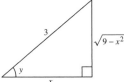

75. Let $y = \arctan\dfrac{x}{\sqrt{2}}$.

$$\tan y = \frac{x}{\sqrt{2}},$$

$$\csc\left(\arctan\frac{x}{\sqrt{2}}\right) = \csc y = \frac{\sqrt{x^2 + 2}}{x}$$

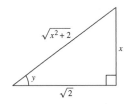

77. $f(x) = \sin(\arctan 2x)$, $g(x) = \dfrac{2x}{\sqrt{1 + 4x^2}}$

They are equal. Let $y = \arctan 2x$,

$$\tan y = 2x = \frac{2x}{1},$$

and $\sin y = \dfrac{2x}{\sqrt{1 + 4x^2}}$.

$$g(x) = \frac{2x}{\sqrt{1 + 4x^2}} = f(x)$$

The graph has horizontal asymptotes at $y = \pm 1$.

79. Let $y = \arctan\dfrac{9}{x}$.

$$\tan y = \frac{9}{x} \text{ and } \sin y = \frac{9}{\sqrt{x^2 + 81}}, x > 0; \frac{-9}{\sqrt{x^2 + 81}}, x < 0$$

So,

$$\arctan\frac{9}{x} = \arcsin\left(\frac{9}{\sqrt{x^2 + 81}}\right), x > 0;$$

$$\arctan\frac{9}{x} = \arcsin\left(\frac{-9}{\sqrt{x^2 + 81}}\right), x < 0.$$

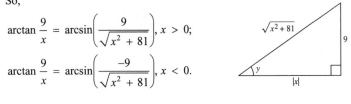

81. Let $y = \arccos \dfrac{3}{\sqrt{x^2 - 2x + 10}}$. Then,

$$\cos y = \dfrac{3}{\sqrt{x^2 - 2x + 10}} = \dfrac{3}{\sqrt{(x-1)^2 + 9}}$$

and $\sin y = \dfrac{|x-1|}{\sqrt{(x-1)^2 + 9}}$.

So, $y = \arcsin \dfrac{|x-1|}{\sqrt{x^2 - 2x + 10}}$.

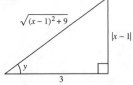

83. $g(x) = \arcsin(x - 1)$

Domain: $0 \le x \le 2$

Range: $-\dfrac{\pi}{2} \le y \le \dfrac{\pi}{2}$

This is the graph of $f(x) = \arcsin(x)$ shifted one unit to the right.

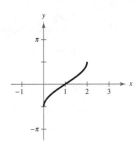

85. $y = 2 \arccos x$

Domain: $-1 \le x \le 1$

Range: $0 \le y \le 2\pi$

This is the graph of $f(x) = \arccos x$ with a factor of 2.

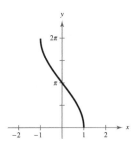

87. $f(x) = \arctan 2x$

Domain: all real numbers

Range: $-\dfrac{\pi}{2} < y < \dfrac{\pi}{2}$

This is the graph of $g(x) = \arctan(x)$ with a horizontal shrink of a factor of 2.

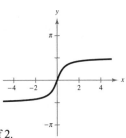

89. $h(v) = \tan(\arccos v) = \dfrac{\sqrt{1 - v^2}}{v}$

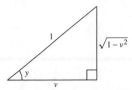

Domain: $-1 \le v \le 1, v \ne 0$

Range: all real numbers

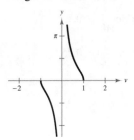

91. $f(x) = 2 \arccos(2x)$

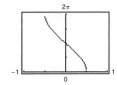

93. $f(x) = \arctan(2x - 3)$

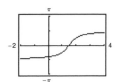

95. $f(x) = \pi - \sin^{-1}\left(\dfrac{2}{3}\right) \approx 2.412$

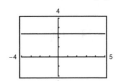

97. $f(t) = 3 \cos 2t + 3 \sin 2t = \sqrt{3^2 + 3^2} \, \sin\left(2t + \arctan \dfrac{3}{3}\right)$

$$= 3\sqrt{2} \, \sin(2t + \arctan 1)$$

$$= 3\sqrt{2} \, \sin\left(2t + \dfrac{\pi}{4}\right)$$

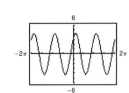

The graph implies that the identity is true.

99. $\dfrac{\pi}{2}$

101. $\dfrac{\pi}{2}$

103. π

105. (a) $\sin\theta = \dfrac{5}{s}$

$\theta = \arcsin\dfrac{5}{s}$

(b) $s = 40$: $\theta = \arcsin\dfrac{5}{40} \approx 0.13$

$s = 20$: $\theta = \arcsin\dfrac{5}{20} \approx 0.25$

107. $\beta = \arctan\dfrac{3x}{x^2+4}$

(a)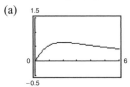

(b) β is maximum when $x = 2$ feet.

(c) The graph has a horizontal asymptote at $\beta = 0$.
As x increases, β decreases.

109.

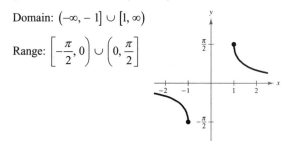

(a) $\tan\theta = \dfrac{20}{41}$

$\theta = \arctan\left(\dfrac{20}{41}\right) \approx 26.0°$

(b) $\tan 26° = \dfrac{h}{50}$

$h = 50\tan 26° \approx 24.4$ feet

111. (a) $\tan\theta = \dfrac{x}{20}$

$\theta = \arctan\dfrac{x}{20}$

(b) $x = 5$: $\theta = \arctan\dfrac{5}{20} \approx 14.0°$

$x = 12$: $\theta = \arctan\dfrac{12}{20} \approx 31.0°$

113. False.

$\dfrac{5\pi}{4}$ is not in the range of the arctangent function.

$\arctan 1 = \dfrac{\pi}{4}$

115. $y = \operatorname{arccot} x$ if and only if $\cot y = x$.

Domain: $(-\infty, \infty)$

Range: $(0, \pi)$

117. $y = \operatorname{arccsc} x$ if and only if $\csc y = x$.

Domain: $(-\infty, -1] \cup [1, \infty)$

Range: $\left[-\dfrac{\pi}{2}, 0\right) \cup \left(0, \dfrac{\pi}{2}\right]$

119. $y = \operatorname{arcsec}\sqrt{2} \Rightarrow \sec y = \sqrt{2}$ and

$0 \le y < \dfrac{\pi}{2} \cup \dfrac{\pi}{2} < y \le \pi \Rightarrow y = \dfrac{\pi}{4}$

121. $y = \operatorname{arccot}(-1) \Rightarrow \cot y = -1$ and

$0 < y < \pi \Rightarrow y = \dfrac{3\pi}{4}$

123. $y = \operatorname{arccsc} 2 \Rightarrow \csc y = 2$ and

$-\dfrac{\pi}{2} \le y < 0 \cup 0 < y \le \dfrac{\pi}{2} \Rightarrow y = \dfrac{\pi}{6}$

125. $y = \operatorname{arccsc}\left(\dfrac{2\sqrt{3}}{3}\right) \Rightarrow \csc y = \dfrac{2\sqrt{3}}{3}$ and

$-\dfrac{\pi}{2} \le y < 0 \cup 0 < y \le \dfrac{\pi}{2} \Rightarrow y = \dfrac{\pi}{3}$

127. $\operatorname{arcsec} 2.54 = \arccos\left(\dfrac{1}{2.54}\right) \approx 1.17$

129. $\operatorname{arccot} 5.25 = \arctan\left(\dfrac{1}{5.25}\right) \approx 0.19$

131. $\operatorname{arccot}\left(\tfrac{5}{3}\right) = \arctan\left(\tfrac{3}{5}\right) \approx 0.54$

133. $\operatorname{arccsc}\left(-\frac{25}{3}\right) = \arcsin\left(-\frac{3}{25}\right) \approx -0.12$

135. Area $= \arctan b - \arctan a$

(a) $a = 0, b = 1$

Area $= \arctan 1 - \arctan 0 = \dfrac{\pi}{4} - 0 = \dfrac{\pi}{4}$

(b) $a = -1, b = 1$

Area $= \arctan 1 - \arctan(-1)$

$= \dfrac{\pi}{4} - \left(-\dfrac{\pi}{4}\right) = \dfrac{\pi}{2}$

(c) $a = 0, b = 3$

Area $= \arctan 3 - \arctan 0$

$\approx 1.25 - 0 = 1.25$

(d) $a = -1, b = 3$

Area $= \arctan 3 - \arctan(-1)$

$\approx 1.25 - \left(-\dfrac{\pi}{4}\right) \approx 2.03$

137. $f(x) = \sin(x), \; f^{-1}(x) = \arcsin(x)$

(a) $f \circ f^{-1} = \sin(\arcsin x)$

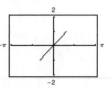

$f^{-1} \circ f = \arcsin(\sin x)$

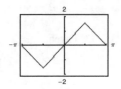

(b) The graphs coincide with the graph of $y = x$ only for certain values of x.

$f \circ f^{-1} = x$ over its entire domain, $-1 \le x \le 1$.

$f^{-1} \circ f = x$ over the region $-\dfrac{\pi}{2} \le x \le \dfrac{\pi}{2}$, corresponding to the region where $\sin x$ is one-to-one and has an inverse.

Section 6.7   Applications and Models

■ You should be able to solve right triangles.

■ You should be able to solve right triangle applications.

■ You should be able to solve applications of simple harmonic motion.

---

1. bearing

3. period

5. Given: $A = 30°, b = 3$

$\tan A = \dfrac{a}{b} \Rightarrow a = b \tan A = 3 \tan 30° \approx 1.73$

$\cos A = \dfrac{b}{c} \Rightarrow c = \dfrac{b}{\cos A} = \dfrac{3}{\cos 30°} \approx 3.46$

$B = 90° - 30° = 60°$

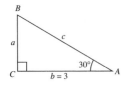

7. Given: $B = 71°, b = 24$

$\tan B = \dfrac{b}{a} \Rightarrow a = \dfrac{b}{\tan B} = \dfrac{24}{\tan 71°} \approx 8.26$

$\sin B = \dfrac{b}{c} \Rightarrow c = \dfrac{b}{\sin B} = \dfrac{24}{\sin 71°} \approx 25.38$

$A = 90° - 71° = 19°$

9. Given: $a = 3, b = 4$

$a^2 + b^2 = c^2 \Rightarrow c^2 = (3)^2 + (4)^2 \Rightarrow c = 5$

$\tan A = \dfrac{a}{b} \Rightarrow A = \tan^{-1}\left(\dfrac{a}{b}\right) = \tan^{-1}\left(\dfrac{3}{4}\right) \approx 36.87°$

$B = 90° - 36.87° = 53.13°$

11. Given: $b = 16, c = 52$

$$a = \sqrt{52^2 - 16^2}$$

$$= \sqrt{2448} = 12\sqrt{17} \approx 49.48$$

$$\cos A = \frac{16}{52}$$

$$A = \arccos \frac{16}{52} \approx 72.80°$$

$$B = 90° - 72.08° \approx 17.92°$$

13. Given: $A = 12°15', c = 430.5$

$$B = 90° - 12°15' = 77°45'$$

$$\sin 12°15' = \frac{a}{430.5}$$

$$a = 430.5 \sin 12°15' \approx 91.34$$

$$\cos 12°15' = \frac{b}{430.5}$$

$$b = 430.5 \cos 12°15' \approx 420.70$$

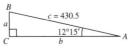

15. $\theta = 45°, b = 6$

$$\tan \theta = \frac{h}{(1/2)b} \Rightarrow h = \frac{1}{2}b \tan \theta$$

$$h = \frac{1}{2}(6) \tan 45° = 3.00 \text{ units}$$

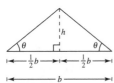

17. $\theta = 32°, b = 8$

$$\tan \theta = \frac{h}{(1/2)b} \Rightarrow h = \frac{1}{2}b \tan \theta$$

$$h = \frac{1}{2}(8) \tan 32° \approx 2.50 \text{ units}$$

19. $\tan 25° = \dfrac{100}{x}$

$$x = \frac{100}{\tan 25°}$$

$$\approx 214.45 \text{ feet}$$

21. $\sin 80° = \dfrac{h}{20}$

$$20 \sin 80° = h$$

$$h \approx 19.7 \text{ feet}$$

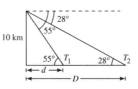

23. Let the height of the church $= x$ and the height of the church and steeple $= y$. Then,

$$\tan 35° = \frac{x}{50} \text{ and } \tan 47°40' = \frac{y}{50}$$

$$x = 50 \tan 35° \text{ and } y = 50 \tan 47°40'$$

$$h = y - x = 50(\tan 47°40' - \tan 35°).$$

$$h \approx 19.9 \text{ feet}$$

25. $\cot 55 = \dfrac{d}{10} \Rightarrow d \approx 7 \text{ kilometers}$

$$\cot 28° = \frac{D}{10} \Rightarrow D \approx 18.8 \text{ kilometers}$$

Distance between towns:

$$D - d = 18.8 - 7 = 11.8 \text{ kilometers}$$

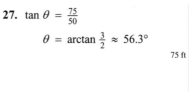

27. $\tan \theta = \dfrac{75}{50}$

$$\theta = \arctan \frac{3}{2} \approx 56.3°$$

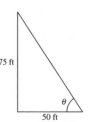

29. $1200 \text{ feet} + 150 \text{ feet} - 400 \text{ feet} = 950 \text{ feet}$

$$5 \text{ miles} = 5 \text{ miles}\left(\frac{5280 \text{ feet}}{1 \text{ mile}}\right) = 26,400 \text{ feet}$$

$$\tan \theta = \frac{950}{26,400}$$

$$\theta = \arctan\left(\frac{950}{26,400}\right) \approx 2.06°$$

Not drawn to scale

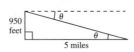

31. (a) $l^2 = (h + 17)^2 + 100^2$

$\quad\quad l = \sqrt{(h + 17)^2 + 10,000}$

$\quad\quad\quad = \sqrt{h^2 + 34h + 10,289}$

(b) $\cos \theta = \dfrac{100}{l}$

$\quad\quad \theta = \arccos\left(\dfrac{100}{l}\right)$

(c) $\cos \theta = \dfrac{100}{l}$

$\quad\quad \cos 35° = \dfrac{100}{l}$

$\quad\quad\quad l \approx 122.077$

$\quad\quad l^2 = 100^2 + (h + 17)^2$

$\quad\quad l^2 = h^2 + 34h + 10.289$

$\quad\quad 0 = h^2 + 34h - 4613.794$

$\quad\quad h \approx 53.02$ feet

33. (a) $l^2 = (200)^2 + (150)^2$

$\quad\quad l = 250$ feet

$\quad\quad \tan A = \dfrac{150}{200} \Rightarrow A = \arctan\left(\dfrac{150}{200}\right) \approx 36.87°$

$\quad\quad \tan B = \dfrac{200}{150} \Rightarrow B = \arctan\left(\dfrac{200}{150}\right) \approx 53.13°$

(b) $250 \text{ ft} \times \dfrac{\text{mile}}{5280 \text{ ft}} \times \dfrac{\text{hour}}{35 \text{ miles}} \times \dfrac{3600 \text{ sec}}{\text{hour}} \approx 4.87$ seconds

35. The plane has traveled $1.5(600) = 900$ miles.

$\quad \sin 38° = \dfrac{a}{900} \Rightarrow a \approx 554$ miles north

$\quad \cos 38° = \dfrac{b}{900} \Rightarrow b \approx 709$ miles east

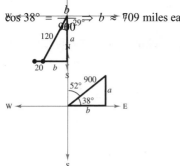

37.

(a) $\cos 29° = \dfrac{a}{120} \Rightarrow a \approx 104.95$ nautical miles south

$\quad \sin 29° = \dfrac{b}{120} \Rightarrow b \approx 58.18$ nautical miles west

(b) $\tan \theta = \dfrac{20 + b}{a} \approx \dfrac{78.18}{104.95} \Rightarrow \theta \approx 36.7°$

Bearing: S 36.7° W

Distance: $d \approx \sqrt{104.95^2 + 78.18^2}$

$\quad\quad\quad \approx 130.9$ nautical miles from port

39. $\tan \theta = \dfrac{45}{30} \Rightarrow \theta \approx 56.3°$

$\quad$ Bearing: N 56.31°

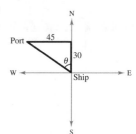

41. $\theta = 32°, \phi = 68°$

(a) $\alpha = 90° - 32° = 58°$

Bearing from A to C: N 58° E

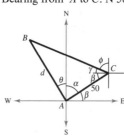

(b) $\beta = \theta = 32°$

$\quad \gamma = 90° - \phi = 22°$

$\quad C = \beta + \gamma = 54°$

$\quad \tan C = \dfrac{d}{50} \Rightarrow \tan 54°$

$\quad\quad = \dfrac{d}{50} \Rightarrow d \approx 68.82$ meters

43. L_1: $3x - 2y = 5 \Rightarrow y = \frac{3}{2}x - \frac{5}{2} \Rightarrow m_1 = \frac{3}{2}$

L_2: $x + y = 1 \Rightarrow y = -x + 1 \Rightarrow m_2 = -1$

$\tan \alpha = \left| \dfrac{-1 - (3/2)}{1 + (-1)(3/2)} \right| = \left| \dfrac{-5/2}{-1/2} \right| = 5$

$\alpha = \arctan 5 \approx 78.7°$

45. The diagonal of the base has a length of
$\sqrt{a^2 + a^2} = \sqrt{2}a$. Now, you have

$\tan \theta = \dfrac{a}{\sqrt{2}a} = \dfrac{1}{\sqrt{2}}$

$\theta = \arctan \dfrac{1}{\sqrt{2}}$

$\theta \approx 35.3°$.

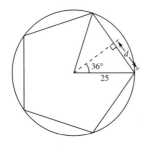

47. $\sin 36° = \dfrac{d}{25} \Rightarrow d \approx 14.69$

Length of side: $2d \approx 29.4$ inches

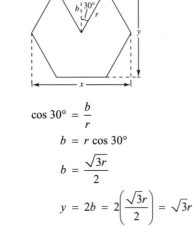

49.

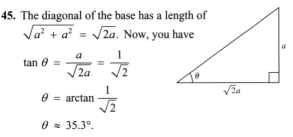

$\cos 30° = \dfrac{b}{r}$

$b = r \cos 30°$

$b = \dfrac{\sqrt{3}r}{2}$

$y = 2b = 2\left(\dfrac{\sqrt{3}r}{2} \right) = \sqrt{3}r$

51.

$\tan 35° = \dfrac{b}{10}$

$b = 10 \tan 35° \approx 7$

$\cos 35° = \dfrac{10}{a}$

$a = \dfrac{10}{\cos 35°} \approx 12.2$

53. Use $d = a \sin \omega t$ because $d = 0$ when $t = 0$.

Period: $\dfrac{2\pi}{\omega} = 2 \Rightarrow \omega = \pi$

So, $d = 4 \sin(\pi t)$.

55. Use $d = a \cos \omega t$ because $d = 3$ when $t = 0$.

Period: $\dfrac{2\pi}{\omega} = 1.5 \Rightarrow \omega = \dfrac{4\pi}{3}$

So, $d = 3 \cos\left(\dfrac{4\pi}{3}t \right) = 3 \cos\left(\dfrac{4\pi t}{3} \right)$.

57. $d = 9 \cos \dfrac{6\pi}{5}t$

(a) Maximum displacement $=$ amplitude $= 9$

(b) Frequency $= \dfrac{\omega}{2\pi} = \dfrac{\frac{6\pi}{5}}{2\pi}$

$= \dfrac{3}{5}$ cycle per unit of time

(c) $d = 9 \cos \dfrac{6\pi}{5}(5) = 9$

(d) $9 \cos \dfrac{6\pi}{5}t = 0$

$\cos \dfrac{6\pi}{5}t = 0$

$\dfrac{6\pi}{5}t = \arccos 0$

$\dfrac{6\pi}{5}t = \dfrac{\pi}{2}$

$t = \dfrac{5}{12}$

59. $d = \dfrac{1}{4}\sin 6\pi t$

(a) Maximum displacement = amplitude = $\dfrac{1}{4}$

(b) Frequency $= \dfrac{\omega}{2\pi} = \dfrac{6\pi}{2\pi}$

$\qquad\qquad = 3$ cycles per unit of time

(c) $d = \dfrac{1}{4}\sin 30\pi \approx 0$

(d) $\dfrac{1}{4}\sin 6\pi t = 0$

$\qquad \sin 6\pi t = 0$

$\qquad 6\pi t = \arcsin 0$

$\qquad 6\pi t = \pi$

$\qquad\quad t = \dfrac{1}{6}$

61. $\qquad d = a\sin \omega t$

Frequency $= \dfrac{\omega}{2\pi}$

$264 = \dfrac{\omega}{2\pi}$

$\omega = 2\pi(264) = 528\pi$

63. $y = \dfrac{1}{4}\cos 16t,\ t > 0$

(a)
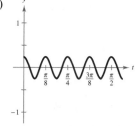

(b) Period: $\dfrac{2\pi}{16} = \dfrac{\pi}{8}$

(c) $\dfrac{1}{4}\cos 16t = 0$ when $16t = \dfrac{\pi}{2} \Rightarrow t = \dfrac{\pi}{32}$

65. (a)

θ	L_1	L_2	$L_1 + L_2$
0.1	$\dfrac{2}{\sin 0.1}$	$\dfrac{3}{\cos 0.1}$	23.0
0.2	$\dfrac{2}{\sin 0.2}$	$\dfrac{3}{\cos 0.2}$	13.1
0.3	$\dfrac{2}{\sin 0.3}$	$\dfrac{3}{\cos 0.3}$	9.9
0.4	$\dfrac{2}{\sin 0.4}$	$\dfrac{3}{\cos 0.4}$	8.4

(b)

θ	L_1	L_2	$L_1 + L_2$
0.5	$\dfrac{2}{\sin 0.5}$	$\dfrac{3}{\cos 0.5}$	7.6
0.6	$\dfrac{2}{\sin 0.6}$	$\dfrac{3}{\cos 0.6}$	7.2
0.7	$\dfrac{2}{\sin 0.7}$	$\dfrac{3}{\cos 0.7}$	7.0
0.8	$\dfrac{2}{\sin 0.8}$	$\dfrac{3}{\cos 0.8}$	7.1

The minimum length of the elevator is 7.0 meters.

(c) $L = L_1 + L_2 = \dfrac{2}{\sin \theta} + \dfrac{3}{\cos \theta}$

(d)
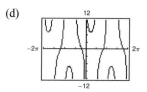

From the graph, it appears that the minimum length is 7.0 meters, which agrees with the estimate of part (b).

67. (a)
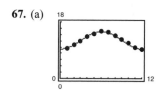

(b) Period $= \dfrac{2\pi}{n} = \dfrac{2\pi}{\pi/6} = 12$

The period is what you expect as the model examines the number of hours of daylight over one year (12 months).

(c) Amplitude $= |2.77| = 2.77$

The amplitude represents the maximum displacement from the average number of hours of daylight.

69. False. The tower isn't vertical and so the triangle formed is not a right triangle.

Review Exercises for Chapter 6

1.

The angle shown is 60°.

3. (a) $\theta = 85°$

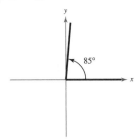

(b) The angle lies in Quadrant I.

(c) Coterminal angles: $85° + 360° = 445°$

$85° - 360° = -275°$

5. $\theta = -110°$

(a)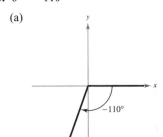

(b) Quadrant III

(c) Coterminal angles:

$-110° + 360° = 250°$

$-110° - 360° = -470°$

7. $\theta = \dfrac{15\pi}{4}$

(a)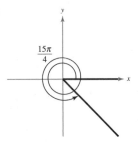

(b) Quadrant IV

(c) $\dfrac{15\pi}{4} - 2\pi = \dfrac{7\pi}{4}$

$\dfrac{7\pi}{4} - 2\pi = -\dfrac{\pi}{4}$

9. $\theta = -\dfrac{4\pi}{3}$

(a)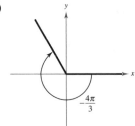

(b) Quadrant II

(c) Coterminal angles:

$-\dfrac{4\pi}{3} + 2\pi = \dfrac{2\pi}{3}$

$-\dfrac{4\pi}{3} - 2\pi = -\dfrac{10\pi}{3}$

11. $450° = 450° \cdot \dfrac{\pi \text{ rad}}{180°} = \dfrac{5\pi}{2} \approx 7.854 \text{ rad}$

13. $-16.5° = -16.5° \cdot \dfrac{\pi \text{ rad}}{180°} \approx -0.288 \text{ rad}$

15. $-33°45' = -33.75° = -33.75° \cdot \dfrac{\pi \text{ rad}}{180°}$

$= -\dfrac{3\pi}{16} \text{ rad} \approx -0.589 \text{ rad}$

17. $-84°15' = 84.25° = 84.25° \cdot \dfrac{\pi \text{ rad}}{180°} \approx 1.470 \text{ rad}$

19. $\dfrac{3\pi}{10} = \dfrac{3\pi}{10} \cdot \dfrac{180°}{\pi \text{ rad}} = 54.000°$

21. $-\dfrac{3\pi}{5} \text{ rad} = -\dfrac{3\pi \text{ rad}}{5} \cdot \dfrac{180°}{\pi \text{ rad}} = -108°$

23. $-3.5 \text{ rad} = -3.5 \text{ rad} \cdot \dfrac{180°}{\pi \text{ rad}} \approx -200.535°$

25. $4.75 \text{ rad} = \dfrac{4.75}{1} \text{ rad} \cdot \dfrac{180°}{\pi \text{ rad}} \approx 272.155°$

27. $138° = \dfrac{138\pi}{180} = \dfrac{23\pi}{30} \text{ radians}$

$s = r\theta = 20\left(\dfrac{23\pi}{30}\right) \approx 48.17 \text{ inches}$

29. $120° = \dfrac{120\pi}{180} = \dfrac{2\pi}{3} \text{ radians}$

$A = \dfrac{1}{2}r^2\theta = \dfrac{1}{2}(18)^2\left(\dfrac{2\pi}{3}\right) \approx 339.29 \text{ square inches}$

31. opp $= 4$, adj $= 5$, hyp $= \sqrt{4^2 + 5^2} = \sqrt{41}$

$$\sin\theta = \frac{\text{opp}}{\text{hyp}} = \frac{4}{\sqrt{41}} = \frac{4\sqrt{41}}{41} \qquad \csc\theta = \frac{\text{hyp}}{\text{opp}} = \frac{\sqrt{41}}{4}$$

$$\cos\theta = \frac{\text{adj}}{\text{hyp}} = \frac{5}{\sqrt{41}} = \frac{5\sqrt{41}}{41} \qquad \sec\theta = \frac{\text{hyp}}{\text{adj}} = \frac{\sqrt{41}}{5}$$

$$\tan\theta = \frac{\text{opp}}{\text{adj}} = \frac{4}{5} \qquad\qquad \cot\theta = \frac{\text{adj}}{\text{opp}} = \frac{5}{4}$$

33. $\sin\theta = \dfrac{1}{3}$

(a) $\csc\theta = \dfrac{1}{\sin\theta} = 3$

(b) $\sin^2\theta + \cos^2\theta = 1$

$$\left(\frac{1}{3}\right)^2 + \cos^2\theta = 1$$

$$\cos^2\theta = 1 - \frac{1}{9}$$

$$\cos^2\theta = \frac{8}{9}$$

$$\cos\theta = \sqrt{\frac{8}{9}}$$

$$\cos\theta = \frac{2\sqrt{2}}{3}$$

(c) $\sec\theta = \dfrac{1}{\cos\theta} = \dfrac{3}{2\sqrt{2}} = \dfrac{3\sqrt{2}}{4}$

(d) $\tan\theta = \dfrac{\sin\theta}{\cos\theta} = \dfrac{1/3}{\left(2\sqrt{2}\right)/3} = \dfrac{1}{2\sqrt{2}} = \dfrac{\sqrt{2}}{4}$

35. $\csc\theta = 4$

(a) $\sin\theta = \dfrac{1}{\csc\theta} = \dfrac{1}{4}$

(b) $\sin^2\theta + \cos^2\theta = 1$

$$\left(\frac{1}{4}\right)^2 + \cos^2\theta = 1$$

$$\cos^2\theta = 1 - \frac{1}{16}$$

$$\cos^2\theta = \frac{15}{16}$$

$$\cos\theta = \sqrt{\frac{15}{16}}$$

$$\cos\theta = \frac{\sqrt{15}}{4}$$

(c) $\sec\theta = \dfrac{1}{\cos\theta} = \dfrac{4}{\sqrt{15}} = \dfrac{4\sqrt{15}}{15}$

(d) $\tan\theta = \dfrac{\sin\theta}{\cos\theta} = \dfrac{1/4}{\sqrt{15}/4} = \dfrac{1}{\sqrt{15}} = \dfrac{\sqrt{15}}{15}$

37. $\tan 41° \approx 0.8693$

39. $\cos 38.9° \approx 0.7782$

41. $\cot 25°13' \approx \cot 25.2167° = \dfrac{1}{\tan 25.2167°} \approx 2.1235$

43. $\cos \dfrac{\pi}{18} \approx 0.9848$

45. $\sin 1°10' = \dfrac{x}{3.5}$

$x = 3.5 \sin 1°10' \approx 0.07$ kilometer or 71.3 meters

Not drawn to scale

47. $x = 12$, $y = 16$, $r = \sqrt{144 + 256} = \sqrt{400} = 20$

$$\sin\theta = \frac{y}{r} = \frac{4}{5} \qquad \csc\theta = \frac{r}{y} = \frac{5}{4}$$

$$\cos\theta = \frac{x}{r} = \frac{3}{5} \qquad \sec\theta = \frac{r}{x} = \frac{5}{3}$$

$$\tan\theta = \frac{y}{x} = \frac{4}{3} \qquad \cot\theta = \frac{x}{y} = \frac{3}{4}$$

49. $x = \dfrac{2}{3}, \; y = \dfrac{5}{2}$

$$r = \sqrt{\left(\dfrac{2}{3}\right)^2 + \left(\dfrac{5}{2}\right)^2} = \dfrac{\sqrt{241}}{6}$$

$$\sin \theta = \dfrac{y}{r} = \dfrac{5/2}{\sqrt{241}/6} = \dfrac{15}{\sqrt{241}} = \dfrac{15\sqrt{241}}{241} \qquad \csc \theta = \dfrac{r}{y} = \dfrac{\sqrt{241}/6}{5/2} = \dfrac{2\sqrt{241}}{30} = \dfrac{\sqrt{241}}{15}$$

$$\cos \theta = \dfrac{x}{r} = \dfrac{2/3}{\sqrt{241}/6} = \dfrac{4}{\sqrt{241}} = \dfrac{4\sqrt{241}}{241} \qquad \sec \theta = \dfrac{r}{x} = \dfrac{\sqrt{241}/6}{2/3} = \dfrac{\sqrt{241}}{4}$$

$$\tan \theta = \dfrac{y}{x} = \dfrac{5/2}{2/3} = \dfrac{15}{4} \qquad\qquad\qquad\qquad \cot \theta = \dfrac{x}{y} = \dfrac{2/3}{5/2} = \dfrac{4}{15}$$

51. $x = -0.5, \; y = 4.5$

$$r = \sqrt{\left(-0.5\right)^2 + \left(4.5\right)^2} = \sqrt{20.5} = \dfrac{\sqrt{82}}{2}$$

$$\sin \theta = \dfrac{y}{r} = \dfrac{4.5}{\sqrt{82}/2} = \dfrac{9\sqrt{82}}{82} \qquad \csc \theta = \dfrac{r}{y} = \dfrac{\sqrt{82}/2}{4.5} = \dfrac{\sqrt{82}}{9}$$

$$\cos \theta = \dfrac{x}{r} = \dfrac{-0.5}{\sqrt{82}/2} = \dfrac{-\sqrt{82}}{82} \qquad \sec \theta = \dfrac{r}{x} = \dfrac{\sqrt{82}/2}{-0.5} = -\sqrt{82}$$

$$\tan \theta = \dfrac{y}{x} = \dfrac{4.5}{-0.5} = -9 \qquad\qquad \cot \theta = \dfrac{x}{y} = \dfrac{-0.5}{4.5} = -\dfrac{1}{9}$$

53. $\left(x, 4x\right), \; x > 0$

$$x' = x, \; y' = 4x$$

$$r = \sqrt{x^2 + \left(4x\right)^2} = \sqrt{17}x$$

$$\sin \theta = \dfrac{y'}{r} = \dfrac{4x}{\sqrt{17}x} = \dfrac{4\sqrt{17}}{17} \qquad \csc \theta = \dfrac{r}{y'} = \dfrac{\sqrt{17}x}{4x} = \dfrac{\sqrt{17}}{4}$$

$$\cos \theta = \dfrac{x'}{r} = \dfrac{x}{\sqrt{17}x} = \dfrac{\sqrt{17}}{17} \qquad \sec \theta = \dfrac{r}{x'} = \dfrac{\sqrt{17}x}{x} = \sqrt{17}$$

$$\tan \theta = \dfrac{y'}{x'} = \dfrac{4x}{x} = 4 \qquad\qquad \cot \theta = \dfrac{x'}{y'} = \dfrac{x}{4x} = \dfrac{1}{4}$$

55. $\sec \theta = \dfrac{6}{5}, \; \tan \theta < 0 \Rightarrow \theta$ is in Quadrant IV.

$$r = 6, \; x = 5, \; y = -\sqrt{36 - 25} = -\sqrt{11}$$

$$\sin \theta = \dfrac{y}{r} = -\dfrac{\sqrt{11}}{6}$$

$$\cos \theta = \dfrac{x}{r} = \dfrac{5}{6}$$

$$\tan \theta = \dfrac{y}{x} = -\dfrac{\sqrt{11}}{5}$$

$$\csc \theta = \dfrac{r}{y} = -\dfrac{6\sqrt{11}}{11}$$

$$\sec \theta = \dfrac{6}{5}$$

$$\cot \theta = -\dfrac{5\sqrt{11}}{11}$$

57. $\tan \theta = \dfrac{7}{3}, \; \cos \theta < 0 \Rightarrow \theta$ is in Quadrant III.

$$y = -7, \; x = -3, \; r = \sqrt{58}$$

$$\sin \theta = \dfrac{y}{r} = -\dfrac{7\sqrt{58}}{58}$$

$$\cos \theta = \dfrac{x}{r} = -\dfrac{3\sqrt{58}}{58}$$

$$\tan \theta = \dfrac{y}{x} = \dfrac{7}{3}$$

$$\csc \theta = \dfrac{r}{y} = -\dfrac{\sqrt{58}}{7}$$

$$\sec \theta = \dfrac{r}{x} = -\dfrac{\sqrt{58}}{3}$$

$$\cot \theta = \dfrac{x}{y} = \dfrac{3}{7}$$

59. $\tan\theta = \dfrac{y}{x} = -\dfrac{40}{9} \Rightarrow r = 41$

$\sin\theta > 0 \Rightarrow \theta$ is in Quandrant II $\Rightarrow x = -9, y = 40$

$\sin\theta = \dfrac{y}{r} = \dfrac{40}{41}$

$\cos\theta = \dfrac{x}{r} = -\dfrac{9}{41}$

$\tan\theta = \dfrac{y}{x} = -\dfrac{40}{9}$

$\csc\theta = \dfrac{r}{y} = \dfrac{41}{40}$

$\sec\theta = \dfrac{r}{x} = -\dfrac{41}{9}$

$\cot\theta = \dfrac{x}{y} = -\dfrac{9}{40}$

61. $\theta = 264°$

$\theta' = 264° - 180° = 84°$

63. $\theta = -\dfrac{6\pi}{5}$

$-\dfrac{6\pi}{5} + 2\pi = \dfrac{4\pi}{5}$

$\theta' = \pi - \dfrac{4\pi}{5} = \dfrac{\pi}{5}$

65. $\sin\dfrac{\pi}{3} = \dfrac{\sqrt{3}}{2}$

$\cos\dfrac{\pi}{3} = \dfrac{1}{2}$

$\tan\dfrac{\pi}{3} = \sqrt{3}$

67. $\sin\dfrac{5\pi}{6} = \sin\left(\pi - \dfrac{5\pi}{6}\right) = \sin\dfrac{\pi}{6} = \dfrac{1}{2}$

$\cos\dfrac{5\pi}{6} = -\cos\left(\pi - \dfrac{5\pi}{6}\right) = -\cos\dfrac{\pi}{6} = -\dfrac{\sqrt{3}}{2}$

$\tan\dfrac{5\pi}{6} = -\tan\left(\pi - \dfrac{5\pi}{6}\right) = -\tan\dfrac{\pi}{6} = -\dfrac{\sqrt{3}}{3}$

69. $\sin\left(-\dfrac{7\pi}{3}\right) = -\sin\dfrac{\pi}{3} = -\dfrac{\sqrt{3}}{2}$

$\cos\left(-\dfrac{7\pi}{3}\right) = \cos\dfrac{\pi}{3} = \dfrac{1}{2}$

$\tan\left(-\dfrac{7\pi}{3}\right) = -\tan\dfrac{\pi}{3} = -\sqrt{3}$

71. $\sin 495° = \sin 45° = \dfrac{\sqrt{2}}{2}$

$\cos 495° = -\cos 45° = -\dfrac{\sqrt{2}}{2}$

$\tan 495° = -\tan 45° = -1$

73. $\sin(-150°) = -\dfrac{1}{2}$

$\cos(-150°) = -\dfrac{\sqrt{3}}{2}$

$\tan(-150°) = \dfrac{-1/2}{-\sqrt{3}/2} = \dfrac{\sqrt{3}}{3}$

75. $\sin 10 \approx -0.5440$

77. $\sec 2.8 = \dfrac{1}{\cos 2.8} \approx -1.0613$

79. $\sin\left(-\dfrac{17\pi}{15}\right) \approx 0.4067$

81. $t = \dfrac{2\pi}{3}, (x, y) = \left(-\dfrac{1}{2}, \dfrac{\sqrt{3}}{2}\right)$

$\sin\dfrac{2\pi}{3} = y = \dfrac{\sqrt{3}}{2}$

$\cos\dfrac{2\pi}{3} = x = -\dfrac{1}{2}$

$\tan\dfrac{2\pi}{3} = \dfrac{y}{x} = \dfrac{\sqrt{3}/2}{-1/2} = -\sqrt{3}$

83. $t = \dfrac{7\pi}{6}, (x, y) = \left(-\dfrac{\sqrt{3}}{2}, -\dfrac{1}{2}\right)$

$\sin\left(\dfrac{7\pi}{6}\right) = y = -\dfrac{1}{2}$

$\cos\left(\dfrac{7\pi}{6}\right) = x = -\dfrac{\sqrt{3}}{2}$

$\tan\left(\dfrac{7\pi}{6}\right) = \dfrac{x}{y} = \dfrac{-1/2}{-\sqrt{3}/2} = \dfrac{1}{\sqrt{3}} = \dfrac{\sqrt{3}}{3}$

85. $y = \sin 6x$

Amplitude: 1

Period: $\dfrac{2\pi}{6} = \dfrac{\pi}{3}$

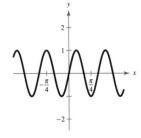

87. $y = 3\cos 2\pi x$

Amplitude: 3

Period: $\dfrac{2\pi}{2\pi} = 1$

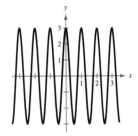

89. $f(x) = 5\sin\dfrac{2x}{5}$

Amplitude: 5

Period: $\dfrac{2\pi}{2/5} = 5\pi$

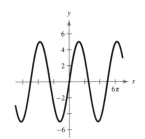

91. $y = 5 + \sin x$

Amplitude: 1

Period: 2π

Shift the graph of $y = \sin x$ 5 units upward

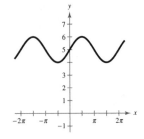

93. $g(t) = \dfrac{5}{2}\sin(t - \pi)$

Amplitude: $\dfrac{5}{2}$

Period: 2π

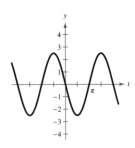

95. $y = a\sin bx$

(a) $a = 2$,

$\dfrac{2\pi}{b} = \dfrac{1}{264} \Rightarrow b = 528\pi$

$y = 2\sin 528\pi x$

(b) $f = \dfrac{1}{1/264} = 264$ cycles per second

97. $f(x) = 3\tan 2x$

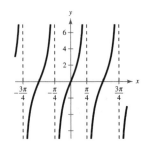

99. $f(x) = \dfrac{1}{2}\cot x$

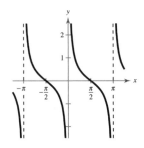

101. $f(x) = 3\sec x$

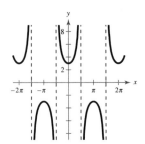

103. $f(x) = \dfrac{1}{2}\csc\dfrac{x}{2}$

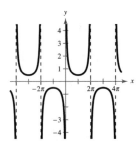

105. $f(x) = x\cos x$

Damping factor: x

As $x \to +\infty$, $f(x)$ oscillates.

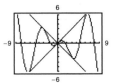

107. $\arcsin\left(-\dfrac{1}{2}\right) = -\arcsin\dfrac{1}{2} = -\dfrac{\pi}{6}$

109. $\arcsin 0.4 \approx 0.41$ radian

111. $\sin^{-1}(-0.44) \approx -0.46$ radian

113. $\arccos\left(-\dfrac{\sqrt{2}}{2}\right) = \dfrac{3\pi}{4}$

115. $\cos^{-1}(-1) = \pi$

117. $\arccos 0.425 \approx 1.13$ radians

119. $\tan^{-1}(-1.5) \approx -0.98$ radian

121. $f(x) = 2 \arcsin \dfrac{x}{2} = 2 \sin^{-1}\left(\dfrac{x}{2}\right)$

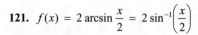
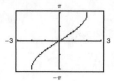

123. $f(x) = \arctan\left(\dfrac{x}{2}\right) = \tan^{-1}\left(\dfrac{x}{2}\right)$

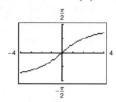

125. Let $u = \arctan \dfrac{3}{4}$ then $\tan u = \dfrac{3}{4}$.

$\cos\left(\arctan \dfrac{3}{4}\right) = \dfrac{4}{5}$

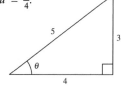

127. Let $u = \arctan \dfrac{12}{5}$

then $\tan u = \dfrac{12}{5}$.

$\sec\left(\arctan \dfrac{12}{5}\right) = \dfrac{13}{5}$

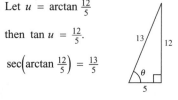

129. Let $y = \arccos\left(\dfrac{x}{2}\right)$. Then

$\cos y = \dfrac{x}{2}$ and $\tan y = \tan\left(\arccos\left(\dfrac{x}{2}\right)\right) = \dfrac{\sqrt{4 - x^2}}{x}$.

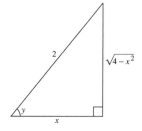

131. $\tan \theta = \dfrac{70}{30}$

$\theta = \arctan\left(\dfrac{70}{30}\right) \approx 66.8°$

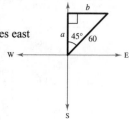

133. $\cos 45° = \dfrac{a}{60} \Rightarrow$

$a \approx 42.43$ nautical miles north

$\sin 45° = \dfrac{b}{60} \Rightarrow$

$b \approx 42.43$ nautical miles east

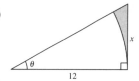

135. False. For each θ there corresponds exactly one value of y.

137. $f(\theta) = \sec \theta$ is undefined at the zeros of

$g(\theta) = \cos \theta$ because $\sec \theta = \dfrac{1}{\cos \theta}$.

139. The ranges for the other four trigonometric functions are not bounded. For $y = \tan x$ and $y = \cot x$, the range is $(-\infty, \infty)$. For $y = \sec x$ and $y = \csc x$, the range is $(-\infty, -1] \cup [1, \infty)$.

141. (a)

$\tan \theta = \dfrac{x}{12}$

$x = 12 \tan \theta$

Area = Area of triangle − Area of sector

$= \left(\dfrac{1}{2}bh\right) - \left(\dfrac{1}{2}r^2\theta\right)$

$= \dfrac{1}{2}(12)(12 \tan \theta) - \dfrac{1}{2}(12^2)(\theta)$

$= 72 \tan \theta - 72\theta$

$= 72(\tan \theta - \theta)$

(b)

As θ approaches $\dfrac{\pi}{2}$, the area increases without bound.

Problem Solving for Chapter 6

1. (a) $8:57 - 6:45 = 2$ hours 12 minutes $= 132$ minutes

$$\frac{132}{48} = \frac{11}{4} \text{ revolutions}$$

$$\theta = \left(\frac{11}{4}\right)(2\pi) = \frac{11\pi}{2} \text{ radians or } 990°$$

(b) $s = r\theta = 47.25(5.5\pi) \approx 816.42$ feet

3. (a) $\sin 39° = \dfrac{3000}{d}$

$$d = \frac{3000}{\sin 39°} \approx 4767 \text{ feet}$$

(b) $\tan 39° = \dfrac{3000}{x}$

$$x = \frac{3000}{\tan 39°} \approx 3705 \text{ feet}$$

(c) $\tan 63° = \dfrac{w + 3705}{3000}$

$$3000 \tan 63° = w + 3705$$

$$w = 3000 \tan 63° - 3705 \approx 2183 \text{ feet}$$

5. (a) $h(x) = \cos^2 x$

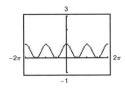

h is even.

(b) $h(x) = \sin^2 x$

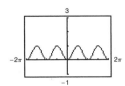

h is even.

7. If you alter the model so that $h = 1$ when $t = 0$, you can use either a sine or a cosine model.

$$a = \frac{1}{2}[\text{max} - \text{min}] = \frac{1}{2}[101 - 1] = 50$$

$$d = \frac{1}{2}[\text{max} + \text{min}] = \frac{1}{2}[101 + 1] = 51$$

$$b = 8\pi$$

Cosine model: $h = 51 - 50 \cos(8\pi t)$

Sine model: $h = 51 - 50 \sin\left(8\pi t + \dfrac{\pi}{2}\right)$

Notice that you needed the horizontal shift so that the sine value was one when $t = 0$.

Another model would be: $h = 51 + 50 \sin\left(8\pi t + \dfrac{3\pi}{2}\right)$

Here you wanted the sine value to be 1 when $t = 0$.

9. Physical (23 days): $P = \sin \dfrac{2\pi t}{23}, t \geq 0$

Emotional (28 days): $E = \sin \dfrac{2\pi t}{28}, t \geq 0$

Intellectual (33 days): $I = \sin \dfrac{2\pi t}{33}, t \geq 0$

(a)

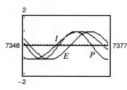

(b) Number of days since birth until September 1, 2008:

$$t = \underbrace{365 \times 20}_{\substack{20 \text{ years}}} + \underbrace{5}_{\substack{\text{leap year}}} + \underbrace{11}_{\substack{\text{remaining} \\ \text{July days}}} + \underbrace{31}_{\substack{\text{August days}}} + \underbrace{1}_{\substack{\text{days in} \\ \text{September}}}$$

$t = 7348$

All three drop early in the month, then peak toward the middle of the month, and drop again toward the latter part of the month.

(c) For September 22, 2008, use $t = 7369$.

$P \approx 0.631$

$E \approx 0.901$

$I \approx 0.945$

11. (a) Both graphs have a period of 2 and intersect when
$x = 5.35$. They should also intersect when
$x = 5.35 - 2 = 3.35$ and $x = 5.35 + 2 = 7.35$.

(b) The graphs intersect when $x = 5.35 - 3(2) = -0.65$.

(c) Because $13.35 = 5.35 + 4(2)$ and $-4.65 = 5.35 - 5(2)$ the graphs will intersect again at these values. So,
$f(13.35) = g(-4.65)$.

13.

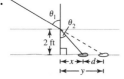

(a) $\dfrac{\sin \theta_1}{\sin \theta_2} = 1.333$

$\sin \theta_2 = \dfrac{\sin \theta_1}{1.333} = \dfrac{\sin 60°}{1.333} \approx 0.6497$

$\theta_2 = 40.5°$

(b) $\tan \theta_2 = \dfrac{x}{2} \Rightarrow x = 2 \tan 40.52° \approx 1.71 \text{ feet}$

$\tan \theta_1 = \dfrac{y}{2} \Rightarrow y = 2 \tan 60° \approx 3.46 \text{ feet}$

(c) $d = y - x = 3.46 - 1.71 = 1.75 \text{ feet}$

(d) As you move closer to the rock, θ_1 decreases, which causes y to decrease, which in turn causes d to decrease.

Chapter 6 Practice Test

1. Express 350° in radian measure.

2. Express $(5\pi)/9$ in degree measure.

3. Convert $135°\,14'\,12''$ to decimal form.

4. Convert $-22.569°$ to $D°\,M'\,S''$ form.

5. If $\cos\theta = \frac{2}{3}$, use the trigonometric identities to find $\tan\theta$.

6. Find θ given $\sin\theta = 0.9063$.

7. Solve for x in the figure below.

8. Find the magnitude of the reference angle for $\theta = (6\pi)/5$.

9. Evaluate $\csc 3.92$.

10. Find $\sec\theta$ given that θ lies in Quadrant III and $\tan\theta = 6$.

11. Graph $y = 3\sin\dfrac{x}{2}$.

12. Graph $y = -2\cos(x - \pi)$.

13. Graph $y = \tan 2x$.

14. Graph $y = -\csc\left(x + \dfrac{\pi}{4}\right)$.

15. Graph $y = 2x + \sin x$, using a graphing calculator.

16. Graph $y = 3x\cos x$, using a graphing calculator.

17. Evaluate $\arcsin 1$.

18. Evaluate $\arctan(-3)$.

19. Evaluate $\sin\left(\arccos\dfrac{4}{\sqrt{35}}\right)$.

20. Write an algebraic expression for $\cos\left(\arcsin\dfrac{x}{4}\right)$.

For Exercises 21–23, solve the right triangle.

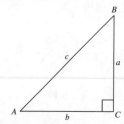

21. $A = 40°, c = 12$

22. $B = 6.84°, a = 21.3$

23. $a = 5, b = 9$

24. A 20-foot ladder leans against the side of a barn. Find the height of the top of the ladder if the angle of elevation of the ladder is 67°.

25. An observer in a lighthouse 250 feet above sea level spots a ship off the shore. If the angle of depression to the ship is 5°, how far out is the ship?

C H A P T E R   7
Analytic Trigonometry

C H A P T E R   7
Analytic Trigonometry

Section 7.1   Using Fundamental Identities

- You should know the fundamental trigonometric identities.

  (a) Reciprocal Identities

  $$\sin u = \frac{1}{\csc u} \qquad\qquad \csc u = \frac{1}{\sin u}$$

  $$\cos u = \frac{1}{\sec u} \qquad\qquad \sec u = \frac{1}{\cos u}$$

  $$\tan u = \frac{1}{\cot u} = \frac{\sin u}{\cos u} \qquad \cot u = \frac{1}{\tan u} = \frac{\cos u}{\sin u}$$

  (b) Pythagorean Identities

  $$\sin^2 u + \cos^2 u = 1$$

  $$1 + \tan^2 u = \sec^2 u$$

  $$1 + \cot^2 u = \csc^2 u$$

  (c) Cofunction Identities

  $$\sin\left(\frac{\pi}{2} - u\right) = \cos u \qquad\qquad \cos\left(\frac{\pi}{2} - u\right) = \sin u$$

  $$\tan\left(\frac{\pi}{2} - u\right) = \cot u \qquad\qquad \cot\left(\frac{\pi}{2} - u\right) = \tan u$$

  $$\sec\left(\frac{\pi}{2} - u\right) = \csc u \qquad\qquad \csc\left(\frac{\pi}{2} - u\right) = \sec u$$

  (d) Even/Odd Identities

  $$\sin(-x) = -\sin x \qquad\qquad \csc(-x) = -\csc x$$

  $$\cos(-x) = \cos x \qquad\qquad \sec(-x) = \sec x$$

  $$\tan(-x) = -\tan x \qquad\qquad \cot(-x) = -\cot x$$

- You should be able to use these fundamental identities to find function values.

- You should be able to convert trigonometric expressions to equivalent forms by using the fundamental identities.

1. $\tan u$

3. $\cot u$

5. $\cot^2 u$

7. $\cos u$

9. $\cos u$

11. $\sin x = \dfrac{1}{2},\ \cos x = \dfrac{\sqrt{3}}{2} \Rightarrow x$ is in Quadrant I.

$$\tan x = \frac{\sin x}{\cos x} = \frac{1/2}{\sqrt{3}/2} = \frac{1}{\sqrt{3}} = \frac{\sqrt{3}}{3}$$

$$\cot x = \frac{1}{\tan x} = \frac{1}{1/\sqrt{3}} = \sqrt{3}$$

$$\sec x = \frac{1}{\cos x} = \frac{1}{\sqrt{3}/2} = \frac{2}{\sqrt{3}} = \frac{2\sqrt{3}}{3}$$

$$\csc x = \frac{1}{\sin x} = \frac{1}{1/2} = 2$$

13. $\sec\theta = \sqrt{2}$, $\sin\theta = -\dfrac{\sqrt{2}}{2} \Rightarrow \theta$ is in Quadrant IV.

$$\cos\theta = \frac{1}{\sec\theta} = \frac{1}{\sqrt{2}} = \frac{\sqrt{2}}{2}$$

$$\tan\theta = \frac{\sin\theta}{\cos\theta} = \frac{-\sqrt{2}/2}{\sqrt{2}/2} = -1$$

$$\cot\theta = \frac{1}{\tan\theta} = -1$$

$$\csc\theta = \frac{1}{\sin\theta} = -\sqrt{2}$$

15. $\tan x = \dfrac{8}{15}$, $\sec x = -\dfrac{17}{15} \Rightarrow x$ is in Quadrant III.

$$\cos x = \frac{1}{\sec x} = \frac{1}{-17/15} = -\frac{15}{17}$$

$$\sin x = -\sqrt{1-\cos^2 x} = -\sqrt{1-\frac{225}{289}} = -\frac{8}{17}$$

$$\cot x = \frac{1}{\tan x} = \frac{1}{8/15} = \frac{15}{8}$$

$$\csc x = \frac{1}{\sin x} = \frac{1}{-8/17} = -\frac{17}{8}$$

17. $\sec\phi = \dfrac{3}{2}$, $\csc\phi = -\dfrac{3\sqrt{5}}{5} \Rightarrow \phi$ is in Quadrant IV.

$$\sin\phi = \frac{1}{\csc\phi} = \frac{1}{-3\sqrt{5}/5} = -\frac{\sqrt{5}}{3}$$

$$\cos\phi = \frac{1}{\sec\phi} = \frac{1}{3/2} = \frac{2}{3}$$

$$\tan\phi = \frac{\sin\phi}{\cos\phi} = \frac{-\sqrt{5}/3}{2/3} = -\frac{\sqrt{5}}{2}$$

$$\cot\phi = \frac{1}{\tan\phi} = \frac{1}{-\sqrt{5}/2} = -\frac{2}{\sqrt{5}} = -\frac{2\sqrt{5}}{5}$$

19. $\sin(-x) = -\dfrac{1}{3} \Rightarrow \sin x = \dfrac{1}{3}$, $\tan x = -\dfrac{\sqrt{2}}{4} \Rightarrow x$ is in Quadrant II.

$$\cos x = -\sqrt{1-\sin^2 x} = -\sqrt{1-\frac{1}{9}} = -\frac{2\sqrt{2}}{3}$$

$$\cot x = \frac{1}{\tan x} = \frac{1}{-\sqrt{2}/4} = -2\sqrt{2}$$

$$\sec x = \frac{1}{\cos x} = \frac{1}{-2\sqrt{2}/3} = -\frac{3\sqrt{2}}{4}$$

$$\csc x = \frac{1}{\sin x} = \frac{1}{1/3} = 3$$

21. $\tan\theta = 2$, $\sin\theta < 0 \Rightarrow \theta$ is in Quadrant III.

$$\sec\theta = -\sqrt{\tan^2\theta + 1} = -\sqrt{4+1} = -\sqrt{5}$$

$$\cos\theta = \frac{1}{\sec\theta} = -\frac{1}{\sqrt{5}} = -\frac{\sqrt{5}}{5}$$

$$\sin\theta = -\sqrt{1-\cos^2\theta}$$
$$= -\sqrt{1-\frac{1}{5}} = -\frac{2}{\sqrt{5}} = -\frac{2\sqrt{5}}{5}$$

$$\csc\theta = \frac{1}{\sin\theta} = -\frac{\sqrt{5}}{2}$$

$$\cot\theta = \frac{1}{\tan\theta} = \frac{1}{2}$$

23. $\sin\theta = -1$, $\cot\theta = 0 \Rightarrow \theta = \dfrac{3\pi}{2}$

$$\cos\theta = \sqrt{1-\sin^2\theta} = 0$$

$\sec\theta$ is undefined.

$\tan\theta$ is undefined.

$$\csc\theta = -1$$

25. $\sec x \cos x = \sec x \cdot \dfrac{1}{\sec x} = 1$

Matches (d).

26. $\tan x \csc x = \dfrac{\sin x}{\cos x} \cdot \dfrac{1}{\sin x} = \dfrac{1}{\cos x} = \sec x$

Matches (a).

27. $\cot^2 x - \csc^2 x = \cot^2 x - \left(1+\cot^2 x\right) = -1$

Matches (b).

28. $\left(1-\cos^2 x\right)(\csc x) = \left(\sin^2 x\right)\dfrac{1}{\sin x} = \sin x$

Matches (f).

29. $\dfrac{\sin(-x)}{\cos(-x)} = \dfrac{-\sin x}{\cos x} = -\tan x$

Matches (e).

30. $\dfrac{\sin\left[(\pi/2)-x\right]}{\cos\left[(\pi/2)-x\right]} = \dfrac{\cos x}{\sin x} = \cot x$

Matches (c).

31. $\sin x \sec x = \sin x \cdot \dfrac{1}{\cos x} = \tan x$

Matches (b).

32. $\cos^2 x\left(\sec^2 x - 1\right) = \cos^2 x\left(\tan^2 x\right)$

$$= \cos^2 x\left(\frac{\sin^2 x}{\cos^2 x}\right)$$

$$= \sin^2 x$$

Matches (c).

33. $\sec^4 x - \tan^4 x = \left(\sec^2 x + \tan^2 x\right)\left(\sec^2 x - \tan^2 x\right)$

$$= \left(\sec^2 x + \tan^2 x\right)(1)$$

$$= \sec^2 x + \tan^2 x$$

Matches (f).

34. $\cot x \sec x = \dfrac{\cos x}{\sin x} \cdot \dfrac{1}{\cos x} = \dfrac{1}{\sin x} = \csc x$

Matches (a).

35. $\dfrac{\sec^2 x - 1}{\sin^2 x} = \dfrac{\tan^2 x}{\sin^2 x} = \dfrac{\sin^2 x}{\cos^2 x} \cdot \dfrac{1}{\sin^2 x} = \sec^2 x$

Matches (e).

36. $\dfrac{\cos^2\left[(\pi/2) - x\right]}{\cos x} = \dfrac{\sin^2 x}{\cos x} = \dfrac{\sin x}{\cos x}\sin x = \tan x \sin x$

Matches (d).

37. $\cot \theta \sec \theta = \dfrac{\cos \theta}{\sin \theta} \cdot \dfrac{1}{\cos \theta} = \dfrac{1}{\sin \theta} = \csc \theta$

39. $\tan(-x)\cos x = -\tan x \cos x$

$$= -\frac{\sin x}{\cos x} \cdot \cos x$$

$$= -\sin x$$

41. $\sin \phi\left(\csc \phi - \sin \phi\right) = \left(\sin \phi\right)\dfrac{1}{\sin \phi} - \sin^2 \phi$

$$= 1 - \sin^2 \phi = \cos^2 \phi$$

43. $\dfrac{\cot x}{\csc x} = \dfrac{\cos x/\sin x}{1/\sin x}$

$$= \frac{\cos x}{\sin x} \cdot \frac{\sin x}{1}$$

$$= \cos x$$

45. $\dfrac{1 - \sin^2 x}{\csc^2 x - 1} = \dfrac{\cos^2 x}{\cot^2 x} = \cos^2 x \tan^2 x = \left(\cos^2 x\right)\dfrac{\sin^2 x}{\cos^2 x}$

$$= \sin^2 x$$

47. $\dfrac{\tan \theta \cot \theta}{\sec \theta} = \dfrac{1}{\sec \theta} = \cos \theta$

49. $\sec \alpha \dfrac{\sin \alpha}{\tan \alpha} = \dfrac{1}{\cos \alpha}\left(\sin \alpha\right)\cot \alpha$

$$= \frac{1}{\cos \alpha}\left(\sin \alpha\right)\left(\frac{\cos \alpha}{\sin \alpha}\right) = 1$$

51. $\cos\left(\dfrac{\pi}{2} - x\right)\sec x = \left(\sin x\right)\left(\sec x\right)$

$$= \left(\sin x\right)\left(\frac{1}{\cos x}\right)$$

$$= \frac{\sin x}{\cos x}$$

$$= \tan x$$

53. $\dfrac{\cos^2 y}{1 - \sin y} = \dfrac{1 - \sin^2 y}{1 - \sin y}$

$$= \frac{\left(1 + \sin y\right)\left(1 - \sin y\right)}{1 - \sin y} = 1 + \sin y$$

55. $\sin \beta \tan \beta + \cos \beta = \left(\sin \beta\right)\dfrac{\sin \beta}{\cos \beta} + \cos \beta$

$$= \frac{\sin^2 \beta}{\cos \beta} + \frac{\cos^2 \beta}{\cos \beta}$$

$$= \frac{\sin^2 \beta + \cos^2 \beta}{\cos \beta}$$

$$= \frac{1}{\cos \beta}$$

$$= \sec \beta$$

57. $\cot u \sin u + \tan u \cos u = \dfrac{\cos u}{\sin u}\left(\sin u\right) + \dfrac{\sin u}{\cos u}\left(\cos u\right)$

$$= \cos u + \sin u$$

59. $\tan^2 x - \tan^2 x \sin^2 x = \tan^2 x\left(1 - \sin^2 x\right)$

$$= \tan^2 x \cos^2 x$$

$$= \frac{\sin^2 x}{\cos^2 x} \cdot \cos^2 x$$

$$= \sin^2 x$$

61. $\sin^2 x \sec^2 x - \sin^2 x = \sin^2 x\left(\sec^2 x - 1\right)$

$$= \sin^2 x \tan^2 x$$

63. $\dfrac{\sec^2 x - 1}{\sec x - 1} = \dfrac{\left(\sec x + 1\right)\left(\sec x - 1\right)}{\sec x - 1}$

$$= \sec x + 1$$

65. $\tan^4 x + 2 \tan^2 x + 1 = \left(\tan^2 x + 1\right)^2$

$$= \left(\sec^2 x\right)^2$$

$$= \sec^4 x$$

67. $\sin^4 x - \cos^4 x = \left(\sin^2 x + \cos^2 x\right)\left(\sin^2 x - \cos^2 x\right)$

$$= (1)\left(\sin^2 x - \cos^2 x\right)$$

$$= \sin^2 x - \cos^2 x$$

69. $\csc^3 x - \csc^2 x - \csc x + 1 = \csc^2 x\left(\csc x - 1\right) - 1\left(\csc x - 1\right)$

$$= \left(\csc^2 x - 1\right)\left(\csc x - 1\right)$$

$$= \cot^2 x\left(\csc x - 1\right)$$

71. $\left(\sin x + \cos x\right)^2 = \sin^2 x + 2\sin x \cos x + \cos^2 x$

$$= \left(\sin^2 x + \cos^2 x\right) + 2\sin x \cos x$$

$$= 1 + 2\sin x \cos x$$

73. $\left(2\csc x + 2\right)\left(2\csc x - 2\right) = 4\csc^2 x - 4$

$$= 4\left(\csc^2 x - 1\right)$$

$$= 4\cot^2 x$$

75. $\dfrac{1}{1+\cos x} + \dfrac{1}{1-\cos x} = \dfrac{1-\cos x + 1 + \cos x}{\left(1+\cos x\right)\left(1-\cos x\right)}$

$$= \dfrac{2}{1-\cos^2 x}$$

$$= \dfrac{2}{\sin^2 x}$$

$$= 2\csc^2 x$$

77. $\dfrac{\cos x}{1+\sin x} + \dfrac{1+\sin x}{\cos x} = \dfrac{\cos^2 x + \left(1+\sin x\right)^2}{\cos x\left(1+\sin x\right)}$

$$= \dfrac{\cos^2 x + 1 + 2\sin x + \sin^2 x}{\cos x\left(1+\sin x\right)}$$

$$= \dfrac{2 + 2\sin x}{\cos x\left(1+\sin x\right)}$$

$$= \dfrac{2\left(1+\sin x\right)}{\cos x\left(1+\sin x\right)}$$

$$= \dfrac{2}{\cos x}$$

$$= 2\sec x$$

79. $\tan x + \dfrac{\cos x}{1+\sin x} = \dfrac{\sin x}{\cos x} + \dfrac{\cos x}{1+\sin x}$

$$= \dfrac{\sin x + \sin^2 x + \cos^2 x}{\cos x\left(1+\sin x\right)}$$

$$= \dfrac{1+\sin x}{\cos x\left(1+\sin x\right)}$$

$$= \dfrac{1}{\cos x}$$

$$= \sec x$$

81. $\dfrac{\sin^2 y}{1-\cos y} = \dfrac{1-\cos^2 y}{1-\cos y}$

$$= \dfrac{\left(1+\cos y\right)\left(1-\cos y\right)}{1-\cos y} = 1 + \cos y$$

83. $\dfrac{3}{\sec x - \tan x} \cdot \dfrac{\sec x + \tan x}{\sec x + \tan x} = \dfrac{3\left(\sec x + \tan x\right)}{\sec^2 x - \tan^2 x}$

$$= \dfrac{3\left(\sec x + \tan x\right)}{1}$$

$$= 3\left(\sec x + \tan x\right)$$

85. $y_1 = \cos\left(\dfrac{\pi}{2} - x\right), y_2 = \sin x$

x	0.2	0.4	0.6	0.8	1.0	1.2	1.4
y_1	0.1987	0.3894	0.5646	0.7174	0.8415	0.9320	0.9854
y_2	0.1987	0.3894	0.5646	0.7174	0.8415	0.9320	0.9854

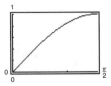

Conclusion: $y_1 = y_2$

87. $y_1 = \dfrac{\cos x}{1 - \sin x},\ y_2 = \dfrac{1 + \sin x}{\cos x}$

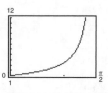

x	0.2	0.4	0.6	0.8	1.0	1.2	1.4
y_1	1.2230	1.5085	1.8958	2.4650	3.4082	5.3319	11.6814
y_2	1.2230	1.5085	1.8958	2.4650	3.4082	5.3319	11.6814

Conclusion: $y_1 = y_2$

89. $y_1 = \cos x \cot x + \sin x = \csc x$

$$\cos x \cot x + \sin x = \cos x\left(\frac{\cos x}{\sin x}\right) + \sin x$$

$$= \frac{\cos^2 x}{\sin x} + \frac{\sin^2 x}{\sin x}$$

$$= \frac{\cos^2 x + \sin^2 x}{\sin x} = \frac{1}{\sin x} = \csc x$$

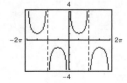

91. $y_1 = \dfrac{1}{\sin x}\left(\dfrac{1}{\cos x} - \cos x\right) = \tan x$

$$\frac{1}{\sin x}\left(\frac{1}{\cos x} - \cos x\right) = \frac{1}{\sin x \cos x} - \frac{\cos x}{\sin x} = \frac{1 - \cos^2 x}{\sin x \cos x} = \frac{\sin^2 x}{\sin x \cos x} = \frac{\sin x}{\cos x} = \tan x$$

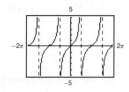

93. Let $x = 3 \cos \theta$.

$$\sqrt{9 - x^2} = \sqrt{9 - (3\cos\theta)^2}$$

$$= \sqrt{9 - 9\cos^2\theta}$$

$$= \sqrt{9(1 - \cos^2\theta)}$$

$$= \sqrt{9\sin^2\theta} = 3\sin\theta$$

95. Let $x = 4 \sin \theta$.

$$\sqrt{16 - x^2} = \sqrt{16 - (4\sin\theta)^2}$$

$$= \sqrt{16 - 16\sin^2\theta}$$

$$= \sqrt{16(1 - \sin^2\theta)}$$

$$= \sqrt{16\cos^2\theta}$$

$$= 4\cos\theta$$

97. Let $x = 3 \sec \theta$.

$$\sqrt{x^2 - 9} = \sqrt{(3\sec\theta)^2 - 9}$$

$$= \sqrt{9\sec^2\theta - 9}$$

$$= \sqrt{9(\sec^2\theta - 1)}$$

$$= \sqrt{9\tan^2\theta}$$

$$= 3\tan\theta$$

99. Let $x = 5 \tan \theta$.

$$\sqrt{x^2 + 25} = \sqrt{(5\tan\theta)^2 + 25}$$

$$= \sqrt{25\tan^2\theta + 25}$$

$$= \sqrt{25(\tan^2\theta + 1)}$$

$$= \sqrt{25\sec^2\theta}$$

$$= 5\sec\theta$$

101. Let $2x = 3 \tan \theta$.

$$\sqrt{4x^2 + 9} = \sqrt{(2x)^2 + 9}$$

$$= \sqrt{(3\tan\theta)^2 + 9}$$

$$= \sqrt{9\tan^2\theta + 9}$$

$$= \sqrt{9(\tan^2\theta + 1)}$$

$$= \sqrt{9\sec^2\theta}$$

$$= 3\sec\theta$$

103. Let $x = \sqrt{2}\,\sin \theta$.

$$\sqrt{2 - x^2} = \sqrt{2 - (\sqrt{2}\sin\theta)^2}$$

$$= \sqrt{2 - 2\sin^2\theta}$$

$$= \sqrt{2(1 - \sin^2\theta)}$$

$$= \sqrt{2\cos^2\theta}$$

$$= \sqrt{2}\,\cos\theta$$

105. Let $x = 3 \sin \theta$.

$$\sqrt{9 - x^2} = 3$$

$$\sqrt{9 - (3 \sin \theta)^2} = 3$$

$$\sqrt{9 - 9 \sin^2 \theta} = 3$$

$$\sqrt{9(1 - \sin^2 \theta)} = 3$$

$$\sqrt{9 \cos^2 \theta} = 3$$

$$3 \cos \theta = 3$$

$$\cos \theta = 1$$

$$\sin \theta = \sqrt{1 - \cos^2 \theta} = \sqrt{1 - (1)^2} = 0$$

107. Let $x = 2 \cos \theta$.

$$\sqrt{16 - 4x^2} = 2\sqrt{2}$$

$$\sqrt{16 - 4(2 \cos \theta)^2} = 2\sqrt{2}$$

$$\sqrt{16 - 16 \cos^2 \theta} = 2\sqrt{2}$$

$$\sqrt{16(1 - \cos^2 \theta)} = 2\sqrt{2}$$

$$\sqrt{16 \sin^2 \theta} = 2\sqrt{2}$$

$$4 \sin \theta = 2\sqrt{2}$$

$$\sin \theta = \frac{\sqrt{2}}{2}$$

$$\cos \theta = \sqrt{1 - \sin^2 \theta}$$

$$= \sqrt{1 - \frac{1}{2}}$$

$$= \sqrt{\frac{1}{2}}$$

$$= \frac{\sqrt{2}}{2}$$

109. $\sin \theta = \sqrt{1 - \cos^2 \theta}$

Let $y_1 = \sin x$ and $y_2 = \sqrt{1 - \cos^2 x}, 0 \le x \le 2\pi$.

$y_1 = y_2$ for $0 \le x \le \pi$.

So, $\sin \theta = \sqrt{1 - \cos^2 \theta}$ for $0 \le \theta \le \pi$.

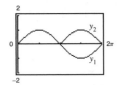

111. $\sec \theta = \sqrt{1 + \tan^2 \theta}$

Let $y_1 = \dfrac{1}{\cos x}$ and $y_2 = \sqrt{1 + \tan^2 x}, 0 \le x \le 2\pi$.

$y_1 = y_2$ for $0 \le x < \dfrac{\pi}{2}$ and $\dfrac{3\pi}{2} < x \le 2\pi$.

So, $\sec \theta = \sqrt{1 + \tan^2 \theta}$ for $0 \le \theta < \dfrac{\pi}{2}$ and

$\dfrac{3\pi}{2} < \theta < 2\pi$.

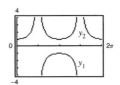

113. $\ln|\cos x| - \ln|\sin x| = \ln \left| \dfrac{\cos x}{\sin x} \right| = \ln|\cot x|$

115. $\ln|\sin x| + \ln|\cot x| = \ln|\sin x \cot x|$

$$= \ln \left| \sin x \cdot \dfrac{\cos x}{\sin x} \right|$$

$$= \ln|\cos x|$$

117. $\ln|\cot t| + \ln(1 + \tan^2 t) = \ln \left[|\cot t| (1 + \tan^2 t) \right]$

$$= \ln|\cot t \sec^2 t|$$

$$= \ln \left| \dfrac{\cot t}{\sin t} \cdot \dfrac{1}{\cos^2 t} \right|$$

$$= \ln \left| \dfrac{1}{\sin t \cos t} \right|$$

$$= \ln|\csc t \sec t|$$

119. (a) $\theta = 132°$

$$\csc^2 132° - \cot^2 132° \approx 1.8107 - 0.8107 = 1$$

(b) $\theta = \dfrac{2\pi}{7}$

$$\csc^2 \dfrac{2\pi}{7} - \cot^2 \dfrac{2\pi}{7} \approx 1.6360 - 0.6360 = 1$$

121. $\cos\left(\dfrac{\pi}{2} - \theta\right) = \sin\theta$

 (a) $\qquad\qquad\theta = 80°$

$\qquad\cos(90° - 80°) = \sin 80°$

$\qquad\qquad 0.9848 = 0.9848$

 (b) $\qquad\qquad\theta = 0.8$

$\qquad\cos\left(\dfrac{\pi}{2} - 0.8\right) = \sin 0.8$

$\qquad\qquad 0.7174 = 0.7174$

123. $\mu W \cos\theta = W \sin\theta$

$\qquad\mu = \dfrac{W \sin\theta}{W \cos\theta} = \tan\theta$

125. $\sec x \tan x - \sin x = \dfrac{1}{\cos x} \cdot \dfrac{\sin x}{\cos x} - \sin x$

$\qquad\qquad = \dfrac{\sin x}{\cos^2 x} - \sin x$

$\qquad\qquad = \dfrac{\sin x - \sin x \cos^2 x}{\cos^2 x}$

$\qquad\qquad = \dfrac{\sin x\left(1 - \cos^2 x\right)}{\cos^2 x}$

$\qquad\qquad = \dfrac{\sin x \sin^2 x}{\cos^2 x}$

$\qquad\qquad = \sin x \tan^2 x$

127. True. For example, $\sin(-x) = -\sin x$ means that the graph of $\sin x$ is symmetric about the origin.

129. As $x \to \dfrac{\pi^-}{2}$, $\sin x \to 1$ and $\csc x \to 1$.

131. As $x \to \dfrac{\pi^-}{2}$, $\tan x \to \infty$ and $\cot x \to 0$.

133. $\cos\theta = \sqrt{1 - \sin^2\theta}$ *is not* an identity.

$\qquad\cos\theta = \pm\sqrt{1 - \sin^2\theta}$

135. $\dfrac{\sin k\theta}{\cos k\theta} = \tan\theta$ *is not* an identity.

$\qquad\dfrac{\sin k\theta}{\cos k\theta} = \tan k\theta$

137. $\sin\theta \csc\theta = 1$ *is* an identity.

$\qquad\sin\theta \cdot \dfrac{1}{\sin\theta} = 1$, provided $\sin\theta \neq 0$.

139. Let $v = a \sin\theta$, then

$\qquad\sqrt{a^2 - v^2} = \sqrt{a^2 - \left(a \sin\theta\right)^2}$

$\qquad\qquad = \sqrt{a^2 - a^2 \sin^2\theta}$

$\qquad\qquad = \sqrt{a^2\left(1 - \sin^2\theta\right)}$

$\qquad\qquad = \sqrt{a^2 \cos^2\theta}$

$\qquad\qquad = a \cos\theta.$

141. Let $u = a \sec\theta$, then

$\qquad\sqrt{u^2 - a^2} = \sqrt{\left(a \sec\theta\right)^2 - a^2}$

$\qquad\qquad = \sqrt{a^2 \sec^2\theta - a^2}$

$\qquad\qquad = \sqrt{a^2\left(\sec^2\theta - 1\right)}$

$\qquad\qquad = \sqrt{a^2 \tan^2\theta}$

$\qquad\qquad = a \tan\theta.$

Section 7.2   Verifying Trigonometric Identities

- You should know the difference between an expression, a conditional equation, and an identity.
- You should be able to solve trigonometric identities, using the following techniques.
  (a) Work with *one* side at a time. Do not "cross" the equal sign.
  (b) Use algebraic techniques such as combining fractions, factoring expressions, rationalizing denominators, and squaring binomials.
  (c) Use the fundamental identities.
  (d) Convert all the terms into sines and cosines.

1. identity

3. $\tan u$

5. $\cos^2 u$

7. $-\csc u$

9. $\tan t \cot t = \dfrac{\sin t}{\cos t} \cdot \dfrac{\cos t}{\sin t} = 1$

11. $\cot^2 y\left(\sec^2 y - 1\right) = \cot^2 y \tan^2 y = 1$

13. $(1 + \sin\alpha)(1 - \sin\alpha) = 1 - \sin^2\alpha = \cos^2\alpha$

15. $\cos^2\beta - \sin^2\beta = (1 - \sin^2\beta) - \sin^2\beta$
$$= 1 - 2\sin^2\beta$$

17. $\dfrac{\tan^2\theta}{\sec\theta} = \dfrac{(\sin\theta/\cos\theta)\tan\theta}{1/\cos\theta} = \sin\theta\tan\theta$

19. $\dfrac{\cot^2 t}{\csc t} = \dfrac{\cos^2 t/\sin^2 t}{1/\sin t} = \dfrac{\cos^2 t}{\sin t} = \dfrac{1 - \sin^2 t}{\sin t}$

21. $\sin^{1/2} x \cos x - \sin^{5/2} x \cos x = \sin^{1/2} x \cos x\left(1 - \sin^2 x\right) = \sin^{1/2} x \cos x \cdot \cos^2 x = \cos^3 x\sqrt{\sin x}$

23. $\dfrac{\cot x}{\sec x} = \dfrac{\cos x/\sin x}{1/\cos x} = \dfrac{\cos^2 x}{\sin x} = \dfrac{1 - \sin^2 x}{\sin x} = \dfrac{1}{\sin x} - \dfrac{\sin^2 x}{\sin x} = \csc x - \sin x$

25. $\csc x - \sin x = \dfrac{1}{\sin x} - \dfrac{\sin^2 x}{\sin x}$
$$= \dfrac{1 - \sin^2 x}{\sin x}$$
$$= \dfrac{\cos^2 x}{\sin x}$$
$$= \dfrac{\cos x}{1} \cdot \dfrac{\cos x}{\sin x}$$
$$= \cos x \cot x$$

27. $\dfrac{1}{\tan x} + \dfrac{1}{\cot x} = \dfrac{\cot x + \tan x}{\tan x \cot x}$
$$= \dfrac{\cot x + \tan x}{1}$$
$$= \tan x + \cot x$$

29. $\dfrac{1 + \sin\theta}{\cos\theta} + \dfrac{\cos\theta}{1 + \sin\theta} = \dfrac{(1 + \sin\theta)^2 + \cos^2\theta}{\cos\theta(1 + \sin\theta)}$
$$= \dfrac{1 + 2\sin\theta + \sin^2\theta + \cos^2\theta}{\cos\theta(1 + \sin\theta)}$$
$$= \dfrac{2 + 2\sin\theta}{\cos\theta(1 + \sin\theta)}$$
$$= \dfrac{2(1 + \sin\theta)}{\cos\theta(1 + \sin\theta)}$$
$$= \dfrac{2}{\cos\theta}$$
$$= 2\sec\theta$$

31. $\dfrac{1}{\cos x + 1} + \dfrac{1}{\cos x - 1} = \dfrac{\cos x - 1 + \cos x + 1}{(\cos x + 1)(\cos x - 1)}$
$$= \dfrac{2\cos x}{\cos^2 x - 1}$$
$$= \dfrac{2\cos x}{-\sin^2 x}$$
$$= -2 \cdot \dfrac{1}{\sin x} \cdot \dfrac{\cos x}{\sin x}$$
$$= -2\csc x \cot x$$

33. $\tan\left(\dfrac{\pi}{2} - \theta\right)\tan\theta = \cot\theta\tan\theta$
$$= \left(\dfrac{1}{\tan\theta}\right)\tan\theta$$
$$= 1$$

35. $\dfrac{\tan x \cot x}{\cos x} = \dfrac{1}{\cos x} = \sec x$

37. $(1 + \sin y)\left[1 + \sin(-y)\right] = (1 + \sin y)(1 - \sin y)$
$$= 1 - \sin^2 y$$
$$= \cos^2 y$$

39. $\dfrac{\tan x + \cot y}{\tan x \cot y} = \dfrac{\dfrac{1}{\cot x} + \dfrac{1}{\tan y}}{\dfrac{1}{\cot x} \cdot \dfrac{1}{\tan y}} \cdot \dfrac{\cot x \tan y}{\cot x \tan y}$
$$= \tan y + \cot x$$

41. $\sqrt{\dfrac{1 + \sin\theta}{1 - \sin\theta}} = \sqrt{\dfrac{1 + \sin\theta}{1 - \sin\theta} \cdot \dfrac{1 + \sin\theta}{1 + \sin\theta}}$
$$= \sqrt{\dfrac{(1 + \sin\theta)^2}{1 - \sin^2\theta}}$$
$$= \sqrt{\dfrac{(1 + \sin\theta)^2}{\cos^2\theta}}$$
$$= \dfrac{1 + \sin\theta}{|\cos\theta|}$$

43. $\cos^2\beta + \cos^2\left(\dfrac{\pi}{2} - \beta\right) = \cos^2\beta + \sin^2\beta = 1$

45. $\sin t \csc\left(\dfrac{\pi}{2} - t\right) = \sin t \sec t = \sin t\left(\dfrac{1}{\cos t}\right)$
$$= \dfrac{\sin t}{\cos t} = \tan t$$

47. Let $\theta = \sin^{-1} x \Rightarrow \sin \theta = x = \dfrac{x}{1}$.

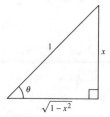

From the diagram,

$$\tan\left(\sin^{-1} x\right) = \tan \theta = \frac{x}{\sqrt{1 - x^2}}.$$

49. Let $\theta = \sin^{-1} \dfrac{x-1}{4} \Rightarrow \sin \theta = \dfrac{x-1}{4}$.

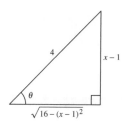

From the diagram,

$$\tan\left(\sin^{-1} \frac{x-1}{4}\right) = \tan \theta = \frac{x-1}{\sqrt{16 - \left(x-1\right)^2}}.$$

51. The first line claims that $\cot(-x) = \cot x$, which is not true. The correct substitution is $\cot(-x) = -\cot x$.

53. (a)

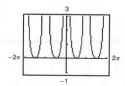

Identity

(b)

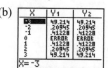

Identity

(c) $\left(1 + \cot^2 x\right)\left(\cos^2 x\right) = \csc^2 x \cos^2 x$

$$= \frac{1}{\sin^2 x} \cdot \cos^2 x$$

$$= \cot^2 x$$

55. (a)

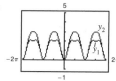

Not an Identity

(b)

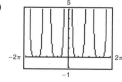

Not an Identity

(c) $2 + \cos^2 x - 3\cos^4 x = \left(1 - \cos^2 x\right)\left(2 + 3\cos^2 x\right)$

$$= \sin^2 x\left(2 + 3\cos^2 x\right)$$

$$\neq \sin^2 x\left(3 + 2\cos^2 x\right)$$

57. (a)

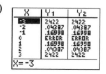

Identity

(b)

Identity

(c) $\csc^4 x - 2\csc^2 x + 1 = \left(\csc^2 x - 1\right)^2$

$$= \left(\cot^2 x\right)^2 = \cot^4 x$$

59. (a)

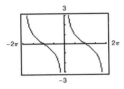

Identity

(b)

Identity

(c)
$$\frac{1 + \cos x}{\sin x} = \frac{(1 + \cos x)(1 - \cos x)}{\sin x(1 - \cos x)}$$

$$= \frac{1 - \cos^2 x}{\sin x(1 - \cos x)}$$

$$= \frac{\sin^2 x}{\sin x(1 - \cos x)}$$

$$= \frac{\sin x}{1 - \cos x}$$

61. $\tan^3 x \sec^2 x - \tan^3 x = \tan^3 x(\sec^2 x - 1)$

$$= \tan^3 x \tan^2 x$$

$$= \tan^5 x$$

63. $(\sin^2 x - \sin^4 x)\cos x = \sin^2 x(1 - \sin^2 x)\cos x$

$$= \sin^2 x \cos^2 x \cos x$$

$$= \sin^2 x \cos^3 x$$

65. $\sin^2 25° + \sin^2 65° = \sin^2 25° + \cos^2(90° - 65°)$

$$= \sin^2 25° + \cos^2 25°$$

$$= 1$$

67. $\cos^2 20° + \cos^2 52° + \cos^2 38° + \cos^2 70° = \cos^2 20° + \cos^2 52° + \sin^2(90° - 38°) + \sin^2(90° - 70°)$

$$= \cos^2 20° + \cos^2 52° + \sin^2 52° + \sin^2 20°$$

$$= (\cos^2 20° + \sin^2 20°) + (\cos^2 52° + \sin^2 52°)$$

$$= 1 + 1$$

$$= 2$$

69. $\cos x - \csc x \cot x = \cos x - \dfrac{1}{\sin x}\dfrac{\cos x}{\sin x}$

$$= \cos x\left(1 - \frac{1}{\sin^2 x}\right)$$

$$= \cos x(1 - \csc^2 x)$$

$$= -\cos x(\csc^2 x - 1)$$

$$= -\cos x \cot^2 x$$

71. True. You can use many different techniques to verify a trigonometric identity.

73. Because $\sin^2 \theta = 1 - \cos^2 \theta$, then

$\sin \theta = \pm\sqrt{1 - \cos^2 \theta}$; $\sin \theta \neq \sqrt{1 - \cos^2 \theta}$ if θ

lies in Quadrant III or IV. One such angle is $\theta = \dfrac{7\pi}{4}$.

75.
$$1 - \cos \theta = \sin \theta$$

$$(1 - \cos \theta)^2 = (\sin \theta)^2$$

$$1 - 2\cos \theta + \cos^2 \theta = \sin^2 \theta$$

$$1 - 2\cos \theta + \cos^2 \theta = 1 - \cos^2 \theta$$

$$2\cos^2 \theta - 2\cos \theta = 0$$

$$2\cos \theta(\cos \theta - 1) = 0$$

The equation is not an identity because it is only true when $\cos \theta = 0$ or $\cos \theta = 1$. So, one angle for which the equation is not true is $-\dfrac{\pi}{2}$.

77.
$$1 + \tan \theta = \sec \theta$$

$$(1 + \tan \theta)^2 = (\sec \theta)^2$$

$$1 + 2\tan \theta + \tan^2 \theta = \sec^2 \theta$$

$$1 + 2\tan \theta + \tan^2 \theta = 1 + \tan^2 \theta$$

$$2\tan \theta = 0$$

$$\tan \theta = 0$$

This equation is not an identity because it is only true when $\tan \theta = 0$. So, one angle for which the equation is not true is $\dfrac{\pi}{6}$.

Section 7.3   Solving Trigonometric Equations

- You should be able to identify and solve trigonometric equations.

- A trigonometric equation is a conditional equation. It is true for a specific set of values.

- To solve trigonometric equations, use algebraic techniques such as collecting like terms, extracting square roots, factoring, squaring, converting to quadratic type, using formulas, and using inverse functions. Study the examples in this section.

---

1. isolate

3. quadratic

5. $2 \cos x - 1 = 0$

   (a) $x = \dfrac{\pi}{3}$

   $$2 \cos \dfrac{\pi}{3} - 1 = 2\left(\dfrac{1}{2}\right) - 1 = 0$$

   (b) $x = \dfrac{5\pi}{3}$

   $$2 \cos \dfrac{5\pi}{3} - 1 = 2\left(\dfrac{1}{2}\right) - 1 = 0$$

7. $3 \tan^2 2x - 1 = 0$

   (a) $x = \dfrac{\pi}{12}$

   $$3 \left[\tan 2\left(\dfrac{\pi}{12}\right) \right]^2 - 1 = 3 \tan^2 \dfrac{\pi}{6} - 1$$
   $$= 3\left(\dfrac{1}{\sqrt{3}}\right)^2 - 1$$
   $$= 0$$

   (b) $x = \dfrac{5\pi}{12}$

   $$3 \left[\tan 2\left(\dfrac{5\pi}{12}\right) \right]^2 - 1 = 3 \tan^2 \dfrac{5\pi}{6} - 1$$
   $$= 3\left(-\dfrac{1}{\sqrt{3}}\right)^2 - 1$$
   $$= 0$$

9. $2 \sin^2 x - \sin x - 1 = 0$

   (a) $x = \dfrac{\pi}{2}$

   $$2 \sin^2 \dfrac{\pi}{2} - \sin \dfrac{\pi}{2} - 1 = 2(1)^2 - 1 - 1$$
   $$= 0$$

   (b) $x = \dfrac{7\pi}{6}$

   $$2 \sin^2 \dfrac{7\pi}{6} - \sin \dfrac{7\pi}{6} - 1 = 2\left(-\dfrac{1}{2}\right)^2 - \left(-\dfrac{1}{2}\right) - 1$$
   $$= \dfrac{1}{2} + \dfrac{1}{2} - 1$$
   $$= 0$$

11. $2 \cos x + 1 = 0$
   $$2 \cos x = -1$$
   $$\cos x = -\dfrac{1}{2}$$
   $$x = \dfrac{2\pi}{3} + 2n\pi$$
   $$\text{or } x = \dfrac{4\pi}{3} + 2n\pi$$

13. $\sqrt{3} \csc x - 2 = 0$
   $$\sqrt{3} \csc x = 2$$
   $$\csc x = \dfrac{2}{\sqrt{3}}$$
   $$x = \dfrac{\pi}{3} + 2n\pi$$
   $$\text{or } x = \dfrac{2\pi}{3} + 2n\pi$$

15. $3 \sec^2 x - 4 = 0$

$$\sec^2 x = \frac{4}{3}$$

$$\sec x = \pm \frac{2}{\sqrt{3}}$$

$$x = \frac{\pi}{6} + n\pi$$

$$\text{or } x = \frac{5\pi}{6} + n\pi$$

17. $\sin x(\sin x + 1) = 0$

$\sin x = 0 \quad$ or $\quad \sin x = -1$

$$x = n\pi \qquad\qquad x = \frac{3\pi}{2} + 2n\pi$$

19. $4 \cos^2 x - 1 = 0$

$$\cos^2 x = \frac{1}{4}$$

$$\cos^2 x = \pm\frac{1}{2}$$

$$x = \frac{\pi}{3} + n\pi \quad \text{or} \quad x = \frac{2\pi}{3} + n\pi$$

21. $2 \sin^2 2x = 1$

$$\sin 2x = \pm\frac{1}{\sqrt{2}} = \pm\frac{\sqrt{2}}{2}$$

$$2x = \frac{\pi}{4} + 2n\pi, \, 2x = \frac{3\pi}{4} + 2n\pi,$$

$$2x = \frac{5\pi}{4} + 2n\pi, \, 2x = \frac{7\pi}{4} + 2n\pi$$

So, $x = \dfrac{\pi}{8} + n\pi, \dfrac{3\pi}{8} + n\pi, \dfrac{5\pi}{8} + n\pi, \dfrac{7\pi}{8} + n\pi.$

You can combine these as follows:

$$x = \frac{\pi}{8} + \frac{n\pi}{2}, x = \frac{3\pi}{8} + \frac{n\pi}{2}$$

23. $\tan 3x(\tan x - 1) = 0$

$\tan 3x = 0 \quad$ or $\quad \tan x - 1 = 0$

$$3x = n\pi \qquad\qquad \tan x = 1$$

$$x = \frac{n\pi}{3} \qquad\qquad x = \frac{\pi}{4} + n\pi$$

25. $\cos^3 x = \cos x$

$$\cos^3 x - \cos x = 0$$

$$\cos x(\cos^2 x - 1) = 0$$

$\cos x = 0 \qquad$ or $\quad \cos^2 x - 1 = 0$

$$x = \frac{\pi}{2}, \frac{3\pi}{2} \qquad\qquad \cos x = \pm 1$$

$$x = 0, \pi$$

27. $3 \tan^3 x - \tan x = 0$

$$\tan x(3 \tan^2 x - 1) = 0$$

$\tan x = 0 \qquad$ or $\quad 3 \tan^2 x - 1 = 0$

$$x = 0, \pi \qquad\qquad \tan x = \pm\frac{\sqrt{3}}{3}$$

$$x = \frac{\pi}{6}, \frac{5\pi}{6}, \frac{7\pi}{6}, \frac{11\pi}{6}$$

29. $\sec^2 x - \sec x - 2 = 0$

$$(\sec x - 2)(\sec x + 1) = 0$$

$\sec x - 2 = 0 \qquad$ or $\quad \sec x + 1 = 0$

$$\sec x = 2 \qquad\qquad \sec x = -1$$

$$x = \frac{\pi}{3}, \frac{5\pi}{3} \qquad\qquad x = \pi$$

31. $2 \sin x + \csc x = 0$

$$2 \sin x + \frac{1}{\sin x} = 0$$

$$2 \sin^2 x + 1 = 0$$

$$\sin^2 x = -\frac{1}{2} \Rightarrow \text{No solution}$$

33. $2 \cos^2 x + \cos x - 1 = 0$

$$(2 \cos x - 1)(\cos x + 1) = 0$$

$2 \cos x - 1 = 0 \qquad$ or $\quad \cos x + 1 = 0$

$$\cos x = \frac{1}{2} \qquad\qquad \cos x = -1$$

$$x = \pi$$

$$x = \frac{\pi}{3}, \frac{5\pi}{3}$$

35. $2 \sec^2 x + \tan^2 x - 3 = 0$

$$2(\tan^2 x + 1) + \tan^2 x - 3 = 0$$

$$3 \tan^2 x - 1 = 0$$

$$\tan x = \pm\frac{\sqrt{3}}{3}$$

$$x = \frac{\pi}{6}, \frac{5\pi}{6}, \frac{7\pi}{6}, \frac{11\pi}{6}$$

37.

$$\csc x + \cot x = 1$$
$$(\csc x + \cot x)^2 = 1^2$$
$$\csc^2 x + 2\csc x \cot x + \cot^2 x = 1$$
$$\cot^2 x + 1 + 2\csc x \cot x + \cot^2 x = 1$$
$$2\cot^2 x + 2\csc x \cot x = 0$$
$$2\cot x(\cot x + \csc x) = 0$$

$$2\cot x = 0 \qquad \text{or} \quad \cot x + \csc x = 0$$

$$x = \frac{\pi}{2}, \frac{3\pi}{2} \qquad\qquad \frac{\cos x}{\sin x} = -\frac{1}{\sin x}$$

$$\left(\frac{3\pi}{2} \text{ is extraneous.}\right) \qquad \cos x = -1$$

$$x = \pi$$

$$(\pi \text{ is extraneous.})$$

$x = \pi/2$ is the only solution.

39. $\cos 2x = \dfrac{1}{2}$

$$2x = \frac{\pi}{3} + 2n\pi \quad \text{or} \quad 2x = \frac{5\pi}{3} + 2n\pi$$

$$x = \frac{\pi}{6} + n\pi \qquad\qquad x = \frac{5\pi}{6} + n\pi$$

41. $\tan 3x = 1$

$$3x = \frac{\pi}{4} + 2n\pi \quad \text{or} \quad 3x = \frac{5\pi}{4} + 2n\pi$$

$$x = \frac{\pi}{12} + \frac{2n\pi}{3} \qquad\qquad x = \frac{5\pi}{12} + \frac{2n\pi}{3}$$

These can be combined as $x = \dfrac{\pi}{12} + \dfrac{n\pi}{3}$.

43. $\cos\left(\dfrac{x}{2}\right) = \dfrac{\sqrt{2}}{2}$

$$\frac{x}{2} = \frac{\pi}{4} + 2n\pi \quad \text{or} \quad \frac{x}{2} = \frac{7\pi}{4} + 2n\pi$$

$$x = \frac{\pi}{2} + 4n\pi \qquad\qquad x = \frac{7\pi}{2} + 4n\pi$$

45. $y = \sin\dfrac{\pi x}{2} + 1$

$$\sin\left(\frac{\pi x}{2}\right) + 1 = 0$$

$$\sin\left(\frac{\pi x}{2}\right) = -1$$

$$\frac{\pi x}{2} = \frac{3\pi}{2} + 2n\pi$$

$$x = 3 + 4n$$

For $-2 < x < 4$, the intercepts are -1 and 3.

47. $y = \tan^2\left(\dfrac{\pi x}{6}\right) - 3$

$$\tan^2\left(\frac{\pi x}{6}\right) - 3 = 0$$

$$\tan^2\left(\frac{\pi x}{6}\right) = 3$$

$$\tan\left(\frac{\pi x}{6}\right) = \pm\sqrt{3}$$

$$\frac{\pi x}{6} = \pm\frac{\pi}{3} + n\pi$$

$$x = \pm 2 + 6n$$

For $-3 < x < 3$, the intercepts are -2 and 2.

49. $2\sin x + \cos x = 0$

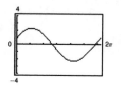

$$x \approx 2.678 \text{ and } x \approx 5.820$$

51. $\dfrac{1 + \sin x}{\cos x} + \dfrac{\cos x}{1 + \sin x} - 4 = 0$

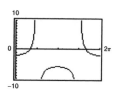

$$x = \frac{\pi}{3} \approx 1.047 \text{ and } x = \frac{5\pi}{3} \approx 5.236$$

53. $x\tan x - 1 = 0$

$$x \approx 0.860 \text{ and } x \approx 3.426$$

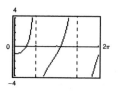

55. $\sec^2 x + 0.5\tan x - 1 = 0$

$$x = 0, x \approx 2.678,$$
$$x = \pi \approx 3.142$$
$$x \approx 5.820$$

57. $2 \tan^2 x + 7 \tan x - 15 = 0$

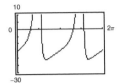

$x \approx 0.983, \ x \approx 1.768, \ x \approx 4.124 \ \text{and} \ x \approx 4.910$

59. $12 \sin^2 x - 13 \sin x + 3 = 0$

$$\sin x = \frac{-(-13) \pm \sqrt{(-13)^2 - 4(12)(3)}}{2(12)} = \frac{13 \pm 5}{24}$$

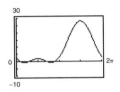

$$\sin x = \frac{1}{3} \quad \text{or} \quad \sin x = \frac{3}{4}$$

$$x \approx 0.3398, 2.8018 \qquad x \approx 0.8481, 2.2935$$

The x-intercepts occur at $x \approx 0.3398$, $x \approx 0.8481$, $x \approx 2.2935$, and $x \approx 2.8018$.

61. $\tan^2 x + 3 \tan x + 1 = 0$

$$\tan x = \frac{-3 \pm \sqrt{3^2 - 4(1)(1)}}{2(1)} = \frac{-3 \pm \sqrt{5}}{2}$$

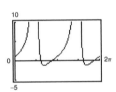

$$\tan x = \frac{-3 - \sqrt{5}}{2} \quad \text{or} \quad \tan x = \frac{-3 + \sqrt{5}}{2}$$

$$x \approx 1.9357, 5.0773 \qquad x \approx 2.7767, 5.9183$$

The x-intercepts occur at $x \approx 1.9357$, $x \approx 2.7767$, $x \approx 5.0773$, and $x \approx 5.9183$.

63. $\tan^2 x + \tan x - 12 = 0$

$(\tan x + 4)(\tan x - 3) = 0$

$\tan x + 4 = 0 \qquad\qquad\qquad \text{or} \quad \tan x - 3 = 0$

$\quad\tan x = -4 \qquad\qquad\qquad\qquad \tan x = 3$

$\qquad x = \arctan(-4) + \pi, \arctan(-4) + 2\pi \qquad x = \arctan 3, \arctan 3 + \pi$

65. $\tan^2 x - 6 \tan x + 5 = 0$

$(\tan x - 1)(\tan x - 5) = 0$

$\tan x - 1 = 0 \quad \text{or} \quad \tan x - 5 = 0$

$\quad \tan x = 1 \qquad\qquad \tan x = 5$

$\qquad x = \dfrac{\pi}{4}, \dfrac{5\pi}{4} \qquad\qquad x = \arctan 5, \arctan 5 + \pi$

67. $2 \cos^2 x - 5 \cos x + 2 = 0$

$(2 \cos x - 1)(\cos x - 2) = 0$

$2 \cos x - 1 = 0 \quad \text{or} \quad \cos x - 2 = 0$

$\quad \cos x = \dfrac{1}{2} \qquad\qquad \cos x = 2$

$\qquad x = \dfrac{\pi}{3}, \dfrac{5\pi}{3} \qquad\qquad \text{No solution}$

69. $\cot^2 x - 9 = 0$

$\cot^2 x = 9$

$\dfrac{1}{9} = \tan^2 x$

$\pm\dfrac{1}{3} = \tan x$

$x = \arctan\dfrac{1}{3},\ \arctan\dfrac{1}{3} + \pi,\ \arctan\left(-\dfrac{1}{3}\right) + \pi,\ \arctan\left(-\dfrac{1}{3}\right) + 2\pi$

71. $\sec^2 x - 4\sec x = 0$

$\sec x(\sec x - 4) = 0$

$\sec x = 0 \qquad \sec x - 4 = 0$

No solution $\qquad \sec x = 4$

$\dfrac{1}{4} = \cos x$

$x = \arccos\dfrac{1}{4},\ -\arccos\dfrac{1}{4} + 2\pi$

73. $\csc^2 x + 3\csc x - 4 = 0$

$(\csc x + 4)(\csc x - 1) = 0$

$\csc x + 4 = 0 \qquad\qquad \text{or} \qquad\qquad \csc x - 1 = 0$

$\csc x = -4 \qquad\qquad\qquad\qquad\qquad \csc x = 1$

$-\dfrac{1}{4} = \sin x \qquad\qquad\qquad\qquad\qquad 1 = \sin x$

$x = \arcsin\left(\dfrac{1}{4}\right) + \pi,\ \arcsin\left(-\dfrac{1}{4}\right) + 2\pi \qquad x = \dfrac{\pi}{2}$

75. $3\tan^2 x + 5\tan x - 4 = 0,\ \left[-\dfrac{\pi}{2}, \dfrac{\pi}{2}\right]$

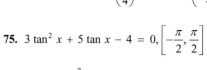

$x \approx -1.154,\ 0.534$

77. $4\cos^2 x - 2\sin x + 1 = 0,\ \left[-\dfrac{\pi}{2}, \dfrac{\pi}{2}\right]$

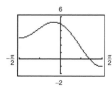

$x \approx 1.110$

79. (a) $f(x) = \sin^2 x + \cos x$

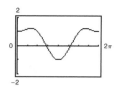

Maximum: $(1.0472, 1.25)$

Maximum: $(5.2360, 1.25)$

Minimum: $(0, 1)$

Minimum: $(3.1416, -1)$

(b) $2\sin x \cos x - \sin x = 0$

$\sin x(2\cos x - 1) = 0$

$\sin x = 0 \quad \text{or} \quad 2\cos x - 1 = 0$

$x = 0, \pi \qquad\qquad\qquad \cos x = \dfrac{1}{2}$

$\approx 0, 3.1416 \qquad\qquad x = \dfrac{\pi}{3}, \dfrac{5\pi}{3}$

$\approx 1.0472, 5.2360$

81. (a) $f(x) = \sin x + \cos x$

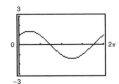

Maximum: $(0.7854, 1.4142)$

Minimum: $(3.9270, -1.4142)$

(b) $\cos x - \sin x = 0$

$$\cos x = \sin x$$

$$1 = \frac{\sin x}{\cos x}$$

$$\tan x = 1$$

$$x = \frac{\pi}{4}, \frac{5\pi}{4}$$

$$\approx 0.7854, 3.9270$$

83. (a) $f(x) = \sin x \cos x$

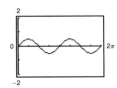

Maximum: $(0.7854, 0.5)$

Maximum: $(3.9270, 0.5)$

Minimum: $(2.3562, -0.5)$

Minimum: $(5.4978, -0.5)$

(b) $-\sin^2 x + \cos^2 x = 0$

$$-\sin^2 x + 1 - \sin^2 x = 0$$

$$-2\sin^2 x + 1 = 0$$

$$\sin^2 x = \frac{1}{2}$$

$$\sin x = \pm\sqrt{\frac{1}{2}} = \pm\frac{\sqrt{2}}{2}$$

$$x = \frac{\pi}{4}, \frac{3\pi}{4}, \frac{5\pi}{4}, \frac{7\pi}{4}$$

$$\approx 0.7854, 2.3562, 3.9270, 5.4978$$

85. $f(x) = \tan \dfrac{\pi x}{4}$

Because $\tan \pi/4 = 1$, $x = 1$ is the smallest nonnegative fixed point.

87. $f(x) = \cos \dfrac{1}{x}$

(a) Domain: all real numbers x except $x = 0$.

(b) The graph has y-axis symmetry and a horizontal asymptote at $y = 1$.

(c) As $x \to 0$, $f(x)$ oscillates between -1 and 1.

(d) There are infinitely many solutions in the interval $[-1, 1]$. They occur at $x = \dfrac{2}{(2n + 1)\pi}$ where n is any integer.

(e) The greatest solution appears to occur at $x \approx 0.6366$.

89. $$y = \frac{1}{12}(\cos 8t - 3\sin 8t)$$

$$\frac{1}{12}(\cos 8t - 3\sin 8t) = 0$$

$$\cos 8t = 3\sin 8t$$

$$\frac{1}{3} = \tan 8t$$

$$8t \approx 0.32175 + n\pi$$

$$t \approx 0.04 + \frac{n\pi}{8}$$

In the interval $0 \le t \le 1$, $t \approx 0.04, 0.43,$ and 0.83.

91. $S = 74.50 + 43.75 \sin \dfrac{\pi t}{6}$

t	1	2	3	4	5	6	7	8	9	10	11	12
S	96.4	112.4	118.3	112.4	96.4	74.5	52.6	36.6	30.8	36.6	52.6	74.5

Sales exceed 100,000 units during February, March, and April.

93.
$$\text{Range} = 300 \text{ feet}$$
$$v_0 = 100 \text{ feet per second}$$
$$r = \tfrac{1}{32}v_0{}^2 \sin 2\theta$$
$$\tfrac{1}{32}(100)^2 \sin 2\theta = 300$$
$$\sin 2\theta = 0.96$$
$$2\theta = \arcsin(0.96) \approx 73.74°$$
$$\theta \approx 36.9°$$

or

$$2\theta = 180° - \arcsin(0.96) \approx 106.26°$$
$$\theta \approx 53.1°$$

95. $h(t) = 53 + 50 \sin\left(\dfrac{\pi}{16}t - \dfrac{\pi}{2}\right)$

(a) $h(t) = 53$ when $50 \sin\left(\dfrac{\pi}{16}t - \dfrac{\pi}{2}\right) = 0.$

$$\dfrac{\pi}{16}t - \dfrac{\pi}{2} = 0 \quad \text{or} \quad \dfrac{\pi}{16}t - \dfrac{\pi}{2} = \pi$$

$$\dfrac{\pi}{16}t = \dfrac{\pi}{2} \qquad\qquad \dfrac{\pi}{16}t = \dfrac{3\pi}{2}$$

$$t = 8 \qquad\qquad\qquad t = 24$$

A person on the Ferris wheel will be 53 feet above ground at 8 seconds and at 24 seconds.

(b) The person will be at the top of the Ferris wheel when

$$\sin\left(\dfrac{\pi}{16}t - \dfrac{\pi}{2}\right) = 1$$

$$\dfrac{\pi}{16}t - \dfrac{\pi}{2} = \dfrac{\pi}{2}$$

$$\dfrac{\pi}{16}t = \pi$$

$$t = 16.$$

The first time this occurs is after 16 seconds. The period of this function is $\dfrac{2\pi}{\pi/16} = 32.$ During 160 seconds, 5 cycles will take place and the person will be at the top of the ride 5 times, spaced 32 seconds apart. The times are: 16 seconds, 48 seconds, 80 seconds, 112 seconds, and 144 seconds.

97. $A = 2x \cos x, 0 < x < \dfrac{\pi}{2}$

(a)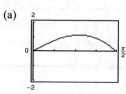

The maximum area of $A \approx 1.12$ occurs when $x \approx 0.86.$

(b) $A \geq 1$ for $0.6 < x < 1.1$

99. True. The period of $2 \sin 4t - 1$ is $\dfrac{\pi}{2}$ and the period of $2 \sin t - 1$ is $2\pi.$

In the interval $[0, 2\pi)$ the first equation has four cycles whereas the second equation has only one cycle, so the first equation has four times the x-intercepts (solutions) as the second equation.

101.
$$\cot x \cos^2 x = 2 \cot x$$
$$\cos^2 x = 2$$
$$\cos x = \pm\sqrt{2}$$

No solution

Because you solved this problem by first dividing by $\cot x,$ you do not get the same solution as Example 3.

When solving equations, you do not want to divide each side by a variable expression that will cancel out because you may accidentally remove one of the solutions.

103. Answers will vary.

Section 7.4 Sum and Difference Formulas

■ You should know the sum and difference formulas.

$$\sin(u \pm v) = \sin u \cos v \pm \cos u \sin v$$

$$\cos(u \pm v) = \cos u \cos v \mp \sin u \sin v$$

$$\tan(u \pm v) = \frac{\tan u \pm \tan v}{1 \mp \tan u \tan v}$$

■ You should be able to use these formulas to find the values of the trigonometric functions of angles whose sums or differences are special angles.

■ You should be able to use these formulas to solve trigonometric equations.

---

1. $\sin u \cos v - \cos u \sin v$

3. $\dfrac{\tan u + \tan v}{1 - \tan u \tan v}$

5. $\cos u \cos v + \sin u \sin v$

7. (a) $\cos\left(\dfrac{\pi}{4} + \dfrac{\pi}{3}\right) = \cos\dfrac{\pi}{4}\cos\dfrac{\pi}{3} - \sin\dfrac{\pi}{4}\sin\dfrac{\pi}{3}$

$$= \frac{\sqrt{2}}{2}\cdot\frac{1}{2} - \frac{\sqrt{2}}{2}\cdot\frac{\sqrt{3}}{2}$$

$$= \frac{\sqrt{2} - \sqrt{6}}{4}$$

(b) $\cos\dfrac{\pi}{4} + \cos\dfrac{\pi}{3} = \dfrac{\sqrt{2}}{2} + \dfrac{1}{2} = \dfrac{\sqrt{2}+1}{2}$

9. (a) $\sin\left(\dfrac{7\pi}{6} - \dfrac{\pi}{3}\right) = \sin\dfrac{5\pi}{6} = \sin\dfrac{\pi}{6} = \dfrac{1}{2}$

(b) $\sin\dfrac{7\pi}{6} - \sin\dfrac{\pi}{3} = -\dfrac{1}{2} - \dfrac{\sqrt{3}}{2} = \dfrac{-1-\sqrt{3}}{2}$

11. (a) $\sin(135° - 30°) = \sin 135° \cos 30° - \cos 135° \sin 30°$

$$= \left(\frac{\sqrt{2}}{2}\right)\left(\frac{\sqrt{3}}{2}\right) - \left(-\frac{\sqrt{2}}{2}\right)\left(\frac{1}{2}\right) = \frac{\sqrt{6}+\sqrt{2}}{4}$$

(b) $\sin 135° - \cos 30° = \dfrac{\sqrt{2}}{2} - \dfrac{\sqrt{3}}{2} = \dfrac{\sqrt{2}-\sqrt{3}}{2}$

13. $\sin\dfrac{11\pi}{12} = \sin\left(\dfrac{3\pi}{4} + \dfrac{\pi}{6}\right)$

$$= \sin\frac{3\pi}{4}\cos\frac{\pi}{6} + \cos\frac{3\pi}{4}\sin\frac{\pi}{6}$$

$$= \frac{\sqrt{2}}{2}\cdot\frac{\sqrt{3}}{2} + \left(-\frac{\sqrt{2}}{2}\right)\frac{1}{2}$$

$$= \frac{\sqrt{2}}{4}\left(\sqrt{3} - 1\right)$$

$\cos\dfrac{11\pi}{12} = \cos\left(\dfrac{3\pi}{4} + \dfrac{\pi}{6}\right)$

$$= \cos\frac{3\pi}{4}\cos\frac{\pi}{6} - \sin\frac{3\pi}{4}\sin\frac{\pi}{6}$$

$$= -\frac{\sqrt{2}}{2}\cdot\frac{\sqrt{3}}{2} - \frac{\sqrt{2}}{2}\cdot\frac{1}{2} = -\frac{\sqrt{2}}{4}\left(\sqrt{3} + 1\right)$$

$\tan\dfrac{11\pi}{4} = \tan\left(\dfrac{3\pi}{4} + \dfrac{\pi}{6}\right)$

$$= \frac{\tan\dfrac{3\pi}{4} + \tan\dfrac{\pi}{6}}{1 - \tan\dfrac{3\pi}{4}\tan\dfrac{\pi}{6}}$$

$$= \frac{-1 + \dfrac{\sqrt{3}}{3}}{1 - (-1)\dfrac{\sqrt{3}}{3}}$$

$$= \frac{-3 + \sqrt{3}}{3 + \sqrt{3}}\cdot\frac{3 - \sqrt{3}}{3 - \sqrt{3}}$$

$$= \frac{-12 + 6\sqrt{3}}{6} = -2 + \sqrt{3}$$

15. $\sin \dfrac{17\pi}{12} = \sin\left(\dfrac{9\pi}{4} - \dfrac{5\pi}{6}\right)$

$\qquad\qquad = \sin\dfrac{9\pi}{4}\cos\dfrac{5\pi}{6} - \cos\dfrac{9\pi}{4}\sin\dfrac{5\pi}{6}$

$\qquad\qquad = \dfrac{\sqrt{2}}{2}\left(-\dfrac{\sqrt{3}}{2}\right) - \left(\dfrac{\sqrt{2}}{2}\right)\left(\dfrac{1}{2}\right)$

$\qquad\qquad = -\dfrac{\sqrt{2}}{4}\left(\sqrt{3}+1\right)$

$\cos\dfrac{17\pi}{12} = \cos\left(\dfrac{9\pi}{4} - \dfrac{5\pi}{6}\right)$

$\qquad\qquad = \cos\dfrac{9\pi}{4}\cos\dfrac{5\pi}{6} + \sin\dfrac{9\pi}{4}\sin\dfrac{5\pi}{6}$

$\qquad\qquad = \dfrac{\sqrt{2}}{2}\left(-\dfrac{\sqrt{3}}{2}\right) + \dfrac{\sqrt{2}}{2}\left(\dfrac{1}{2}\right)$

$\qquad\qquad = \dfrac{\sqrt{2}}{4}\left(1-\sqrt{3}\right)$

$\tan\dfrac{17\pi}{12} = \tan\left(\dfrac{9\pi}{4} - \dfrac{5\pi}{6}\right)$

$\qquad\qquad = \dfrac{\tan(9\pi/4) - \tan(5\pi/6)}{1 + \tan(9\pi/4)\tan(5\pi/6)}$

$\qquad\qquad = \dfrac{1 - \left(-\sqrt{3}/3\right)}{1 + \left(-\sqrt{3}/3\right)}$

$\qquad\qquad = \dfrac{3+\sqrt{3}}{3-\sqrt{3}} \cdot \dfrac{3+\sqrt{3}}{3+\sqrt{3}}$

$\qquad\qquad = \dfrac{12+6\sqrt{3}}{6} = 2+\sqrt{3}$

17. $\sin 105° = \sin(60° + 45°)$

$\qquad\qquad = \sin 60° \cos 45° + \cos 60° \sin 45°$

$\qquad\qquad = \dfrac{\sqrt{3}}{2} \cdot \dfrac{\sqrt{2}}{2} + \dfrac{1}{2} \cdot \dfrac{\sqrt{2}}{2}$

$\qquad\qquad = \dfrac{\sqrt{2}}{4}\left(\sqrt{3}+1\right)$

$\cos 105° = \cos(60° + 45°)$

$\qquad\qquad = \cos 60° \cos 45° - \sin 60° \sin 45°$

$\qquad\qquad = \dfrac{1}{2} \cdot \dfrac{\sqrt{2}}{2} - \dfrac{\sqrt{3}}{2} \cdot \dfrac{\sqrt{2}}{2}$

$\qquad\qquad = \dfrac{\sqrt{2}}{4}\left(1-\sqrt{3}\right)$

$\tan 105° = \tan(60° + 45°)$

$\qquad\qquad = \dfrac{\tan 60° + \tan 45°}{1 - \tan 60° \tan 45°}$

$\qquad\qquad = \dfrac{\sqrt{3}+1}{1-\sqrt{3}} = \dfrac{\sqrt{3}+1}{1-\sqrt{3}} \cdot \dfrac{1+\sqrt{3}}{1+\sqrt{3}}$

$\qquad\qquad = \dfrac{4+2\sqrt{3}}{-2} = -2-\sqrt{3}$

19. $\sin 195° = \sin(225° - 30°)$

$\qquad\qquad = \sin 225° \cos 30° - \cos 225° \sin 30°$

$\qquad\qquad = -\sin 45° \cos 30° + \cos 45° \sin 30°$

$\qquad\qquad = -\dfrac{\sqrt{2}}{2} \cdot \dfrac{\sqrt{3}}{2} + \dfrac{\sqrt{2}}{2} \cdot \dfrac{1}{2}$

$\qquad\qquad = \dfrac{\sqrt{2}}{4}\left(1-\sqrt{3}\right)$

$\cos 195° = \cos(225° - 30°)$

$\qquad\qquad = \cos 225° \cos 30° + \sin 225° \sin 30°$

$\qquad\qquad = -\cos 45° \cos 30° - \sin 45° \sin 30°$

$\qquad\qquad = -\dfrac{\sqrt{2}}{2} \cdot \dfrac{\sqrt{3}}{2} - \dfrac{\sqrt{2}}{2} \cdot \dfrac{1}{2}$

$\qquad\qquad = -\dfrac{\sqrt{2}}{4}\left(\sqrt{3}+1\right)$

$\tan 195° = \tan(225° - 30°)$

$\qquad\qquad = \dfrac{\tan 225° - \tan 30°}{1 + \tan 225° \tan 30°}$

$\qquad\qquad = \dfrac{\tan 45° - \tan 30°}{1 + \tan 45° \tan 30°}$

$\qquad\qquad = \dfrac{1 - \left(\dfrac{\sqrt{3}}{3}\right)}{1 + \left(\dfrac{\sqrt{3}}{3}\right)} = \dfrac{3-\sqrt{3}}{3+\sqrt{3}} \cdot \dfrac{3-\sqrt{3}}{3-\sqrt{3}}$

$\qquad\qquad = \dfrac{12-6\sqrt{3}}{6} = 2-\sqrt{3}$

21. $\dfrac{13\pi}{12} = \dfrac{3\pi}{4} + \dfrac{\pi}{3}$

$\begin{aligned} \sin\dfrac{13\pi}{12} &= \sin\left(\dfrac{3\pi}{4} + \dfrac{\pi}{3}\right)\\ &= \sin\dfrac{3\pi}{4}\cos\dfrac{\pi}{3} + \cos\dfrac{3\pi}{4}\sin\dfrac{\pi}{3}\\ &= \dfrac{\sqrt{2}}{2}\cdot\dfrac{1}{2} + \left(-\dfrac{\sqrt{2}}{2}\right)\left(\dfrac{\sqrt{3}}{2}\right)\\ &= \dfrac{\sqrt{2}}{4}\left(1 - \sqrt{3}\right) \end{aligned}$

$\begin{aligned} \cos\dfrac{13\pi}{12} &= \cos\left(\dfrac{3\pi}{4} + \dfrac{\pi}{3}\right)\\ &= \cos\dfrac{3\pi}{4}\cos\dfrac{\pi}{3} - \sin\dfrac{3\pi}{4}\sin\dfrac{\pi}{3}\\ &= -\dfrac{\sqrt{2}}{2}\cdot\dfrac{1}{2} - \dfrac{\sqrt{2}}{2}\cdot\dfrac{\sqrt{3}}{2} = -\dfrac{\sqrt{2}}{4}\left(1 + \sqrt{3}\right) \end{aligned}$

$\begin{aligned} \tan\dfrac{13\pi}{12} &= \tan\left(\dfrac{3\pi}{4} + \dfrac{\pi}{3}\right)\\ &= \dfrac{\tan\left(\dfrac{3\pi}{4}\right) + \tan\left(\dfrac{\pi}{3}\right)}{1 - \tan\left(\dfrac{3\pi}{4}\right)\tan\left(\dfrac{\pi}{3}\right)}\\ &= \dfrac{-1 + \sqrt{3}}{1 - (-1)\left(\sqrt{3}\right)}\\ &= \dfrac{1 - \sqrt{3}}{1 + \sqrt{3}}\cdot\dfrac{1 - \sqrt{3}}{1 - \sqrt{3}}\\ &= -\dfrac{4 - 2\sqrt{3}}{-2}\\ &= 2 - \sqrt{3} \end{aligned}$

23. $-\dfrac{13\pi}{12} = -\left(\dfrac{3\pi}{4} + \dfrac{\pi}{3}\right)$

$\begin{aligned} \sin\left[-\left(\dfrac{3\pi}{4} + \dfrac{\pi}{3}\right)\right] &= -\sin\left(\dfrac{3\pi}{4} + \dfrac{\pi}{3}\right) = -\left[\sin\dfrac{3\pi}{4}\cos\dfrac{\pi}{3} + \cos\dfrac{3\pi}{4}\sin\dfrac{\pi}{3}\right]\\ &= -\left[\dfrac{\sqrt{2}}{2}\left(\dfrac{1}{2}\right) + \left(-\dfrac{\sqrt{2}}{2}\right)\left(\dfrac{\sqrt{3}}{2}\right)\right] = -\dfrac{\sqrt{2}}{4}\left(1 - \sqrt{3}\right) = \dfrac{\sqrt{2}}{4}\left(\sqrt{3} - 1\right) \end{aligned}$

$\begin{aligned} \cos\left[-\left(\dfrac{3\pi}{4} + \dfrac{\pi}{3}\right)\right] &= \cos\left(\dfrac{3\pi}{4} + \dfrac{\pi}{3}\right) = \cos\dfrac{3\pi}{4}\cos\dfrac{\pi}{3} - \sin\dfrac{3\pi}{4}\sin\dfrac{\pi}{3}\\ &= -\dfrac{\sqrt{2}}{2}\left(\dfrac{1}{2}\right) - \dfrac{\sqrt{2}}{2}\left(\dfrac{\sqrt{3}}{2}\right) = -\dfrac{\sqrt{2}}{4}\left(\sqrt{3} + 1\right) \end{aligned}$

$\begin{aligned} \tan\left[-\left(\dfrac{3\pi}{4} + \dfrac{\pi}{3}\right)\right] &= -\tan\left(\dfrac{3\pi}{4} + \dfrac{\pi}{3}\right) = -\dfrac{\tan\dfrac{3\pi}{4} + \tan\dfrac{\pi}{3}}{1 - \tan\dfrac{3\pi}{4}\tan\dfrac{\pi}{3}} = -\dfrac{-1 + \sqrt{3}}{1 - \left(-\sqrt{3}\right)}\\ &= \dfrac{1 - \sqrt{3}}{1 + \sqrt{3}}\cdot\dfrac{1 - \sqrt{3}}{1 - \sqrt{3}} = \dfrac{4 - 2\sqrt{3}}{-2} = -2 + \sqrt{3} \end{aligned}$

25. $285° = 225° + 60°$

$\begin{aligned} \sin 285° &= \sin(225° + 60°) = \sin 225°\cos 60° + \cos 225°\sin 60°\\ &= -\dfrac{\sqrt{2}}{2}\left(\dfrac{1}{2}\right) - \dfrac{\sqrt{2}}{2}\left(\dfrac{\sqrt{3}}{2}\right) = -\dfrac{\sqrt{2}}{4}\left(\sqrt{3} + 1\right) \end{aligned}$

$\begin{aligned} \cos 285° &= \cos(225° + 60°) = \cos 225°\cos 60° - \sin 225°\sin 60°\\ &= -\dfrac{\sqrt{2}}{2}\left(\dfrac{1}{2}\right) - \left(-\dfrac{\sqrt{2}}{2}\right)\left(\dfrac{\sqrt{3}}{2}\right) = \dfrac{\sqrt{2}}{4}\left(\sqrt{3} - 1\right) \end{aligned}$

$\begin{aligned} \tan 285° &= \tan(225° + 60°) = \dfrac{\tan 225° + \tan 60°}{1 - \tan 225°\tan 60°}\\ &= \dfrac{1 + \sqrt{3}}{1 - \sqrt{3}}\cdot\dfrac{1 + \sqrt{3}}{1 + \sqrt{3}} = \dfrac{4 + 2\sqrt{3}}{-2} = -2 - \sqrt{3} = -\left(2 + \sqrt{3}\right) \end{aligned}$

27. $-165° = -(120° + 45°)$

$\sin(-165°) = \sin\left[-(120° + 45°)\right] = -\sin(120° + 45°) = -\left[\sin 120° \cos 45° + \cos 120° \sin 45°\right]$

$$= -\left[\frac{\sqrt{3}}{2} \cdot \frac{\sqrt{2}}{2} - \frac{1}{2} \cdot \frac{\sqrt{2}}{2}\right] = -\frac{\sqrt{2}}{4}\left(\sqrt{3} - 1\right)$$

$\cos(-165°) = \cos\left[-(120° + 45°)\right] = \cos(120° + 45°) = \cos 120° \cos 45° - \sin 120° \sin 45°$

$$= -\frac{1}{2} \cdot \frac{\sqrt{2}}{2} - \frac{\sqrt{3}}{2} \cdot \frac{\sqrt{2}}{2} = -\frac{\sqrt{2}}{4}\left(1 + \sqrt{3}\right)$$

$\tan(-165°) = \tan\left[-(120° + 45°)\right] = -\tan(120° + \tan 45°) = -\dfrac{\tan 120° + \tan 45°}{1 - \tan 120° \tan 45°}$

$$= -\frac{-\sqrt{3} + 1}{1 - \left(-\sqrt{3}\right)(1)} = -\frac{1 - \sqrt{3}}{1 + \sqrt{3}} \cdot \frac{1 - \sqrt{3}}{1 - \sqrt{3}} = -\frac{4 - 2\sqrt{3}}{-2} = 2 - \sqrt{3}$$

29. $\sin 3 \cos 1.2 - \cos 3 \sin 1.2 = \sin(3 - 1.2) = \sin 1.8$

31. $\sin 60° \cos 15° + \cos 60° \sin 15° = \sin(60° + 15°)$
$$= \sin 75°$$

33. $\dfrac{\tan 45° - \tan 30°}{1 + \tan 45° \tan 30°} = \tan(45° - 30°)$
$$= \tan 15°$$

35. $\dfrac{\tan 2x + \tan x}{1 - \tan 2x \tan x} = \tan(2x + x) = \tan 3x$

37. $\sin \dfrac{\pi}{12} \cos \dfrac{\pi}{4} + \cos \dfrac{\pi}{12} \sin \dfrac{\pi}{4} = \sin\left(\dfrac{\pi}{12} + \dfrac{\pi}{4}\right)$
$$= \sin \frac{\pi}{3}$$
$$= \frac{\sqrt{3}}{2}$$

39. $\sin 120° \cos 60° - \cos 120° \sin 60° = \sin(120° - 60°)$
$$= \sin 60°$$
$$= \frac{\sqrt{3}}{2}$$

41. $\dfrac{\tan(5\pi/6) - \tan(\pi/6)}{1 + \tan(5\pi/6) \tan(\pi/6)} = \tan\left(\dfrac{5\pi}{6} - \dfrac{\pi}{6}\right)$
$$= \tan \frac{2\pi}{3}$$
$$= -\sqrt{3}$$

For Exercises 43–49, you have:

$\sin u = \frac{5}{13}$, u **in Quadrant II** $\Rightarrow \cos u = -\frac{12}{13}$, $\tan u = -\frac{5}{12}$

$\cos v = -\frac{3}{5}$, v **in Quadrant II** $\Rightarrow \sin v = \frac{4}{5}$, $\tan v = -\frac{4}{3}$

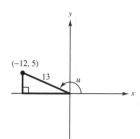

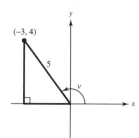

Figures for Exercises 43–49

43. $\sin(u + v) = \sin u \cos v + \cos u \sin v$
$$= \left(\tfrac{5}{13}\right)\left(-\tfrac{3}{5}\right) + \left(-\tfrac{12}{13}\right)\left(\tfrac{4}{5}\right)$$
$$= -\frac{63}{65}$$

45. $\cos(u + v) = \cos u \cos v - \sin u \sin v$
$$= \left(-\tfrac{12}{13}\right)\left(-\tfrac{3}{5}\right) - \left(\tfrac{5}{13}\right)\left(\tfrac{4}{5}\right) = \frac{16}{65}$$

47. $\tan(u + v) = \dfrac{\tan u + \tan v}{1 - \tan u \tan v} = \dfrac{-\dfrac{5}{12} + \left(-\dfrac{4}{3}\right)}{1 - \left(-\dfrac{5}{12}\right)\left(-\dfrac{4}{3}\right)}$

$= \dfrac{-\dfrac{21}{12}}{1 - \dfrac{5}{9}} = \left(-\dfrac{7}{4}\right)\left(\dfrac{9}{4}\right) = -\dfrac{63}{16}$

49. $\sec(v - u) = \dfrac{1}{\cos(v - u)} = \dfrac{1}{\cos v \cos u + \sin v \sin u}$

$= \dfrac{1}{\left(-\dfrac{3}{5}\right)\left(-\dfrac{12}{13}\right) + \left(\dfrac{4}{5}\right)\left(\dfrac{5}{13}\right)} = \dfrac{1}{\left(\dfrac{36}{65}\right) + \left(\dfrac{20}{65}\right)}$

$= \dfrac{1}{\dfrac{56}{65}} = \dfrac{65}{56}$

For Exercises 51–55, you have:

$\sin u = -\dfrac{7}{25}$, u **in Quadrant III** $\Rightarrow \cos u = -\dfrac{24}{25}$, $\tan u = \dfrac{7}{24}$

$\cos v = -\dfrac{4}{5}$, v **in Quadrant III** $\Rightarrow \sin v = -\dfrac{3}{5}$, $\tan v = \dfrac{3}{4}$

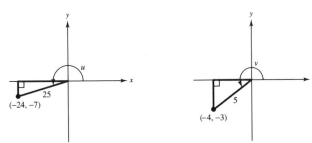

Figures for Exercises 51–55

51. $\cos(u + v) = \cos u \cos v - \sin u \sin v$

$= \left(-\dfrac{24}{25}\right)\left(-\dfrac{4}{5}\right) - \left(-\dfrac{7}{25}\right)\left(-\dfrac{3}{5}\right)$

$= \dfrac{3}{5}$

53. $\tan(u - v) = \dfrac{\tan u - \tan v}{1 + \tan u \tan v}$

$= \dfrac{\dfrac{7}{24} - \dfrac{3}{4}}{1 + \left(\dfrac{7}{24}\right)\left(\dfrac{3}{4}\right)} = \dfrac{-\dfrac{11}{24}}{\dfrac{39}{32}} = -\dfrac{44}{117}$

55. $\csc(u - v) = \dfrac{1}{\sin(u - v)} = \dfrac{1}{\sin u \cos v - \cos u \sin v}$

$= \dfrac{1}{\left(-\dfrac{7}{25}\right)\left(-\dfrac{4}{5}\right) - \left(-\dfrac{24}{25}\right)\left(-\dfrac{3}{5}\right)}$

$= \dfrac{1}{-\dfrac{44}{125}}$

$= -\dfrac{125}{44}$

57. $\sin(\arcsin x + \arccos x) = \sin(\arcsin x)\cos(\arccos x) + \sin(\arccos x)\cos(\arcsin x)$

$= x \cdot x + \sqrt{1 - x^2} \cdot \sqrt{1 - x^2}$

$= x^2 + 1 - x^2$

$= 1$

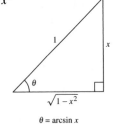

$\theta = \arcsin x$

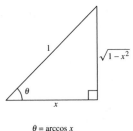

$\theta = \arccos x$

59. $\cos(\arccos x + \arcsin x) = \cos(\arccos x)\cos(\arcsin x) - \sin(\arccos x)\sin(\arcsin x)$

$= x \cdot \sqrt{1 - x^2} - \sqrt{1 - x^2} \cdot x$

$= 0$

(Use the triangles in Exercise 57.)

61. $\sin\left(\dfrac{\pi}{2} - x\right) = \sin\dfrac{\pi}{2}\cos x - \cos\dfrac{\pi}{2}\sin x$

$\qquad\qquad = (1)(\cos x) - (0)(\sin x)$

$\qquad\qquad = \cos x$

63. $\sin\left(\dfrac{\pi}{6} + x\right) = \sin\dfrac{\pi}{6}\cos x + \cos\dfrac{\pi}{6}\sin x$

$\qquad\qquad = \dfrac{1}{2}\left(\cos x + \sqrt{3}\sin x\right)$

65. $\cos(\pi - \theta) + \sin\left(\dfrac{\pi}{2} + \theta\right) = \cos\pi\cos\theta + \sin\pi\sin\theta + \sin\dfrac{\pi}{2}\cos\theta + \cos\dfrac{\pi}{2}\sin\theta$

$\qquad\qquad\qquad\qquad = (-1)(\cos\theta) + (0)(\sin\theta) + (1)(\cos\theta) + (\sin\theta)(0)$

$\qquad\qquad\qquad\qquad = -\cos\theta + \cos\theta$

$\qquad\qquad\qquad\qquad = 0$

67. $\cos(x + y)\cos(x - y) = (\cos x\cos y - \sin x\sin y)(\cos x\cos y + \sin x\sin y)$

$\qquad\qquad\qquad = \cos^2 x\cos^2 y - \sin^2 x\sin^2 y$

$\qquad\qquad\qquad = \cos^2 x(1 - \sin^2 y) - \sin^2 x\sin^2 y$

$\qquad\qquad\qquad = \cos^2 x - \cos^2 x\sin^2 y - \sin^2 x\sin^2 y$

$\qquad\qquad\qquad = \cos^2 x - \sin^2 y(\cos^2 x + \sin^2 x)$

$\qquad\qquad\qquad = \cos^2 x - \sin^2 y$

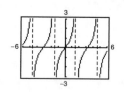

69. $\sin(x + y) + \sin(x - y) = \sin x\cos y + \cos x\sin y + \sin x\cos y - \cos x\sin y$

$\qquad\qquad\qquad = 2\sin x\cos y$

71. $\cos\left(\dfrac{3\pi}{2} - x\right) = \cos\dfrac{3\pi}{2}\cos x + \sin\dfrac{3\pi}{2}\sin x$

$\qquad\qquad = (0)(\cos x) + (-1)(\sin x)$

$\qquad\qquad = -\sin x$

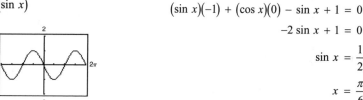

75.

$\qquad\qquad \sin(x + \pi) - \sin x + 1 = 0$

$\qquad\sin x\cos\pi + \cos x\sin\pi - \sin x + 1 = 0$

$\qquad(\sin x)(-1) + (\cos x)(0) - \sin x + 1 = 0$

$\qquad\qquad\qquad\qquad -2\sin x + 1 = 0$

$\qquad\qquad\qquad\qquad \sin x = \dfrac{1}{2}$

$\qquad\qquad\qquad\qquad x = \dfrac{\pi}{6}, \dfrac{5\pi}{6}$

73. $\sin\left(\dfrac{3\pi}{2} + \theta\right) = \sin\dfrac{3\pi}{2}\cos\theta + \cos\dfrac{3\pi}{2}\sin\theta$

$\qquad\qquad = (-1)(\cos\theta) + (0)(\sin\theta)$

$\qquad\qquad = -\cos\theta$

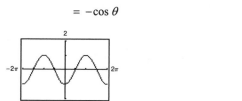

77.

$\qquad\qquad \cos(x + \pi) - \cos x - 1 = 0$

$\qquad\cos x\cos\pi - \sin x\sin\pi - \cos x - 1 = 0$

$\qquad(\cos x)(-1) - (\sin x)(0) - \cos x - 1 = 0$

$\qquad\qquad\qquad\qquad -2\cos x - 1 = 0$

$\qquad\qquad\qquad\qquad \cos x = -\dfrac{1}{2}$

$\qquad\qquad\qquad\qquad x = \dfrac{2\pi}{3}, \dfrac{4\pi}{3}$

79.
$$\sin\left(x + \frac{\pi}{6}\right) - \sin\left(x - \frac{\pi}{6}\right) = \frac{1}{2}$$

$$\sin x \cos \frac{\pi}{6} + \cos x \sin \frac{\pi}{6} - \left(\sin x \cos \frac{\pi}{6} - \cos x \sin \frac{\pi}{6}\right) = \frac{1}{2}$$

$$2 \cos x (0.5) = \frac{1}{2}$$

$$\cos x = \frac{1}{2}$$

$$x = \frac{\pi}{3}, \frac{5\pi}{3}$$

81.
$$\cos\left(x + \frac{\pi}{4}\right) - \cos\left(x - \frac{\pi}{4}\right) = 1$$

$$\cos x \cos \frac{\pi}{4} - \sin x \sin \frac{\pi}{4} - \left(\cos x \cos \frac{\pi}{4} + \sin x \sin \frac{\pi}{4}\right) = 1$$

$$-2 \sin x \left(\frac{\sqrt{2}}{2}\right) = 1$$

$$-\sqrt{2} \sin x = 1$$

$$\sin x = -\frac{1}{\sqrt{2}}$$

$$\sin x = -\frac{\sqrt{2}}{2}$$

$$x = \frac{5\pi}{4}, \frac{7\pi}{4}$$

83.
$$\sin\left(x + \frac{\pi}{2}\right) - \cos^2 x = 0$$

$$\sin x \cos \frac{\pi}{2} + \cos x \sin \frac{\pi}{2} - \cos^2 x = 0$$

$$(\sin x)(0) + (\cos x)(1) - \cos^2 x = 0$$

$$\cos x - \cos^2 x = 0$$

$$\cos x(1 - \cos x) = 0$$

$$\cos x = 0 \qquad \text{or} \quad 1 - \cos x = 0$$

$$x = \frac{\pi}{2}, \frac{3\pi}{2} \qquad\qquad \cos x = 1$$

$$x = 0$$

85. $\cos\left(x + \frac{\pi}{4}\right) + \cos\left(x - \frac{\pi}{4}\right) = 1$

Graph $y_1 = \cos\left(x + \frac{\pi}{4}\right) + \cos\left(x - \frac{\pi}{4}\right)$ and $y_2 = 1$.

$x = \frac{\pi}{4}, \frac{7\pi}{4}$

87. $\sin\left(x + \frac{\pi}{2}\right) + \cos^2 x = 0$

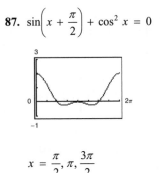

$x = \frac{\pi}{2}, \pi, \frac{3\pi}{2}$

89. $y = \dfrac{1}{3}\sin 2t + \dfrac{1}{4}\cos 2t$

   (a) $a = \dfrac{1}{3}, b = \dfrac{1}{4}, B = 2$

$$C = \arctan\dfrac{b}{a} = \arctan\dfrac{3}{4} \approx 0.6435$$

$$y \approx \sqrt{\left(\dfrac{1}{3}\right)^2 + \left(\dfrac{1}{4}\right)^2}\, \sin(2t + 0.6435) = \dfrac{5}{12}\sin(2t + 0.6435)$$

   (b) Amplitude: $\dfrac{5}{12}$ feet

   (c) Frequency: $\dfrac{1}{\text{period}} = \dfrac{B}{2\pi} = \dfrac{2}{2\pi} = \dfrac{1}{\pi}$ cycle per second

91. True.

$$\sin(u + v) = \sin u \cos v + \cos u \sin v$$

$$\sin(u - v) = \sin u \cos v - \cos u \sin v$$

So, $\sin(u \pm v) = \sin u \cos v \pm \cos u \sin v$.

93. False.

$$\tan\left(x - \dfrac{\pi}{4}\right) = \dfrac{\tan x - \tan(\pi/4)}{1 + \tan x \tan(\pi/4)}$$

$$= \dfrac{\tan x - 1}{1 + \tan x}$$

95. $\cos(n\pi + \theta) = \cos n\pi \cos\theta - \sin n\pi \sin\theta$

$$= (-1)^n(\cos\theta) - (0)(\sin\theta)$$

$$= (-1)^n(\cos\theta),\text{ where }n\text{ is an integer.}$$

97. $C = \arctan\dfrac{b}{a} \Rightarrow \sin C = \dfrac{b}{\sqrt{a^2 + b^2}},\ \cos C = \dfrac{a}{\sqrt{a^2 + b^2}}$

$$\sqrt{a^2 + b^2}\, \sin(B\theta + C) = \sqrt{a^2 + b^2}\left(\sin B\theta \cdot \dfrac{a}{\sqrt{a^2 + b^2}} + \dfrac{b}{\sqrt{a^2 + b^2}} \cdot \cos B\theta\right) = a\sin B\theta + b\cos B\theta$$

99. $\sin\theta + \cos\theta$

$a = 1, b = 1, B = 1$

   (a) $C = \arctan\dfrac{b}{a} = \arctan 1 = \dfrac{\pi}{4}$

$$\sin\theta + \cos\theta = \sqrt{a^2 + b^2}\, \sin(B\theta + C)$$

$$= \sqrt{2}\,\sin\left(\theta + \dfrac{\pi}{4}\right)$$

   (b) $C = \arctan\dfrac{a}{b} = \arctan 1 = \dfrac{\pi}{4}$

$$\sin\theta + \cos\theta = \sqrt{a^2 + b^2}\, \cos(B\theta - C)$$

$$= \sqrt{2}\,\cos\left(\theta - \dfrac{\pi}{4}\right)$$

101. $12\sin 3\theta + 5\cos 3\theta$

$a = 12, b = 5, B = 3$

   (a) $C = \arctan\dfrac{b}{a} = \arctan\dfrac{5}{12} \approx 0.3948$

$$12\sin 3\theta + 5\cos 3\theta = \sqrt{a^2 + b^2}\, \sin(B\theta + C)$$

$$\approx 13\sin(3\theta + 0.3948)$$

   (b) $C = \arctan\dfrac{a}{b} = \arctan\dfrac{12}{5} \approx 1.1760$

$$12\sin 3\theta + 5\cos 3\theta = \sqrt{a^2 + b^2}\, \cos(B\theta - C)$$

$$\approx 13\cos(3\theta - 1.1760)$$

103. $C = \arctan \dfrac{b}{a} = \dfrac{\pi}{4} \Rightarrow a = b, a > 0, b > 0$

$\sqrt{a^2 + b^2} = 2 \Rightarrow a = b = \sqrt{2}$

$B = 1$

$2 \sin\left(\theta + \dfrac{\pi}{4}\right) = \sqrt{2} \sin\theta + \sqrt{2} \cos\theta$

105.

$$\dfrac{\cos(x + h) - \cos x}{h} = \dfrac{\cos x \cos h - \sin x \sin h - \cos x}{h}$$

$$= \dfrac{\cos x \cos h - \cos x - \sin x \sin h}{h}$$

$$= \dfrac{\cos x(\cos h - 1) - \sin x \sin h}{h}$$

$$= \dfrac{\cos x(\cos h - 1)}{h} - \dfrac{\sin x \sin h}{h}$$

107.

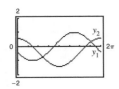

$m_1 = \tan\alpha$ and $m_2 = \tan\beta$

$\beta + \delta = 90° \Rightarrow \delta = 90° - \beta$

$\alpha + \theta + \delta = 90° \Rightarrow \alpha + \theta + (90° - \beta) = 90° \Rightarrow \theta = \beta - \alpha$

So, $\theta = \arctan m_2 - \arctan m_1$.

For $y = x$ and $y = \sqrt{3}x$ you have $m_1 = 1$ and $m_2 = \sqrt{3}$.

$\theta = \arctan\sqrt{3} - \arctan 1 = 60° - 45° = 15°$

109. $y_1 = \cos(x + 2)$, $y_2 = \cos x + \cos 2$

No, $y_1 \ne y_2$ because their graphs are different.

111. (a) To prove the identity for $\sin(u + v)$ you first need to prove the identity for $\cos(u - v)$.

Assume $0 < v < u < 2\pi$ and locate u, v, and $u - v$ on the unit circle.

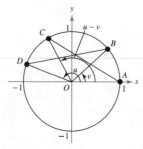

The coordinates of the points on the circle are:

$A = (1, 0)$, $B = (\cos v, \sin v)$, $C = (\cos(u - v), \sin(u - v))$, and $D = (\cos u, \sin u)$.

Because $\angle DOB = \angle COA$, chords AC and BD are equal. By the Distance Formula:

$$\sqrt{\left[\cos(u - v) - 1\right]^2 + \left[\sin(u - v) - 0\right]^2} = \sqrt{\left(\cos u - \cos v\right)^2 + \left(\sin u - \sin v\right)^2}$$

$$\cos^2(u - v) - 2\cos(u - v) + 1 + \sin^2(u - v) = \cos^2 u - 2\cos u \cos v + \cos^2 v + \sin^2 u - 2\sin u \sin v + \sin^2 v$$

$$\left[\cos^2(u - v) + \sin^2(u - v)\right] + 1 - 2\cos(u - v) = \left(\cos^2 u + \sin^2 u\right) + \left(\cos^2 v + \sin^2 v\right) - 2\cos u \cos v - 2\sin u \sin v$$

$$2 - 2\cos(u - v) = 2 - 2\cos u \cos v - 2\sin u \sin v$$

$$-2\cos(u - v) = -2\left(\cos u \cos v + \sin u \sin v\right)$$

$$\cos(u - v) = \cos u \cos v + \sin u \sin v$$

Now, to prove the identity for $\sin(u + v)$, use cofunction identities.

$$\sin(u + v) = \cos\left[\frac{\pi}{2} - (u + v)\right] = \cos\left[\left(\frac{\pi}{2} - u\right) - v\right]$$

$$= \cos\left(\frac{\pi}{2} - u\right)\cos v + \sin\left(\frac{\pi}{2} - u\right)\sin v$$

$$= \sin u \cos v + \cos u \sin v$$

(b) First, prove $\cos(u - v) = \cos u \cos v + \sin u \sin v$ using the figure containing points

$A(1, 0)$

$B(\cos(u - v), \sin(u - v))$

$C(\cos v, \sin v)$

$D(\cos u, \sin u)$

on the unit circle.

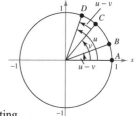

Because chords AB and CD are each subtended by angle $u - v$, their lengths are equal. Equating

$\left[d(A, B)\right]^2 = \left[d(C, D)\right]^2$ you have $\left(\cos(u - v) - 1\right)^2 + \sin^2(u - v) = \left(\cos u - \cos v\right)^2 + \left(\sin u - \sin v\right)^2$.

Simplifying and solving for $\cos(u - v)$, you have $\cos(u - v) = \cos u \cos v + \sin u \sin v$.

Using $\sin \theta = \cos\left(\frac{\pi}{2} - \theta\right)$,

$$\sin(u - v) = \cos\left[\frac{\pi}{2} - (u - v)\right] = \cos\left[\left(\frac{\pi}{2} - u\right) - (-v)\right] = \cos\left(\frac{\pi}{2} - u\right)\cos(-v) + \sin\left(\frac{\pi}{2} - u\right)\sin(-v)$$

$$= \sin u \cos v - \cos u \sin v$$

Section 7.5 Multiple-Angle and Product-to-Sum Formulas

- You should know the following double-angle formulas.

  (a) $\sin 2u = 2 \sin u \cos u$

  (b) $\cos 2u = \cos^2 u - \sin^2 u$

  $= 2 \cos^2 u - 1$

  $= 1 - 2 \sin^2 u$

  (c) $\tan 2u = \dfrac{2 \tan u}{1 - \tan^2 u}$

- You should be able to reduce the power of a trigonometric function.

  (a) $\sin^2 u = \dfrac{1 - \cos 2u}{2}$ (b) $\cos^2 u = \dfrac{1 + \cos 2u}{2}$ (c) $\tan^2 u = \dfrac{1 - \cos 2u}{1 + \cos 2u}$

- You should be able to use the half-angle formulas. The signs of $\sin \dfrac{u}{2}$ and $\cos \dfrac{u}{2}$ depend on the quadrant in which $\dfrac{u}{2}$ lies.

  (a) $\sin \dfrac{u}{2} = \pm \sqrt{\dfrac{1 - \cos u}{2}}$

  (b) $\cos \dfrac{u}{2} = \pm \sqrt{\dfrac{1 + \cos u}{2}}$

  (c) $\tan \dfrac{u}{2} = \dfrac{1 - \cos u}{\sin u} = \dfrac{\sin u}{1 + \cos u}$

- You should be able to use the product-sum formulas.

  (a) $\sin u \sin v = \dfrac{1}{2} \big[\cos(u - v) - \cos(u + v) \big]$ (b) $\cos u \cos v = \dfrac{1}{2} \big[\cos(u - v) + \cos(u + v) \big]$

  (c) $\sin u \cos v = \dfrac{1}{2} \big[\sin(u + v) + \sin(u - v) \big]$ (d) $\cos u \sin v = \dfrac{1}{2} \big[\sin(u + v) - \sin(u - v) \big]$

- You should be able to use the sum-product formulas.

  (a) $\sin x + \sin y = 2 \sin \left(\dfrac{x + y}{2} \right) \cos \left(\dfrac{x - y}{2} \right)$ (b) $\sin x - \sin y = 2 \cos \left(\dfrac{x + y}{2} \right) \sin \left(\dfrac{x - y}{2} \right)$

  (c) $\cos x + \cos y = 2 \cos \left(\dfrac{x + y}{2} \right) \cos \left(\dfrac{x - y}{2} \right)$ (d) $\cos x - \cos y = -2 \sin \left(\dfrac{x + y}{2} \right) \sin \left(\dfrac{x - y}{2} \right)$

1. $2 \sin u \cos u$

3. $\cos^2 u - \sin^2 u = 2 \cos^2 u - 1 = 1 - 2 \sin^2 u$

5. $\pm \sqrt{\dfrac{1 - \cos u}{2}}$

7. $\dfrac{1}{2} \big[\cos(u - v) + \cos(u + v) \big]$

9. $2 \sin \left(\dfrac{u + v}{2} \right) \cos \left(\dfrac{u - v}{2} \right)$

Figure for Exercises 11–17

$$\sin \theta = \frac{\sqrt{17}}{17}$$

$$\cos \theta = \frac{4\sqrt{17}}{17}$$

$$\tan \theta = \frac{1}{4}$$

11. $\cos 2\theta = 2 \cos^2 \theta - 1$

$$= 2 \left(\frac{4\sqrt{17}}{17} \right)^2 - 1$$

$$= \frac{32}{17} - 1$$

$$= \frac{15}{17}$$

13. $\tan 2\theta = \dfrac{2 \tan \theta}{1 - \tan^2 \theta}$

$$= \frac{2 \left(\frac{1}{4} \right)}{1 - \left(\frac{1}{4} \right)^2}$$

$$= \frac{\frac{1}{2}}{1 - \frac{1}{16}}$$

$$= \frac{1}{2} \cdot \frac{16}{15}$$

$$= \frac{8}{15}$$

15. $\csc 2\theta = \dfrac{1}{\sin 2\theta} = \dfrac{1}{2 \sin \theta \cos \theta} = \dfrac{1}{2 \left(\frac{\sqrt{17}}{17} \right) \left(\frac{4\sqrt{17}}{17} \right)}$

$$= \frac{17}{8}$$

17. From Exercise 11, $\cos 2\theta = \frac{15}{17}$.

From Exercise 12, $\sin 2\theta = \frac{8}{17}$.

$\sin 4\theta = 2 \sin 2\theta \cos 2\theta$

$$= 2 \left(\tfrac{8}{17} \right) \left(\tfrac{15}{17} \right)$$

$$= \tfrac{240}{289}$$

19.
$$\sin 2x - \sin x = 0$$
$$2 \sin x \cos x - \sin x = 0$$
$$\sin x (2 \cos x - 1) = 0$$

$\sin x = 0 \quad$ or $\quad 2 \cos x - 1 = 0$

$x = 0, \pi \qquad\qquad \cos x = \dfrac{1}{2}$

$$x = \frac{\pi}{3}, \frac{5\pi}{3}$$

$$x = 0, \frac{\pi}{3}, \pi, \frac{5\pi}{3}$$

21. $4 \sin x \cos x = 1$

$$2 \sin 2x = 1$$

$$\sin 2x = \frac{1}{2}$$

$2x = \dfrac{\pi}{6} + 2n\pi \quad$ or $\quad 2x = \dfrac{5\pi}{6} + 2n\pi$

$x = \dfrac{\pi}{12} + n\pi \qquad\qquad x = \dfrac{5\pi}{12} + n\pi$

$x = \dfrac{\pi}{12}, \dfrac{13\pi}{12} \qquad\qquad x = \dfrac{5\pi}{12}, \dfrac{17\pi}{12}$

23.
$$\cos 2x - \cos x = 0$$
$$\cos 2x = \cos x$$
$$\cos^2 x - \sin^2 x = \cos x$$
$$\cos^2 x - \left(1 - \cos^2 x \right) - \cos x = 0$$
$$2 \cos^2 x - \cos x - 1 = 0$$
$$\left(2 \cos x + 1 \right)\left(\cos x - 1 \right) = 0$$

$2 \cos x + 1 = 0 \quad$ or $\quad \cos x - 1 = 0$

$\cos x = -\dfrac{1}{2} \qquad\qquad \cos x = 1$

$x = \dfrac{2\pi}{3}, \dfrac{4\pi}{3} \qquad\qquad x = 0$

25.
$$\sin 4x = -2 \sin 2x$$
$$\sin 4x + 2 \sin 2x = 0$$
$$2 \sin 2x \cos 2x + 2 \sin 2x = 0$$
$$2 \sin 2x (\cos 2x + 1) = 0$$

$2 \sin 2x = 0 \quad$ or $\quad \cos 2x + 1 = 0$

$\sin 2x = 0 \qquad\qquad \cos 2x = -1$

$2x = n\pi \qquad\qquad 2x = \pi + 2n\pi$

$x = \dfrac{n}{2}\pi \qquad\qquad x = \dfrac{\pi}{2} + n\pi$

$x = 0, \dfrac{\pi}{2}, \pi, \dfrac{3\pi}{2} \qquad\qquad x = \dfrac{\pi}{2}, \dfrac{3\pi}{2}$

27. $\tan 2x - \cot x = 0$

$$\frac{2 \tan x}{1 - \tan^2 x} = \cot x$$

$$2 \tan x = \cot x \left(1 - \tan^2 x\right)$$

$$2 \tan x = \cot x - \cot x \tan^2 x$$

$$2 \tan x = \cot x - \tan x$$

$$3 \tan x = \cot x$$

$$3 \tan x - \cot x = 0$$

$$3 \tan x - \frac{1}{\tan x} = 0$$

$$\frac{3 \tan^2 x - 1}{\tan x} = 0$$

$$\frac{1}{\tan x} \left(3 \tan^2 x - 1\right) = 0$$

$$\cot x \left(3 \tan^2 x - 1\right) = 0$$

$$\cot x = 0 \quad \text{or} \quad 3 \tan^2 x - 1 = 0$$

$$x = \frac{\pi}{2}, \frac{3\pi}{2} \qquad \tan^2 x = \frac{1}{3}$$

$$\tan x = \pm \frac{\sqrt{3}}{3}$$

$$x = \frac{\pi}{6}, \frac{5\pi}{6}, \frac{7\pi}{6}, \frac{11\pi}{6}$$

$$x = \frac{\pi}{6}, \frac{\pi}{2}, \frac{5\pi}{6}, \frac{7\pi}{6}, \frac{3\pi}{2}, \frac{11\pi}{6}$$

29. $6 \sin x \cos x = 3\left(2 \sin x \cos x\right)$

$$= 3 \sin 2x$$

31. $6 \cos^2 x - 3 = 3\left(2 \cos^2 x - 1\right)$

$$= 3 \cos 2x$$

33. $4 - 8 \sin^2 x = 4\left(1 - 2 \sin^2 x\right)$

$$= 4 \cos 2x$$

35. $\left(\cos x + \sin x\right)\left(\cos x - \sin x\right) = \cos^2 x - \sin^2 x$

$$= \cos 2x$$

37. $\sin u = -\frac{3}{5}, \dfrac{3\pi}{2} < u < 2\pi$

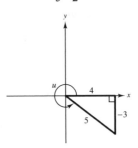

$$\sin 2u = 2 \sin u \cos u = 2\left(-\frac{3}{5}\right)\left(\frac{4}{5}\right) = -\frac{24}{25}$$

$$\cos 2u = \cos^2 u - \sin^2 u = \frac{16}{25} - \frac{9}{25} = \frac{7}{25}$$

$$\tan 2u = \frac{2 \tan u}{1 - \tan^2 u} = \frac{2\left(-\dfrac{3}{4}\right)}{1 - \dfrac{9}{16}} = -\frac{3}{2}\left(\frac{16}{7}\right) = -\frac{24}{7}$$

39. $\tan u = \dfrac{3}{5}, 0 < u < \dfrac{\pi}{2}$

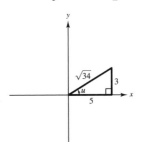

$$\sin 2u = 2 \sin u \cos u = 2\left(\frac{3}{\sqrt{34}}\right)\left(\frac{5}{\sqrt{34}}\right) = \frac{15}{17}$$

$$\cos 2u = \cos^2 u - \sin^2 u = \frac{25}{34} - \frac{9}{34} = \frac{8}{17}$$

$$\tan 2u = \frac{2 \tan u}{1 - \tan^2 u} = \frac{2\left(\dfrac{3}{5}\right)}{1 - \dfrac{9}{25}} = \frac{6}{5}\left(\frac{25}{16}\right) = \frac{15}{8}$$

41. $\sec u = -2, \dfrac{\pi}{2} < u < \pi$

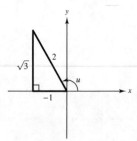

$$\sin 2u = 2 \sin u \cos u = 2\left(\frac{\sqrt{3}}{2}\right)\left(-\frac{1}{2}\right) = -\frac{\sqrt{3}}{2}$$

$$\cos 2u = \cos^2 u - \sin^2 u = \frac{1}{4} - \frac{3}{4} = -\frac{1}{2}$$

$$\tan 2u = \frac{2 \tan u}{1 - \tan^2 u} = \frac{2(-\sqrt{3})}{1 - 3} = \sqrt{3}$$

43. $\cos^4 x = (\cos^2 x)(\cos^2 x) = \left(\dfrac{1 + \cos 2x}{2}\right)\left(\dfrac{1 + \cos 2x}{2}\right) = \dfrac{1 + 2\cos 2x + \cos^2 2x}{4}$

$$= \frac{1 + 2\cos 2x + \dfrac{1 + \cos 4x}{2}}{4}$$

$$= \frac{2 + 4\cos 2x + 1 + \cos 4x}{8}$$

$$= \frac{3 + 4\cos 2x + \cos 4x}{8}$$

$$= \frac{1}{8}(3 + 4\cos 2x + \cos 4x)$$

45. $\cos^4 2x = (\cos^2 2x)^2$

$$= \left(\frac{1 + \cos 4x}{2}\right)^2$$

$$= \frac{1}{4}(1 + 2\cos 4x + \cos^2 4x)$$

$$= \frac{1}{4}\left(1 + 2\cos 4x + \frac{1 + \cos 8x}{2}\right)$$

$$= \frac{1}{4} + \frac{1}{2}\cos 4x + \frac{1}{8} + \frac{1}{8}\cos 8x$$

$$= \frac{3}{8} + \frac{1}{2}\cos 4x + \frac{1}{8}\cos 8x$$

$$= \frac{1}{8}(3 + 4\cos 4x + \cos 8x)$$

47. $\tan^4 2x = (\tan^2 2x)^2$

$$= \left(\frac{1 - \cos 4x}{1 + \cos 4x}\right)^2$$

$$= \frac{1 - 2\cos 4x + \cos^2 4x}{1 + 2\cos 4x + \cos^2 4x}$$

$$= \frac{1 - 2\cos 4x + \dfrac{1 + \cos 8x}{2}}{1 + 2\cos 4x + \dfrac{1 + \cos 8x}{2}}$$

$$= \frac{\dfrac{1}{2}(2 - 4\cos 4x + 1 + \cos 8x)}{\dfrac{1}{2}(2 + 4\cos 4x + 1 + \cos 8x)}$$

$$= \frac{3 - 4\cos 4x + \cos 8x}{3 + 4\cos 4x + \cos 8x}$$

49. $\sin^2 2x \cos^2 2x = \left(\dfrac{1 - \cos 4x}{2}\right)\left(\dfrac{1 + \cos 4x}{2}\right)$

$\qquad\qquad = \dfrac{1}{4}\left(1 - \cos^2 4x\right)$

$\qquad\qquad = \dfrac{1}{4}\left(1 - \dfrac{1 + \cos 8x}{2}\right)$

$\qquad\qquad = \dfrac{1}{4} - \dfrac{1}{8} - \dfrac{1}{8}\cos 8x$

$\qquad\qquad = \dfrac{1}{8} - \dfrac{1}{8}\cos 8x$

$\qquad\qquad = \dfrac{1}{8}\left(1 - \cos 8x\right)$

51. $\sin^4 x \cos^2 x = \sin^2 x \sin^2 x \cos^2 x$

$\qquad\qquad = \left(\dfrac{1 - \cos 2x}{2}\right)\left(\dfrac{1 - \cos 2x}{2}\right)\left(\dfrac{1 + \cos 2x}{2}\right)$

$\qquad\qquad = \dfrac{1}{8}\left(1 - \cos 2x\right)\left(1 - \cos^2 2x\right)$

$\qquad\qquad = \dfrac{1}{8}\left(1 - \cos 2x - \cos^2 2x + \cos^3 2x\right)$

$\qquad\qquad = \dfrac{1}{8}\left[1 - \cos 2x - \left(\dfrac{1 + \cos 4x}{2}\right) + \cos 2x\left(\dfrac{1 + \cos 4x}{2}\right)\right]$

$\qquad\qquad = \dfrac{1}{16}\left[2 - 2\cos 2x - 1 - \cos 4x + \cos 2x + \cos 2x \cos 4x\right]$

$\qquad\qquad = \dfrac{1}{16}\left[1 - \cos 2x - \cos 4x + \cos 2x \cos 4x\right]$

Figure for Exercises 53–57

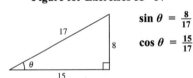

$\sin\theta = \dfrac{8}{17}$

$\cos\theta = \dfrac{15}{17}$

53. $\cos\dfrac{\theta}{2} = \sqrt{\dfrac{1 + \cos\theta}{2}} = \sqrt{\dfrac{1 + \dfrac{15}{17}}{2}} = \sqrt{\dfrac{32}{34}} = \sqrt{\dfrac{16}{17}} = \dfrac{4\sqrt{17}}{17}$

55. $\tan\dfrac{\theta}{2} = \dfrac{\sin\theta}{1 + \cos\theta} = \dfrac{8/17}{1 + (15/17)} = \dfrac{8}{17}\cdot\dfrac{17}{32} = \dfrac{1}{4}$

57. $\csc\dfrac{\theta}{2} = \dfrac{1}{\sin(\theta/2)} = \dfrac{1}{\sqrt{(1 - \cos\theta)/2}} = \dfrac{1}{\sqrt{[1 - (15/17)]/2}} = \dfrac{1}{\sqrt{1/17}} = \sqrt{17}$

59. $\sin 75° = \sin\left(\dfrac{1}{2} \cdot 150°\right) = \sqrt{\dfrac{1 - \cos 150°}{2}} = \sqrt{\dfrac{1 + \left(\sqrt{3}/2\right)}{2}}$

$= \dfrac{1}{2}\sqrt{2 + \sqrt{3}}$

$\cos 75° = \cos\left(\dfrac{1}{2} \cdot 150°\right) = \sqrt{\dfrac{1 + \cos 150°}{2}} = \sqrt{\dfrac{1 - \left(\sqrt{3}/2\right)}{2}}$

$= \dfrac{1}{2}\sqrt{2 - \sqrt{3}}$

$\tan 75° = \tan\left(\dfrac{1}{2} \cdot 150°\right) = \dfrac{\sin 150°}{1 + \cos 150°} = \dfrac{1/2}{1 - \left(\sqrt{3}/2\right)}$

$= \dfrac{1}{2 - \sqrt{3}} \cdot \dfrac{2 + \sqrt{3}}{2 + \sqrt{3}} = \dfrac{2 + \sqrt{3}}{4 - 3} = 2 + \sqrt{3}$

61. $\sin 112° 30' = \sin\left(\dfrac{1}{2} \cdot 225°\right) = \sqrt{\dfrac{1 - \cos 225°}{2}} = \sqrt{\dfrac{1 + \left(\sqrt{2}/2\right)}{2}} = \dfrac{1}{2}\sqrt{2 + \sqrt{2}}$

$\cos 112° 30' = \cos\left(\dfrac{1}{2} \cdot 225°\right) = -\sqrt{\dfrac{1 + \cos 225°}{2}} = -\sqrt{\dfrac{1 - \left(\sqrt{2}/2\right)}{2}} = -\dfrac{1}{2}\sqrt{2 - \sqrt{2}}$

$\tan 112° 30' = \tan\left(\dfrac{1}{2} \cdot 225°\right) = \dfrac{\sin 225°}{1 + \cos 225°} = \dfrac{-\sqrt{2}/2}{1 - \left(\sqrt{2}/2\right)} = -1 - \sqrt{2}$

63. $\sin \dfrac{\pi}{8} = \sin\left[\dfrac{1}{2}\left(\dfrac{\pi}{4}\right)\right] = \sqrt{\dfrac{1 - \cos \dfrac{\pi}{4}}{2}} = \dfrac{1}{2}\sqrt{2 - \sqrt{2}}$

$\cos \dfrac{\pi}{8} = \cos\left[\dfrac{1}{2}\left(\dfrac{\pi}{4}\right)\right] = \sqrt{\dfrac{1 + \cos \dfrac{\pi}{4}}{2}} = \dfrac{1}{2}\sqrt{2 + \sqrt{2}}$

$\tan \dfrac{\pi}{8} = \tan\left[\dfrac{1}{2}\left(\dfrac{\pi}{4}\right)\right] = \dfrac{\sin \dfrac{\pi}{4}}{1 + \cos \dfrac{\pi}{4}} = \dfrac{\dfrac{\sqrt{2}}{2}}{1 + \dfrac{\sqrt{2}}{2}} = \sqrt{2} - 1$

65. $\sin \dfrac{3\pi}{8} = \sin\left(\dfrac{1}{2} \cdot \dfrac{3\pi}{4}\right) = \sqrt{\dfrac{1 - \cos \dfrac{3\pi}{4}}{2}} = \sqrt{\dfrac{1 + \dfrac{\sqrt{2}}{2}}{2}} = \dfrac{1}{2}\sqrt{2 + \sqrt{2}}$

$\cos \dfrac{3\pi}{8} = \cos\left(\dfrac{1}{2} \cdot \dfrac{3\pi}{4}\right) = \sqrt{\dfrac{1 + \cos \dfrac{3\pi}{4}}{2}} = \sqrt{\dfrac{1 - \dfrac{\sqrt{2}}{2}}{2}} = \dfrac{1}{2}\sqrt{2 - \sqrt{2}}$

$\tan \dfrac{3\pi}{8} = \tan\left(\dfrac{1}{2} \cdot \dfrac{3\pi}{4}\right) = \dfrac{\sin \dfrac{3\pi}{4}}{1 + \cos \dfrac{3\pi}{4}} = \dfrac{\dfrac{\sqrt{2}}{2}}{1 - \dfrac{\sqrt{2}}{2}} = \dfrac{\dfrac{\sqrt{2}}{2}}{\dfrac{\left(2 - \sqrt{2}\right)}{2}} = \dfrac{\sqrt{2}}{2 - \sqrt{2}} = \sqrt{2} + 1$

67. $\cos u = \dfrac{7}{25}, 0 < u < \dfrac{\pi}{2}$

    (a) Because u is in Quadrant I, $\dfrac{u}{2}$ is also in Quadrant I.

    (b) $\sin \dfrac{u}{2} = \sqrt{\dfrac{1 - \cos u}{2}} = \sqrt{\dfrac{1 - \dfrac{7}{25}}{2}} = \sqrt{\dfrac{9}{25}} = \dfrac{3}{5}$

        $\cos \dfrac{u}{2} = \sqrt{\dfrac{1 + \cos u}{2}} = \sqrt{\dfrac{1 + \dfrac{7}{25}}{2}} = \sqrt{\dfrac{16}{25}} = \dfrac{4}{5}$

        $\tan \dfrac{u}{2} = \dfrac{1 - \cos u}{\sin u} = \dfrac{1 - \dfrac{7}{25}}{\dfrac{24}{25}} = \dfrac{3}{4}$

69. $\tan u = -\dfrac{5}{12}, \dfrac{3\pi}{2} < u < 2\pi$

    (a) Because u is in Quadrant IV, $\dfrac{u}{2}$ is in Quadrant II.

    (b) $\sin \dfrac{u}{2} = \sqrt{\dfrac{1 - \cos u}{2}} = \sqrt{\dfrac{1 - \dfrac{12}{13}}{2}} = \sqrt{\dfrac{1}{26}} = \dfrac{\sqrt{26}}{26}$

        $\cos \dfrac{u}{2} = -\sqrt{\dfrac{1 + \cos u}{2}} = -\sqrt{\dfrac{1 + \dfrac{12}{13}}{2}} = -\sqrt{\dfrac{25}{26}} = -\dfrac{5\sqrt{26}}{26}$

        $\tan \dfrac{u}{2} = \dfrac{1 - \cos u}{\sin u} = \dfrac{1 - \dfrac{12}{13}}{\left(-\dfrac{5}{13}\right)} = -\dfrac{1}{5}$

71. $\csc u = -\dfrac{5}{3}, \pi < u < \dfrac{3\pi}{2} \Rightarrow \sin u = -\dfrac{3}{5}$ and $\cos u = -\dfrac{4}{5}$

    (a) Because u is in Quadrant III, $\dfrac{u}{2}$ is in Quadrant II.

    (b) $\sin\left(\dfrac{u}{2}\right) = \sqrt{\dfrac{1 - \cos u}{2}} = \sqrt{\dfrac{1 + \dfrac{4}{5}}{2}} = \dfrac{3\sqrt{10}}{10}$

        $\cos\left(\dfrac{u}{2}\right) = -\sqrt{\dfrac{1 + \cos u}{2}} = -\sqrt{\dfrac{1 - \dfrac{4}{5}}{2}} = -\dfrac{\sqrt{10}}{10}$

        $\tan\left(\dfrac{u}{2}\right) = \dfrac{1 - \cos u}{\sin u} = \dfrac{1 + \dfrac{4}{5}}{-\dfrac{3}{5}} = -3$

73. $\sqrt{\dfrac{1 - \cos 6x}{2}} = \left|\sin 3x\right|$

75. $-\sqrt{\dfrac{1 - \cos 8x}{1 + \cos 8x}} = -\dfrac{\sqrt{\dfrac{1 - \cos 8x}{2}}}{\sqrt{\dfrac{1 + \cos 8x}{2}}}$

                               $= -\left|\dfrac{\sin 4x}{\cos 4x}\right|$

                               $= -\left|\tan 4x\right|$

77. $\sin \dfrac{x}{2} + \cos x = 0$

$\pm\sqrt{\dfrac{1 - \cos x}{2}} = -\cos x$

$\dfrac{1 - \cos x}{2} = \cos^2 x$

$0 = 2\cos^2 x + \cos x - 1$

$= (2\cos x - 1)(\cos x + 1)$

$\cos x = \dfrac{1}{2} \quad \text{or} \quad \cos x = -1$

$x = \dfrac{\pi}{3}, \dfrac{5\pi}{3} \qquad x = \pi$

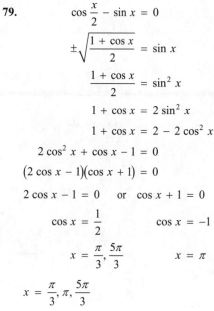

By checking these values in the original equation,
$x = \pi/3$ and $x = 5\pi/3$ are extraneous, and $x = \pi$ is
the only solution.

79. $\cos \dfrac{x}{2} - \sin x = 0$

$\pm\sqrt{\dfrac{1 + \cos x}{2}} = \sin x$

$\dfrac{1 + \cos x}{2} = \sin^2 x$

$1 + \cos x = 2\sin^2 x$

$1 + \cos x = 2 - 2\cos^2 x$

$2\cos^2 x + \cos x - 1 = 0$

$(2\cos x - 1)(\cos x + 1) = 0$

$2\cos x - 1 = 0 \quad \text{or} \quad \cos x + 1 = 0$

$\cos x = \dfrac{1}{2} \qquad\qquad \cos x = -1$

$x = \dfrac{\pi}{3}, \dfrac{5\pi}{3} \qquad\qquad x = \pi$

$x = \dfrac{\pi}{3}, \pi, \dfrac{5\pi}{3}$

$\pi/3, \pi,$ and $5\pi/3$ are all solutions to the equation.

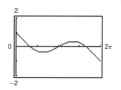

81. $\sin \dfrac{\pi}{3} \cos \dfrac{\pi}{6} = \dfrac{1}{2}\left[\sin\left(\dfrac{\pi}{3} + \dfrac{\pi}{6}\right) + \sin\left(\dfrac{\pi}{3} - \dfrac{\pi}{6}\right)\right]$

$= \dfrac{1}{2}\left(\sin \dfrac{\pi}{2} + \sin \dfrac{\pi}{6}\right)$

83. $10 \cos 75° \cos 15° = 10\left(\dfrac{1}{2}\right)\left[\cos(75° - 15°) + \cos(75° + 15°)\right] = 5\left[\cos 60° + \cos 90°\right]$

85. $\sin 5\theta \sin 3\theta = \dfrac{1}{2}\left[\cos(5\theta - 3\theta) - \cos(5\theta + 3\theta)\right]$

$= \dfrac{1}{2}\left(\cos 2\theta - \cos 8\theta\right)$

87. $7 \cos(-5\beta) \sin 3\beta = 7 \cdot \dfrac{1}{2}\left[\sin(-5\beta + 3\beta) - \sin(-5\beta - 3\beta)\right]$

$= \dfrac{7}{2}\left(\sin(-2\beta) - \sin(-8\beta)\right)$

89. $\sin(x + y)\sin(x - y) = \dfrac{1}{2}(\cos 2y - \cos 2x)$

91. $\sin 3\theta + \sin \theta = 2\sin\left(\dfrac{3\theta + \theta}{2}\right)\cos\left(\dfrac{3\theta - \theta}{2}\right)$

$= 2\sin 2\theta \cos \theta$

93. $\cos 6x + \cos 2x = 2\cos\left(\dfrac{6x + 2x}{2}\right)\cos\left(\dfrac{6x - 2x}{2}\right)$

$= 2\cos 4x \cos 2x$

95. $\sin(\alpha + \beta) - \sin(\alpha - \beta) = 2\cos\left(\dfrac{\alpha + \beta + \alpha - \beta}{2}\right)\sin\left(\dfrac{\alpha + \beta - \alpha + \beta}{2}\right) = 2\cos \alpha \sin \beta$

97. $\cos\left(\theta + \dfrac{\pi}{2}\right) - \cos\left(\theta - \dfrac{\pi}{2}\right) = -2\sin\left[\dfrac{\left(\theta + \dfrac{\pi}{2}\right) + \left(\theta - \dfrac{\pi}{2}\right)}{2}\right]\sin\left[\dfrac{\left(\theta + \dfrac{\pi}{2}\right) - \left(\theta - \dfrac{\pi}{2}\right)}{2}\right] = -2\sin\theta\sin\dfrac{\pi}{2} = -2\sin\theta$

99. $\sin 75° + \sin 15° = 2\sin\left(\dfrac{75° + 15°}{2}\right)\cos\left(\dfrac{75° - 15°}{2}\right)$

$\qquad\qquad\qquad\quad = 2\sin 45° \cos 30°$

$\qquad\qquad\qquad\quad = 2\left(\dfrac{\sqrt{2}}{2}\right)\left(\dfrac{\sqrt{3}}{2}\right)$

$\qquad\qquad\qquad\quad = \dfrac{\sqrt{6}}{2}$

101. $\cos\dfrac{3\pi}{4} - \cos\dfrac{\pi}{4} = -2\sin\left(\dfrac{\dfrac{3\pi}{4} + \dfrac{\pi}{4}}{2}\right)\sin\left(\dfrac{\dfrac{3\pi}{4} - \dfrac{\pi}{4}}{2}\right) = -2\sin\dfrac{\pi}{2}\sin\dfrac{\pi}{4}$

$\cos\dfrac{3\pi}{4} - \cos\dfrac{\pi}{4} = -\dfrac{\sqrt{2}}{2} - \dfrac{\sqrt{2}}{2} = -\sqrt{2}$

103. $\qquad\qquad\sin 6x + \sin 2x = 0$

$2\sin\left(\dfrac{6x + 2x}{2}\right)\cos\left(\dfrac{6x - 2x}{2}\right) = 0$

$\qquad\qquad 2(\sin 4x)\cos 2x = 0$

$\sin 4x = 0 \quad \text{or} \quad \cos 2x = 0$

$\quad 4x = n\pi \qquad\qquad 2x = \dfrac{\pi}{2} + n\pi$

$\quad\ x = \dfrac{n\pi}{4} \qquad\qquad x = \dfrac{\pi}{4} + \dfrac{n\pi}{2}$

In the interval $[0, 2\pi)$

$x = 0, \dfrac{\pi}{4}, \dfrac{\pi}{2}, \dfrac{3\pi}{4}, \pi, \dfrac{5\pi}{4}, \dfrac{3\pi}{2}, \dfrac{7\pi}{4}.$

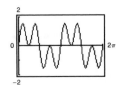

105. $\dfrac{\cos 2x}{\sin 3x - \sin x} - 1 = 0$

$\qquad \dfrac{\cos 2x}{\sin 3x - \sin x} = 1$

$\qquad \dfrac{\cos 2x}{2\cos 2x \sin x} = 1$

$\qquad\qquad 2\sin x = 1$

$\qquad\qquad\ \ \sin x = \dfrac{1}{2}$

$\qquad\qquad\qquad x = \dfrac{\pi}{6}, \dfrac{5\pi}{6}$

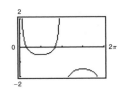

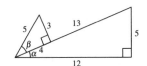

Figure for Exercises 107–109

107. $\sin 2\alpha = 2\sin\alpha\cos\alpha$

$\qquad\quad\ = 2\left(\dfrac{5}{13}\right)\left(\dfrac{12}{13}\right)$

$\qquad\quad\ = \dfrac{120}{169}$

109. $\cos\dfrac{\beta}{2} = \sqrt{\dfrac{1 + \cos\beta}{2}} = \sqrt{\dfrac{1 + \dfrac{4}{5}}{2}} = \sqrt{\dfrac{9}{10}} = \dfrac{3\sqrt{10}}{10}$

111. $\csc 2\theta = \dfrac{1}{\sin 2\theta}$

$$= \dfrac{1}{2\sin\theta\cos\theta}$$

$$= \dfrac{1}{\sin\theta} \cdot \dfrac{1}{2\cos\theta}$$

$$= \dfrac{\csc\theta}{2\cos\theta}$$

113. $\sin\left(\dfrac{\alpha}{3}\right)\cos\left(\dfrac{\alpha}{3}\right) = \dfrac{1}{2}\left[2\left(\sin\left(\dfrac{\alpha}{3}\right)\cos\left(\dfrac{\alpha}{3}\right)\right)\right]$

$$= \dfrac{1}{2}\sin\dfrac{2\alpha}{3}$$

115. $1 + \cos 10y = 1 + \cos^2 5y - \sin^2 5y$

$$= 1 + \cos^2 5y - \left(1 - \cos^2 5y\right)$$

$$= 2\cos^2 5y$$

117. $\cos 4\alpha = \cos 2(2\alpha) = \cos^2 2\alpha - \sin^2 2\alpha$

119. $\tan\dfrac{u}{2} = \dfrac{1 - \cos u}{\sin u}$

$$= \dfrac{1}{\sin u} - \dfrac{\cos u}{\sin u}$$

$$= \csc u - \cot u$$

121. $\dfrac{\cos 4x + \cos 2x}{\sin 4x + \sin 2x} = \dfrac{2\cos\left(\dfrac{4x + 2x}{2}\right)\cos\left(\dfrac{4x - 2x}{2}\right)}{2\sin\left(\dfrac{4x + 2x}{2}\right)\cos\left(\dfrac{4x - 2x}{2}\right)}$

$$= \dfrac{2\cos 3x\cos x}{2\sin 3x\cos x} = \cot 3x$$

123. $\sin\left(\dfrac{\pi}{6} + x\right) + \sin\left(\dfrac{\pi}{6} - x\right) = 2\sin\dfrac{\pi}{6}\cos x$

$$= 2 \cdot \dfrac{1}{2}\cos x$$

$$= \cos x$$

125.

Let $y_1 = \cos(3x)$ and

$y_2 = (\cos x)^3 - 3(\sin x)^2\cos x.$

$\cos 3\beta = \cos(2\beta + \beta)$

$$= \cos 2\beta\cos\beta - \sin 2\beta\sin\beta$$

$$= \left(\cos^2\beta - \sin^2\beta\right)\cos\beta - 2\sin\beta\cos\beta\sin\beta$$

$$= \cos^3\beta - \sin^2\beta\cos\beta - 2\sin^2\beta\cos\beta$$

$$= \cos^3\beta - 3\sin^2\beta\cos\beta$$

127.

Let $y_1 = \dfrac{\left(\cos 4x - \cos 2x\right)}{\left(2\sin 3x\right)}$

and $y_2 = -\sin x.$

$\dfrac{\cos 4x - \cos 2x}{2\sin 3x} = \dfrac{-2\sin\left(\dfrac{4x + 2x}{2}\right)\sin\left(\dfrac{4x - 2x}{2}\right)}{2\sin 3x}$

$$= \dfrac{-2\sin 3x\sin x}{2\sin 3x} = -\sin x$$

129. $\sin^2 x = \dfrac{1 - \cos 2x}{2} = \dfrac{1}{2} - \dfrac{\cos 2x}{2}$

131. $\sin(2\arcsin x) = 2\sin(\arcsin x)\cos(\arcsin x)$

$$= 2x\sqrt{1 - x^2}$$

133. $\cos(2\arcsin x) = \cos^2(\arcsin x) - \sin^2(\arcsin x)$

$$= \left(\sqrt{1 - x^2}\right)^2 - \left(x^2\right)$$

$$= 1 - 2x^2$$

135. $\dfrac{1}{32}(75)^2 \sin 2\theta = 130$

$$\sin 2\theta = \dfrac{130(32)}{75^2}$$

$$\theta = \dfrac{1}{2}\sin^{-1}\left(\dfrac{130(32)}{75^2}\right)$$

$$\theta \approx 23.85°$$

137. $\sin \dfrac{\theta}{2} = \dfrac{1}{M}$

(a) $\sin \dfrac{\theta}{2} = 1$

$\dfrac{\theta}{2} = \arcsin 1$

$\dfrac{\theta}{2} = \dfrac{\pi}{2}$

$\theta = \pi$

(b) $\sin \dfrac{\theta}{2} = \dfrac{1}{4.5}$

$\dfrac{\theta}{2} = \arcsin\left(\dfrac{1}{4.5}\right)$

$\theta = 2\arcsin\left(\dfrac{1}{4.5}\right)$

$\theta \approx 0.4482$

(c) $\dfrac{S}{760} = 1$

$S = 760$ miles per hour

$\dfrac{S}{760} = 4.5$

$S = 3420$ miles per hour

(d) $\sin \dfrac{\theta}{2} = \dfrac{1}{M}$

$\dfrac{\theta}{2} = \arcsin\left(\dfrac{1}{M}\right)$

$\theta = 2\arcsin\left(\dfrac{1}{M}\right)$

139. False. For $u < 0$,

$$\sin 2u = -\sin(-2u)$$

$$= -2\sin(-u)\cos(-u)$$

$$= -2(-\sin u)\cos u$$

$$= 2\sin u \cos u.$$

Review Exercises for Chapter 7

1. $\dfrac{\sin x}{\cos x} = \tan x$

3. $\dfrac{1}{\sec x} = \cos x$

5. $\sqrt{\cot^2 x + 1} = \sqrt{\csc^2 x} = \left|\csc x\right|$

7. $\sin x = \dfrac{5}{13}, \cos x = \dfrac{12}{13} \Rightarrow x$ is in Quadrant I.

$$\tan x = \dfrac{\sin x}{\cos x} = \dfrac{5/13}{12/13} = \dfrac{5}{12}$$

$$\cot x = \dfrac{1}{\tan x} = \dfrac{1}{5/12} = \dfrac{12}{5}$$

$$\sec x = \dfrac{1}{\cos x} = \dfrac{1}{12/13} = \dfrac{13}{12}$$

$$\csc x = \dfrac{1}{\sin x} = \dfrac{1}{5/13} = \dfrac{13}{5}$$

9. $\sin\left(\dfrac{\pi}{2} - x\right) = \dfrac{\sqrt{2}}{2} \Rightarrow \cos x = \dfrac{1}{\sqrt{2}} = \dfrac{\sqrt{2}}{2}$

$\sin x = -\dfrac{\sqrt{2}}{2}$

$\tan x = \dfrac{\sin x}{\cos x} = \dfrac{-\dfrac{1}{\sqrt{2}}}{\dfrac{1}{\sqrt{2}}} = -1$

$\cot x = \dfrac{1}{\tan x} = -1$

$\sec x = \dfrac{1}{\cos x} = \sqrt{2}$

$\csc x = \dfrac{1}{\sin x} = -\sqrt{2}$

11. $\dfrac{1}{\cot^2 x + 1} = \dfrac{1}{\csc^2 x} = \sin^2 x$

13. $\tan^2 x\left(\csc^2 x - 1\right) = \tan^2 x\left(\cot^2 x\right)$

$\qquad = \tan^2 x\left(\dfrac{1}{\tan^2 x}\right)$

$\qquad = 1$

15. $\dfrac{\sin\left(\dfrac{\pi}{2} - \theta\right)}{\sin \theta} = \dfrac{\cos \theta}{\sin \theta} = \cot \theta$

17. $\dfrac{\sin^2 \theta + \cos^2 \theta}{\sin \theta} = \dfrac{1}{\sin \theta} = \csc \theta$

19. $\cos^2 x + \cos^2 x \cot^2 x = \cos^2 x\left(1 + \cot^2 x\right)$

$\qquad = \cos^2 x\left(\csc^2 x\right)$

$\qquad = \cos^2 x\left(\dfrac{1}{\sin^2 x}\right)$

$\qquad = \dfrac{\cos^2 x}{\sin^2 x}$

$\qquad = \cot^2 x$

21. $\left(\tan x + 1\right)^2 \cos x = \left(\tan^2 x + 2 \tan x + 1\right)\cos x$

$\qquad = \left(\sec^2 x + 2 \tan x\right)\cos x$

$\qquad = \sec^2 x \cos x + 2\left(\dfrac{\sin x}{\cos x}\right)\cos x$

$\qquad = \sec x + 2 \sin x$

23. $\dfrac{1}{\csc \theta + 1} - \dfrac{1}{\csc \theta - 1} = \dfrac{\left(\csc \theta - 1\right) - \left(\csc \theta + 1\right)}{\left(\csc \theta + 1\right)\left(\csc \theta - 1\right)}$

$\qquad = \dfrac{-2}{\csc^2 \theta - 1}$

$\qquad = \dfrac{-2}{\cot^2 \theta}$

$\qquad = -2 \tan^2 \theta$

25. Let $x = 5 \sin \theta$, then

$\sqrt{25 - x^2} = \sqrt{25 - \left(5 \sin \theta\right)^2}$

$\qquad = \sqrt{25 - 25 \sin^2 \theta}$

$\qquad = \sqrt{25\left(1 - \sin^2 \theta\right)}$

$\qquad = \sqrt{25 \cos^2 \theta}$

$\qquad = 5 \cos \theta.$

27. $\csc^2 x - \csc x \cot x = \dfrac{1}{\sin^2 x} - \left(\dfrac{1}{\sin x}\right)\left(\dfrac{\cos x}{\sin x}\right)$

$\qquad = \dfrac{1 - \cos x}{\sin^2 x}$

29. $\cos x\left(\tan^2 x + 1\right) = \cos x \sec^2 x$

$\qquad = \dfrac{1}{\sec x} \sec^2 x$

$\qquad = \sec x$

31. $\sec\left(\dfrac{\pi}{2} - \theta\right) = \csc \theta$

33. $\dfrac{1}{\tan \theta \csc \theta} = \dfrac{1}{\dfrac{\sin \theta}{\cos \theta} \cdot \dfrac{1}{\sin \theta}} = \cos \theta$

35. $\sin^5 x \cos^2 x = \sin^4 x \cos^2 x \sin x$

$\qquad = \left(1 - \cos^2 x\right)^2 \cos^2 x \sin x$

$\qquad = \left(1 - 2\cos^2 x + \cos^4 x\right)\cos^2 x \sin x$

$\qquad = \left(\cos^2 x - 2\cos^4 x + \cos^6 x\right)\sin x$

37. $\sin x = \sqrt{3} - \sin x$

$\sin x = \dfrac{\sqrt{3}}{2}$

$x = \dfrac{\pi}{3} + 2\pi n, \dfrac{2\pi}{3} + 2\pi n$

39. $3\sqrt{3}\,\tan u = 3$

$$\tan u = \frac{1}{\sqrt{3}}$$

$$u = \frac{\pi}{6} + n\pi$$

41. $3\csc^2 x = 4$

$$\csc^2 x = \frac{4}{3}$$

$$\sin x = \pm\frac{\sqrt{3}}{2}$$

$$x = \frac{\pi}{3} + 2\pi n, \frac{2\pi}{3} + 2\pi n, \frac{4\pi}{3} + 2\pi n, \frac{5\pi}{3} + 2\pi n$$

These can be combined as:

$$x = \frac{\pi}{3} + n\pi \quad \text{or} \quad x = \frac{2\pi}{3} + n\pi$$

43.
$$2\cos^2 x - \cos x = 1$$
$$2\cos^2 x - \cos x - 1 = 0$$
$$(2\cos x + 1)(\cos x - 1) = 0$$

$$2\cos x + 1 = 0 \qquad \cos x - 1 = 0$$

$$\cos x = -\frac{1}{2} \qquad \cos x = 1$$

$$x = \frac{2\pi}{3}, \frac{4\pi}{3} \qquad x = 0$$

45.
$$\cos^2 x + \sin x = 1$$
$$1 - \sin^2 x + \sin x - 1 = 0$$
$$-\sin x(\sin x - 1) = 0$$

$$\sin x = 0 \qquad \sin x - 1 = 0$$
$$x = 0, \pi \qquad \sin x = 1$$
$$\qquad x = \frac{\pi}{2}$$

47. $2\sin 2x - \sqrt{2} = 0$

$$\sin 2x = \frac{\sqrt{2}}{2}$$

$$2x = \frac{\pi}{4} + 2\pi n, \frac{3\pi}{4} + 2\pi n$$

$$x = \frac{\pi}{8} + \pi n, \frac{3\pi}{8} + \pi n$$

$$x = \frac{\pi}{8}, \frac{3\pi}{8}, \frac{9\pi}{8}, \frac{11\pi}{8}$$

49. $3\tan^2\left(\dfrac{x}{3}\right) - 1 = 0$

$$\tan^2\left(\frac{x}{3}\right) = \frac{1}{3}$$

$$\tan\frac{x}{3} = \pm\sqrt{\frac{1}{3}}$$

$$\tan\frac{x}{3} = \pm\frac{\sqrt{3}}{3}$$

$$\frac{x}{3} = \frac{\pi}{6}, \frac{5\pi}{6}, \frac{7\pi}{6}$$

$$x = \frac{\pi}{2}, \frac{5\pi}{2}, \frac{7\pi}{2}$$

$\dfrac{5\pi}{2}$ and $\dfrac{7\pi}{2}$ are greater than 2π, so they are not

solutions. The solution is $x = \dfrac{\pi}{2}$.

51. $\cos 4x(\cos x - 1) = 0$

$$\cos 4x = 0 \qquad\qquad \cos x - 1 = 0$$

$$4x = \frac{\pi}{2} + 2\pi n, \frac{3\pi}{2} + 2\pi n \qquad \cos x = 1$$

$$x = \frac{\pi}{8} + \frac{\pi}{2}n, \frac{3\pi}{8} + \frac{\pi}{2}n \qquad x = 0$$

$$x = 0, \frac{\pi}{8}, \frac{3\pi}{8}, \frac{5\pi}{8}, \frac{7\pi}{8}, \frac{9\pi}{8}, \frac{11\pi}{8}, \frac{13\pi}{8}, \frac{15\pi}{8}$$

53. $\sin^2 x - 2\sin x = 0$

$$\sin x(\sin x - 2) = 0$$

$$\sin x = 0 \qquad \sin x - 2 = 0$$
$$x = 0, \pi \qquad \text{No solution}$$

55.
$$\tan^2\theta + \tan\theta - 6 = 0$$
$$(\tan\theta + 3)(\tan\theta - 2) = 0$$

$$\tan\theta + 3 = 0 \qquad\qquad \text{or} \qquad\qquad \tan\theta - 2 = 0$$
$$\tan\theta = -3 \qquad\qquad\qquad\qquad \tan\theta = 2$$
$$\theta = \arctan(-3) + \pi, \arctan(-3) + 2\pi \qquad \theta = \arctan 2, \arctan 2 + \pi$$

57. $\sin 285° = \sin(315° - 30°)$

$$= \sin 315° \cos 30° - \cos 315° \sin 30°$$

$$= \left(-\frac{\sqrt{2}}{2}\right)\left(\frac{\sqrt{3}}{2}\right) - \left(\frac{\sqrt{2}}{2}\right)\left(\frac{1}{2}\right)$$

$$= -\frac{\sqrt{2}}{4}\left(\sqrt{3} + 1\right)$$

$\cos 285° = \cos(315° - 30°)$

$$= \cos 315° \cos 30° + \sin 315° \sin 30°$$

$$= \left(\frac{\sqrt{2}}{2}\right)\left(\frac{\sqrt{3}}{2}\right) + \left(-\frac{\sqrt{2}}{2}\right)\left(\frac{1}{2}\right)$$

$$= \frac{\sqrt{2}}{4}\left(\sqrt{3} - 1\right)$$

$\tan 285° = \tan(315° - 30°) = \dfrac{\tan 315° - \tan 30°}{1 + \tan 315° \tan 30°}$

$$= \frac{(-1) - \left(\dfrac{\sqrt{3}}{3}\right)}{1 + (-1)\left(\dfrac{\sqrt{3}}{3}\right)} = -2 - \sqrt{3}$$

59. $\sin \dfrac{25\pi}{12} = \sin\left(\dfrac{11\pi}{6} + \dfrac{\pi}{4}\right)$

$$= \sin \frac{11\pi}{6} \cos \frac{\pi}{4} + \cos \frac{11\pi}{6} \sin \frac{\pi}{4}$$

$$= \left(-\frac{1}{2}\right)\left(\frac{\sqrt{2}}{2}\right) + \left(\frac{\sqrt{3}}{2}\right)\left(\frac{\sqrt{2}}{2}\right)$$

$$= \frac{\sqrt{2}}{4}\left(\sqrt{3} - 1\right)$$

$\cos \dfrac{25\pi}{12} = \cos\left(\dfrac{11\pi}{6} + \dfrac{\pi}{4}\right)$

$$= \cos \frac{11\pi}{6} \cos \frac{\pi}{4} - \sin \frac{11\pi}{6} \sin \frac{\pi}{4}$$

$$= \left(\frac{\sqrt{3}}{2}\right)\left(\frac{\sqrt{2}}{2}\right) - \left(-\frac{1}{2}\right)\left(\frac{\sqrt{2}}{2}\right)$$

$$= \frac{\sqrt{2}}{4}\left(\sqrt{3} + 1\right)$$

$\tan \dfrac{25\pi}{12} = \tan\left(\dfrac{11\pi}{6} + \dfrac{\pi}{4}\right)$

$$= \frac{\tan \dfrac{11\pi}{6} + \tan \dfrac{\pi}{4}}{1 - \tan \dfrac{11\pi}{6} \tan \dfrac{\pi}{4}}$$

$$= \frac{\left(-\dfrac{\sqrt{3}}{3}\right) + 1}{1 - \left(-\dfrac{\sqrt{3}}{3}\right)(1)}$$

$$= 2 - \sqrt{3}$$

61. $\sin 60° \cos 45° - \cos 60° \sin 45° = \sin(60° - 45°)$

$$= \sin 15°$$

63. $\dfrac{\tan 25° + \tan 10°}{1 - \tan 25° \tan 10°} = \tan(25° + 10°)$

$$= \tan 35°$$

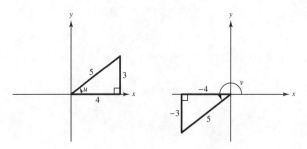

Figures for Exercises 65–69

65. $\sin(u + v) = \sin u \cos v + \cos u \sin v$

$$= \tfrac{3}{5}\left(-\tfrac{4}{5}\right) + \tfrac{4}{5}\left(-\tfrac{3}{5}\right)$$

$$= -\tfrac{24}{25}$$

67. $\cos(u - v) = \cos u \cos v + \sin u \sin v$

$$= \tfrac{4}{5}\left(-\tfrac{4}{5}\right) + \tfrac{3}{5}\left(-\tfrac{3}{5}\right)$$

$$= -1$$

69. $\cos(u + v) = \cos u \cos v - \sin u \sin v$

$$= \tfrac{4}{5}\left(-\tfrac{4}{5}\right) - \tfrac{3}{5}\left(-\tfrac{3}{5}\right)$$

$$= -\tfrac{7}{25}$$

71. $\cos\left(x + \dfrac{\pi}{2}\right) = \cos x \cos \dfrac{\pi}{2} - \sin x \sin \dfrac{\pi}{2}$

$$= \cos x(0) - \sin x(1)$$

$$= -\sin x$$

73. $\tan\left(x - \dfrac{\pi}{2}\right) = -\tan\left(\dfrac{\pi}{2} - x\right) = -\cot x$

75. $\cos 3x = \cos(2x + x)$

$$= \cos 2x \cos x - \sin 2x \sin x$$

$$= \left(\cos^2 x - \sin^2 x\right)\cos x - (2 \sin x \cos x)\sin x$$

$$= \cos^3 x - \sin^2 x \cos x - 2 \sin^2 x \cos x$$

$$= \cos^3 x - 3 \sin^2 x \cos x$$

$$= \cos^3 x - 3\left(1 - \cos^2 x\right)\cos x$$

$$= \cos^3 x - 3 \cos x + 3 \cos^3 x$$

$$= 4 \cos^3 x - 3 \cos x$$

77. $\sin\left(x + \dfrac{\pi}{4}\right) - \sin\left(x - \dfrac{\pi}{4}\right) = 1$

$$2\cos x \sin \dfrac{\pi}{4} = 1$$

$$\cos x = \dfrac{\sqrt{2}}{2}$$

$$x = \dfrac{\pi}{4}, \dfrac{7\pi}{4}$$

79. $\sin\left(x + \dfrac{\pi}{2}\right) - \sin\left(x - \dfrac{\pi}{2}\right) = \sqrt{3}$

$$2\cos x \sin \dfrac{\pi}{2} = \sqrt{3}$$

$$\cos x = \dfrac{\sqrt{3}}{2}$$

$$x = \dfrac{\pi}{6}, \dfrac{11\pi}{6}$$

81. $\sin u = -\dfrac{4}{5}, \ \pi < u < \dfrac{3\pi}{2}$

$$\cos u = -\sqrt{1 - \sin^2 u} = \dfrac{-3}{5}$$

$$\tan u = \dfrac{\sin u}{\cos u} = \dfrac{4}{3}$$

$$\sin 2u = 2 \sin u \cos u = 2\left(-\dfrac{4}{5}\right)\left(-\dfrac{3}{5}\right) = \dfrac{24}{25}$$

$$\cos 2u = \cos^2 u - \sin^2 u = \left(-\dfrac{3}{5}\right)^2 - \left(-\dfrac{4}{5}\right)^2 = -\dfrac{7}{25}$$

$$\tan 2u = \dfrac{2 \tan u}{1 - \tan^2 u} = \dfrac{2\left(\dfrac{4}{3}\right)}{1 - \left(\dfrac{4}{3}\right)^2} = -\dfrac{24}{7}$$

83. $\sec u = -3, \ \dfrac{\pi}{2} < u < \pi$

$$\cos u = \dfrac{1}{\sec u} = \dfrac{1}{-3} = -\dfrac{1}{3}$$

$$\sin u = \sqrt{1 - \cos^2 u} = \sqrt{1 - \left(-\dfrac{1}{3}\right)^2} = \sqrt{\dfrac{8}{9}} = \dfrac{2\sqrt{2}}{3}$$

$$\tan u = \dfrac{\sin u}{\cos u} = -2\sqrt{2}$$

$$\sin 2u = 2 \sin u \cos u = 2\left(\dfrac{2\sqrt{2}}{3}\right)\left(-\dfrac{1}{3}\right) = \dfrac{-4\sqrt{2}}{9}$$

$$\cos 2u = \cos^2 u - \sin^2 u = \left(-\dfrac{1}{3}\right)^2 - \left(\dfrac{2\sqrt{2}}{3}\right)^2 = -\dfrac{7}{9}$$

$$\tan 2u = \dfrac{2 \tan u}{1 - \tan^2 u} = \dfrac{2\left(-2\sqrt{2}\right)}{1 - \left(-2\sqrt{2}\right)^2} = \dfrac{4\sqrt{2}}{7}$$

85. $\sin 4x = 2 \sin 2x \cos 2x$

$$= 2\left[2 \sin x \cos x\left(\cos^2 x - \sin^2 x\right)\right]$$

$$= 4 \sin x \cos x\left(2\cos^2 x - 1\right)$$

$$= 8\cos^3 x \sin x - 4\cos x \sin x$$

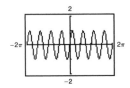

87. $\tan^2 2x = \dfrac{\sin^2 2x}{\cos^2 2x} = \dfrac{\dfrac{1 - \cos 4x}{2}}{\dfrac{1 + \cos 4x}{2}} = \dfrac{1 - \cos 4x}{1 + \cos 4x}$

89. $\sin^2 x \tan^2 x = \sin^2 x\left(\dfrac{\sin^2 x}{\cos^2 x}\right) = \dfrac{\sin^4 x}{\cos^2 x}$

$$= \dfrac{\left(\dfrac{1 - \cos 2x}{2}\right)^2}{\dfrac{1 + \cos 2x}{2}}$$

$$= \dfrac{\dfrac{1 - 2\cos 2x + \cos^2 2x}{4}}{\dfrac{1 + \cos 2x}{2}}$$

$$= \dfrac{1 - 2\cos 2x + \dfrac{1 + \cos 4x}{2}}{2(1 + \cos 2x)}$$

$$= \dfrac{2 - 4\cos 2x + 1 + \cos 4x}{4(1 + \cos 2x)}$$

$$= \dfrac{3 - 4\cos 2x + \cos 4x}{4(1 + \cos 2x)}$$

91. $\sin(-75°) = -\sqrt{\dfrac{1 - \cos 150°}{2}} = -\sqrt{\dfrac{1 - \left(-\dfrac{\sqrt{3}}{2}\right)}{2}} = -\dfrac{\sqrt{2 + \sqrt{3}}}{2} = -\dfrac{1}{2}\sqrt{2 + \sqrt{3}}$

$\cos(-75°) = -\sqrt{\dfrac{1 + \cos 150°}{2}} = \sqrt{\dfrac{1 + \left(-\dfrac{\sqrt{3}}{2}\right)}{2}} = \dfrac{\sqrt{2 - \sqrt{3}}}{2} = \dfrac{1}{2}\sqrt{2 - \sqrt{3}}$

$\tan(-75°) = -\left(\dfrac{1 - \cos 150°}{\sin 150°}\right) = -\left(\dfrac{1 - \left(-\dfrac{\sqrt{3}}{2}\right)}{\dfrac{1}{2}}\right) = -\left(2 + \sqrt{3}\right) = -2 - \sqrt{3}$

93. $\sin\left(\dfrac{19\pi}{12}\right) = -\sqrt{\dfrac{1 - \cos\dfrac{19\pi}{6}}{2}} = -\sqrt{\dfrac{1 - \left(-\dfrac{\sqrt{3}}{2}\right)}{2}} = -\dfrac{\sqrt{2 + \sqrt{3}}}{2} = -\dfrac{1}{2}\sqrt{2 + \sqrt{3}}$

$\cos\left(\dfrac{19\pi}{12}\right) = \sqrt{\dfrac{1 + \cos\dfrac{19\pi}{6}}{2}} = \sqrt{\dfrac{1 + \left(-\dfrac{\sqrt{3}}{2}\right)}{2}} = \dfrac{\sqrt{2 - \sqrt{3}}}{2} = \dfrac{1}{2}\sqrt{2 - \sqrt{3}}$

$\tan\left(\dfrac{19\pi}{12}\right) = \dfrac{1 - \cos\dfrac{19\pi}{6}}{\sin\dfrac{19\pi}{6}} = \dfrac{1 - \left(-\dfrac{\sqrt{3}}{2}\right)}{-\dfrac{1}{2}} = -2 - \sqrt{3}$

95. $\sin u = \dfrac{7}{25}, \; 0 < u < \dfrac{\pi}{2}$

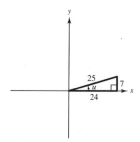

(a) Because u is in Quadrant I, $\dfrac{u}{2}$ is also in Quadrant I.

(b) $\sin\dfrac{u}{2} = \sqrt{\dfrac{1 - \cos u}{2}} = \sqrt{\dfrac{1 - \dfrac{24}{25}}{2}} = \sqrt{\dfrac{1}{50}} = \dfrac{\sqrt{2}}{10}$

$\cos\dfrac{u}{2} = \sqrt{\dfrac{1 + \cos u}{2}} = \sqrt{\dfrac{1 + \dfrac{24}{25}}{2}} = \sqrt{\dfrac{49}{50}} = \dfrac{7\sqrt{2}}{10}$

$\tan\dfrac{u}{2} = \dfrac{1 - \cos u}{\sin u} = \dfrac{1 - \dfrac{24}{25}}{\dfrac{7}{25}} = \dfrac{1}{7}$

97. $\cos u = -\dfrac{2}{7}, \dfrac{\pi}{2} < u < \pi$

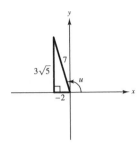

(a) Because u is in Quadrant II, $\dfrac{u}{2}$ is in Quadrant I.

(b) $\sin \dfrac{u}{2} = \sqrt{\dfrac{1 - \cos u}{2}} = \sqrt{\dfrac{1 - \left(-\dfrac{2}{7}\right)}{2}} = \sqrt{\dfrac{9}{14}}$

$= \dfrac{3\sqrt{14}}{14}$

$\cos \dfrac{u}{2} = \sqrt{\dfrac{1 + \cos u}{2}} = \sqrt{\dfrac{1 + \left(-\dfrac{2}{7}\right)}{2}} = \sqrt{\dfrac{5}{14}}$

$= \dfrac{\sqrt{70}}{14}$

$\tan \dfrac{u}{2} = \dfrac{1 - \cos u}{\sin u} = \dfrac{1 - \left(-\dfrac{2}{7}\right)}{\dfrac{3\sqrt{5}}{7}} = \dfrac{3\sqrt{5}}{5}$

99. $-\sqrt{\dfrac{1 + \cos 10x}{2}} = -\left|\cos \dfrac{10x}{2}\right| = -\left|\cos 5x\right|$

101. $\cos \dfrac{\pi}{6} \sin \dfrac{\pi}{6} = \dfrac{1}{2}\left[\sin \dfrac{\pi}{3} - \sin 0\right] = \dfrac{1}{2}\sin \dfrac{\pi}{3}$

103. $\cos 4\theta \sin 6\theta = \dfrac{1}{2}\left[\sin(4\theta + 6\theta) - \sin(4\theta - 6\theta)\right]$

$= \dfrac{1}{2}(\sin 10\theta - \sin(-2\theta))$

105. $\sin 4\theta - \sin 8\theta = 2 \cos\left(\dfrac{4\theta + 8\theta}{2}\right) \sin\left(\dfrac{4\theta - 8\theta}{2}\right)$

$= 2 \cos 6\theta \sin(-2\theta)$

107. $\cos\left(x + \dfrac{\pi}{6}\right) - \cos\left(x - \dfrac{\pi}{6}\right) = -2 \sin x \sin \dfrac{\pi}{6}$

109. $r = \dfrac{1}{32}v_0^2 \sin 2\theta$

range $= 100$ feet

$v_0 = 80$ feet per second

$r = \dfrac{1}{32}(80)^2 \sin 2\theta = 100$

$\sin 2\theta = 0.5$

$2\theta = 30°$

$\theta = 15°$ or $\dfrac{\pi}{12}$

111. $y = 1.5 \sin 8t - 0.5 \cos 8t$

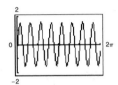

113. Amplitude $= \dfrac{\sqrt{10}}{2}$ feet

115. False. If $\dfrac{\pi}{2} < \theta < \pi$, then $\dfrac{\pi}{4} < \dfrac{\theta}{2} < \dfrac{\pi}{2}$ and $\dfrac{\theta}{2}$ is in Quadrant I.

$\cos \dfrac{\theta}{2} > 0$

117. True. $4 \sin(-x)\cos(-x) = 4(-\sin x) \cos x$

$= -4 \sin x \cos x$

$= -2(2 \sin x \cos x)$

$= -2 \sin 2x$

119. Reciprocal Identities:

$\sin \theta = \dfrac{1}{\csc \theta}$ $\qquad$ $\csc \theta = \dfrac{1}{\sin \theta}$

$\cos \theta = \dfrac{1}{\sec \theta}$ $\qquad$ $\sec \theta = \dfrac{1}{\cos \theta}$

$\tan \theta = \dfrac{1}{\cot \theta}$ $\qquad$ $\cot \theta = \dfrac{1}{\tan \theta}$

Quotient Identities:

$\tan \theta = \dfrac{\sin \theta}{\cos \theta}$ $\qquad$ $\cot \theta = \dfrac{\cos \theta}{\sin \theta}$

Pythagorean Identities:

$\sin^2 \theta + \cos^2 \theta = 1$

$1 + \tan^2 \theta = \sec^2 \theta$

$1 + \cot^2 \theta = \csc^2 \theta$

121. $a \sin x - b = 0$

$$\sin x = \frac{b}{a}$$

If $|b| > |a|$, then $\left|\dfrac{b}{a}\right| > 1$ and there is no solution

because $|\sin x| \le 1$ for all x.

123. The graph of y_1 is a vertical shift of the graph of y_2 one unit upward so $y_1 = y_2 + 1$.

125. $y = \sqrt{x + 3} + 4 \cos x$

Zeros: $x \approx -1.8431, 2.1758,$ $3.9903, 8.8935, 9.8820$

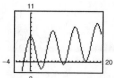

Problem Solving for Chapter 7

1. (a) Because $\sin^2 \theta + \cos^2 \theta = 1$ and
$\cos^2 \theta = 1 - \sin^2 \theta$:

$$\cos \theta = \pm\sqrt{1 - \sin^2 \theta}$$

$$\tan \theta = \frac{\sin \theta}{\cos \theta} = \pm\frac{\sin \theta}{\sqrt{1 - \sin^2 \theta}}$$

$$\cot \theta = \frac{1}{\tan \theta} = \pm\frac{\sqrt{1 - \sin^2 \theta}}{\sin \theta}$$

$$\sec \theta = \frac{1}{\cos \theta} = \pm\frac{1}{\sqrt{1 - \sin^2 \theta}}$$

$$\csc \theta = \frac{1}{\sin \theta}$$

You also have the following relationships:

$$\cos \theta = \sin\left(\frac{\pi}{2} - \theta\right)$$

$$\tan \theta = \frac{\sin \theta}{\sin\left(\dfrac{\pi}{2} - \theta\right)}$$

$$\cot \theta = \frac{\sin\left(\dfrac{\pi}{2} - \theta\right)}{\sin \theta}$$

$$\sec \theta = \frac{1}{\sin\left(\dfrac{\pi}{2} - \theta\right)}$$

$$\csc \theta = \frac{1}{\sin \theta}$$

(b) $\sin \theta = \pm\sqrt{1 - \cos^2 \theta}$

$$\tan \theta = \frac{\sin \theta}{\cos \theta} = \pm\frac{\sqrt{1 - \cos^2 \theta}}{\cos \theta}$$

$$\csc \theta = \frac{1}{\sin \theta} = \pm\frac{1}{\sqrt{1 - \cos^2 \theta}}$$

$$\sec \theta = \frac{1}{\cos \theta}$$

$$\cot \theta = \frac{1}{\tan \theta} = \pm\frac{\cos \theta}{\sqrt{1 - \cos^2 \theta}}$$

You also have the following relationships:

$$\sin \theta = \cos\left(\frac{\pi}{2} - \theta\right)$$

$$\tan \theta = \frac{\cos\left[(\pi/2) - \theta\right]}{\cos \theta}$$

$$\csc \theta = \frac{1}{\cos\left[(\pi/2) - \theta\right]}$$

$$\sec \theta = \frac{1}{\cos \theta}$$

$$\cot \theta = \frac{\cos \theta}{\cos\left[(\pi/2) - \theta\right]}$$

3. $\sin\left[\dfrac{(12n + 1)\pi}{6}\right] = \sin\left[\dfrac{1}{6}(12n\pi + \pi)\right]$

$$= \sin\left(2n\pi + \frac{\pi}{6}\right)$$

$$= \sin\frac{\pi}{6} = \frac{1}{2}$$

So, $\sin\left[\dfrac{(12n + 1)\pi}{6}\right] = \dfrac{1}{2}$ for all integers n.

394 *Chapter 7 Analytic Trigonometry*

5. From the figure, it appears that $u + v = w$. Assume that u, v, and w are all in Quadrant I.

From the figure:

$$\tan u = \frac{s}{3s} = \frac{1}{3}$$

$$\tan v = \frac{s}{2s} = \frac{1}{2}$$

$$\tan w = \frac{s}{s} = 1$$

$$\tan(u + v) = \frac{\tan u + \tan v}{1 - \tan u \tan v} = \frac{1/3 + 1/2}{1 - (1/3)(1/2)} = \frac{5/6}{1 - (1/6)} = 1 = \tan w.$$

So, $\tan(u + v) = \tan w$. Because u, v, and w are all in Quadrant I, you have

$$\arctan\left[\tan(u + v)\right] = \arctan[\tan w] \quad u + v = w.$$

7.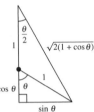

The hypotenuse of the larger right triangle is:

$$\sqrt{\sin^2 \theta + (1 + \cos \theta)^2} = \sqrt{\sin^2 \theta + 1 + 2 \cos \theta + \cos^2 \theta}$$

$$= \sqrt{2 + 2 \cos \theta}$$

$$= \sqrt{2(1 + \cos \theta)}$$

$$\sin\left(\frac{\theta}{2}\right) = \frac{\sin \theta}{\sqrt{2(1 + \cos \theta)}} = \frac{\sin \theta}{\sqrt{2(1 + \cos \theta)}} \cdot \frac{\sqrt{1 - \cos \theta}}{\sqrt{1 - \cos \theta}} = \frac{\sin \theta \sqrt{1 - \cos \theta}}{\sqrt{2(1 - \cos^2 \theta)}} = \frac{\sin \theta \sqrt{1 - \cos \theta}}{\sqrt{2} \sin \theta} = \sqrt{\frac{1 - \cos \theta}{2}}$$

$$\cos\left(\frac{\theta}{2}\right) = \frac{1 + \cos \theta}{\sqrt{2(1 + \cos \theta)}} = \frac{\sqrt{(1 + \cos \theta)^2}}{\sqrt{2(1 + \cos \theta)}} = \sqrt{\frac{1 + \cos \theta}{2}}$$

$$\tan\left(\frac{\theta}{2}\right) = \frac{\sin \theta}{1 + \cos \theta}$$

9. Seward: $D = 12.2 - 6.4 \cos\left[\dfrac{\pi(t + 0.2)}{182.6}\right]$

New Orleans: $D = 12.2 - 1.9 \cos\left[\dfrac{\pi(t + 0.2)}{182.6}\right]$

(a)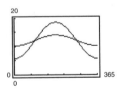

(b) The graphs intersect when $t \approx 91$ and when $t \approx 274$. These values correspond to April 1 and October 1, the spring equinox and the fall equinox.

(c) Seward has the greater variation in the number of daylight hours. This is determined by the amplitudes, 6.4 and 1.9.

(d) Period: $\dfrac{2\pi}{\pi/182.6} = 365.2$ days

11. (a) Let $y_1 = \sin x$ and $y_2 = 0.5$.

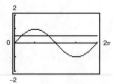

$\sin x \geq 0.5$ on the interval $\left[\dfrac{\pi}{6}, \dfrac{5\pi}{6}\right]$.

(b) Let $y_1 = \cos x$ and $y_2 = -0.5$.

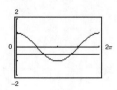

$\cos x \leq -0.5$ on the interval $\left[\dfrac{2\pi}{3}, \dfrac{4\pi}{3}\right]$.

(c) Let $y_1 = \tan x$ and $y_2 = \sin x$.

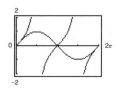

$\tan x < \sin x$ on the intervals $\left(\dfrac{\pi}{2}, \pi\right)$ and $\left(\dfrac{3\pi}{2}, 2\pi\right)$.

(d) Let $y_1 = \cos x$ and $y_2 = \sin x$.

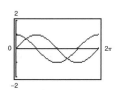

$\cos x \geq \sin x$ on the intervals $\left[0, \dfrac{\pi}{4}\right]$ and $\left[\dfrac{5\pi}{4}, 2\pi\right]$.

13. (a) $\sin(u + v + w) = \sin\left[(u + v) + w\right]$

$\qquad = \sin(u + v)\cos w + \cos(u + v)\sin w$

$\qquad = \left[\sin u \cos v + \cos u \sin v\right]\cos w + \left[\cos u \cos v - \sin u \sin v\right]\sin w$

$\qquad = \sin u \cos v \cos w + \cos u \sin v \cos w + \cos u \cos v \sin w - \sin u \sin v \sin w$

(b) $\tan(u + v + w) = \tan\big[(u + v) + w\big]$

$$= \frac{\tan(u + v) + \tan w}{1 - \tan(u + v) \tan w}$$

$$= \frac{\left[\dfrac{\tan u + \tan v}{1 - \tan u \tan v}\right] + \tan w}{1 - \left[\dfrac{\tan u + \tan v}{1 - \tan u \tan v}\right] \tan w} \cdot \frac{(1 - \tan u \tan v)}{(1 - \tan u \tan v)}$$

$$= \frac{\tan u + \tan v + (1 - \tan u \tan v) \tan w}{(1 - \tan u \tan v) - (\tan u + \tan v) \tan w}$$

$$= \frac{\tan u + \tan v + \tan w - \tan u \tan v \tan w}{1 - \tan u \tan v - \tan u \tan w - \tan v \tan w}$$

15. $h_1 = 3.75 \sin 733t + 7.5$

$h_2 = 3.75 \sin 733\left(t + \dfrac{4\pi}{3}\right) + 7.5$

(a)

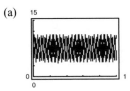

(b) The period for h_1 and h_2 is $\dfrac{2\pi}{733} \approx 0.0086$.

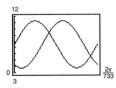

The graphs intersect twice per cycle.

There are $\dfrac{1}{2\pi/733} \approx 116.66$ cycles in the interval $[0, 1]$, so the graphs intersect approximately 233.3 times.

Chapter 7   Practice Test

1. Find the value of the other five trigonometric functions, given $\tan x = \frac{4}{11}$, $\sec x < 0$.

2. Simplify $\dfrac{\sec^2 x + \csc^2 x}{\csc^2 x\left(1 + \tan^2 x\right)}$.

3. Rewrite as a single logarithm and simplify $\ln|\tan\theta| - \ln|\cot\theta|$.

4. True or false:

   $$\cos\left(\frac{\pi}{2} - x\right) = \frac{1}{\csc x}$$

5. Factor and simplify: $\sin^4 x + \left(\sin^2 x\right)\cos^2 x$

6. Multiply and simplify: $\left(\csc x + 1\right)\left(\csc x - 1\right)$

7. Rationalize the denominator and simplify:

   $$\frac{\cos^2 x}{1 - \sin x}$$

8. Verify:

   $$\frac{1 + \cos\theta}{\sin\theta} + \frac{\sin\theta}{1 + \cos\theta} = 2\csc\theta$$

9. Verify:

   $$\tan^4 x + 2\tan^2 x + 1 = \sec^4 x$$

10. Use the sum or difference formulas to determine:
    (a) $\sin 105°$
    (b) $\tan 15°$

11. Simplify: $\left(\sin 42°\right)\cos 38° - \left(\cos 42°\right)\sin 38°$

12. Verify $\tan\left(\theta + \dfrac{\pi}{4}\right) = \dfrac{1 + \tan\theta}{1 - \tan\theta}$.

13. Write $\sin\left(\arcsin x - \arccos x\right)$ as an algebraic expression in x.

14. Use the double-angle formulas to determine:
    (a) $\cos 120°$
    (b) $\tan 300°$

15. Use the half-angle formulas to determine:
    (a) $\sin 22.5°$
    (b) $\tan \dfrac{\pi}{12}$

16. Given $\sin\theta = 4/5$, θ lies in Quadrant II, find $\cos(\theta/2)$.

17. Use the power-reducing identities to write $\left(\sin^2 x\right)\cos^2 x$ in terms of the first power of cosine.

18. Rewrite as a sum: $6(\sin 5\theta)\cos 2\theta$.

19. Rewrite as a product: $\sin(x + \pi) + \sin(x - \pi)$.

20. Verify $\dfrac{\sin 9x + \sin 5x}{\cos 9x - \cos 5x} = -\cot 2x$.

21. Verify:

$$(\cos u)\sin v = \tfrac{1}{2}\left[\sin(u + v) - \sin(u - v)\right].$$

22. Find all solutions in the interval $[0, 2\pi)$:

$$4\sin^2 x = 1$$

23. Find all solutions in the interval $[0, 2\pi)$:

$$\tan^2 \theta + \left(\sqrt{3} - 1\right)\tan \theta - \sqrt{3} = 0$$

24. Find all solutions in the interval $[0, 2\pi)$:

$$\sin 2x = \cos x$$

25. Use the quadratic formula to find all solutions in the interval $[0, 2\pi)$:

$$\tan^2 x - 6\tan x + 4 = 0$$

CHAPTER 8
Additional Topics in Trigonometry

C H A P T E R 8
Additional Topics in Trigonometry

Section 8.1   Law of Sines

■ If ABC is any oblique triangle with sides a, b, and c, then

$$\frac{a}{\sin A} = \frac{b}{\sin B} = \frac{c}{\sin C}.$$

■ You should be able to use the Law of Sines to solve an oblique triangle for the remaining three parts, given:

(a) Two angles and any side (AAS or ASA)

(b) Two sides and an angle opposite one of them (SSA)

   1. If A is acute and $h = b \sin A$:

   (a) $a < h$, no triangle is possible.

   (b) $a = h$ or $a > b$, one triangle is possible.

   (c) $h < a < b$, two triangles are possible.

   2. If A is obtuse and $h = b \sin A$:

   (a) $a \leq b$, no triangle is possible.

   (b) $a > b$, one triangle is possible.

■ The area of any triangle equals one-half the product of the lengths of two sides and the sine of their included angle.

$$A = \tfrac{1}{2}ab \sin C = \tfrac{1}{2}ac \sin B = \tfrac{1}{2}bc \sin A$$

1. oblique

3. angles; side

5.

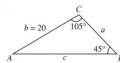

Given: $B = 45°$, $C = 105°$, $b = 20$

$A = 180° - B - C = 30°$

$a = \dfrac{b}{\sin B}(\sin A) = \dfrac{20 \sin 30°}{\sin 45°} = 10\sqrt{2} \approx 14.14$

$C = \dfrac{b}{\sin B}(\sin C) = \dfrac{20 \sin 105°}{\sin 45°} \approx 27.32$

7.

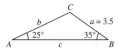

Given: $A = 25°$, $B = 35°$, $a = 3.5$

$C = 180° - A - B = 120°$

$b = \dfrac{a}{\sin A}(\sin B) = \dfrac{3.5}{\sin 25°}(\sin 35°) \approx 4.75$

$c = \dfrac{a}{\sin A}(\sin C) = \dfrac{3.5}{\sin 25°}(\sin 120°) \approx 7.17$

9. Given: $A = 102.4°$, $C = 16.7°$, $a = 21.6$

$B = 180° - A - C = 60.9°$

$b = \dfrac{a}{\sin A}(\sin B) = \dfrac{21.6}{\sin 102.4°}(\sin 60.9°) \approx 19.32$

$c = \dfrac{a}{\sin A}(\sin C) = \dfrac{21.6}{\sin 102.4°}(\sin 16.7°) \approx 6.36$

11. Given: $A = 83°20'$, $C = 54.6°$, $c = 18.1$

$B = 180° - A - C = 180° - 83°20' - 54°36' = 42°4'$

$a = \dfrac{c}{\sin C}(\sin A) = \dfrac{18.1}{\sin 54.6°}(\sin 83°20') \approx 22.05$

$b = \dfrac{c}{\sin C}(\sin B) = \dfrac{18.1}{\sin 54.6°}(\sin 42°4') \approx 14.88$

13. Given: $A = 35°$, $B = 65°$, $c = 10$

$C = 180° - A - B = 80°$

$a = \dfrac{c}{\sin C}(\sin A) = \dfrac{10 \sin 35°}{\sin 80°} \approx 5.82$

$b = \dfrac{c}{\sin C}(\sin B) = \dfrac{10 \sin 65°}{\sin 80°} \approx 9.20$

15. Given: $A = 55°$, $B = 42°$, $c = \dfrac{3}{4}$

$$C = 180° - A - B = 83°$$

$$a = \frac{c}{\sin C}(\sin A) = \frac{0.75}{\sin 83°}(\sin 55°) \approx 0.62$$

$$b = \frac{c}{\sin C}(\sin B) = \frac{0.75}{\sin 83°}(\sin 42°) \approx 0.51$$

17. Given: $A = 36°$, $a = 8$, $b = 5$

$$\sin B = \frac{b \sin A}{a} = \frac{5 \sin 36°}{8} \approx 0.36737 \Rightarrow B \approx 21.55°$$

$$C = 180° - A - B \approx 180° - 36° - 21.55 = 122.45°$$

$$c = \frac{a}{\sin A}(\sin C) = \frac{8}{\sin 36°}(\sin 122.45°) \approx 11.49$$

19. Given: $B = 15°30'$, $a = 4.5$, $b = 6.8$

$$\sin A = \frac{a \sin B}{b} = \frac{4.5 \sin 15°30'}{6.8} \approx 0.17685 \Rightarrow A \approx 10°11'$$

$$C = 180° - A - B \approx 180° - 10°11' - 15°30' = 154°19'$$

$$c = \frac{b}{\sin B}(\sin C) = \frac{6.8}{\sin 15°30'}(\sin 154°19') \approx 11.03$$

21. Given: $A = 145°$, $a = 14$, $b = 4$

$$\sin B = \frac{b \sin A}{a} = \frac{4 \sin 145°}{14} \approx 0.1639 \Rightarrow B \approx 9.43°$$

$$C = 180° - A - B \approx 25.57°$$

$$c = \frac{a}{\sin A}(\sin C) \approx \frac{14 \sin 25.57°}{\sin 145°} \approx 10.53$$

23. Given: $A = 110°15'$, $a = 48$, $b = 16$

$$\sin B = \frac{b \sin A}{a} = \frac{16 \sin 110°15'}{48} \approx 0.31273 \Rightarrow B \approx 18°13'$$

$$C = 180° - A - B \approx 180° - 110°15' - 18°13' = 51°32'$$

$$c = \frac{a}{\sin A}(\sin C) = \frac{48}{\sin 110°15'}(\sin 15°32') \approx 40.06$$

25. Given: $A = 110°$, $a = 125$, $b = 100$

$$\sin B = \frac{b \sin A}{a} = \frac{100 \sin 110°}{125} \approx 0.75175 \Rightarrow B \approx 48.74°$$

$$C = 180° - A - B \approx 21.26°$$

$$c = \frac{a \sin C}{\sin A} \approx \frac{125 \sin 21.26°}{\sin 110°} \approx 48.23$$

27. Given: $a = 18$, $b = 20$, $A = 76°$

$$h = 20 \sin 76° \approx 19.41$$

Because $a < h$, no triangle is formed.

29. Given: $A = 58°, a = 11.4, c = 12.8$

$$\sin B = \frac{b \sin A}{a} = \frac{12.8 \sin 58°}{11.4} \approx 0.9522 \Rightarrow B \approx 72.21° \text{ or } B \approx 107.79°$$

Case 1

$B \approx 72.21°$

$C = 180° - A - B \approx 49.79°$

$c = \frac{a}{\sin A}(\sin C) \approx \frac{11.4 \sin 49.79°}{\sin 58°} \approx 10.27$

Case 2

$B \approx 107.79°$

$C = 180° - A - B \approx 14.21°$

$c = \frac{a}{\sin A}(\sin C) \approx \frac{11.4 \sin 14.21°}{\sin 58°} \approx 3.30$

31. Given: $A = 120°, a = b = 25$

No triangle is formed because A is obtuse and $a = b$.

33. Given: $A = 45°, a = b = 1$

Because $a = b = 1, B = 45°$.

$C = 180° - A - B = 90°$

$c = \frac{a}{\sin A}(\sin C) = \frac{1 \sin 90°}{\sin 45°} = \sqrt{2} \approx 1.41$

35. Given: $A = 36°, a = 5$

(a) One solution if $b \le 5$ or $b = \dfrac{5}{\sin 36°}$

(b) Two solutions if $5 < b < \dfrac{5}{\sin 36°}$

(c) No solution if $b > \dfrac{5}{\sin 36°}$

37. Given: $A = 10°, a = 10.8$

(a) One solution if $b \le 10.8$ or $b = \dfrac{10.8}{\sin 10°}$

(b) Two solutions if $10.8 < b < \dfrac{10.8}{\sin 10°}$

(c) No solution if $b > \dfrac{10.8}{\sin 10°}$

39. Area $= \frac{1}{2}ab \sin C = \frac{1}{2}(4)(6) \sin 120° \approx 10.4$

41. Area $= \frac{1}{2}bc \sin A = \frac{1}{2}(57)(85) \sin 43°45' \approx 1675.2$

43. Area $= \frac{1}{2}ac \sin B = \frac{1}{2}(105)(64) \sin(72°30') \approx 3204.5$

45. $C = 180° - 94° - 30° = 56°$

$h = \frac{40}{\sin 56°}(\sin 30°) \approx 24.1$ meters

47. $\dfrac{\sin(42° - \theta)}{10} = \dfrac{\sin 48°}{17}$

$\sin(42° - \theta) \approx 0.43714$

$42° - \theta \approx 25.9°$

$\theta \approx 16.1°$

49. Given: $c = 100$

$A = 74° - 28° = 46°,$

$B = 180° - 41° - 74° = 65°,$

$C = 180° - 46° - 65° = 69°$

$a = \frac{c}{\sin C}(\sin A) = \frac{100}{\sin 69°}(\sin 46°) \approx 77$ meters

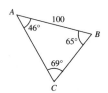

51. (a)

Not drawn to scale

(b) $\dfrac{x}{\sin 17.5°} = \dfrac{9000}{\sin 1.3°}$

$x \approx 119,289.1261$ feet ≈ 22.6 miles

(c) $\dfrac{y}{\sin 71.2°} = \dfrac{x}{\sin 90°}$

$y = x \sin 71.2° \approx 119,289.1261 \sin 71.2°$

$\approx 112,924.963$ feet ≈ 21.4 miles

(d) $z = x \sin 18.8° \approx 119,289.1261 \sin 18.8°$

$\approx 38,443$ feet ≈ 7.3 miles

53.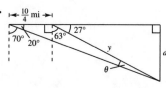

In 15 minutes the boat has traveled

$$(10 \text{ mph})\left(\frac{1}{4} \text{ hr}\right) = \frac{10}{4} \text{ miles.}$$

$$\theta = 180° - 20° - (90° + 63°)$$

$$\theta = 7°$$

$$\frac{10/4}{\sin 7°} = \frac{y}{\sin 20°}$$

$$y \approx 7.0161$$

$$\sin 27° = \frac{d}{7.0161}$$

$$d \approx 3.2 \text{ miles}$$

55. Given: $A = 55°, c = 2.2$

$$B = 180° - 72° = 108°$$

$$C = 180° - 55° - 108° = 17°$$

$$a = \frac{c}{\sin c}(\sin A) = \frac{2.2 \sin 55°}{\sin 17°} \approx 6.16$$

$$h = a \sin 72° \approx 6.16 \sin 72° \approx 5.86 \text{ miles}$$

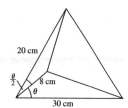

57. True. If one angle of a triangle is obtuse, then there is less than 90° left for the other two angles, so it cannot contain a right angle. It must be oblique.

59. False. To solve an oblique triangle using the Law of Sines, you need to know two angles and any side, or two sides and an angle opposite one of them.

61. (a) $A = \frac{1}{2}(30)(20) \sin\left(\theta + \frac{\theta}{2}\right) - \frac{1}{2}(8)(20) \sin \frac{\theta}{2} - \frac{1}{2}(8)(30) \sin \theta$

$$= 300 \sin \frac{3\theta}{2} - 80 \sin \frac{\theta}{2} - 120 \sin \theta$$

$$= 20\left[15 \sin \frac{3\theta}{2} - 4 \sin \frac{\theta}{2} - 6 \sin \theta\right]$$

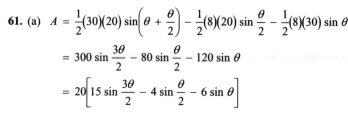

(b)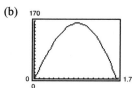

(c) Domain: $0 \le \theta \le 1.6690$

The domain would increase in length and the area would have a greater maximum value if the 8-centimeter line segment were decreased.

Section 8.2   Law of Cosines

- If *ABC* is any oblique triangle with sides *a*, *b*, and *c*, the following equations are valid.

  (a) $a^2 = b^2 + c^2 - 2bc \cos A$ or $\cos A = \dfrac{b^2 + c^2 - a^2}{2bc}$

  (b) $b^2 = a^2 + c^2 - 2ac \cos B$ or $\cos B = \dfrac{a^2 + c^2 - b^2}{2ac}$

  (c) $c^2 = a^2 + b^2 - 2ab \cos C$ or $\cos C = \dfrac{a^2 + b^2 - c^2}{2ab}$

- You should be able to use the Law of Cosines to solve an oblique triangle for the remaining three parts, given:

  (a) Three sides (SSS)

  (b) Two sides and their included angle (SAS)

- Given any triangle with sides of length *a*, *b*, and *c*, the area of the triangle is

  $$\text{Area} = \sqrt{s(s - a)(s - b)(s - c)}, \text{ where } s = \frac{a + b + c}{2}. \quad \text{(Heron's Formula)}$$

1. Cosines

3. $b^2 = a^2 + c^2 - 2ac \cos B$

5. Given: $a = 10, b = 12, c = 16$

$$\cos C = \frac{a^2 + b^2 - c^2}{2ab} = \frac{100 + 144 - 256}{2(10)(12)} = -0.05 \Rightarrow C \approx 92.87°$$

$$\sin B = \frac{b \sin C}{c} \approx \frac{12 \sin 92.87°}{16} \approx 0.7491 \Rightarrow B \approx 48.51°$$

$$A \approx 180° - 48.51° - 92.87° \approx 38.62°$$

7. Given: $A = 30°, b = 15, c = 30$

$$a^2 = b^2 + c^2 - 2bc \cos A$$

$$= 225 + 900 - 2(15)(30) \cos 30° \approx 345.5771$$

$$a \approx 18.59$$

$$\cos C = \frac{a^2 + b^2 - c^2}{2ab} \approx \frac{(18.59)^2 + 15^2 - 30^2}{2(18.59)(15)} \approx -0.5907 \Rightarrow C \approx 126.21°$$

$$B \approx 180° - 30° - 126.21° \approx 23.79°$$

9. Given: $a = 11, b = 15, c = 21$

$$\cos C = \frac{a^2 + b^2 - c^2}{2ab} = \frac{121 + 225 - 441}{2(11)(15)} \approx -0.2879 \Rightarrow C \approx 106.73°$$

$$\sin B = \frac{b \sin C}{c} = \frac{15 \sin 106.73°}{21} \approx 0.6841 \Rightarrow B \approx 43.16°$$

$$A \approx 180° - 43.16° - 106.73° \approx 30.11°$$

11. Given: $a = 75.4, b = 52, c = 52$

$$\cos A = \frac{b^2 + c^2 - a^2}{2bc} = \frac{52^2 + 52^2 - 75.4^2}{2(52)(52)} = -0.05125 \Rightarrow A \approx 92.94°$$

$$\sin B = \frac{b \sin A}{a} \approx \frac{52(0.9987)}{75.4} \approx 0.68876 \Rightarrow B \approx 43.53°$$

$$C = B \approx 43.53°$$

13. Given: $A = 120°, b = 6, c = 7$

$$a^2 = b^2 + c^2 - 2bc \cos A = 36 + 49 - 2(6)(7) \cos 120° = 127 \Rightarrow a \approx 11.27$$

$$\sin B = \frac{b \sin A}{a} \approx \frac{6 \sin 120°}{11.27} \approx 0.4611 \Rightarrow B \approx 27.46°$$

$$C \approx 180° - 120° - 27.46° \approx 32.54°$$

15. Given: $B = 10° 35', a = 40, c = 30$

$$b^2 = a^2 + c^2 - 2ac \cos B = 1600 + 900 - 2(40)(30) \cos 10° 35' \approx 140.8268 \Rightarrow b \approx 11.87$$

$$\sin C = \frac{c \sin B}{b} = \frac{30 \sin 10° 35'}{11.87} \approx 0.4642 \Rightarrow C \approx 27.66° \approx 27° 40'$$

$$A \approx 180° - 10° 35' - 27° 40' = 141° 45'$$

17. Given: $B = 125° 40', a = 37, c = 37$

$$b^2 = a^2 + c^2 - 2ac \cos B = 1369 + 1369 - 2(37)(37) \cos 125° 40' \approx 4334.4420 \Rightarrow b \approx 65.84$$

$$A = C \Rightarrow 2A = 180 - 125° 40' = 54° 20' \Rightarrow A = C = 27° 10'$$

19. $C = 43°, a = \dfrac{4}{9}, b = \dfrac{7}{9}$

$$c^2 = a^2 + b^2 - 2ab\cos C = \left(\dfrac{4}{9}\right)^2 + \left(\dfrac{7}{9}\right)^2 - 2\left(\dfrac{4}{9}\right)\left(\dfrac{7}{9}\right)\cos 43° \approx 0.2968 \Rightarrow c \approx 0.54$$

$$\sin A = \dfrac{a\sin C}{c} = \dfrac{(4/9)\sin 43°}{0.5448} \approx 0.5564 \Rightarrow A \approx 33.80°$$

$$B \approx 180° - 43° - 33.8° \approx 103.20°$$

21. $d^2 = 5^2 + 8^2 - 2(5)(8)\cos 45° \approx 32.4315 \Rightarrow d \approx 5.69$

$2\phi = 360° - 2(45°) = 270° \Rightarrow \phi = 135°$

$c^2 = 5^2 + 8^2 - 2(5)(8)\cos 135° \approx 145.5685 \Rightarrow c \approx 12.07$

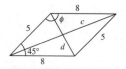

23. $\cos\phi = \dfrac{10^2 + 14^2 - 20^2}{2(10)(14)}$

$\phi \approx 111.8°$

$2\theta \approx 360° - 2(111.8°)$

$\theta = 68.2°$

$d^2 = 10^2 + 14^2 - 2(10)(14)\cos 68.2°$

$d \approx 13.86$

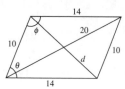

25. $\cos\alpha = \dfrac{(12.5)^2 + (15)^2 - 10^2}{2(12.5)(15)} = 0.75 \Rightarrow \alpha \approx 41.41°$

$\cos\beta = \dfrac{10^2 + 15^2 - (12.5)^2}{2(10)(15)} = 0.5625 \Rightarrow \beta \approx 55.77°$

$z = 180° - \alpha - \beta = 82.82°$

$u = 180° - z = 97.18°$

$b^2 = 12.5^2 + 10^2 - 2(12.5)(10)\cos 97.18° \approx 287.4967 \Rightarrow b \approx 16.96$

$\cos\delta = \dfrac{12.5^2 + 16.96^2 - 10^2}{2(12.5)(16.96)} \approx 0.8111 \Rightarrow \delta \approx 35.80°$

$\theta = \alpha + \delta = 41.41° + 35.80° \approx 77.2°$

$2\phi = 360° - 2\theta \Rightarrow \phi = \dfrac{360° - 2(77.21°)}{2} = 102.8°$

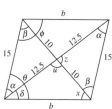

27. Given: $a = 8, c = 5, B = 40°$

Given two sides and included angle, use the Law of Cosines.

$$b^2 = a^2 + c^2 - 2ac\cos B = 64 + 25 - 2(8)(5)\cos 40° \approx 27.7164 \Rightarrow b \approx 5.26$$

$$\cos A = \dfrac{b^2 + c^2 - a^2}{2bc} \approx \dfrac{(5.26)^2 + 25 - 64}{2(5.26)(5)} \approx -0.2154 \Rightarrow A \approx 102.44°$$

$$C \approx 180° - 102.44° - 40° \approx 37.56°$$

29. Given: $A = 24°, a = 4, b = 18$

Given two sides and an angle opposite one of them, use the Law of Sines.

$h = b\sin A = 18\sin 24° \approx 7.32$

Because $a < h$, no triangle is formed.

31. Given: $A = 42°, B = 35°, c = 1.2$

Given two angles and a side, use the Law of Sines.

$C = 180° - 42° - 35° = 103°$

$a = \dfrac{c \sin A}{\sin C} = \dfrac{1.2 \sin 42°}{\sin 103°} \approx 0.82$

$b = \dfrac{c \sin B}{\sin C} = \dfrac{1.2 \sin 35°}{\sin 103°} \approx 0.71$

33. $a = 8, b = 12, c = 17$

$s = \dfrac{a + b + c}{2} = \dfrac{8 + 12 + 17}{2} = 18.5$

Area $= \sqrt{s(s - a)(s - b)(s - c)} = \sqrt{18.5(10.5)(6.5)(1.5)} \approx 43.52$

35. $a = 2.5, b = 10.2, c = 9$

$s = \dfrac{a + b + c}{2} = 10.85$

Area $= \sqrt{s(s - a)(s - b)(s - c)} = \sqrt{10.85(8.35)(0.65)(1.85)} \approx 10.4$

37. $a = 12.32, b = 8.46, c = 15.05$

$s = \dfrac{a + b + c}{2} = 17.915$

Area $= \sqrt{s(s - a)(s - b)(s - c)} = \sqrt{17.915(5.595)(9.455)(2.865)} \approx 52.11$

39. $a = 1, b = \dfrac{1}{2}, c = \dfrac{3}{4}$

$s = \dfrac{a + b + c}{2} = \dfrac{1 + \dfrac{1}{2} + \dfrac{3}{4}}{2} = \dfrac{9}{8}$

Area $= \sqrt{s(s - a)(s - b)(s - c)} = \sqrt{\dfrac{9}{8}\left(\dfrac{1}{8}\right)\left(\dfrac{5}{8}\right)\left(\dfrac{3}{8}\right)} \approx 0.18$

41. $\cos B = \dfrac{1700^2 + 3700^2 - 3000^2}{2(1700)(3700)} \Rightarrow B \approx 52.9°$

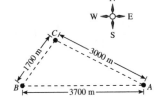

Bearing: $90° - 52.9° = $ N $37.1°$ E

$\cos C = \dfrac{1700^2 + 3000^2 - 3700^2}{2(1700)(3000)} \Rightarrow C \approx 100.2°$

Bearing: $90° - 26.9° = $ S $63.1°$ E

43. $b^2 = 220^2 + 250^2 - 2(220)(250) \cos 105° \Rightarrow b \approx 373.3$ meters

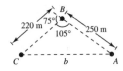

45. The largest angle is across from the largest side.

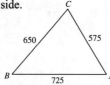

$$\cos C = \frac{650^2 + 575^2 - 725^2}{2(650)(575)}$$

$$C \approx 72.3°$$

47. $C = 180° - 53° - 67° = 60°$

$c^2 = a^2 + b^2 - 2ab \cos C$

$\quad = 36^2 + 48^2 - 2(36)(48)(0.5)$

$\quad = 1872$

$c \approx 43.3 \text{ mi}$

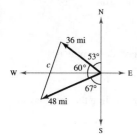

49. (a) $\cos \theta = \dfrac{273^2 + 178^2 - 235^2}{2(273)(178)}$

$\theta \approx 58.4°$

Bearing: N 58.4° W

(b) $\cos \phi = \dfrac{235^2 + 178^2 - 273^2}{2(235)(178)}$

$\phi \approx 81.5°$

Bearing: S 81.5° W

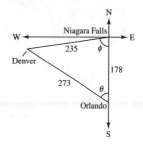

51. $d^2 = 60.5^2 + 90^2 - 2(60.5)(90) \cos 45° \approx 4059.8572 \Rightarrow d \approx 63.7 \text{ ft}$

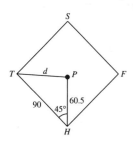

53. $a^2 = 35^2 + 20^2 - 2(35)(20) \cos 42°$

$\quad \approx 24.2 \text{ miles}$

55. $\overline{RS} = \sqrt{8^2 + 10^2} = \sqrt{164} = 2\sqrt{41} \approx 12.8 \text{ ft}$

$\overline{PQ} = \dfrac{1}{2}\sqrt{16^2 + 10^2} = \dfrac{1}{2}\sqrt{356} = \sqrt{89} \approx 9.4 \text{ ft}$

$\tan P = \dfrac{10}{16} = \dfrac{\overline{QS}}{\overline{PS}} = \dfrac{\overline{QS}}{8} \Rightarrow \overline{QS} = 5$

57. $d^2 = 10^2 + 7^2 - 2(10)(7) \cos \theta$

$\theta = \arccos\left[\dfrac{10^2 + 7^2 - d^2}{2(10)(7)}\right]$

$s = \dfrac{360° - \theta}{360°}(2\pi r) = \dfrac{(360° - \theta)\pi}{45°}$

d (inches)	9	10	12	13	14	15	16
θ (degrees)	60.9°	69.5°	88.0°	98.2°	109.6°	122.9°	139.8°
s (inches)	20.88	20.28	18.99	18.28	17.48	16.55	15.37

59. $a = 200$

$b = 500$

$c = 600 \Rightarrow s = \dfrac{200 + 500 + 600}{2} = 650$

Area $= \sqrt{650(450)(150)(50)} \approx 46{,}837.5$ square feet

61. $s = \dfrac{510 + 840 + 1120}{2} = 1235$

Area $= \sqrt{1235(1235 - 510)(1235 - 840)(1235 - 1120)}$

$\approx 201{,}674$ square yards

Cost $\approx \left(\dfrac{201{,}674}{4840}\right)(2000) \approx \$83{,}336.36$

63. False. The average of the three sides of a triangle is

$\dfrac{a + b + c}{3}$, not $\dfrac{a + b + c}{2} = s$.

65. No. If $a = 10$, $b = 16$, and $c = 5$, then by the Law of Cosines,

$\cos A = \dfrac{16^2 + 5^2 - 10^2}{2(16)(5)} = 1.13125 > 1$.

This is not possible. In general, if the sum of any two sides is less than the third side, then they cannot form a triangle.

Here $10 + 5$ is less than 16.

67. (a) Working with $\triangle ODC$, we have $\cos \alpha = \dfrac{a/2}{R}$.

This implies that $2R = \dfrac{a}{\cos \alpha}$.

Because

$\dfrac{a}{\sin A} = \dfrac{b}{\sin B} = \dfrac{c}{\sin C}$,

the proof can be completed by showing that $\cos \alpha = \sin A$. The solution of the system

$A + B + C = 180°$

$\alpha - C + A = \beta$

$\alpha + \beta = B$

is $\alpha = 90° - A$. So,

$2R = \dfrac{a}{\cos \alpha} = \dfrac{a}{\cos(90° - A)} = \dfrac{a}{\sin A}$.

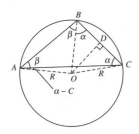

(b) By Heron's Formula, the area of the triangle is

Area $= \sqrt{s(s - a)(s - b)(s - c)}$.

You can also find the area by dividing the area into six triangles and using the fact that the area is $\dfrac{1}{2}$ the base times the height. Using the figure as given,

Area $= \dfrac{1}{2}xr + \dfrac{1}{2}xr + \dfrac{1}{2}yr + \dfrac{1}{2}yr + \dfrac{1}{2}zr + \dfrac{1}{2}zr$

$= r(x + y + z)$

$= rs$.

So, $rs = \sqrt{s(s - a)(s - b)(s - c)} \Rightarrow$

$r = \sqrt{\dfrac{(s - a)(s - b)(s - c)}{s}}$.

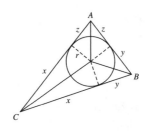

69. Given: $a = 200$ ft, $b = 250$ ft, $c = 325$ ft

$s = \dfrac{200 + 250 + 325}{2} \approx 387.5$

Radius of the inscribed circle: $r = \sqrt{\dfrac{(s - a)(s - b)(s - c)}{s}} = \sqrt{\dfrac{(187.5)(137.5)(62.5)}{387.5}} \approx 64.5$ ft (see #67)

Circumference of an inscribed circle: $C = 2\pi r \approx 2\pi(64.5) \approx 405.2$ ft

71. You could solve for either angle, B or C. Because A is obtuse, there is only one solution for B or C.

73. There is no method that can be used to solve the no-solution case of SSA.

The Law of Cosines can be used to solve the single-solution case of SSA. You can substitute values into $a^2 = b^2 + c^2 - 2bc \cos A$. The simplified quadratic equation in terms of c can be solved, with one positive solution and one negative solution. The negative solution can be discarded because length is positive. You can use the positive solution to solve the triangle.

75. $\dfrac{1}{2}bc(1 + \cos A) = \dfrac{1}{2}bc\left[1 + \dfrac{b^2 + c^2 - a^2}{2bc}\right]$

$\qquad = \dfrac{1}{2}bc\left[\dfrac{2bc + b^2 + c^2 - a^2}{2bc}\right]$

$\qquad = \dfrac{1}{4}\left[(b + c)^2 - a^2\right]$

$\qquad = \dfrac{1}{4}\left[(b + c) + a\right]\left[(b + c) - a\right]$

$\qquad = \dfrac{b + c + a}{2} \cdot \dfrac{b + c - a}{2}$

$\qquad = \dfrac{a + b + c}{2} \cdot \dfrac{-a + b + c}{2}$

Section 8.3   Vectors in the Plane

- A vector **v** is the collection of all directed line segments that are equivalent to a given directed line segment $\overline{PQ}$.

- You should be able to *geometrically* perform the operations of vector addition and scalar multiplication.

- The component form of the vector with initial point $P = (p_1, p_2)$ and terminal point $Q = (q_1, q_2)$ is
  $\overline{PQ} = \langle q_1 - p_1, q_2 - p_2 \rangle = \langle v_1, v_2 \rangle = \mathbf{v}.$

- The magnitude of $\mathbf{v} = \langle v_1, v_2 \rangle$ is given by $\|\mathbf{v}\| = \sqrt{v_1{}^2 + v_2{}^2}$.

- If $\|\mathbf{v}\| = 1$, **v** is a unit vector.

- You should be able to perform the operations of scalar multiplication and vector addition in component form.
  (a) $\mathbf{u} + \mathbf{v} = \langle u_1 + v_1, u_2 + v_2 \rangle$
  (b) $k\mathbf{u} = \langle ku_1, ku_2 \rangle$

- You should know the following properties of vector addition and scalar multiplication.
  (a) $\mathbf{u} + \mathbf{v} = \mathbf{v} + \mathbf{u}$
  (b) $(\mathbf{u} + \mathbf{v}) + \mathbf{w} = \mathbf{u} + (\mathbf{v} + \mathbf{w})$
  (c) $\mathbf{u} + \mathbf{0} = \mathbf{u}$
  (d) $\mathbf{u} + (-\mathbf{u}) = \mathbf{0}$
  (e) $c(d\mathbf{u}) = (cd)\mathbf{u}$
  (f) $(c + d)\mathbf{u} = c\mathbf{u} + d\mathbf{u}$
  (g) $c(\mathbf{u} + \mathbf{v}) = c\mathbf{u} + c\mathbf{v}$
  (h) $1(\mathbf{u}) = \mathbf{u}, 0\mathbf{u} = \mathbf{0}$
  (i) $\|c\mathbf{v}\| = |c| \|\mathbf{v}\|$

- A unit vector in the direction of **v** is $\mathbf{u} = \dfrac{\mathbf{v}}{\|\mathbf{v}\|}$.

- The standard unit vectors are $\mathbf{i} = \langle 1, 0 \rangle$ and $\mathbf{j} = \langle 0, 1 \rangle$. $\mathbf{v} = \langle v_1, v_2 \rangle$ can be written as $\mathbf{v} = v_1\mathbf{i} + v_2\mathbf{j}$.

- A vector **v** with magnitude $\|\mathbf{v}\|$ and direction θ can be written as $\mathbf{v} = a\mathbf{i} + b\mathbf{j} = \|\mathbf{v}\|(\cos \theta)\mathbf{i} + \|\mathbf{v}\|(\sin \theta)\mathbf{j}$, where $\tan \theta = b/a$.

1. directed line segment

3. magnitude

5. magnitude; direction

7. unit vector

9. resultant

11. $\|\mathbf{u}\| = \sqrt{(6-2)^2 + (5-4)^2} = \sqrt{17}$

$\|\mathbf{v}\| = \sqrt{(4-0)^2 + (1-0)^2} = \sqrt{17}$

$\text{slope}_{\mathbf{u}} = \dfrac{5-4}{6-2} = \dfrac{1}{4}$

$\text{slope}_{\mathbf{v}} = \dfrac{1-0}{4-0} = \dfrac{1}{4}$

u and **v** have the same magnitude and direction so they are equal.

13. Initial point: $(0, 0)$

Terminal point: $(1, 3)$

$\mathbf{v} = \langle 1-0, 3-0 \rangle = \langle 1, 3 \rangle$

$\|\mathbf{v}\| = \sqrt{1^2 + 3^2} = \sqrt{10}$

15. Initial point: $(-1, -1)$

Terminal point: $(3, 5)$

$\mathbf{v} = \langle 3-(-1), 5-(-1) \rangle = \langle 4, 6 \rangle$

$\|\mathbf{v}\| = \sqrt{4^2 + 6^2} = \sqrt{52} = 2\sqrt{13}$

17. Initial point: $(3, -2)$

Terminal point: $(3, 3)$

$\mathbf{v} = \langle 3-3, 3-(-2) \rangle = \langle 0, 5 \rangle$

$\|\mathbf{v}\| = \sqrt{0^2 + 5^2} = \sqrt{25} = 5$

19. Initial point: $(-3, -5)$

Terminal point: $(5, 1)$

$\mathbf{v} = \langle 5-(-3), 1-(-5) \rangle = \langle 8, 6 \rangle$

$\|\mathbf{v}\| = \sqrt{8^2 + 6^2} = \sqrt{100} = 10$

21. Initial point: $(1, 3)$

Terminal point: $(-8, -9)$

$\mathbf{v} = \langle -8-1, -9-3 \rangle = \langle -9, -12 \rangle$

$\|\mathbf{v}\| = \sqrt{(-9)^2 + (-12)^2} = \sqrt{225} = 15$

23. Initial point: $(-1, 5)$

Terminal point: $(15, 12)$

$\mathbf{v} = \langle 15-(-1), 12-5 \rangle = \langle 16, 7 \rangle$

$\|\mathbf{v}\| = \sqrt{16^2 + 7^2} = \sqrt{305}$

25. $-\mathbf{v}$

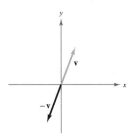

27. $\mathbf{u} + \mathbf{v}$

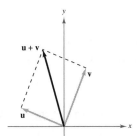

29. $\mathbf{u} - \mathbf{v}$

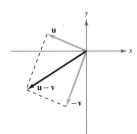

31. $\mathbf{u} = \langle 2, 1 \rangle$, $\mathbf{v} = \langle 1, 3 \rangle$

(a) $\mathbf{u} + \mathbf{v} = \langle 3, 4 \rangle$

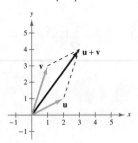

(b) $\mathbf{u} - \mathbf{v} = \langle 1, -2 \rangle$

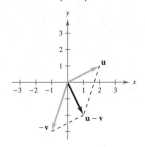

(c) $2\mathbf{u} - 3\mathbf{v} = \langle 4, 2 \rangle - \langle 3, 9 \rangle = \langle 1, -7 \rangle$

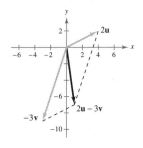

33. $\mathbf{u} = \langle -5, 3 \rangle$, $\mathbf{v} = \langle 0, 0 \rangle$

(a) $\mathbf{u} + \mathbf{v} = \langle -5, 3 \rangle = \mathbf{u}$

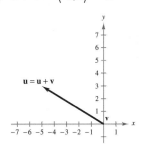

(b) $\mathbf{u} - \mathbf{v} = \langle -5, 3 \rangle = \mathbf{u}$

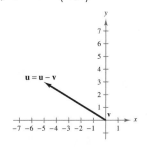

(c) $2\mathbf{u} - 3\mathbf{v} = \langle -10, 6 \rangle = 2\mathbf{u}$

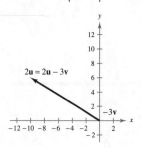

35. $\mathbf{u} = \mathbf{i} + \mathbf{j}$, $\mathbf{v} = 2\mathbf{i} - 3\mathbf{j}$

(a) $\mathbf{u} + \mathbf{v} = 3\mathbf{i} - 2\mathbf{j}$

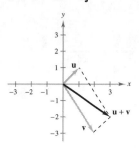

(b) $\mathbf{u} - \mathbf{v} = -\mathbf{i} + 4\mathbf{j}$

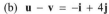
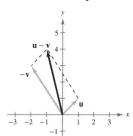

(c) $2\mathbf{u} - 3\mathbf{v} = (2\mathbf{i} + 2\mathbf{j}) - (6\mathbf{i} - 9\mathbf{j})$
$= -4\mathbf{i} + 11\mathbf{j}$

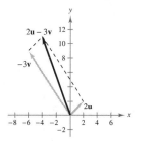

37. $\mathbf{u} = 2\mathbf{i}$, $\mathbf{v} = \mathbf{j}$

(a) $\mathbf{u} + \mathbf{v} = 2\mathbf{i} + \mathbf{j}$

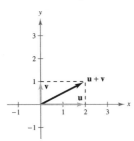

(b) $\mathbf{u} - \mathbf{v} = 2\mathbf{i} - \mathbf{j}$

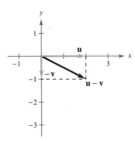

(c) $2\mathbf{u} - 3\mathbf{v} = 4\mathbf{i} - 3\mathbf{j}$

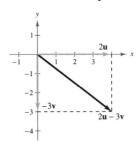

39. $\mathbf{u} = \langle 3, 0 \rangle$

$$\mathbf{v} = \frac{1}{\|\mathbf{u}\|}\mathbf{u} = \frac{1}{\sqrt{3^2 + 0^2}}\langle 3, 0 \rangle = \frac{1}{3}\langle 3, 0 \rangle = \langle 1, 0 \rangle$$

$$\|\mathbf{v}\| = \sqrt{1^2 + 0^2} = 1$$

41. $\mathbf{v} = \langle -2, 2 \rangle$

$$\mathbf{u} = \frac{1}{\|\mathbf{v}\|}\mathbf{v} = \frac{1}{\sqrt{(-2)^2 + 2^2}}\langle -2, 2 \rangle = \frac{1}{2\sqrt{2}}\langle -2, 2 \rangle$$

$$= \left\langle -\frac{1}{\sqrt{2}}, -\frac{1}{\sqrt{2}} \right\rangle$$

$$= \left\langle -\frac{\sqrt{2}}{2}, \frac{\sqrt{2}}{2} \right\rangle$$

$$\|\mathbf{v}\| = \sqrt{\left(\frac{-\sqrt{2}}{2}\right)^2 + \left(\frac{\sqrt{2}}{2}\right)^2} = 1$$

43. $\mathbf{v} = \mathbf{i} + \mathbf{j}$

$$\mathbf{u} = \frac{1}{\|\mathbf{v}\|}\mathbf{v}$$

$$= \frac{1}{\sqrt{1^2 + 1^2}}(\mathbf{i} + \mathbf{j}) = \frac{1}{\sqrt{2}}(\mathbf{i} + \mathbf{j}) = \frac{\sqrt{2}}{2}\mathbf{i} + \frac{\sqrt{2}}{2}\mathbf{j}$$

$$\|\mathbf{u}\| = \sqrt{\left(\frac{\sqrt{2}}{2}\right)^2 + \left(\frac{\sqrt{2}}{2}\right)^2} = 1$$

45. $\mathbf{w} = 4\mathbf{j}$

$$\mathbf{u} = \frac{1}{\|\mathbf{w}\|}\mathbf{w} = \frac{1}{4}(4\mathbf{j}) = \mathbf{j}$$

$$\|\mathbf{u}\| = \sqrt{0^2 + 1^2} = 1$$

47. $\mathbf{w} = \mathbf{i} - 2\mathbf{j}$

$$\mathbf{u} = \frac{1}{\|\mathbf{w}\|}\mathbf{w} = \frac{1}{\sqrt{1^2 + (-2)^2}}(\mathbf{i} - 2\mathbf{j}) = \frac{1}{\sqrt{5}}(\mathbf{i} - 2\mathbf{j})$$

$$= \frac{1}{\sqrt{5}}\mathbf{i} - \frac{2}{\sqrt{5}}\mathbf{j} = \frac{\sqrt{5}}{5}\mathbf{i} - \frac{2\sqrt{5}}{5}\mathbf{j}$$

$$\|\mathbf{u}\| = \sqrt{\left(\frac{\sqrt{5}}{5}\right)^2 + \left(-\frac{2\sqrt{5}}{5}\right)^2} = 1$$

49. $\mathbf{v} = 10\left(\frac{1}{\|\mathbf{u}\|}\mathbf{u}\right) = 10\left(\frac{1}{\sqrt{(-3)^2 + 4^2}}\langle -3, 4 \rangle\right)$

$$= 2\langle -3, 4 \rangle$$

$$= \langle -6, 8 \rangle$$

51. $9\left(\frac{1}{\|\mathbf{u}\|}\mathbf{u}\right) = 9\left(\frac{1}{\sqrt{2^2 + 5^2}}\langle 2, 5 \rangle\right) = \frac{9}{\sqrt{29}}\langle 2, 5 \rangle$

$$= \left\langle \frac{18}{\sqrt{29}}, \frac{45}{\sqrt{29}} \right\rangle = \left\langle \frac{18\sqrt{29}}{29}, \frac{45\sqrt{29}}{29} \right\rangle$$

53. $\mathbf{u} = \langle 3 - (-2), -2 - 1 \rangle$

$$= \langle 5, -3 \rangle$$

$$= 5\mathbf{i} - 3\mathbf{j}$$

55. $\mathbf{u} = \langle 0 - (-6), 1 - 4 \rangle$

$$\mathbf{u} = \langle 6, -3 \rangle$$

$$\mathbf{u} = 6\mathbf{i} - 3\mathbf{j}$$

57. $\mathbf{v} = \frac{3}{2}\mathbf{u}$

$\quad\quad = \frac{3}{2}(2\mathbf{i} - \mathbf{j})$

$\quad\quad = 3\mathbf{i} - \frac{3}{2}\mathbf{j} = \left\langle 3, -\frac{3}{2} \right\rangle$

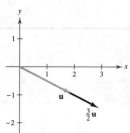

59. $\mathbf{v} = \mathbf{u} + 2\mathbf{w}$

$\quad\quad = (2\mathbf{i} - \mathbf{j}) + 2(\mathbf{i} + 2\mathbf{j})$

$\quad\quad = 4\mathbf{i} + 3\mathbf{j} = \langle 4, 3 \rangle$

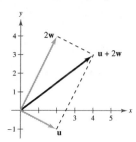

61. $\mathbf{v} = \frac{1}{2}(3\mathbf{u} + \mathbf{w})$

$\quad\quad = \frac{1}{2}(6\mathbf{i} - 3\mathbf{j} + \mathbf{i} + 2\mathbf{j})$

$\quad\quad = \frac{7}{2}\mathbf{i} - \frac{1}{2}\mathbf{j} = \left\langle \frac{7}{2}, -\frac{1}{2} \right\rangle$

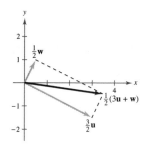

63. $\mathbf{v} = 6\mathbf{i} - 6\mathbf{j}$

$\quad \|\mathbf{v}\| = \sqrt{6^2 + (-6)^2} = \sqrt{72} = 6\sqrt{2}$

$\quad \tan\theta = \dfrac{-6}{6} = -1$

Since $\mathbf{v}$ lies in Quadrant IV, $\theta = 315°$.

65. $\mathbf{v} = 3(\cos 60°\mathbf{i} + \sin 60°\mathbf{j})$

$\quad \|\mathbf{v}\| = 3,\ \theta = 60°$

67. $\mathbf{v} = \langle 3\cos 0°, 3\sin 0° \rangle$

$\quad\quad = \langle 3, 0 \rangle$

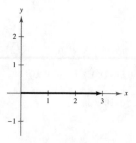

69. $\mathbf{v} = \left\langle \frac{7}{2}\cos 150°, \frac{7}{2}\sin 150° \right\rangle$

$\quad\quad = \left\langle -\frac{7\sqrt{3}}{4}, \frac{7}{4} \right\rangle$

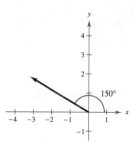

71. $\mathbf{v} = \left\langle 2\sqrt{3}\cos 45°, 2\sqrt{3}\sin 45° \right\rangle$

$\quad\quad = \left\langle \sqrt{6}, \sqrt{6} \right\rangle$

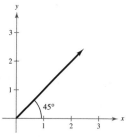

73. $\mathbf{v} = 3\left(\dfrac{1}{\sqrt{3^2 + 4^2}}\right)(3\mathbf{i} + 4\mathbf{j})$

$\quad\quad = \frac{3}{5}(3\mathbf{i} + 4\mathbf{j})$

$\quad\quad = \frac{9}{5}\mathbf{i} + \frac{12}{5}\mathbf{j} = \left\langle \frac{9}{5}, \frac{12}{5} \right\rangle$

75. $\mathbf{u} = \langle 5\cos 0°, 5\sin 0° \rangle = \langle 5, 0 \rangle$

$\mathbf{v} = \langle 5\cos 90°, 5\sin 90° \rangle = \langle 0, 5 \rangle$

$\mathbf{u} + \mathbf{v} = \langle 5, 5 \rangle$

77. $\mathbf{u} = \langle 20\cos 45°, 20\sin 45° \rangle = \langle 10\sqrt{2}, 10\sqrt{2} \rangle$

$\mathbf{v} = \langle 50\cos 180°, 50\sin 180° \rangle = \langle -50, 0 \rangle$

$\mathbf{u} + \mathbf{v} = \langle 10\sqrt{2} - 50, 10\sqrt{2} \rangle$

79. $\mathbf{v} = \mathbf{i} + \mathbf{j}$

$\mathbf{w} = 2\mathbf{i} - 2\mathbf{j}$

$\mathbf{u} = \mathbf{v} - \mathbf{w} = -\mathbf{i} + 3\mathbf{j}$

$\|\mathbf{v}\| = \sqrt{2}$

$\|\mathbf{w}\| = 2\sqrt{2}$

$\|\mathbf{v} - \mathbf{w}\| = \sqrt{10}$

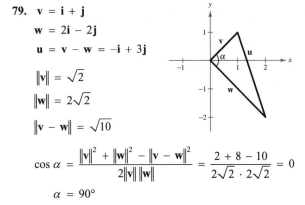

$\cos \alpha = \dfrac{\|\mathbf{v}\|^2 + \|\mathbf{w}\|^2 - \|\mathbf{v} - \mathbf{w}\|^2}{2\|\mathbf{v}\|\,\|\mathbf{w}\|} = \dfrac{2 + 8 - 10}{2\sqrt{2} \cdot 2\sqrt{2}} = 0$

$\alpha = 90°$

81. Force One: $\mathbf{u} = 45\mathbf{i}$

Force Two: $\mathbf{v} = 60\cos\theta\,\mathbf{i} + 60\sin\theta\,\mathbf{j}$

Resultant Force: $\mathbf{u} + \mathbf{v} = (45 + 60\cos\theta)\mathbf{i} + 60\sin\theta\,\mathbf{j}$

$\|\mathbf{u} + \mathbf{v}\| = \sqrt{(45 + 60\cos\theta)^2 + (60\sin\theta)^2} = 90$

$\qquad\qquad 2025 + 5400\cos\theta + 3600 = 8100$

$\qquad\qquad\qquad\qquad 5400\cos\theta = 2475$

$\qquad\qquad\qquad\qquad\qquad \cos\theta = \tfrac{2475}{5400} \approx 0.4583$

$\qquad\qquad\qquad\qquad\qquad\quad \theta \approx 62.7°$

83. Horizontal component of velocity: $1200\cos 6° \approx 1193.4$ ft/sec

Vertical component of velocity: $1200\sin 6° \approx 125.4$ ft/sec

85. $\qquad \mathbf{u} = 300\mathbf{i}$

$\qquad \mathbf{v} = (125\cos 45°)\mathbf{i} + (125\sin 45°)\mathbf{j} = \dfrac{125}{\sqrt{2}}\mathbf{i} + \dfrac{125}{\sqrt{2}}\mathbf{j}$

$\quad \mathbf{u} + \mathbf{v} = \left(300 + \dfrac{125}{\sqrt{2}}\right)\mathbf{i} + \dfrac{125}{\sqrt{2}}\mathbf{j}$

$\|\mathbf{u} + \mathbf{v}\| = \sqrt{\left(300 + \dfrac{125}{\sqrt{2}}\right)^2 + \left(\dfrac{125}{\sqrt{2}}\right)^2} \approx 398.32$ newtons

$\tan\theta = \dfrac{\dfrac{125}{\sqrt{2}}}{300 + \left(\dfrac{125}{\sqrt{2}}\right)} \Rightarrow \theta \approx 12.8°$

87. $\qquad \mathbf{u} = (75\cos 30°)\mathbf{i} + (75\sin 30°)\mathbf{j} \approx 64.95\mathbf{i} + 37.5\mathbf{j}$

$\qquad \mathbf{v} = (100\cos 45°)\mathbf{i} + (100\sin 45°)\mathbf{j} \approx 70.71\mathbf{i} + 70.71\mathbf{j}$

$\qquad \mathbf{w} = (125\cos 120°)\mathbf{i} + (125\sin 120°)\mathbf{j} \approx -62.5\mathbf{i} + 108.3\mathbf{j}$

$\mathbf{u} + \mathbf{v} + \mathbf{w} \approx 73.16\mathbf{i} + 216.5\mathbf{j}$

$\|\mathbf{u} + \mathbf{v} + \mathbf{w}\| \approx 228.5$ pounds

$\tan\theta \approx \dfrac{216.5}{73.16} \approx 2.9593$

$\theta \approx 71.3°$

89. Left cable: $\mathbf{u} = \|\mathbf{u}\|(\cos 20°\mathbf{i} - \sin 20°\mathbf{j})$

Right cable: $\mathbf{v} = \|\mathbf{v}\|(-\cos 20°\mathbf{i} - \sin 20°\mathbf{j})$

Resultant: $\mathbf{u} + \mathbf{v} = -12\mathbf{j}$

$\|\mathbf{u}\|\cos 20° - \|\mathbf{v}\|\cos 20° = 0$

$-\|\mathbf{u}\|\sin 20° - \|\mathbf{v}\|\sin 20° = -12$

Solving this system of equations yields:

$T_L = \|\mathbf{u}\| \approx 17.5$ pounds

$T_R = \|\mathbf{v}\| \approx 17.5$ pounds

91. Cable $\overrightarrow{AC}$: $\mathbf{u} = \|\mathbf{u}\|(\cos 50°\mathbf{i} - \sin 50°\mathbf{j})$

Cable $\overrightarrow{BC}$: $\mathbf{v} = \|\mathbf{v}\|(-\cos 30°\mathbf{i} - \sin 30°\mathbf{j})$

Resultant: $\mathbf{u} + \mathbf{v} = -2000\mathbf{j}$

$\|\mathbf{u}\|\cos 50° - \|\mathbf{v}\|\cos 30° = 0$

$-\|\mathbf{u}\|\sin 50° - \|\mathbf{v}\|\sin 30° = -2000$

Solving this system of equations yields:

$T_{AC} = \|\mathbf{u}\| \approx 1758.8$ pounds

$T_{BC} = \|\mathbf{v}\| \approx 1305.4$ pounds

93. Towline 1: $\mathbf{u} = \|\mathbf{u}\|(\cos 18°\mathbf{i} + \sin 18°\mathbf{j})$

Towline 2: $\mathbf{v} = \|\mathbf{u}\|(\cos 18°\mathbf{i} - \sin 18°\mathbf{j})$

Resultant: $\mathbf{u} + \mathbf{v} = 6000\mathbf{i}$

$\|\mathbf{u}\|\cos 18° + \|\mathbf{u}\|\cos 18° = 6000$

$\|\mathbf{u}\| \approx 3154.4$

So, the tension on each towline is $\|\mathbf{u}\| \approx 3154.4$ pounds.

95. $W = 100, \theta = 12°$

$\sin\theta = \dfrac{F}{W}$

$F = W\sin\theta = 100\sin 12° \approx 20.8$ pounds

97. $F = 5000, W = 15{,}000$

$\sin\theta = \dfrac{F}{W}$

$\sin\theta = \dfrac{5000}{15{,}000}$

$\theta = \sin^{-1}\dfrac{1}{3} \approx 19.5°$

99. $W = FD = (100\cos 50°)(30) = 1928.4$ foot-pounds

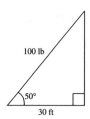

101. Airspeed: $\mathbf{u} = (875\cos 58°)\mathbf{i} - (875\sin 58°)\mathbf{j}$

Groundspeed: $\mathbf{v} = (800\cos 50°)\mathbf{i} - (800\sin 50°)\mathbf{j}$

Wind: $\mathbf{w} = \mathbf{v} - \mathbf{u} = (800\cos 50° - 875\cos 58°)\mathbf{i} + (-800\sin 50° + 875\sin 58°)\mathbf{j}$

$\approx 50.5507\mathbf{i} + 129.2065\mathbf{j}$

Wind speed: $\|\mathbf{w}\| \approx \sqrt{(50.5507)^2 + (129.2065)^2} \approx 138.7$ kilometers per hour

Wind direction: $\tan\theta \approx \dfrac{129.2065}{50.5507}$

$\theta \approx 68.6°; 90° - \theta = 21.4°$

Bearing: N 21.4° E

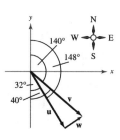

103. True. **a** and **d** have the same magnitude, are parallel, and are pointing in opposite directions.

105. True. By definition of vector addition.

107. True. $\mathbf{a} = -\mathbf{d}, \mathbf{w} = -\mathbf{d}, \mathbf{a} + \mathbf{w} = -\mathbf{d} + (-\mathbf{d}) = -2\mathbf{d}$

109. False. $\mathbf{u} - \mathbf{v} = 2\mathbf{u}$ and $-2(\mathbf{b} + \mathbf{t}) = -2(-2\mathbf{u}) = 4\mathbf{u}$

111. Let $\mathbf{v} = (\cos\theta)\mathbf{i} + (\sin\theta)\mathbf{j}$.

$\|\mathbf{v}\| = \sqrt{\cos^2\theta + \sin^2\theta} = \sqrt{1} = 1$

So, **v** is a unit vector for any value of θ.

113. $\mathbf{F}_1 = \langle 10, 0 \rangle$, $\mathbf{F}_2 = 5\langle \cos\theta, \sin\theta \rangle$

(a) $\mathbf{F}_1 + \mathbf{F}_2 = \langle 10 + 5\cos\theta, 5\sin\theta \rangle$

$$\|\mathbf{F}_1 + \mathbf{F}_2\| = \sqrt{(10 + 5\cos\theta)^2 + (5\sin\theta)^2}$$
$$= \sqrt{100 + 100\cos\theta + 25\cos^2\theta + 25\sin^2\theta}$$
$$= 5\sqrt{4 + 4\cos\theta + \cos^2\theta + \sin^2\theta}$$
$$= 5\sqrt{4 + 4\cos\theta + 1}$$
$$= 5\sqrt{5 + 4\cos\theta}$$

(b)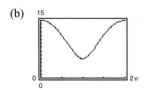

(c) Range: $[5, 15]$

Maximum is 15 when $\theta = 0$.

Minimum is 5 when $\theta = \pi$.

(d) The magnitude of the resultant is never 0 because the magnitudes of $\mathbf{F}_1$ and $\mathbf{F}_2$ are not the same.

115.
$$\mathbf{u} = \langle 5 - 1, 2 - 6 \rangle = \langle 4, -4 \rangle$$
$$\mathbf{v} = \langle 9 - 4, 4 - 5 \rangle = \langle 5, -1 \rangle$$
$$\mathbf{u} - \mathbf{v} = \langle -1, -3 \rangle \text{ or } \mathbf{v} - \mathbf{u} = \langle 1, 3 \rangle$$

117. Answers will vary. *Sample answer*: A scalar is a real number such as 2. A vector is represented by a directed line segment. A vector has both magnitude and direction.

For example $\langle \sqrt{3}, 1 \rangle$ has direction $\dfrac{\pi}{6}$ and a magnitude of 2.

119. (a) Vector. The velocity has both magnitude and direction.

(b) Scalar. The price has only magnitude.

(c) Scalar. The temperature has only magnitude.

(d) Vector. The weight has magnitude and direction.

Section 8.4   Vectors and Dot Products

- Know the definition of the dot product of $\mathbf{u} = \langle u_1, u_2 \rangle$ and $\mathbf{v} = \langle v_1, v_2 \rangle$.

  $\mathbf{u} \cdot \mathbf{v} = u_1 v_1 + u_2 v_2$

- Know the following properties of the dot product:

  1. $\mathbf{u} \cdot \mathbf{v} = \mathbf{v} \cdot \mathbf{u}$

  2. $\mathbf{0} \cdot \mathbf{v} = 0$

  3. $\mathbf{u} \cdot (\mathbf{v} + \mathbf{w}) = \mathbf{u} \cdot \mathbf{v} + \mathbf{u} \cdot \mathbf{w}$

  4. $\mathbf{v} \cdot \mathbf{v} = \|\mathbf{v}\|^2$

  5. $c(\mathbf{u} \cdot \mathbf{v}) = c\mathbf{u} \cdot \mathbf{v} = \mathbf{u} \cdot c\mathbf{v}$

- If θ is the angle between two nonzero vectors $\mathbf{u}$ and $\mathbf{v}$, then

  $\cos\theta = \dfrac{\mathbf{u} \cdot \mathbf{v}}{\|\mathbf{u}\| \|\mathbf{v}\|}$.

- The vectors $\mathbf{u}$ and $\mathbf{v}$ are orthogonal if $\mathbf{u} \cdot \mathbf{v} = 0$.

- Know the definition of vector components.

  $\mathbf{u} = \mathbf{w}_1 + \mathbf{w}_2$ where $\mathbf{w}_1$ and $\mathbf{w}_2$ are orthogonal, and $\mathbf{w}_1$ is parallel to $\mathbf{v}$. $\mathbf{w}_1$ is called the projection of $\mathbf{u}$ onto $\mathbf{v}$ and is

  denoted by $\mathbf{w}_1 = \text{proj}_{\mathbf{v}}\mathbf{u} = \left(\dfrac{\mathbf{u} \cdot \mathbf{v}}{\|\mathbf{v}\|^2} \right)\mathbf{v}$. Then we have $\mathbf{w}_2 = \mathbf{u} - \mathbf{w}_1$.

- Know the definition of work.

  1. Projection form: $W = \left\| \text{proj}_{\overrightarrow{PQ}}\mathbf{F} \right\| \left\| \overrightarrow{PQ} \right\|$ 2. Dot product form: $W = \mathbf{F} \cdot \overrightarrow{PQ}$

1. dot product

3. $\dfrac{\mathbf{u}\cdot\mathbf{v}}{\|\mathbf{u}\|\,\|\mathbf{v}\|}$

5. $\left(\dfrac{\mathbf{u}\cdot\mathbf{v}}{\|\mathbf{v}\|^{2}}\right)\mathbf{v}$

7. $\mathbf{u}=\langle 7,1\rangle,\ \mathbf{v}=\langle -3,2\rangle$

$\mathbf{u}\cdot\mathbf{v}=7(-3)+1(2)=-19$

9. $\mathbf{u}=\langle -4,1\rangle,\ \mathbf{v}=\langle 2,-3\rangle$

$\mathbf{u}\cdot\mathbf{v}=-4(2)+1(-3)=-11$

11. $\mathbf{u}=4\mathbf{i}-2\mathbf{j},\ \mathbf{v}=\mathbf{i}-\mathbf{j}$

$\mathbf{u}\cdot\mathbf{v}=4(1)+(-2)(-1)=6$

13. $\mathbf{u}=3\mathbf{i}+2\mathbf{j},\ \mathbf{v}=-2\mathbf{i}-3\mathbf{j}$

$\mathbf{u}\cdot\mathbf{v}=3(-2)+2(-3)=-12$

15. $\mathbf{u}=\langle 3,3\rangle$

$\mathbf{u}\cdot\mathbf{u}=3(3)+3(3)=18$

The result is a scalar.

17. $\mathbf{u}=\langle 3,3\rangle,\ \mathbf{v}=\langle -4,2\rangle$

$(\mathbf{u}\cdot\mathbf{v})\mathbf{v}=\bigl[3(-4)+3(2)\bigr]\langle -4,2\rangle$

$=-6\langle -4,2\rangle$

$=\langle 24,-12\rangle$

The result is a vector.

19. $\mathbf{u}=\langle 3,3\rangle,\ \mathbf{v}=\langle -4,2\rangle,\ \mathbf{w}=\langle 3,-1\rangle$

$(3\mathbf{w}\cdot\mathbf{v})\mathbf{u}=3\bigl[3(-4)+(-1)(2)\bigr]\langle 3,3\rangle$

$=3(-14)\langle 3,3\rangle$

$=-42\langle 3,3\rangle$

$=\langle -126,-126\rangle$

The result is a vector.

21. $\mathbf{w}=\langle 3,-1\rangle$

$\|\mathbf{w}\|-1=\sqrt{3^{2}+(-1)^{2}}-1=\sqrt{10}-1$

The result is a scalar.

23. $\mathbf{u}=\langle 3,3\rangle,\ \mathbf{v}=\langle -4,2\rangle,\ \mathbf{w}=\langle 3,-1\rangle$

$(\mathbf{u}\cdot\mathbf{v})-(\mathbf{u}\cdot\mathbf{w})=\bigl[3(-4)+3(2)\bigr]-\bigl[3(3)+3(-1)\bigr]$

$=-6-6$

$=-12$

The result is a scalar.

25. $\mathbf{u}=\langle -8,15\rangle$

$\|\mathbf{u}\|=\sqrt{\mathbf{u}\cdot\mathbf{u}}=\sqrt{(-8)(-8)+15(15)}=\sqrt{289}=17$

27. $\mathbf{u}=20\mathbf{i}+25\mathbf{j}$

$\|\mathbf{u}\|=\sqrt{\mathbf{u}\cdot\mathbf{u}}=\sqrt{(20)^{2}+(25)^{2}}=\sqrt{1025}=5\sqrt{41}$

29. $\mathbf{u}=6\mathbf{j}$

$\|\mathbf{u}\|=\sqrt{\mathbf{u}\cdot\mathbf{u}}=\sqrt{(0)^{2}+(6)^{2}}=\sqrt{36}=6$

31. $\mathbf{u}=\langle 1,0\rangle,\ \mathbf{v}=\langle 0,-2\rangle$

$\cos\theta=\dfrac{\mathbf{u}\cdot\mathbf{v}}{\|\mathbf{u}\|\,\|\mathbf{v}\|}=\dfrac{0}{(1)(2)}=0$

$\theta=90°$

33. $\mathbf{u}=3\mathbf{i}+4\mathbf{j},\ \mathbf{v}=-2\mathbf{j}$

$\cos\theta=\dfrac{\mathbf{u}\cdot\mathbf{v}}{\|\mathbf{u}\|\,\|\mathbf{v}\|}=-\dfrac{8}{(5)(2)}$

$\theta=\arccos\left(-\dfrac{4}{5}\right)$

$\theta\approx 143.13°$

35. $\mathbf{u}=2\mathbf{i}-\mathbf{j},\ \mathbf{v}=6\mathbf{i}+4\mathbf{j}$

$\cos\theta=\dfrac{\mathbf{u}\cdot\mathbf{v}}{\|\mathbf{u}\|\,\|\mathbf{v}\|}=\dfrac{8}{\sqrt{5}\sqrt{52}}=0.4961$

$\theta=60.26°$

37. $\mathbf{u}=5\mathbf{i}+5\mathbf{j},\ \mathbf{v}=-6\mathbf{i}+6\mathbf{j}$

$\cos\theta=\dfrac{\mathbf{u}\cdot\mathbf{v}}{\|\mathbf{u}\|\,\|\mathbf{v}\|}=0$

$\theta=90°$

39. $\mathbf{u}=\left(\cos\dfrac{\pi}{3}\right)\mathbf{i}+\left(\sin\dfrac{\pi}{3}\right)\mathbf{j}=\dfrac{1}{2}\mathbf{i}+\dfrac{\sqrt{3}}{2}\mathbf{j}$

$\mathbf{v}=\left(\cos\dfrac{3\pi}{4}\right)\mathbf{i}+\left(\sin\dfrac{3\pi}{4}\right)\mathbf{j}=-\dfrac{\sqrt{2}}{2}\mathbf{i}+\dfrac{\sqrt{2}}{2}\mathbf{j}$

$\|\mathbf{u}\|=\|\mathbf{v}\|=1$

$\cos\theta=\dfrac{\mathbf{u}\cdot\mathbf{v}}{\|\mathbf{u}\|\,\|\mathbf{v}\|}=\mathbf{u}\cdot\mathbf{v}$

$=\left(\dfrac{1}{2}\right)\left(-\dfrac{\sqrt{2}}{2}\right)+\left(\dfrac{\sqrt{3}}{2}\right)\left(\dfrac{\sqrt{2}}{2}\right)=\dfrac{-\sqrt{2}+\sqrt{6}}{4}$

$\theta=\arccos\left(\dfrac{-\sqrt{2}+\sqrt{6}}{4}\right)=75°=\dfrac{5\pi}{12}$

41. $\mathbf{u} = 3\mathbf{i} + 4\mathbf{j}$

$\mathbf{v} = -7\mathbf{i} + 5\mathbf{j}$

$\cos \theta = \dfrac{\mathbf{u} \cdot \mathbf{v}}{\|\mathbf{u}\| \|\mathbf{v}\|}$

$= \dfrac{3(-7) + 4(5)}{3\sqrt{74}}$

$= \dfrac{-1}{5\sqrt{74}} \approx -0.0232$

$\theta = 91.33°$

43. $\mathbf{u} = 5\mathbf{i} + 5\mathbf{j}$

$\mathbf{v} = -8\mathbf{i} + 8\mathbf{j}$

$\cos \theta = \dfrac{\mathbf{u} \cdot \mathbf{v}}{\|\mathbf{u}\| \|\mathbf{v}\|}$

$= \dfrac{5(-8) + 5(8)}{\sqrt{50}\sqrt{128}}$

$= 0$

$\theta = 90°$

45. $P = (1, 2), Q = (3, 4), R = (2, 5)$

$\overline{PQ} = \langle 2, 2 \rangle, \overline{PR} = \langle 1, 3 \rangle, \overline{QR} = \langle -1, 1 \rangle$

$\cos \alpha = \dfrac{\overline{PQ} \cdot \overline{PR}}{\|\overline{PQ}\| \|\overline{PR}\|} = \dfrac{8}{(2\sqrt{2})(\sqrt{10})} \Rightarrow \alpha = \arccos \dfrac{2}{\sqrt{5}} \approx 26.57°$

$\cos \beta = \dfrac{\overline{PQ} \cdot \overline{QR}}{\|\overline{PQ}\| \|\overline{QR}\|} = 0 \Rightarrow \beta = 90°$

$\gamma = 180° - 26.57° - 90° = 63.43°$

47. $P = (-3, 0), Q = (2, 2), R = (0, 6)$

$\overline{QP} = \langle -5, -2 \rangle, \overline{PR} = \langle 3, 6 \rangle, \overline{QR} = \langle -2, 4 \rangle, \overline{PQ} = \langle 5, 2 \rangle$

$\cos \alpha = \dfrac{\overline{PQ} \cdot \overline{PR}}{\|\overline{PQ}\| \|\overline{PR}\|} = \dfrac{27}{\sqrt{29}\sqrt{45}} \Rightarrow \alpha \approx 41.63°$

$\cos \beta = \dfrac{\overline{QP} \cdot \overline{QR}}{\|\overline{QP}\| \|\overline{PR}\|} = \dfrac{2}{\sqrt{29}\sqrt{20}} \Rightarrow \beta \approx 85.24°$

$\delta = 180° - 41.63° - 85.24° = 53.13°$

49. $\|\mathbf{u}\| = 4, \|\mathbf{v}\| = 10, \theta = \dfrac{2\pi}{3}$

$\mathbf{u} \cdot \mathbf{v} = \|\mathbf{u}\| \|\mathbf{v}\| \cos \theta$

$= (4)(10) \cos \dfrac{2\pi}{3}$

$= 40\left(-\dfrac{1}{2}\right)$

$= -20$

51. $\|\mathbf{u}\| = 9, \|\mathbf{v}\| = 36, \theta = \dfrac{3\pi}{4}$

$\mathbf{u} \cdot \mathbf{v} = \|\mathbf{u}\| \|\mathbf{v}\| \cos \theta$

$= (9)(36) \cos \dfrac{3\pi}{4}$

$= 324\left(-\dfrac{\sqrt{2}}{2}\right)$

$= -162\sqrt{2} \approx -229.1$

53. $\mathbf{u} = \langle -12, 30 \rangle, \|\mathbf{v}\| = \left\langle \dfrac{1}{2}, -\dfrac{5}{4} \right\rangle$

$\mathbf{u} = -24\mathbf{v} \Rightarrow \mathbf{u}$ and $\mathbf{v}$ are parallel.

55. $\mathbf{u} = \dfrac{1}{4}(3\mathbf{i} - \mathbf{j}), \mathbf{v} = 5\mathbf{i} + 6\mathbf{j}$

$\mathbf{u} \neq k\mathbf{v} \Rightarrow$ Not parallel

$\mathbf{u} \cdot \mathbf{v} \neq 0 \Rightarrow$ Not orthogonal

Neither

57. $\mathbf{u} = 2\mathbf{i} - 2\mathbf{j}, \mathbf{v} = -\mathbf{i} - \mathbf{j}$

$\mathbf{u} \cdot \mathbf{v} = 0 \Rightarrow \mathbf{u}$ and $\mathbf{v}$ are orthogonal.

59. $\mathbf{u} = \langle 2, 2 \rangle, \mathbf{v} = \langle 6, 1 \rangle$

$\mathbf{w}_1 = \text{proj}_\mathbf{v} \mathbf{u} = \left(\dfrac{\mathbf{u} \cdot \mathbf{v}}{\|\mathbf{v}\|^2}\right)\mathbf{v} = \dfrac{14}{37}\langle 6, 1 \rangle = \dfrac{1}{37}\langle 84, 14 \rangle$

$\mathbf{w}_2 = \mathbf{u} - \mathbf{w}_1 = \langle 2, 2 \rangle - \dfrac{14}{37}\langle 6, 1 \rangle = \left\langle -\dfrac{10}{37}, \dfrac{60}{37} \right\rangle = \dfrac{10}{37}\langle -1, 6 \rangle = \dfrac{1}{37}\langle -10, 60 \rangle$

$\mathbf{u} = \dfrac{1}{37}\langle 84, 14 \rangle + \dfrac{1}{37}\langle -10, 60 \rangle = \langle 2, 2 \rangle$

61. $\mathbf{u} = \langle 0, 3 \rangle$, $\mathbf{v} = \langle 2, 15 \rangle$

$$\mathbf{w}_1 = \text{proj}_{\mathbf{v}}\mathbf{u} = \left(\frac{\mathbf{u} \cdot \mathbf{v}}{\|\mathbf{v}\|^2} \right)\mathbf{v} = \frac{45}{229}\langle 2, 15 \rangle$$

$$\mathbf{w}_2 = \mathbf{u} - \mathbf{w}_1 = \langle 0, 3 \rangle - \frac{45}{229}\langle 2, 15 \rangle = \left\langle -\frac{90}{229}, \frac{12}{229} \right\rangle = \frac{6}{229}\langle -15, 2 \rangle$$

$$\mathbf{u} = \frac{45}{229}\langle 2, 15 \rangle + \frac{6}{229}\langle -15, 2 \rangle = \langle 0, 3 \rangle$$

63. $\text{proj}_{\mathbf{v}}\mathbf{u} = \mathbf{u}$ because they are parallel.

$$\text{proj}_{\mathbf{v}}\mathbf{u} = \frac{\mathbf{u} \cdot \mathbf{v}}{\|\mathbf{v}\|^2}\mathbf{v} = \frac{3(6) + 2(4)}{\left(\sqrt{6^2 + 4^2} \right)^2}\langle 6, 4 \rangle = \frac{1}{2}\langle 6, 4 \rangle = \langle 3, 2 \rangle = \mathbf{u}$$

65. $\text{proj}_{\mathbf{v}}\mathbf{u} = 0$ because they are perpendicular.

Because $\mathbf{u}$ and $\mathbf{v}$ are orthogonal,
$\mathbf{u} \cdot \mathbf{v} = 0$ and $\text{proj}_{\mathbf{v}}\mathbf{u} = 0$.

$$\text{proj}_{\mathbf{v}}\mathbf{u} = \frac{\mathbf{u} \cdot \mathbf{v}}{\|\mathbf{v}\|^2}\mathbf{v} = 0, \text{ because } \mathbf{u} \cdot \mathbf{v} = 0.$$

67. $\mathbf{u} = \langle 3, 5 \rangle$

For $\mathbf{v}$ to be orthogonal to $\mathbf{u}$, $\mathbf{u} \cdot \mathbf{v}$ must equal 0.

Two possibilities: $\langle -5, 3 \rangle$ and $\langle 5, -3 \rangle$

69. $\mathbf{u} = \frac{1}{2}\mathbf{i} - \frac{2}{3}\mathbf{j}$

For $\mathbf{u}$ and $\mathbf{v}$ to be orthogonal, $\mathbf{u} \cdot \mathbf{v}$ must equal 0.

Two possibilities: $\mathbf{v} = \frac{2}{3}\mathbf{i} + \frac{1}{2}\mathbf{j}$ and $\mathbf{v} = -\frac{2}{3}\mathbf{i} - \frac{1}{2}\mathbf{j}$

71. Work $= \left\| \text{proj}_{\overrightarrow{PQ}}\mathbf{v} \right\| \left\| \overrightarrow{PQ} \right\|$ where $\overrightarrow{PQ} = \langle 4, 7 \rangle$ and $\mathbf{v} = \langle 1, 4 \rangle$.

$$\text{proj}_{\overrightarrow{PQ}}\mathbf{v} = \left(\frac{\mathbf{v} \cdot \overrightarrow{PQ}}{\left\| \overrightarrow{PQ} \right\|^2} \right)\overrightarrow{PQ} = \left(\frac{32}{65} \right)\langle 4, 7 \rangle$$

$$\text{Work} = \left\| \text{proj}_{\overrightarrow{PQ}}\mathbf{v} \right\| \left\| \overrightarrow{PQ} \right\| = \left(\frac{32\sqrt{65}}{65} \right)\left(\sqrt{65} \right) = 32$$

73. $\mathbf{u} = \langle 4600, 5250 \rangle$, $\mathbf{v} = \langle 79.99, 99.99 \rangle$

(a) $\mathbf{u} \cdot \mathbf{v} = 4600(79.99) + 5250(99.99) = 892{,}901.5$

The total revenue that can be earned by selling all of the cellular phones is \$892,901.50.

(b) Increase prices by 5%: $1.05\mathbf{v}$

The operation is scalar multiplication.

75. (a) Force due to gravity:

$$\mathbf{F} = -30{,}000\mathbf{j}$$

Unit vector along hill:

$$\mathbf{v} = (\cos d)\mathbf{i} + (\sin d)\mathbf{j}$$

Projection of $\mathbf{F}$ onto $\mathbf{v}$:

$$\mathbf{w}_1 = \text{proj}_{\mathbf{v}}\mathbf{F} = \left(\frac{\mathbf{F} \cdot \mathbf{v}}{\|\mathbf{v}\|^2} \right)\mathbf{v} = (\mathbf{F} \cdot \mathbf{v})\mathbf{v} = -30{,}000 \sin d\mathbf{v}$$

The magnitude of the force is $30{,}000 \sin d$.

(b)

d	0°	1°	2°	3°	4°	5°	6°	7°	8°	9°	10°
Force	0	523.6	1047.0	1570.1	2092.7	2614.7	3135.9	3656.1	4175.2	4693.0	5209.4

(c) Force perpendicular to the hill when $d = 5°$:

$$\text{Force} = \sqrt{(30{,}000)^2 - (2614.7)^2} \approx 29{,}885.8 \text{ pounds}$$

77. Work $= (245)(3) = 735$ newton-meters

79. Work $= (\cos 30°)(45)(20) \approx 779.4$ foot-pounds

81. Work $= (\cos 30°)(250)(100) \approx 21{,}650.64$ foot-pounds

83. (a)–(c) Programs will vary.

85. Programs will vary.

87. False. Work is represented by a scalar.

89. In a rhombus, $\|\mathbf{u}\| = \|\mathbf{v}\|$. The diagonals are $\mathbf{u} + \mathbf{v}$ and
$\mathbf{u} - \mathbf{v}$.

$$(\mathbf{u} + \mathbf{v}) \cdot (\mathbf{u} - \mathbf{v}) = (\mathbf{u} + \mathbf{v}) \cdot \mathbf{u} - (\mathbf{u} + \mathbf{v}) \cdot \mathbf{v}$$
$$= \mathbf{u} \cdot \mathbf{u} + \mathbf{v} \cdot \mathbf{u} - \mathbf{u} \cdot \mathbf{v} - \mathbf{v} \cdot \mathbf{v}$$
$$= \|\mathbf{u}\|^2 - \|\mathbf{v}\|^2 = 0$$

So, the diagonals are orthogonal.

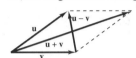

91. (a) $\text{proj}_\mathbf{v}\mathbf{u} = \mathbf{u} \Rightarrow \mathbf{u}$ and $\mathbf{v}$ are parallel.

(b) $\text{proj}_\mathbf{v}\mathbf{u} = 0 \Rightarrow \mathbf{u}$ and $\mathbf{v}$ are orthogonal.

93.

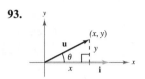

$$\cos \theta = \frac{x}{\|\mathbf{u}\|}, \ \sin \theta = \frac{y}{\|\mathbf{u}\|}$$

Because $\mathbf{u}$ is a unit vector, $\|\mathbf{u}\| = 1$. So, $x = \cos \theta$ and
$y = \sin \theta$.

So, $\mathbf{u} = \langle x, y \rangle = \cos \theta\mathbf{i} + \sin \theta\mathbf{j}$

Section 8.5   Trigonometric Form of a Complex Number

- You should be able to graphically represent complex numbers and know the following facts and about them.

- The absolute value of the complex number $z = a + bi$ is $|z| = \sqrt{a^2 + b^2}$.

- The trigonometric form of the complex number $z = a + bi$ is $z = r(\cos \theta + i \sin \theta)$ where

  (a) $a = r \cos \theta$.

  (b) $b = r \sin \theta$.

  (c) $r = \sqrt{a^2 + b^2}$; r is called the modulus of z.

  (d) $\tan \theta = \dfrac{b}{a}$; θ is called the argument of z.

- Given $z_1 = r_1(\cos \theta_1 + i \sin \theta_1)$ and $z_2 = r_2(\cos \theta_2 + i \sin \theta_2)$:

  (a) $z_1 z_2 = r_1 r_2 \left[\cos(\theta_1 + \theta_2) + i \sin(\theta_1 + \theta_2) \right]$

  (b) $\dfrac{z_1}{z_2} = \dfrac{r_1}{r_2} \left[\cos(\theta_1 - \theta_2) + i \sin(\theta_1 - \theta_2) \right], z_2 \neq 0$

- You should know DeMoivre's Theorem: If $z = r(\cos \theta + i \sin \theta)$, then for any positive integer n,

  $z^n = r^n(\cos n\theta + i \sin n\theta)$.

- You should know that for any positive integer n, $z = r(\cos \theta + i \sin \theta)$ has n distinct nth roots given by

  $$\sqrt[n]{r} \left[\cos\left(\frac{\theta + 2\pi k}{n} \right) + i \sin\left(\frac{\theta + 2\pi k}{n} \right) \right]$$

  where $k = 0, 1, 2, \ldots, n - 1$.

1. absolute value

3. DeMoivre's

5. $|-6 + 8i| = \sqrt{(-6)^2 + 8^2}$

$\qquad = \sqrt{100} = 10$

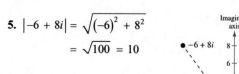

7. $|-7i| = \sqrt{0^2 + (-7)^2}$

$\qquad = \sqrt{49} = 7$

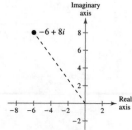

9. $|4 - 6i| = \sqrt{4^2 + (-6)^2}$

$\qquad = \sqrt{52} = 2\sqrt{13}$

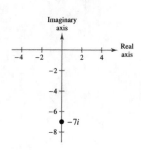

11. $z = 3i$

$\qquad r = \sqrt{0^2 + 3^2} = \sqrt{9} = 3$

$\qquad \tan\theta = \dfrac{3}{0}, \text{ undefined} \Rightarrow \theta = \dfrac{\pi}{2}$

$\qquad z = 3\left(\cos\dfrac{\pi}{2} + i\sin\dfrac{\pi}{2}\right)$

13. $z = -3 - 3i$

$\qquad r = \sqrt{(-3)^2 + (-3)^2} = \sqrt{18} = 3\sqrt{2}$

$\qquad \tan\theta = \dfrac{-3}{-3} = 1, \theta \text{ is in Quadrant III} \Rightarrow \theta = \dfrac{5\pi}{4}.$

$\qquad z = 3\sqrt{2}\left(\cos\dfrac{5\pi}{4} + i\sin\dfrac{5\pi}{4}\right)$

15. $z = 1 + i$

$\qquad r = \sqrt{1^2 + 1^2} = \sqrt{2}$

$\qquad \tan\theta = 1, \theta \text{ is in Quadrant I} \Rightarrow \theta = \dfrac{\pi}{4}.$

$\qquad z = \sqrt{2}\left(\cos\dfrac{\pi}{4} + i\sin\dfrac{\pi}{4}\right)$

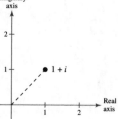

17. $z = 1 - \sqrt{3}i$

$\qquad r = \sqrt{1^2 + (-\sqrt{3})^2} = \sqrt{4} = 2$

$\qquad \tan\theta = -\sqrt{3}, \theta \text{ is in Quadrant IV} \Rightarrow \theta = \dfrac{5\pi}{3}.$

$\qquad z = 2\left(\cos\dfrac{5\pi}{3} + i\sin\dfrac{5\pi}{3}\right)$

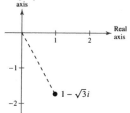

19. $z = -2\left(1 + \sqrt{3}i\right)$

$\qquad r = \sqrt{(-2)^2 + (-2\sqrt{3})^2} = \sqrt{16} = 4$

$\qquad \tan\theta = \dfrac{\sqrt{3}}{1} = \sqrt{3}, \theta \text{ is in Quadrant III} \Rightarrow \theta = \dfrac{4\pi}{3}.$

$\qquad z = 4\left(\cos\dfrac{4\pi}{3} + i\sin\dfrac{4\pi}{3}\right)$

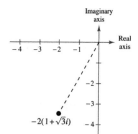

21. $z = -5i$

$r = \sqrt{0^2 + (-5)^2} = \sqrt{25} = 5$

$\tan \theta = \dfrac{-5}{0}$, undefined $\Rightarrow \theta = \dfrac{3\pi}{2}$

$z = 5\left(\cos \dfrac{3\pi}{2} + i \sin \dfrac{3\pi}{2}\right)$

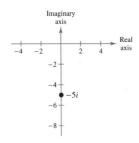

23. $z = -7 + 4i$

$r = \sqrt{(-7)^2 + (4)^2} = \sqrt{65}$

$\tan \theta = \dfrac{4}{-7}$, θ is in Quadrant II $\Rightarrow \theta \approx 2.62$.

$z \approx \sqrt{65}\left(\cos 2.62 + i \sin 2.62\right)$

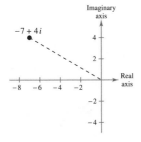

25. $z = 2$

$r = \sqrt{2^2 + 0^2} = \sqrt{4} = 2$

$\tan \theta = 0 \Rightarrow \theta = 0$

$z = 2(\cos 0 + i \sin 0)$

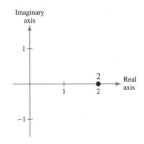

27. $z = 2\sqrt{2} - i$

$r = \sqrt{\left(2\sqrt{2}\right)^2 + (-1)^2} = \sqrt{9} = 3$

$\tan \theta = \dfrac{-1}{2\sqrt{2}} = -\dfrac{\sqrt{2}}{4} \Rightarrow \theta \approx 5.94$ radians

$z = 3(\cos 5.94 + i \sin 5.94)$

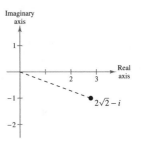

29. $z = 5 + 2i$

$r = \sqrt{5^2 + 2^2} = \sqrt{29}$

$\tan \theta = \dfrac{2}{5}$

$\theta \approx 0.38$

$z \approx \sqrt{29}(\cos 0.38 + i \sin 0.38)$

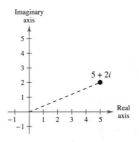

31. $z = -8 - 5\sqrt{3}i$

$r = \sqrt{(-8)^2 + \left(-5\sqrt{3}\right)^2} = \sqrt{139}$

$\tan \theta = \dfrac{5\sqrt{3}}{8}$

$\theta \approx 3.97$

$z \approx \sqrt{139}(\cos 3.97 + i \sin 3.97)$

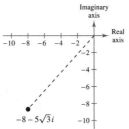

33. $2(\cos 60° + i \sin 60°) = 2\left(\dfrac{1}{2} + \dfrac{\sqrt{3}}{2}i\right)$

$$= 1 + \sqrt{3}i$$

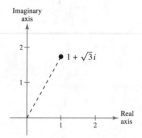

35. $\sqrt{48}\left[\cos(-30°) + i \sin(-30°)\right] = 4\sqrt{3}\left(\dfrac{\sqrt{3}}{2} - \dfrac{1}{2}i\right)$

$$= 6 - 2\sqrt{3}i$$

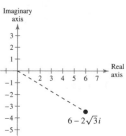

37. $\dfrac{9}{4}\left(\cos \dfrac{3\pi}{4} + i \sin \dfrac{3\pi}{4}\right) = \dfrac{9}{4}\left(-\dfrac{\sqrt{2}}{2} + \dfrac{\sqrt{2}}{2}i\right)$

$$= -\dfrac{9\sqrt{2}}{8} + \dfrac{9\sqrt{2}}{8}i$$

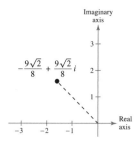

39. $7(\cos 0° + i \sin 0°) = 7$

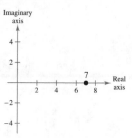

41. $5\left[\cos(198°45') + i \sin(198°45')\right] \approx -4.7347 - 1.6072i$

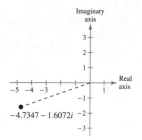

43. $5\left(\cos \dfrac{\pi}{9} + i \sin \dfrac{\pi}{9}\right) \approx 4.6985 + 1.7101i$

45. $2(\cos 155° + i \sin 155°) \approx -1.8126 + 0.8452i$

47. $\left[2\left(\cos \dfrac{\pi}{4} + i \sin \dfrac{\pi}{4}\right)\right]\left[6\left(\cos \dfrac{\pi}{12} + i \sin \dfrac{\pi}{12}\right)\right] = (2)(6)\left[\cos\left(\dfrac{\pi}{4} + \dfrac{\pi}{12}\right) + i \sin\left(\dfrac{\pi}{4} + \dfrac{\pi}{12}\right)\right]$

$$= 12\left(\cos \dfrac{\pi}{3} + i \sin \dfrac{\pi}{3}\right)$$

49. $\left[\dfrac{5}{3}(\cos 120° + i \sin 120°)\right]\left[\dfrac{2}{3}(\cos 30° + i \sin 30°)\right] = \dfrac{5}{3}\left(\dfrac{2}{3}\right)\left[\cos(120° + 30°) + i \sin(120° + 30°)\right]$

$$= \dfrac{10}{9}(\cos 150° + i \sin 150°)$$

51. $\left(\cos 80° + i \sin 80°\right)\left(\cos 330° + i \sin 330°\right) = \cos\left(80° + 330°\right) + i \sin\left(80° + 330°\right)$

$$= \cos 410° + i \sin 410°$$

$$= \cos 50° + i \sin 50°$$

53. $\dfrac{3\left(\cos 50° + i \sin 50°\right)}{9\left(\cos 20° + i \sin 20°\right)} = \dfrac{1}{3}\left[\cos\left(50° - 20°\right) + i \sin\left(50° - 20°\right)\right] = \dfrac{1}{3}\left(\cos 30° + i \sin 30°\right)$

55. $\dfrac{\cos \pi + i \sin \pi}{\cos\left(\pi/3\right) + i \sin\left(\pi/3\right)} = \cos\left(\pi - \dfrac{\pi}{3}\right) + i \sin\left(\pi - \dfrac{\pi}{3}\right) = \cos\dfrac{2\pi}{3} + i \sin\dfrac{2\pi}{3}$

57. $\dfrac{12\left(\cos 92° + i \sin 92°\right)}{2\left(\cos 122° + i \sin 122°\right)} = 6\left[\cos\left(92° - 122°\right) + i \sin\left(92° - 122°\right)\right]$

$$= 6\left[\cos\left(-30°\right) + i \sin\left(-30°\right)\right]$$

$$= 6\left(\cos 330° + i \sin 330°\right)$$

59. (a) $2 + 2i = 2\sqrt{2}\left(\cos\dfrac{\pi}{4} + i \sin\dfrac{\pi}{4}\right)$

$1 - i = \sqrt{2}\left[\cos\left(-\dfrac{\pi}{4}\right) + i \sin\left(-\dfrac{\pi}{4}\right)\right] = \sqrt{2}\left(\cos\dfrac{7\pi}{4} + i \sin\dfrac{7\pi}{4}\right)$

(b) $\left(2 + 2i\right)\left(1 - i\right) = \left[2\sqrt{2}\left(\cos\dfrac{\pi}{4} + i \sin\dfrac{\pi}{4}\right)\right]\left[\sqrt{2}\left(\cos\left(\dfrac{7\pi}{4}\right) + i \sin\left(\dfrac{7\pi}{4}\right)\right)\right] = 4\left(\cos 2\pi + i \sin 2\pi\right)$

$$= 4\left(\cos 0 + i \sin 0\right) = 4$$

(c) $\left(2 + 2i\right)\left(1 - i\right) = 2 - 2i + 2i - 2i^2 = 2 + 2 = 4$

61. (a) $-2i = 2\left[\cos\left(-\dfrac{\pi}{2}\right) + i \sin\left(-\dfrac{\pi}{2}\right)\right] = 2\left(\cos\dfrac{3\pi}{2} + i \sin\dfrac{3\pi}{2}\right)$

$1 + i = \sqrt{2}\left(\cos\dfrac{\pi}{4} + i \sin\dfrac{\pi}{4}\right)$

(b) $-2i\left(1 + i\right) = 2\left[\cos\left(\dfrac{3\pi}{2}\right) + i \sin\left(\dfrac{3\pi}{2}\right)\right]\left[\sqrt{2}\left(\cos\dfrac{\pi}{4} + i \sin\dfrac{\pi}{4}\right)\right]$

$$= 2\sqrt{2}\left[\cos\left(\dfrac{7\pi}{4}\right) + i \sin\left(\dfrac{7\pi}{4}\right)\right]$$

$$= 2\sqrt{2}\left[\dfrac{1}{\sqrt{2}} - \dfrac{1}{\sqrt{2}}i\right] = 2 - 2i$$

(c) $-2i\left(1 + i\right) = -2i - 2i^2 = -2i + 2 = 2 - 2i$

63. (a) $3 + 4i \approx 5\left(\cos 0.93 + i \sin 0.93\right)$

$1 - \sqrt{3}i = 2\left(\cos\dfrac{5\pi}{3} + i \sin\dfrac{5\pi}{3}\right)$

(b) $\dfrac{3 + 4i}{1 - \sqrt{3}i} \approx \dfrac{5\left(\cos 0.93 + i \sin 0.93\right)}{2\left(\cos\dfrac{5\pi}{3} + i \sin\dfrac{5\pi}{3}\right)} \approx 2.5\left[\cos\left(-4.31\right) + i \sin\left(-4.31\right)\right] = \dfrac{5}{2}\left(\cos 1.97 + i \sin 1.97\right) \approx -0.982 + 2.299i$

(c) $\dfrac{3 + 4i}{1 - \sqrt{3}i} = \dfrac{3 + 4i}{1 - \sqrt{3}i} \cdot \dfrac{1 + \sqrt{3}i}{1 + \sqrt{3}i} = \dfrac{3 + \left(4 + 3\sqrt{3}\right)i + 4\sqrt{3}i^2}{1 + 3} = \dfrac{3 - 4\sqrt{3}}{4} + \dfrac{4 + 3\sqrt{3}}{4}i \approx -0.982 + 2.299i$

65. $z = \dfrac{\sqrt{2}}{2}(1 + i) = \cos 45° + i \sin 45°$

$z^2 = \cos 90° + i \sin 90° = i$

$z^3 = \cos 135° + i \sin 135° = \dfrac{\sqrt{2}}{2}(-1 + i)$

$z^4 = \cos 180° + i \sin 180° = -1$

The absolute value of each is 1, and consecutive powers of z are each $45°$ apart.

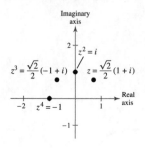

67. $(1 + i)^5 = \left[\sqrt{2} \left(\cos \dfrac{\pi}{4} + i \sin \dfrac{\pi}{4} \right) \right]^5$

$= \left(\sqrt{2} \right)^5 \left(\cos \dfrac{5\pi}{4} + i \sin \dfrac{5\pi}{4} \right)$

$= 4\sqrt{2} \left(-\dfrac{\sqrt{2}}{2} - \dfrac{\sqrt{2}}{2} i \right)$

$= -4 - 4i$

69. $(-1 + i)^6 = \left[\sqrt{2} \left(\cos \dfrac{3\pi}{4} + i \sin \dfrac{3\pi}{4} \right) \right]^6$

$= \left(\sqrt{2} \right)^6 \left(\cos \dfrac{18\pi}{4} + i \sin \dfrac{18\pi}{4} \right)$

$= 8 \left(\cos \dfrac{9\pi}{2} + i \sin \dfrac{9\pi}{2} \right)$

$= 8(0 + i)$

$= 8i$

71. $2\left(\sqrt{3} + i \right)^{10} = 2 \left[2 \left(\cos \dfrac{\pi}{6} + i \sin \dfrac{\pi}{6} \right) \right]^{16}$

$= 2 \left[2^{10} \left(\cos \dfrac{10\pi}{6} + i \sin \dfrac{10\pi}{6} \right) \right]$

$= 2048 \left(\cos \dfrac{5\pi}{3} + i \sin \dfrac{5\pi}{3} \right)$

$= 2048 \left(\dfrac{1}{2} - \dfrac{\sqrt{3}}{2} i \right)$

$= 1024 - 1024\sqrt{3} i$

73. $\left[5 \left(\cos 20° + i \sin 20° \right) \right]^3 = 5^3 \left(\cos 60° + i \sin 60° \right)$

$= \dfrac{125}{2} + \dfrac{125\sqrt{3}}{2} i$

75. $\left(\cos \dfrac{\pi}{4} + i \sin \dfrac{\pi}{4} \right)^{12} = \cos \dfrac{12\pi}{4} + i \sin \dfrac{12\pi}{4}$

$= \cos 3\pi + i \sin 3\pi$

$= -1$

77. $\left[5 \left(\cos 3.2 + i \sin 3.2 \right) \right]^4 = 5^4 \left(\cos 12.8 + i \sin 12.8 \right)$

$\approx 608.0 + 144.7 i$

79. $(3 - 2i)^5 \approx \left[3.6056 \left[\cos(-0.588) + i \sin(-0.588) \right] \right]^5$

$\approx (3.6056)^5 \left[\cos(-2.94) + i \sin(-2.94) \right]$

$\approx -597 - 122 i$

81. $\left[3 \left(\cos 15° + i \sin 15° \right) \right]^4 = 81 \left(\cos 60° + i \sin 60° \right)$

$= \dfrac{81}{2} + \dfrac{81\sqrt{3}}{2} i$

83. (a) Square roots of $5 \left(\cos 120° + i \sin 120° \right)$:

$\sqrt{5} \left[\cos \left(\dfrac{120° + 360°k}{2} \right) + i \sin \left(\dfrac{120° + 360°k}{2} \right) \right], k = 0, 1$

$k = 0$: $\sqrt{5} \left(\cos 60° + i \sin 60° \right)$

$k = 1$: $\sqrt{5} \left(\cos 240° + i \sin 240° \right)$

(c) $\dfrac{\sqrt{5}}{2} + \dfrac{\sqrt{15}}{2} i, -\dfrac{\sqrt{5}}{2} - \dfrac{\sqrt{15}}{2} i$

(b)

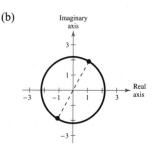

85. (a) Cube roots of $8\left(\cos\dfrac{2\pi}{3} + i\sin\dfrac{2\pi}{3}\right)$:

(b)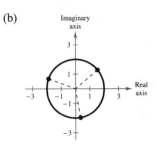

$$\sqrt[3]{8}\left[\cos\left(\frac{(2\pi/3) + 2\pi k}{3}\right) + i\sin\left(\frac{(2\pi/3) + 2\pi k}{3}\right)\right], k = 0, 1, 2$$

$k = 0:\ 2\left(\cos\dfrac{2\pi}{9} + i\sin\dfrac{2\pi}{9}\right)$

$k = 1:\ 2\left(\cos\dfrac{8\pi}{9} + i\sin\dfrac{8\pi}{9}\right)$

$k = 2:\ 2\left(\cos\dfrac{14\pi}{9} + i\sin\dfrac{14\pi}{9}\right)$

(c) $1.5321 + 1.2856i,\ -1.8794 + 0.6840i,\ 0.3473 - 1.9696i$

87. (a) Cube roots of $-\dfrac{125}{2}\left(1 + \sqrt{3}i\right) = 125\left(\cos\dfrac{4\pi}{3} + i\sin\dfrac{4\pi}{3}\right)$:

(b)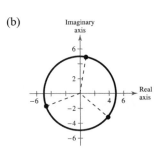

$$\sqrt[3]{125}\left[\cos\left(\frac{\dfrac{4\pi}{3} + 2k\pi}{3}\right) + i\sin\left(\frac{\dfrac{4\pi}{3} + 2k\pi}{3}\right)\right], k = 0, 1, 2$$

$k = 0:\ 5\left(\cos\dfrac{4\pi}{9} + i\sin\dfrac{4\pi}{9}\right)$

$k = 1:\ 5\left(\cos\dfrac{10\pi}{9} + i\sin\dfrac{10\pi}{9}\right)$

$k = 2:\ 5\left(\cos\dfrac{16\pi}{9} + i\sin\dfrac{16\pi}{9}\right)$

(c) $0.8682 + 4.9240i,\ -4.6985 - 1.7101i,\ 3.8302 - 3.2140i$

89. (a) Square roots of $-25i = 25\left(\cos\dfrac{3\pi}{2} + i\sin\dfrac{3\pi}{2}\right)$:

(b)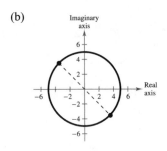

$$\sqrt{25}\left[\cos\left(\frac{\dfrac{3\pi}{2} + 2k\pi}{2}\right) + i\sin\left(\frac{\dfrac{3\pi}{2} + 2k\pi}{2}\right)\right], k = 0, 1$$

$k = 0:\ 5\left(\cos\dfrac{3\pi}{4} + i\sin\dfrac{3\pi}{4}\right)$

$k = 1:\ 5\left(\cos\dfrac{7\pi}{4} + i\sin\dfrac{7\pi}{4}\right)$

(c) $-\dfrac{5\sqrt{2}}{2} + \dfrac{5\sqrt{2}}{2}i,\ \dfrac{5\sqrt{2}}{2} - \dfrac{5\sqrt{2}}{2}i$

91. (a) Fourth roots of $16 = 16(\cos 0 + i \sin 0)$:

$$\sqrt[4]{16}\left[\cos \frac{0 + 2\pi k}{4} + i \sin \frac{0 + 2\pi k}{4}\right], k = 0, 1, 2, 3$$

$k = 0$: $2(\cos 0 + i \sin 0)$

$k = 1$: $2\left(\cos \frac{\pi}{2} + i \sin \frac{\pi}{2}\right)$

$k = 2$: $2(\cos \pi + i \sin \pi)$

$k = 3$: $2\left(\cos \frac{3\pi}{2} + i \sin \frac{3\pi}{2}\right)$

(c) $2, 2i, -2, -2i$

(b)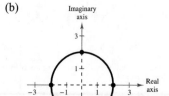

93. (a) Fifth roots of $1 = \cos 0 + i \sin 0$:

$$\cos\left(\frac{2k\pi}{5}\right) + i \sin\left(\frac{2k\pi}{5}\right), k = 0, 1, 2, 3, 4$$

$k = 0$: $\cos 0 + i \sin 0$

$k = 1$: $\cos \frac{2\pi}{5} + i \sin \frac{2\pi}{5}$

$k = 2$: $\cos \frac{4\pi}{5} + i \sin \frac{4\pi}{5}$

$k = 3$: $\cos \frac{6\pi}{5} + i \sin \frac{6\pi}{5}$

$k = 4$: $\cos \frac{8\pi}{5} + i \sin \frac{8\pi}{5}$

(c) $1, 0.3090 + 0.9511i, -0.8090 + 0.5878i, -0.8090 - 0.5878i, 0.3090 - 0.9511i$

(b)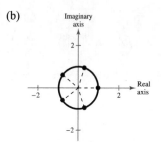

95. (a) Cube roots of $-125 = 125(\cos \pi + i \sin \pi)$:

$$\sqrt[3]{125}\left[\cos\left(\frac{\pi + 2\pi k}{3}\right) + i \sin\left(\frac{\pi + 2\pi k}{3}\right)\right], k = 0, 1, 2$$

$k = 0$: $5\left(\cos \frac{\pi}{3} + i \sin \frac{\pi}{3}\right)$

$k = 1$: $5(\cos \pi + i \sin \pi)$

$k = 2$: $5\left(\cos \frac{5\pi}{3} + i \sin \frac{5\pi}{3}\right)$

(b)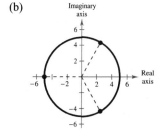

(c) $\dfrac{5}{2} + \dfrac{5\sqrt{3}}{2}i, -5, \dfrac{5}{2} - \dfrac{5\sqrt{3}}{2}i$

97. (a) Fifth roots of $4(1 - i) = 4\sqrt{2}\left(\cos\dfrac{7\pi}{4} + i\sin\dfrac{7\pi}{4}\right)$:

(b)

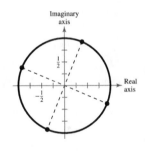

$$\sqrt[5]{4\sqrt{2}}\left[\cos\left(\dfrac{\dfrac{7\pi}{4} + 2\pi k}{5}\right) + i\sin\left(\dfrac{\dfrac{7\pi}{4} + 2\pi k}{5}\right)\right], k = 0, 1, 2, 3, 4$$

$$k = 0: \sqrt{2}\left(\cos\dfrac{7\pi}{20} + i\sin\dfrac{7\pi}{20}\right)$$

$$k = 1: \sqrt{2}\left(\cos\dfrac{3\pi}{4} + i\sin\dfrac{3\pi}{4}\right)$$

$$k = 2: \sqrt{2}\left(\cos\dfrac{23\pi}{20} + i\sin\dfrac{23\pi}{20}\right)$$

$$k = 3: \sqrt{2}\left(\cos\dfrac{31\pi}{20} + i\sin\dfrac{31\pi}{20}\right)$$

$$k = 4: \sqrt{2}\left(\cos\dfrac{39\pi}{20} + i\sin\dfrac{39\pi}{20}\right)$$

(c) $0.6420 + 1.2601i, -1 + 1i, -1.2601 - 0.6420i, 0.2212 - 1.3968i, 1.3968 - 0.2212i$

99. $x^4 + i = 0$

$$x^4 = -i$$

The solutions are the fourth roots of $i = \cos\dfrac{3\pi}{2} + i\sin\dfrac{3\pi}{2}$:

$$\sqrt[4]{1}\left[\cos\left(\dfrac{\dfrac{3\pi}{2} + 2k\pi}{4}\right) + i\sin\left(\dfrac{\dfrac{3\pi}{2} + 2k\pi}{4}\right)\right], k = 0, 1, 2, 3$$

$$k = 0: \cos\dfrac{3\pi}{8} + i\sin\dfrac{3\pi}{8} \approx 0.3827 + 0.9239i$$

$$k = 1: \cos\dfrac{7\pi}{8} + i\sin\dfrac{7\pi}{8} \approx -0.9239 + 0.3827i$$

$$k = 2: \cos\dfrac{11\pi}{8} + i\sin\dfrac{11\pi}{8} \approx -0.3827 - 0.9239i$$

$$k = 3: \cos\dfrac{15\pi}{8} + i\sin\dfrac{15\pi}{8} \approx 0.9239 - 0.3827i$$

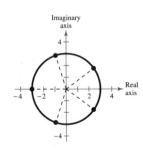

101. $x^5 + 243 = 0$

$$x^5 = -243$$

The solutions are the fifth roots of $-243 = 243(\cos\pi + i\sin\pi)$:

$$\sqrt[5]{243}\left[\cos\left(\dfrac{\pi + 2k\pi}{5}\right) + i\sin\left(\dfrac{\pi + 2k\pi}{5}\right)\right], k = 0, 1, 2, 3, 4$$

$$k = 0: 3\left(\cos\dfrac{\pi}{5} + i\sin\dfrac{\pi}{5}\right) \approx 2.4271 + 1.7634i$$

$$k = 1: 3\left(\cos\dfrac{3\pi}{5} + i\sin\dfrac{3\pi}{5}\right) \approx -0.9271 + 2.8532i$$

$$k = 2: 3(\cos\pi + i\sin\pi) = -3$$

$$k = 3: 3\left(\cos\dfrac{7\pi}{5} + i\sin\dfrac{7\pi}{5}\right) \approx -0.9271 - 2.8532i$$

$$k = 4: 3\left(\cos\dfrac{9\pi}{5} + i\sin\dfrac{9\pi}{5}\right) \approx 2.4271 - 1.7634i$$

103. $x^4 + 16i = 0$

$$x^4 = -16i$$

The solutions are the fourth roots of $-16i = 16\left(\cos\dfrac{3\pi}{2} + i\sin\dfrac{3\pi}{2}\right)$:

$$\sqrt[4]{16}\left[\cos\dfrac{\dfrac{3\pi}{2} + 2\pi k}{4} + i\sin\dfrac{\dfrac{3\pi}{2} + 2\pi k}{4}\right], k = 0, 1, 2, 3$$

$k = 0$: $2\left(\cos\dfrac{3\pi}{8} + i\sin\dfrac{3\pi}{8}\right) \approx 0.7654 + 1.8478i$

$k = 1$: $2\left(\cos\dfrac{7\pi}{8} + i\sin\dfrac{7\pi}{8}\right) \approx -1.8478 + 0.7654i$

$k = 2$: $2\left(\cos\dfrac{11\pi}{8} + i\sin\dfrac{11\pi}{8}\right) \approx -0.7654 - 1.8478i$

$k = 3$: $2\left(\cos\dfrac{15\pi}{8} + i\sin\dfrac{15\pi}{8}\right) \approx 1.8478 - 0.7654i$

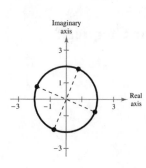

105. $x^3 - (1 - i) = 0$

$$x^3 = 1 - i = \sqrt{2}\left(\cos\dfrac{7\pi}{4} + i\sin\dfrac{7\pi}{4}\right)$$

The solutions are the cube roots of $1 - i$:

$$\sqrt[3]{\sqrt{2}}\left[\cos\left(\dfrac{(7\pi/4) + 2\pi k}{3}\right) + i\sin\left(\dfrac{(7\pi/4) + 2\pi k}{3}\right)\right], k = 0, 1, 2$$

$k = 0$: $\sqrt[6]{2}\left(\cos\dfrac{7\pi}{12} + i\sin\dfrac{7\pi}{12}\right) \approx -0.2905 + 1.0842i$

$k = 1$: $\sqrt[6]{2}\left(\cos\dfrac{5\pi}{4} + i\sin\dfrac{5\pi}{4}\right) \approx -0.7937 - 0.7937i$

$k = 2$: $\sqrt[6]{2}\left(\cos\dfrac{23\pi}{12} + i\sin\dfrac{23\pi}{12}\right) \approx 1.0842 - 0.2905i$

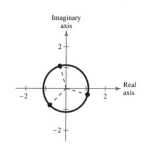

107. False. They are equally spaced along the circle centered at the origin with radius $\sqrt[n]{r}$.

109. $\dfrac{z_1}{z_2} = \dfrac{r_1(\cos\theta_1 + i\sin\theta_1)}{r_2(\cos\theta_2 + i\sin\theta_2)} \cdot \dfrac{\cos\theta_2 - i\sin\theta_2}{\cos\theta_2 - i\sin\theta_2}$

$$= \dfrac{r_1}{r_2(\cos^2\theta_2 + \sin^2\theta_2)}\left[\cos\theta_1\cos\theta_2 + \sin\theta_1\sin\theta_2 + i(\sin\theta_1\cos\theta_2 - \sin\theta_2\cos\theta_1)\right]$$

$$= \dfrac{r_1}{r_2}\left[\cos(\theta_1 - \theta_2) + i\sin(\theta_1 - \theta_2)\right]$$

111. (a) $z\bar{z} = \left[r(\cos\theta + i\sin\theta)\right]\left[r(\cos(-\theta) + i\sin(-\theta))\right]$

$= r^2\left[\cos(\theta - \theta) + i\sin(\theta - \theta)\right]$

$= r^2\left[\cos 0 + i\sin 0\right]$

$= r^2$

(b) $\dfrac{z}{\bar{z}} = \dfrac{r(\cos\theta + i\sin\theta)}{r\left[\cos(-\theta) + i\sin(-\theta)\right]}$

$= \dfrac{r}{r}\left[\cos(\theta - (-\theta)) + i\sin(\theta - (-\theta))\right]$

$= \cos 2\theta + i\sin 2\theta$

113. $\dfrac{1}{2}\left(1 - \sqrt{3}i\right) = \cos\dfrac{5\pi}{3} + i\sin\dfrac{5\pi}{3}$

$$\left[\dfrac{1}{2}\left(1 - \sqrt{3}i\right)\right]^{9} = \left(\cos\dfrac{5\pi}{3} + i\sin\dfrac{5\pi}{3}\right)^{9}$$

$$= \cos 15\pi + i\sin 15\pi$$

$$= -1$$

115. The solution of $x^4 + 16 = 0$ is $x = \sqrt[4]{-16}$. By writing -16 as $16\left(\cos \pi + i\sin \pi\right)$, you can find the fourth roots of -16 by using Demoivre's Theorem and the formula

$$\sqrt[n]{r}\left(\cos\dfrac{\theta + 2\pi k}{n} + i\sin\dfrac{\theta + 2\pi k}{n}\right), k = 0, 1, \ldots n - 1,$$

where $n = 4$.

Review Exercises for Chapter 8

1. Given: $A = 38°, B = 70°, a = 8$

$C = 180° - 38° - 70° = 72°$

$b = \dfrac{a \sin B}{\sin A} = \dfrac{8 \sin 70°}{\sin 38°} \approx 12.21$

$c = \dfrac{a \sin C}{\sin A} = \dfrac{8 \sin 72°}{\sin 38°} \approx 12.36$

3. Given: $B = 72°, C = 82°, b = 54$

$A = 180° - 72° - 82° = 26°$

$a = \dfrac{b \sin A}{\sin B} = \dfrac{54 \sin 26°}{\sin 72°} \approx 24.89$

$c = \dfrac{b \sin C}{\sin B} = \dfrac{54 \sin 82°}{\sin 72°} \approx 56.23$

9. Given: $B = 150°, b = 30, c = 10$

$\sin C = \dfrac{c \sin B}{b} = \dfrac{10 \sin 150°}{30} \approx 0.1667 \Rightarrow C \approx 9.59°$

$A \approx 180° - 150° - 9.59° = 20.41°$

$a = \dfrac{b \sin A}{\sin B} = \dfrac{30 \sin 20.41°}{\sin 150°} \approx 20.92$

11. $A = 75°, a = 51.2, b = 33.7$

$\sin B = \dfrac{b \sin A}{a} = \dfrac{33.7 \sin 75°}{51.2} \approx 0.6358 \Rightarrow B \approx 39.48°$

$C \approx 180° - 75° - 39.48° = 65.52°$

$c = \dfrac{a \sin C}{\sin A} = \dfrac{51.2 \sin 65.52°}{\sin 75°} \approx 48.24$

13. $A = 33°, b = 7, c = 10$

Area $= \dfrac{1}{2}bc \sin A = \dfrac{1}{2}(7)(10) \sin 33° \approx 19.06$

5. Given: $A = 16°, B = 98°, c = 8.4$

$C = 180° - 16° - 98° = 66°$

$a = \dfrac{c \sin A}{\sin C} = \dfrac{8.4 \sin 16°}{\sin 66°} \approx 2.53$

$b = \dfrac{c \sin B}{\sin C} = \dfrac{8.4 \sin 98°}{\sin 66°} \approx 9.11$

7. Given: $A = 24°, C = 48°, b = 27.5$

$B = 180° - 24° - 48° = 108°$

$a = \dfrac{b \sin A}{\sin B} = \dfrac{27.5 \sin 24°}{\sin 108°} \approx 11.76$

$c = \dfrac{b \sin C}{\sin B} = \dfrac{27.5 \sin 48°}{\sin 108°} \approx 21.49$

15. $C = 119°, a = 18, b = 6$

Area $= \dfrac{1}{2}ab \sin C = \dfrac{1}{2}(18)(6) \sin 119° \approx 47.23$

17. $\tan 17° = \dfrac{h}{x + 50} \Rightarrow h = (x + 50)\tan 17°$

$$h = x\tan 17° + 50\tan 17°$$

$\tan 31° = \dfrac{h}{x} \Rightarrow h = x\tan 31°$

$x\tan 17° + 50\tan 17° = x\tan 31°$

$$50\tan 17° = x(\tan 31° - \tan 17°)$$

$$\dfrac{50\tan 17°}{\tan 31° - \tan 17°} = x$$

$$x \approx 51.7959$$

$h = x\tan 31° \approx 51.7959\tan 31° \approx 31.1 \text{ meters}$

The height of the building is approximately 31.1 meters.

19. $\dfrac{h}{\sin 17°} = \dfrac{75}{\sin 45°}$

$$h = \dfrac{75\sin 17°}{\sin 45°}$$

$$h \approx 31.01 \text{ feet}$$

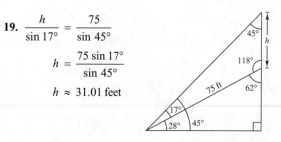

21. Given: $a = 8, b = 14, c = 17$

$\cos C = \dfrac{a^2 + b^2 - c^2}{2ab} = \dfrac{64 + 196 - 289}{2(8)(14)} \approx -0.1295 \Rightarrow C \approx 97.44°$

$\sin B = \dfrac{b\sin C}{c} \approx \dfrac{14\sin 97.44}{17} \approx 0.8166 \Rightarrow B \approx 54.75°$

$A \approx 180° - 54.75° - 97.44° = 27.81°$

23. Given: $a = 6, b = 9, c = 14$

$\cos C = \dfrac{a^2 + b^2 - c^2}{2ab} = \dfrac{36 + 81 - 196}{2(6)(9)} \approx -0.7315 \Rightarrow C \approx 137.01°$

$\sin B = \dfrac{b\sin C}{c} \approx \dfrac{9\sin 137.01°}{14} \approx 0.4383 \Rightarrow B \approx 26.00°$

$A \approx 180° - 26.00° - 137.01° = 16.99°$

25. Given: $a = 2.5, b = 5.0, c = 4.5$

$\cos B = \dfrac{a^2 + c^2 - b^2}{2ac} = 0.0667 \Rightarrow B \approx 86.18°$

$\cos C = \dfrac{a^2 + b^2 - c^2}{2ab} = 0.44 \Rightarrow C \approx 63.90°$

$A = 180° - B - C \approx 29.92°$

27. Given: $B = 108°, a = 11, c = 11$

$b^2 = a^2 + c^2 - 2ac\cos B = 11^2 + 11^2 - 2(11)(11)\cos 108° \Rightarrow b \approx 17.80$

$A = C = \frac{1}{2}(180° - 108°) = 36°$

29. Given: $C = 43°, a = 22.5, b = 31.4$

$c = \sqrt{a^2 + b^2 - 2ab\cos C} \approx 21.42$

$\cos B = \dfrac{a^2 + c^2 - b^2}{2ac} \approx -0.02169 \Rightarrow B \approx 91.24°$

$A = 180° - B - C \approx 45.76°$

31. Given: $b = 9, c = 13, C = 64°$

Given two sides and an angle opposite one of them, use the Law of Sines.

$$\sin B = \frac{b \sin C}{c} = \frac{9 \sin 64°}{13} \approx 0.6222 \Rightarrow B \approx 38.48°$$

$$A \approx 180° - 38.48° - 64° = 77.52°$$

$$a = \frac{c \sin A}{\sin C} \approx \frac{13 \sin 77.52°}{\sin 64°} \approx 14.12$$

33. Given: $a = 13, b = 15, c = 24$

Given three sides, use the Law of Cosines.

$$\cos C = \frac{a^2 + b^2 - c^2}{2ab} = \frac{169 + 225 - 576}{2(13)(15)} \approx -0.4667 \Rightarrow C \approx 117.82°$$

$$\sin A = \frac{a \sin C}{c} \approx \frac{13 \sin 117.82°}{24} \approx 0.4791 \Rightarrow A \approx 28.62°$$

$$B \approx 180° - 28.62° - 117.82° = 33.56°$$

35.

$$a^2 = 5^2 + 8^2 - 2(5)(8) \cos 28° \approx 18.364$$

$$a \approx 4.3 \text{ feet}$$

$$b^2 = 8^2 + 5^2 - 2(8)(5) \cos 152° \approx 159.636$$

$$b \approx 12.6 \text{ feet}$$

37. Length of $AC = \sqrt{300^2 + 425^2 - 2(300)(425) \cos 115°}$

$$\approx 615.1 \text{ meters}$$

39. $a = 3, b = 6, c = 8$

$$s = \frac{a + b + c}{2} = \frac{3 + 6 + 8}{2} = 8.5$$

$$\text{Area} = \sqrt{s(s - a)(s - b)(s - c)}$$

$$= \sqrt{8.5(5.5)(2.5)(0.5)}$$

$$\approx 7.64$$

41. $a = 12.3, b = 15.8, c = 3.7$

$$s = \frac{a + b + c}{2} = \frac{12.3 + 15.8 + 3.7}{2} = 15.9$$

$$\text{Area} = \sqrt{s(s - a)(s - b)(s - c)}$$

$$= \sqrt{15.9(3.6)(0.1)(12.2)} = 8.36$$

43. $\|\mathbf{u}\| = \sqrt{(4 - (-2))^2 + (6 - 1)^2} = \sqrt{61}$

$\|\mathbf{v}\| = \sqrt{(6 - 0)^2 + (3 - (-2))^2} = \sqrt{61}$

$\mathbf{u}$ is directed along a line with a slope of $\frac{6 - 1}{4 - (-2)} = \frac{5}{6}$.

$\mathbf{v}$ is directed along a line with a slope of $\frac{3 - (-2)}{6 - 0} = \frac{5}{6}$.

Because $\mathbf{u}$ and $\mathbf{v}$ have identical magnitudes and directions, $\mathbf{u} = \mathbf{v}$.

45. Initial point: $(-5, 4)$

Terminal point: $(2, -1)$

$$\mathbf{v} = \langle 2 - (-5), -1 - 4 \rangle = \langle 7, -5 \rangle$$

47. Initial point: $(0, 10)$

Terminal point: $(7, 3)$

$$\mathbf{v} = \langle 7 - 0, 3 - 10 \rangle = \langle 7, -7 \rangle$$

49. $\|\mathbf{v}\| = 8, \theta = 120°$

$$\langle 8 \cos 120°, 8 \sin 120° \rangle = \langle -4, 4\sqrt{3} \rangle$$

51. $\mathbf{u} = \langle -1, -3 \rangle, \mathbf{v} = \langle -3, 6 \rangle$

(a) $\mathbf{u} + \mathbf{v} = \langle -1, -3 \rangle + \langle -3, 6 \rangle = \langle -4, 3 \rangle$

(b) $\mathbf{u} - \mathbf{v} = \langle -1, -3 \rangle - \langle -3, 6 \rangle = \langle 2, -9 \rangle$

(c) $4\mathbf{u} = 4\langle -1, -3 \rangle = \langle -4, -12 \rangle$

(d) $3\mathbf{v} + 5\mathbf{u} = 3\langle -3, 6 \rangle + 5\langle -1, -3 \rangle = \langle -9, 18 \rangle + \langle -5, -15 \rangle = \langle -14, 3 \rangle$

53. $\mathbf{u} = \langle -5, 2 \rangle$, $\mathbf{v} = \langle 4, 4 \rangle$

(a) $\mathbf{u} + \mathbf{v} = \langle -5, 2 \rangle + \langle 4, 4 \rangle = \langle -1, 6 \rangle$

(b) $\mathbf{u} - \mathbf{v} = \langle -5, 2 \rangle - \langle 4, 4 \rangle = \langle -9, -2 \rangle$

(c) $4\mathbf{u} = 4\langle -5, 2 \rangle = \langle -20, 8 \rangle$

(d) $3\mathbf{v} + 5\mathbf{u} = 3\langle 4, 4 \rangle + 5\langle -5, 2 \rangle = \langle 12, 12 \rangle + \langle -25, 10 \rangle = \langle -13, 22 \rangle$

55. $\mathbf{u} = 2\mathbf{i} - \mathbf{j}$, $\mathbf{v} = 5\mathbf{i} + 3\mathbf{j}$

(a) $\mathbf{u} + \mathbf{v} = (2\mathbf{i} - \mathbf{j}) + (5\mathbf{i} + 3\mathbf{j}) = 7\mathbf{i} + 2\mathbf{j}$

(b) $\mathbf{u} - \mathbf{v} = (2\mathbf{i} - \mathbf{j}) - (5\mathbf{i} + 3\mathbf{j}) = -3\mathbf{i} - 4\mathbf{j}$

(c) $4\mathbf{u} = 4(2\mathbf{i} - \mathbf{j}) = 8\mathbf{i} - 4\mathbf{j}$

(d) $3\mathbf{v} + 5\mathbf{u} = 3(5\mathbf{i} + 3\mathbf{j}) + 5(2\mathbf{i} - \mathbf{j}) = 15\mathbf{i} + 9\mathbf{j} + 10\mathbf{i} - 5\mathbf{j} = 25\mathbf{i} + 4\mathbf{j}$

57. $\mathbf{u} = 4\mathbf{i}$, $\mathbf{v} = -\mathbf{i} + 6\mathbf{j}$

(a) $\mathbf{u} + \mathbf{v} = 4\mathbf{i} + (-\mathbf{i} + 6\mathbf{j}) = 3\mathbf{i} + 6\mathbf{j}$

(b) $\mathbf{u} - \mathbf{v} = 4\mathbf{i} - (-\mathbf{i} + 6\mathbf{j}) = 5\mathbf{i} - 6\mathbf{j}$

(c) $4\mathbf{u} = 4(4\mathbf{i}) = 16\mathbf{i}$

(d) $3\mathbf{v} + 5\mathbf{u} = 3(-\mathbf{i} + 6\mathbf{j}) + 5(4\mathbf{i}) = -3\mathbf{i} + 18\mathbf{j} + 20\mathbf{i} = 17\mathbf{i} + 18\mathbf{j}$

59. $\mathbf{u} = 6\mathbf{i} - 5\mathbf{j}$, $\mathbf{v} = 10\mathbf{i} + 3\mathbf{j}$

$$2\mathbf{u} + \mathbf{v} = 2(6\mathbf{i} - 5\mathbf{j}) + (10\mathbf{i} + 3\mathbf{j})$$
$$= 22\mathbf{i} - 7\mathbf{j}$$
$$= \langle 22, -7 \rangle$$

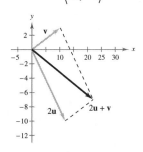

61. $\mathbf{v} = 10\mathbf{i} + 3\mathbf{j}$

$$3\mathbf{v} = 3(10\mathbf{i} + 3\mathbf{j})$$
$$= 30\mathbf{i} + 9\mathbf{j}$$
$$= \langle 30, 9 \rangle$$

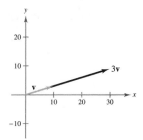

63. $\mathbf{u} = \langle -1, 5 \rangle = -\mathbf{i} + 5\mathbf{j}$

65. Initial point: $(3, 4)$

Terminal point: $(9, 8)$

$\mathbf{u} = (9 - 3)\mathbf{i} + (8 - 4)\mathbf{j} = 6\mathbf{i} + 4\mathbf{j}$

67. $\mathbf{v} = -10\mathbf{i} + 10\mathbf{j}$

$\|\mathbf{v}\| = \sqrt{(-10)^2 + (10)^2} = \sqrt{200} = 10\sqrt{2}$

$\tan \theta = \dfrac{10}{-10} = -1 \Rightarrow \theta = 135°$ because $\mathbf{v}$ is in Quadrant II.

$\mathbf{v} = 10\sqrt{2}\langle \mathbf{i} \cos 135° + \mathbf{j} \sin 135° \rangle$

69. $\mathbf{v} = 7(\cos 60° \, \mathbf{i} + \sin 60° \, \mathbf{j})$

$\|\mathbf{v}\| = 7$

$\theta = 60°$

71. $\mathbf{v} = 5\mathbf{i} + 4\mathbf{j}$

$\|\mathbf{v}\| = \sqrt{5^2 + 4^2} = \sqrt{41}$

$\tan \theta = \dfrac{4}{5} \Rightarrow \theta \approx 38.7°$

73. $\mathbf{v} = -3\mathbf{i} - 3\mathbf{j}$

$\|\mathbf{v}\| = \sqrt{(-3)^2 + (-3)^2} = 3\sqrt{2}$

$\tan \theta = \dfrac{-3}{-3} = 1 \Rightarrow \theta = 225°$

75. Magnitude of resultant:

$$c = \sqrt{85^2 + 50^2 - 2(85)(50)\cos 165°} \approx 133.92 \text{ pounds}$$

Let θ be the angle between the resultant and the 85-pound force.

$$\cos \theta \approx \frac{(133.92)^2 + 85^2 - 50^2}{2(133.92)(85)} \approx 0.9953 \Rightarrow \theta \approx 5.6°$$

77. Airspeed: $\mathbf{u} = 430(\cos 45°\mathbf{i} - \sin 45°\mathbf{j}) = 215\sqrt{2}(\mathbf{i} - \mathbf{j})$

Wind: $\mathbf{w} = 35(\cos 60°\mathbf{i} + \sin 60°\mathbf{j}) = \frac{35}{2}(\mathbf{i} + \sqrt{3}\mathbf{j})$

Groundspeed: $\mathbf{u} + \mathbf{w} = \left(215\sqrt{2} + \frac{35}{2}\right)\mathbf{i} + \left(\frac{35\sqrt{3}}{2} - 215\sqrt{2}\right)\mathbf{j}$

$$\|\mathbf{u} + \mathbf{w}\| = \sqrt{\left(215\sqrt{2} + \frac{35}{2}\right)^2 + \left(\frac{35\sqrt{3}}{2} - 215\sqrt{2}\right)^2} \approx 422.30 \text{ miles per hour}$$

Bearing: $\tan \theta' = \dfrac{17.5\sqrt{3} - 215\sqrt{2}}{215\sqrt{2} + 17.5}$

$\theta' \approx -40.4°$

$\theta = 90° + |\theta'| = 130.4°$

79. $\mathbf{u} = \langle 6, 7 \rangle, \mathbf{v} = \langle -3, 9 \rangle$

$\mathbf{u} \cdot \mathbf{v} = 6(-3) + 7(9) = 45$

81. $\mathbf{u} = 3\mathbf{i} + 7\mathbf{j}, \mathbf{v} = 11\mathbf{i} - 5\mathbf{j}$

$\mathbf{u} \cdot \mathbf{v} = 3(11) + 7(-5) = -2$

83. $\mathbf{u} = \langle -4, 2 \rangle$

$2\mathbf{u} = \langle -8, 4 \rangle$

$2\mathbf{u} \cdot \mathbf{u} = -8(-4) + 4(2) = 40$

The result is a scalar.

85. $\mathbf{u} = \langle -4, 2 \rangle$

$4 - \|\mathbf{u}\| = 4 - \sqrt{(-4)^2 + 2^2} = 4 - \sqrt{20} = 4 - 2\sqrt{5}$

The result is a scalar.

87. $\mathbf{u} = \langle -4, 2 \rangle, \mathbf{v} = \langle 5, 1 \rangle$

$\mathbf{u}(\mathbf{u} \cdot \mathbf{v}) = \langle -4, 2 \rangle [-4(5) + 2(1)]$

$= -18\langle -4, 2 \rangle$

$= \langle 72, -36 \rangle$

The result is a vector.

89. $\mathbf{u} = \langle -4, 2 \rangle, \mathbf{v} = \langle 5, 1 \rangle$

$(\mathbf{u} \cdot \mathbf{u}) - (\mathbf{u} \cdot \mathbf{v}) = [-4(-4) + 2(2)] - [-4(5) + 2(1)]$

$= 20 - (-18)$

$= 38$

The result is a scalar.

91. $\mathbf{u} = \cos \frac{7\pi}{4}\mathbf{i} + \sin \frac{7\pi}{4}\mathbf{j} = \left\langle \frac{1}{\sqrt{2}}, -\frac{1}{\sqrt{2}} \right\rangle$

$\mathbf{v} = \cos \frac{5\pi}{6}\mathbf{i} + \sin \frac{5\pi}{6}\mathbf{j} = \left\langle -\frac{\sqrt{3}}{2}, \frac{1}{2} \right\rangle$

$\cos \theta = \dfrac{\mathbf{u} \cdot \mathbf{v}}{\|\mathbf{u}\|\|\mathbf{v}\|} = \dfrac{-\sqrt{3} - 1}{2\sqrt{2}} \Rightarrow \theta = \dfrac{11\pi}{12}$

93. $\mathbf{u} = \langle 2\sqrt{2}, -4 \rangle, \mathbf{v} = \langle -\sqrt{2}, 1 \rangle$

$\cos \theta = \dfrac{\mathbf{u} \cdot \mathbf{v}}{\|\mathbf{u}\|\|\mathbf{v}\|} = \dfrac{-8}{(\sqrt{24})(\sqrt{3})} \Rightarrow \theta \approx 160.5°$

95. $\mathbf{u} = \langle -3, 8 \rangle$

$\mathbf{v} = \langle 8, 3 \rangle$

$\mathbf{u} \cdot \mathbf{v} = -3(8) + 8(3) = 0$

$\mathbf{u}$ and $\mathbf{v}$ are orthogonal.

97. $\mathbf{u} = -\mathbf{i}$

$\mathbf{v} = \mathbf{i} + 2\mathbf{j}$

$\mathbf{u} \cdot \mathbf{v} \neq 0 \Rightarrow$ Not orthogonal

$\mathbf{v} \neq k\mathbf{u} \Rightarrow$ Not parallel

Neither

99. $\mathbf{u} = \langle -4, 3 \rangle$, $\mathbf{v} = \langle -8, -2 \rangle$

$\mathbf{w}_1 = \text{proj}_\mathbf{v}\mathbf{u} = \left(\dfrac{\mathbf{u} \cdot \mathbf{v}}{\|\mathbf{v}\|^2} \right)\mathbf{v} = \left(\dfrac{26}{68} \right)\langle -8, -2 \rangle = -\dfrac{13}{17}\langle 4, 1 \rangle$

$\mathbf{w}_2 = \mathbf{u} - \mathbf{w}_1 = \langle -4, 3 \rangle - \left(-\dfrac{13}{17} \right)\langle 4, 1 \rangle = \dfrac{16}{17}\langle -1, 4 \rangle$

$\mathbf{u} = \mathbf{w}_1 + \mathbf{w}_2 = -\dfrac{13}{17}\langle 4, 1 \rangle + \dfrac{16}{17}\langle -1, 4 \rangle$

101. $\mathbf{u} = \langle 2, 7 \rangle$, $\mathbf{v} = \langle 1, -1 \rangle$

$\mathbf{w}_1 = \text{proj}_\mathbf{v}\mathbf{u} = \left(\dfrac{\mathbf{u} \cdot \mathbf{v}}{\|\mathbf{v}\|^2} \right)\mathbf{v} = -\dfrac{5}{2}\langle 1, -1 \rangle = \dfrac{5}{2}\langle -1, 1 \rangle$

$\mathbf{w}_2 = \mathbf{u} - \mathbf{w}_1 = \langle 2, 7 \rangle - \left(\dfrac{5}{2} \right)\langle -1, 1 \rangle = \dfrac{9}{2}\langle 1, 1 \rangle$

$\mathbf{u} = \mathbf{w}_1 + \mathbf{w}_2 = \dfrac{5}{2}\langle -1, 1 \rangle + \dfrac{9}{2}\langle 1, 1 \rangle$

103. $P = (5, 3)$, $Q = (8, 9) \Rightarrow \overrightarrow{PQ} = \langle 3, 6 \rangle$

Work $= \mathbf{v} \cdot \overrightarrow{PQ} = \langle 2, 7 \rangle \cdot \langle 3, 6 \rangle = 48$

105. Work $= (18{,}000)\left(\dfrac{48}{12} \right) = 72{,}000$ foot-pounds

107. $|7i| = \sqrt{0^2 + 7^2} = 7$

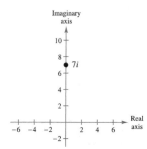

109. $|5 + 3i| = \sqrt{5^2 + 3^2}$

$\qquad\quad = \sqrt{34}$

111. $\left| \sqrt{2} - \sqrt{2}i \right| = \sqrt{\left(\sqrt{2} \right)^2 + \left(\sqrt{2} \right)^2} = 2$

113. $z = 4i$

$r = \sqrt{0^2 + 4^2} = \sqrt{16} = 4$

$\tan \theta = \dfrac{4}{0}$, undefined $\Rightarrow \theta = \dfrac{\pi}{2}$

$z = 4\left(\cos \dfrac{\pi}{2} + i \sin \dfrac{\pi}{2} \right)$

115. $5 - 5i$

$r = \sqrt{5^2 + (-5)^2} = \sqrt{50} = 5\sqrt{2}$

$\tan \theta = \dfrac{-5}{5} = -1 \Rightarrow \theta = \dfrac{7\pi}{4}$ because the complex number is in Quadrant IV.

$5 - 5i = 5\sqrt{2}\left(\cos \dfrac{7\pi}{4} + i \sin \dfrac{7\pi}{4} \right)$

117. $z = -5 - 12i$

$r = \sqrt{(-5)^2 + (-12)^2} = \sqrt{169} = 13$

$\tan \theta = \dfrac{12}{5}$, θ is in Quadrant III $\Rightarrow \theta \approx 4.32$

$z = 13(\cos 4.32 + i \sin 4.32)$

119. (a) $z_1 = 2\sqrt{3} - 2i = 4\left(\cos\dfrac{11\pi}{6} + i\sin\dfrac{11\pi}{6}\right)$

$z_2 = -10i = 10\left(\cos\dfrac{3\pi}{2} + i\sin\dfrac{3\pi}{2}\right)$

(b) $z_1z_2 = \left[4\left(\cos\dfrac{11\pi}{6} + i\sin\dfrac{11\pi}{6}\right)\right]\left[10\left(\cos\dfrac{3\pi}{2} + i\sin\dfrac{3\pi}{2}\right)\right]$

$= 40\left(\cos\dfrac{10\pi}{3} + i\sin\dfrac{10\pi}{3}\right)$

$\dfrac{z_1}{z_2} = \dfrac{4\left(\cos\dfrac{11\pi}{6} + i\sin\dfrac{11\pi}{6}\right)}{10\left(\cos\dfrac{3\pi}{2} + i\sin\dfrac{3\pi}{2}\right)} = \dfrac{2}{5}\left(\cos\dfrac{\pi}{3} + i\sin\dfrac{\pi}{3}\right)$

121. $\left[5\left(\cos\dfrac{\pi}{12} + i\sin\dfrac{\pi}{12}\right)\right]^4 = 5^4\left(\cos\dfrac{4\pi}{12} + i\sin\dfrac{4\pi}{12}\right)$

$= 625\left(\cos\dfrac{\pi}{3} + i\sin\dfrac{\pi}{3}\right)$

$= 625\left(\dfrac{1}{2} + \dfrac{\sqrt{3}}{2}i\right)$

$= \dfrac{625}{2} + \dfrac{625\sqrt{3}}{2}i$

123. $(2 + 3i)^6 \approx \left[\sqrt{13}(\cos 56.3° + i\sin 56.3°)\right]^6$

$= 13^3(\cos 337.9° + i\sin 337.9°)$

$\approx 13^3(0.9263 - 0.3769i)$

$\approx 2035 - 828i$

125. Sixth roots of $-729i = 729\left(\cos\dfrac{3\pi}{2} + i\sin\dfrac{3\pi}{2}\right)$:

(a) $\sqrt[6]{729}\left[\cos\left(\dfrac{\dfrac{3\pi}{2} + 2k\pi}{6}\right) + i\sin\left(\dfrac{\dfrac{3\pi}{2} + 2k\pi}{6}\right)\right], k = 0, 1, 2, 3, 4, 5$

$k = 0: 3\left(\cos\dfrac{\pi}{4} + i\sin\dfrac{\pi}{4}\right)$

$k = 1: 3\left(\cos\dfrac{7\pi}{12} + i\sin\dfrac{7\pi}{12}\right)$

$k = 2: 3\left(\cos\dfrac{11\pi}{12} + i\sin\dfrac{11\pi}{12}\right)$

$k = 3: 3\left(\cos\dfrac{5\pi}{4} + i\sin\dfrac{5\pi}{4}\right)$

$k = 4: 3\left(\cos\dfrac{19\pi}{12} + i\sin\dfrac{19\pi}{12}\right)$

$k = 5: 3\left(\cos\dfrac{23\pi}{12} + i\sin\dfrac{23\pi}{12}\right)$

(b)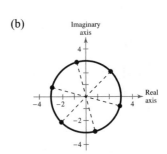

(c) $\dfrac{3\sqrt{2}}{2} + \dfrac{3\sqrt{2}}{2}i$

$-0.776 + 2.898i$

$-2.898 + 0.776i$

$\dfrac{-3\sqrt{2}}{2} - \dfrac{3\sqrt{2}}{2}i$

$0.776 - 2.898i$

$2.898 - 0.776i$

127. Cube roots of $8 = 8(\cos 0 + i \sin 0)$, $k = 0, 1, 2$

(a) $\sqrt[3]{8}\left[\cos\left(\dfrac{0 + 2\pi k}{3}\right) + i \sin\left(\dfrac{0 + 2\pi k}{3}\right)\right]$

$k = 0:\ 2(\cos 0 + i \sin 0)$

$k = 1:\ 2\left(\cos\dfrac{2\pi}{3} + i \sin\dfrac{2\pi}{3}\right)$

$k = 2:\ 2\left(\cos\dfrac{4\pi}{3} + i \sin\dfrac{4\pi}{3}\right)$

(b)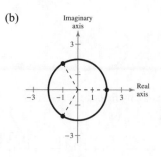

(c) 2

$-1 + \sqrt{3}i$

$-1 - \sqrt{3}i$

129. $x^4 + 81 = 0$

$x^4 = -81$ Solve by finding the fourth roots of -81.

$-81 = 81(\cos \pi + i \sin \pi)$

$\sqrt[4]{-81} = \sqrt[4]{81}\left[\cos\left(\dfrac{\pi + 2\pi k}{4}\right) + i \sin\left(\dfrac{\pi + 2\pi k}{4}\right)\right],\ k = 0, 1, 2, 3$

$k = 0:\ 3\left(\cos\dfrac{\pi}{4} + i \sin\dfrac{\pi}{4}\right) = \dfrac{3\sqrt{2}}{2} + \dfrac{3\sqrt{2}}{2}i$

$k = 1:\ 3\left(\cos\dfrac{3\pi}{4} + i \sin\dfrac{3\pi}{4}\right) = -\dfrac{3\sqrt{2}}{2} + \dfrac{3\sqrt{2}}{2}i$

$k = 2:\ 3\left(\cos\dfrac{5\pi}{4} + i \sin\dfrac{5\pi}{4}\right) = -\dfrac{3\sqrt{2}}{2} - \dfrac{3\sqrt{2}}{2}i$

$k = 3:\ 3\left(\cos\dfrac{7\pi}{4} + i \sin\dfrac{7\pi}{4}\right) = \dfrac{3\sqrt{2}}{2} - \dfrac{3\sqrt{2}}{2}i$

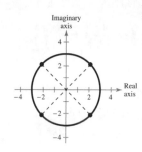

131. $x^3 + 8i = 0$

$x^3 = -8i$ Solve by finding the cube roots of $-8i$.

$-8i = 8\left(\cos\dfrac{3\pi}{2} + i \sin\dfrac{3\pi}{2}\right)$

$\sqrt[3]{-8i} = \sqrt[3]{8}\left[\cos\left(\dfrac{\dfrac{3\pi}{2} + 2\pi k}{3}\right) + i \sin\left(\dfrac{\dfrac{3\pi}{2} + 2\pi k}{3}\right)\right],\ k = 0, 1, 2$

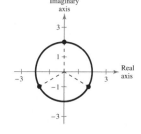

$k = 0:\ 2\left(\cos\dfrac{\pi}{2} + i \sin\dfrac{\pi}{2}\right) = 2i$

$k = 1:\ 2\left(\cos\dfrac{7\pi}{6} + i \sin\dfrac{7\pi}{6}\right) = -\sqrt{3} - i$

$k = 2:\ 2\left(\cos\dfrac{11\pi}{6} + i \sin\dfrac{11\pi}{6}\right) = \sqrt{3} - i$

133.
$$x^5 + x^3 - x^2 - 1 = 0$$
$$x^3(x^2 + 1) - 1(x^2 + 1) = 0$$
$$(x^3 - 1)(x^2 + 1) = 0$$
$$x^3 - 1 = 0 \text{ or } x^2 + 1 = 0$$

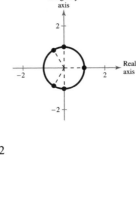

Case 1: $x^3 - 1 = 0$

$x^3 = 1$ Solve by finding the cube roots of 1.

$1 = 1(\cos 0 + i \sin 0)$

$\sqrt[3]{1} = \sqrt[3]{1}\left[\cos\left(\dfrac{0 + 2\pi k}{3}\right) + i \sin\left(\dfrac{0 + 2\pi k}{3}\right)\right], k = 0, 1, 2$

$k = 0$: $1(\cos 0 + i \sin 0) = 1$

$k = 1$: $1\left(\cos\dfrac{2\pi}{3} + i \sin\dfrac{2\pi}{3}\right) = -\dfrac{1}{2} + \dfrac{\sqrt{3}}{2}i$

$k = 2$: $1\left(\cos\dfrac{4\pi}{3} + i \sin\dfrac{4\pi}{3}\right) = -\dfrac{1}{2} - \dfrac{\sqrt{3}}{2}i$

Case 2: $x^2 + 1 = 0$

$x^2 = -1$ Solve by finding the square roots of -1.

$-1 = 1(\cos \pi + i \sin \pi)$

$\sqrt{-1} = \sqrt{1}\left[\cos\left(\dfrac{\pi + 2\pi k}{2}\right) + i \sin\left(\dfrac{\pi + 2\pi k}{2}\right)\right], k = 0, 1$

$k = 0$: $1\left(\cos\dfrac{\pi}{2} + i \sin\dfrac{\pi}{2}\right) = i$

$k = 1$: $1\left(\cos\dfrac{3\pi}{2} + i \sin\dfrac{3\pi}{2}\right) = -i$

135. True. $\sin 90°$ is defined in the Law of Sines.

137. True, by the definition of a unit vector.

$$\mathbf{u} = \dfrac{\mathbf{v}}{\|\mathbf{v}\|} \text{ so } \mathbf{v} = \|\mathbf{v}\|\mathbf{u}$$

139. False. $x = \sqrt{3} + i$ is a solution to $x^3 - 8i = 0$, not $x^2 - 8i = 0$.

Also, $\left(\sqrt{3} + i\right)^2 - 8i = 2 + \left(2\sqrt{3} - 8\right)i \neq 0$.

141. $a^2 = b^2 + c^2 - 2bc \cos A$

$b^2 = a^2 + c^2 - 2ac \cos B$

$c^2 = a^2 + b^2 - 2ab \cos C$

143. **A** and **C** appear to have the same magnitude and direction.

145. If $k > 0$, the direction of $k\mathbf{u}$ is the same, and the magnitude is $k\|\mathbf{u}\|$.

If $k < 0$, the direction of $k\mathbf{u}$ is the opposite direction of $\mathbf{u}$, and the magnitude is $|k|\,\|\mathbf{u}\|$.

147. (a) The trigonometric form of the three roots shown is:

$4(\cos 60° + i \sin 60°)$

$4(\cos 180° + i \sin 180°)$

$4(\cos 300° + i \sin 300°)$

(b) Because there are three evenly spaced roots on the circle of radius 4, they are cube roots of a complex number of modulus $4^3 = 64$.

Cubing them yields -64.

$\left[4(\cos 60° + i \sin 60°)\right]^3 = -64$

$\left[4(\cos 180° + i \sin 180°)\right]^3 = -64$

$\left[4(\cos 300° + i \sin 300°)\right]^3 = -64$

149. $z_1 = 2(\cos\theta + i\sin\theta)$

$z_2 = 2(\cos(\pi - \theta) + i\sin(\pi - \theta))$

$z_1 z_2 = (2)(2)\left[\cos(\theta + (\pi - \theta)) + i\sin(\theta + (\pi - \theta))\right]$

$\qquad = 4(\cos\pi + i\sin\pi)$

$\qquad = -4$

$\dfrac{z_1}{z_2} = \dfrac{2(\cos\theta + i\sin\theta)}{2(\cos(\pi - \theta) + i\sin(\pi - \theta))}$

$\qquad = 1\left[\cos(\theta - (\pi - \theta)) + i\sin(\theta - (\pi - \theta))\right]$

$\qquad = \cos(2\theta - \pi) + i\sin(2\theta - \pi)$

$\qquad = \cos 2\theta \cos\pi + \sin 2\theta \sin\pi + i(\sin 2\theta \cos\pi - \cos 2\theta \sin\pi)$

$\qquad = -\cos 2\theta - i\sin 2\theta$

Problem Solving for Chapter 8

1. $\left(\overrightarrow{PQ}\right)^2 = 4.7^2 + 6^2 - 2(4.7)(6)\cos 25°$

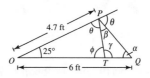

$\overrightarrow{PQ} \approx 2.6409$ feet

$\dfrac{\sin\alpha}{4.7} = \dfrac{\sin 25°}{2.6409} \Rightarrow \alpha \approx 48.78°$

$\theta + \beta = 180° - 25° - 48.78° = 106.22°$

$(\theta + \beta) + \theta = 180° \Rightarrow \theta = 180° - 106.22° = 73.78°$

$\beta = 106.22° - 73.78° = 32.44°$

$\gamma = 180° - \alpha - \beta = 180° - 48.78° - 32.44° = 98.78°$

$\phi = 180° - \gamma = 180° - 98.78° = 81.22°$

$\dfrac{\overrightarrow{PT}}{\sin 25°} = \dfrac{4.7}{\sin 81.22°}$

$\overrightarrow{PT} \approx 2.01$ feet

3. (a)

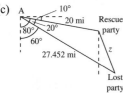

(b) $\dfrac{x}{\sin 15°} = \dfrac{75}{\sin 135°}$ and $\dfrac{y}{\sin 30°} = \dfrac{75}{\sin 135°}$

$\qquad\qquad x \approx 27.45$ miles $\qquad\qquad y \approx 53.03$ miles

(c)

$z^2 = (27.45)^2 + (20)^2 - 2(27.45)(20)\cos 20°$

$z \approx 11.03$ miles

$\dfrac{\sin\theta}{27.45} = \dfrac{\sin 20°}{11.03}$

$\sin\theta \approx 0.8511$

$\theta = 180° - \sin^{-1}(0.8511)$

$\theta \approx 121.7°$

To find the bearing, we have $\theta - 10° - 90° \approx 21.7°$.

Bearing: S 21.7° E

5. If $\mathbf{u} \neq 0$, $\mathbf{v} \neq 0$, and $\mathbf{u} + \mathbf{v} \neq 0$, then $\left\| \dfrac{\mathbf{u}}{\|\mathbf{u}\|} \right\| = \left\| \dfrac{\mathbf{v}}{\|\mathbf{v}\|} \right\| = \left\| \dfrac{\mathbf{u} + \mathbf{v}}{\|\mathbf{u} + \mathbf{v}\|} \right\| = 1$ because all of these are magnitudes of unit vectors.

(a) $\mathbf{u} = \langle 1, -1 \rangle$, $\quad \mathbf{v} = \langle -1, 2 \rangle$, $\quad \mathbf{u} + \mathbf{v} = \langle 0, 1 \rangle$

   (i) $\|\mathbf{u}\| = \sqrt{2}$ (ii) $\|\mathbf{v}\| = \sqrt{5}$ (iii) $\|\mathbf{u} + \mathbf{v}\| = 1$ (iv) $\left\| \dfrac{\mathbf{u}}{\|\mathbf{u}\|} \right\| = 1$ (v) $\left\| \dfrac{\mathbf{v}}{\|\mathbf{v}\|} \right\| = 1$ (vi) $\left\| \dfrac{\mathbf{u} + \mathbf{v}}{\|\mathbf{u} + \mathbf{v}\|} \right\| = 1$

(b) $\mathbf{u} = \langle 0, 1 \rangle$, $\quad \mathbf{v} = \langle 3, -3 \rangle$, $\quad \mathbf{u} + \mathbf{v} = \langle 3, -2 \rangle$

   (i) $\|\mathbf{u}\| = 1$ (ii) $\|\mathbf{v}\| = \sqrt{18} = 3\sqrt{2}$ (iii) $\|\mathbf{u} + \mathbf{v}\| = \sqrt{13}$ (iv) $\left\| \dfrac{\mathbf{u}}{\|\mathbf{u}\|} \right\| = 1$ (v) $\left\| \dfrac{\mathbf{v}}{\|\mathbf{v}\|} \right\| = 1$ (vi) $\left\| \dfrac{\mathbf{u} + \mathbf{v}}{\|\mathbf{u} + \mathbf{v}\|} \right\| = 1$

(c) $\mathbf{u} = \left\langle 1, \dfrac{1}{2} \right\rangle$, $\mathbf{v} = \langle 2, 3 \rangle$, $\mathbf{u} + \mathbf{v} = \left\langle 3, \dfrac{7}{2} \right\rangle$

   (i) $\|\mathbf{u}\| = \dfrac{\sqrt{5}}{2}$ (ii) $\|\mathbf{v}\| = \sqrt{13}$ (iii) $\|\mathbf{u} + \mathbf{v}\| = \sqrt{9 + \dfrac{49}{4}} = \dfrac{\sqrt{85}}{2}$ (iv) $\left\| \dfrac{\mathbf{u}}{\|\mathbf{u}\|} \right\| = 1$

   (v) $\left\| \dfrac{\mathbf{v}}{\|\mathbf{v}\|} \right\| = 1$ (vi) $\left\| \dfrac{\mathbf{u} + \mathbf{v}}{\|\mathbf{u} + \mathbf{v}\|} \right\| = 1$

(d) $\mathbf{u} = \langle 2, -4 \rangle$, $\quad \mathbf{v} = \langle 5, 5 \rangle$, $\quad \mathbf{u} + \mathbf{v} = \langle 7, 1 \rangle$

   (i) $\|\mathbf{u}\| = \sqrt{20} = 2\sqrt{5}$ (ii) $\|\mathbf{v}\| = \sqrt{50} = 5\sqrt{2}$ (iii) $\|\mathbf{u} + \mathbf{v}\| = \sqrt{50} = 5\sqrt{2}$ (iv) $\left\| \dfrac{\mathbf{u}}{\|\mathbf{u}\|} \right\| = 1$

   (v) $\left\| \dfrac{\mathbf{v}}{\|\mathbf{v}\|} \right\| = 1$ (vi) $\left\| \dfrac{\mathbf{u} + \mathbf{v}}{\|\mathbf{u} + \mathbf{v}\|} \right\| = 1$

---

7. Initial point: $(0, 0)$

Terminal point: $\left(\dfrac{\mathbf{u}_1 + \mathbf{v}_1}{2}, \dfrac{\mathbf{u}_2 + \mathbf{v}_2}{2} \right)$

$\mathbf{w} = \left\langle \dfrac{\mathbf{u}_1 + \mathbf{v}_1}{2}, \dfrac{\mathbf{u}_2 + \mathbf{v}_2}{2} \right\rangle = \dfrac{1}{2}(\mathbf{u} + \mathbf{v})$

Initial point: $(\mathbf{u}_1, \mathbf{u}_2)$

Terminal point: $\dfrac{1}{2}(\mathbf{u}_1 + \mathbf{v}_1, \mathbf{u}_2 + \mathbf{v}_2)$

$\mathbf{w} = \left\langle \dfrac{\mathbf{u}_1 + \mathbf{v}_1}{2} - \mathbf{u}_1, \dfrac{\mathbf{u}_2 + \mathbf{v}_2}{2} - \mathbf{u}_2 \right\rangle$

$= \left\langle \dfrac{\mathbf{v}_1 - \mathbf{u}_1}{2}, \dfrac{\mathbf{v}_2 - \mathbf{u}_2}{2} \right\rangle = \dfrac{1}{2}(\mathbf{v} - \mathbf{u})$

9. $W = (\cos \theta)\|\mathbf{F}\| \left\| \overrightarrow{PQ} \right\|$ and $\|\mathbf{F}_1\| = \|\mathbf{F}_2\|$

(a)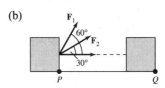

If $\theta_1 = -\theta_2$ then the work is the same because $\cos(-\theta) = \cos \theta$.

(b)

If $\theta_1 = 60°$ then $W_1 = \dfrac{1}{2}\|\mathbf{F}_1\| \left\| \overrightarrow{PQ} \right\|$.

If $\theta_2 = 30°$ then $W_2 = \dfrac{\sqrt{3}}{2}\|\mathbf{F}_2\| \left\| \overrightarrow{PQ} \right\|$.

$W_2 = \sqrt{3}W_1$

The amount of work done by $\mathbf{F}_2$ is $\sqrt{3}$ times as great as the amount of work done by $\mathbf{F}_1$.

Chapter 8 Practice Test

For Exercises 1 and 2, use the Law of Sines to find the remaining sides and angles of the triangle.

1. $A = 40°, B = 12°, b = 100$

2. $C = 150°, a = 5, c = 20$

3. Find the area of the triangle: $a = 3, b = 6, C = 130°$.

4. Determine the number of solutions to the triangle: $a = 10, b = 35, A = 22.5°$.

For Exercises 5 and 6, use the Law of Cosines to find the remaining sides and angles of the triangle.

5. $a = 49, b = 53, c = 38$

6. $C = 29°, a = 100, c = 300$

7. Use Heron's Formula to find the area of the triangle: $a = 4.1, b = 6.8, c = 5.5$.

8. A ship travels 40 miles due east, then adjusts its course 12° southward. After traveling 70 miles in that direction, how far is the ship from its point of departure?

9. $\mathbf{w} = 4\mathbf{u} - 7\mathbf{v}$ where $\mathbf{u} = 3\mathbf{i} + \mathbf{j}$ and $\mathbf{v} = -\mathbf{i} + 2\mathbf{j}$. Find $\mathbf{w}$.

10. Find a unit vector in the direction of $\mathbf{v} = 5\mathbf{i} - 3\mathbf{j}$.

11. Find the dot product and the angle between $\mathbf{u} = 6\mathbf{i} + 5\mathbf{j}$ and $\mathbf{v} = 2\mathbf{i} - 3\mathbf{j}$.

12. $\mathbf{v}$ is a vector of magnitude 4 making an angle of 30° with the positive x-axis. Find $\mathbf{v}$ in component form.

13. Find the projection of $\mathbf{u}$ onto $\mathbf{v}$ given $\mathbf{u} = \langle 3, -1 \rangle$ and $\mathbf{v} = \langle -2, 4 \rangle$.

14. Give the trigonometric form of $z = 5 - 5i$.

15. Give the standard form of $z = 6(\cos 225° + i \sin 225°)$.

16. Multiply $\left[7(\cos 23° + i \sin 23°) \right]\left[4(\cos 7° + i \sin 7°) \right]$.

17. Divide $\dfrac{9\left(\cos \dfrac{5\pi}{4} + i \sin \dfrac{5\pi}{4} \right)}{3(\cos \pi + i \sin \pi)}$.

18. Find $(2 + 2i)^8$.

19. Find the cube roots of $8\left(\cos \dfrac{\pi}{3} + i \sin \dfrac{\pi}{3} \right)$.

20. Find all the solutions to $x^4 + i = 0$.

CHAPTER 9
Systems of Equations and Inequalities

C H A P T E R   9
Systems of Equations and Inequalities

Section 9.1 Linear and Nonlinear Systems of Equations

- You should be able to solve systems of equations by the method of substitution.
  1. Solve one of the equations for one of the variables.
  2. Substitute this expression into the other equation and solve.
  3. Back-substitute into the first equation to find the value of the other variable.
  4. Check your answer in each of the original equations.

- You should be able to find solutions graphically. (See Example 5 in textbook.)

1. system; equations

3. solving

5. point of intersection

7. $\begin{cases} 2x - y = 4 \\ 8x + y = -9 \end{cases}$

  (a) $(0, -4)$

    $8(0) - 4 \neq -9$

    $(0, -4)$ *is not* a solution.

  (b) $(-2, 7)$

    $2(-2) - 7 \neq 4$

    $(-2, 7)$ *is not* a solution.

  (c) $\left(\frac{3}{2}, -1\right)$

    $8\left(\frac{3}{2}\right) - 1 \neq -9$

    $\left(\frac{3}{2}, -1\right)$ *is not* a solution.

  (d) $\left(-\frac{1}{2}, -5\right)$

    $2\left(-\frac{1}{2}\right) + 5 \overset{?}{=} 4$

    $-1 + 5 = 4$

    $8\left(-\frac{1}{2}\right) - 5 \overset{?}{=} -9$

    $-4 - 5 = -9$

    $\left(-\frac{1}{2}, -5\right)$ *is* a solution.

9. $\begin{cases} y = -4e^x \\ 7x - y = 4 \end{cases}$

  (a) $(-4, 0)$

    $7(-4) - 0 \neq 4$

    $(-4, 0)$ *is not* a solution.

  (b) $(0, -4)$

    $-4 \overset{?}{=} -4e^0$

    $-4 = -4(1)$

    $7(0) - (-4) \overset{?}{=} 4$

    $4 = 4$

    $(0, -4)$ *is* a solution.

  (c) $(0, -2)$

    $7(0) - (-2) \neq 4$

    $(0, -2)$ *is not* a solution.

  (d) $(-1, -3)$

    $7(-1) - (-3) \neq 4$

    $(-1, -3)$ *is not* a solution.

11. $\begin{cases} 2x + y = 6 & \text{Equation 1} \\ -x + y = 0 & \text{Equation 2} \end{cases}$

  Solve for y in Equation 1: $y = 6 - 2x$

  Substitute for y in Equation 2: $-x + (6 - 2x) = 0$

  Solve for x: $-3x + 6 = 0 \Rightarrow x = 2$

  Back-substitute $x = 2$: $y = 6 - 2(2) = 2$

  Solution: $(2, 2)$

13. $\begin{cases} x - y = -4 & \text{Equation 1} \\ x^2 - y = -2 & \text{Equation 2} \end{cases}$

Solve for y in Equation 1: $y = x + 4$

Substitute for y in Equation 2: $x^2 - (x + 4) = -2$

Solve for x: $x^2 - x - 2 = 0 \Rightarrow (x + 1)(x - 2) = 0 \Rightarrow x = -1, 2$

Back-substitute $x = -1$: $y = -1 + 4 = 3$

Back-substitute $x = 2$: $y = 2 + 4 = 6$

Solutions: $(-1, 3), (2, 6)$

15. $\begin{cases} -\frac{1}{2}x + y = -\frac{5}{2} & \text{Equation 1} \\ x^2 + y^2 = 25 & \text{Equation 2} \end{cases}$

Solve for x in Equation 1: $-\frac{1}{2}x = -y - \frac{5}{2} \Rightarrow x = 2y + 5$

Substitute for x in Equation 2: $(2y + 5)^2 + y^2 = 25$

Solve for $4y^2 + 20y + 25 + y^2 = 25 \Rightarrow 5y^2 + 20y = 0 \Rightarrow 5y(y + 4) = 0 \Rightarrow y = 0, y = -4$

Back-substitute $y = 0$: $-\frac{1}{2}x + 0 = -\frac{5}{2} \Rightarrow x = 5$

Back-substitute $y = -4$: $-\frac{1}{2}x - 4 = -\frac{5}{2} \Rightarrow x = -3$

Solutions: $(-3, -4), (5, 0)$

17. $\begin{cases} x^2 + y = 0 & \text{Equation 1} \\ x^2 - 4x - y = 0 & \text{Equation 2} \end{cases}$

Solve for y in Equation 1: $y = -x^2$

Substitute for y in Equation 2: $x^2 - 4x - (-x^2) = 0$

Solve for x:

$2x^2 - 4x = 0 \Rightarrow 2x(x - 2) = 0 \Rightarrow x = 0, 2$

Back-substitute $x = 0$: $y = -0^2 = 0$

Back-substitute $x = 2$: $y = -2^2 = -4$

Solutions: $(0, 0), (2, -4)$

19. $\begin{cases} y = x^3 - 3x^2 + 1 & \text{Equation 1} \\ y = x^2 - 3x + 1 & \text{Equation 2} \end{cases}$

Substitute for y in Equation 2:

$$x^3 - 3x^2 + 1 = x^2 - 3x + 1$$
$$x^3 - 4x^2 + 3x = 0$$
$$x(x - 1)(x - 3) = 0 \Rightarrow x = 0, 1, 3$$

Back-substitute $x = 0$: $y = 0^3 - 3(0)^2 + 1 = 1$

Back-substitute $x = 1$: $y = 1^3 - 3(1)^2 + 1 = -1$

Back-substitute $x = 3$: $y = 3^3 - 3(3)^2 + 1 = 1$

Solutions: $(0, 1), (1, -1), (3, 1)$

21. $\begin{cases} x - y = 2 & \text{Equation 1} \\ 6x - 5y = 16 & \text{Equation 2} \end{cases}$

Solve for x in Equation 1: $x = y + 2$

Substitute for x in Equation 2: $6(y + 2) - 5y = 16 \Rightarrow 6y + 12 - 5y = 16 \Rightarrow y = 4$

Back-substitute $y = 4$: $x - 4 = 2 \Rightarrow x = 6$

Solution: $(6, 4)$

23. $\begin{cases} 2x - y + 2 = 0 & \text{Equation 1} \\ 4x + y - 5 = 0 & \text{Equation 2} \end{cases}$

Solve for y in Equation 1: $y = 2x + 2$

Substitute for y in Equation 2: $4x + (2x + 2) - 5 = 0$

Solve for x: $6x - 3 = 0 \Rightarrow x = \frac{1}{2}$

Back-substitute $x = \frac{1}{2}$: $y = 2x + 2 = 2\left(\frac{1}{2}\right) + 2 = 3$

Solution: $\left(\frac{1}{2}, 3\right)$

25. $\begin{cases} 1.5x + 0.8y = 2.3 & \text{Equation 1} \\ 0.3x - 0.2y = 0.1 & \text{Equation 2} \end{cases}$

Multiply the equations by 10.

$\quad 15x + 8y = 23 \qquad$ Revised Equation 1

$\quad\;\; 3x - 2y = 1 \qquad$ Revised Equation 2

Solve for y in revised Equation 2: $y = \frac{3}{2}x - \frac{1}{2}$

Substitute for y in revised Equation 1: $15x + 8\left(\frac{3}{2}x - \frac{1}{2}\right) = 23$

Solve for x: $15x + 12x - 4 = 23 \Rightarrow 27x = 27 \Rightarrow x = 1$

Back-substitute $x = 1$: $y = \frac{3}{2}(1) - \frac{1}{2} = 1$

Solution: $(1, 1)$

27. $\begin{cases} \frac{1}{5}x + \frac{1}{2}y = 8 & \text{Equation 1} \\ x + y = 20 & \text{Equation 2} \end{cases}$

Solve for x in Equation 2: $x = 20 - y$

Substitute for x in Equation 1: $\frac{1}{5}(20 - y) + \frac{1}{2}y = 8$

Solve for y: $4 + \frac{3}{10}y = 8 \Rightarrow y = \frac{40}{3}$

Back-substitute $y = \frac{40}{3}$: $x = 20 - y = 20 - \frac{40}{3} = \frac{20}{3}$

Solution: $\left(\frac{20}{3}, \frac{40}{3}\right)$

29. $\begin{cases} 6x + 5y = -3 & \text{Equation 1} \\ -x - \frac{5}{6}y = -7 & \text{Equation 2} \end{cases}$

Solve for x in Equation 2: $x = 7 - \frac{5}{6}y$

Substitute for x in Equation 1: $6\left(7 - \frac{5}{6}y\right) + 5y = -3$

Solve for y: $42 - 5y + 5y = -3 \Rightarrow 42 = -3$ (False)

No solution

31. $\begin{cases} x^2 - y = 0 & \text{Equation 1} \\ 2x + y = 0 & \text{Equation 2} \end{cases}$

Solve for y in Equation 2: $y = -2x$

Substitute for y in Equation 1: $x^2 - (-2x) = 0$

Solve for x:

$\quad x^2 + 2x = 0 \Rightarrow x(x + 2) = 0 \Rightarrow x = 0, -2$

Back-substitute $x = 0$: $y = -2(0) = 0$

Back-substitute $x = -2$: $y = -2(-2) = 4$

Solutions: $(0, 0), (-2, 4)$

33. $\begin{cases} x - y = -1 & \text{Equation 1} \\ x^2 - y = -4 & \text{Equation 2} \end{cases}$

Solve for y in Equation 1: $y = x + 1$

Substitute for y in Equation 2: $x^2 - (x + 1) = -4$

Solve for x: $x^2 - x - 1 = -4 \Rightarrow x^2 - x + 3 = 0$

The Quadratic Formula yields no real solutions.

35. $\begin{cases} -x + 2y = -2 \\ 3x + y = 20 \end{cases}$

Point of intersection:

$(6, 2)$

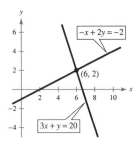

37. $\begin{cases} x - 3y = -3 \\ 5x + 3y = -6 \end{cases}$

Point of intersection:

$\left(-\frac{3}{2}, \frac{1}{2}\right)$

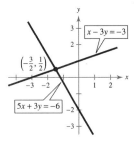

39. $\begin{cases} x + y = 4 \\ x^2 + y^2 - 4x = 0 \end{cases}$

Points of intersection:

$(2, 2), (4, 0)$

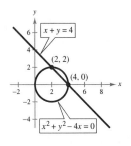

41. $\begin{cases} x - y + 3 = 0 \\ y = x^2 - 4x + 7 \end{cases}$

Points of intersection:

$(1, 4), (4, 7)$

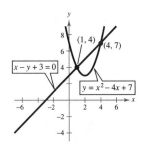

43. $\begin{cases} 7x + 8y = 24 \\ x - 8y = 8 \end{cases}$

Point of intersection:

$\left(4, -\frac{1}{2}\right)$

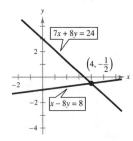

45. $\begin{cases} 3x - 2y = 0 \\ x^2 - y^2 = 4 \end{cases}$

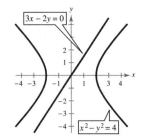

No points of intersection $\Rightarrow$ No solution

47. $\begin{cases} x^2 + y^2 = 25 \\ 3x^2 - 16y = 0 \end{cases}$

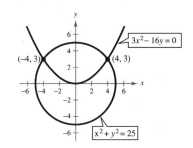

Points of intersection: $(-4, 3), (4, 3)$

49. $\begin{cases} y = e^x \\ x - y + 1 = 0 \end{cases} \Rightarrow y = x + 1$

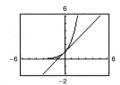

Point of intersection: $(0, 1)$

51. $\begin{cases} x + 2y = 8 \quad \Rightarrow y = -\frac{1}{2}x + 4 \\ y = \log_2 x \Rightarrow y = \frac{\ln x}{\ln 2} \end{cases}$

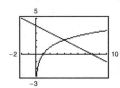

Point of intersection: $(4, 2)$

53. $\begin{cases} x^2 + y^2 = 169 \Rightarrow y_1 = \sqrt{169 - x^2} \text{ and } y_2 = -\sqrt{169 - x^2} \\ x^2 - 8y = 104 \Rightarrow y_3 = \frac{1}{8}x^2 - 13 \end{cases}$

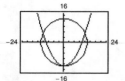

Points of intersection: $(0, -13), (12, 5), (-12, 5)$

55. $\begin{cases} y = 2x & \text{Equation 1} \\ y = x^2 + 1 & \text{Equation 2} \end{cases}$

Substitute for y in Equation 2: $2x = x^2 + 1$

Solve for x: $x^2 - 2x + 1 = (x - 1)^2 = 0 \Rightarrow x = 1$

Back-substitute $x = 1$ in Equation 1: $y = 2x = 2$

Solution: $(1, 2)$

57. $\begin{cases} x - 2y = 4 & \text{Equation 1} \\ x^2 - y = 0 & \text{Equation 2} \end{cases}$

Solve for y in Equation 2: $y = x^2$

Substitute for y in Equation 1: $x - 2x^2 = 4$

Solve for x: $0 = 2x^2 - x + 4 \Rightarrow x = \dfrac{1 \pm \sqrt{1 - 4(2)(4)}}{2(2)} \Rightarrow x = \dfrac{1 \pm \sqrt{-31}}{4}$

The discriminant in the Quadratic Formula is negative.

No real solution

59. $\begin{cases} y - e^{-x} = 1 \Rightarrow y = e^{-x} + 1 \\ y - \ln x = 3 \Rightarrow y = \ln x + 3 \end{cases}$

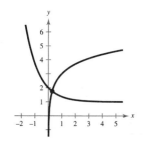

Point of intersection: approximately $(0.287, 1.751)$

61. $\begin{cases} y = x^4 - 2x^2 + 1 & \text{Equation 1} \\ y = 1 - x^2 & \text{Equation 2} \end{cases}$

Substitute for y in Equation 1: $1 - x^2 = x^4 - 2x^2 + 1$

Solve for x: $x^4 - x^2 = 0 \Rightarrow x^2(x^2 - 1) = 0$

$\Rightarrow x = 0, \pm 1$

Back-substitute $x = 0$: $y = 1 - x^2 = 1 - 0^2 = 1$

Back-substitute $x = 1$: $y = 1 - x^2 = 1 - 1^2 = 0$

Back-substitute $x = -1$: $y = 1 - x^2 = 1 - (-1)^2 = 0$

Solutions: $(0, 1), (1, 0), (-1, 0)$

63. $\begin{cases} xy - 1 = 0 & \text{Equation 1} \\ 2x - 4y + 7 = 0 & \text{Equation 2} \end{cases}$

Solve for y in Equation 1: $y = \dfrac{1}{x}$

Substitute for y in Equation 2: $2x - 4\left(\dfrac{1}{x}\right) + 7 = 0$

Solve for x: $2x^2 - 4 + 7x = 0 \Rightarrow (2x - 1)(x + 4) = 0 \Rightarrow x = \dfrac{1}{2}, -4$

Back-substitute $x = \dfrac{1}{2}$: $y = \dfrac{1}{1/2} = 2$

Back-substitute $x = -4$: $y = \dfrac{1}{-4} = -\dfrac{1}{4}$

Solutions: $\left(\dfrac{1}{2}, 2\right), \left(-4, -\dfrac{1}{4}\right)$

65. $C = 8650x + 250{,}000,\ R = 9950x$

$$R = C$$
$$9950x = 8650x + 250{,}000$$
$$1300x = 250{,}000$$
$$x \approx 192 \text{ units}$$

67. $C = 45.25x + 25{,}000,\ R = 69.95x$

(a)
$$R = C$$
$$69.95x = 45.25x + 25{,}000$$
$$24.70x = 25{,}000$$
$$x \approx 1013$$

About 1013 units must be sold to break even.

(b)
$$P = R - C$$
$$100{,}000 = 69.95x - (45.25x + 25{,}000)$$
$$100{,}000 = 69.95x - 45.25x - 25{,}000$$
$$125{,}000 = 24.70x$$
$$x \approx 5061$$

About 5061 units must be sold to earn a $100,000 profit.

69. $\begin{cases} R = 360 - 24x & \text{Equation 1} \\ R = 24 + 18x & \text{Equation 2} \end{cases}$

(a) Substitute for R in Equation 2: $360 - 24x = 24 + 18x$

Solve for x: $336 = 42x \Rightarrow x = 8$ weeks

(b)

Weeks, x	1	2	3	4	5	6	7	8	9	10
$R = 360 - 24x$	336	312	288	264	240	216	192	168	144	120
$R = 24 + 18x$	42	60	78	96	114	132	150	168	186	204

The rentals are equal when $x = 8$ weeks.

71. $\begin{cases} S = 0.06x & \text{Straight comission} \\ S = 500 + 0.03x & \text{Salary plus comission} \end{cases}$

$$0.06x > 500 + 0.03x$$
$$0.03x > 500$$
$$x > \$16{,}666.67$$

To make straight commission the better offer, you would have to sell more than $16,666.67 per week.

73. (a) $\begin{cases} x + y = 25,000 \\ 0.06x + 0.085y = 2000 \end{cases}$

(b) $y_1 = 25,000 - x$

$y_2 = \dfrac{2000 - 0.06x}{0.085}$

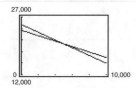

As the amount at 6% increases, the amount at 8.5% decreases. The amount of interest is fixed at $2000.

(c) The point of intersection occurs when $x = 5000$, so the most that can be invested at 6% and still earn $2000 per year in interest is $5000.

75. (a) Solar: $C \approx 0.0598t^3 - 1.719t^2 + 14.66t + 32.2$

Wind: $C \approx 3.237t^2 - 51.97t + 247.9$

(b) and (c)

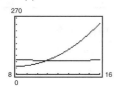

Point of intersection: $(10.9, 65.26)$

During the year 2000, the consumption of solar energy was equal to the consumption of wind energy (in trillions of Btus).

(d) Sample answer: No. As time increases, each model tends to infinity.

(e) Answers will vary.

77. (a) Public: $T_1 \approx 26.560t^2 + 85.54t + 2468.5$

Private: $T_2 \approx 794.14t + 14,124.6$

(b)

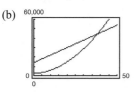

(c) From the graph, it appears that the tuition for public universities will exceed the tuition for private universities when $t > 38$, or sometime during the year 2038.

(d) $T_1 > T_2$

$26.560t^2 + 85.54t + 2468.5 > 794.14t + 14.125$

$26.560t^2 - 708.6t - 11,656.5 > 0$

Key Numbers: $t \approx 38.18, t \approx -11.50$

When $t > 38.18$, or during the year 2038, tuition for public universities will exceed tuition for private universities.

79. $2l + 2w = 280 \quad \Rightarrow \quad l + w = 140$

$w = l - 20 \Rightarrow l + (l - 20) = 140$

$2l = 160$

$l = 80$

$w = l - 20 = 80 - 20 = 60$

Dimensions: 60 centimeters $\times$ 80 centimeters

81. $2l + 2w = 484 \quad \Rightarrow \quad l + w = 242$

$l = 4.5w \Rightarrow (4.5w) + w = 242$

$5.5w = 242$

$w = 44$ feet

$l = 4.5(44) = 198$ feet

Dimensions: 44 feet $\times$ 198 feet

83. $44 = 2l + 2w$

$22 = l + w \Rightarrow l = 22 - w$

$A = lw$

$120 = lw$

$120 = (22 - w)w$

$120 = 22w - w^2$

$w^2 - 22w + 120 = 0$

$(w - 10)(w - 12) = 0$

$w = 10, \quad w = 12$

When $w = 10, l = 22 - 10 = 12$.

When $w = 12, l = 22 - 12 = 10$.

Dimensions: 10 kilometers $\times$ 12 kilometers

85. False. To solve a system of equations by substitution, you can solve for either variable in one of the two equations and then back-substitute.

87.

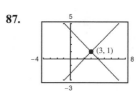

Point of intersection: $(3, 1)$

The point of intersection is equal to the solution found in Example 1.

89. For a linear system, the result will be a contradictory equation such as $0 = N$, where N is a nonzero real number. For a nonlinear system, there may be an equation with imaginary solutions.

91. $y = x^2$

    (a) Line with two points of intersection   (b) Line with one point of intersection   (c) Line with no points of intersection

        $y = 2x$ $y = 0$ $y = x - 2$

        $(0, 0)$ and $(2, 4)$ $(0, 0)$

Section 9.2   Two-Variable Linear Systems

■ You should be able to solve a linear system by the method of elimination.

1. Obtain coefficients for either x or y that differ only in sign. This is done by multiplying all the terms of one or both equations by appropriate constants.

2. Add the equations to eliminate one of the variables and then solve for the remaining variable.

3. Use back-substitution into either original equation and solve for the other variable.

4. Check your answer.

■ You should know that for a system of two linear equations, one of the following is true.

1. There are infinitely many solutions; the lines are identical. The system is consistent. The slopes are equal.

2. There is no solution; the lines are parallel. The system is inconsistent. The slopes are equal.

3. There is one solution; the lines intersect at one point. The system is consistent. The slopes are not equal.

---

1. elimination

3. consistent; inconsistent

5. $\begin{cases} 2x + y = 5 & \text{Equation 1} \\ x - y = 1 & \text{Equation 2} \end{cases}$

Add to eliminate y: $2x + y = 5$

$$\underline{\begin{array}{r} x - y = 1 \\ 3x \quad\quad = 6 \end{array}} \Rightarrow x = 2$$

Substitute $x = 2$ in Equation 2: $2 - y = 1 \Rightarrow y = 1$

Solution: $(2, 1)$

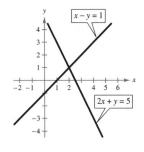

7. $\begin{cases} x + y = 0 & \text{Equation 1} \\ 3x + 2y = 1 & \text{Equation 2} \end{cases}$

Multiply Equation 1 by -2: $-2x - 2y = 0$

Add this to Equation 2 to eliminate y:

$$\begin{array}{r} -2x - 2y = 0 \\ \underline{3x + 2y = 1} \\ x \quad\quad = 1 \end{array}$$

Substitute $x = 1$ in Equation 1: $1 + y = 0 \Rightarrow y = -1$

Solution: $(1, -1)$

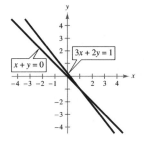

9. $\begin{cases} x - y = 2 & \text{Equation 1} \\ -2x + 2y = 5 & \text{Equation 2} \end{cases}$

Multiply Equation 1 by 2: $2x - 2y = 4$

Add this to Equation 2: $\begin{array}{r} 2x - 2y = 4 \\ \underline{-2x + 2y = 5} \\ 0 = 9 \end{array}$

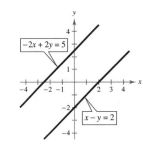

There are no solutions.

11. $\begin{cases} 3x - 2y = 5 & \text{Equation 1} \\ -6x + 4y = -10 & \text{Equation 2} \end{cases}$

Multiply Equation 1 by 2: $6x - 4y = 10$

Add this to Equation 2: $\begin{aligned} 6x - 4y &= 10 \\ -6x + 4y &= -10 \\ \hline 0 &= 0 \end{aligned}$

The equations are dependent. There are infinitely many solutions.

Let $x = a$, then $y = \dfrac{3a - 5}{2} = \dfrac{3}{2}a - \dfrac{5}{2}$.

Solution: $\left(a, \dfrac{3}{2}a - \dfrac{5}{2}\right)$, where a is any real number.

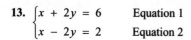

13. $\begin{cases} x + 2y = 6 & \text{Equation 1} \\ x - 2y = 2 & \text{Equation 2} \end{cases}$

Add the equations to eliminate y: $\begin{aligned} x + 2y &= 6 \\ x - 2y &= 2 \\ \hline 2x &= 8 \Rightarrow x = 4 \end{aligned}$

Substitute $x = 4$ into Equation 1: $4 + 2y = 6 \Rightarrow y = 1$

Solution: $(4, 1)$

15. $\begin{cases} 5x + 3y = 6 & \text{Equation 1} \\ 3x - y = 5 & \text{Equation 2} \end{cases}$

Multiply Equation 2 by 3: $9x - 3y = 15$

Add this to Equation 1 to eliminate y: $\begin{aligned} 5x + 3y &= 6 \\ 9x - 3y &= 15 \\ \hline 14x &= 21 \Rightarrow x = \dfrac{3}{2} \end{aligned}$

Substitute $x = \dfrac{3}{2}$ into Equation 1: $5\left(\dfrac{3}{2}\right) + 3y = 6 \Rightarrow y = -\dfrac{1}{2}$

Solution: $\left(\dfrac{3}{2}, -\dfrac{1}{2}\right)$

17. $\begin{cases} 3x + 2y = 10 & \text{Equation 1} \\ 2x + 5y = 3 & \text{Equation 2} \end{cases}$

Multiply Equation 1 by 2 and Equation 2 by -3:

$\begin{cases} 6x + 4y = 20 \\ -6x - 15y = -9 \end{cases}$

Add to eliminate x: $-11y = 11 \Rightarrow y = -1$

Substitute $y = -1$ in Equation 1:

$3x - 2 = 10 \Rightarrow x = 4$

Solution: $(4, -1)$

19. $\begin{cases} 5u + 6v = 24 & \text{Equation 1} \\ 3u + 5v = 18 & \text{Equation 2} \end{cases}$

Multiply Equation 1 by 5 and Equation 2 by -6:

$\begin{cases} 25u + 30v = 120 \\ -18u - 30v = -108 \end{cases}$

Add to eliminate v: $7u = 12 \Rightarrow u = \dfrac{12}{7}$

Substitute $u = \dfrac{12}{7}$ in Equation 1:

$5\left(\dfrac{12}{7}\right) + 6v = 24 \Rightarrow 6v = \dfrac{108}{7} \Rightarrow v = \dfrac{18}{7}$

Solution: $\left(\dfrac{12}{7}, \dfrac{18}{7}\right)$

21. $\begin{cases} \frac{9}{5}x + \frac{6}{5}y = 4 & \text{Equation 1} \\ 9x + 6y = 3 & \text{Equation 2} \end{cases}$

Multiply Equation 1 by 10 and Equation 2 by -2:

$\begin{cases} 18x + 12y = 40 \\ -18x - 12y = -6 \end{cases}$

Add these two together: $0 = 34$

No solution

23. $\begin{cases} -5x + 6y = -3 & \text{Equation 1} \\ 20x - 24y = 12 & \text{Equation 2} \end{cases}$

Multiply Equation 1 by 4:

$\begin{cases} -20x + 24y = -12 \\ 20x - 24y = 12 \end{cases}$

Add these two together: $0 = 0$

The equations are dependent. There are infinitely many solutions.

Let $x = a$, then

$-5a + 6y = -3 \Rightarrow y = \dfrac{5a - 3}{6} = \dfrac{5}{6}a - \dfrac{1}{2}.$

Solution: $\left(a, \dfrac{5}{6}a - \dfrac{1}{2}\right)$, where a is any real number

25. $\begin{cases} 0.2x - 0.5y = -27.8 & \text{Equation 1} \\ 0.3x + 0.4y = 68.7 & \text{Equation 2} \end{cases}$

Multiply Equation 1 by 4 and Equation 2 by 5:

$\begin{cases} 0.8x - 2y = -111.2 \\ 1.5x + 2y = 343.5 \end{cases}$

Add these to eliminate y: $\begin{aligned} 0.8x - 2y &= -111.2 \\ 1.5x + 2y &= 343.5 \\ \hline 2.3x &= 232.3 \\ x &= 101 \end{aligned}$

Substitute $x = 101$ in Equation 1:

$0.2(101) - 0.5y = -27.8 \Rightarrow y = 96$

Solution: $(101, 96)$

27. $\begin{cases} 4b + 3m = 3 & \text{Equation 1} \\ 3b + 11m = 13 & \text{Equation 2} \end{cases}$

Multiply Equation 1 by 3 and Equation 2 by -4:

$\begin{cases} 12b + 9m = 9 \\ -12b - 44m = -52 \end{cases}$

Add to eliminate b: $-35m = -43 \Rightarrow m = \frac{43}{35}$

Substitute $m = \frac{43}{35}$ in Equation 1:

$4b + 3\left(\frac{43}{35}\right) = 3 \Rightarrow b = -\frac{6}{35}$

Solution: $\left(-\frac{6}{35}, \frac{43}{35}\right)$

29. $\begin{cases} \dfrac{x + 3}{4} + \dfrac{y - 1}{3} = 1 & \text{Equation 1} \\ 2x - y = 12 & \text{Equation 2} \end{cases}$

Multiply Equation 1 by 12 and Equation 2 by 4:

$\begin{cases} 3x + 4y = 7 \\ 8x - 4y = 48 \end{cases}$

Add to eliminate y: $11x = 55 \Rightarrow x = 5$

Substitute $x = 5$ into Equation 2:

$2(5) - y = 12 \Rightarrow y = -2$

Solution: $(5, -2)$

31. $\begin{cases} 2x - 5y = 0 \\ x - y = 3 \end{cases}$

Multiply Equation 2 by -5:

$\begin{cases} 2x - 5y = 0 \\ -5x + 5y = -15 \end{cases}$

Add to eliminate y: $-3x = -15 \Rightarrow x = 5$

Matches graph (b).

Number of solutions: One

Consistent

33. $\begin{cases} -7x + 6y = -4 \\ 14x - 12y = 8 \end{cases}$

Multiply Equation 1 by 2:

$\begin{cases} -14x + 12y = -8 \\ 14x - 12y = 8 \end{cases}$

Add this to Equation 2: $0 = 0$

The original equations are dependent.

Matches graph (a).

Number of solutions: Infinite

Consistent

35. $\begin{cases} 3x - 5y = 7 & \text{Equation 1} \\ 2x + y = 9 & \text{Equation 2} \end{cases}$

Multiply Equation 2 by 5:

$10x + 5y = 45$

Add this to Equation 1:

$13x = 52 \Rightarrow x = 4$

Back-substitute $x = 4$ into Equation 2:

$2(4) + y = 9 \Rightarrow y = 1$

Solution: $(4, 1)$

37. $\begin{cases} y = 2x - 5 & \text{Equation 1} \\ y = 5x - 11 & \text{Equation 2} \end{cases}$

Because both equations are solved for y, set them equal to one another and solve for x.

$$2x - 5 = 5x - 11$$
$$6 = 3x$$
$$2 = x$$

Back-substitute $x = 2$ into Equation 1:

$$y = 2(2) - 5 = -1$$

Solution: $(2, -1)$

39. $\begin{cases} x - 5y = 21 & \text{Equation 1} \\ 6x + 5y = 21 & \text{Equation 2} \end{cases}$

Add the equations: $7x = 42 \Rightarrow x = 6$

Back-substitute $x = 6$ into Equation 1:

$$6 - 5y = 21 \Rightarrow -5y = 15 \Rightarrow y = -3$$

Solution: $(6, -3)$

41. $\begin{cases} -5x + 9y = 13 & \text{Equation 1} \\ y = x - 4 & \text{Equation 2} \end{cases}$

Substitute the expression for y from Equation 2 into Equation 1.

$$-5x + 9(x - 4) = 13$$
$$-5x + 9x - 36 = 13$$
$$4x = 49$$
$$x = \frac{49}{4}$$

Back-substitute $x = \frac{49}{4}$ into Equation 2:

$$y = \frac{49}{4} - 4 = \frac{33}{4}$$

Solution: $\left(\frac{49}{4}, \frac{33}{4}\right)$

43. Let $r_1 = $ the air speed of the plane and $r_2 = $ the wind air speed.

$$3.6(r_1 - r_2) = 1800 \quad \text{Equation 1} \Rightarrow \quad r_1 - r_2 = 500$$
$$3(r_1 + r_2) = 1800 \quad \text{Equation 2} \Rightarrow \quad \underline{r_1 + r_2 = 600}$$
$$2r_1 \quad = 1100 \quad \text{Add the equations.}$$
$$r_1 \quad = 550$$
$$550 + r_2 = 600$$
$$r_2 = 50$$

The air speed of the plane is 550 miles per hour and the speed of the wind is 50 miles per hour.

45.
$$500 - 0.4x = 380 + 0.1x$$
$$120 = 0.5x$$
$$x = 240 \text{ units}$$
$$p = \$404$$

Equilibrium point: $(240, 404)$

47.
$$140 - 0.00002x = 80 + 0.00001x$$
$$60 = 0.00003x$$
$$x = 2{,}000{,}000 \text{ units}$$
$$p = \$100.00$$

Equilibrium point: $(2{,}000{,}000, 100)$

49. Let $x = $ number of calories in a cheeseburger. Let $y = $ number of calories in a small order of french fries.

$$\begin{cases} 2x + y = 830 & \text{Equation 1} \\ 3x + 2y = 1360 & \text{Equation 2} \end{cases}$$

Multiply Equation 1 by -2: $-4x - 2y = -1660$

Add this to Equation 2 to eliminate y:

$$-4x - 2y = -1660$$
$$\underline{3x + 2y = 1360}$$
$$-x = -300$$
$$x = 300 \text{ calories}$$

Back-substitute $x = 300$ into Equation 2:

$$3(300) + 2y = 1360$$
$$2y = 460$$
$$y = 230 \text{ calories}$$

The cheeseburger contains 300 calories and the fries contain 230 calories.

51. (a) Let x = the number of liters at 25%.

Let y = the number of liters at 50%.

$$\begin{cases} 0.25x + 0.50y = 12 \\ x + y = 30 \end{cases}$$

(b)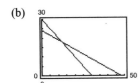

As the amount of 25% solution increases, the amount of 50% solution decreases.

(c) $$\begin{cases} 0.25x + 0.50y = 12 & \text{Equation 1} \\ x + y = 30 & \text{Equation 2} \end{cases}$$

Solve Equation 2 for y: $y = 30 - x$

Substitute this into Equation 1 to eliminate y:

$$0.25x + 0.50(30 - x) = 12$$

$$0.25x + 15 - 0.50x = 12$$

$$-0.25x = -3$$

$$x = 12 \text{ liters}$$

Back-substitute $x = 12$ into Equation 2:

$$12 + y = 30 \Rightarrow y = 18 \text{ liters}$$

The final mixture should contain 12 liters of the 25% solution and 18 liters of the 50% solution.

53. Let x = amount of money invested at 3.5%.

Let y = amount of money invested at 5%.

$$\begin{cases} x + y = 24{,}000 & \text{Equation 1} \\ 0.035x + 0.05y = 930 & \text{Equation 2} \end{cases}$$

Solve Equation 1 for x: $x = 24{,}000 - y$

Substitute this into Equation 2 to eliminate x:

$$0.035(24{,}000 - y) + 0.05y = 930$$

$$840 + 0.015y = 930$$

$$y = \$6000$$

Back-substitute $y = 6000$ into Equation 1:

$$x + 6000 = 24{,}000$$

$$x = \$18{,}000$$

$18,000 should be invested in the 3.5% bond.

55. (a)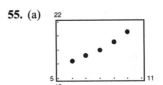

Pharmacy A: $P = 0.52t + 16.0$

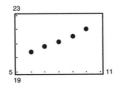

Pharmacy B: $P = 0.39t + 18.0$

(b) Pharmacy A $\overset{?}{>}$ Pharmacy B

$$0.52t + 16.0 > 0.39t + 18.0$$

$$0.13t > 2.0$$

$$t > 15.4$$

Yes. During the year 2015, the number of prescriptions filled at Pharmacy A will exceed the number of prescriptions filled at Pharmacy B.

57. $$\begin{cases} 5b + 10a = 20.2 \Rightarrow -10b - 20a = -40.4 \\ 10b + 30a = 50.1 \Rightarrow \underline{10b + 30a = 50.1} \end{cases}$$

$$\begin{aligned} 10a &= 9.7 \\ a &= 0.97 \\ b &= 2.1 \end{aligned}$$

Least squares regression line: $y = 0.97x + 2.1$

59. $(0, 8), (1, 6), (2, 4), (3, 2)$

$$4b + \left(\sum_{i=1}^{4} x_i\right) a = \sum_{i=1}^{4} y_i$$

$$4b + (0 + 1 + 2 + 3)a = (8 + 6 + 4 + 2)$$

$$4b + 6a = 20$$

$$2b + 3a = 10$$

$$\left(\sum_{i=1}^{4} x_i\right) b + \left(\sum_{i=1}^{4} x_i^2\right) a = \left(\sum_{i=1}^{4} x_i y_i\right)$$

$$(0 + 1 + 2 + 3) b + (0^2 + 1^2 + 2^2 + 3^2) a = (0 \cdot 8 + 1 \cdot 6 + 2 \cdot 4 + 3 \cdot 2)$$

$$6b + 14a = 20$$

$$3b + 7a = 10$$

$$\begin{cases} 2b + 3a = 10 & \text{Equation 1} \\ 3b + 7a = 10 & \text{Equation 2} \end{cases}$$

Multiply Equation 1 by -3: $-6b - 9a = -30$

Multiply Equation 2 by 2: $6b + 14a = 20$

Add these together to eliminate b:
$$\begin{aligned} -6b - 9a &= -30 \\ 6b + 14a &= 20 \\ \hline 5a &= -10 \\ a &= -2 \end{aligned}$$

Back-substitute $a = -2$ into Equation 1:

$$2b + 3(-2) = 10$$
$$2b = 16$$
$$b = 8$$

Least squares regression line: $y = -2x + 8$

61. (a) $(1.0, 32), (1.5, 41), (2.0, 48), (2.5, 53)$

$$\begin{cases} 4b + 7a = 174 \Rightarrow -7b - 12.25a = -304.5 \\ 7b + 13.5a = 322 \Rightarrow 7b + 13.5a = 322 \end{cases}$$
$$\begin{aligned} \hline 1.25a &= 17.5 \\ a &= 14 \\ 4b + 98 &= 174 \\ b &= 19 \end{aligned}$$

Least squares regression line: $y = 14x + 19$

(b) When $x = 1.6$: $y = 14(1.6) + 19 = 41.4$ bushels per acre.

63. False. Two lines that coincide have infinitely many points of intersection.

65. No, it is not possible for a consistent system of linear equations to have exactly two solutions. Either the lines will intersect once or they will coincide and then the system would have infinite solutions.

67. The method of elimination is much easier.

69. $\begin{cases} 100y - x = 200 & \text{Equation 1} \\ 99y - x = -198 & \text{Equation 2} \end{cases}$

Subtract Equation 2 from Equation 1 to eliminate x:

$$\begin{aligned} 100y - x &= 200 \\ -99y + x &= 198 \\ \hline y &= 398 \end{aligned}$$

Substitute $y = 398$ into Equation 1:

$$100(398) - x = 200 \Rightarrow x = 39{,}600$$

Solution: $(39{,}600, 398)$

The lines are not parallel. The scale on the axes must be changed to see the point of intersection.

71. $\begin{cases} 4x - 8y = -3 & \text{Equation 1} \\ 2x + ky = 16 & \text{Equation 2} \end{cases}$

Multiply Equation 2 by -2: $-4x - 2ky = -32$

Add this to Equation 1:

$$\begin{aligned} 4x - 8y &= -3 \\ -4x - 2ky &= -32 \\ \hline -8y - 2ky &= -35 \end{aligned}$$

The system is inconsistent if $-8y - 2ky = 0$.

This occurs when $k = -4$.

Section 9.3 Multivariable Linear Systems

- You should know the operations that lead to equivalent systems of equations:

  (a) Interchange any two equations.

  (b) Multiply all terms of an equation by a nonzero constant.

  (c) Replace an equation by the sum of itself and a constant multiple of any other equation in the system.

- You should be able to use the method of Gaussian elimination with back-substitution.

---

1. row-echelon

3. Gaussian

5. nonsquare

7. $\begin{cases} 6x - y + z = -1 \\ 4x \quad\;\; - 3z = -19 \\ \quad\;\; 2y + 5z = 25 \end{cases}$

(a) $(2, 0, -2)$

$$4(2) - 3(-2) \neq -19$$

$(2, 0, -2)$ *is not* a solution.

(b) $(-3, 0, 5)$

$$6(-3) - 0 + 5 \neq -1$$

$(-3, 0, 5)$ *is not* a solution.

(c) $(0, -1, 4)$

$$4(0) - 3(4) \neq -19$$

$(0, -1, 4)$ *is not* a solution.

(d) $(-1, 0, 5)$

$$6(-1) - 0 + 5 = -1$$
$$4(-1) - 3(5) = -19$$
$$2(0) + 5(5) = 25$$

$(-1, 0, 5)$ *is a solution.*

9. $\begin{cases} 4x + y - z = 0 \\ -8x - 6y + z = -\frac{7}{4} \\ 3x - y \quad\;\; = -\frac{9}{4} \end{cases}$

(a) $4\left(\frac{1}{2}\right) + \left(-\frac{3}{4}\right) - \left(-\frac{7}{4}\right) \neq 0$

$\left(\frac{1}{2}, -\frac{3}{4}, -\frac{7}{4}\right)$ *is not* a solution.

(b) $4\left(-\frac{3}{2}\right) + \left(\frac{5}{4}\right) - \left(-\frac{5}{4}\right) \neq 0$

$\left(-\frac{3}{2}, \frac{5}{4}, -\frac{5}{4}\right)$ *is not* a solution.

(c) $4\left(-\frac{1}{2}\right) + \left(\frac{3}{4}\right) - \left(-\frac{5}{4}\right) = 0$

$-8\left(-\frac{1}{2}\right) - 6\left(\frac{3}{4}\right) + \left(-\frac{5}{4}\right) = -\frac{7}{4}$

$3\left(-\frac{1}{2}\right) - \left(\frac{3}{4}\right) \quad\;\; = -\frac{9}{4}$

$\left(-\frac{1}{2}, \frac{3}{4}, -\frac{5}{4}\right)$ *is a solution.*

(d) $4\left(-\frac{1}{2}\right) + \left(\frac{1}{6}\right) - \left(-\frac{3}{4}\right) \neq 0$

$\left(-\frac{1}{2}, \frac{1}{6}, -\frac{3}{4}\right)$ *is not* a solution.

11. $\begin{cases} 2x - y + 5z = 24 & \text{Equation 1} \\ \quad\; y + 2z = 6 & \text{Equation 2} \\ \qquad\quad z = 8 & \text{Equation 3} \end{cases}$

Back-substitute $z = 8$ into Equation 2:

$$y + 2(8) = 6$$
$$y = -10$$

Back-substitute $y = -10$ and $z = 8$ into Equation 1:

$$2x + 10 + 5(8) = 24$$
$$2x = -26$$
$$x = -13$$

Solution: $(-13, -10, 8)$

13. $\begin{cases} 2x + y - 3z = 10 & \text{Equation 1} \\ \quad\; y + z = 12 & \text{Equation 2} \\ \qquad\quad z = 2 & \text{Equation 3} \end{cases}$

Back-substitute $z = 2$ into Equation 2:
$$y + 2 = 12 \Rightarrow y = 10$$

Back-substitute $y = 10$ and $z = 2$ into Equation 1:

$$2x + 10 - 3(2) = 10$$
$$2x + 4 = 10$$
$$2x = 6$$
$$x = 3$$

Solution: $(3, 10, 2)$

15. $\begin{cases} 4x - 2y + z = 8 & \text{Equation 1} \\ \quad -y + z = 4 & \text{Equation 2} \\ \qquad\quad z = 11 & \text{Equation 3} \end{cases}$

Back-substitute $z = 11$ into Equation 2:

$$-y + 11 = 4$$
$$y = 7$$

Back-substitute $y = 7$ and $z = 11$ into Equation 1:

$$4x - 2(7) + 11 = 8$$
$$4x = 11$$
$$x = \frac{11}{4}$$

Solution: $\left(\frac{11}{4}, 7, 11\right)$

17. $\begin{cases} x - 2y + 3z = 5 & \text{Equation 1} \\ -x + 3y - 5z = 4 & \text{Equation 2} \\ 2x \qquad - 3z = 0 & \text{Equation 3} \end{cases}$

Add Equation 1 to Equation 2:

$$\begin{cases} x - 2y + 3z = 5 \\ \quad\; y - 2z = 9 \\ 2x \qquad - 3z = 0 \end{cases}$$

This is the first step in putting the system in row-echelon form.

19. $\begin{cases} x + y + z = 7 & \text{Equation 1} \\ 2x - y + z = 9 & \text{Equation 2} \\ 3x \quad - z = 10 & \text{Equation 3} \end{cases}$

$\begin{cases} x + y + z = 7 \\ 3x \quad + 2z = 16 & \text{Eq.2 + Eq.1} \\ 3x \quad - z = 10 \end{cases}$

$\begin{cases} x + y + z = 7 \\ 3x \quad + 2z = 16 \\ 9x \qquad\quad = 36 & \text{Eq.2 + 2Eq.3} \end{cases}$

$\begin{cases} x + y + z = 7 \\ 3x \quad + 2z = 16 \\ x \qquad\quad = 4 & \frac{1}{4}\,\text{Eq.3} \end{cases}$

$$3(4) + 2z = 16$$
$$2z = 4$$
$$z = 2$$
$$4 + y + 2 = 7$$
$$y = 1$$

Solution: $(4, 1, 2)$

21. $\begin{cases} 2x \qquad + 2z = 2 & \text{Equation 1} \\ 5x + 3y \qquad = 4 & \text{Equation 2} \\ \quad\; 3y - 4z = 4 & \text{Equation 3} \end{cases}$

$\begin{cases} x \qquad + z = 1 & \frac{1}{2}\,\text{Eq.1} \\ 5x + 3y \qquad = 4 \\ \quad\; 3y - 4z = 4 \end{cases}$

$\begin{cases} x \qquad + z = 1 \\ \quad\; 3y - 5z = -1 & -5\,\text{Eq.1 + Eq.2} \\ \quad\; 3y - 4z = 4 \end{cases}$

$\begin{cases} x \qquad + z = 1 \\ \quad\; 3y - 5z = -1 \\ \qquad\quad z = 5 & -\text{Eq.2 + Eq.3} \end{cases}$

$$3y - 5(5) = -1 \Rightarrow y = 8$$
$$x + 5 = 1 \Rightarrow x = -4$$

Solution: $(-4, 8, 5)$

23. $\begin{cases} 3x + 3y & = 9 \\ 2x & - 3z = 10 \\ 6y + 4z = -12 \end{cases}$ Interchange equations.

$\begin{cases} x + y & = 3 \\ 2x & - 3z = 10 \\ 6y + 4z = -12 \end{cases}$ $\frac{1}{3}$Eq.1

$\begin{cases} x + y & = 3 \\ -2y - 3z = 4 \\ 6y + 4z = -12 \end{cases}$ -2Eq.1 + Eq.2

$\begin{cases} x + y & = 3 \\ -2y - 3z = 4 \\ -5z = 0 \end{cases}$ 3Eq.2 + Eq.3

$\begin{cases} x + y & = 3 \\ -2y - 3z = 4 \\ z = 0 \end{cases}$ $-\frac{1}{5}$Eq.3

$-2y - 3(0) = 4 \Rightarrow y = -2$
$x - 2 = 3 \Rightarrow x = 5$

Solution: $(5, -2, 0)$

25. $\begin{cases} x - 2y + 2z = -9 \\ 2x + y - z = 7 \\ 3x - y + z = 5 \end{cases}$ Interchange equations.

$\begin{cases} x - 2y + 2z = -9 \\ 5y - 5z = 25 \\ 5y - 5z = 32 \end{cases}$ -2Eq.1 + Eq.2
    -3Eq.1 + Eq.3

$\begin{cases} x - 2y + 2z = -9 \\ 5y - 5z = 25 \\ 0 = 7 \end{cases}$ $-$Eq.2 + Eq.3

Inconsistent, no solution

27. $\begin{cases} 3x - 5y + 5z = 1 \\ 5x - 2y + 3z = 0 \\ 7x - y + 3z = 0 \end{cases}$ Equation 1
    Equation 2
    Equation 3

$\begin{cases} 6x - 10y + 10z = 2 \\ 5x - 2y + 3z = 0 \\ 7x - y + 3z = 0 \end{cases}$ 2Eq.1

$\begin{cases} x - 8y + 7z = 2 \\ 5x - 2y + 3z = 0 \\ 7x - y + 3z = 0 \end{cases}$ $-$Eq.2 + Eq.1

$\begin{cases} x - 8y + 7z = 2 \\ 38y - 32z = -10 \\ 55y - 46z = -14 \end{cases}$ -5Eq.1 + Eq.2
    -7Eq.1 + Eq.3

$\begin{cases} x - 8y + 7z = 2 \\ 2090y - 1760z = -550 \\ -2090y + 1748z = 532 \end{cases}$ 55Eq.2
    -38Eq.3

$\begin{cases} x - 8y + 7z = 2 \\ 2090y - 1760z = -550 \\ -12z = -18 \end{cases}$ Eq.2 + Eq.3

$-12z = -18 \Rightarrow z = \frac{3}{2}$

$38y - 32\left(\frac{3}{2}\right) = -10 \Rightarrow y = 1$

$x - 8(1) + 7\left(\frac{3}{2}\right) = 2 \Rightarrow x = -\frac{1}{2}$

Solution: $\left(-\frac{1}{2}, 1, \frac{3}{2}\right)$

29. $\begin{cases} x + 2y - 7z = -4 \\ 2x + y + z = 13 \\ 3x + 9y - 36z = -33 \end{cases}$ Equation 1
    Equation 2
    Equation 3

$\begin{cases} x + 2y - 7z = -4 \\ -3y + 15z = 21 \\ 3y - 15z = -21 \end{cases}$ -2Eq.1 + Eq.2
    -3Eq.1 + Eq.3

$\begin{cases} x + 2y - 7z = -4 \\ -3y + 15z = 21 \\ 0 = 0 \end{cases}$ Eq.2 + Eq.3

$\begin{cases} x + 2y - 7z = -4 \\ y - 5z = -7 \end{cases}$ $-\frac{1}{3}$Eq.2

$\begin{cases} x + 3z = 10 \\ y - 5z = -7 \end{cases}$ -2Eq.2 + Eq.1

Let $z = a$, then:

$y = 5a - 7$

$x = -3a + 10$

Solution: $(-3a + 10, 5a - 7, a)$

31. $\begin{cases} 3x - 3y + 6z = 6 & \text{Equation 1} \\ x + 2y - z = 5 & \text{Equation 2} \\ 5x - 8y + 13z = 7 & \text{Equation 3} \end{cases}$

$\begin{cases} x - y + 2z = 2 & \tfrac{1}{3}\text{Eq.1} \\ x + 2y - z = 5 \\ 5x - 8y + 13z = 7 \end{cases}$

$\begin{cases} x - y + 2z = 2 \\ 3y - 3z = 3 & -\text{Eq.1} + \text{Eq.2} \\ -3y + 3z = -3 & -5\text{Eq.1} + \text{Eq.3} \end{cases}$

$\begin{cases} x - y + 2z = 2 \\ y - z = 1 & \tfrac{1}{3}\text{Eq.2} \\ 0 = 0 & \text{Eq.2} + \text{Eq.3} \end{cases}$

$\begin{cases} x + z = 3 & \text{Eq.2} + \text{Eq.1} \\ y - z = 1 \end{cases}$

Let $z = a$, then:

$y = a + 1$

$x = -a + 3$

Solution: $\left(-a + 3, a + 1, a\right)$

33. $\begin{cases} x - 2y + 5z = 2 \\ 4x - z = 0 \end{cases}$

Let $z = a$, then: $x = \tfrac{1}{4}a$.

$\tfrac{1}{4}a - 2y + 5a = 2$

$a - 8y + 20a = 8$

$-8y = -21a + 8$

$y = \tfrac{21}{8}a - 1$

Answer: $\left(\tfrac{1}{4}a, \tfrac{21}{8}a - 1, a\right)$

To avoid fractions, we could go back and let $z = 8a$, then $4x - 8a = 0 \Rightarrow x = 2a$.

$2a - 2y + 5(8a) = 2$

$-2y + 42a = 2$

$y = 21a - 1$

Solution: $\left(2a, 21a - 1, 8a\right)$

35. $\begin{cases} 2x - 3y + z = -2 & \text{Equation 1} \\ -4x + 9y = 7 & \text{Equation 2} \end{cases}$

$\begin{cases} 2x - 3y + z = -2 \\ 3y + 2z = 3 & 2\text{Eq.1} + \text{Eq.2} \end{cases}$

$\begin{cases} 2x + 3z = 1 & \text{Eq.2} + \text{Eq.1} \\ 3y + 2z = 3 \end{cases}$

Let $z = a$, then:

$y = -\tfrac{2}{3}a + 1$

$x = -\tfrac{3}{2}a + \tfrac{1}{2}$

Solution: $\left(-\tfrac{3}{2}a + \tfrac{1}{2}, -\tfrac{2}{3}a + 1, a\right)$

37. $\begin{cases} x + 3w = 4 & \text{Equation 1} \\ 2y - z - w = 0 & \text{Equation 2} \\ 3y - 2w = 1 & \text{Equation 3} \\ 2x - y + 4z = 5 & \text{Equation 4} \end{cases}$

$\begin{cases} x + 3w = 4 \\ 2y - z - w = 0 \\ 3y - 2w = 1 \\ -y + 4z - 6w = -3 & -2\text{Eq.1} + \text{Eq.4} \end{cases}$

$\begin{cases} x + 3w = 4 \\ y - 4z + 6w = 3 & -\text{Eq.4 and interchange} \\ 2y - z - w = 0 & \text{the equations.} \\ 3y - 2w = 1 \end{cases}$

$\begin{cases} x + 3w = 4 \\ y - 4z + 6w = 3 \\ 7z - 13w = -6 & -\text{Eq.2} + \text{Eq.3} \\ 12z - 20w = -8 & -3\text{Eq.2} + \text{Eq.4} \end{cases}$

$\begin{cases} x + 3w = 4 \\ y - 4z + 6w = 3 \\ z - 3w = -2 & -\tfrac{1}{2}\text{Eq.4} + \text{Eq.3} \\ 12z - 20w = -8 \end{cases}$

$\begin{cases} x + 3w = 4 \\ y - 4z + 6w = 3 \\ z - 3w = -2 \\ 16w = 16 & -12\text{Eq.3} + \text{Eq.4} \end{cases}$

$16w = 16 \Rightarrow w = 1$

$z - 3(1) = -2 \Rightarrow z = 1$

$y - 4(1) + 6(1) = 3 \Rightarrow y = 1$

$x + 3(1) = 4 \Rightarrow x = 1$

Solution: $(1, 1, 1, 1)$

39. $\begin{cases} x \quad\quad + 4z = 1 & \text{Equation 1} \\ x + y + 10z = 10 & \text{Equation 2} \\ 2x - y + 2z = -5 & \text{Equation 3} \end{cases}$

$\begin{cases} x \quad\quad + 4z = 1 \\ \quad y + 6z = 9 & -\text{Eq.1} + \text{Eq.2} \\ \quad -y - 6z = -7 & -2\text{Eq.1} + \text{Eq.3} \end{cases}$

$\begin{cases} x \quad\quad + 4z = 1 \\ \quad y + 6z = 9 \\ \quad\quad\quad 0 = 2 & \text{Eq.2} + \text{Eq.3} \end{cases}$

No solution, inconsistent

41. $\begin{cases} 2x + 3y \quad\quad = 0 & \text{Equation 1} \\ 4x + 3y - z = 0 & \text{Equation 2} \\ 8x + 3y + 3z = 0 & \text{Equation 3} \end{cases}$

$\begin{cases} 2x + 3y \quad\quad = 0 \\ \quad -3y - z = 0 & -2\text{Eq.1} + \text{Eq.2} \\ \quad -9y + 3z = 0 & -4\text{Eq.1} + \text{Eq.3} \end{cases}$

$\begin{cases} 2x + 3y \quad\quad = 0 \\ \quad -3y - z = 0 \\ \quad\quad\quad 6z = 0 & -3\text{Eq.2} + \text{Eq.3} \end{cases}$

$6z = 0 \Rightarrow z = 0$

$-3y - 0 = 0 \Rightarrow y = 0$

$2x + 3(0) = 0 \Rightarrow x = 0$

Solution: $(0, 0, 0)$

43. $\begin{cases} 12x + 5y + z = 0 & \text{Equation 1} \\ 23x + 4y - z = 0 & \text{Equation 2} \end{cases}$

$\begin{cases} 24x + 10y + 2z = 0 & 2\text{Eq. 1} \\ 23x + 4y - z = 0 \end{cases}$

$\begin{cases} x + 6y + 3z = 0 & -\text{Eq.2} + \text{Eq.1} \\ 23x + 4y - z = 0 \end{cases}$

$\begin{cases} x + 6y + 3z = 0 \\ \quad -134y - 70z = 0 & -23\text{Eq.1} + \text{Eq.2} \end{cases}$

$\begin{cases} x + 6y + 3z = 0 \\ \quad -67y - 35z = 0 & \frac{1}{2}\text{Eq.2} \end{cases}$

To avoid fractions, let $z = 67a$, then:

$-67y - 35(67a) = 0$

$y = -35a$

$x + 6(-35a) + 3(67a) = 0$

$x = 9a$

Solution: $(9a, -35a, 67a)$

45. $s = \frac{1}{2}at^2 + v_0t + s_0$

$(1, 128), (2, 80), (3, 0)$

$128 = \frac{1}{2}a + v_0 + s_0 \Rightarrow a + 2v_0 + 2s_0 = 256$

$80 = 2a + 2v_0 + s_0 \Rightarrow 2a + 2v_0 + s_0 = 80$

$0 = \frac{9}{2}a + 3v_0 + s_0 \Rightarrow 9a + 6v_0 + 2s_0 = 0$

Solving this system yields $a = -32, v_0 = 0, s_0 = 144$.

So, $s = \frac{1}{2}(-32)t^2 + (0)t + 144 = -16t^2 + 144$.

47. $s = \frac{1}{2}at^2 + v_0t + s_0$

$(1, 352), (2, 272), (3, 160)$

$352 = \frac{1}{2}a(1)^2 + v_0(1) + s_0 \Rightarrow a + 2v_0 + 2s_0 = 704$

$272 = \frac{1}{2}a(2)^2 + v_0(2) + s_0 \Rightarrow 2a + 2v_0 + s_0 = 272$

$160 = \frac{1}{2}a(3)^2 + v_0(3) + s_0 \Rightarrow 9a + 6v_0 + 2s_0 = 320$

Solving this system yields $a = -32, v_0 = -32,$ and $s_0 = 400$.

So, $s = \frac{1}{2}(-32)t^2 - 32t + 400 = -16t^2 - 32t + 400$.

49. $y = ax^2 + bx + c$ passing through $(0, 0), (2, -2), (4, 0)$

$(0, 0): 0 = c$

$(2, -2): -2 = 4a + 2b + c \Rightarrow -1 = 2a + b$

$(4, 0): 0 = 16a + 4b + c \Rightarrow 0 = 4a + b$

Solution: $a = \frac{1}{2}, b = -2, c = 0$

The equation of the parabola is $y = \frac{1}{2}x^2 - 2x$.

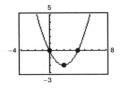

51. $y = ax^2 + bx + c$ passing through $(2, 0), (3, -1), (4, 0)$

$(2, 0)$: $0 = 4a + 2b + c$

$(3, -1)$: $-1 = 9a + 3b + c$

$(4, 0)$: $0 = 16a + 4b + c$

$$\begin{cases} 0 = 4a + 2b + c \\ -1 = 5a + b \qquad\qquad -\text{Eq.1} + \text{Eq.2} \\ 0 = 12a + 2b \qquad\quad -\text{Eq.1} + \text{Eq.3} \end{cases}$$

$$\begin{cases} 0 = 4a + 2b + c \\ -1 = 5a + b \\ 2 = 2a \qquad\qquad\quad -2\text{Eq.2} + \text{Eq.3} \end{cases}$$

Solution: $a = 1, b = -6, c = 8$

The equation of the parabola is $y = x^2 - 6x + 8$.

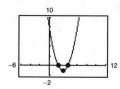

53. $y = ax^2 + bx + c$ passing through $\left(\frac{1}{2}, 1\right), (1, 3), (2, 13)$

$\left(\frac{1}{2}, 1\right)$: $1 = a\left(\frac{1}{2}\right)^2 + b\left(\frac{1}{2}\right) + c$

$(1, 3)$: $3 = a(1)^2 + b(1) + c$

$(2, 13)$: $13 = a(2)^2 + b(2) + c$

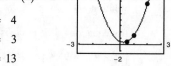

$$\begin{cases} a + 2b + 4c = 4 \\ a + b + c = 3 \\ 4a + 2b + c = 13 \end{cases}$$

Solution: $a = 4, b = -2, c = 1$

The equation of the parabola is $y = 4x^2 - 2x + 1$.

55. $x^2 + y^2 + Dx + Ey + F = 0$ passing through $(0, 0), (5, 5), (10, 0)$

$(0, 0)$: $0^2 + 0^2 + D(0) + E(0) + F = 0 \Rightarrow F = 0$

$(5, 5)$: $5^2 + 5^2 + D(5) + E(5) + F = 0 \Rightarrow 5D + 5E + F = -50$

$(10, 0)$: $10^2 + 0^2 + D(10) + E(0) + F = 0 \Rightarrow 10D + F = -100$

Solution: $D = -10, E = 0, F = 0$

The equation of the circle is $x^2 + y^2 - 10x = 0$. To graph, complete the square first, then solve for y.

$$\left(x^2 - 10x + 25\right) + y^2 = 25$$

$$\left(x - 5\right)^2 + y^2 = 25$$

$$y^2 = 25 - \left(x - 5\right)^2$$

$$y = \pm\sqrt{25 - \left(x - 5\right)^2}$$

Let $y_1 = \sqrt{25 - \left(x - 5\right)^2}$ and $y_2 = -\sqrt{25 - \left(x - 5\right)^2}$.

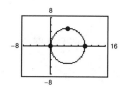

57. $x^2 + y^2 + Dx + Ey + F = 0$ passing through $(-3, -1), (2, 4), (-6, 8)$

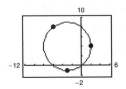

$(-3, -1)$: $10 - 3D - E + F = 0 \Rightarrow 10 = 3D + E - F$

$(2, 4)$: $20 + 2D + 4E + F = 0 \Rightarrow 20 = -2D - 4E - F$

$(-6, 8)$: $100 - 6D + 8E + F = 0 \Rightarrow 100 = 6D - 8E - F$

Solution: $D = 6, E = -8, F = 0$

The equation of the circle is $x^2 + y^2 + 6x - 8y = 0$. To graph, complete the squares first, then solve for y.

$$\left(x^2 + 6x + 9\right) + \left(y^2 - 8y + 16\right) = 0 + 9 + 16$$

$$\left(x + 3\right)^2 + \left(y - 4\right)^2 = 25$$

$$\left(y - 4\right)^2 = 25 - \left(x + 3\right)^2$$

$$y - 4 = \pm\sqrt{25 - \left(x + 3\right)^2}$$

$$y = 4 \pm \sqrt{25 - \left(x + 3\right)^2}$$

Let $y_1 = 4 + \sqrt{25 - \left(x + 3\right)^2}$ and $y_2 = 4 - \sqrt{25 - \left(x + 3\right)^2}$.

59. Let x = number of touchdowns.

Let y = number of extra-point kicks.

Let z = number of field goals.

$$\begin{cases} x + y + z = 13 \\ 6x + y + 3z = 45 \\ x - y = 0 \\ x - 6z = 0 \end{cases}$$

$$\begin{cases} x + y + z = 13 \\ -5y - 3z = -33 \quad -6\text{Eq.1} + \text{Eq.2} \\ -2y - z = -13 \quad -\text{Eq.1} + \text{Eq.3} \\ -y - 7z = -13 \quad -\text{Eq.1} + \text{Eq.4} \end{cases}$$

$$\begin{cases} x + y + z = 13 \\ -y - 7z = -13 \quad \text{Interchange Eq.2 and Eq.4.} \\ -2y - z = -13 \\ -5y - 3z = -33 \end{cases}$$

$$\begin{cases} x + y + z = 13 \\ y + 7z = 13 \quad -\text{Eq.2} \\ -2y - z = -13 \\ -5y - 3z = -33 \end{cases}$$

$$\begin{cases} x + y + z = 13 \\ y + 7z = 13 \\ 13z = 13 \quad 2\text{Eq.2} + \text{Eq.3} \\ 32z = 32 \quad 5\text{Eq.2} + \text{Eq.4} \end{cases}$$

$z = 1$

$y + 7(1) = 13 \Rightarrow y = 6$

$x + 6 + 1 = 13 \Rightarrow x = 6$

So, 6 touchdowns, 6 extra-point kicks, and 1 field goal were scored.

61. Let x = amount at 8%.

Let y = amount at 9%.

Let z = amount at 10%.

$$\begin{cases} x + y + z = 775{,}000 \\ 0.08x + 0.09y + 0.10z = 67{,}500 \\ x = 4z \end{cases}$$

$$\begin{cases} y + 5z = 775{,}000 \\ 0.09y + 0.42z = 67{,}500 \end{cases}$$

$z = 75{,}000$

$y = 775{,}000 - 5z = 400{,}000$

$x = 4z = 300{,}000$

$\$300{,}000$ was borrowed at 8%.

$\$400{,}000$ was borrowed at 9%.

$\$75{,}000$ was borrowed at 10%.

63. Let C = amount in certificates of deposit.

Let M = amount in municipal bonds.

Let B = amount in blue chip stocks.

Let G = amount in growth stocks.

$$\begin{cases} C + M + B + G = 500{,}000 \\ 0.03C + 0.05M + 0.08B + 0.10G = 0.05(500{,}000) \\ B + G = \tfrac{1}{4}(500{,}000) \end{cases}$$

The system has infinitely many solutions.

Let $G = s$, then:

$187{,}500 + s$ in certificates of deposit,

$187{,}500 - s$ in municipal bonds,

$125{,}000 - s$ in blue-chip stocks,

s in growth stocks.

65. Let x = pounds of brand X.

Let y = pounds of brand Y.

Let z = pounds of brand Z.

Fertilizer A: $\frac{1}{3}y + \frac{2}{9}z = 5$

Fertilizer B: $\frac{1}{2}x + \frac{2}{3}y + \frac{5}{9}z = 13$

Fertilizer C: $\frac{1}{2}x \qquad + \frac{2}{9}z = 4$

$\begin{cases} \frac{1}{2}x + \frac{2}{3}y + \frac{5}{9}z = 13 & \text{Interchange Eq.1 and Eq.2.} \\ \quad\ \ \frac{1}{3}y + \frac{2}{9}z = 5 \\ \frac{1}{2}x \qquad + \frac{2}{9}z = 4 \end{cases}$

$\begin{cases} \frac{1}{2}x + \frac{2}{3}y + \frac{5}{9}z = 13 \\ \quad\ \ \frac{1}{3}y + \frac{2}{9}z = 5 \\ \quad -\frac{2}{3}y - \frac{1}{3}z = -9 & -\text{Eq.1} + \text{Eq.3} \end{cases}$

$\begin{cases} \frac{1}{2}x + \frac{2}{3}y + \frac{5}{9}z = 13 \\ \quad\ \ \frac{1}{3}y + \frac{2}{9}z = 5 \\ \qquad\quad \frac{1}{9}z = 1 & 2\text{Eq.2} + \text{Eq.3} \end{cases}$

$z = 9$

$\frac{1}{3}y + \frac{2}{9}(9) = 5 \Rightarrow y = 9$

$\frac{1}{2}x + \frac{2}{3}(9) + \frac{5}{9}(9) = 13 \Rightarrow x = 4$

4 pounds of brand X, 9 pounds of brand Y, and 9 pounds of brand Z are needed to obtain the desired mixture.

67. Let x = the longest side (hypotenuse).

Let y = leg.

Let z = shortest leg.

$\begin{cases} x + y + z = 110 \\ \qquad\quad x = z + 21 \\ \quad y + z = 14 + x \end{cases}$

$\begin{cases} x + y + z = 110 \\ x \quad\ - z = 21 \\ -x + y + z = 14 \end{cases}$

$\begin{cases} x + y + z = 110 \\ x \quad\ - z = 21 \\ \quad y \qquad = 35 & \text{Eq.2} + \text{Eq.3} \end{cases}$

$\begin{cases} 2x + y \qquad = 1 & \text{Eq.2} + \text{Eq.1} \\ x \quad\ - z = 21 \\ \quad y \qquad = 35 \end{cases}$

$y = 35$

$2x + 35 = 131 \Rightarrow x = 48$

$48 - z = 21 \Rightarrow z = 27$

So, the longest side measures 48 feet, the shortest side measures 27 feet, and the third side measures 35 feet.

69. $\begin{cases} x + y + z = 180 \\ 2x + 7 + z = 180 \\ y + 2x - 7 = 180 \end{cases}$

$\begin{cases} x + y + z = 180 \\ 2x \quad\ + z = 173 \\ 2x + y \qquad = 187 \end{cases}$

$\begin{cases} -x + y \qquad = 7 & -\text{Eq.2} + \text{Eq.1} \\ 2x \quad\ + z = 173 \\ 2x + y \qquad = 187 \end{cases}$

$\begin{cases} -x + y \qquad = 7 \\ 2x \quad\ + z = 173 \\ 3x \qquad\quad = 180 & -\text{Eq.1} + \text{Eq.3} \end{cases}$

$x = 60°$

$2(60) + z = 173 \Rightarrow z = 53°$

$-60 + y = 7 \Rightarrow y = 67°$

71. Let x = number of television ads.

Let y = number of radio ads.

Let z = number of local newspaper ads.

$\begin{cases} x + y + z = 60 \\ 1000x + 200y + 500z = 42{,}000 \\ x - y - z = 0 \end{cases}$

$\begin{cases} x + y + z = 60 \\ -800y - 500z = -18{,}000 & -1000\text{Eq.1} + \text{Eq.2} \\ -2y - 2z = -60 & -\text{Eq.1} + \text{Eq.3.} \end{cases}$

$\begin{cases} x + y + z = 60 \\ -2y - 2z = -60 & \text{Interchange} \\ -800y - 500z = -18{,}000 & \text{Eq.2 and Eq.3} \end{cases}$

$\begin{cases} x + y + z = 60 \\ -2y - 2z = -60 \\ 300z = 6000 & -400\text{Eq.2} + \text{Eq.3} \end{cases}$

$z = 20$

$-2y - 2(20) = -60 \Rightarrow y = 10$

$x + 10 + 20 = 60 \Rightarrow x = 30$

30 television ads, 10 radio ads, and 20 newspaper ads can be run each month.

73. (a) To use 2 liters of the 50% solution:

Let x = amount of 10% solution.

Let y = amount of 20% solution.

$$x + y = 8 \Rightarrow y = 8 - x$$

$$x(0.10) + y(0.20) + 2(0.50) = 10(0.25)$$

$$0.10x + 0.20(8 - x) + 1 = 2.5$$

$$0.10x + 1.6 - 0.20x + 1 = 2.5$$

$$-0.10x = -0.1$$

$$x = 1 \text{ liter of 10\% solution}$$

$$y = 7 \text{ liters of 20\% solution}$$

Given: 2 liters of 50% solution

(b) To use as little of the 50% solution as possible, the chemist should use no 10% solution.

Let x = amount of 20% solution.

Let y = amount of 50% solution.

$$x + y = 10 \Rightarrow y = 10 - x$$

$$x(0.20) + y(0.50) = 10(0.25)$$

$$x(0.20) + (10 - x)(0.50) = 10(0.25)$$

$$x(0.20) + 5 - 0.50x = 2.5$$

$$-0.30x = -2.5$$

$$x = 8\tfrac{1}{3} \text{ liters of 20\% solution}$$

$$y = 1\tfrac{2}{3} \text{ liters of 50\% solution}$$

(c) To use as much of the 50% solution as possible, the chemist should use no 20% solution.

Let x = amount of 10% solution.

Let y = amount of 50% solution.

$$x + y = 10 \Rightarrow y = 10 - x$$

$$x(0.10) + y(0.50) = 10(0.25)$$

$$0.10x + 0.50(10 - x) = 2.5$$

$$0.10x + 5 - 0.50x = 2.5$$

$$-0.40x = -2.5$$

$$x = 6\tfrac{1}{4} \text{ liters of 10\% solution}$$

$$y = 3\tfrac{3}{4} \text{ liters of 50\% solution}$$

75. $\begin{cases} I_1 - I_2 + I_3 = 0 & \text{Equation 1} \\ 3I_1 + 2I_2 \quad\quad = 7 & \text{Equation 2} \\ \quad\quad 2I_2 + 4I_3 = 8 & \text{Equation 3} \end{cases}$

$\begin{cases} I_1 - I_2 + I_3 = 0 \\ \quad 5I_2 - 3I_3 = 7 \quad (-3)\text{Eq.1} + \text{Eq.2} \\ \quad 2I_2 + 4I_3 = 8 \end{cases}$

$\begin{cases} I_1 - I_2 + I_3 = 0 \\ \quad 10I_2 - 6I_3 = 14 \quad 2\text{Eq.2} \\ \quad 10I_2 + 20I_3 = 40 \quad 5\text{Eq.3} \end{cases}$

$\begin{cases} I_1 - I_2 + I_3 = 0 \\ \quad 10I_2 - 6I_3 = 14 \\ \quad\quad 26I_3 = 26 \quad (-1)\text{Eq.2} + \text{Eq.3} \end{cases}$

$$26I_3 = 26 \Rightarrow I_3 = 1$$

$$10I_2 - 6(1) = 14 \Rightarrow I_2 = 2$$

$$I_1 - 2 + 1 = 0 \Rightarrow I_1 = 1$$

Solution: $I_1 = 1, I_2 = 2, I_3 = 1$

77. $(-4, 5), (-2, 6), (2, 6), (4, 2)$

$$n = 4, \sum_{i=1}^{4} x_i = 0, \sum_{i=1}^{4} x_i^2 = 40, \sum_{i=1}^{4} x_i^3 = 0, \sum_{i=1}^{4} x_i^4 = 544,$$

$$\sum_{i=1}^{4} y_i = 19, \sum_{i=1}^{4} x_i y_i = -12, \sum_{i=1}^{4} x_i^2 y_i = 160$$

$\begin{cases} 4c \quad\quad + 40a = 19 \\ \quad 40b \quad\quad = -12 \\ 40c \quad\quad + 544a = 160 \end{cases}$

$\begin{cases} 4c \quad\quad + 40a = 19 \\ \quad 40b \quad\quad = -12 \\ \quad\quad 144a = -30 \quad -10\text{Eq.1} + \text{Eq.3} \end{cases}$

$$144a = -30 \Rightarrow a = -\tfrac{5}{24}$$

$$40b = -12 \Rightarrow b = -\tfrac{3}{10}$$

$$4c + 40\left(-\tfrac{5}{24}\right) = 19 \Rightarrow c = \tfrac{41}{6}$$

Least squares regression parabola:

$$y = -\tfrac{5}{24}x^2 - \tfrac{3}{10}x + \tfrac{41}{6}$$

79. $(0, 0), (2, 2), (3, 6), (4, 12)$

$$n = 4, \sum_{i=1}^{4} x_i = 9, \sum_{i=1}^{4} x_i^2 = 29, \sum_{i=1}^{4} x_i^3 = 99, \sum_{i=1}^{4} x_i^4 = 353, \sum_{i=1}^{4} y_i = 20, \sum_{i=1}^{4} x_i y_i = 70, \sum_{i=1}^{4} x_i^2 y_i = 254$$

$$\begin{cases} 4c + 9b + 29a = 20 \\ 9c + 29b + 99a = 70 \\ 29c + 99b + 353a = 254 \end{cases}$$

$$\begin{cases} 9c + 29b + 99a = 70 \\ 4c + 9b + 29a = 20 \\ 29c + 99b + 353a = 254 \end{cases} \quad \text{Interchange equations.}$$

$$\begin{cases} c + 11b + 41a = 30 & -2\text{Eq.2} + \text{Eq.1} \\ -35b - 135a = -100 & -4\text{Eq.1} + \text{Eq.2} \\ -220b - 836a = -616 & -29\text{Eq.1} + \text{Eq.3} \end{cases}$$

$$\begin{cases} c + 11b + 41a = 30 \\ 1540b + 5940a = 4400 & -44\text{Eq.2} \\ -1540b - 5852a = -4312 & 7\text{Eq.3} \end{cases}$$

$$\begin{cases} c + 11b + 41a = 30 \\ 1540b + 5940a = 4400 \\ 88a = 88 & \text{Eq.2} + \text{Eq.3} \end{cases}$$

$$88a = 88 \Rightarrow a = 1$$
$$1540b + 5940(1) = 4400 \Rightarrow b = -1$$
$$c + 11(-1) + 41(1) = 30 \Rightarrow c = 0$$

Least squares regression parabola: $y = x^2 - x$

81. (a) $(100, 75), (120, 68), (140, 55)$

$$n = 3, \sum_{i=1}^{3} x_i = 360, \sum_{i=1}^{3} x_i^2 = 44{,}000, \sum_{i=1}^{3} x_i^3 = 5{,}472{,}000, \sum_{i=1}^{3} x_i^4 = 691{,}520{,}000, \sum_{i=1}^{3} y_i = 198,$$

$$\sum_{i=1}^{3} x_i y_i = 23{,}360, \sum_{i=1}^{3} x_i^2 y_i = 2{,}807{,}200$$

$$\begin{cases} 3c + 360b + 44{,}000a = 198 \\ 360c + 44{,}000b + 5{,}472{,}000a = 23{,}360 \\ 44{,}000c + 5{,}472{,}000b + 691{,}520{,}000a = 2{,}807{,}200 \end{cases}$$

Solving this system yields $a = -0.0075, b = 1.3,$ and $c = 20.$

Least squares regression parabola: $y = -0.0075x^2 + 1.3x + 20$

(b)

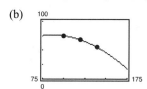

(c)

x	Actual Percent, y	Model Approximation, y
100	75	75
120	68	68
140	55	55

The model is a good fit to the actual data.
The values are the same.

   (d) For $x = 170$: $y = -0.0075(170)^2 + 1.3(170) + 20 = 24.25\%$

   (e) For $y = 40$:

$$40 = -0.0075x^2 + 1.3x + 20$$

$$0.0075x^2 - 1.3x + 20 = 0$$

By the Quadratic Formula, $x \approx 17$ or $x \approx 156$.

Choosing the value that fits with our data, you have 156 females.

83. Let x = number of touchdowns.

Let y = number of extra-point kicks.

Let z = number of field goals.

Let s = number of safeties.

$$\begin{cases} 6x + y + 3z + 2s = 50 \\ x + y + z + s = 15 \\ x = 3z \\ y = x \end{cases}$$

$$\begin{cases} 6x + y + 3z + 2s = 50 \\ x + y + z + s = 15 \\ x \quad\quad - 3z \quad\quad = 0 \Rightarrow z = \frac{1}{3}x \\ x - y \quad\quad\quad = 0 \Rightarrow y = x \end{cases}$$

$$\begin{cases} 6x + x + 3\left(\frac{1}{3}x\right) + 2s = 50 \\ x + x + \frac{1}{3}x + s = 15 \end{cases}$$

$$\begin{cases} 8x + 2s = 50 \\ 2\frac{1}{3}x + s = 15 \end{cases}$$

$$\begin{array}{r} 8x + 2s = 50 \\ -4\frac{2}{3}x - 2s = -30 \\ \hline 3\frac{1}{3}x \quad\quad = 20 \end{array}$$

$$x = 6$$
$$s = 1$$
$$z = \tfrac{1}{3}x = 2$$
$$y = x = 6$$

So, 6 touchdowns, 6 extra-point kicks, 2 field goals, and 1 safety were scored during the game.

85. $\begin{cases} y + \lambda = 0 \\ x \quad\quad + \lambda = 0 \\ x + y - 10 = 0 \end{cases} \Rightarrow \begin{aligned} & x = y = -\lambda \\ & \Rightarrow 2x - 10 = 0 \end{aligned}$

$$x = 5$$
$$y = 5$$
$$\lambda = -5$$

87. $\begin{cases} 2x - 2x\lambda = 0 \Rightarrow 2x(1 - \lambda) = 0 \Rightarrow \lambda = 1 \text{ or } x = 0 \\ -2y + \lambda = 0 \\ y - x^2 = 0 \end{cases}$

If $\lambda = 1$: $2y = \lambda \Rightarrow y = \dfrac{1}{2}$

$$x^2 = y \Rightarrow x = \pm\sqrt{\frac{1}{2}} = \pm\frac{\sqrt{2}}{2}$$

If $x = 0$: $x^2 = y \Rightarrow y = 0$

$$2y = \lambda \Rightarrow \lambda = 0$$

Solution: $x = \pm\dfrac{\sqrt{2}}{2}$ or $x = 0$

$$y = \frac{1}{2} \quad\quad\quad y = 0$$
$$\lambda = 1 \quad\quad\quad \lambda = 0$$

89. False. Equation 2 does not have a leading coefficient of 1.

91. No, they are not equivalent. There are two arithmetic errors. The constant in the second equation should be -11 and the coefficient of z in the third equation should be 2.

93. Sample answer: There are an infinite number of linear systems that have $(3, -4, 2)$ as their solution. Two systems are:

$$\begin{cases} 2x + y - z = 0 \\ 3x + 2y - \frac{1}{2}z = 0 \\ -x + 2y + z = -9 \end{cases}$$

$$\begin{cases} 4x + 3y = 0 \\ 4y + 8z = 0 \\ 2x - z = 4 \end{cases}$$

95. Sample answer: There are an infinite number of linear systems that have $\left(-6, -\frac{1}{2}, -\frac{7}{4}\right)$ as their solution. Two systems are:

$$\begin{cases} x - 12y - 4z = 7 \\ x + 2y - 4z = 0 \\ -2x + 18y + 4z = -4 \end{cases}$$

$$\begin{cases} -8y - 4z = 11 \\ x - 12y = 0 \\ x - 8y + 4z = -9 \end{cases}$$

Section 9.4   Partial Fractions

- You should know how to decompose a rational function $\dfrac{N(x)}{D(x)}$ into partial fractions.

  (a) If the fraction is improper, divide to obtain

  $$\frac{N(x)}{D(x)} = p(x) + \frac{N_1(x)}{D(x)}$$

  where $p(x)$ is a polynomial.

  (b) Factor the denominator completely into linear and irreducible quadratic factors.

  (c) For each factor of the form $(px + q)^m$, the partial fraction decomposition includes the terms

  $$\frac{A_1}{(px + q)} + \frac{A_2}{(px + q)^2} + \cdots + \frac{A_m}{(px + q)^m}.$$

  (d) For each factor of the form $\left(ax^2 + bx + c\right)^n$, the partial fraction decomposition includes the terms

  $$\frac{B_1x + C_1}{ax^2 + bx + c} + \frac{B_2x + C_2}{\left(ax^2 + bx + c\right)^2} + \cdots + \frac{B_nx + C_n}{\left(ax^2 + bx + c\right)^n}.$$

- You should know how to determine the values of the constants in the numerators.

  (a) Set $\dfrac{N_1(x)}{D(x)}$ = partial fraction decomposition.

  (b) Multiply both sides by $D(x)$ to obtain the basic equation.

  (c) For distinct linear factors, substitute the zeros of the distinct linear factors into the basic equation.

  (d) For repeated linear factors, use the coefficients found in part (c) to rewrite the basic equation. Then use other values of x to solve for the remaining coefficients.

  (e) For quadratic factors, expand the basic equation, collect like terms, and then equate the coefficients of like terms.

1. partial fraction decomposition

3. partial fraction

5. $\dfrac{3x - 1}{x(x - 4)} = \dfrac{A}{x} + \dfrac{B}{x - 4}$

Matches (b).

6. $\dfrac{3x - 1}{x^2(x - 4)} = \dfrac{A}{x} + \dfrac{B}{x^2} + \dfrac{C}{x - 4}$

Matches (c).

7. $\dfrac{3x - 1}{x(x^2 + 4)} = \dfrac{A}{x} + \dfrac{Bx + C}{x^2 + 4}$

Matches (d).

8. $\dfrac{3x - 1}{x(x^2 - 4)} = \dfrac{3x - 1}{x(x - 2)(x + 2)}$

$$= \dfrac{A}{x} + \dfrac{B}{x - 2} + \dfrac{C}{x + 2}$$

Matches (a).

9. $\dfrac{3}{x^2 - 2x} = \dfrac{3}{x(x-2)} = \dfrac{A}{x} + \dfrac{B}{x-2}$

11. $\dfrac{9}{x^3 - 7x^2} = \dfrac{9}{x^2(x-7)} = \dfrac{A}{x} + \dfrac{B}{x^2} + \dfrac{C}{x-7}$

13. $\dfrac{4x^2 + 3}{(x-5)^3} = \dfrac{A}{x-5} + \dfrac{B}{(x-5)^2} + \dfrac{C}{(x-5)^3}$

15. $\dfrac{2x-3}{x^3 + 10x} = \dfrac{2x-3}{x(x^2+10)} = \dfrac{A}{x} + \dfrac{Bx + C}{x^2 + 10}$

17. $\dfrac{x-1}{x(x^2+1)^2} = \dfrac{A}{x} + \dfrac{Bx+C}{x^2+1} + \dfrac{Dx+E}{(x^2+1)^2}$

19. $\dfrac{1}{x^2 + x} = \dfrac{A}{x} + \dfrac{B}{x+1}$

$1 = A(x+1) + Bx$

Let $x = 0$: $1 = A$

Let $x = -1$: $1 = -B \Rightarrow B = -1$

$\dfrac{1}{x^2 + x} = \dfrac{1}{x} - \dfrac{1}{x+1}$

21. $\dfrac{1}{2x^2 + x} = \dfrac{A}{2x+1} + \dfrac{B}{x}$

$1 = Ax + B(2x+1)$

Let $x = -\dfrac{1}{2}$: $1 = -\dfrac{1}{2}A \Rightarrow A = -2$

Let $x = 0$: $1 = B$

$\dfrac{1}{2x^2 + x} = \dfrac{1}{x} - \dfrac{2}{2x+1}$

23. $\dfrac{3}{x^2 + x - 2} = \dfrac{A}{x-1} + \dfrac{B}{x+2}$

$3 = A(x+2) + B(x-1)$

Let $x = 1$: $3 = 3A \Rightarrow A = 1$

Let $x = -2$: $3 = -3B \Rightarrow B = -1$

$\dfrac{3}{x^2 + x - 2} = \dfrac{1}{x-1} - \dfrac{1}{x+2}$

25. $\dfrac{1}{x^2 - 1} = \dfrac{A}{x+1} + \dfrac{B}{x-1}$

$1 = A(x-1) + B(x+1)$

Let $x = -1$: $1 = -2A \Rightarrow A = -\dfrac{1}{2}$

Let $x = 1$: $1 = 2B \Rightarrow B = \dfrac{1}{2}$

$\dfrac{1}{x^2 - 1} = \dfrac{1/2}{x-1} - \dfrac{1/2}{x+1} = \dfrac{1}{2}\left(\dfrac{1}{x-1} - \dfrac{1}{x+1}\right)$

27. $\dfrac{x^2 + 12x + 12}{x^3 - 4x} = \dfrac{A}{x} + \dfrac{B}{x+2} + \dfrac{C}{x-2}$

$x^2 + 12x + 12 = A(x+2)(x-2) + Bx(x-2) + Cx(x+2)$

Let $x = 0$: $12 = -4A \Rightarrow A = -3$

Let $x = -2$: $-8 = 8B \Rightarrow B = -1$

Let $x = 2$: $40 = 8C \Rightarrow C = 5$

$\dfrac{x^2 + 12x + 12}{x^3 - 4x} = -\dfrac{3}{x} - \dfrac{1}{x+2} + \dfrac{5}{x-2}$

29. $\dfrac{3x}{(x-3)^2} = \dfrac{A}{x-3} + \dfrac{B}{(x-3)^2}$

$3x = A(x-3) + B$

Let $x = 3$: $9 = B$

Let $x = 0$: $0 = -3A + B$

$0 = -3A + 9$

$3 = A$

$\dfrac{3x}{(x-3)^2} = \dfrac{3}{x-3} + \dfrac{9}{(x-3)^2}$

31. $\dfrac{4x^2 + 2x - 1}{x^2(x+1)} = \dfrac{A}{x} + \dfrac{B}{x^2} + \dfrac{C}{x+1}$

$4x^2 + 2x - 1 = Ax(x+1) + B(x+1) + Cx^2$

Let $x = 0$: $-1 = B$

Let $x = -1$: $1 = C$

Let $x = 1$: $5 = 2A + 2B + C$

$5 = 2A - 2 + 1$

$6 = 2A$

$3 = A$

$\dfrac{4x^2 + 2x - 1}{x^2(x+1)} = \dfrac{3}{x} - \dfrac{1}{x^2} + \dfrac{1}{x+1}$

33. $\dfrac{x^2 + 2x + 3}{x^3 + x} = \dfrac{A}{x} + \dfrac{Bx + C}{x^2 + 1}$

$x^2 + 2x + 3 = A(x^2 + 1) + (Bx + C)(x)$

$x^2 + 2x + 3 = x^2(A + B) + Cx + A$

Equating coefficients of like terms gives $A + B = 1, C = 2$, and $A = 3$.

So, $A = 3, B = -2$, and $C = 2$.

$\dfrac{x^2 + 2x + 3}{x^3 + x} = \dfrac{3}{x} - \dfrac{2x - 2}{x^2 + 1}$

35. $\dfrac{x}{x^3 - x^2 - 2x + 2} = \dfrac{x}{(x - 1)(x^2 - 2)} = \dfrac{A}{x - 1} + \dfrac{Bx + C}{x^2 - 2}$

$x = A(x^2 - 2) + (Bx + C)(x - 1) = Ax^2 - 2A + Bx^2 - Bx + Cx - C$

$= (A + B)x^2 + (C - B)x - (2A + C)$

Equating coefficients of like terms gives $0 = A + B, 1 = C - B$, and $0 = 2A + C$. So, $A = -1, B = 1$, and $C = 2$.

$\dfrac{x}{x^3 - x^2 - 2x + 2} = -\dfrac{1}{x - 1} + \dfrac{x + 2}{x^2 - 2}$

37. $\dfrac{2x^2 + x + 8}{(x^2 + 4)^2} = \dfrac{Ax + B}{x^2 + 4} + \dfrac{Cx + D}{(x^2 + 4)^2}$

$2x^2 + x + 8 = (Ax + B)(x^2 + 4) + Cx + D$

$2x^2 + x + 8 = Ax^3 + Bx^2 + (4A + C)x + (4B + D)$

Equating coefficients of like terms gives $0 = A$

$2 = B$

$1 = 4A + C \Rightarrow C = 1$

$8 = 4B + D \Rightarrow D = 0$

$\dfrac{2x^2 + x + 8}{(x^2 + 4)^2} = \dfrac{2}{x^2 + 4} + \dfrac{x}{(x^2 + 4)^2}$

39. $\dfrac{x}{16x^4 - 1} = \dfrac{x}{(4x^2 - 1)(4x^2 + 1)} = \dfrac{x}{(2x + 1)(2x - 1)(4x^2 + 1)} = \dfrac{A}{2x + 1} + \dfrac{B}{2x - 1} + \dfrac{Cx + D}{4x^2 + 1}$

$x = A(2x - 1)(4x^2 + 1) + B(2x + 1)(4x^2 + 1) + (Cx + D)(2x + 1)(2x - 1)$

$= A(8x^3 - 4x^2 + 2x - 1) + B(8x^3 + 4x^2 + 2x + 1) + (Cx + D)(4x^2 - 1)$

$= 8Ax^3 - 4Ax^2 + 2Ax - A + 8Bx^3 + 4Bx^2 + 2Bx + B + 4Cx^3 + 4Dx^2 - Cx - D$

$= (8A + 8B + 4C)x^3 + (-4A + 4B + 4D)x^2 + (2A + 2B - C)x + (-A + B - D)$

Equating coefficients of like terms gives $0 = 8A + 8B + 4C, 0 = -4A + 4B + 4D, 1 = 2A + 2B - C$, and $0 = -A + B - D$.

Using the first and third equations, $2A + 2B + C = 0$ and $2A + 2B - C = 1$; by subtraction, $2C = -1$, so $C = -\dfrac{1}{2}$.

Using the second and fourth equations, $-A + B + D = 0$ and $-A + B - D = 0$; by subtraction $2D = 0$, so $D = 0$.

Substituting $-\dfrac{1}{2}$ for C and 0 for D in the first and second equations, $8A + 8B = 2$ and $-4A + 4B = 0$, so $A = \dfrac{1}{8}$ and $B = \dfrac{1}{8}$.

$\dfrac{x}{16x^4 - 1} = \dfrac{\frac{1}{8}}{2x + 1} + \dfrac{\frac{1}{8}}{2x - 1} + \dfrac{\left(-\frac{1}{2}\right)x}{4x^2 + 1} = \dfrac{1}{8(2x + 1)} + \dfrac{1}{8(2x - 1)} - \dfrac{x}{2(4x^2 + 1)} = \dfrac{1}{8}\left(\dfrac{1}{2x + 1} + \dfrac{1}{2x - 1} - \dfrac{4x}{4x^2 + 1}\right)$

41. $\dfrac{x^2 + 5}{(x + 1)(x^2 - 2x + 3)} = \dfrac{A}{x + 1} + \dfrac{Bx + C}{x^2 - 2x + 3}$

$$x^2 + 5 = A(x^2 - 2x + 3) + (Bx + C)(x + 1)$$
$$= Ax^2 - 2Ax + 3A + Bx^2 + Bx + Cx + C$$
$$= (A + B)x^2 + (-2A + B + C)x + (3A + C)$$

Equating coefficients of like terms gives $1 = A + B, 0 = -2A + B + C,$ and $5 = 3A + C.$

Subtracting both sides of the second equation from the first gives $1 = 3A - C$; combining this with the third equation gives $A = 1$ and $C = 2.$ Because $A + B = 1, B = 0.$

$$\dfrac{x^2 + 5}{(x + 1)(x^2 - 2x + 3)} = \dfrac{1}{x + 1} + \dfrac{2}{x^2 - 2x + 3}$$

43. $\dfrac{x^2 - x}{x^2 + x + 1} = 1 + \dfrac{-2x - 1}{x^2 + x + 1} = 1 - \dfrac{2x + 1}{x^2 + x + 1}$

45. $\dfrac{2x^3 - x^2 + x + 5}{x^2 + 3x + 2} = 2x - 7 + \dfrac{18x + 19}{(x + 1)(x + 2)}$

$\dfrac{18x + 19}{(x + 1)(x + 2)} = \dfrac{A}{x + 1} + \dfrac{B}{x + 2}$

$18x + 19 = A(x + 2) + B(x + 1)$

Let $x = -1$: $1 = A$

Let $x = -2$: $-17 = -B \Rightarrow B = 17$

$\dfrac{2x^3 - x^2 + x + 5}{x^2 + 3x + 2} = 2x - 7 + \dfrac{1}{x + 1} + \dfrac{17}{x + 2}$

47. $\dfrac{x^4}{(x - 1)^3} = \dfrac{x^4}{x^3 - 3x^2 + 3x - 1}$

$= x + 3 + \dfrac{6x^2 - 8x + 3}{(x - 1)^3}$

$\dfrac{6x^2 - 8x + 3}{(x - 1)^3} = \dfrac{A}{x - 1} + \dfrac{B}{(x - 1)^2} + \dfrac{C}{(x - 1)^3}$

$6x^2 - 8x + 3 = A(x - 1)^2 + B(x - 1) + C$

Let $x = 1$: $1 = C$

$\left.\begin{array}{l} \text{Let } x = 0: \; 3 = A - B + 1 \\ \text{Let } x = 2: \; 11 = A + B + 1 \end{array}\right\}$ $\begin{array}{l} A - B = 2 \\ A + B = 10 \end{array}$

So, $A = 6$ and $B = 4.$

$\dfrac{x^4}{(x - 1)^3} = x + 3 + \dfrac{6}{x - 1} + \dfrac{4}{(x - 1)^2} + \dfrac{1}{(x - 1)^3}$

49. $\dfrac{x^4 + 2x^3 + 4x^2 + 8x + 2}{x^3 + 2x^2 + x} = x + \dfrac{3x^2 + 8x + 2}{x^3 + 2x^2 + x}$

$= x + \dfrac{3x^2 + 8x + 2}{x(x + 1)^2}$

$\dfrac{3x^2 + 8x + 2}{x(x + 1)^2} = \dfrac{A}{x} + \dfrac{B}{x + 1} + \dfrac{C}{(x + 1)^2}$

$3x^2 + 8x + 2 = A(x + 1)^2 + B(x)(x + 1) + C(x)$

$3x^2 + 8x + 2 = Ax^2 + 2Ax + A + Bx^2 + Bx + Cx$

$3x^2 + 8x + 2 = (A + B)x^2 + (2A + B + C)x + A$

Equating coefficients of like terms gives

$A + B = 3, 2A + B + C = 8,$ and $A = 2.$

So, $A = 2, B = 1,$ and $C = 3.$

$\dfrac{x^4 + 2x^3 + 4x^2 + 8x + 2}{x^3 + 2x^2 + x} = x + \dfrac{2}{x} + \dfrac{1}{x + 1} + \dfrac{3}{(x + 1)^2}$

51. $\dfrac{5 - x}{2x^2 + x - 1} = \dfrac{A}{2x - 1} + \dfrac{B}{x + 1}$

$-x + 5 = A(x + 1) + B(2x - 1)$

Let $x = \dfrac{1}{2}$: $\dfrac{9}{2} = \dfrac{3}{2}A \Rightarrow A = 3$

Let $x = -1$: $6 = -3B \Rightarrow B = -2$

$\dfrac{5 - x}{2x^2 + x - 1} = \dfrac{3}{2x - 1} - \dfrac{2}{x + 1}$

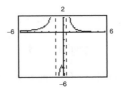

53. $\dfrac{4x^2 - 1}{2x(x+1)^2} = \dfrac{A}{2x} + \dfrac{B}{x+1} + \dfrac{C}{(x+1)^2}$

$4x^2 - 1 = A(x+1)^2 + 2Bx(x+1) + 2Cx$

Let $x = 0$: $-1 = A$

Let $x = -1$: $3 = -2C \Rightarrow C = -\dfrac{3}{2}$

Let $x = 1$:

$3 = 4A + 4B + 2C$

$3 = -4 + 4B - 3$

$\dfrac{5}{2} = B$

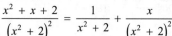

$\dfrac{4x^2 - 1}{2x(x+1)^2} = \dfrac{1}{2}\left[-\dfrac{1}{x} + \dfrac{5}{x+1} - \dfrac{3}{(x+1)^2}\right]$

55. $\dfrac{x^2 + x + 2}{(x^2+2)^2} = \dfrac{Ax + B}{x^2+2} + \dfrac{Cx + D}{(x^2+2)^2}$

$x^2 + x + 2 = (Ax + B)(x^2 + 2) + Cx + D$

$x^2 + x + 2 = Ax^3 + Bx^2 + (2A + C)x + (2B + D)$

Equating coefficients of like terms gives

$0 = A$

$1 = B$

$1 = 2A + C \Rightarrow C = 1$

$2 = 2B + D \Rightarrow D = 0$

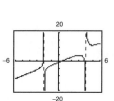

$\dfrac{x^2 + x + 2}{(x^2+2)^2} = \dfrac{1}{x^2+2} + \dfrac{x}{(x^2+2)^2}$

57. $\dfrac{2x^3 - 4x^2 - 15x + 5}{x^2 - 2x - 8} = 2x + \dfrac{x + 5}{(x+2)(x-4)}$

$\dfrac{x + 5}{(x+2)(x-4)} = \dfrac{A}{x+2} + \dfrac{B}{x-4}$

$x + 5 = A(x - 4) + B(x + 2)$

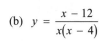

Let $x = -2$: $3 = -6A \Rightarrow A = -\dfrac{1}{2}$

Let $x = 4$: $9 = 6B \Rightarrow B = \dfrac{3}{2}$

$\dfrac{2x^3 - 4x^2 - 15x + 5}{x^2 - 2x - 8} = 2x + \dfrac{1}{2}\left(\dfrac{3}{x-4} - \dfrac{1}{x+2}\right)$

59. (a) $\dfrac{x - 12}{x(x-4)} = \dfrac{A}{x} + \dfrac{B}{x-4}$

$x - 12 = A(x - 4) + Bx$

Let $x = 0$: $-12 = -4A \Rightarrow A = 3$

Let $x = 4$: $-8 = 4B \Rightarrow B = -2$

$\dfrac{x - 12}{x(x-4)} = \dfrac{3}{x} - \dfrac{2}{x-4}$

(b) $y = \dfrac{x - 12}{x(x-4)}$ $y = \dfrac{3}{x}$ $y = -\dfrac{2}{x-4}$

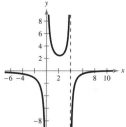

Vertical asymptotes: $x = 0$
and $x = 4$

Vertical asymptote: $x = 0$

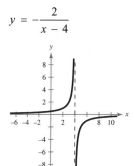

Vertical asymptote: $x = 4$

(c) The combination of the vertical asymptotes of the terms of the decomposition are the same as the vertical asymptotes of the rational function.

61. $C = \dfrac{120p}{10,000 - p^2} = \dfrac{120p}{(100 + p)(100 - p)}$

$\qquad = \dfrac{A}{100 + p} + \dfrac{B}{100 - p}$

$\qquad 120p = A(100 - p) + B(100 + p)$

Let $p = 100$: $200B = 12,000$

$\qquad\qquad\qquad B = 60$

Let $p = -100$: $200A = -12,000$

$\qquad\qquad\qquad A = -60$

$C = \dfrac{120p}{10,000 - p^2} = -\dfrac{60}{100 + p} + \dfrac{60}{100 - p}$

Let

$y_1 = \dfrac{120p}{10,000 - p^2}$ and $y_2 = -\dfrac{60}{100 + p} + \dfrac{60}{100 - p}.$

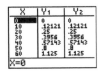

63. False. The partial fraction decomposition is

$\dfrac{A}{x + 10} + \dfrac{B}{x - 10} + \dfrac{C}{(x - 10)^2}.$

65. True. The expression is an improper rational expression.

67. $\dfrac{1}{a^2 - x^2} = \dfrac{A}{a + x} + \dfrac{B}{a - x},$ a is a constant.

$\qquad 1 = A(a - x) + B(a + x)$

Let $x = -a$: $1 = 2aA \Rightarrow A = \dfrac{1}{2a}$

Let $x = a$: $1 = 2aB \Rightarrow B = \dfrac{1}{2a}$

$\dfrac{1}{a^2 - x^2} = \dfrac{1}{2a}\left(\dfrac{1}{a + x} + \dfrac{1}{a - x}\right)$

69. $\dfrac{1}{y(a - y)} = \dfrac{A}{y} + \dfrac{B}{a - y}$

$\qquad 1 = A(a - y) + By$

Let $y = 0$: $1 = aA \Rightarrow A = \dfrac{1}{a}$

Let $y = a$: $1 = aB \Rightarrow B = \dfrac{1}{a}$

$\dfrac{1}{y(a - y)} = \dfrac{1}{a}\left(\dfrac{1}{y} + \dfrac{1}{a - y}\right)$

71. One way to find the constants is to choose values of the variable that eliminate one or more of the constants in the basic equation so that you can solve for another constant. If necessary, you can then use these constants with other chosen values of the variable to solve for any remaining constants. Another way is to expand the basic equation and collect like terms. Then you can equate coefficients of the like terms on each side of the equation to obtain simple equations involving the constants. If necessary, you can solve these equations using substitution.

Section 9.5   Systems of Inequalities

- You should be able to sketch the graph of an inequality in two variables.
  - (a) Replace the inequality with an equal sign and graph the equation. Use a dashed line for < or >, a solid line for ≤ or ≥.
  - (b) Test a point in each region formed by the graph. If the point satisfies the inequality, shade the whole region.
- You should be able to sketch systems of inequalities.

1. solution

3. linear

5. solution set

7. $y < 5 - x^2$

Using a dashed line, graph $y = 5 - x^2$, and shade the region inside the parabola.

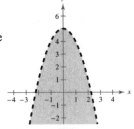

9. $x \geq 6$

Using a solid line, graph the vertical line $x = 6$, and shade to the right of this line.

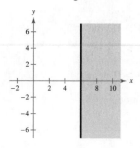

11. $y > -7$

Using a dashed line, graph the horizontal line $y = -7$, and shade above the line.

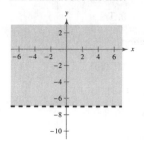

13. $y < 2 - x$

Using a dashed line, graph $y = 2 - x$, and then shade below the line. (Use $(0, 0)$ as a test point.)

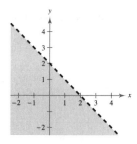

15. $2y - x \geq 4$

Using a solid line, graph $2y - x = 4$, and then shade above the line. (Use $(0, 0)$ as a test point.)

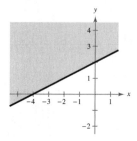

17. $(x + 1)^2 + (y - 2)^2 < 9$

Using a dashed line, sketch the circle $(x + 1)^2 + (y - 2)^2 = 9$.

Center: $(-1, 2)$

Radius: 3

Test point: $(0, 0)$

Shade the inside of the circle.

19. $y \leq \dfrac{1}{1 + x^2}$

Using a solid line, graph $y = \dfrac{1}{1 + x^2}$, and then shade below the curve. (Use $(0, 0)$ as a test point.)

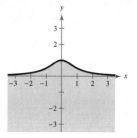

21. $y < \ln x$

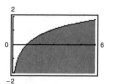

23. $y < 4^{-x-5}$

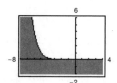

25. $y \geq \frac{5}{9}x - 2$

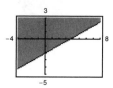

27. $y < -3.8x + 1.1$

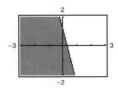

29. $x^2 + 5y - 10 \leq 0$

$$y \leq 2 - \frac{x^2}{5}$$

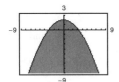

31. $\frac{5}{2}y - 3x^2 - 6 \geq 0$

$$y \geq \frac{2}{5}(3x^2 + 6)$$

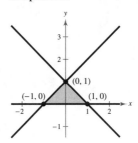

33. The line through $(-5, 0)$ and $(-1, 0)$ is $y = 5x + 5$.
The shaded region below the line gives $y < 5x + 5$.

35. The line through $(0, 2)$ and $(3, 0)$ is $y = -\frac{2}{3}x + 2$.
The shaded region above the line gives $y \geq -\frac{2}{3}x + 2$.

37. $\begin{cases} x \geq -4 \\ y > -3 \\ y \leq -8x - 3 \end{cases}$

(a) $0 \leq -8(0) - 3$, False

$(0, 0)$ *is not* a solution.

(b) $-3 > -3$, False

$(-1, -3)$ *is not* a solution.

(c) $-4 \geq -4$, True

$0 > -3$, True

$0 \leq -8(-4) - 3$, True

$(-4, 0)$ *is* a solution.

(d) $-3 \geq -4$, True

$11 > -3$, True

$11 < -8(-3) - 3$, True

$(-3, 11)$ *is* a solution.

39. $\begin{cases} 3x + y > 1 \\ -y - \frac{1}{2}x^2 \leq -4 \\ -15x + 4y > 0 \end{cases}$

(a) $3(0) + 10 > 1$, True

$-10 - \frac{1}{2}(0)^2 \leq -4$, True

$-15(0) + 4(10) > 0$, True

$(0, 10)$ *is* a solution.

(b) $3(0) + (-1) > 1$, False

$(0, -1)$ *is not* a solution.

(c) $3(2) + 9 > 1$, True

$-9 - \frac{1}{2}(2)^2 \leq -4$, True

$-15(2) + 4(9) > 0$, True

$(2, 9)$ *is* a solution.

(d) $3(-1) + 6 > 1$, True

$-6 - \frac{1}{2}(-1)^2 \leq -4$, True

$-15(-1) + 4(6) > 0$, True

$(-1, 6)$ *is* a solution.

41. $\begin{cases} x + y \leq 1 \\ -x + y \leq 1 \\ y \geq 0 \end{cases}$

First, find the points of intersection of each pair of equations.

Vertex A	Vertex B	Vertex C
$x + y = 1$	$x + y = 1$	$-x + y = 1$
$-x + y = 1$	$y = 0$	$y = 0$
$(0, 1)$	$(1, 0)$	$(-1, 0)$

43. $\begin{cases} x^2 + y \le 7 \\ x \ \ \ge -2 \\ \ \ \ y \ge 0 \end{cases}$

First, find the points of intersection of each pair of equations.

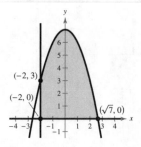

$(-2, 3)$

$(-2, 0)$

$(\sqrt{7}, 0)$

Vertex A

$x^2 + y = 7, x = -2$

$4 + y = 7$

$y = 3$

$(-2, 3)$

Vertex B

$x^2 + y = 7, y = 0$

$x^2 = 7$

$x = \sqrt{7}$

$(\sqrt{7}, 0)$

Vertex C

$x = -2, y = 0$

$(-2, 0)$

45. $\begin{cases} 2x + y > 2 \\ 6x + 3y < 2 \end{cases}$

The graphs of $2x + y = 2$ and $6x + 3y = 2$ are parallel lines. The first inequality has the region above the line shaded. The second inequality has the region below the line shaded. There are no points that satisfy both inequalities.

No solution

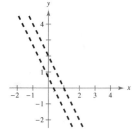

47. $\begin{cases} -3x + 2y < 6 \\ x - 4y > -2 \\ 2x + y < 3 \end{cases}$

First, find the points of intersection of each pair of equations.

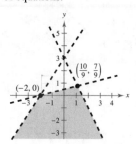

$\left(\frac{10}{9}, \frac{7}{9}\right)$

$(-2, 0)$

Vertex A

$-3x + 2y = 6$

$x - 4y = -2$

$(-2, 0)$

Vertex B

$-3x + 2y = 6$

$2x + y = 3$

$(0, 3)$

Vertex C

$x - 4y = -2$

$2x + y = 3$

$\left(\frac{10}{9}, \frac{7}{9}\right)$

Note that B is not a vertex of the solution region.

49. $\begin{cases} x > y^2 \\ x < y + 2 \end{cases}$

Points of intersection:

$$y^2 = y + 2$$
$$y^2 - y - 2 = 0$$
$$(y + 1)(y - 2) = 0$$
$$y = -1, 2$$

$(1, -1), (4, 2)$

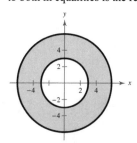

$(4, 2)$

$(1, -1)$

51. $\begin{cases} x^2 + y^2 \le 36 \\ x^2 + y^2 \ge 9 \end{cases}$

There are no points of intersection. The region common to both in equalities is the region between the circles.

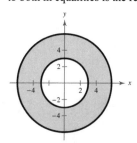

53. $3x + 4 \geq y^2$

$x - y < 0$

Points of intersection:

$x - y = 0 \Rightarrow y = x$

$3y + 4 = y^2$

$0 = y^2 - 3y - 4$

$0 = (y - 4)(y + 1)$

$y = 4$ or $y = -1$

$x = 4 \quad x = -1$

$(4, 4)$ and $(-1, -1)$

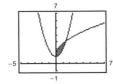

55. $\begin{cases} y \leq \sqrt{3x} + 1 \\ y \geq x^2 + 1 \end{cases}$

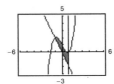

57. $\begin{cases} y < x^3 - 2x + 1 \\ y > -2x \\ x \leq 1 \end{cases}$

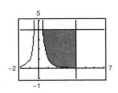

59. $\begin{cases} x^2y \geq 1 \Rightarrow y \geq \dfrac{1}{x^2} \\ 0 < x \leq 4 \\ y \leq 4 \end{cases}$

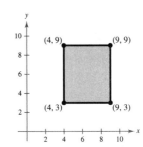

61. Line through points $(6, 0)$ and $(0, 6)$: $y = 6 - x$

$\begin{cases} x \geq 0 \\ y \geq 0 \\ y \leq 6 - x \end{cases}$

63. Line through points $(0, 4)$ and $(4, 0)$: $y = 4 - x$

Line through points $(0, 2)$ and $(8, 0)$: $y = 2 - \frac{1}{4}x$

$\begin{cases} y \geq 4 - x \\ y \geq 2 - \frac{1}{4}x \\ x \geq 0 \\ y \geq 0 \end{cases}$

65. $(8, 0), (0, 8)$

$\begin{cases} x \geq 0 \\ y \geq 0 \\ x^2 + y^2 < 64 \end{cases}$

67. Rectangular region with vertices at

$(4, 3), (9, 3), (9, 9), (4, 9)$

$\begin{cases} x \geq 4 \\ x \leq 9 \\ y \geq 3 \\ y \leq 9 \end{cases}$

This system may be written as:

$\begin{cases} 4 \leq x \leq 9 \\ 3 \leq y \leq 9 \end{cases}$

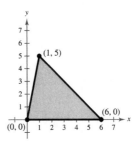

69. Triangle with vertices at $(0, 0), (6, 0), (1, 5)$

$(0, 0), (6, 0)$: $y = 0$

$(0, 0), (1, 5)$: $y = 5x$

$(6, 0), (1, 5)$: $y = -x + 6$

$\begin{cases} y \geq 0 \\ y \leq 5x \\ y \leq -x + 6 \end{cases}$

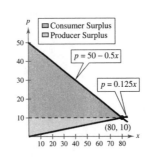

71. (a) Demand = Supply

$50 - 0.5x = 0.125x$

$50 = 0.625x$

$80 = x$

$10 = p$

Point of equilibrium: $(80, 10)$

(b) The consumer surplus is the area of the triangular region defined by

$$\begin{cases} p \le 50 - 0.5x \\ p \ge 10 \\ x \ge 0. \end{cases}$$

Consumer surplus $= \frac{1}{2}(\text{base})(\text{height}) = \frac{1}{2}(80)(40) = \$1600$

The producer surplus is the area of the triangular region defined by

$$\begin{cases} p \ge 0.125x \\ p \le 10 \\ x \ge 0. \end{cases}$$

Producer surplus $= \frac{1}{2}(\text{base})(\text{height}) = \frac{1}{2}(80)(10) = \$400$

73. (a)

$$\text{Demand} = \text{Supply}$$
$$140 - 0.00002x = 80 + 0.00001x$$
$$60 = 0.00003x$$
$$2,000,000 = x$$
$$100 = p$$

Point of equilibrium: $(2,000,000, 100)$

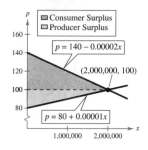

(b) The consumer surplus is the area of the triangular region defined by

$$\begin{cases} p \le 140 - 0.00002x \\ p \ge 100 \\ x \ge 0. \end{cases}$$

Consumer surplus $= \frac{1}{2}(\text{base})(\text{height})$
$$= \frac{1}{2}(2,000,000)(40)$$
$$= \$40,000,000$$

The producer surplus is the area of the triangular region defined by

$$\begin{cases} p \ge 80 + 0.00001x \\ p \le 100 \\ x \ge 0. \end{cases}$$

Producer surplus $= \frac{1}{2}(\text{base})(\text{height})$
$$= \frac{1}{2}(2,000,000)(20)$$
$$= \$20,000,000$$

75. $x =$ number of tables

$y =$ number of chairs

$$\begin{cases} x + \frac{3}{2}y \le 12 & \text{Assembly center} \\ \frac{4}{3}x + \frac{3}{2}y \le 15 & \text{Finishing center} \\ x \ge 0 \\ y \ge 0 \end{cases}$$

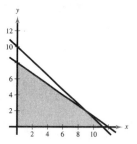

77. $x =$ amount in smaller account

$y =$ amount in larger account

Account constraints:

$$\begin{cases} x + y \le 20,000 \\ y \ge 2x \\ x \ge 5,000 \\ y \ge 5,000 \end{cases}$$

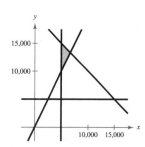

79. $x = $ number of packages of gravel

$y = $ number of bags of stone

$$\begin{cases} 55x + 70y \leq 7500 & \text{Weight} \\ x \geq 50 \\ y \geq 40 \end{cases}$$

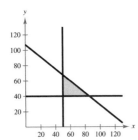

81. (a) $x = $ number of ounces of food X

$y = $ number of ounces of food Y

$$\begin{cases} 20x + 10y \geq 300 & \text{(calcium)} \\ 15x + 10y \geq 150 & \text{(iron)} \\ 10x + 20y \geq 200 & \text{(vitamin B)} \\ x \geq 0 \\ y \geq 0 \end{cases}$$

(b)

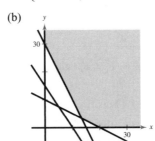

(c) Answers will vary. Some possible solutions which would satisfy the minimum daily requirements for calcium, iron, and vitamin B:

$(0, 30) \Rightarrow 30$ ounces of food Y

$(20, 0) \Rightarrow 20$ ounces of food X

$\left(13\frac{1}{3}, 3\frac{1}{3}\right) \Rightarrow 13\frac{1}{3}$ ounces of food X and

$3\frac{1}{3}$ ounces of food Y

83. (a) $y = 16.75t + 148.4$

(b)

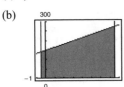

(c) Area of a trapezoid: $A = \dfrac{h}{2}(B + b)$

$h = 7.5 + 0.5 = 8$

$b = 16.75(-0.5) + 148.4 = 140.025$

$B = 16.75(7.5) + 148.4 = 274.025$

85. (a)

$$\begin{cases} xy \geq 500 & \text{Body-building space} \\ 2x + \pi y \geq 125 & \text{Track (Two semi-circles and two lengths)} \\ x \geq 0 \\ y \geq 0 \end{cases}$$

(b)

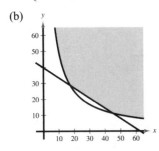

87. False. The graph shows the solution of the system

$$\begin{cases} y < 6 \\ -4x - 9y < 6 \\ 3x + y^2 \geq 2. \end{cases}$$

89. x = radius of smaller circle

y = radius of larger circle

(a) Constraints on circles:

$$\begin{cases} \pi y^2 - \pi x^2 \geq 10 \\ \quad\quad y > x \\ \quad\quad x > 0 \end{cases}$$

(b)

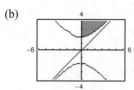

(c) The line is an asymptote to the boundary. The larger the circles, the closer the radii can be and the constraint still be satisfied.

91. $\begin{cases} x^2 + y^2 \leq 16 & \Rightarrow \text{ region inside the circle} \\ x + y \geq 4 & \Rightarrow \text{ region above the line} \end{cases}$

Matches graph (d).

93. $\begin{cases} x^2 + y^2 \geq 16 & \Rightarrow \text{ region outside the circle} \\ x + y \geq 4 & \Rightarrow \text{ region above the line} \end{cases}$

Matches graph (c).

Section 9.6   Linear Programming

> ■ To solve a linear programming problem:
> 1. Sketch the solution set for the system of constraints.
> 2. Find the vertices of the region.
> 3. Test the objective function at each of the vertices.

1. optimization

3. objective

5. inside; on

7. $z = 4x + 3y$

At $(0, 5)$: $z = 4(0) + 3(5) = 15$

At $(0, 0)$: $z = 4(0) + 3(0) = 0$

At $(5, 0)$: $z = 4(5) + 3(0) = 20$

The minimum value is 0 at $(0, 0)$.

The maximum value is 20 at $(5, 0)$.

9. $z = 2x + 5y$

At $(0, 0)$: $z = 2(0) + 5(0) = 0$

At $(4, 0)$: $z = 2(4) + 5(0) = 8$

At $(3, 4)$: $z = 2(3) + 5(4) = 26$

At $(0, 5)$: $z = 2(0) + 5(5) = 25$

The minimum value is 0 at $(0, 0)$.

The maximum value is 26 at $(3, 4)$.

11. $z = 10x + 7y$

At $(0, 45)$: $z = 10(0) + 7(45) = 315$

At $(30, 45)$: $z = 10(30) + 7(45) = 615$

At $(60, 20)$: $z = 10(60) + 7(20) = 740$

At $(60, 0)$: $z = 10(60) + 7(0) = 600$

At $(0, 0)$: $z = 10(0) + 7(0) = 0$

The minimum value is 0 at $(0, 0)$.

The maximum value is 740 at $(60, 20)$.

13. $z = 3x + 2y$

At $(0, 10)$: $z = 3(0) + 2(10) = 20$

At $(4, 0)$: $z = 3(4) + 2(0) = 12$

At $(2, 0)$: $z = 3(2) + 2(0) = 6$

The minimum value is 6 at $(2, 0)$.

The maximum value is 20 at $(0, 10)$.

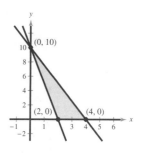

15. $z = 4x + 5y$

At $(10, 0)$: $z = 4(10) + 5(0) = 40$

At $(5, 3)$: $z = 4(5) + 5(3) = 35$

At $(0, 8)$: $z = 4(0) + 5(8) = 40$

The minimum value is 35 at $(5, 3)$.

The region is unbounded. There is no maximum.

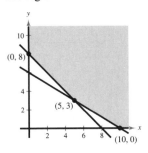

17. $z = 3x + y$

At $(16, 0)$: $z = 3(16) + 0 = 48$

At $(60, 0)$: $z = 3(60) + 0 = 180$

At $(7.2, 13.2)$: $z = 3(7.2) + 13.2 = 34.8$

The minimum value is 34.8 at $(7.2, 13.2)$.

The maximum value is 180 at $(60, 0)$.

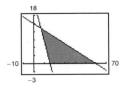

19. $z = x + 4y$

At $(16, 0)$: $z = 16 + 4(0) = 16$

At $(60, 0)$: $z = 60 + 4(0) = 60$

At $(7.2, 13.2)$: $z = 7.2 + 4(13.2) = 60$

The minimum value is 16 at $(16, 0)$.

The maximum value is 60 at any point along the line segment connecting $(60, 0)$ and $(7.2, 13.2)$.

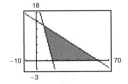

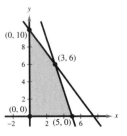

Figure for Exercises 21 and 23

21. $z = 2x + y$

At $(0, 10)$: $z = 2(0) + 10 = 10$

At $(3, 6)$: $z = 2(3) + 6 = 12$

At $(5, 0)$: $z = 2(5) + 0 = 10$

At $(0, 0)$: $z = 2(0) + 0 = 0$

The minimum value is 0 at $(0, 0)$.

The maximum value is 12 at $(3, 6)$.

23. $z = x + y$

At $(0, 10)$: $z = 0 + 10 = 10$

At $(3, 6)$: $z = 3 + 6 = 9$

At $(5, 0)$: $z = 5 + 0 = 5$

At $(0, 0)$: $z = 0 + 0 = 0$

The minimum value is 0 at $(0, 0)$.

The maximum value is 10 at $(0, 10)$.

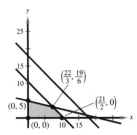

Figure for Exercises 25 and 27

25. $z = x + 5y$

At $(0, 5)$: $z = 0 + 5(5) = 25$

At $\left(\frac{22}{3}, \frac{19}{6}\right)$: $z = \frac{22}{3} + 5\left(\frac{19}{6}\right) = \frac{139}{6}$

At $\left(\frac{21}{2}, 0\right)$: $z = \frac{21}{2} + 5(0) = \frac{21}{2}$

At $(0, 0)$: $z = 0 + 5(0) = 0$

The minimum value is 0 at $(0, 0)$.

The maximum value is 25 at $(0, 5)$.

27. $z = 4x + 5y$

At $(0, 5)$: $z = 4(0) + 5(5) = 25$

At $\left(\frac{22}{3}, \frac{19}{6}\right)$: $z = 4\left(\frac{22}{3}\right) + 5\left(\frac{19}{6}\right) = \frac{271}{6}$

At $\left(\frac{21}{2}, 0\right)$: $z = 4\left(\frac{21}{2}\right) + 5(0) = 42$

At $(0, 0)$: $z = 4(0) + 5(0) = 0$

The minimum value is 0 at $(0, 0)$.

The maximum value is $\frac{271}{6}$ at $\left(\frac{22}{3}, \frac{19}{6}\right)$.

29. Objective function: $z = 2.5x + y$

Constraints:
$x \geq 0, y \geq 0, 3x + 5y \leq 15, 5x + 2y \leq 10$

At $(0, 0)$: $z = 0$

At $(2, 0)$: $z = 5$

At $\left(\frac{20}{19}, \frac{45}{19}\right)$: $z = \frac{95}{19} = 5$

At $(0, 3)$: $z = 3$

The minimum value is 0 at $(0, 0)$.

The maximum value of 5 occurs at any point on the line segment connecting $(2, 0)$ and $\left(\frac{20}{19}, \frac{45}{19}\right)$.

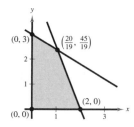

31. Objective function: $z = -x + 2y$

Constraints: $x \geq 0, y \geq 0, x \leq 10, x + y \leq 7$

At $(0, 0)$: $z = -0 + 2(0) = 0$

At $(0, 7)$: $z = -0 + 2(7) = 14$

At $(7, 0)$: $z = -7 + 2(0) = -7$

The constraint $x \leq 10$ is extraneous.

The minimum value is -7 at $(7, 0)$.

The maximum value is 14 at $(0, 7)$.

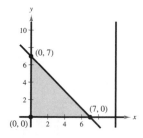

33. Objective function: $z = 3x + 4y$

Constraints: $x \geq 0, y \geq 0, x + y \leq 1, 2x + y \leq 4$

At $(0, 0)$: $z = 3(0) + 4(0) = 0$

At $(0, 1)$: $z = 3(0) + 4(1) = 4$

At $(1, 0)$: $z = 3(1) + 4(0) = 3$

The constraint $2x + y \leq 4$ is extraneous.

The minimum value is 0 at $(0, 0)$.

The maximum value is 4 at $(0, 1)$.

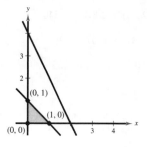

35. $x =$ number of \$225 models

$y =$ number of \$250 models

Constraints:
$$225x + 250y \leq 63,000$$
$$x + y \leq 275$$
$$x \geq 0$$
$$y \geq 0$$

Objective function: $P = 30x + 31y$

Vertices: $(0, 0), (0, 252), (230, 45)$ and $(275, 0)$

At $(0, 0)$: $P = 30(0) + 31(0) = 0$

At $(0, 252)$: $P = 30(0) + 31(252) = 7812$

At $(230, 45)$: $P = 30(230) + 31(45) = 8295$

At $(275, 0)$: $P = 30(275) + 31(0) = 8250$

An optimal profit of \$8295 occurs when 230 units of the \$225 model and 45 units of the \$250 model are stocked in inventory.

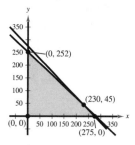

37. x = number of bags of Brand X

   y = number of bags of Brand Y

   Constraints: $2x + \ y \geq 12$

   $\qquad\qquad 2x + 9y \geq 36$

   $\qquad\qquad 2x + 3y \geq 24$

   $\qquad\qquad\qquad x \geq 0$

   $\qquad\qquad\qquad y \geq 0$

   Objective function: $C = 25x + 20y$

   Vertices: $(0, 12), (3, 6), (9, 2), (18, 0)$

   At $(0, 12)$: $C = 25(0) + 20(12) = 240$

   At $(3, 6)$: $C = 25(3) + 20(6) = 195$

   At $(9, 2)$: $C = 25(9) + 20(2) = 265$

   At $(18, 0)$: $C = 25(18) + 20(0) = 450$

   To optimize cost, use three bags of Brand X and six bags of Brand Y for an optimal cost of $195.

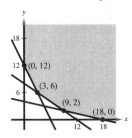

39. x = number of audits

   y = number of tax returns

   Constraints: $60x + 10y \leq 780$

   $\qquad\qquad 16x + \ 4y \leq 272$

   $\qquad\qquad\qquad x \geq 0$

   $\qquad\qquad\qquad y \geq 0$

   Objective function: $R = 1600x + 250y$

   Vertices: $(0, 0), (13, 0), (5, 48), (0, 68)$

   At $(0, 0)$: $R = 1600(0) + 250(0) = 0$

   At $(13, 0)$: $R = 1600(13) + 250(0) = 20{,}800$

   At $(5, 48)$: $R = 1600(5) + 250(48) = 20{,}000$

   At $(0, 68)$: $R = 1600(0) + 250(68) = 17{,}000$

   A maximum revenue of $20,800 occurs when the firm conducts 13 audits and 0 tax returns.

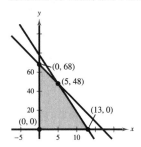

41. x = number of TV ads

   y = number of newspaper ads

   Constraints: $100{,}000x + \ 20{,}000y \leq 1{,}000{,}000$

   $\qquad\qquad\quad 100{,}000x \leq 800{,}000$

   $\qquad\qquad\qquad\qquad x \geq 0$

   $\qquad\qquad\qquad\qquad y \geq 0$

   Objective function: $A = 20x + 5y$ (A in millions)

   Vertices: $(0, 0), (0, 50), (8, 10), (8, 0)$

   At $(0, 0)$: $A = 20(0) + 5(0) = 0$

   At $(0, 50)$: $A = 20(0) + 5(50) = 250$ million

   At $(8, 10)$: $A = 20(8) + 5(10) = 210$ million

   At $(8, 0)$: $A = 20(8) + 5(0) = 160$ million

The company should spend $0 on television ads and $1,000,000 on newspaper ads. The optimal total audience is 250 million people.

43. x = amount of type A

y = amount of type B

Constraints: $x + y \le 250{,}000$

$x \ge \frac{1}{4}(250{,}000)$

$y \ge \frac{1}{4}(250{,}000)$

Objective function: $P = 0.08x + 0.10y$

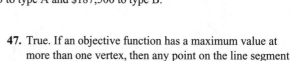

Vertices: $(62{,}500, 62{,}500), (62{,}500, 187{,}500), (187{,}500, 62{,}500)$

At $(62{,}500, 62{,}500)$: $P = 0.08(62{,}500) + 0.10(62{,}500) = 11{,}250$

At $(62{,}500, 187{,}500)$: $P = 0.08(62{,}500) + 0.10(187{,}500) = 23{,}750$

At $(187{,}500, 62{,}500)$: $P = 0.08(187{,}500) + 0.10(62{,}500) = 21{,}250$

To obtain an optimal return, the investor should allocate \$62,500 to type A and \$187,500 to type B. The optimal return is \$23,750.

45. True. The objective function has a maximum value at any point on the line segment connecting the two vertices. Both of these points are on the line $y = -x + 11$ and lie between $(4, 7)$ and $(8, 3)$.

47. True. If an objective function has a maximum value at more than one vertex, then any point on the line segment connecting the points will produce the maximum value.

Review Exercises for Chapter 9

1. $\begin{cases} x + y = 2 \\ x - y = 0 \end{cases} \Rightarrow x = y$

$x + x = 2$

$2x = 2$

$x = 1$

$y = 1$

Solution: $(1, 1)$

3. $\begin{cases} 4x - y - 1 = 0 \Rightarrow y = 4x - 1 \\ 8x + y - 17 = 0 \end{cases}$

$8x + (4x - 1) - 17 = 0$

$12x = 18$

$x = \frac{3}{2}$

$4\left(\frac{3}{2}\right) - y - 1 = 0$

$-y + 5 = 0$

$y = 5$

Solution: $\left(\frac{3}{2}, 5\right)$

5. $\begin{cases} 0.5x + y = 0.75 \Rightarrow y = 0.75 - 0.5x \\ 1.25x - 4.5y = -2.5 \end{cases}$

$1.25x - 4.5(0.75 - 0.5x) = -2.5$

$1.25x - 3.375 + 2.25x = -2.5$

$3.50x = 0.875$

$x = 0.25$

$y = 0.625$

Solution: $(0.25, 0.625)$

7. $\begin{cases} x^2 - y^2 = 9 \\ x - y = 1 \end{cases} \Rightarrow x = y + 1$

$(y + 1)^2 - y^2 = 9$

$2y + 1 = 9$

$y = 4$

$x = 5$

Solution: $(5, 4)$

9. $\begin{cases} y = 2x^2 \\ y = x^4 - 2x^2 \end{cases} \Rightarrow 2x^2 = x^4 - 2x^2$

$0 = x^4 - 4x^2$

$0 = x^2(x^2 - 4)$

$0 = x^2(x + 2)(x - 2) \Rightarrow x = 0, -2, 2$

$x = 0: y = 2(0)^2 = 0$

$x = -2: y = 2(-2)^2 = 8$

$x = 2: y = 2(2)^2 = 8$

Solutions: $(0, 0), (-2, 8), (2, 8)$

11. $\begin{cases} 2x - y = 10 \\ x + 5y = -6 \end{cases}$

Point of intersection: $(4, -2)$

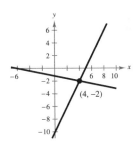

13. $\begin{cases} y = 2x^2 - 4x + 1 \\ y = x^2 - 4x + 3 \end{cases}$

Points of intersection: $(1.41, -0.66), (-1.41, 10.66)$

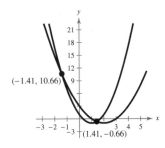

15. $\begin{cases} y = -2e^{-x} \\ 2e^x + y = 0 \Rightarrow y = -2e^x \end{cases}$

Point of intersection: $(0, -2)$

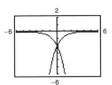

17. $\begin{cases} y = 2 + \log x \\ y = \frac{3}{4}x + 5 \end{cases}$

No Solution

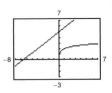

19. Let x = number of kits.

$C = 12x + 50{,}000$

$R = 25x$

Break-even: $R = C$

$$25x = 12x + 50{,}000$$
$$13x = 50{,}000$$
$$x \approx 3846.15$$

You would need to sell 3847 kits to cover your costs.

21. $\begin{cases} 2l + 2w = 480 \\ l = 1.50w \end{cases}$

$$2(1.50w) + 2w = 480$$
$$5w = 480$$
$$w = 96$$
$$l = 144$$

The dimensions are 96 meters $\times$ 144 meters.

23. $\begin{cases} 20 = l + w \Rightarrow w = 20 - l \\ 96 = lw \end{cases}$

$$96 = l(20 - l)$$
$$0 = 20l - l^2 - 96$$
$$0 = l^2 - 20l + 96$$
$$0 = (l - 8)(l - 12)$$
$$l = 8, l = 12$$

When $l = 8$, $w = 20 - 8 = 12$.

When $l = 12$, $w = 20 - 12 = 8$.

Dimensions: 12 inches $\times$ 8 inches

25. $\begin{cases} 2x - y = 2 \Rightarrow 16x - 8y = 16 \\ 6x + 8y = 39 \Rightarrow 6x + 8y = 39 \end{cases}$

$$\begin{aligned} 22x &= 55 \\ x &= \frac{55}{22} = \frac{5}{2} \end{aligned}$$

Back-substitute $x = \frac{5}{2}$ into Equation 1.

$$2\left(\frac{5}{2}\right) - y = 2$$
$$y = 3$$

Solution: $\left(\frac{5}{2}, 3\right)$

27. $\begin{cases} 0.2x + 0.3y = 0.14 \Rightarrow 20x + 30y = 14 \Rightarrow 20x + 30y = 14 \\ 0.4x + 0.5y = 0.20 \Rightarrow 4x + 5y = 2 \Rightarrow \underline{-20x - 25y = -10} \end{cases}$

$$5y = 4$$
$$y = \tfrac{4}{5}$$

Back-substitute $y = \tfrac{4}{5}$ into Equation 2.

$$4x + 5\left(\tfrac{4}{5}\right) = 2$$
$$4x = -2$$
$$x = -\tfrac{1}{2}$$

Solution: $\left(-\tfrac{1}{2}, \tfrac{4}{5}\right) = (-0.5, 0.8)$

29. $\begin{cases} 3x - 2y = 0 \Rightarrow 3x - 2y = 0 \\ 3x + 2(y + 5) = 10 \Rightarrow \underline{3x + 2y = 0} \end{cases}$

$$6x = 0$$
$$x = 0$$

Back-substitute $x = 0$ into Equation 1.

$$3(0) - 2y = 0$$
$$2y = 0$$
$$y = 0$$

Solution: $(0, 0)$

31. $\begin{cases} 1.25x - 2y = 3.5 \Rightarrow 5x - 8y = 14 \\ 5x - 8y = 14 \Rightarrow \underline{-5x + 8y = -14} \end{cases}$

$$0 = 0$$

There are infinitely many solutions.

Let $y = a$, then $5x - 8a = 14 \Rightarrow x = \tfrac{8}{5}a + \tfrac{14}{5}$.

Solution: $\left(\tfrac{8}{5}a + \tfrac{14}{5}, a\right)$ where a is any real number.

33. $\begin{cases} x + 5y = 4 \Rightarrow x + 5y = 4 \\ x - 3y = 6 \Rightarrow \underline{-x + 3y = -6} \end{cases}$

$$8y = -2 \Rightarrow y = -\tfrac{1}{4}$$

Matches graph (d). The system has one solution and is consistent.

35. $\begin{cases} 3x - y = 7 \Rightarrow 6x - 2y = 14 \\ -6x + 2y = 8 \Rightarrow \underline{-6x + 2y = 8} \end{cases}$

$$0 \neq 22$$

Matches graph (b). The system has no solution and is inconsistent.

37. $37 - 0.0002x = 22 + 0.00001x$

$$15 = 0.00021x$$
$$x = \frac{500{,}000}{7}, \; p = \frac{159}{7}$$

Point of equilibrium: $\left(\dfrac{500{,}000}{7}, \dfrac{159}{7}\right)$

39. $\begin{cases} x - 4y + 3z = 3 \\ -y + z = -1 \\ z = -5 \end{cases}$

$$-y + (-5) = -1 \Rightarrow y = -4$$
$$x - 4(-4) + 3(-5) = 3 \Rightarrow x = 2$$

Solution: $(2, -4, -5)$

41. $\begin{cases} 4x - 3y - 2z = -65 \\ 8y - 7z = -14 \\ z = 10 \end{cases}$

$$8y - 7(10) = -14 \Rightarrow y = 7$$
$$4x - 3(7) - 2(10) = -65 \Rightarrow x = -6$$

Solution: $(-6, 7, 10)$

43. $\begin{cases} x + 2y + 6z = 4 & \text{Equation 1} \\ -3x + 2y - z = -4 & \text{Equation 2} \\ 4x + 2z = 16 & \text{Equation 3} \end{cases}$

$\begin{cases} x + 2y + 6z = 4 \\ 8y + 17z = 8 & 3\text{Eq.1} + \text{Eq.2} \\ -8y - 22z = 0 & -4\text{Eq.1} + \text{Eq.3} \end{cases}$

$\begin{cases} x + 2y + 6z = 4 \\ 8y + 17z = 8 \\ -5z = 8 & \text{Eq.2} + \text{Eq.3} \end{cases}$

$\begin{cases} x + 2y + 6z = 4 \\ 8y + 17z = 8 \\ z = -\frac{8}{5} & -\frac{1}{5}\text{Eq.3} \end{cases}$

$8y + 17\left(-\frac{8}{5}\right) = 8 \Rightarrow y = \frac{22}{5}$

$x + 2\left(\frac{22}{5}\right) + 6\left(-\frac{8}{5}\right) = 4 \Rightarrow x = \frac{24}{5}$

Solution: $\left(\frac{24}{5}, \frac{22}{5}, -\frac{8}{5}\right)$

45. $\begin{cases} x - 2y + z = -6 & \text{Equation 1} \\ 2x - 3y = -7 & \text{Equation 2} \\ -x + 3y - 3z = 11 & \text{Equation 3} \end{cases}$

$\begin{cases} x - 2y + z = -6 \\ y - 2z = 5 & -2\text{Eq.1} + \text{Eq.2} \\ y - 2z = 5 & \text{Eq.1} + \text{Eq.3} \end{cases}$

$\begin{cases} x - 2y + z = -6 \\ y - 2z = 5 \\ 0 = 0 & -\text{Eq.2} + \text{Eq.3} \end{cases}$

Let $z = a$, then:

$y = 2a + 5$

$x - 2(2a + 5) + a = -6$

$x - 3a - 10 = -6$

$x = 3a + 4$

Solution: $(3a + 4, 2a + 5, a)$ where a is any real number.

49. $\begin{cases} 5x - 12y + 7z = 16 \Rightarrow \\ 3x - 7y + 4z = 9 \Rightarrow \end{cases} \begin{cases} 15x - 36y + 21z = 48 \\ -15x + 35y - 20z = -45 \\ \hline -y + z = 3 \end{cases}$

Let $z = a$. Then $y = a - 3$ and $5x - 12(a - 3) + 7a = 16 \Rightarrow x = a - 4$.

Solution: $(a - 4, a - 3, a)$ where a is any real number.

47. $\begin{cases} x + 4w = 1 & \text{Equation 1} \\ 3y + z - w = 4 & \text{Equation 2} \\ 2y - 3w = 2 & \text{Equation 3} \\ 4x - y + 2z = 5 & \text{Equation 4} \end{cases}$

$\begin{cases} x + 4w = 1 \\ 3y + z - w = 4 \\ 2y - 3w = 2 \\ -y + 2z - 16w = 1 & -4\text{Eq.1} + \text{Eq.4} \end{cases}$

$\begin{cases} x + 4w = 1 \\ 3y + z - w = 4 \\ 2y - 3w = 2 \\ 4z - 35w = 4 & \text{Eq.3} + 2\text{Eq.4} \end{cases}$

$\begin{cases} x + 4w = 1 \\ 3y + z - w = 4 \\ -2z - 7w = -2 & -2\text{Eq.2} + 3\text{Eq.3} \\ 4z - 35w = 4 \end{cases}$

$\begin{cases} x + 4w = 1 \\ 3y + z - w = 4 \\ -2z - 7w = -2 \\ -49w = 0 & 2\text{Eq.3} + \text{Eq.4} \end{cases}$

$w = 0$

$-2z - 7(0) = -2 \Rightarrow z = 1$

$3y + 1 - 0 = 4 \Rightarrow y = 1$

$x + 4(0) = 1 \Rightarrow x = 1$

Solution: $(1, 1, 1, 0)$

51. $y = ax^2 + bx + c$ through $(0, -5), (1, -2),$ and $(2, 5)$.

$(0, -5):\ -5 = \qquad\quad c \Rightarrow \qquad c = -5$

$(1, -2):\ -2 = \ a + \ b + c \Rightarrow \begin{cases} a + b = \quad 3 \end{cases}$

$(2,\ 5):\ \ 5 = 4a + 2b + c \Rightarrow \begin{cases} 2a + b = \quad 5 \end{cases}$

$$\begin{cases} 2a + b = \ \ 5 \\ \underline{-a - b = -3} \\ \ a \qquad\quad = \ \ 2 \\ \ \ \ \ \quad b = \ \ 1 \end{cases}$$

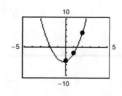

The equation of the parabola is $y = 2x^2 + x - 5$.

53. $x^2 + y^2 + Dx + Ey + F = 0$ through $(-1, -2), (5, -2),$ and $(2, 1)$.

$(-1, -2):\ \ 5 - \ \ D - 2E + F = 0 \Rightarrow \begin{cases} D + 2E - F = \quad 5 \end{cases}$

$(5, -2):\ \ 29 + 5D - 2E + F = 0 \Rightarrow \begin{cases} 5D - 2E + F = -29 \end{cases}$

$(2, 1):\ \quad 5 + 2D + \ \ E + F = 0 \Rightarrow \begin{cases} 2D + \ \ E + F = \ \ -5 \end{cases}$

From the first two equations

$6D = -24$

$D = \ -4.$

Substituting $D = -4$ into the second and third equations yields:

$-20 - 2E + F = -29 \Rightarrow \begin{cases} -2E + F = \ \ -9 \end{cases}$

$-8 + \ \ E + F = \ \ -5 \Rightarrow \begin{cases} \underline{-E - F = \ \ -3} \end{cases}$

$$\begin{aligned} -3E \qquad &= -12 \\ E \qquad &= \quad 4 \\ F &= \ \ -1 \end{aligned}$$

The equation of the circle is $x^2 + y^2 - 4x + 4y - 1 = 0$.

To verify the result using a graphing utility, solve the equation for y.

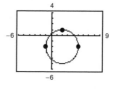

$\left(x^2 - 4x + 4\right) + \left(y^2 + 4y + 4\right) = 1 + 4 + 4$

$(x - 2)^2 + (y + 2)^2 = 9$

$(y + 2)^2 = 9 - (x - 2)^2$

$y = -2 \pm \sqrt{9 - (x - 2)^2}$

Let $y_1 = -2 + \sqrt{9 - (x - 2)^2}$ and $y_2 = -2 - \sqrt{9 - (x - 2)^2}$.

55. $(10, 267.8), (11, 301.0), (12, 334.7)$

(a) $n = 3, \sum_{i=1}^{3} x_i = 33, \sum_{i=1}^{3} x_i^2 = 365, \sum_{i=1}^{3} x_i^3 = 4059, \sum_{i=1}^{3} x_i^4 = 45,377, \sum_{i=1}^{3} y_i = 903.5,$

$\sum_{i=1}^{3} x_i y_i = 10,005.4, \sum_{i=1}^{3} x_i^2 y_i = 111,397.8$

$$\begin{cases} 3c + \quad 33b + \quad 365a = \quad 903.5 \\ 33c + \quad 365b + \quad 4059a = \quad 10,005.4 \\ 365c + 4059b + 45,377a = 111,397.8 \end{cases}$$

Solving this system yields $a = 0.25, b = 27.95,$ and $c = -36.7$.

Quadratic model: $y = 0.25x^2 + 27.95x - 36.7$

(b)

The model is a good fit.

(c) $2015 \rightarrow x = 15$

$$y = 0.25(15)^2 + 27.95(15) - 36.7 = 438.8$$

$438.8 billion

This answer seems reasonable.

57. Let x = amount invested at 7%

y = amount invested at 9%

z = amount invested at 11%.

$y = x - 3000$ and $z = x - 5000 \Rightarrow y + z = 2x - 8000$

$$\begin{cases} x + y + z = 40{,}000 \\ 0.07x + 0.09y + 0.11z = 3500 \\ y + z = 2x - 8000 \end{cases}$$

$x + (2x - 8000) = 40{,}000 \Rightarrow x = 16{,}000$

$y = 16{,}000 - 3000 \Rightarrow y = 13{,}000$

$z = 16{,}000 - 5000 \Rightarrow z = 11{,}000$

So, $16,000 was invested at 7%, $13,000 at 9%, and $11,000 at 11%.

59. Let x = number of par-3 holes

y = number of par-4 holes

z = number of par-5 holes.

$$\begin{cases} x + y + z = 18 & \text{Equation 1} \\ y = 2z + 2 & \text{Equation 2} \\ x = z & \text{Equation 3} \end{cases}$$

Back substitute $x = z$ into Equation 2: $y = 2x + 2$

Back substitute $x = z$ and $y = 2x + 2$ into Equation 1:

$$x + (2x + 2) + x = 18$$

$$4x = 16$$

$$x = 4$$

$$x = z \Rightarrow z = 4$$

$$y = 2(4) + 2 = 10$$

So, there are 4 par-3 holes, 10 par-4 holes, and 4 par-5 holes.

61. $\dfrac{3}{x^2 + 20x} = \dfrac{3}{x(x + 20)} = \dfrac{A}{x} + \dfrac{B}{x + 20}$

63. $\dfrac{3x - 4}{x^3 - 5x^2} = \dfrac{3x - 4}{x^2(x - 5)} = \dfrac{A}{x} + \dfrac{B}{x^2} + \dfrac{C}{x - 5}$

65. $\dfrac{4 - x}{x^2 + 6x + 8} = \dfrac{A}{x + 2} + \dfrac{B}{x + 4}$

$4 - x = A(x + 4) + B(x + 2)$

Let $x = -2$: $6 = 2A \Rightarrow A = 3$

Let $x = -4$: $8 = -2B \Rightarrow B = -4$

$\dfrac{4 - x}{x^2 + 6x + 8} = \dfrac{3}{x + 2} - \dfrac{4}{x + 4}$

67. $\dfrac{x^2}{x^2 + 2x - 15} = 1 - \dfrac{2x - 15}{x^2 + 2x - 15}$

$\dfrac{-2x + 15}{(x + 5)(x - 3)} = \dfrac{A}{x + 5} + \dfrac{B}{x - 3}$

$-2x + 15 = A(x - 3) + B(x + 5)$

Let $x = -5$: $25 = -8A \Rightarrow A = -\dfrac{25}{8}$

Let $x = 3$: $9 = 8B \Rightarrow B = \dfrac{9}{8}$

$\dfrac{x^2}{x^2 + 2x - 15} = 1 - \dfrac{25}{8(x + 5)} + \dfrac{9}{8(x - 3)}$

69. $\dfrac{x^2 + 2x}{x^3 - x^2 + x - 1} = \dfrac{x^2 + 2x}{(x-1)(x^2+1)} = \dfrac{A}{x-1} + \dfrac{Bx + C}{x^2+1}$

$x^2 + 2x = A(x^2 + 1) + (Bx + C)(x - 1) = Ax^2 + A + Bx^2 - Bx + Cx - C = (A + B)x^2 + (-B + C)x + (A - C)$

Equating coefficients of like terms gives $1 = A + B$, $2 = -B + C$, and $0 = A - C$.

Adding both sides of all three equations gives $3 = 2A$. So, $A = \dfrac{3}{2}, B = -\dfrac{1}{2}$, and $C = \dfrac{3}{2}$.

$\dfrac{x^2 + 2x}{x^3 - x^2 + x - 1} = \dfrac{\dfrac{3}{2}}{x-1} + \dfrac{-\dfrac{1}{2}x + \dfrac{3}{2}}{x^2+1} = \dfrac{1}{2}\left(\dfrac{3}{x-1} - \dfrac{x-3}{x^2+1}\right)$

71. $\dfrac{3x^2 + 4x}{(x^2 + 1)^2} = \dfrac{Ax + B}{x^2 + 1} + \dfrac{Cx + D}{(x^2 + 1)^2}$

$3x^2 + 4x = (Ax + B)(x^2 + 1) + Cx + D$

$\quad = Ax^3 + Bx^2 + (A + C)x + (B + D)$

Equating coefficients of like terms gives

$0 = A$

$3 = B$

$4 = 0 + C \Rightarrow C = 4$

$0 = B + D \Rightarrow D = -3$

$\dfrac{3x^2 + 4x}{(x^2 + 1)^2} = \dfrac{3}{x^2 + 1} + \dfrac{4x - 3}{(x^2 + 1)^2}$

73. $y \le 5 - \dfrac{1}{2}x$

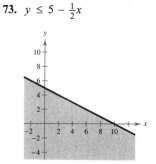

75. $y - 4x^2 > -1$

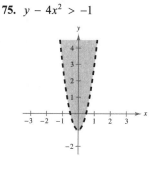

77. $(x - 1)^2 + (y - 3)^2 < 16$

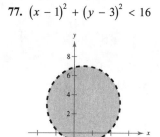

79. $\begin{cases} x + 2y \le 160 \\ 3x + y \le 180 \\ \quad\ x \ge 0 \\ \quad\ y \ge 0 \end{cases}$

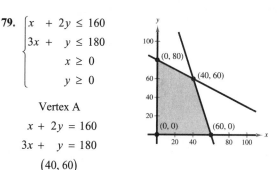

Vertex A

$x + 2y = 160$

$3x + y = 180$

$(40, 60)$

Vertex B	Vertex C	Vertex D
$x + 2y = 160$	$3x + y = 180$	$x = 0$
$x = 0$	$y = 0$	$y = 0$
$(0, 80)$	$(60, 0)$	$(0, 0)$

81. $\begin{cases} 3x + 2y \geq 24 \\ x + 2y \geq 12 \\ 2 \leq x \leq 15 \\ y \leq 15 \end{cases}$

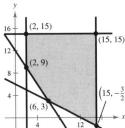

Vertex A	Vertex B	Vertex C
$3x + 2y = 24$	$3x + 2y = 24$	$x = 2$
$x + 2y = 12$	$x = 2$	$y = 15$
$(6, 3)$	$(2, 9)$	$(2, 15)$

Vertex D	Vertex E
$x = 15$	$x + 2y = 12$
$y = 15$	$x = 15$
$(15, 15)$	$\left(15, -\frac{3}{2}\right)$

83. $\begin{cases} y < x + 1 \\ y > x^2 - 1 \end{cases}$

Vertices:

$x + 1 = x^2 - 1$

$0 = x^2 - x - 2 = (x + 1)(x - 2)$

$x = -1 \text{ or } x = 2$

$y = 0 \qquad y = 3$

$(-1, 0) \qquad (2, 3)$

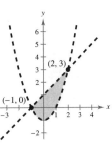

85. $\begin{cases} 2x - 3y \geq 0 \\ 2x - y \leq 8 \\ y \geq 0 \end{cases}$

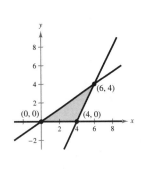

Vertex A

$2x - 3y = 0$

$2x - y = 8$

$(6, 4)$

Vertex B	Vertex C
$2x - 3y = 0$	$2x - y = 8$
$y = 0$	$y = 0$
$(0, 0)$	$(4, 0)$

87. x = number of units of Product I

y = number of units of Product II

$\begin{cases} 20x + 30y \leq 24{,}000 \\ 12x + 8y \leq 12{,}400 \\ x \geq 0 \\ y \geq 0 \end{cases}$

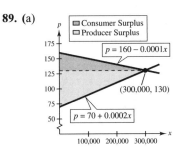

89. (a)

$160 - 0.0001x = 70 + 0.0002x$

$90 = 0.0003x$

$x = 300{,}000 \text{ units}$

$p = \$130$

Point of equilibrium: $(300{,}000, 130)$

(b) Consumer surplus: $\frac{1}{2}(300{,}000)(30) = \$4{,}500{,}000$

Producer surplus: $\frac{1}{2}(300{,}000)(60) = \$9{,}000{,}000$

91. Rectangular region with vertices at:

$(3, 1), (7, 1), (7, 10), \text{ and } (3, 10)$

$\begin{cases} x \geq 3 \\ x \leq 7 \\ y \geq 1 \\ y \leq 10 \end{cases}$

This system may be written as:

$\begin{cases} 3 \leq x \leq 7 \\ 1 \leq y \leq 10 \end{cases}$

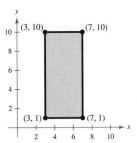

93. Objective function: $z = 3x + 4y$

Constraints: $\begin{cases} x \geq 0 \\ y \geq 0 \\ 2x + 5y \leq 50 \\ 4x + y \leq 28 \end{cases}$

At $(0, 0)$: $z = 0$

At $(0, 10)$: $z = 40$

At $(5, 8)$: $z = 47$

At $(7, 0)$: $z = 21$

The minimum value is 0 at $(0, 0)$.

The maximum value is 47 at $(5, 8)$.

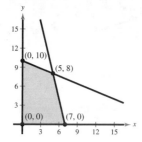

95. Objective function: $z = 1.75x + 2.25y$

Constraints: $\begin{cases} x \geq 0 \\ y \geq 0 \\ 2x + y \geq 25 \\ 3x + 2y \geq 45 \end{cases}$

At $(0, 25)$: $z = 56.25$

At $(5, 15)$: $z = 42.5$

At $(15, 0)$: $z = 26.25$

The minimum value is 26.25 at $(15, 0)$.

Because the region is unbounded, there is no maximum value.

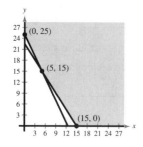

97. Objective function: $z = 5x + 11y$

Constraints: $\begin{cases} x \geq 0 \\ y \geq 0 \\ x + 3y \leq 12 \\ 3x + 2y \leq 15 \end{cases}$

At $(0, 0)$: $z = 0$

At $(5, 0)$: $z = 25$

At $(3, 3)$: $z = 48$

At $(0, 4)$: $z = 44$

The minimum value is 0 at $(0, 0)$.

The maximum value is 48 at $(3, 3)$.

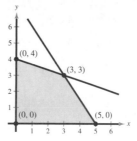

99. x = number of haircuts

y = number of permanents

Objective function: Optimize $R = 25x + 70y$ subject to the following constraints:

$\begin{cases} x \geq 0 \\ y \geq 0 \\ \left(\frac{20}{60}\right)x + \left(\frac{70}{60}\right)y \leq 24 \Rightarrow 2x + 7y \leq 144 \end{cases}$

At $(0, 0)$: $R = 0$

At $(72, 0)$: $R = 1800$

At $\left(0, \frac{144}{7}\right)$: $R = 1440$

The revenue is optimal if the student does 72 haircuts and no permanents. The maximum revenue is \$1800.

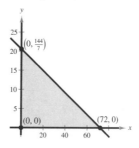

101. $x =$ number of model A

$y =$ number of model B

Constraints: $\begin{cases} 2x + 2.5y \le 4000 \\ 4x + y \le 4800 \\ x + 0.75y \le 1500 \\ x \ge 0 \\ y \ge 0 \end{cases}$

Objective function: $P = 45x + 50y$

Vertices: $(0, 0), (0, 1600), (750, 1000), (1050, 600), (1200, 0)$

At $(0, 0)$: $P = 45(0) + 50(0) = 0$

At $(0, 1600)$: $P = 45(0) + 50(1600) = 80{,}000$

At $(750, 1000)$: $P = 45(750) + 50(1000) = 83{,}750$

At $(1050, 600)$: $P = 45(1050) + 50(600) = 77{,}250$

At $(1200, 0)$: $P = 45(1200) + 50(0) = 54{,}000$

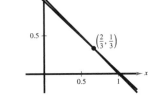

The optimal profit of \$83,750 occurs when 750 units of model A and 1000 units of model B are produced.

103. $x =$ fraction of regular gasoline

$y =$ fraction of premium gasoline

Constraints: $\begin{cases} 87x + 93y = 89 \\ x + y = 1 \\ x \ge 0 \\ y \ge 0 \end{cases}$

Objective function: $C = 1.85x + 2.10y$

Note that the "region" defined by the constraints is actually the point of intersection, $\left(\frac{2}{3}, \frac{1}{3}\right)$.

At $\left(\frac{2}{3}, \frac{1}{3}\right)$: $C = 1.85\left(\frac{2}{3}\right) + 2.10\left(\frac{1}{3}\right) = \$1.93$

The minimum cost is \$1.93 an occurs with a mixture that is $\frac{2}{3}$ gallon regular and $\frac{1}{3}$ gallon premium gasoline.

105. False. The system $y \le 5$, $y \ge -2$, $y \le \frac{7}{2}x - 9$, and $y \le -\frac{7}{2}x + 26$ represents the region covered by an isosceles trapezoid.

107. There are an infinite number of linear systems with the solution $(-8, 10)$. One possible system is:

$\begin{cases} 4x + y = -22 \\ \frac{1}{2}x + y = 6 \end{cases}$

109. There are infinite linear systems with the solution $\left(\frac{4}{3}, 3\right)$.

One possible system is:

$\begin{cases} 3x + y = 7 \\ -6x + 3y = 1 \end{cases}$

111. There are an infinite number of linear systems with the solution $(4, -1, 3)$. One possible system is as follows:

$\begin{cases} x + y + z = 6 \\ x + y - z = 0 \\ x - y - z = 2 \end{cases}$

113. There are an infinite number of linear systems with the solution $\left(5, \frac{3}{2}, 2\right)$. One possible system is:

$\begin{cases} 2x + 2y - 3z = 7 \\ x - 2y + z = 4 \\ -x + 4y - z = -1 \end{cases}$

115. A system of linear equations is inconsistent if it has no solution.

Problem Solving for Chapter 9

1. The longest side of the triangle is a diameter of the circle and has a length of 20.

The lines $y = \frac{1}{2}x + 5$ and $y = -2x + 20$ intersect at the point $(6, 8)$.

The distance between $(-10, 0)$ and $(6, 8)$ is:

$$d_1 = \sqrt{\left(6 - (-10)\right)^2 + (8 - 0)^2} = \sqrt{320} = 8\sqrt{5}$$

The distance between $(6, 8)$ and $(10, 0)$ is:

$$d_2 = \sqrt{(10 - 6)^2 + (0 - 8)^2} = \sqrt{80} = 4\sqrt{5}$$

Because $\left(\sqrt{320}\right)^2 + \left(\sqrt{80}\right)^2 = (20)^2$

$$400 = 400,$$

the sides of the triangle satisfy the Pythagorean Theorem. So, the triangle is a right triangle.

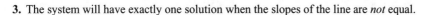

3. The system will have exactly one solution when the slopes of the line are *not* equal.

$$\begin{cases} ax + by = e \Rightarrow y = -\dfrac{a}{b}x + \dfrac{e}{b} \\ cx + dy = f \Rightarrow y = -\dfrac{c}{d}x + \dfrac{f}{d} \end{cases}$$

$$-\frac{a}{b} \neq -\frac{c}{d}$$

$$\frac{a}{b} \neq \frac{c}{d}$$

$$ad \neq bc$$

5. There are a finite number of solutions.

(a) If both equations are linear, then the maximum number of solutions to a finite system is *one*.

(b) If one equation is linear and the other is quadratic, then the maximum number of solutions is *two*.

(c) If both equations are quadratic, then the maximum number of solutions to a finite system is *four*.

7. The point where the two sections meet is at a depth of 10.1 feet. The distance between $(0, -10.1)$ and $(252.5, 0)$ is:

$$d = \sqrt{(252.5 - 0)^2 + \left(0 - (-10.1)\right)^2} = \sqrt{63,858.26}$$

$$d \approx 252.7$$

Each section is approximately 252.7 feet long.

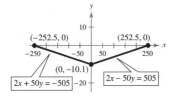

9. Let x = cost of the cable, per foot.

Let y = cost of a connector.

$$\begin{cases} 6x + 2y = 15.50 \Rightarrow 6x + 2y = 15.50 \\ 3x + 2y = 10.25 \Rightarrow -3x - 2y = -10.25 \end{cases}$$

$$\begin{aligned} 3x &= 5.25 \\ x &= 1.75 \\ y &= 2.50 \end{aligned}$$

For a four-foot cable with a connector on each end, the cost should be $4(1.75) + 2(2.50) = \$12.00$.

11. Let $X = \dfrac{1}{x}, Y = \dfrac{1}{y},$ and $Z = \dfrac{1}{z}$.

(a)
$$\begin{cases} \dfrac{12}{x} - \dfrac{12}{y} = 7 \Rightarrow 12X - 12Y = 7 \Rightarrow 12X - 12Y = 7 \\[2mm] \dfrac{3}{x} - \dfrac{4}{y} = 0 \Rightarrow 3X + 4Y = 0 \Rightarrow \underline{9X + 12Y = 0} \end{cases}$$

$$\begin{aligned} 21X &= 7 \\ X &= \frac{1}{3} \\ Y &= -\frac{1}{4} \end{aligned}$$

So, $\dfrac{1}{x} = \dfrac{1}{3} \Rightarrow x = 3$ and $\dfrac{1}{y} = -\dfrac{1}{4} \Rightarrow y = -4$.

Solution: $(3, -4)$

(b)
$$\begin{cases} \dfrac{2}{x} + \dfrac{1}{y} - \dfrac{3}{z} = 4 \Rightarrow 2X + Y - 3Z = 4 \quad \text{Eq.1} \\[2mm] \dfrac{4}{x} \phantom{+ \dfrac{1}{y}} + \dfrac{2}{z} = 10 \Rightarrow 4X + 2Z = 10 \quad \text{Eq.2} \\[2mm] -\dfrac{2}{x} + \dfrac{3}{y} - \dfrac{13}{z} = -8 \Rightarrow -2X + 3Y - 13Z = -8 \quad \text{Eq.3} \end{cases}$$

$$\begin{cases} 2X + Y - 3Z = 4 \\ -2Y + 8Z = 2 \qquad -2\text{Eq.1} + \text{Eq.2} \\ 4Y - 16Z = -4 \qquad \text{Eq.1} + \text{Eq.3} \end{cases}$$

$$\begin{cases} 2X + Y - 3Z = 4 \\ -2Y + 8Z = 2 \\ 0 = 0 \qquad 2\text{Eq.2} + \text{Eq.3} \end{cases}$$

The system has infinite solutions.

Let $Z = a$, then $Y = 4a - 1$ and $X = \dfrac{-a + 5}{2}$.

Then $\dfrac{1}{z} = a \Rightarrow z = \dfrac{1}{a}, \dfrac{1}{y} = 4a - 1 \Rightarrow y = \dfrac{1}{4a - 1},$ and $\dfrac{1}{x} = \dfrac{-a + 5}{2} \Rightarrow x = \dfrac{2}{-a + 5}$.

Solution: $\left(\dfrac{2}{-a + 5}, \dfrac{1}{4a - 1}, \dfrac{1}{a} \right), a \neq 5, \dfrac{1}{4}, 0$

13. Solution: $(1, -1, 2)$

$$\begin{cases} 4x - 2y + 5z = 16 & \text{Equation 1} \\ x + y = 0 & \text{Equation 2} \\ -x - 3y + 2z = 6 & \text{Equation 3} \end{cases}$$

(a) $\begin{cases} 4x - 2y + 5z = 16 \\ x + y = 0 \end{cases}$

$\begin{cases} x + y = 0 & \text{Interchange the equations.} \\ 4x - 2y + 5z = 16 \end{cases}$

$\begin{cases} x + y = 0 \\ -6y + 5z = 16 & -4\text{Eq.1} + \text{Eq.2} \end{cases}$

Let $z = a$, then $y = \dfrac{5a - 16}{6}$ and $x = \dfrac{-5a + 16}{6}$.

Solution: $\left(\dfrac{-5a + 16}{6}, \dfrac{5a - 16}{6}, a \right)$

When $a = 2$, we have the original solution.

(b) $\begin{cases} 4x - 2y + 5z = 16 \\ -x - 3y + 2z = 6 \end{cases}$

$\begin{cases} -x - 2y + 2z = 6 & \text{Interchange the equations.} \\ 4x - 3y + 5z = 16 \end{cases}$

$\begin{cases} -x - 3y + 2z = 6 & 4\text{Eq.1} + \text{Eq.2} \\ -14y + 13z = 40 \end{cases}$

Let $z = a$, then $y = \dfrac{13a - 40}{14}$ and $x = \dfrac{-11a + 36}{14}$.

Solution: $\left(\dfrac{-11a + 36}{14}, \dfrac{13a - 40}{14}, a \right)$

When $a = 2$, we have the original solution.

(c) $\begin{cases} x + y = 0 \\ -x - 3y + 2z = 6 \end{cases}$

$\begin{cases} x + y = 0 \\ -2y + 2z = 6 & \text{Eq.1} + \text{Eq.2} \end{cases}$

Let $z = a$, then $y = a - 3$ and $x = -a + 3$.

Solution: $(-a + 3, a - 3, a)$

When $a = 2$, we have the original solution.

(d) Each of these systems has infinite solutions.

15. $t =$ amount of terrestrial vegetation in kilograms

$a =$ amount of aquatic vegetation in kilograms

$$\begin{cases} a + t \le 32 \\ 0.15a \ge 1.9 \\ 193a + 772t \ge 11{,}000 \end{cases}$$

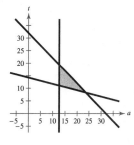

17. x = milligrams of HDL cholesterol

y = milligrams of LDL cholesterol

(a) $\begin{cases} 0 < y \le 130 \\ x \ge 60 \\ x + y \le 200 \end{cases}$

(b)

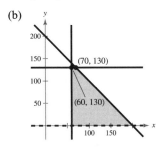

(c) $y = 120$ is in the region because $0 < y \le 130$.

$x = 90$ is in the region because $60 \le x \le 200$.

$x + y = 210$ is not the region because $x + y \le 200$.

(d) If the LDL reading is 135 and the HDL reading is 65, then $x \ge 60$ and $x + y \le 200$, but $y \not\le 130$.

(e) $\dfrac{x + y}{x} < 5$

$x + y < 5x$

$y < 4x$

The point $(75, 105)$ is in the region, $105 < 4(75)$, and $\dfrac{180}{75} = 2.4 < 5$.

Chapter 9   Practice Test

For Exercises 1–3, solve the given system by the method of substitution.

1. $\begin{cases} x + y = 1 \\ 3x - y = 15 \end{cases}$

2. $\begin{cases} x - 3y = -3 \\ x^2 + 6y = 5 \end{cases}$

3. $\begin{cases} x + y + z = 6 \\ 2x - y + 3z = 0 \\ 5x + 2y - z = -3 \end{cases}$

4. Find the two numbers whose sum is 110 and product is 2800.

5. Find the dimensions of a rectangle if its perimeter is 170 feet and its area is 1500 square feet.

For Exercises 6–8, solve the linear system by elimination.

6. $\begin{cases} 2x + 15y = 4 \\ x - 3y = 23 \end{cases}$

7. $\begin{cases} x + y = 2 \\ 38x - 19y = 7 \end{cases}$

8. $\begin{cases} 0.4x + 0.5y = 0.112 \\ 0.3x - 0.7y = -0.131 \end{cases}$

9. Herbert invests $17,000 in two funds that pay 11% and 13% simple interest, respectively. If he receives $2080 in yearly interest, how much is invested in each fund?

10. Find the least squares regression line for the points $(4, 3), (1, 1), (-1, -2),$ and $(-2, -1)$.

For Exercises 11–12, solve the system of equations.

11. $\begin{cases} x + y = -2 \\ 2x - y + z = 11 \\ 4y - 3z = -20 \end{cases}$

12. $\begin{cases} 3x + 2y - z = 5 \\ 6x - y + 5z = 2 \end{cases}$

13. Find the equation of the parabola $y = ax^2 + bx + c$ passing through the points $(0, -1), (1, 4)$ and $(2, 13)$.

For Exercises 14–15, write the partial fraction decomposition of the rational functions.

14. $\dfrac{10x - 17}{x^2 - 7x - 8}$

15. $\dfrac{x^2 + 4}{x^4 + x^2}$

16. Graph $x^2 + y^2 \geq 9$.

17. Graph the solution of the system.

$$\begin{cases} x + y \leq 6 \\ \quad\; x \geq 2 \\ \quad\; y \geq 0 \end{cases}$$

18. Derive a set of inequalities to describe the triangle with vertices $(0, 0), (0, 7),$ and $(2, 3)$.

19. Find the maximum value of the objective function, $z = 30x + 26y$, subject to the following constraints.

$$\begin{cases} \quad\quad\; x \geq 0 \\ \quad\quad\; y \geq 0 \\ 2x + 3y \leq 21 \\ 5x + 3y \leq 30 \end{cases}$$

20. Graph the system of inequalities.

$$\begin{cases} \quad\quad\; x^2 + y^2 \leq 4 \\ (x - 2)^2 + y^2 \geq 4 \end{cases}$$

For Exercises 21–22, write the partial fraction decomposition for the rational expression.

21. $\dfrac{1 - 2x}{x^2 + x}$

22. $\dfrac{6x - 17}{(x - 3)^2}$

CHAPTER 10
Matrices and Determinants

C H A P T E R   1 0
Matrices and Determinants

Section 10.1   Matrices and Systems of Equations

1. matrix

3. main diagonal

5. augmented

7. row-equivalent

9. Because the matrix has one row and two columns, its order is 1×2.

11. Because the matrix has three rows and one column, its order is 3×1.

13. Because the matrix has two rows and two columns, its order is 2×2.

15. $\begin{cases} 4x - 3y = -5 \\ -x + 3y = 12 \end{cases}$

$\begin{bmatrix} 4 & -3 & \vdots & -5 \\ -1 & 3 & \vdots & 12 \end{bmatrix}$

17. $\begin{cases} x + 10y - 2z = 2 \\ 5x - 3y + 4z = 0 \\ 2x + y = 6 \end{cases}$

$\begin{bmatrix} 1 & 10 & -2 & \vdots & 2 \\ 5 & -3 & 4 & \vdots & 0 \\ 2 & 1 & 0 & \vdots & 6 \end{bmatrix}$

19. $\begin{cases} 7x - 5y + z = 13 \\ 19x - 8z = 10 \end{cases}$

$\begin{bmatrix} 7 & -5 & 1 & \vdots & 13 \\ 19 & 0 & -8 & \vdots & 10 \end{bmatrix}$

21. $\begin{bmatrix} 1 & 2 & \vdots & 7 \\ 2 & -3 & \vdots & 4 \end{bmatrix}$

$\begin{cases} x + 2y = 7 \\ 2x - 3y = 4 \end{cases}$

23. $\begin{bmatrix} 2 & 0 & 5 & \vdots & -12 \\ 0 & 1 & -2 & \vdots & 7 \\ 6 & 3 & 0 & \vdots & 2 \end{bmatrix}$

$\begin{cases} 2x + 5z = -12 \\ y - 2z = 7 \\ 6x + 3y = 2 \end{cases}$

25. $\begin{bmatrix} 9 & 12 & 3 & 0 & \vdots & 0 \\ -2 & 18 & 5 & 2 & \vdots & 10 \\ 1 & 7 & -8 & 0 & \vdots & -4 \\ 3 & 0 & 2 & 0 & \vdots & -10 \end{bmatrix}$

$\begin{cases} 9x + 12y + 3z = 0 \\ -2x + 18y + 5z + 2w = 10 \\ x + 7y - 8z = -4 \\ 3x + 2z = -10 \end{cases}$

27. $\begin{bmatrix} 1 & 4 & 3 \\ 2 & 10 & 5 \end{bmatrix}$

$-2R_1 + R_2 \rightarrow \begin{bmatrix} 1 & 4 & 3 \\ 0 & \boxed{2} & -1 \end{bmatrix}$

29. $\begin{bmatrix} 1 & 1 & 1 \\ 5 & -2 & 4 \end{bmatrix}$

$-5R_1 + R_2 \rightarrow \begin{bmatrix} 1 & 1 & 1 \\ 0 & \boxed{-7} & -1 \end{bmatrix}$

31.
$$\begin{bmatrix} 1 & 5 & 4 & -1 \\ 0 & 1 & -2 & 2 \\ 0 & 0 & 1 & -7 \end{bmatrix}$$

$$-5R_2 + R_1 \rightarrow \begin{bmatrix} 1 & 0 & \boxed{14} & \boxed{-11} \\ 0 & 1 & -2 & 2 \\ 0 & 0 & 1 & -7 \end{bmatrix}$$

33.
$$\begin{bmatrix} 1 & 1 & 4 & -1 \\ 3 & 8 & 10 & 3 \\ -2 & 1 & 12 & 6 \end{bmatrix}$$

$$\begin{array}{c} -3R_1 + R_2 \rightarrow \\ 2R_1 + R_3 \rightarrow \end{array} \begin{bmatrix} 1 & 1 & 4 & -1 \\ 0 & 5 & \boxed{-2} & \boxed{6} \\ 0 & 3 & \boxed{20} & \boxed{4} \end{bmatrix}$$

$$\tfrac{1}{5}R_2 \rightarrow \begin{bmatrix} 1 & 1 & 4 & -1 \\ 0 & 1 & -\tfrac{2}{5} & \tfrac{6}{5} \\ 0 & 3 & \boxed{20} & \boxed{4} \end{bmatrix}$$

35.
$$\begin{bmatrix} -2 & 5 & 1 \\ 3 & -1 & -8 \end{bmatrix} \rightarrow \begin{bmatrix} 13 & 0 & -39 \\ 3 & -1 & -8 \end{bmatrix}$$

Add 5 times Row 2 to Row 1.

37.
$$\begin{bmatrix} 0 & -1 & -5 & 5 \\ -1 & 3 & -7 & 6 \\ 4 & -5 & 1 & 3 \end{bmatrix} \rightarrow \begin{bmatrix} -1 & 3 & -7 & 6 \\ 0 & -1 & -5 & 5 \\ 0 & 7 & -27 & 27 \end{bmatrix}$$

Interchange Row 1 and Row 2. Then add 4 times the new Row 1 to Row 3.

39.
$$\begin{bmatrix} 1 & 2 & 3 \\ 2 & -1 & -4 \\ 3 & 1 & -1 \end{bmatrix}$$

(a) $\begin{bmatrix} 1 & 2 & 3 \\ 0 & -5 & -10 \\ 3 & 1 & -1 \end{bmatrix}$ (b) $\begin{bmatrix} 1 & 2 & 3 \\ 0 & -5 & -10 \\ 0 & -5 & -10 \end{bmatrix}$ (c) $\begin{bmatrix} 1 & 2 & 3 \\ 0 & -5 & -10 \\ 0 & 0 & 0 \end{bmatrix}$ (d) $\begin{bmatrix} 1 & 2 & 3 \\ 0 & 1 & 2 \\ 0 & 0 & 0 \end{bmatrix}$ (e) $\begin{bmatrix} 1 & 0 & -1 \\ 0 & 1 & 2 \\ 0 & 0 & 0 \end{bmatrix}$ This matrix is in reduced row-echelon form.

41.
$$\begin{bmatrix} 1 & 0 & 0 & 0 \\ 0 & 1 & 1 & 5 \\ 0 & 0 & 0 & 0 \end{bmatrix}$$

This matrix is in reduced row-echelon form.

43.
$$\begin{bmatrix} 1 & 0 & 0 & 1 \\ 0 & 1 & 0 & -1 \\ 0 & 0 & 0 & 2 \end{bmatrix}$$

This matrix is not in row-echelon form.

45.
$$\begin{bmatrix} 1 & 1 & 0 & 5 \\ -2 & -1 & 2 & -10 \\ 3 & 6 & 7 & 14 \end{bmatrix}$$

$$\begin{array}{c} 2R_1 + R_2 \rightarrow \\ -3R_1 + R_3 \rightarrow \end{array} \begin{bmatrix} 1 & 1 & 0 & 5 \\ 0 & 1 & 2 & 0 \\ 0 & 3 & 7 & -1 \end{bmatrix}$$

$$-3R_2 + R_3 \rightarrow \begin{bmatrix} 1 & 1 & 0 & 5 \\ 0 & 1 & 2 & 0 \\ 0 & 0 & 1 & -1 \end{bmatrix}$$

47.
$$\begin{bmatrix} 1 & -1 & -1 & 1 \\ 5 & -4 & 1 & 8 \\ -6 & 8 & 18 & 0 \end{bmatrix}$$

$$\begin{array}{c} -5R_1 + R_2 \rightarrow \\ 6R_1 + R_3 \rightarrow \end{array} \begin{bmatrix} 1 & -1 & -1 & 1 \\ 0 & 1 & 6 & 3 \\ 0 & 2 & 12 & 6 \end{bmatrix}$$

$$-2R_2 + R_3 \rightarrow \begin{bmatrix} 1 & -1 & -1 & 1 \\ 0 & 1 & 6 & 3 \\ 0 & 0 & 0 & 0 \end{bmatrix}$$

49. Use the reduced row-echelon form feature of a graphing utility.

$$\begin{bmatrix} 3 & 3 & 3 \\ -1 & 0 & -4 \\ 2 & 4 & -2 \end{bmatrix} \Rightarrow \begin{bmatrix} 1 & 0 & 0 \\ 0 & 1 & 0 \\ 0 & 0 & 1 \end{bmatrix}$$

51. Use the reduced row-echelon form feature of a graphing utility.

$$\begin{bmatrix} 1 & 2 & 3 & -5 \\ 1 & 2 & 4 & -9 \\ -2 & -4 & -4 & 3 \\ 4 & 8 & 11 & -14 \end{bmatrix} \Rightarrow \begin{bmatrix} 1 & 2 & 0 & 0 \\ 0 & 0 & 1 & 0 \\ 0 & 0 & 0 & 1 \\ 0 & 0 & 0 & 0 \end{bmatrix}$$

53. Use the reduced row-echelon form feature of a graphing utility.

$$\begin{bmatrix} -3 & 5 & 1 & 12 \\ 1 & -1 & 1 & 4 \end{bmatrix} \Rightarrow \begin{bmatrix} 1 & 0 & 3 & 16 \\ 0 & 1 & 2 & 12 \end{bmatrix}$$

55. $\begin{cases} x - 2y = 4 \\ \quad\quad y = -3 \end{cases}$

$x - 2(-3) = 4$

$x = -2$

Solution: $(-2, -3)$

57. $\begin{cases} x - y + 2z = 4 \\ \quad\quad y - z = 2 \\ \quad\quad\quad z = -2 \end{cases}$

$y - (-2) = 2$

$y = 0$

$x - 0 + 2(-2) = 4$

$x = 8$

Solution: $(8, 0, -2)$

59. $\begin{bmatrix} 1 & 0 & \vdots & 3 \\ 0 & 1 & \vdots & -4 \end{bmatrix}$

$x = 3$

$y = -4$

Solution: $(3, -4)$

61. $\begin{bmatrix} 1 & 0 & 0 & \vdots & -4 \\ 0 & 1 & 0 & \vdots & -10 \\ 0 & 0 & 1 & \vdots & 4 \end{bmatrix}$

$x = -4$

$y = -10$

$z = 4$

Solution: $(-4, -10, 4)$

63. $\begin{cases} x + 2y = 7 \\ 2x + y = 8 \end{cases}$

$$\begin{bmatrix} 1 & 2 & \vdots & 7 \\ 2 & 1 & \vdots & 8 \end{bmatrix}$$

$-2R_1 + R_2 \rightarrow \begin{bmatrix} 1 & 2 & \vdots & 7 \\ 0 & -3 & \vdots & -6 \end{bmatrix}$

$-\frac{1}{3}R_2 \rightarrow \begin{bmatrix} 1 & 2 & \vdots & 7 \\ 0 & 1 & \vdots & 2 \end{bmatrix}$

$\begin{cases} x + 2y = 7 \\ \quad\quad y = 2 \end{cases}$

$y = 2$

$x + 2(2) = 7 \Rightarrow x = 3$

Solution: $(3, 2)$

65. $\begin{cases} 3x - 2y = -27 \\ x + 3y = 13 \end{cases}$

$$\begin{bmatrix} 3 & -2 & \vdots & -27 \\ 1 & 3 & \vdots & 13 \end{bmatrix}$$

$\begin{matrix} R_1 \\ R_2 \end{matrix} \begin{bmatrix} 1 & 3 & \vdots & 13 \\ 3 & -2 & \vdots & -27 \end{bmatrix}$

$-3R_1 + R_2 \rightarrow \begin{bmatrix} 1 & 3 & \vdots & 13 \\ 0 & -11 & \vdots & -66 \end{bmatrix}$

$-\frac{1}{11}R_2 \rightarrow \begin{bmatrix} 1 & 3 & \vdots & 13 \\ 0 & 1 & \vdots & 6 \end{bmatrix}$

$\begin{cases} x + 3y = 13 \\ \quad\quad y = 6 \end{cases}$

$y = 6$

$x + 3(6) = 13 \Rightarrow x = -5$

Solution: $(-5, 6)$

67. $\begin{cases} -2x + 6y = -22 \\ x + 2y = -9 \end{cases}$

$$\begin{bmatrix} -2 & 6 & \vdots & -22 \\ 1 & 2 & \vdots & -9 \end{bmatrix}$$

$\begin{matrix} R_1 \\ R_2 \end{matrix} \begin{bmatrix} 1 & 2 & \vdots & -9 \\ -2 & 6 & \vdots & -22 \end{bmatrix}$

$2R_1 + R_2 \rightarrow \begin{bmatrix} 1 & 2 & \vdots & -9 \\ 0 & 10 & \vdots & -40 \end{bmatrix}$

$\frac{1}{10}R_2 \rightarrow \begin{bmatrix} 1 & 2 & \vdots & -9 \\ 0 & 1 & \vdots & -4 \end{bmatrix}$

$\begin{cases} x + 2y = -9 \\ \quad\quad y = -4 \end{cases}$

$y = -4$

$x + 2(-4) = -9 \Rightarrow x = -1$

Solution: $(-1, -4)$

69. $\begin{cases} 8x - 4y = 7 \\ 5x + 2y = 1 \end{cases}$

$$\begin{bmatrix} 8 & -4 & \vdots & 7 \\ 5 & 2 & \vdots & 1 \end{bmatrix}$$

$$-5R_1 + 8R_2 \rightarrow \begin{bmatrix} 8 & -4 & \vdots & 7 \\ 0 & 36 & \vdots & -27 \end{bmatrix}$$

$$\tfrac{1}{36}R_2 \rightarrow \begin{bmatrix} 8 & -4 & \vdots & 7 \\ 0 & 1 & \vdots & -\tfrac{3}{4} \end{bmatrix}$$

$$\begin{cases} 8x - 4y = 7 \\ \quad\quad y = -\tfrac{3}{4} \end{cases}$$

$$y = -\tfrac{3}{4}$$

$$8x - 4\left(-\tfrac{3}{4}\right) = 7 \Rightarrow x = \tfrac{1}{2}$$

Solution: $\left(\tfrac{1}{2}, -\tfrac{3}{4}\right)$

71. $\begin{cases} x \quad\quad - 3z = -2 \\ 3x + y - 2z = 5 \\ 2x + 2y + z = 4 \end{cases}$

$$\begin{bmatrix} 1 & 0 & -3 & \vdots & -2 \\ 3 & 1 & -2 & \vdots & 5 \\ 2 & 2 & 1 & \vdots & 4 \end{bmatrix}$$

$$\begin{matrix} -3R_1 + R_2 \rightarrow \\ -2R_1 + R_3 \rightarrow \end{matrix} \begin{bmatrix} 1 & 0 & -3 & \vdots & -2 \\ 0 & 1 & 7 & \vdots & 11 \\ 0 & 2 & 7 & \vdots & 8 \end{bmatrix}$$

$$-2R_2 + R_3 \rightarrow \begin{bmatrix} 1 & 0 & -3 & \vdots & -2 \\ 0 & 1 & 7 & \vdots & 11 \\ 0 & 0 & -7 & \vdots & -14 \end{bmatrix}$$

$$-\tfrac{1}{7}R_3 \rightarrow \begin{bmatrix} 1 & 0 & -3 & \vdots & -2 \\ 0 & 1 & 7 & \vdots & 11 \\ 0 & 0 & 1 & \vdots & 2 \end{bmatrix}$$

$$\begin{cases} x - 3z = -2 \\ y + 7z = 11 \\ \quad\quad z = 2 \end{cases}$$

$$z = 2$$

$$y + 7(2) = 11 \Rightarrow y = -3$$

$$x - 3(2) = -2 \Rightarrow x = 4$$

Solution: $(4, -3, 2)$

73. $\begin{cases} -x + y - z = -14 \\ 2x - y + z = 21 \\ 3x + 2y + z = 19 \end{cases}$

$$\begin{bmatrix} -1 & 1 & -1 & \vdots & -14 \\ 2 & -1 & 1 & \vdots & 21 \\ 3 & 2 & 1 & \vdots & 19 \end{bmatrix}$$

$$-R_1 \rightarrow \begin{bmatrix} 1 & -1 & 1 & \vdots & 14 \\ 2 & -1 & 1 & \vdots & 21 \\ 3 & 2 & 1 & \vdots & 19 \end{bmatrix}$$

$$\begin{matrix} -2R_1 + R_2 \rightarrow \\ -3R_1 + R_3 \rightarrow \end{matrix} \begin{bmatrix} 1 & -1 & 1 & \vdots & 14 \\ 0 & 1 & -1 & \vdots & -7 \\ 0 & 5 & -2 & \vdots & -23 \end{bmatrix}$$

$$-5R_2 + R_3 \rightarrow \begin{bmatrix} 1 & -1 & 1 & \vdots & 14 \\ 0 & 1 & -1 & \vdots & -7 \\ 0 & 0 & 3 & \vdots & 12 \end{bmatrix}$$

$$\tfrac{1}{3}R_3 \rightarrow \begin{bmatrix} 1 & -1 & 1 & \vdots & 14 \\ 0 & 1 & -1 & \vdots & -7 \\ 0 & 0 & 1 & \vdots & 4 \end{bmatrix}$$

$$\begin{cases} x - y + z = 14 \\ y - z = -7 \\ \quad\quad z = 4 \end{cases}$$

$$z = 4$$

$$y - 4 = -7 \Rightarrow y = -3$$

$$x - (-3) + 4 = 14 \Rightarrow x = 7$$

Solution: $(7, -3, 4)$

75. $\begin{cases} x + 2y - 3z = -28 \\ \quad\;\; 4y + 2z = \;\;\;0 \\ -x + \;\; y - \;\; z = -5 \end{cases}$

$$\begin{bmatrix} 1 & 2 & -3 & \vdots & -28 \\ 0 & 4 & 2 & \vdots & 0 \\ -1 & 1 & -1 & \vdots & -5 \end{bmatrix}$$

$$\begin{matrix} \frac{1}{4}R_2 \to \\ R_1 + R_3 \to \end{matrix} \begin{bmatrix} 1 & 2 & -3 & \vdots & -28 \\ 0 & 1 & \frac{1}{2} & \vdots & 0 \\ 0 & 3 & -4 & \vdots & -33 \end{bmatrix}$$

$$-3R_2 + R_3 \to \begin{bmatrix} 1 & 2 & -3 & \vdots & -28 \\ 0 & 1 & \frac{1}{2} & \vdots & 0 \\ 0 & 0 & -\frac{11}{2} & \vdots & -33 \end{bmatrix}$$

$$-\frac{2}{11}R_3 \to \begin{bmatrix} 1 & 2 & -3 & \vdots & -28 \\ 0 & 1 & \frac{1}{2} & \vdots & 0 \\ 0 & 0 & 1 & \vdots & 6 \end{bmatrix}$$

$$\begin{cases} x + 2y - 3z = -28 \\ \quad\;\; y + \frac{1}{2}z = \;\;\;0 \\ \qquad\qquad z = \;\;\;6 \end{cases}$$

$$z = 6$$

$$y + \frac{1}{2}(6) = 0 \Rightarrow y = -3$$

$$x + 2(-3) - 3(6) = -28 \Rightarrow x = -4$$

Solution: $(-4, -3, 6)$

77. $\begin{cases} x + 2y = 0 \\ -x - \;\; y = 0 \end{cases}$

$$\begin{bmatrix} 1 & 2 & \vdots & 0 \\ -1 & -1 & \vdots & 0 \end{bmatrix}$$

$$R_1 + R_2 \to \begin{bmatrix} 1 & 2 & \vdots & 0 \\ 0 & 1 & \vdots & 0 \end{bmatrix}$$

$$\begin{cases} x + 2y = 0 \\ \quad\;\; y = 0 \end{cases}$$

$$y = 0$$

$$x + 2(0) = 0 \Rightarrow x = 0$$

Solution: $(0, 0)$

79. $\begin{cases} x + 2y + \;\; z = \;\; 8 \\ 3x + 7y + 6z = 26 \end{cases}$

$$\begin{bmatrix} 1 & 2 & 1 & \vdots & 8 \\ 3 & 7 & 6 & \vdots & 26 \end{bmatrix}$$

$$-3R_1 + R_2 \to \begin{bmatrix} 1 & 2 & 1 & \vdots & 8 \\ 0 & 1 & 3 & \vdots & 2 \end{bmatrix}$$

$$\begin{cases} x + 2y + \;\; z = 8 \\ \quad\;\; y + 3z = 2 \end{cases}$$

Let $z = a$.

$$y + 3a = 2 \Rightarrow y = -3a + 2$$

$$x + 2(-3a + 2) + a = 8 \Rightarrow x = \;\; 5a + 4$$

Solution: $(5a + 4, -3a + 2, a)$ where a is a real number

81. $\begin{cases} -x + \;\; y = -22 \\ 3x + 4y = \;\;\;4 \\ 4x - 8y = \;\;32 \end{cases}$

$$\begin{bmatrix} -1 & 1 & \vdots & -22 \\ 3 & 4 & \vdots & 4 \\ 4 & -8 & \vdots & 32 \end{bmatrix}$$

$$-R_1 \to \begin{bmatrix} 1 & -1 & \vdots & 22 \\ 3 & 4 & \vdots & 4 \\ 4 & -8 & \vdots & 32 \end{bmatrix}$$

$$\begin{matrix} -3R_1 + R_2 \to \\ -4R_1 + R_3 \to \end{matrix} \begin{bmatrix} 1 & -1 & \vdots & 22 \\ 0 & 7 & \vdots & -62 \\ 0 & -4 & \vdots & -56 \end{bmatrix}$$

$$\begin{matrix} \frac{1}{7}R_2 \to \\ -\frac{1}{4}R_3 \to \end{matrix} \begin{bmatrix} 1 & -1 & \vdots & 22 \\ 0 & 1 & \vdots & -\frac{62}{7} \\ 0 & 1 & \vdots & 14 \end{bmatrix}$$

$$-R_2 + R_3 \to \begin{bmatrix} 1 & -1 & \vdots & 22 \\ 0 & 1 & \vdots & -\frac{62}{7} \\ 0 & 0 & \vdots & \frac{160}{7} \end{bmatrix}$$

The system is inconsistent and there is no solution.

83. Use the reduced row-echelon form feature of a graphing utility.

$$\begin{cases} 3x + 2y - z + w = 0 \\ x - y + 4z + 2w = 25 \\ -2x + y + 2z - w = 2 \\ x + y + z + w = 6 \end{cases}$$

$$\begin{bmatrix} 3 & 2 & -1 & 1 & \vdots & 0 \\ 1 & -1 & 4 & 2 & \vdots & 25 \\ -2 & 1 & 2 & -1 & \vdots & 2 \\ 1 & 1 & 1 & 1 & \vdots & 6 \end{bmatrix} \Rightarrow \begin{bmatrix} 1 & 0 & 0 & 0 & \vdots & 3 \\ 0 & 1 & 0 & 0 & \vdots & -2 \\ 0 & 0 & 1 & 0 & \vdots & 5 \\ 0 & 0 & 0 & 1 & \vdots & 0 \end{bmatrix}$$

$x = 3$

$y = -2$

$z = 5$

$w = 0$

Solution: $(3, -2, 5, 0)$

85. Use the reduced row-echelon form feature of a graphic utility.

$$\begin{cases} 3x + 3y + 12z = 6 \\ x + y + 4z = 2 \\ 2x + 5y + 20z = 10 \\ -x + 2y + 8z = 4 \end{cases} \begin{bmatrix} 3 & 3 & 12 & \vdots & 6 \\ 1 & 1 & 4 & \vdots & 2 \\ 2 & 5 & 20 & \vdots & 10 \\ -1 & 2 & 8 & \vdots & 4 \end{bmatrix} \Rightarrow \begin{bmatrix} 1 & 0 & 0 & \vdots & 0 \\ 0 & 1 & 4 & \vdots & 2 \\ 0 & 0 & 0 & \vdots & 0 \\ 0 & 0 & 0 & \vdots & 0 \end{bmatrix} \Rightarrow \begin{cases} x = 0 \\ y + 4z = 2 \end{cases}$$

Let $z = a$.

$y = 2 - 4a$

$x = 0$

Solution: $(0, 2 - 4a, a)$ where a is any real number

87. Use the reduced row-echelon form feature of a graphing utility.

$$\begin{cases} 2x + y - z + 2w = -6 \\ 3x + 4y + w = 1 \\ x + 5y + 2z + 6w = -3 \\ 5x + 2y - z - w = 3 \end{cases} \begin{bmatrix} 2 & 1 & -1 & 2 & \vdots & -6 \\ 3 & 4 & 0 & 1 & \vdots & 1 \\ 1 & 5 & 2 & 6 & \vdots & -3 \\ 5 & 2 & -1 & -1 & \vdots & 3 \end{bmatrix} \Rightarrow \begin{bmatrix} 1 & 0 & 0 & 0 & \vdots & 1 \\ 0 & 1 & 0 & 0 & \vdots & 0 \\ 0 & 0 & 1 & 0 & \vdots & 4 \\ 0 & 0 & 0 & 1 & \vdots & -2 \end{bmatrix}$$

$x = 1$

$y = 0$

$z = 4$

$w = -2$

Solution: $(1, 0, 4, -2)$

89. Use the reduced row-echelon form feature of a graphing utility.

$$\begin{cases} x + y + z + w = 0 \\ 2x + 3y + z - 2w = 0 \\ 3x + 5y + z = 0 \end{cases} \begin{bmatrix} 1 & 1 & 1 & 1 & \vdots & 0 \\ 2 & 3 & 1 & -2 & \vdots & 0 \\ 3 & 5 & 1 & 0 & \vdots & 0 \end{bmatrix} \Rightarrow \begin{bmatrix} 1 & 0 & 2 & 0 & \vdots & 0 \\ 0 & 1 & -1 & 0 & \vdots & 0 \\ 0 & 0 & 0 & 1 & \vdots & 0 \end{bmatrix}$$

$$\begin{cases} x + 2z = 0 \\ y - z = 0 \\ w = 0 \end{cases}$$

Let $z = a$. Then $x = -2a$ and $y = a$.

Solution: $(-2a, a, a, 0)$ where a is a real number

91. (a) $\begin{cases} x - 2y + z = -6 \\ y - 5z = 16 \\ z = -3 \end{cases}$

$$y - 5(-3) = 16$$
$$y = 1$$
$$x - 2(1) + (-3) = -6$$
$$x = -1$$

Solution: $(-1, 1, -3)$

(b) $\begin{cases} x + y - 2z = 6 \\ y + 3z = -8 \\ z = -3 \end{cases}$

$$y + 3(-3) = -8$$
$$y = 1$$
$$x + 1 - 2(-3) = 6$$
$$x = -1$$

Solution: $(-1, 1, -3)$

Both systems yield the same solution, $(-1, 1, -3)$.

93. (a) $\begin{cases} x - 4y + 5z = 27 \\ y - 7z = -54 \\ z = 8 \end{cases}$

$$y - 7(8) = -54$$
$$y = 2$$
$$x - 4(2) + 5(8) = 27$$
$$x = -5$$

Solution: $(-5, 2, 8)$

(b) $\begin{cases} x - 6y + z = 15 \\ y + 5z = 42 \\ z = 8 \end{cases}$

$$y + 5(8) = 42$$
$$y = 2$$
$$x - 6(2) + 8 = 15$$
$$x = 19$$

Solution: $(19, 2, 8)$

The systems do *not* yield the same solution.

95. $\begin{cases} a + b + c = 1 \\ 4a + 2b + c = -1 \\ 9a + 3b + c = -5 \end{cases}$

$$\begin{bmatrix} 1 & 1 & 1 & \vdots & 1 \\ 4 & 2 & 1 & \vdots & -1 \\ 9 & 3 & 1 & \vdots & -5 \end{bmatrix} \Rightarrow \begin{bmatrix} 1 & 0 & 0 & \vdots & -1 \\ 0 & 1 & 0 & \vdots & 1 \\ 0 & 0 & 1 & \vdots & 1 \end{bmatrix}$$

$$a = 1$$
$$b = 1$$
$$c = 1$$

So, $f(x) = -x^2 + x + 1$.

97. $\begin{cases} 4a - 2b + c = -15 \\ a - b + c = 7 \\ a + b + c = -3 \end{cases}$

$$\begin{bmatrix} 4 & -2 & 1 & \vdots & -15 \\ 4 & -1 & 1 & \vdots & 7 \\ 1 & 1 & 1 & \vdots & -3 \end{bmatrix} \Rightarrow \begin{bmatrix} 1 & 0 & 0 & \vdots & -9 \\ 0 & 1 & 0 & \vdots & -5 \\ 0 & 0 & 1 & \vdots & 11 \end{bmatrix}$$

$$a = -9$$
$$b = -5$$
$$c = 11$$

So, $f(x) = -9x^2 - 5x + 11$.

99. $\begin{cases} -a + b - c + d = -5 \\ a + b + c + d = -1 \\ 8a + 4b + 2c + d = 1 \\ 27a + 9b + 3c + d = 11 \end{cases}$

$$\begin{bmatrix} -1 & 1 & -1 & 1 & \vdots & -5 \\ 1 & 1 & 1 & 1 & \vdots & -1 \\ 8 & 4 & 2 & 1 & \vdots & 1 \\ 27 & 9 & 3 & 1 & \vdots & 11 \end{bmatrix} \Rightarrow \begin{bmatrix} 1 & 0 & 0 & 0 & \vdots & 1 \\ 0 & 1 & 0 & 0 & \vdots & -2 \\ 0 & 0 & 1 & 0 & \vdots & 1 \\ 0 & 0 & 0 & 1 & \vdots & -1 \end{bmatrix}$$

$$a = 1$$
$$b = -2$$
$$c = 1$$
$$d = -1$$

So, $f(x) = x^3 - 2x^2 + x - 1$.

101. $\begin{cases} -8a + 4b - 2c + d = -7 \\ -a + b - c + d = 2 \\ a + b + c + d = -4 \\ 8a + 4b + 2c + d = -7 \end{cases}$

$$\begin{bmatrix} -8 & 4 & -2 & 1 & \vdots & -7 \\ -1 & 1 & -1 & 1 & \vdots & 2 \\ 1 & 1 & 1 & 1 & \vdots & -4 \\ 8 & 4 & 2 & 1 & \vdots & -7 \end{bmatrix} \Rightarrow \begin{bmatrix} 1 & 0 & 0 & 0 & \vdots & 1 \\ 0 & 1 & 0 & 0 & \vdots & -2 \\ 0 & 0 & 1 & 0 & \vdots & -4 \\ 0 & 0 & 0 & 1 & \vdots & 1 \end{bmatrix}$$

$a = 1$

$b = -2$

$c = -4$

$d = 1$

So, $f(x) = x^3 - 2x^2 - 4x + 1$.

103. $\begin{cases} x + 3y + z = 3 \\ x + 5y + 5z = 1 \\ 2x + 6y + 3z = 8 \end{cases}$

$$\begin{bmatrix} 1 & 3 & 1 & \vdots & 3 \\ 1 & 5 & 5 & \vdots & 1 \\ 2 & 6 & 3 & \vdots & 8 \end{bmatrix}$$

$\begin{matrix} \\ -R_1 + R_2 \to \\ -2R_1 + R_3 \to \end{matrix} \begin{bmatrix} 1 & 3 & 1 & \vdots & 3 \\ 0 & 2 & 4 & \vdots & -2 \\ 0 & 0 & 1 & \vdots & 2 \end{bmatrix}$

$\frac{1}{2}R_2 \to \begin{bmatrix} 1 & 3 & 1 & \vdots & 3 \\ 0 & 1 & 2 & \vdots & -1 \\ 0 & 0 & 1 & \vdots & 2 \end{bmatrix}$ This is a matrix in row echelon form

$\begin{bmatrix} 1 & 3 & \frac{3}{2} & \vdots & 4 \\ 0 & 1 & \frac{7}{4} & \vdots & -\frac{3}{2} \\ 0 & 0 & 1 & \vdots & 2 \end{bmatrix}$ The row-echelon form feature of a graphing utility yields this form.

There are infinitely many matrices in row-echelon form that correspond to the original system of equations. All such matrices will yield the same solution, $(16, -5, 2)$.

105. $\dfrac{4x^2}{(x+1)^2(x-1)} = \dfrac{A}{x-1} + \dfrac{B}{x+1} + \dfrac{C}{(x+1)^2}$

$4x^2 = A(x+1)^2 + B(x+1)(x-1) + C(x-1)$

$4x^2 = A(x^2 + 2x + 1) + B(x^2 - 1) + C(x-1)$

$4x^2 = (A+B)x^2 + (2A+C)x + (A - B - C)$

System of equations: $\begin{cases} A + B = 4 \\ 2A + C = 0 \\ A - B - C = 0 \end{cases}$

$$\begin{bmatrix} 1 & 1 & 0 & \vdots & 4 \\ 2 & 0 & 1 & \vdots & 0 \\ 1 & -1 & -1 & \vdots & 0 \end{bmatrix} \underset{\to}{\text{rref}} \begin{bmatrix} 1 & 0 & 0 & \vdots & 1 \\ 0 & 1 & 0 & \vdots & 3 \\ 0 & 0 & 1 & \vdots & -2 \end{bmatrix}$$

$A = 1, B = 3, C = -2$

So, $\dfrac{4x^2}{(x+1)^2(x-1)} = \dfrac{1}{x-1} + \dfrac{3}{x+1} - \dfrac{2}{(x+1)^2}$.

107. $x = $ amount at 7%

$y = $ amount at 8%

$z = $ amount at 10%

$z = 4x \Rightarrow -4x + z = 0$

$\begin{cases} x + y + z = 1{,}5000{,}000 \\ 0.07x + 0.08y + 0.10z = 130{,}500 \\ -4x + z = 0 \end{cases}$

$$\begin{bmatrix} 1 & 1 & 1 & \vdots & 1{,}500{,}000 \\ 0.07 & 0.08 & 0.10 & \vdots & 130{,}500 \\ -4 & 0 & 1 & \vdots & 0 \end{bmatrix}$$

$\begin{matrix} -0.07R_1 + R_2 \to \\ 4R_1 + R_3 \to \end{matrix} \begin{bmatrix} 1 & 1 & 1 & \vdots & 1{,}500{,}000 \\ 0 & 0.01 & 0.03 & \vdots & 25{,}500 \\ 0 & 4 & 5 & \vdots & 6{,}000{,}000 \end{bmatrix}$

$100R_2 \to \begin{bmatrix} 1 & 1 & 1 & \vdots & 1{,}500{,}000 \\ 0 & 1 & 3 & \vdots & 2{,}550{,}000 \\ 0 & 4 & 5 & \vdots & 6{,}000{,}000 \end{bmatrix}$

$-4R_2 + R_3 \to \begin{bmatrix} 1 & 1 & 1 & \vdots & 1{,}500{,}000 \\ 0 & 1 & 3 & \vdots & 2{,}550{,}000 \\ 0 & 0 & -7 & \vdots & -4{,}200{,}000 \end{bmatrix}$

$-\frac{1}{7}R_3 \to \begin{bmatrix} 1 & 1 & 1 & \vdots & 1{,}500{,}000 \\ 0 & 1 & 3 & \vdots & 2{,}550{,}000 \\ 0 & 0 & 1 & \vdots & 600{,}000 \end{bmatrix}$

$\begin{cases} x + y + z = 1{,}500{,}000 \\ y + 3z = 2{,}550{,}000 \\ z = 600{,}000 \end{cases}$

$y + 3(600{,}000) = 2{,}550{,}000 \Rightarrow y = 750{,}000$

$x + 750{,}000 + 600{,}000 = 1{,}5000{,}000 \Rightarrow x = 150{,}000$

Solution: \$150,000 at 7%, \$750,000 at 8%, \$600,000 at 10%

109. x = number of \$1 bills

y = number of \$5 bills

z = number of \$10 bills

w = number of \$20 bills

$$\begin{cases} x + 5y + 10z + 20w = 95 \\ x + y + z + w = 26 \\ y - 4z = 0 \\ x - 2y = -1 \end{cases}$$

$$\begin{bmatrix} 1 & 5 & 10 & 20 & \vdots & 95 \\ 1 & 1 & 1 & 1 & \vdots & 26 \\ 0 & 1 & -4 & 0 & \vdots & 0 \\ 1 & -2 & 0 & 0 & \vdots & -1 \end{bmatrix} \Rightarrow \begin{bmatrix} 1 & 0 & 0 & 0 & \vdots & 15 \\ 0 & 1 & 0 & 0 & \vdots & 8 \\ 0 & 0 & 1 & 0 & \vdots & 2 \\ 0 & 0 & 0 & 1 & \vdots & 1 \end{bmatrix}$$

$x = 15$

$y = 8$

$z = 2$

$w = 1$

The server has 15 \$1 bills, 8 \$5 bills, 2 \$10 bills and one \$20 bill.

111. $y = ax^2 + bx + c$

$$\begin{cases} a + b + c = 8 \\ 4a + 2b + c = 13 \\ 9a + 3b + c = 20 \end{cases}$$

$$\begin{bmatrix} 1 & 1 & 1 & \vdots & 8 \\ 4 & 2 & 1 & \vdots & 13 \\ 9 & 3 & 1 & \vdots & 20 \end{bmatrix}$$

$$\begin{matrix} -4R_1 + R_2 \to \\ -9R_1 + R_3 \to \end{matrix} \begin{bmatrix} 1 & 1 & 1 & \vdots & 8 \\ 0 & -2 & -3 & \vdots & -19 \\ 0 & -6 & -8 & \vdots & -52 \end{bmatrix}$$

$$\begin{matrix} -\frac{1}{2}R_2 \to \\ -3R_2 + R_3 \to \end{matrix} \begin{bmatrix} 1 & 1 & 1 & \vdots & 8 \\ 0 & 1 & \frac{3}{2} & \vdots & \frac{19}{2} \\ 0 & 0 & 1 & \vdots & 5 \end{bmatrix}$$

$$\begin{cases} a + b + c = 8 \\ b + \frac{3}{2}c = \frac{19}{2} \\ c = 5 \end{cases}$$

$c = 5$

$b + \frac{3}{2}(5) = \frac{19}{2} \Rightarrow b = 2$

$a + 2 + 5 = 8 \Rightarrow a = 1$

Equation of parabola: $y = x^2 + 2x + 5$

113. (a) $(0, 5.0), (15, 9.6), (30, 12.4)$

$y = ax^2 + bx + c$

$$\begin{cases} c = 5 \\ 225a + 15b + c = 9.6 \Rightarrow 225a + 15b = 4.6 \\ 900a + 30b + c = 12.4 \Rightarrow 900a + 30b = 7.4 \end{cases}$$

$$\begin{bmatrix} 225 & 15 & \vdots & 4.6 \\ 900 & 30 & \vdots & 7.4 \end{bmatrix}$$

$$-4R_1 + R_2 \to \begin{bmatrix} 225 & 15 & \vdots & 4.6 \\ 0 & -30 & \vdots & -11 \end{bmatrix}$$

$$\begin{matrix} \frac{1}{225}R_1 \to \\ \left(-\frac{1}{30}\right)R_2 \to \end{matrix} \begin{bmatrix} 1 & \frac{1}{15} & \vdots & \frac{23}{1125} \\ 0 & 1 & \vdots & \frac{11}{30} \end{bmatrix}$$

$$\begin{cases} a + \frac{1}{15}b = \frac{23}{1125} \\ b = \frac{11}{30} \end{cases}$$

$a + \frac{1}{15}\left(\frac{11}{30}\right) = \frac{23}{1125} \Rightarrow a = -\frac{1}{250} = -0.004$

Equation of parabola: $y = -0.004x^2 + 0.367x + 5$

(b)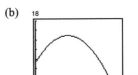

(c) The maximum height is approximately 13 feet and the ball strikes the ground at approximately 104 feet.

(d) The maximum height occurs at the vertex.

$x = -\dfrac{b}{2a} = \dfrac{-0.367}{2(-0.004)} = 45.875$

$y = -0.004(45.875)^2 + 0.367(45.875) + 5$

$= 13.418$ feet

The ball strikes the ground when $y = 0$.

$-0.004x^2 + 0.367x + 5 = 0$

By the Quadratic Formula and using the positive value for x, you have $x \approx 103.793$ feet.

(e) The values found in part (d) are more accurate, but still very close to the estimates found in part (c).

115. (a) $x_1 + x_3 = 600$

$x_1 = x_2 + x_4 \Rightarrow x_1 - x_2 - x_4 = 0$

$x_2 + x_5 = 500$

$x_3 + x_6 = 600$

$x_4 + x_7 = x_6 \Rightarrow x_4 - x_6 + x_7 = 0$

$x_5 + x_7 = 500$

$$\left[\begin{array}{ccccccc|c} 1 & 0 & 1 & 0 & 0 & 0 & 0 & 600 \\ 1 & -1 & 0 & -1 & 0 & 0 & 0 & 0 \\ 0 & 1 & 0 & 0 & 1 & 0 & 0 & 500 \\ 0 & 0 & 1 & 0 & 0 & 1 & 0 & 600 \\ 0 & 0 & 0 & 1 & 0 & -1 & 1 & 0 \\ 0 & 0 & 0 & 0 & 1 & 0 & 1 & 500 \end{array}\right]$$

$$\begin{array}{c} \\ -R_1 + R_2 \rightarrow \\ R_2 + R_3 \rightarrow \\ R_3 + R_4 \rightarrow \\ R_4 + R_5 \rightarrow \\ -R_5 + R_6 \rightarrow \end{array}\left[\begin{array}{ccccccc|c} 1 & 0 & 1 & 0 & 0 & 0 & 0 & 600 \\ 0 & -1 & -1 & -1 & 0 & 0 & 0 & -600 \\ 0 & 0 & -1 & -1 & 1 & 0 & 0 & -100 \\ 0 & 0 & 0 & -1 & 1 & 1 & 0 & 500 \\ 0 & 0 & 0 & 0 & 1 & 0 & 1 & 500 \\ 0 & 0 & 0 & 0 & 0 & 0 & 0 & 0 \end{array}\right]$$

$$\begin{array}{c} \\ -R_3 + R_2 \rightarrow \\ -R_4 + R_3 \rightarrow \\ -R_4 \rightarrow \\ \\ \end{array}\left[\begin{array}{ccccccc|c} 1 & 0 & 1 & 0 & 0 & 0 & 0 & 600 \\ 0 & -1 & 0 & 0 & -1 & 0 & 0 & -500 \\ 0 & 0 & -1 & 0 & 0 & -1 & 0 & -600 \\ 0 & 0 & 0 & 1 & -1 & -1 & 0 & -500 \\ 0 & 0 & 0 & 0 & 1 & 0 & 1 & 500 \\ 0 & 0 & 0 & 0 & 0 & 0 & 1 & 0 \end{array}\right]$$

$$\begin{array}{c} \\ -R_2 \rightarrow \\ -R_3 \rightarrow \\ \\ \\ \end{array}\left[\begin{array}{ccccccc|c} 1 & 0 & 1 & 0 & 0 & 0 & 0 & 600 \\ 0 & 1 & 0 & 0 & 1 & 0 & 0 & 500 \\ 0 & 0 & 1 & 0 & 0 & 1 & 0 & 600 \\ 0 & 0 & 0 & 1 & -1 & -1 & 0 & -500 \\ 0 & 0 & 0 & 0 & 1 & 0 & 1 & 500 \\ 0 & 0 & 0 & 0 & 0 & 0 & 0 & 0 \end{array}\right]$$

$$\begin{cases} x_1 + x_3 = 600 \\ x_2 + x_5 = 500 \\ x_3 + x_6 = 600 \\ x_4 - x_5 - x_6 = -500 \\ x_5 + x_7 = 500 \end{cases}$$

Let $x_7 = t$ and $x_6 = s$, then $x_5 = 500 - t$,

$x_4 = -500 + s + (500 - t) = s - t$,

$x_3 = 600 - s, x_2 = 500 - (500 - t) = t$,

$x_1 = 600 - (600 - s) = s$.

Solution: $(s, t, 600 - s, s - t, 500 - t, s, t)$

(b) When $x_6 = 0$ and $x_7 = 0$, $s = 0$ and $t = 0$: $x_1 = 0, x_2 = 0, x_3 = 600, x_4 = 0, x_5 = 500, x_6 = 0, x_7 = 0$

(c) When $x_5 = 400$ and $x_6 = 500$, $s = 500$ and $t = 100$: $x_1 = 500, x_2 = 100, x_3 = 100, x_4 = 400, x_5 = 400,$ $x_6 = 500, x_7 = 100$

117. False. It is a 2×4 matrix.

119. $z = a$

$y = -4a + 1$

$x = -3a - 2$

One possible system is:

$$\begin{cases} x + y + 7z = (-3a - 2) + (-4a + 1) + 7a = -1 \\ x + 2y + 11z = (-3a - 2) + 2(-4a + 1) + 11a = 0 \\ 2x + y + 10z = 2(-3a - 2) + (-4a + 1) + 10a = -3 \end{cases} \text{ or } \begin{cases} x + y + 7z = -1 \\ x + 2y + 11z = 0 \\ 2x + y + 10z = -3 \end{cases}$$

(Note that the coefficients of $x, y,$ and z have been chosen so that the a-terms cancel.)

121. 1. Interchange two rows.

2. Multiply a row by a nonzero constant.

3. Add a multiple of one row to another row.

123. They are the same.

Section 10.2   Operations with Matrices

- $A = B$ if and only if they have the same order and $a_{ij} = b_{ij}$.

- You should be able to perform the operations of matrix addition, scalar multiplication, and matrix multiplication.

- Some properties of matrix addition and scalar multiplication are:

  (a) $A + B = B + A$

  (b) $A + (B + C) = (A + B) + C$

  (c) $(cd)A = c(dA)$

  (d) $1A = A$

  (e) $c(A + B) = cA + cB$

  (f) $(c + d)A = cA + dA$

- You should remember that $AB \neq BA$ in general.

- Some properties of matrix multiplication are:

  (a) $A(BC) = (AB)C$ (b) $A(B + C) = AB + AC$

  (c) $(A + B)C = AC + BC$ (d) $c(AB) = (cA)B = A(cB)$

- You should know that I_n, the identity matrix of order n, is an $n \times n$ matrix consisting of ones on its main diagonal and zeros elsewhere. If A is an $n \times n$ matrix, then $AI_n = I_nA = A$.

1. equal

3. zero; O

5. (a) (iii)

   (b) (iv)

   (c) (i)

   (d) (v)

   (e) (ii)

7. $x = -4, y = 22$

9. $2x + 1 = 5, 3x = 6, 3y - 5 = 4$

   $x = 2, y = 3$

11. (a) $A + B = \begin{bmatrix} 1 & -1 \\ 2 & -1 \end{bmatrix} + \begin{bmatrix} 2 & -1 \\ -1 & 8 \end{bmatrix} = \begin{bmatrix} 1+2 & -1-1 \\ 2-1 & -1+8 \end{bmatrix} = \begin{bmatrix} 3 & -2 \\ 1 & 7 \end{bmatrix}$

(b) $A - B = \begin{bmatrix} 1 & -1 \\ 2 & -1 \end{bmatrix} - \begin{bmatrix} 2 & -1 \\ -1 & 8 \end{bmatrix} = \begin{bmatrix} 1-2 & -1+1 \\ 2+1 & -1-8 \end{bmatrix} = \begin{bmatrix} -1 & 0 \\ 3 & -9 \end{bmatrix}$

(c) $3A = 3\begin{bmatrix} 1 & -1 \\ 2 & -1 \end{bmatrix} = \begin{bmatrix} 3(1) & 3(-1) \\ 3(2) & 3(-1) \end{bmatrix} = \begin{bmatrix} 3 & -3 \\ 6 & -3 \end{bmatrix}$

(d) $3A - 2B = \begin{bmatrix} 3 & -3 \\ 6 & -3 \end{bmatrix} - 2\begin{bmatrix} 2 & -1 \\ -1 & 8 \end{bmatrix} = \begin{bmatrix} 3 & -3 \\ 6 & -3 \end{bmatrix} + \begin{bmatrix} -4 & 2 \\ 2 & -16 \end{bmatrix} = \begin{bmatrix} -1 & -1 \\ 8 & -19 \end{bmatrix}$

13. $A = \begin{bmatrix} 8 & -1 \\ 2 & 3 \\ -4 & 5 \end{bmatrix}, B = \begin{bmatrix} 1 & 6 \\ -1 & -5 \\ 1 & 10 \end{bmatrix}$

(a) $A + B = \begin{bmatrix} 8 & -1 \\ 2 & 3 \\ -4 & 5 \end{bmatrix} + \begin{bmatrix} 1 & 6 \\ -1 & -5 \\ 1 & 10 \end{bmatrix} = \begin{bmatrix} 8+1 & -1+6 \\ 2-1 & 3-5 \\ -4+1 & 5+10 \end{bmatrix} = \begin{bmatrix} 9 & 5 \\ 1 & -2 \\ -3 & 15 \end{bmatrix}$

(b) $A - B = \begin{bmatrix} 8 & -1 \\ 2 & 3 \\ -4 & 5 \end{bmatrix} - \begin{bmatrix} 1 & 6 \\ -1 & -5 \\ 1 & 10 \end{bmatrix} = \begin{bmatrix} 8-1 & -1-6 \\ 2-(-1) & 3-(-5) \\ -4-1 & 5-10 \end{bmatrix} = \begin{bmatrix} 7 & -7 \\ 3 & 8 \\ -5 & -5 \end{bmatrix}$

(c) $3A = 3\begin{bmatrix} 8 & -1 \\ 2 & 3 \\ -4 & 5 \end{bmatrix} = \begin{bmatrix} 3(8) & 3(-1) \\ 3(2) & 3(3) \\ 3(-4) & 3(5) \end{bmatrix} = \begin{bmatrix} 24 & -3 \\ 6 & 9 \\ -12 & 15 \end{bmatrix}$

(d) $3A - 2B = \begin{bmatrix} 24 & -3 \\ 6 & 9 \\ -12 & 15 \end{bmatrix} - 2\begin{bmatrix} 1 & 6 \\ -1 & -5 \\ 1 & 10 \end{bmatrix} = \begin{bmatrix} 24-2 & -3-12 \\ 6+2 & 9+10 \\ -12-2 & 15-20 \end{bmatrix} = \begin{bmatrix} 22 & -15 \\ 8 & 19 \\ -14 & -5 \end{bmatrix}$

15. $A = \begin{bmatrix} 4 & 5 & -1 & 3 & 4 \\ 1 & 2 & -2 & -1 & 0 \end{bmatrix}, B = \begin{bmatrix} 1 & 0 & -1 & 1 & 0 \\ -6 & 8 & 2 & -3 & -7 \end{bmatrix}$

(a) $A + B = \begin{bmatrix} 4 & 5 & -1 & 3 & 4 \\ 1 & 2 & -2 & -1 & 0 \end{bmatrix} + \begin{bmatrix} 1 & 0 & -1 & 1 & 0 \\ -6 & 8 & 2 & -3 & -7 \end{bmatrix} = \begin{bmatrix} 4+1 & 5+0 & -1-1 & 3+1 & 4+0 \\ 1-6 & 2+8 & -2+2 & -1-3 & 0-7 \end{bmatrix}$

$= \begin{bmatrix} 5 & 5 & -2 & 4 & 4 \\ -5 & 10 & 0 & -4 & -7 \end{bmatrix}$

(b) $A - B = \begin{bmatrix} 4 & 5 & -1 & 3 & 4 \\ 1 & 2 & -2 & -1 & 0 \end{bmatrix} - \begin{bmatrix} 1 & 0 & -1 & 1 & 0 \\ -6 & 8 & 2 & -3 & -7 \end{bmatrix} = \begin{bmatrix} 4-1 & 5-0 & -1-(-1) & 3-1 & 4-0 \\ 1-(-6) & 2-8 & -2-2 & -1-(-3) & 0-(-7) \end{bmatrix}$

$= \begin{bmatrix} 3 & 5 & 0 & 2 & 4 \\ 7 & -6 & -4 & 2 & 7 \end{bmatrix}$

(c) $3A = 3\begin{bmatrix} 4 & 5 & -1 & 3 & 4 \\ 1 & 2 & -2 & -1 & 0 \end{bmatrix} = \begin{bmatrix} 3(4) & 3(5) & 3(-1) & 3(3) & 3(4) \\ 3(1) & 3(2) & 3(-2) & 3(-1) & 3(0) \end{bmatrix} = \begin{bmatrix} 12 & 15 & -3 & 9 & 12 \\ 3 & 6 & -6 & -3 & 0 \end{bmatrix}$

(d) $3A - 2B = \begin{bmatrix} 12 & 15 & -3 & 9 & 12 \\ 3 & 6 & -6 & -3 & 0 \end{bmatrix} - 2\begin{bmatrix} 1 & 0 & -1 & 1 & 0 \\ -6 & 8 & 2 & -3 & -7 \end{bmatrix} = \begin{bmatrix} 12-2 & 15+0 & -3+2 & 9-2 & 12-0 \\ 3+12 & 6-16 & -6-4 & -3+6 & 0+14 \end{bmatrix}$

$= \begin{bmatrix} 10 & 15 & -1 & 7 & 12 \\ 15 & -10 & -10 & 3 & 14 \end{bmatrix}$

17. $A = \begin{bmatrix} 6 & 0 & 3 \\ -1 & -4 & 0 \end{bmatrix}$, $B = \begin{bmatrix} 8 & -1 \\ 4 & -3 \end{bmatrix}$

    (a) $A + B$ is not possible. A and B do not have the same order.

    (b) $A - B$ is not possible. A and B do not have the same order.

    (c) $3A = \begin{bmatrix} 18 & 0 & 9 \\ -3 & -12 & 0 \end{bmatrix}$

    (d) $3A - 2B$ is not possible. A and B do not have the same order.

19. $\begin{bmatrix} -5 & 0 \\ 3 & -6 \end{bmatrix} + \begin{bmatrix} 7 & 1 \\ -2 & -1 \end{bmatrix} + \begin{bmatrix} -10 & -8 \\ 14 & 6 \end{bmatrix} = \begin{bmatrix} -5 + 7 + (-10) & 0 + 1 + (-8) \\ 3 + (-2) + 14 & -6 + (-1) + 6 \end{bmatrix} = \begin{bmatrix} -8 & -7 \\ 15 & -1 \end{bmatrix}$

21. $4\left(\begin{bmatrix} -4 & 0 & 1 \\ 0 & 2 & 3 \end{bmatrix} - \begin{bmatrix} 2 & 1 & -2 \\ 3 & -6 & 0 \end{bmatrix} \right) = 4\begin{bmatrix} -6 & -1 & 3 \\ -3 & 8 & 3 \end{bmatrix} = \begin{bmatrix} -24 & -4 & 12 \\ -12 & 32 & 12 \end{bmatrix}$

23. $-3\left(\begin{bmatrix} 0 & -3 \\ 7 & 2 \end{bmatrix} + \begin{bmatrix} -6 & 3 \\ 8 & 1 \end{bmatrix} \right) - 2\begin{bmatrix} 4 & -4 \\ 7 & -9 \end{bmatrix} = -3\begin{bmatrix} -6 & 0 \\ 15 & 3 \end{bmatrix} - \begin{bmatrix} 8 & -8 \\ 14 & -18 \end{bmatrix} = \begin{bmatrix} 18 & 0 \\ -45 & -9 \end{bmatrix} - \begin{bmatrix} 8 & -8 \\ 14 & -18 \end{bmatrix} = \begin{bmatrix} 10 & 8 \\ -59 & 9 \end{bmatrix}$

25. $\frac{3}{7}\begin{bmatrix} 2 & 5 \\ -1 & -4 \end{bmatrix} + 6\begin{bmatrix} -3 & 0 \\ 2 & 2 \end{bmatrix} \approx \begin{bmatrix} -17.143 & 2.143 \\ 11.571 & 10.286 \end{bmatrix}$

27. $-\begin{bmatrix} 3.211 & 6.829 \\ -1.004 & 4.914 \\ 0.055 & -3.889 \end{bmatrix} - \begin{bmatrix} -1.630 & -3.090 \\ 5.256 & 8.335 \\ -9.768 & 4.251 \end{bmatrix} = \begin{bmatrix} -1.581 & -3.739 \\ -4.252 & -13.249 \\ 9.713 & -0.362 \end{bmatrix}$

29. $X = 3\begin{bmatrix} -2 & -1 \\ 1 & 0 \\ 3 & -4 \end{bmatrix} - 2\begin{bmatrix} 0 & 3 \\ 2 & 0 \\ -4 & -1 \end{bmatrix} = \begin{bmatrix} -6 & -3 \\ 3 & 0 \\ 9 & -12 \end{bmatrix} - \begin{bmatrix} 0 & 6 \\ 4 & 0 \\ -8 & -2 \end{bmatrix} = \begin{bmatrix} -6 & -9 \\ -1 & 0 \\ 17 & -10 \end{bmatrix}$

31. $X = -\frac{3}{2}A + \frac{1}{2}B = -\frac{3}{2}\begin{bmatrix} -2 & -1 \\ 1 & 0 \\ 3 & -4 \end{bmatrix} + \frac{1}{2}\begin{bmatrix} 0 & 3 \\ 2 & 0 \\ -4 & -1 \end{bmatrix} = \begin{bmatrix} 3 & \frac{3}{2} \\ -\frac{3}{2} & 0 \\ -\frac{9}{2} & 6 \end{bmatrix} + \begin{bmatrix} 0 & \frac{3}{2} \\ 1 & 0 \\ -2 & -\frac{1}{2} \end{bmatrix} = \begin{bmatrix} 3 & 3 \\ -\frac{1}{2} & 0 \\ -\frac{13}{2} & \frac{11}{2} \end{bmatrix}$

33. A is 3×2 and B is 3×3. AB is not possible.

35. A is 3×2, B is $2 \times 2 \Rightarrow AB$ is 3×2.

$A = \begin{bmatrix} -1 & 6 \\ -4 & 5 \\ 0 & 3 \end{bmatrix}$, $B = \begin{bmatrix} 2 & 3 \\ 0 & 9 \end{bmatrix}$

$AB = \begin{bmatrix} -1 & 6 \\ -4 & 5 \\ 0 & 3 \end{bmatrix}\begin{bmatrix} 2 & 3 \\ 0 & 9 \end{bmatrix} = \begin{bmatrix} (-1)(2) + (6)(0) & (-1)(3) + (6)(9) \\ (-4)(2) + (5)(0) & (-4)(3) + (5)(9) \\ (0)(2) + (3)(0) & (0)(3) + (3)(9) \end{bmatrix} = \begin{bmatrix} -2 & 51 \\ -8 & 33 \\ 0 & 27 \end{bmatrix}$

37. A is 3×3, B is $3 \times 3 \Rightarrow AB$ is 3×3.

$AB = \begin{bmatrix} 5 & 0 & 0 \\ 0 & -8 & 0 \\ 0 & 0 & 7 \end{bmatrix}\begin{bmatrix} \frac{1}{5} & 0 & 0 \\ 0 & -\frac{1}{8} & 0 \\ 0 & 0 & \frac{1}{2} \end{bmatrix} = \begin{bmatrix} 1 & 0 & 0 \\ 0 & 1 & 0 \\ 0 & 0 & \frac{7}{2} \end{bmatrix}$

39. A is 2×1, B is $1 \times 4 \Rightarrow AB$ is 2×4.

$$\begin{bmatrix} 10 \\ 12 \end{bmatrix} \begin{bmatrix} 6 & -2 & 1 & 6 \end{bmatrix} = \begin{bmatrix} 60 & -20 & 10 & 60 \\ 72 & -24 & 12 & 72 \end{bmatrix}$$

41. $A = \begin{bmatrix} 7 & 5 & -4 \\ -2 & 5 & 1 \\ 10 & -4 & -7 \end{bmatrix}$, $B = \begin{bmatrix} 2 & -2 & 3 \\ 8 & 1 & 4 \\ -4 & 2 & -8 \end{bmatrix}$

$$AB = \begin{bmatrix} 7 & 5 & -4 \\ -2 & 5 & 1 \\ 10 & -4 & -7 \end{bmatrix} \begin{bmatrix} 2 & -2 & 3 \\ 8 & 1 & 4 \\ -4 & 2 & -8 \end{bmatrix} = \begin{bmatrix} 70 & -17 & 73 \\ 32 & 11 & 6 \\ 16 & -38 & 70 \end{bmatrix}$$

43. $\begin{bmatrix} -3 & 8 & -6 & 8 \\ -12 & 15 & 9 & 6 \\ 5 & -1 & 1 & 5 \end{bmatrix} \begin{bmatrix} 3 & 1 & 6 \\ 24 & 15 & 14 \\ 16 & 10 & 21 \\ 8 & -4 & 10 \end{bmatrix} = \begin{bmatrix} 151 & 25 & 48 \\ 516 & 279 & 387 \\ 47 & -20 & 87 \end{bmatrix}$

45. A is 2×4, B is 2×4. AB is not possible.

47. (a) $AB = \begin{bmatrix} 1 & 2 \\ 4 & 2 \end{bmatrix} \begin{bmatrix} 2 & -1 \\ -1 & 8 \end{bmatrix} = \begin{bmatrix} (1)(2) + (2)(-1) & (1)(-1) + (2)(8) \\ (4)(2) + (2)(-1) & (4)(-1) + (2)(8) \end{bmatrix} = \begin{bmatrix} 0 & 15 \\ 6 & 12 \end{bmatrix}$

(b) $BA = \begin{bmatrix} 2 & -1 \\ -1 & 8 \end{bmatrix} \begin{bmatrix} 1 & 2 \\ 4 & 2 \end{bmatrix} = \begin{bmatrix} (2)(1) + (-1)(4) & (2)(2) + (-1)(2) \\ (-1)(1) + (8)(4) & (-1)(2) + (8)(2) \end{bmatrix} = \begin{bmatrix} -2 & 2 \\ 31 & 14 \end{bmatrix}$

(c) $A^2 = \begin{bmatrix} 1 & 2 \\ 4 & 2 \end{bmatrix} \begin{bmatrix} 1 & 2 \\ 4 & 2 \end{bmatrix} = \begin{bmatrix} (1)(1) + (2)(4) & (1)(2) + (2)(2) \\ (4)(1) + (2)(4) & (4)(2) + (2)(2) \end{bmatrix} = \begin{bmatrix} 9 & 6 \\ 12 & 12 \end{bmatrix}$

49. (a) $AB = \begin{bmatrix} 3 & -1 \\ 1 & 3 \end{bmatrix} \begin{bmatrix} 1 & -3 \\ 3 & 1 \end{bmatrix} = \begin{bmatrix} (3)(1) + (-1)(3) & (3)(-3) + (-1)(1) \\ (1)(1) + (3)(3) & (1)(-3) + (3)(1) \end{bmatrix} = \begin{bmatrix} 0 & -10 \\ 10 & 0 \end{bmatrix}$

(b) $BA = \begin{bmatrix} 1 & -3 \\ 3 & 1 \end{bmatrix} \begin{bmatrix} 3 & -1 \\ 1 & 3 \end{bmatrix} = \begin{bmatrix} (1)(3) + (-3)(1) & (1)(-1) + (-3)(3) \\ (3)(3) + (1)(1) & (3)(-1) + (1)(3) \end{bmatrix} = \begin{bmatrix} 0 & -10 \\ 10 & 0 \end{bmatrix}$

(c) $A^2 = \begin{bmatrix} 3 & -1 \\ 1 & 3 \end{bmatrix} \begin{bmatrix} 3 & -1 \\ 1 & 3 \end{bmatrix} = \begin{bmatrix} (3)(3) + (-1)(1) & (3)(-1) + (-1)(3) \\ (1)(3) + (3)(1) & (1)(-1) + (3)(3) \end{bmatrix} = \begin{bmatrix} 8 & -6 \\ 6 & 8 \end{bmatrix}$

51. (a) $AB = \begin{bmatrix} 7 \\ 8 \\ -1 \end{bmatrix} \begin{bmatrix} 1 & 1 & 2 \end{bmatrix} = \begin{bmatrix} 7(1) & 7(1) & 7(2) \\ 8(1) & 8(1) & 8(2) \\ -1(1) & -1(1) & -1(2) \end{bmatrix} = \begin{bmatrix} 7 & 7 & 4 \\ 8 & 8 & 16 \\ -1 & -1 & -2 \end{bmatrix}$

(b) $BA = \begin{bmatrix} 1 & 1 & 2 \end{bmatrix} \begin{bmatrix} 7 \\ 8 \\ -1 \end{bmatrix} = \begin{bmatrix} (1)(7) + (1)(8) + (2)(-1) \end{bmatrix} = \begin{bmatrix} 13 \end{bmatrix}$

(c) A^2 is not possible.

53. $\begin{bmatrix} 3 & 1 \\ 0 & -2 \end{bmatrix} \begin{bmatrix} 1 & 0 \\ -2 & 2 \end{bmatrix} \begin{bmatrix} 1 & 0 \\ 2 & 4 \end{bmatrix} = \begin{bmatrix} 1 & 2 \\ 4 & -4 \end{bmatrix} \begin{bmatrix} 1 & 0 \\ 2 & 4 \end{bmatrix} = \begin{bmatrix} 5 & 8 \\ -4 & -16 \end{bmatrix}$

55. $\begin{bmatrix} 0 & 2 & -2 \\ 4 & 1 & 2 \end{bmatrix} \left(\begin{bmatrix} 4 & 0 \\ 0 & -1 \\ -1 & 2 \end{bmatrix} + \begin{bmatrix} -2 & 3 \\ -3 & 5 \\ 0 & -3 \end{bmatrix} \right) = \begin{bmatrix} 0 & 2 & -2 \\ 4 & 1 & 2 \end{bmatrix} \begin{bmatrix} 2 & 3 \\ -3 & 4 \\ -1 & -1 \end{bmatrix} = \begin{bmatrix} -4 & 10 \\ 3 & 14 \end{bmatrix}$

57. (a) $\begin{bmatrix} -1 & 1 \\ -2 & 1 \end{bmatrix}\begin{bmatrix} x_1 \\ x_2 \end{bmatrix} = \begin{bmatrix} 4 \\ 0 \end{bmatrix}$

(b) $\begin{bmatrix} -1 & 1 & \vdots & 4 \\ -2 & 1 & \vdots & 0 \end{bmatrix}$

$-R_2 + R_1 \rightarrow \begin{bmatrix} 1 & 0 & \vdots & 4 \\ -2 & 1 & \vdots & 0 \end{bmatrix}$

$2R_1 + R_2 \rightarrow \begin{bmatrix} 1 & 0 & \vdots & 4 \\ 0 & 1 & \vdots & 8 \end{bmatrix}$

$X = \begin{bmatrix} 4 \\ 8 \end{bmatrix}$

59. (a) $\begin{bmatrix} -2 & -3 \\ 6 & 1 \end{bmatrix}\begin{bmatrix} x_1 \\ x_2 \end{bmatrix} = \begin{bmatrix} -4 \\ -36 \end{bmatrix}$

(b) $\begin{bmatrix} -2 & -3 & \vdots & -4 \\ 6 & 1 & \vdots & -36 \end{bmatrix}$

$3R_1 + R_2 \rightarrow \begin{bmatrix} -2 & -3 & \vdots & -4 \\ 0 & -8 & \vdots & -48 \end{bmatrix}$

$\begin{matrix} -\frac{1}{2}R_1 \rightarrow \\ -\frac{1}{8}R_2 \rightarrow \end{matrix} \begin{bmatrix} 1 & \frac{3}{2} & \vdots & 2 \\ 0 & 1 & \vdots & 6 \end{bmatrix}$

$-\frac{3}{2}R_2 + R_1 \rightarrow \begin{bmatrix} 1 & 0 & \vdots & -7 \\ 0 & 1 & \vdots & 6 \end{bmatrix}$

$X = \begin{bmatrix} -7 \\ 6 \end{bmatrix}$

61. (a) $\begin{bmatrix} 1 & -2 & 3 \\ -1 & 3 & -1 \\ 2 & -5 & 5 \end{bmatrix}\begin{bmatrix} x_1 \\ x_2 \\ x_3 \end{bmatrix} = \begin{bmatrix} 9 \\ -6 \\ 17 \end{bmatrix}$

(b) $\begin{bmatrix} 1 & -2 & 3 & \vdots & 9 \\ -1 & 3 & -1 & \vdots & -6 \\ 2 & -5 & 5 & \vdots & 17 \end{bmatrix}$

$\begin{matrix} R_1 + R_2 \rightarrow \\ -2R_2 + R_3 \rightarrow \end{matrix} \begin{bmatrix} 1 & -2 & 3 & \vdots & 9 \\ 0 & 1 & 2 & \vdots & 3 \\ 0 & -1 & -1 & \vdots & -1 \end{bmatrix}$

$\begin{matrix} 2R_2 + R_1 \rightarrow \\ \\ R_2 + R_3 \rightarrow \end{matrix} \begin{bmatrix} 1 & 0 & 7 & \vdots & 15 \\ 0 & 1 & 2 & \vdots & 3 \\ 0 & 0 & 1 & \vdots & 2 \end{bmatrix}$

$\begin{matrix} -7R_3 + R_1 \rightarrow \\ -2R_3 + R_2 \rightarrow \end{matrix} \begin{bmatrix} 1 & 0 & 0 & \vdots & 1 \\ 0 & 1 & 0 & \vdots & -1 \\ 0 & 0 & 1 & \vdots & 2 \end{bmatrix}$

$X = \begin{bmatrix} 1 \\ -1 \\ 2 \end{bmatrix}$

63. (a) $\begin{bmatrix} 1 & -5 & 2 \\ -3 & 1 & -1 \\ 0 & -2 & 5 \end{bmatrix}\begin{bmatrix} x_1 \\ x_2 \\ x_3 \end{bmatrix} = \begin{bmatrix} -20 \\ 8 \\ -16 \end{bmatrix}$

(b) $\begin{bmatrix} 1 & -5 & 2 & \vdots & -20 \\ -3 & 1 & -1 & \vdots & 8 \\ 0 & -2 & 5 & \vdots & -16 \end{bmatrix}$

$3R_1 + R_2 \rightarrow \begin{bmatrix} 1 & -5 & 2 & \vdots & -20 \\ 0 & -14 & 5 & \vdots & -52 \\ 0 & -2 & 5 & \vdots & -16 \end{bmatrix}$

$-R_3 + R_2 \rightarrow \begin{bmatrix} 1 & -5 & 2 & \vdots & -20 \\ 0 & -12 & 0 & \vdots & -36 \\ 0 & -2 & 5 & \vdots & -16 \end{bmatrix}$

$-\frac{1}{12}R_2 \rightarrow \begin{bmatrix} 1 & -5 & 2 & \vdots & -20 \\ 0 & 1 & 0 & \vdots & 3 \\ 0 & -2 & 5 & \vdots & -16 \end{bmatrix}$

$\begin{matrix} 5R_2 + R_1 \rightarrow \\ \\ 2R_2 + R_3 \rightarrow \end{matrix} \begin{bmatrix} 1 & 0 & 2 & \vdots & -5 \\ 0 & 1 & 0 & \vdots & 3 \\ 0 & 0 & 5 & \vdots & -10 \end{bmatrix}$

$\frac{1}{5}R_3 \rightarrow \begin{bmatrix} 1 & 0 & 2 & \vdots & -5 \\ 0 & 1 & 0 & \vdots & 3 \\ 0 & 0 & 1 & \vdots & -2 \end{bmatrix}$

$-2R_3 + R_1 \rightarrow \begin{bmatrix} 1 & 0 & 0 & \vdots & -1 \\ 0 & 1 & 0 & \vdots & 3 \\ 0 & 0 & 1 & \vdots & -2 \end{bmatrix}$

$X = \begin{bmatrix} -1 \\ 3 \\ -2 \end{bmatrix}$

65. $1.2\begin{bmatrix} 70 & 50 & 25 \\ 35 & 100 & 70 \end{bmatrix} = \begin{bmatrix} 84 & 60 & 30 \\ 42 & 120 & 84 \end{bmatrix}$

67. (a)

	Farmer's Market	Fruit Stand	Fruit Farm	
$A =$	125	100	75	Apples
	100	175	125	Peaches

The entries represent the number of bushels of each crop that are shipped to each outlet.

(b) $B = \begin{bmatrix} \$3.50 & \$6.00 \end{bmatrix}$

The entries represent the profit per bushel of each crop.

(c) $BA = \begin{bmatrix} \$3.50 & \$6.00 \end{bmatrix}\begin{bmatrix} 125 & 100 & 75 \\ 100 & 175 & 125 \end{bmatrix}$

$= \begin{bmatrix} \$1037.50 & \$1400.00 & \$1012.50 \end{bmatrix}$

The entries represent the profits from both crops at each of the three outlets.

69. $ST = \begin{bmatrix} 3 & 2 & 2 & 3 & 0 \\ 0 & 2 & 3 & 4 & 3 \\ 4 & 2 & 1 & 3 & 2 \end{bmatrix} \begin{bmatrix} 840 & 1100 \\ 1200 & 1350 \\ 1450 & 1650 \\ 2650 & 3000 \\ 3050 & 3200 \end{bmatrix} = \begin{bmatrix} \$15,770 & \$18,300 \\ \$26,500 & \$29,250 \\ \$21,260 & \$24,150 \end{bmatrix}$

The entries represent the wholesale and retail values of the inventories at the three outlets.

71. $P^3 = P^2 P = \begin{bmatrix} 0.40 & 0.15 & 0.15 \\ 0.28 & 0.53 & 0.17 \\ 0.32 & 0.32 & 0.68 \end{bmatrix} \begin{bmatrix} 0.6 & 0.1 & 0.1 \\ 0.2 & 0.7 & 0.1 \\ 0.2 & 0.2 & 0.8 \end{bmatrix} = \begin{bmatrix} 0.300 & 0.175 & 0.175 \\ 0.308 & 0.433 & 0.217 \\ 0.392 & 0.392 & 0.608 \end{bmatrix}$

$P^4 = P^3 P = \begin{bmatrix} 0.300 & 0.175 & 0.175 \\ 0.308 & 0.433 & 0.217 \\ 0.392 & 0.392 & 0.608 \end{bmatrix} \begin{bmatrix} 0.6 & 0.1 & 0.1 \\ 0.2 & 0.7 & 0.1 \\ 0.2 & 0.2 & 0.8 \end{bmatrix} = \begin{bmatrix} 0.250 & 0.188 & 0.188 \\ 0.315 & 0.377 & 0.248 \\ 0.435 & 0.435 & 0.565 \end{bmatrix}$

$P^5 = P^4 P = \begin{bmatrix} 0.250 & 0.188 & 0.188 \\ 0.315 & 0.377 & 0.248 \\ 0.435 & 0.435 & 0.565 \end{bmatrix} \begin{bmatrix} 0.6 & 0.1 & 0.1 \\ 0.2 & 0.7 & 0.1 \\ 0.2 & 0.2 & 0.8 \end{bmatrix} = \begin{bmatrix} 0.225 & 0.194 & 0.194 \\ 0.314 & 0.345 & 0.267 \\ 0.461 & 0.461 & 0.539 \end{bmatrix}$

$P^6 = \begin{bmatrix} 0.213 & 0.197 & 0.197 \\ 0.311 & 0.326 & 0.280 \\ 0.477 & 0.477 & 0.523 \end{bmatrix}$

$P^7 = \begin{bmatrix} 0.206 & 0.198 & 0.198 \\ 0.308 & 0.316 & 0.288 \\ 0.486 & 0.486 & 0.514 \end{bmatrix}$

$P^8 = \begin{bmatrix} 0.203 & 0.199 & 0.199 \\ 0.305 & 0.309 & 0.292 \\ 0.492 & 0.492 & 0.508 \end{bmatrix}$

As *P* is raised to higher and higher powers, the resulting matrices appear to be approaching the matrix

$\begin{bmatrix} 0.2 & 0.2 & 0.2 \\ 0.3 & 0.3 & 0.3 \\ 0.5 & 0.5 & 0.5 \end{bmatrix}$.

73. (a)

$AB = \begin{bmatrix} 40 & 64 & 52 \\ 60 & 82 & 76 \\ 76 & 96 & 84 \end{bmatrix} \begin{bmatrix} 3.45 & 1.20 \\ 3.65 & 1.30 \\ 3.85 & 1.45 \end{bmatrix} = \begin{matrix} \text{Sales \$} & \text{Profit} \\ \begin{bmatrix} 571.8 & 206.6 \\ 798.9 & 288.8 \\ 936 & 337.8 \end{bmatrix} \end{matrix}$

The entries represent the total sales (in dollars) and the profit (in dollars) for milk on Friday, Saturday, and Sunday.

(b) Total profit = \$206.60 + \$288.80 + \$337.80 = \$833.20

75. (a)   Bicycled   Jogged   Walked

$B = \begin{bmatrix} 2 & 0.5 & 3 \end{bmatrix}$ 20-minute time periods

(b) $BA = \begin{bmatrix} 2 & 0.5 & 3 \end{bmatrix} \begin{bmatrix} 109 & 136 \\ 127 & 159 \\ 64 & 79 \end{bmatrix} = \begin{matrix} \text{120-pound} & \text{150-pound} \\ \text{person} & \text{person} \\ \begin{bmatrix} 473.5 & 588.5 \end{bmatrix} \end{matrix}$ Calories burned

The entries represent the total calories burned.

77. True.

The sum of two matrices of different orders is undefined.

For 79–85, *A* is of order 2 × 3, *B* is of order 2 × 3, *C* is of order 3 × 2, and *D* is of order 2 × 2.

79. $A + 2C$ is not possible. *A* and *C* are not of the same order.

81. AB is not possible. The number of columns of *A* does not equal the number of rows of *B*.

83. $BC - D$ is possible. The resulting order is 2 × 2.

85. $D(A - 3B)$ is possible. The resulting order is 2 × 3.

87. $A = \begin{bmatrix} 3 & -1 \\ 4 & 7 \end{bmatrix}, B = \begin{bmatrix} -2 & 0 \\ 8 & 1 \end{bmatrix}, C = \begin{bmatrix} 5 & 2 \\ 2 & -6 \end{bmatrix}$

(a) $A + B = \begin{bmatrix} 3 & -1 \\ 4 & 7 \end{bmatrix} + \begin{bmatrix} -2 & 0 \\ 8 & 1 \end{bmatrix} = \begin{bmatrix} 3-2 & -1+0 \\ 4+8 & 7+1 \end{bmatrix} = \begin{bmatrix} 1 & -1 \\ 12 & 8 \end{bmatrix}$

$B + A = \begin{bmatrix} -2 & 0 \\ 8 & 1 \end{bmatrix} + \begin{bmatrix} 3 & -1 \\ 4 & 7 \end{bmatrix} = \begin{bmatrix} -2+3 & 0-1 \\ 8+4 & 1+7 \end{bmatrix} = \begin{bmatrix} 1 & -1 \\ 12 & 8 \end{bmatrix}$

(b) $(A + B) + C = \begin{bmatrix} 1 & -1 \\ 12 & 8 \end{bmatrix} + \begin{bmatrix} 5 & 2 \\ 2 & -6 \end{bmatrix} = \begin{bmatrix} 1+5 & -1+2 \\ 12+2 & 8-6 \end{bmatrix} = \begin{bmatrix} 6 & 1 \\ 14 & 2 \end{bmatrix}$

$B + C = \begin{bmatrix} -2 & 0 \\ 8 & 1 \end{bmatrix} + \begin{bmatrix} 5 & 2 \\ 2 & -6 \end{bmatrix} = \begin{bmatrix} -2+5 & 0+2 \\ 8+2 & 1-6 \end{bmatrix} = \begin{bmatrix} 3 & 2 \\ 10 & -5 \end{bmatrix}$

$(B + C) + A = \begin{bmatrix} 3 & 2 \\ 10 & -5 \end{bmatrix} + \begin{bmatrix} 3 & -1 \\ 4 & 7 \end{bmatrix} = \begin{bmatrix} 6 & 1 \\ 14 & 2 \end{bmatrix}$

(c) $2A = 2\begin{bmatrix} 3 & -1 \\ 4 & 7 \end{bmatrix} = \begin{bmatrix} 2(3) & 2(-1) \\ 2(4) & 2(7) \end{bmatrix} = \begin{bmatrix} 6 & -2 \\ 8 & 14 \end{bmatrix}$

$2B = 2\begin{bmatrix} -2 & 0 \\ 8 & 1 \end{bmatrix} = \begin{bmatrix} 2(-2) & 2(0) \\ 2(8) & 2(1) \end{bmatrix} = \begin{bmatrix} -4 & 0 \\ 16 & 2 \end{bmatrix}$

$2A + 2B = \begin{bmatrix} 6 & -2 \\ 8 & 14 \end{bmatrix} + \begin{bmatrix} -4 & 0 \\ 16 & 2 \end{bmatrix} = \begin{bmatrix} 6+(-4) & -2+0 \\ 8+16 & 14+2 \end{bmatrix} = \begin{bmatrix} 2 & -2 \\ 24 & 16 \end{bmatrix}$

$2(A + B) = 2\begin{bmatrix} 1 & -1 \\ 12 & 8 \end{bmatrix} = \begin{bmatrix} 2(1) & 2(-1) \\ 2(12) & 2(8) \end{bmatrix} = \begin{bmatrix} 2 & -2 \\ 24 & 16 \end{bmatrix}$

89. $AC = \begin{bmatrix} 0 & 1 \\ 0 & 1 \end{bmatrix}\begin{bmatrix} 2 & 3 \\ 2 & 3 \end{bmatrix} = \begin{bmatrix} 2 & 3 \\ 2 & 3 \end{bmatrix}$

$BC = \begin{bmatrix} 1 & 0 \\ 1 & 0 \end{bmatrix}\begin{bmatrix} 2 & 3 \\ 2 & 3 \end{bmatrix} = \begin{bmatrix} 2 & 3 \\ 2 & 3 \end{bmatrix}$

So, $AC = BC$ even though $A \neq B$.

91. The product of two diagonal matrices of the same order is a diagonal matrix whose entries are the products of the corresponding diagonal entries of *A* and *B*.

93. Sample answer:

$A = \begin{bmatrix} 1 & 0 \\ 0 & 1 \end{bmatrix}, B = \begin{bmatrix} 0 & 1 \\ 1 & 0 \end{bmatrix}$

$AB = \begin{bmatrix} 1 & 0 \\ 0 & 1 \end{bmatrix}\begin{bmatrix} 0 & 1 \\ 1 & 0 \end{bmatrix} = \begin{bmatrix} (1)(0)+(0)(1) & (1)(1)+(0)(0) \\ (0)(0)+(1)(1) & (0)(1)+(1)(0) \end{bmatrix} = \begin{bmatrix} 0 & 1 \\ 1 & 0 \end{bmatrix}$

$BA = \begin{bmatrix} 0 & 1 \\ 1 & 0 \end{bmatrix}\begin{bmatrix} 1 & 0 \\ 0 & 1 \end{bmatrix} = \begin{bmatrix} (0)(1)+(1)(0) & (0)(0)+(1)(1) \\ (1)(1)+(0)(0) & (1)(0)+(0)(1) \end{bmatrix} = \begin{bmatrix} 0 & 1 \\ 1 & 0 \end{bmatrix}$

So, $AB = BA$.

Section 10.3   The Inverse of a Square Matrix

- You should know that the inverse of an $n \times n$ matrix A is the $n \times n$ matrix A^{-1}, if is exists, such that $AA^{-1} = A^{-1}A = I$, where I is the $n \times n$ identity matrix.

- You should be able to find the inverse, if it exists, of a square matrix.

  (a) Write the $n \times 2n$ matrix that consists of the given matrix A on the left and the $n \times n$ identity matrix I on the right to obtain $\begin{bmatrix} A & \vdots & I \end{bmatrix}$. Note that we separate the matrices A and I by a dotted line. We call this process adjoining the matrices A and I.

  (b) If possible, row reduce A to I using elementary row operations on the *entire* matrix $\begin{bmatrix} A & \vdots & I \end{bmatrix}$. The result will be the matrix $\begin{bmatrix} I & \vdots & A^{-1} \end{bmatrix}$. If this is not possible, then A is not invertible.

  (c) Check your work by multiplying to see that $AA^{-1} = I = A^{-1}A$.

- The inverse of $A = \begin{bmatrix} a & b \\ c & d \end{bmatrix}$ is $A^{-1} = \dfrac{1}{ad - bc}\begin{bmatrix} d & -b \\ -c & a \end{bmatrix}$ if $ad - cb \neq 0$.

- You should be able to use inverse matrices to solve systems of linear equations if the coefficient matrix is square and invertible.

---

1. square

3. nonsingular; singular

5. $AB = \begin{bmatrix} 2 & 1 \\ 5 & 3 \end{bmatrix}\begin{bmatrix} 3 & -1 \\ -5 & 2 \end{bmatrix} = \begin{bmatrix} 6-5 & -2+2 \\ 15-15 & -5+6 \end{bmatrix} = \begin{bmatrix} 1 & 0 \\ 0 & 1 \end{bmatrix}$

   $BA = \begin{bmatrix} 3 & -1 \\ -5 & 2 \end{bmatrix}\begin{bmatrix} 2 & 1 \\ 5 & 3 \end{bmatrix} = \begin{bmatrix} 6-5 & 3-3 \\ -10+10 & -5+6 \end{bmatrix} = \begin{bmatrix} 1 & 0 \\ 0 & 1 \end{bmatrix}$

7. $AB = \begin{bmatrix} 1 & 2 \\ 3 & 4 \end{bmatrix}\begin{bmatrix} -2 & 1 \\ \frac{3}{2} & -\frac{1}{2} \end{bmatrix} = \begin{bmatrix} -2+3 & 1-1 \\ -6+6 & 3-2 \end{bmatrix} = \begin{bmatrix} 1 & 0 \\ 0 & 1 \end{bmatrix}$

   $BA = \begin{bmatrix} -2 & 1 \\ \frac{3}{2} & -\frac{1}{2} \end{bmatrix}\begin{bmatrix} 1 & 2 \\ 3 & 4 \end{bmatrix} = \begin{bmatrix} -2+3 & -4+4 \\ \frac{3}{2}-\frac{3}{2} & 3-2 \end{bmatrix} = \begin{bmatrix} 1 & 0 \\ 0 & 1 \end{bmatrix}$

9. $AB = \begin{bmatrix} 2 & -17 & 11 \\ -1 & 11 & -7 \\ 0 & 3 & -2 \end{bmatrix}\begin{bmatrix} 1 & 1 & 2 \\ 2 & 4 & -3 \\ 3 & 6 & -5 \end{bmatrix} = \begin{bmatrix} 2-34+33 & 2-68+66 & 4+51-55 \\ -1+22-21 & -1+44-42 & -2-33+35 \\ 6-6 & 12-12 & -9+10 \end{bmatrix} = \begin{bmatrix} 1 & 0 & 0 \\ 0 & 1 & 0 \\ 0 & 0 & 1 \end{bmatrix}$

   $BA = \begin{bmatrix} 1 & 1 & 2 \\ 2 & 4 & -3 \\ 3 & 6 & -5 \end{bmatrix}\begin{bmatrix} 2 & -17 & 11 \\ -1 & 11 & -7 \\ 0 & 3 & -2 \end{bmatrix} = \begin{bmatrix} 2-1 & -17+11+6 & 11-7-4 \\ 4-4 & -34+44-9 & 22-28+6 \\ 6-6 & -51+66-15 & 33-42+10 \end{bmatrix} = \begin{bmatrix} 1 & 0 & 0 \\ 0 & 1 & 0 \\ 0 & 0 & 1 \end{bmatrix}$

11. $AB = \frac{1}{3}\begin{bmatrix} -2 & 2 & 3 \\ 1 & -1 & 0 \\ 0 & 1 & 4 \end{bmatrix}\begin{bmatrix} -4 & -5 & 3 \\ -4 & -8 & 3 \\ 1 & 2 & 0 \end{bmatrix} = \frac{1}{3}\begin{bmatrix} 8-8+3 & 10-16+6 & -6+6 \\ -4+4 & -5+8 & 3-3 \\ -4+4 & -8+8 & 3 \end{bmatrix} = \frac{1}{3}\begin{bmatrix} 3 & 0 & 0 \\ 0 & 3 & 0 \\ 0 & 0 & 3 \end{bmatrix} = \begin{bmatrix} 1 & 0 & 0 \\ 0 & 1 & 0 \\ 0 & 0 & 1 \end{bmatrix}$

   $BA = \frac{1}{3}\begin{bmatrix} -4 & -5 & 3 \\ -4 & -8 & 3 \\ 1 & 2 & 0 \end{bmatrix}\begin{bmatrix} -2 & 2 & 3 \\ 1 & -1 & 0 \\ 0 & 1 & 4 \end{bmatrix} = \frac{1}{3}\begin{bmatrix} 8-5 & -8+5+3 & -12+12 \\ 8-8 & -8+8+3 & -12+12 \\ -2+2 & 2-2 & 3 \end{bmatrix} = \begin{bmatrix} 1 & 0 & 0 \\ 0 & 1 & 0 \\ 0 & 0 & 1 \end{bmatrix}$

13. $[A \; \vdots \; I] = \begin{bmatrix} 2 & 0 & \vdots & 1 & 0 \\ 0 & 3 & \vdots & 0 & 1 \end{bmatrix}$

$\begin{matrix} \frac{1}{2}R_1 \to \\ \frac{1}{3}R_2 \to \end{matrix} \begin{bmatrix} 1 & 0 & \vdots & \frac{1}{2} & 0 \\ 0 & 1 & \vdots & 0 & \frac{1}{3} \end{bmatrix} = \begin{bmatrix} I & \vdots & A^{-1} \end{bmatrix}$

$A^{-1} = \begin{bmatrix} \frac{1}{2} & 0 \\ 0 & \frac{1}{3} \end{bmatrix}$

15. $[A \; \vdots \; I] = \begin{bmatrix} 1 & -2 & \vdots & 1 & 0 \\ 2 & -3 & \vdots & 0 & 1 \end{bmatrix}$

$-2R_1 + R_2 \to \begin{bmatrix} 1 & -2 & \vdots & 1 & 0 \\ 0 & 1 & \vdots & -2 & 1 \end{bmatrix}$

$2R_2 + R_1 \to \begin{bmatrix} 1 & 0 & \vdots & -3 & 2 \\ 0 & 1 & \vdots & -2 & 1 \end{bmatrix} = \begin{bmatrix} I & \vdots & A^{-1} \end{bmatrix}$

$A^{-1} = \begin{bmatrix} -3 & 2 \\ -2 & 1 \end{bmatrix}$

17. $[A \; \vdots \; I] = \begin{bmatrix} 3 & 1 & \vdots & 1 & 0 \\ 4 & 2 & \vdots & 0 & 1 \end{bmatrix}$

$\frac{1}{2}R_2 \to \begin{bmatrix} 3 & 1 & \vdots & 1 & 0 \\ 2 & 1 & \vdots & 0 & \frac{1}{2} \end{bmatrix}$

$-R_2 + R_1 \to \begin{bmatrix} 1 & 0 & \vdots & 1 & -\frac{1}{2} \\ 2 & 1 & \vdots & 0 & \frac{1}{2} \end{bmatrix}$

$-2R_1 + R_2 \to \begin{bmatrix} 1 & 0 & \vdots & 1 & -\frac{1}{2} \\ 0 & 1 & \vdots & -2 & \frac{3}{2} \end{bmatrix} = \begin{bmatrix} I & \vdots & A^{-1} \end{bmatrix}$

$A^{-1} = \begin{bmatrix} 1 & -\frac{1}{2} \\ -2 & \frac{3}{2} \end{bmatrix}$

19. $[A \; \vdots \; I] = \begin{bmatrix} 1 & 1 & 1 & \vdots & 1 & 0 & 0 \\ 3 & 5 & 4 & \vdots & 0 & 1 & 0 \\ 3 & 6 & 5 & \vdots & 0 & 0 & 1 \end{bmatrix}$

$\begin{matrix} -3R_1 + R_2 \to \\ -3R_1 + R_3 \to \end{matrix} \begin{bmatrix} 1 & 1 & 1 & \vdots & 1 & 0 & 0 \\ 0 & 2 & 1 & \vdots & -3 & 1 & 0 \\ 0 & 3 & 2 & \vdots & -3 & 0 & 1 \end{bmatrix}$

$\frac{1}{2}R_2 \to \begin{bmatrix} 1 & 1 & 1 & \vdots & 1 & 0 & 0 \\ 0 & 1 & \frac{1}{2} & \vdots & -\frac{3}{2} & \frac{1}{2} & 0 \\ 0 & 3 & 2 & \vdots & -3 & 0 & 1 \end{bmatrix}$

$\begin{matrix} -R_2 + R_1 \to \\ \\ -3R_2 + R_3 \to \end{matrix} \begin{bmatrix} 1 & 0 & \frac{1}{2} & \vdots & \frac{5}{2} & -\frac{1}{2} & 0 \\ 0 & 1 & \frac{1}{2} & \vdots & -\frac{3}{2} & \frac{1}{2} & 0 \\ 0 & 0 & \frac{1}{2} & \vdots & \frac{3}{2} & -\frac{3}{2} & 1 \end{bmatrix}$

$\begin{matrix} -R_3 + R_1 \to \\ -R_3 + R_2 \to \\ \\ \end{matrix} \begin{bmatrix} 1 & 0 & 0 & \vdots & 1 & 1 & -1 \\ 0 & 1 & 0 & \vdots & -3 & 2 & -1 \\ 0 & 0 & \frac{1}{2} & \vdots & \frac{3}{2} & -\frac{3}{2} & 1 \end{bmatrix}$

$\begin{matrix} \\ \\ 2R_3 \to \end{matrix} \begin{bmatrix} 1 & 0 & 0 & \vdots & 1 & 1 & -1 \\ 0 & 1 & 0 & \vdots & -3 & 2 & -1 \\ 0 & 0 & 1 & \vdots & 3 & -3 & 2 \end{bmatrix} = \begin{bmatrix} I & \vdots & A^{-1} \end{bmatrix}$

$A^{-1} = \begin{bmatrix} 1 & 1 & -1 \\ -3 & 2 & -1 \\ 3 & -3 & 2 \end{bmatrix}$

21. $[A \; \vdots \; I] = \begin{bmatrix} -5 & 0 & 0 & \vdots & 1 & 0 & 0 \\ 2 & 0 & 0 & \vdots & 0 & 1 & 0 \\ -1 & 5 & 7 & \vdots & 0 & 0 & 1 \end{bmatrix} \begin{matrix} \\ \\ R_2 + 2R_3 \to \end{matrix} \begin{bmatrix} -5 & 0 & 0 & \vdots & 1 & 0 & 0 \\ 2 & 0 & 0 & \vdots & 0 & 1 & 0 \\ 0 & 10 & 14 & \vdots & 0 & 1 & 2 \end{bmatrix} 2R_1 + 5R_2 \to \begin{bmatrix} -5 & 0 & 0 & \vdots & 1 & 0 & 0 \\ 0 & 0 & 0 & \vdots & 2 & 5 & 0 \\ 0 & 10 & 14 & \vdots & 0 & 1 & 2 \end{bmatrix}$

Because the first three entries of row 2 are all zeros, the inverse of A does not exist.

23. $[A \; \vdots \; I] = \begin{bmatrix} -8 & 0 & 0 & 0 & \vdots & 1 & 0 & 0 & 0 \\ 0 & 1 & 0 & 0 & \vdots & 0 & 1 & 0 & 0 \\ 0 & 0 & 4 & 0 & \vdots & 0 & 0 & 1 & 0 \\ 0 & 0 & 0 & -5 & \vdots & 0 & 0 & 0 & 1 \end{bmatrix} \begin{matrix} -\frac{1}{8}R_1 \to \\ \\ \frac{1}{4}R_3 \to \\ -\frac{1}{5}R_4 \to \end{matrix} \begin{bmatrix} 1 & 0 & 0 & 0 & \vdots & -\frac{1}{8} & 0 & 0 & 0 \\ 0 & 1 & 0 & 0 & \vdots & 0 & 1 & 0 & 0 \\ 0 & 0 & 1 & 0 & \vdots & 0 & 0 & \frac{1}{4} & 0 \\ 0 & 0 & 0 & 1 & \vdots & 0 & 0 & 0 & -\frac{1}{5} \end{bmatrix} = \begin{bmatrix} I & \vdots & A^{-1} \end{bmatrix}$

$A^{-1} = \begin{bmatrix} -\frac{1}{8} & 0 & 0 & 0 \\ 0 & 1 & 0 & 0 \\ 0 & 0 & \frac{1}{4} & 0 \\ 0 & 0 & 0 & -\frac{1}{5} \end{bmatrix}$

25. $A = \begin{bmatrix} 1 & 2 & -1 \\ 3 & 7 & -10 \\ -5 & -7 & -15 \end{bmatrix}$

$A^{-1} = \begin{bmatrix} -175 & 37 & -13 \\ 95 & -20 & 7 \\ 14 & -3 & 1 \end{bmatrix}$

27. $A = \begin{bmatrix} 1 & 1 & 2 \\ 3 & 1 & 0 \\ -2 & 0 & 3 \end{bmatrix}$

$A^{-1} = \begin{bmatrix} -1.5 & 1.5 & 1 \\ 4.5 & -3.5 & -3 \\ -1 & 1 & 1 \end{bmatrix}$

29. $A = \begin{bmatrix} -\frac{1}{2} & \frac{3}{4} & \frac{1}{4} \\ 1 & 0 & -\frac{3}{2} \\ 0 & -1 & \frac{1}{2} \end{bmatrix}$

$A^{-1} = \begin{bmatrix} -12 & -5 & -9 \\ -4 & -2 & -4 \\ -8 & -4 & -6 \end{bmatrix}$

31. $A = \begin{bmatrix} 0.1 & 0.2 & 0.3 \\ -0.3 & 0.2 & 0.2 \\ 0.5 & 0.4 & 0.4 \end{bmatrix}$

$A^{-1} = \begin{bmatrix} 0 & -1.\overline{81} & 0.\overline{90} \\ -10 & 5 & 5 \\ 10 & -2.\overline{72} & -3.\overline{63} \end{bmatrix}$

33. $A = \begin{bmatrix} -1 & 0 & 1 & 0 \\ 0 & 2 & 0 & -1 \\ 2 & 0 & -1 & 0 \\ 0 & -1 & 0 & 1 \end{bmatrix}$

$A^{-1} = \begin{bmatrix} 1 & 0 & 1 & 0 \\ 0 & 1 & 0 & 1 \\ 2 & 0 & 1 & 0 \\ 0 & 1 & 0 & 2 \end{bmatrix}$

35. $A = \begin{bmatrix} a & b \\ c & d \end{bmatrix}$, $A^{-1} = \dfrac{1}{ad - bc}\begin{bmatrix} d & -b \\ -c & a \end{bmatrix}$

$A = \begin{bmatrix} 2 & 3 \\ -1 & 5 \end{bmatrix}$

$ad - bc = (2)(5) - (3)(-1) = 13$

$A^{-1} = \dfrac{1}{13}\begin{bmatrix} 5 & -3 \\ 1 & 2 \end{bmatrix} = \begin{bmatrix} \frac{5}{13} & -\frac{3}{13} \\ \frac{1}{13} & \frac{2}{13} \end{bmatrix}$

37. $A = \begin{bmatrix} -4 & -6 \\ 2 & 3 \end{bmatrix}$

$ad - bc = (-4)(3) - (-2)(-6) = 0$

Because $ad - bc = 0$, A^{-1} does not exist.

39. $A = \begin{bmatrix} \frac{7}{2} & -\frac{3}{4} \\ \frac{1}{5} & \frac{4}{5} \end{bmatrix}$

$ad - bc = \left(\dfrac{7}{2}\right)\left(\dfrac{4}{5}\right) - \left(-\dfrac{3}{4}\right)\left(\dfrac{1}{5}\right) = \dfrac{28}{10} + \dfrac{3}{20} = \dfrac{59}{20}$

$A^{-1} = \dfrac{20}{59}\begin{bmatrix} \frac{4}{5} & \frac{3}{4} \\ -\frac{1}{5} & \frac{7}{2} \end{bmatrix} = \begin{bmatrix} \frac{16}{59} & \frac{15}{59} \\ -\frac{4}{59} & \frac{70}{59} \end{bmatrix}$

41. $\begin{bmatrix} x \\ y \end{bmatrix} = \begin{bmatrix} -3 & 2 \\ -2 & 1 \end{bmatrix}\begin{bmatrix} 5 \\ 10 \end{bmatrix} = \begin{bmatrix} 5 \\ 0 \end{bmatrix}$

Solution: $(5, 0)$

43. $\begin{bmatrix} x \\ y \end{bmatrix} = \begin{bmatrix} -3 & 2 \\ -2 & 1 \end{bmatrix}\begin{bmatrix} 4 \\ 2 \end{bmatrix} = \begin{bmatrix} -8 \\ -6 \end{bmatrix}$

Solution: $(-8, -6)$

45. $\begin{bmatrix} x \\ y \\ z \end{bmatrix} = \begin{bmatrix} 1 & 1 & -1 \\ -3 & 2 & -1 \\ 3 & -3 & 2 \end{bmatrix}\begin{bmatrix} 0 \\ 5 \\ 2 \end{bmatrix} = \begin{bmatrix} 3 \\ 8 \\ -11 \end{bmatrix}$

Solution: $(3, 8, -11)$

47. $\begin{bmatrix} x_1 \\ x_2 \\ x_3 \\ x_4 \end{bmatrix} = \begin{bmatrix} -24 & 7 & 1 & -2 \\ -10 & 3 & 0 & -1 \\ -29 & 7 & 3 & -2 \\ 12 & -3 & -1 & 1 \end{bmatrix}\begin{bmatrix} 0 \\ 1 \\ -1 \\ 2 \end{bmatrix} = \begin{bmatrix} 2 \\ 1 \\ 0 \\ 0 \end{bmatrix}$

Solution: $(2, 1, 0, 0)$

49.
$$\begin{cases} x_1 + 2x_2 - x_3 + 3x_4 - x_5 = -3 \\ x_1 - 3x_2 + x_3 + 2x_4 - x_5 = -3 \\ 2x_1 + x_2 + x_3 - 3x_4 + x_5 = 6 \\ x_1 - x_2 + 2x_3 + x_4 - x_5 = 2 \\ 2x_1 + x_2 - x_3 + 2x_4 + x_5 = -3 \end{cases}$$

$$AX = B = \begin{bmatrix} 1 & 2 & -1 & 3 & -1 \\ 1 & -3 & 1 & 2 & -1 \\ 2 & 1 & 1 & -3 & 1 \\ 1 & -1 & 2 & 1 & -1 \\ 2 & 1 & -1 & 2 & 1 \end{bmatrix} \begin{bmatrix} x_1 \\ x_2 \\ x_3 \\ x_4 \\ x_5 \end{bmatrix} = \begin{bmatrix} -3 \\ -3 \\ 6 \\ 2 \\ -3 \end{bmatrix}$$

Using a graphing utility:

$$X = A^{-1}B = \begin{bmatrix} 0 \\ 1 \\ 2 \\ -1 \\ 0 \end{bmatrix}$$

Solution: $(0, 1, 2, -1, 0)$

51. $A = \begin{bmatrix} 3 & 4 \\ 5 & 3 \end{bmatrix}$

$$A^{-1} = \frac{1}{9 - 20} \begin{bmatrix} 3 & -4 \\ -5 & 3 \end{bmatrix}$$

$$\begin{bmatrix} x \\ y \end{bmatrix} = -\frac{1}{11} \begin{bmatrix} 3 & -4 \\ -5 & 3 \end{bmatrix} \begin{bmatrix} -2 \\ 4 \end{bmatrix} = -\frac{1}{11} \begin{bmatrix} -22 \\ 22 \end{bmatrix} = \begin{bmatrix} 2 \\ -2 \end{bmatrix}$$

Solution: $(2, -2)$

53. $A = \begin{bmatrix} -0.4 & 0.8 \\ 2 & -4 \end{bmatrix}$

$$A^{-1} = \frac{1}{1.6 - 1.6} \begin{bmatrix} -4 & -0.8 \\ -2 & -0.4 \end{bmatrix}$$

A^{-1} does not exist.

This implies that there is no unique solution; that is, either the system is inconsistent *or* there are infinitely many solutions.

Find the reduced row-echelon form of the matrix corresponding to the system.

$$\begin{bmatrix} -0.4 & 0.8 & \vdots & 1.6 \\ 2 & -4 & \vdots & 5 \end{bmatrix}$$

$$-2.5R_1 \rightarrow \begin{bmatrix} 1 & -2 & \vdots & -4 \\ 2 & -4 & \vdots & 5 \end{bmatrix}$$

$$-2R_1 + R_2 \rightarrow \begin{bmatrix} 1 & -2 & \vdots & -4 \\ 0 & 0 & \vdots & 13 \end{bmatrix}$$

The given system is inconsistent and there is no solution.

55. $A = \begin{bmatrix} -\dfrac{1}{4} & \dfrac{3}{8} \\ \dfrac{3}{2} & \dfrac{3}{4} \end{bmatrix}$

$$A^{-1} = \frac{1}{-\dfrac{3}{16} - \dfrac{9}{16}} \begin{bmatrix} \dfrac{3}{4} & -\dfrac{3}{8} \\ -\dfrac{3}{2} & -\dfrac{1}{4} \end{bmatrix}$$

$$= -\frac{4}{3} \begin{bmatrix} \dfrac{3}{4} & -\dfrac{3}{8} \\ -\dfrac{3}{2} & -\dfrac{1}{4} \end{bmatrix}$$

$$= \begin{bmatrix} -1 & \dfrac{1}{2} \\ 2 & \dfrac{1}{3} \end{bmatrix}$$

$$\begin{bmatrix} x \\ y \end{bmatrix} = \begin{bmatrix} -1 & \dfrac{1}{2} \\ 2 & \dfrac{1}{3} \end{bmatrix} \begin{bmatrix} -2 \\ -12 \end{bmatrix} = \begin{bmatrix} -4 \\ -8 \end{bmatrix}$$

Solution: $(-4, -8)$

57. $A = \begin{bmatrix} 4 & -1 & 1 \\ 2 & 2 & 3 \\ 5 & -2 & 6 \end{bmatrix}$

Find A^{-1}.

$$[A \;\vdots\; I] = \begin{bmatrix} 4 & -1 & 1 & \vdots & 1 & 0 & 0 \\ 2 & 2 & 3 & \vdots & 0 & 1 & 0 \\ 5 & -2 & 6 & \vdots & 0 & 0 & 1 \end{bmatrix}$$

$$\begin{matrix} {\scriptstyle R_1} \\ \\ {\scriptstyle R_3} \end{matrix} \begin{bmatrix} 5 & -2 & 6 & \vdots & 0 & 0 & 1 \\ 2 & 2 & 3 & \vdots & 0 & 1 & 0 \\ 4 & -1 & 1 & \vdots & 1 & 0 & 0 \end{bmatrix}$$

$$-R_3 + R_1 \rightarrow \begin{bmatrix} 1 & -1 & 5 & \vdots & -1 & 0 & 1 \\ 2 & 2 & 3 & \vdots & 0 & 1 & 0 \\ 4 & -1 & 1 & \vdots & 1 & 0 & 0 \end{bmatrix}$$

$$\begin{matrix} -2R_1 + R_2 \rightarrow \\ -4R_1 + R_3 \rightarrow \end{matrix} \begin{bmatrix} 1 & -1 & 5 & \vdots & -1 & 0 & 1 \\ 0 & 4 & -7 & \vdots & 2 & 1 & -2 \\ 0 & 3 & -19 & \vdots & 5 & 0 & -4 \end{bmatrix}$$

$$-R_3 + R_2 \rightarrow \begin{bmatrix} 1 & -1 & 5 & \vdots & -1 & 0 & 1 \\ 0 & 1 & 12 & \vdots & -3 & 1 & 2 \\ 0 & 3 & -19 & \vdots & 5 & 0 & -4 \end{bmatrix}$$

$$\begin{matrix} R_2 + R_1 \rightarrow \\ \\ -3R_2 + R_3 \rightarrow \end{matrix} \begin{bmatrix} 1 & 0 & 17 & \vdots & -4 & 1 & 3 \\ 0 & 1 & 12 & \vdots & -3 & 1 & 2 \\ 0 & 0 & -55 & \vdots & 14 & -3 & -10 \end{bmatrix}$$

$$-\tfrac{1}{55}R_3 \rightarrow \begin{bmatrix} 1 & 0 & 17 & \vdots & -4 & 1 & 3 \\ 0 & 1 & 12 & \vdots & -3 & 1 & 2 \\ 0 & 0 & 1 & \vdots & -\tfrac{14}{55} & \tfrac{3}{55} & \tfrac{2}{11} \end{bmatrix}$$

$$\begin{matrix} -17R_3 + R_1 \rightarrow \\ -12R_3 + R_2 \rightarrow \\ \\ \end{matrix} \begin{bmatrix} 1 & 0 & 0 & \vdots & \tfrac{18}{55} & \tfrac{4}{55} & -\tfrac{1}{11} \\ 0 & 1 & 0 & \vdots & \tfrac{3}{55} & \tfrac{19}{55} & -\tfrac{2}{11} \\ 0 & 0 & 1 & \vdots & -\tfrac{14}{55} & \tfrac{3}{55} & \tfrac{2}{11} \end{bmatrix} = [I \;\vdots\; A^{-1}]$$

$$A^{-1} = \tfrac{1}{55} \begin{bmatrix} 18 & 4 & -5 \\ 3 & 19 & -10 \\ -14 & 3 & 10 \end{bmatrix}$$

$$\begin{bmatrix} x \\ y \\ z \end{bmatrix} = \tfrac{1}{55} \begin{bmatrix} 18 & 4 & -5 \\ 3 & 19 & -10 \\ -14 & 3 & 10 \end{bmatrix} \begin{bmatrix} -5 \\ 10 \\ 1 \end{bmatrix} = \tfrac{1}{55} \begin{bmatrix} -55 \\ 165 \\ 110 \end{bmatrix} = \begin{bmatrix} -1 \\ 3 \\ 2 \end{bmatrix}$$

Solution: $(-1, 3, 2)$

59. $A = \begin{bmatrix} 5 & -3 & 2 \\ 2 & 2 & -3 \\ 1 & -7 & 8 \end{bmatrix}$

A^{-1} does not exist. This implies that there is no unique solution; that is, either the system is inconsistent *or* the system has infinitely many solutions. Use a graphing utility to find the reduced row-echelon form of the matrix corresponding to the system.

$$\begin{bmatrix} 5 & -3 & 2 & \vdots & 2 \\ 2 & 2 & -3 & \vdots & 3 \\ 1 & -7 & 8 & \vdots & -4 \end{bmatrix}$$

$$\begin{bmatrix} 1 & 0 & -\frac{5}{16} & \vdots & \frac{13}{16} \\ 0 & 1 & -\frac{19}{16} & \vdots & \frac{11}{16} \\ 0 & 0 & 0 & \vdots & 0 \end{bmatrix}$$

$$\begin{cases} x - \frac{5}{16}z = \frac{13}{16} \\ y - \frac{19}{16}z = \frac{11}{16} \end{cases}$$

Let $z = a$. Then $x = \frac{5}{16}a + \frac{13}{16}$ and $y = \frac{19}{16}a + \frac{11}{16}$.

Solution: $\left(\frac{5}{16}a + \frac{13}{16}, \frac{19}{16}a + \frac{11}{16}, a\right)$ where a is a real number

65. $A = \begin{bmatrix} 1 & 1 & 1 \\ 0.065 & 0.07 & 0.09 \\ 0 & 2 & -1 \end{bmatrix}$

$$[A \ \vdots \ I] = \begin{bmatrix} 1 & 1 & 1 & \vdots & 1 & 0 & 0 \\ 0.065 & 0.07 & 0.09 & \vdots & 0 & 1 & 0 \\ 0 & 2 & -1 & \vdots & 0 & 0 & 1 \end{bmatrix}$$

$$200R_2 \rightarrow \begin{bmatrix} 1 & 1 & 1 & \vdots & 1 & 0 & 0 \\ 13 & 14 & 18 & \vdots & 0 & 200 & 0 \\ 0 & 2 & -1 & \vdots & 0 & 0 & 1 \end{bmatrix}$$

$$-13R_1 + R_2 \rightarrow \begin{bmatrix} 1 & 1 & 1 & \vdots & 1 & 0 & 0 \\ 0 & 1 & 5 & \vdots & -13 & 200 & 0 \\ 0 & 2 & -1 & \vdots & 0 & 0 & 1 \end{bmatrix}$$

$$\begin{matrix} -R_2 + R_1 \rightarrow \\ \\ -2R_2 + R_3 \rightarrow \end{matrix} \begin{bmatrix} 1 & 0 & -4 & \vdots & 14 & -200 & 0 \\ 0 & 1 & 5 & \vdots & -13 & 200 & 0 \\ 0 & 0 & -11 & \vdots & 26 & -400 & 1 \end{bmatrix}$$

$$-\frac{1}{11}R_3 \rightarrow \begin{bmatrix} 1 & 0 & -4 & \vdots & 14 & -200 & 0 \\ 0 & 1 & 5 & \vdots & -13 & 200 & 0 \\ 0 & 0 & 1 & \vdots & -\frac{26}{11} & \frac{400}{11} & -\frac{1}{11} \end{bmatrix}$$

$$\begin{matrix} 4R_3 + R_1 \rightarrow \\ \\ -5R_3 + R_2 \rightarrow \\ \\ \\ \end{matrix} \begin{bmatrix} 1 & 0 & 0 & \vdots & \frac{50}{11} & -\frac{600}{11} & -\frac{4}{11} \\ 0 & 1 & 0 & \vdots & -\frac{13}{11} & \frac{200}{11} & \frac{5}{11} \\ 0 & 0 & 1 & \vdots & -\frac{26}{11} & \frac{400}{11} & -\frac{1}{11} \end{bmatrix} = [I \ \vdots \ A^{-1}]$$

$$X = A^{-1}B = \frac{1}{11}\begin{bmatrix} 50 & -600 & -4 \\ -13 & 200 & 5 \\ -26 & 400 & -1 \end{bmatrix}\begin{bmatrix} 10,000 \\ 705 \\ 0 \end{bmatrix} = \begin{bmatrix} 7000 \\ 1000 \\ 2000 \end{bmatrix}$$

Solution: $7000 in AAA-rated bonds, $1000 in A-rated bonds, $2000 in B-rated bonds

61. $A = \begin{bmatrix} 3 & -2 & 1 \\ -4 & 1 & -3 \\ 1 & -5 & 1 \end{bmatrix}$

$$A^{-1} = \begin{bmatrix} 0.56 & 0.12 & -0.2 \\ -0.04 & -0.08 & -0.2 \\ -0.76 & -0.52 & 0.2 \end{bmatrix}$$

$$\begin{bmatrix} x \\ y \\ z \end{bmatrix} = \begin{bmatrix} 0.56 & 0.12 & -0.2 \\ -0.04 & -0.08 & -0.2 \\ -0.76 & -0.52 & 0.2 \end{bmatrix}\begin{bmatrix} -29 \\ 37 \\ -24 \end{bmatrix} = \begin{bmatrix} -7 \\ 3 \\ -2 \end{bmatrix}$$

Solution: $(-7, 3, -2)$

63. $A = \begin{bmatrix} a & b \\ c & d \end{bmatrix}$, $A^{-1} = \dfrac{1}{ad - bc}\begin{bmatrix} d & -b \\ -c & a \end{bmatrix}$

$$A = \begin{bmatrix} \sin\theta & \cos\theta \\ -\cos\theta & \sin\theta \end{bmatrix}$$

$$ad - bc = (\sin\theta)(\sin\theta) - (\cos\theta)(-\cos\theta)$$

$$= \sin^2\theta + \cos^2\theta = 1$$

$$A^{-1} = \frac{1}{1}\begin{bmatrix} \sin\theta & -\cos\theta \\ \cos\theta & \sin\theta \end{bmatrix} = \begin{bmatrix} \sin\theta & -\cos\theta \\ \cos\theta & \sin\theta \end{bmatrix}$$

67. Use the inverse matrix A^{-1} from Exercise 65.

$$X = A^{-1}B = \frac{1}{11}\begin{bmatrix} 50 & -600 & -4 \\ -13 & 200 & 5 \\ -26 & 400 & -1 \end{bmatrix}\begin{bmatrix} 12{,}000 \\ 835 \\ 0 \end{bmatrix} = \begin{bmatrix} 9000 \\ 1000 \\ 2000 \end{bmatrix}$$

Solution: $9000 in AAA-rated bonds, $1000 in A-rated bonds, $2000 in B-rated bonds

In Exercises 69 and 71, use the following:

Let x = number of muffins, y = number of bones, and z = number of cookies.

$$AX = B = \begin{bmatrix} 2 & 1 & 2 \\ 3 & 1 & 1 \\ 2 & 1 & 1.5 \end{bmatrix}\begin{bmatrix} x \\ y \\ z \end{bmatrix} = \begin{bmatrix} \text{Beef} \\ \text{Chicken} \\ \text{Liver} \end{bmatrix}$$

$$A^{-1} = \begin{bmatrix} 1 & 1 & -2 \\ -5 & -2 & 8 \\ 2 & 0 & -2 \end{bmatrix}$$

69. $A^{-1}\begin{bmatrix} 700 \\ 500 \\ 600 \end{bmatrix} = \begin{bmatrix} 1 & 1 & -2 \\ -5 & -2 & 8 \\ 2 & 0 & -2 \end{bmatrix}\begin{bmatrix} 700 \\ 500 \\ 600 \end{bmatrix} = \begin{bmatrix} 0 \\ 300 \\ 200 \end{bmatrix}$

Solution: 0 muffins, 300 bones, 200 cookies

71. $A^{-1}\begin{bmatrix} 800 \\ 750 \\ 725 \end{bmatrix} = \begin{bmatrix} 1 & 1 & -2 \\ -5 & 2 & 8 \\ 2 & 0 & -2 \end{bmatrix}\begin{bmatrix} 800 \\ 750 \\ 725 \end{bmatrix} = \begin{bmatrix} 100 \\ 300 \\ 150 \end{bmatrix}$

Solution: 100 muffins, 300 bones, 150 cookies

73. (a) $\begin{cases} 2f + 2.5h + 3s = 26 \\ f + h + s = 10 \\ h - s = 0 \end{cases}$

(b) $\begin{bmatrix} 2 & 2.5 & 3 \\ 1 & 1 & 1 \\ 0 & 1 & -1 \end{bmatrix}\begin{bmatrix} f \\ h \\ s \end{bmatrix} = \begin{bmatrix} 26 \\ 10 \\ 0 \end{bmatrix}$

$\qquad A \qquad X = B$

(c) $A^{-1} = \begin{bmatrix} -\frac{4}{3} & \frac{11}{3} & -\frac{1}{3} \\ \frac{2}{3} & -\frac{4}{3} & \frac{2}{3} \\ \frac{2}{3} & -\frac{4}{3} & -\frac{1}{3} \end{bmatrix}$

$X = A^{-1}B = \begin{bmatrix} -\frac{4}{3} & \frac{11}{3} & -\frac{1}{3} \\ \frac{2}{3} & -\frac{4}{3} & \frac{2}{3} \\ \frac{2}{3} & -\frac{4}{3} & -\frac{1}{3} \end{bmatrix}\begin{bmatrix} 26 \\ 10 \\ 0 \end{bmatrix} = \begin{bmatrix} 2 \\ 4 \\ 4 \end{bmatrix}$

So, there are 2 pounds of French vanilla, 4 pounds of hazelnut, and 4 pounds of Swiss chocolate flavored coffee in the 10 pound package.

75. $(10, 13.89), (11, 14.04), (12, 14.20)$

(a) $\begin{cases} 100a + 10b + c = 13.89 \\ 121a + 11b + c = 14.04 \\ 144a + 12b + c = 14.20 \end{cases}$

(b) $AX = B = \begin{bmatrix} 100 & 10 & 1 \\ 121 & 11 & 1 \\ 144 & 12 & 1 \end{bmatrix}\begin{bmatrix} a \\ b \\ c \end{bmatrix} = \begin{bmatrix} 13.89 \\ 14.04 \\ 14.20 \end{bmatrix}$

$A^{-1} = \begin{bmatrix} 0.5 & -1 & 0.5 \\ -11.5 & 22 & -10.5 \\ 66 & -120 & 55 \end{bmatrix}$

$X = A^{-1}B = \begin{bmatrix} a \\ b \\ c \end{bmatrix} = \begin{bmatrix} 0.5 & -1 & 0.5 \\ -11.5 & 22 & -10.5 \\ 66 & -120 & 55 \end{bmatrix}\begin{bmatrix} 13.89 \\ 14.04 \\ 14.20 \end{bmatrix} = \begin{bmatrix} 0.005 \\ 0.45 \\ 12.94 \end{bmatrix}$

So, $y = 0.005t^2 + 0.45t + 12.94$.

(c)

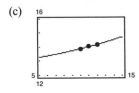

(d) For the immediate future, the model provides a good prediction.
In the long-term, the model does not provide good predictions.

77. True. If B is the inverse of A, then $AB = I = BA$.

79. If the determinant of a 2×2 matrix is not equal to 0, then the inverse exists.

To find the inverse, take 1 divided by the determinant and multiply it by the matrix which has a diagonal from top left to bottom right that has the terms from the original matrix flipped and the other diagonal is the negative of the terms from the original matrix.

81. (a) Given $A = \begin{bmatrix} a_{11} & 0 \\ 0 & a_{22} \end{bmatrix}$, $A^{-1} = \begin{bmatrix} \frac{1}{a_{11}} & 0 \\ 0 & \frac{1}{a_{22}} \end{bmatrix}$.

Given $A = \begin{bmatrix} a_{11} & 0 & 0 \\ 0 & a_{22} & 0 \\ 0 & 0 & a_{33} \end{bmatrix}$, $A^{-1} = \begin{bmatrix} \frac{1}{a_{11}} & 0 & 0 \\ 0 & \frac{1}{a_{22}} & 0 \\ 0 & 0 & \frac{1}{a_{33}} \end{bmatrix}$.

(b) In general, the inverse of a matrix in the form of A is

$$\begin{bmatrix} \frac{1}{a_{11}} & 0 & 0 & \cdots & 0 \\ 0 & \frac{1}{a_{22}} & 0 & \cdots & 0 \\ 0 & 0 & \frac{1}{a_{33}} & \cdots & 0 \\ \vdots & \vdots & \vdots & \cdots & \vdots \\ 0 & 0 & 0 & \cdots & \frac{1}{a_{nn}} \end{bmatrix}.$$

Section 10.4   The Determinant of a Square Matrix

- You should be able to determine the determinant of a matrix of order 2×2 by using the difference of the products of the diagonals.

- You should be able to use expansion by cofactors to find the determinant of a matrix of order 3×3 or greater.

- The determinant of a triangular matrix equals the product of the entries on the main diagonal.

1. determinant

3. cofactor

5. 4

7. $\begin{vmatrix} 8 & 4 \\ 2 & 3 \end{vmatrix} = (8)(3) - (4)(2) = 16$

9. $\begin{vmatrix} 6 & 2 \\ -5 & 3 \end{vmatrix} = (6)(3) - (2)(-5) = 28$

11. $\begin{vmatrix} -7 & 0 \\ 3 & 0 \end{vmatrix} = -7(0) - 0(3) = 0$

13. $\begin{vmatrix} 2 & 6 \\ 0 & 3 \end{vmatrix} = 2(3) - 6(0) = 6$

15. $\begin{vmatrix} -3 & -2 \\ -6 & -1 \end{vmatrix} = (-3)(-1) - (-2)(-6) = 3 - 12 = -9$

17. $\begin{vmatrix} -7 & 6 \\ \frac{1}{2} & 3 \end{vmatrix} = (-7)(3) - (6)(\frac{1}{2}) = -24$

19. $\begin{vmatrix} -\frac{1}{2} & \frac{1}{3} \\ -6 & \frac{1}{3} \end{vmatrix} = -\frac{1}{2}(\frac{1}{3}) - \frac{1}{3}(-6) = -\frac{1}{2} + 2 = \frac{11}{6}$

21. $\begin{vmatrix} 0.3 & 0.2 & 0.2 \\ 0.2 & 0.2 & 0.2 \\ -0.4 & 0.4 & 0.3 \end{vmatrix} = -0.002$

23. $\begin{vmatrix} 0.9 & 0.7 & 0 \\ -0.1 & 0.3 & 1.3 \\ -2.2 & 4.2 & 6.1 \end{vmatrix} = -4.842$

25. $\begin{bmatrix} 4 & 5 \\ 3 & -6 \end{bmatrix}$

(a) $M_{11} = -6$ (b) $C_{11} = M_{11} = -6$

     $M_{12} = 3$ $C_{12} = -M_{12} = -3$

     $M_{21} = 5$ $C_{21} = -M_{21} = -5$

     $M_{22} = 4$ $C_{22} = M_{22} = 4$

27. $\begin{bmatrix} 3 & 1 \\ -2 & -4 \end{bmatrix}$

(a) $M_{11} = -4$ (b) $C_{11} = M_{11} = -4$

     $M_{12} = -2$ $C_{12} = -M_{12} = 2$

     $M_{21} = 1$ $C_{21} = -M_{21} = -1$

     $M_{22} = 3$ $C_{22} = M_{22} = 3$

29. $\begin{bmatrix} 4 & 0 & 2 \\ -3 & 2 & 1 \\ 1 & -1 & 1 \end{bmatrix}$

(a) $M_{11} = \begin{vmatrix} 2 & 1 \\ -1 & 1 \end{vmatrix} = 2 - (-1) = 3$

     $M_{12} = \begin{vmatrix} -3 & 1 \\ 1 & 1 \end{vmatrix} = -3 - 1 = -4$

     $M_{13} = \begin{vmatrix} -3 & 2 \\ 1 & -1 \end{vmatrix} = 3 - 2 = 1$

     $M_{21} = \begin{vmatrix} 0 & 2 \\ -1 & 1 \end{vmatrix} = 0 - (-2) = 2$

     $M_{22} = \begin{vmatrix} 4 & 2 \\ 1 & 1 \end{vmatrix} = 4 - 2 = 2$

     $M_{23} = \begin{vmatrix} 4 & 0 \\ 1 & -1 \end{vmatrix} = -4 - 0 = -4$

     $M_{31} = \begin{vmatrix} 0 & 2 \\ 2 & 1 \end{vmatrix} = 0 - 4 = -4$

     $M_{32} = \begin{vmatrix} 4 & 2 \\ -3 & 1 \end{vmatrix} = 4 - (-6) = 10$

     $M_{33} = \begin{vmatrix} 4 & 0 \\ -3 & 2 \end{vmatrix} = 8 - 0 = 8$

(b) $C_{11} = (-1)^2 M_{11} = 3$

     $C_{12} = (-1)^3 M_{12} = 4$

     $C_{13} = (-1)^4 M_{13} = 1$

     $C_{21} = (-1)^3 M_{21} = -2$

     $C_{22} = (-1)^4 M_{22} = 2$

     $C_{23} = (-1)^5 M_{23} = 4$

     $C_{31} = (-1)^4 M_{31} = -4$

     $C_{32} = (-1)^5 M_{32} = -10$

     $C_{33} = (-1)^6 M_{33} = 8$

31. $\begin{bmatrix} -4 & 6 & 3 \\ 7 & -2 & 8 \\ 1 & 0 & -5 \end{bmatrix}$

(a) $M_{11} = \begin{vmatrix} -2 & 8 \\ 0 & -5 \end{vmatrix} = (-2)(-5) - (8)(0) = 10$

     $M_{12} = \begin{vmatrix} 7 & 8 \\ 1 & -5 \end{vmatrix} = (7)(-5) - (8)(1) = -43$

     $M_{13} = \begin{vmatrix} 7 & -2 \\ 1 & 0 \end{vmatrix} = (7)(0) - (-2)(1) = 2$

     $M_{21} = \begin{vmatrix} 6 & 3 \\ 0 & -5 \end{vmatrix} = (6)(-5) - (3)(0) = -30$

     $M_{22} = \begin{vmatrix} -4 & 3 \\ 1 & -5 \end{vmatrix} = (-4)(-5) - (3)(1) = 17$

     $M_{23} = \begin{vmatrix} -4 & 6 \\ 1 & 0 \end{vmatrix} = (-4)(0) - (6)(1) = -6$

     $M_{31} = \begin{vmatrix} 6 & 3 \\ -2 & 8 \end{vmatrix} = (6)(8) - (3)(-2) = 54$

     $M_{32} = \begin{vmatrix} -4 & 3 \\ 7 & 8 \end{vmatrix} = (-4)(8) - (3)(7) = -53$

     $M_{33} = \begin{vmatrix} -4 & 6 \\ 7 & -2 \end{vmatrix} = (-4)(-2) - (6)(7) = -34$

(b) $C_{11} = (-1)^2 M_{11} = 10$

     $C_{12} = (-1)^3 M_{12} = 43$

     $C_{13} = (-1)^4 M_{13} = 2$

     $C_{21} = (-1)^3 M_{21} = 30$

     $C_{22} = (-1)^4 M_{22} = 17$

     $C_{23} = (-1)^5 M_{23} = 6$

     $C_{31} = (-1)^4 M_{31} = 54$

     $C_{32} = (-1)^5 M_{32} = 53$

     $C_{33} = (-1)^6 M_{33} = -34$

33. (a) $\begin{vmatrix} -3 & 2 & 1 \\ 4 & 5 & 6 \\ 2 & -3 & 1 \end{vmatrix} = -3\begin{vmatrix} 5 & 6 \\ -3 & 1 \end{vmatrix} - 2\begin{vmatrix} 4 & 6 \\ 2 & 1 \end{vmatrix} + \begin{vmatrix} 4 & 5 \\ 2 & -3 \end{vmatrix} = -3(23) - 2(-8) - 22 = -75$

(b) $\begin{vmatrix} -3 & 2 & 1 \\ 4 & 5 & 6 \\ 2 & -3 & 1 \end{vmatrix} = -2\begin{vmatrix} 4 & 6 \\ 2 & 1 \end{vmatrix} + 5\begin{vmatrix} -3 & 1 \\ 2 & 1 \end{vmatrix} + 3\begin{vmatrix} -3 & 1 \\ 4 & 6 \end{vmatrix} = -2(-8) + 5(-5) + 3(-22) = -75$

35. (a) $\begin{vmatrix} 5 & 0 & -3 \\ 0 & 12 & 4 \\ 1 & 6 & 3 \end{vmatrix} = 0\begin{vmatrix} 0 & -3 \\ 6 & 3 \end{vmatrix} + 12\begin{vmatrix} 5 & -3 \\ 1 & 3 \end{vmatrix} - 4\begin{vmatrix} 5 & 0 \\ 1 & 6 \end{vmatrix} = 0(18) + 12(18) - 4(30) = 96$

(b) $\begin{vmatrix} 5 & 0 & -3 \\ 0 & 12 & 4 \\ 1 & 6 & 3 \end{vmatrix} = 0\begin{vmatrix} 0 & 4 \\ 1 & 3 \end{vmatrix} + 12\begin{vmatrix} 5 & -3 \\ 1 & 3 \end{vmatrix} - 6\begin{vmatrix} 5 & -3 \\ 0 & 4 \end{vmatrix} = 0(-4) + 12(18) - 6(20) = 96$

37. (a) $\begin{vmatrix} 6 & 0 & -3 & 5 \\ 4 & 13 & 6 & -8 \\ -1 & 0 & 7 & 4 \\ 8 & 6 & 0 & 2 \end{vmatrix} = -4\begin{vmatrix} 0 & -3 & 5 \\ 0 & 7 & 4 \\ 6 & 0 & 2 \end{vmatrix} + 13\begin{vmatrix} 6 & -3 & 5 \\ -1 & 7 & 4 \\ 8 & 0 & 2 \end{vmatrix} - 6\begin{vmatrix} 6 & 0 & 5 \\ -1 & 0 & 4 \\ 8 & 6 & 2 \end{vmatrix} - 8\begin{vmatrix} 6 & 0 & -3 \\ -1 & 0 & 7 \\ 8 & 6 & 0 \end{vmatrix}$

$= -4(-282) + 13(-298) - 6(-174) - 8(-234) = 170$

(b) $\begin{vmatrix} 6 & 0 & -3 & 5 \\ 4 & 13 & 6 & -8 \\ -1 & 0 & 7 & 4 \\ 8 & 6 & 0 & 2 \end{vmatrix} = 0\begin{vmatrix} 4 & 6 & -8 \\ -1 & 7 & 4 \\ 8 & 0 & 2 \end{vmatrix} + 13\begin{vmatrix} 6 & -3 & 5 \\ -1 & 7 & 4 \\ 8 & 0 & 2 \end{vmatrix} + 0\begin{vmatrix} 6 & -3 & 5 \\ 4 & 6 & -8 \\ 8 & 0 & 2 \end{vmatrix} + 6\begin{vmatrix} 6 & -3 & 5 \\ 4 & 6 & -8 \\ -1 & 7 & 4 \end{vmatrix}$

$= 0 + 13(-298) + 0 + 6(674) = 170$

39. Expand along Column 1.

$\begin{vmatrix} 2 & -1 & 0 \\ 4 & 2 & 1 \\ 4 & 2 & 1 \end{vmatrix} = 2\begin{vmatrix} 2 & 1 \\ 2 & 1 \end{vmatrix} - 4\begin{vmatrix} -1 & 0 \\ 2 & 1 \end{vmatrix} + 4\begin{vmatrix} -1 & 0 \\ 2 & 1 \end{vmatrix}$

$= 2(0) - 4(-1) + 4(-1) = 0$

41. Expand along Row 2.

$\begin{vmatrix} 6 & 3 & -7 \\ 0 & 0 & 0 \\ 4 & -6 & 3 \end{vmatrix} = 0\begin{vmatrix} 3 & -7 \\ -6 & 3 \end{vmatrix} - 0\begin{vmatrix} 6 & -7 \\ 4 & 3 \end{vmatrix} + 0\begin{vmatrix} 6 & 3 \\ 4 & -6 \end{vmatrix} = 0$

43. Expand along Column 1.

$\begin{vmatrix} -1 & 2 & -5 \\ 0 & 3 & 4 \\ 0 & 0 & 3 \end{vmatrix} = -1\begin{vmatrix} 3 & 4 \\ 0 & 3 \end{vmatrix} - 0\begin{vmatrix} 2 & -5 \\ 0 & 3 \end{vmatrix} + 0\begin{vmatrix} 2 & -5 \\ 3 & 4 \end{vmatrix}$

$= -1(9) - 0(6) + 0(23) = -9$

45. Expand along Column 3.

$\begin{vmatrix} 1 & 4 & -2 \\ 3 & 2 & 0 \\ -1 & 4 & 3 \end{vmatrix} = -2\begin{vmatrix} 3 & 2 \\ -1 & 4 \end{vmatrix} + 3\begin{vmatrix} 1 & 4 \\ 3 & 2 \end{vmatrix}$

$= -2(14) + 3(-10) = -58$

47. Expand along Column 1.

$\begin{vmatrix} 2 & 4 & 6 \\ 0 & 3 & 1 \\ 0 & 0 & -5 \end{vmatrix} = 2\begin{vmatrix} 3 & 1 \\ 0 & -5 \end{vmatrix} - 0\begin{vmatrix} 4 & 6 \\ 0 & -5 \end{vmatrix} + 0\begin{vmatrix} 4 & 6 \\ 3 & 1 \end{vmatrix}$

$= 2(-15) - 0(-20) + 0(-14) = -30$

49. Expand along Column 3.

$\begin{vmatrix} 2 & 6 & 6 & 2 \\ 2 & 7 & 3 & 6 \\ 1 & 5 & 0 & 1 \\ 3 & 7 & 0 & 7 \end{vmatrix} = 6\begin{vmatrix} 2 & 7 & 6 \\ 1 & 5 & 1 \\ 3 & 7 & 7 \end{vmatrix} - 3\begin{vmatrix} 2 & 6 & 2 \\ 1 & 5 & 1 \\ 3 & 7 & 7 \end{vmatrix}$

$= 6(-20) - 3(16) = -168$

51. Expand along Column 1.

$$\begin{vmatrix} 5 & 3 & 0 & 6 \\ 4 & 6 & 4 & 12 \\ 0 & 2 & -3 & 4 \\ 0 & 1 & -2 & 2 \end{vmatrix} = 5\begin{vmatrix} 6 & 4 & 12 \\ 2 & -3 & 4 \\ 1 & -2 & 2 \end{vmatrix} - 4\begin{vmatrix} 3 & 0 & 6 \\ 2 & -3 & 4 \\ 1 & -2 & 2 \end{vmatrix} = 5(0) - 4(0) = 0$$

53. Expand along Column 2, then along Column 4.

$$\begin{vmatrix} 3 & 2 & 4 & -1 & 5 \\ -2 & 0 & 1 & 3 & 2 \\ 1 & 0 & 0 & 4 & 0 \\ 6 & 0 & 2 & -1 & 0 \\ 3 & 0 & 5 & 1 & 0 \end{vmatrix} = -2\begin{vmatrix} -2 & 1 & 3 & 2 \\ 1 & 0 & 4 & 0 \\ 6 & 2 & -1 & 0 \\ 3 & 5 & 1 & 0 \end{vmatrix} = (-2)(-2)\begin{vmatrix} 1 & 0 & 4 \\ 6 & 2 & -1 \\ 3 & 5 & 1 \end{vmatrix} = 4(103) = 412$$

55. $\begin{vmatrix} 3 & 8 & -7 \\ 0 & -5 & 4 \\ 8 & 1 & 6 \end{vmatrix} = -126$

57. $\begin{vmatrix} 7 & 0 & -14 \\ -2 & 5 & 4 \\ -6 & 2 & 12 \end{vmatrix} = 0$

59. $\begin{vmatrix} 1 & -1 & 8 & 4 \\ 2 & 6 & 0 & -4 \\ 2 & 0 & 2 & 6 \\ 0 & 2 & 8 & 0 \end{vmatrix} = -336$

61. $\begin{vmatrix} 3 & -2 & 4 & 3 & 1 \\ -1 & 0 & 2 & 1 & 0 \\ 5 & -1 & 0 & 3 & 2 \\ 4 & 7 & -8 & 0 & 0 \\ 1 & 2 & 3 & 0 & 2 \end{vmatrix} = 410$

63. (a) $\begin{vmatrix} -1 & 0 \\ 0 & 3 \end{vmatrix} = -3$

(b) $\begin{vmatrix} 2 & 0 \\ 0 & -1 \end{vmatrix} = -2$

(c) $\begin{bmatrix} -1 & 0 \\ 0 & 3 \end{bmatrix}\begin{bmatrix} 2 & 0 \\ 0 & -1 \end{bmatrix} = \begin{bmatrix} -2 & 0 \\ 0 & -3 \end{bmatrix}$

(d) $\begin{vmatrix} -2 & 0 \\ 0 & -3 \end{vmatrix} = 6$

65. (a) $\begin{vmatrix} 4 & 0 \\ 3 & -2 \end{vmatrix} = -8$

(b) $\begin{vmatrix} -1 & 1 \\ -2 & 2 \end{vmatrix} = 0$

(c) $\begin{bmatrix} 4 & 0 \\ 3 & -2 \end{bmatrix}\begin{bmatrix} -1 & 1 \\ -2 & 2 \end{bmatrix} = \begin{bmatrix} -4 & 4 \\ 1 & -1 \end{bmatrix}$

(d) $\begin{vmatrix} -4 & 4 \\ 1 & -1 \end{vmatrix} = 0$

67. (a) $\begin{vmatrix} 0 & 1 & 2 \\ -3 & -2 & 1 \\ 0 & 4 & 1 \end{vmatrix} = -21$

(b) $\begin{vmatrix} 3 & -2 & 0 \\ 1 & -1 & 2 \\ 3 & 1 & 1 \end{vmatrix} = -19$

(c) $\begin{bmatrix} 0 & 1 & 2 \\ -3 & -2 & 1 \\ 0 & 4 & 1 \end{bmatrix}\begin{bmatrix} 3 & -2 & 0 \\ 1 & -1 & 2 \\ 3 & 1 & 1 \end{bmatrix} = \begin{bmatrix} 7 & 1 & 4 \\ -8 & 9 & -3 \\ 7 & -3 & 9 \end{bmatrix}$

(d) $\begin{vmatrix} 7 & 1 & 4 \\ -8 & 9 & -3 \\ 7 & -3 & 9 \end{vmatrix} = 399$

69. (a) $\begin{vmatrix} -1 & 2 & 1 \\ 1 & 0 & 1 \\ 0 & 1 & 0 \end{vmatrix} = 2$

(b) $\begin{vmatrix} -1 & 0 & 0 \\ 0 & 2 & 0 \\ 0 & 0 & 3 \end{vmatrix} = -6$

(c) $\begin{bmatrix} -1 & 2 & 1 \\ 1 & 0 & 1 \\ 0 & 1 & 0 \end{bmatrix}\begin{bmatrix} -1 & 0 & 0 \\ 0 & 2 & 0 \\ 0 & 0 & 3 \end{bmatrix} = \begin{bmatrix} 1 & 4 & 3 \\ -1 & 0 & 3 \\ 0 & 2 & 0 \end{bmatrix}$

(d) $\begin{vmatrix} 1 & 4 & 3 \\ -1 & 0 & 3 \\ 0 & 2 & 0 \end{vmatrix} = -12$

71. $\begin{vmatrix} w & x \\ y & z \end{vmatrix} = wz - xy$

$-\begin{vmatrix} y & z \\ w & x \end{vmatrix} = -(xy - wz) = wz - xy$

So, $\begin{vmatrix} w & x \\ y & z \end{vmatrix} = -\begin{vmatrix} y & z \\ w & x \end{vmatrix}$.

73. $\begin{vmatrix} w & x \\ y & z \end{vmatrix} = wz - xy$

$\begin{vmatrix} w & x+cw \\ y & z+cy \end{vmatrix} = w(z+cy) - y(x+cw) = wz - xy$

So, $\begin{vmatrix} w & x \\ y & z \end{vmatrix} = \begin{vmatrix} w & x+cw \\ y & z+cy \end{vmatrix}$.

75. $\begin{vmatrix} 1 & x & x^2 \\ 1 & y & y^2 \\ 1 & z & z^2 \end{vmatrix} = \begin{vmatrix} y & y^2 \\ z & z^2 \end{vmatrix} - \begin{vmatrix} x & x^2 \\ z & z^2 \end{vmatrix} + \begin{vmatrix} x & x^2 \\ y & y^2 \end{vmatrix}$

$= (yz^2 - y^2z) - (xz^2 - x^2z) + (xy^2 - x^2y)$

$= yz^2 - xz^2 - y^2z + x^2z + xy(y - x)$

$= z^2(y - x) - z(y^2 - x^2) + xy(y - x)$

$= z^2(y - x) - z(y - x)(y + x) + xy(y - x)$

$= (y - x)[z^2 - z(y + x) + xy]$

$= (y - x)[z^2 - zy - zx + xy]$

$= (y - x)[z^2 - zx - zy + xy]$

$= (y - x)[z(z - x) - y(z - x)]$

$= (y - x)(z - x)(z - y)$

77. $\begin{vmatrix} x & 2 \\ 1 & x \end{vmatrix} = 2$

$x^2 - 2 = 2$

$x^2 = 4$

$x = \pm 2$

79. $\begin{vmatrix} x & 1 \\ 2 & x-2 \end{vmatrix} = -1$

$x(x - 2) - 2 = -1$

$x^2 - 2x - 1 = 0$

Using the Quadratic Formula:

$x = \dfrac{2 \pm \sqrt{4 - 4(1)(-1)}}{2}$

$x = \dfrac{2 \pm 2\sqrt{2}}{2}$

$x = 1 \pm \sqrt{2}$

81. $\begin{vmatrix} x-1 & 2 \\ 3 & x-2 \end{vmatrix} = 0$

$(x - 1)(x - 2) - 6 = 0$

$x^2 - 3x - 4 = 0$

$(x + 1)(x - 4) = 0$

$x = -1 \text{ or } x = 4$

83. $\begin{vmatrix} x+3 & 2 \\ 1 & x+2 \end{vmatrix} = 0$

$(x + 3)(x + 2) - 2 = 0$

$x^2 + 5x + 4 = 0$

$(x + 1)(x + 4) = 0$

$x = -1 \text{ or } x = -4$

85. $\begin{vmatrix} 4u & -1 \\ -1 & 2v \end{vmatrix} = 8uv - 1$

87. $\begin{vmatrix} e^{2x} & e^{3x} \\ 2e^{2x} & 3e^{3x} \end{vmatrix} = 3e^{5x} - 2e^{5x} = e^{5x}$

89. $\begin{vmatrix} x & \ln x \\ 1 & \dfrac{1}{x} \end{vmatrix} = 1 - \ln x$

91. True. If an entire row is zero, then each cofactor in the expansion is multiplied by zero.

93. Sample answer: Let $A = \begin{bmatrix} 1 & 3 \\ -2 & 4 \end{bmatrix}$ and $B = \begin{bmatrix} -4 & 0 \\ 3 & 5 \end{bmatrix}$.

$|A| = \begin{vmatrix} 1 & 3 \\ -2 & 4 \end{vmatrix} = 10, \; |B| = \begin{vmatrix} -4 & 0 \\ 3 & 5 \end{vmatrix} = -20$

$|A| + |B| = -10$

$A + B = \begin{bmatrix} -3 & 3 \\ 1 & 9 \end{bmatrix}, \; |A + B| = \begin{vmatrix} -3 & 3 \\ 1 & 9 \end{vmatrix} = -30$

So, $|A + B| \neq |A| + |B|$.

95. A square matrix is a square array of numbers. The determinant of a square matrix is a real number.

97. (a) $\begin{vmatrix} 1 & 3 & 4 \\ -7 & 2 & -5 \\ 6 & 1 & 2 \end{vmatrix} = -115$

(b) $\begin{vmatrix} 1 & 3 & 4 \\ -2 & 2 & 0 \\ 1 & 6 & 2 \end{vmatrix} = -40$

$-\begin{vmatrix} 1 & 4 & 3 \\ -7 & -5 & 2 \\ 6 & 2 & 1 \end{vmatrix} = -115$

$-\begin{vmatrix} 1 & 6 & 2 \\ -2 & 2 & 0 \\ 1 & 3 & 4 \end{vmatrix} = -40$

Column 2 and Column 3 were interchanged.
Row 1 and Row 3 were interchanged.

99. (a) $A = \begin{bmatrix} 1 & 2 \\ 2 & -3 \end{bmatrix}, B = \begin{bmatrix} 5 & 10 \\ 2 & -3 \end{bmatrix}$

$|B| = \begin{vmatrix} 5 & 10 \\ 2 & -3 \end{vmatrix} = -35$

$5|A| = 5\begin{vmatrix} 1 & 2 \\ 2 & -3 \end{vmatrix} = -35$

Row 1 was multiplied by 5.

$|B| = 5|A|$

(b) $A = \begin{bmatrix} 1 & 2 & -1 \\ 3 & -3 & 2 \\ 7 & 1 & 3 \end{bmatrix}, B = \begin{bmatrix} 1 & 8 & -3 \\ 3 & -12 & 6 \\ 7 & 4 & 9 \end{bmatrix}$

$|B| = \begin{vmatrix} 1 & 8 & -3 \\ 3 & -12 & 6 \\ 7 & 4 & 9 \end{vmatrix} = -300$

$12|A| = 12\begin{vmatrix} 1 & 2 & -1 \\ 3 & -3 & 2 \\ 7 & 1 & 3 \end{vmatrix} = -300$

Column 2 was multiplied by 4 and Column 3 was multiplied by 3.

$|B| = (4)(3)|A| = 12|A|$

101. $\begin{vmatrix} 1 & 0 & 0 \\ 0 & 5 & 0 \\ 0 & 0 & 2 \end{vmatrix} = 1\begin{vmatrix} 5 & 0 \\ 0 & 2 \end{vmatrix} - 0\begin{vmatrix} 0 & 0 \\ 0 & 2 \end{vmatrix} + 0\begin{vmatrix} 0 & 5 \\ 0 & 0 \end{vmatrix}$

$= 1(10) - 0(0) + 0(0)$

$= 10$

103. $\begin{vmatrix} -1 & 2 & -5 \\ 0 & 3 & 4 \\ 0 & 0 & 3 \end{vmatrix} = -1\begin{vmatrix} 3 & 4 \\ 0 & 3 \end{vmatrix} - 2\begin{vmatrix} 0 & 4 \\ 0 & 3 \end{vmatrix} - 5\begin{vmatrix} 0 & 3 \\ 0 & 0 \end{vmatrix}$

$= -1(9) - 2(0) - 5(0)$

$= -9$

105. The determinant of a triangular matrix is the product of the terms in the diagonal.

Section 10.5 Applications of Matrices and Determinants

- You should be able to use Cramer's Rule to solve a system of linear equations.

- Now you should be able to solve a system of linear equations by graphing, substitution, elimination, elementary row operations on an augmented matrix, using the inverse matrix, or Cramer's Rule.

- You should be able to find the area of a triangle with vertices (x_1, y_1), (x_2, y_2), and (x_3, y_3).

$$\text{Area} = \pm \frac{1}{2} \begin{vmatrix} x_1 & y_1 & 1 \\ x_2 & y_2 & 1 \\ x_3 & y_3 & 1 \end{vmatrix}$$

  The $\pm$ symbol indicates that the appropriate sign should be chosen so that the area is positive.

- You should be able to test to see if three points, (x_1, y_1), (x_2, y_2), and (x_3, y_3), are collinear.

$$\begin{vmatrix} x_1 & y_1 & 1 \\ x_2 & y_2 & 1 \\ x_3 & y_3 & 1 \end{vmatrix} = 0, \text{ if and only if they are collinear.}$$

- You should be able to find the equation of the line through (x_1, y_1) and (x_2, y_2) by evaluating.

$$\begin{vmatrix} x & y & 1 \\ x_1 & y_1 & 1 \\ x_2 & y_2 & 1 \end{vmatrix} = 0$$

- You should be able to encode and decode messages by using an invertible $n \times n$ matrix.

---

1. Cramer's Rule

3. $A = \pm \dfrac{1}{2} \begin{vmatrix} x_1 & y_1 & 1 \\ x_2 & y_2 & 1 \\ x_3 & y_3 & 1 \end{vmatrix}$

5. uncoded; coded

7. $\begin{cases} -7x + 11y = -1 \\ 3x - 9y = 9 \end{cases}$

$$x = \frac{\begin{vmatrix} -1 & 11 \\ 9 & -9 \end{vmatrix}}{\begin{vmatrix} -7 & 11 \\ 3 & -9 \end{vmatrix}} = \frac{-90}{30} = -3$$

$$y = \frac{\begin{vmatrix} -7 & -1 \\ 3 & 9 \end{vmatrix}}{\begin{vmatrix} -7 & 11 \\ 3 & -9 \end{vmatrix}} = \frac{-60}{30} = -2$$

Solution: $(-3, -2)$

9. $\begin{cases} 3x + 2y = -2 \\ 6x + 4y = 4 \end{cases}$

Because $\begin{vmatrix} 3 & 2 \\ 6 & 4 \end{vmatrix} = 0$, Cramer's Rule does not apply.

The system is inconsistent in this case and has no solution.

11. $\begin{cases} -0.4x + 0.8y = 1.6 \\ 0.2x + 0.3y = 2.2 \end{cases}$

$$x = \frac{\begin{vmatrix} 1.6 & 0.8 \\ 2.2 & 0.3 \end{vmatrix}}{\begin{vmatrix} -0.4 & 0.8 \\ 0.2 & 0.3 \end{vmatrix}} = \frac{-1.28}{-0.28} = \frac{32}{7}$$

$$y = \frac{\begin{vmatrix} -0.4 & 1.6 \\ 0.2 & 2.2 \end{vmatrix}}{\begin{vmatrix} -0.4 & 0.8 \\ 0.2 & 0.3 \end{vmatrix}} = \frac{-1.20}{-0.28} = \frac{30}{7}$$

Solution: $\left(\dfrac{32}{7}, \dfrac{30}{7} \right)$

13. $\begin{cases} 4x - y + z = -5 \\ 2x + 2y + 3z = 10, \\ 5x - 2y + 6z = 1 \end{cases}$ $D = \begin{vmatrix} 4 & -1 & 1 \\ 2 & 2 & 3 \\ 5 & -1 & 6 \end{vmatrix} = 55$

$$x = \dfrac{\begin{vmatrix} -5 & -1 & 1 \\ 10 & 2 & 3 \\ 1 & -2 & 6 \end{vmatrix}}{55} = \dfrac{-55}{55} = -1, \; y = \dfrac{\begin{vmatrix} 4 & -5 & 1 \\ 2 & 10 & 3 \\ 5 & 1 & 6 \end{vmatrix}}{55} = \dfrac{165}{55} = 3, \; z = \dfrac{\begin{vmatrix} 4 & -1 & -5 \\ 2 & 2 & 10 \\ 5 & -2 & 1 \end{vmatrix}}{55} = \dfrac{110}{55} = 2$$

Solution: $(-1, 3, 2)$

15. $\begin{cases} x + 2y + 3z = -3 \\ -2x + y - z = 6, \\ 3x - 3y + 2z = -11 \end{cases}$ $D = \begin{vmatrix} 1 & 2 & 3 \\ -2 & 1 & -1 \\ 3 & -3 & 2 \end{vmatrix} = 10$

$$x = \dfrac{\begin{vmatrix} -3 & 2 & 3 \\ 6 & 1 & -1 \\ -11 & -3 & 2 \end{vmatrix}}{10} = \dfrac{-20}{10} = -2$$

$$y = \dfrac{\begin{vmatrix} 1 & -3 & 3 \\ -2 & 6 & -1 \\ 3 & -11 & 2 \end{vmatrix}}{10} = \dfrac{10}{10} = 1$$

$$z = \dfrac{\begin{vmatrix} 1 & 2 & -3 \\ -2 & 1 & 6 \\ 3 & -3 & -11 \end{vmatrix}}{10} = \dfrac{-10}{10} = -1$$

Solution: $(-2, 1, -1)$

17. $\begin{cases} 3x + 3y + 5z = 1 \\ 3x + 5y + 9z = 2, \\ 5x + 9y + 17z = 4 \end{cases}$ $D = \begin{vmatrix} 3 & 3 & 5 \\ 3 & 5 & 9 \\ 5 & 9 & 17 \end{vmatrix} = 4$

$$x = \dfrac{\begin{vmatrix} 1 & 3 & 5 \\ 2 & 5 & 9 \\ 4 & 9 & 17 \end{vmatrix}}{4} = \dfrac{0}{4} = 0$$

$$y = \dfrac{\begin{vmatrix} 3 & 1 & 5 \\ 3 & 2 & 9 \\ 5 & 4 & 17 \end{vmatrix}}{4} = \dfrac{-2}{4} = -\dfrac{1}{2}$$

$$z = \dfrac{\begin{vmatrix} 3 & 3 & 1 \\ 3 & 5 & 2 \\ 5 & 9 & 4 \end{vmatrix}}{4} = \dfrac{2}{4} = \dfrac{1}{2}$$

Solution: $\left(0, -\dfrac{1}{2}, \dfrac{1}{2}\right)$

19. $\begin{cases} 2x - y + z = 5 \\ x - 2y - z = 1, \\ 3x + y + z = 4 \end{cases}$ $D = \begin{vmatrix} 2 & -1 & 1 \\ 1 & -2 & -1 \\ 3 & 1 & 1 \end{vmatrix} = 9$

$$x = \dfrac{\begin{vmatrix} 5 & -1 & 1 \\ 1 & -2 & -1 \\ 4 & 1 & 1 \end{vmatrix}}{9} = \dfrac{9}{9} = 1$$

$$y = \dfrac{\begin{vmatrix} 2 & 5 & 1 \\ 1 & 1 & -1 \\ 3 & 4 & 1 \end{vmatrix}}{9} = \dfrac{-9}{9} = -1$$

$$z = \dfrac{\begin{vmatrix} 2 & -1 & 5 \\ 1 & -2 & 1 \\ 3 & 1 & 4 \end{vmatrix}}{9} = \dfrac{18}{9} = 2$$

Solution: $(1, -1, 2)$

21. Vertices: $(0, 0)\, (3, 1), (1, 5)$

$$\text{Area} = \dfrac{1}{2}\begin{vmatrix} 0 & 0 & 1 \\ 3 & 1 & 1 \\ 1 & 5 & 1 \end{vmatrix} = \dfrac{1}{2}\begin{vmatrix} 3 & 1 \\ 1 & 5 \end{vmatrix} = 7 \text{ square units}$$

23. Vertices: $(-2, -3), (2, -3), (0, 4)$

$$\text{Area} = \frac{1}{2}\begin{vmatrix} -2 & -3 & 1 \\ 2 & -3 & 1 \\ 0 & 4 & 1 \end{vmatrix} = \frac{1}{2}\left(-2\begin{vmatrix} -3 & 1 \\ 4 & 1 \end{vmatrix} - 2\begin{vmatrix} -3 & 1 \\ 4 & 1 \end{vmatrix}\right) = \frac{1}{2}(14 + 14) = 14 \text{ square units}$$

25. Vertices: $\left(0, \frac{1}{2}\right), \left(\frac{5}{2}, 0\right), (4, 3)$

$$\text{Area} = \frac{1}{2}\begin{vmatrix} 0 & \frac{1}{2} & 1 \\ \frac{5}{2} & 0 & 1 \\ 4 & 3 & 1 \end{vmatrix} = \frac{1}{2}\left(-\frac{1}{2}\begin{vmatrix} \frac{5}{2} & 1 \\ 4 & 1 \end{vmatrix} + 1\begin{vmatrix} \frac{5}{2} & 0 \\ 4 & 3 \end{vmatrix}\right) = \frac{1}{2}\left(\frac{3}{4} + \frac{15}{2}\right) = \frac{33}{8} \text{ square units}$$

27. Vertices: $(-2, 4), (2, 3), (-1, 5)$

$$\text{Area} = \frac{1}{2}\begin{vmatrix} -2 & 4 & 1 \\ 2 & 3 & 1 \\ -1 & 5 & 1 \end{vmatrix} = \frac{1}{2}\left[\begin{vmatrix} 2 & 3 \\ -1 & 5 \end{vmatrix} - \begin{vmatrix} -2 & 4 \\ -1 & 5 \end{vmatrix} + \begin{vmatrix} -2 & 4 \\ 2 & 3 \end{vmatrix}\right] = \frac{1}{2}(13 + 6 - 14) = \frac{5}{2} \text{ square units}$$

29. Vertices: $(-3, 5), (2, 6), (3, -5)$

$$\text{Area} = -\frac{1}{2}\begin{vmatrix} -3 & 5 & 1 \\ 2 & 6 & 1 \\ 3 & -5 & 1 \end{vmatrix} = -\frac{1}{2}\left[\begin{vmatrix} 2 & 6 \\ 3 & -5 \end{vmatrix} - \begin{vmatrix} -3 & 5 \\ 3 & -5 \end{vmatrix} + \begin{vmatrix} -3 & 5 \\ 2 & 6 \end{vmatrix}\right] = -\frac{1}{2}(-28 + 0 - 28) = 28 \text{ square units}$$

31. Vertices: $(-4, 2), \left(0, \frac{7}{2}\right), \left(3, -\frac{1}{2}\right)$

$$\text{Area} = -\frac{1}{2}\begin{vmatrix} -4 & 2 & 1 \\ 0 & \frac{7}{2} & 1 \\ 3 & -\frac{1}{2} & 1 \end{vmatrix} = -\frac{1}{2}\left(-4\begin{vmatrix} \frac{7}{2} & 1 \\ -\frac{1}{2} & 1 \end{vmatrix} - 2\begin{vmatrix} 0 & 1 \\ 3 & 1 \end{vmatrix} + 1\begin{vmatrix} 0 & \frac{7}{2} \\ 3 & -\frac{1}{2} \end{vmatrix}\right) = -\frac{1}{2}\left(-16 + 6 - \frac{21}{2}\right) = \frac{41}{4} \text{ square units}$$

33. $4 = \pm\frac{1}{2}\begin{vmatrix} -5 & 1 & 1 \\ 0 & 2 & 1 \\ -2 & y & 1 \end{vmatrix}$

$\pm 8 = -5\begin{vmatrix} 2 & 1 \\ y & 1 \end{vmatrix} - 2\begin{vmatrix} 1 & 1 \\ 2 & 1 \end{vmatrix}$

$\pm 8 = -5(2 - y) - 2(-1)$

$\pm 8 = 5y - 8$

$y = \dfrac{8 \pm 8}{5}$

$y = \dfrac{16}{5} \text{ or } y = 0$

35. $6 = \pm\frac{1}{2}\begin{vmatrix} -2 & -3 & 1 \\ 1 & -1 & 1 \\ -8 & y & 1 \end{vmatrix}$

$\pm 12 = \begin{vmatrix} 1 & -1 \\ -8 & y \end{vmatrix} - \begin{vmatrix} -2 & -3 \\ -8 & y \end{vmatrix} + \begin{vmatrix} -2 & -3 \\ 1 & -1 \end{vmatrix}$

$\pm 12 = (y - 8) - (-2y - 24) + 5$

$\pm 12 = 3y + 21$

$y = \dfrac{-21 \pm 12}{3} = -7 \pm 4$

$y = -3 \text{ or } y = -11$

37. Vertices: $(0, 25), (10, 0), (28, 5)$

$$\text{Area} = \frac{1}{2}\begin{vmatrix} 0 & 25 & 1 \\ 10 & 0 & 1 \\ 28 & 5 & 1 \end{vmatrix} = 250 \text{ square miles}$$

39. Points: $(3, -1), (0, -3), (12, 5)$

$$\begin{vmatrix} 3 & -1 & 1 \\ 0 & -3 & 1 \\ 12 & 5 & 1 \end{vmatrix} = 3\begin{vmatrix} -3 & 1 \\ 5 & 1 \end{vmatrix} + 12\begin{vmatrix} -1 & 1 \\ -3 & 1 \end{vmatrix} = 3(-8) + 12(2) = 0$$

The points are collinear.

41. Points: $\left(2, -\frac{1}{2}\right), (-4, 4), (6, -3)$

$$\begin{vmatrix} 2 & -\frac{1}{2} & 1 \\ -4 & 4 & 1 \\ 6 & -3 & 1 \end{vmatrix} = \begin{vmatrix} -4 & 4 \\ 6 & -3 \end{vmatrix} - \begin{vmatrix} 2 & -\frac{1}{2} \\ 6 & -3 \end{vmatrix} + \begin{vmatrix} 2 & -\frac{1}{2} \\ -4 & 4 \end{vmatrix} = -12 + 3 + 6 = -3 \neq 0$$

The points are not collinear.

43. Points: $(0, 2), (1, 2.4), (-1, 1.6)$

$$\begin{vmatrix} 0 & 2 & 1 \\ 1 & 2.4 & 1 \\ -1 & 1.6 & 1 \end{vmatrix} = -2\begin{vmatrix} 1 & 1 \\ -1 & 1 \end{vmatrix} + \begin{vmatrix} 1 & 2.4 \\ -1 & 1.6 \end{vmatrix} = -2(2) + 4 = 0$$

The points are collinear.

45.

$$\begin{vmatrix} 2 & -5 & 1 \\ 4 & y & 1 \\ 5 & -2 & 1 \end{vmatrix} = 0$$

$$2\begin{vmatrix} y & 1 \\ -2 & 1 \end{vmatrix} + 5\begin{vmatrix} 4 & 1 \\ 5 & 1 \end{vmatrix} + \begin{vmatrix} 4 & y \\ 5 & -2 \end{vmatrix} = 0$$

$$2(y + 2) + 5(-1) + (-8 - 5y) = 0$$

$$-3y - 9 = 0$$

$$y = -3$$

47. Points: $(0, 0), (5, 3)$

Equation: $\begin{vmatrix} x & y & 1 \\ 0 & 0 & 1 \\ 5 & 3 & 1 \end{vmatrix} = -\begin{vmatrix} x & y \\ 5 & 3 \end{vmatrix} = 5y - 3x = 0 \Rightarrow 3x - 5y = 0$

49. Points: $(-4, 3), (2, 1)$

Equation: $\begin{vmatrix} x & y & 1 \\ -4 & 3 & 1 \\ 2 & 1 & 1 \end{vmatrix} = x\begin{vmatrix} 3 & 1 \\ 1 & 1 \end{vmatrix} - y\begin{vmatrix} -4 & 1 \\ 2 & 1 \end{vmatrix} + \begin{vmatrix} -4 & 3 \\ 2 & 1 \end{vmatrix} = 2x + 6y - 10 = 0 \Rightarrow x + 3y - 5 = 0$

51. Points: $\left(-\frac{1}{2}, 3\right), \left(\frac{5}{2}, 1\right)$

Equation: $\begin{vmatrix} x & y & 1 \\ -\frac{1}{2} & 3 & 1 \\ \frac{5}{2} & 1 & 1 \end{vmatrix} = x\begin{vmatrix} 3 & 1 \\ 1 & 1 \end{vmatrix} - y\begin{vmatrix} -\frac{1}{2} & 1 \\ \frac{5}{2} & 1 \end{vmatrix} + \begin{vmatrix} -\frac{1}{2} & 3 \\ \frac{5}{2} & 1 \end{vmatrix} = 2x + 3y - 8 = 0$

53. (a) Uncoded: C   O   M   E   _   H   O   M   E   _   S   O   O   N
$$[3 \quad 15][13 \quad 5][0 \quad 8][15 \quad 13][5 \quad 0] \ [19 \quad 15][15 \quad 14]$$

(b) $[3 \quad 15]\begin{bmatrix} 1 & 2 \\ 3 & 5 \end{bmatrix} = [48 \quad 81]$

$[13 \quad 5]\begin{bmatrix} 1 & 2 \\ 3 & 5 \end{bmatrix} = [28 \quad 51]$

$[0 \quad 8]\begin{bmatrix} 1 & 2 \\ 3 & 5 \end{bmatrix} = [24 \quad 40]$

$[15 \quad 13]\begin{bmatrix} 1 & 2 \\ 3 & 5 \end{bmatrix} = [54 \quad 95]$

$[5 \quad 0]\begin{bmatrix} 1 & 2 \\ 3 & 5 \end{bmatrix} = [5 \quad 10]$

$[19 \quad 15]\begin{bmatrix} 1 & 2 \\ 3 & 5 \end{bmatrix} = [64 \quad 113]$

$[15 \quad 14]\begin{bmatrix} 1 & 2 \\ 3 & 5 \end{bmatrix} = [57 \quad 100]$

Encoded: 48   81   28   51   24   40   54   95   5   10   64   113   57   100

55. (a) Uncoded: C   A   L   L   _   M   E   _   T   O   M   O   R   R   O   W   _   _
$$[3 \quad 1 \quad 12][12 \quad 0 \quad 13][5 \quad 0 \quad 20][15 \quad 13 \quad 15] \ [18 \quad 18 \quad 15][23 \quad 0 \quad 0]$$

(b) $[3 \quad 1 \quad 12]\begin{bmatrix} 1 & -1 & 0 \\ 1 & 0 & -1 \\ -6 & 2 & 3 \end{bmatrix} = [-68 \quad 21 \quad 35]$

$[12 \quad 0 \quad 13]\begin{bmatrix} 1 & -1 & 0 \\ 1 & 0 & -1 \\ -6 & 2 & 3 \end{bmatrix} = [-66 \quad 14 \quad 39]$

$[5 \quad 0 \quad 20]\begin{bmatrix} 1 & -1 & 0 \\ 1 & 0 & -1 \\ -6 & 2 & 3 \end{bmatrix} = [-115 \quad 35 \quad 60]$

$[15 \quad 13 \quad 15]\begin{bmatrix} 1 & -1 & 0 \\ 1 & 0 & -1 \\ -6 & 2 & 3 \end{bmatrix} = [-62 \quad 15 \quad 32]$

$[18 \quad 18 \quad 15]\begin{bmatrix} 1 & -1 & 0 \\ 1 & 0 & -1 \\ -6 & 2 & 3 \end{bmatrix} = [-54 \quad 12 \quad 27]$

$[23 \quad 0 \quad 0]\begin{bmatrix} 1 & -1 & 0 \\ 1 & 0 & -1 \\ -6 & 2 & 3 \end{bmatrix} = [23 \quad -23 \quad 0]$

Encoded: −68   21   35   −66   14   39   −115   35   60   −62   15   32   −54   12   27   23   −23   0

In Exercises 57 and 59, use the matrix $A = \begin{bmatrix} 1 & 2 & 2 \\ 3 & 7 & 9 \\ -1 & -4 & -7 \end{bmatrix}$.

57. L A N D I N G _ S U C C E S S F U L
$[12 \ \ 1 \ \ 14][4 \ \ 9 \ \ 14][7 \ \ 0 \ \ 19][21 \ \ 3 \ \ 3][5 \ \ 19 \ \ 19][6 \ \ 21 \ \ 12]$

$[12 \ \ 1 \ \ 14] \begin{bmatrix} 1 & 2 & 2 \\ 3 & 7 & 9 \\ -1 & -4 & -7 \end{bmatrix} = [1 \ \ -25 \ \ -65]$

$[4 \ \ 9 \ \ 14] \begin{bmatrix} 1 & 2 & 2 \\ 3 & 7 & 9 \\ -1 & -4 & -7 \end{bmatrix} = [17 \ \ 15 \ \ -9]$

$[7 \ \ 0 \ \ 19] \begin{bmatrix} 1 & 2 & 2 \\ 3 & 7 & 9 \\ -1 & -4 & -7 \end{bmatrix} = [-12 \ \ -62 \ \ -119]$

$[21 \ \ 3 \ \ 3] \begin{bmatrix} 1 & 2 & 2 \\ 3 & 7 & 9 \\ -1 & -4 & -7 \end{bmatrix} = [27 \ \ 51 \ \ 48]$

$[5 \ \ 19 \ \ 19] \begin{bmatrix} 1 & 2 & 2 \\ 3 & 7 & 9 \\ -1 & -4 & -7 \end{bmatrix} = [43 \ \ 67 \ \ 48]$

$[6 \ \ 21 \ \ 12] \begin{bmatrix} 1 & 2 & 2 \\ 3 & 7 & 9 \\ -1 & -4 & -7 \end{bmatrix} = [57 \ \ 111 \ \ 117]$

Cryptogram: 1 -25 -65 17 15 -9 -12 -62 -119 27 51 48 43 67 48 57 111 117

59. H A P P Y _ B I R T H D A Y _
$[8 \ \ 1 \ \ 16][16 \ \ 25 \ \ 0][2 \ \ 9 \ \ 18][20 \ \ 8 \ \ 4][1 \ \ 25 \ \ 0]$

$[8 \ \ 1 \ \ 16] \quad A = [-5 \ \ -41 \ \ -87]$

$[16 \ \ 25 \ \ 0] \quad A = [91 \ \ 207 \ \ 257]$

$[2 \ \ 9 \ \ 18] \quad A = [11 \ \ -5 \ \ -41]$

$[20 \ \ 8 \ \ 4] \quad A = [40 \ \ 80 \ \ 84]$

$[1 \ \ 25 \ \ 0] \quad A = [76 \ \ 177 \ \ 227]$

Cryptogram: -5 -41 -87 91 207 257 11 -5 -41 40 80 84 76 177 227

61. $A^{-1} = \begin{bmatrix} 1 & 2 \\ 3 & 5 \end{bmatrix}^{-1} = \begin{bmatrix} -5 & 2 \\ 3 & -1 \end{bmatrix}$

$\begin{bmatrix} 11 & 21 \\ 64 & 112 \\ 25 & 50 \\ 29 & 53 \\ 23 & 46 \\ 40 & 75 \\ 55 & 92 \end{bmatrix} \begin{bmatrix} -5 & 2 \\ 3 & -1 \end{bmatrix} = \begin{bmatrix} 8 & 1 \\ 16 & 16 \\ 25 & 0 \\ 14 & 5 \\ 23 & 0 \\ 25 & 5 \\ 1 & 18 \end{bmatrix} \begin{matrix} \text{H} & \text{A} \\ \text{P} & \text{P} \\ \text{Y} & \_ \\ \text{N} & \text{E} \\ \text{W} & \_ \\ \text{Y} & \text{E} \\ \text{A} & \text{R} \end{matrix}$

Message: HAPPY NEW YEAR

63. $A^{-1} = \begin{bmatrix} 1 & -1 & 0 \\ 1 & 0 & -1 \\ -6 & 2 & 3 \end{bmatrix}^{-1} = \begin{bmatrix} -2 & -3 & -1 \\ -3 & -3 & -1 \\ -2 & -4 & -1 \end{bmatrix}$

$\begin{bmatrix} 9 & -1 & -9 \\ 38 & -19 & -19 \\ 28 & -9 & -19 \\ -80 & 25 & 41 \\ -64 & 21 & 31 \\ 9 & -5 & -4 \end{bmatrix} \begin{bmatrix} -2 & -3 & -1 \\ -3 & -3 & -1 \\ -2 & -4 & -1 \end{bmatrix} = \begin{bmatrix} 3 & 12 & 1 \\ 19 & 19 & 0 \\ 9 & 19 & 0 \\ 3 & 1 & 14 \\ 3 & 5 & 12 \\ 5 & 4 & 0 \end{bmatrix} \begin{matrix} \text{C L A} \\ \text{S S \_} \\ \text{I S \_} \\ \text{C A N} \\ \text{C E L} \\ \text{E D \_} \end{matrix}$

Message: CLASS IS CANCELED

65. $A^{-1} = \begin{bmatrix} 1 & 2 & 2 \\ 3 & 7 & 9 \\ -1 & -4 & -7 \end{bmatrix}^{-1} = \begin{bmatrix} -13 & 6 & 4 \\ 12 & -5 & -3 \\ -5 & 2 & 1 \end{bmatrix}$

$\begin{bmatrix} 20 & 17 & -15 \\ -12 & -56 & -104 \\ 1 & -25 & -65 \\ 62 & 143 & 181 \end{bmatrix} \begin{bmatrix} -13 & 6 & 4 \\ 12 & -5 & -3 \\ -5 & 2 & 1 \end{bmatrix} = \begin{bmatrix} 19 & 5 & 14 \\ 4 & 0 & 16 \\ 12 & 1 & 14 \\ 5 & 19 & 0 \end{bmatrix} \begin{matrix} \text{S E N} \\ \text{D \_ P} \\ \text{L A N} \\ \text{E S \_} \end{matrix}$

Message: SEND PLANES

67. Let A be the 2×2 matrix needed to decode the message.

$\begin{bmatrix} -18 & -18 \\ 1 & 16 \end{bmatrix} A = \begin{bmatrix} 0 & 18 \\ 15 & 14 \end{bmatrix} \begin{matrix} \text{\_ R} \\ \text{O N} \end{matrix}$

$A = \begin{bmatrix} -18 & -18 \\ 1 & 16 \end{bmatrix}^{-1} \begin{bmatrix} 0 & 18 \\ 15 & 14 \end{bmatrix} = \begin{bmatrix} -\frac{8}{135} & -\frac{1}{15} \\ \frac{1}{270} & \frac{1}{15} \end{bmatrix} \begin{bmatrix} 0 & 18 \\ 15 & 14 \end{bmatrix} = \begin{bmatrix} -1 & -2 \\ 1 & 1 \end{bmatrix}$

$\begin{bmatrix} 8 & 21 \\ -15 & -10 \\ -13 & -13 \\ 5 & 10 \\ 5 & 25 \\ 5 & 19 \\ -1 & 6 \\ 20 & 40 \\ -18 & -18 \\ 1 & 16 \end{bmatrix} \begin{bmatrix} -1 & -2 \\ 1 & 1 \end{bmatrix} = \begin{bmatrix} 13 & 5 \\ 5 & 20 \\ 0 & 13 \\ 5 & 0 \\ 20 & 15 \\ 14 & 9 \\ 7 & 8 \\ 20 & 0 \\ 0 & 18 \\ 15 & 14 \end{bmatrix} \begin{matrix} \text{M E} \\ \text{E T} \\ \text{\_ M} \\ \text{E \_} \\ \text{T O} \\ \text{N I} \\ \text{G H} \\ \text{T \_} \\ \text{\_ R} \\ \text{O N} \end{matrix}$

Message: MEET ME TONIGHT RON

69. $(0, 16.7), (1, 18.2), (2, 20.1), (3, 21.6),$
$(4, 23.2), (5, 25.5), (6, 27.7), (7, 29.1)$

(a) $n = 8, \sum_{i=1}^{n} x_i = 28, \sum_{i=1}^{n} x_i^2 = 140, \sum_{i=1}^{n} x_i^3 = 784, \sum_{i=1}^{n} x_i^4 = 4676$

$\sum_{i=1}^{n} y_i = 182.1, \sum_{i=1}^{n} x_i y_i = 713.4, \sum_{i=1}^{n} x_i^2 y_i = 3724.8$

System: $\begin{cases} 8c + 28b + 140a = 182.1 \\ 28c + 140b + 784a = 713.4 \\ 140c + 784b + 4676a = 3724.8 \end{cases}$

(b) $D = \begin{vmatrix} 8 & 28 & 140 \\ 28 & 140 & 784 \\ 140 & 784 & 4676 \end{vmatrix} = 56{,}448$

$c = \dfrac{\begin{vmatrix} 182.1 & 28 & 140 \\ 713.4 & 140 & 784 \\ 3724.8 & 784 & 4676 \end{vmatrix}}{56{,}448} = \dfrac{1333}{80} \approx 16.66$

$b = \dfrac{\begin{vmatrix} 8 & 182.1 & 140 \\ 28 & 713.4 & 784 \\ 140 & 3724.8 & 4676 \end{vmatrix}}{56{,}448} = \dfrac{881}{560} \approx 1.57$

$a = \dfrac{\begin{vmatrix} 8 & 28 & 182.1 \\ 28 & 140 & 713.4 \\ 140 & 784 & 3724.8 \end{vmatrix}}{56{,}448} = \dfrac{19}{560} \approx 0.034$

The least squares regression parabola is $y = 0.034t^2 + 1.57t + 16.66$.

(c)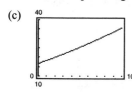

(d) The intersection of the regression parabola and the line $y = 35$ is about $t = 9$, so the consumption of bottled water will exceed 35 gallons in 2009.

71. False. In Cramer's Rule, the denominator is the determinant of the coefficient matrix.

73. False. If the determinant of the coefficient matrix is zero, the system has either no solution or infinitely many solutions.

75. Answers will vary.

77. Area $= \dfrac{1}{2}\begin{vmatrix} 3 & -1 & 1 \\ 7 & -1 & 1 \\ 7 & 5 & 1 \end{vmatrix} = \dfrac{1}{2}\left(3\begin{vmatrix} -1 & 1 \\ 5 & 1 \end{vmatrix} + 1\begin{vmatrix} 7 & 1 \\ 7 & 1 \end{vmatrix} + 1\begin{vmatrix} 7 & -1 \\ 7 & 5 \end{vmatrix}\right) = \dfrac{1}{2}(-18 + 0 + 42) = 12$ square units

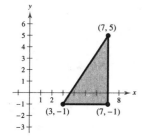

Area $= \dfrac{1}{2}(\text{base})(\text{height}) = \dfrac{1}{2}(7 - 3)(5 - (-1)) = \dfrac{1}{2}(4)(6) = 12$ square units

Review Exercises for Chapter 10

1. $\begin{bmatrix} -4 \\ 0 \\ 5 \end{bmatrix}$

Order: 3×1

3. $\begin{bmatrix} 3 \end{bmatrix}$

Order: 1×1

5. $\begin{cases} 3x - 10y = 15 \\ 5x + 4y = 22 \end{cases}$

$\begin{bmatrix} 3 & -10 & \vdots & 15 \\ 5 & 4 & \vdots & 22 \end{bmatrix}$

7. $\begin{bmatrix} 5 & 1 & 7 & \vdots & -9 \\ 4 & 2 & 0 & \vdots & 10 \\ 9 & 4 & 2 & \vdots & 3 \end{bmatrix}$

$\begin{cases} 5x + y + 7z = -9 \\ 4x + 2y = 10 \\ 9x + 4y + 2z = 3 \end{cases}$

9. $\begin{bmatrix} 0 & 1 & 1 \\ 1 & 2 & 3 \\ 2 & 2 & 2 \end{bmatrix}$

$\begin{matrix} R_1 \\ R_2 \end{matrix}$ $\begin{bmatrix} 1 & 2 & 3 \\ 0 & 1 & 1 \\ 2 & 2 & 2 \end{bmatrix}$

$-2R_1 + R_3 \rightarrow \begin{bmatrix} 1 & 2 & 3 \\ 0 & 1 & 1 \\ 0 & -2 & -4 \end{bmatrix}$

$2R_2 + R_3 \rightarrow \begin{bmatrix} 1 & 2 & 3 \\ 0 & 1 & 1 \\ 0 & 0 & -2 \end{bmatrix}$

$-\frac{1}{2}R_3 \rightarrow \begin{bmatrix} 1 & 2 & 3 \\ 0 & 1 & 1 \\ 0 & 0 & 1 \end{bmatrix}$

11. $\begin{bmatrix} 1 & 2 & 3 & \vdots & 9 \\ 0 & 1 & -2 & \vdots & 2 \\ 0 & 0 & 1 & \vdots & 0 \end{bmatrix} \Rightarrow \begin{cases} x + 2y + 3z = 9 \\ y - 2z = 2 \\ z = 0 \end{cases}$

$y - 2(0) = 2 \Rightarrow y = 2$

$x + 2(2) + 3(0) = 9 \Rightarrow x = 5$

Solution: $(5, 2, 0)$

13. $\begin{bmatrix} 1 & -5 & 4 & \vdots & 1 \\ 0 & 1 & 2 & \vdots & 3 \\ 0 & 0 & 1 & \vdots & 4 \end{bmatrix} \Rightarrow \begin{cases} x - 5y + 4z = 1 \\ y + 2z = 3 \\ z = 4 \end{cases}$

$y + 2(4) = 3 \Rightarrow y = -5$

$x - 5(-5) + 4(4) = 1 \Rightarrow x = -40$

Solution: $(-40, -5, 4)$

15. $\begin{bmatrix} 5 & 4 & \vdots & 2 \\ -1 & 1 & \vdots & -22 \end{bmatrix}$

$4R_2 + R_1 \rightarrow \begin{bmatrix} 1 & 8 & \vdots & -86 \\ -1 & 1 & \vdots & -22 \end{bmatrix}$

$R_1 + R_2 \rightarrow \begin{bmatrix} 1 & 8 & \vdots & -86 \\ 0 & 9 & \vdots & -108 \end{bmatrix}$

$\frac{1}{9}R_2 \rightarrow \begin{bmatrix} 1 & 8 & \vdots & -86 \\ 0 & 1 & \vdots & -12 \end{bmatrix}$

$\begin{cases} x + 8y = -86 \\ y = -12 \end{cases}$

$y = -12$

$x + 8(-12) = -86 \Rightarrow x = 10$

Solution: $(10, -12)$

17. $\begin{bmatrix} 0.3 & -0.1 & \vdots & -0.13 \\ 0.2 & -0.3 & \vdots & -0.25 \end{bmatrix}$

$\begin{matrix} 10R_1 \rightarrow \\ 10R_2 \rightarrow \end{matrix} \begin{bmatrix} 3 & -1 & \vdots & -1.3 \\ 2 & -3 & \vdots & -2.5 \end{bmatrix}$

$-R_2 + R_1 \rightarrow \begin{bmatrix} 1 & 2 & \vdots & 1.2 \\ 2 & -3 & \vdots & -2.5 \end{bmatrix}$

$-2R_1 + R_2 \rightarrow \begin{bmatrix} 1 & 2 & \vdots & 1.2 \\ 0 & -7 & \vdots & -4.9 \end{bmatrix}$

$-\frac{1}{7}R_2 \rightarrow \begin{bmatrix} 1 & 2 & \vdots & 1.2 \\ 0 & 1 & \vdots & 0.7 \end{bmatrix}$

$\begin{cases} x + 2y = 1.2 \\ y = 0.7 \end{cases}$

$y = 0.7$

$x + 2(0.7) = 1.2 \Rightarrow x = -0.2$

Solution: $(-0.2, 0.7) = \left(-\frac{1}{5}, \frac{7}{10}\right)$

19. $\begin{cases} -x + 2y = 3 \\ 2x - 4y = 6 \end{cases}$

$$\begin{bmatrix} -1 & 2 & \vdots & 3 \\ 2 & -4 & \vdots & 6 \end{bmatrix}$$

$2R_1 + R_2 \rightarrow \begin{bmatrix} -1 & 2 & \vdots & 3 \\ 0 & 0 & \vdots & 12 \end{bmatrix}$

Because the last row consists of all zeros except for the last entry, the system is inconsistent and there is no solution.

21. $\begin{cases} x - 2y + z = 7 \\ 2x + y - 2z = -4 \\ -x + 3y + 2z = -3 \end{cases}$

$$\begin{bmatrix} 1 & -2 & 1 & \vdots & 7 \\ 2 & 1 & -2 & \vdots & -4 \\ -1 & 3 & 2 & \vdots & -3 \end{bmatrix}$$

$\begin{matrix} -2R_1 + R_2 \rightarrow \\ R_1 + R_3 \rightarrow \end{matrix} \begin{bmatrix} 1 & -2 & 1 & \vdots & 7 \\ 0 & 5 & -4 & \vdots & -18 \\ 0 & 1 & 3 & \vdots & 4 \end{bmatrix}$

$R_2 + (-5)R_3 \rightarrow \begin{bmatrix} 1 & -2 & 1 & \vdots & 7 \\ 0 & 0 & -19 & \vdots & -38 \\ 0 & 1 & 3 & \vdots & 4 \end{bmatrix}$

$$-19z = -38$$
$$z = 2$$
$$y + 3(2) = 4 \Rightarrow y = -2$$
$$x - 2(-2) + 2 = 7 \Rightarrow x = 1$$

Solution: $(1, -2, 2)$

23. $\begin{bmatrix} 2 & 1 & 2 & \vdots & 4 \\ 2 & 2 & 0 & \vdots & 5 \\ 2 & -1 & 6 & \vdots & 2 \end{bmatrix}$

$\begin{matrix} -R_1 + R_2 \rightarrow \\ -R_1 + R_3 \rightarrow \end{matrix} \begin{bmatrix} 2 & 1 & 2 & \vdots & 4 \\ 0 & 1 & -2 & \vdots & 1 \\ 0 & -2 & 4 & \vdots & -2 \end{bmatrix}$

$\begin{matrix} -R_2 + R_1 \rightarrow \\ \\ 2R_2 + R_3 \rightarrow \end{matrix} \begin{bmatrix} 2 & 0 & 4 & \vdots & 3 \\ 0 & 1 & -2 & \vdots & 1 \\ 0 & 0 & 0 & \vdots & 0 \end{bmatrix}$

$\frac{1}{2}R_1 \rightarrow \begin{bmatrix} 1 & 0 & 2 & \vdots & \frac{3}{2} \\ 0 & 1 & -2 & \vdots & 1 \\ 0 & 0 & 0 & \vdots & 0 \end{bmatrix}$

Let $z = a$, then:

$$y - 2a = 1 \Rightarrow y = 2a + 1$$
$$x + 2a = \frac{3}{2} \Rightarrow x = -2a + \frac{3}{2}$$

Solution: $\left(-2a + \frac{3}{2}, 2a + 1, a\right)$ where a is any real number

25. $\begin{bmatrix} 2 & 3 & 1 & \vdots & 10 \\ 2 & -3 & -3 & \vdots & 22 \\ 4 & -2 & 3 & \vdots & -2 \end{bmatrix}$

$\begin{matrix} -R_1 + R_2 \rightarrow \\ -2R_1 + R_3 \rightarrow \end{matrix} \begin{bmatrix} 2 & 3 & 1 & \vdots & 10 \\ 0 & -6 & -4 & \vdots & 12 \\ 0 & -8 & 1 & \vdots & -22 \end{bmatrix}$

$\begin{matrix} \frac{1}{2}R_1 \rightarrow \\ -\frac{1}{6}R_2 \rightarrow \end{matrix} \begin{bmatrix} 1 & \frac{3}{2} & \frac{1}{2} & \vdots & 5 \\ 0 & 1 & \frac{2}{3} & \vdots & -2 \\ 0 & -8 & 1 & \vdots & -22 \end{bmatrix}$

$8R_2 + R_3 \rightarrow \begin{bmatrix} 1 & \frac{3}{2} & \frac{1}{2} & \vdots & 5 \\ 0 & 1 & \frac{2}{3} & \vdots & -2 \\ 0 & 0 & \frac{19}{3} & \vdots & -38 \end{bmatrix}$

$\frac{3}{19}R_3 \rightarrow \begin{bmatrix} 1 & \frac{3}{2} & \frac{1}{2} & \vdots & 5 \\ 0 & 1 & \frac{2}{3} & \vdots & -2 \\ 0 & 0 & 1 & \vdots & -6 \end{bmatrix}$

$$z = -6$$
$$y + \frac{2}{3}(-6) = -2 \Rightarrow y = 2$$
$$x + \frac{3}{2}(2) + \frac{1}{2}(-6) = 5 \Rightarrow x = 5$$

Solution: $(5, 2, -6)$

27.

$$\begin{bmatrix} 2 & 1 & 1 & 0 & \vdots & 6 \\ 0 & -2 & 3 & -1 & \vdots & 9 \\ 3 & 3 & -2 & -2 & \vdots & -11 \\ 1 & 0 & 1 & 3 & \vdots & 14 \end{bmatrix}$$

$-R_4 + R_1 \to \begin{bmatrix} 1 & 1 & 0 & -3 & \vdots & -8 \\ 0 & -2 & 3 & -1 & \vdots & 9 \\ 3 & 3 & -2 & -2 & \vdots & -11 \\ 1 & 0 & 1 & 3 & \vdots & 14 \end{bmatrix}$

$\begin{matrix} \\ \\ -3R_1 + R_3 \to \\ -R_1 + R_4 \to \end{matrix} \begin{bmatrix} 1 & 1 & 0 & -3 & \vdots & -8 \\ 0 & -2 & 3 & -1 & \vdots & 9 \\ 0 & 0 & -2 & 7 & \vdots & 13 \\ 0 & -1 & 1 & 6 & \vdots & 22 \end{bmatrix}$

$-3R_4 + R_2 \to \begin{bmatrix} 1 & 1 & 0 & -3 & \vdots & -8 \\ 0 & 1 & 0 & -19 & \vdots & -57 \\ 0 & 0 & -2 & 7 & \vdots & 13 \\ 0 & -1 & 1 & 6 & \vdots & 22 \end{bmatrix}$

$R_2 + R_4 \to \begin{bmatrix} 1 & 1 & 0 & -3 & \vdots & -8 \\ 0 & 1 & 0 & -19 & \vdots & -57 \\ 0 & 0 & -2 & 7 & \vdots & 13 \\ 0 & 0 & 1 & -13 & \vdots & -35 \end{bmatrix}$

$\begin{matrix} \\ \\ R_4 \\ R_3 \end{matrix} \begin{bmatrix} 1 & 1 & 0 & -3 & \vdots & -8 \\ 0 & 1 & 0 & -19 & \vdots & -57 \\ 0 & 0 & 1 & -13 & \vdots & -35 \\ 0 & 0 & -2 & 7 & \vdots & 13 \end{bmatrix}$

$2R_3 + R_4 \to \begin{bmatrix} 1 & 1 & 0 & -3 & \vdots & -8 \\ 0 & 1 & 0 & -19 & \vdots & -57 \\ 0 & 0 & 1 & -13 & \vdots & -35 \\ 0 & 0 & 0 & -19 & \vdots & -57 \end{bmatrix}$

$\frac{1}{19}R_4 \to \begin{bmatrix} 1 & 1 & 0 & -3 & \vdots & -8 \\ 0 & 1 & 0 & -19 & \vdots & -57 \\ 0 & 0 & 1 & -13 & \vdots & -35 \\ 0 & 0 & 0 & 1 & \vdots & 3 \end{bmatrix}$

$w = 3$

$z - 13(3) = -35 \Rightarrow z = 4$

$y - 19(3) = -57 \Rightarrow y = 0$

$x + 0 - 3(3) = -8 \Rightarrow x = 1$

Solution: $(1, 0, 4, 3)$

29. $\begin{cases} x + 2y - z = 3 \\ x - y - z = -3 \\ 2x + y + 3z = 10 \end{cases}$

$$\begin{bmatrix} 1 & 2 & -1 & \vdots & 3 \\ 1 & -1 & -1 & \vdots & -3 \\ 2 & 1 & 3 & \vdots & 10 \end{bmatrix}$$

$\begin{matrix} -R_1 + R_2 \to \\ -2R_2 + R_3 \to \end{matrix} \begin{bmatrix} 1 & 2 & -1 & \vdots & 3 \\ 0 & -3 & 0 & \vdots & -6 \\ 0 & 3 & 5 & \vdots & 16 \end{bmatrix}$

$R_2 + R_3 \to \begin{bmatrix} 1 & 2 & -1 & \vdots & 3 \\ 0 & -3 & 0 & \vdots & -6 \\ 0 & 0 & 5 & \vdots & 10 \end{bmatrix}$

$3R_1 + 2R_2 \to \begin{bmatrix} 3 & 0 & -3 & \vdots & -3 \\ 0 & -3 & 0 & \vdots & -6 \\ 0 & 0 & 5 & \vdots & 10 \end{bmatrix}$

$5R_1 + 3R_3 \to \begin{bmatrix} 15 & 0 & 0 & \vdots & 15 \\ 0 & -3 & 0 & \vdots & -6 \\ 0 & 0 & 5 & \vdots & 10 \end{bmatrix}$

$\begin{matrix} \frac{1}{15}R_1 \to \\ \frac{1}{3}R_2 \to \\ \frac{1}{5}R_3 \to \end{matrix} \begin{bmatrix} 1 & 0 & 0 & \vdots & 1 \\ 0 & 1 & 0 & \vdots & 2 \\ 0 & 0 & 1 & \vdots & 2 \end{bmatrix}$

$x = 1$

$y = 2$

$z = 2$

Solution: $(1, 2, 2)$

31.
$$\begin{bmatrix} -1 & 1 & 2 & \vdots & 1 \\ 2 & 3 & 1 & \vdots & -2 \\ 5 & 4 & 2 & \vdots & 4 \end{bmatrix}$$

$$-R_1 \rightarrow \begin{bmatrix} 1 & -1 & -2 & \vdots & -1 \\ 2 & 3 & 1 & \vdots & -2 \\ 5 & 4 & 2 & \vdots & 4 \end{bmatrix}$$

$$\begin{matrix} \\ -2R_1 + R_2 \rightarrow \\ -5R_1 + R_3 \rightarrow \end{matrix} \begin{bmatrix} 1 & -1 & -2 & \vdots & -1 \\ 0 & 5 & 5 & \vdots & 0 \\ 0 & 9 & 12 & \vdots & 9 \end{bmatrix}$$

$$\tfrac{1}{5}R_2 \rightarrow \begin{bmatrix} 1 & -1 & -2 & \vdots & -1 \\ 0 & 1 & 1 & \vdots & 0 \\ 0 & 9 & 12 & \vdots & 9 \end{bmatrix}$$

$$\begin{matrix} R_2 + R_1 \rightarrow \\ \\ -9R_2 + R_3 \rightarrow \end{matrix} \begin{bmatrix} 1 & 0 & -1 & \vdots & -1 \\ 0 & 1 & 1 & \vdots & 0 \\ 0 & 0 & 3 & \vdots & 9 \end{bmatrix}$$

$$\tfrac{1}{3}R_3 \rightarrow \begin{bmatrix} 1 & 0 & -1 & \vdots & -1 \\ 0 & 1 & 1 & \vdots & 0 \\ 0 & 0 & 1 & \vdots & 3 \end{bmatrix}$$

$$\begin{matrix} R_3 + R_1 \rightarrow \\ -R_3 + R_2 \rightarrow \\ \end{matrix} \begin{bmatrix} 1 & 0 & 0 & \vdots & 2 \\ 0 & 1 & 0 & \vdots & -3 \\ 0 & 0 & 1 & \vdots & 3 \end{bmatrix}$$

$x = 2, y = -3, z = 3$

Solution: $(2, -3, 3)$

33.
$$\begin{bmatrix} 2 & -1 & 9 & \vdots & -8 \\ -1 & -3 & 4 & \vdots & -15 \\ 5 & 2 & -1 & \vdots & 17 \end{bmatrix}$$

$$R_2 + R_1 \rightarrow \begin{bmatrix} 1 & -4 & 13 & \vdots & -23 \\ -1 & -3 & 4 & \vdots & -15 \\ 5 & 2 & -1 & \vdots & 17 \end{bmatrix}$$

$$\begin{matrix} R_1 + R_2 \rightarrow \\ -5R_1 + R_3 \rightarrow \end{matrix} \begin{bmatrix} 1 & -4 & 13 & \vdots & -23 \\ 0 & -7 & 17 & \vdots & -38 \\ 0 & 22 & -66 & \vdots & 132 \end{bmatrix}$$

$$\begin{matrix} R_3 \\ R_2 \end{matrix} \begin{bmatrix} 1 & -4 & 13 & \vdots & -23 \\ 0 & 22 & -66 & \vdots & 132 \\ 0 & -7 & 17 & \vdots & -38 \end{bmatrix}$$

$$\tfrac{1}{22}R_2 \rightarrow \begin{bmatrix} 1 & -4 & 13 & \vdots & -23 \\ 0 & 1 & -3 & \vdots & 6 \\ 0 & -7 & 17 & \vdots & -38 \end{bmatrix}$$

$$7R_2 + R_3 \rightarrow \begin{bmatrix} 1 & -4 & 13 & \vdots & -23 \\ 0 & 1 & -3 & \vdots & 6 \\ 0 & 0 & -4 & \vdots & 4 \end{bmatrix}$$

$$-\tfrac{1}{4}R_3 \rightarrow \begin{bmatrix} 1 & -4 & 13 & \vdots & -23 \\ 0 & 1 & -3 & \vdots & 6 \\ 0 & 0 & 1 & \vdots & -1 \end{bmatrix}$$

$$4R_2 + R_1 \rightarrow \begin{bmatrix} 1 & 0 & 1 & \vdots & 1 \\ 0 & 1 & -3 & \vdots & 6 \\ 0 & 0 & 1 & \vdots & -1 \end{bmatrix}$$

$$\begin{matrix} -R_3 + R_1 \rightarrow \\ 3R_3 + R_2 \rightarrow \\ \end{matrix} \begin{bmatrix} 1 & 0 & 0 & \vdots & 2 \\ 0 & 1 & 0 & \vdots & 3 \\ 0 & 0 & 1 & \vdots & -1 \end{bmatrix}$$

$x = 2, y = 3, z = -1$

Solution: $(2, 3, -1)$

35. Use the reduced row-echelon form feature of a graphing utility.

$$\begin{bmatrix} 3 & -1 & 5 & -2 & \vdots & -44 \\ 1 & 6 & 4 & -1 & \vdots & 1 \\ 5 & -1 & 1 & 3 & \vdots & -15 \\ 0 & 4 & -1 & -8 & \vdots & 58 \end{bmatrix} \Rightarrow \begin{bmatrix} 1 & 0 & 0 & 0 & \vdots & 2 \\ 0 & 1 & 0 & 0 & \vdots & 6 \\ 0 & 0 & 1 & 0 & \vdots & -10 \\ 0 & 0 & 0 & 1 & \vdots & -3 \end{bmatrix}$$

$x = 2, y = 6, z = -10, w = -3$

Solution: $(2, 6, -10, -3)$

37. $\begin{bmatrix} -1 & x \\ y & 9 \end{bmatrix} = \begin{bmatrix} -1 & 12 \\ -7 & 9 \end{bmatrix} \Rightarrow x = 12$ and $y = -7$

39. $\begin{bmatrix} x+3 & -4 & 4y \\ 0 & -3 & 2 \\ -2 & y+5 & 6x \end{bmatrix} = \begin{bmatrix} 5x-1 & -4 & 44 \\ 0 & -3 & 2 \\ -2 & 16 & 6 \end{bmatrix}$

$\left.\begin{array}{l} x+3 = 5x-1 \\ 4y = 44 \\ y+5 = 16 \\ 6x = 6 \end{array}\right\}$ $x = 1$ and $y = 11$

41. (a) $A+B = \begin{bmatrix} 2 & -2 \\ 3 & 5 \end{bmatrix} + \begin{bmatrix} -3 & 10 \\ 12 & 8 \end{bmatrix} = \begin{bmatrix} -1 & 8 \\ 15 & 13 \end{bmatrix}$

(b) $A-B = \begin{bmatrix} 2 & -2 \\ 3 & 5 \end{bmatrix} - \begin{bmatrix} -3 & 10 \\ 12 & 8 \end{bmatrix} = \begin{bmatrix} 5 & -12 \\ -9 & -3 \end{bmatrix}$

(c) $4A = 4\begin{bmatrix} 2 & -2 \\ 3 & 5 \end{bmatrix} = \begin{bmatrix} 8 & -8 \\ 12 & 20 \end{bmatrix}$

(d) $A+3B = \begin{bmatrix} 2 & -2 \\ 3 & 5 \end{bmatrix} + 3\begin{bmatrix} -3 & 10 \\ 12 & 8 \end{bmatrix} = \begin{bmatrix} 2 & -2 \\ 3 & 5 \end{bmatrix} + \begin{bmatrix} -9 & 30 \\ 36 & 24 \end{bmatrix} = \begin{bmatrix} -7 & 28 \\ 39 & 29 \end{bmatrix}$

43. (a) $A+B = \begin{bmatrix} 5 & 4 \\ -7 & 2 \\ 11 & 2 \end{bmatrix} + \begin{bmatrix} 0 & 3 \\ 4 & 12 \\ 20 & 40 \end{bmatrix} = \begin{bmatrix} 5 & 7 \\ -3 & 14 \\ 31 & 42 \end{bmatrix}$

(b) $A-B = \begin{bmatrix} 5 & 4 \\ -7 & 2 \\ 11 & 2 \end{bmatrix} - \begin{bmatrix} 0 & 3 \\ 4 & 12 \\ 20 & 40 \end{bmatrix} = \begin{bmatrix} 5 & 1 \\ -11 & -10 \\ -9 & -38 \end{bmatrix}$

(c) $4A = 4\begin{bmatrix} 5 & 4 \\ -7 & 2 \\ 11 & 2 \end{bmatrix} = \begin{bmatrix} 20 & 16 \\ -28 & 8 \\ 44 & 8 \end{bmatrix}$

(d) $A+3B = \begin{bmatrix} 5 & 4 \\ -7 & 2 \\ 11 & 2 \end{bmatrix} + 3\begin{bmatrix} 0 & 3 \\ 4 & 12 \\ 20 & 40 \end{bmatrix} = \begin{bmatrix} 5 & 4 \\ -7 & 2 \\ 11 & 2 \end{bmatrix} + \begin{bmatrix} 0 & 9 \\ 12 & 36 \\ 60 & 120 \end{bmatrix} = \begin{bmatrix} 5 & 13 \\ 5 & 38 \\ 71 & 122 \end{bmatrix}$

45. $\begin{bmatrix} 7 & 3 \\ -1 & 5 \end{bmatrix} + \begin{bmatrix} 10 & -20 \\ 14 & -3 \end{bmatrix} = \begin{bmatrix} 7+10 & 3-20 \\ -1+14 & 5-3 \end{bmatrix} = \begin{bmatrix} 17 & -17 \\ 13 & 2 \end{bmatrix}$

47. $-2\begin{bmatrix} 1 & 2 \\ 5 & -4 \\ 6 & 0 \end{bmatrix} + 8\begin{bmatrix} 7 & 1 \\ 1 & 2 \\ 1 & 4 \end{bmatrix} = \begin{bmatrix} -2 & -4 \\ -10 & 8 \\ -12 & 0 \end{bmatrix} + \begin{bmatrix} 56 & 8 \\ 8 & 16 \\ 8 & 32 \end{bmatrix} = \begin{bmatrix} 54 & 4 \\ -2 & 24 \\ -4 & 32 \end{bmatrix}$

49. $3\begin{bmatrix} 8 & -2 & 5 \\ 1 & 3 & -1 \end{bmatrix} + 6\begin{bmatrix} 4 & -2 & -3 \\ 2 & 7 & 6 \end{bmatrix} = \begin{bmatrix} 24 & -6 & 15 \\ 3 & 9 & -3 \end{bmatrix} + \begin{bmatrix} 24 & -12 & -18 \\ 12 & 42 & 36 \end{bmatrix} = \begin{bmatrix} 48 & -18 & -3 \\ 15 & 51 & 33 \end{bmatrix}$

51. $X = 2A - 3B = 2\begin{bmatrix} -4 & 0 \\ 1 & -5 \\ -3 & 2 \end{bmatrix} - 3\begin{bmatrix} 1 & 2 \\ -2 & 1 \\ 4 & 4 \end{bmatrix} = \begin{bmatrix} -8 & 0 \\ 2 & -10 \\ -6 & 4 \end{bmatrix} + \begin{bmatrix} -3 & -6 \\ 6 & -3 \\ -12 & -12 \end{bmatrix} = \begin{bmatrix} -11 & -6 \\ 8 & -13 \\ -18 & -8 \end{bmatrix}$

53. $X = \frac{1}{3}[B - 2A] = \frac{1}{3}\left(\begin{bmatrix} 1 & 2 \\ -2 & 1 \\ 4 & 4 \end{bmatrix} - 2\begin{bmatrix} -4 & 0 \\ 1 & -5 \\ -3 & 2 \end{bmatrix}\right) = \frac{1}{3}\begin{bmatrix} 9 & 2 \\ -4 & 11 \\ 10 & 0 \end{bmatrix} = \begin{bmatrix} 3 & \frac{2}{3} \\ -\frac{4}{3} & \frac{11}{3} \\ \frac{10}{3} & 0 \end{bmatrix}$

55. A and B are both 2×2, so AB exists.

$$AB = \begin{bmatrix} 2 & -2 \\ 3 & 5 \end{bmatrix}\begin{bmatrix} -3 & 10 \\ 12 & 8 \end{bmatrix} = \begin{bmatrix} 2(-3) + (-2)(12) & 2(10) + (-2)(8) \\ 3(-3) + 5(12) & 3(10) + 5(8) \end{bmatrix} = \begin{bmatrix} -30 & 4 \\ 51 & 70 \end{bmatrix}$$

57. Because A is 3×2 and B is 2×2, AB exists.

$$AB = \begin{bmatrix} 5 & 4 \\ -7 & 2 \\ 11 & 2 \end{bmatrix}\begin{bmatrix} 4 & 12 \\ 20 & 40 \end{bmatrix} = \begin{bmatrix} 5(4) + 4(20) & 5(12) + 4(40) \\ -7(4) + 2(20) & -7(12) + 2(40) \\ 11(4) + 2(20) & 11(12) + 2(40) \end{bmatrix} = \begin{bmatrix} 100 & 220 \\ 12 & -4 \\ 84 & 212 \end{bmatrix}$$

59. $\begin{bmatrix} 1 & 2 \\ 5 & -4 \\ 6 & 0 \end{bmatrix}\begin{bmatrix} 6 & -2 & 8 \\ 4 & 0 & 0 \end{bmatrix} = \begin{bmatrix} 1(6) + 2(4) & 1(-2) + 2(0) & 1(8) + 2(0) \\ 5(6) + (-4)(4) & 5(-2) + (-4)(0) & 5(8) + (-4)(0) \\ 6(6) + (0)(4) & 6(-2) + (0)(0) & 6(8) + (0)(0) \end{bmatrix} = \begin{bmatrix} 14 & -2 & 8 \\ 14 & -10 & 40 \\ 36 & -12 & 48 \end{bmatrix}$

61. $\begin{bmatrix} 1 & 5 & 6 \\ 2 & -4 & 0 \end{bmatrix}\begin{bmatrix} 6 & 4 \\ -2 & 0 \\ 8 & 0 \end{bmatrix} = \begin{bmatrix} 1(6) + 5(-2) + 6(8) & 1(4) + 5(0) + 6(0) \\ 2(6) - 4(-2) + 0(8) & 2(4) - 4(0) + 0(0) \end{bmatrix} = \begin{bmatrix} 44 & 4 \\ 20 & 8 \end{bmatrix}$

63. Not possible. The number of columns of the first matrix does not equal the number of rows of the second matrix.

65. $\begin{bmatrix} 2 & 1 \\ 6 & 0 \end{bmatrix}\left(\begin{bmatrix} 4 & 2 \\ -3 & 1 \end{bmatrix} + \begin{bmatrix} -2 & 4 \\ 0 & 4 \end{bmatrix}\right) = \begin{bmatrix} 2 & 1 \\ 6 & 0 \end{bmatrix}\begin{bmatrix} 2 & 6 \\ -3 & 5 \end{bmatrix} = \begin{bmatrix} 2(2) + 1(-3) & 2(6) + 1(5) \\ 6(2) + 0 & 6(6) + 0 \end{bmatrix} = \begin{bmatrix} 1 & 17 \\ 12 & 36 \end{bmatrix}$

67. $\begin{bmatrix} 4 & 1 \\ 11 & -7 \\ 12 & 3 \end{bmatrix}\begin{bmatrix} 3 & -5 & 6 \\ 2 & -2 & -2 \end{bmatrix} = \begin{bmatrix} 14 & -22 & 22 \\ 19 & -41 & 80 \\ 42 & -66 & 66 \end{bmatrix}$

69. $0.95A = 0.95\begin{bmatrix} 80 & 120 & 140 \\ 40 & 100 & 80 \end{bmatrix} = \begin{bmatrix} 76 & 114 & 133 \\ 38 & 95 & 76 \end{bmatrix}$

71. $BA = \begin{bmatrix} \$79.99 & \$109.95 & \$189.99 \end{bmatrix}\begin{bmatrix} 8200 & 7400 \\ 6500 & 9800 \\ 5400 & 4800 \end{bmatrix} = \begin{bmatrix} \$2,396,539 & \$2,581,388 \end{bmatrix}$

The merchandise shipped to warehouse 1 is worth \$2,396,539 and the merchandise shipped to warehouse 2 is worth \$2,581,388.

73. $AB = \begin{bmatrix} -4 & -1 \\ 7 & 2 \end{bmatrix}\begin{bmatrix} -2 & -1 \\ 7 & 4 \end{bmatrix} = \begin{bmatrix} -4(-2) + (-1)(7) & -4(-1) + (-1)(4) \\ 7(-2) + 2(7) & 7(-1) + 2(4) \end{bmatrix} = \begin{bmatrix} 1 & 0 \\ 0 & 1 \end{bmatrix} = I$

$BA = \begin{bmatrix} -2 & -1 \\ 7 & 4 \end{bmatrix}\begin{bmatrix} -4 & -1 \\ 7 & 2 \end{bmatrix} = \begin{bmatrix} -2(-4) + (-1)(7) & -2(-1) + (-1)(2) \\ 7(-4) + 4(7) & 7(-1) + 4(2) \end{bmatrix} = \begin{bmatrix} 1 & 0 \\ 0 & 1 \end{bmatrix} = I$

75. $AB = \begin{bmatrix} 1 & 1 & 0 \\ 1 & 0 & 1 \\ 6 & 2 & 3 \end{bmatrix} \begin{bmatrix} -2 & -3 & 1 \\ 3 & 3 & -1 \\ 2 & 4 & -1 \end{bmatrix} = \begin{bmatrix} 1(-2)+1(3)+0(2) & 1(-3)+1(3)+0(4) & 1(1)+1(-1)+0(-1) \\ 1(-2)+0(3)+1(2) & 1(-3)+0(3)+1(4) & 1(1)+0(-1)+1(-1) \\ 6(-2)+2(3)+3(2) & 6(-3)+2(3)+3(4) & 6(1)+2(-1)+3(-1) \end{bmatrix} = \begin{bmatrix} 1 & 0 & 0 \\ 0 & 1 & 0 \\ 0 & 0 & 1 \end{bmatrix} = I$

$BA = \begin{bmatrix} -2 & -3 & 1 \\ 3 & 3 & -1 \\ 2 & 4 & -1 \end{bmatrix} \begin{bmatrix} 1 & 1 & 0 \\ 1 & 0 & 1 \\ 6 & 2 & 3 \end{bmatrix} = \begin{bmatrix} -2(1)+(-3)(1)+1(6) & -2(1)+(-3)(0)+1(2) & -2(0)+(-3)(1)+1(3) \\ 3(1)+3(1)+(-1)(6) & 3(1)+3(0)+(-1)(2) & 3(0)+3(1)+(-1)(3) \\ 2(1)+4(1)+(-1)(6) & 2(1)+4(0)+(-1)(2) & 2(0)+4(1)+(-1)(3) \end{bmatrix}$

$= \begin{bmatrix} 1 & 0 & 0 \\ 0 & 1 & 0 \\ 0 & 0 & 1 \end{bmatrix} = I$

77. $[A \;\vdots\; I] = \begin{bmatrix} -6 & 5 & \vdots & 1 & 0 \\ -5 & 4 & \vdots & 0 & 1 \end{bmatrix}$

$-\frac{1}{6}R_1 \rightarrow \begin{bmatrix} 1 & -\frac{5}{6} & \vdots & -\frac{1}{6} & 0 \\ -5 & 4 & \vdots & 0 & 1 \end{bmatrix}$

$5R_1 + R_2 \rightarrow \begin{bmatrix} 1 & -\frac{5}{6} & \vdots & -\frac{1}{6} & 0 \\ 0 & -\frac{1}{6} & \vdots & -\frac{5}{6} & 1 \end{bmatrix}$

$-6R_2 \rightarrow \begin{bmatrix} 1 & -\frac{5}{6} & \vdots & -\frac{1}{6} & 0 \\ 0 & 1 & \vdots & 5 & -6 \end{bmatrix}$

$\frac{5}{6}R_2 + R_1 \rightarrow \begin{bmatrix} 1 & 0 & \vdots & 4 & -5 \\ 0 & 1 & \vdots & 5 & -6 \end{bmatrix} = [I \;\vdots\; A^{-1}]$

$A^{-1} = \begin{bmatrix} 4 & -5 \\ 5 & -6 \end{bmatrix}$

79. $[A \;\vdots\; I] = \begin{bmatrix} 2 & 0 & 3 & \vdots & 1 & 0 & 0 \\ -1 & 1 & 1 & \vdots & 0 & 1 & 0 \\ 2 & -2 & 1 & \vdots & 0 & 0 & 1 \end{bmatrix}$

$2R_2 + R_3 \rightarrow \begin{bmatrix} 2 & 0 & 3 & \vdots & 1 & 0 & 0 \\ -1 & 1 & 1 & \vdots & 0 & 1 & 0 \\ 0 & 0 & 3 & \vdots & 0 & 2 & 1 \end{bmatrix}$

$-R_3 + R_1 \rightarrow \begin{bmatrix} 2 & 0 & 0 & \vdots & 1 & -2 & -1 \\ -1 & 1 & 1 & \vdots & 0 & 1 & 0 \\ 0 & 0 & 3 & \vdots & 0 & 2 & 1 \end{bmatrix}$

$\frac{1}{2}R_1 \rightarrow \atop \frac{1}{3}R_3 \rightarrow \begin{bmatrix} 1 & 0 & 0 & \vdots & \frac{1}{2} & -1 & -\frac{1}{2} \\ -1 & 1 & 1 & \vdots & 0 & 1 & 0 \\ 0 & 0 & 1 & \vdots & 0 & \frac{2}{3} & \frac{1}{3} \end{bmatrix}$

$R_1 + R_2 \rightarrow \begin{bmatrix} 1 & 0 & 0 & \vdots & \frac{1}{2} & -1 & -\frac{1}{2} \\ 0 & 1 & 1 & \vdots & \frac{1}{2} & 0 & -\frac{1}{2} \\ 0 & 0 & 1 & \vdots & 0 & \frac{2}{3} & \frac{1}{3} \end{bmatrix}$

$-R_3 + R_2 \rightarrow \begin{bmatrix} 1 & 0 & 0 & \vdots & \frac{1}{2} & -1 & -\frac{1}{2} \\ 0 & 1 & 0 & \vdots & \frac{1}{2} & -\frac{2}{3} & -\frac{5}{6} \\ 0 & 0 & 1 & \vdots & 0 & \frac{2}{3} & \frac{1}{3} \end{bmatrix} = [I \;\vdots\; A^{-1}]$

$A^{-1} = \begin{bmatrix} \frac{1}{2} & -1 & -\frac{1}{2} \\ \frac{1}{2} & -\frac{2}{3} & -\frac{5}{6} \\ 0 & \frac{2}{3} & \frac{1}{3} \end{bmatrix}$

81. $\begin{bmatrix} -1 & -2 & -2 \\ 3 & 7 & 9 \\ 1 & 4 & 7 \end{bmatrix}^{-1} = \begin{bmatrix} 13 & 6 & -4 \\ -12 & -5 & 3 \\ 5 & 2 & -1 \end{bmatrix}$

83. $\begin{bmatrix} 1 & 3 & 1 & 6 \\ 4 & 4 & 2 & 6 \\ 3 & 4 & 1 & 2 \\ -1 & 2 & -1 & -2 \end{bmatrix}^{-1} = \begin{bmatrix} -3 & 6 & -5.5 & 3.5 \\ 1 & -2 & 2 & -1 \\ 7 & -15 & 14.5 & -9.5 \\ -1 & 2.5 & -2.5 & 1.5 \end{bmatrix}$

85. $A = \begin{bmatrix} -7 & 2 \\ -8 & 2 \end{bmatrix}$

$A^{-1} = \dfrac{1}{-7(2)-2(-8)}\begin{bmatrix} 2 & -2 \\ 8 & -7 \end{bmatrix} = \dfrac{1}{2}\begin{bmatrix} 2 & -2 \\ 8 & -7 \end{bmatrix} = \begin{bmatrix} 1 & -1 \\ 4 & -\dfrac{7}{2} \end{bmatrix}$

87. $A = \begin{bmatrix} -12 & 6 \\ 10 & -5 \end{bmatrix}$

$ad - bc = (-12)(-5) - (6)(10) = 0$

A^{-1} does not exist.

89. $A = \begin{bmatrix} -\dfrac{1}{2} & 20 \\ \dfrac{3}{10} & -6 \end{bmatrix}$

$A^{-1} = \dfrac{1}{-\dfrac{1}{2}(-6) - 20\left(\dfrac{3}{10}\right)}\begin{bmatrix} -6 & -20 \\ -\dfrac{3}{10} & -\dfrac{1}{2} \end{bmatrix} = -\dfrac{1}{3}\begin{bmatrix} -6 & -20 \\ -\dfrac{3}{10} & -\dfrac{1}{2} \end{bmatrix}$

$= \begin{bmatrix} 2 & \dfrac{20}{3} \\ \dfrac{1}{10} & \dfrac{1}{6} \end{bmatrix}$

91. $A = \begin{bmatrix} 0.5 & 0.1 \\ -0.2 & -0.4 \end{bmatrix}$

$ad - bc = (0.5)(-0.4) - (0.1)(-0.2) = -0.18$

$A^{-1} = \dfrac{1}{-0.18}\begin{bmatrix} -0.4 & -0.1 \\ 0.2 & 0.5 \end{bmatrix} = -\dfrac{50}{9}\begin{bmatrix} -0.4 & -0.1 \\ 0.2 & 0.5 \end{bmatrix} = \begin{bmatrix} \dfrac{20}{9} & \dfrac{5}{9} \\ -\dfrac{10}{9} & -\dfrac{25}{9} \end{bmatrix}$

93. $\begin{cases} -x + 4y = 8 \\ 2x - 7y = -5 \end{cases}$

$\begin{bmatrix} x \\ y \end{bmatrix} = \begin{bmatrix} -1 & 4 \\ 2 & -7 \end{bmatrix}^{-1}\begin{bmatrix} 8 \\ -5 \end{bmatrix} = \begin{bmatrix} 7 & 4 \\ 2 & 1 \end{bmatrix}\begin{bmatrix} 8 \\ -5 \end{bmatrix}$

$= \begin{bmatrix} 7(8)+4(-5) \\ 2(8)+1(-5) \end{bmatrix} = \begin{bmatrix} 36 \\ 11 \end{bmatrix}$

Solution: $(36, 11)$

95. $\begin{cases} -3x + 10y = 8 \\ 5x - 17y = -13 \end{cases}$

$\begin{bmatrix} x \\ y \end{bmatrix} = \begin{bmatrix} -3 & 10 \\ 5 & -17 \end{bmatrix}^{-1}\begin{bmatrix} 8 \\ -13 \end{bmatrix} = \begin{bmatrix} -17 & -10 \\ -5 & -3 \end{bmatrix}\begin{bmatrix} 8 \\ -13 \end{bmatrix}$

$= \begin{bmatrix} -17(8)+(-10)(-13) \\ -5(8)+(-3)(-13) \end{bmatrix} = \begin{bmatrix} -6 \\ -1 \end{bmatrix}$

Solution: $(-6, -1)$

97. $\begin{cases} \frac{1}{2}x + \frac{1}{3}y = 2 \\ -3x + 2y = 0 \end{cases}$

$\begin{bmatrix} x \\ y \end{bmatrix} = \begin{bmatrix} \frac{1}{2} & \frac{1}{3} \\ -3 & 2 \end{bmatrix}^{-1}\begin{bmatrix} 2 \\ 0 \end{bmatrix} = \begin{bmatrix} 1 & -\frac{1}{6} \\ \frac{3}{2} & \frac{1}{4} \end{bmatrix}\begin{bmatrix} 2 \\ 0 \end{bmatrix} = \begin{bmatrix} 2 \\ 3 \end{bmatrix}$

Solution: $(2, 3)$

99. $\begin{cases} 0.3x + 0.7y = 10.2 \\ 0.4x + 0.6y = 7.6 \end{cases}$

$\begin{bmatrix} x \\ y \end{bmatrix} = \begin{bmatrix} 0.3 & 0.7 \\ 0.4 & 0.6 \end{bmatrix}^{-1}\begin{bmatrix} 10.2 \\ 7.6 \end{bmatrix} = \begin{bmatrix} -6 & 7 \\ 4 & -3 \end{bmatrix}\begin{bmatrix} 10.2 \\ 7.6 \end{bmatrix} = \begin{bmatrix} -8 \\ 18 \end{bmatrix}$

Solution: $(-8, 18)$

101. $\begin{cases} 3x + 2y - z = 6 \\ x - y + 2z = -1 \\ 5x + y + z = 7 \end{cases}$

$$\begin{bmatrix} x \\ y \\ z \end{bmatrix} = \begin{bmatrix} 3 & 2 & -1 \\ 1 & -1 & 2 \\ 5 & 1 & 1 \end{bmatrix}^{-1} \begin{bmatrix} 6 \\ -1 \\ 7 \end{bmatrix} = \begin{bmatrix} -1 & -1 & 1 \\ 3 & \frac{8}{3} & -\frac{7}{3} \\ 2 & \frac{7}{3} & -\frac{5}{3} \end{bmatrix} \begin{bmatrix} 6 \\ -1 \\ 7 \end{bmatrix}$$

$$= \begin{bmatrix} -1(6) - 1(-1) + 1(7) \\ 3(6) + \frac{8}{3}(-1) - \frac{7}{3}(7) \\ 2(6) + \frac{7}{3}(-1) - \frac{5}{3}(7) \end{bmatrix} = \begin{bmatrix} 2 \\ -1 \\ -2 \end{bmatrix}$$

Solution: $(2, -1, -2)$

103. $\begin{cases} -2x + y + 2z = -13 \\ -x - 4y + z = -11 \\ -y - z = 0 \end{cases}$

$$\begin{bmatrix} x \\ y \\ z \end{bmatrix} = \begin{bmatrix} -2 & 1 & 2 \\ -1 & -4 & 1 \\ 0 & -1 & -1 \end{bmatrix}^{-1} \begin{bmatrix} -13 \\ -11 \\ 0 \end{bmatrix} = \begin{bmatrix} -\frac{5}{9} & \frac{1}{9} & -1 \\ \frac{1}{9} & -\frac{2}{9} & 0 \\ -\frac{1}{9} & \frac{2}{9} & -1 \end{bmatrix} \begin{bmatrix} -13 \\ -11 \\ 0 \end{bmatrix}$$

$$= \begin{bmatrix} -\frac{5}{9}(-13) + \frac{1}{9}(-11) - 1(0) \\ \frac{1}{9}(-13) - \frac{2}{9}(-11) + 0(0) \\ -\frac{1}{9}(-13) + \frac{2}{9}(-11) - 1(0) \end{bmatrix} = \begin{bmatrix} 6 \\ 1 \\ -1 \end{bmatrix}$$

Solution: $(6, 1, -1)$

109. $\begin{cases} -3x - 3y - 4z = 2 \\ y + z = -1 \\ 4x + 3y + 4z = -1 \end{cases}$

$$\begin{bmatrix} x \\ y \\ z \end{bmatrix} = \begin{bmatrix} -3 & -3 & -4 \\ 0 & 1 & 1 \\ 4 & 3 & 4 \end{bmatrix}^{-1} \begin{bmatrix} 2 \\ -1 \\ -1 \end{bmatrix} = \begin{bmatrix} 1 & 0 & 1 \\ 4 & 4 & 3 \\ -4 & -3 & -3 \end{bmatrix} \begin{bmatrix} 2 \\ -1 \\ -1 \end{bmatrix} = \begin{bmatrix} 1 \\ 1 \\ -2 \end{bmatrix}$$

Solution: $(1, 1, -2)$

111. $\begin{vmatrix} 8 & 5 \\ 2 & -4 \end{vmatrix} = 8(-4) - 5(2) = -42$

113. $\begin{vmatrix} 50 & -30 \\ 10 & 5 \end{vmatrix} = 50(5) - (-30)(10) = 550$

105. $\begin{cases} x + 2y = -1 \\ 3x + 4y = -5 \end{cases}$

$$\begin{bmatrix} x \\ y \end{bmatrix} = \begin{bmatrix} 1 & 2 \\ 3 & 4 \end{bmatrix}^{-1} \begin{bmatrix} -1 \\ -5 \end{bmatrix} = \begin{bmatrix} -2 & 1 \\ \frac{3}{2} & -\frac{1}{2} \end{bmatrix} \begin{bmatrix} -1 \\ -5 \end{bmatrix} = \begin{bmatrix} -3 \\ 1 \end{bmatrix}$$

Solution: $(-3, 1)$

107. $\begin{cases} \frac{6}{5}x - \frac{4}{7}y = \frac{6}{5} \\ -\frac{12}{5}x + \frac{12}{7}y = -\frac{17}{5} \end{cases}$

$$\begin{bmatrix} x \\ y \end{bmatrix} = \begin{bmatrix} \frac{6}{5} & -\frac{4}{7} \\ -\frac{12}{5} & \frac{12}{7} \end{bmatrix}^{-1} \begin{bmatrix} \frac{6}{5} \\ -\frac{17}{5} \end{bmatrix} = \begin{bmatrix} \frac{5}{2} & \frac{5}{6} \\ \frac{7}{2} & \frac{7}{4} \end{bmatrix} \begin{bmatrix} \frac{6}{5} \\ -\frac{17}{5} \end{bmatrix} = \begin{bmatrix} \frac{1}{6} \\ -\frac{7}{4} \end{bmatrix}$$

Solution: $\left(\frac{1}{6}, -\frac{7}{4} \right)$

115. $\begin{bmatrix} 2 & -1 \\ 7 & 4 \end{bmatrix}$

(a) $M_{11} = 4$

$M_{12} = 7$

$M_{21} = -1$

$M_{22} = 2$

(b) $C_{11} = M_{11} = 4$

$C_{12} = -M_{12} = -7$

$C_{21} = -M_{21} = 1$

$C_{22} = M_{22} = 2$

117. $\begin{bmatrix} 3 & 2 & -1 \\ -2 & 5 & 0 \\ 1 & 8 & 6 \end{bmatrix}$

(a) $M_{11} = \begin{vmatrix} 5 & 0 \\ 8 & 6 \end{vmatrix} = 30$

$M_{12} = \begin{vmatrix} -2 & 0 \\ 1 & 6 \end{vmatrix} = -12$

$M_{13} = \begin{vmatrix} -2 & 5 \\ 1 & 8 \end{vmatrix} = -21$

$M_{21} = \begin{vmatrix} 2 & -1 \\ 8 & 6 \end{vmatrix} = 20$

$M_{22} = \begin{vmatrix} 3 & -1 \\ 1 & 6 \end{vmatrix} = 19$

$M_{23} = \begin{vmatrix} 3 & 2 \\ 1 & 8 \end{vmatrix} = 22$

$M_{31} = \begin{vmatrix} 2 & -1 \\ 5 & 0 \end{vmatrix} = 5$

$M_{32} = \begin{vmatrix} 3 & -1 \\ -2 & 0 \end{vmatrix} = -2$

$M_{33} = \begin{vmatrix} 3 & 2 \\ -2 & 5 \end{vmatrix} = 19$

(b) $C_{11} = M_{11} = 30$

$C_{12} = -M_{12} = 12$

$C_{13} = M_{13} = -21$

$C_{21} = -M_{21} = -20$

$C_{22} = M_{22} = 19$

$C_{23} = -M_{23} = -22$

$C_{31} = M_{31} = 5$

$C_{32} = -M_{32} = 2$

$C_{33} = M_{33} = 19$

119. Expand using Row 1.

$\begin{vmatrix} -2 & 0 & 0 \\ 2 & -1 & 0 \\ -1 & 1 & -3 \end{vmatrix} = -2 \begin{vmatrix} -1 & 0 \\ 1 & -3 \end{vmatrix} - 0 \begin{vmatrix} 2 & 0 \\ -1 & 3 \end{vmatrix} + 0 \begin{vmatrix} 2 & -1 \\ -1 & 1 \end{vmatrix} = -2(3) - 0(6) + 0(1) = -6$

121. Expand using Row 3.

$\begin{vmatrix} 4 & 1 & -1 \\ 2 & 3 & 2 \\ 1 & -1 & 0 \end{vmatrix} = 1 \begin{vmatrix} 1 & -1 \\ 3 & 2 \end{vmatrix} + 1 \begin{vmatrix} 4 & -1 \\ 2 & 2 \end{vmatrix} + 0 \begin{vmatrix} 4 & 1 \\ 2 & 3 \end{vmatrix} = 1(5) + 1(10) + 0(10) = 15$

123. Expand using Column 2.

$\begin{vmatrix} -2 & 4 & 1 \\ -6 & 0 & 2 \\ 5 & 3 & 4 \end{vmatrix} = -4 \begin{vmatrix} -6 & 2 \\ 5 & 4 \end{vmatrix} - 3 \begin{vmatrix} -2 & 1 \\ -6 & 2 \end{vmatrix} = -4(-34) - 3(2) = 130$

125. Expand using Row 4.

$\begin{vmatrix} 1 & 2 & -1 & 0 \\ 1 & 2 & -4 & 1 \\ 2 & -4 & -3 & 1 \\ 2 & 0 & 0 & 0 \end{vmatrix} = -2 \begin{vmatrix} 2 & -1 & 0 \\ 2 & -4 & 1 \\ -4 & -3 & 1 \end{vmatrix} + 0 \begin{vmatrix} 1 & -1 & 0 \\ 1 & -4 & 1 \\ 2 & -3 & 1 \end{vmatrix} - 0 \begin{vmatrix} 1 & 2 & 0 \\ 1 & 2 & 1 \\ 2 & -4 & 1 \end{vmatrix} + 0 \begin{vmatrix} 1 & 2 & -1 \\ 1 & 2 & 4 \\ 2 & -4 & -3 \end{vmatrix}$

$= -2 \big[0(-22) - 1(-10) + 1(-6) \big]$

$= -2(4)$

$= -8$

127. Expand along Row 1.

$$\begin{vmatrix} 3 & 0 & -4 & 0 \\ 0 & 8 & 1 & 2 \\ 6 & 1 & 8 & 2 \\ 0 & 3 & -4 & 1 \end{vmatrix} = 3\begin{vmatrix} 8 & 1 & 2 \\ 1 & 8 & 2 \\ 3 & -4 & 1 \end{vmatrix} + (-4)\begin{vmatrix} 0 & 8 & 2 \\ 6 & 1 & 2 \\ 0 & 3 & 1 \end{vmatrix}$$

$$= 3\big[8(8 - (-8)) - 1(1 - 6) + 2(-4 - 24)\big] - 4\big[0 - 6(8 - 6) + 0\big]$$

$$= 3[128 + 5 - 56] - 4[-12]$$

$$= 279$$

129. $\begin{cases} 5x - 2y = 6 \\ -11x + 3y = -23 \end{cases}$

$$x = \frac{\begin{vmatrix} 6 & -2 \\ -23 & 3 \end{vmatrix}}{\begin{vmatrix} 5 & -2 \\ -11 & 3 \end{vmatrix}} = \frac{-28}{-7} = 4, \qquad y = \frac{\begin{vmatrix} 5 & 6 \\ -11 & -23 \end{vmatrix}}{\begin{vmatrix} 5 & -2 \\ -11 & 3 \end{vmatrix}} = \frac{-49}{-7} = 7$$

Solution: $(4, 7)$

131. $\begin{cases} -2x + 3y - 5z = -11 \\ 4x - y + z = -3 \\ -x - 4y + 6z = 15 \end{cases}$

$$D = \begin{vmatrix} -2 & 3 & -5 \\ 4 & -1 & 1 \\ -1 & -4 & 6 \end{vmatrix} = -2(-1)^2\begin{vmatrix} -1 & 1 \\ -4 & 6 \end{vmatrix} + 4(-1)^3\begin{vmatrix} 3 & -5 \\ -4 & 6 \end{vmatrix} - 1(-1)^4\begin{vmatrix} 3 & -5 \\ -1 & 1 \end{vmatrix} = -2(-2) - 4(-2) - (-2) = 14$$

$$x = \frac{\begin{vmatrix} -11 & 3 & -5 \\ -3 & -1 & 1 \\ 15 & -4 & 6 \end{vmatrix}}{14} = \frac{-11(-1)^2\begin{vmatrix} -1 & 1 \\ -4 & 6 \end{vmatrix} - 3(-1)^3\begin{vmatrix} 3 & -5 \\ -4 & 6 \end{vmatrix} + 15(-1)^4\begin{vmatrix} 3 & -5 \\ -1 & 1 \end{vmatrix}}{14} = \frac{-11(-2) + 3(-2) + 15(-2)}{14} = \frac{-14}{14} = -1$$

$$y = \frac{\begin{vmatrix} -2 & -11 & -5 \\ 4 & -3 & 1 \\ -1 & 15 & 6 \end{vmatrix}}{14} = \frac{-2(-1)^2\begin{vmatrix} -3 & 1 \\ 15 & 6 \end{vmatrix} + 4(-1)^3\begin{vmatrix} -11 & -5 \\ 15 & 6 \end{vmatrix} - 1(-1)^4\begin{vmatrix} -11 & -5 \\ -3 & 1 \end{vmatrix}}{14} = \frac{-2(-33) - 4(9) - 1(-26)}{14} = \frac{56}{14} = 4$$

$$z = \frac{\begin{vmatrix} -2 & 3 & -11 \\ 4 & -1 & -3 \\ -1 & -4 & 15 \end{vmatrix}}{14} = \frac{-2(-1)^2\begin{vmatrix} -1 & -3 \\ -4 & 15 \end{vmatrix} + 4(-1)^3\begin{vmatrix} 3 & -11 \\ -4 & 15 \end{vmatrix} - 1(-1)^4\begin{vmatrix} 3 & -11 \\ -1 & -3 \end{vmatrix}}{14} = \frac{-2(-27) - 4(1) - 1(-20)}{14} = \frac{70}{14} = 5$$

Solution: $(-1, 4, 5)$

133. $(1, 0), (5, 0), (5, 8)$

$$\text{Area} = \frac{1}{2}\begin{vmatrix} 1 & 0 & 1 \\ 5 & 0 & 1 \\ 5 & 8 & 1 \end{vmatrix} = \frac{1}{2}\left(1\begin{vmatrix} 0 & 1 \\ 8 & 1 \end{vmatrix} + 1\begin{vmatrix} 5 & 0 \\ 5 & 8 \end{vmatrix}\right) = \frac{1}{2}(-8 + 40) = \frac{1}{2}(32) = 16 \text{ square units}$$

135. $(1, -4), (-2, 3), (0, 5)$

$$\text{Area} = -\frac{1}{2}\begin{vmatrix} 1 & -4 & 1 \\ -2 & 3 & 1 \\ 0 & 5 & 1 \end{vmatrix}$$

$$= -\frac{1}{2}\left(-5\begin{vmatrix} 1 & 1 \\ -2 & 1 \end{vmatrix} + \begin{vmatrix} 1 & -4 \\ -2 & 3 \end{vmatrix}\right)$$

$$= -\frac{1}{2}(-5(3) + (-5)) = 10 \text{ square units}$$

137. $(-1, 7), (3, -9), (-3, 15)$

$$\begin{vmatrix} -1 & 7 & 1 \\ 3 & -9 & 1 \\ -3 & 15 & 1 \end{vmatrix} = \begin{vmatrix} 3 & -9 \\ -3 & 15 \end{vmatrix} - \begin{vmatrix} -1 & 7 \\ -3 & 15 \end{vmatrix} + \begin{vmatrix} -1 & 7 \\ 3 & -9 \end{vmatrix}$$

$$= 18 - 6 - 12 = 0$$

The points are collinear.

139. $(-4, 0), (4, 4)$

$$\begin{vmatrix} x & y & 1 \\ -4 & 0 & 1 \\ 4 & 4 & 1 \end{vmatrix} = 0$$

$$1\begin{vmatrix} -4 & 0 \\ 4 & 4 \end{vmatrix} - 1\begin{vmatrix} x & y \\ 4 & 4 \end{vmatrix} + 1\begin{vmatrix} x & y \\ -4 & 0 \end{vmatrix} = 0$$

$$-16 - (4x - 4y) + 4y = 0$$

$$-4x + 8y - 16 = 0$$

$$x - 2y + 4 = 0$$

141. $\left(-\frac{5}{2}, 3\right), \left(\frac{7}{2}, 1\right)$

$$\begin{vmatrix} x & y & 1 \\ -\frac{5}{2} & 3 & 1 \\ \frac{7}{2} & 1 & 1 \end{vmatrix} = 0$$

$$1\begin{vmatrix} -\frac{5}{2} & 3 \\ \frac{7}{2} & 1 \end{vmatrix} - 1\begin{vmatrix} x & y \\ \frac{7}{2} & 1 \end{vmatrix} + 1\begin{vmatrix} x & y \\ -\frac{5}{2} & 3 \end{vmatrix} = 0$$

$$-13 - \left(x - \frac{7}{2}y\right) + \left(3x + \frac{5}{2}y\right) = 0$$

$$2x + 6y - 13 = 0$$

143. (a) Uncoded: L O O K _ O U T _ B E L O W _

$$\begin{bmatrix} 12 & 15 & 15 \end{bmatrix} \begin{bmatrix} 11 & 0 & 15 \end{bmatrix} \begin{bmatrix} 21 & 20 & 0 \end{bmatrix} \begin{bmatrix} 2 & 5 & 12 \end{bmatrix} \begin{bmatrix} 15 & 23 & 0 \end{bmatrix}$$

(b) $\begin{bmatrix} 12 & 15 & 15 \end{bmatrix} \begin{bmatrix} 2 & -2 & 0 \\ 3 & 0 & -3 \\ -6 & 2 & 3 \end{bmatrix} = \begin{bmatrix} -21 & 6 & 0 \end{bmatrix}$

$\begin{bmatrix} 11 & 0 & 15 \end{bmatrix} \begin{bmatrix} 2 & -2 & 0 \\ 3 & 0 & -3 \\ -6 & 2 & 3 \end{bmatrix} = \begin{bmatrix} -68 & 8 & 45 \end{bmatrix}$

$\begin{bmatrix} 21 & 20 & 0 \end{bmatrix} \begin{bmatrix} 2 & -2 & 0 \\ 3 & 0 & -3 \\ -6 & 2 & 3 \end{bmatrix} = \begin{bmatrix} 102 & -42 & -60 \end{bmatrix}$

$\begin{bmatrix} 2 & 5 & 12 \end{bmatrix} \begin{bmatrix} 2 & -2 & 0 \\ 3 & 0 & -3 \\ -6 & 2 & 3 \end{bmatrix} = \begin{bmatrix} -53 & 20 & 21 \end{bmatrix}$

$\begin{bmatrix} 15 & 23 & 0 \end{bmatrix} \begin{bmatrix} 2 & -2 & 0 \\ 3 & 0 & -3 \\ -6 & 2 & 3 \end{bmatrix} = \begin{bmatrix} 99 & -30 & -69 \end{bmatrix}$

Encoded: $-21 \quad 6 \quad 0 \quad -68 \quad 8 \quad 45 \quad 102 \quad -42 \quad -42 \quad -60 \quad -53 \quad 20 \quad 21 \quad 99 \quad -30 \quad -69$

145. $A^{-1} = \begin{bmatrix} -1 & 2 & -3 \\ 2 & 1 & 0 \\ 4 & -2 & 5 \end{bmatrix}$

$\begin{bmatrix} -5 & 11 & -2 \end{bmatrix} \begin{bmatrix} -1 & 2 & -3 \\ 2 & 1 & 0 \\ 4 & -2 & 5 \end{bmatrix} = \begin{bmatrix} 19 & 5 & 5 \end{bmatrix}$ S E E

$\begin{bmatrix} 370 & -265 & 225 \end{bmatrix} \begin{bmatrix} -1 & 2 & -3 \\ 2 & 1 & 0 \\ 4 & -2 & 5 \end{bmatrix} = \begin{bmatrix} 0 & 25 & 15 \end{bmatrix}$ _ Y O

$\begin{bmatrix} -57 & 48 & -33 \end{bmatrix} \begin{bmatrix} -1 & 2 & -3 \\ 2 & 1 & 0 \\ 4 & -2 & 5 \end{bmatrix} = \begin{bmatrix} 21 & 0 & 6 \end{bmatrix}$ U _ F

$\begin{bmatrix} 32 & -15 & 20 \end{bmatrix} \begin{bmatrix} -1 & 2 & -3 \\ 2 & 1 & 0 \\ 4 & -2 & 5 \end{bmatrix} = \begin{bmatrix} 18 & 9 & 4 \end{bmatrix}$ R I D

$\begin{bmatrix} 245 & -171 & 147 \end{bmatrix} \begin{bmatrix} -1 & 2 & -3 \\ 2 & 1 & 0 \\ 4 & -2 & 5 \end{bmatrix} = \begin{bmatrix} 1 & 25 & 0 \end{bmatrix}$ A Y _

Message: SEE YOU FRIDAY

147. False. The matrix must be square

149. An error message appears because $1(6) - (-3)(-2) = 0$.

151. If A is a square matrix, the cofactor C_{ij} of the entry a_{ij} is $(-1)^{i+j} M_{ij}$, where M_{ij} is the determinant obtained by deleting the ith row and jth column of A. The determinant of A is the sum of the entries of any row or column of A multiplied by their respective cofactors.

153. The part of the matrix corresponding to the coefficients of the system reduces to a matrix in which the number of rows with nonzero entries is the same as the number of variables.

Problem Solving for Chapter 10

1. $A = \begin{bmatrix} 0 & -1 \\ 1 & 0 \end{bmatrix}$ $T = \begin{bmatrix} 1 & 2 & 3 \\ 1 & 4 & 2 \end{bmatrix}$

(a) $AT = \begin{bmatrix} -1 & -4 & -2 \\ 1 & 2 & 3 \end{bmatrix}$ $AAT = \begin{bmatrix} -1 & -2 & -3 \\ -1 & -4 & -2 \end{bmatrix}$

Original Triangle AT Triangle AAT Triangle

The transformation A interchanges the x and y coordinates and then takes the negative of the x coordinate. A represents a counterclockwise rotation by $90°$.

(b) AAT is rotated clockwise $90°$ to obtain AT. AT is then rotated clockwise $90°$ to obtain T.

3. (a) $A^2 = \begin{bmatrix} 1 & 0 \\ 0 & 0 \end{bmatrix}\begin{bmatrix} 1 & 0 \\ 0 & 0 \end{bmatrix} = \begin{bmatrix} 1 & 0 \\ 0 & 0 \end{bmatrix} = A$

A *is* idempotent.

(c) $A^2 = \begin{bmatrix} 2 & 3 \\ -1 & -2 \end{bmatrix}\begin{bmatrix} 2 & 3 \\ -1 & -2 \end{bmatrix} = \begin{bmatrix} 1 & 0 \\ 0 & 1 \end{bmatrix} \neq A$

A is *not* idempotent.

(b) $A^2 = \begin{bmatrix} 0 & 1 \\ 1 & 0 \end{bmatrix}\begin{bmatrix} 0 & 1 \\ 1 & 0 \end{bmatrix} = \begin{bmatrix} 1 & 0 \\ 0 & 1 \end{bmatrix} \neq A$

A is *not* idempotent.

(d) $A^2 = \begin{bmatrix} 2 & 3 \\ 1 & 2 \end{bmatrix}\begin{bmatrix} 2 & 3 \\ 1 & 2 \end{bmatrix} = \begin{bmatrix} 7 & 12 \\ 4 & 7 \end{bmatrix} \neq A$

A is *not* idempotent.

5. (a) $\begin{bmatrix} 0.70 & 0.15 & 0.15 \\ 0.20 & 0.80 & 0.15 \\ 0.10 & 0.05 & 0.70 \end{bmatrix}\begin{bmatrix} 25{,}000 \\ 30{,}000 \\ 45{,}000 \end{bmatrix} = \begin{bmatrix} 28{,}750 \\ 35{,}750 \\ 35{,}500 \end{bmatrix}$

Gold Satellite System: 28,750 subscribers
Galaxy Satellite Network: 35,750 subscribers
Nonsubscribers: 35,500

(c) $\begin{bmatrix} 0.70 & 0.15 & 0.15 \\ 0.20 & 0.80 & 0.15 \\ 0.10 & 0.05 & 0.70 \end{bmatrix}\begin{bmatrix} 30{,}812.5 \\ 39{,}675 \\ 29{,}512.5 \end{bmatrix} \approx \begin{bmatrix} 31{,}947 \\ 42{,}329 \\ 25{,}724 \end{bmatrix}$

Gold Satellite System: 31,947 subscribers
Galaxy Satellite Network: 42,329 subscribers
Nonsubscribers: 25,724

(b) $\begin{bmatrix} 0.70 & 0.15 & 0.15 \\ 0.20 & 0.80 & 0.15 \\ 0.10 & 0.05 & 0.70 \end{bmatrix}\begin{bmatrix} 28{,}750 \\ 35{,}750 \\ 35{,}500 \end{bmatrix} \approx \begin{bmatrix} 30{,}813 \\ 39{,}675 \\ 29{,}513 \end{bmatrix}$

Gold Satellite System: 30,813 subscribers
Galaxy Satellite Network: 39,675 subscribers
Nonsubscribers: 29,513

(d) Both satellite companies are increasing the number of subscribers, while the number of nonsubscribers is decreasing each year.

7. If $A = \begin{bmatrix} 4 & x \\ -2 & -3 \end{bmatrix}$ is singular then

$ad - bc = -12 + 2x = 0.$

So, $x = 6.$

9. $(a - b)(b - c)(c - a) = -a^2b + a^2c + ab^2 - ac^2 - b^2c + bc^2$

$\begin{vmatrix} 1 & 1 & 1 \\ a & b & c \\ a^2 & b^2 & c^2 \end{vmatrix} = \begin{vmatrix} b & c \\ b^2 & c^2 \end{vmatrix} - \begin{vmatrix} a & c \\ a^2 & c^2 \end{vmatrix} + \begin{vmatrix} a & b \\ a^2 & b^2 \end{vmatrix} = bc^2 - b^2c - ac^2 + a^2c + ab^2 - a^2b$

So, $\begin{vmatrix} 1 & 1 & 1 \\ a & b & c \\ a^2 & b^2 & c^2 \end{vmatrix} = (a - b)(b - c)(c - a).$

11. $\begin{vmatrix} x & 0 & c \\ -1 & x & b \\ 0 & -1 & a \end{vmatrix} = x\begin{vmatrix} x & b \\ -1 & a \end{vmatrix} + c\begin{vmatrix} -1 & x \\ 0 & -1 \end{vmatrix} = x(ax + b) + c(1 - 0) = ax^2 + bx + c$

13.
$$
\begin{aligned}
4S + 4N \qquad &= 184 \\
S \qquad\quad + 6F &= 146 \\
2N + 4F &= 104
\end{aligned}
$$

$$
D = \begin{vmatrix} 4 & 4 & 0 \\ 1 & 0 & 6 \\ 0 & 2 & 4 \end{vmatrix} = -64
$$

$$
S = \dfrac{\begin{vmatrix} 184 & 4 & 0 \\ 146 & 0 & 6 \\ 104 & 2 & 4 \end{vmatrix}}{-64} = \dfrac{-2048}{-64} = 32
$$

$$
N = \dfrac{\begin{vmatrix} 4 & 184 & 0 \\ 1 & 146 & 6 \\ 0 & 104 & 4 \end{vmatrix}}{-64} = \dfrac{-896}{-64} = 14
$$

$$
F = \dfrac{\begin{vmatrix} 4 & 4 & 184 \\ 1 & 0 & 146 \\ 0 & 2 & 104 \end{vmatrix}}{-64} = \dfrac{-1216}{-64} = 19
$$

Element	Atomic mass
Sulfur	32
Nitrogen	14
Fluoride	19

15.
$$
A = \begin{bmatrix} -1 & 1 & -2 \\ 2 & 0 & 1 \end{bmatrix}, \qquad B = \begin{bmatrix} -3 & 0 \\ 1 & 2 \\ 1 & -1 \end{bmatrix}
$$

$$
A^T = \begin{bmatrix} -1 & 2 \\ 1 & 0 \\ -2 & 1 \end{bmatrix}, \qquad B^T = \begin{bmatrix} -3 & 1 & 1 \\ 0 & 2 & -1 \end{bmatrix}
$$

$$
AB = \begin{bmatrix} 2 & 4 \\ -5 & -1 \end{bmatrix}, \qquad (AB)^T = \begin{bmatrix} 2 & -5 \\ 4 & -1 \end{bmatrix}
$$

$$
B^T A^T = \begin{bmatrix} -3 & 1 & 1 \\ 0 & 2 & -1 \end{bmatrix} \begin{bmatrix} -1 & 2 \\ 1 & 0 \\ -2 & 1 \end{bmatrix} = \begin{bmatrix} 2 & -5 \\ 4 & -1 \end{bmatrix}
$$

So, $(AB)^T = B^T A^T$.

17. (a)
$$
\begin{bmatrix} 45 & -35 \end{bmatrix} \begin{bmatrix} w & x \\ y & z \end{bmatrix} = \begin{bmatrix} 10 & 15 \end{bmatrix}
$$

$$
\begin{bmatrix} 38 & -30 \end{bmatrix} \begin{bmatrix} w & x \\ y & z \end{bmatrix} = \begin{bmatrix} 8 & 14 \end{bmatrix}
$$

$$
\begin{aligned}
45w - 35y &= 10 \\
45x - 35z &= 15 \\
38w - 30y &= 8 \\
38x - 30z &= 14
\end{aligned}
$$

$$
\left. \begin{aligned} 45w - 35y &= 10 \\ 38w - 30y &= 8 \end{aligned} \right\} \Rightarrow w = 1,\ y = 1
$$

$$
\left. \begin{aligned} 45x - 35z &= 15 \\ 38x - 30z &= 14 \end{aligned} \right\} \Rightarrow x = -2,\ z = -3
$$

$$
A^{-1} = \begin{bmatrix} 1 & -2 \\ 1 & -3 \end{bmatrix}
$$

(b)
$$
\begin{bmatrix} 45 & -35 \\ 38 & -30 \\ 18 & -18 \\ 35 & -30 \\ 81 & -60 \\ 42 & -28 \\ 75 & -55 \\ 2 & -2 \\ 22 & -21 \\ 15 & -10 \end{bmatrix}
\begin{bmatrix} 1 & -2 \\ 1 & -3 \end{bmatrix}
=
\begin{bmatrix} 10 & 15 \\ 8 & 14 \\ 0 & 18 \\ 5 & 20 \\ 21 & 18 \\ 14 & 0 \\ 20 & 15 \\ 0 & 2 \\ 1 & 19 \\ 5 & 0 \end{bmatrix}
\begin{matrix} J & O \\ H & N \\ \_ & R \\ E & T \\ U & R \\ N & \_ \\ T & O \\ \_ & B \\ A & S \\ E & \_ \end{matrix}
$$

Message: JOHN RETURN TO BASE

19. Let $A = \begin{bmatrix} 3 & -3 \\ 5 & -5 \end{bmatrix}$, then $|A| = 0$.

Let $A = \begin{bmatrix} 2 & 4 & -6 \\ -3 & 1 & 2 \\ 5 & -8 & 3 \end{bmatrix}$, then $|A| = 0$.

Let $A = \begin{bmatrix} 3 & -7 & 5 & -1 \\ -6 & 4 & 0 & 2 \\ 5 & 8 & -6 & -7 \\ 9 & 11 & -4 & -16 \end{bmatrix}$, then $|A| = 0$.

Conjecture: If A is an $n \times n$ matrix, each of whose rows add up to zero, then $|A| = 0$.

Chapter 10 Practice Test

1. Put the matrix in reduced row-echelon form.

$$\begin{bmatrix} 1 & -2 & 4 \\ 3 & -5 & 9 \end{bmatrix}$$

For Exercises 2–4, use matrices to solve the system of equations.

2. $\begin{cases} 3x + 5y = 3 \\ 2x - y = -11 \end{cases}$

3. $\begin{cases} 2x + 3y = -3 \\ 3x + 2y = 8 \\ x + y = 1 \end{cases}$

4. $\begin{cases} x + 3z = -5 \\ 2x + y = 0 \\ 3x + y - z = 3 \end{cases}$

5. Multiply $\begin{bmatrix} 1 & 4 & 5 \\ 2 & 0 & -3 \end{bmatrix} \begin{bmatrix} 1 & 6 \\ 0 & -7 \\ -1 & 2 \end{bmatrix}$.

6. Given $A = \begin{bmatrix} 9 & 1 \\ -4 & 8 \end{bmatrix}$ and $B = \begin{bmatrix} 6 & -2 \\ 3 & 5 \end{bmatrix}$, find $3A - 5B$.

7. Find $f(A)$.

$$f(x) = x^2 - 7x + 8, \ A = \begin{bmatrix} 3 & 0 \\ 7 & 1 \end{bmatrix}$$

8. True or false:

$(A + B)(A + 3B) = A^2 + 4AB + 3B^2$ where A and B are matrices.

(Assume that A^2, AB, and B^2 exist.)

For Exercises 9–10, find the inverse of the matrix, if it exists.

9. $\begin{bmatrix} 1 & 2 \\ 3 & 5 \end{bmatrix}$

10. $\begin{bmatrix} 1 & 1 & 1 \\ 3 & 6 & 5 \\ 6 & 10 & 8 \end{bmatrix}$

11. Use an inverse matrix to solve the systems.

(a) $\begin{cases} x + 2y = 4 \\ 3x + 5y = 1 \end{cases}$

(b) $\begin{cases} x + 2y = 3 \\ 3x + 5y = -2 \end{cases}$

For Exercises 12–14, find the determinant of the matrix.

12. $\begin{bmatrix} 6 & -1 \\ 3 & 4 \end{bmatrix}$

13. $\begin{bmatrix} 1 & 3 & -1 \\ 5 & 9 & 0 \\ 6 & 2 & -5 \end{bmatrix}$

14. $\begin{bmatrix} 1 & 4 & 2 & 3 \\ 0 & 1 & -2 & 0 \\ 3 & 5 & -1 & 1 \\ 2 & 0 & 6 & 1 \end{bmatrix}$

15. Evaluate $\begin{vmatrix} 6 & 4 & 3 & 0 & 6 \\ 0 & 5 & 1 & 4 & 8 \\ 0 & 0 & 2 & 7 & 3 \\ 0 & 0 & 0 & 9 & 2 \\ 0 & 0 & 0 & 0 & 1 \end{vmatrix}$.

16. Use a determinant to find the area of the triangle with vertices $(0, 7), (5, 0),$ and $(3, 9)$.

17. Use a determinant to find the equation of the line passing through $(2, 7)$ and $(-1, 4)$.

For Exercises 18–20, use Cramer's Rule to find the indicated value.

18. Find x.
$$\begin{cases} 6x - 7y = 4 \\ 2x + 5y = 11 \end{cases}$$

19. Find z.
$$\begin{cases} 3x \quad\quad + z = 1 \\ \quad\quad y + 4z = 3 \\ x - y \quad\quad = 2 \end{cases}$$

20. Find y.
$$\begin{cases} 721.4x - 29.1y = 33.77 \\ 45.9x + 105.6y = 19.85 \end{cases}$$

CHAPTER 11
Sequences, Series, and Probability

CHAPTER 11
Sequences, Series, and Probability

Section 11.1   Sequences and Series

- Given the general nth term in a sequence, you should be able to find, or list, some of the terms.

- You should be able to find an expression for the apparent nth term of a sequence.

- You should be able to use and evaluate factorials.

- You should be able to use summation notation for a sum.

- You should know that the sum of the terms of a sequence is a series.

1. infinite sequence

3. finite

5. factorial

7. index; upper; lower

9. $a_n = 2n + 5$

$a_1 = 2(1) + 5 = 7$

$a_2 = 2(2) + 5 = 9$

$a_3 = 2(3) + 5 = 11$

$a_4 = 2(4) + 5 = 13$

$a_5 = 2(5) + 5 = 15$

11. $a_n = 2^n$

$a_1 = 2^1 = 2$

$a_2 = 2^2 = 4$

$a_3 = 2^3 = 8$

$a_4 = 2^4 = 16$

$a_5 = 2^5 = 32$

13. $a_n = (-2)^n$

$a_1 = (-2)^1 = -2$

$a_2 = (-2)^2 = 4$

$a_3 = (-2)^3 = -8$

$a_4 = (-2)^4 = 16$

$a_5 = (-2)^5 = -32$

15. $a_n = \dfrac{n + 2}{n}$

$a_1 = \dfrac{1 + 2}{1} = 3$

$a_2 = \dfrac{4}{2} = 2$

$a_3 = \dfrac{5}{3}$

$a_4 = \dfrac{6}{4} = \dfrac{3}{2}$

$a_5 = \dfrac{7}{5}$

17. $a_n = \dfrac{6n}{3n^2 - 1}$

$a_1 = \dfrac{6(1)}{3(1)^2 - 1} = 3$

$a_2 = \dfrac{6(2)}{3(2)^2 - 1} = \dfrac{12}{11}$

$a_3 = \dfrac{6(3)}{3(3)^2 - 1} = \dfrac{9}{13}$

$a_4 = \dfrac{6(4)}{3(4)^2 - 1} = \dfrac{24}{47}$

$a_5 = \dfrac{6(5)}{3(5)^2 - 1} = \dfrac{15}{37}$

19. $a_n = \dfrac{1 + (-1)^n}{n}$

$a_1 = 0$

$a_2 = \dfrac{2}{2} = 1$

$a_3 = 0$

$a_4 = \dfrac{2}{4} = \dfrac{1}{2}$

$a_5 = 0$

21. $a_n = 2 - \dfrac{1}{3^n}$

$a_1 = 2 - \dfrac{1}{3} = \dfrac{5}{3}$

$a_2 = 2 - \dfrac{1}{9} = \dfrac{17}{9}$

$a_3 = 2 - \dfrac{1}{27} = \dfrac{53}{27}$

$a_4 = 2 - \dfrac{1}{81} = \dfrac{161}{81}$

$a_5 = 2 - \dfrac{1}{243} = \dfrac{485}{243}$

23. $a_n = \dfrac{1}{n^{3/2}}$

$a_1 = \dfrac{1}{1} = 1$

$a_2 = \dfrac{1}{2^{3/2}}$

$a_3 = \dfrac{1}{3^{3/2}}$

$a_4 = \dfrac{1}{4^{3/2}} = \dfrac{1}{8}$

$a_5 = \dfrac{1}{5^{3/2}}$

25. $a_n = \dfrac{(-1)^n}{n^2}$

$a_1 = -\dfrac{1}{1} = -1$

$a_2 = \dfrac{1}{4}$

$a_3 = -\dfrac{1}{9}$

$a_4 = \dfrac{1}{16}$

$a_5 = -\dfrac{1}{25}$

27. $a_n = \dfrac{2}{3}$

$a_1 = \dfrac{2}{3}$

$a_2 = \dfrac{2}{3}$

$a_3 = \dfrac{2}{3}$

$a_4 = \dfrac{2}{3}$

$a_5 = \dfrac{2}{3}$

29. $a_n = n(n-1)(n-2)$

$a_1 = (1)(0)(-1) = 0$

$a_2 = (2)(1)(0) = 0$

$a_3 = (3)(2)(1) = 6$

$a_4 = (4)(3)(2) = 24$

$a_5 = (5)(4)(3) = 60$

31. $a_n = \dfrac{(-1)^{n+1}}{n^2 + 1}$

$a_1 = \dfrac{(-1)^{1+1}}{1^2 + 1} = \dfrac{(-1)^2}{2} = \dfrac{1}{2}$

$a_2 = \dfrac{(-1)^{2+1}}{2^2 + 1} = \dfrac{(-1)^3}{5} = -\dfrac{1}{5}$

$a_3 = \dfrac{(-1)^{3+1}}{3^2 + 1} = \dfrac{(-1)^4}{10} = \dfrac{1}{10}$

$a_4 = \dfrac{(-1)^{4+1}}{4^2 + 1} = \dfrac{(-1)^5}{17} = -\dfrac{1}{17}$

$a_5 = \dfrac{(-1)^{5+1}}{5^2 + 1} = \dfrac{(-1)^6}{26} = \dfrac{1}{26}$

33. $a_{25} = (-1)^{25}(3(25) - 2) = -73$

35. $a_{11} = \dfrac{4(11)}{2(11)^2 - 3} = \dfrac{44}{239}$

37. $a_n = \dfrac{2}{3}n$

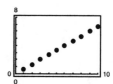

39. $a_n = 16(-0.5)^{n-1}$

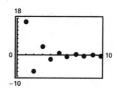

41. $a_n = \dfrac{2n}{n+1}$

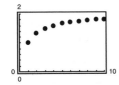

43. $a_n = \dfrac{8}{n+1}$

$a_1 = 4, a_{10} = \dfrac{8}{11}$

The sequence decreases.

Matches graph (c).

45. $a_n = 4(0.5)^{n-1}$

$a_1 = 4, a_{10} = \dfrac{1}{128}$

The sequence decreases.

Matches graph (d).

47. $1, 4, 7, 10, 13, \ldots$

$a_n = 1 + (n-1)3 = 3n - 2$

49. $0, 3, 8, 15, 24, \ldots$

$a_n = n^2 - 1$

51. $-\dfrac{2}{3}, \dfrac{3}{4}, -\dfrac{4}{5}, \dfrac{5}{6}, -\dfrac{6}{7}, \ldots$

$a_n = (-1)^n \left(\dfrac{n+1}{n+2} \right)$

53. $\dfrac{2}{1}, \dfrac{3}{3}, \dfrac{4}{5}, \dfrac{5}{7}, \dfrac{6}{9}, \ldots$

$a_n = \dfrac{n+1}{2n-1}$

55. $1, \dfrac{1}{4}, \dfrac{1}{9}, \dfrac{1}{16}, \dfrac{1}{25}, \ldots$

$a_n = \dfrac{1}{n^2}$

57. $1, -1, 1, -1, 1, \ldots$

$\quad n: 1 \quad 2 \quad 3 \quad 4 \quad 5 \ldots \quad n$

Terms: $1 \quad -1 \quad 1 \quad -1 \quad 1 \ldots a_n$

Apparent pattern:

Each term is either 1 or -1 which implies that

$a_n = (-1)^{n+1}$.

59. $1, 3, 1, 3, 1, \ldots$

$\quad n: 1 \quad 2 \quad 3 \quad 4 \quad 5 \ldots \quad n$

Terms: $1 \quad 3 \quad 1 \quad 3 \quad 1 \ldots a_n$

Apparent pattern:

Each term is either 1 or 3 which implies that

$a_n = (-1)^n + 2$.

61. $1 + \dfrac{1}{1}, 1 + \dfrac{1}{2}, 1 + \dfrac{1}{3}, 1 + \dfrac{1}{4}, 1 + \dfrac{1}{5}, \ldots$

$a_n = 1 + \dfrac{1}{n}$

63. $a_1 = 28$ and $a_{k+1} = a_k - 4$

$a_1 = 28$

$a_2 = a_1 - 4 = 28 - 4 = 24$

$a_3 = a_2 - 4 = 24 - 4 = 20$

$a_4 = a_3 - 4 = 20 - 4 = 16$

$a_5 = a_4 - 4 = 16 - 4 = 12$

65. $a_1 = 3$, and $a_{k+1} = 2(a_k - 1)$

$a_1 = 3$

$a_2 = 2(a_1 - 1) = 2(3 - 1) = 4$

$a_3 = 2(a_2 - 1) = 2(4 - 1) = 6$

$a_4 = 2(a_3 - 1) = 2(6 - 1) = 10$

$a_5 = 2(a_4 - 1) = 2(10 - 1) = 18$

67. $a_1 = 6$ and $a_{k+1} = a_k + 2$

$a_1 = 6$

$a_2 = a_1 + 2 = 6 + 2 = 8$

$a_3 = a_2 + 2 = 8 + 2 = 10$

$a_4 = a_3 + 2 = 10 + 2 = 12$

$a_5 = a_4 + 2 = 12 + 2 = 14$

In general, $a_n = 2n + 4$.

69. $a_1 = 81$ and $a_{k+1} = \dfrac{1}{3} a_k$

$a_1 = 81$

$a_2 = \dfrac{1}{3} a_1 = \dfrac{1}{3}(81) = 27$

$a_3 = \dfrac{1}{3} a_2 = \dfrac{1}{3}(27) = 9$

$a_4 = \dfrac{1}{3} a_3 = \dfrac{1}{3}(9) = 3$

$a_5 = \dfrac{1}{3} a_4 = \dfrac{1}{3}(3) = 1$

In general,

$a_n = 81 \left(\dfrac{1}{3} \right)^{n-1} = 81(3) \left(\dfrac{1}{3} \right)^n = \dfrac{243}{3^n}$.

71. $a_n = \dfrac{1}{n!}$

$a_0 = \dfrac{1}{0!} = \dfrac{1}{1} = 1$

$a_1 = \dfrac{1}{1!} = \dfrac{1}{1} = 1$

$a_2 = \dfrac{1}{2!} = \dfrac{1}{2 \cdot 1} = \dfrac{1}{2}$

$a_3 = \dfrac{1}{3!} = \dfrac{1}{3 \cdot 2 \cdot 1} = \dfrac{1}{6}$

$a_4 = \dfrac{1}{4!} = \dfrac{1}{4 \cdot 3 \cdot 2 \cdot 1} = \dfrac{1}{24}$

73. $a_n = \dfrac{1}{(n+1)!}$

$a_0 = \dfrac{1}{1!} = 1$

$a_1 = \dfrac{1}{2!} = \dfrac{1}{2}$

$a_2 = \dfrac{1}{3!} = \dfrac{1}{6}$

$a_3 = \dfrac{1}{4!} = \dfrac{1}{24}$

$a_4 = \dfrac{1}{5!} = \dfrac{1}{120}$

75. $a_n = \dfrac{(-1)^{2n}}{(2n)!} = \dfrac{1}{(2n)!}$

$a_0 = \dfrac{1}{0!} = 1$

$a_1 = \dfrac{1}{2!} = \dfrac{1}{2}$

$a_2 = \dfrac{1}{4!} = \dfrac{1}{24}$

$a_3 = \dfrac{1}{6!} = \dfrac{1}{720}$

$a_4 = \dfrac{1}{8!} = \dfrac{1}{40,320}$

77. $\dfrac{4!}{6!} = \dfrac{1 \cdot 2 \cdot 3 \cdot 4}{1 \cdot 2 \cdot 3 \cdot 4 \cdot 5 \cdot 6} = \dfrac{1}{5 \cdot 6} = \dfrac{1}{30}$

79. $\dfrac{12!}{4! \cdot 8!} = \dfrac{1 \cdot 2 \cdot 3 \cdot 4 \cdot 5 \cdot 6 \cdot 7 \cdot 8 \cdot 9 \cdot 10 \cdot 11 \cdot 12}{1 \cdot 2 \cdot 3 \cdot 4 \cdot 1 \cdot 2 \cdot 3 \cdot 4 \cdot 5 \cdot 6 \cdot 7 \cdot 8} = \dfrac{9 \cdot \overset{5}{10} \cdot 11 \cdot 12}{1 \cdot 2 \cdot 3 \cdot 4} = 495$

81. $\dfrac{(n+1)!}{n!} = \dfrac{1 \cdot 2 \cdot 3 \cdots n \cdot (n+1)}{1 \cdot 2 \cdot 3 \cdots n} = \dfrac{n+1}{1}$

$\qquad = n + 1$

83. $\dfrac{(2n-1)!}{(2n+1!)} = \dfrac{1 \cdot 2 \cdot 3 \cdots (2n-1)}{1 \cdot 2 \cdot 3 \cdots (2n-1) \cdot (2n) \cdot (2n+1)}$

$\qquad = \dfrac{1}{2n(2n+1)}$

85. $\displaystyle\sum_{i=1}^{5} (2i+1) = (2+1) + (4+1) + (6+1) + (8+1) + (10+1) = 35$

87. $\displaystyle\sum_{k=1}^{4} 10 = 10 + 10 + 10 + 10 = 40$

89. $\displaystyle\sum_{i=0}^{4} i^2 = 0^2 + 1^2 + 2^2 + 3^2 + 4^2 = 30$

91. $\displaystyle\sum_{k=0}^{3} \dfrac{1}{k^2+1} = \dfrac{1}{1} + \dfrac{1}{1+1} + \dfrac{1}{4+1} + \dfrac{1}{9+1} = \dfrac{9}{5}$

93. $\displaystyle\sum_{k=2}^{5} (k+1)^2(k-3) = (3)^2(-1) + (4)^2(0) + (5)^2(1) + (6)^2(2) = 88$

95. $\displaystyle\sum_{i=1}^{4} 2^i = 2^1 + 2^2 + 2^3 + 2^4 = 30$

99. $\displaystyle\sum_{k=0}^{4} \dfrac{(-1)^k}{k+1} = \dfrac{47}{60}$

97. $\displaystyle\sum_{n=0}^{5} \dfrac{1}{2n+1} = \dfrac{6508}{3465}$

101. $\displaystyle\sum_{n=0}^{25} \dfrac{1}{4^n} \approx 1.33$

103. $\dfrac{1}{3(1)} + \dfrac{1}{3(2)} + \dfrac{1}{3(3)} + \cdots + \dfrac{1}{3(9)} = \displaystyle\sum_{i=1}^{9} \dfrac{1}{3i}$

105. $\left[2\left(\dfrac{1}{8}\right) + 3\right] + \left[2\left(\dfrac{2}{8}\right) + 3\right] + \left[2\left(\dfrac{3}{8}\right) + 3\right] + \cdots + \left[2\left(\dfrac{8}{8}\right) + 3\right] = \displaystyle\sum_{i=1}^{8}\left[2\left(\dfrac{i}{8}\right) + 3\right]$

107. $3 - 9 + 27 - 81 + 243 - 729 = \displaystyle\sum_{i=1}^{6} (-1)^{i+1}3i$

109. $\dfrac{1}{1^2} - \dfrac{1}{2^2} + \dfrac{1}{3^2} - \dfrac{1}{4^2} + \cdots - \dfrac{1}{20^2} = \displaystyle\sum_{i=1}^{20} \dfrac{(-1)^{i+1}}{i^2}$

111. $\dfrac{1}{4} + \dfrac{3}{8} + \dfrac{7}{16} + \dfrac{15}{32} + \dfrac{31}{64} = \displaystyle\sum_{i=1}^{5} \dfrac{2^i - 1}{2^{i+1}}$

113. $\displaystyle\sum_{i=1}^{4} 5\left(\dfrac{1}{2}\right)^i = 5\left(\dfrac{1}{2}\right) + 5\left(\dfrac{1}{2}\right)^2 + 5\left(\dfrac{1}{2}\right)^3 + 5\left(\dfrac{1}{2}\right)^4 = \dfrac{75}{16}$

115. $\displaystyle\sum_{n=1}^{3} 4\left(-\dfrac{1}{2}\right)^n = 4\left(-\dfrac{1}{2}\right) + 4\left(-\dfrac{1}{2}\right)^2 + 4\left(-\dfrac{1}{2}\right)^3 = -\dfrac{3}{2}$

117. $\displaystyle\sum_{i=1}^{\infty} 6\left(\dfrac{1}{10}\right)^i = 0.6 + 0.06 + 0.006 + 0.0006 + \cdots = \dfrac{2}{3}$

119. By using a calculator,

$$\displaystyle\sum_{k=1}^{10} 7\left(\dfrac{1}{10}\right)^k \approx 0.7777777777$$

$$\displaystyle\sum_{k=1}^{50} 7\left(\dfrac{1}{10}\right)^k \approx 0.7777777778$$

$$\displaystyle\sum_{k=1}^{100} 7\left(\dfrac{1}{10}\right)^k \approx \dfrac{7}{9}.$$

The terms approach zero as $n \to \infty$.

So, $\displaystyle\sum_{k=1}^{\infty} 7\left(\dfrac{1}{10}\right)^k = \dfrac{7}{9}.$

121. $A_n = 25{,}000\left(1 + \dfrac{0.07}{12}\right)^n$, $n = 1, 2, 3, \ldots$

  (a) $A_1 = \$25{,}145.83$
      $A_2 = \$25{,}292.52$
      $A_3 = \$25{,}440.06$
      $A_4 = \$25{,}586.46$
      $A_5 = \$25{,}737.72$
      $A_6 = \$25{,}887.86$

  (b) $A_{60} = \$35{,}440.63$

  (c) No

      $A_{120} = \$50{,}241.53 \neq 2A_{60}$

123. (a) $y = 76.4n + 380$

  (b) $y = 2.18n^2 + 56.8n + 418$

  (c)

n	2	3	4	5	6	7
Actual data	548	595	668	786	822	923
Linear model	533	609	686	762	838	915
Quadratic model	540	608	680	757	837	922

      The quadratic model is a better fit.

  (d) For the year 2013 $(n = 13)$, we have the following predictions:

      Linear model: 1373 stores

      Quadratic model: 1524 stores

      The quadratic model provides the better prediction.

125. (a) $a_n = 1.0904n^3 - 6.348n^2 + 41.76n + 4871.3$

$a_5 \approx \$5057.7$ $a_{12} \approx \$6342.5$

$a_6 \approx \$5128.9$ $a_{13} \approx \$6737.0$

$a_7 \approx \$5226.6$ $a_{14} \approx \$7203.8$

$a_8 \approx \$5357.4$ $a_{15} \approx \$7749.5$

$a_9 \approx \$5527.9$ $a_{16} \approx \$8380.7$

$a_{10} \approx \$5744.5$ $a_{17} \approx \$9103.8$

$a_{11} \approx \$6013.9$

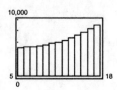

(b) The federal debt is increasing.

127. True, $\displaystyle\sum_{i=1}^{4}\left(i^2 + 2i\right) = \sum_{i=1}^{4} i^2 + 2\sum_{i=1}^{4} i$ by the Properties of Sums.

129. $a_1 = 1, a_2 = 1, a_{k+2} + a_{k+1} + a_k, k \geq 1$

$a_1 = 1$ $b_1 = \frac{1}{1} = 1$

$a_2 = 1$ $b_2 = \frac{2}{1} = 2$

$a_3 = 1 + 1 = 2$ $b_3 = \frac{3}{2}$

$a_4 = 2 + 1 = 3$ $b_4 = \frac{5}{3}$

$a_5 = 3 + 2 = 5$ $b_5 = \frac{8}{5}$

$a_6 = 5 + 3 = 8$ $b_6 = \frac{13}{8}$

$a_7 = 8 + 5 = 13$ $b_7 = \frac{21}{13}$

$a_8 = 13 + 8 = 21$ $b_8 = \frac{34}{21}$

$a_9 = 21 + 13 = 34$ $b_9 = \frac{55}{34}$

$a_{10} = 34 + 21 = 55$ $b_{10} = \frac{89}{55}$

$a_{11} = 55 + 34 = 89$

$a_{12} = 89 + 55 = 144$

131. $\dfrac{327.15 + 785.69 + 433.04 + 265.38 + 604.12 + 590.30}{6} \approx \$500.95$

133. $\displaystyle\sum_{i=1}^{n}\left(x_1 - \bar{x}\right) = \sum_{i=1}^{n} x_i - \sum_{i=1}^{n}\bar{x} = \left(\sum_{i=1}^{n} x_i\right) - n\bar{x} = \left(\sum_{i=1}^{n} x_i\right) - n\left(\frac{1}{n}\sum_{i=1}^{n} x_i\right) = 0$

135. $a_n = \dfrac{x^n}{n!}$

$a_1 = \dfrac{x^1}{1!} = x$

$a_2 = \dfrac{x^2}{2!} = \dfrac{x^2}{2}$

$a_3 = \dfrac{x^3}{3!} = \dfrac{x^3}{6}$

$a_4 = \dfrac{x^4}{4!} = \dfrac{x^4}{24}$

$a_5 = \dfrac{x^5}{5!} = \dfrac{x^5}{120}$

137. $a_n = \dfrac{(-1)^n x^{2n}}{(2n)!}$

$a_1 = \dfrac{-x^2}{2!} = -\dfrac{x^2}{2}$

$a_2 = \dfrac{x^4}{4!} = \dfrac{x^4}{24}$

$a_3 = \dfrac{-x^6}{6!} = -\dfrac{x^6}{720}$

$a_4 = \dfrac{x^8}{8!} = \dfrac{x^8}{40,320}$

$a_5 = \dfrac{-x^{10}}{10!} = -\dfrac{x^{10}}{3,628,800}$

139. $a_n = \dfrac{(-1)^{n+1}}{2n + 1}$

$a_1 = \dfrac{(-1)^{1+1}}{2(1) + 1} = \dfrac{(-1)^2}{2 + 1} = \dfrac{1}{3}$

$a_2 = \dfrac{(-1)^{2+1}}{2(2) + 1} = \dfrac{(-1)^3}{4 + 1} = \dfrac{-1}{5}$

$a_3 = \dfrac{(-1)^{3+1}}{2(3) + 1} = \dfrac{(-1)^4}{6 + 1} = \dfrac{1}{7}$

$a_4 = \dfrac{(-1)^{4+1}}{2(4) + 1} = \dfrac{(-1)^5}{8 + 1} = \dfrac{-1}{9}$

$a_5 = \dfrac{(-1)^{5+1}}{2(5) + 1} = \dfrac{(-1)^6}{10 + 1} = \dfrac{1}{11}$

No. The signs are opposite.

141. (a) and (b)

Number of blue cube faces	0	1	2	3
$3 \times 3 \times 3$	1	6	12	8
$4 \times 4 \times 4$	8	24	24	8
$5 \times 5 \times 5$	27	54	36	8
$6 \times 6 \times 6$	64	96	48	8

(c) The columns change at different rates

(d)

Number of blue cube faces	0	1	2	3
$n \times n \times n$	$(n-2)^3$	$6(n-2)^2$	$12(n-2)$	8

Section 11.2 Arithmetic Sequences and Partial Sums

- You should be able to recognize an arithmetic sequence, find its common difference, and find its nth term.

- You should be able to find the nth partial sum of an arithmetic sequence by using the formula

  $$S_n = \frac{n}{2}(a_1 + a_n).$$

1. arithmetic; common

3. recursion

5. $10, 8, 6, 4, 2, \ldots$

   Arithmetic sequence, $d = -2$

7. $1, 2, 4, 8, 16, \ldots$

   Not an arithmetic sequence

9. $\frac{9}{4}, 2, \frac{7}{4}, \frac{3}{2}, \frac{5}{4}, \ldots$

   Arithmetic sequence, $d = -\frac{1}{4}$

11. $3.7, 4.3, 4.9, 5.5, 6.1, \ldots$

   Arithmetic sequence, $d = 0.6$

13. $\ln 1, \ln 2, \ln 3, \ln 4, \ln 5, \ldots$

   Not an arithmetic sequence

15. $a_n = 5 + 3n$

   $8, 11, 14, 17, 20$

   Arithmetic sequence, $d = 3$

17. $a_n = 3 - 4(n - 2)$

   $7, 3, -1, -5, -9$

   Arithmetic sequence, $d = -4$

19. $a_n = (-1)^n$

   $-1, 1, -1, 1, -1$

   Not an arithmetic sequence

21. $a_n = \dfrac{(-1)^n 3}{n}$

   $-3, \dfrac{3}{2}, -1, \dfrac{3}{4}, -\dfrac{3}{5}$

   Not an arithmetic sequence

23. $a_1 = 1, d = 3$

   $a_n = a_1 + (n-1)d = 1 + (n-1)(3) = 3n - 2$

25. $a_1 = 100, d = -8$

   $a_n = a_1 + (n-1)d = 100 + (n-1)(-8)$

   $\qquad\qquad = -8n + 108$

27. $4, \frac{3}{2}, -1, -\frac{7}{2}, \ldots$

   $d = -\frac{5}{2}$

   $a_n = a_1 + (n-1)d = 4 + (n-1)\left(-\frac{5}{2}\right) = -\frac{5}{2}n + \frac{13}{2}$

29. $a_1 = 5, a_4 = 15$

   $a_4 = a_1 + 3d \Rightarrow 15 = 5 + 3d \Rightarrow d = \frac{10}{3}$

   $a_n = a_1 + (n-1)d = 5 + (n-1)\left(\frac{10}{3}\right) = \frac{10}{3}n + \frac{5}{3}$

31. $a_3 = 94, a_6 = 85$

$a_6 = a_3 + 3d \Rightarrow 85 = 94 + 3d \Rightarrow d = -3$

$a_1 = a_3 - 2d \Rightarrow a_1 = 94 - 2(-3) = 100$

$a_n = a_1 + (n-1)d = 100 + (n-1)(-3)$

$\qquad\qquad = -3n + 103$

33. $a_1 = 5, d = 6$

$a_1 = 5$

$a_2 = 5 + 6 = 11$

$a_3 = 11 + 6 = 17$

$a_4 = 17 + 6 = 23$

$a_5 = 23 + 6 = 29$

35. $a_1 = -2.6, d = -0.4$

$a_1 = -2.6$

$a_2 = -2.6 + (-0.4) = -3.0$

$a_3 = -3.0 + (-0.4) = -3.4$

$a_4 = -3.4 + (-0.4) = -3.8$

$a_5 = -3.8 + (-0.4) = -4.2$

37. $a_1 = 2, a_{12} = 46$

$46 = 2 + (12 - 1)d$

$44 = 11d$

$4 = d$

$a_1 = 2$

$a_2 = 2 + 4 = 6$

$a_3 = 6 + 4 = 10$

$a_4 = 10 + 4 = 14$

$a_5 = 14 + 4 = 18$

39. $a_8 = 26, a_{12} = 42$

$a_{12} = a_8 + 4d$

$42 = 26 + 4d \Rightarrow d = 4$

$a_8 = a_1 + 7d$

$26 = a_1 + 28 \Rightarrow a_1 = -2$

$a_1 = -2$

$a_2 = -2 + 4 = 2$

$a_3 = 2 + 4 = 6$

$a_4 = 6 + 4 = 10$

$a_5 = 10 + 4 = 14$

41. $a_1 = 15, a_{n+1} = a_n + 4$

$a_2 = 15 + 4 = 19$

$a_3 = 19 + 4 = 23$

$a_4 = 23 + 4 = 27$

$a_5 = 27 + 4 = 31$

43. $a_1 = 200, a_{n+1} = a_n - 10$

$a_2 = 200 - 10 = 190$

$a_3 = 190 - 10 = 180$

$a_4 = 180 - 10 = 170$

$a_5 = 170 - 10 = 160$

45. $a_1 = \frac{5}{8}, a_{n+1} = a_n - \frac{1}{8}$

$a_1 = \frac{5}{8}$

$a_2 = \frac{5}{8} - \frac{1}{8} = \frac{1}{2}$

$a_3 = \frac{1}{2} - \frac{1}{8} = \frac{3}{8}$

$a_4 = \frac{3}{8} - \frac{1}{8} = \frac{1}{4}$

$a_5 = \frac{1}{4} - \frac{1}{8} = \frac{1}{8}$

47. $a_1 = 5, a_2 = 11 \Rightarrow d = 11 - 5 = 6$

$a_n = a_1 + (n-1)d \Rightarrow a_{10} = 5 + 9(6) = 59$

49. $a_1 = 4.2, a_2 = 6.6 \Rightarrow d = 6.6 - 4.2 = 2.4$

$a_n = a_1 + (n-1)d \Rightarrow a_7 = 4.2 + 6(2.4) = 18.6$

51. $S_{10} = \frac{10}{2}(2 + 20) = 110$

53. $S_5 = \frac{5}{2}(-1 + (-9)) = -25$

55. $S_{50} = \frac{50}{2}(2 + 100) = 2550$

57. $S_{131} = \frac{131}{2}(-100 + 30) = -4585$

59. $8, 20, 32, 44, \ldots$

$a_1 = 8, d = 12, n = 10$

$a_{10} = 8 + 9(12) = 116$

$S_{10} = \frac{10}{2}(8 + 116) = 620$

61. $4.2, 3.7, 3.2, 2.7, \ldots$

$a_1 = 4.2, d = -0.5, n = 12$

$a_{12} = 4.2 + 11(-0.5) = -1.3$

$S_{12} = \frac{12}{2}[4.2 + (-1.3)] = 17.4$

63. $40, 37, 34, 31, \ldots$

$a_1 = 40, d = -3, n = 10$

$a_{10} = 40 + 9(-3) = 13$

$S_{10} = \frac{10}{2}(40 + 13) = 265$

65. $a_1 = 100, a_{25} = 220, n = 25$

$S_n = \frac{n}{2}[a_1 + a_n]$

$S_{25} = \frac{25}{2}(100 + 220) = 4000$

67. $a_1 = 1, a_{50} = 50, n = 50$

$$\sum_{n=1}^{50} n = \frac{50}{2}(1 + 50) = 1275$$

69. $a_{10} = 60, a_{100} = 600, n = 91$

$$\sum_{n=10}^{100} 6n = \frac{91}{2}(60 + 600) = 30{,}030$$

71. $\displaystyle\sum_{n=11}^{30} n - \sum_{n=1}^{10} n = \frac{20}{2}(11 + 30) - \frac{10}{2}(1 + 10) = 355$

73. $a_1 = 9, a_{500} = 508, n = 500$

$$\sum_{n=1}^{500} (n + 8) = \frac{500}{2}(9 + 508) = 129{,}250$$

75. $a_n = -\frac{3}{4}n + 8$

$d = -\frac{3}{4}$ so the sequence is decreasing and $a_1 = 7\frac{1}{4}$.

Matches (b).

77. $a_n = 2 + \frac{3}{4}n$

$d = \frac{3}{4}$ so the sequence is increasing and $a_1 = 2\frac{3}{4}$.

Matches (c).

79. $a_n = 15 - \frac{3}{2}n$

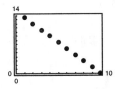

81. $a_n = 0.2n + 3$

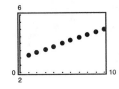

83. $\displaystyle\sum_{n=1}^{20} (2n + 1) = 440$

85. $\displaystyle\sum_{n=1}^{100} \left(\frac{n+1}{2}\right) = 2575$

87. $\displaystyle\sum_{i=1}^{60} \left(250 - \frac{2}{5}i\right) = 14{,}268$

89. (a) $a_1 = 32{,}500, d = 1500$

$a_6 = a_1 + 5d = 32{,}500 + 5(1500) = \$40{,}000$

(b) $S_6 = \frac{6}{2}[32{,}500 + 40{,}000] = \$217{,}500$

91. $a_1 = 20, d = 4, n = 30$

$a_{30} = 20 + 29(4) = 136$

$S_{30} = \frac{30}{2}(20 + 136) = 2340$ seats

93. $a_1 = 14, a_{18} = 31$

$S_{18} = \frac{18}{2}(14 + 31) = 405$ bricks

95. $4.9, 14.7, 24.5, 34.3, \ldots$

$d = 9.8$

$a_{10} = 4.9 + 9(9.8) = 93.1$ meters

$S_{10} = \frac{10}{2}(4.9 + 93.1) = 490$ meters

97. (a) $a_1 = 200, a_2 = 175 \Rightarrow d = -25$

$c = 200 - (-25) = 225$

$a_n = -25n + 225$

(b) $a_8 = -25(8) + 225 = 25$

$S_8 = \frac{8}{2}(200 + 25) = \$900$

99. $a_n = 1500n + 6500$

$a_1 = 8000, a_6 = 15{,}500$

$S_6 = \frac{6}{2}(8000 + 15{,}500) = \$70{,}500$

The cost of gasoline, labor, equipment, insurance, and maintenance are a few economic factors that could prevent the company from meeting its goals, but the biggest unknown variable is the amount of annual snowfall.

101. (a)

Monthly Payment	Unpaid Balance
$a_1 = 200 + 0.01(2000) = \$220$	\$1800
$a_2 = 200 + 0.01(1800) = \$218$	\$1600
$a_3 = 200 + 0.01(1600) = \$216$	\$1400
$a_4 = 200 + 0.01(1400) = \$214$	\$1200
$a_5 = 200 + 0.01(1200) = \$212$	\$1000
$a_6 = 200 + 0.01(1000) = \$210$	\$800

(b) $a_n = -2n + 222 \Rightarrow a_{10} = 202$

$S_{10} = \frac{10}{2}(220 + 202) = \$2110$

Interest paid: \$110

103. (a) Using a graphing utility, $a_n = 1594n + 27{,}087$.

(b)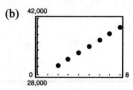

(c) $2009 \to n = 9$

$a_n = 1594(9) + 27{,}087 = \$41{,}433$

(d) Answers will vary.

105. True; given a_1 and a_2 then $d = a_2 - a_1$ and

$a_n = a_1 + (n - 1)d$.

107. $a_1 = x, d = 2x$

$a_n = x + (n - 1)2x$

$a_n = 2xn - x$

$a_1 = 2x(1) - x = x$ $\qquad a_6 = 2x(6) - x = 11x$

$a_2 = 2x(2) - x = 3x$ $\qquad a_7 = 2x(7) - x = 13x$

$a_3 = 2x(3) - x = 5x$ $\qquad a_8 = 2x(8) - x = 15x$

$a_4 = 2x(4) - x = 7x$ $\qquad a_9 = 2x(9) - x = 17x$

$a_5 = 2x(5) - x = 9x$ $\qquad a_{10} = 2x(10) - x = 19x$

109. First term plus $(n - 1)$ times the common difference

111. (a) $a_n = 2 + 3n$

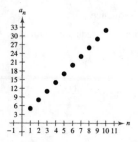

(b) $y = 3x + 2$

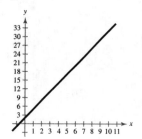

(c) The graph of $a_n = 2 + 3n$ contains only points at the positive integers. The graph of $y = 3x + 2$ is a solid line which contains these points.

(d) The slope $m = 3$ is equal to the common difference $d = 3$. In general, these should be equal.

113. $S_{20} = \frac{20}{2}\{a_1 + [a_1 + (20 - 1)(3)]\} = 650$

$10(2a_1 + 57) = 650$

$2a_1 + 57 = 65$

$2a_1 = 8$

$a_1 = 4$

Section 11.3   Geometric Sequences and Series

- You should be able to identify a geometric sequence, find its common ratio, and find the nth term.

- You should know that the nth term of a geometric sequence with common ratio r is given by $a_n = a_1 r^{n-1}$.

- You should know that the nth partial sum of a geometric sequence with common ratio $r \neq 1$ is given by

$$S_n = a_1\left(\frac{1 - r^n}{1 - r}\right).$$

- You should know that if $|r| < 1$, then

$$\sum_{n=1}^{\infty} a_1 r^{n-1} = \sum_{n=0}^{\infty} a_1 r^n = \frac{a_1}{1 - r}.$$

1. geometric; common

3. $S_n = a_1\left(\frac{1 - r^n}{1 - r}\right)$

5. $2, 10, 50, 250, \ldots$

Geometric sequence, $r = 5$

7. $3, 12, 21, 30, \ldots$

Not a geometric sequence

Note: It is an arithmetic sequence with $d = 9$.

9. $1, -\frac{1}{2}, \frac{1}{4}, -\frac{1}{8}, \ldots$

Geometric sequence, $r = -\frac{1}{2}$

11. $\frac{1}{8}, \frac{1}{4}, \frac{1}{2}, 1, \ldots$

Geometric sequence, $r = 2$

13. $1, \frac{1}{2}, \frac{1}{3}, \frac{1}{4}, \ldots$

Not a geometric sequence

15. $1, -\sqrt{7}, 7, -7\sqrt{7}, \ldots$

Geometric sequence, $r = -\sqrt{7}$

17. $a_1 = 4, r = 3$

$a_1 = 4$

$a_2 = 4(3) = 12$

$a_3 = 12(3) = 36$

$a_4 = 36(3) = 108$

$a_5 = 108(3) = 324$

19. $a_1 = 1, r = \frac{1}{2}$

$a_1 = 1$

$a_2 = 1\left(\frac{1}{2}\right) = \frac{1}{2}$

$a_3 = \frac{1}{2}\left(\frac{1}{2}\right) = \frac{1}{4}$

$a_4 = \frac{1}{4}\left(\frac{1}{2}\right) = \frac{1}{8}$

$a_5 = \frac{1}{8}\left(\frac{1}{2}\right) = \frac{1}{16}$

21. $a_1 = 5, r = -\frac{1}{10}$

$a_1 = 5$

$a_2 = 5\left(-\frac{1}{10}\right) = -\frac{1}{2}$

$a_3 = \left(-\frac{1}{2}\right)\left(-\frac{1}{10}\right) = \frac{1}{20}$

$a_4 = \frac{1}{20}\left(-\frac{1}{10}\right) = -\frac{1}{200}$

$a_5 = \left(-\frac{1}{200}\right)\left(-\frac{1}{10}\right) = \frac{1}{2000}$

23. $a_1 = 1, r = e$

$a_1 = 1$

$a_2 = 1(e) = e$

$a_3 = (e)(e) = e^2$

$a_4 = (e^2)(e) = e^3$

$a_5 = (e^3)(e) = e^4$

25. $a_1 = 3, r = \sqrt{5}$

$a_1 = 3$

$a_2 = 3\left(\sqrt{5}\right)^1 = 3\sqrt{5}$

$a_3 = 3\left(\sqrt{5}\right)^2 = 15$

$a_4 = 3\left(\sqrt{5}\right)^3 = 15\sqrt{5}$

$a_5 = 3\left(\sqrt{5}\right)^4 = 75$

27. $a_1 = 2, r = \frac{x}{4}$

$a_1 = 2$

$a_2 = 2\left(\frac{x}{4}\right) = \frac{x}{2}$

$a_3 = \left(\frac{x}{2}\right)\left(\frac{x}{4}\right) = \frac{x^2}{8}$

$a_4 = \left(\frac{x^2}{8}\right)\left(\frac{x}{4}\right) = \frac{x^3}{32}$

$a_5 = \left(\frac{x^3}{32}\right)\left(\frac{x}{4}\right) = \frac{x^4}{128}$

29. $a_1 = 64, a_{k+1} = \frac{1}{2}a_k$

$a_1 = 64$

$a_2 = \frac{1}{2}(64) = 32$

$a_3 = \frac{1}{2}(32) = 16$

$a_4 = \frac{1}{2}(16) = 8$

$a_5 = \frac{1}{2}(8) = 4$

$r = \frac{1}{2}$

$a_n = 64\left(\frac{1}{2}\right)^{n-1} = 128\left(\frac{1}{2}\right)^n$

31. $a_1 = 9, a_{k+1} = 2a_k$

$a_1 = 9$

$a_2 = 2(9) = 18$

$a_3 = 2(18) = 36$

$a_4 = 2(36) = 72$

$a_5 = 2(72) = 144$

$r = 2$

$a_n = \frac{9}{2}(2)^n$

33. $a_1 = 6, a_{k+1} = -\frac{3}{2}a_k$

$a_1 = 6$

$a_2 = -\frac{3}{2}(6) = -9$

$a_3 = -\frac{3}{2}(-9) = \frac{27}{2}$

$a_4 = -\frac{3}{2}\left(\frac{27}{2}\right) = -\frac{81}{4}$

$a_5 = -\frac{3}{2}\left(-\frac{81}{4}\right) = \frac{243}{8}$

$r = -\frac{3}{2}$

$a_n = 6\left(-\frac{3}{2}\right)^{n-1}$ or $a_n = -4\left(-\frac{3}{2}\right)^n$

35. $a_1 = 4, r = \frac{1}{2}, n = 10$

$a_n = a_1 r^{n-1} = 4\left(\frac{1}{2}\right)^{n-1}$

$a_{10} = 4\left(\frac{1}{2}\right)^9 = \left(\frac{1}{2}\right)^7 = \frac{1}{128}$

37. $a_1 = 6, r = -\frac{1}{3}, n = 12$

$a_n = a_1 r^{n-1} = 6\left(-\frac{1}{3}\right)^{n-1}$

$a_{12} = 6\left(-\frac{1}{3}\right)^{11} = -\frac{2}{3^{10}} = -\frac{2}{59,049}$

39. $a_1 = 100, r = e^x, n = 9$

$a_n = a_1 r^{n-1} = 100\left(e^x\right)^{n-1}$

$a_9 = 100\left(e^x\right)^8 = 100e^{8x}$

41. $a_1 = 1, r = \sqrt{2}, n = 12$

$a_n = 1\left(\sqrt{2}\right)^{n-1} = \left(\sqrt{2}\right)^{n-1}$

$a_{12} = \left(\sqrt{2}\right)^{12-1} = 32\sqrt{2}$

43. $a_1 = 500, r = 1.02, n = 40$

$a_n = a_1 r^{n-1} = 500(1.02)^{n-1}$

$a_{40} = 500(1.02)^{39} \approx 1082.372$

45. $11, 33, 99, \ldots \Rightarrow r = 3$

$a_n = 11(3)^{n-1}$

$a_9 = 11(3)^{9-1} = 72{,}171$

47. $5, 30, 180, \ldots \Rightarrow r = 6$

$a_n = 5(6)^{n-1}$

$a_{10} = 5(6)^9 = 50{,}388{,}480$

49. $\frac{1}{2}, -\frac{1}{8}, \frac{1}{32}, -\frac{1}{128}, \ldots \Rightarrow r = -\frac{1}{4}$

$a_n = \frac{1}{2}\left(-\frac{1}{4}\right)^{n-1}$

$a_8 = \frac{1}{2}\left(-\frac{1}{4}\right)^{8-1} = -\frac{1}{32{,}768}$

51. $a_1 = 16, a_4 = \frac{27}{4}$

$a_4 = a_1 r^3$

$\frac{27}{4} = 16r^3$

$\frac{27}{64} = r^3$

$\frac{3}{4} = r$

$a_n = 16\left(\frac{3}{4}\right)^{n-1}$

$a_3 = 16\left(\frac{3}{4}\right)^2 = 9$

53. $a_4 = -18, a_7 = \frac{2}{3}$

$a_7 = a_4 r^3$

$\frac{2}{3} = -18r^3$

$-\frac{1}{27} = r^3$

$-\frac{1}{3} = r$

$a_6 = \frac{a_7}{r} = \frac{2/3}{-1/3} = -2$

55. $a_2 = 2, a_3 = -\sqrt{2} \Rightarrow r = -\frac{1}{\sqrt{2}} = -\frac{\sqrt{2}}{2}$

$a_5 = -\sqrt{2}\left(-\frac{\sqrt{2}}{2}\right)^{5-3}$

$a_5 = -\frac{\sqrt{2}}{2}$

57. $a_n = 18\left(\frac{2}{3}\right)^{n-1}$

$a_1 = 18$ and $r = \frac{2}{3}$

Because $0 < r < 1$, the sequence is decreasing.

Matches (a).

59. $a_n = 18\left(\frac{3}{2}\right)^{n-1}$

Because $a_1 = 18$ and $r = \frac{3}{2} > 1$, the sequence is increasing.

Matches (b).

61. $a_n = 12(-0.75)^{n-1}$

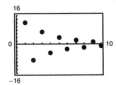

63. $a_n = 12(-0.4)^{n-1}$

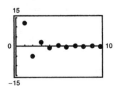

65. $a_n = 2(1.3)^{n-1}$

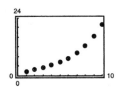

67. $\displaystyle\sum_{n=1}^{7} 4^{n-1} = 1 + 4^1 + 4^2 + 4^3 + 4^4 + 4^5 + 4^6 \Rightarrow a_1 = 1, r = 4$

$$S_7 = \frac{1(1 - 4^7)}{1 - 4} = 5461$$

69. $\displaystyle\sum_{n=1}^{6} (-7)^{n-1} = 1 + (-7) + (-7)^2 + \cdots + (-7)^5 \Rightarrow a_1 = 1, r = -7$

$$S_6 = \frac{1\left(1 - (-7)^6\right)}{1 - (-7)} = -14{,}706$$

71. $\displaystyle\sum_{i=1}^{7} 64\left(-\frac{1}{2}\right)^{i-1} = 64 + 64\left(-\frac{1}{2}\right)^1 + 64\left(-\frac{1}{2}\right)^2 + \cdots + 64\left(-\frac{1}{2}\right)^6 \Rightarrow a_1 = 64, r = -\frac{1}{2}$

$$S_7 = 64\left[\frac{1 - \left(-\frac{1}{2}\right)^7}{1 - \left(-\frac{1}{2}\right)}\right] = \frac{128}{3}\left[1 - \left(-\frac{1}{2}\right)^7\right] = 43$$

73. $\displaystyle\sum_{i=1}^{6} 32\left(\frac{1}{4}\right)^{i-1} = 32 + 32\left(\frac{1}{4}\right)^1 + 32\left(\frac{1}{4}\right)^2 + 32\left(\frac{1}{4}\right)^3 + 32\left(\frac{1}{4}\right)^4 + 32\left(\frac{1}{4}\right)^5 \Rightarrow a_1 = 32, r = \frac{1}{4}, n = 6$

$$S_6 = 32\left[\frac{1 - \left(\frac{1}{4}\right)^6}{1 - \frac{1}{4}}\right] = \frac{1365}{32}$$

75. $\displaystyle\sum_{n=0}^{20} 3\left(\frac{3}{2}\right)^n = \sum_{n=1}^{21} 3\left(\frac{3}{2}\right)^{n-1} = 3 + 3\left(\frac{3}{2}\right)^1 + 3\left(\frac{3}{2}\right)^2 + \cdots + 3\left(\frac{3}{2}\right)^{20} \Rightarrow a_1 = 3, r = \frac{3}{2}$

$$S_{21} = 3\left[\frac{1 - \left(\frac{3}{2}\right)^{21}}{1 - \frac{3}{2}}\right] = -6\left[1 - \left(\frac{3}{2}\right)^{21}\right] \approx 29{,}921.311$$

77. $\displaystyle\sum_{n=0}^{15} 2\left(\frac{4}{3}\right)^n = \sum_{n=1}^{16} 2\left(\frac{4}{3}\right)^{n-1} = 2 + 2\left(\frac{4}{3}\right)^1 + 2\left(\frac{4}{3}\right)^2 + \cdots + 2\left(\frac{4}{3}\right)^{15} \Rightarrow a_1 = 2, r = \frac{4}{3}, n = 16$

$S_{16} = 2\left[\dfrac{1 - \left(\frac{4}{3}\right)^{16}}{1 - \frac{4}{3}}\right] \approx 592.647$

79. $\displaystyle\sum_{n=0}^{5} 300(1.06)^n = \sum_{n=1}^{6} 300(1.06)^{n-1}$

$= 300 + 300(1.06)^1 + 300(1.06)^2 + 300(1.06)^3 + 300(1.06)^4 + 300(1.06)^5 \Rightarrow a_1 = 300, r = 1.06$

$S_6 = 300\left[\dfrac{1 - (1.06)^6}{1 - 1.06}\right] \approx 2092.596$

81. $\displaystyle\sum_{n=0}^{40} 2\left(-\frac{1}{4}\right)^n = 2 + 2\left(-\frac{1}{4}\right) + 2\left(-\frac{1}{4}\right)^2 + \cdots + 2\left(-\frac{1}{4}\right)^{40} \Rightarrow a_1 = 2, r = -\frac{1}{4}, n = 41$

$S_{41} = 2\left[\dfrac{1 - \left(-\frac{1}{4}\right)^{41}}{1 - \left(-\frac{1}{4}\right)}\right] = \frac{8}{5}\left[1 - \left(-\frac{1}{4}\right)^{41}\right] \approx 1.6 = \frac{8}{5}$

83. $\displaystyle\sum_{i=1}^{10} 8\left(-\frac{1}{4}\right)^{i-1} = 8 + 8\left(-\frac{1}{4}\right)^1 + 8\left(-\frac{1}{4}\right)^2 + \cdots + 8\left(-\frac{1}{4}\right)^9 \Rightarrow a_1 = 8, r = -\frac{1}{4}$

$S_{10} = 8\left[\dfrac{1 - \left(-\frac{1}{4}\right)^{10}}{1 - \left(-\frac{1}{4}\right)}\right] = \frac{32}{5}\left[1 - \left(-\frac{1}{4}\right)^{10}\right] \approx 6.400$

85. $\displaystyle\sum_{i=1}^{10} 5\left(-\frac{1}{3}\right)^{i-1} = 5 + 5\left(-\frac{1}{3}\right)^1 + 5\left(-\frac{1}{3}\right)^2 + \cdots + 5\left(-\frac{1}{3}\right)^9 \Rightarrow a_1 = 5, r = -\frac{1}{3}, n = 10$

$S_{10} = 5\left[\dfrac{1 - \left(-\frac{1}{3}\right)^{10}}{1 - \left(-\frac{1}{3}\right)}\right] \approx 3.750$

87. $10 + 30 + 90 + \cdots + 7290$

$r = 3$ and $7290 = 10(3)^{n-1}$

$729 = 3^{n-1}$

$6 = n - 1 \Rightarrow n = 7$

So, the sum can be written as $\displaystyle\sum_{n=1}^{7} 10(3)^{n-1}$.

89. $2 - \frac{1}{2} + \frac{1}{8} - \cdots + \frac{1}{2048}$

$r = -\frac{1}{4}$ and $\frac{1}{2048} = 2\left(-\frac{1}{4}\right)^{n-1}$

By trial and error, we find that $n = 7$.

So, the sum can be written as $\displaystyle\sum_{n=1}^{7} 2\left(-\frac{1}{4}\right)^{n-1}$.

91. $0.1 + 0.4 + 1.6 + \cdots + 102.4$

$r = 4$ and $102.4 = 0.1(4)^{n-1}$

$1024 = 4^{n-1} \Rightarrow 5 = n - 1 \Rightarrow n = 6$

So, the sum can be written as $\displaystyle\sum_{n=1}^{6} 0.1(4)^{n-1}$.

93. $\displaystyle\sum_{n=0}^{\infty} \left(\frac{1}{2}\right)^n = 1 + \left(\frac{1}{2}\right)^1 + \left(\frac{1}{2}\right)^2 + \cdots$

$a_1 = 1, r = \frac{1}{2}$

$\displaystyle\sum_{n=0}^{\infty} \left(\frac{1}{2}\right)^n = \frac{a_1}{1 - r} = \frac{1}{1 - \left(\frac{1}{2}\right)} = 2$

95. $\sum_{n=0}^{\infty} \left(-\frac{1}{2}\right)^n = 1 + \left(-\frac{1}{2}\right)^1 + \left(-\frac{1}{2}\right)^2 + \cdots$

$a_1 = 1, r = -\frac{1}{2}$

$\sum_{n=0}^{\infty} \left(-\frac{1}{2}\right)^n = \frac{a_1}{1-r} = \frac{1}{1-\left(-\frac{1}{2}\right)} = \frac{2}{3}$

97. $\sum_{n=0}^{\infty} 4\left(\frac{1}{4}\right)^n = 4 + 4\left(\frac{1}{4}\right)^1 + 4\left(\frac{1}{4}\right)^2 + \cdots$

$a_1 = 4, r = \frac{1}{4}$

$\sum_{n=0}^{\infty} 4\left(\frac{1}{4}\right)^n = \frac{a_1}{1-r} = \frac{4}{1-\left(\frac{1}{4}\right)} = \frac{16}{3}$

99. $\sum_{n=0}^{\infty} (0.4)^n = 1 + (0.4)^1 + (0.4)^2 + \cdots$

$a_1 = 1, r = 0.4$

$\sum_{n=0}^{\infty} (0.4)^n = \frac{1}{1 - 0.4} = \frac{5}{3}$

101. $\sum_{n=0}^{\infty} -3(0.9)^n = -3 - 3(0.9)^1 - 3(0.9)^2 - \cdots$

$a_1 = -3, r = 0.9$

$\sum_{n=0}^{\infty} -3(0.9)^n = \frac{-3}{1 - 0.9} = -30$

103. $8 + 6 + \frac{9}{2} + \frac{27}{8} + \cdots = \sum_{n=0}^{\infty} 8\left(\frac{3}{4}\right)^n = \frac{8}{1 - \frac{3}{4}} = 32$

105. $\frac{1}{9} - \frac{1}{3} + 1 - 3 + \cdots = \sum_{n=0}^{\infty} \frac{1}{9}(-3)^n$

The sum is undefined because

$|r| = |-3| = 3 > 1.$

107. $0.\overline{36} = \sum_{n=0}^{\infty} 0.36(0.01)^n = \frac{0.36}{1 - 0.01} = \frac{0.36}{0.99} = \frac{36}{99} = \frac{4}{11}$

109. $0.3\overline{18} = 0.3 + \sum_{n=0}^{\infty} 0.018(0.01)^n = \frac{3}{10} + \frac{0.018}{1 - 0.01}$

$= \frac{3}{10} + \frac{0.018}{0.99} = \frac{3}{10} + \frac{18}{990} = \frac{3}{10} + \frac{2}{110}$

$= \frac{35}{110} = \frac{7}{22}$

111. $f(x) = 6\left[\frac{1 - (0.5)^x}{1 - (0.5)}\right], \sum_{n=0}^{\infty} 6\left(\frac{1}{2}\right)^n = \frac{6}{1 - \frac{1}{2}} = 12$

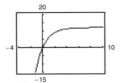

The horizontal asymptote of $f(x)$ is $y = 12$.

This corresponds to the sum of the series.

113. (a) $a_n \approx 1269.10(1.006)^n$

(b) The population of China is growing at a rate of about 0.6% per year

(c) $2015 \to n = 15$

$a_{15} = 1269.10(1.006)^{15} \approx 1388.2$ million.

This value is close to the value predicted by the Census Bureau.

(d) $a_n = 1.35$ billion $\to a_n = 1350$ million

$1350 = 1269.10(1.006)^n$

$\frac{1350}{1269.10} = (1.006)^n$

$\ln\left(\frac{1350}{1269.10}\right) = n \ln(1.006)$

$\frac{\ln\left(\frac{1350}{1269.10}\right)}{\ln 1.006} = n$

$n \approx 10.33$

So, sometime during the year 2010 China's population will reach 1.35 billion people.

115. $A = P\left(1 + \frac{r}{n}\right)^{nt} = 2500\left(1 + \frac{0.02}{n}\right)^{n(20)}$

(a) $n = 1: A = 2500\left(1 + \frac{0.02}{1}\right)^{(1)(20)} \approx \$3714.87$

(b) $n = 2: A = 2500\left(1 + \frac{0.02}{2}\right)^{(2)(20)} \approx \$3722.16$

(c) $n = 4: A = 2500\left(1 + \frac{0.02}{2}\right)^{(4)(20)} \approx \$3725.85$

(d) $n = 12: A = 2500\left(1 + \frac{0.02}{12}\right)^{(12)(20)} \approx \$3728.32$

(e) $n = 365: A = 2500\left(1 + \frac{0.02}{365}\right)^{(365)(20)} \approx \$3729.52$

117. $A = \displaystyle\sum_{n=1}^{60} 100\left(1 + \frac{0.06}{12}\right)^n = \sum_{n=1}^{60} 100(1.005)^n = 100(1.005) \cdot \frac{\left[1 - 1.005^{60}\right]}{\left[1 - 1.005\right]} \approx \$7011.89$

119. Let $N = 12t$ be the total number of deposits.

$$A = P\left(1 + \frac{r}{12}\right) + P\left(1 + \frac{r}{12}\right)^2 + \cdots + P\left(1 + \frac{r}{12}\right)^N$$

$$= \left(1 + \frac{r}{12}\right)\left[P + P\left(1 + \frac{r}{12}\right) + \cdots + P\left(1 + \frac{r}{12}\right)^{N-1}\right]$$

$$= P\left(1 + \frac{r}{12}\right)\sum_{n=1}^{N}\left(1 + \frac{r}{12}\right)^{n-1}$$

$$= P\left(1 + \frac{r}{12}\right)\left[\frac{1 - \left(1 + \frac{r}{12}\right)^N}{1 - \left(1 + \frac{r}{12}\right)}\right]$$

$$= P\left(1 + \frac{r}{12}\right)\left(-\frac{12}{r}\right)\left[1 - \left(1 + \frac{r}{12}\right)^N\right]$$

$$= P\left(\frac{12}{r} + 1\right)\left[-1 + \left(1 + \frac{r}{12}\right)^N\right]$$

$$= P\left[\left(1 + \frac{r}{12}\right)^N - 1\right]\left(1 + \frac{12}{r}\right)$$

$$= P\left[\left(1 + \frac{r}{12}\right)^{12t} - 1\right]\left(1 + \frac{12}{r}\right)$$

121. $P = \$50, r = 5\%, t = 20$ years

(a) Compounded monthly:

$$A = 50\left[\left(1 + \frac{0.05}{12}\right)^{12(20)} - 1\right]\left(1 + \frac{12}{0.05}\right)$$

$$\approx \$20,637.32$$

(b) Compounded continuously:

$$A = \frac{50e^{0.05/12}\left(e^{0.05(20)} - 1\right)}{e^{0.05/12} - 1} \approx \$20,662.37$$

123. $P = \$100, r = 2\%, t = 40$ years

(a) Compounded monthly:

$$A = 100\left[\left(1 + \frac{0.02}{12}\right)^{12(40)} - 1\right]\left(1 + \frac{12}{0.02}\right)$$

$$\approx \$73,565.97$$

(b) Compounded continuously:

$$A = \frac{100e^{0.02/12}\left(e^{0.02(40)} - 1\right)}{e^{0.02/12} - 1} \approx \$73,593.75$$

125. $P = W\displaystyle\sum_{n=1}^{12t}\left[\left(1 + \frac{r}{12}\right)^{-1}\right]^n$

$$= W\left(1 + \frac{r}{12}\right)^{-1}\left[\frac{1 - \left(1 + \frac{r}{12}\right)^{-12t}}{1 - \left(1 + \frac{r}{12}\right)^{-1}}\right]$$

$$= W\left(\frac{1}{1 + \frac{r}{12}}\right)\left[\frac{1 - \left(1 + \frac{r}{12}\right)^{-12t}}{1 - \frac{1}{\left(1 + \frac{r}{12}\right)}}\right]$$

$$= W\frac{\left[1 - \left(1 + \frac{r}{12}\right)^{-12t}\right]}{\left(1 + \frac{r}{12}\right) - 1}$$

$$= W\left(\frac{12}{r}\right)\left[1 - \left(1 + \frac{r}{12}\right)^{-12t}\right]$$

127. $\displaystyle\sum_{n=0}^{\infty} 400(0.75)^n = \frac{400}{1 - 0.75} = \$1600$

129. $\displaystyle\sum_{n=0}^{\infty} 600(0.725)^n = \frac{600}{1 - 0.725} \approx \$2181.82$

131. $64 + 32 + 16 + 8 + 4 + 2 = 126$

Total area of shaded region is approximately 126 square inches.

133. $a_n = 45,000(1.05)^{n-1}$

Total compensation $= T = \displaystyle\sum_{n=1}^{40} 45,000(1.05)^{n-1} = 45,000\frac{\left(1 - 1.05^{40}\right)}{\left(1 - 1.05\right)} \approx \$5,435,989.84$

135. (a) Downward: $850 + 0.75(850) + (0.75)^2(850) + \cdots + (0.75)^9(850) = \displaystyle\sum_{n=0}^{9} 850(0.75)^n$

$$\approx 3208.53 \text{ feet}$$

Upward: $0.75(850) + (0.75)^2(850) + \cdots + (0.75)^{10}(850) = \displaystyle\sum_{n=0}^{9} (0.75)(850)(0.75)^n$

$$= \displaystyle\sum_{n=0}^{9} 637.5(0.75)^n \approx 2406.4 \text{ feet}$$

Total distance: $3208.53 + 2406.4 = 5614.93$ feet

(b) $\displaystyle\sum_{n=0}^{\infty} 850(0.75)^n + \displaystyle\sum_{n=0}^{\infty} 637.5(0.75)^n = \dfrac{850}{1 - 0.75} + \dfrac{637.5}{1 - 0.75} = 5950$ feet

137. False. A sequence is geometric if the ratios of consecutive terms are the same.

139. $y = \left(\dfrac{1 - r^x}{1 - r} \right)$

(a)

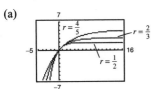

As $x \to \infty$, $y \to \dfrac{1}{1 - r}$.

(b)

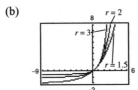

As $x \to \infty$, $y \to \infty$.

141. Given a real number r between -1 and 1, as the exponent n increases, r^n approaches zero.

Section 11.4 Mathematical Induction

- You should be sure that you understand the principle of mathematical induction. If P_n is a statement involving the positive integer n, where P_1 is true and the truth of P_k implies the truth of P_{k+1} for every positive k, then P_n is true for all positive integers n.

- You should be able to verify (by induction) the formulas for the sums of powers of integers and be able to use these formulas.

- You should be able to calculate the first and second differences of a sequence.

- You should be able to find the quadratic model for a sequence, when it exists.

1. mathematical induction

3. arithmetic

5. $P_k = \dfrac{5}{k(k + 1)}$

$P_{k+1} = \dfrac{5}{(k + 1)[(k + 1) + 1]} = \dfrac{5}{(k + 1)(k + 2)}$

7. $P_k = \dfrac{k^2(k + 3)^2}{6}$

$P_{k+1} = \dfrac{(k + 1)^2[(k + 1) + 3]^2}{6} = \dfrac{(k + 1)^2(k + 4)^2}{6}$

9. $P_k = \dfrac{3}{(k + 2)(k + 3)}$

$P_{k+1} = \dfrac{3}{[(k + 1) + 2][(k + 1) + 3]} = \dfrac{3}{(k + 3)(k + 4)}$

11. 1. When $n = 1$, $S_1 = 2 = 1(1 + 1)$.

    2. Assume that

$$S_k = 2 + 4 + 6 + 8 + \cdots + 2k = k(k + 1).$$

    Then,

$$S_{k+1} = 2 + 4 + 6 + 8 + \cdots + 2k + 2(k + 1)$$
$$= S_k + 2(k + 1) = k(k + 1) + 2(k + 1) = (k + 1)(k + 2).$$

So, we conclude that the formula is valid for all positive integer values of n.

13. 1. When $n = 1$, $S_1 = 2 = \dfrac{1}{2}(5(1) - 1)$.

    2. Assume that

$$S_k = 2 + 7 + 12 + 17 + \cdots + (5k - 3) = \frac{k}{2}(5k - 1).$$

    Then,

$$S_{k+1} = 2 + 7 + 12 + 17 + \cdots + (5k - 3) + \left[5(k + 1) - 3 \right]$$

$$= S_k + (5k + 5 - 3) = \frac{k}{2}(5k - 1) + 5k + 2$$

$$= \frac{5k^2 - k + 10k + 4}{2} = \frac{5k^2 + 9k + 4}{2}$$

$$= \frac{(k + 1)(5k + 4)}{2} = \frac{(k + 1)}{2}\left[5(k + 1) - 1 \right].$$

So, we conclude that this formula is valid for all positive integer values of n.

15. 1. When $n = 1$, $S_1 = 1 = 2^1 - 1$.

    2. Assume that

$$S_k = 1 + 2 + 2^2 + 2^3 + \cdots + 2^{k-1} = 2^k - 1.$$

    Then,

$$S_{k+1} = 1 + 2 + 2^2 + 2^3 + \cdots + 2^{k-1} + 2^k = S_k + 2^k = 2^k - 1 + 2^k = 2(2^k) - 1 = 2^{k+1} - 1.$$

So, we conclude that this formula is valid for all positive integer values of n.

17. 1. When $n = 1$, $S_1 = 1 = \dfrac{1(1 + 1)}{2}$.

    2. Assume that

$$S_k = 1 + 2 + 3 + 4 + \cdots + k = \frac{k(k + 1)}{2}.$$

    Then,

$$S_{k+1} = 1 + 2 + 3 + 4 + \cdots + k + (k + 1) = S_k + (k + 1) = \frac{k(k + 1)}{2} + \frac{2(k + 1)}{2} = \frac{(k + 1)(k + 2)}{2}.$$

So, we conclude that this formula is valid for all positive integer values of n.

19. 1. When $n = 1$, $S_1 = 1^2 = \dfrac{1(2(1) - 1)(2(1) + 1)}{3}$

2. Assume that

$$S_k = 1^2 + 3^2 + \cdots + (2k - 1)^2 = \frac{k(2k - 1)(2k + 1)}{3}$$

Then,

$$S_{k+1} = 1^2 + 3^2 + \cdots + (2k - 1)^2 + (2k + 1)^2$$

$$= S_k + (2k + 1)^2 = \frac{k(2k - 1)(2k + 1)}{3} + (2k + 1)^2$$

$$= (2k + 1)\left[\frac{k(2k - 1)}{3} + (2k + 1)\right] = \frac{2k + 1}{3}\left[2k^2 - k + 6k + 3\right]$$

$$= \frac{2k + 1}{3}(2k + 3)(k + 1) = \frac{(k + 1)(2(k + 1) - 1)(2(k + 1) + 1)}{3}$$

So, we conclude that this formula is valid for all positive integer values of n.

21. 1. When $n = 1$, $S_1 = 1 = \dfrac{(1)^2(1 + 1)^2(2(1)^2 + 2(1) - 1)}{12}$.

2. Assume that

$$S_k = \sum_{i=1}^{k} i^5 = \frac{k^2(k + 1)^2(2k^2 + 2k - 1)}{12}.$$

Then,

$$S_{k+1} = \sum_{i=1}^{k+1} i^5 = \left(\sum_{i=1}^{k} i^5\right) + (k + 1)^5$$

$$= \frac{k^2(k + 1)^2(2k^2 + 2k - 1)}{12} + \frac{12(k + 1)^5}{12}$$

$$= \frac{(k + 1)^2\left[k^2(2k^2 + 2k - 1) + 12(k + 1)^3\right]}{12}$$

$$= \frac{(k + 1)^2\left[2k^4 + 2k^3 - k^2 + 12(k^3 + 3k^2 + 3k + 1)\right]}{12}$$

$$= \frac{(k + 1)^2\left[2k^4 + 14k^3 + 35k^2 + 36k + 12\right]}{12}$$

$$= \frac{(k + 1)^2(k^2 + 4k + 4)(2k^2 + 6k + 3)}{12}$$

$$= \frac{(k + 1)^2(k + 2)^2\left[2(k + 1)^2 + 2(k + 1) - 1\right]}{12}.$$

So, we conclude that this formula is valid for all positive integer values of n.

Note: The easiest way to complete the last two steps is to "work backwards." Start with the desired expression for S_{k+1} and multiply out to show that it is equal to the expression you found for $S_k + (k + 1)^5$.

23. 1. When $n = 1$, $S_1 = 2 = \dfrac{1(2)(3)}{3}$.

2. Assume that

$$S_k = 1(2) + 2(3) + 3(4) + \cdots + k(k + 1) = \frac{k(k + 1)(k + 2)}{3}.$$

Then,

$$S_{k+1} = 1(2) + 2(3) + 3(4) + \cdots + k(k + 1) + (k + 1)(k + 2)$$

$$= S_k + (k + 1)(k + 2) = \frac{k(k + 1)(k + 2)}{3} + \frac{3(k + 1)(k + 2)}{3} = \frac{(k + 1)(k + 2)(k + 3)}{3}.$$

So, we conclude that this formula is valid for all positive integer values of n.

25. 1. When $n = 4$, $4! = 24$ and $2^4 = 16$, thus $4! > 2^4$.

2. Assume

$$k! > 2^k, k > 4.$$

Then,

$$(k + 1)! = k!(k + 1) > 2^k(2) \text{ since } k! > 2^k \text{ and } k + 1 > 2.$$

Thus, $(k + 1)! > 2^{k+1}$.

So, by extended mathematical induction, the inequality is valid for all integers n such that $n \geq 4$.

27. 1. When $n = 2$, $\dfrac{1}{\sqrt{1}} + \dfrac{1}{\sqrt{2}} \approx 1.707$ and $\sqrt{2} \approx 1.414$, thus $\dfrac{1}{\sqrt{1}} + \dfrac{1}{\sqrt{2}} > \sqrt{2}$.

2. Assume that

$$\frac{1}{\sqrt{1}} + \frac{1}{\sqrt{2}} + \frac{1}{\sqrt{3}} + \cdots + \frac{1}{\sqrt{k}} > \sqrt{k}, k > 2.$$

Then,

$$\frac{1}{\sqrt{1}} + \frac{1}{\sqrt{2}} + \frac{1}{\sqrt{3}} + \cdots + \frac{1}{\sqrt{k}} + \frac{1}{\sqrt{k + 1}} > \sqrt{k} + \frac{1}{\sqrt{k + 1}}.$$

Now it is sufficient to show that

$$\sqrt{k} + \frac{1}{\sqrt{k + 1}} > \sqrt{k + 1}, k > 2,$$

or equivalently $\left(\text{multiplying by } \sqrt{k + 1}\right)$,

$$\sqrt{k}\sqrt{k + 1} + 1 > k + 1.$$

This is true because

$$\sqrt{k}\sqrt{k + 1} + 1 > \sqrt{k}\sqrt{k} + 1 = k + 1.$$

Therefore,

$$\frac{1}{\sqrt{1}} + \frac{1}{\sqrt{2}} + \frac{1}{\sqrt{3}} + \cdots + \frac{1}{\sqrt{k}} + \frac{1}{\sqrt{k + 1}} > \sqrt{k + 1}.$$

So, by extended mathematical induction, the inequality is valid for all integers n such that $n \geq 2$.

29. $(1 + a)^n \geq na, n \geq 1$ and $a > 0$

Because a is positive, then all of the terms in the binomial expansion are positive.

$$(1 + a)^n = 1 + na + \cdots + na^{n-1} + a^n > na$$

31. 1. When $n = 1$, $(ab)^1 = a^1 b^1 = ab$.

   2. Assume that $(ab)^k = a^k b^k$.

   Then, $(ab)^{k+1} = (ab)^k (ab)$
   $$= a^k b^k ab$$
   $$= a^{k+1} b^{k+1}.$$

   So, $(ab)^n = a^n b^n$.

33. 1. When $n = 2$, $(x_1 x_2)^{-1} = \dfrac{1}{x_1 x_2} = \dfrac{1}{x_1} \cdot \dfrac{1}{x_2} = x_1^{-1} x_2^{-1}$.

   2. Assume that
   $$(x_1 x_2 x_3 \cdots x_k)^{-1} = x_1^{-1} x_2^{-1} x_3^{-1} \cdots x_k^{-1}.$$

   Then,
   $$(x_1 x_2 x_3 \cdots x_k x_{k+1})^{-1} = \left[(x_1 x_2 x_3 \cdots x_k) x_{k+1} \right]^{-1}$$
   $$= (x_1 x_2 x_3 \cdots x_k)^{-1} x_{k+1}^{-1}$$
   $$= x_1^{-1} x_2^{-1} x_3^{-1} \cdots x_k^{-1} x_{k+1}^{-1}.$$

   So, the formula is valid.

35. 1. When $n = 1$, $x(y_1) = xy_1$.

   2. Assume that $x(y_1 + y_2 + \cdots + y_k) = xy_1 + xy_2 + \cdots + xy_k$.

   Then, $xy_1 + xy_2 + \cdots + xy_k + xy_{k+1} = x(y_1 + y_2 + \cdots + y_k) + xy_{k+1}$
   $$= x\left[(y_1 + y_2 + \cdots + y_k) + y_{k+1} \right]$$
   $$= x(y_1 + y_2 + \cdots + y_k + y_{k+1}).$$

   So, the formula holds.

37. 1. When $n = 1$, $\left[1^3 + 3(1)^2 + 2(1) \right] = 6$ and 3 is a factor.

   2. Assume that 3 is a factor of $k^3 + 3k^2 + 2k$.

   Then, $(k + 1)^3 + 3(k + 1)^2 + 2(k + 1) = k^3 + 3k^2 + 3k + 1 + 3k^2 + 6k + 3 + 2k + 2$
   $$= (k^3 + 3k^2 + 2k) + (3k^2 + 9k + 6)$$
   $$= (k^3 + 3k^2 + 2k) + 3(k^2 + 3k + 2).$$

   Because 3 is a factor of each term, 3 is a factor of the sum.

   So, 3 is a factor of $(n^3 + 3n^2 + 2n)$ for every positive integer n.

39. A factor of $n^4 - n + 4$ is 2.

   1. When $n = 1$, $1^4 - 1 + 4 = 4$ and 2 is a factor.

   2. Assume that 2 is a factor of $k^4 - k + 4$.

   Then, $(k + 1)^4 - (k + 1) + 4 = k^4 + 4k^3 + 6k^2 + 4k + 1 - k - 1 + 4$
   $$= (k^4 - k + 4) + (4k^3 + 6k^2 + 4k)$$
   $$= (k^4 - k + 4) + 2(2k^3 + 3k^2 + 2k).$$

   Because 2 is a factor of each term, 2 is a factor of the sum.

   So, 2 is a factor of $n^4 - n + 4$ for every positive integer n.

41. A factor of $2^{4n-2} + 1$ is 5.

   1. When $n = 1$, $2^{4(1)-2} + 1 = 5$ and 5 is a factor.

   2. Assume that 5 is a factor of $2^{4k-2} + 1$.

   Then, $2^{4(k+1)-2} + 1 = 2^{4k+4-2} + 1 = 2^{4k-2} \cdot 2^4 + 1 = 2^{4k-2} \cdot 16 + 1 = \left(2^{4k-2} + 1 \right) + 15 \cdot 2^{4k-2}$.

   Because 5 is a factor of each term, 5 is a factor of the sum.

   So, 5 is a factor of $2^{4n-2} + 1$ for every positive integer n.

43. $S_n = 1 + 5 + 9 + 13 + \cdots + (4n - 3)$

$S_1 = 1 = 1 \cdot 1$

$S_2 = 1 + 5 = 6 = 2 \cdot 3$

$S_3 = 1 + 5 + 9 = 15 = 3 \cdot 5$

$S_4 = 1 + 5 + 9 + 13 = 28 = 4 \cdot 7$

From this sequence, it appears that $S_n = n(2n - 1)$.

This can be verified by mathematical induction. The formula has already been verified for $n = 1$. Assume that the formula is valid for $n = k$.

Then, $S_{k+1} = \left[1 + 5 + 9 + 13 + \cdots + (4k - 3)\right] + \left[4(k + 1) - 3\right]$

$= k(2k - 1) + (4k + 1)$

$= 2k^2 + 3k + 1$

$= (k + 1)(2k + 1)$

$= (k + 1)\left[2(k + 1) - 1\right].$

So, the formula is valid.

45. $S_n = 1 + \dfrac{9}{10} + \dfrac{81}{100} + \dfrac{729}{1000} + \cdots + \left(\dfrac{9}{10}\right)^{n-1}$

Because this series is geometric,

$$S_n = \sum_{i=1}^{n}\left(\frac{9}{10}\right)^{i-1} = \frac{1 - \left(\frac{9}{10}\right)^n}{1 - \frac{9}{10}} = 10\left[1 - \left(\frac{9}{10}\right)^n\right] = 10 - 10\left(\frac{9}{10}\right)^n.$$

47. $S_n = \dfrac{1}{4} + \dfrac{1}{12} + \dfrac{1}{24} + \dfrac{1}{40} + \cdots + \dfrac{1}{2n(n + 1)}$

$S_1 = \dfrac{1}{4} = \dfrac{1}{2(2)}$

$S_2 = \dfrac{1}{4} + \dfrac{1}{12} = \dfrac{4}{12} = \dfrac{2}{6} = \dfrac{2}{2(3)}$

$S_3 = \dfrac{1}{4} + \dfrac{1}{12} + \dfrac{1}{24} = \dfrac{9}{24} = \dfrac{3}{8} = \dfrac{3}{2(4)}$

$S_4 = \dfrac{1}{4} + \dfrac{1}{12} + \dfrac{1}{24} + \dfrac{1}{40} = \dfrac{16}{40} = \dfrac{4}{10} = \dfrac{4}{2(5)}$

From the sequence, it appears that $S_n = \dfrac{n}{2(n + 1)}$.

This can be verified by mathematical induction. The formula has already been verified for $n = 1$. Assume that the formula is valid for $n = k$. Then,

$$S_{k+1} = \left[\frac{1}{4} + \frac{1}{12} + \frac{1}{40} + \cdots + \frac{1}{2k(k + 1)}\right] + \frac{1}{2(k + 1)(k + 2)} = \frac{k}{2(k + 1)} + \frac{1}{2(k + 1)(k + 2)}$$

$$= \frac{k(k + 2) + 1}{2(k + 1)(k + 2)} = \frac{k^2 + 2k + 1}{2(k + 1)(k + 2)} = \frac{(k + 1)^2}{2(k + 1)(k + 2)} = \frac{k + 1}{2(k + 2)}.$$

So, the formula is valid.

49. $\displaystyle\sum_{n=1}^{15} n = \dfrac{15(15 + 1)}{2} = 120$

51. $\displaystyle\sum_{n=1}^{6} n^2 = \dfrac{6(6 + 1)\left[2(6) + 1\right]}{6} = 91$

53. $\displaystyle\sum_{n=1}^{5} n^4 = \frac{5(5+1)\left[2(5)+1\right]\left[3(5)^2+3(5)-1\right]}{30} = 979$

55. $\displaystyle\sum_{n=1}^{6}(n^2-n) = \sum_{n=1}^{6} n^2 - \sum_{n=1}^{6} n$

$$= \frac{6(6+1)\left[2(6)+1\right]}{6} - \frac{6(6+1)}{2}$$

$$= 91 - 21 = 70$$

57. $\displaystyle\sum_{i=1}^{6}(6i-8i^3) = 6\sum_{i=1}^{6} i - 8\sum_{i=1}^{6} i^3 = 6\left[\frac{6(6+1)}{2}\right] - 8\left[\frac{(6)^2(6+1)^2}{4}\right] = 6(21) - 8(441) = -3402$

59. $5, 13, 21, 29, 37, 45, \ldots$

Linear

Note: This is an arithmetic sequence.

$a_1 = 5, d = 8$

$a_n = 5 + (n-1)8$

$a_n = 8n - 3$

61. $6, 15, 30, 51, 78, 111, \ldots$

Quadratic

$\begin{cases} a+b+c = 6 \\ 4a+2b+c = 15 \\ 9a+3b+c = 30 \end{cases}$

Solving this system yields $a = 3$, $b = 0$, and $c = 3$.

So, $a_n = 3n^2 + 3$.

63. $-2, 1, 6, 13, 22, 33, \ldots$

Quadratic

$\begin{cases} a+b+c = -2 \\ 4a+2b+c = 1 \\ 9a+3b+c = 6 \end{cases}$

Solving this system yields $a = 1$, $b = 0$, and $c = -3$.

So, $a_n = n^2 - 3$.

65. $a_1 = 0$, $a_n = a_{n-1} + 3$

$a_1 = a_1 = 0$

$a_2 = a_1 + 3 = 0 + 3 = 3$

$a_3 = a_2 + 3 = 3 + 3 = 6$

$a_4 = a_3 + 3 = 6 + 3 = 9$

$a_5 = a_4 + 3 = 9 + 3 = 12$

$a_6 = a_5 + 3 = 12 + 3 = 15$

a_n: $\quad 0 \quad\;\; 3 \quad\;\; 6 \quad\;\; 9 \quad\;\; 12 \quad\;\; 15$

First differences: $\quad 3 \quad\;\; 3 \quad\;\; 3 \quad\;\; 3 \quad\;\; 3$

Second differences: $\quad\;\; 0 \quad\;\; 0 \quad\;\; 0 \quad\;\; 0$

Because the first differences are equal, the sequence has a linear model.

67. $a_1 = 3$, $a_n = a_{n-1} - n$

$a_1 = a_1 = 3$

$a_2 = a_1 - 2 = 3 - 2 = 1$

$a_3 = a_2 - 3 = 1 - 3 = -2$

$a_4 = a_3 - 4 = -2 - 4 = -6$

$a_5 = a_4 - 5 = -6 - 5 = -11$

$a_6 = a_5 - 6 = -11 - 6 = -17$

a_n: $\quad 3 \quad\;\; 1 \quad\; -2 \quad\; -6 \quad\; -11 \quad\; -17$

First differences: $\; -2 \quad -3 \quad -4 \quad -5 \quad -6$

Second differences: $\quad\;\; -1 \quad -1 \quad -1 \quad -1$

Because the second differences are all the same, the sequence has a quadratic model.

69. $a_0 = 2$, $a_n = \left(a_{n-1}\right)^2$

$a_0 = 2$

$a_1 = a_0^2 = 2^2 = 4$

$a_2 = a_1^2 = 4^2 = 16$

$a_3 = a_2^2 = 16^2 = 256$

$a_4 = a_3^2 = 256^2 = 65,536$

$a_5 = a_4^2 = 65,536^2 = 4,294,967,296$

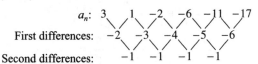

a_n: $\quad 2 \quad\;\; 4 \quad\;\; 16 \quad\;\; 256 \quad\; 65,536 \quad 4,294,967,296$

First differences: $\quad 2 \quad\;\; 12 \quad\;\; 240 \quad 65,280 \quad 4,294,901,760$

Second differences: $\quad\;\; 10 \quad 228 \quad 65,040 \quad 4,294,836,480$

Because neither the first differences nor the second differences are equal, the sequence does not have a linear or quadratic model.

71. $a_1 = 2, a_n = n - a_{n-1}$

$a_2 = 0$

$a_3 = 3$

$a_4 = 1$

$a_5 = 4$

$a_6 = 2$

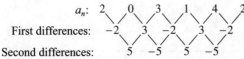

a_n: 2   0   3   1   4   2

First differences: -2 3 -2 3 -2

Second differences: 5 -5 5 -5

Because neither the first differences nor the second differences are equal, the sequence does not have a linear or quadratic model.

73. $a_0 = 3, a_1 = 3, a_4 = 15$

Let $a_n = an^2 + bn + c$.

Then:

$a_0 = a(0)^2 + b(0) + c = 3 \Rightarrow \qquad c = 3$

$a_1 = a(1)^2 + b(1) + c = 3 \Rightarrow a + b + c = 3$

$\qquad\qquad\qquad\qquad\qquad a + b \quad = 0$

$a_4 = a(4)^2 + b(4) + c = 15 \Rightarrow 16a + 4b + c = 15$

$\qquad\qquad\qquad\qquad\qquad 16a + 4b \quad = 12$

$\qquad\qquad\qquad\qquad\qquad 4a + b \quad = 3$

By elimination: $-a - b = 0$

$\qquad\qquad\qquad \underline{4a + b = 3}$

$\qquad\qquad\qquad 3a \quad = 3$

$\qquad\qquad\qquad a = 1 \Rightarrow b = -1$

So, $a_n = n^2 - n + 3$.

75. $a_0 = -3, a_2 = 1, a_4 = 9$

Let $a_n = an^2 + bn + c$.

Then: $a_0 = a(0)^2 + b(0) + c = -3 \Rightarrow c = -3$

$a_2 = a(2)^2 + b(2) + c = 1 \Rightarrow 4a + 2b + c = 1$

$\qquad\qquad\qquad\qquad\qquad 4a + 2b \quad = 4$

$\qquad\qquad\qquad\qquad\qquad 2a + b \quad = 2$

$a_4 = a(4)^2 + b(4) + c = 9 \Rightarrow 16a + 4b + c = 9$

$\qquad\qquad\qquad\qquad\qquad 16a + 4b \quad = 12$

$\qquad\qquad\qquad\qquad\qquad 4a + b \quad = 3$

By elimination: $-2a - b = -2$

$\qquad\qquad\qquad \underline{4a + b = \quad 3}$

$\qquad\qquad\qquad 2a \quad = 1$

$\qquad\qquad\qquad a = \frac{1}{2} \Rightarrow b = 1$

So, $a_n = \frac{1}{2}n^2 + n - 3$.

77. $a_1 = 0, a_2 = 8, a_4 = 30$

Let $a_n = an^2 + bn + c$. Then:

$a_1 = a(1)^2 + b(1) + c = \quad 0 \Rightarrow \quad a + b + c = 0$

$a_2 = a(2)^2 + b(2) + c = \quad 8 \Rightarrow \quad 4a + 2b + c = 8$

$a_4 = a(4)^2 + b(4) + c = 30 \Rightarrow 16a + 4b + c = 30$

$$\begin{cases} a + b + c = 0 \\ 4a + 2b + c = 8 \\ 16a + 4b + c = 30 \end{cases}$$

Solving this system yields $a = 1, b = 5$, and $c = -6$.

So, $a_n = n^2 + 5n - 6$.

79. (a) 8, 11, 7, 8, 6

(b) Although the first differences are not exactly equal, for the case of real life data the differences are close enough to use a linear model.

$a_2 = 643$ average common difference, $d = 8$

Because the model starts at a_2, use the formula

$a_n = a_2 + d(n - 2)$.

$a_n = 643 + 8(n - 2)$

$a_n = 8n + 627$

(c) $a_n \approx 8.1n + 628$

(d) $2013 \rightarrow n = 13$

Part (b): $a_{13} = 8(13) + 627 = 731$

Part (c): $a_{13} = 8.1(13) + 628 = 733.3$

The values are very similar.

81. True. P_7 may be false.

83. True. If the second differences are all zero, then the first differences are all the same, so the sequence is arithmetic.

85. False. A sequence that is arithmetic has second differences equal to zero.

Sample answer: The sequence 2, 5, 8, 11, 14, … is arithmetic with first differences of 3 and second differences equal to zero.

Section 11.5   The Binomial Theorem

- You should be able to use the formula

$$(x + y)^n = x^n + nx^{n-1}y + \frac{n(n-1)}{2!}x^{n-2}y^2 + \cdots + {}_nC_r x^{n-r} y^r + \cdots + y^n$$

  where ${}_nC_r = \dfrac{n!}{(n-r)!r!}$, to expand $(x+y)^n$. Also, ${}_nC_r = \begin{pmatrix} n \\ r \end{pmatrix}$.

- You should be able to use Pascal's Triangle in binomial expansion.

1. binomial coefficients

3. $\begin{pmatrix} n \\ r \end{pmatrix}$ or ${}_nC_r$

5. ${}_5C_3 = \dfrac{5!}{3! \cdot 2!} = \dfrac{5 \cdot 4}{2 \cdot 1} = 10$

7. ${}_{12}C_0 = \dfrac{12!}{0! \cdot 12!} = 1$

9. ${}_{20}C_{15} = \dfrac{20!}{15! \cdot 5!} = \dfrac{20 \cdot 19 \cdot 18 \cdot 17 \cdot 16}{5 \cdot 4 \cdot 3 \cdot 2 \cdot 1} = 15{,}504$

11. $\begin{pmatrix} 10 \\ 4 \end{pmatrix} = \dfrac{10!}{6! \cdot 4!} = \dfrac{10 \cdot 9 \cdot 8 \cdot 7 \cdot 6!}{6!(24)} = 210$

13. $\begin{pmatrix} 100 \\ 98 \end{pmatrix} = \dfrac{100!}{2! \cdot 98!} = \dfrac{100 \cdot 99}{2 \cdot 1} = 4950$

15.
```
            1
          1   1
        1   2   1
      1   3   3   1
    1   4   6   4   1
  1   5  10  10   5   1
1   6  15  20  15 (6)  1
```

$\begin{pmatrix} 6 \\ 5 \end{pmatrix} = 6$, the 6th entry in the 6th row

17.
```
              1
            1   1
          1   2   1
        1   3   3   1
      1   4   6   4   1
    1   5  10  10   5   1
  1   6  15  20  15   6   1
1   7  21  35 (35) 21   7   1
```

${}_7C_4 = 35$, the 5th entry in the 7th row

19. $(x + 1)^4 = {}_4C_0 x^4 + {}_4C_1 x^3(1) + {}_4C_2 x^2(1)^2 + {}_4C_3 x(1)^3 + {}_4C_4(1)^4$

   $= x^4 + 4x^3 + 6x^2 + 4x + 1$

21. $(a + 6)^4 = {}_4C_0 a^4 + {}_4C_1 a^3(6) + {}_4C_2 a^2(6)^2 + {}_4C_3 a(6)^3 + {}_4C_4(6)^4$

   $= 1a^4 + 4a^3(6) + 6a^2(6)^2 + 4a(6)^3 + 1(6)^4$

   $= a^4 + 24a^3 + 216a^2 + 864a + 1296$

23. $(y - 4)^3 = {}_3C_0 y^3 - {}_3C_1 y^2(4) + {}_3C_2 y(4)^2 - {}_3C_3(4)^3$

   $= 1y^3 - 3y^2(4) + 3y(4)^2 - 1(4)^3$

   $= y^3 - 12y^2 + 48y - 64$

25. $(x + y)^5 = {}_5C_0 x^5 + {}_5C_1 x^4 y + {}_5C_2 x^3 y^2 + {}_5C_3 x^2 y^3 + {}_5C_4 xy^4 + {}_5C_5 y^5$

   $= x^5 + 5x^4 y + 10x^3 y^2 + 10x^2 y^3 + 5xy^4 + y^5$

27. $(2x + y)^3 = {}_3C_0(2x)^3 + {}_3C_1(2x)^2(y) + {}_3C_2(2x)(y^2) + {}_3C_3(y^3)$

   $= (1)(8x^3) + (3)(4x^2)(y) + (3)(2x)(y^2) + (1)(y^3)$

   $= 8x^3 + 12x^2 y + 6xy^2 + y^3$

29. $(r + 3s)^6 = {}_6C_0r^6 + {}_6C_1r^5(3s) + {}_6C_2r^4(3s)^2 + {}_6C_3r^3(3s)^3 + {}_6C_4r^2(3s)^4 + {}_6C_5r(3s)^5 + {}_6C_6(3s)^6$

$\qquad = 1r^6 + 6r^5(3s) + 15r^4(3s)^2 + 20r^3(3s)^3 + 15r^2(3s)^4 + 6r(3s)^5 + 1(3s)^6$

$\qquad = r^6 + 18r^5s + 135r^4s^2 + 540r^3s^3 + 1215r^2s^4 + 1458rs^5 + 729s^6$

31. $(3a - 4b)^5 = {}_5C_0(3a)^5 - {}_5C_1(3a)^4(4b) + {}_5C_2(3a)^3(4b)^2 - {}_5C_3(3a)^2(4b)^3 + {}_5C_4(3a)(4b)^4 - {}_5C_5(4b)^5$

$\qquad = (1)(243a^5) - 5(81a^4)(4b) + 10(27a^3)(16b^2) - 10(9a^2)(64b^3) + 5(3a)(256b^4) - (1)(1024b^5)$

$\qquad = 243a^5 - 1620a^4b + 4320a^3b^2 - 5760a^2b^3 + 3840ab^4 - 1024b^5$

33. $(x^2 + y^2)^4 = {}_4C_0(x^2)^4 + {}_4C_1(x^2)^3(y^2) + {}_4C_2(x^2)^2(y^2)^2 + {}_4C_3(x^2)(y^2)^3 + {}_4C_4(y^2)^4$

$\qquad = (1)(x^8) + (4)(x^6y^2) + (6)(x^4y^4) + (4)(x^2y^6) + (1)(y^8)$

$\qquad = x^8 + 4x^6y^2 + 6x^4y^4 + 4x^2y^6 + y^8$

35. $\left(\dfrac{1}{x} + y\right)^5 = {}_5C_0\left(\dfrac{1}{x}\right)^5 + {}_5C_1\left(\dfrac{1}{x}\right)^4 y + {}_5C_2\left(\dfrac{1}{x}\right)^3 y^2 + {}_5C_3\left(\dfrac{1}{x}\right)^2 y^3 + {}_5C_4\left(\dfrac{1}{x}\right)y^4 + {}_5C_5y^5$

$\qquad = \dfrac{1}{x^5} + \dfrac{5y}{x^4} + \dfrac{10y^2}{x^3} + \dfrac{10y^3}{x^2} + \dfrac{5y^4}{x} + y^5$

37. $\left(\dfrac{2}{x} - y\right)^4 = {}_4C_0\left(\dfrac{2}{x}\right)^4 - {}_4C_1\left(\dfrac{2}{x}\right)^3 y + {}_4C_2\left(\dfrac{2}{x}\right)^2 y^2 - {}_4C_3\left(\dfrac{2}{x}\right)y^3 + {}_4C_4y^4$

$\qquad = \dfrac{16}{x^4} - \dfrac{32y}{x^3} + \dfrac{24y^2}{x^2} - \dfrac{8y^3}{x} + y^4$

39. $2(x - 3)^4 + 5(x - 3)^2 = 2\left[x^4 - 4(x^3)(3) + 6(x^2)(3^2) - 4(x)(3^3) + 3^4\right] + 5\left[x^2 - 2(x)(3) + 3^2\right]$

$\qquad = 2(x^4 - 12x^3 + 54x^2 - 108x + 81) + 5(x^2 - 6x + 9)$

$\qquad = 2x^4 - 24x^3 + 113x^2 - 246x + 207$

41. 5th Row of Pascal's Triangle: 1   5   10   10   5   1

$\qquad (2t - s)^5 = 1(2t)^5 - 5(2t)^4(s) + 10(2t)^3(s)^2 - 10(2t)^2(s)^3 + 5(2t)(s)^4 - 1(s)^5$

$\qquad\qquad = 32t^5 - 80t^4s + 80t^3s^2 - 40t^2s^3 + 10ts^4 - s^5$

43. 5th Row of Pascal's Triangle: 1   5   10   10   5   1

$\qquad (x + 2y)^5 = 1x^5 + 5x^4(2y) + 10x^3(2y)^2 + 10x^2(2y)^3 + 5x(2y)^4 + 1(2y)^5$

$\qquad\qquad = x^5 + 10x^4y + 40x^3y^2 + 80x^2y^3 + 80xy^4 + 32y^5$

45. The 4th term in the expansion of $(x + y)^{10}$ is

$\qquad {}_{10}C_3x^{10-3}y^3 = 120x^7y^3.$

47. The 3rd term in the expansion of $(x - 6y)^5$ is

$\qquad {}_5C_2x^{5-2}(-6y)^2 = 10x^3(36y^2) = 360x^3y^2.$

49. The 8th term in the expansion of $(4x + 3y)^9$ is

$\qquad {}_9C_7(4x)^{9-7}(3y)^7 = 36(16x^2)(2187y^7)$

$\qquad\qquad = 1,259,712x^2y^7.$

51. The 10th term in the expansion of $(10x - 3y)^{12}$ is

$\qquad {}_{12}C_9(10x)^{12-9}(-3y)^9 = 220(1000x^3)(-19,683y^9)$

$\qquad\qquad = -4,330,260,000x^3y^9.$

53. The term involving x^5 in the expansion of $(x + 3)^{12}$ is

$\qquad {}_{12}C_7x^5(3)^7 = \dfrac{12!}{7! \cdot 5!} \cdot 3^7x^5 = 1,732,104x^5.$

The coefficient is 1,732,104.

55. The term involving $x^2 y^8$ in the expansion of $(4x - y)^{10}$

is $_{10}C_8(4x)^2(-y)^8 = \dfrac{10!}{(10-8)!8!} \cdot 16x^2 y^8 = 720x^2 y^8$.

The coefficient is 720.

57. The term involving $x^4 y^5$ in the expansion of $(2x - 5x)^9$

is $_9C_5(2x)^4(-5y)^5 = 126(16x^4)(-3125y^5)$

$= -6,300,000x^4 y^5$.

The coefficient is $-6,300,000$.

59. The term involving $x^8 y^6 = \left(x^2\right)^4 y^6$ in the expansion of

$\left(x^2 + y\right)^{10}$ is $_{10}C_6\left(x^2\right)^4 y^6 = \dfrac{10!}{4!6!}\left(x^2\right)^4 y^6 = 210x^8 y^6$.

The coefficient is 210.

61. $\left(\sqrt{x} + 5\right)^3 = \left(\sqrt{x}\right)^3 + 3\left(\sqrt{x}\right)^2(5) + 3\left(\sqrt{x}\right)(5^2) + 5^3$

$= x^{3/2} + 15x + 75x^{1/2} + 125$

63. $\left(x^{2/3} - y^{1/3}\right)^3 = \left(x^{2/3}\right)^3 - 3\left(x^{2/3}\right)^2\left(y^{1/3}\right) + 3\left(x^{2/3}\right)\left(y^{1/3}\right)^2 - \left(y^{1/3}\right)^3 = x^2 - 3x^{4/3}y^{1/3} + 3x^{2/3}y^{2/3} - y$

65. $\left(3\sqrt{t} + \sqrt[4]{t}\right)^4 = \left(3\sqrt{t}\right)^4 + 4\left(3\sqrt{t}\right)^3\left(\sqrt[4]{t}\right) + 6\left(3\sqrt{t}\right)^2\left(\sqrt[4]{t}\right)^2 + 4\left(3\sqrt{t}\right)\left(\sqrt[4]{t}\right)^3 + \left(\sqrt[4]{t}\right)^4$

$= 81t^2 + 108t^{7/4} + 54t^{3/2} + 12t^{5/4} + t$

67. $\dfrac{f(x+h) - f(x)}{h} = \dfrac{(x+h)^3 - x^3}{h}$

$= \dfrac{x^3 + 3x^2 h + 3xh^2 + h^3 - x^3}{h}$

$= \dfrac{h(3x^2 + 3xh + h^2)}{h}$

$= 3x^2 + 3xh + h^2, h \neq 0$

69. $\dfrac{f(x+h) - f(x)}{h} = \dfrac{(x+h)^6 - x^6}{h}$

$= \dfrac{x^6 + 6x^5 h + 15x^4 h^2 + 20x^3 h^3 + 15x^2 h^4 + 6xh^5 + h^6 - x^6}{h}$

$= \dfrac{h(6x^5 + 15x^4 h + 20x^3 h^2 + 15x^2 h^3 + 6xh^4 + h^5)}{h}$

$= 6x^5 + 15x^4 h + 20x^3 h^2 + 15x^2 h^3 + 6xh^4 + h^5, h \neq 0$

71. $\dfrac{f(x+h) - f(x)}{h} = \dfrac{\sqrt{x+h} - \sqrt{x}}{h}$

$= \dfrac{\sqrt{x+h} - \sqrt{x}}{h} \cdot \dfrac{\sqrt{x+h} + \sqrt{x}}{\sqrt{x+h} + \sqrt{x}}$

$= \dfrac{(x+h) - x}{h\left(\sqrt{x+h} + \sqrt{x}\right)}$

$= \dfrac{1}{\sqrt{x+h} + \sqrt{x}}, h \neq 0$

73. $(1 + i)^4 = {}_4C_0(1)^4 + {}_4C_1(1)^3 i + {}_4C_2(1)^2 i^2 + {}_4C_3(1)i^3 + {}_4C_4 i^4$

$= 1 + 4i - 6 - 4i + 1$

$= -4$

75. $(2 - 3i)^6 = {}_6C_0 2^6 - {}_6C_1 2^5(3i) + {}_6C_2 2^4(3i)^2 - {}_6C_3 2^3(3i)^3 + {}_6C_4 2^2(3i)^4 - {}_6C_5 2(3i)^5 + {}_6C_6(3i)^6$

$= (1)(64) - (6)(32)(3i) + 15(16)(-9) - 20(8)(-27i) + 15(4)(81) - 6(2)(243i) + (1)(-729)$

$= 64 - 576i - 2160 + 4320i + 4860 - 2916i - 729$

$= 2035 + 828i$

77. $\left(-\dfrac{1}{2} + \dfrac{\sqrt{3}}{2}i\right)^3 = \dfrac{1}{8}\left[(-1)^3 + 3(-1)^2\left(\sqrt{3}i\right) + 3(-1)\left(\sqrt{3}i\right)^2 + \left(\sqrt{3}i\right)^3\right]$

$= \dfrac{1}{8}\left[-1 + 3\sqrt{3}i + 9 - 3\sqrt{3}i\right]$

$= 1$

79. $(1.02)^8 = (1 + 0.02)^8$

$= 1 + 8(0.02) + 28(0.02)^2 + 56(0.02)^3 + 70(0.02)^4 + 56(0.02)^5 + 28(0.02)^6 + 8(0.02)^7 + (0.02)^8$

$= 1 + 0.16 + 0.0112 + 0.000448 + \cdots$

≈ 1.172

81. $(2.99)^{12} = (3 - 0.01)^{12}$

$= 3^{12} - 12(3)^{11}(0.01) + 66(3)^{10}(0.01)^2 - 220(3)^9(0.01)^3 + 495(3)^8(0.01)^4$

$\quad - 792(3)^7(0.01)^5 + 924(3)^6(0.01)^6 - 792(3)^5(0.01)^7 + 495(3)^4(0.01)^8$

$\quad - 220(3)^3(0.01)^9 + 66(3)^2(0.01)^{10} - 12(3)(0.01)^{11} + (0.01)^{12}$

$\approx 531,441 - 21,257.64 + 389.7234 - 4.3303 + 0.0325 - 0.0002 + \cdots \approx 510,568.785$

83. $f(x) = x^2 - 4x$

$g(x) = f(x + 4)$

$= (x + 4)^3 - 4(x + 4)$

$= x^3 + 3x^2(4) + 3x(4)^2 + (4)^3 - 4x - 16$

$= x^3 + 12x^2 + 48x + 64 - 4x - 16$

$= x^3 + 12x^2 + 44x + 48$

The graph of g is the same as the graph of f shifted four units to the left.

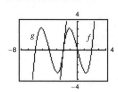

85. ${}_7C_4\left(\dfrac{1}{2}\right)^4\left(\dfrac{1}{2}\right)^3 = \dfrac{7!}{3!4!}\left(\dfrac{1}{16}\right)\left(\dfrac{1}{8}\right) = 35\left(\dfrac{1}{16}\right)\left(\dfrac{1}{8}\right) \approx 0.273$

87. ${}_8C_4\left(\dfrac{1}{3}\right)^4\left(\dfrac{2}{3}\right)^4 = \dfrac{8!}{4!4!}\left(\dfrac{1}{81}\right)\left(\dfrac{16}{81}\right) = 70\left(\dfrac{1}{81}\right)\left(\dfrac{16}{81}\right) \approx 0.171$

89.

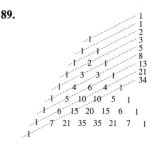

The first nine terms of the sequence are 1, 1, 2, 3, 5, 8, 13, 21, 34, …

After the first two terms, the next terms are formed by adding the previous two terms.

$a_1 = 1, a_2 = 1$

$a_3 = a_1 + a_2 = 1 + 1 = 2$

$a_4 = a_2 + a_3 = 1 + 2 = 3$

$a_5 = a_3 + a_4 = 2 + 3 = 5$

$a_6 = a_4 + a_5 = 3 + 5 = 8$

$a_7 = a_5 + a_6 = 5 + 8 = 13$

This is called the Fibonacci sequence.

91. $f(t) = -4.702t^2 + 110.18t + 1026.7$

   (a) $g(t) = f(t + 5)$

$$= -4.702(t + 5)^2 + 110.18(t + 5) + 1026.7$$

$$= -4.702(t^2 + 10t + 25) + 110.18(t + 5) + 1026.7$$

$$= -4.702t^2 + 63.16t + 1460.05$$

   (b)

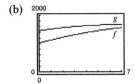

   (c) For 2006, $f(6) = g(1) = \$1518.51$

      For 2007, $f(7) = g(2) = \$1567.56$

      So, the average child support collections will exceed $1525 in the year 2007.

93. True. The coefficients from the Binomial Theorem can be used to find the numbers in Pascal's Triangle.

95. False.

   The coefficient of the x^{10}-term is $_{12}C_7(3)^7 = 1,732,104$.

   The coefficient of the x^{14}-term is $_{12}C_5(3)^5 = 192,456$.

97.

```
                  1
               1     1
             1    2    1
           1    3    3    1
         1    4    6    4    1
       1    5   10   10    5    1
     1    6   15   20   15    6    1
    1   7   21   35   35   21   7   1
  1   8   28   56   70   56   28   8   1
 1   9  36   84  126  126  84   36   9   1
1  10  45  120  210  252  210  120  45  10  1
```

99. The functions $f(x) = (1 - x)^3$ and

   $k(x) = 1 - 3x + 3x^2 + x^3$ (choices (a) and (d)) have

   identical graphs, because $k(x)$ is the expansion of $f(x)$.

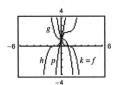

101. $_{n}C_{n-r} = \dfrac{n!}{(n - (n - r))!(n - r)!}$

$$= \dfrac{n!}{r!(n - r)!}$$

$$= \dfrac{n!}{(n - r)!r!}$$

$$= \,_{n}C_{r}$$

103. $_{n}C_{r} + \,_{n}C_{r-1} = \dfrac{n!}{(n - r)!r!} + \dfrac{n!}{(n - r + 1)!(r - 1)!}$

$$= \dfrac{n!(n - r + 1)!(r - 1)! + n!(n - r)!r!}{(n - r)!r!(n - r + 1)!(r - 1)!}$$

$$= \dfrac{n![(n - r + 1)!(r - 1)! + r!(n - r)!]}{(n - r)!r!(n - r + 1)!(r - 1)!}$$

$$= \dfrac{n!(r - 1)![(n - r + 1)! + r(n - r)!]}{(n - r)!r!(n - r + 1)!(r - 1)!}$$

$$= \dfrac{n!(n - r)![(n - r + 1) + r]}{(n - r)!r!(n - r + 1)!}$$

$$= \dfrac{n![n + 1]}{r!(n - r + 1)!}$$

$$= \dfrac{(n + 1)!}{[(n + 1) - r]!r!}$$

$$= \,_{n+1}C_{r}$$

105.

n	r	$_nC_r$	$_nC_{n-r}$
9	5	126	126
7	1	7	7
12	4	495	495
6	0	1	1
10	7	120	120

$$_nC_r = {}_nC_{n-r}$$

The table illustrates the symmetry of Pascal's Triangle.

Section 11.6   Counting Principles

- You should know The Fundamental Counting Principle.

- $_nP_r = \dfrac{n!}{(n-r)!}$ is the number of permutations of n elements taken r at a time.

- Given a set of n objects that has n_1 of one kind, n_2 of a second kind, and so on, the number of distinguishable permutations is

  $$\frac{n!}{n_1!n_2!\ldots n_k!}.$$

- $_nC_r = \dfrac{n!}{(n-r)!r!}$ is the number of combinations of n elements taken r at a time.

---

1. Fundamental Counting Principle

3. $_nP_r = \dfrac{n!}{(n-r)!}$

5. combinations

7. Odd integers: 1, 3, 5, 7, 9, 11

   6 ways

9. Prime integers: 2, 3, 5, 7, 11

   5 ways

11. Divisible by 4: 4, 8, 12

   3 ways

13. Sum is 9: $1 + 8, 2 + 7, 3 + 6, 4 + 5, 5 + 4,$

          $6 + 3, 7 + 2, 8 + 1$

   8 ways

15. Amplifiers: 3 choices

   Compact disc players: 2 choices

   Speakers: 5 choices

   Total: $3 \cdot 2 \cdot 5 = 30$ ways

17. Math courses: 2

   Science courses: 3

   Social sciences and humanities courses: 5

   Total: $2 \cdot 3 \cdot 5 = 30$ schedules

19. $2^6 = 64$

21. $26 \cdot 26 \cdot 26 \cdot 10 \cdot 10 \cdot 10 \cdot 10 = 175,760,000$

   distinct license plate numbers

23. (a) $9 \cdot 10 \cdot 10 = 900$

   (b) $9 \cdot 9 \cdot 8 = 648$

   (c) $9 \cdot 10 \cdot 2 = 180$

   (d) $6 \cdot 10 \cdot 10 = 600$

25. $40^3 = 64,000$

27. (a) $8 \cdot 7 \cdot 6 \cdot 5 \cdot 4 \cdot 3 \cdot 2 \cdot 1 = 40,320$

   (b) $8 \cdot 1 \cdot 6 \cdot 1 \cdot 4 \cdot 1 \cdot 2 \cdot 1 = 384$

29. $_4P_4 = \dfrac{4!}{0!} = 4! = 24.$

31. $_8P_3 = \dfrac{8!}{5!} = 8 \cdot 7 \cdot 6 = 336$

---

33. $_5P_4 = \dfrac{5!}{1!} = 120$

35. $_{20}P_5 = 1,860,480$

37. $_{100}P_3 = 970,200$

39. $5! = 120$ ways

41. $_{12}P_4 = \dfrac{12!}{8!} = 12 \cdot 11 \cdot 10 \cdot 9 = 11,880$ ways

43. $\dfrac{7!}{2!1!3!1!} = \dfrac{7!}{2!3!} = 420$

45. $\dfrac{7!}{2!1!1!1!1!1!} = \dfrac{7!}{2!} = 7 \cdot 6 \cdot 5 \cdot 4 \cdot 3 = 2520$

47.

ABCD	BACD	CABD	DABC
ABDC	BADC	CADB	DACB
ACBD	BCAD	CBAD	DBAC
ACDB	BCDA	CBDA	DBCA
ADBC	BDAC	CDAB	DCAB
ADCB	BDCA	CDBA	DCBA

49. $_{15}P_9 = \dfrac{15!}{6!} = 1,816,214,400$

different batting orders

51. $_5C_2 = \dfrac{5!}{2!3!} = 10$

53. $_4C_1 = \dfrac{4!}{1!3!} = 4$

55. $_{25}C_0 = \dfrac{25!}{0!25!} = 1$

57. $_{20}C_4 = 4845$

59. $_{42}C_5 = 850,668$

61. $_6C_2 = 15$

The 15 ways are listed below.

AB, AC, AD, AE, AF, BC, BD, BE, BF, CD, CE, CF, DE, DF, FF

63. $_{40}C_{12} = \dfrac{40!}{28!12!} = 5,586,853,480$ ways

65. $_{35}C_5 = \dfrac{35!}{30!5!} = 324,632$ ways

67. There are 22 good units and 3 defective units.

(a) $_{22}C_4 = \dfrac{22!}{4!8!} = 7315$ ways

(b) $_{22}C_2 \cdot _3C_2 = \dfrac{22!}{2!20!} \cdot \dfrac{3!}{2!1!} = 231 \cdot 3 = 693$ ways

(c) $_{22}C_4 \cdot _{22}C_3 \cdot _3C_1 + _{22}C_2 \cdot _3C_2 = \dfrac{22!}{4!18!} + \dfrac{22!}{3!19!} \cdot \dfrac{3!}{1!2!} + \dfrac{22!}{2!20!} \cdot \dfrac{3!}{2!1!}$

$= 7315 + 1540 \cdot 3 + 231 \cdot 3$

$= 12,628$ ways

69. (a) Select type of card for three of a kind: $_{13}C_1$

Select three of four cards for three of a kind: $_4C_3$

Select type of card for pair: $_{12}C_1$

Select two of four cards for pair: $_4C_2$

$_{13}C_1 \cdot _4C_3 \cdot _{12}C_1 \cdot _4C_2 = \dfrac{13!}{(13-1)!1!} \cdot \dfrac{4!}{(4-3)!3!} \cdot \dfrac{12!}{(12-1)!1!} \cdot \dfrac{4!}{(4-2)!2!} = 3744$

(b) Select two jacks: $_4C_2$

Select three aces: $_4C_3$

$_4C_2 \cdot _4C_3 = \dfrac{4!}{(4-2)!2!} \cdot \dfrac{4!}{(4-3)!3!} = 24$

71. $_7C_1 \cdot {}_{12}C_3 \cdot {}_{20}C_2 = \dfrac{7!}{(7-1)!1!} \cdot \dfrac{12!}{(12-3)!3!} \cdot \dfrac{20!}{(20-2)!2!} = 292{,}600$

73. $_5C_2 - 5 = 10 - 5 = 5$ diagonals

75. $_8C_2 - 8 = 28 - 8 = 20$ diagonals

77. $_9C_2 = \dfrac{9!}{2!7!} = 36$ lines

79. $14 \cdot {}_nP_3 = {}_{n+2}P_4$ **Note:** $n \geq 3$ for this to be defined.

$$14\left(\frac{n!}{(n-3)!}\right) = \frac{(n+2)!}{(n-2)!}$$

$14n(n-1)(n-2) = (n+2)(n+1)n(n-1)$ (We can divide here by $n(n-1)$ because $n \neq 0, n \neq 1$.)

$$14(n-2) = (n+2)(n+1)$$
$$14n - 28 = n^2 + 3n + 2$$
$$0 = n^2 - 11n + 30$$
$$0 = (n-5)(n-6)$$
$$n = 5 \text{ or } n = 6$$

81. $_nP_4 = 10 \cdot {}_{n-1}P_3$ **Note:** $n \geq 4$ for this to be defined.

$$\frac{n!}{(n-4)!} = 10 \cdot \frac{(n-1)!}{(n-4)!}$$
$$n(n-1)(n-2)(n-3) = 10(n-1)(n-2)(n-3) \quad \begin{pmatrix} \text{We can divide by } (n-1)(n-2)(n-3) \text{ because} \\ n \neq 1, n \neq 2, \text{ and } n \neq 3. \end{pmatrix}$$
$$n = 10$$

83. $_{n+1}P_3 = 4 \cdot {}_nP_2$ **Note:** $n \geq 2$ for this to be defined.

$$\frac{(n+1)!}{(n-2)!} = 4 \cdot \frac{n!}{(n-2)!}$$
$$(n+1)(n)(n-1) = 4(n)(n-1) \quad \left(\text{We can divide by } n(n-1) \text{ because } n \neq 0, \text{ and } n \neq 1.\right)$$
$$n + 1 = 4$$
$$n = 3$$

85. $4 \cdot {}_{n+1}P_2 = {}_{n+2}P_3$ **Note:** $n \geq 1$ for this to be defined.

$$\frac{(n+1)!}{(n-1)!} = \frac{(n+2)!}{(n-1)!}$$
$$4(n+1)(n) = (n+2)(n+1)n \quad \left(\text{We can divide by } (n+1)n \text{ because } n \neq 1 \text{ and } n \neq 0.\right)$$
$$4 = n + 2$$
$$2 = n$$

87. False.

It is an example of a combination.

89. $_nC_r = {}_nC_{n-r}$ They are the same.

91. $_nP_{n-1} = \dfrac{n!}{(n-(n-1))!} = \dfrac{n!}{1!} = \dfrac{n!}{0!} = {}_nP_n$

93. $_nC_{n-1} = \dfrac{n!}{(n-(n-1))!(n-1)!} = \dfrac{n!}{(1)!(n-1)!}$

$\qquad = \dfrac{n!}{(n-1)!1!} = {}_nC_1$

95. $_{100}P_{80} \approx 3.836 \times 10^{139}$

This number is too large for some calculators to evaluate.

97. The symbol $_nP_r$ denotes the number of ways to choose and order r elements out of a collection of n elements.

Section 11.7 Probability

You should know the following basic principles of probability.

- If an event E has $n(E)$ equally likely outcomes and its sample space has $n(S)$ equally likely outcomes, then the probability of event E is

$$P(E) = \frac{n(E)}{n(S)}, \text{ where } 0 \le P(E) \le 1.$$

- If A and B are mutually exclusive events, then $P(A \cup B) = P(A) + P(B)$.

  If A and B are not mutually exclusive events, then $P(A \cup B) = P(A) + P(B) - P(A \cap B)$.

- If A and B are independent events, then the probability that both A and B will occur is $P(A)P(B)$.

- The complement of an event A is denoted by A' and its probability is $P(A') = 1 - P(A)$.

---

1. experiment; outcomes

3. probability

5. mutually exclusive

7. complement

9. $\{(H, 1), (H, 2), (H, 3), (H, 4), (H, 5), (H, 6),$
   $(T, 1), (T, 2), (T, 3), (T, 4), (T, 5), (T, 6)\}$

11. $\{ABC, ACB, BAC, BCA, CAB, CBA\}$

13. $\{AB, AC, AD, AE, BC, BD, BE, CD, CE, DE\}$

21. $E = \{K\clubsuit, K\diamondsuit, K\heartsuit, K\spadesuit, Q\clubsuit, Q\diamondsuit, Q\heartsuit, Q\spadesuit, J\clubsuit, J\diamondsuit, J\heartsuit, J\spadesuit\}$

$$P(E) = \frac{n(E)}{n(S)} = \frac{12}{52} = \frac{3}{13}$$

23. $E = \{K\diamondsuit, K\heartsuit, Q\diamondsuit, Q\heartsuit, J\diamondsuit, J\heartsuit\}$

$$P(E) = \frac{n(E)}{n(S)} = \frac{6}{52} = \frac{3}{26}$$

25. $E = \{(1, 5), (2, 4), (3, 3), (4, 2), (5, 1)\}$

$$P(E) = \frac{n(E)}{n(S)} = \frac{5}{36}$$

29. $E_3 = \{(1, 2), (2, 1)\}, n(E_3) = 2$

$E_5 = \{(1, 4), (2, 3), (3, 2), (4, 1)\}, n(E_5) = 4$

$E_7 = \{(1, 6), (2, 5), (3, 4), (4, 3), (5, 2), (6, 1)\}, n(E_7) = 6$

$E = E_3 \cup E_5 \cup E_7$

$n(E) = 2 + 4 + 6 = 12$

$$P(E) = \frac{n(E)}{n(S)} = \frac{12}{36} = \frac{1}{3}$$

15. $E = \{HHT, HTH, THH\}$

$$P(E) = \frac{n(E)}{n(S)} = \frac{3}{8}$$

17. $E = \{HHH, HHT, HTH, HTT\}$

$$P(E) = \frac{n(E)}{n(S)} = \frac{4}{8} = \frac{1}{2}$$

19. $E = \{HHH, HHT, HTH, HTT, THH, THT, TTH\}$

$$P(E) = \frac{n(E)}{n(S)} = \frac{7}{8}$$

27. Use the complement.

$$E' = \{(5, 6), (6, 5), (6, 6)\}$$

$$P(E') = \frac{n(E')}{n(S)} = \frac{3}{36} = \frac{1}{12}$$

$$P(E) = 1 - P(E') = 1 - \frac{1}{12} = \frac{11}{12}$$

31. $P(E) = \dfrac{_3C_2}{_6C_2} = \dfrac{3}{15} = \dfrac{1}{5}$

33. $P(E) = \dfrac{_4C_2}{_6C_2} = \dfrac{6}{15} = \dfrac{2}{5}$

35. $P(E') = 1 - P(E) = 1 - 0.87 = 0.13$

37. $P(E') = 1 - P(E) = 1 - \dfrac{1}{4} = \dfrac{3}{4}$

39. $P(E) = 1 - P(E') = 1 - 0.23 = 0.77$

41. $P(E) = 1 - P(E') = 1 - \dfrac{17}{35} = \dfrac{18}{35}$

43. (a) $0.14(8.92) \approx 1.25$ million

    (b) $40\% = \dfrac{2}{5}$

    (c) $26\% = \dfrac{13}{50}$

    (d) $26\% + 3\% = 29\% = \dfrac{29}{100}$

45. (a) $0.24(1011) \approx 243$ adults

    (b) $2\% = \dfrac{1}{50}$

    (c) $52\% + 12\% = 64\% = \dfrac{16}{25}$

47. (a) $\dfrac{290}{500} = 0.58 = 58\%$

    (b) $\dfrac{478}{500} = 0.956 = 95.6\%$

    (c) $\dfrac{2}{500} = 0.004 = 0.4\%$

49. (a) $\dfrac{672}{1254} = \dfrac{112}{209}$

    (b) $\dfrac{582}{1254} = \dfrac{97}{209}$

    (c) $\dfrac{672 - 124}{1254} = \dfrac{548}{1254} = \dfrac{274}{627}$

51. $1 - 0.37 - 0.44 = 0.19 = 19\%$

53. (a) $\dfrac{_{15}C_{10}}{_{20}C_{10}} = \dfrac{3003}{184{,}756} = \dfrac{21}{1292} \approx 0.016$

    (b) $\dfrac{_{15}C_8 \cdot _5C_2}{_{20}C_{10}} = \dfrac{64{,}350}{184{,}756} = \dfrac{225}{646} \approx 0.348$

    (c) $\dfrac{_{15}C_9 \cdot _5C_1}{_{20}C_{10}} + \dfrac{_{15}C_{10}}{_{20}C_{10}} + \dfrac{25{,}025 + 3003}{184{,}756} = \dfrac{28{,}028}{184{,}756}$

$$= \dfrac{49}{323}$$

$$\approx 0.152$$

55. (a) $\dfrac{1}{_5P_5} = \dfrac{1}{120}$

    (b) $\dfrac{1}{_4P_4} = \dfrac{1}{24}$

57. (a) $\dfrac{20}{52} = \dfrac{5}{13}$

    (b) $\dfrac{26}{52} = \dfrac{1}{2}$

    (c) $\dfrac{16}{52} = \dfrac{4}{13}$

59. (a) $\dfrac{_9C_4}{_{12}C_4} = \dfrac{126}{495} = \dfrac{14}{55}$ (4 good units)

    (b) $\dfrac{_9C_2 \cdot _3C_2}{_{12}C_4} = \dfrac{108}{495} = \dfrac{12}{55}$ (2 good units)

    (c) $\dfrac{_9C_3 \cdot _3C_1}{_{12}C_4} = \dfrac{252}{495} = \dfrac{28}{55}$ (3 good units)

At least 2 good units: $\dfrac{12}{55} + \dfrac{28}{55} + \dfrac{14}{55} + \dfrac{54}{55}$

61. (a) $P(EE) = \dfrac{20}{40} \cdot \dfrac{20}{40} = \dfrac{1}{4}$

    (b) $P(EO \text{ or } OE) = 2\left(\dfrac{20}{40}\right)\left(\dfrac{20}{40}\right) = \dfrac{1}{2}$

    (c) $P(N_1 < 30, N_2 < 30) = \dfrac{29}{40} \cdot \dfrac{29}{40} = \dfrac{841}{1600}$

    (d) $P(N_1 N_1) = \dfrac{40}{40} \cdot \dfrac{1}{40} = \dfrac{1}{40}$

63. $(0.78)^3 \approx 0.4746$

65. (a) $P(SS) = (0.985)^2 \approx 0.9702$

    (b) $P(S) = 1 - P(FF) = 1 - (0.015)^2 \approx 0.9998$

    (c) $P(FF) = (0.015)^2 \approx 0.0002$

67. (a) $\dfrac{1}{38}$

    (b) $\dfrac{18}{38} = \dfrac{9}{19}$

    (c) $\dfrac{2}{38} + \dfrac{18}{38} = \dfrac{20}{38} = \dfrac{10}{19}$

    (d) $\dfrac{1}{38} \cdot \dfrac{1}{38} = \dfrac{1}{1444}$

    (e) $\dfrac{18}{38} \cdot \dfrac{18}{38} \cdot \dfrac{18}{38} = \dfrac{5832}{54{,}872} = \dfrac{729}{6859}$

69. $1 - \dfrac{(45)^2}{(60)^2} = 1 - \left(\dfrac{45}{60}\right)^2 = 1 - \left(\dfrac{3}{4}\right)^2 = 1 - \dfrac{9}{16} = \dfrac{7}{16}$

71. True. Two events are independent if the occurrence of one has no effect on the occurrence of the other.

73. (a) As you consider successive people with distinct birthdays, the probabilities must decrease to take into account the birth dates already used. Because the birth dates of people are independent events, multiply the respective probabilities of distinct birthdays.

(b) $\dfrac{365}{365} \cdot \dfrac{364}{365} \cdot \dfrac{363}{365} \cdot \dfrac{362}{365}$

(c) $P_1 = \dfrac{365}{365} = 1$

$P_2 = \dfrac{365}{365} \cdot \dfrac{364}{365} = \dfrac{364}{365}P_1 = \dfrac{365 - (2-1)}{365}P_1$

$P_3 = \dfrac{365}{365} \cdot \dfrac{364}{365} \cdot \dfrac{363}{365} = \dfrac{363}{365}P_2 = \dfrac{365 - (3-1)}{365}P_2$

$P_n = \dfrac{365}{365} \cdot \dfrac{364}{365} \cdot \dfrac{363}{365} \cdots \dfrac{365 - (n-1)}{365} = \dfrac{365 - (n-1)}{365}P_{n-1}$

(d) Q_n is the probability that the birthdays are not distinct which is equivalent to at least two people having the same birthday.

(e)

n	10	15	20	23	30	40	50
P_n	0.88	0.75	0.59	0.49	0.29	0.11	0.03
Q_n	0.12	0.25	0.41	0.51	0.71	0.89	0.97

(f) 23; $Q_n > 0.5$ for $n \ge 23$.

75. If a weather forecast indicates that the probability of rain is 40%, this means the meteorological records indicate that over an extended period of time with similar weather conditions it will rain 40% of the time.

Review Exercises for Chapter 11

1. $a_n = 2 + \dfrac{6}{n}$

$a_1 = 2 + \dfrac{6}{1} = 8$

$a_2 = 2 + \dfrac{6}{2} = 5$

$a_3 = 2 + \dfrac{6}{3} = 4$

$a_4 = 2 + \dfrac{6}{4} = \dfrac{7}{2}$

$a_5 = 2 + \dfrac{6}{5} = \dfrac{16}{5}$

3. $a_n = \dfrac{72}{n!}$

$a_1 = \dfrac{72}{1!} = 72$

$a_2 = \dfrac{72}{2!} = 36$

$a_3 = \dfrac{72}{3!} = 12$

$a_4 = \dfrac{72}{4!} = 3$

$a_5 = \dfrac{72}{5!} = \dfrac{3}{5}$

5. $-2, 2, -2, 2, -2, \ldots$

$a_n = 2(-1)^n$

7. $4, 2, \dfrac{4}{3}, 1, \dfrac{4}{5}, \ldots$

$a_n = \dfrac{4}{n}$

9. $9! = 9 \cdot 8 \cdot 7 \cdot 6 \cdot 5 \cdot 4 \cdot 3 \cdot 2 \cdot 1 = 362{,}880$

11. $\dfrac{3!\,5!}{6!} = \dfrac{(3 \cdot 2 \cdot 1)5!}{6 \cdot 5!} = 1$

13. $\displaystyle\sum_{i=1}^{6} 8 = 8 + 8 + 8 + 8 + 8 + 8 = 48$

15. $\displaystyle\sum_{j=1}^{4} \dfrac{6}{j^2} = \dfrac{6}{1^2} + \dfrac{6}{2^2} + \dfrac{6}{3^2} + \dfrac{6}{4^2}$

$\quad = 6 + \dfrac{3}{2} + \dfrac{2}{3} + \dfrac{3}{8}$

$\quad = \dfrac{205}{24}$

17. $\displaystyle\sum_{k=1}^{10} 2k^3 = 2(1)^3 + 2(2)^3 + 2(3)^3 + \cdots + 2(10)^3 = 6050$

19. $\dfrac{1}{2(1)} + \dfrac{1}{2(2)} + \dfrac{1}{2(3)} + \cdots + \dfrac{1}{2(20)} = \displaystyle\sum_{k=1}^{20} \dfrac{1}{2k}$

21. $\displaystyle\sum_{i=1}^{\infty} \dfrac{4}{10^i} = \sum_{i=1}^{\infty} 4\left(\dfrac{1}{10^i}\right) = \dfrac{\dfrac{4}{10}}{1 - \dfrac{1}{10}} = \dfrac{4}{9}$

23. $A_n = 10{,}000\left(1 + \dfrac{0.08}{12}\right)^n$

(a) $A_1 \approx \$10{,}066.67$

$A_2 \approx \$10{,}133.78$

$A_3 \approx \$10{,}201.34$

$A_4 \approx \$10{,}269.35$

$A_5 \approx \$10{,}337.81$

$A_6 \approx \$10{,}406.73$

$A_7 \approx \$10{,}476.10$

$A_8 \approx \$10{,}545.95$

$A_9 \approx \$10{,}616.25$

$A_{10} \approx \$10{,}687.03$

(b) $A_{120} \approx \$22{,}196.40$

25. $6, -1, -8, -15, -22, \ldots$

Arithmetic sequence, $d = -7$

27. $\dfrac{1}{2}, 1, \dfrac{3}{2}, 2, \dfrac{5}{2}, \ldots$

Arithmetic sequence, $d = \dfrac{1}{2}$

29. $a_1 = 3, d = 11$

$a_1 = 3$

$a_2 = 3 + 11 = 14$

$a_3 = 14 + 11 = 25$

$a_4 = 25 + 11 = 36$

$a_5 = 36 + 11 = 47$

31. $a_1 = 25, a_{k+1} = a_k + 3$

$a_1 = 25$

$a_2 = 25 + 3 = 28$

$a_3 = 28 + 3 = 31$

$a_4 = 31 + 3 = 34$

$a_5 = 34 + 3 = 37$

33. $a_1 = 7, d = 12$

$a_n = 7 + (n - 1)12$

$\quad = 7 + 12n - 12$

$\quad = 12n - 5$

35. $a_1 = y, d = 3y$

$a_n = y + (n - 1)3y$

$\quad = y + 3ny - 3y$

$\quad = 3ny - 2y$

37. $a_2 = 93, a_6 = 65$

$a_6 = a_2 + 4d \Rightarrow 65 = 93 + 4d \Rightarrow -28 = 4d \Rightarrow d = -7$

$a_1 = a_2 - d \Rightarrow a_1 = 93 - (-7) = 100$

$a_n = a_1 + (n - 1)d = 100 + (n - 1)(-7) = -7n + 107$

39. $\displaystyle\sum_{k=1}^{100} 7k$ is arithmetic. Therefore, $a_1 = 7, a_{100} = 700, S_{700} = \dfrac{100}{2}(7 + 700) = 35{,}350$.

41. $\displaystyle\sum_{j=1}^{10} (2j - 3)$ is arithmetic. Therefore, $a_1 = -1, a_{10} = 17, S_{10} = \dfrac{10}{2}[-1 + 17] = 80$.

43. $\sum_{k=1}^{11} \left(\frac{2}{3}k + 4 \right)$ is arithmetic. Therefore, $a_1 = \frac{14}{3}$, $a_{11} = \frac{34}{3}$, $S_{11} = \frac{11}{2}\left[\frac{14}{3} + \frac{34}{3} \right] = 88$.

45. $a_n = 43{,}800 + (n - 1)(1950)$

    (a) $a_5 = 43{,}800 + 4(1950) = \$51{,}600$

    (b) $S_5 = \frac{5}{2}(43{,}800 + 51{,}600) = \$238{,}500$

47. $6, 12, 24, 48, \ldots$

    Geometric sequence, $r = 2$

49. $\frac{1}{5}, -\frac{3}{5}, \frac{9}{5}, -\frac{27}{5}, \ldots$

    Geometric sequence, $r = -3$

51. $a_1 = 4$, $r = -\frac{1}{4}$

    $a_1 = 4$

    $a_2 = 4\left(-\frac{1}{4}\right) = -1$

    $a_3 = -1\left(-\frac{1}{4}\right) = \frac{1}{4}$

    $a_4 = \frac{1}{4}\left(-\frac{1}{4}\right) = -\frac{1}{16}$

    $a_5 = -\frac{1}{16}\left(-\frac{1}{4}\right) = \frac{1}{64}$

53. $a_1 = 9$, $a_3 = 4$

    $a_3 = a_1 r^2$

    $4 = 9r^2$

    $\frac{4}{9} = r^2 \Rightarrow r = \pm\frac{2}{3}$

    $a_1 = 9$ $a_1 = 9$

    $a_2 = 9\left(\frac{2}{3}\right) = 6$ $a_2 = 9\left(-\frac{2}{3}\right) = -6$

    $a_3 = 6\left(\frac{2}{3}\right) = 4$ or $a_3 = -6\left(-\frac{2}{3}\right) = 4$

    $a_4 = 4\left(\frac{2}{3}\right) = \frac{8}{3}$ $a_4 = 4\left(-\frac{2}{3}\right) = -\frac{8}{3}$

    $a_5 = \frac{8}{3}\left(\frac{2}{3}\right) = \frac{16}{9}$ $a_5 = -\frac{8}{3}\left(-\frac{2}{3}\right) = \frac{16}{9}$

55. $a_1 = 18$, $a_2 = -9$

    $a_2 = a_1 r$

    $-9 = 18r$

    $-\frac{1}{2} = r$

    $a_n = 18\left(-\frac{1}{2}\right)^{n-1}$

    $a_{10} = 18\left(-\frac{1}{2}\right)^9 = \frac{9}{256}$

57. $a_1 = 100$, $r = 1.05$

    $a_n = 100(1.05)^{n-1}$

    $a_{10} = 100(1.05)^9 \approx 155.133$

59. $\sum_{i=1}^{7} 2^{i-1} = \frac{1 - 2^7}{1 - 2} = 127$

61. $\sum_{i=1}^{4} \left(\frac{1}{2}\right)^i = \frac{1}{2} + \frac{1}{4} + \frac{1}{8} + \frac{1}{16} = \frac{15}{16}$

63. $\sum_{i=1}^{5} (2)^{i-1} = 1 + 2 + 4 + 8 + 16 = 31$

65. $\sum_{i=1}^{10} 10\left(\frac{3}{5}\right)^{i-1} \approx 24.85$

67. $\sum_{i=1}^{\infty} \left(\frac{7}{8}\right)^{i-1} = \frac{1}{1 - \frac{7}{8}} = 8$

69. $\sum_{k=1}^{\infty} 4\left(\frac{2}{3}\right)^{k-1} = \frac{4}{1 - \frac{2}{3}} = 12$

71. (a) $a_n = 120{,}000(0.7)^n$

    (b) $a_5 = 120{,}000(0.7)^5$

            $= \$20{,}168.40$

73. 1. When $n = 1$, $3 = 1(1 + 2)$.

    2. Assume that $S_k = 3 + 5 + 7 + \cdots + (2k + 1) = k(k + 2)$.

        Then, $S_{k+1} = 3 + 5 + 7 + \cdots + (2k + 1) + \left[2(k + 1) + 1 \right] = S_k + (2k + 3)$

                 $= k(k + 2) + 2k + 3$

                 $= k^2 + 4k + 3$

                 $= (k + 1)(k + 3)$

                 $= (k + 1)\left[(k + 1) + 2 \right]$.

    So, by mathematical induction, the formula is valid for all positive integer values of n.

75. 1. When $n = 1$, $a = a\left(\dfrac{1-r}{1-r}\right)$.

    2. Assume that $S_k = \displaystyle\sum_{i=0}^{k-1} ar^i = \dfrac{a(1-r^k)}{1-r}$.

      Then $S_{k+1} = \displaystyle\sum_{i=0}^{k} ar^i = \left(\sum_{i=0}^{k-1} ar^i\right) + ar^k = \dfrac{a(1-r^k)}{1-r} + ar^k$

      $= \dfrac{a(1-r^k + r^k - r^{k+1})}{1-r} = \dfrac{a(1-r^{k+1})}{1-r}$.

      So, by mathematical induction, the formula is valid for all positive integer values of n.

77. $S_1 = 9 = 1(9) = 1\big[2(1) + 7\big]$

    $S_2 = 9 + 13 = 22 = 2(11) = 2\big[2(2) + 7\big]$

    $S_3 = 9 + 13 + 17 = 39 = 3(13) = 3\big[2(3) + 7\big]$

    $S_4 = 9 + 13 + 17 + 21 = 60 = 4(15) = 4\big[2(4) + 7\big]$

    $S_n = n(2n + 7)$

79. $S_1 = 1$

    $S_2 = 1 + \dfrac{3}{5} = \dfrac{8}{5}$

    $S_3 = 1 + \dfrac{3}{5} = \dfrac{9}{25} = \dfrac{49}{25}$

    $S_4 = 1 + \dfrac{3}{5} + \dfrac{9}{25} + \dfrac{27}{125} = \dfrac{272}{125}$

    Because the series is geometric,

$$S_n = \dfrac{1 - \left(\dfrac{3}{5}\right)^n}{1 - \dfrac{3}{5}} = \dfrac{5}{2}\left[1 - \left(\dfrac{3}{5}\right)^n\right].$$

81. $\displaystyle\sum_{n=1}^{50} n = \dfrac{50(51)}{2} = 1275$

83. $a_1 = f(1) = 5$, $a_n = a_{n-1} + 5$

    $a_1 = 5$

    $a_2 = 5 + 5 = 10$

    $a_3 = 10 + 5 = 15$

    $a_4 = 15 + 5 = 20$

    $a_5 = 20 + 5 = 25$

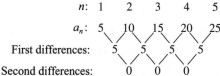

    n: 1   2   3   4   5

    a_n: 5 10 15 20 25

First differences: 5 5 5 5

Second differences: 0 0 0

Because the first differences are all the same, the sequence has a linear model.

85. $a_1 = f(1) = 16$, $a_n = a_{n-1} - 1$

    $a_1 = 16$

    $a_2 = 16 - 1 = 15$

    $a_3 = 15 - 1 = 14$

    $a_4 = 14 - 1 = 13$

    $a_5 = 13 - 1 = 12$

    n: 1   2   3   4   5

    a_n: 16 15 14 13 12

First differences: -1 -1 -1 -1

Second differences: 0 0 0

Because the first differences are all the same, the sequence has a linear model.

87. $_6C_4 = \dfrac{6!}{2!4!} = 15$

89. $\dbinom{8}{6} = 28$

$$
\begin{array}{c}
1 \\
1 \quad 1 \\
1 \quad 2 \quad 1 \\
1 \quad 3 \quad 3 \quad 1 \\
1 \quad 4 \quad 6 \quad 4 \quad 1 \\
1 \quad 5 \quad 10 \quad 10 \quad 5 \quad 1 \\
1 \quad 6 \quad 15 \quad 20 \quad 15 \quad 6 \quad 1 \\
1 \quad 7 \quad 21 \quad 35 \quad 35 \quad 21 \quad 7 \quad 1 \\
1 \quad 8 \quad 28 \quad 56 \quad 70 \quad 56 \quad \boxed{28} \quad 8 \quad 1
\end{array}
$$

$\dbinom{8}{6} = 28$, the 7th entry in the 8th row

91. $(x + 4)^4 = x^4 + 4x^3(4) + 6x^2(4)^2 + 4x(4)^3 + 4^4 = x^4 + 16x^3 + 96x^2 + 256x + 256$

93. $(a - 3b)^5 = a^5 - 5a^4(3b) + 10a^3(3b)^2 - 10a^2(3b)^3 + 5a(3b)^4 - (3b)^5$

$\qquad = a^5 - 15a^4b + 90a^3b^2 - 270a^2b^3 + 405ab^4 - 243b^5$

95. $(5 + 2i)^4 = (5)^4 + 4(5)^3(2i) + 6(5)^2(2i)^2 + 4(5)(2i)^3 + (2i)^4$

$\qquad = 625 + 1000i + 600i^2 + 160i^3 + 16i^4$

$\qquad = 625 + 1000i - 600 - 160i + 16 = 41 + 840i$

97.   First number: 1   2   3   4   5   6   7   8   9   10   11
   Second number: 11   10   9   8   7   6   5   4   3   2   1

From this list, you can see that a total of 12 occurs 11 different ways.

99. $(10)(10)(10)(10) = 10{,}000$ different telephone numbers

101. $_{10}P_3 = \dfrac{10!}{7!} = \dfrac{10 \cdot 9 \cdot 8 \cdot 7!}{7!}$

$\qquad = 10 \cdot 9 \cdot 8 = 720$ different ways

103. $_8C_3 = \dfrac{8!}{5!3!} = 56$

105. $(1)\left(\dfrac{1}{9}\right) = \dfrac{1}{9}$

107. (a) $25\% + 18\% = 43\%$

(b) $100\% - 18\% = 82\%$

109. $\left(\frac{1}{6}\right)\left(\frac{1}{6}\right)\left(\frac{1}{6}\right)\left(\frac{1}{6}\right) = \frac{1}{1296}$

111. $1 - \frac{13}{52} = 1 - \frac{1}{4} = \frac{3}{4}$

113. True. $\dfrac{(n + 2)!}{n!} = \dfrac{(n + 2)(n + 1)n!}{n!} = (n + 2)(n + 1)$

115. True. $\displaystyle\sum_{k=1}^{8} 3k = 3\sum_{k=1}^{8} k$ by the Properties of Sums.

117. The domain of an infinite sequence is the set of natural numbers.

119. Each term of the sequence is defined in terms of preceding terms.

Problem Solving for Chapter 11

1. $x_0 = 1$ and $x_n = \dfrac{1}{2}x_{n-1} + \dfrac{1}{x_{n-1}}, n = 1, 2, \ldots$

$x_0 = 1$

$x_1 = \dfrac{1}{2}(1) + \dfrac{1}{1} = \dfrac{3}{2} = 1.5$

$x_2 = \dfrac{1}{2}\left(\dfrac{3}{2}\right) + \dfrac{1}{3/2} = \dfrac{17}{12} = 1.41\overline{6}$

$x_3 = \dfrac{1}{2}\left(\dfrac{17}{12}\right) + \dfrac{1}{17/12} = \dfrac{577}{408} \approx 1.414215686$

$x_4 = \dfrac{1}{2}\left(\dfrac{577}{408}\right) + \dfrac{1}{577/408} \approx 1.414213562$

$x_5 = \dfrac{1}{2}x_4 + \dfrac{1}{x_4} \approx 1.414213562$

$x_6 \approx x_7 \approx x_8 \approx x_9 \approx 1.414213562$

Conjecture: $x_n \to \sqrt{2}$ as $n \to \infty$

3. $a_n = 3 + (-1)^n$

(a)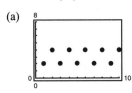

(b) $a_n = \begin{cases} 2, & \text{if } n \text{ is odd} \\ 4, & \text{if } n \text{ is even} \end{cases}$

(c)

n	1	10	101	1000	10,001
a_n	2	4	2	4	2

(d) As $n \to \infty$, a_n oscillates between 2 and 4 and does not approach a fixed value.

5. (a)

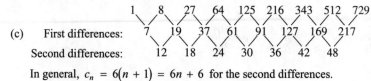

First differences: 3 5 7 9 11 13 15 17

In general, $b_n = 2n + 1$ for the first differences.

(b) Find the second differences of the perfect cubes.

1   8   27   64   125   216   343   512   729

(c)   First differences: 7 19 37 61 91 127 169 217

Second differences: 12 18 24 30 36 42 48

In general, $c_n = 6(n + 1) = 6n + 6$ for the second differences.

(d) Find the third differences of the perfect fourth powers.

(e)

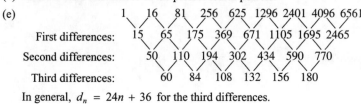

1   16   81   256   625   1296   2401   4096   6561

First differences: 15 65 175 369 671 1105 1695 2465

Second differences: 50 110 194 302 434 590 770

Third differences: 60 84 108 132 156 180

In general, $d_n = 24n + 36$ for the third differences.

7. Side lengths: $1, \frac{1}{2}, \frac{1}{4}, \frac{1}{8}, \dots$

$$S_n = \left(\frac{1}{2}\right)^{n-1} \text{ for } n \geq 1$$

Areas: $\dfrac{\sqrt{3}}{4}, \dfrac{\sqrt{3}}{4}\left(\dfrac{1}{2}\right)^2, \dfrac{\sqrt{3}}{4}\left(\dfrac{1}{4}\right)^2, \dfrac{\sqrt{3}}{4}\left(\dfrac{1}{8}\right)^2, \dots$

$$A_n = \frac{\sqrt{3}}{4}\left[\left(\frac{1}{2}\right)^{n-1}\right]^2 = \frac{\sqrt{3}}{4}\left(\frac{1}{2}\right)^{2n-2} = \frac{\sqrt{3}}{4}S_n^2$$

9. The numbers $1, 5, 12, 22, 35, 51, \dots$ can be written recursively as $P_n = P_{n-1} + (3n - 2)$. Show that $P_n = n(3n - 1)/2$.

1. For $n = 1$: $1 = \dfrac{1(3 - 1)}{2}$

2. Assume $P_k = \dfrac{k(3k - 1)}{2}$.

Then, $P_{k+1} = P_k + \left[3(k + 1) - 2\right]$

$\qquad = \dfrac{k(3k - 1)}{2} + (3k + 1) = \dfrac{k(3k - 1) + 2(3k + 1)}{2}$

$\qquad = \dfrac{3k^2 + 5k + 2}{2} = \dfrac{(k + 1)(3k + 2)}{2}$

$\qquad = \dfrac{(k + 1)\left[3(k + 1) - 1\right]}{2}$.

So, by mathematical induction, the formula is valid for all integers $n \geq 1$.

11. (a) The Fibonacci sequence is defined as follows: $f_1 = 1, f_2 = 1, f_n = f_{n-2} + f_{n-1}$ for $n \geq 3$.

By this definition $f_3 = f_1 + f_2 = 2, f_4 = f_2 + f_3 = 3, f_5 = f_4 + f_3 = 5, f_6 = f_5 + f_4 = 8, \dots$

1. For $n = 2$: $f_1 + f_2 = 2$ and $f_4 - 1 = 2$

2. Assume $f_1 + f_2 + \cdots + f_k = f_{k+2} - 1$.

Then, $f_1 + f_2 + f_3 + \cdots + f_k + f_{k+1} = f_{k+2} - 1 + f_{k+1} = (f_{k+2} + f_{k+1}) - 1 = f_{k+3} - 1 = f_{(k+1)+2} - 1$.

So, by mathematical induction, the formula is valid for all integers $n \geq 2$.

(b) $S_{20} = f_{22} - 1 = 17{,}711 - 1 = 17{,}710$

13. $\frac{1}{3}$

15. (a) $V = \left(\dfrac{1}{_{47}C_5(27)}\right)(12{,}000{,}000) + \left(1 - \dfrac{1}{_{47}C_5(27)}\right)(-1)$

$\approx -\$0.71$

(b) $V = \dfrac{1}{36}(1) + \dfrac{1}{36}(4) + \dfrac{1}{36}(9) + \dfrac{1}{36}(16) + \dfrac{1}{36}(25) + \dfrac{1}{36}(36) + \dfrac{30}{36}(0) \approx 2.53$

$\dfrac{60}{2.53} \approx 24$ turns

Chapter 11   Practice Test

1. Write out the first five terms of the sequence $a_n = \dfrac{2n}{(n+2)!}$.

2. Write an expression for the nth term of the sequence $\dfrac{4}{3}, \dfrac{5}{9}, \dfrac{6}{27}, \dfrac{7}{81}, \dfrac{8}{243}, \dots$.

3. Find the sum $\displaystyle\sum_{i=1}^{6} (2i - 1)$.

4. Write out the first five terms of the arithmetic sequence where $a_1 = 23$ and $d = -2$.

5. Find a_n for the arithmetic sequence with $a_1 = 12$, $d = 3$, and $n = 50$.

6. Find the sum of the first 200 positive integers.

7. Write out the first five terms of the geometric sequence with $a_1 = 7$ and $r = 2$.

8. Evaluate $\displaystyle\sum_{n=1}^{10} 6\left(\dfrac{2}{3}\right)^{n-1}$.

9. Evaluate $\displaystyle\sum_{n=0}^{\infty} (0.03)^n$.

10. Use mathematical induction to prove that $1 + 2 + 3 + 4 + \cdots + n = \dfrac{n(n+1)}{2}$.

11. Use mathematical induction to prove that $n! > 2^n$, $n \geq 4$.

12. Evaluate $_{13}C_4$.

13. Expand $(x + 3)^5$.

14. Find the term involving x^7 in $(x - 2)^{12}$.

15. Evaluate $_{30}P_4$.

16. How many ways can six people sit at a table with six chairs?

17. Twelve cars run in a race. How many different ways can they come in first, second, and third place? (Assume that there are no ties.)

18. Two six-sided dice are tossed. Find the probability that the total of the two dice is less than 5.

19. Two cards are selected at random from a deck of 52 playing cards without replacement. Find the probability that the first card is a King and the second card is a black ten.

20. A manufacturer has determined that for every 1000 units it produces, 3 will be faulty. What is the probability that an order of 50 units will have one or more faulty units?

Appendix A   Errors and the Algebra of Calculus

- You should be able to recognize and avoid the common algebraic errors involving parentheses, fractions, exponents, radicals, and cancellation.

- You should be able to "unsimplify" algebraic expressions by the following methods.
  - (a) Unusual Factoring
  - (b) Rewriting with Negative Exponents
  - (c) Writing a Fraction as a Sum of Terms
  - (d) Inserting Factors or Terms

---

1. numerator

3. $2x - (3y + 4) \neq 2x - 3y + 4$

Change all signs when distributing the minus sign.

$2x - (3y + 4) = 2x - 3y - 4$

5. $\dfrac{4}{16x - (2x + 1)} \neq \dfrac{4}{14x + 1}$

Change all signs when distributing the minus sign.

$\dfrac{4}{16x - (2x + 1)} = \dfrac{4}{16x - 2x - 1} = \dfrac{4}{14x - 1}$

7. $(5z)(6z) \neq 30z$

z occurs twice as a factor.

$(5z)(6z) = 30z^2$

9. $a\left(\dfrac{x}{y}\right) \neq \dfrac{ax}{ay}$

The fraction as a whole is multiplied by a, not the numerator and denominator separately.

$a\left(\dfrac{x}{y}\right) = \dfrac{a}{1} \cdot \dfrac{x}{y} = \dfrac{ax}{y}$

11. $\sqrt{x + 9} \neq \sqrt{x} + 3$

Do not apply the radical to the terms.

$\sqrt{x + 9}$ does not simplify.

13. $\dfrac{2x^2 + 1}{5x} \neq \dfrac{2x + 1}{5}$

Divide out common factors not common terms.

$\dfrac{2x^2 + 1}{5x}$ cannot be simplified.

15. $\dfrac{1}{a^{-1} + b^{-1}} \neq \left(\dfrac{1}{a + b}\right)^{-1}$

To get rid of negative exponents:

$\dfrac{1}{a^{-1} + b^{-1}} = \dfrac{1}{a^{-1} + b^{-1}} \cdot \dfrac{ab}{ab} = \dfrac{ab}{b + a}$

17. $(x^2 + 5x)^{1/2} \neq x(x + 5)^{1/2}$

Factor within grouping symbols before applying the exponent to each factor.

$(x^2 + 5x)^{1/2} = [x(x + 5)]^{1/2} = x^{1/2}(x + 5)^{1/2}$

19. $\dfrac{3}{x} + \dfrac{4}{y} = \dfrac{3}{x} \cdot \dfrac{y}{y} + \dfrac{4}{y} \cdot \dfrac{x}{x} = \dfrac{3y + 4x}{xy}$

To add fractions, they must have a common denominator.

21. To add fractions, first find a common denominator.

$\dfrac{x}{2y} + \dfrac{y}{3} = \dfrac{3x}{6y} + \dfrac{2y^2}{6y} = \dfrac{3x + 2y^2}{6y}$

23. $\dfrac{5x + 3}{4} = \dfrac{1}{4}(5x + 3)$

The required factor is $5x + 3$.

25. $\frac{2}{3}x^2 + \frac{1}{3}x + 5 = \frac{2}{3}x^2 + \frac{1}{3}x + \frac{15}{3} = \frac{1}{3}(2x^2 + x + 15)$

The required factor is $2x^2 + x + 15$.

27. $x^2(x^3 - 1)^4 = \frac{1}{3}(x^3 - 1)^4(3x^2)$

The required factor is $\frac{1}{3}$.

29. $2(y - 5)^{1/2} + y(y - 5)^{-1/2} = (y - 5)^{-1/2}(3y - 10)$

The required factor is $3y - 10$.

31. $\dfrac{4x + 6}{(x^2 + 3x + 7)^3} = \dfrac{2(2x + 3)}{(x^2 + 3x + 7)^3} = \dfrac{2}{1} \cdot \dfrac{(2x + 3)}{1} \cdot \dfrac{1}{(x^2 + 3x + 7)^3} = (2)\dfrac{1}{(x^2 + 3x + 7)^3}(2x + 3)$

The required factor is 2.

33. $\dfrac{3}{x} + \dfrac{5}{2x^2} - \dfrac{3}{2}x = \dfrac{6x}{2x^2} + \dfrac{5}{2x^2} - \dfrac{3x^3}{2x^2}$

$\qquad\qquad\qquad = \left(\dfrac{1}{2x^2}\right)(6x + 5 - 3x^3)$

The required factor is $\dfrac{1}{2x^2}$.

35. $\dfrac{25x^2}{36} + \dfrac{4y^2}{9} = \dfrac{x^2}{\dfrac{36}{25}} + \dfrac{y^2}{\dfrac{9}{4}}$

The required factors are $\dfrac{36}{25}$ and $\dfrac{9}{4}$.

37. $\dfrac{x^2}{\dfrac{3}{10}} - \dfrac{y^2}{\dfrac{4}{5}} = \dfrac{10x^2}{3} - \dfrac{5y^2}{4}$

The required factors are 3 and 4.

39. $x^{1/3} - 5x^{4/3} = x^{1/3}\left(1 - 5x^{3/3}\right) = x^{1/3}(1 - 5x)$

The required factor is $1 - 5x$.

41. $(1 - 3x)^{4/3} - 4x(1 - 3x)^{1/3} = (1 - 3x)^{1/3}\left[(1 - 3x)^1 - 4x\right]$

$\qquad\qquad\qquad\qquad\qquad = (1 - 3x)^{1/3}(1 - 7x)$

The required factor is $1 - 7x$.

43. $\tfrac{1}{10}(2x + 1)^{5/2} - \tfrac{1}{6}(2x + 1)^{3/2} = \tfrac{3}{30}(2x + 1)^{3/2}(2x + 1)^1 - \tfrac{5}{30}(2x + 1)^{3/2}$

$\qquad\qquad\qquad\qquad\qquad\quad = \tfrac{1}{30}(2x + 1)^{3/2}\left[3(2x + 1) - 5\right]$

$\qquad\qquad\qquad\qquad\qquad\quad = \tfrac{1}{30}(2x + 1)^{3/2}(6x - 2)$

$\qquad\qquad\qquad\qquad\qquad\quad = \tfrac{1}{30}(2x + 1)^{3/2}\,2(3x - 1)$

$\qquad\qquad\qquad\qquad\qquad\quad = \tfrac{1}{15}(2x + 1)^{3/2}(3x - 1)$

The required factor is $3x - 1$.

45. $\dfrac{7}{(x + 3)^5} = 7(x + 3)^{-5}$

47. $\dfrac{2x^5}{(3x + 5)^4} = 2x^5(3x + 5)^{-4}$

49. $\dfrac{4}{3x} + \dfrac{4}{x^4} - \dfrac{7x}{\sqrt[3]{2x}} = 4(3x)^{-1} + 4x^{-4} - 7x(2x)^{-1/3}$

51. $\dfrac{x^2 + 6x + 12}{3x} = \dfrac{x^2}{3x} + \dfrac{6x}{3x} + \dfrac{12}{3x}$

$\qquad\qquad\qquad\; = \dfrac{x}{3} + 2 + \dfrac{4}{x}$

53. $\dfrac{4x^3 - 7x^2 + 1}{x^{1/3}} = \dfrac{4x^3}{x^{1/3}} - \dfrac{7x^2}{x^{1/3}} + \dfrac{1}{x^{1/3}}$

$\qquad\qquad\qquad\quad = 4x^{3-1/3} - 7x^{2-1/3} + \dfrac{1}{x^{1/3}}$

$\qquad\qquad\qquad\quad = 4x^{8/3} - 7x^{5/3} + \dfrac{1}{x^{1/3}}$

55. $\dfrac{3 - 5x^2 - x^4}{\sqrt{x}} = \dfrac{3}{\sqrt{x}} - \dfrac{5x^2}{\sqrt{x}} - \dfrac{x^4}{\sqrt{x}}$

$\qquad\qquad\qquad\quad = \dfrac{3}{\sqrt{x}} - 5x^{2-1/2} - x^{4-1/2}$

$\qquad\qquad\qquad\quad = \dfrac{3}{x^{1/2}} - 5x^{3/2} - x^{7/2}$

57. $\dfrac{-2(x^2 - 3)^{-3}(2x)(x + 1)^3 - 3(x + 1)^2(x^2 - 3)^{-2}}{\left[(x + 1)^3\right]^2} = \dfrac{(x^2 - 3)^{-3}(x + 1)^2\left[-4x(x + 1) - 3(x^2 - 3)\right]}{(x + 1)^6}$

$\qquad\qquad\qquad\qquad\qquad\qquad\qquad\qquad\qquad = \dfrac{-4x^2 - 4x - 3x^2 + 9}{(x^2 - 3)^3(x + 1)^4}$

$\qquad\qquad\qquad\qquad\qquad\qquad\qquad\qquad\qquad = \dfrac{-7x^2 - 4x + 9}{(x^2 - 3)^3(x + 1)^4}$

59. $\dfrac{(6x + 1)^3(27x^2 + 2) - (9x^3 + 2x)(3)(6x + 1)^2(6)}{\left[(6x + 1)^3\right]^2} = \dfrac{(6x + 1)^2\left[(6x + 1)(27x^2 + 2) - 18(9x^3 + 2x)\right]}{(6x + 1)^6}$

$$= \dfrac{162x^3 + 12x + 27x^2 + 2 - 162x^3 - 36x}{(6x + 1)^4}$$

$$= \dfrac{27x^2 - 24x + 2}{(6x + 1)^4}$$

61. $\dfrac{(x + 2)^{3/4}(x + 3)^{-2/3} - (x + 3)^{1/3}(x + 2)^{-1/4}}{\left[(x + 2)^{3/4}\right]^2} = \dfrac{(x + 2)^{-1/4}(x + 3)^{-2/3}\left[(x + 2) - (x + 3)\right]}{(x + 2)^{6/4}}$

$$= \dfrac{x + 2 - x - 3}{(x + 2)^{1/4}(x + 3)^{2/3}(x + 2)^{6/4}}$$

$$= -\dfrac{1}{(x + 3)^{2/3}(x + 2)^{7/4}}$$

63. $\dfrac{2(3x - 1)^{1/3} - (2x + 1)(1/3)(3x - 1)^{-2/3}(3)}{(3x - 1)^{2/3}} = \dfrac{(3x - 1)^{-2/3}\left[2(3x - 1) - (2x + 1)\right]}{(3x - 1)^{2/3}}$

$$= \dfrac{6x - 2 - 2x - 1}{(3x - 1)^{2/3}(3x - 1)^{2/3}}$$

$$= \dfrac{4x - 3}{(3x - 1)^{4/3}}$$

65. $\dfrac{1}{(x^2 + 4)^{1/2}} \cdot \dfrac{1}{2}(x^2 + 4)^{-1/2}(2x) = \dfrac{1}{(x^2 + 4)^{1/2}} \cdot \dfrac{1}{(x^2 + 4)^{1/2}} \cdot \dfrac{1}{2}(2x)$

$$= \dfrac{1}{(x^2 + 4)^1}(x)$$

$$= \dfrac{x}{x^2 + 4}$$

67. $(x^2 + 5)^{1/2}\left(\dfrac{3}{2}\right)(3x - 2)^{1/2}(3) + (3x - 2)^{3/2}\left(\dfrac{1}{2}\right)(x^2 + 5)^{-1/2}(2x) = \dfrac{9}{2}(x^2 + 5)^{1/2}(3x - 2)^{1/2} + x(x^2 + 5)^{-1/2}(3x - 2)^{3/2}$

$$= \dfrac{9}{2}(x^2 + 5)^{1/2}(3x - 2)^{1/2} + \dfrac{2}{2}x(x^2 + 5)^{-1/2}(3x - 2)^{3/2}$$

$$= \dfrac{1}{2}(x^2 + 5)^{-1/2}(3x - 2)^{1/2}\left[9(x^2 + 5)^1 + 2x(3x - 2)^1\right]$$

$$= \dfrac{1}{2}(x^2 + 5)^{-1/2}(3x - 2)^{1/2}(9x^2 + 45 + 6x^2 - 4x)$$

$$= \dfrac{(3x - 2)^{1/2}(15x^2 - 4x + 45)}{2(x^2 + 5)^{1/2}}$$

69. $t = \dfrac{\sqrt{x^2 + 4}}{2} + \dfrac{\sqrt{(4 - x)^2 + 4}}{6}$

(a)

x	t
0.5	1.70
1.0	1.72
1.5	1.78
2.0	1.89
2.5	2.02
3.0	2.18
3.5	2.36
4.0	2.57

(b) She should swim to a point about $\dfrac{1}{2}$ mile down the coast to minimize the time required to reach the finish line.

(c) $\dfrac{1}{2}x(x^2 + 4)^{-1/2} + \dfrac{1}{6}(x - 4)(x^2 - 8x + 20)^{-1/2} = \dfrac{3}{6}x(x^2 + 4)^{-1/2} + \dfrac{1}{6}(x - 4)(x^2 - 8x + 20)^{-1/2}$

$$= \dfrac{1}{6}\left[3x(x^2 + 4)^{-1/2} + (x - 4)(x^2 - 8x + 20)^{-1/2}\right]$$

$$= \dfrac{1}{6}\left[\dfrac{3x}{(x^2 + 4)^{1/2}} + \dfrac{x - 4}{(x^2 - 8x + 20)^{1/2}}\right]$$

$$= \dfrac{3x\sqrt{x^2 - 8x + 20} + (x - 4)\sqrt{x^2 + 4}}{6\sqrt{x^2 + 4}\sqrt{x^2 - 8x + 20}}$$

71. You cannot move term-by-term from the denominator to the numerator.

Chapter P   Practice Test Solutions

1. $\dfrac{|-42| - 20}{15 - |-4|} = \dfrac{42 - 20}{15 - 4} = \dfrac{22}{11} = 2$

2. $\dfrac{x}{z} - \dfrac{z}{y} = \dfrac{x}{z} \cdot \dfrac{y}{y} - \dfrac{z}{y} \cdot \dfrac{z}{z} = \dfrac{xy - z^2}{yz}$

3. $|x - 7| \le 4$

4. $10(-5)^3 = 10(-125) = -1250$

5. $\left(-4x^3\right)\left(-2x^{-5}\right)\left(\dfrac{1}{16}x\right) = (-4)(-2)\left(\dfrac{1}{16}\right)x^{3+(-5)+1}$

$\qquad = \dfrac{8}{16}x^{-1}$

$\qquad = \dfrac{1}{2x}$

6. $0.0000412 = 4.12 \times 10^{-5}$

7. $125^{2/3} = \left(\sqrt[3]{125}\right)^2 = (5)^2 = 25$

8. $\sqrt[4]{64x^7y^9} = \sqrt[4]{16 \cdot 4x^4x^3y^8y} = 2xy^2\sqrt[4]{4x^3y}$

9. $\dfrac{6}{\sqrt{12}} = \dfrac{6}{2\sqrt{3}} \cdot \dfrac{\sqrt{3}}{\sqrt{3}} = \dfrac{6\sqrt{3}}{6} = \sqrt{3}$

10. $3\sqrt{80} - 7\sqrt{500} = 3\left(4\sqrt{5}\right) - 7\left(10\sqrt{5}\right)$

$\qquad = 12\sqrt{5} - 70\sqrt{5}$

$\qquad = -58\sqrt{5}$

11. $\left(8x^4 - 9x^2 + 2x - 1\right) - \left(3x^3 + 5x + 4\right) = 8x^4 - 3x^3 - 9x^2 - 3x - 5$

12. $(x - 3)\left(x^2 + x - 7\right) = x^3 + x^2 - 7x - 3x^2 - 3x + 21 = x^3 - 2x^2 - 10x + 21$

13. $\left[(x - 2) - y\right]^2 = (x - 2)^2 - 2y(x - 2) + y^2 = x^2 - 4x + 4 - 2xy + 4y + y^2 = x^2 + y^2 - 2xy - 4x + 4y + 4$

14. $16x^4 - 1 = \left(4x^2 + 1\right)\left(4x^2 - 1\right)$

$\qquad = \left(4x^2 + 1\right)(2x + 1)(2x - 1)$

15. $6x^2 + 5x - 4 = (2x - 1)(3x + 4)$

16. $x^3 - 64 = x^3 - 4^3 = (x - 4)\left(x^2 + 4x + 16\right)$

17. $-\dfrac{3}{x} + \dfrac{x}{x^2 + 2} = \dfrac{-3\left(x^2 + 2\right) + x^2}{x\left(x^2 + 2\right)} = \dfrac{-2x^2 - 6}{x\left(x^2 + 2\right)} = -\dfrac{2\left(x^2 + 3\right)}{x\left(x^2 + 2\right)}$

18. $\dfrac{x - 3}{4x} \div \dfrac{x^2 - 9}{x^2} = \dfrac{x - 3}{4x} \cdot \dfrac{x^2}{(x + 3)(x - 3)} = \dfrac{x}{4(x + 3)}, \; x \ne 0, 3$

19. $\dfrac{1 - \dfrac{1}{x}}{1 - \dfrac{1}{1 - (1/x)}} = \dfrac{\dfrac{x - 1}{x}}{1 - \dfrac{1}{(x-1)/x}} = \dfrac{\dfrac{x - 1}{x}}{1 - \dfrac{x}{x - 1}} = \dfrac{\dfrac{x - 1}{x}}{\dfrac{-1}{x - 1}} = \dfrac{x - 1}{x} \cdot \dfrac{x - 1}{-1} = \dfrac{-(x - 1)^2}{x}, \; x \ne 1$

20. (a)

(b) $d = \sqrt{\left[5 - (-3)\right]^2 + (-1 - 7)^2}$

$\qquad = \sqrt{(8)^2 + (-8)^2}$

$\qquad = \sqrt{64 + 64}$

$\qquad = \sqrt{128}$

$\qquad = 8\sqrt{2}$

(c) $\left(\dfrac{-3 + 5}{2}, \dfrac{7 + (-1)}{2}\right) = (1, 3)$

Chapter 1   Practice Test Solutions

1. $3x - 5y = 15$

Line

x-intercept: $(5, 0)$

y-intercept: $(0, -3)$

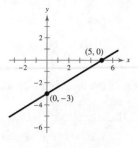

2. $y = \sqrt{9 - x}$

Domain: $(-\infty, 9]$

x-intercept: $(9, 0)$

y-intercept: $(0, 3)$

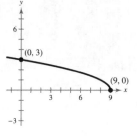

3. $5x + 4 = 7x - 8$

$\quad 4 + 8 = 7x - 5x$

$\quad\quad 12 = 2x$

$\quad\quad\, x = 6$

4. $\dfrac{x}{3} - 5 = \dfrac{x}{5} + 1$

$15\left(\dfrac{x}{3} - 5\right) = 15\left(\dfrac{x}{5} + 1\right)$

$\quad 5x - 75 = 3x + 15$

$\quad\quad 2x = 90$

$\quad\quad\, x = 45$

5. $\dfrac{3x + 1}{6x - 7} = \dfrac{2}{5}$

$5(3x + 1) = 2(6x - 7)$

$15x + 5 = 12x - 14$

$\quad\quad 3x = -19$

$\quad\quad\, x = -\dfrac{19}{3}$

6. $(x - 3)^2 + 4 = (x + 1)^2$

$x^2 - 6x + 9 + 4 = x^2 + 2x + 1$

$\quad\quad\quad -8x = -12$

$\quad\quad\quad\quad x = \dfrac{-12}{-8}$

$\quad\quad\quad\quad x = \dfrac{3}{2}$

7. $\quad A = \dfrac{1}{2}(a + b)h$

$\quad 2A = ah + bh$

$2A - bh = ah$

$\dfrac{2A - bh}{h} = a$

8. Percent $= \dfrac{301}{4300} = 0.07 = 7\%$

9. Let $x = $ number of quarters.

Then $53 - x = $ number of nickels.

$25x + 5(53 - x) = 605$

$\quad 20x + 265 = 605$

$\quad\quad\quad 20x = 340$

$\quad\quad\quad\quad x = 17$ quarters

$\quad\quad 53 - x = 36$ nickels

10. Let $x = $ amount in the $9\frac{1}{2}\%$ fund.

Then $15{,}000 - x = $ amount in 11% fund.

$0.095x + 0.11(15{,}000 - x) = 1582.50$

$\quad\quad -0.015x + 1650 = 1582.50$

$\quad\quad\quad\quad -0.015x = -67.5$

$\quad\quad\quad\quad\quad\quad x = \$4500$ at $9\frac{1}{2}\%$

$\quad\quad 15{,}000 - x = \$10{,}500$ at 11%

11. $28 + 5x - 3x^2 = 0$

$(4 - x)(7 + 3x) = 0$

$4 - x = 0 \Rightarrow x = 4$

$7 + 3x = 0 \Rightarrow x = -\dfrac{7}{3}$

12. $(x - 2)^2 = 24$

$x - 2 = \pm\sqrt{24}$

$x - 2 = \pm 2\sqrt{6}$

$\quad\; x = 2 \pm 2\sqrt{6}$

13. $x^2 - 4x - 9 = 0$

$x^2 - 4x + 2^2 = 9 + 2^2$

$\quad (x - 2)^2 = 13$

$\quad\quad x - 2 = \pm\sqrt{13}$

$\quad\quad\quad\; x = 2 \pm \sqrt{13}$

14. $x^2 + 5x - 1 = 0$

$a = 1, b = 5, c = -1$

$x = \dfrac{-5 \pm \sqrt{(5)^2 - 4(1)(-1)}}{2(1)}$

$ = \dfrac{-5 \pm \sqrt{25 + 4}}{2}$

$ = \dfrac{-5 \pm \sqrt{29}}{2}$

15. $3x^2 - 2x + 4 = 0$

$a = 3, b = -2, c = 4$

$x = \dfrac{-(-2) \pm \sqrt{(-2)^2 - 4(3)(4)}}{2(3)}$

$ = \dfrac{2 \pm \sqrt{4 - 48}}{6}$

$ = \dfrac{2 \pm \sqrt{-44}}{6}$

$ = \dfrac{2 \pm 2i\sqrt{11}}{6}$

$ = \dfrac{1 \pm i\sqrt{11}}{3} = \dfrac{1}{3} \pm \dfrac{\sqrt{11}}{3}i$

16.

$60{,}000 = xy$

$y = \dfrac{60{,}000}{x}$

$2x + 2y = 1100$

$2x + 2\left(\dfrac{60{,}000}{x}\right) = 1100$

$x + \dfrac{60{,}000}{x} = 550$

$x^2 + 60{,}000 = 550x$

$x^2 - 550x + 60{,}000 = 0$

$(x - 150)(x - 400) = 0$

$x = 150 \quad \text{or} \quad x = 400$

$y = 400 \qquad\quad y = 150$

Length: 400 feet; width: 150 feet

17.

$x(x + 2) = 624$

$x^2 + 2x - 624 = 0$

$(x - 24)(x + 26) = 0$

$x = 24 \quad \text{or} \quad x = -26, \text{ (extraneous solution)}$

$x + 2 = 26$

The integers are 24 and 26.

18. $x^2 - 10x^2 + 24x = 0$

$x(x^2 - 10x + 24) = 0$

$x(x - 4)(x - 6) = 0$

$x = 0, x = 4, x = 6$

19. $\sqrt[3]{6 - x} = 4$

$6 - x = 64$

$-x = 58$

$x = -58$

20. $(x^2 - 8)^{2/5} = 4$

$x^2 - 8 = \pm 4^{5/2}$

$x^2 - 8 = 32 \qquad \text{or} \quad x^2 - 8 = -32$

$x^2 = 40 \qquad\qquad\quad x^2 = -24$

$x = \pm\sqrt{40} \qquad\qquad x = \pm\sqrt{-24}$

$x = \pm 2\sqrt{10} \qquad\qquad x = \pm 2\sqrt{6}i$

21. $x^4 - x^2 - 12 = 0$

$(x^2 - 4)(x^2 + 3) = 0$

$x^2 = 4 \quad \text{or} \quad x^2 = -3$

$x^2 = \pm 2 \qquad\quad x = \pm\sqrt{3}i$

22. $4 - 3x > 16$

$-3x > 12$

$x < -4$

23. $\left|\dfrac{x - 3}{2}\right| < 5$

$-5 < \dfrac{x - 3}{2} < 5$

$-10 < x - 3 < 10$

$-7 < x < 13$

24.
$$\frac{x+1}{x-3} < 2$$

$$\frac{x+1}{x-3} - 2 < 0$$

$$\frac{x+1-2(x-3)}{x-3} < 0$$

$$\frac{7-x}{x-3} < 0$$

Critical numbers: $x = 7$ and $x = 3$

Test intervals: $(-\infty, 3), (3, 7), (7, \infty)$

Test: Is $\dfrac{7-x}{x-3} < 0$?

Solution intervals: $(-\infty, 3) \cup (7, \infty)$

25. $|3x - 4| \geq 9$

$$3x - 4 \leq -9 \quad \text{or} \quad 3x - 4 \geq 9$$

$$3x \leq -5 \qquad\qquad 3x \geq 13$$

$$x \leq -\tfrac{5}{3} \qquad\qquad x \geq \tfrac{13}{3}$$

Chapter 2   Practice Test Solutions

1.
$$m = \frac{-1-4}{3-2} = -5$$

$$y - 4 = -5(x - 2)$$

$$y - 4 = -5x + 10$$

$$y = -5x + 14$$

2. $y = \tfrac{4}{3}x - 3$

3. $2x + 3y = 0$

$$y = -\tfrac{2}{3}x$$

$$m_1 = -\tfrac{2}{3}$$

$$\perp m_2 = \tfrac{3}{2} \text{ through } (4, 1)$$

$$y - 1 = \tfrac{3}{2}(x - 4)$$

$$y - 1 = \tfrac{3}{2}x - 6$$

$$y = \tfrac{3}{2}x - 5$$

4. $(5, 32)$ and $(9, 44)$

$$m = \frac{44 - 32}{9 - 5} = \frac{12}{4} = 3$$

$$y - 32 = 3(x - 5)$$

$$y - 32 = 3x - 15$$

$$y = 3x + 17$$

When $x = 20$, $y = 3(20) + 17$

$$y = \$77.$$

5. $f(x - 3) = (x - 3)^2 - 2(x - 3) + 1$

$$= x^2 - 6x + 9 - 2x + 6 + 1$$

$$= x^2 - 8x + 16$$

6.
$$f(3) = 12 - 11 = 1$$

$$\frac{f(x) - f(3)}{x - 3} = \frac{(4x - 11) - 1}{x - 3}$$

$$= \frac{4x - 12}{x - 3}$$

$$= \frac{4(x - 3)}{x - 3} = 4, \; x \neq 3$$

7. $f(x) = \sqrt{36 - x^2} = \sqrt{(6 + x)(6 - x)}$

Domain: $[-6, 6]$

Range: $[0, 6]$, because

$$(6 + x)(6 - x) \geq 0 \text{ on this interval.}$$

8. (a) $6x - 5y + 4 = 0$

$$y = \frac{6x + 4}{5} \text{ is a function of } x.$$

(b) $x^2 + y^2 = 9$

$$y = \pm\sqrt{9 - x^2} \text{ is not a function of } x.$$

(c) $y^3 = x^2 + 6$

$$y = \sqrt[3]{x^2 + 6} \text{ is a function of } x.$$

9. Parabola

Vertex: $(0, -5)$

Intercepts:

$(0, -5), (\pm\sqrt{5}, 0)$

y-axis symmetry

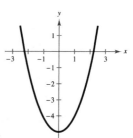

10. Intercepts: $(0, 3), (-3, 0)$

x	-4	-3	-2	-1	0	1	2
y	1	0	1	2	3	4	5

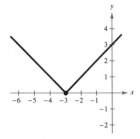

11.

x	0	1	2	3
y	1	3	5	7

x	-1	-2	-3
y	2	6	12

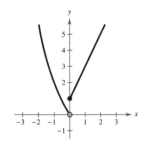

12. (a) $f(x + 2)$

Horizontal shift two units to the left

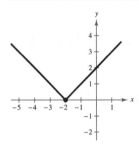

(b) $-f(x) + 2$

Reflection in the x-axis and a vertical shift two units upward

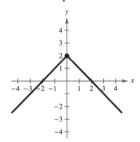

13. (a) $(g - f)(x) = g(x) - f(x)$

$$= (2x^2 - 5) - (3x + 7)$$

$$= 2x^2 - 3x - 12$$

(b) $(fg)(x) = f(x)g(x)$

$$= (3x + 7)(2x^2 - 5)$$

$$= 6x^3 + 14x^2 - 15x - 35$$

14. $f(g(x)) = f(2x + 3)$

$$= (2x + 3)^2 - 2(2x + 3) + 16$$

$$= 4x^2 + 12x + 9 - 4x - 6 + 16$$

$$= 4x^2 + 8x + 19$$

15. $f(x) = x^3 + 7$

$$y = x^3 + 7$$

$$x = y^3 + 7$$

$$x - 7 = y^3$$

$$\sqrt[3]{x - 7} = y$$

$$f^{-1}(x) = \sqrt[3]{x - 7}$$

16. (a) $f(x) = |x - 6|$ does not have an inverse. Its graph does not pass the Horizontal Line Test.

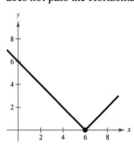

(b) $f(x) = ax + b, a \neq 0$ does have an inverse.

$$y = ax + b$$

$$x = ay + b$$

$$\frac{x - b}{a} = y$$

$$f^{-1}(x) = \frac{x - b}{a}$$

(c) $f(x) = x^3 - 19$ does have an inverse.

$$y = x^3 - 19$$

$$x = y^3 - 19$$

$$x + 19 = y^3$$

$$\sqrt[3]{x + 19} = y$$

$$f^{-1}(x) = \sqrt[3]{x + 19}$$

17. $f(x) = \sqrt{\dfrac{3-x}{x}}, \ 0 < x \le 3, \ y \ge 0$

$y = \sqrt{\dfrac{3-x}{x}}$

$x = \sqrt{\dfrac{3-y}{y}}, \ 0 < y \le 3, \ x \ge 0$

$x^2 = \dfrac{3-y}{y}$

$x^2 y = 3 - y$

$x^2 y + y = 3$

$y(x^2 + 1) = 3$

$y = \dfrac{3}{x^2 + 1}$

$f^{-1}(x) = \dfrac{3}{x^2 + 1}, \ x \ge 0$

18. False. The slopes of 3 and $\tfrac{1}{3}$ are not **negative** reciprocals.

19. True. Let $y = (f \circ g)(x)$. Then $x = (f \circ g)^{-1}(y)$.
Also,

$(f \circ g)(x) = y$

$f(g(x)) = y$

$g(x) = f^{-1}(y)$

$x = g^{-1}(f^{-1}(y))$

$x = (g^{-1} \circ f^{-1})(y).$

Since $x = x$, we have $(f \circ g)^{-1}(y) = (g^{-1} \circ f^{-1})(y).$

20. True. It must pass the Vertical Line Test to be a function and it must pass the Horizontal Line Test to have an inverse.

Chapter 3   Practice Test Solutions

1. x-intercepts: $(1, 0), (5, 0)$

y-intercept: $(0, 5)$

Vertex: $(3, -4)$

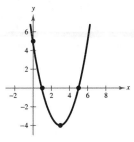

2. $a = 0.01, \ b = -90$

$\dfrac{-b}{2a} = \dfrac{90}{2(0.01)} = 4500$ units

3. Vertex $(1, 7)$ opening downward through $(2, 5)$

$y = a(x - 1)^2 + 7$ Standard form

$5 = a(2 - 1)^2 + 7$

$5 = a + 7$

$a = -2$

$y = -2(x - 1)^2 + 7$

$= -2(x^2 - 2x + 1) + 7$

$= -2x^2 + 4x + 5$

4. $y = \pm a(x - 2)(3x - 4)$ where a is any real nonzero number.

$y = \pm(3x^2 - 10x + 8)$

5. Leading coefficient: -3

Degree: 5 (odd)

Falls to the right.

Rises to the left.

6. $0 = x^5 - 5x^3 + 4x$

$= x(x^4 - 5x^2 + 4)$

$= x(x^2 - 1)(x^2 - 4)$

$= x(x + 1)(x - 1)(x + 2)(x - 2)$

$x = 0, \ x = \pm 1, \ x = \pm 2$

7. $f(x) = x(x - 3)(x + 2)$

$= x(x^2 - x - 6)$

$= x^3 - x^2 - 6x$

8. $f(x) = x^3 - 12x$

Intercepts: $(0, 0), \left(\pm 2\sqrt{3}, 0\right)$

Rises to the right.

Falls to the left.

x	-2	-1	0	1	2
y	16	11	0	-11	-16

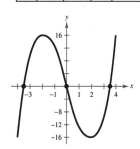

9.

$$
\begin{array}{r}
3x^3 + 9x^2 + 20x + 62 \\
x - 3\overline{\smash{\big)}\, 3x^4 + 0x^3 - 7x^2 + 2x - 10} \\
\underline{3x^4 - 9x^3} \\
9x^3 - 7x^2 \\
\underline{9x^3 - 27x^2} \\
20x^2 + 2x \\
\underline{20x^2 - 60x} \\
62x - 10 \\
\underline{62x - 186} \\
176
\end{array}
$$

$$\frac{3x^4 - 7x^2 + 2x - 10}{x - 3} = 3x^3 + 9x^2 + 20x + 6 + \frac{176}{x - 3}$$

10.

$$
\begin{array}{r}
x - 2 \\
x^2 + 2x - 1\overline{\smash{\big)}\, x^3 + 0x^2 + 0x - 11} \\
\underline{x^3 + 2x^2 - x} \\
-2x^2 + x - 11 \\
\underline{-2x^2 - 4x + 2} \\
5x - 13
\end{array}
$$

$$\frac{x^3 - 11}{x^2 + 2x - 1} = x - 2 + \frac{5x - 13}{x^2 + 2x - 1}$$

11.

$$
\begin{array}{r|rrrrrr}
-5 & 3 & 13 & 0 & 0 & 12 & -1 \\
   &   & -15 & 10 & -50 & 250 & -1310 \\
\hline
   & 3 & -2 & 10 & -50 & 262 & -1311
\end{array}
$$

$$\frac{3x^5 + 13x^4 + 12x - 1}{x + 5} = 3x^4 - 2x^3 + 10x^2 - 50x + 262 - \frac{1311}{x + 5}$$

12.

$$
\begin{array}{r|rrrr}
-6 & 7 & 40 & -12 & 15 \\
   &   & -42 & 12 & 0 \\
\hline
   & 7 & -2 & 0 & 15
\end{array}
$$

$$f(-6) = 15$$

13. $0 = x^3 - 19x - 30$

Possible rational zeros: $\pm 1, \pm 2, \pm 3, \pm 5, \pm 6, \pm 10, \pm 15, \pm 30$

$$
\begin{array}{r|rrrr}
-2 & 1 & 0 & -19 & -30 \\
   &   & -2 & 4 & 30 \\
\hline
   & 1 & -2 & -15 & 0
\end{array}
$$

$$0 = (x + 2)(x^2 - 2x - 15)$$
$$0 = (x + 2)(x + 3)(x - 5)$$

Zeros: $x = -2,\ x = -3,\ x = 5$

14. $0 = x^4 + x^3 - 8x^2 - 9x - 9$

Possible rational zeros: $\pm 1, \pm 3, \pm 9$

$$
\begin{array}{r|rrrrr}
3 & 1 & 1 & -8 & -9 & -9 \\
  &   & 3 & 12 & 12 & 9 \\
\hline
  & 1 & 4 & 4 & 3 & 0
\end{array}
$$

$0 = (x - 3)(x^3 + 4x^2 + 4x + 3)$

Possible rational zeros of $x^3 + 4x^2 + 4x + 3$: $\pm 1, \pm 3$

$$
\begin{array}{r|rrrr}
-3 & 1 & 4 & 4 & 3 \\
   &   & -3 & -3 & -3 \\
\hline
   & 1 & 1 & 1 & 0
\end{array}
$$

$0 = (x - 3)(x + 3)(x^2 + x + 1)$

Use the Quadratic Formula. The zeros of $x^2 + x + 1$
are $x = \dfrac{-1 \pm \sqrt{3}i}{2}$.

Zeros:

$x = 3, \; x = -3, \; x = -\dfrac{1}{2} + \dfrac{\sqrt{3}}{2}i, \; x = -\dfrac{1}{2} - \dfrac{\sqrt{3}}{2}i$

The real zeros are $x = 3$ and $x = -3$.

15. $0 = 6x^3 - 5x^2 + 4x - 15$

Possible rational zeros:

$\pm 1, \pm 3, \pm 5, \pm 15, \pm\frac{1}{2}, \pm\frac{3}{2}, \pm\frac{5}{2}, \pm\frac{15}{2}, \pm\frac{1}{3}, \pm\frac{5}{3}, \pm\frac{1}{6}, \pm\frac{5}{6}$

16. $0 = x^3 - \frac{20}{3}x^2 + 9x - \frac{10}{3}$

$0 = 3x^3 - 20x^2 + 27x - 10$

Possible rational zeros:

$\pm 1, \pm 2, \pm 5, \pm 10, \pm\frac{1}{3}, \pm\frac{2}{3}, \pm\frac{5}{3}, \pm\frac{10}{3}$

$$
\begin{array}{r|rrrr}
1 & 3 & -20 & 27 & -10 \\
  &   & 3 & -17 & 10 \\
\hline
  & 3 & -17 & 10 & 0
\end{array}
$$

$0 = (x - 1)(3x^2 - 17x + 10)$

$0 = (x - 1)(3x - 2)(x - 5)$

Zeros: $x = 1, \; x = \frac{2}{3}, \; x = 5$

17. Possible rational zeros: $\pm 1, \pm 2, \pm 5, \pm 10$

$$
\begin{array}{r|rrrrr}
1 & 1 & 1 & 3 & 5 & -10 \\
  &   & 1 & 2 & 5 & 10 \\
\hline
  & 1 & 2 & 5 & 10 & 0
\end{array}
$$

$$
\begin{array}{r|rrrr}
-2 & 1 & 2 & 5 & 10 \\
   &   & -2 & 0 & -10 \\
\hline
   & 1 & 0 & 5 & 0
\end{array}
$$

$f(x) = (x - 1)(x + 2)(x^2 + 5)$

$ = (x - 1)(x + 2)(x + \sqrt{5}i)(x - \sqrt{5}i)$

18. $f(x) = (x - 2)[x - (3 + i)][x - (3 - i)]$

$ = (x - 2)[(x - 3) - i][(x - 3) + i]$

$ = (x - 2)[(x - 3)^2 - (i)^2]$

$ = (x - 2)[x^2 - 6x + 10]$

$ = x^3 - 8x^2 + 22x - 20$

19.
$$
\begin{array}{r|rrrr}
3i & 1 & 4 & 9 & 36 \\
   &   & 3i & 12i - 9 & -36 \\
\hline
   & 1 & 4 + 3i & 12i & 0
\end{array}
$$

Thus, $f(3i) = 0$.

20. $z = \dfrac{kx^2}{\sqrt{y}}$

Chapter 4   Practice Test Solutions

1. Vertical asymptote: $x = 0$

Horizontal asymptote: $y = \frac{1}{2}$

x-intercept: $(1, 0)$

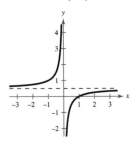

2. Vertical asymptote: $x = 0$

Slant asymptote: $y = 3x$

x-intercepts: $\left(\pm\dfrac{2}{\sqrt{3}}, 0\right)$

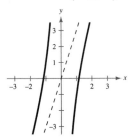

3. $y = 8$ is a horizontal asymptote since the degree of the numerator equals the degree of the denominator. There are no vertical asymptotes.

4. $x = 1$ is a vertical asymptote.

$$\frac{4x^2 - 2x + 7}{x - 1} = 4x + 2 + \frac{9}{x - 1}$$

so $y = 4x + 2$ is a slant asymptote.

5. $f(x) = \dfrac{x - 5}{(x - 5)^2} = \dfrac{1}{x - 5}$

Vertical asymptote: $x = 5$

Horizontal asymptote: $y = 0$

y-intercept: $\left(0, -\dfrac{1}{5}\right)$

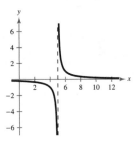

6. $(x - 0)^2 = 4(5)(y - 0)$

Vertex: $(0, 0)$

Focus: $(0, 5)$

Directrix: $y = -5$

7. $(y - 0)^2 = 4(7)(x - 0)$

$y^2 = 28x$

8. $a = 12, b = 5, h = k = 0,$

$c = \sqrt{144 - 25} = \sqrt{119}$

Center: $(0, 0)$

Foci: $\left(\pm\sqrt{119}, 0\right)$

Vertices: $(\pm 12, 0)$

9. Center: $(0, 0)$

$c = 4, 2b = 6 \Rightarrow b = 3,$

$a = \sqrt{16 + 9} = 5$

$\dfrac{x^2}{25} + \dfrac{y^2}{9} = 1$

10. $a = 12, b = 13, c = \sqrt{144 + 169} = \sqrt{313}$

Center: $(0, 0)$

Foci: $\left(0, \pm\sqrt{313}\right)$

Vertices: $(0, \pm 12)$

Asymptotes: $y = \pm\frac{12}{13}x$

11. Center: $(0, 0)$

$a = 4, \pm\dfrac{1}{2} = \pm\dfrac{b}{4} \Rightarrow b = 2$

$\dfrac{x^2}{16} - \dfrac{y^2}{4} = 1$

12. $p = 4$

$(x - 6)^2 = 4(4)(y + 1)$

$(x - 6)^2 = 16(y + 1)$

13.
$$16x^2 - 96x + 9y^2 + 36y = -36$$
$$16(x^2 - 6x + 9) + 9(y^2 + 4y + 4) = -36 + 144 + 36$$
$$16(x-3)^2 + 9(y+2)^2 = 144$$
$$\frac{(x-3)^2}{9} + \frac{(y+2)^2}{16} = 1$$

$a = 4, b = 3, c = \sqrt{16-9} = \sqrt{7}$

Center: $(3, -2)$

Foci: $\left(3, -2 \pm \sqrt{7}\right)$

Vertices: $(3, -2 \pm 4)$ OR $(3, 2)$ and $(3, -6)$

14. Center: $(3, 1)$

$a = 4, 2b = 2 \Rightarrow b = 1$

$$\frac{(x-3)^2}{16} + \frac{(y-1)^2}{1} = 1$$

15. $\dfrac{(x+3)^2}{1/4} - \dfrac{(y-1)^2}{1/9} = 1$

$a = \dfrac{1}{2}, b = \dfrac{1}{3}, c = \sqrt{\dfrac{1}{4} + \dfrac{1}{9}} = \dfrac{\sqrt{13}}{6}$

Center: $(-3, 1)$

Vertices: $\left(-3 \pm \dfrac{1}{2}, 1\right)$ OR $\left(-\dfrac{5}{2}, 1\right)$ and $\left(-\dfrac{7}{2}, 1\right)$

Foci: $\left(-3 \pm \dfrac{\sqrt{13}}{6}, 1\right)$

Asymptotes: $y = \pm\dfrac{1/3}{1/2}(x+3) + 1 = \pm\dfrac{2}{3}(x+3) + 1$

16. Center: $(3, 0)$

$a = 4, c = 7, b = \sqrt{49-16} = \sqrt{33}$

$$\frac{y^2}{16} - \frac{(x-3)^2}{33} = 1$$

Chapter 5   Practice Test Solutions

1. $x^{3/5} = 8$

$x = 8^{5/3} = \left(\sqrt[3]{8}\right)^5 = 2^5 = 32$

2. $3^{x-1} = \dfrac{1}{81}$

$3^{x-1} = 3^{-4}$

$x - 1 = -4$

$x = -3$

3. $f(x) = 2^{-x} = \left(\frac{1}{2}\right)^x$

x	-2	-1	0	1	2
$f(x)$	4	2	1	$\frac{1}{2}$	$\frac{1}{4}$

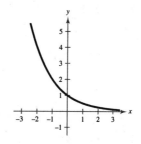

4. $g(x) = e^x + 1$

x	-2	-1	0	1	2
$g(x)$	1.14	1.37	2	3.72	8.39

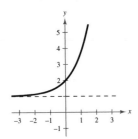

5. (a) $A = P\left(1 + \dfrac{r}{n}\right)^{nt}$

$A = 5000\left(1 + \dfrac{0.09}{12}\right)^{12(3)} \approx \$6543.23$

(b) $A = P\left(1 + \dfrac{r}{n}\right)^{nt}$

$A = 5000\left(1 + \dfrac{0.09}{4}\right)^{4(3)} \approx \$6530.25$

(c) $A = Pe^{rt}$

$A = 5000e^{(0.09)(3)} \approx \$6549.82$

6. $7^{-2} = \dfrac{1}{49}$

$\log_7 \dfrac{1}{49} = -2$

7. $x - 4 = \log_2 \dfrac{1}{64}$

$2^{x-4} = \dfrac{1}{64}$

$2^{x-4} = 2^{-6}$

$x - 4 = -6$

$x = -2$

8. $\log_b \sqrt[4]{\dfrac{8}{25}} = \dfrac{1}{4} \log_b \dfrac{8}{25}$

$= \dfrac{1}{4}\left[\log_b 8 - \log_b 25\right]$

$= \dfrac{1}{4}\left[\log_b 2^3 - \log_b 5^2\right]$

$= \dfrac{1}{4}\left[3 \log_b 2 - 2 \log_b 5\right]$

$= \dfrac{1}{4}\left[3(0.3562) - 2(0.8271)\right]$

$= -0.1464$

9. $5 \ln x - \dfrac{1}{2} \ln y + 6 \ln z = \ln x^5 - \ln \sqrt{y} + \ln z^6$

$= \ln\left(\dfrac{x^5 z^6}{\sqrt{y}}\right), z > 0$

10. $\log_9 28 = \dfrac{\log 28}{\log 9} \approx 1.5166$

11. $\log N = 0.6646$

$N = 10^{0.6646} \approx 4.62$

12.

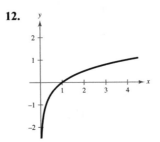

13. Domain:

$x^2 - 9 > 0$

$(x + 3)(x - 3) > 0$

$x < -3 \text{ or } x > 3$

14.

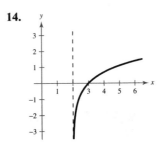

15. False. $\dfrac{\ln x}{\ln y} \neq \ln(x - y)$ because $\dfrac{\ln x}{\ln y} = \log_y x.$

16. $5^3 = 41$

$x = \log_5 41 = \dfrac{\ln 41}{\ln 5} \approx 2.3074$

17. $x - x^2 = \log_5 \dfrac{1}{25}$

$5^{x-x^2} = \dfrac{1}{25}$

$5^{x-x^2} = 5^{-2}$

$x - x^2 = -2$

$0 = x^2 - x - 2$

$0 = (x + 1)(x - 2)$

$x = -1 \text{ or } x = 2$

18. $\log_2 x + \log_2(x - 3) = 2$

$\log_2\left[x(x - 3)\right] = 2$

$x(x - 3) = 2^2$

$x^2 - 3x = 4$

$x^2 - 3x - 4 = 0$

$(x + 1)(x - 4) = 0$

$x = 4$

$x = -1\ (\text{extraneous})$

$x = 4$ is the only solution.

19. $\dfrac{e^x + e^{-x}}{3} = 4$

$e^x\left(e^x + e^{-x}\right) = 12e^x$

$e^{2x} + 1 = 12e^x$

$e^{2x} - 12e^x + 1 = 0$

$e^x = \dfrac{12 \pm \sqrt{144 - 4}}{2}$

$e^x \approx 11.9161$	or	$e^x \approx 0.0839$
$x = \ln 11.9161$		$x = \ln 0.0839$
$x \approx 2.478$		$x \approx -2.478$

20. $A = Pe^{rt}$

$12{,}000 = 6000e^{0.13t}$

$2 = e^{0.13t}$

$0.13t = \ln 2$

$t = \dfrac{\ln 2}{0.13}$

$t \approx 5.3319$ years or 5 years 4 months

Chapter 6   Practice Test Solutions

1. $350° = 350\left(\dfrac{\pi}{180}\right) = \dfrac{35\pi}{18}$

2. $\dfrac{5\pi}{9} = \dfrac{5\pi}{9} \cdot \dfrac{180}{\pi} = 100°$

3. $135° \, 14' \, 12'' = \left(135 + \dfrac{14}{60} + \dfrac{12}{3600}\right)°$

$\approx 135.2367°$

4. $-22.569° = -\left(22° + 0.569(60)'\right)$

$= -22° \, 34.14'$

$= -\left(22° \, 34' + 0.14(60)''\right)$

$\approx -22° \, 34' \, 8''$

5. $\cos\theta = \dfrac{2}{3}$

$x = 2, r = 3, y = \pm\sqrt{9 - 4} = \pm\sqrt{5}$

$\tan\theta = \dfrac{y}{x} = \pm\dfrac{\sqrt{5}}{2}$

6. $\sin\theta = 0.9063$

$\theta = \arcsin(0.9063)$

$\theta = 65° = \dfrac{13\pi}{36}$ or $\theta = 180° - 65° = 115° = \dfrac{23\pi}{36}$

7. $\tan 20° = \dfrac{35}{x}$

$x = \dfrac{35}{\tan 20°}$

≈ 96.1617

8. $\theta = \dfrac{6\pi}{5}, \theta$ is in Quadrant III.

Reference angle: $\dfrac{6\pi}{5} - \pi = \dfrac{\pi}{5}$ or $36°$

9. $\csc 3.92 = \dfrac{1}{\sin 3.92} \approx -1.4242$

10. $\tan \theta = 6 = \dfrac{6}{1}$, θ lies in Quandrant III.

$y = -6$, $x = -1$, $r = \sqrt{36 + 1} = \sqrt{37}$, so

$\sec \theta = \dfrac{\sqrt{37}}{-1} \approx -6.0828$.

11. Period: 4π

Amplitude: 3

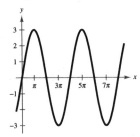

12. Period: 2π

Amplitude: 2

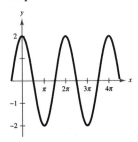

13. Period: $\dfrac{\pi}{2}$

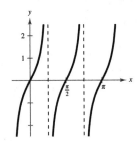

14. Period: 2π

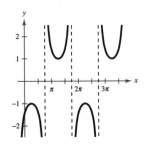

15.

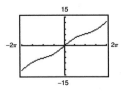

16.

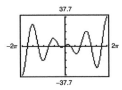

17. $\theta = \arcsin 1$

$\sin \theta = 1$

$\theta = \dfrac{\pi}{2} = 90°$

18. $\theta = \arctan(-3)$

$\tan \theta = -3$

$\theta \approx -1.249 \approx -71.565°$

19. $\sin\left(\arccos \dfrac{4}{\sqrt{35}}\right)$

$\sin \theta = \dfrac{\sqrt{19}}{\sqrt{35}} \approx 0.7368$

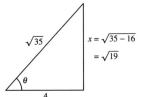

20. $\cos\left(\arcsin \dfrac{x}{4}\right)$

$\cos \theta = \dfrac{\sqrt{16 - x^2}}{4}$

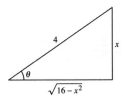

21. Given $A = 40°$, $c = 12$

$B = 90° - 40° = 50°$

$\sin 40° = \dfrac{a}{12}$

$a = 12 \sin 40° \approx 7.713$

$\cos 40° = \dfrac{b}{12}$

$b = 12 \cos 40° \approx 9.193$

22. Given $B = 6.84°$, $a = 21.3$

$$A = 90° - 6.84° = 83.16°$$

$$\sin 83.16° = \frac{21.3}{c}$$

$$c = \frac{21.3}{\sin 83.16°} \approx 21.453$$

$$\tan 83.16° = \frac{21.3}{b}$$

$$b = \frac{21.3}{\tan 83.16°} \approx 2.555$$

23. Given $a = 5, b = 9$

$$c = \sqrt{25 + 81} = \sqrt{106} \approx 10.296$$

$$\tan A = \tfrac{5}{9}$$

$$A = \arctan \tfrac{5}{9} \approx 29.055°$$

$$B \approx 90° - 29.055° = 60.945°$$

24. $\sin 67° = \dfrac{x}{20}$

$$x = 20 \sin 67° \approx 18.41 \text{ feet}$$

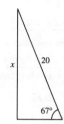

25. $\tan 5° = \dfrac{250}{x}$

$$x = \frac{250}{\tan 5°}$$

$$\approx 2857.513 \text{ feet}$$

$$\approx 0.541 \text{ mi}$$

Chapter 7   Practice Test Solutions

1. $\tan x = \dfrac{4}{11}$, $\sec x < 0 \Rightarrow x$ is in Quadrant III.

$$y = -4, x = -11, r = \sqrt{16 + 121} = \sqrt{137}$$

$$\sin x = -\frac{4}{\sqrt{137}} = -\frac{4\sqrt{137}}{137} \qquad \csc x = -\frac{\sqrt{137}}{4}$$

$$\cos x = -\frac{11}{\sqrt{137}} = -\frac{11\sqrt{137}}{137} \qquad \sec x = -\frac{\sqrt{137}}{11}$$

$$\tan x = \frac{4}{11} \qquad \cot x = \frac{11}{4}$$

2.
$$\frac{\sec^2 x + \csc^2 x}{\csc^2 x \left(1 + \tan^2 x\right)} = \frac{\sec^2 x + \csc^2 x}{\csc^2 x + \left(\csc^2 x\right) \tan^2 x}$$

$$= \frac{\sec^2 x + \csc^2 x}{\csc^2 x + \dfrac{1}{\sin^2 x} \cdot \dfrac{\sin^2 x}{\cos^2 x}}$$

$$= \frac{\sec^2 x + \csc^2 x}{\csc^2 x + \dfrac{1}{\cos^2 x}}$$

$$= \frac{\sec^2 x + \csc^2 x}{\csc^2 x + \sec^2 x} = 1$$

3. $\ln|\tan \theta| - \ln|\cot \theta| = \ln \left|\dfrac{\tan \theta}{\cot \theta}\right| = \ln \left|\dfrac{\sin \theta/\cos \theta}{\cos \theta/\sin \theta}\right| = \ln \left|\dfrac{\sin^2 \theta}{\cos^2 \theta}\right| = \ln|\tan^2 \theta| = 2 \ln|\tan \theta|$

4. $\cos\left(\dfrac{\pi}{2} - x\right) = \dfrac{1}{\csc x}$ is true since $\cos\left(\dfrac{\pi}{2} - x\right) = \sin x = \dfrac{1}{\csc x}$.

5. $\sin^4 x + \left(\sin^2 x\right) \cos^2 x = \sin^2 x\left(\sin^2 x + \cos^2 x\right)$

$$= \sin^2 x (1) = \sin^2 x$$

6. $\left(\csc x + 1\right)\left(\csc x - 1\right) = \csc^2 x - 1 = \cot^2 x$

7. $\dfrac{\cos^2 x}{1 - \sin x} \cdot \dfrac{1 + \sin x}{1 + \sin x} = \dfrac{\cos^2 x(1 + \sin x)}{1 - \sin^2 x} = \dfrac{\cos^2 x(1 + \sin x)}{\cos^2 x} = 1 + \sin x$

8. $\dfrac{1 + \cos \theta}{\sin \theta} + \dfrac{\sin \theta}{1 + \cos \theta} = \dfrac{\left(1 + \cos \theta\right)^2 + \sin^2 \theta}{\sin \theta(1 + \cos \theta)}$

$$= \frac{1 + 2 \cos \theta + \cos^2 \theta + \sin^2 \theta}{\sin \theta(1 + \cos \theta)} = \frac{2 + 2 \cos \theta}{\sin \theta(1 + \cos \theta)} = \frac{2}{\sin \theta} = 2 \csc \theta$$

9. $\tan^4 x + 2 \tan^2 x + 1 = \left(\tan^2 x + 1\right)^2 = \left(\sec^2 x\right)^2 = \sec^4 x$

10. (a) $\sin 105° = \sin(60° + 45°) = \sin 60° \cos 45° + \cos 60° \sin 45°$

$$= \frac{\sqrt{3}}{2} \cdot \frac{\sqrt{2}}{2} + \frac{1}{2} \cdot \frac{\sqrt{2}}{2} = \frac{\sqrt{2}}{4}\left(\sqrt{3} + 1\right)$$

    (b) $\tan 15° = \tan(60° - 45°) = \dfrac{\tan 60° - \tan 45°}{1 + \tan 60° \tan 45°}$

$$= \frac{\sqrt{3} - 1}{1 + \sqrt{3}} \cdot \frac{1 - \sqrt{3}}{1 - \sqrt{3}} = \frac{2\sqrt{3} - 1 - 3}{1 - 3} = \frac{2\sqrt{3} - 4}{-2} = 2 - \sqrt{3}$$

11. $(\sin 42°) \cos 38° - (\cos 42°) \sin 38° = \sin(42° - 38°) = \sin 4°$

12. $\tan\left(\theta + \dfrac{\pi}{4}\right) = \dfrac{\tan \theta + \tan\left(\dfrac{\pi}{4}\right)}{1 - (\tan \theta) \tan\left(\dfrac{\pi}{4}\right)} = \dfrac{\tan \theta + 1}{1 - \tan \theta(1)} = \dfrac{1 + \tan \theta}{1 - \tan \theta}$

13. $\sin(\arcsin x - \arccos x) = \sin(\arcsin x) \cos(\arccos x) - \cos(\arcsin x) \sin(\arccos x)$

$$= (x)(x) - \left(\sqrt{1 - x^2}\right)\left(\sqrt{1 - x^2}\right) = x^2 - \left(1 - x^2\right) = 2x^2 - 1$$

14. (a) $\cos(120°) = \cos\left[2(60°)\right] = 2 \cos^2 60° - 1 = 2\left(\dfrac{1}{2}\right)^2 - 1 = -\dfrac{1}{2}$

    (b) $\tan(300°) = \tan\left[2(150°)\right] = \dfrac{2 \tan 150°}{1 - \tan^2 150°} = \dfrac{-\dfrac{2\sqrt{3}}{3}}{1 - \left(\dfrac{1}{3}\right)} = -\sqrt{3}$

15. (a) $\sin 22.5° = \sin \dfrac{45°}{2} = \sqrt{\dfrac{1 - \cos 45°}{2}} = \sqrt{\dfrac{1 - \dfrac{\sqrt{2}}{2}}{2}} = \dfrac{\sqrt{2 - \sqrt{2}}}{2}$

    (b) $\tan \dfrac{\pi}{12} = \tan \dfrac{\dfrac{\pi}{6}}{2} = \dfrac{\sin \dfrac{\pi}{6}}{1 + \cos\left(\dfrac{\pi}{6}\right)} = \dfrac{\dfrac{1}{2}}{1 + \dfrac{\sqrt{3}}{2}} = \dfrac{1}{2 + \sqrt{3}} = 2 - \sqrt{3}$

16. $\sin \theta = \dfrac{4}{5}$, θ lies in Quadrant II $\Rightarrow \cos \theta = -\dfrac{3}{5}$.

$$\cos \frac{\theta}{2} = \sqrt{\frac{1 + \cos \theta}{2}} = \sqrt{\frac{1 - \dfrac{3}{5}}{2}} = \sqrt{\frac{2}{10}} = \frac{1}{\sqrt{5}} = \frac{\sqrt{5}}{5}$$

17. $\left(\sin^2 x\right) \cos^2 x = \dfrac{1 - \cos 2x}{2} \cdot \dfrac{1 + \cos 2x}{2} = \dfrac{1}{4}\left[1 - \cos^2 2x\right] = \dfrac{1}{4}\left[1 - \dfrac{1 + \cos 4x}{2}\right]$

$$= \frac{1}{8}\left[2 - (1 + \cos 4x)\right] = \frac{1}{8}\left[1 - \cos 4x\right]$$

18. $6(\sin 5\theta) \cos 2\theta = 6\left\{\dfrac{1}{2}\left[\sin(5\theta + 2\theta) + \sin(5\theta - 2\theta)\right]\right\} = 3\left[\sin 7\theta + \sin 3\theta\right]$

19. $\sin(x + \pi) + \sin(x - \pi) = 2\left(\sin \dfrac{[(x + \pi) + (x - \pi)]}{2} \right) \cos \dfrac{[(x + \pi) - (x - \pi)]}{2}$

$$= 2 \sin x \cos \pi = -2 \sin x$$

20. $\dfrac{\sin 9x + \sin 5x}{\cos 9x - \cos 5x} = \dfrac{2 \sin 7x \cos 2x}{-2 \sin 7x \sin 2x} = -\dfrac{\cos 2x}{\sin 2x} = -\cot 2x$

21. $\frac{1}{2}[\sin(u + v) - \sin(u - v)] = \frac{1}{2}\{(\sin u) \cos v + (\cos u) \sin v - [(\sin u) \cos v - (\cos u) \sin v]\}$

$$= \frac{1}{2}[2(\cos u) \sin v] = (\cos u) \sin v$$

22. $4 \sin^2 x = 1$

$\sin^2 x = \dfrac{1}{4}$

$\sin x = \pm \dfrac{1}{2}$

$\sin x = \dfrac{1}{2}$ or $\sin x = -\dfrac{1}{2}$

$x = \dfrac{\pi}{6}$ or $\dfrac{5\pi}{6}$ $x = \dfrac{7\pi}{6}$ or $\dfrac{11\pi}{6}$

23. $\tan^2 \theta + \left(\sqrt{3} - 1\right) \tan \theta - \sqrt{3} = 0$

$\left(\tan \theta - 1\right)\left(\tan \theta + \sqrt{3}\right) = 0$

$\tan \theta = 1$ or $\tan \theta = -\sqrt{3}$

$\theta = \dfrac{\pi}{4}$ or $\dfrac{5\pi}{4}$ $\theta = \dfrac{2\pi}{3}$ or $\dfrac{5\pi}{3}$

24. $\sin 2x = \cos x$

$2(\sin x) \cos x - \cos x = 0$

$\cos x(2 \sin x - 1) = 0$

$\cos x = 0$ or $\sin x = \dfrac{1}{2}$

$x = \dfrac{\pi}{2}$ or $\dfrac{3\pi}{2}$ $x = \dfrac{\pi}{6}$ or $\dfrac{5\pi}{6}$

25. $\tan^2 x - 6 \tan x + 4 = 0$

$\tan x = \dfrac{-(-6) \pm \sqrt{(-6)^2 - 4(1)(4)}}{2(1)}$

$\tan x = \dfrac{6 \pm \sqrt{20}}{2} = 3 \pm \sqrt{5}$

$\tan x = 3 + \sqrt{5}$ or $\tan x = 3 - \sqrt{5}$

$x \approx 1.3821$ or 4.5237 $x \approx 0.6524$ or 3.7940

Chapter 8   Practice Test Solutions

1. $C = 180° - \left(40° + 12°\right) = 128°$

$a = \sin 40°\left(\dfrac{100}{\sin 12°} \right) \approx 309.164$

$c = \sin 128°\left(\dfrac{100}{\sin 12°} \right) \approx 379.012$

2. $\sin A = 5\left(\dfrac{\sin 150°}{20} \right) = 0.125$

$A \approx 7.181°$

$B \approx 180° - \left(150° + 7.181°\right) = 22.819°$

$b = \sin 22.819°\left(\dfrac{20}{\sin 150°} \right) \approx 15.513$

3. Area $= \frac{1}{2}ab \sin C = \frac{1}{2}(3)(6) \sin 130° \approx 6.894$ square units

4. $h = b \sin A = 35 \sin 22.5° \approx 13.394$

$a = 10$

Since $a < h$ and A is acute, the triangle has no solution.

5. $\cos A = \dfrac{(53)^2 + (38)^2 - (49)^2}{2(53)(38)} \approx 0.4598$

$A \approx 62.627°$

$\cos B = \dfrac{(49)^2 + (38)^2 - (53)^2}{2(49)(38)} \approx 0.2782$

$B \approx 73.847°$

$C \approx 180° - \left(62.627° + 73.847°\right)$

$= 43.526°$

6. $c^2 = (100)^2 + (300)^2 - 2(100)(300)\cos 29°$

$\approx 47{,}522.8176$

$c \approx 218$

$\cos A = \dfrac{(300)^2 + (218)^2 - (100)^2}{2(300)(218)} \approx 0.97495$

$A \approx 12.85°$

$B \approx 180° - (12.85° + 29°) = 138.15°$

7. $s = \dfrac{a + b + c}{2} = \dfrac{4.1 + 6.8 + 5.5}{2} = 8.2$

Area $= \sqrt{s(s - a)(s - b)(s - c)}$

$= \sqrt{8.2(8.2 - 4.1)(8.2 - 6.8)(8.2 - 5.5)}$

≈ 11.273 square units

8. $x^2 = (40)^2 + (70)^2 - 2(40)(70)\cos 168°$

$\approx 11{,}977.6266$

$x \approx 190.442$ miles

9. $\mathbf{w} = 4(3\mathbf{i} + \mathbf{j}) - 7(-\mathbf{i} + 2\mathbf{j})$

$= 19\mathbf{i} - 10\mathbf{j}$

10. $\dfrac{\mathbf{v}}{\|\mathbf{v}\|} = \dfrac{5\mathbf{i} - 3\mathbf{j}}{\sqrt{25 + 9}} = \dfrac{5}{\sqrt{34}}\mathbf{i} - \dfrac{3}{\sqrt{34}}\mathbf{j}$

$= \dfrac{5\sqrt{34}}{34}\mathbf{i} - \dfrac{3\sqrt{34}}{34}\mathbf{j}$

11. $\mathbf{u} = 6\mathbf{i} + 5\mathbf{j}, \ \mathbf{v} = 2\mathbf{i} - 3\mathbf{j}$

$\mathbf{u} \cdot \mathbf{v} = 6(2) + 5(-3) = -3$

$\|\mathbf{u}\| = \sqrt{61}, \qquad \|\mathbf{v}\| = \sqrt{13}$

$\cos \theta = \dfrac{-3}{\sqrt{61}\sqrt{13}}$

$\theta \approx 96.116°$

12. $4(\mathbf{i} \cos 30° + \mathbf{j} \sin 30°) = 4\left(\dfrac{\sqrt{3}}{2}\mathbf{i} + \dfrac{1}{2}\mathbf{j}\right)$

$= \langle 2\sqrt{3}, 2\rangle$

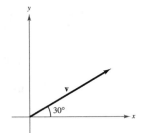

13. $\text{proj}_{\mathbf{v}}\mathbf{u} = \left(\dfrac{\mathbf{u} \cdot \mathbf{v}}{\|\mathbf{v}\|^2}\right)\mathbf{v} = \dfrac{-10}{20}\langle -2, 4\rangle = \langle 1, -2\rangle$

14. $r = \sqrt{25 + 25} = \sqrt{50} = 5\sqrt{2}$

$\tan \theta = \dfrac{-5}{5} = -1$

Because z is in Quadrant IV, $\theta = 315°$.

$z = 5\sqrt{2}(\cos 315° + i \sin 315°)$

15. $\cos 225° = -\dfrac{\sqrt{2}}{2}, \sin 225° = -\dfrac{\sqrt{2}}{2}$

$z = 6\left(-\dfrac{\sqrt{2}}{2} - i\dfrac{\sqrt{2}}{2}\right) = -3\sqrt{2} - 3\sqrt{2}i$

16. $\left[7(\cos 23° + i \sin 23°)\right]\left[4(\cos 7° + i \sin 7°)\right] = 7(4)\left[\cos(23° + 7°) + i \sin(23° + 7°)\right]$

$= 28(\cos 30° + i \sin 30°)$

17. $\dfrac{9\left(\cos \dfrac{5\pi}{4} + i \sin \dfrac{5\pi}{4}\right)}{3(\cos \pi + i \sin \pi)} = \dfrac{9}{3}\left[\cos\left(\dfrac{5\pi}{4} - \pi\right) + i \sin\left(\dfrac{5\pi}{4} - \pi\right)\right] = 3\left(\cos \dfrac{\pi}{4} + i \sin \dfrac{\pi}{4}\right)$

18. $(2 + 2i)^8 = \left[2\sqrt{2}(\cos 45° + i \sin 45°)\right]^8 = \left(2\sqrt{2}\right)^8\left[\cos(8)(45°) + i \sin(8)(45°)\right]$

$= 4096[\cos 360° + i \sin 360°] = 4096$

19. $z = 8\left(\cos\dfrac{\pi}{3} + i\sin\dfrac{\pi}{3}\right), n = 3$

The cube roots of z are: $\sqrt[3]{8}\left[\cos\dfrac{\left(\frac{\pi}{3}\right) + 2\pi k}{3} + i\sin\dfrac{\left(\frac{\pi}{3}\right) + 2\pi k}{3}\right], k = 0, 1, 2$

For $k = 0$: $\sqrt[3]{8}\left[\cos\dfrac{\frac{\pi}{3}}{3} + i\sin\dfrac{\frac{\pi}{3}}{3}\right] = 2\left(\cos\dfrac{\pi}{9} + i\sin\dfrac{\pi}{9}\right)$

For $k = 1$: $\sqrt[3]{8}\left[\cos\dfrac{\left(\frac{\pi}{3}\right) + 2\pi}{3} + i\sin\dfrac{\left(\frac{\pi}{3}\right) + 2\pi}{3}\right] = 2\left(\cos\dfrac{7\pi}{9} + i\sin\dfrac{7\pi}{9}\right)$

For $k = 2$: $\sqrt[3]{8}\left[\cos\dfrac{\left(\frac{\pi}{3}\right) + 4\pi}{3} + i\sin\dfrac{\left(\frac{\pi}{3}\right) + 4\pi}{3}\right] = 2\left(\cos\dfrac{13\pi}{9} + i\sin\dfrac{13\pi}{9}\right)$

20. $x^4 = -i = 1\left(\cos\dfrac{3\pi}{2} + i\sin\dfrac{3\pi}{2}\right)$

The fourth roots are: $\sqrt[4]{1}\left[\cos\dfrac{\left(\frac{3\pi}{2}\right) + 2\pi k}{4} + i\sin\dfrac{\left(\frac{3\pi}{2}\right) + 2\pi k}{4}\right], k = 0, 1, 2, 3$

For $k = 0$: $\cos\dfrac{\frac{3\pi}{2}}{4} + i\sin\dfrac{\frac{3\pi}{2}}{4} = \cos\dfrac{3\pi}{8} + i\sin\dfrac{3\pi}{8}$

For $k = 1$: $\cos\dfrac{\left(\frac{3\pi}{2}\right) + 2\pi}{4} + i\sin\dfrac{\left(\frac{3\pi}{2}\right) + 2\pi}{4} = \cos\dfrac{7\pi}{8} + i\sin\dfrac{7\pi}{8}$

For $k = 2$: $\cos\dfrac{\left(\frac{3\pi}{2}\right) + 4\pi}{4} + i\sin\dfrac{\left(\frac{3\pi}{2}\right) + 4\pi}{4} = \cos\dfrac{11\pi}{8} + i\sin\dfrac{11\pi}{8}$

For $k = 3$: $\cos\dfrac{\left(\frac{3\pi}{2}\right) + 6\pi}{4} + i\sin\dfrac{\left(\frac{3\pi}{2}\right) + 6\pi}{4} = \cos\dfrac{15\pi}{8} + i\sin\dfrac{15\pi}{8}$

Chapter 9   Practice Test Solutions

1. $\begin{cases} x + y = 1 \\ 3x - y = 15 \Rightarrow y = 3x - 15 \end{cases}$

$x + (3x - 15) = 1$

$4x = 16$

$x = 4$

$y = -3$

Solution: $(4, -3)$

2. $\begin{cases} x - 3y = -3 \Rightarrow x = 3y - 3 \\ x^2 + 6y = 5 \end{cases}$

$(3y - 3)^2 + 6y = 5$

$9y^2 - 18y + 9 + 6y = 5$

$9y^2 - 12y + 4 = 0$

$(3y - 2)^2 = 0$

$y = \frac{2}{3}$

$x = -1$

Solution: $\left(-1, \frac{2}{3}\right)$

3. $\begin{cases} x + y + z = 6 \Rightarrow z = 6 - x - y \\ 2x - y + 3z = 0 \Rightarrow 2x - y + 3(6 - x - y) = 0 \Rightarrow -x - 4y = -18 \Rightarrow x = 18 - 4y \\ 5x + 2y - z = -3 \Rightarrow 5x + 2y - (6 - x - y) = -3 \Rightarrow 6x + 3y = 3 \end{cases}$

$$6(18 - 4y) + 3y = 3$$
$$-21y = -105$$
$$y = 5$$
$$x = 18 - 4y = -2$$
$$z = 6 - x - y = 3$$

Solution: $(-2, 5, 3)$

4. $x + y = 110 \Rightarrow y = 110 - x$

$$xy = 2800$$

$$x(110 - x) = 2800$$
$$0 = x^2 - 110x + 2800$$
$$0 = (x - 40)(x - 70)$$

$x = 40$ or $x = 70$
$y = 70 \qquad y = 40$

Solution: The two numbers are 40 and 70.

5. $2x + 2y = 170 \Rightarrow y = \dfrac{170 - 2x}{2} = 85 - x$

$$xy = 1500$$
$$x(85 - x) = 1500$$
$$0 = x^2 - 85x + 1500$$
$$0 = (x - 25)(x - 60)$$

$x = 25$ or $x = 60$
$y = 60 \qquad y = 25$

Dimensions: $60 \text{ ft} \times 25 \text{ ft}$

6. $\begin{cases} 2x + 15y = 4 \Rightarrow 2x + 15y = 4 \\ x - 3y = 23 \Rightarrow 5x - 15y = 115 \end{cases}$
$$\overline{ 7x = 119}$$
$$x = 17$$
$$y = \frac{x - 23}{3}$$
$$= -2$$

Solution: $(17, -2)$

7. $\begin{cases} x + y = 2 \Rightarrow 19x + 19y = 38 \\ 38x - 19y = 7 \Rightarrow 38x - 19y = 7 \end{cases}$
$$\overline{ 57x = 45}$$
$$x = \frac{15}{19}$$
$$y = 2 - x$$
$$= \frac{38}{19} - \frac{15}{19}$$
$$= \frac{23}{19}$$

Solution: $\left(\frac{15}{19}, \frac{23}{19} \right)$

8. $\begin{cases} 0.4x + 0.5y = 0.112 \Rightarrow 0.28x + 0.35y = 0.0784 \\ 0.3x - 0.7y = -0.131 \Rightarrow 0.15x - 0.35y = -0.0655 \end{cases}$
$$\overline{ 0.43x = 0.0129}$$

$$x = \frac{0.0129}{0.43} = 0.03$$
$$y = \frac{0.112 - 0.4x}{0.5} = 0.20$$

Solution: $(0.03, 0.20)$

9. Let $x =$ amount in 11% fund and $y =$ amount in 13% fund.

$$x + y = 17{,}000 \Rightarrow y = 17{,}000 - x$$
$$0.11x + 0.13y = 2080$$
$$0.11x + 0.13(17{,}000 - x) = 2080$$
$$-0.02x = -130$$
$$x = \$6500 \quad \text{at } 11\%$$
$$y = \$10{,}500 \text{ at } 13\%$$

10. $(4, 3), (1, 1), (-1, -2), (-2, -1)$

Use a calculator.

$$y = ax + b = \tfrac{11}{14}x - \tfrac{1}{7}$$

11. $\begin{cases} x + y = -2 \\ 2x - y + z = 11 \\ 4y - 3z = -20 \end{cases}$

$\begin{cases} x + y = -2 \\ -3y + z = 15 \\ 4y - 3z = -20 \end{cases}$ $-2\text{Eq.1} + \text{Eq.2}$

$\begin{cases} x + y = -2 \\ y - 2z = -5 \\ 4y - 3z = -20 \end{cases}$ $\text{Eq.3} + \text{Eq.2}$

$\begin{cases} x + y = -2 \\ y - 2z = -5 \\ 5z = 0 \end{cases}$ $-4\text{Eq.2} + \text{Eq.3}$

$\begin{cases} x + y = -2 \\ y - 2z = -5 \\ z = 0 \end{cases}$

$y - 2(0) = -5 \Rightarrow y = -5$

$x + (-5) = -2 \Rightarrow x = 3$

Solution: $(3, -5, 0)$

12. $\begin{cases} 4x - y + 5z = 4 \\ 2x + y - z = 0 \\ 2x + 4y + 8z = 0 \end{cases}$

$\begin{cases} 2x + 4y + 8z = 0 \\ 2x + y - z = 0 \\ 4x - y + 5z = 4 \end{cases}$ Interchange equations.

$\begin{cases} 2x + 4y + 8z = 0 \\ -3y - 9z = 0 \\ -9y - 11z = 4 \end{cases}$ $\begin{aligned} &-\text{Eq.1} + \text{Eq.2} \\ &-2\text{Eq.1} + \text{Eq.3} \end{aligned}$

$\begin{cases} 2x + 4y + 8z = 0 \\ -3y - 9z = 0 \\ 16z = 4 \end{cases}$ $-3\text{Eq.2} + \text{Eq.3}$

$\begin{cases} x + 2y + 4z = 0 \\ y + 3z = 0 \\ z = \frac{1}{4} \end{cases}$ $\begin{aligned} &\tfrac{1}{2}\text{Eq.1} \\ &-\tfrac{1}{3}\text{Eq.2} \\ &\tfrac{1}{16}\text{Eq.3} \end{aligned}$

$y + 3\left(\frac{1}{4}\right) = 0 \Rightarrow y = -\frac{3}{4}$

$x + 2\left(-\frac{3}{4}\right) + 4\left(\frac{1}{4}\right) = 0 \Rightarrow x = \frac{1}{2}$

Solution: $\left(-\frac{1}{2}, -\frac{3}{4}, \frac{1}{4}\right)$

13. $\begin{cases} 3x + 2y - z = 5 \\ 6x - y + 5z = 2 \end{cases}$

$\begin{cases} 3x + 2y - z = 5 \\ -5y + 7z = -8 \end{cases}$ $-2\text{Eq.1} + \text{Eq.2}$

$\begin{cases} x + \frac{2}{3}y - \frac{1}{3}z = \frac{5}{3} \\ y - \frac{7}{5}z = \frac{8}{5} \end{cases}$ $\begin{aligned} &\tfrac{1}{3}\text{Eq.1} \\ &-\tfrac{1}{5}\text{Eq.2} \end{aligned}$

Let $a = z$.

Then $y = \frac{7}{5}a + \frac{8}{5}$, and $x + \frac{2}{3}\left(\frac{7}{5}a + \frac{8}{5}\right) - \frac{1}{3}a = \frac{5}{3}$

$x + \frac{3}{5}a = \frac{3}{5}$

$x = -\frac{3}{5}a + \frac{3}{5}$.

Solution: $\left(-\frac{3}{5}a + \frac{3}{5}, \frac{7}{5}a + \frac{8}{5}, a\right)$ where a is any real number

14. $y = ax^2 + bx + c$ passes through $(0, -1)$, $(1, 4)$, and $(2, 13)$.

At $(0, -1)$: $-1 = a(0)^2 + b(0) + c \Rightarrow c = -1$

At $(1, 4)$: $4 = a(1)^2 + b(1) - 1 \Rightarrow 5 = a + b \Rightarrow 5 = a + b$

At $(2, 13)$: $13 = a(2)^2 + b(2) - 1 \Rightarrow 14 = 4a + 2b \Rightarrow \underline{-7 = -2a - b}$

$-2 = -a$

$a = 2$

$b = 3$

So, the equation of the parabola is $y = 2x^2 + 3x - 1$.

15. $s = \frac{1}{2}at^2 + v_0t + s_0$ passes through $(1, 12), (2, 5),$ and $(3, 4)$.

At $(1, 12)$: $12 = \frac{1}{2}a + v_0 + s_0$
At $(2, 5)$: $5 = 2a + 2v_0 + s_0$
At $(3, 4)$: $4 = \frac{9}{2}a + 3v_0 + s_0$

$$\begin{cases} a + 2v_0 + 2s_0 = 24 \\ 2a + 2v_0 + s_0 = 5 \\ 9a + 6v_0 + 2s_0 = 8 \end{cases}$$

$$\begin{cases} a + 2v_0 + 2s_0 = 24 \\ -2v_0 - 3s_0 = -43 \qquad -2\text{Eq.1} + \text{Eq.2} \\ -12v_0 - 16s_0 = -208 \qquad -9\text{Eq.1} + \text{Eq.3} \end{cases}$$

$$\begin{cases} a + 2v_0 + 2s_0 = 24 \\ -2v_0 - 3s_0 = -43 \\ 2s_0 = 50 \qquad -6\text{Eq.2} + \text{Eq.3} \end{cases}$$

$$\begin{cases} a + 2v_0 + 2s_0 = 24 \\ v_0 + \frac{3}{2}s_0 = \frac{43}{2} \qquad -\frac{1}{2}\text{Eq.2} \\ s_0 = 25 \qquad \frac{1}{2}\text{Eq.3} \end{cases}$$

$s_0 = 25$

$v_0 + \frac{3}{2}(25) = \frac{43}{2} \Rightarrow v_0 = 16$

$a + 2(-16) + 2(25) = 24 \Rightarrow a = 6$

So, $s = \frac{1}{2}(6)t^2 - 16t + 25 = 3t^2 - 16t + 25$.

16. $x^2 + y^2 \geq 9$

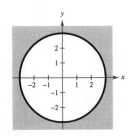

17. $\begin{cases} x + y \leq 6 \\ x \geq 2 \\ y \geq 0 \end{cases}$

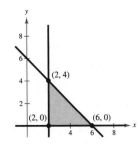

18. Line through $(0, 0)$ and $(0, 7)$: $x = 0$

Line through $(0, 0)$ and $(2, 3)$:

$y = \frac{3}{2}x$ or $3x - 2y = 0$

Line through $(0, 7)$ and $(2, 3)$:

$y = -2x + 7$ or $2x + y = 7$

Inequalities: $\begin{cases} x \geq 0 \\ 3x - 2y \leq 0 \\ 2x + y \leq 7 \end{cases}$

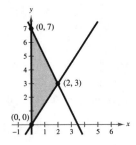

19. Vertices $(0, 0), (0, 7), (6, 0), (3, 5)$

$z = 30x + 26y$

At $(0, 0): z = 0$

At $(0, 7): z = 182$

At $(6, 0): z = 180$

At $(3, 5): z = 220$

The maximum value of z occurs at $(3, 5)$ and is 220.

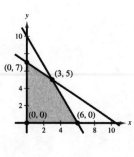

20. $x^2 + y^2 \le 4$

$(x - 2)^2 + y^2 \ge 4$

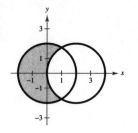

21. $\dfrac{1 - 2x}{x^2 + x} = \dfrac{1 - 2x}{x(x + 1)} = \dfrac{A}{x} + \dfrac{B}{x + 1}$

$1 - 2x = A(x + 1) + Bx$

When $x = 0, 1 = A.$

When $x = -1, 3 = -B \Rightarrow B = -3.$

$\dfrac{1 - 2x}{x^2 + x} = \dfrac{1}{x} - \dfrac{3}{x + 1}$

22. $\dfrac{6x - 17}{(x - 3)^2} = \dfrac{A}{x - 3} + \dfrac{B}{(x - 3)^2}$

$6x - 17 = A(x - 3) + B$

When $x = 3, 1 = B.$

When $x = 0, -17 = -3A + B \Rightarrow A = 6.$

$\dfrac{6x - 17}{(x - 3)^2} = \dfrac{6}{x - 3} + \dfrac{1}{(x - 3)^2}$

Chapter 10   Practice Test Solutions

1.
$$\begin{bmatrix} 1 & -2 & 4 \\ 3 & -5 & 9 \end{bmatrix}$$

$-3R_1 + R_2 \rightarrow \begin{bmatrix} 1 & -2 & 4 \\ 0 & 1 & -3 \end{bmatrix}$

$2R_2 + R_1 \rightarrow \begin{bmatrix} 1 & 0 & -2 \\ 0 & 1 & -3 \end{bmatrix}$

2. $\begin{cases} 3x + 5y = 3 \\ 2x - y = -11 \end{cases}$

$$\begin{bmatrix} 3 & 5 & \vdots & 3 \\ 2 & -1 & \vdots & -11 \end{bmatrix}$$

$-R_2 + R_1 \rightarrow \begin{bmatrix} 1 & 6 & \vdots & 14 \\ 2 & -1 & \vdots & -11 \end{bmatrix}$

$-2R_1 + R_2 \rightarrow \begin{bmatrix} 1 & 6 & \vdots & 14 \\ 0 & -13 & \vdots & -39 \end{bmatrix}$

$-\frac{1}{13}R_2 \rightarrow \begin{bmatrix} 1 & 6 & \vdots & 14 \\ 0 & 1 & \vdots & 3 \end{bmatrix}$

$-6R_2 + R_1 \rightarrow \begin{bmatrix} 1 & 0 & \vdots & -4 \\ 0 & 1 & \vdots & 3 \end{bmatrix}$

$x = -4, y = 3$

Solution: $(-4, 3)$

3. $\begin{cases} 2x + 3y = -3 \\ 3x - 2y = 8 \\ x + y = 1 \end{cases}$

$$\begin{bmatrix} 2 & 3 & \vdots & -3 \\ 3 & 2 & \vdots & 8 \\ 1 & 1 & \vdots & 1 \end{bmatrix}$$

$\begin{matrix} R_3 \\ \\ R_1 \end{matrix} \begin{bmatrix} 1 & 1 & \vdots & 1 \\ 3 & 2 & \vdots & 8 \\ 2 & 3 & \vdots & -3 \end{bmatrix}$

$\begin{matrix} -3R_1 + R_2 \rightarrow \\ -2R_1 + R_3 \rightarrow \end{matrix} \begin{bmatrix} 1 & 1 & \vdots & 1 \\ 0 & -1 & \vdots & 5 \\ 0 & 1 & \vdots & -5 \end{bmatrix}$

$-R_2 \rightarrow \begin{bmatrix} 1 & 1 & \vdots & 1 \\ 0 & 1 & \vdots & -5 \\ 0 & 1 & \vdots & -5 \end{bmatrix}$

$\begin{matrix} -R_2 + R_1 \rightarrow \\ \\ -R_2 + R_3 \rightarrow \end{matrix} \begin{bmatrix} 1 & 0 & \vdots & 6 \\ 0 & 1 & \vdots & -5 \\ 0 & 0 & \vdots & 0 \end{bmatrix}$

$x = 6, y = -5$

Solution: $(6, -5)$

4. $\begin{cases} x & + 3z = -5 \\ 2x + y & = 0 \\ 3x + y - z = -3 \end{cases}$

$$\begin{bmatrix} 1 & 0 & 3 & \vdots & -5 \\ 2 & 1 & 0 & \vdots & 0 \\ 3 & 1 & -1 & \vdots & 3 \end{bmatrix}$$

$\begin{matrix} \\ -2R_1 + R_2 \to \\ -3R_1 + R_3 \to \end{matrix} \begin{bmatrix} 1 & 0 & 3 & \vdots & -5 \\ 0 & 1 & -6 & \vdots & 10 \\ 0 & 1 & -10 & \vdots & 18 \end{bmatrix}$

$\begin{matrix} \\ \\ -R_2 + R_3 \to \end{matrix} \begin{bmatrix} 1 & 0 & 3 & \vdots & -5 \\ 0 & 1 & -6 & \vdots & 10 \\ 0 & 0 & -4 & \vdots & 8 \end{bmatrix}$

$\begin{matrix} \\ \\ -\frac{1}{4}R_3 \to \end{matrix} \begin{bmatrix} 1 & 0 & 3 & \vdots & -5 \\ 0 & 1 & -6 & \vdots & 10 \\ 0 & 0 & 1 & \vdots & -2 \end{bmatrix}$

$\begin{matrix} -3R_3 + R_1 \to \\ 6R_3 + R_2 \to \\ \\ \end{matrix} \begin{bmatrix} 1 & 0 & 0 & \vdots & 1 \\ 0 & 1 & 0 & \vdots & -2 \\ 0 & 0 & 1 & \vdots & -2 \end{bmatrix}$

$x = 1, y = -2, z = -2$

Solution: $(1, -2, -2)$

5. $\begin{bmatrix} 1 & 4 & 5 \\ 2 & 0 & -3 \end{bmatrix}\begin{bmatrix} 1 & 6 \\ 0 & -7 \\ -1 & 2 \end{bmatrix} = \begin{bmatrix} (1)(1) + (4)(0) + (5)(-1) & (1)(6) + (4)(-7) + (5)(2) \\ (2)(1) + (0)(0) + (-3)(-1) & (2)(6) + (0)(-7) + (-3)(2) \end{bmatrix} = \begin{bmatrix} -4 & -12 \\ 5 & 6 \end{bmatrix}$

6. $3A - 5B = 3\begin{bmatrix} 9 & 1 \\ -4 & 8 \end{bmatrix} - 5\begin{bmatrix} 6 & -2 \\ 3 & 5 \end{bmatrix}$

$= \begin{bmatrix} 27 & 3 \\ -12 & 24 \end{bmatrix} - \begin{bmatrix} 30 & -10 \\ 15 & 25 \end{bmatrix}$

$= \begin{bmatrix} -3 & 13 \\ -27 & -1 \end{bmatrix}$

7. $f(A) = \begin{bmatrix} 3 & 0 \\ 7 & 1 \end{bmatrix}^2 - 7\begin{bmatrix} 3 & 0 \\ 7 & 1 \end{bmatrix} + 8\begin{bmatrix} 1 & 0 \\ 0 & 1 \end{bmatrix}$

$= \begin{bmatrix} 3 & 0 \\ 7 & 1 \end{bmatrix}\begin{bmatrix} 3 & 0 \\ 7 & 1 \end{bmatrix} - \begin{bmatrix} 21 & 0 \\ 49 & 7 \end{bmatrix} + \begin{bmatrix} 8 & 0 \\ 0 & 8 \end{bmatrix}$

$= \begin{bmatrix} 9 & 0 \\ 28 & 1 \end{bmatrix} - \begin{bmatrix} 21 & 0 \\ 49 & 7 \end{bmatrix} + \begin{bmatrix} 8 & 0 \\ 0 & 8 \end{bmatrix}$

$= \begin{bmatrix} -4 & 0 \\ -21 & 2 \end{bmatrix}$

8. False.

$(A + B)(A + 3B) = A(A + 3B) + B(A + 3B)$

$\qquad\qquad\qquad = A^2 + 3AB + BA + 3B^2$ and, in general, $AB \neq BA$.

9.

$$\begin{bmatrix} 1 & 2 & \vdots & 1 & 0 \\ 3 & 5 & \vdots & 0 & 1 \end{bmatrix}$$

$$-3R_1 + R_2 \rightarrow \begin{bmatrix} 1 & 2 & \vdots & 1 & 0 \\ 0 & -1 & \vdots & -3 & 1 \end{bmatrix}$$

$$2R_2 + R_1 \rightarrow \begin{bmatrix} 1 & 0 & \vdots & -5 & 2 \\ 0 & -1 & \vdots & -3 & 1 \end{bmatrix}$$

$$-R_2 \rightarrow \begin{bmatrix} 1 & 0 & \vdots & -5 & 2 \\ 0 & 1 & \vdots & 3 & -1 \end{bmatrix}$$

$$A^{-1} = \begin{bmatrix} -5 & 2 \\ 3 & -1 \end{bmatrix}$$

10.

$$\begin{bmatrix} 1 & 1 & 1 & \vdots & 1 & 0 & 0 \\ 3 & 6 & 5 & \vdots & 0 & 1 & 0 \\ 6 & 10 & 8 & \vdots & 0 & 0 & 1 \end{bmatrix}$$

$$\begin{matrix} -3R_1 + R_2 \rightarrow \\ -6R_1 + R_3 \rightarrow \end{matrix} \begin{bmatrix} 1 & 1 & 1 & \vdots & 1 & 0 & 0 \\ 0 & 3 & 2 & \vdots & -3 & 1 & 0 \\ 0 & 4 & 2 & \vdots & -6 & 0 & 1 \end{bmatrix}$$

$$-R_3 + R_2 \rightarrow \begin{bmatrix} 1 & 1 & 1 & \vdots & 1 & 0 & 0 \\ 0 & -1 & 0 & \vdots & 3 & 1 & -1 \\ 0 & 4 & 2 & \vdots & -6 & 0 & 1 \end{bmatrix}$$

$$\begin{matrix} R_2 + R_1 \rightarrow \\ \\ 4R_2 + R_3 \rightarrow \end{matrix} \begin{bmatrix} 1 & 0 & 1 & \vdots & 4 & 1 & -1 \\ 0 & -1 & 0 & \vdots & 3 & 1 & -1 \\ 0 & 0 & 2 & \vdots & 6 & 4 & -3 \end{bmatrix}$$

$$\begin{matrix} -R_2 \rightarrow \\ \frac{1}{2}R_3 \rightarrow \end{matrix} \begin{bmatrix} 1 & 0 & 1 & \vdots & 4 & 1 & -1 \\ 0 & 1 & 0 & \vdots & -3 & -1 & 1 \\ 0 & 0 & 1 & \vdots & 3 & 2 & -\frac{3}{2} \end{bmatrix}$$

$$-R_3 + R_1 \rightarrow \begin{bmatrix} 1 & 0 & 0 & \vdots & 1 & -1 & \frac{1}{2} \\ 0 & 1 & 0 & \vdots & -3 & -1 & 1 \\ 0 & 0 & 1 & \vdots & 3 & 2 & -\frac{3}{2} \end{bmatrix}$$

$$A^{-1} = \begin{bmatrix} 1 & -1 & \frac{1}{2} \\ -3 & -1 & 1 \\ 3 & 2 & -\frac{3}{2} \end{bmatrix}$$

11. (a) $\begin{cases} x + 2y = 4 \\ 3x + 5y = 1 \end{cases}$

$$A = \begin{bmatrix} 1 & 2 \\ 3 & 5 \end{bmatrix}$$

$$A^{-1} = \frac{1}{5 - 6}\begin{bmatrix} 5 & -2 \\ -3 & 1 \end{bmatrix} = \begin{bmatrix} -5 & 2 \\ 3 & -1 \end{bmatrix}$$

$$\begin{bmatrix} x \\ y \end{bmatrix} = A^{-1}B = \begin{bmatrix} -5 & 2 \\ 3 & -1 \end{bmatrix}\begin{bmatrix} 4 \\ 1 \end{bmatrix} = \begin{bmatrix} -18 \\ 11 \end{bmatrix}$$

$$x = -18, \, y = 11$$

Solution: $(-18, 11)$

(b) $\begin{cases} x + 2y = 3 \\ 3x + 5y = -2 \end{cases}$

Again, $A^{-1} = \begin{bmatrix} -5 & 2 \\ 3 & -1 \end{bmatrix}$.

$$\begin{bmatrix} x \\ y \end{bmatrix} = A^{-1}B = \begin{bmatrix} -5 & 2 \\ 3 & -1 \end{bmatrix}\begin{bmatrix} 3 \\ -2 \end{bmatrix} = \begin{bmatrix} -19 \\ 11 \end{bmatrix}$$

$$x = -19, \, y = 11$$

Solution: $(-19, 11)$

12. $\begin{vmatrix} 6 & -1 \\ 3 & 4 \end{vmatrix} = 24 - (-3) = 27$

13. $\begin{vmatrix} 1 & 3 & -1 \\ 5 & 9 & 0 \\ 6 & 2 & -5 \end{vmatrix} = -1\begin{vmatrix} 5 & 9 \\ 6 & 2 \end{vmatrix} - 5\begin{vmatrix} 1 & 3 \\ 5 & 9 \end{vmatrix}$

$$= -(-44) - 5(-6) = 74$$

14. Expand along Row 2.

$$\begin{vmatrix} 1 & 4 & 2 & 3 \\ 0 & 1 & -2 & 0 \\ 3 & 5 & -1 & 1 \\ 2 & 0 & 6 & 1 \end{vmatrix} = \begin{vmatrix} 1 & 2 & 3 \\ 3 & -1 & 1 \\ 2 & 6 & 1 \end{vmatrix} + 2\begin{vmatrix} 1 & 4 & 3 \\ 3 & 5 & 1 \\ 2 & 0 & 1 \end{vmatrix}$$

$$= 51 + 2(-29) = -7$$

15. $\begin{vmatrix} 6 & 4 & 3 & 0 & 6 \\ 0 & 5 & 1 & 4 & 8 \\ 0 & 0 & 2 & 7 & 3 \\ 0 & 0 & 0 & 9 & 2 \\ 0 & 0 & 0 & 0 & 1 \end{vmatrix} = 6\begin{vmatrix} 5 & 1 & 4 & 8 \\ 0 & 2 & 7 & 3 \\ 0 & 0 & 9 & 2 \\ 0 & 0 & 0 & 1 \end{vmatrix} = 6(5)\begin{vmatrix} 2 & 7 & 3 \\ 0 & 9 & 2 \\ 0 & 0 & 1 \end{vmatrix} = 6(5)(2)\begin{vmatrix} 9 & 2 \\ 0 & 1 \end{vmatrix} = 6(5)(2)(9) = 540$

16. Area $= \dfrac{1}{2}\begin{vmatrix} 0 & 7 & 1 \\ 5 & 0 & 1 \\ 3 & 9 & 1 \end{vmatrix} = \dfrac{1}{2}(31) = \dfrac{31}{2}$

17. $\begin{vmatrix} x & y & 1 \\ 2 & 7 & 1 \\ -1 & 4 & 1 \end{vmatrix} = 3x - 3y + 15 = 0$ or $x - y + 5 = 0$

18. $x = \dfrac{\begin{vmatrix} 4 & -7 \\ 11 & 5 \end{vmatrix}}{\begin{vmatrix} 6 & -7 \\ 2 & 5 \end{vmatrix}} = \dfrac{97}{44}$

20. $y = \dfrac{\begin{vmatrix} 721.4 & 33.77 \\ 45.9 & 19.85 \end{vmatrix}}{\begin{vmatrix} 721.4 & -29.1 \\ 45.9 & 105.6 \end{vmatrix}} = \dfrac{12{,}769.747}{77{,}515.530} \approx 0.1647$

19. $z = \dfrac{\begin{vmatrix} 3 & 0 & 1 \\ 0 & 1 & 3 \\ 1 & -1 & 2 \end{vmatrix}}{\begin{vmatrix} 3 & 0 & 1 \\ 0 & 1 & 4 \\ 1 & -1 & 0 \end{vmatrix}} = \dfrac{14}{11}$

Chapter 11   Practice Test Solutions

1. $a_n = \dfrac{2n}{(n+2)!}$

$a_1 = \dfrac{2(1)}{3!} = \dfrac{2}{6} = \dfrac{1}{3}$

$a_2 = \dfrac{2(2)}{4!} = \dfrac{4}{24} = \dfrac{1}{6}$

$a_3 = \dfrac{2(3)}{5!} = \dfrac{6}{120} = \dfrac{1}{20}$

$a_4 = \dfrac{2(4)}{6!} = \dfrac{8}{720} = \dfrac{1}{90}$

$a_5 = \dfrac{2(5)}{7!} = \dfrac{10}{5040} = \dfrac{1}{504}$

Terms: $\dfrac{1}{3}, \dfrac{1}{6}, \dfrac{1}{20}, \dfrac{1}{90}, \dfrac{1}{504}$

2. $a_n = \dfrac{n+3}{3^n}$

3. $\displaystyle\sum_{i=1}^{6}(2i-1) = 1+3+5+7+9+11 = 36$

4. $a_1 = 23, d = -2$

$a_2 = 23 + (-2) = 21$

$a_3 = 21 + (-2) = 19$

$a_4 = 19 + (-2) = 17$

$a_5 = 17 + (-2) = 15$

Terms: 23, 21, 19, 17, 15

5. $a_1 = 12, d = 3, n = 50$

$a_n = a_1 + (n-1)d$

$a_{50} = 12 + (50-1)3 = 159$

6. $a_1 = 1$

$a_{200} = 200$

$S_n = \dfrac{n}{2}(a_1 + a_n)$

$S_{200} = \dfrac{200}{2}(1 + 200) = 20{,}100$

7. $a_1 = 7, r = 2$

$a_2 = 7(2) = 14$

$a_3 = 7(2)^2 = 28$

$a_4 = 7(2)^3 = 56$

$a_5 = 7(2)^4 = 112$

Terms: 7, 14, 28, 56, 112

8. $\displaystyle\sum_{n=1}^{10} 6\left(\dfrac{2}{3}\right)^{n-1}, a_1 = 6, r = \dfrac{2}{3}, n = 10$

$S_n = \dfrac{a_1(1-r^n)}{1-r} = \dfrac{6\left[1 - \left(\dfrac{2}{3}\right)^{10}\right]}{1 - \dfrac{2}{3}} = 18\left(1 - \dfrac{1024}{59{,}049}\right) = \dfrac{116{,}050}{6561} \approx 17.6879$

9. $\sum_{n=0}^{\infty} (0.03)^n = \sum_{n=1}^{\infty} (0.03)^{n-1}, a_1 = 1, r = 0.03$

$S = \dfrac{a_1}{1-r} = \dfrac{1}{1-0.03} = \dfrac{1}{0.97} = \dfrac{100}{97} \approx 1.0309$

10. For $n = 1, 1 = \dfrac{1(1+1)}{2}$.

Assume that $S_k = 1 + 2 + 3 + 4 + \cdots + k = \dfrac{k(k+1)}{2}$.

Then $S_{k+1} = 1 + 2 + 3 + 4 + \cdots + k + (k+1) = \dfrac{k(k+1)}{2} + k + 1$

$= \dfrac{k(k+1)}{2} + \dfrac{2(k+1)}{2}$

$= \dfrac{(k+1)(k+2)}{2}$.

Thus, by the principle of mathematical induction, $1 + 2 + 3 + 4 + \cdots + n = \dfrac{n(n+1)}{2}$ for all integers $n \geq 1$.

11. For $n = 4, 4! > 2^4$. Assume that $k! > 2^k$.

Then $(k+1)! = (k+1)(k!) > (k+1)2^k > 2 \cdot 2^k = 2^{k+1}$.

Thus, by the extended principle of mathematical induction, $n! > 2^n$ for all integers $n \geq 4$.

12. $_{13}C_4 = \dfrac{13!}{(13-4)!4!} = 715$

13. $(x+3)^5 = x^5 + 5x^4(3) + 10x^3(3)^2 + 10x^2(3)^3 + 5x(3)^4 + (3)^5$

$= x^5 + 15x^4 + 90x^3 + 270x^2 + 405x + 243$

14. $-_{12}C_5x^7(2)^5 = -25,344x^7$

15. $_{30}P_4 = \dfrac{30!}{(30-4)!} = 657,720$

16. $6! = 720$ ways

17. $_{12}P_3 = 1320$

18. $P(2) + P(3) + P(4) = \frac{1}{36} + \frac{2}{36} + \frac{3}{36}$

$= \frac{6}{36} = \frac{1}{6}$

19. $P(\text{K, B10}) = \frac{4}{52} \cdot \frac{2}{51} = \frac{2}{663}$

20. Let A = probability of no faulty units.

$P(A) = \left(\dfrac{997}{1000}\right)^{50} \approx 0.8605$

$P(A') = 1 - P(A) \approx 0.1395$

PART II

Chapter P   Chapter Test Solutions

1. $-\frac{10}{3} = -3\frac{1}{3}$

$-|-4| = -4$

$-\frac{10}{3} > -|-4|$

2. $\left|-5.4 - 3\frac{3}{4}\right| = 9.15$

3. $(5 - x) + 0 = 5 - x$

Additive Identity Property

4. (a) $27\left(-\frac{2}{3}\right) = -18$

(b) $\frac{5}{18} \div \frac{5}{8} = \frac{5}{18} \cdot \frac{8}{5} = \frac{4}{9}$

(c) $\left(-\frac{3}{5}\right)^3 = -\frac{27}{125}$

(d) $\left(\frac{3^2}{2}\right)^{-3} = \left(\frac{2}{9}\right)^3 = \frac{8}{729}$

5. (a) $\sqrt{5} \cdot \sqrt{125} = \sqrt{625} = 25$

(b) $\frac{\sqrt{27}}{\sqrt{2}} = \frac{3\sqrt{3}}{\sqrt{2}} \cdot \frac{\sqrt{2}}{\sqrt{2}} = \frac{3\sqrt{6}}{2}$

(c) $\frac{5.4 \times 10^8}{3 \times 10^3} = \frac{5.4}{3} \times 10^{8-3} = 1.8 \times 10^5$

(d) $\left(3 \times 10^4\right)^3 = 27 \times 10^{12} = 2.7 \times 10^{13}$

6. (a) $3z^2\left(2z^3\right)^2 = 3z^2\left(4z^6\right) = 12z^8$

(b) $(u - 2)^{-4}(u - 2)^{-3} = (u - 2)^{-7} = \frac{1}{(u - 2)^7}$

(c) $\left(\frac{x^{-2}y^2}{3}\right)^{-1} = \frac{x^2 y^{-2}}{3^{-1}} = \frac{3x^2}{y^2}$

7. (a) $9z\sqrt{8z} - 3\sqrt{2z^3} = 18z\sqrt{2z} - 3z\sqrt{2z}$

$= 15z\sqrt{2z}$

Since $\sqrt{8z}$ appears in the expression, we may assume that $z \geq 0$. It is not necessary to use an absolute value when simplifying $\sqrt{2z^3}$.

(b) $\left(4x^{3/5}\right)\left(x^{1/3}\right) = 4x^{(3/5)+(1/3)} = 4x^{14/15}$

(c) $\sqrt[3]{\frac{16}{v^5}} = \sqrt[3]{\frac{8}{v^6} \cdot 2v} = \frac{2}{v^2}\sqrt[3]{2v}$

8. Standard form: $-2x^5 - x^4 + 3x^3 + 3$

Degree: 5

Leading coefficient: -2

9. $\left(x^2 + 3\right) - \left[3x + \left(8 - x^2\right)\right] = x^2 + 3 - 3x - 8 + x^2$

$= 2x^2 - 3x - 5$

10. $\left(x + \sqrt{5}\right)\left(x - \sqrt{5}\right) = x^2 - \left(\sqrt{5}\right)^2 = x^2 - 5$

11. $\frac{5x}{x - 4} + \frac{20}{4 - x} = \frac{5x}{x - 4} - \frac{20}{x - 4}$

$= \frac{5x - 20}{x - 4}$

$= \frac{5(x - 4)}{x - 4}$

$= 5, \ x \neq 4$

12. $\dfrac{\left(\dfrac{2}{x} - \dfrac{2}{x + 1}\right)}{\left(\dfrac{4}{x^2 - 1}\right)} = \dfrac{2(x + 1) - 2x}{x(x + 1)} \cdot \dfrac{x^2 - 1}{4}$

$= \dfrac{2}{x(x + 1)} \cdot \dfrac{(x + 1)(x - 1)}{4}$

$= \dfrac{x - 1}{2x}, \ x \neq \pm 1$

13. (a) $2x^4 - 3x^3 - 2x^2 = x^2\left(2x^2 - 3x - 2\right)$

$= x^2(2x + 1)(x - 2)$

(b) $x^3 + 2x^2 - 4x - 8 = x^2(x + 2) - 4(x + 2)$

$= (x + 2)\left(x^2 - 4\right)$

$= (x + 2)(x + 2)(x - 2)$

$= (x + 2)^2(x - 2)$

14. (a) $\frac{16}{\sqrt[3]{16}} = \frac{16}{\sqrt[3]{16}} \cdot \frac{\sqrt[3]{4}}{\sqrt[3]{4}} = \frac{16\sqrt[3]{4}}{\sqrt[3]{64}} = \frac{16\sqrt[3]{4}}{4} = 4\sqrt[3]{4}$

(b) $\frac{4}{1 - \sqrt{2}} = \frac{4}{1 - \sqrt{2}} \cdot \frac{1 + \sqrt{2}}{1 + \sqrt{2}}$

$= \frac{4\left(1 + \sqrt{2}\right)}{1 - 2}$

$= -4\left(1 + \sqrt{2}\right)$

15. The domain of $\dfrac{6-x}{1-x}$ is all real numbers x except $x = 1$.

16. $\dfrac{y^2 + 8y + 16}{2y - 4} \cdot \dfrac{8y - 16}{(y+4)^3} = \dfrac{(y+4)^2}{2(y-2)} \cdot \dfrac{8(y-2)}{(y+4)^3} = \dfrac{4}{y+4}, \; y \neq 2$

17. $P = R - C$

$= 15x - (1480 + 6x)$

$= 9x - 1480$

When $x = 225$,

$P = 9(225) - 1480$

$= \$545.$

18.

Midpoint: $\left(\dfrac{-2+6}{2}, \dfrac{5+0}{2}\right) = \left(2, \dfrac{5}{2}\right)$

Distance: $d = \sqrt{(-2-6)^2 + (5-0)^2}$

$= \sqrt{64 + 25}$

$= \sqrt{89}$

19. Area = Area of large triangle − Area of small triangle

$A = \dfrac{1}{2}(3x)(\sqrt{3}x) - \dfrac{1}{2}(2x)\left(\dfrac{2}{3}\sqrt{3}x\right)$

$= \dfrac{3\sqrt{3}x^2}{2} - \dfrac{2\sqrt{3}x^2}{3}$

$= \dfrac{9\sqrt{3}x^2 - 4\sqrt{3}x^2}{6}$

$= \dfrac{5\sqrt{3}x^2}{6}$

$= \dfrac{5}{6}\sqrt{3}x^2$

Chapter 1   Chapter Test Solutions

1. $y = 4 - \dfrac{3}{4}x$

No symmetry

x-intercept: $\left(\dfrac{16}{3}, 0\right)$

y-intercept: $(0, 4)$

2. $y = 4 - \dfrac{3}{4}|x|$

y-axis symmetry

x-intercepts: $\left(\pm\dfrac{16}{3}, 0\right)$

y-intercept: $(0, 4)$

3. $y = 4 - (x - 2)^2$

Parabola; vertex: $(2, 4)$

No x-axis, y-axis, or origin symmetry

x-intercepts: $(0, 0)$ and $(4, 0)$

$$0 = 4 - (x - 2)^2$$
$$(x - 2)^2 = 4$$
$$x - 2 = \pm 2$$
$$x = 2 \pm 2$$
$$x = 4 \quad \text{or} \quad x = 0$$

y-intercept: $(0, 0)$

4. $y = x - x^3$

Origin symmetry

x-intercepts: $(0, 0), (1, 0), (-1, 0)$

$$0 = x - x^3$$
$$0 = x(1 + x)(1 - x)$$
$$x = 0, x = \pm 1$$

y-intercept: $(0, 0)$

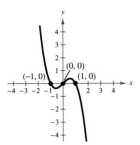

5. $y = \sqrt{5 - x}$

Domain: $x \le 5$

No symmetry

x-intercept: $(5, 0)$

y-intercept: $(0, \sqrt{5})$

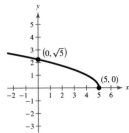

6. $(x - 3)^2 + y^2 = 9$

Circle

Center: $(3, 0)$

Radius: 3

x-axis symmetry

x-intercepts: $(0, 0), (6, 0)$

y-intercept: $(0, 0)$

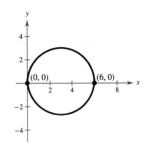

7. $\frac{2}{3}(x - 1) + \frac{1}{4}x = 10$

$$12\left[\frac{2}{3}(x - 1) + \frac{1}{4}x\right] = 12(10)$$
$$8(x - 1) + 3x = 120$$
$$8x - 8 + 3x = 120$$
$$11x = 128$$
$$x = \frac{128}{11}$$

8. $(x - 4)(x + 2) = 7$
$$x^2 - 2x - 8 = 7$$
$$x^2 - 2x - 15 = 0$$
$$(x + 3)(x - 5) = 0$$
$$x = -3 \quad \text{or} \quad x = 5$$

9. $\dfrac{x - 2}{x + 2} + \dfrac{4}{x + 2} + 4 = 0, x \ne -2$
$$\frac{x + 2}{x + 2} = -4$$
$$1 \ne -4 \Rightarrow \text{No solution}$$

10. $x^4 + x^2 - 6 = 0$
$$(x^2 - 2)(x^2 + 3) = 0$$
$$x^2 = 2 \Rightarrow x = \pm\sqrt{2}$$
$$x^2 = -3 \Rightarrow x = \pm\sqrt{3}i$$

11. $2\sqrt{x} - \sqrt{2x+1} = 1$

$$-\sqrt{2x+1} = 1 - 2\sqrt{x}$$

$$\left(-\sqrt{2x+1}\right)^2 = \left(1 - 2\sqrt{x}\right)^2$$

$$2x + 1 = 1 - 4\sqrt{x} + 4x$$

$$-2x = -4\sqrt{x}$$

$$x = 2\sqrt{x}$$

$$x^2 = 4x$$

$$x^2 - 4x = 0$$

$$x(x - 4) = 0$$

$$x = 0 \quad \text{or} \quad x = 4$$

Only $x = 4$ is a solution to the original equation. $x = 0$ is extraneous.

12. $|3x - 1| = 7$

$$3x - 1 = 7 \quad \text{or} \quad 3x - 1 = -7$$

$$3x = 8 \qquad\qquad 3x = -6$$

$$x = \tfrac{8}{3} \qquad\qquad x = -2$$

13. $-3 \le 2(x + 4) < 14$

$$-3 \le 2x + 8 < 14$$

$$-11 \le 2x < 6$$

$$-\tfrac{11}{2} \le x < 3$$

14. $\dfrac{2}{x} > \dfrac{5}{x + 6}$

$$\frac{2}{x} - \frac{5}{x + 6} > 0$$

$$\frac{2(x + 6) - 5x}{x(x + 6)} > 0$$

$$\frac{-3x + 12}{x(x + 6)} > 0$$

$$\frac{-3(x - 4)}{x(x + 6)} > 0$$

Critical numbers: $x = 4$, $x = 0$, $x = -6$

Test intervals: $(-\infty, -6), (-6, 0), (0, 4), (4, \infty)$

Test: Is $\dfrac{-3(x - 4)}{x(x + 6)} > 0$?

Solution set: $(-\infty, -6) \cup (0, 4)$

In inequality notation: $x < -6 \quad \text{or} \quad 0 < x < 4$

15. $2x^2 + 5x > 12$

$$2x^2 + 5x - 12 > 0$$

$$(2x - 3)(x + 4) > 0$$

Critical numbers: $x = \tfrac{3}{2}$, $x = -4$

Test intervals: $(-\infty, -4), \left(-4, \tfrac{3}{2}\right), \left(\tfrac{3}{2}, \infty\right)$

Test: Is $(2x - 3)(x + 4) > 0$?

Solution set: $(-\infty, -4) \cup \left(\tfrac{3}{2}, \infty\right)$

In inequality notation: $x < -4 \quad \text{or} \quad x > \tfrac{3}{2}$

16. $|3x + 5| \ge 10$

$$3x + 5 \le -10 \quad \text{or} \quad 3x + 5 \ge 10$$

$$3x \le -15 \qquad\qquad 3x \ge 5$$

$$x \le -5 \qquad\qquad x \ge \tfrac{5}{3}$$

17. (a) $10i - \left(3 + \sqrt{-25}\right) = 10i - (3 + 5i) = -3 + 5i$

(b) $(-1 - 5i)(-1 + 5i) = 1 - 25i^2 = 1 + 25 = 26$

18. $\dfrac{5}{2 + i} = \dfrac{5}{2 + i} \cdot \dfrac{2 - i}{2 - i} = \dfrac{5(2 - i)}{4 + 1} = 2 - i$

19. $y = 4.41t - 14.6$

(a)
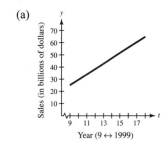
Year ($9 \leftrightarrow 1999$)

(b) For 2013, let $t = 23$. The estimated sales in 2013 are $86.83 billion.

(c) $y(23) = 4.41(23) - 14.6 = 86.83$

20. $V = \dfrac{4}{3}\pi r^3$

$$\frac{4}{3}\pi r^3 = 455.9$$

$$r = \sqrt[3]{\frac{455.9(3)}{4\pi}} \approx 4.774 \text{ inches}$$

21. $\left(100 \text{ km/hr}\right)\left(2\frac{1}{4} \text{ hr}\right) + \left(x \text{ km/hr}\right)\left(1\frac{1}{3} \text{ hr}\right) = 350 \text{ km}$

$$225 + \tfrac{4}{3}x = 350$$

$$\tfrac{4}{3}x = 125$$

$$x = \tfrac{375}{4} = 93\tfrac{3}{4} \text{ km/hr}$$

22. $a + b = 100 \Rightarrow b = 100 - a$

Area of ellipse = Area of circle

$$\pi ab = \pi (40)^2$$

$$a(100 - a) = 1600$$

$$0 = a^2 - 100a + 1600$$

$$0 = (a - 80)(a - 20)$$

$$a = 80 \Rightarrow b = 20$$

or

$$a = 20 \Rightarrow b = 80$$

Because $a > b$, choose $a = 80$ and $b = 20$.

Chapter 2   Chapter Test Solutions

1. $(4, -5), (-2, 7)$

$$m = \frac{7 - (-5)}{-2 - 4} = \frac{12}{-6} = -2$$

$$y - 7 = -2(x - (-2))$$

$$y - 7 = -2x - 4$$

$$y = -2x + 3$$

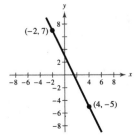

2. $(3, 0.8), (7, -6)$

$$m = \frac{-6 - 0.8}{7 - 3} = \frac{-6.8}{4} = -1.7$$

$$y - 0.8 = -1.7(x - 3)$$

$$y - 0.8 = -1.7x + 5.1$$

$$y = -1.7x + 5.9$$

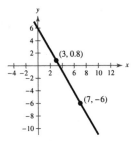

3. $5x + 2y = 3$

$$2y = -5x + 3$$

$$y = -\tfrac{5}{2}x + \tfrac{3}{2}$$

(a) Parallel line:

$$m = -\tfrac{5}{2}$$

$$y - 4 = -\tfrac{5}{2}(x - 0)$$

$$y - 4 = -\tfrac{5}{2}x$$

$$y = -\tfrac{5}{2}x + 4$$

(b) Perpendicular line:

$$m = \tfrac{2}{5}$$

$$y - 4 = \tfrac{2}{5}(x - 0)$$

$$y - 4 = \tfrac{2}{5}x$$

$$y = \tfrac{2}{5}x + 4$$

4. $f(x) = |x + 2| - 15$

(a) $f(-8) = -9$

(b) $f(14) = 1$

(c) $f(x - 6) = |x - 4| - 15$

5. $f(x) = \dfrac{\sqrt{x+9}}{x^2 - 81}$

   (a) $f(7) = \dfrac{4}{-32} = -\dfrac{1}{8}$

   (b) $f(-5) = \dfrac{2}{-56} = -\dfrac{1}{28}$

   (c) $f(x-9) = \dfrac{\sqrt{x}}{(x-9)^2 - 81} = \dfrac{\sqrt{x}}{x^2 - 18x}$

6. $f(x) = |-x + 6| + 2$

   Domain: All real numbers x

7. $f(x) = 10\sqrt{3-x}$

   $3 - x \geq 0$

      $3 \geq x$

      $x \leq 3$

   Domain: All real numbers x such that $x \leq 3$

8. $f(x) = 2x^6 + 5x^4 - x^2$

   (a)
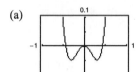

   (b) Increasing on $(-0.31, 0), (0.31, \infty)$

      Decreasing on $(-\infty, -0.31), (0, 0.31)$

   (c) y-axis symmetry $\Rightarrow$ the function is even.

9. $f(x) = 4x\sqrt{3-x}$

   (a)
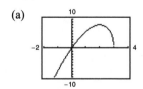

   (b) Increasing on $(-\infty, 2)$

      Decreasing on $(2, 3)$

   (c) The function is neither odd nor even.

10. $f(x) = |x + 5|$

   (a)
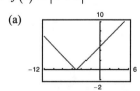

   (b) Increasing on $(-5, \infty)$

      Decreasing on $(-\infty, -5)$

   (c) The function is neither odd nor even.

11. $f(x) = -x^3 + 2x - 1$

   Relative minimum: $(-0.816, -2.089)$

   Relative maximum: $(0.816, 0.089)$

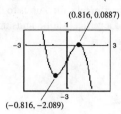

12. $f(x) = -2x^2 + 5x - 3$

   $\dfrac{f(3) - f(1)}{3 - 1} = \dfrac{-6 - 0}{2} = -3$

   The average rate of change of f from $x_1 = 1$ to $x_2 = 3$ is -3.

13. $f(x) = \begin{cases} 3x + 7 & x \leq -3 \\ 4x^2 - 1, & x > -3 \end{cases}$

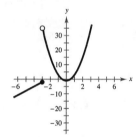

14. (a) Parent function: $f(x) = [\![x]\!]$

   (b) $h(x) = 3[\![x]\!]$

      Vertical stretch

   (c)
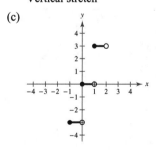

15. $h(x) = -\sqrt{x+5} + 8$

    (a) Parent function: $f(x) = x\sqrt{x}$

    (b) Transformation: Reflection in the x-axis,
           a horizontal shift 5 units to the left,
           and a vertical shift 8 units upward

    (c)

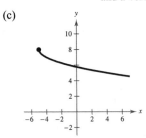

16. (a) Parent function: $f(x) = x^3$

    (b) $h(x) = -2(x-5)^3 + 3$

        Vertical stretch, reflection in x-axis, horizontal shift
        5 units to the right, vertical shift 3 units upward

    (c)

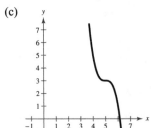

17. $f(x) = 3x^2 - 7,\ g(x) = -x^2 - 4x + 5$

    (a) $(f+g)(x) = (3x^2 - 7) + (-x^2 - 4x + 5) = 2x^2 - 4x - 2$

    (b) $(f-g)(x) = (3x^2 - 7) - (-x^2 - 4x + 5) = 4x^2 + 4x - 12$

    (c) $(fg)(x) = (3x^2 - 7)(-x^2 - 4x + 5) = -3x^4 - 12x^3 + 22x^2 + 28x - 35$

    (d) $\left(\dfrac{f}{g}\right)(x) = \dfrac{3x^2 - 7}{-x^2 - 4x + 5},\ x \neq -5, 1$

    (e) $(f \circ g)(x) = f(g(x)) = f(-x^2 - 4x + 5) = 3(-x^2 - 4x + 5)^2 - 7 = 3x^4 + 24x^3 + 18x^2 - 120x + 68$

    (f) $(g \circ f)(x) = g(f(x)) = g(3x^2 - 7) = -(3x^2 - 7)^2 - 4(3x^2 - 7) + 5 = -9x^4 + 30x^2 - 16$

18. $f(x) = \dfrac{1}{x},\ g(x) = 2\sqrt{x}$

    (a) $(f+g)(x) = \dfrac{1}{x} + 2\sqrt{x} = \dfrac{1 + 2x^{3/2}}{x},\ x > 0$

    (b) $(f-g)(x) = \dfrac{1}{x} - 2\sqrt{x} = \dfrac{1 - 2x^{3/2}}{x},\ x > 0$

    (c) $(fg)(x) = \left(\dfrac{1}{x}\right)(2\sqrt{x}) = \dfrac{2\sqrt{x}}{x},\ x > 0$

    (d) $\left(\dfrac{f}{g}\right)(x) = \dfrac{1/x}{2\sqrt{x}} = \dfrac{1}{2x\sqrt{x}} = \dfrac{1}{2x^{3/2}},\ x > 0$

    (e) $(f \circ g)(x) = f(g(x)) = f(2\sqrt{x}) = \dfrac{1}{2\sqrt{x}} = \dfrac{\sqrt{x}}{2x},\ x > 0$

    (f) $(g \circ f)(x) = g(f(x)) = g\left(\dfrac{1}{x}\right) = 2\sqrt{\dfrac{1}{x}} = \dfrac{2}{\sqrt{x}} = \dfrac{2\sqrt{x}}{x},\ x > 0$

19. $f(x) = x^3 + 8$

Since f is one-to-one, f has an inverse.

$$y = x^3 + 8$$
$$x = y^3 + 8$$
$$x - 8 = y^3$$
$$\sqrt[3]{x - 8} = y$$
$$f^{-1}(x) = \sqrt[3]{x - 8}$$

20. $f(x) = \left|x^2 - 3\right| + 6$

Since f is not one-to-one, f does not have an inverse.

21. $f(x) = 3x\sqrt{x} = 3x^{3/2}, x \geq 0$

Because f is one-to-one, f has an inverse.

$$y = 3x^{3/2}$$
$$x = 3y^{3/2}$$
$$\tfrac{1}{3}x = y^{3/2}$$
$$\left(\tfrac{1}{3}x\right)^{2/3} = y, x \geq 0$$
$$f^{-1}(x) = \left(\tfrac{1}{3}x\right)^{2/3}, x \geq 0$$

22. $(6, 58)$ and $(10, 78)$

$$m = \frac{78 - 58}{10 - 6} = 5$$
$$C - 58 = 5(x - 6)$$
$$C = 5x + 28$$

When $x = 25$: $C = 5(25) + 28 = \$153$

Chapters P–2   Cumulative Test Solutions

1. $\dfrac{8x^2y^{-3}}{30x^{-1}y^2} = \dfrac{8x^2x}{30y^2y^3} = \dfrac{4x^3}{15y^5}, x \neq 0$

2. $\sqrt{18x^3y^4} = \sqrt{9x^2y^4 2x} = 3xy^2\sqrt{2x}$

3. $4x - \left[2x + 3(2 - x)\right] = 4x - \left[2x + 6 - 3x\right]$
$$= 4x - \left[-x + 6\right]$$
$$= 5x - 6$$

4. $(x - 2)(x^2 + x - 3) = x^3 + x^2 - 3x - 2x^2 - 2x + 6$
$$= x^3 - x^2 - 5x + 6$$

5. $\dfrac{2}{s + 3} - \dfrac{1}{s + 1} = \dfrac{2(s + 1) - (s + 3)}{(s + 3)(s + 1)}$
$$= \dfrac{2s + 2 - s - 3}{(s + 3)(s + 1)}$$
$$= \dfrac{s - 1}{(s + 3)(s + 1)}$$

6. $25 - (x - 2)^2 = \left[5 + (x - 2)\right]\left[5 - (x - 2)\right]$
$$= (3 + x)(7 - x)$$

7. $x - 5x^2 - 6x^3 = x\left(1 - 5x - 6x^2\right)$
$$= x(1 + x)(1 - 6x)$$

8. $54x^3 + 16 = 2\left(27x^3 + 8\right)$
$$= 2\left((3x)^3 + 2^3\right)$$
$$= 2(3x + 2)\left(9x^2 - 6x + 4\right)$$

9. $x(2x + 4) + 2x(x + 4) = 2x^2 + 4x + 2x^2 + 8x$
$$= 4x^2 + 12x$$

10. $\tfrac{1}{2}(x + 5)\left[(x - 1) + 2(x + 1)\right] = \tfrac{1}{2}(x + 5)\left[x - 1 + 2x + 2\right]$
$$= \tfrac{1}{2}(x + 5)(3x + 1)$$
$$= \tfrac{3}{2}x^2 + 8x + \tfrac{5}{2}$$

11. $x - 3y + 12 = 0$

Line

x-intercept: $(-12, 0)$

y-intercept: $(0, 4)$

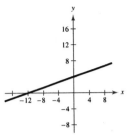

12. $y = x^2 - 9$

Parabola

x-intercept: $(\pm 3, 0)$

y-intercept: $(0, -9)$

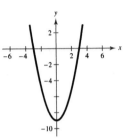

13. $y = \sqrt{4 - x}$

Domain: $x \le 4$

x-intercept: $(4, 0)$

y-intercept: $(0, 2)$

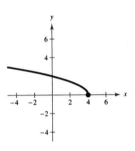

14. $3x - 5 = 6x + 8$

$-3x = 13$

$x = -\dfrac{13}{3}$

15. $-(x + 3) = 14(x - 6)$

$-x - 3 = 14x - 84$

$-15x = -81$

$x = \dfrac{-81}{-15} = \dfrac{27}{5}$

16. $\dfrac{1}{x - 2} = \dfrac{10}{4x + 3}$

$4x + 3 = 10(x - 2)$

$4x + 3 = 10x - 20$

$-6x = -23$

$x = \dfrac{23}{6}$

17. Factoring

$x^2 - 4x + 3 = 0$

$(x - 1)(x - 3) = 0$

$x - 1 = 0 \Rightarrow x = 1$

$x - 3 = 0 \Rightarrow x = 3$

18. Completing the Square

$-2x^2 + 8x + 12 = 0$

$-2(x^2 - 4x - 6) = 0$

$x^2 - 4x - 6 = 0$

$x^2 - 4x = 6$

$x^2 - 4x + 4 = 6 + 4$

$(x - 2)^2 = 10$

$x - 2 = \pm\sqrt{10}$

$x = 2 \pm \sqrt{10}$

19. Extracting Square Roots

$\frac{2}{3}x^2 = 24$

$x^2 = 36$

$x = \pm 6$

20. Quadratic Formula

$3x^2 + 5x - 6 = 0$

$a = 3, b = 5, c = -6$

$x = \dfrac{-5 \pm \sqrt{5^2 - 4(3)(-6)}}{2(3)}$

$= \dfrac{-5 \pm \sqrt{25 + 72}}{6}$

$= \dfrac{-5 \pm \sqrt{97}}{6}$

21. Quadratic Formula

$3x^2 + 9x + 1 = 0$

$a = 3, b = 9, c = 1$

$x = \dfrac{-9 \pm \sqrt{9^2 - 4(3)(1)}}{2(3)}$

$= \dfrac{-9 \pm \sqrt{81 - 12}}{6}$

$= \dfrac{-9 \pm \sqrt{69}}{6}$

$= -\dfrac{3}{2} \pm \dfrac{\sqrt{69}}{6}$

22. Extracting Square Roots

$\frac{1}{2}x^2 - 7 = 25$

$\frac{1}{2}x^2 = 32$

$x^2 = 64$

$x = \pm\sqrt{64}$

$x = \pm 8$

23. $x^4 + 12x^3 + 4x^2 + 48x = 0$

$x^3(x + 12) + 4x(x + 12) = 0$

$(x^3 + 4x)(x + 12) = 0$

$x(x^2 + 4)(x + 12) = 0$

$x = 0$

$x^2 + 4 = 0 \Rightarrow x = \pm 2i$

$x + 12 = 0 \Rightarrow x = -12$

24. $8x^3 - 48x^2 + 72x = 0$

$8x(x^2 - 6x + 9) = 0$

$8x(x - 3)^2 = 0$

$8x = 0 \Rightarrow x = 0$

$x - 3 = 0 \Rightarrow x = 3$

25. $x^{2/3} + 13 = 17$

$\sqrt[3]{x^2} = 4$

$x^2 = 4^3$

$x = \pm\sqrt{64}$

$x = \pm 8$

26. $\sqrt{x + 10} = x - 2$

$x + 10 = x^2 - 4x + 4$

$0 = x^2 - 5x - 6$

$0 = (x - 6)(x + 1)$

$x = 6 \text{ or } x = -1$

Only $x = 6$ is a solution to the original equation. $x = -1$ is extraneous.

27. $|3(x - 4)| = 27$

$3(x - 4) = -27 \text{ or } 3(x - 4) = 27$

$x - 4 = -9 \qquad x - 4 = 9$

$x = -5 \qquad\quad x = 13$

28. $|x - 12| = -2$

No solution. The absolute value of a number cannot be negative.

29. $4x + 2 > 7$

(a) $4(-1) + 2 \not> 7$

$x = -1$ *is not* a solution.

(b) $4\left(\frac{1}{2}\right) + 2 \not> 7$

$x = \frac{1}{2}$ *is not* a solution.

(c) $4\left(\frac{3}{2}\right) + 2 > 7$

$x = \frac{3}{2}$ *is* a solution.

(d) $4(2) + 2 > 7$

$x = 2$ *is* a solution.

30. $|5x - 1| < 4$

(a) $|5(-1) - 1| \not< 4$

$x = -1$ *is not* a solution.

(b) $\left|5\left(-\frac{1}{2}\right) - 1\right| < 4$

$x = -\frac{1}{2}$ *is* a solution.

(c) $|5(1) - 1| \not< 4$

$x = 1$ *is not* a solution.

(d) $|5(2) - 1| \not< 4$

$x = 2$ *is not* a solution.

31. $|x + 1| \le 6$

$-6 \le x + 1 \le 6$

$-7 \le \quad x \quad \le 5$

32. $|5 + 6x| > 3$

$5 + 6x < -3 \text{ or } 5 + 6x > 3$

$6x < -8 \qquad 6x > -2$

$x < -\frac{4}{3} \qquad x > -\frac{1}{3}$

33. $5x^2 + 12x + 7 \ge 0$

$(5x + 7)(x + 1) \ge 0$

Critical numbers: $x = -\frac{7}{5}, -1$

Test intervals: $\left(-\infty, -\frac{7}{5}\right), \left(-\frac{7}{5}, -1\right), (-1, \infty)$

Test: Is $5x^2 + 12x + 7 \ge 0$?

Solution: $x \le -\frac{7}{5}, x \ge -1$

34. $-x^2 + x + 4 < 0$

$x^2 - x - 4 > 0$

Critical numbers: $x = \dfrac{1 \pm \sqrt{17}}{2}$ (by the Quadratic Formula)

Test intervals: $\left(-\infty, \dfrac{1 - \sqrt{17}}{2}\right), \left(\dfrac{1 - \sqrt{17}}{2}, \dfrac{1 + \sqrt{17}}{2}\right),$

$\left(\dfrac{1 + \sqrt{17}}{2}, \infty\right)$

Test: Is $-x^2 + x + 4 < 0$?

Solution: $x < \dfrac{1 - \sqrt{17}}{2}, x > \dfrac{1 + \sqrt{17}}{2}$

35. $\left(-\dfrac{1}{2}, 1\right)$ and $(3, 8)$

$m = \dfrac{8 - 1}{3 - (-1/2)} = \dfrac{7}{7/2} = 2$

$y - 8 = 2(x - 3)$

$y - 8 = 2x - 6$

$y = 2x + 2$

36. It fails the Vertical Line Test. For some values of x there correspond two values of y.

37. $f(x) = \dfrac{x}{x - 2}$

(a) $f(6) = \dfrac{6}{4} = \dfrac{3}{2}$

(b) $f(2)$ is undefined because division by zero is undefined.

(c) $f(s + 2) = \dfrac{s + 2}{(s + 2) - 2} = \dfrac{s + 2}{s}$

38. $f(x) = 5 + \sqrt{4 - x}$

$f(-x) = 5 + \sqrt{4 - (-x)} = 5 + \sqrt{4 + x}$

$-f(x) = -5 - \sqrt{4 - x}$

$f(-x) \neq f(x)$ and $f(-x) \neq -f(x)$

The function is neither even nor odd.

39. $f(x) = x^5 - x^3 + 2$

$f(-x) = (-x)^5 - (-x)^3 + 2 = -x^5 + x^3 + 2$

$f(-x) \neq f(x)$ and $f(-x) \neq -f(x)$

The function is neither even nor odd.

40. $f(x) = 2x^4 - 4$

$f(-x) = 2(-x)^4 - 4 = 2x^4 - 4 = f(x)$

The function is even.

41. $y = \sqrt[3]{x}$

(a) $r(x) = \frac{1}{2}\sqrt[3]{x}$ is a vertical shrink by a factor of $\frac{1}{2}$.

(b) $h(x) = \sqrt[3]{x} + 2$ is a vertical shift two units upward.

(c) $g(x) = \sqrt[3]{x + 2}$ is a horizontal shift two units to the left.

42. $f(x) = x - 4, g(x) = 3x + 1$

(a) $(f + g)(x) = f(x) + g(x)$

$= (x - 4) + (3x + 1)$

$= 4x - 3$

(b) $(f - g)(x) = f(x) - g(x)$

$= (x - 4) - (3x + 1)$

$= -2x - 5$

(c) $(fg)(x) = f(x)g(x)$

$= (x - 4)(3x + 1)$

$= 3x^2 - 11x - 4$

(d) $\left(\dfrac{f}{g}\right)(x) = \dfrac{f(x)}{g(x)} = \dfrac{x - 4}{3x + 1}$

Domain: All real numbers x except $x = -\dfrac{1}{3}$

43. $f(x) = \sqrt{x - 1}, g(x) = x^2 + 1$

(a) $(f + g)(x) = f(x) + g(x)$

$= \sqrt{x - 1} + x^2 + 1$

(b) $(f - g)(x) = f(x) - g(x)$

$= \sqrt{x - 1} - x^2 - 1$

(c) $(fg)(x) = f(x)g(x)$

$= \sqrt{x - 1}(x^2 + 1)$

$= x^2\sqrt{x - 1} + \sqrt{x - 1}$

(d) $\left(\dfrac{f}{g}\right)(x) = \dfrac{f(x)}{g(x)} = \dfrac{\sqrt{x - 1}}{x^2 + 1}$

Domain: all real numbers x such that $x \geq 1$

44. $f(x) = 2x^2$, $g(x) = \sqrt{x+6}$

  (a) $(f \circ g)(x) = f(g(x))$

$$= f\left(\sqrt{x+6}\right)$$

$$= 2\left(\sqrt{x+6}\right)^2$$

$$= 2(x+6)$$

$$= 2x + 12$$

Domain: all real numbers x such that $x \geq -6$

  (b) $(g \circ f)(x) = g(f(x))$

$$= g\left(2x^2\right)$$

$$= \sqrt{2x^2 + 6}$$

Domain: all real numbers x

45. $f(x) = x - 2$, $g(x) = |x|$

  (a) $(f \circ g)(x) = f(g(x))$

$$= f(|x|)$$

$$= |x| - 2$$

Domain: all real numbers x

  (b) $(g \circ f)(x) = g(f(x))$

$$= g(x - 2)$$

$$= |x - 2|$$

Domain: all real numbers x

46. $h(x) = 3x - 4$

Because h is one-to-one, h has an inverse.

$$y = 3x - 4$$

$$x = 3y - 4$$

$$x + 4 = 3y$$

$$\tfrac{1}{3}(x + 4) = y$$

$$h^{-1}(x) = \tfrac{1}{3}(x + 4)$$

Chapter 3   Chapter Test Solutions

1. $f(x) = x^2$

  (a) $g(x) = 2 - x^2$

Reflection in the x-axis followed by a vertical translation two units upward

  (b) $g(x) = \left(x - \tfrac{3}{2}\right)^2$

Horizontal translation $\tfrac{3}{2}$ units to the right

47. Cost per person: $\dfrac{36{,}000}{n}$

If three additional people join the group, the cost per person is $\dfrac{36{,}000}{n+3}$.

$$\frac{36{,}000}{n} = \frac{36{,}000}{n+3} + 1000$$

$$36{,}000(n+3) = 36{,}000n + 1000n(n+3)$$

$$36(n+3) = 36n + n(n+3)$$

$$36n + 108 = 36n + n^2 + 3n$$

$$0 = n^2 + 3n - 108$$

$$0 = (n+12)(n-9)$$

Choosing the positive value, the group has $n = 9$ people.

48. Rate $= 10 - 0.05(n - 60)$, $n \geq 60$

  (a) Revenue $=$ (number of people)(rate per person)

$$R(n) = n\left[10 - 0.05(n - 60)\right]$$

$$= 10n - 0.05n(n - 60)$$

$$= 10n - 0.05n^2 + 3n$$

$$= -0.05n^2 + 13n, \ n \geq 60$$

  (b)

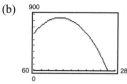

The revenue is maximum when $n = 130$ passengers.

49. $s(t) = -16t^2 + 36t + 8$

$$\frac{s(2) - s(0)}{2 - 0} = \frac{16 - 8}{2} = 4$$

The average rate of change in the height of the object from $t_1 = 0$ to $t_2 = 2$ seconds is 4 feet per second.

2. $y = x^2 + 4x + 3$

$$= x^2 + 4x + 4 - 4 + 3$$

$$= (x + 2)^2 - 1$$

Vertex: $(-2, -1)$

x-intercepts: $0 = x^2 + 4x + 3$

$$0 = (x + 3)(x + 1)$$

$$x = -3 \text{ or } x = -1$$

$$(-3, 0) \text{ or } (-1, 0)$$

y-intercept: $(0, 3)$

3. Vertex: $(3, -6)$

$$y = a(x - 3)^2 - 6$$

Point on the graph: $(0, 3)$

$$3 = a(0 - 3)^2 - 6$$

$$9 = 9a \Rightarrow a = 1$$

So, $y = (x - 3)^2 - 6$.

4. (a) $y = -\frac{1}{20}x^2 + 3x + 5$

$$= -\frac{1}{20}(x^2 - 60x + 900 - 900) + 5$$

$$= -\frac{1}{20}\left[(x - 30)^2 - 900\right] + 5$$

$$= -\frac{1}{20}(x - 30)^2 + 50$$

Vertex: $(30, 50)$

The maximum height is 50 feet.

(b) The constant term, $c = 5$, determines the height at which the ball was thrown. Changing this constant results in a vertical translation of the graph, and therefore, changes the maximum height.

5. $h(t) = -\frac{3}{4}t^5 + 2t^2$

The degree is odd and the leading coefficient is negative. The graph rises to the left and falls to the right.

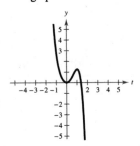

6. $x^2 + 0x + 1\overline{\smash{\big)}3x^3 + 0x^2 + 4x - 1}$ with quotient $3x + \dfrac{x - 1}{x^2 + 1}$

$$\underline{3x^3 + 0x^2 + 3x}$$
$$x - 1$$

Thus, $\dfrac{3x^3 + 4x - 1}{x^2 + 1} = 3x + \dfrac{x - 1}{x^2 + 1}.$

7.

$$\begin{array}{r|rrrrr}
2 & 2 & 0 & -5 & 0 & -3 \\
  &   & 4 & 8 & 6 & 12 \\
\hline
  & 2 & 4 & 3 & 6 & 9
\end{array}$$

Thus,

$$\frac{2x^4 - 5x^2 - 3}{x - 2} = 2x^3 + 4x^2 + 3x + 6 + \frac{9}{x - 2}.$$

8.

$$\begin{array}{r|rrrr}
\sqrt{3} & 2 & -5 & -6 & 15 \\
 & & 2\sqrt{3} & 6 - 5\sqrt{3} & -15 \\
\hline
 & 2 & 2\sqrt{3} - 5 & -5\sqrt{3} & 0
\end{array}$$

$$\begin{array}{r|rrr}
-\sqrt{3} & 2 & 2\sqrt{3} - 5 & -5\sqrt{3} \\
 & & -2\sqrt{3} & 5\sqrt{3} \\
\hline
 & 2 & -5 & 0
\end{array}$$

$$2x^3 - 5x^2 - 6x + 15 = \left(x - \sqrt{3}\right)\left(x + \sqrt{3}\right)(2x - 5)$$

The real zeros of $f(x)$ are $x = \pm\sqrt{3}$ and $x = \frac{5}{2}$.

9. $g(t) = 2t^4 - 3t^3 + 16t - 24$

Possible rational zeros:

$$\pm 1, \pm 2, \pm 3, \pm 4, \pm 6, \pm 8, \pm 12, \pm 24, \pm\tfrac{1}{2}, \pm\tfrac{3}{2}$$

From the graph, we have $t = -2$ and $t = \frac{3}{2}$.

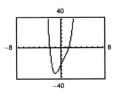

10. $h(x) = 3x^5 + 2x^4 - 3x - 2$

Possible rational zeros: $\pm 1, \pm 2, \pm\frac{1}{3}, \pm\frac{2}{3}$

From the graph, we have $x = \pm 1$ and $x = -\frac{2}{3}$.

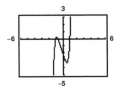

11. $f(x) = x(x - 3)\big(x - (2 + i)\big)\big(x - (2 - i)\big)$

$$= \left(x^2 - 3x\right)\left[(x - 2) - i\right]\left[(x - 2) + i\right]$$

$$= \left(x^2 - 3x\right)\left[(x - 2)^2 - i^2\right]$$

$$= \left(x^2 - 3x\right)\left(x^2 - 4x + 5\right)$$

$$= x^4 - 7x^3 + 17x^2 - 15x$$

12. Because $1 - \sqrt{3}i$ is a zero, $1 + \sqrt{3}i$ is also a zero.

$$\begin{aligned} f(x) &= (x-2)(x-2)\left[x-\left(1-\sqrt{3}i\right)\right]\left[x-\left(1+\sqrt{3}i\right)\right] \\ &= \left(x^2-4x+4\right)\left[(x-1)+\sqrt{3}i\right]\left[(x-1)-\sqrt{3}i\right] \\ &= \left(x^2-4x+4\right)\left[(x-1)^2-\left(\sqrt{3}i\right)^2\right] \\ &= \left(x^2-4x+4\right)\left(x^2-2x+4\right) \\ &= x^4-6x^3+16x^2-24x+16 \end{aligned}$$

13. $f(x) = 3x^3 + 14x^2 - 7x - 10$

Possible rational zeros:
$\pm 1, \pm 2, \pm 5, \pm 10, \pm\frac{1}{3}, \pm\frac{2}{3}, \pm\frac{5}{3}, \pm\frac{10}{3}$

$$\begin{array}{r|rrrr} 1 & 3 & 14 & -7 & -10 \\ & & 3 & 17 & 10 \\ \hline & 3 & 17 & 10 & 0 \end{array}$$

$$\begin{aligned} f(x) &= (x-1)(3x^2+17x+10) \\ &= (x-1)(3x+2)(x+5) \end{aligned}$$

The zeros of $f(x)$ are $x=-5$, $x=-\frac{2}{3}$, and $x=1$.

14. $f(x) = x^4 - 9x^2 - 22x - 24$

Possible rational zeros:
$\pm 1, \pm 2, \pm 3, \pm 4, \pm 6, \pm 8, \pm 12, \pm 24$

$$\begin{array}{r|rrrrr} -2 & 1 & 0 & -9 & -22 & -24 \\ & & -2 & 4 & 10 & 24 \\ \hline & 1 & -2 & -5 & -12 & 0 \end{array}$$

$$\begin{array}{r|rrrr} 4 & 1 & -2 & -5 & -12 \\ & & 4 & 8 & 12 \\ \hline & 1 & 2 & 3 & 0 \end{array}$$

$$f(x) = (x+2)(x-4)(x^2+2x+3)$$

By the Quadratic Formula the zeros of x^2+2x+3 are $x=-1\pm\sqrt{2}i$. The zeros of f are: $x=-2, 4$, $-1\pm\sqrt{2}i$.

15. $v = k\sqrt{s}$

$24 = k\sqrt{16}$

$6 = k$

$v = 6\sqrt{s}$

16. $A = kxy$

$500 = k(15)(8)$

$500 = k(120)$

$\frac{25}{6} = k$

$A = \frac{25}{6}xy$

17. $b = \dfrac{k}{a}$

$32 = \dfrac{k}{1.5}$

$48 = k$

$b = \dfrac{48}{a}$

18. $S = 385t + 115$

Year, t	Salary, S	Model
4	1550	1655
5	2150	2040
6	2500	2425
7	2750	2810
8	3175	3195

The model is a fairly good fit for the actual data.

Chapter 4   Chapter Test Solutions

1. $y = \dfrac{3x}{x+1}$

Domain: all real numbers x except $x=-1$

Vertical asymptote: $x=-1$

Horizontal asymptote: $y=3$

2. $f(x) = \dfrac{3-x^2}{3+x^2} = \dfrac{-x^2+3}{x^2+3}$

Domain: all real numbers x

Vertical asymptote: None

Horizontal asymptote: $y = \dfrac{-1}{1} = -1$

3. $g(x) = \dfrac{x^2 - 7x + 12}{x - 3}$

$ = \dfrac{(x - 3)(x - 4)}{(x - 3)}$

$ = x - 4, \, x \neq 3$

Domain: all real numbers x except $x = 3$

No asymptotes

4. $h(x) = \dfrac{4}{x^2} - 1 = \dfrac{4 - x^2}{x^2} = \dfrac{(2 - x)(2 + x)}{x^2}$

x-intercepts: $(\pm 2, 0)$

Vertical asymptote: $x = 0$

Horizontal asymptote: $y = -1$

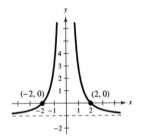

5. $g(x) = \dfrac{x^2 + 2}{x - 1} = x + 1 + \dfrac{3}{x - 1}$

y-intercept: $(0, -2)$

Vertical asymptote: $x = 1$

Slant asymptote: $y = x + 1$

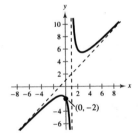

6. $f(x) = \dfrac{x + 1}{x^2 + x - 12} = \dfrac{x + 1}{(x + 4)(x - 3)}$

x-intercept: $(-1, 0)$

y-intercept: $\left(0, -\dfrac{1}{12}\right)$

Vertical asymptotes: $x = -4, \, x = 3$

Horizontal asymptote: $y = 0$

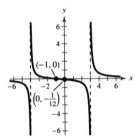

7. $f(x) = \dfrac{2x^2 - 5x - 12}{x^2 - 16}$

$ = \dfrac{(2x + 3)(x - 4)}{(x + 4)(x - 4)}$

$ = \dfrac{2x + 3}{x + 4}, \, x \neq 4$

x-intercept: $\left(-\dfrac{3}{2}, 0\right)$

y-intercept: $\left(0, \dfrac{3}{4}\right)$

Vertical asymptote: $x = -4$

Horizontal asymptote: $y = 2$

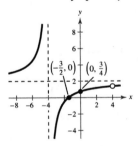

8. $f(x) = \dfrac{2x^2 + 9}{5x^2 + 9}$

y-intercept: $(0, 1)$

Horizontal asymptote: $y = \dfrac{2}{5}$

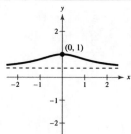

9. $g(x) = \dfrac{2x^3 - 7x^2 + 4x + 4}{x^2 - x - 2}$

$= 2x - 5 + \dfrac{3(x - 2)}{(x - 2)(x + 1)}$

$= 2x - 5 + \dfrac{3}{x + 1}$

$= \dfrac{2x^2 - 3x - 2}{x + 1}$

$= \dfrac{(2x + 1)(x - 2)}{x + 1}, x \neq 2$

x-intercept: $\left(-\dfrac{1}{2}, 0\right)$

y-intercept: $(0, -2)$

Vertical asymptote: $x = -1$

Slant asymptote: $y = 2x - 5$

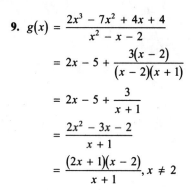

10.

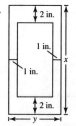

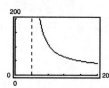

Minimize $A = xy$.

Given: $(x - 4)(y - 2) = 36$

$y = \dfrac{36}{x - 4} + 2$

$A = x\left(\dfrac{36}{x - 4} + 2\right) = x\left(\dfrac{2x + 28}{x - 4}\right) = \dfrac{2x(x + 14)}{x - 4}$

Domain: $x > 4$

From the graph of A we see that the minimum occurs when $x \approx 12.49$ inches and

$y = \dfrac{36}{x - 4} + 2 \approx 6.24$ inches. The dimensions are 6.24 inches by 12.49 inches.

Note: The exact values are $x = 4 + 6\sqrt{2}$ and $y = 2 + 3\sqrt{2}$.

11. (a) Equate the slopes.

$\dfrac{y - 1}{0 - 2} = \dfrac{1 - 0}{2 - x}$

$\dfrac{y - 1}{-2} = \dfrac{1}{2 - x}$

$y - 1 = -2\left(\dfrac{1}{2 - x}\right)$

$y = 1 + \dfrac{2}{x - 2}$

(b) $A = \dfrac{1}{2}xy$

$= \dfrac{1}{2}x\left[1 + \dfrac{2}{x - 2}\right]$

$= \dfrac{x}{2} + \dfrac{x}{x - 2}$

$= \dfrac{x^2}{2(x - 2)}$

In context, we have $x > 2$ for the domain.

(c)

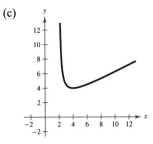

The minimum area occurs at $x = 4$ and is $A = 4$.

12. $y^2 - 4x = 0$

$$y^2 = 4x$$

$$y^2 = 4(1)x \Rightarrow p = 1$$

Parabola

Vertex: $(0, 0)$

Focus: $(1, 0)$

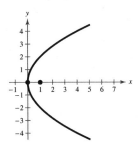

13. $\qquad x^2 + y^2 - 10x + 4y + 4 = 0$

$$\left(x^2 - 10x\right) + \left(y^2 + 4y\right) = -4$$

$$\left(x^2 - 10x + 25\right) + \left(y^2 + 4y + 4\right) = -4 + 25 + 4$$

$$(x - 5)^2 + (y + 2)^2 = 25$$

Circle

Center: $(5, -2)$

Radius: 5

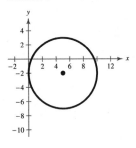

14. $x^2 - 10x - 2y + 19 = 0$

$$x^2 - 10x = 2y - 19$$

$$x^2 - 10x + 25 = 2y - 19 + 25$$

$$(x - 5)^2 = 2(y + 3)$$

Parabola

Vertex: $(5, -3)$

Focus: $\left(5, -\dfrac{5}{2}\right)$

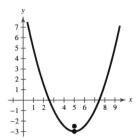

15. $\dfrac{x^2}{1} - \dfrac{y^2}{4} = 1$

Hyperbola

Center: $(0, 0)$

$$a = 1, b = 2, c = \sqrt{5}$$

Horizontal transverse axis

Vertices: $(\pm 1, 0)$

Foci: $\left(\pm\sqrt{5}, 0\right)$

Asymptotes: $y = \pm 2x$

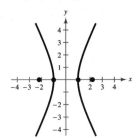

16. $\dfrac{y^2}{4} - x^2 = 1$

Hyperbola

Center: $(0, 0)$

$a = 2, b = 1, c = \sqrt{5}$

Vertical transverse axis

Vertices: $(0, \pm 2)$

Foci: $\left(0, \pm\sqrt{5}\right)$

Asymptotes: $y = \pm 2x$

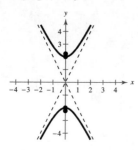

17.
$$x^2 + 3y^2 - 2x + 36y + 100 = 0$$
$$\left(x^2 - 2x\right) + 3\left(y^2 + 12y\right) = -100$$
$$\left(x^2 - 2x + 1\right) + 3\left(y^2 + 12y + 36\right) = -100 + 1 + 108$$
$$(x - 1)^2 + 3(y + 6)^2 = 9$$
$$\dfrac{(x - 1)^2}{9} + \dfrac{(y + 6)^2}{3} = 1$$

Ellipse

Center: $(1, -6)$

$a = 3, b = \sqrt{3}, c = \sqrt{6}$

Horizontal major axis

Vertices: $(-2, -6), (4, -6)$

Foci: $\left(1 \pm \sqrt{6}, -6\right)$

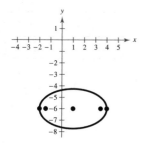

18. Ellipse

Vertices: $(0, 2)$ and $(8, 2)$

Center: $(4, 2)$

Horizontal major axis: $a = 4$

Minor axis of length 4: $2b = 4 \Rightarrow b = 2$

$$\dfrac{(x - h)^2}{a^2} + \dfrac{(y - k)^2}{b^2} = 1$$
$$\dfrac{(x - 4)^2}{16} + \dfrac{(y - 2)^2}{4} = 1$$

19. Hyperbola

Vertices: $(0, \pm 3)$

Center: $(0, 0)$

Vertical transverse axis: $a = 3$

Asymptotes: $y = \pm\dfrac{3}{2}x$

$$\pm\dfrac{a}{b} = \pm\dfrac{3}{2} \Rightarrow b = 2$$
$$\dfrac{(y - k)^2}{a^2} - \dfrac{(x - h)^2}{b^2} = 1$$
$$\dfrac{y^2}{9} - \dfrac{x^2}{4} = 1$$

20.

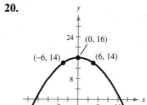

$$x^2 = 4p(y - 16)$$
$$36 = 4p(14 - 16)$$
$$-\dfrac{9}{2} = p$$
$$x^2 = -18(y - 16)$$

When $y = 0$: $x^2 = -18(-16) \Rightarrow x \approx 17 \Rightarrow 2x \approx 34$

At ground level, the archway is approximately 34 meters.

21. $a = \frac{1}{2}(768,800) = 384,400$

$b = \frac{1}{2}(767,640) = 383,820$

$c = \sqrt{384,400^2 - 383,820^2} \approx 21,108$

Smallest distance (perigee): $a - c \approx 363,292$ km

Greatest distance (apogee): $a + c \approx 405,508$ km

Chapter 5 Chapter Test Solutions

1. $4.2^{0.6} \approx 2.366$

2. $4^{3\pi/2} \approx 687.291$

3. $e^{-7/10} \approx 0.497$

4. $e^{3.1} \approx 22.198$

5. $f(x) = 10^{-x}$

x	-1	$-\frac{1}{2}$	0	$\frac{1}{2}$	1
$f(x)$	10	3.162	1	0.316	0.1

Horizontal asymptote: $y = 0$

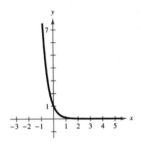

6. $f(x) = -6^{x-2}$

x	-1	0	1	2	3
$f(x)$	-0.005	-0.028	-0.167	-1	-6

Horizontal asymptote: $y = 0$

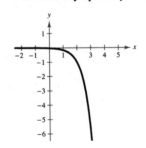

7. $f(x) = 1 - e^{2x}$

x	-1	$-\frac{1}{2}$	0	$\frac{1}{2}$	1
$f(x)$	0.865	0.632	0	-1.718	-6.389

Horizontal asymptote: $y = 1$

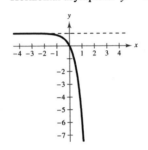

8. (a) $\log_7 7^{-0.89} = -0.89$

(b) $4.6 \ln e^2 = 4.6(2) = 9.2$

9. $f(x) = -\log x - 6$

x	$\frac{1}{2}$	1	$\frac{3}{2}$	2	4
$f(x)$	-5.699	-6	-6.176	-6.301	-6.602

Vertical asymptote: $x = 0$

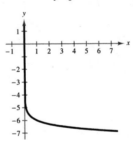
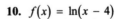

10. $f(x) = \ln(x - 4)$

x	5	7	9	11	13
$f(x)$	0	1.099	1.609	1.946	2.197

Vertical asymptote: $x = 4$

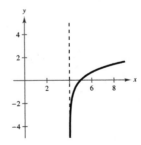

11. $f(x) = 1 + \ln(x + 6)$

x	-5	-3	-1	0	1
$f(x)$	1	2.099	2.609	2.792	2.946

Vertical asymptote: $x = -6$

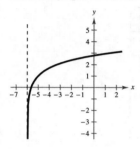

12. $\log_7 44 = \dfrac{\ln 44}{\ln 7} = \dfrac{\log 44}{\log 7} \approx 1.945$

13. $\log_{16} 0.63 = \dfrac{\log 0.63}{\log 16} \approx -0.167$

14. $\log_{3/4} 24 = \dfrac{\log 24}{\log (3/4)} \approx -11.047$

15. $\log_2 3a^4 = \log_2 3 + \log_2 a^4 = \log_2 3 + 4 \log_2 |a|$

16. $\ln \dfrac{5\sqrt{x}}{6} = \ln\left(5\sqrt{x}\right) - \ln 6$

$\qquad = \ln 5 + \ln \sqrt{x} - \ln 6$

$\qquad = \ln 5 + \dfrac{1}{2} \ln x - \ln 6$

17. $\log \dfrac{(x-1)^3}{y^2 z} = \log(x-1)^3 - \log y^2 z$

$\qquad = 3 \log(x-1) - \left(\log y^2 + \log z\right)$

$\qquad = 3 \log(x-1) - 2 \log y - \log z$

18. $\log_3 13 + \log_3 y = \log_3 13y$

19. $4 \ln x - 4 \ln y = \ln x^4 - \ln y^4 = \ln \dfrac{x^4}{y^4}$

20. $3 \ln x - \ln(x + 3) + 2 \ln y = \ln x^3 - \ln(x + 3) + \ln y^2 = \ln \dfrac{x^3 y^2}{x + 3}$

21. $5^x = \dfrac{1}{25}$

$\qquad 5^x = 5^{-2}$

$\qquad x = -2$

22. $3e^{-5x} = 132$

$\qquad e^{-5x} = 44$

$\qquad -5x = \ln 44$

$\qquad x = \dfrac{\ln 44}{-5} \approx -0.757$

23. $\dfrac{1025}{8 + e^{4x}} = 5$

$\qquad 1025 = 5\left(8 + e^{4x}\right)$

$\qquad 205 = 8 + e^{4x}$

$\qquad 197 = e^{4x}$

$\qquad \ln 197 = 4x$

$\qquad x = \dfrac{\ln 197}{4} \approx 1.321$

24. $\ln x = \dfrac{1}{2}$

$\qquad x = e^{1/2} \approx 1.649$

25. $18 + 4 \ln x = 7$

$\qquad 4 \ln x = -11$

$\qquad \ln x = -\dfrac{11}{4}$

$\qquad x = e^{-11/4} \approx 0.0639$

26. $\log x + \log(x - 15) = 2$

$\qquad \log\left[x(x - 15)\right] = 2$

$\qquad x(x - 15) = 10^2$

$\qquad x^2 - 15x - 100 = 0$

$\qquad (x - 20)(x + 5) = 0$

$\qquad x - 20 = 0 \quad \text{or} \quad x + 5 = 0$

$\qquad x = 20 \qquad\qquad x = -5$

The value $x = -5$ is extraneous. The only solution is $x = 20$.

27. $y = ae^{bt}$

$(0, 2745):\ 2745 = ae^{b(0)} \Rightarrow a = 2745$

$\qquad y = 2745e^{bt}$

$(9, 11{,}277):\qquad 11{,}277 = 2745e^{b(9)}$

$\qquad\qquad \dfrac{11{,}277}{2745} = e^{9b}$

$\qquad\qquad \ln\left(\dfrac{11{,}277}{2745}\right) = 9b$

$\qquad\qquad \dfrac{1}{9} \ln\left(\dfrac{11{,}277}{2745}\right) = b \Rightarrow b \approx 0.1570$

So, $y = 2745e^{0.1570t}$.

28. $y = ae^{bt}$

$$\frac{1}{2}a = ae^{b(21.77)}$$

$$\frac{1}{2} = e^{21.77b}$$

$$\ln\left(\frac{1}{2}\right) = 21.77b$$

$$b = \frac{\ln(1/2)}{21.77} \approx -0.0318$$

$$y = ae^{-0.0318t}$$

When $t = 19$: $y = ae^{-0.0318(19)} \approx 0.55a$

So, 55% will remain after 19 years.

29. $H = 70.228 + 5.104x + 9.222 \ln x,\; \frac{1}{4} \leq x \leq 6$

(a)

x	H (cm)
$\frac{1}{4}$	58.720
$\frac{1}{2}$	66.388
1	75.332
2	86.828
3	95.671
4	103.43
5	110.59
6	117.38

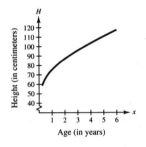

(b) Estimate: 103

When $x = 4$, $H \approx 103.43$ cm.

Chapters 3–5 Cumulative Test Solutions

1. Vertex: $(-8, 5)$

Point: $(-4, -7)$

$$y - k = a(x - h)^2$$

$$y - 5 = a(x + 8)^2$$

$$-7 - 5 = a(-4 + 8)^2$$

$$-12 = 16a$$

$$-\frac{3}{4} = a$$

$$y = -\frac{3}{4}(x + 8)^2 + 5$$

2. $h(x) = -\left(x^2 + 4x\right)$

$$= -\left(x^2 + 4x + 4 - 4\right)$$

$$= -(x + 2)^2 + 4$$

Parabola

Vertex: $(-2, 4)$

Intercepts: $(-4, 0), (0, 0)$

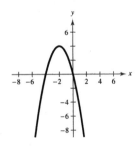

3. $f(t) = \frac{1}{4}t(t - 2)^2$

Cubic

Falls to the left

Rises to the right

Intercepts: $(0, 0), (2, 0)$

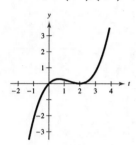

4. $g(s) = s^2 + 2s + 9$

$= (s^2 + 2s + 1) - 1 + 9$

$= (s + 1)^2 + 8$

Parabola

Vertex: $(-1, 8)$

Intercept: $(0, 9)$

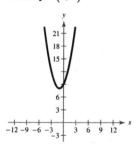

8.

$$\begin{array}{r|rrrrr}
2 & 3 & 0 & 2 & -5 & 3 \\
  &   & 6 & 12 & 28 & 46 \\
\hline
  & 3 & 6 & 14 & 23 & 49
\end{array}$$

Thus, $\dfrac{3x^4 + 2x^2 - 5x + 3}{x - 2} = 3x^3 + 6x^2 + 14x + 23 + \dfrac{49}{x - 2}$.

9. $g(x) = x^3 + 3x^2 - 6$

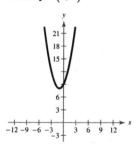

$x \approx 1.20$

5. $f(x) = x^3 + 2x^2 + 4x + 8$

$= x^2(x + 2) + 4(x + 2)$

$= (x + 2)(x^2 + 4)$

$x + 2 = 0 \Rightarrow x = -2$

$x^2 + 4 = 0 \Rightarrow x = \pm 2i$

The zeros of $f(x)$ are -2 and $\pm 2i$.

6. $f(x) = x^4 + 4x^3 - 21x^2$

$= x^2(x^2 + 4x - 21)$

$= x^2(x + 7)(x - 3)$

The zeros of $f(x)$ are $0, -7,$ and 3.

7.

$$\require{enclose}
\begin{array}{r}
3x - 2 + \dfrac{-3x + 2}{2x^2 + 1} \\[2mm]
2x^2 + 0x + 1 \enclose{longdiv}{6x^3 - 4x^2 + 0x + 0} \\[1mm]
\underline{6x^3 + 0x^2 + 3x} \\
-4x^2 - 3x + 0 \\
\underline{-4x^2 + 0x - 2} \\
-3x + 2
\end{array}$$

Thus, $\dfrac{6x^3 - 4x^2}{2x^2 + 1} = 3x - 2 - \dfrac{3x - 2}{2x^2 + 1}$.

10. Because $2 + \sqrt{3}i$ is a zero, so is $2 - \sqrt{3}i$.

$$
\begin{aligned}
f(x) &= (x + 5)(x + 2)\left[x - \left(2 + \sqrt{3}i\right)\right]\left[x - \left(2 - \sqrt{3}i\right)\right] \\
&= \left(x^2 + 7x + 10\right)\left[(x - 2) - \sqrt{3}i\right]\left[(x - 2) + \sqrt{3}i\right] \\
&= \left(x^2 + 7x + 10\right)\left[(x - 2)^2 + 3\right] \\
&= \left(x^2 + 7x + 10\right)\left(x^2 - 4x + 7\right) \\
&= x^4 + 3x^3 - 11x^2 + 9x + 70
\end{aligned}
$$

11. $f(x) = \dfrac{2x}{x - 3}$

Domain: all real numbers x except $x = 3$

Vertical asymptote: $x = 3$

Horizontal asymptote: $y = 2$

Intercept: $(0, 0)$

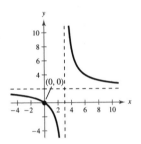

12. $f(x) = \dfrac{4x^2}{x - 5} = 4x + 20 + \dfrac{100}{x - 5}$

Domain: all real numbers x except $x = 5$

Vertical asymptote: $x = 5$

Slant asymptote: $y = 4x + 20$

Intercept: $(0, 0)$

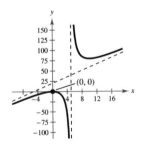

13. $f(x) = \dfrac{2x}{x^2 + 2x - 3}$

$ = \dfrac{2x}{(x + 3)(x - 1)}$

Intercept: $(0, 0)$

Vertical asymptotes: $x = -3, x = 1$

Horizontal asymptote: $y = 0$

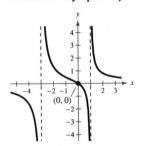

14. $f(x) = \dfrac{x^2 - 4}{x^2 + x - 2}$

$ = \dfrac{(x + 2)(x - 2)}{(x + 2)(x - 1)}$

$ = \dfrac{x - 2}{x - 1}, x \neq -2$

Vertical asymptote: $x = 1$

Horizontal asymptote: $y = 1$

x-intercept: $(2, 0)$

y-intercept: $(0, 2)$

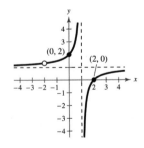

15. $f(x) = \dfrac{x^3 - 2x^2 - 9x + 18}{x^2 + 4x + 3}$

$= \dfrac{x^2(x - 2) - 9(x - 2)}{(x + 1)(x + 3)}$

$= \dfrac{(x - 2)(x^2 - 9)}{(x + 1)(x + 3)}$

$= \dfrac{(x - 2)(x + 3)(x - 3)}{(x + 1)(x + 3)}$

$= \dfrac{(x - 2)(x - 3)}{x + 1}$

$= \dfrac{x^2 - 5x + 6}{x + 1}$

$= x - 6 + \dfrac{12}{x + 1}, x \neq -3$

Vertical asymptote: $x = -1$

Slant asymptote: $y = x - 6$

x-intercepts: $(2, 0), (3, 0)$

y-intercept: $(0, 6)$

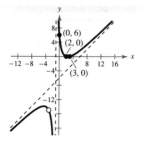

16. $\dfrac{(x + 3)^2}{16} - \dfrac{(y + 4)^2}{25} = 1$

Hyperbola

Center: $(-3, -4)$

Vertices: $(-7, -4), (1, -4)$

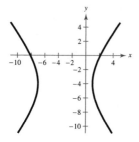

17. $\dfrac{(x - 2)^2}{4} + \dfrac{(y + 1)^2}{9} = 1$

Ellipse

Center: $(2, -1)$

Vertices: $(2, -4), (2, 2)$

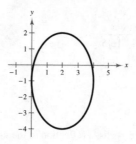

18. Parabola

Vertex: $(3, -2) \Rightarrow y = a(x - 3)^2 - 2$

Point: $(0, 4) \Rightarrow 4 = a(0 - 3)^2 - 2$

$\qquad\qquad\qquad 6 = 9a \Rightarrow a = \tfrac{2}{3}$

Equation: $\qquad y = \tfrac{2}{3}(x - 3)^2 - 2$

$y + 2 = \tfrac{2}{3}(x - 3)^2$

$\tfrac{3}{2}(y + 2) = (x - 3)^2$

$(x - 3)^2 = \tfrac{3}{2}(y + 2)$

19. Hyperbola

Foci: $(0, 0)$ and $(0, 4) \Rightarrow$ Center: $(0, 2)$ and vertical transverse axis

Asymptotes: $y = \pm\dfrac{1}{2}x + 2 \Rightarrow \dfrac{a}{b} = \dfrac{1}{2} \Rightarrow 2a = b$

$c^2 = a^2 + b^2 \Rightarrow 4 = a^2 + 4a^2 \Rightarrow a^2 = \dfrac{4}{5}$ and $b^2 = \dfrac{16}{5}$

Equation: $\dfrac{(y - 2)^2}{4/5} - \dfrac{x^2}{16/5} = 1$

20. $f(x) = \left(\dfrac{2}{5}\right)^x$

$g(x) = -\left(\dfrac{2}{5}\right)^{-x+3}$

g is a reflection in the x-axis, a reflection in the y-axis, and a horizontal shift three units to the right of the graph of f.

21. $f(x) = 2.2^x$

$g(x) = -2.2^x + 4$

g is a reflection in the x-axis, and a vertical shift four units upward of the graph of f.

22. $\log 98 \approx 1.991$

23. $\log\left(\dfrac{6}{7}\right) \approx -0.067$

24. $\ln\sqrt{31} \approx 1.717$

25. $\ln\left(\sqrt{30} - 4\right) \approx 0.390$

26. $\log_5 4.3 = \dfrac{\log_{10} 4.3}{\log_{10} 5} = \dfrac{\ln 4.3}{\ln 5} \approx 0.906$

27. $\log_3 0.149 = \dfrac{\log_{10} 0.149}{\log_{10} 3} = \dfrac{\ln 0.149}{\ln 3} \approx -1.733$

28. $\log_{1/2} 17 = \dfrac{\log_{10} 17}{\log_{10}(1/2)} = \dfrac{\ln 17}{\ln(1/2)} \approx -4.087$

29. $\ln\left(\dfrac{x^2 - 16}{x^4}\right) = \ln(x^2 - 16) - \ln x^4$

$= \ln(x + 4)(x - 4) - 4\ln x$

$= \ln(x + 4) + \ln(x - 4) - 4\ln x, \ x > 4$

30. $2\ln x - \dfrac{1}{2}\ln(x + 5) = \ln x^2 - \ln\sqrt{x + 5}$

$= \ln\dfrac{x^2}{\sqrt{x + 5}}, \ x > 0$

31. $6e^{2x} = 72$

$e^{2x} = 12$

$2x = \ln 12$

$x = \dfrac{\ln 12}{2} \approx 1.242$

32. $4^{x-5} + 21 = 30$

$4^{x-5} = 9$

$x - 5 = \log_4 9$

$x = 5 + \log_4 9$

$x = 5 + \dfrac{\ln 9}{\ln 4}$

$x \approx 6.585$

33. $e^{2x} - 13e^x + 42 = 0$

$(e^x - 6)(e^x - 7) = 0$

$e^x - 6 = 0 \Rightarrow e^x = 6 \Rightarrow x = \ln 6 \approx 1.792$

$e^x - 7 = 0 \Rightarrow e^x = 7 \Rightarrow x = \ln 7 \approx 1.946$

34. $\log_2 x + \log_2 5 = 6$

$\log_2 5x = 6$

$5x = 2^6$

$x = \dfrac{64}{5} = 12.8$

35. $\ln 4x - \ln 2 = 8$

$$\ln \frac{4x}{2} = 8$$

$$\ln 2x = 8$$

$$2x = e^8$$

$$x = \frac{e^8}{2} \approx 1490.479$$

36. $\ln \sqrt{x + 2} = 3$

$$\tfrac{1}{2}\ln(x + 2) = 3$$

$$\ln(x + 2) = 6$$

$$x + 2 = e^6$$

$$x = e^6 - 2 \approx 401.429$$

37. $f(x) = \dfrac{1000}{1 + 4e^{-0.2x}}$

Horizontal asymptotes: $y = 0$, $y = 1000$

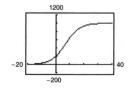

38. $P = -\tfrac{1}{2}x^2 + 20x + 230$

$$= -\tfrac{1}{2}\left(x^2 - 40x + 400 - 400\right) + 230$$

$$= -\tfrac{1}{2}(x - 20)^2 + 430$$

The maximum occurs at the vertex when $x = 20$, which corresponds to spending $(20)(100) = \$2000$ on advertising.

39. (a) and (c)

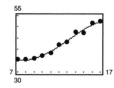

The model is a good fit for the data.

(b) $S = -0.0297t^3 + 1.175t^2 - 12.96t + 79.0$

(d) For 2015, use $t = 25$: $S(25) \approx \$25.3$ billion.

No, this does not seem reasonable. You would expect the sales to continue to increase, but after $t = 18$ the model predicts sales will decrease.

40. $A = 2500e^{(0.075)(25)} \approx \$16,302.05$

41. $N = 175e^{kt}$

$$420 = 175e^{k(8)}$$

$$2.4 = e^{8k}$$

$$\ln 2.4 = 8k$$

$$\frac{\ln 2.4}{8} = k$$

$$k \approx 0.1094$$

$$N = 175e^{0.1094t}$$

$$350 = 175e^{0.1094t}$$

$$2 = e^{0.1094t}$$

$$\ln 2 = 0.1094t$$

$$t = \frac{\ln 2}{0.1094} \approx 6.3 \text{ hours to double}$$

42. 28 million $= 28,000$ thousand

$$P = 20,879e^{0.0189t}$$

$$28,000 = 20,879e^{0.0189t}$$

$$\frac{28,000}{20,879} = e^{0.0189t}$$

$$\ln\left(\frac{28,000}{20,879}\right) = 0.0189t$$

$$t = \frac{\ln\left(\dfrac{28,000}{20,879}\right)}{0.0189} \approx 15.53$$

According to the model, the population will reach 28 million during the year 2015.

43. $p = \dfrac{1200}{1 + 3e^{-t/5}}$

(a) $p(0) = \dfrac{1200}{1 + 3e^0} = \dfrac{1200}{4} = 300$ birds

(b) $p(5) = \dfrac{1200}{1 + 3e^{-1}} \approx 570$ birds

(c) $800 = \dfrac{1200}{1 + 3e^{-t/5}}$

$$800\left(1 + 3e^{-t/5}\right) = 1200$$

$$1 + 3e^{-t/5} = 1.5$$

$$3e^{-t/5} = 0.5$$

$$e^{-t/5} = \frac{1}{6}$$

$$-\frac{t}{5} = \ln\left(\frac{1}{6}\right)$$

$$t = -5\ln\left(\frac{1}{6}\right) \approx 9 \text{ years}$$

Chapter 6   Chapter Test Solutions

1. $\theta = \dfrac{5\pi}{4}$

(a)

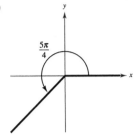

(b) $\dfrac{5\pi}{4} + 2\pi = \dfrac{13\pi}{4}$

$\dfrac{5\pi}{4} - 2\pi = -\dfrac{3\pi}{4}$

(c) $\dfrac{5\pi}{4}\left(\dfrac{180°}{\pi}\right) = 225°$

2. $\dfrac{105 \text{ km}}{\text{hr}} \times \dfrac{1 \text{ hr}}{60 \text{ min}} = 1.75 \text{ km per min}$

diameter $= 1$ meter $= 0.001$ km

radius $= \dfrac{1}{2}$ diameter $= 0.0005$ km

Angular speed $= \dfrac{\theta}{t}$

$= \dfrac{1.75}{2\pi(0.0005)} \cdot 2\pi$

$= 3500$ radians per minute

3. $130° = \dfrac{130\pi}{180} = \dfrac{13\pi}{18}$ radians

$A = \dfrac{1}{2}r^2\theta = \dfrac{1}{2}(25)^2\left(\dfrac{13\pi}{18}\right) \approx 709.04$ square feet

4. $x = -2, y = 6$

$r = \sqrt{(-2)^2 + (6)^2} = 2\sqrt{10}$

$\sin\theta = \dfrac{y}{r} = \dfrac{6}{2\sqrt{10}} = \dfrac{3}{\sqrt{10}} = \dfrac{3\sqrt{10}}{10}$

$\cos\theta = \dfrac{x}{r} = \dfrac{-2}{2\sqrt{10}} = -\dfrac{1}{\sqrt{10}} = -\dfrac{\sqrt{10}}{10}$

$\tan\theta = \dfrac{y}{x} = \dfrac{6}{-2} = -3$

$\csc\theta = \dfrac{r}{y} = \dfrac{2\sqrt{10}}{6} = \dfrac{\sqrt{10}}{3}$

$\sec\theta = \dfrac{r}{x} = \dfrac{2\sqrt{10}}{-2} = -\sqrt{10}$

$\cot\theta = \dfrac{x}{y} = \dfrac{-2}{6} = -\dfrac{1}{3}$

5.

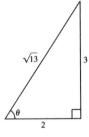

For $0 \le \theta < \dfrac{\pi}{2}$:

$\sin\theta = \dfrac{\text{opp}}{\text{hyp}} = \dfrac{3}{\sqrt{13}} = \dfrac{3\sqrt{13}}{13}$

$\cos\theta = \dfrac{\text{adj}}{\text{hyp}} = \dfrac{2}{\sqrt{13}} = \dfrac{2\sqrt{13}}{13}$

$\csc\theta = \dfrac{\text{hyp}}{\text{opp}} = \dfrac{\sqrt{13}}{3}$

$\sec\theta = \dfrac{\text{hyp}}{\text{adj}} = \dfrac{\sqrt{13}}{2}$

$\cot\theta = \dfrac{\text{adj}}{\text{opp}} = \dfrac{2}{3}$

For $\pi \le \theta < \dfrac{3\pi}{2}$:

$\sin\theta = -\dfrac{3\sqrt{13}}{13}$

$\cos\theta = -\dfrac{2\sqrt{13}}{13}$

$\csc\theta = -\dfrac{\sqrt{13}}{3}$

$\sec\theta = -\dfrac{\sqrt{13}}{2}$

$\cot\theta = \dfrac{2}{3}$

6. $\theta = 205°$

$\theta' = 205° - 180° = 25°$

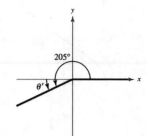

7. $\sec \theta < 0$ and $\tan \theta > 0$

$\dfrac{r}{x} < 0$ and $\dfrac{y}{x} > 0$

Quadrant III

8. $\cos \theta = -\dfrac{\sqrt{3}}{2}$

Reference angle is 30° and θ is in Quadrant II or III.

$\theta = 150°$ or $210°$

9. $\csc \theta = 1.030$

$\dfrac{1}{\sin \theta} = 1.030$

$\sin \theta = \dfrac{1}{1.030}$

$\theta = \arcsin \dfrac{1}{1.030}$

$\theta \approx 1.33$ and $\pi - 1.33 \approx 1.81$

10. $\cos \theta = \frac{3}{5}$, $\tan \theta < 0 \Rightarrow \theta$ lies in Quadrant IV.

Let $x = 3, r = 5 \Rightarrow y = -4$.

$\sin \theta = -\dfrac{4}{5}$

$\cos \theta = \dfrac{3}{5}$

$\tan \theta = -\dfrac{4}{3}$

$\csc \theta = -\dfrac{5}{4}$

$\sec \theta = \dfrac{5}{3}$

$\cot \theta = -\dfrac{3}{4}$

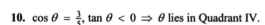

11. $\sec \theta = -\dfrac{29}{20}$, $\sin \theta > 0 \Rightarrow \theta$ lies in Quadrant II.

Let $r = 29, x = -20 \Rightarrow y = 21$.

$\sin \theta = \dfrac{21}{29}$

$\cos \theta = -\dfrac{20}{29}$

$\tan \theta = -\dfrac{21}{20}$

$\csc \theta = \dfrac{29}{21}$

$\cot \theta = -\dfrac{20}{21}$

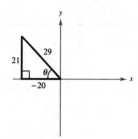

12. $g(x) = -2 \sin\left(x - \dfrac{\pi}{4}\right)$

Period: 2π

Amplitude: $|-2| = 2$

Shifted to the right by $\dfrac{\pi}{4}$ units and reflected in the x-axis.

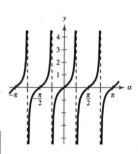

x	0	$\dfrac{\pi}{4}$	$\dfrac{3\pi}{4}$	$\dfrac{5\pi}{4}$	$\dfrac{7\pi}{4}$
y	$\sqrt{2}$	0	-2	0	2

13. $f(\alpha) = \dfrac{1}{2} \tan 2\alpha$

Period: $\dfrac{\pi}{2}$

Asymptotes:

$x = -\dfrac{\pi}{4}, x = \dfrac{\pi}{4}$

α	$-\dfrac{\pi}{8}$	0	$\dfrac{\pi}{8}$
$f(\alpha)$	$-\dfrac{1}{2}$	0	$\dfrac{1}{2}$

14. $y = \sin 2\pi x + 2 \cos \pi x$

Periodic: period $= 2$

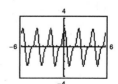

15. $y = 6e^{-0.12t}\cos(0.25t), 0 \le t \le 32$

Not periodic

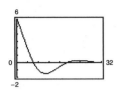

16. $f(x) = a\sin(bx + c)$

Amplitude: $2 \Rightarrow |a| = 2$

Reflected in the x-axis: $a = -2$

Period: $4\pi = \dfrac{2\pi}{b} \Rightarrow b = \dfrac{1}{2}$

Phase shift: $\dfrac{c}{b} = -\dfrac{\pi}{2} \Rightarrow c = -\dfrac{\pi}{4}$

$f(x) = -2\sin\left(\dfrac{x}{2} - \dfrac{\pi}{4}\right)$

17. $\cot\left(\arcsin\dfrac{3}{8}\right)$

Let $y = \arcsin\dfrac{3}{8}$. Then $\sin y = \dfrac{3}{8}$ and

$\cot\left(\arcsin\dfrac{3}{8}\right) = \cot y = \dfrac{\sqrt{55}}{3}$.

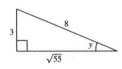

18. $f(x) = 2\arcsin\left(\dfrac{1}{2}x\right)$

Domain: $[-2, 2]$

Range: $[-\pi, \pi]$

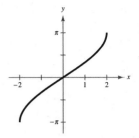

19. $\tan\theta = -\dfrac{110}{90}$

$\theta = \arctan\left(-\dfrac{110}{90}\right)$

$\theta \approx -50.7$

$\theta \approx 309.3°$

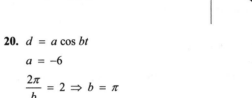

20. $d = a\cos bt$

$a = -6$

$\dfrac{2\pi}{b} = 2 \Rightarrow b = \pi$

$d = -6\cos\pi t$

Chapter 7   Chapter Test Solutions

1. $\tan\theta = \dfrac{6}{5}, \cos\theta < 0$

θ is in Quadrant III.

$\sec\theta = -\sqrt{1 + \tan^2\theta} = -\sqrt{1 + \left(\dfrac{6}{5}\right)^2} = -\dfrac{\sqrt{61}}{5}$

$\cos\theta = \dfrac{1}{\sec\theta} = -\dfrac{5}{\sqrt{61}} = -\dfrac{5\sqrt{61}}{61}$

$\sin\theta = \tan\theta\cos\theta = \left(\dfrac{6}{5}\right)\left(-\dfrac{5\sqrt{61}}{61}\right) = -\dfrac{6\sqrt{61}}{61}$

$\csc\theta = \dfrac{1}{\sin\theta} = -\dfrac{\sqrt{61}}{6}$

$\cot\theta = \dfrac{1}{\tan\theta} = \dfrac{5}{6}$

2. $\csc^2\beta(1 - \cos^2\beta) = \dfrac{1}{\sin^2\beta}(\sin^2\beta) = 1$

3. $\dfrac{\sec^4 x - \tan^4 x}{\sec^2 x + \tan^2 x} = \dfrac{(\sec^2 x + \tan^2 x)(\sec^2 x - \tan^2 x)}{\sec^2 x + \tan^2 x}$

$\qquad\qquad = \sec^2 x - \tan^2 x = 1$

4. $\dfrac{\cos\theta}{\sin\theta} + \dfrac{\sin\theta}{\cos\theta} = \dfrac{\cos^2\theta + \sin^2\theta}{\sin\theta\cos\theta} = \dfrac{1}{\sin\theta\cos\theta}$

$\qquad\qquad = \csc\theta\sec\theta$

5. $y = \tan\theta,\ y = -\sqrt{\sec^2\theta - 1}$

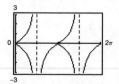

$$\tan\theta = -\sqrt{\sec^2\theta - 1} \text{ on}$$

$$\theta = 0, \frac{\pi}{2} < \theta \le \pi, \frac{3\pi}{2} < \theta < 2\pi.$$

6. $y_1 = \cos x + \sin x \tan x,\ y_2 = \sec x$

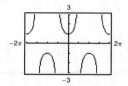

It appears that $y_1 = y_2$.

$$\cos x + \sin x \tan x = \cos + \sin x \frac{\sin x}{\cos x}$$

$$= \cos + \frac{\sin^2 x}{\cos x}$$

$$= \frac{\cos^2 x + \sin^2 x}{\cos x}$$

$$= \frac{1}{\cos x} = \sec x$$

7. $\sin\theta \sec\theta = \sin\theta \dfrac{1}{\cos\theta} = \dfrac{\sin\theta}{\cos\theta} = \tan\theta$

8. $\sec^2 x \tan^2 x + \sec^2 x = \sec^2 x(\sec^2 x - 1) + \sec^2 x = \sec^4 x - \sec^2 x + \sec^2 x = \sec^4 x$

9. $\dfrac{\csc\alpha + \sec\alpha}{\sin\alpha + \cos\alpha} = \dfrac{\dfrac{1}{\sin\alpha} + \dfrac{1}{\cos\alpha}}{\sin\alpha + \cos\alpha} = \dfrac{\dfrac{\cos\alpha + \sin\alpha}{\sin\alpha\cos\alpha}}{\sin\alpha + \cos\alpha} = \dfrac{1}{\sin\alpha\cos\alpha}$

$$= \frac{\cos^2\alpha + \sin^2\alpha}{\sin\alpha\cos\alpha} = \frac{\cos^2\alpha}{\sin\alpha\cos\alpha} + \frac{\sin^2\alpha}{\sin\alpha\cos\alpha}$$

$$= \frac{\cos\alpha}{\sin\alpha} + \frac{\sin\alpha}{\cos\alpha} = \cot\alpha + \tan\alpha$$

10. $\tan\left(x + \dfrac{\pi}{2}\right) = \tan\left(\dfrac{\pi}{2} - (-x)\right) = \cot(-x) = -\cot x$

11. $\sin(n\pi + \theta) = (-1)^n \sin\theta$, n is an integer.

For n odd:

$\sin(n\pi + \theta) = \sin n\pi \cos\theta + \cos n\pi \sin\theta$

$\qquad\qquad = (0)\cos\theta + (-1)\sin\theta = -\sin\theta$

For n even:

$\sin(n\pi + \theta) = \sin n\pi \cos\theta + \cos n\pi \sin\theta$

$\qquad\qquad = (0)\cos\theta + (1)\sin\theta = \sin\theta$

When n is odd, $(-1)^n = -1$. When n is even $(-1)^n = 1$.

So, $\sin(n\pi + \theta) = (-1)^n \sin\theta$ for any integer n.

12. $(\sin x + \cos x)^2 = \sin^2 x + 2\sin x \cos x + \cos^2 x$

$\qquad\qquad\qquad = 1 + 2\sin x \cos x$

$\qquad\qquad\qquad = 1 + \sin 2x$

13. $\sin^4 \dfrac{x}{2} = \left(\sin^2 \dfrac{x}{2}\right)^2$

$$= \left(\frac{1 - \cos 2\left(\dfrac{x}{2}\right)}{2}\right)^2$$

$$= \left(\frac{1 - \cos x}{2}\right)^2$$

$$= \frac{1}{4}(1 - 2\cos x + \cos^2 x)$$

$$= \frac{1}{4}\left(1 - 2\cos x + \frac{1 + \cos 2x}{2}\right)$$

$$= \frac{1}{8}(3 - 4\cos x + \cos 2x)$$

14. $\dfrac{\sin 4\theta}{1 + \cos 4\theta} = \tan \dfrac{4\theta}{2} = \tan 2\theta$

15. $4 \sin 3\theta \cos 2\theta = 4 \cdot \frac{1}{2}\left[\sin(3\theta + 2\theta) + \sin(3\theta - 2\theta)\right]$

$\qquad = 2(\sin 5\theta + \sin \theta)$

16. $\cos 3\theta - \cos \theta = -2 \sin\left(\dfrac{3\theta + \theta}{2}\right) \sin\left(\dfrac{3\theta - \theta}{2}\right)$

$\qquad = -2 \sin 2\theta \sin \theta$

17. $\tan^2 x + \tan x = 0$

$\tan x(\tan x + 1) = 0$

$\qquad \tan x = 0 \qquad \text{or} \quad \tan x + 1 = 0$

$\qquad\qquad x = 0, \pi \qquad\qquad \tan x = -1$

$\qquad\qquad\qquad\qquad\qquad\qquad x = \dfrac{3\pi}{4}, \dfrac{7\pi}{4}$

18. $\qquad \sin 2\alpha - \cos \alpha = 0$

$2 \sin \alpha \cos \alpha - \cos \alpha = 0$

$\qquad \cos \alpha(2 \sin \alpha - 1) = 0$

$\cos \alpha = 0 \quad \text{or} \quad 2 \sin \alpha - 1 = 0$

$\qquad \alpha = \dfrac{\pi}{2}, \dfrac{3\pi}{2} \qquad \sin \alpha = \dfrac{1}{2}$

$\qquad\qquad\qquad\qquad\qquad \alpha = \dfrac{\pi}{6}, \dfrac{5\pi}{6}$

19. $4 \cos^2 x - 3 = 0$

$\qquad \cos^2 x = \dfrac{3}{4}$

$\qquad \cos x = \pm\sqrt{\dfrac{3}{4}} = \pm\dfrac{\sqrt{3}}{2}$

$\qquad\qquad x = \dfrac{\pi}{6}, \dfrac{5\pi}{6}, \dfrac{7\pi}{6}, \dfrac{11\pi}{6}$

20. $\qquad \csc^2 x - \csc x - 2 = 0$

$(\csc x - 2)(\csc x + 1) = 0$

$\csc x - 2 = 0 \qquad \text{or} \quad \csc x + 1 = 0$

$\qquad \csc x = 2 \qquad\qquad\qquad \csc = -1$

$\qquad \dfrac{1}{\sin x} = 2 \qquad\qquad\qquad \dfrac{1}{\sin x} = -1$

$\qquad \sin x = \dfrac{1}{2} \qquad\qquad\qquad \sin x = -1$

$\qquad x = \dfrac{\pi}{6}, \dfrac{5\pi}{6} \qquad\qquad\qquad x = \dfrac{3\pi}{2}$

21. $5 \sin x - x = 0$

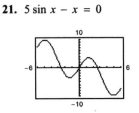

$x \approx -2.596, 0, 2.596$

22. $\qquad 105° = 135° - 30°$

$\cos 105° = \cos(135° - 30°)$

$\qquad = \cos 135° \cos 30° + \sin 135° \sin 30°$

$\qquad = -\cos 45° \cos 30° + \sin 45° \sin 30°$

$\qquad = \left(-\dfrac{\sqrt{2}}{2}\right)\left(\dfrac{\sqrt{3}}{2}\right) + \left(\dfrac{\sqrt{2}}{2}\right)\left(\dfrac{1}{2}\right)$

$\qquad = \dfrac{-\sqrt{6} + \sqrt{2}}{4} = \dfrac{\sqrt{2} - \sqrt{6}}{4}$

23. $x = 2, y = -5, r = \sqrt{29}$

$\sin 2u = 2 \sin u \cos u = 2\left(-\dfrac{5}{\sqrt{29}}\right)\left(\dfrac{2}{\sqrt{29}}\right) = -\dfrac{20}{29}$

$\cos 2u = \cos^2 u - \sin^2 u = \left(\dfrac{2}{\sqrt{29}}\right)^2 - \left(-\dfrac{5}{\sqrt{29}}\right)^2 = -\dfrac{21}{29}$

$\tan 2u = \dfrac{2 \tan u}{1 - \tan^2 u} = \dfrac{2\left(-\dfrac{5}{2}\right)}{1 - \left(-\dfrac{5}{2}\right)^2} = \dfrac{20}{21}$

24. Let $y_1 = 31 \sin\left(\dfrac{2\pi t}{365} - 1.4\right)$ and $y_2 = 20$.

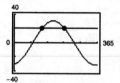

The points of intersection occur when $t \approx 123$ and $t \approx 223$.
The number of days that $D > 20°$ is 100, from day 123 to day 223.

25.
$$28 \cos 10t + 38 = 28 \cos\left[10\left(t - \frac{\pi}{6}\right)\right] + 38$$

$$\cos 10t = \cos\left[10\left(t - \frac{\pi}{6}\right)\right]$$

$$0 = \cos\left[10\left(t - \frac{\pi}{6}\right)\right] - \cos 10t$$

$$= -2 \sin\left(\frac{10\left(t - (\pi/6)\right) + 10t}{2}\right) \sin\left(\frac{10\left(t - (\pi/6)\right) - 10t}{2}\right)$$

$$= -2 \sin\left(10t - \frac{5\pi}{6}\right) \sin\left(-\frac{5\pi}{6}\right)$$

$$= -2 \sin\left(10t - \frac{5\pi}{6}\right)\left(-\frac{1}{2}\right)$$

$$= \sin\left(10t - \frac{5\pi}{6}\right)$$

$$10t - \frac{5\pi}{6} = n\pi \text{ where } n \text{ is any integer.}$$

$$t = \frac{n\pi}{10} + \frac{\pi}{12} \text{ where } n \text{ is any integer.}$$

The first six times the two people are at the same height are: 0.26 minutes,
0.58 minutes, 0.89 minutes, 1.20 minutes, 1.52 minutes, 1.83 minutes.

Chapter 8   Chapter Test Solutions

1. $A = 24°, B = 68°, a = 12.2$

$C = 180° - 24° - 68° = 88°$

$b = \dfrac{a \sin B}{\sin A} = \dfrac{12.2 \sin 68°}{\sin 24°} \approx 27.81$

$c = \dfrac{a \sin C}{\sin A} = \dfrac{12.2 \sin 88°}{\sin 24°} \approx 29.98$

2. $B = 110°, C = 28°, a = 15.6$

$A = 180° - 110° - 28° = 42°$

$b = \dfrac{a \sin B}{\sin A} = \dfrac{15.6 \sin 110°}{\sin 42°} \approx 21.91$

$c = \dfrac{a \sin C}{\sin A} = \dfrac{15.6 \sin 28°}{\sin 42°} \approx 10.95$

3. $A = 24°, a = 11.2, b = 13.4$

$\sin B = \dfrac{b \sin A}{a} = \dfrac{13.4 \sin 24°}{11.2} \approx 0.4866$

Two Solutions

$B \approx 29.12°$	or	$B \approx 150.88°$
$C \approx 126.88°$		$C \approx 5.12°$

$c = \dfrac{a \sin C}{\sin A} = \dfrac{11.2 \sin 126.88°}{\sin 24°}$ $c = \dfrac{11.2 \sin 5.12°}{\sin 24°}$

$c \approx 22.03$ $c \approx 2.46$

4. $a = 4.0, b = 7.3, c = 12.4$

$$\cos C = \frac{a^2 + b^2 - c^2}{2ab}$$

$$= \frac{4^2 + 7.3^2 - 12.4^2}{2(4)(7.3)}$$

$$\approx -1.4464 < -1$$

No solution

5. $B = 100°, a = 15, b = 23$

$$\sin A = \frac{a \sin B}{b} = \frac{15 \sin 100°}{23} \Rightarrow A \approx 39.96°$$

$$C \approx 180° - 100° - 39.96° = 40.04°$$

$$c \approx \frac{b \sin C}{\sin B} = \frac{23 \sin 40.04°}{\sin 100°} \approx 15.02$$

6. $C = 121°, a = 34, b = 55$

$$c^2 = a^2 + b^2 - 2ab \cos C = 34^2 + 55^2 - 2(34)(55) \cos 121° \Rightarrow c \approx 78.15$$

$$\sin A = \frac{a \sin C}{c} \approx \frac{34 \sin 121°}{78.15} \approx 0.3729 \Rightarrow A \approx 21.90°$$

$$B \approx 180° - 21.90° - 121° = 37.10°$$

7. $a = 60, b = 70, c = 82$

$$s = \frac{a + b + c}{2} = \frac{60 + 70 + 82}{2} = 106$$

$$\text{Area} = \sqrt{s(s - a)(s - b)(s - c)} = \sqrt{106(46)(36)(24)} \approx 2052.5 \text{ square meters}$$

8. $b^2 = 370^2 + 240^2 - 2(370)(240)\cos 167°$

$$b \approx 606.3 \text{ miles}$$

$$\sin A = \frac{a \sin B}{b} = \frac{240 \sin 167°}{606.3}$$

$$A \approx 5.1°$$

Bearing: $24° + 5.1° = 29.1°$

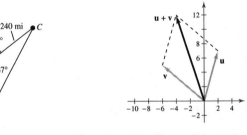

11. $\mathbf{u} = \langle 2, 7 \rangle, \mathbf{v} = \langle -6, 5 \rangle$

$$\mathbf{u} + \mathbf{v} = \langle 2, 7 \rangle + \langle -6, 5 \rangle = \langle -4, 12 \rangle$$

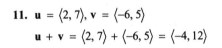

9. Initial point: $(-3, 7)$

Terminal point: $(11, -16)$

$$\mathbf{v} = \langle 11 - (-3), -16 - 7 \rangle = \langle 14, -23 \rangle$$

10. $\mathbf{v} = 12\left(\frac{\mathbf{u}}{\|\mathbf{u}\|}\right) = 12\left(\frac{\langle 3, -5 \rangle}{\sqrt{3^2 + (-5)^2}}\right) = \frac{12}{\sqrt{34}}\langle 3, -5 \rangle$

$$= \frac{6\sqrt{34}}{17}\langle 3, -5 \rangle = \left\langle \frac{18\sqrt{34}}{17}, -\frac{30\sqrt{34}}{17} \right\rangle$$

12. $\mathbf{u} = \langle 2, 7 \rangle, \mathbf{v} = \langle -6, 5 \rangle$

$$\mathbf{u} - \mathbf{v} = \langle 2, 7 \rangle - \langle -6, 5 \rangle = \langle 8, 2 \rangle$$

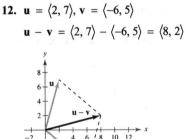

13. $\mathbf{u} = \langle 2, 7 \rangle$, $\mathbf{v} = \langle -6, 5 \rangle$

$$5\mathbf{u} - 3\mathbf{v} = 5\langle 2, 7 \rangle - 3\langle -6, 5 \rangle$$
$$= \langle 10, 35 \rangle - \langle -18, 15 \rangle$$
$$= \langle 28, 20 \rangle$$

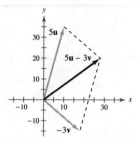

14. $\mathbf{u} = \langle 2, 7 \rangle$, $\mathbf{v} = \langle -6, 5 \rangle$

$$4\mathbf{u} + 2\mathbf{v} = 4\langle 2, 7 \rangle + 2\langle -6, 5 \rangle$$
$$= \langle 8, 28 \rangle + \langle -12, 10 \rangle$$
$$= \langle -4, 38 \rangle$$

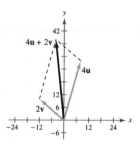

15. $\dfrac{\mathbf{u}}{\|\mathbf{u}\|} = \dfrac{\langle 24, -7 \rangle}{\sqrt{24^2 + (-7)^2}} = \dfrac{1}{25}\langle 24, -7 \rangle = \left\langle \dfrac{24}{25}, -\dfrac{7}{25} \right\rangle$

16. $\mathbf{u} = 250(\cos 45° \, \mathbf{i} + \sin 45° \, \mathbf{j})$

$\mathbf{v} = 130(\cos(-60°)\mathbf{i} + \sin(-60°)\mathbf{j})$

$\mathbf{R} = \mathbf{u} + \mathbf{v} \approx 241.7767 \, \mathbf{i} + 64.1934 \, \mathbf{j}$

$\|\mathbf{R}\| \approx \sqrt{241.7767^2 + 64.1934^2} \approx 250.15$ pounds

$\tan \theta \approx \dfrac{64.1934}{241.7767} \Rightarrow \theta \approx 14.9°$

17. $\mathbf{u} = \langle -1, 5 \rangle$, $\mathbf{v} = \langle 3, -2 \rangle$

$\cos \theta = \dfrac{\mathbf{u} \cdot \mathbf{v}}{\|\mathbf{u}\|\|\mathbf{v}\|} = \dfrac{-13}{\sqrt{26}\sqrt{13}} \Rightarrow \theta = 135°$

18. $\mathbf{u} = \langle 6, -10 \rangle$, $\mathbf{v} = \langle 5, 3 \rangle$

$\mathbf{u} \cdot \mathbf{v} = 6(5) + (-10)(3) = 0$

$\mathbf{u}$ and $\mathbf{v}$ are orthogonal.

19. $\mathbf{u} = \langle 6, 7 \rangle$, $\mathbf{v} = \langle -5, -1 \rangle$

$\mathbf{w}_1 = \text{proj}_{\mathbf{v}}\,\mathbf{u} = \left(\dfrac{\mathbf{u} \cdot \mathbf{v}}{\|\mathbf{v}\|^2} \right)\mathbf{v} = -\dfrac{37}{26}\langle -5, -1 \rangle = \dfrac{37}{26}\langle 5, 1 \rangle$

$\mathbf{w}_2 = \mathbf{u} - \mathbf{w}_1 = \langle 6, 7 \rangle - \dfrac{37}{26}\langle 5, 1 \rangle$

$$= \left\langle -\dfrac{29}{26}, \dfrac{145}{26} \right\rangle$$

$$= \dfrac{29}{26}\langle -1, 5 \rangle$$

$\mathbf{u} = \mathbf{w}_1 + \mathbf{w}_2 = \dfrac{37}{26}\langle 5, 1 \rangle + \dfrac{29}{26}\langle -1, 5 \rangle$

20. $\mathbf{F} = -500\mathbf{j}$, $\mathbf{v} = (\cos 12°)\mathbf{i} + (\sin 12°)\mathbf{j}$

$\mathbf{w}_1 = \text{proj}_{\mathbf{v}}\,\mathbf{F} = \left(\dfrac{\mathbf{F} \cdot \mathbf{v}}{\|\mathbf{v}\|^2} \right)\mathbf{v} = (\mathbf{F} \cdot \mathbf{v})\mathbf{v}$

$= (-500 \sin 12°)\mathbf{v}$

The magnitude of the force is $500 \sin 12° \approx 104$ pounds.

21. $z = 5 - 5i$

$|z| = \sqrt{5^2 + (-5)^2} = \sqrt{50} = 5\sqrt{2}$

$\tan \theta = \dfrac{-5}{5} = -1$ and θ is in Quadrant IV $\Rightarrow \theta = \dfrac{7\pi}{4}$

$z = 5\sqrt{2}\left(\cos \dfrac{7\pi}{4} + i \sin \dfrac{7\pi}{4} \right)$

22. $z = 6(\cos 120° + i \sin 120°)$

$= 6\left(-\dfrac{1}{2} + \dfrac{\sqrt{3}}{2}i \right) = -3 + 3\sqrt{3}i$

23. $\left[3\left(\cos \dfrac{7\pi}{6} + i \sin \dfrac{7\pi}{6} \right) \right]^8 = 3^8\left(\cos \dfrac{28\pi}{3} + i \sin \dfrac{28\pi}{3} \right)$

$$= 6561\left(-\dfrac{1}{2} - \dfrac{\sqrt{3}}{2}i \right)$$

$$= -\dfrac{6561}{2} - \dfrac{6561\sqrt{3}}{2}i$$

24. $(3 - 3i)^6 = \left[3\sqrt{2}\left(\cos \dfrac{7\pi}{4} + i \sin \dfrac{7\pi}{4} \right) \right]^6$

$= (3\sqrt{2})^6\left(\cos \dfrac{21\pi}{2} + i \sin \dfrac{21\pi}{2} \right)$

$= 5832(0 + i)$

$= 5832i$

25. $z = 256\left(1 + \sqrt{3}i\right)$

$|z| = 256\sqrt{1^2 + \left(\sqrt{3}\right)^2} = 256\sqrt{4} = 512$

$\tan\theta = \dfrac{\sqrt{3}}{1} \Rightarrow \theta = \dfrac{\pi}{3}$

$z = 512\left(\cos\dfrac{\pi}{3} + i\sin\dfrac{\pi}{3}\right)$

Fourth roots of $z = \sqrt[4]{512}\left[\cos\dfrac{\dfrac{\pi}{3} + 2\pi k}{4} + i\sin\dfrac{\dfrac{\pi}{3} + 2\pi k}{4}\right]$, $k = 0, 1, 2, 3$

$k = 0: \ 4\sqrt[4]{2}\left(\cos\dfrac{\pi}{12} + i\sin\dfrac{\pi}{12}\right)$

$k = 1: \ 4\sqrt[4]{2}\left(\cos\dfrac{7\pi}{12} + i\sin\dfrac{7\pi}{12}\right)$

$k = 2: \ 4\sqrt[4]{2}\left(\cos\dfrac{13\pi}{12} + i\sin\dfrac{13\pi}{12}\right)$

$k = 3: \ 4\sqrt[4]{2}\left(\cos\dfrac{19\pi}{12} + i\sin\dfrac{19\pi}{12}\right)$

26. $x^3 - 27i = 0 \Rightarrow x^3 = 27i$

The solutions to the equation are the cube roots of $27i = 27\left(\cos\dfrac{\pi}{2} + i\sin\dfrac{\pi}{2}\right)$.

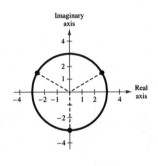

Cube roots: $\sqrt[3]{27}\left[\cos\dfrac{\dfrac{\pi}{2} + 2\pi k}{3} + i\sin\dfrac{\dfrac{\pi}{2} + 2\pi k}{3}\right]$, $k = 0, 1, 2$

$k = 0: \ 3\left(\cos\dfrac{\pi}{6} + i\sin\dfrac{\pi}{6}\right) = 3\left(\dfrac{\sqrt{3}}{2} + \dfrac{1}{2}i\right) = \dfrac{3\sqrt{3}}{2} + \dfrac{3}{2}i$

$k = 1: \ 3\left(\cos\dfrac{5\pi}{6} + i\sin\dfrac{5\pi}{6}\right) = 3\left(-\dfrac{\sqrt{3}}{2} + \dfrac{1}{2}i\right) = -\dfrac{3\sqrt{3}}{2} + \dfrac{3}{2}i$

$k = 2: \ 3\left(\cos\dfrac{3\pi}{2} + i\sin\dfrac{3\pi}{2}\right) = 3(0 - i) = -3i$

Chapters 6–8   Cumulative Test Solutions

1. (a)

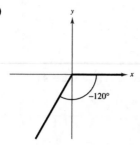

(b) $-120° + 360° = 240°$

(c) $-120\left(\dfrac{\pi}{180°}\right) = -\dfrac{2\pi}{3}$

(d) $-120° + 360° = 240°$

$\theta' = 240° - 180° = 60°$

(e) $\sin(-120°) = -\sin 60° = -\dfrac{\sqrt{3}}{2}$

$\cos(-120°) = -\cos 60° = -\dfrac{1}{2}$

$\tan(-120°) = \tan 60° = \sqrt{3}$

$\csc(-120°) = \dfrac{1}{-\sin 60°} = -\dfrac{2\sqrt{3}}{3}$

$\sec(-120°) = \dfrac{1}{-\cos 60°} = -2$

$\cot(-120°) = \dfrac{1}{\tan 60°} = \dfrac{\sqrt{3}}{3}$

2. $-1.45\left(\dfrac{180}{\pi}\right) \approx -83.1°$

3. $\tan\theta = \dfrac{y}{x} = -\dfrac{21}{20} \Rightarrow r = 29$

Because $\sin\theta < 0$, θ is in Quadrant IV. $\Rightarrow x = 20$

$\cos\theta = \dfrac{x}{r} = \dfrac{20}{29}$

4. $f(x) = 3 - 2\sin\pi x$

Period: $\dfrac{2\pi}{\pi} = 2$

Amplitude: $|a| = |-2| = 2$

Upward shift of 3 units (reflected in x-axis prior to shift)

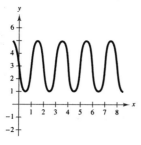

5. $g(x) = \dfrac{1}{2}\tan\left(x - \dfrac{\pi}{2}\right)$

Period: π

Asymptotes: $x = 0, x = \pi$

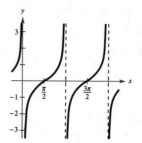

6. $h(x) = -\sec(x + \pi)$

Graph $y = -\cos(x + \pi)$ first.

Period: 2π

Amplitude: 1

Set $x + \pi = 0$ and $x + \pi = 2\pi$ for one cycle.

$\qquad x = -\pi \qquad\qquad x = \pi$

The asymptotes of $h(x)$ corresponds to the x-intercepts of $y = -\cos(x + \pi)$.

$x + \pi = \dfrac{(2n + 1)\pi}{2}$

$x = \dfrac{(2n - 1)\pi}{2}$ where n is any integer

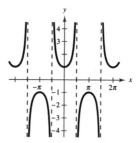

7. $h(x) = a\cos(bx + c)$

Graph is reflected in x-axis.

Amplitude: $a = -3$

Period: $2 = \dfrac{2\pi}{\pi} \Rightarrow b = \pi$

No phase shift: $c = 0$

$h(x) = -3\cos(\pi x)$

8. $f(x) = \dfrac{x}{2}\sin x,\ -3\pi \le x \le 3\pi$

$$-\dfrac{x}{2} \le f(x) \le \dfrac{x}{2}$$

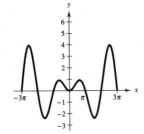

9. $\tan(\arctan 4.9) = 4.9$

10. $\tan\!\left(\arcsin\dfrac{3}{5}\right) = \dfrac{3}{4}$

11. $y = \arccos(2x)$

$$\sin y = \sin(\arccos(2x)) = \sqrt{1 - 4x^2}$$

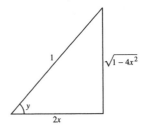

12. $\cos\!\left(\dfrac{\pi}{2} - x\right)\csc x = \sin x\!\left(\dfrac{1}{\sin x}\right) = 1$

13.
$$\dfrac{\sin\theta - 1}{\cos\theta} - \dfrac{\cos\theta}{\sin\theta - 1} = \dfrac{\sin\theta - 1}{\cos\theta} - \dfrac{\cos\theta(\sin\theta + 1)}{\sin^2\theta - 1}$$

$$= \dfrac{\sin\theta - 1}{\cos\theta} + \dfrac{\cos\theta(\sin\theta + 1)}{\cos^2\theta} = \dfrac{\sin\theta - 1}{\cos\theta} + \dfrac{\sin\theta + 1}{\cos\theta} = \dfrac{2\sin\theta}{\cos\theta} = 2\tan\theta$$

14. $\cot^2\alpha(\sec^2\alpha - 1) = \cot^2\alpha\,\tan^2\alpha = 1$

15.
$$\sin(x + y)\sin(x - y) = \tfrac{1}{2}\big[\cos(x + y - (x - y)) - \cos(x + y + x - y)\big]$$

$$= \tfrac{1}{2}[\cos 2y - \cos 2x] = \tfrac{1}{2}\big[1 - 2\sin^2 y - (1 - 2\sin^2 x)\big] = \sin^2 x - \sin^2 y$$

16.
$$\sin^2 x \cos^2 x = \left(\dfrac{1 - \cos 2x}{2}\right)\!\left(\dfrac{1 + \cos 2x}{2}\right)$$

$$= \dfrac{1}{4}(1 - \cos 2x)(1 + \cos 2x)$$

$$= \dfrac{1}{4}(1 - \cos^2 2x)$$

$$= \dfrac{1}{4}\left(1 - \dfrac{1 + \cos 4x}{2}\right)$$

$$= \dfrac{1}{8}(2 - (1 + \cos 4x))$$

$$= \dfrac{1}{8}(1 - \cos 4x)$$

17. $2\cos^2\beta - \cos\beta = 0$

$\cos\beta(2\cos\beta - 1) = 0$

$\cos\beta = 0$ or $2\cos\beta - 1 = 0$

$\beta = \dfrac{\pi}{2}, \dfrac{3\pi}{2}$ $\cos\beta = \dfrac{1}{2}$

$\beta = \dfrac{\pi}{3}, \dfrac{5\pi}{3}$

Answer: $\dfrac{\pi}{3}, \dfrac{\pi}{2}, \dfrac{3\pi}{2}, \dfrac{5\pi}{3}$

18. $3\tan\theta - \cot\theta = 0$

$$3\tan\theta - \dfrac{1}{\tan\theta} = 0$$

$$\dfrac{3\tan^2\theta - 1}{\tan\theta} = 0$$

$$3\tan^2\theta - 1 = 0$$

$$\tan^2\theta = \dfrac{1}{3}$$

$$\tan\theta = \pm\dfrac{\sqrt{3}}{3}$$

$$\theta = \dfrac{\pi}{6}, \dfrac{5\pi}{6}, \dfrac{7\pi}{6}, \dfrac{11\pi}{6}$$

19. $\sin^2 x + 2\sin x + 1 = 0$

$(\sin x + 1)(\sin x + 1) = 0$

$\sin x + 1 = 0$

$\sin x = -1$

$x = \dfrac{3\pi}{2}$

20. $\sin u = \frac{12}{13} \Rightarrow \cos u = \frac{5}{13}$ and $\tan u = \frac{12}{5}$ because u is in Quadrant I.

$\cos v = \frac{3}{5} \Rightarrow \sin v = \frac{4}{5}$ and $\tan v = \frac{4}{3}$ because v is in Quadrant I.

$$\tan(u - v) = \frac{\tan u - \tan v}{1 + \tan u \tan v} = \frac{\dfrac{12}{5} - \dfrac{4}{3}}{1 + \left(\dfrac{12}{5}\right)\left(\dfrac{4}{3}\right)} = \frac{16}{63}$$

21. $\tan \theta = \dfrac{1}{2}$

$$\tan 2\theta = \frac{2 \tan \theta}{1 - \tan^2 \theta} = \frac{2\left(\dfrac{1}{2}\right)}{1 - \left(\dfrac{1}{2}\right)^2} = \frac{4}{3}$$

22. $\tan \theta = \dfrac{4}{3} \Rightarrow \cos \theta = \pm\dfrac{3}{5}$

$$\sin \frac{\theta}{2} = \sqrt{\frac{1 - \cos \theta}{2}} = \sqrt{\frac{1 - \dfrac{3}{5}}{2}} = \frac{\sqrt{5}}{5}$$

$$\text{or } = \sqrt{\frac{1 + \dfrac{3}{5}}{2}} = \frac{2\sqrt{5}}{5}$$

23. $5 \sin \dfrac{3\pi}{4} \cos \dfrac{7\pi}{4} = \dfrac{5}{2}\left[\sin\left(\dfrac{3\pi}{4} + \dfrac{7\pi}{4}\right) + \sin\left(\dfrac{3\pi}{4} - \dfrac{7\pi}{4}\right)\right]$

$\qquad\qquad\qquad\quad = \dfrac{5}{2}\left[\sin\dfrac{5\pi}{2} + \sin(-\pi)\right]$

$\qquad\qquad\qquad\quad = \dfrac{5}{2}\left(\sin\dfrac{5\pi}{2} - \sin\pi\right)$

24. $\cos 9x - \cos 7x = -2 \sin\left(\dfrac{9x + 7x}{2}\right)\sin\left(\dfrac{9x - 7x}{2}\right)$

$\qquad\qquad\qquad\quad = -2 \sin 8x \sin x$

25. Given: $A = 30°, a = 9, b = 8$

$$\frac{\sin B}{8} = \frac{\sin 30°}{9}$$

$$\sin B = \frac{8\left(\dfrac{1}{2}\right)}{9}$$

$$B = \arcsin\left(\frac{4}{9}\right)$$

$$B \approx 26.39°$$

$$C = 180° - A - B \approx 123.61°$$

$$\frac{c}{\sin 123.61°} = \frac{9}{\sin 30°}$$

$$c \approx 14.99$$

26. Given: $A = 30°, b = 8, c = 10$

$$a^2 = 8^2 + 10^2 - 2(8)(10)\cos 30°$$

$$a^2 \approx 25.4359$$

$$a \approx 5.04$$

$$\cos B = \frac{5.04^2 + 10^2 - 8^2}{2(5.04)(10)}$$

$$\cos B \approx 0.6091$$

$$B \approx 52.48°$$

$$C = 180° - A - B \approx 97.52°$$

27. Given: $A = 30°, C = 90°, b = 10$

$$B = 180° - 30° - 90° = 60°$$

$$\tan 30° = \frac{a}{10} \Rightarrow a = 10 \tan 30° \approx 5.77$$

$$\cos 30° = \frac{10}{c} \Rightarrow c = \frac{10}{\cos 30°} \approx 11.55$$

28. Given: $a = 4.7, b = 8.1, c = 10.3$

$$\cos C = \frac{a^2 + b^2 - c^2}{2ab} = \frac{4.7^2 + 8.1^2 + 10.3^2}{2(4.7)(8.1)} \approx -0.2415 \Rightarrow C \approx 103.98°$$

$$\sin A = \frac{a \sin C}{c} \approx \frac{4.7 \sin 103.98°}{10.3} \approx 0.4428 \Rightarrow A \approx 26.28°$$

$$B \approx 180° - 26.28° - 103.98° = 49.74°$$

29. Given: $A = 45°, B = 26°, c = 20$

Given two angles and a side, use the Law of Sines.

$C = 180° - 45° - 26° = 109°$

$a = \dfrac{c \sin A}{\sin C} = \dfrac{20 \sin 45°}{\sin 109°} \approx 14.96$

$b = \dfrac{c \sin B}{\sin C} = \dfrac{20 \sin 26°}{\sin 109°} \approx 9.27$

30. Given: $a = 1.2, b = 10, C = 80°$

Given two sides and the included angle, use the Law of Cosines.

$c^2 = a^2 + b^2 - 2ab \cos C = 1.2^2 + 10^2 - 2(1.2)(10)\cos 80° \Rightarrow c \approx 9.86$

$\cos B = \dfrac{a^2 + c^2 - b^2}{2ac} = \dfrac{1.2^2 + 9.86^2 - 10^2}{2(1.2)(9.86)} \approx -0.0566 \Rightarrow B \approx 93.25°$

$A \approx 180° - 93.25° - 80° = 6.75°$

31. Area $= \dfrac{1}{2}(7)(12)\sin 99° = 41.48$ in.^{2}

32. $a = 30, b = 41, c = 45$

$s = \dfrac{a + b + c}{2} = \dfrac{30 + 41 + 45}{2} = 58$

Area $= \sqrt{s(s - a)(s - b)(s - c)}$

$\qquad = \sqrt{58(28)(17)(13)}$

$\qquad \approx 599.09$ m^2

33. $\mathbf{u} = \langle 7, 8 \rangle = 7\mathbf{i} + 8\mathbf{j}$

34. $\mathbf{v} = \mathbf{i} + \mathbf{j}$

$\|\mathbf{v}\| = \sqrt{1^2 + 1^2} = \sqrt{2}$

$\mathbf{u} = \dfrac{\mathbf{v}}{\|\mathbf{v}\|} = \dfrac{1}{\sqrt{2}}(\mathbf{i} + \mathbf{j}) = \dfrac{\sqrt{2}}{2}(\mathbf{i} + \mathbf{j})$

35. $\mathbf{u} = 3\mathbf{i} + 4\mathbf{j}, \mathbf{v} = \mathbf{i} - 2\mathbf{j}$

$\mathbf{u} \cdot \mathbf{v} = 3(1) + 4(-2) = -5$

36. $\mathbf{u} = \langle 8, -2 \rangle, \mathbf{v} = \langle 1, 5 \rangle$

$\mathbf{w}_1 = \text{proj}_\mathbf{v}\, \mathbf{u} = \left(\dfrac{\mathbf{u} \cdot \mathbf{v}}{\|\mathbf{v}\|^2} \right)\mathbf{v} = \dfrac{-2}{26}\langle 1, 5 \rangle = -\dfrac{1}{13}\langle 1, 5 \rangle$

$\mathbf{w}_2 = \mathbf{u} - \mathbf{w}_1 = \langle 8, -2 \rangle - \left\langle -\dfrac{1}{13}, -\dfrac{5}{13} \right\rangle = \left\langle \dfrac{105}{13}, -\dfrac{21}{13} \right\rangle$

$\qquad = \dfrac{21}{13}\langle 5, -1 \rangle$

$\mathbf{u} = \mathbf{w}_1 + \mathbf{w}_2 = -\dfrac{1}{13}\langle 1, 5 \rangle + \dfrac{21}{13}\langle 5, -1 \rangle$

37. $r = |-2 + 2i| = \sqrt{(-2)^2 + (2)^2} = 2\sqrt{2}$

$\tan \theta = \dfrac{2}{-2} = -1$

Because $\tan \theta = -1$ and $-2 + 2i$ lies in Quadrant II,

$\theta = \dfrac{3\pi}{4}$. So, $-2 + 2i = 2\sqrt{2}\left(\cos\dfrac{3\pi}{4} + i \sin\dfrac{3\pi}{4} \right)$.

38. $\left[4(\cos 30° + i \sin 30°) \right]\left[6(\cos 120° + i \sin 120°) \right] = (4)(6)\left[\cos(30° + 120°) + i \sin(30° + 120°) \right]$

$\qquad\qquad = 24(\cos 150° + i \sin 150°)$

$\qquad\qquad = 24\left(-\dfrac{\sqrt{3}}{2} + \dfrac{1}{2}i \right)$

$\qquad\qquad = -12\sqrt{3} + 12i$

39. $1 = 1(\cos 0 + i \sin 0)$

$$\sqrt[3]{1} = \sqrt[3]{1}\left[\cos\left(\frac{0 + 2\pi k}{3}\right) + i \sin\left(\frac{0 + 2\pi k}{3}\right)\right], k = 0, 1, 2$$

$$k = 0: \sqrt[3]{1}\left[\left(\cos\left(\frac{0 + 2\pi(0)}{3}\right) + i \sin\left(\frac{0 + 2\pi(0)}{3}\right)\right)\right] = \cos 0 + i \sin 0 = 1$$

$$k = 1: \sqrt[3]{1}\left[\left(\cos\left(\frac{0 + 2\pi(1)}{3}\right) + i \sin\left(\frac{0 + 2\pi(1)}{3}\right)\right)\right] = \cos \frac{2\pi}{3} + i \sin \frac{2\pi}{3} = -\frac{1}{2} + \frac{\sqrt{3}}{2}i$$

$$k = 2: \sqrt[3]{1}\left[\left(\cos\left(\frac{0 + 2\pi(2)}{3}\right) + i \sin\left(\frac{0 + 2\pi(2)}{3}\right)\right)\right] = \cos \frac{4\pi}{3} + i \sin \frac{4\pi}{3} = -\frac{1}{2} - \frac{\sqrt{3}}{2}i$$

40. $x^5 + 243 = 0 \Rightarrow x^5 = -243$

The solutions to the equation are the fifth roots of $-243 = 243(\cos \pi + i \sin \pi)$, which are:

$$\sqrt[5]{243}\left[\cos\left(\frac{\pi + 2\pi k}{5}\right) + i \sin\left(\frac{\pi + 2\pi k}{5}\right)\right], k = 0, 1, 2, 3, 4$$

$$k = 0: 3\left(\cos \frac{\pi}{5} + i \sin \frac{\pi}{5}\right)$$

$$k = 1: 3\left(\cos \frac{3\pi}{5} + i \sin \frac{3\pi}{5}\right)$$

$$k = 2: 3(\cos \pi + i \sin \pi)$$

$$k = 3: 3\left(\cos \frac{7\pi}{5} + i \sin \frac{7\pi}{5}\right)$$

$$k = 4: 3\left(\cos \frac{9\pi}{5} + i \sin \frac{9\pi}{5}\right)$$

41. Angular speed $= \dfrac{\theta}{t} = \dfrac{2\pi(63)}{1} \approx 395.8$ radians per minute

Linear speed $= \dfrac{s}{t} = \dfrac{42\pi(63)}{1} \approx 8312.7$ inches per minute

42. Area $= \dfrac{\theta r^2}{2} = \dfrac{(105°)\left(\dfrac{\pi}{180°}\right)(12)^2}{2} = 42\pi \approx 131.95$ yd^2

43. Height of smaller triangle:

$$\tan 16° \, 45' = \frac{h_1}{200}$$

$$h_1 = 200 \tan 16.75°$$

$$\approx 60.2 \text{ feet}$$

Height of larger triangle:

$$\tan 18° = \frac{h_2}{200}$$

$$h_2 = 200 \tan 18° \approx 65.0 \text{ feet}$$

Height of flag: $h_2 - h_1 = 65.0 - 60.2 \approx 5$ feet

Not drawn to scale

44. $\tan \theta = \dfrac{5}{12} \Rightarrow \theta \approx 22.6°$

45. $d = a \cos bt$

$$|a| = 4 \Rightarrow a = 4$$

$$\frac{2\pi}{b} = 8 \Rightarrow b = \frac{\pi}{4}$$

$$d = 4 \cos \frac{\pi}{4} t$$

46. $\mathbf{v}_1 = 500\langle\cos 60°, \sin 60°\rangle = \langle 250, 250\sqrt{3}\rangle$

$\mathbf{v}_2 = 50\langle\cos 30°, \sin 30°\rangle = \langle 25\sqrt{3}, 25\rangle$

$\mathbf{v} = \mathbf{v}_1 + \mathbf{v}_2 = \langle 250 + 25\sqrt{3}, 250\sqrt{3} + 25\rangle \approx \langle 293.3, 458.0\rangle$

$\|\mathbf{v}\| = \sqrt{(293.3)^2 + (458.0)^2} \approx 543.9$

$\tan\theta = \dfrac{458.0}{293.3} \approx 1.56 \Rightarrow \theta \approx 57.4°$

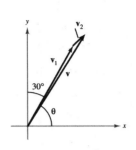

Bearing: $90° - 57.4° = 32.6°$

The plane is traveling on a bearing of $32.6°$ at 543.9 kilometers per hour.

47. $\mathbf{w} = (85)(10)\cos 60° = 425$ foot-pounds

Chapter 9   Chapter Test Solutions

1. $\begin{cases} x + y = -9 \Rightarrow x = -y - 9 \\ 5x - 8y = 20 \end{cases}$

$5(-y - 9) - 8y = 20$

$-13y = 65$

$y = -5$

$x - 5 = -9 \Rightarrow x = -4$

Solution: $(-4, -5)$

2. $\begin{cases} y = x - 1 \\ y = (x - 1)^3 \end{cases}$

$x - 1 = (x - 1)^3$

$x - 1 = x^3 - 3x^2 + 3x - 1$

$0 = x^3 - 3x^2 + 2x$

$0 = x(x - 1)(x - 2)$

$x = 0$ or $x = 1$ or $x = 2$

$y = -1$ $y = 0$ $y = 1$

Solutions: $(0, -1), (1, 0), (2, 1)$

3. $\begin{cases} x - y = 4 \Rightarrow x = y + 4 \\ 2x - y^2 = 0 \Rightarrow 2(y + 4) - y^2 = 0 \end{cases}$

$0 = y^2 - 2y - 8$

$0 = (y + 2)(y - 4)$

$y = -2$ or $y = 4$

$x = 2$ $x = 8$

Solutions: $(2, -2), (8, 4)$

4. $\begin{cases} 3x - 6y = 0 \Rightarrow y = \dfrac{1}{2}x \\ 3x + 6y = 18 \Rightarrow y = -\dfrac{1}{2}x + 3 \end{cases}$

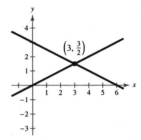

Solution: $\left(3, \dfrac{3}{2}\right)$

5. $\begin{cases} y = 9 - x^2 \\ y = x + 3 \end{cases}$

Solutions: $(-3, 0), (2, 5)$

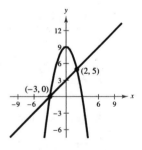

6. $\begin{cases} y - \ln x = 12 \Rightarrow y = 12 + \ln x \\ 7x - 2y + 11 = -6 \Rightarrow y = \dfrac{7}{2}x + \dfrac{17}{2} \end{cases}$

Solutions:

$(1, 12), (0.034, 8.619)$

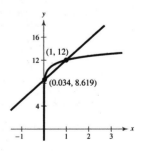

7. $\begin{cases} 3x + 4y = -26 & \text{Equation 1} \\ 7x - 5y = 11 & \text{Equation 2} \end{cases}$

Multiply Equation 1 by 5: $15x + 20y = -130$

Multiply Equation 2 by 4: $28x - 20y = 44$

Add the equations to eliminate y:
$$\begin{array}{r} 15x + 20y = -130 \\ 28x - 20y = 44 \\ \hline 43x \quad\quad = -86 \\ x = -2 \end{array}$$

Back-substitute $x = -2$ into Equation 1:
$3(-2) + 4y = -26$
$y = -5$

Solution: $(-2, -5)$

8. $\begin{cases} 1.4x - y = 17 & \text{Equation 1} \\ 0.8x + 6y = -10 & \text{Equation 2} \end{cases}$

Multiply Equation 1 by 6: $8.4x - 6y = 102$

Add this to Equation 2 to eliminate y:
$$\begin{array}{r} 8.4x - 6y = 102 \\ 0.8x + 6y = -10 \\ \hline 9.2x \quad\quad = 92 \\ x = 10 \end{array}$$

Back-substitute $x = 10$ into Equation 2:
$0.8(10) + 6y = -10$
$6y = -18$
$y = -3$

Solution: $(10, -3)$

9. $\begin{cases} x - 2y + 3z = 11 \\ 2x \quad\quad - z = 3 \\ \quad 3y + z = -8 \end{cases}$

$\begin{cases} x - 2y + 3z = 11 \\ 4y - 7z = -19 & -2\text{Eq.1} + \text{Eq.2} \\ 3y + z = -8 \end{cases}$

$\begin{cases} x - 2y + 3z = 11 \\ y - 8z = -11 & -\text{Eq.3} + \text{Eq.2} \\ 3y + z = -8 \end{cases}$

$\begin{cases} x - 2y + 3z = 11 \\ y - 8z = -11 \\ 25z = 25 & -3\text{Eq.2} + \text{Eq.3} \end{cases}$

$\begin{cases} x - 2y + 3z = 11 \\ y - 8z = -11 \\ z = 1 & \frac{1}{25}\text{Eq.3} \end{cases}$

$y - 8(1) = -11 \Rightarrow y = -3$
$x - 2(-3) + 3(1) = 11 \Rightarrow x = 2$

Solution: $(2, -3, 1)$

10. $\begin{cases} 3x + 2y + z = 17 & \text{Equation 1} \\ -x + y + z = 4 & \text{Equation 2} \\ x - y - z = 3 & \text{Equation 3} \end{cases}$

Interchange Equations 1 and 3

$\begin{cases} x - y - z = 3 \\ -x + y + z = 4 \\ 3x + 2y + z = 17 \end{cases}$

$\begin{cases} x - y - z = 3 \\ 0 \neq 7 & \text{Eq. 1} + \text{Eq. 2} \\ 3x + 2y + z = 17 \end{cases}$

Inconsistent

No solution

11. $\dfrac{2x + 5}{x^2 - x - 2} = \dfrac{2x + 5}{(x - 2)(x + 1)} = \dfrac{A}{x - 2} + \dfrac{B}{x + 1}$

$2x + 5 = A(x + 1) + B(x - 2)$

Let $x = 2$: $9 = 3A \Rightarrow A = 3$

Let $x = -1$: $3 = -3B \Rightarrow B = -1$

$\dfrac{2x + 5}{x^2 - x - 2} = \dfrac{3}{x - 2} - \dfrac{1}{x + 1}$

12. $\dfrac{3x^2 - 2x + 4}{x^2(2 - x)} = \dfrac{A}{x} + \dfrac{B}{x^2} + \dfrac{C}{2 - x}$

$3x^2 - 2x + 4 = Ax(2 - x) + B(2 - x) + Cx^2$

Let $x = 0$: $4 = 2B \Rightarrow B = 2$

Let $x = 2$: $12 = 4C \Rightarrow C = 3$

Let $x = 1$: $5 = A + B + C = A + 2 + 3 \Rightarrow A = 0$

$\dfrac{3x^2 - 2x + 4}{x^2(2 - x)} = \dfrac{2}{x^2} + \dfrac{3}{2 - x}$

13. $\dfrac{x^2 + 5}{x^3 - x} = \dfrac{x^2 + 5}{x(x + 1)(x - 1)} = \dfrac{A}{x} + \dfrac{B}{x + 1} + \dfrac{C}{x - 1}$

$x^2 + 5 = A(x + 1)(x - 1) + Bx(x - 1) + Cx(x + 1)$

Let $x = 0$: $5 = -A \Rightarrow A = -5$

Let $x = -1$: $6 = 2B \Rightarrow B = 3$

Let $x = 1$: $6 = 2C \Rightarrow C = 3$

$\dfrac{x^2 + 5}{x^3 - x} = -\dfrac{5}{x} + \dfrac{3}{x + 1} + \dfrac{3}{x - 1}$

14. $\dfrac{x^2 - 4}{x^3 + 2x} = \dfrac{x^2 - 4}{x(x^2 + 2)} = \dfrac{A}{x} + \dfrac{Bx + C}{x^2 + 2}$

$x^2 - 4 = A(x^2 + 2) + (Bx + C)x$

$\qquad = Ax^2 + 2A + Bx^2 + Cx$

$\qquad = (A + B)x^2 + Cx + 2A$

Equate the coefficients of like terms:

$1 = A + B, 0 = C, -4 = 2A$

So, $A = -2, B = 3, C = 0$.

$\dfrac{x^2 - 4}{x^3 + 2x} = -\dfrac{2}{x} + \dfrac{3x}{x^2 + 2}$

15. $\begin{cases} 2x + y \le 4 \\ 2x - y \ge 0 \\ \qquad x \ge 0 \end{cases}$

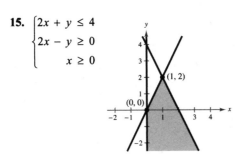

16. $\begin{cases} y < -x^2 + x + 4 \\ y > 4x \end{cases}$

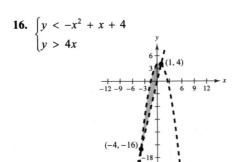

17. $\begin{cases} x^2 + y^2 \le 36 \\ \qquad x \ge 2 \\ \qquad y \ge -4 \end{cases}$

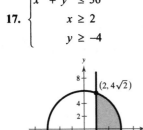

18. Maximize $z = 20x + 12y$ subject to:

$\begin{cases} x \ge 0, y \ge 0 \\ x + 4y \le 32 \\ 3x + 2y \le 36 \end{cases}$

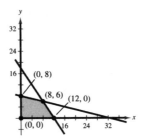

At $(0, 0)$ we have $z = 0$.

At $(0, 8)$ we have $z = 96$.

At $(8, 6)$ we have $z = 232$.

At $(12, 0)$ we have $z = 240$.

The maximum value, $z = 240$, occurs at $(12, 0)$.

The minimum value, $z = 0$ occurs at $(0, 0)$.

19. Let x = amount of money invested at 4%.

Let y = amount of money invested at 5.5%.

$\begin{cases} x + y = 50{,}000 \qquad \text{Equation 1} \\ 0.04x + 0.055y = 2390 \qquad \text{Equation 2} \end{cases}$

Multiply Equation 1 by -4: $-4x - 4y = -200{,}000$

Multiply Equation 2 by 100: $4x + 5.5y = 239{,}000$

Add these two equations to eliminate x:

$\begin{aligned} -4x - 4y &= -200{,}000 \\ 4x + 5.5y &= 239{,}000 \\ \hline 1.5y &= 39{,}000 \\ y &= 26{,}000 \end{aligned}$

Back-substitute $y = 26{,}000$ into Equation 1:

$x + 26{,}000 = 50{,}000$

$x = 24{,}000$

So, $24{,}000$ should be invested at 4% and $26{,}000$ should be invested at 5.5%.

20. $y = ax^2 + bx + c$

$(0, 6): 6 = c$

$(-2, 2): 2 = 4a - 2b + c$

$\left(3, \frac{9}{2}\right): \frac{9}{2} = 9a + 3b + c$

Solving this system yields: $a = -\frac{1}{2}, b = 1$, and $c = 6$.

So, $y = -\frac{1}{2}x^2 + x + 6$.

21. Optimize $P = 30x + 40y$ subject to:

$$\begin{cases} x \geq 0, \ y \geq 0 \\ 0.5x + 0.75y \leq 4000 \\ 2.0x + 1.5y \leq 8950 \\ 0.5x + 0.5y \leq 2650 \end{cases}$$

At $(0, 0)$: $P = 0$

At $(0, 5300)$: $P = 212,000$

At $(2000, 3300)$: $P = 192,000$

At $(4475, 0)$: $P = 134,250$

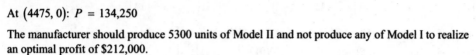

The manufacturer should produce 5300 units of Model II and not produce any of Model I to realize an optimal profit of \$212,000.

Chapter 10   Chapter Test Solutions

1.
$$\begin{bmatrix} 1 & -1 & 5 \\ 6 & 2 & 3 \\ 5 & 3 & -3 \end{bmatrix}$$

$$\begin{matrix} -6R_1 + R_2 \rightarrow \\ -5R_1 + R_3 \rightarrow \end{matrix} \begin{bmatrix} 1 & -1 & 5 \\ 0 & 8 & -27 \\ 0 & 8 & -28 \end{bmatrix}$$

$$-R_2 + R_3 \rightarrow \begin{bmatrix} 1 & -1 & 5 \\ 0 & 8 & -27 \\ 0 & 0 & -1 \end{bmatrix}$$

$$\begin{matrix} \frac{1}{8}R_2 \rightarrow \\ -R_3 \rightarrow \end{matrix} \begin{bmatrix} 1 & -1 & 5 \\ 0 & 1 & -\frac{27}{8} \\ 0 & 0 & 1 \end{bmatrix}$$

$$R_2 + R_1 \rightarrow \begin{bmatrix} 1 & 0 & \frac{13}{8} \\ 0 & 1 & -\frac{27}{8} \\ 0 & 0 & 1 \end{bmatrix}$$

$$\begin{matrix} -\frac{13}{8}R_3 + R_1 \rightarrow \\ \frac{27}{8}R_3 + R_2 \rightarrow \end{matrix} \begin{bmatrix} 1 & 0 & 0 \\ 0 & 1 & 0 \\ 0 & 0 & 1 \end{bmatrix}$$

2.
$$\begin{bmatrix} 1 & 0 & -1 & 2 \\ -1 & 1 & 1 & -3 \\ 1 & 1 & -1 & 1 \\ 3 & 2 & -3 & 4 \end{bmatrix}$$

$$\begin{matrix} R_1 + R_2 \rightarrow \\ -R_1 + R_3 \rightarrow \\ -3R_1 + R_4 \rightarrow \end{matrix} \begin{bmatrix} 1 & 0 & -1 & 2 \\ 0 & 1 & 0 & -1 \\ 0 & 1 & 0 & -1 \\ 0 & 2 & 0 & -2 \end{bmatrix}$$

$$\begin{matrix} -R_2 + R_3 \rightarrow \\ -2R_2 + R_4 \rightarrow \end{matrix} \begin{bmatrix} 1 & 0 & -1 & 2 \\ 0 & 1 & 0 & -1 \\ 0 & 0 & 0 & 0 \\ 0 & 0 & 0 & 0 \end{bmatrix}$$

3.
$$\begin{bmatrix} 4 & 3 & -2 & \vdots & 14 \\ -1 & -1 & 2 & \vdots & -5 \\ 3 & 1 & -4 & \vdots & 8 \end{bmatrix}$$

$$3R_2 + R_1 \rightarrow \begin{bmatrix} 1 & 0 & 4 & \vdots & -1 \\ -1 & -1 & 2 & \vdots & -5 \\ 3 & 1 & -4 & \vdots & 8 \end{bmatrix}$$

$$\begin{matrix} R_1 + R_2 \rightarrow \\ -3R_1 + R_3 \rightarrow \end{matrix} \begin{bmatrix} 1 & 0 & 4 & \vdots & -1 \\ 0 & -1 & 6 & \vdots & -6 \\ 0 & 1 & -16 & \vdots & 11 \end{bmatrix}$$

$$R_2 + R_3 \rightarrow \begin{bmatrix} 1 & 0 & 4 & \vdots & -1 \\ 0 & -1 & 6 & \vdots & -6 \\ 0 & 0 & -10 & \vdots & 5 \end{bmatrix}$$

$$\begin{matrix} -R_2 \rightarrow \\ -\frac{1}{10}R_3 \rightarrow \end{matrix} \begin{bmatrix} 1 & 0 & 4 & \vdots & -1 \\ 0 & 1 & -6 & \vdots & 6 \\ 0 & 0 & 1 & \vdots & -\frac{1}{2} \end{bmatrix}$$

$$\begin{matrix} -4R_3 + R_1 \rightarrow \\ 6R_3 + R_2 \rightarrow \end{matrix} \begin{bmatrix} 1 & 0 & 0 & \vdots & 1 \\ 0 & 1 & 0 & \vdots & 3 \\ 0 & 0 & 1 & \vdots & -\frac{1}{2} \end{bmatrix}$$

Solution: $\left(1, 3, -\frac{1}{2}\right)$

4. $A = \begin{bmatrix} 6 & 5 \\ -5 & -5 \end{bmatrix}$, $\quad B = \begin{bmatrix} 5 & 0 \\ -5 & -1 \end{bmatrix}$

(a) $A - B = \begin{bmatrix} 6 & 5 \\ -5 & -5 \end{bmatrix} - \begin{bmatrix} 5 & 0 \\ -5 & -1 \end{bmatrix} = \begin{bmatrix} 6 - 5 & 5 - 0 \\ -5 - (-5) & -5 - (-1) \end{bmatrix} = \begin{bmatrix} 1 & 5 \\ 0 & -4 \end{bmatrix}$

(b) $3A = 3\begin{bmatrix} 6 & 5 \\ -5 & -5 \end{bmatrix} = \begin{bmatrix} 3(6) & 3(5) \\ 3(-5) & 3(-5) \end{bmatrix} = \begin{bmatrix} 18 & 15 \\ -15 & -15 \end{bmatrix}$

(c) $3A - 2B = \begin{bmatrix} 18 & 15 \\ -15 & -15 \end{bmatrix} - 2\begin{bmatrix} 5 & 0 \\ -5 & -1 \end{bmatrix} = \begin{bmatrix} 18 - 2(5) & 15 - 2(0) \\ -15 - 2(-5) & -15 - 2(-1) \end{bmatrix} = \begin{bmatrix} 8 & 15 \\ -5 & -13 \end{bmatrix}$

(d) $AB = \begin{bmatrix} 6 & 5 \\ -5 & -5 \end{bmatrix}\begin{bmatrix} 5 & 0 \\ -5 & -1 \end{bmatrix} = \begin{bmatrix} (6)(5) + (5)(-5) & (6)(0) + (5)(-1) \\ (-5)(5) + (-5)(-5) & (-5)(0) + (-5)(-1) \end{bmatrix} = \begin{bmatrix} 5 & -5 \\ 0 & 5 \end{bmatrix}$

5. $A = \begin{bmatrix} a & b \\ c & d \end{bmatrix}$, $\quad A^{-1} = \dfrac{1}{ad - bc}\begin{bmatrix} d & -b \\ -c & a \end{bmatrix}$

$A = \begin{bmatrix} -4 & 3 \\ 5 & -2 \end{bmatrix}$

$ad - bc = (-4)(-2) - (3)(5) = -7$

$A^{-1} = -\dfrac{1}{7}\begin{bmatrix} -2 & -3 \\ -5 & -4 \end{bmatrix} = \begin{bmatrix} \frac{2}{7} & \frac{3}{7} \\ \frac{5}{7} & \frac{4}{7} \end{bmatrix}$

6.

$\begin{bmatrix} -2 & 4 & -6 & \vdots & 1 & 0 & 0 \\ 2 & 1 & 0 & \vdots & 0 & 1 & 0 \\ 4 & -2 & 5 & \vdots & 0 & 0 & 1 \end{bmatrix}$

$\begin{matrix} \\ R_1 + R_2 \rightarrow \\ 2R_1 + R_3 \rightarrow \end{matrix}\begin{bmatrix} -2 & 4 & -6 & \vdots & 1 & 0 & 0 \\ 0 & 5 & -6 & \vdots & 1 & 1 & 0 \\ 0 & 6 & -7 & \vdots & 2 & 0 & 1 \end{bmatrix}$

$\begin{matrix} -\frac{1}{2}R_1 \rightarrow \\ -R_3 + R_2 \rightarrow \\ \\ \end{matrix}\begin{bmatrix} 1 & -2 & 3 & \vdots & -\frac{1}{2} & 0 & 0 \\ 0 & -1 & 1 & \vdots & -1 & 1 & -1 \\ 0 & 6 & -7 & \vdots & 2 & 0 & 1 \end{bmatrix}$

$\begin{matrix} -2R_2 + R_1 \rightarrow \\ \\ 6R_2 + R_3 \rightarrow \end{matrix}\begin{bmatrix} 1 & 0 & 1 & \vdots & \frac{3}{2} & -2 & 2 \\ 0 & -1 & 1 & \vdots & -1 & 1 & -1 \\ 0 & 0 & -1 & \vdots & -4 & 6 & -5 \end{bmatrix}$

$\begin{matrix} \\ -R_2 \rightarrow \\ -R_3 \rightarrow \end{matrix}\begin{bmatrix} 1 & 0 & 1 & \vdots & \frac{3}{2} & -2 & 2 \\ 0 & 1 & -1 & \vdots & 1 & -1 & 1 \\ 0 & 0 & 1 & \vdots & 4 & -6 & 5 \end{bmatrix}$

$\begin{matrix} -R_3 + R_1 \rightarrow \\ R_3 + R_2 \rightarrow \\ \\ \end{matrix}\begin{bmatrix} 1 & 0 & 0 & \vdots & -\frac{5}{2} & 4 & -3 \\ 0 & 1 & 0 & \vdots & 5 & -7 & 6 \\ 0 & 0 & 1 & \vdots & 4 & -6 & 5 \end{bmatrix}$

$A^{-1} = \begin{bmatrix} -\frac{5}{2} & 4 & -3 \\ 5 & -7 & 6 \\ 4 & -6 & 5 \end{bmatrix}$

7. $\begin{cases} -4x + 3y = 6 \\ 5x - 2y = 24 \end{cases}$

$\begin{bmatrix} -4 & 3 \\ 5 & -2 \end{bmatrix}\begin{bmatrix} x \\ y \end{bmatrix} = \begin{bmatrix} 6 \\ 24 \end{bmatrix}$

$\begin{bmatrix} x \\ y \end{bmatrix} = \begin{bmatrix} -4 & 3 \\ 5 & -2 \end{bmatrix}^{-1}\begin{bmatrix} 6 \\ 24 \end{bmatrix} = \begin{bmatrix} \frac{2}{7} & \frac{3}{7} \\ \frac{5}{7} & \frac{4}{7} \end{bmatrix}\begin{bmatrix} 6 \\ 24 \end{bmatrix} = \begin{bmatrix} 12 \\ 18 \end{bmatrix}$

Solution: $(12, 18)$

8. $\begin{vmatrix} -6 & 4 \\ 10 & 12 \end{vmatrix} = (-6)(12) - (4)(10) = -112$

9. $\begin{vmatrix} \frac{5}{2} & \frac{13}{4} \\ -8 & \frac{6}{5} \end{vmatrix} = \left(\frac{5}{2}\right)\left(\frac{6}{5}\right) - \left(\frac{13}{4}\right)(-8) = 29$

10. Expand along Column 3.

$\begin{vmatrix} 6 & -7 & 2 \\ 3 & -2 & 0 \\ 1 & 5 & 1 \end{vmatrix} = 2\begin{vmatrix} 3 & -2 \\ 1 & 5 \end{vmatrix} + \begin{vmatrix} 6 & -7 \\ 3 & -2 \end{vmatrix} = 2(17) + 9 = 43$

11. $\begin{cases} 7x + 6y = 9 \\ -2x - 11y = -49 \end{cases}$ $D = \begin{vmatrix} 7 & 6 \\ -2 & -11 \end{vmatrix} = -65$

$$x = \dfrac{\begin{vmatrix} 9 & 6 \\ -49 & -11 \end{vmatrix}}{-65} = \dfrac{195}{-65} = -3$$

$$y = \dfrac{\begin{vmatrix} 7 & 9 \\ -2 & -49 \end{vmatrix}}{-65} = \dfrac{-325}{-65} = 5$$

Solution: $(-3, 5)$

12. $\begin{cases} 6x - y + 2z = -4 \\ -2x + 3y - z = 10 \\ 4x - 4y + z = -18 \end{cases}$ $D = \begin{vmatrix} 6 & -1 & 2 \\ -2 & 3 & -1 \\ 4 & -4 & 1 \end{vmatrix} = -12$

$$x = \dfrac{\begin{vmatrix} -4 & -1 & 2 \\ 10 & 3 & -1 \\ -18 & -4 & 1 \end{vmatrix}}{-12} = \dfrac{24}{-12} = -2$$

$$y = \dfrac{\begin{vmatrix} 6 & -4 & 2 \\ -2 & 10 & -1 \\ 4 & -18 & 1 \end{vmatrix}}{-12} = \dfrac{-48}{-12} = 4$$

$$z = \dfrac{\begin{vmatrix} 6 & -1 & -4 \\ -2 & 3 & 10 \\ 4 & -4 & -18 \end{vmatrix}}{-12} = \dfrac{-72}{-12} = 6$$

Solution: $(-2, 4, 6)$

13. $A = -\dfrac{1}{2} \begin{vmatrix} -5 & 0 & 1 \\ 4 & 4 & 1 \\ 3 & 2 & 1 \end{vmatrix} = -\dfrac{1}{2}(-14) = 7$

14.
$$\begin{matrix} K & N & O \\ C & K & - \\ O & N & - \\ W & O & O \\ D & - & - \end{matrix} \begin{bmatrix} 11 & 14 & 15 \\ 3 & 11 & 0 \\ 15 & 14 & 0 \\ 23 & 15 & 15 \\ 4 & 0 & 0 \end{bmatrix} \begin{bmatrix} 1 & -1 & 0 \\ 1 & 0 & -1 \\ 6 & -2 & -3 \end{bmatrix} = \begin{bmatrix} 115 & -41 & -59 \\ 14 & -3 & -11 \\ 29 & -15 & -14 \\ 128 & -53 & -60 \\ 4 & -4 & 0 \end{bmatrix}$$

Message: $[11 \; 14 \; 15], [3 \; 11 \; 0], [15 \; 14 \; 0], [23 \; 15 \; 15], [4 \; 0 \; 0]$

Encoded Message: 115 −41 −59 14 −3 −11 29 −15 −14 128 −53 −60 4 −4 0

15. Let x = amount of 60% solution and y = amount of 20% solution.

$$\begin{cases} x + y = 100 \Rightarrow y = 100 - x \\ 0.60x + 0.20y = 0.50(100) \Rightarrow 6x + 2y = 500 \end{cases}$$

By substitution,

$$6x + 2(100 - x) = 500$$
$$6x + 200 - 2x = 500$$
$$4x = 300$$
$$x = 75$$
$$y = 100 - x = 25.$$

75 liters of 60% solution and 25 liters of 20% solution.

Chapter 11   Chapter Test Solutions

1. $a_n = \dfrac{(-1)^n}{3n+2}$

$a_1 = -\dfrac{1}{5}$

$a_2 = \dfrac{1}{8}$

$a_3 = -\dfrac{1}{11}$

$a_4 = \dfrac{1}{14}$

$a_5 = -\dfrac{1}{17}$

2. $\dfrac{3}{1!}, \dfrac{4}{2!}, \dfrac{5}{3!}, \dfrac{6}{4!}, \dfrac{7}{5!}, \dots$

$a_n = \dfrac{n+2}{n!}$

3. $8 + 21 + 34 + 47 + \dots$

$a_5 = 60, a_6 = 73, a_7 = 86$

$S_6 = 8 + 21 + 34 + 47 + 60 + 73 = 243$

4. $a_5 = 5.4, a_{12} = 11.0$

$a_{12} = a_5 + 7d$

$11.0 = 5.4 + 7d$

$5.6 = 7d$

$0.8 = d$

$a_1 = a_5 - 4d$

$a_1 = 5.4 - 4(0.8)$

$\quad = 2.2$

$a_n = a_1 + (n-1)d$

$\quad = 2.2 + (n-1)(0.8)$

$\quad = 0.8n + 1.4$

5. $a_2 = 28, a_6 = 7168$

$a_6 = a_2 r^4$

$7168 = 28r^4$

$256 = r^4$

$4 = r$

$a_2 = a_1 r$

$28 = a_1(4)$

$7 = a_1$

$a_n = 7(4)^{n-1}$

6. $a_n = 5(2)^{n-1}$

$a_1 = 5$

$a_2 = 10$

$a_3 = 20$

$a_4 = 40$

$a_5 = 80$

7. $\displaystyle\sum_{i=1}^{50} \left(2i^2 + 5\right) = 2\sum_{i=1}^{50} i^2 + \sum_{i=1}^{50} 5$

$\qquad\qquad = 2\left[\dfrac{50(51)(101)}{6}\right] + 50(5)$

$\qquad\qquad = 86{,}100$

8. $\displaystyle\sum_{n=1}^{9} (12n - 7) = 12\sum_{n=1}^{9} n - \sum_{n=1}^{9} 7$

$\qquad\qquad = 12\left[\dfrac{9(10)}{2}\right] - 9(7)$

$\qquad\qquad = 477$

9. $\displaystyle\sum_{i=1}^{\infty} 4\left(\dfrac{1}{2}\right)^i = \dfrac{2}{1 - \dfrac{1}{2}} = 4$

10. $5 + 10 + 15 + \dots + 5n = \dfrac{5n(n+1)}{2}$

When $n = 1, S_1 = 5 = \dfrac{5(1)(2)}{2}$, so the formula is valid.

Assume that

$S_k = 5 + 10 + 15 + \dots + 5k = \dfrac{5k(k+1)}{2}$, then

$S_{k+1} = S_k + a_{k+1}$

$\qquad = \dfrac{5k(k+1)}{2} + 5(k+1)$

$\qquad = \dfrac{5k(k+1)}{2} + \dfrac{10(k+1)}{2}$

$\qquad = \dfrac{5k(k+1) + 10(k+1)}{2}$

$\qquad = \dfrac{5(k+1)(k+2)}{2}$

$\qquad = \dfrac{5(k+1)\left[(k+1)+1\right]}{2}.$

So, the formula is valid for all integers $n \geq 1$.

11. (a) $(x + 6y)^4 = x^4 + {}_4C_1x^3(6y) + {}_4C_2x^2(6y)^2 + {}_4C_3x(6y)^3 + {}_4C_4(6y)^4$

$$= x^4 + 24x^3y + 216x^2y^2 + 864xy^3 + 1296y^4$$

(b) $3(x - 2)^5 + 4(x - 2)^3 = 3\left[x^5 + {}_5C_1x^4(-2) + {}_5C_2x^3(-2)^2 + {}_5C_3x^2(-2)^3 + {}_5C_4x(-2)^4 + {}_5C_5(-2)^5\right]$

$$+ 4\left[x^3 + {}_3C_1x^2(-2) + {}_3C_2x(-2)^2 + {}_3C_3(-2)^3\right]$$

$$= 3(x^5 - 10x^4 + 40x^3 - 80x^2 + 80x - 32) + 4(x^3 - 6x^2 + 12x - 8)$$

$$= 3x^5 - 30x^4 + 124x^3 - 264x^2 + 288x - 128$$

12. ${}_nC_rx^{n-r}y^r = {}_7C_3(3a)^4(-2b)^3$

$$= 35(81a^4)(-8b^3)$$

$$= -22,680a^4b^3$$

So, the coefficient of a^4b^3 is $-22,680$.

13. (a) ${}_9P_2 = \dfrac{9!}{7!} = 72$

(b) ${}_{70}P_3 = \dfrac{70!}{67!} = 328,440$

14. (a) ${}_{11}C_4 = \dfrac{11!}{7!4!} = 330$

(b) ${}_{66}C_4 = \dfrac{66!}{62!4!} = 720,720$

15. $(26)(10)(10)(10) = 26,000$ distinct license plates

16. $\underbrace{(1)}_{\substack{\text{owner}}} \cdot \underbrace{(3)(2)}_{\substack{\text{bow}\\\text{seats}}} \cdot \underbrace{(5)(4)(3)(2)(1)}_{\substack{\text{remaining}\\\text{seats}}} = 720$ seating arrangements

17. $\dfrac{20}{300} = \dfrac{1}{15} \approx 0.0667$

18. $\dfrac{1}{{}_{30}C_4} = \dfrac{1}{27,405}$

19. $P(E') = 1 - P(E)$

$$= 1 - 0.90$$

$$= 0.10 \text{ or } 10\%$$

Chapters 9–11 Cumulative Test Solutions

1. $\begin{cases} y = 3 - x^2 \\ 2(y - 2) = x - 1 \Rightarrow 2(3 - x^2 - 2) = x - 1 \end{cases}$

$$2(1 - x^2) = x - 1$$

$$2 - 2x^2 = x - 1$$

$$0 = 2x^2 + x - 3$$

$$0 = (2x + 3)(x - 1)$$

$$x = -\tfrac{3}{2} \text{ or } x = 1$$

$$y = \tfrac{3}{4} \qquad y = 2$$

Solutions: $\left(-\tfrac{3}{2}, \tfrac{3}{4}\right), (1, 2)$

2. $\begin{cases} x + 3y = -6 \Rightarrow 4x + 12y = -24 \\ 2x + 4y = -10 \Rightarrow \underline{-6x - 12y = 30} \end{cases}$

$$-2x \qquad = 6$$

$$x = -3 \Rightarrow y = -1$$

Solution: $(-3, -1)$

3. $\begin{cases} -2x + 4y - z = -16 \\ x - 2y + 2z = 5 \\ x - 3y - z = 13 \end{cases}$

Interchange equations.

$\begin{cases} x - 2y + 2z = 5 & \text{Eq.1} \\ -2x + 4y - z = -16 & \text{Eq.2} \\ x - 3y - z = 13 & \text{Eq.3} \end{cases}$

$\begin{cases} x - 2y + 2z = 5 \\ \quad\quad 3z = -6 & 2\text{Eq.1} + \text{Eq.2} \\ \quad -y - 3z = 8 & -\text{Eq.1} + \text{Eq.3} \end{cases}$

From Equation 2, $z = -2$. Substituting this into Equation 3 yields $y = -2$. Using these in Equation 1 yields $x = 5$.

Solution: $(5, -2, -2)$

4. $\begin{cases} x + 3y - 2z = -7 \\ -2x + y - z = -5 \\ 4x + y + z = 3 \end{cases}$

$\begin{cases} x + 3y - 2z = -7 \\ 7y - 5z = -19 \qquad \text{2Eq.1 + Eq.2} \\ -11y + 9z = 31 \qquad \text{−4Eq.1 + Eq.3} \end{cases}$

$\begin{cases} x + 3y - 2z = -7 \\ y - \frac{5}{7}z = -\frac{19}{7} \qquad \frac{1}{7}\text{Eq.2} \\ -11y + 9z = 31 \end{cases}$

$\begin{cases} x + \frac{1}{7}z = \frac{8}{7} \qquad \text{−3Eq.2 + Eq.1} \\ y - \frac{5}{7}z = -\frac{19}{7} \\ \frac{8}{7}z = \frac{8}{7} \qquad \text{11Eq.2 + Eq.3} \end{cases}$

$\begin{cases} x + \frac{1}{7}z = \frac{8}{7} \\ y - \frac{5}{7}z = -\frac{19}{7} \\ z = 1 \qquad \frac{7}{8}\text{Eq.3} \end{cases}$

$\begin{cases} x = 1 \qquad -\frac{1}{7}\text{Eq.3 + Eq.1} \\ y = -2 \qquad \frac{5}{7}\text{Eq.3 + Eq.2} \\ z = 1 \end{cases}$

Solution: $(1, -2, 1)$

5. $\begin{cases} 2x + y \geq -3 \\ x - 3y \leq 2 \end{cases}$

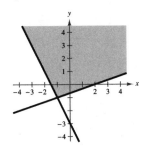

6. $\begin{cases} x - y > 6 \\ 5x + 2y < 10 \end{cases}$

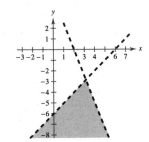

7. Objective function: $z = 3x + 2y$

Subject to: $x + 4y \leq 20$

$ 2x + y \leq 12$

$ x \geq 0, y \geq 0$

At $(0, 0)$: $z = 0$

At $(0, 5)$: $z = 10$

At $(4, 4)$: $z = 30$

At $(6, 0)$: $z = 18$

Minimum of $z = 0$ at $(0, 0)$

Maximum of $z = 20$ at $(4, 4)$

8. $\begin{cases} x + y = 200 \Rightarrow y = 200 - x \\ 0.75x + 1.25y = 0.95(200) \end{cases}$

$0.75x + 1.25(200 - x) = 190$

$0.75x + 250 - 1.25x = 190$

$-0.50x = -60$

$x = 120$

$y = 200 - x = 80$

120 pounds of $0.75 seed and 80 pounds of $1.25 seed.

9. $y = ax^2 + bx + c$

$(0, 6)$: $6 = a(0)^2 + b(0) + c \Rightarrow c = 6$

$(2, 3)$: $3 = a(2)^2 + b(2) + 6 \Rightarrow 4a + 2b = -3$

$ 2a + b = -\frac{3}{2}$

$(4, 2)$: $2 = a(4)^2 + b(4) + 6 \Rightarrow 16a + 4b = -4$

$ 4a + b = -1$

Solving the system:

$\begin{cases} 2a + b = -\frac{3}{2} \\ 4a + b = -1 \end{cases}$ yields $a = \frac{1}{4}$ and $b = -2$.

So, the equation of the parabola is $y = \frac{1}{4}x^2 - 2x + 6$.

10. $\begin{cases} -x + 2y - z = 9 \\ 2x - y + 2z = -9 \\ 3x + 3y - 4z = 7 \end{cases}$

$\begin{bmatrix} -1 & 2 & -1 & \vdots & 9 \\ 2 & -1 & 2 & \vdots & -9 \\ 3 & 3 & -4 & \vdots & 7 \end{bmatrix}$

11.
$$\begin{bmatrix} -1 & 2 & -1 & \vdots & 9 \\ 2 & -1 & 2 & \vdots & -9 \\ 3 & 3 & -4 & \vdots & 7 \end{bmatrix}$$

$$\begin{matrix} \\ 2R_1 + R_2 \to \\ 3R_1 + R_3 \to \end{matrix} \begin{bmatrix} -1 & 2 & -1 & \vdots & 9 \\ 0 & 3 & 0 & \vdots & 9 \\ 0 & 9 & -7 & \vdots & 34 \end{bmatrix}$$

$$\begin{matrix} -R_1 \to \\ \\ -3R_2 + R_3 \to \end{matrix} \begin{bmatrix} 1 & -2 & 1 & \vdots & -9 \\ 0 & 3 & 0 & \vdots & 3 \\ 0 & 0 & -7 & \vdots & 7 \end{bmatrix}$$

$$\begin{matrix} \\ \frac{1}{3}R_2 \to \\ -\frac{1}{7}R_2 \to \end{matrix} \begin{bmatrix} 1 & -2 & 1 & \vdots & -9 \\ 0 & 1 & 0 & \vdots & 3 \\ 0 & 0 & 1 & \vdots & -1 \end{bmatrix}$$

$$\begin{matrix} 2R_2 + R_1 \to \\ \\ \\ \end{matrix} \begin{bmatrix} 1 & 0 & 1 & \vdots & -3 \\ 0 & 1 & 0 & \vdots & 3 \\ 0 & 0 & 1 & \vdots & -1 \end{bmatrix}$$

$$\begin{matrix} -R_3 + R_1 \to \\ \\ \\ \end{matrix} \begin{bmatrix} 1 & 0 & 0 & \vdots & -2 \\ 0 & 1 & 0 & \vdots & 3 \\ 0 & 0 & 1 & \vdots & -1 \end{bmatrix}$$

Solution: $(-2, 3, -1)$

12. $A + B = \begin{bmatrix} 3 & 0 \\ -1 & 4 \end{bmatrix} + \begin{bmatrix} -2 & 5 \\ 0 & -1 \end{bmatrix} = \begin{bmatrix} 1 & 5 \\ -1 & 3 \end{bmatrix}$

13. $-8B = -8\begin{bmatrix} -2 & 5 \\ 0 & -1 \end{bmatrix} = \begin{bmatrix} 16 & -40 \\ 0 & 8 \end{bmatrix}$

14. $2A - 5B = 2A + (-5)B = 2\begin{bmatrix} 3 & 0 \\ -1 & 4 \end{bmatrix} + (-5)\begin{bmatrix} -2 & 5 \\ 0 & -1 \end{bmatrix} = \begin{bmatrix} 6 & 0 \\ -2 & 8 \end{bmatrix} + \begin{bmatrix} 10 & -25 \\ 0 & 5 \end{bmatrix} = \begin{bmatrix} 16 & -25 \\ -2 & 13 \end{bmatrix}$

15. $AB = \begin{bmatrix} 3 & 0 \\ -1 & 4 \end{bmatrix}\begin{bmatrix} -2 & 5 \\ 0 & -1 \end{bmatrix} = \begin{bmatrix} 3(-2) + 0(0) & 3(5) + 0(-1) \\ -1(-2) + 4(0) & -1(5) + 4(-1) \end{bmatrix} = \begin{bmatrix} -6 & 15 \\ 2 & -9 \end{bmatrix}$

16. $A^2 = \begin{bmatrix} 3 & 0 \\ -1 & 4 \end{bmatrix}\begin{bmatrix} 3 & 0 \\ -1 & 4 \end{bmatrix} = \begin{bmatrix} 3(3) + 0(-1) & 3(0) + 0(4) \\ -1(3) + 4(-1) & -1(0) + 4(4) \end{bmatrix} = \begin{bmatrix} 9 & 0 \\ -7 & 16 \end{bmatrix}$

17. $BA - B^2 = \begin{bmatrix} -2 & 5 \\ 0 & -1 \end{bmatrix}\begin{bmatrix} 3 & 0 \\ -1 & 4 \end{bmatrix} - \begin{bmatrix} -2 & 5 \\ 0 & -1 \end{bmatrix}\begin{bmatrix} -2 & 5 \\ 0 & -1 \end{bmatrix}$

$$= \begin{bmatrix} -2(3) + 5(-1) & -2(0) + 5(4) \\ 0(3) + (-1)(-1) & 0(0) + (-1)(4) \end{bmatrix} - \begin{bmatrix} -2(-2) + 5(0) & -2(5) + 5(-1) \\ 0(-2) + (-1)(0) & 0(5) + (-1)(-1) \end{bmatrix}$$

$$= \begin{bmatrix} -11 & 20 \\ 1 & -4 \end{bmatrix} - \begin{bmatrix} 4 & -15 \\ 0 & 1 \end{bmatrix}$$

$$= \begin{bmatrix} -15 & 35 \\ 1 & -5 \end{bmatrix}$$

18. Expand along Row 1.

$$\begin{vmatrix} 7 & 1 & 0 \\ -2 & 4 & -1 \\ 3 & 8 & 5 \end{vmatrix} = 7\begin{vmatrix} 4 & -1 \\ 8 & 5 \end{vmatrix} - 1\begin{vmatrix} -2 & -1 \\ 3 & 5 \end{vmatrix} = 7(28) - 1(-7) = 203$$

19.
$$\left[\begin{array}{rrr:rrr} 1 & 2 & -1 & 1 & 0 & 0 \\ 3 & 7 & -10 & 0 & 1 & 0 \\ -5 & -7 & -15 & 0 & 0 & 1 \end{array}\right]$$

$$\begin{array}{c} \\ -3R_1 + R_2 \to \\ 5R_1 + R_3 \to \end{array} \left[\begin{array}{rrr:rrr} 1 & 2 & -1 & 1 & 0 & 0 \\ 0 & 1 & -7 & -3 & 1 & 0 \\ 0 & 3 & -20 & 5 & 0 & 1 \end{array}\right]$$

$$\begin{array}{c} -2R_2 + R_1 \to \\ \\ -3R_2 + R_3 \to \end{array} \left[\begin{array}{rrr:rrr} 1 & 0 & 13 & 7 & -2 & 0 \\ 0 & 1 & -7 & -3 & 1 & 0 \\ 0 & 0 & 1 & 14 & -3 & 1 \end{array}\right]$$

$$\begin{array}{c} -13R_3 + R_1 \to \\ 7R_3 + R_2 \to \\ \end{array} \left[\begin{array}{rrr:rrr} 1 & 0 & 0 & -175 & 37 & -13 \\ 0 & 1 & 0 & 95 & -20 & 7 \\ 0 & 0 & 1 & 14 & -3 & 1 \end{array}\right]$$

$$\left[\begin{array}{rrr} 1 & 2 & -1 \\ 3 & 7 & -10 \\ -5 & -7 & -15 \end{array}\right]^{-1} = \left[\begin{array}{rrr} -175 & 37 & -13 \\ 95 & -20 & 7 \\ 14 & -3 & 1 \end{array}\right]$$

20. Let x = total sales of gym shoes (in millions),

$\quad\quad y$ = total sales of jogging shoes (in millions),

$\quad\quad z$ = total sales of walking shoes (in millions).

$$\left[\begin{array}{rrr} 0.09 & 0.09 & 0.03 \\ 0.06 & 0.10 & 0.05 \\ 0.12 & 0.25 & 0.12 \end{array}\right]\left[\begin{array}{c} x \\ y \\ z \end{array}\right] = \left[\begin{array}{c} 442.20 \\ 466.57 \\ 1088.09 \end{array}\right]$$

$$\left[\begin{array}{c} x \\ y \\ z \end{array}\right] = \left[\begin{array}{rrr} 0.09 & 0.09 & 0.03 \\ 0.06 & 0.10 & 0.05 \\ 0.12 & 0.25 & 0.12 \end{array}\right]^{-1}\left[\begin{array}{c} 442.20 \\ 466.57 \\ 1088.09 \end{array}\right] \approx \left[\begin{array}{c} 2042 \\ 1733 \\ 3415 \end{array}\right]$$

So, sales for each type of shoe amounted to:

Gym shoes: $2042 million

Jogging shoes: $1733 million

Walking shoes: $ 3415 million

21. $\begin{cases} 8x - 3y = -52 \\ 3x + 5y = 5 \end{cases}, \quad D = \begin{vmatrix} 8 & -3 \\ 3 & 5 \end{vmatrix} = 49$

$$x = \frac{\begin{vmatrix} -52 & -3 \\ 5 & 5 \end{vmatrix}}{49} = \frac{-245}{49} = -5$$

$$y = \frac{\begin{vmatrix} 8 & -52 \\ 3 & 5 \end{vmatrix}}{49} = \frac{196}{49} = 4$$

Solution: $(-5, 4)$

22. $\begin{cases} 5x + 4y + 3z = 7 \\ -3x - 8y + 7z = -9, \\ 7x - 5y - 6z = -53 \end{cases} \quad D = \begin{vmatrix} 5 & 4 & 3 \\ -3 & -8 & 7 \\ 7 & -5 & -6 \end{vmatrix} = 752$

$$x = \frac{\begin{vmatrix} 7 & 4 & 3 \\ -9 & -8 & 7 \\ -53 & -5 & -6 \end{vmatrix}}{752} = \frac{-2256}{752} = -3$$

$$y = \frac{\begin{vmatrix} 5 & 7 & 3 \\ -3 & -9 & 7 \\ 7 & -53 & -6 \end{vmatrix}}{752} = \frac{3008}{752} = 4$$

$$z = \frac{\begin{vmatrix} 5 & 4 & 7 \\ -3 & -8 & -9 \\ 7 & -5 & -53 \end{vmatrix}}{752} = \frac{1504}{752} = 2$$

Solution: $(-3, 4, 2)$

23. $A = \pm\frac{1}{2}\begin{vmatrix} -2 & 3 & 1 \\ 1 & 5 & 1 \\ 4 & 1 & 1 \end{vmatrix} = -\frac{1}{2}(-18) = 9$

24. $a_n = \dfrac{(-1)^{n+1}}{2n+3}$

$a_1 = \dfrac{1}{5}$

$a_2 = -\dfrac{1}{7}$

$a_3 = \dfrac{1}{9}$

$a_4 = -\dfrac{1}{11}$

$a_5 = \dfrac{1}{13}$

25. $\dfrac{2!}{4}, \dfrac{3!}{5}, \dfrac{4!}{6}, \dfrac{5!}{7}, \dfrac{6!}{8}, \ldots$

$a_n = \dfrac{(n+1)!}{n+3}$

26. $6, 18, 30, 42, \ldots$

$a_n = 12n - 6$

$a_1 = 6, a_{16} = 186$

$S_{16} = \dfrac{16}{2}(6 + 186) = 1536$

27. (a) $a_6 = 20.6$

$a_9 = 30.2$

$a_9 = a_6 + 3d$

$30.2 = 20.6 + 3d$

$9.6 = 3d$

$3.2 = d$

$a_{20} = a_9 + 11d = 30.2 + 11(3.2) = 65.4$

(b) $a_1 = a_6 - 5d$

$a_1 = 20.6 - 5(3.2)$

$\quad = 4.6$

$a_n = a_1 + (n-1)d$

$\quad = 4.6 + (n-1)(3.2)$

$\quad = 3.2n + 1.4$

28. $a_n = 3(2)^{n-1}$

$a_1 = 3$

$a_2 = 6$

$a_3 = 12$

$a_4 = 24$

$a_5 = 48$

29. $\displaystyle\sum_{i=0}^{\infty} 1.3\left(\frac{1}{10}\right)^{i-1} = \sum_{i=0}^{\infty} 13\left(\frac{1}{10}\right)^{i} = \frac{13}{1 - \dfrac{1}{10}} = 13\left(\frac{10}{9}\right) = \frac{130}{9}$

30. $S_1 = 3 = 1\big[2(1) + 1\big]$

Assume that $S_k = 3 + 7 + 11 + 15 + \cdots + (4k - 1) = k(2k + 1)$.

Then, $S_{k+1} = 3 + 7 + 11 + 15 + \cdots + (4k - 1) + \big[4(k + 1) - 1\big]$

$= S_k + (4k + 3)$

$= k(2k + 1) + (4k + 3)$

$= 2k^2 + 5k + 3$

$= (k + 1)(2k + 3)$

$= (k + 1)\big[2(k + 1) + 1\big].$

So, the formula is valid for all integers $n \geq 1$.

31. $(w - 9)^4 = w^4 + {_4C_1}w^3(-9) + {_4C_2}w^2(-9)^2 + {_4C_3}w(-9)^3 + (-9)^4$

$= w^4 - 36w^3 + 486w^2 - 2916w + 6561$

32. ${_{14}P_3} = \dfrac{14!}{(14 - 3)!} = \dfrac{14!}{11!} = 2184$

33. ${_{25}P_2} = \dfrac{25!}{(25 - 2)!} = \dfrac{25!}{23!} = 600$

34. $\begin{pmatrix} 8 \\ 4 \end{pmatrix} = {}_8C_4 = \dfrac{8!}{(8-4)!4!} = \dfrac{8!}{4!4!} = 70$

35. ${}_{11}C_6 = \dfrac{11!}{(11-6)!6!} = \dfrac{11!}{5!6!} = 462$

36. B A S K E T B A L L

$\dfrac{10!}{2!2!2!1!1!1!1!} = 453{,}600$ distinguishable permutations

37. A N T A R C T I C A

$\dfrac{10!}{3!2!2!1!1!1!1!} = 151{,}200$ distinguishable permutations

38. ${}_{10}P_3 = \dfrac{10!}{(10-3)!} = \dfrac{10!}{7!} = 720$

39. The first digit is 4 or 5, so the probability of picking it correctly is $\frac{1}{2}$. Then there are two numbers left for the second digit so its probability is also $\frac{1}{2}$. If these two are correct, then the third digit must be the remaining number. The probability of winning is

$\left(\frac{1}{2}\right)\left(\frac{1}{2}\right)(1) = \frac{1}{4}$.